ATLAS OF

Human Brain Connections

"The *Atlas of Human Brain Connections* is intended to help neuroscientists and clinicians in the process of correlating structure with function, and lesions with symptoms...." With those spare words, the authors introduce beginners and experts alike to their wondrous, 500-page treasure-trove of fiber tracts, beautifully illustrated in all three planes of serial sections (axial, coronal, and sagittal) as well as in two-dimensional reconstructions. Faced with the reality of its myriad highways of communication, we are led to think anew about how the human brain performs its miracles. To that end, this uncommon view of brain structure will soon become indispensable.

Dr Mortimer Mishkin, Chief of the Section on Cognitive Neuroscience in the Laboratory of Neuropsychology, NIMH

Cognitive neuroscience is at a crossroads. On one hand, it sits on a rich data set of cortical connectivity in the monkey, an animal that lacks the complicated behaviors of interest. On the other hand, it is amassing an even richer set of facts on the functional map of the human brain, but with relatively little information on underlying structural connectivity. This lavishly illustrated volume by Catani and Thiebaut de Schotten represents a major step in closing this gap. The authors have combined the science of diffusion tensor imaging with the art of tractography in a comprehensive work slated to become a standard reference for exploring the structural foundations of human brain function.

Marsel Mesulam, MD, Director, Cognitive Neurology and Alzheimer's Disease Center

ATLAS OF
Human Brain Connections

Marco Catani
Senior Lecturer and Honorary Consultant Psychiatrist
Head of the Natbrainlab
Department of Forensic and Neurodevelopmental Sciences
Institute of Psychiatry
King's College London
London

Michel Thiebaut de Schotten
Research Fellow
Natbrainlab
Department of Forensic and Neurodevelopmental Sciences
Institute of Psychiatry
King's College London
London

OXFORD
UNIVERSITY PRESS

Great Clarendon Street, Oxford, OX2 6DP,
United Kingdom

Oxford University Press is a department of the University of Oxford.
It furthers the University's objective of excellence in research, scholarship,
and education by publishing worldwide. Oxford is a registered trade mark of
Oxford University Press in the UK and in certain other countries

First published 2012
First published in paperback 2015

Impression: 3

Published in the United States of America by Oxford University Press
198 Madison Avenue, New York, NY 10016, United States of America

British Library Cataloguing in Publication Data

Data available

Library of Congress Cataloging in Publication Data

Data available

ISBN 978–0–19–954116–4 (Hbk.)
ISBN 978–0–19–872937–2 (Pbk.)

Printed and bound in Great Britain by
Ashford Colour Press Ltd, Gosport, Hampshire

Preface

Interest in the anatomy of connections has had phenomenal growth in the last decade, spurred in particular by the advent of diffusion tensor magnetic resonance imaging tractography. The development of this method of analysis has proven invaluable. There is, however, a tendency to equate 'virtual in vivo tractographic trajectories' with axonal pathways, and this is not necessarily true. Other methods are indeed able to visual axonal connections but these are not suitable for the human brain. In the *Atlas of the Human Brain Connections* we attempt to provide an overview of connections derived from virtual in vivo tractography dissections of the human brain.

The first five 'introductory' chapters of the book are dedicated to the 'pioneers' of descriptive anatomy. These chapters underscore the historical salience of animal and human lesion research, in conjunction with older and newer structural imaging methods, as a framework in which to develop theoretical models.

The stereotaxic brain sections provided in Chapter 6 are intended to stimulate clinicians to develop a critical approach to clinico-anatomical correlations, and broaden their view of clinical anatomy beyond the cortical surface in order to encompass the dysfunction related to connecting pathways. By allowing a precise localization of white matter lesions and associated symptoms, the atlas will facilitate future work on the functional correlates of human neural networks as derived from the study of clinical populations. The hodological approach to clinico-anatomical correlation is however, still a work in progress. A quick glance at the figures of our atlas reveals that more than half of the white matter regions of the human brain remain unmapped. These regions correspond to the most superficial short connections that compose the intricate connectivity of neighbouring areas. The delineation of those connections remains the greatest challenge ahead.

Finally the aim of chapters 7-11 is twofold: first, to emphasize the importance of the anatomy of connections as a means of anchoring functional and clinical studies; second, to compare, whenever possible, tractography findings with monkey axonal tracing studies in an attempt to indirectly validate our virtual reconstructions.

A comprehensive review of tractography studies would have been an impossible and certainly daunting task. The publication rate of tractography research advances with a staggering amount of more than 300 papers a year, and it is likely to continue with a similar progression. For this reason, the Atlas is inevitably biased towards our own work. We apologize to those researchers whose papers would have deserved greater attention.

The book has absorbed much of our time and energy for the last five years. Throughout this project we have found inspiration from our mentors and colleagues and support from our families and friends.

Marco Catani would like to thank in particular Flavio Dell'Acqua, Dominic ffytche, Derek Jones, Steve Williams, Declan Murphy, Michael Craig, Robert Howard, and Rosario Donato. A special 'grazie' goes to his beloved wife Raffaela and his beautiful children Giulia and Matteo. He dedicates this book to them.

Michel Thiebaut de Schotten would like to thank colleagues and mentors who encouraged and supported him and his fascination for white matter connections; in particular Paolo Bartolomeo, Hugues Duffau, Bruno Dubois, Richard Levy, Emmanuelle Volle, Flavio Dell'Acqua, Declan Murphy, and Steve Williams. He dedicates this book to his father, Igor de Schotten who passed away during its creation, and his beloved wife, Lauren Sakuma who he met and married during this same period.

Many of our students (Stephanie Forkel, Henrietta Howells, Jamie Kwadler, Sean Froudist Walsh, Christine Longinotti) have read early drafts of some of the chapters and gave precious advice. Only two people had the tenacity to read the entire proof of the book; to Professor Lawrence Bannister and Dr Alberto Bizzi goes our deepest gratitude. We are also delighted for the advance praises that Professors Marsel Mesulam and Mortimer Mishkin have kindly provided. Working with them in the most recent years has been an inspiration and a privilege.

Writing this book has been difficult at times, but immensely enriching. Despite the authors coming from a different background, we have found perfect synchrony throughout the years. Our work has been fuelled by a common passion for anatomy, admiration for the pioneers and a great interest for contemporary technologies. We are also aware that the realization of this Atlas would have not been possible without reciprocal fondness and a profound friendship. Our hope is to transmit to the reader a sense of that curiosity, clinical dedication and scientific enthusiasm that has accompanied us throughout this fascinating exploration of the brain.

Marco Catani and Michel Thiebaut de Schotten

Contents

List of illustrations, *ix*

Chapter 1 Introduction to Descriptive Neuroanatomy, 1
Why study anatomy? 2
Descriptive neuroanatomy, 3
References, 5

Chapter 2 Surface Neuroanatomy, 7
Introduction, 8
Dorso-lateral surface, 8
Medial surface, 9
Ventral surface, 10
The cerebral lobes, 11
From surface neuroanatomy to function, 12
References, 19

Chapter 3 Sectional Neuroanatomy, 21
Introduction, 22
Ventricular system, 22
Subcortical telencephalic and diencephalic nuclei, 23
- The diencephalon, 23
- Basal ganglia, 24
The cerebral cortex, 24
From sectional neuroanatomy to function, 25
References, 35

Chapter 4 Connectional Neuroanatomy, 37
Introduction, 38
Association pathways, 38
Commissural pathways, 40
Projection pathways, 40
From connectional neuroanatomy to function, 41
References, 52

Chapter 5 The Clinico-Anatomical Correlation Method, 55
Introduction, 56
General considerations, 57
- Localization of symptoms or functions? 57
- Cortical or subcortical localization of functions? 58
What is a disconnection syndrome? 59
A hodotopic framework for clinico-anatomical correlation, 61
- Cortical pathology, 62
- Subcortical pathology, 62
- Cortico-subcortical pathology, 62
Synopsis of cerebral lobe syndromes, 64
- Frontal lobe, 64
- Parietal lobe, 66
- Occipital lobe, 68
- Temporal lobe, 69
- Limbic lobe, 70
Conclusions, 72
References, 72

Chapter 6 Atlas of Human Brain Connections (all tracts), 75
Axial maps, 76
Coronal maps, 124
Sagittal maps, 184

Chapter 7 Perisylvian Pathways, 239
Introduction, 240
Functional anatomy of the perisylvian cortical territories, 241
Lateralization of perisylvian pathways, 242
Comparative anatomy of the perisylvian networks, 245
Perisylvian disorders, 246
- Aphasias, 247
- Disorders of writing and reading, 250
- Memory disorders, 250
- Apraxias and discaculia, 250
- Right hemisphere syndromes, 251
- Neurodevelopmental disorders, 252
Superior longitudinal fasciculus system, 254
References, 255
Arcuate fasciculus (atlas), 259
Long segment (atlas), 269
Anterior segment (atlas), 278
Posterior segment (atlas), 287

Chapter 8 Occipital Visual Pathways, 297
Introduction, 298
The optic radiation, 298
Autochthonous connections, 299
Long association fibres, 299
Functional and comparative anatomy of cortical visual areas, 301
Functional and comparative anatomy of visual pathways, 303
Disorders of visual perception, 306
References, 310
Optic radiations (atlas), 313
Inferior longitudinal fasciculus (atlas), 320

Inferior fronto-occipital fasciculus (atlas), 331

Chapter 9 Commissural Pathways, 343
Introduction, 344
Anterior commissure, 344
Corpus callosum, 345
In vivo callosal anatomy, 346
Comparative anatomy, 347
Functional anatomy and tractography studies, 349
Callosal disorders and clinical syndromes, 349
- Dysgenesis of the corpus callosum, 349
- Split-brain patients, 351
- Acquired neurological disorders, 352
- Psychiatric disorders, 352
References, 352
Anterior commissure (atlas), 355
Corpus callosum (atlas), 365

Chapter 10 Projection Systems, 379
Introduction, 380
Corticospinal and corticobulbar projections, 381
Ascending thalamic projections, 384
Cortico-basal ganglia circuit, 387
Cerebellar pathways, 389
- Long-range cerebellar loops and intracerebellar circuits, 390
- Clinical manifestations of cerebellar lesions, 391
References, 394
Internal capsule (atlas), 397
Cortico-ponto-cerebellar tract (atlas), 413
Inferior cerebellar peduncle (atlas), 425
Superior cerebellar peduncle (atlas), 431

Chapter 11 Limbic System, 439
Introduction, 440
Fornix, 441
Mammillo-thalamic tract, 441
Anterior thalamic projections, 441
Cingulum, 442
Uncinate fasciculus, 442
Functional anatomy of the limbic system, 443
Limbic syndromes, 445
- Amnesias and dementias, 445
- Temporal lobe epilepsy, 446
- Antisocial behaviour, 446
- Schizophrenia, 448
- Depression and bipolar disorder, 448
- Obsessive-compulsive disorder, 448
- Autism spectrum disorder, 448
References, 449
Cingulum (atlas), 453
Uncinate fasciculus (atlas), 467
Fornix (atlas), 476

Appendix I, 489
Appendix II, 497

Index, 515

List of illustrations

We are grateful for permission to reproduce the following material in this volume.

Figure 1.1A	Image copyright © The Metropolitan Museum of Art/Art Resouce, NY.
Figures 1.1B,C	The Royal Collection © 2011 Her Majesty Queen Elizabeth II.
Figure 2.5B	Reproduced from Alberto Debernardi, Elena Sala, Giuseppe D'Aliberti, et al, Alcmaeon of Croton, *Neurosurgery*, **66**(2) with kind permission from Wolters Kluwer/Lippincott, Williams and Wilkins.
Figure 2.6	Wellcome Library, London.
Figure 2.7	Wellcome Library, London.
Figure 2.8	Wellcome Library, London.
Figure 2.9A	Reproduced with kind permission from Musée Dupuytren, Université Pierre et Marie Curie, Paris
Figure 2.9	Image of Joseph Jules Dejerine from the Wellcome Library, London.
Figure 2.10A	From *Science*, Norman Geschwind and Walter Levitsky, Human Brain: Left-Right Asymmetries in Temporal Speech Region © 1968 The American Association for the Advancement of Science. Reprinted with permission from AAAS.
Figure 3.4A	Wellcome Library, London.
Figure 3.4C	Reproduced with kind permission from the Klassik Stiftung Weimar.
Figure 3.5C,D, and the image of Arcangelo Piccolomini	Wellcome Library, London.
Figure 3.6	Wellcome Library, London.
Figure 3.7, image of Camillo Golgi	Wellcome Library, London.
Figure 3.8, image of Rudolph Albert von Kölliker	Museum für Naturkunde Berlin Historische Bild- u. Schriftgutsammlungen (Sigel: MfN, HBSB), Bestand: Zool. Mus. Signatur: B I/268.
Figure 3.8, image of Theodore Meynert	Wellcome Library, London.
Figure 3.8C	Reproduced from A.W. Campbell, *Histological studies on the localisation of cerebral function*, 1905, with permission from Cambridge University Press.
Figure 3.9, image of Oskart Vogt	Wellcome Library, London.
Figure 3.10	Reprinted from *The Lancet*, **158**(4077), Dr. Paul Flechsig, Developmental (myelogenetic) localisation of the cerebral cortex in the human subject, pp. 1027-1030, Copyright (1901), with permission from Elsevier.
Figure 3.11	With kind permission from Springer Science+Business Media: *Atlas of Cytoarchitectonics of the Adult Human Cerebral Cortex*, 1925: Constantin von Economo and Georg N. Koskinas.
Figure 3.12A	Reprinted from *NeuroImage*, **28**(3), David C. Van Essen, A Population-Average, Landmark- and Surface-based (PALS) atlas of human cerebral cortex, pp. 635-662, Copyright (2005), with permission from Elsevier.
Figure 3.12B	Reprinted from *NeuroImage*, **11**(1), Katrin Amunts, Aleksandar Malikovic, Hartmut Mohlberg, Thorsten Schormann, and Karl Zilles, Brodmann's Areas 17 and 18 Brought into Stereotaxic Space—Where and How Variable?, pp. 66-84, Copyright (2000), with permission from Elsevier.
Figure 4.2	Reproduced from *Brain*, **128**, Marco Catani and Dominic H. ffytche, The rises and falls of disconnection syndromes, copyright 2005, with permission from Oxford University Press.
Figure 4.3	Reproduced from *Brain*, **128**, Marco Catani and Dominic H. ffytche, The rises and falls of disconnection syndromes, copyright 2005, with permission from Oxford University Press.
Figure 4.4, portraits	Wellcome Library, London.

Figure 4.8, image of Hugo Liepmann	Reproduced with kind permission from Professor Bernd Holdorff.
Figure 4.8B	Reproduced from *Brain*, **7**, 433-484, On aphasia By Lichtheim, L, © 1885 with permission from Oxford University Press.
Figure 4.8D	With kind permission from Springer Science+Business Media: *Lehrbuch der Nervenkrankheiten*, 2nd edn, Apraktische Störungen, 1925, H. Liepmann.
Figure 4.9A	Courtesy of Columbia University Press.
Figure 4.9B	Reproduced here by courtesy of Orell Füssli Verlag AG, Zürich, Switzerland.
Figure 4.10, image of Paul Emile Flechsig	© Quelle: Archiv für Leipziger Psychiatriegeschichte, Universität Leipzig.
Figure 4.10, image of James Papez	Courtesy of the Archives, California Institute of Technology.
Figure 4.10C	The figure was first published in © The Psychoanalytic Quarterly, 1955, *The Psychoanalytic* Quarterly, Volume XXIV, Number 3, page 418.
Figure 4.12	Image of Roger Sperry courtesy of the Archives, California Institute of Technology.
Figure 4.12C	Illustration from the Nobel Committee for Physiology or Medicine, based on "Impact of Science on Society" published by UNESCO. Copyright the Nobel Committee.
Figure 4.13, image of Norman Geschwind	Reprinted from *Cortex*, **44**(8), Marco Catani and Marsel Mesulam, The arcuate fasciculus and the disconnection theme in language and aphasia: History and current state, pp. 953-961, Copyright (2008), with permission from Elsevier.
Figure 4.13A	This article was published in *Principles of Neural Science*, third edition, Kandel, E.R., Schwartz J. and Jessel T. Copyright Elsevier (1991).
Figure 4.13C	Ingle, David, M. Goodale and R. Mansfield, *Analysis of Visual Behavior*, figure from pages 549-86, © 1982 Massachusetts Institute of Technology, by permission of The MIT Press.
Figure 5.1A	Levelt, Willem J. M., *Speaking: From Intention to Articulation*, figure 1.1 page 9, © 1989 Massachusetts Institute of Technology, by permission of The MIT Press.
Figure 5.1C	Reprinted from *Cortex*, **46**(1), Marco Catani and Dominic H. ffytche, On 'the study of the nervous system and behaviour', pp. 106-9, Copyright (2010), with permission from Elsevier.
Figure 5.4D	From *Science*, **170**(3961), Norman Geschwind, The Organization of Language and the Brain: Language disorders after brain damage help in elucidating the neural basis of verbal behavior, © 1970, The American Association for the Advancement of Science. Reprinted with permission from AAAS.
Figure 5.5	Reprinted from *Brain*, **128**(10), Marco Catani and Dominic H. ffytche, The rises and falls of disconnection syndromes, Copyright (2005), with permission from Oxford University Press.
Figure 5.9A	Reprinted from, *NeuroImage*, **41**(4),Marco Catani, Derek K. Jones, Eileen Daly, Nitzia Embiricos, Quinton Deeley, Luca Pugliese, Sarah Curran, Dene Robertson, and Declan G.M. Murphy, Altered cerebellar feedback projections in Asperger syndrome, pp. 1184-1191, Copyright (2008), with permission from Elsevier.
Figure 5.9B	Reprinted from, *Cortex*, **4**(8), Marco Catani and Marsel Mesulam, The arcuate fasciculus and the disconnection theme in language and aphasia: History and current state, pp. 953-961, Copyright (2008), with permission from Elsevier.
Figure 5.9C,D	Reprinted from *Brain*, **130**(5), N. F. Dronkers, O. Plaisant, M. T. Iba-Zizen, and E. A. Cabanis, Paul Broca's historic cases: high resolution MR imaging of the brains of Leborgne and Lelong, Copyright (2007), with permission from Oxford University Press.
Figure 7.1	Reprinted from *NeuroImage*, **17**(1), Marco Catani, Robert J. Howard, Sinisa Pajevic, and Derek K. Jones, Virtual in Vivo Interactive Dissection of White Matter Fasciculi in the Human Brain, pp. 77-94, Copyright (2002), with permission from Elsevier.
Figure 7.2	Reprinted from *Annals of Neurology*, Marco Catani, Derek K. Jones, and Dominic H. ffytche, Virtual in Vivo Interactive Dissection of White Matter Fasciculi in the Human Brain, pp. 8-16, Copyright (2005), with permission from John Wiley and Sons.
Figure 7.3A	Modified from Catani et al. Symmetries in human brain language pathways correlate with verbal recall. Proceedings of the National Academy of Sciences of the United States of America (2007) vol. 104 (43) pp. 17163-8 Copyright (2007) National Academy of Sciences, U.S.A.
Figure 7.4	Modified from Catani et al. Symmetries in human brain language pathways correlate with verbal recall. Proceedings of the National Academy of Sciences of the United States of America (2007) vol. 104 (43) pp. 17163-8 Copyright (2007) National Academy of Sciences, U.S.A.
Figure 7.6	These figures are copyright to University of Wisconsin and Michigan State Comparative Mammalian Brain Collections, and the National Museum of Health and Medicine. Preparation of the images and specimens within these collections have been funded by the National Science Foundation, as well as by the National Institutes of Health (http://brainmuseum.org/index.html).

Figure 7.7A Reproduced from *Brain*, **7**, 433-484, On aphasia By Lichtheim, L, © 1885 with permission from Oxford University Press.

Figure 7.7B Reprinted from *Brain*, **107**, Rosaleen McCarthy and Elizabeth K. Warrington, A two route model of speech production: Evidence from aphasia, Copyright (1984), with permission from Oxford University Press.

Figure 7.10A Reproduced from Baddeley, A.D, *Working memory, thought and action*, © 2007 with permission from Oxford University Press

Figure 7.11A Reprinted by permission from Macmillan Publishers Ltd: *Nature Reviews Neuroscience*, Maurizio Corbetta and Gordon L. Shulman, Control of goal-directed and stimulus-driven attention in the brain, **3**(3) copyright (2002)

Figure 7.13C,D Reprinted by permission from John Wiley and Sons: *Journal of Comparative Neurology*, M. Petrides, D. N. Pandya, Projections to the frontal cortex from the posterior parietal region in the rhesus monkey, **228**(1) pp. 105-116 copyright (1984).

Figure 7.13E Reprinted from *NeuroImage*, **49**(2), Flavio Dell'Acqua, Paola Scifo, Giovanna Rizzo, Marco Catani, Andrew Simmons, Giuseppe Scotti, and Ferruccio Fazio, A modified damped Richardson–Lucy algorithm to reduce isotropic background effects in spherical deconvolution, pp. 1446-1458, Copyright (2010), with permission from Elsevier.

Figure 8.1 Reprinted from *Brain*, **126**, Marco Catani, Derek K. Jones, Rosario Donato, and Dominic H. ffytche, Occipito-temporal connections in the human brain, Copyright (2003), with permission from Elsevier.

Figure 8.2 Reprinted from *NeuroImage*, **17**(1), Marco Catani, Robert J. Howard, Sinisa Pajevic, and Derek K. Jones, Virtual in Vivo Interactive Dissection of White Matter Fasciculi in the Human Brain, pp. 77-94, Copyright (2002), with permission from Oxford University Press.

Figure 8.3 Reproduced from *Brain*, **126**, Marco Catani, Derek K. Jones, Rosario Donato, and Dominic H. ffytche, Occipito-temporal connections in the human brain, Copyright (2003), pp. 2093-107 with permission from Oxford University Press.

Figure 8.5B Reproduced from *British Journal of Opthalmology*, Gordon Holmes, **2**, 353-84, copyright 1918 with permission from BMJ Publishing Group Ltd.

Figure 8.7A Copyright © 1992 Society for Industrial and Applied Mathematics. Reprinted with permission. All rights reserved.

Figure 8.7B David, M. Goodale and R. Mansfield, *Analysis of Visual Behavior*, figure from pages 549-586, © Massachusetts Institute of Technology, by permission of the MIT Press.

Figure 8.8 Reprinted from *Brain*, **128**(10), Marco Catani and Dominic H. ffytche, The rises and falls of disconnection syndromes, Copyright (2005), with permission from Oxford University Press.

Figure 9.1 Reprinted from *NeuroImage*, **17**(1), Marco Catani, Robert J. Howard, Sinisa Pajevic, and Derek K. Jones, Virtual in Vivo Interactive Dissection of White Matter Fasciculi in the Human Brain, pp. 77-94, Copyright (2002), with permission from Elsevier.

Figure 9.2A Reprinted from *Brain*, **126**, Marco Catani, Derek K. Jones, Rosario Donato, and Dominic H. ffytche, Occipito-temporal connections in the human brain, Copyright (2003), with permission from Oxford University Press.

Figure 9.2B Reprinted from *Brain*, **112**, Sandra F. Witelson, Hand and Sex Differences in the isthmus and genu of the human corpus callosum: A postmortem morphological study, Copyright (1989), with permission from Oxford University Press.

Figure 9.2C Reprinted from *Brain Research*, **598**(1-2), Francisco Aboitiz, Arnold B. Scheibel, Robin S. Fisher, and Eran Zaidel, Fiber composition of the human corpus callosum, pp. 143-153, Copyright (1992), with permission from Oxford University Press.

Figure 9.3A-D Reprinted from *Brain*, **132**, Daniel Barazany, Peter J. Basser, and Yaniv Assaf, In vivo measurement of axon diameter distribution in the corpus callosum of rat brain, Copyright (2009), with permission from Oxford University Press.

Figure 9.3F Reprinted from *NeuroImage*, **49**(2), Flavio Dell'Acqua, Paola Scifo, Giovanna Rizzo, Marco Catani, Andrew Simmons, Giuseppe Scotti, and Ferruccio Fazio, A modified damped Richardson–Lucy algorithm to reduce isotropic background effects in spherical deconvolution,pp. 1446-1458, Copyright (2010), with permission from Elsevier.

Figure 9.4A Reproduced from Schmahmann, J.D., and Pandya, D.N., *Fiber Pathways of the Brain*, 2006 with permission from Oxford University Press.

Figure 9.4B Reprinted from *NeuroImage*, **49**(2), Flavio Dell'Acqua, Paola Scifo, Giovanna Rizzo, Marco Catani, Andrew Simmons, Giuseppe Scotti, and Ferruccio Fazio, A modified damped Richardson–Lucy algorithm to reduce isotropic background effects in spherical deconvolution, pp. 1446-1458, Copyright (2010), with permission from Elsevier.

Figure 10.7A,B Reprinted from *NeuroImage*, **25**(4), Nikos Makris, John E. Schlerf, Steven M. Hodge, Christian Haselgrove, Matthew D. Albaugh, Larry J. Seidman, Scott L. Rauch, Gordon Harris, Joseph Biederman, Verne S. Caviness Jr., David N. Kennedy, and Jeremy D. Schmahmann, MRI-based surface-assisted parcellation of human cerebellar cortex: an anatomically specified method with estimate of reliability, pp. 1146-1160, Copyright (2005), with permission from Elsevier.

Figure 10.7C	Reprinted from *NeuroImage*, **37**(4), Sean C.L. Deoni and Marco Catani, Visualization of the deep cerebellar nuclei using quantitative T1 and ρ magnetic resonance imaging at 3 Tesla, pp.1260-1266, Copyright (2007), with permission from Elsevier.
Figure 10.7D	Reproduced with permission from University of Minnesota Press.
Figure 10.8	Reprinted from *NeuroImage*, **41**(4), Marco Catani, Derek K. Jones, Eileen Daly, Nitzia Embiricos, Quinton Deeley, Luca Pugliese, Sarah Curran, Dene Robertson, and Declan G.M. Murphy, Altered cerebellar feedback projections in Asperger syndrome, pp. 1181-1194, Copyright (2008), with permission from Elsevier.
Figure 10.10	Reprinted from *NeuroImage*, **41**(4), Marco Catani, Derek K. Jones, Eileen Daly, Nitzia Embiricos, Quinton Deeley, Luca Pugliese, Sarah Curran, Dene Robertson, and Declan G.M. Murphy, Altered cerebellar feedback projections in Asperger syndrome, pp. 1181-1194, Copyright (2008), with permission from Elsevier.
Figure 11.2	Reprinted from *NeuroImage*, **17**(1), Marco Catani, Robert J. Howard, Sinisa Pajevic, and Derek K. Jones, Virtual in Vivo Interactive Dissection of White Matter Fasciculi in the Human Brain, pp.77-94, Copyright (2002), with permission from Elsevier.
Figure 11.3	Reprinted from *NeuroImage*, **17**(1), Marco Catani, Robert J. Howard, Sinisa Pajevic, and Derek K. Jones, Virtual in Vivo Interactive Dissection of White Matter Fasciculi in the Human Brain, pp.77-94, Copyright (2002), with permission from Elsevier.
Figure 11.6A	Reprinted with permission from the Society for Neuroscience.
Figure 11.6B	Understanding anterograde amnesia: Disconnections and hidden lesions, John P. Aggleton, *The Quarterly Journal of Experimental Psychology*, 2008, Taylor & Francis, reprinted by permission of the publisher (Taylor & Francis Group, http://www.informaworld.com).
Figure 11.6C	Reprinted with permission from The Journal of Neuroscience.
Figure 11.7	Reprinted by permission from Macmillan Publishers Ltd: *Molecular Psychiatry*, M C Craig, M Catani, Q Deeley, R Latham, E Daly et al, Altered connections on the road to psychopathy, **14**(10), 946-53, copyright (2009).
Figure A1.2	With kind permission from Springer Science+Business Media: Brain Mapping. From Neural Basis of Cognition to Surgical Applications, DTI tractography and subcortical fasciculi, 2011, Duffau and Hugues (Ed.).
Figure A1.5	Reprinted from *Brain*, **130**(3), Marco Catani, From hodology to function, Copyright (2007), with permission from Oxford University Press.

CHAPTER 1

Introduction to Descriptive Neuroanatomy

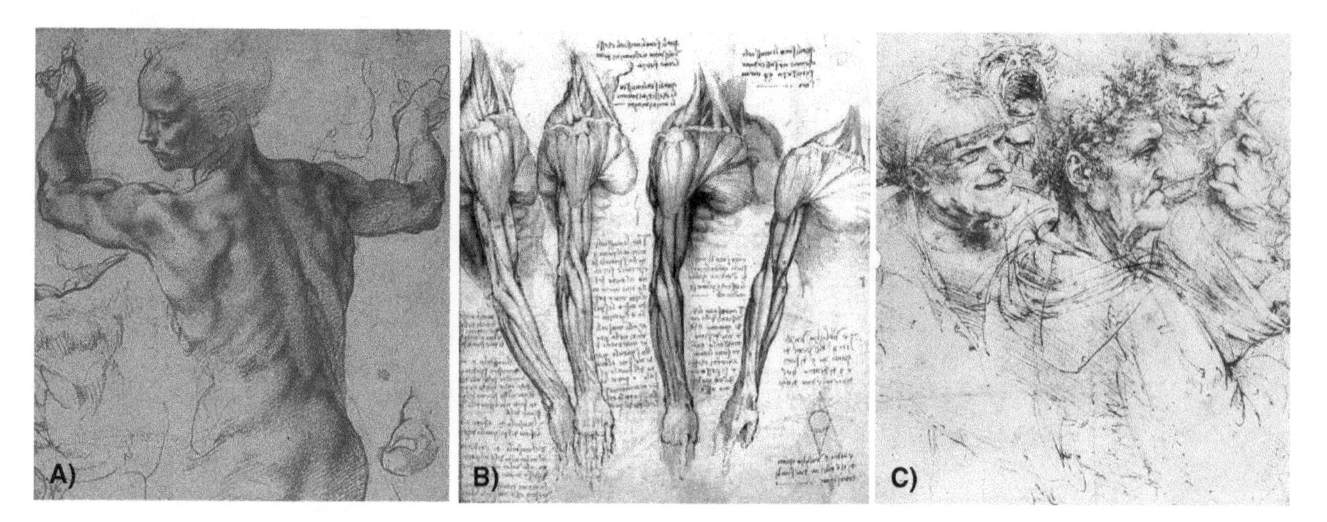

Figure 1.1 The Aristotelian principle of form and function correlation inspired Michelangelo Buonarroti and Leonardo da Vinci. A) Michelangelo's (1508) study for the Libyan Sibyl shows his deep understanding of human anatomy. (Image copyright © The Metropolitan Museum of Art/Art Resource, NY) B) Leonardo's study of the muscles of the shoulder and arm and annotations on possible functional correlates. C) Leonardo's study of the faces according to his principles of physiognomy. Leonardo advice for his students was 'represent your figures in such action as may be fitted to express what purpose is in the mind of each; otherwise your art will not be admirable'. Physiognomy was the first attempt to apply the Aristotelian principle to understand the human mind. ((B) and (C) from The Royal Collection © 2011 Her Majesty Queen Elizabeth II)

Why study anatomy?

The *Atlas of Human Brain Connections* is intended to encourage reflection on two important issues in neurological sciences: the relevance of neuroanatomy to the understanding of brain function and the importance of methodological developments to the study of neuroanatomy. Aristotle, the father of comparative anatomy, was the first to tell us why it is important to study anatomy. For example, in relation to the bodies of birds he wrote *'each bodily part is for the sake of something... Birds' beaks... differ according to their different [ways] of life. Some beaks are... straight if they are used simply for feeding, curved if the bird eats raw meat, because a curved beak is useful for overpowering their prey...'* (Aristotle, 2004). Hence, 'form and function go together for Aristotle. Anatomy and physiology are integral components of the same science' (Blits, 1999). In teleological terms we might say that a unique form subserves a specific function; in evolutionary terms that the function shapes the form. It follows that modification of the form results in dysfunction, a corollary that applies in nature at any level. Scientists have exploited this logic to work out, in a reverse order (from functional deficits to structural modifications), the causes of human diseases. Thus, in molecular biology, the study of linear sequences of bases forming the DNA or amino acids composing proteins, and their ability to fold into relatively stable three-dimensional structures with unique transcriptional or enzymatic activities has led to important insights into normal cellular processes and pathological conditions. At a macroscopical level, the Aristotelian principle has dominated Western culture since the Renaissance (Figure 1.1). Michelangelo's obsession with the human body was driven by the hope of capturing the meaning of human existence through the representation of a perfect body. Leonardo's physiognomy attempted to reveal the many facets of 'human passions' through the representation of facial expressions. Both Michelangelo and Leonardo practised human body dissection, a method of scientific enquiry that grew stronger over the centuries and culminated in the work of the greatest neuroanatomists of the 18th and 19th centuries, all of whom shared a common faith in the clinico-anatomical correlation to explore disorders of the brain in both animals and humans (see Chapter 5) (Catani, 2007).

Anatomy teaching became one of the pillars of medical education. Old anatomists used to introduce their courses with the saying '*sine anatomia non sciemus*' ('without anatomy there is no knowledge') and students would pay costly fees to attend extracurricular dissecting classes. With the move towards holism that took place between the two world wars and the development of more sophisticated psychological methods in the second half of the 20th century, the neuroanatomy understanding of that time became insufficient to capture the complexity of psychological functions. For many, anatomy became largely irrelevant to the development of psychological models of function and dysfunction (Catani and ffytche, 2010). This position is put in a nutshell by the American philosopher and cognitive scientist Jerry Fodor: *'... if the mind happens in space at all, it happens somewhere north of the neck. What exactly turns on knowing how far north?'* (Fodor, 1999).

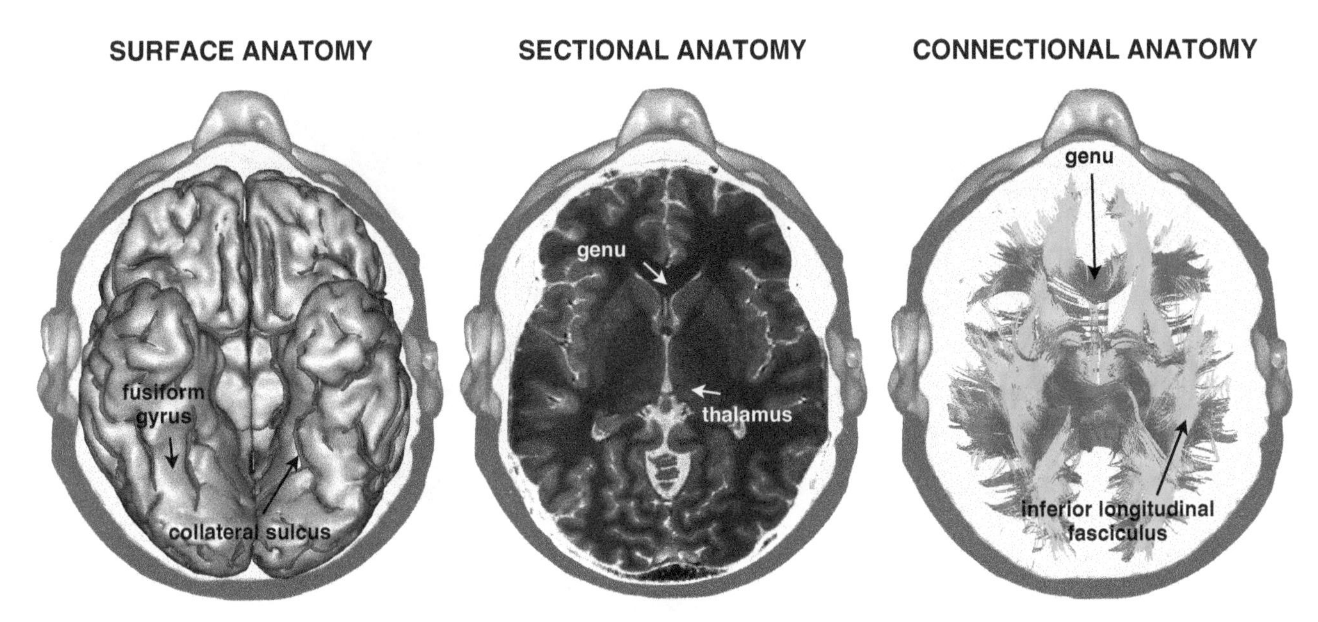

Figure 1.2 Magnetic resonance reconstructions of the surface, sectional, and connectional anatomy of the human brain. Left, 3D reconstruction and automatic segmentation of the sulci and gyri of the ventral aspect of the cerebral hemispheres. Centre, axial section passing through the deep grey nuclei. Right, tractography reconstruction of the tracts passing through an axial plane (different colours correspond to tracts with different courses and orientations).

In the last decade the advent of magnetic resonance imaging (MRI) and the flourishing of techniques for structural and functional imaging are promising to reinstate anatomy to its former position. However, the importance of studying anatomy with 'expensive' neuroimaging techniques continues to be harshly criticized and often people wonder whether much more than a few footnotes have been added to the work of the great neuroanatomists of the past centuries. There is no doubt that neuroimaging is a very powerful tool that is contributing to our understanding of brain anatomy and function. It is also true that many findings derived from neuroimaging research are difficult to translate into benefits for patients.

In contemporary neuroscience the anatomy of the central nervous system has become essentially the anatomy of the connections between distant regions. Understanding the pattern of connectivity and how neurons communicate is probably the next frontier in neurosciences (Mesulam, 2005). Investments on a large scale for projects aimed at mapping the 'connectome' testify to the great interest of the scientific community in the descriptive anatomy of human brain connections. We, however, hold the view that neuroimaging has to prove its clinical validity beyond group-level analysis to stand the test of time. Unfortunately, insights from advanced imaging methods often fail to reach those clinicians who might benefit most from them. The *Atlas of Human Brain Connections* is intended to help neuroscientists and clinicians in the processes of correlating structure with function, and lesions with symptoms, tasks that for a long time have been within the remit of descriptive anatomy.

Descriptive neuroanatomy

Descriptive anatomy of the nervous system can be defined as the process of identifying and labelling different parts of the brain and spinal cord. The ultimate objective of descriptive neuroanatomy is to systematize, classify, and derive general principles of organization and function of the nervous system. Here we focus only on the brain, whose anatomy can be described from its surface, and by means of orthogonal sections and tract dissections (Figure 1.2).

Surface neuroanatomy is the subject matter of Chapter 2, where the general appearance of grooves (*sulci*) and folds (*gyri* or *convolutions*) of the brain is described in detail. The surface of the brain

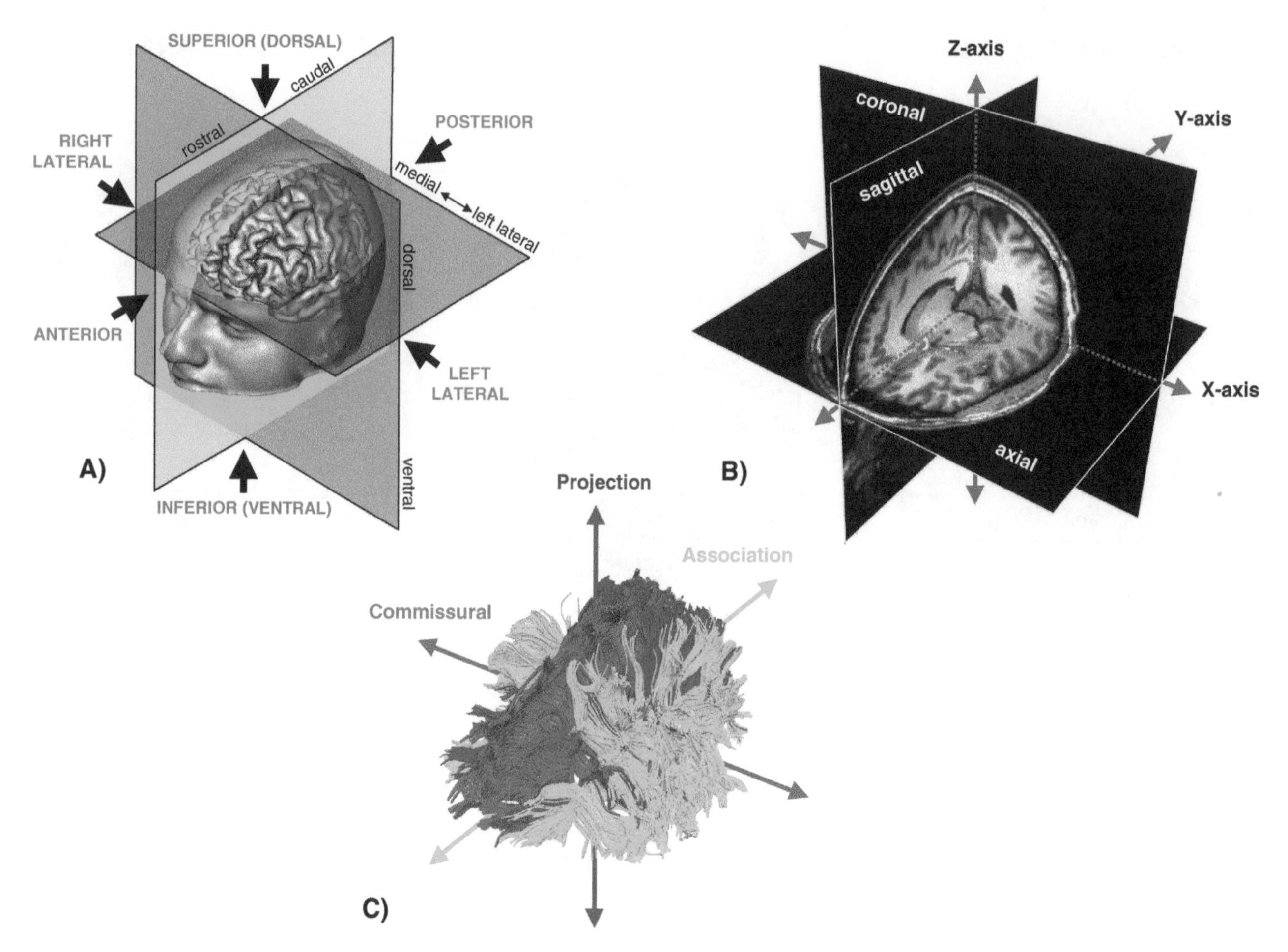

Figure 1.3 Terms commonly used to describe the orientation of the brain in surface (A), sectional (B), and connectional anatomy (C) representations.

can be viewed from the side (lateral view), the middle (medial view), the front (anterior or frontal view), and the back (posterior or occipital view) (Figure 1.3A). The same terminology is used to indicate different regions of the brain surface (e.g. dorso-lateral prefrontal cortex, etc.).

Sectional neuroanatomy describes the relationship between cortical and subcortical structures, most commonly visualized along orthogonal axial, coronal, and sagittal planes (Figure 1.3B). Axial planes divide the brain into an upper and a lower part. In radiological convention, the axial slices are viewed from the feet towards the head. Hence, on axial slices the left side of the brain is on the right side of the page (see axial slices on Chapter 6). Axial planes are also sometimes indicated as horizontal or transverse planes. In neuroimaging studies brains are often displayed within a space of reference (e.g. Talairach and Tournoux Atlas or the Montreal Neurological Institute [MNI] reference) and the axial coordinates move in a z-direction (negative values indicate slices lower than the anterior commissure, while positive values indicate slices above the anterior commissure) (Talairach and Tournoux 1988; Evans et al., 1992). Coronal planes cut through the brain along the left-to-right and top-down directions. The coronal planes are conventionally oriented with the left side of the brain on the right side of the page (frontal view) (see coronal slices on Chapter 6). Slices anterior to the anterior commissure are indicated with positive values along the y-axis. Finally, the sagittal plane divides the brain into two halves. Parasagittal slices through the left hemisphere have negative numbers along the x-axis. Various aspects of sectional anatomy of the human brain, including the cortical cytoarchitectonic maps and subcortical nuclei are described in Chapter 3.

Connectional neuroanatomy delineates the origin, course, and termination of connecting pathways. Post-mortem dissections of white matter tracts require special preparation and are particularly difficult to perform. Recent developments in diffusion MRI tractography have revitalized the field. The tracts are classified according to their course and terminal projections (Figure 1.3C). Commissural pathways run along a horizontal axis and connect the two hemispheres. The majority of the projection pathways have a perpendicular course along a dorso-ventral (descending) or ventro-dorsal (ascending) axis and connect the cerebral cortex to subcortical nuclei, cerebellum, and the spinal cord. The association tracts run longitudinally along an antero-posterior axis from front to back and *vice versa* and connect cortical areas within the same hemisphere. Chapter 4 gives a general overview of the major white matter pathways of the human brain.

An important question to ask is to what extent the study of the anatomical features of the brain can help in the understanding of cerebral function and its disorders. This is addressed in Chapter 5, where the advantages and limitations of the clinico-anatomical correlation method applied to patients with neurological and psychiatric disorders are discussed. The chapter concludes with a synopsis of the most frequent neurological and psychiatric disorders affecting the cerebral lobes.

To facilitate the localization of white matter tracts we have produced composite maps of the major white matter tracts of the human brain using diffusion MRI tractography. The axial, coronal, and sagittal maps of the atlas of human brain connections are normalized in the MNI space and reproduced in Chapter 6.

The detailed anatomy of individual tracts, their functions, and the clinical manifestations commonly associated with their lesions are illustrated separately in the remaining chapters of the atlas. The tracts are grouped together according to their function or anatomical proximity, but there is clearly some degree of arbitrariness in grouping different tracts under the same umbrella.

The functional anatomy of a group of tracts connecting regions around the lateral fissure, the perisylvian pathways, is described in Chapter 7.

Chapter 8 is dedicated to the ventral occipital pathways, which include the optic radiations, the inferior longitudinal fasciculus (ILF), and the inferior fronto-occipital fasciculus (IFOF).

Chapter 9 considers the two major telencephalic commissural pathways of the brain, namely the anterior commissure and the corpus callosum.

Chapter 10 examines the projection pathways. These include the cortico-spinal tract, the thalamic projections, and the connections to the basal ganglia and cerebellum.

Finally the extended limbic network, which includes the uncinate and the Papez circuit formed by the fornix, mamillo-thalamic tract, anterior thalamic projections, and the cingulum is described in Chapter 11.

Finally the methods used for the atlas are reported in the appendices where advantages and limitations of current diffusion tensor imaging methods are discussed.

References

Aristotle. (2004). *On the parts of animals.* Whitefish, MT: Kessinger Publishing.

Blits, K.C. (1999). Aristotle: form, function, and comparative anatomy. *Anat Rec (New Anat)* **257**(2), 58–63.

Catani, M. (2007). From hodology to function. *Brain* **130**(3), 602–5.

Catani, M. and ffytche, D.H. (2010). On 'the study of the nervous system and behaviour'. *Cortex* **46**(1), 106–9.

Evans, A.C., Marrett, S., Neelin, P., et al. (1992) Anatomical mapping of functional activation in stereotactic coordinate space. *NeuroImage* **1**(1), 43-53.

Fodor, J. (1999). Let your brain alone. *LRB* 21, 68–9. Available at: http://www.lrb.co.uk/v21/n19/fodo01_.html.

Mesulam, M.M. (2005). Imaging connectivity in the human cerebral cortex: the next frontier? *Ann Neurol* **57**(1), 5–7.

Talairach, J. and Tournoux, P. (1988). *Co-planar stereotaxic atlas of the human brain*. Thieme: New York.

CHAPTER 2

Surface Neuroanatomy

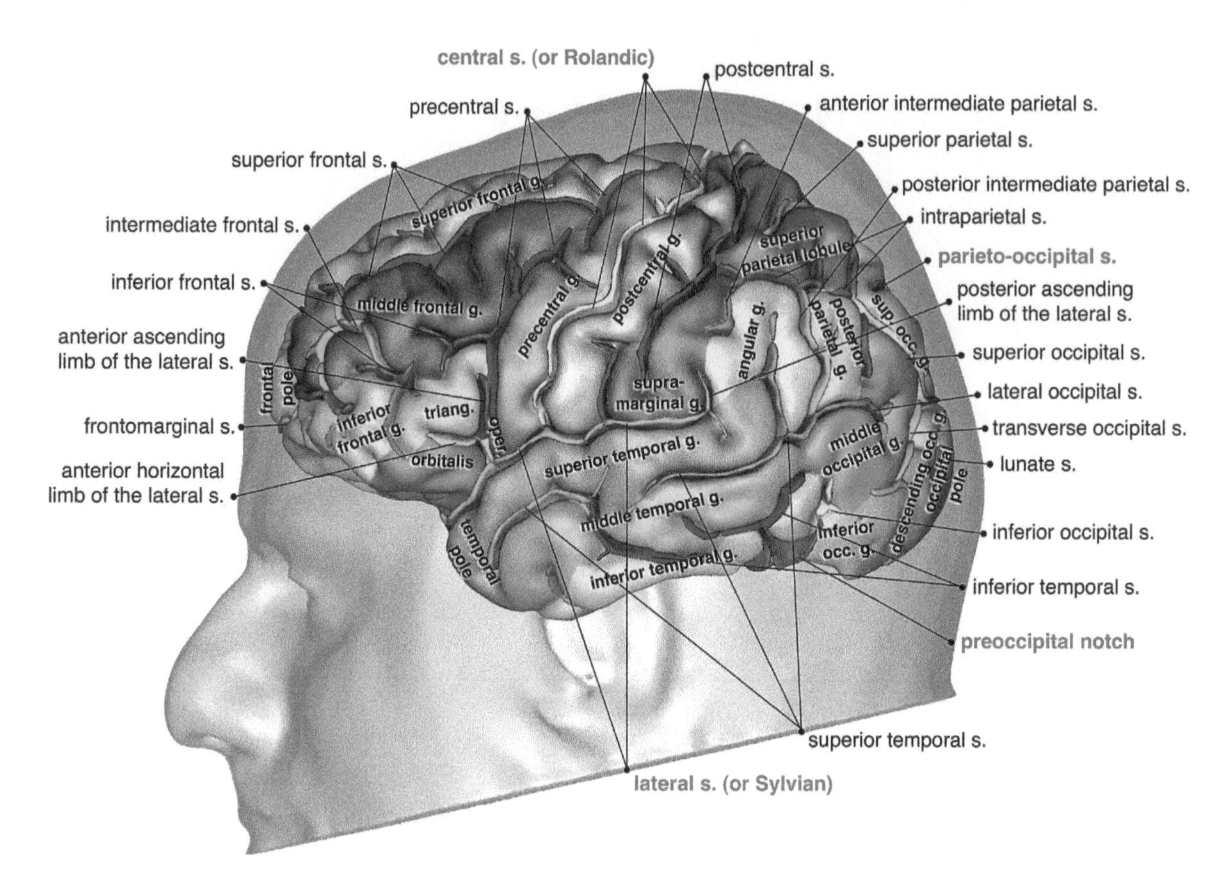

Figure 2.1 3D reconstruction of the brain showing the principal gyri (g.) and sulci (s.) of the dorsolateral surface of the left hemisphere. Interlobar sulci are written in red, intralobar sulci in black.

Introduction

Surface anatomy of the cerebral hemispheres describes the general appearance of grooves and folds, generally referred to as sulci and gyri (or convolutions), respectively. Sulci vary in depth and the term fissure is used by some anatomists for the deepest sulci (e.g. lateral fissure, calcarine fissure). Intralobar sulci separate adjacent gyri within the same lobe, whereas interlobar sulci demarcate the boundaries between lobes. Gyri differ in length and width and are named according to their location (e.g. inferior frontal gyrus) or their shape (e.g. fusiform gyrus). For each gyrus we can distinguish two lateral walls and an external surface. It is estimated that about two-thirds of the cerebral cortex lies along the walls of the sulci and one-third on the visible surface (Henneberg, 1910). Often gyri merge into each other and their boundaries are then rather arbitrary.

Following separation of the two hemispheres along the longitudinal sulcus, one can distinguish three surfaces (dorso-lateral, medial, and ventral or basal) and three poles (frontal, temporal, and occipital) in each hemisphere. On inspection, the three surfaces present characteristic patterns of sulci and gyri.

Dorso-lateral surface

The main gyri of the dorso-lateral surface have either longitudinal (i.e. anterior–posterior) or vertical (i.e. dorsal-ventral) orientations (Figure 2.1). The deep central (or Rolandic) sulcus extends from the dorsal to the ventrolateral surface of the hemisphere and separates the frontal and parietal lobes. The precentral (or ascending frontal) gyrus runs anterior and parallel to the central sulcus and is delimited anteriorly by the precentral sulcus. Similarly the postcentral (or ascending parietal) gyrus lies posterior to the central sulcus and is delimited posteriorly by the postcentral sulcus. The superior and inferior frontal sulci extend horizontally from the precentral sulcus and separate the superior, middle, and inferior frontal gyri. In some subjects the superior and inferior frontal sulci are interconnected through the intermediate frontal sulcus, which cuts through the middle frontal gyrus (Ono et al., 1990). The inferior frontal gyrus lies beneath the inferior frontal sulcus and its posterior part is divided into three regions by the anterior ascending and anterior horizontal limbs of the lateral sulcus. The pars opercularis is behind the anterior ascending limb of the lateral sulcus and anterior to the precentral sulcus. The pars triangularis is between the anterior ascending and anterior horizontal limb of the lateral sulcus. The pars orbitalis is beneath the anterior horizontal limb. The superior, middle, and inferior frontal gyri converge anteriorly to form the frontal pole, which is separated from the orbital gyri by the frontomarginal sulcus.

The lateral (or Sylvian) sulcus extends longitudinally along the lateral surface of the hemisphere and separates the superior temporal gyrus from the inferior margin of the frontal and parietal lobes. The inferior wall of the lateral sulcus is created by the temporal operculum, while the superior wall is formed anteriorly by the frontal operculum and posteriorly by the parietal operculum. The temporal, frontal, and parietal opercula cover the underlying gyri of the insula

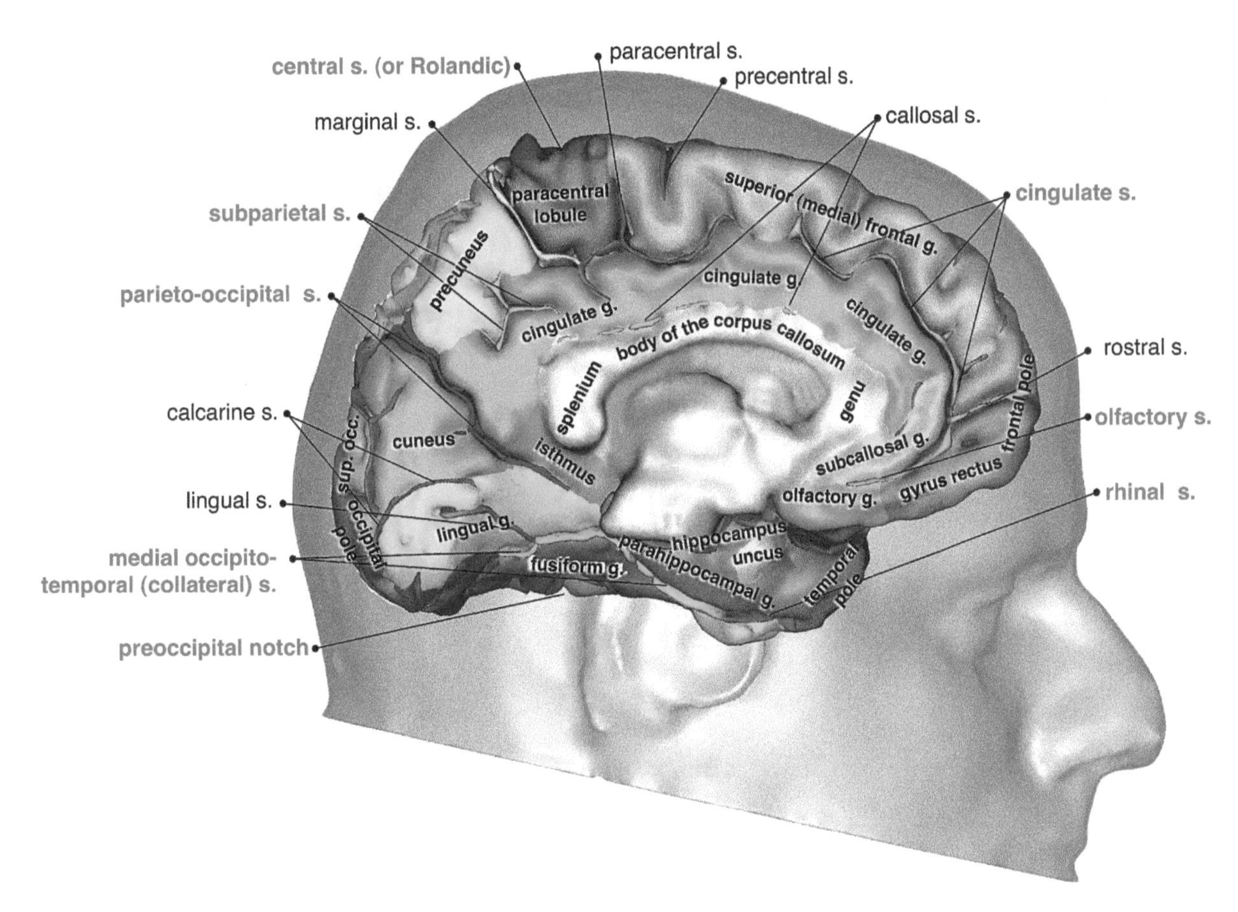

Figure 2.2 3D reconstruction of the brain showing the principal gyri (g.) and sulci (s.) of the medial surface of the left hemisphere. Interlobar sulci are written in red, intralobar sulci in black.

of Reil. Parallel to the lateral sulcus are the superior and inferior temporal sulci, which separate the superior, middle, and inferior temporal gyri. Along the ventrolateral edge of the inferior temporal gyrus a small indentation, known as the preoccipital notch, demarcates the border between the temporal and the occipital lobe.

The superior temporal gyrus continues posteriorly into the angular gyrus, which arches around the terminal branch of the superior temporal sulcus. Anterior to the angular gyrus is the supramarginal gyrus. The angular and supramarginal gyri form the inferior parietal lobule, which is separated from the superior parietal lobule by the intraparietal sulcus. The anterior intermediate parietal sulcus delimitates the borders between the supramarginal and angular gyri. Posterior to the angular gyrus is the posterior parietal gyrus, lying just beneath the intraparietal sulcus. The posterior intermediate parietal sulcus, which separates the angular and the posterior parietal gyri, often continues into the superior temporal sulcus.

In the occipital lobe, the superior occipital sulcus separates the superior from the middle occipital gyrus, whilst the inferior occipital sulcus separates the middle from the inferior occipital gyrus. Often the superior occipital sulcus continues into the intraparietal sulcus. Similarly, the superior occipital gyrus extends into the superior parietal lobule, the middle occipital gyrus into the posterior parietal and middle temporal gyri and the inferior occipital gyrus into the inferior temporal gyrus. The three longitudinal occipital gyri are delimited posteriorly by the transverse occipital sulcus, which, with the lunate sulcus, delimits the descending occipital gyrus. The occipital pole lies posterior to the lunate sulcus.

Medial surface

The gyri and sulci of the medial surface are concentrically distributed around the mid-sagittal portion of the corpus callosum (Figure 2.2). The callosal sulcus is the most internal of all sulci and separates the corpus callosum from the cingulate gyrus. Rostrally, the cingulate gyrus continues into the subcallosal gyrus, a small convolution that is separated from the olfactory gyrus and the gyrus rectus by the olfactory sulcus. Posteriorly, the cingulate gyrus arches around and beneath the splenium of the corpus callosum and continues into the parahippocampal gyrus and hippocampus through the isthmus. The cingulate sulcus separates the cingulate gyrus from the medial aspect of the superior frontal gyrus and paracentral lobule. Posteriorly, the subparietal and parieto-occipital sulci separate the cingulate gyrus from the precuneus and cuneus, respectively. Short sulci radiate out from the cingulate sulcus and separate the outermost gyri of the medial surface. Among them are the rostral sulcus between the gyrus rectus and the superior frontal gyrus, the paracentral sulcus between the superior frontal gyrus and the paracentral lobule, and the marginal sulcus between the paracentral lobule and the precuneus.

The paracentral lobule caps the central sulcus medially and continues into the precentral and postcentral gyri on the dorsal aspect of the hemisphere. The caudal part of the medial hemisphere is occupied by the cuneus, which is delimitated dorsally and anteriorly by the parieto-occipital sulcus and ventrally by the calcarine sulcus.

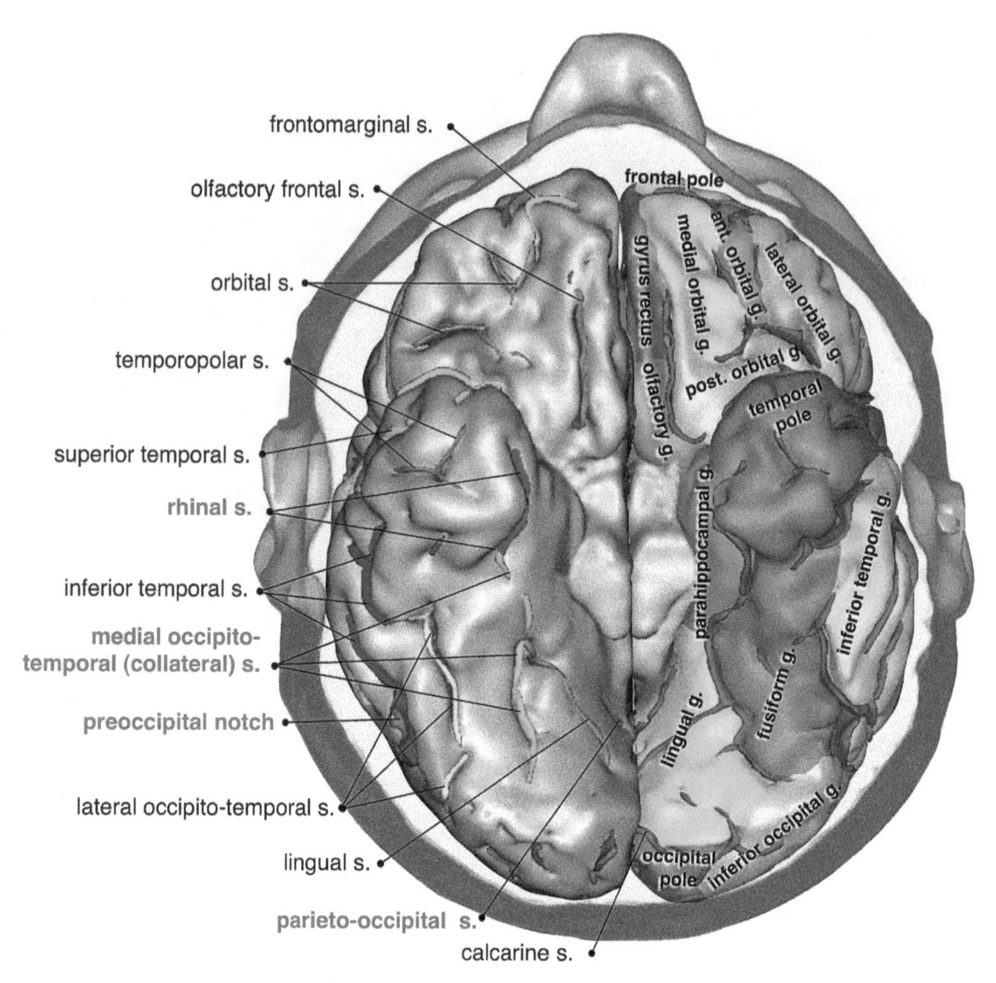

Figure 2.3 3D reconstruction of the brain showing the principal gyri (g.) and sulci (s.) of the ventral surface of the left and right hemispheres. Interlobar sulci are written in red, intralobar sulci in black.

The lingual gyrus, which is located beneath the cuneus and posterior cingulate gyrus, is separated from the fusiform gyrus by the medial occipito-temporal (collateral) sulcus. The lingual gyrus merges anteriorly with the isthmus and continues into the parahippocampal gyrus. The hippocampal and parahippocampal gyri and the temporal pole surround the uncus, a protuberance containing the amygdala.

Ventral surface

The gyri of the ventral surface have a predominantly longitudinal orientation (Figure 2.3). The most anterior aspect of the ventral surface is occupied by the frontal orbital and olfactory gyri. In the orbitofrontal region the orbital sulcus is often in the form of a 'H' or 'X', allowing a separation between anterior, posterior, medial and lateral orbital gyri. Medially, the olfactory frontal sulcus separates the medial orbital gyrus from the gyrus rectus and the olfactory gyrus. The gyrus rectus is a longitudinal convolution that demarcates the border between the medial and ventral surface of the frontal lobe. The frontomarginal sulcus separates the orbital gyri from the polar frontal region.

The posterior two-thirds of the ventral surface is occupied by three longitudinal occipito-temporal gyri. Medially, the parahippocampal gyrus continues posteriorly into the lingual gyrus. Laterally, the inferior temporal gyrus continues into the inferior occipital gyrus. The parahippocampal-lingual and the lateral occipito-temporal gyri merge to form anteriorly the temporal pole and posteriorly the occipital pole. The third convolution, the fusiform gyrus, lies between these two gyral formations. The medial occipito-temporal (or collateral) sulcus separates the lingual from the fusiform gyrus, and frequently continues anteriorly into the rhinal sulcus, between the parahippocampal gyrus and the temporopolar region. The lateral occipito-temporal sulcus separates the fusiform gyrus from the ventral surface of the inferior temporal gyrus. Anteriorly, the temporopolar sulcus forms the border between the fusiform gyrus and the temporal polar region. On the ventral surface only the tip of the calcarine sulcus is visible. In some subjects the calcarine sulcus merges with the lingual sulcus. The preoccipital (or occipito-temporal) notch is visible on the most lateral aspect of the occipital region and, demarcates the border between the inferior temporal gyrus and the ventral surface of the inferior occipital gyrus. Occasionally the preoccipital notch continues into the inferior temporal sulcus.

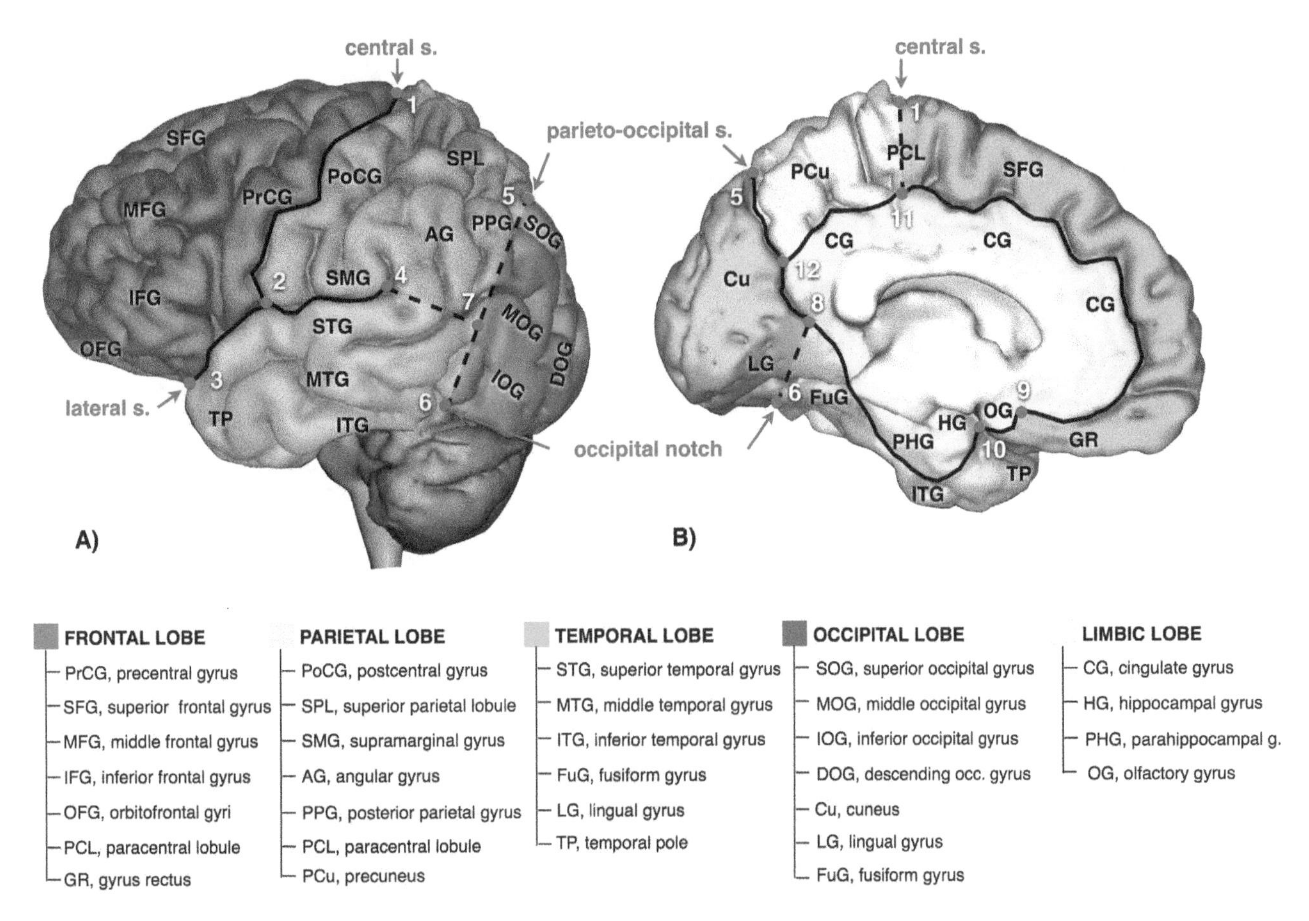

Figure 2.4 Lobar division of the dorsolateral (A) and medial (B) surface of the left cerebral hemisphere. Solid lines represent the interlobar sulci conventionally adopted as anatomical boundaries between lobes. Dashed lines are arbitrary lines of separation that complete the boundaries not made by the sulci (see text). The gyri of each lobe are also indicated.

The cerebral lobes

The cerebral hemispheres can be divided into five lobes according to surface anatomical landmarks (Figure 2.4). These are the frontal, parietal, occipital, temporal, and limbic lobes. Hidden from surface view is a sixth lobe, the insula, which is considered by some authors as belonging to the limbic lobe.

The frontal lobe is the most anterior and is the largest of the human cerebrum. The parietal lobe occupies the space between the frontal and the occipital lobes and lies above the temporal lobe. The occipital lobe is relatively smaller in humans compared to the monkey brain. The limbic lobe, as defined by some authors, consists of a group of structures on the medial surface surrounding the corpus callosum.

On the lateral surface, the central sulcus (from point 1 to point 2 in Figure 2.4A) separates the frontal lobe from the parietal lobe. In some cases the postcentral gyrus continues into the precentral gyrus, and the lower limb of the central sulcus terminates before it reaches the lateral sulcus. In these cases an arbitrary line from the central sulcus perpendicular to the lateral sulcus is used to demarcate the inferior fronto-parietal border.

The lateral sulcus separates the temporal lobe from the frontal lobe (from point 2 to point 3) and, in part, the temporal lobe from the parietal lobe (from point 2 to point 4). The occipital lobe is separated from the temporal and parietal lobes by an arbitrary line between the dorsal tip of the parieto-occipital sulcus and the preoccipital notch (a straight line from point 5 to point 6). The ventral surface of the posterior parietal lobe is separated from the temporal lobe by an arbitrary line that extends from the anterior margin of the occipital lobe to the posterior tip of the lateral sulcus (from point 7 to point 4). This temporo-parietal line is perpendicular to the line that separates the occipital from the parietal and temporal lobes.

On the medial surface (Figure 2.4B), the parieto-occipital, subparietal, cingulate and olfactory sulci demarcate the posterior, dorsal, and anterior borders of the limbic lobe (from point 8 to point 9), whereas the collateral and rhinal sulci delimitate its ventral border (from point 8 to point 10). A vertical line extending from the central sulcus to the cingulate sulcus separates the frontal from the parietal lobe (from point 1 to point 11). The parieto-occipital sulcus divides the parietal from the occipital lobe (from point 5 to point 12). An arbitrary line from the preoccipital notch to the lower tip of the parieto-occipital sulcus demarcates the border between the occipital and temporal lobes (from point 6 to point 8).

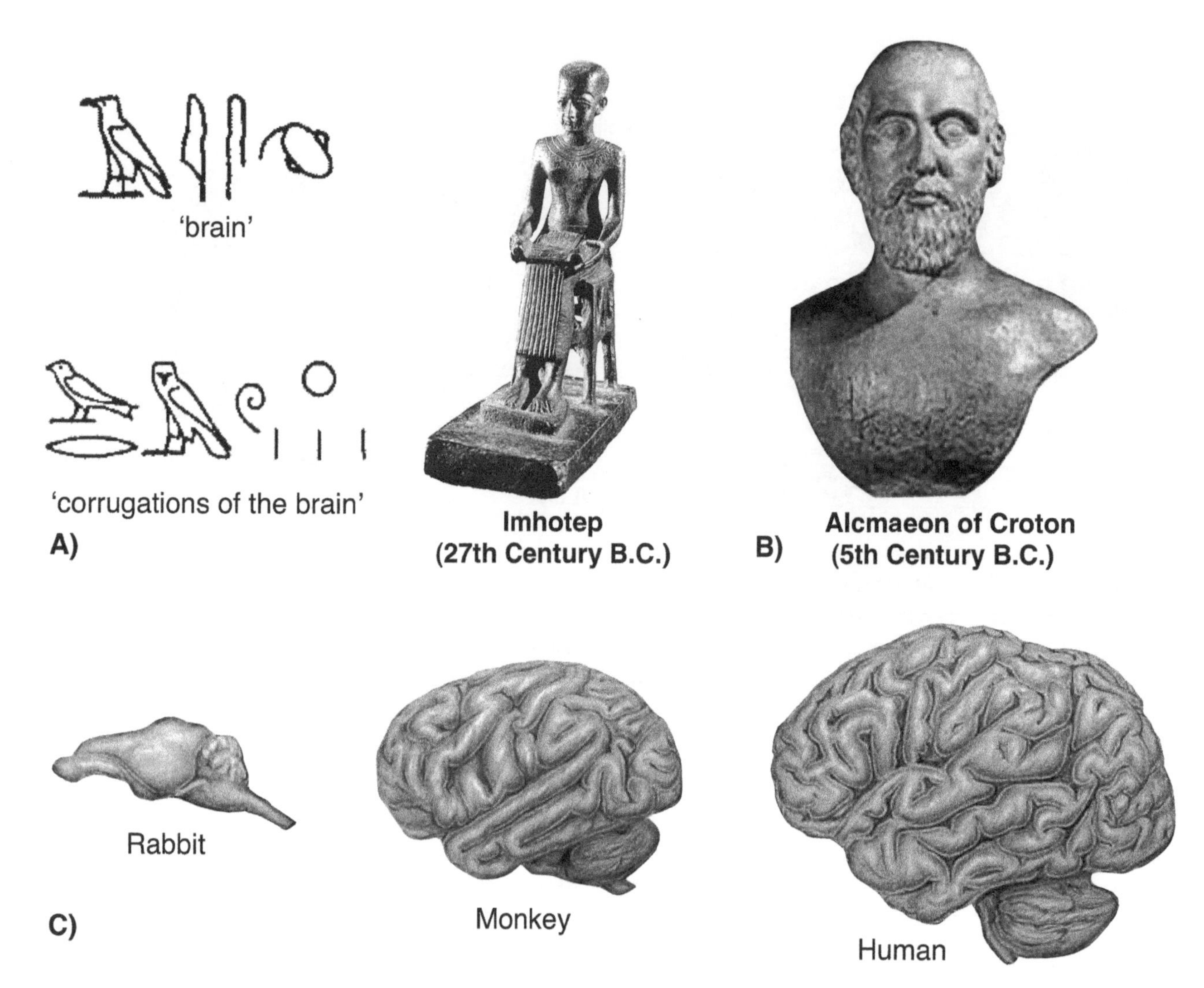

Figure 2.5 The origin of neuroscience and the development of surface neuroanatomy in Egyptian and Greek antiquity. A) Hieroglyphics for 'brain' and 'corrugations of the brain' (i.e. gyri) appeared for the first time in human history in the Edwin Smith Surgical Papyrus, an ancient Egyptian document dated around 17th century BC reporting the teaching of the highest priest Imhotep who lived a thousand years before that time. B) Alcmaeon of Croton was among the first to propose a cephalocentric theory of mind. C) Erasistratus of Alexandria (c.310–250 B.C.) suggested that the superior intellect of humans compared to other animals was related to the complexity of the cerebral convolutions (images of the brains kindly produced by Luca Santanicchia).

From surface neuroanatomy to function

The Edwin Smith surgical papyrus, estimated to have been written in the 17th century BC, is an ancient book on trauma surgery reporting the anatomical observations of numerous injuries, and notes on their treatment and prognosis according to the teaching of the highest priest Imhotep (Breasted, 1930). The hieroglyphics for the word 'brain' and 'corrugations of the brain' appear for the first time in human history in this document (Figure 2.5A). Patient number 20 is of particular interest as he presents with a penetrating wound in the skull and becomes 'speechless' during the palpation of the brain beneath the open fractured bone. This case is probably the first documentation of a correlation between speech arrest and a circumscribed region of the brain that, many centuries later, became recognized as Broca's area (Broca, 1861a). Because Egyptians looked upon the heart, not the brain, as the seat of 'spirit, intellect, passions, motor control, and sensation', Imhotep failed to deduce a general theory of brain–behaviour from his remarkable clinical observations.

The shift from the heart to the brain as the central organ for human cognition and behaviour occurred only in the 6th and 5th century BC and can be found in the writings of the pre-Socratic philosophers. Among them, Alcmaeon of Croton, arguably the first to have performed body dissections, recognized that the sense organs are connected to the brain through the nerves (Figure 2.5B). Alcmaeon's cephalocentric ideas are known to have influenced later philosophers such as Pythagoras, Plato, Herophilus, Erasistratus, Rufus, and Galen, who chose the brain as being central to cognition. Others such as Empedocles, Democritus, Aristotle, Diocles, the Stoics, and the Epicureans continued to favour the heart (Clarke and O'Malley, 1996). In the 3rd century BC Erasistratus of Alexandria (c. 310-250 BC) drew attention to the cerebral convolutions and speculated on possible functional correlations (Figure 2.5C). He observed that the brain is:

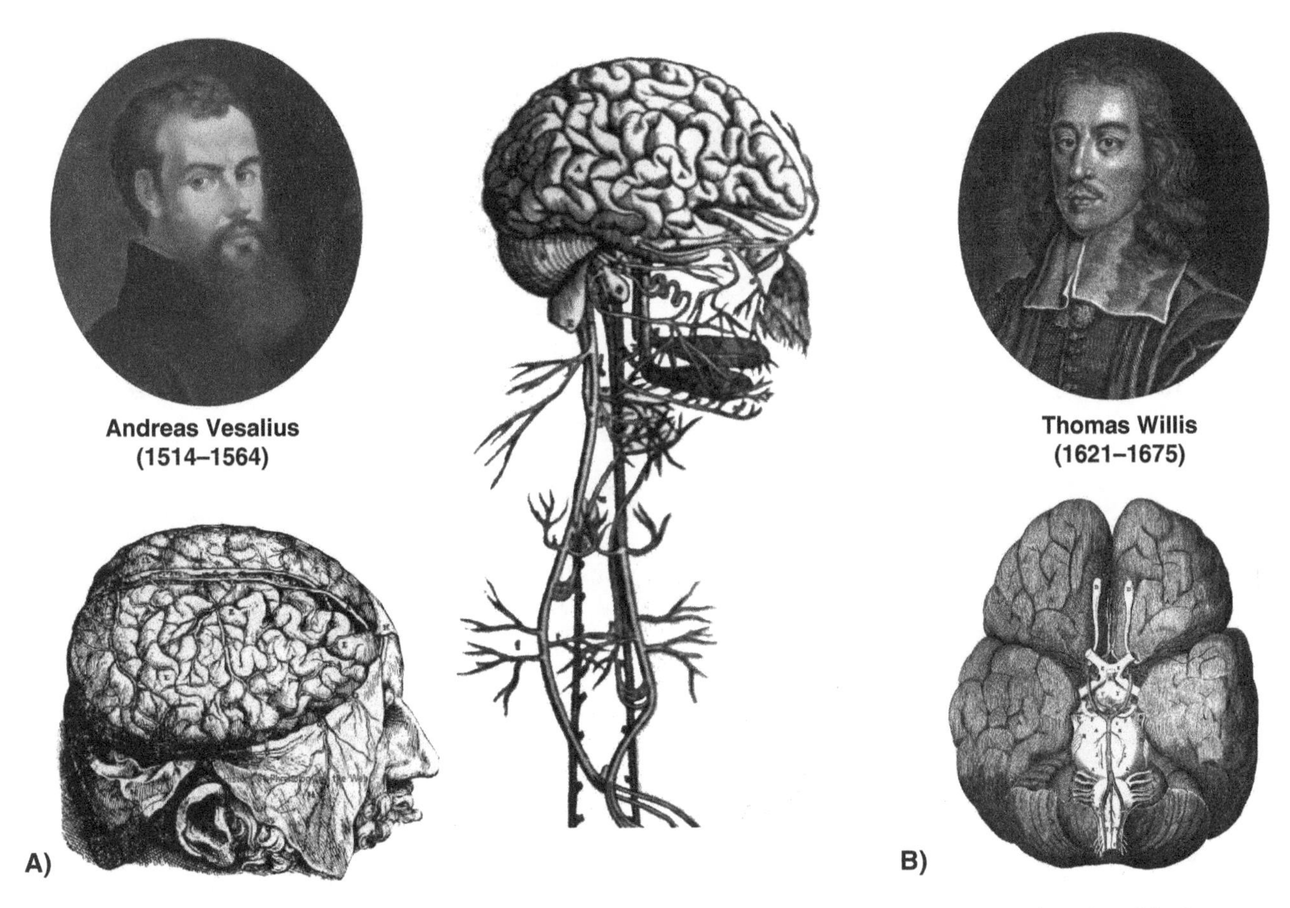

Figure 2.6 Convolutions as a mass of irregular foldings without any distinct pattern in the illustrations of Renaissance anatomists. A) Drawings of the dorsolateral gyri of the human brain from *De humani corporis fabrica* (1543) by Andreas Vesalius (portrait adapted with permission of Musée des Beaux-Arts, Orleans). B) Drawing of the ventral convolutions of the human brain by Christopher Wren for Thomas Willis' *Cerebri Anatome* (1664). (Images from the Wellcome Library, London)

...like the jejunum and very much folded. From this the observer may learn that in those animals that surpass the others in speed of running, such as the stag and hare, well constructed with muscles and nerves for this, so also, since man greatly surpasses other beings in intelligence, his brain is greatly convoluted. (Translated in Clarke and O'Malley, 1996)

This idea represents the first attempt to correlate some general features of brain morphology with complexity of animal behaviour and intelligence. Unfortunately, Erasistratus' idea was rejected by Galen (129–199 AD) in favour of the ventricular theory. Galen observed:

...even donkeys have a very complex brain, although it would seem that, relative to their lack of intelligence, it ought to have been wholly simple and uniform. It would be better to consider intelligence (whatever that may be) as dependent upon a proper portion of the substance, not upon complexity of composition. (Translated in Clarke and O'Malley, 1996)

Galen's doctrine became a dogma, which anatomists believed without feeling the need to prove its validity experimentally (Manzoni, 1998). It is only in the Renaissance that people began to perform dissections and return to the anatomical study of the brain. In Italy, Andreas Vesalius (Figure 2.6A) was among the most active dissectors and critics of Galen's teaching. Vesalius attributed a mechanistic role to convolutions, that of anchoring the vessels to the surface of the brain for a deeper penetration of the nourishing substances into the central regions:

...the convolutions reveal the great ingenuity of the Creator who formed them not otherwise than for the nourishment of the substance of the brain. If the substance of the brain had been continuous without all that folding of membranous fibres, it would not have been sufficiently firm for the distribution through it of veins and arteries as in the other parts of the body... Foreseeing this, nature impressed those sinuous foldings throughout the substance of the brain, so that the thin membrane, filled with numerous vessels, could insert itself into the substance of the brain and so very dexterously administer nourishment. (Vesalius, 1543)

Another great Renaissance anatomist, Thomas Willis (Figure 2.6B), postulated even more complex functions for the cerebral convolutions:

In addition, these twistings of the brain more suitably hide and preserve all the blood vessels that are very slender and smooth and twisted together in various plexuses; if they were distributed externally and uncovered, they would be exposed to very frequent injury... But a no less important reason and necessity for the twistings in the brain arises from the distribution of the animal spirits. Since for the various act of imagination and memory the animal

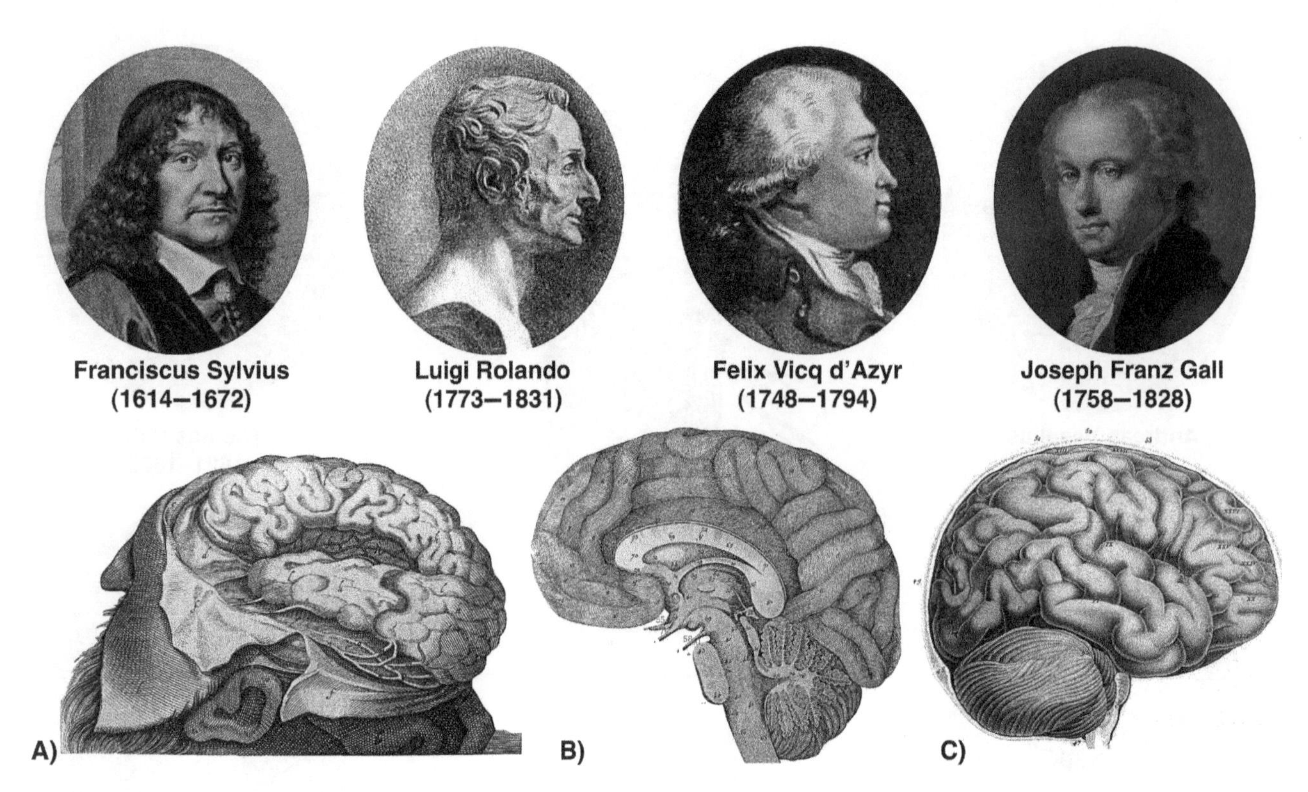

Figure 2.7 After the Renaissance, anatomists recognized distinct sulci and convolutions of the human brain. Franciscus Sylvius described the lateral fissure and Luigi Rolando (1829) the central sulcus (also known as Rolandic sulcus). Others such as Felix Vicq d'Azyr and Joseph Franz Gall hypothesized functional specialization for distinct groups of gyri. A) Figure from *Institutiones Anatomicae* (1641) by Caspar Bartholin, who named the lateral fissure after Sylvius (Sylvius was teaching his students about the lateral fissure but decided to write on it in 1663 only after being credited for the discovery). B) Drawing from Vicq d'Azyr (1786) showing the labelling of different gyri of the brain surface according to a numerical order. C) Gyral division of the lateral surface of the human brain from the *Anatomie et Physiologie du Systeme Nerveux* by Gall and Spurzheim (1810). (Images from the Wellcome Library, London)

spirits must be moved back and forth repeatedly within certain distinct limits and through the same tracts or pathways, therefore numerous folds and convolutions of the brain are required for these various arrangements of the animal spirits; that is, the appearance of perceptible things are stored in them, just as in various storerooms and warehouses, and at given times can be called forth from them. Hence these folds or convolutions are far more numerous and larger in man than in any other animal because of the variety and number of acts of the higher faculties… (Willis, 1664)

Willis combined Vesalius' mechanistic view with Erasistratus' comparative reasoning. Most importantly, his idea of the foldings as 'storerooms' and 'warehouses' of memories can be seen as the forerunner of later localizationist theories of cognitive functions. Willis' anatomical descriptions of the surface anatomy were, however, very general and he failed to identify possible distinct features of the cerebral convolutions.

The first separation of the cerebral gyri into a superior and inferior group was made by a contemporary of Willis, Franciscus Sylvius (François Deleboe), who in 1663 described the lateral fissure that later became know as the Sylvian fissure (Figure 2.7A):

All the surface of the cerebrum is deeply marked by gyri similar to convolutions of small intestine and especially be a distinct fissure or hiatus that begins at the roots of the eyes. It runs posteriorly above the temples as far as the origins of the brain stem. It divides the cerebrum into an upper, larger part and a lower smaller part. (Sylvius, 1663)

After Sylvius, anatomists continued to subdivide the convolutions further into smaller groups and identify anatomical landmarks on the cerebral surface. Vicq d'Azyr, for example, concerned himself with the size of convolutions and their interhemispheric asymmetry, defining anterior, middle, posterior, and inferior groups of convolutions (Figure 2.7B). Vicq d'Azyr, also described the central sulcus, the precentral and postcentral convolutions and the gyri of the insula (Vicq d'Azyr, 1786). The identification of distinct cerebral convolutions was the foundation of one of the most influential and controversial theories in the history of neuroscience: organology. In the early 18th century, Franz Joseph Gall (1758–1828) developed the idea that the brain is the organ of the mind and itself is made up of multiple 'organs', to be identified with the convolutions (Figure 2.7C):

The convolutions, as far as they constitute an organ, receive their fibers from different regions... These fibres or fibre bundles have a constant and uniform direction, different however in each region; they form their own expansions and their own convolutions; they develop at different stages of life; their number varies greatly in different kinds of animal. It is true that all these parts are connected to each other, but do these connections prove that each is not an independent organ?... This is by no means the case; each organ is independent and acts by itself by the virtue of its own powers and it contains directly within itself the proximate cause of the phenomena which it offers. (Gall and Spurzheim, 1810)

Gall then argued that mental faculties or 'propensities' are located within the cerebral organs, and that a particularly well-developed

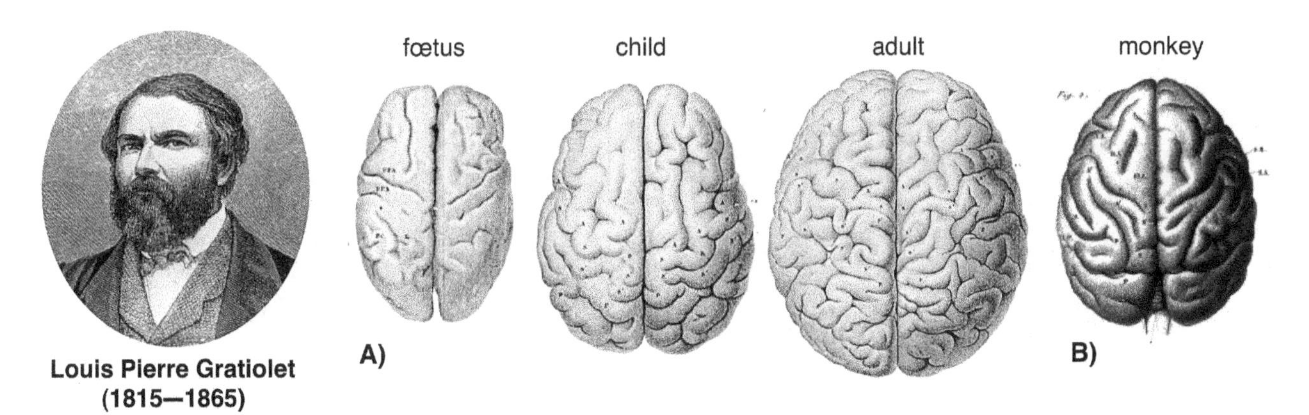

Giuseppe Marco Fieschi (1790–1836) was considered as a famous anarchist by his contemporaries but he was probably best known for his criminal activities. He was among the conspirators that attempted on the life of King Louis-Philippe of France. In Paris together with two members of the Société des Droits de l'Homme, he built a "machine infernale", an unorthodox weapon consisting of twenty gun barrels to be fired simultaneously. On July 28, 1835 Fieschi discharged his machine while Louis-Phillipe was passing from Place de la Republique to the Bastille. A ball grazed the king's forehead but he survived miraculously. Eighteen people were killed and many were wounded; Fieschi himself was severely injured and tried to escape. He was condemned and executed by guillotine the 19th February 1836. Leuret and Gratiolet were asked to perform the post-mortem examination of Fieschi's brain and they observed: *"[...] an advanced dolichocephalic brain where the length is one fourth longer than the width[...] The regions beneath the Sylvian sulcus are larger than the regions above it; the convolutions, though large enough, are less complex and overall less sinuous [...] These findings confirm and discredit the phrenological theories at the same time."*

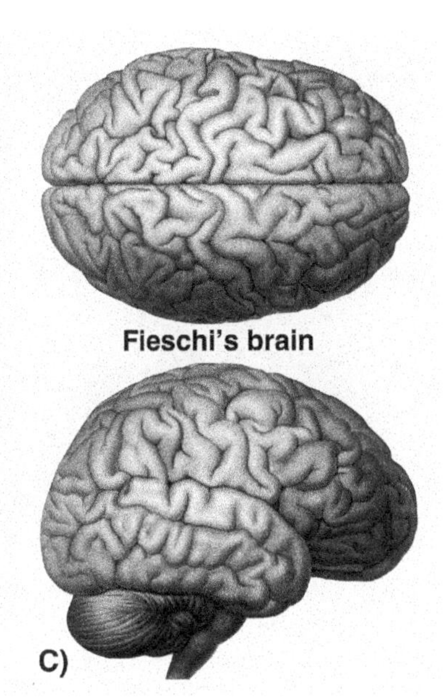

Figure 2.8 Louis Pierre Gratiolet, together with his teacher François Leuret, proposed the first division of the cerebral hemispheres into lobes and promoted comparative surface anatomy studies of the brain, (A) across the lifespan and (B) between species. C) They also performed the autopsy of the famous criminal Fieschi and tried to identify possible anatomical explanations for his 'psychopathy'. This marked the beginning of forensic neuroanatomy in the modern era. Figures from Leuret and Gratiolet (1857). (Images from the Wellcome Library, London)

faculty requires a well-developed cortical organ. He concluded that well-developed organs leave an impression on the skull and, therefore, are indirectly measurable as protrusions on the external cranium:

The gyri or convolutions do not all adopt the same direction. Some have a straight course from before backwards, others run transversely from above to the side, and others again have an oblique course... All the forms of the main divisions, when these latter have developed considerably, declare their presence on the skull under the same type... This explains the relation or the correspondence, which exists between craniology [i.e. palpation of the head] and organology [the localized organs on the cerebral surfaces], or the doctrine of the functions of the brain (cerebral physiology). (Gall and Spurzheim, 1810)

Gall, like his predecessors, used comparative anatomy to support his theory, but took a step forward by identifying a subgroup of cerebral convolutions, those of the frontal lobe, as the most evolved part of the brain in the phylogeny scale:

...until one reaches the brain of man which, in the anterior-superior, and superior-anterior frontal region is endowed with parts of the brain which are denied to the other animals; by which man enjoys higher qualities and faculties of judgement and moral sense. (Gall and Spurzheim, 1810)

Of course, his use of external cranial features as an indirect measure of organology and 'cerebral physiology' was flawed and the system fell into disrepute. Nevertheless Gall's ideas had far-reaching influence and later anatomists began to realize that cerebral convolutions are not distributed in an indiscriminate manner and that they ought to be studied in detail across individuals and species. Furthermore, Gall put forward specific hypotheses about localization of certain mental faculties that clinicians could investigate, this time with a more scientifically sound approach, the clinico-anatomical correlation method. In France, Francois Leuret (1797–1851) and his successor Louis Pierre Gratiolet (Figure 2.8), building on Gall's legacy made an outstanding contribution to the nomenclature of sulci and gyri. They separated those gyri that in the phylogenetic scale appear first (primary gyri) from those which are

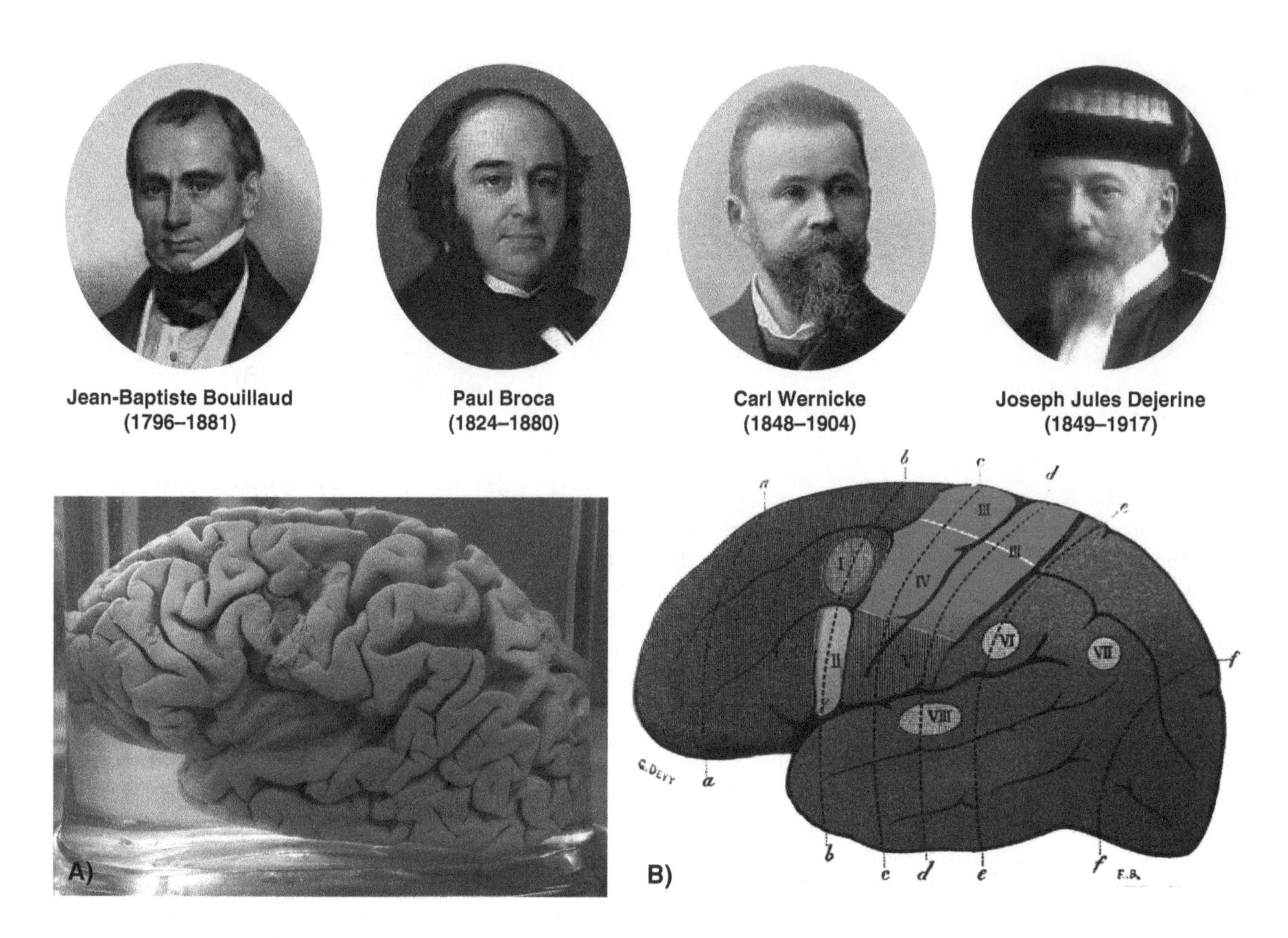

Figure 2.9 Surface anatomy in the 19th century and the rise of cortical localizationism. A) The brain of Monsieur Lelong, the second patient of Paul Broca, who presented with speech difficulties due to a lesion in the posterior third of the left inferior frontal convolution. The brain is preserved at the Dupuytren museum in Paris (image reproduced with kind permission from Musée Dupuytren, Université Pierre et Marie Curie, Paris). B) Drawing of the cerebral centres in the human brain dedicated to higher cognitive functions (from Testut, 1897): I) writing centre of Exner; II) Broca's centre for speech; III) motor centre, lower limb; IV) motor centre, upper limb; V) motor centre, face and tongue; VI–VII) centre for reading proposed by Dejerine; VIII) Wernicke's acoustic centre for verbal comprehension. Blue and purple areas are zones of the association centres according to Flechsig (1901). (Image of Joseph Jules Dejerine from the Wellcome Library, London)

secondary in order of appearance. They also subdivided the brain into five lobes and named them according to the corresponding overlying bones. Their nomenclature of convolutions became very popular and widely used as it enabled to make intra- and interspecies comparisons (Leuret and Gratiolet, 1857). They conducted comparative studies of the gyri following three distinct approaches: i) comparative anatomy of the human brain at different stages of development, from foetal life to senescence; ii) comparative phylogeny studies across species; iii) comparative studies of subjects with moral (e.g. criminals) and intellectual (e.g. congenital brain disorders) deviation from the norm. Gratiolet had even gone so far as to classify races according to what he thought was the dominant anatomical feature (i.e. 'Caucasian is the frontal race', 'Mongoloid the parietal race', and 'Negroid the occipital race'). His ideas on the asymmetrical development of the cerebral hemispheres (earlier maturation of the left) have certainly had a longer lasting validity.

Other neuroanatomists concerned themselves with the morphology of individual sulci. Emil Huschke (1797–1858), professor of anatomy and embryology in Jena, for example, observed that:

> *…the position of the Sylvian fissure is not uniformly oblique in all mammals… where in general a vertical position seems to indicate a smaller frontal and a greater parietal brain, and vice versa. (Huschke, 1854)*

William Turner (1832–1916), professor of anatomy in Edinburgh, described for the first time the intraparietal sulcus and established that the fissure of Rolando was to be considered the posterior limit of the frontal lobe, and not the precentral sulcus as originally proposed by Gratiolet (Turner, 1866). Others focused their activity on Gall's most significant insight into the correlation between convolutions of the frontal lobes and the faculty of language (Gall and Spurzheim, 1810). This idea had a deep influence on Jean-Baptiste Bouillaud (Figure 2.9) and his son-in-law Simon Alexandre Ernest Auburtin (1825–1865), who were among the most fervent supporters of speech localization (Bouillaud, 1825). Bouillaud and Auburtin had the great merit of turning away from cranioscopy and using the clinico-pathological correlation method. They not only collected a considerable number of brains of aphasic patients but they were able to provide the most extraordinary evidence of

OUTSTANDING BRAINS OF CRIMINALS AND GENIUSES

In 1882 Edward Charles Spitzka, a young american neurologist testified as expert witness at the trial of Charles J. Guiteau, the assassin of the American President James A. Garfield. Spitzka was asked to examine the prisoner and formulate his judgement on the mental state of Guiteau. Spitzka gave a vigorous and passionate testimony of the insanity of Guiteau, who nevertheless was convicted and hanged. Despite the presence of pathological signs indicative of syphilis at the autopsy, the opinion of the public and the experts converged towards a diagnosis of hereditary insanity and the sentence to death was deemed as unjust. In 1901 the American President William McKinley was assassinated by the anarchic Leon F. Czolgosz. Czolgosz was condemned to death and electrocuted. Spitzka's son (Edward Anthony Spitzka) was asked this time to conduct the autopsy on Czolgosz. Unlike for Guiteau the autopsy revealed no abnormalities of structure and insanity was ruled out (Haines 1995). Edward Anthony Spitzka continued publishing a long series of papers containing detailed drawings of the brain of distinguished men of philosophy, art and science. Among them Rene Descartes, Karl Friedrich Gauss, William Osler, and Hermann Helmholtz. Spitzka became the editor of three american editions of the Gray's anatomy and director of the Daniel Baugh Institute of Anatomy. His work revitalized and popularized in the United States the anatomical approach to 'outstanding' minds, whose brains were studied to understand deviation from the norm.

Leon F. Czolgosz

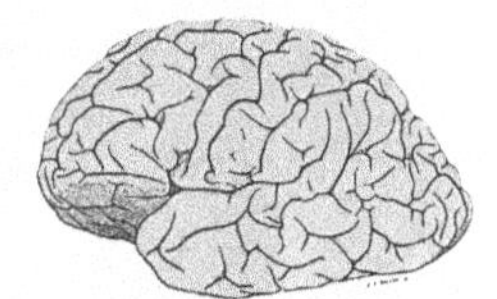

Czolgosz's brain

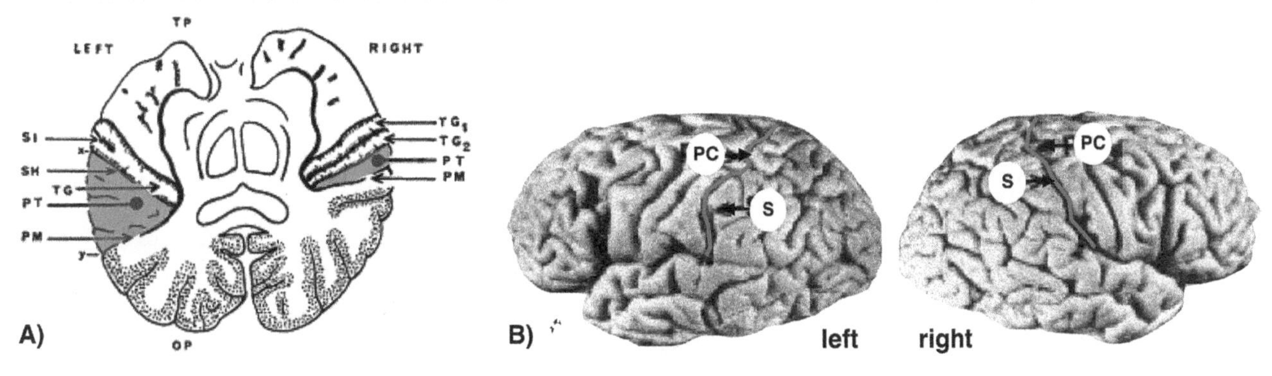

Figure 2.10 The call for outstanding brains in the 20th century and anatomical studies of asymmetry. A) Asymmetry of the planum temporale (red area) was demonstrated by Geschwind and Levitsky in 1968 and associated with language lateralization in man (Geschwind and Levitsky, 1968). B) Albert Einstein died in 1955 and donated his brain for post-mortem studies. The examination showed the confluence of the post-central sulcus (PC) with the posterior ascending limb of the lateral suclus (S), the anterior location of the posterior ascending limb of the lateral suclus and the symmetry between the two hemispheres. These anatomical features of the parietal lobes are not frequently found in the general population and have been linked to Einstein's exceptional visuospatial and mathematical abilities. (Modified from Witelson et al., 1999)

speech localization in a living human brain. At the meeting of the Société d'Anthropologie in April 1861, Auburtin described the case of Monsieur Cullerier who had shot himself in the head and was admitted to hospital with an open wound in his left forehead. The man remained conscious and possessed normal speech, but his anterior brain was exposed and Auburtin had the audacity to apply a light pressure with a blade to the wounded man's frontal lobe:

...the frontal bone was completely removed. The anterior lobes of the brain were exposed, but they were not damaged. Intelligence was intact as well as speech. This unfortunate survived several hours, and we could carry out the following observation. While he was interviewed the blade of a large spatula was placed on the anterior lobes; a light pressure was applied and speech suddenly terminated; a word that had been commenced was cut in two. The faculty of speech reappeared as soon as the compression ceased... (Auburtin, 1861)

Among the audience a young student of Leuret, Paul Broca (Figure 2.9), felt that speech localization promoted by Bouillaud and Auburtin was correct in principle but it needed further experimental support. Broca was an anthropologist, anatomist, and surgeon with access to patients. Just a few days after Auburtin's presentation, Broca had admitted to his service Monsieur Leborgne, a 51-year-old man with a long history of epilepsy and speech difficulties. He presented with gangrene of the right leg and died a week later. Broca took the opportunity of testing the frontal localization of speech at the autopsy. Indeed he found a lesion in the posterior third of the left inferior frontal gyrus. Broca presented his work to the Société d'Anthropologie the same year and he continued to observe cases of patients with verbal fluency deficits associated with lesions to the region of the brain that subsequently was named after him (i.e. Broca's area) (Figure 2.9A) (Broca, 1861a).

Soon after other centres for language were identified in the left hemisphere. Carl Wernicke described the centre for understanding words (Wernicke, 1874), Joseph Jules Dejerine the 'reading centre' (Dejerine, 1892), and Sigmund Exner (1846–1926) the center for writing (Figure 2.9B) (Exner, 1881). In England, John Hughlings Jackson (1835–1911) drew attention to the functions of the right hemisphere (Jackson, 1876) and others argued for the existence of right localized centres, such as the 'emotion centre' (Luys, 1881) or the 'geography centre' (Dunn, 1895).

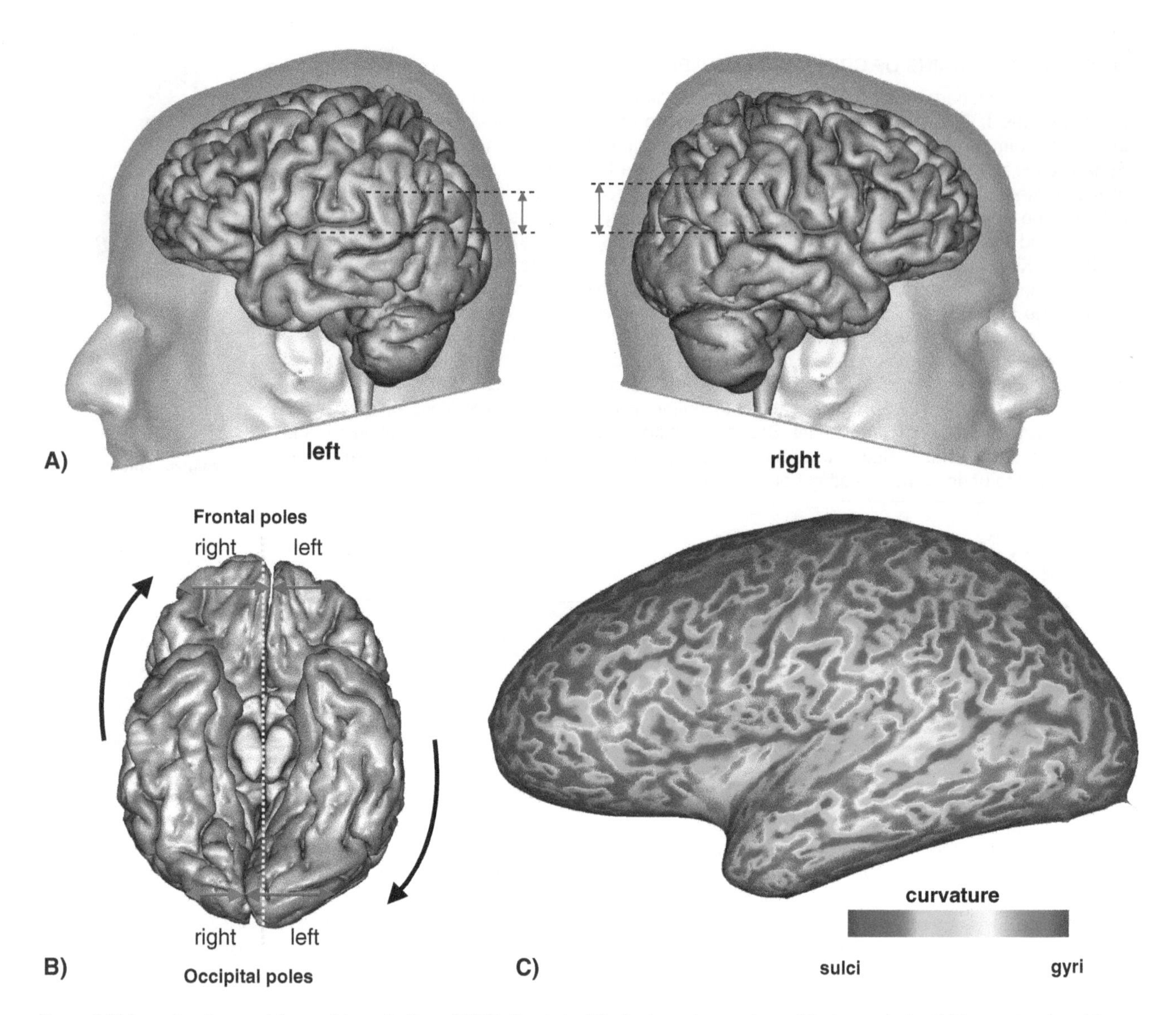

Figure 2.11 Examples of some of the possible applications of MRI to the study of the *in vivo* surface anatomy of the human brain. A) 3D reconstruction of the living human brain for quantitative morphometric studies of sulci and gyri. In this figure the course of the posterior branch of the lateral fissure is reconstructed and measured to assess interhemispheric differences. The asymmetry of the lateral fissure has been associated with language lateralization (LeMay, 1976). B) MRI reconstruction of the ventral aspect of the brain showing left–right asymmetry of the occipital and frontal lobes, whose protrusions produce imprints in the inner surface of the skull (e.g. petalia). C) Flat map of the gyral and sulcal patterns of the human brain.

Although localizationist theories received further support from other disciplines, in particular from cortical electrophysiology, the majority of scientists remained less prone to embrace it and concerned themselves with the study of the relationship between general features of the brain (e.g. weight, or complexity of gyral pattern) and intelligence, broadly defined. This idea was championed in France by Marie-Jean Pierre Flourens (1794–1867), who believed that the brain was equipotential (i.e. any part of the cerebral cortex participates in and is able to perform all the functions of the whole) and that there were differences in overall brain size not only across species, but also within humans (Flourens, 1824). To test this theory, neuroanatomists begun to compare brain sizes among individuals from different races, genders, social backgrounds, and cultures. In Europe, Rudolph Wagner (1805–1864) examined the brain of the great astronomist and mathematician Carl Friedrich Gauss and concluded that despite its average weight the brain was truly exceptional in its intricate pattern of sulci and convolutions (Wagner, 1860). However, Wagner could not find the same features in other brains of professors at Göttingen and concluded that neither weight nor complexity of convolutions are correlated with intellectual abilities. Broca himself contributed to the debate and in reply to Wagner's conclusions stated that *'a professorial robe is not necessarily a certificate of genius; there may be, even at Göttingen, some chairs occupied by not very remarkable men'* (Broca 1861b). On the other side of the spectrum, the brains of criminals were considered by anthropologists to be smaller and closer to simian brains. The Italian alienist and criminologist Cesare Lombroso (1836–1909), believed in the existence of two groups of criminals. Those not born criminals, who, by committing a crime, acted against their good hereditary traits, and those born with adverse hereditary traits who are destined to become criminals. He then argued that only the latter group would show distinct anatomical features, including smaller brains. Furthermore, on the basis of this distinction, Lombroso promulgated the idea that people should be judged not just on the basis of their crime but also on the basis of the hereditary traits of their mind (Lombroso, 1911).

These anatomically-rooted debates were of public interest, especially on those occasions where an entire nation was shaken by the most heinous crimes, such as the assassination of an American President. Often neuroanatomists (mostly alienists) were invited to judge on the mental state of the persons accused and examine the possible anatomical abnormalities of their brains after the execution. The members of the American Anthropometric Society were particularly active at the turn of the 19th century. Among them Edward Anthony Spitzka (1876–1922) collected one of the largest series of eminent brains and conducted autopsies on famous criminals (Spitzka, 1907). Spitzka, found anatomical peculiarities in the brains of most criminals but he also warned that it would be dangerous to make broad generalizations from his observations (Figure 2.10).

As the 20th century progressed, the anatomical approach to the human mind continued to fascinate generations of clinicians, and progressively objective quantitative measurements based on post-mortem examination and neuroimaging morphometry replaced anecdotal reports. Norman Geschwind's report on the asymmetry of the planum temporale of the human brain (Figure 2.10A) (Geschwind and Levitsky, 1968) revitalized the field and stimulated the use of neuroimaging methods to study the anatomy of the brain, in particular its asymmetry with regard to the development of language and its disorders (e.g. dyslexia) (LeMay, 1976; Galaburda et al., 1978). In contemporary neuroscience the search for the anatomical basis of cognitive functions is pursued using sophisticated magnetic resonance imaging methods for image acquisition and processing (Figure 2.11). These methods have the advantage of being suitable for studying the living human brain in large groups. Examples of this approach are the finding of a reduced gyrification index in patients with schizophrenia (Kulynich et al., 1997), or bipolar disorder (Penttila et al., 2009). However, the finding that increased gyrification in the parietal lobe correlates with autistic symptoms in children with the disorder (Kates et al., 2009) raises some doubts on whether 'more' or 'bigger' is always 'better'. Today as two thousand years ago the path that may take us from anatomy to function is obviously more complex than it appears at first glance on the surface.

References

Auburtin, S.A.E. (1861). Reprise de la discussion sur la forme et le volume du cerveau. *Bull Soc Anthropol*, **2**, 209–20.

Bartholin, C. (1641). *Institutiones anatomicae, novis recentiorum opinionibus et observationibus, quarum innumerae hactenus editae non sunt, figurisque auctae ab auctoris filio Thoma Bartholino*. Leiden: Lug. Batavorum, Apud Franciscum Hackium.

Bouillaud, J.B. (1825). Recherches cliniques propres à démontrer que la perte de la parole correspond à la lésion des lobules antérieurs du cerveau, et à confirmer l'opinion de M. Gall, sur le siège de l'organe du langage articulé. *Archs Gén Méd*, **8**, 25–45.

Breasted, J.H. (1930). *The Edwin Smith Surgical Papyrus*. Chicago, IL: University of Chicago Press.

Broca, P. (1861a). Perte de la parole, ramollissement chronique et destruction partielle du lobe antérieur gauche du cerveau. *Bull Soc Anthropol*, **2**, 235–8.

Broca, P. (1861b). Sur le volume et la forme du cerveau suivant les individus et suivant les races. *Bull Soc Anthropol*, **2**, 139–207.

Clarke, E. and O'Malley, C.D. (1996). *The Human Brain and Spinal Cord: A Historical Study Illustrated by Writings from Antiquity to the Twentieth Century*. San Francisco, CA: Norman Publishing.

Debernardi, A., Sala, E., D'Aliberti, G., Talamonti, G., Franchini, A.F., Collice, M.(2010). Alcmaeon of Croton. *Neurosurgery*, **66**(2), 247–52.

Dejerine, J. (1892). Contribution a l'étude anatomo-pathologique et clinique des differentes variétés de cécité-verbale. *Mém Soc Biol*, **4**, 61–90.

Dunn, T.D. (1895). Double hemiplegia with double hemianopsia and loss of geographical center. *Trans Coll Physic Philad*, **17**, 45–55.

Exner, S. (1881). *Untersuchungen über die Localization der Functionen in der Grosshirnrinde des Menschen*. Wien: Wilhelm Braumüller.

Flechsig, P.E. (1901). Developmental (myelogenetic) localisation of the cerebral cortex in the human brain. *Lancet*, **2**, 1027–9.

Flourens, M-.J.P. (1824) Recherches expérimentales sur les propriétés et les fonctions du système nerveux dans les animaux vertébrés. Paris: J.B. Ballière.

Galaburda, A.M., LeMay, M., Kemper, T.L., and Geschwind, N. (1978). Right-left asymmetries in the brain. *Science*, **199**(4331), 852–6.

Gall, F. J. and Spurzheim, J.C. (1810–1819). *Anatomie et physiologie du système nerveux en général, et du cerveau en particulier*. Paris: Schoell.

Geschwind, N., and Levitsky, W. (1968). Human brain: left-right asymmetries in temporal speech region. *Science*, **161**(837), 186–7.

Haines, D.E. (1995). Spitzka and Spitzka on the brains of the assassins of presidents. *J History Neurosci*, **4**(3–4), 236–66.

Henneberg, R. (1910). Messung der Obergflachenausdehnung der Grosshirnrinde. *J Psychol Neurol*, **17**, 144–58.

Huschke, E. (1854). *Schädel, Hirn und Seele des Menshen und der Thiere nach Alter, Geschlecht und Raçe*. Jena: Mauke.

Jackson, J.H. (1876). Case of large cerebral tumour without optic neuritis and with left hemiplegia and imperception. *Ophthal Hosp Rep*, **8**, 434–44.

Kates, W.R., Ikuta, I., and Burnette, C.P. (2009). Gyrification patterns in monozygotic twin pairs varying in discordance for autism. *Autism Res*, **2**(5), 267–78.

Kulynich, J.J., Luevano L.F., Jones D.W., and Weinberger, D.R. (1997) Cortical abnormality in schizophrenia: an in vivo application of the gyrification index. *Biol Psychiatry*, **41**(10), 995–9.

LeMay, M. (1976). Morphological cerebral asymmetries of modern man, fossil man, and nonhuman primate. *Ann N Y Acad Sci*, **280**, 349–66.

Leuret, F. and Gratiolet, P. (1857). *Anatomie comparée du système nerveux considéré dans ses rapports avec l'intelligence (Vol. 2)*. Paris: Ballière.

Lombroso, C. (1911). *Crime: Its Causes and Remedies*. Boston, MA: Little, Brown.

Luys, J.B. (1881). *Traité clinique et pratique des maladies mentales*. Paris: Delahaye et Lecrosnier.

Manzoni, T. (1998). The cerebral ventricles, the animal spirits and the dawn of brain localization of function. *Arch Ital Biol*, **136**(2), 103–52.

Ono, M., Kubik, S., and Abernathey, C.D. (1990). *Atlas of the cerebral sulci*. Stuttgart: Thieme.

Penttila, J., Cachia, A., Martinot, J.L., et al. (2009). Cortical folding difference between patients with early-onset and patients with intermediate-onset bipolar disorder. *Bipolar Disord*, **11**(4), 361–70.

Rolando, L. (1829). *Della Struttura degli Emisferi Cerebrali*. Memorie della Regia Accademia delle Scienze di Torino, **35**, 103–45.

Spitzka, E.A. (1907). A study of the brains of six eminent scientists and scholars belonging to the American Anthropometric Society together with a description of the skull of Professor E. D. Cope. *Trans Am Philosoph Soc, New series*, **21**, 175–308.

Sylvius, F. (1663) Disputationes medicarum pars prima, primarias corporis humani functiones naturales ex anatomicis, practicis et chymicis experimentiis deductas complectens. Amsterdam: J. van den Bergh.

Testut, L. (1897). *Traité d'anatomie humaine*. Paris: O. Doin.

Turner, W. (1866). *The convolutions of the human cerebrum topographically considered*. Edinburgh: Maclachlan & Stewart.

Vesalius (1543). *De humani corporis fabrica libri septem*. Basel: Oporinus.

Vicq d'Azyr, F. (1786). *Traité d'anatomie et de physiologie*. Paris: Didot l'Aine.

Wagner, R. (1860). *Vorstudien zu einer wissenschaftlichen Morphologie und Physiologie des menschlichen Gehirns als Seelenorgan*. Göttingen: Dieterichschen Buchhandlung.

Wernicke, C. (1874). *Der Aphasische Symptomencomplex. Ein psychologische Studie auf anatomischer Basis*. Breslau: Cohn & Weigert.

Willis, T. (1664). *Cerebri anatome*. London: Martyn & Allestry.

Witelson, S.F., Kigar, D.L., and Harvey, T. (1999). The exceptional brain of Albert Einstein. *Lancet*, **353**(9170), 2149–53.

CHAPTER 3

Sectional Neuroanatomy

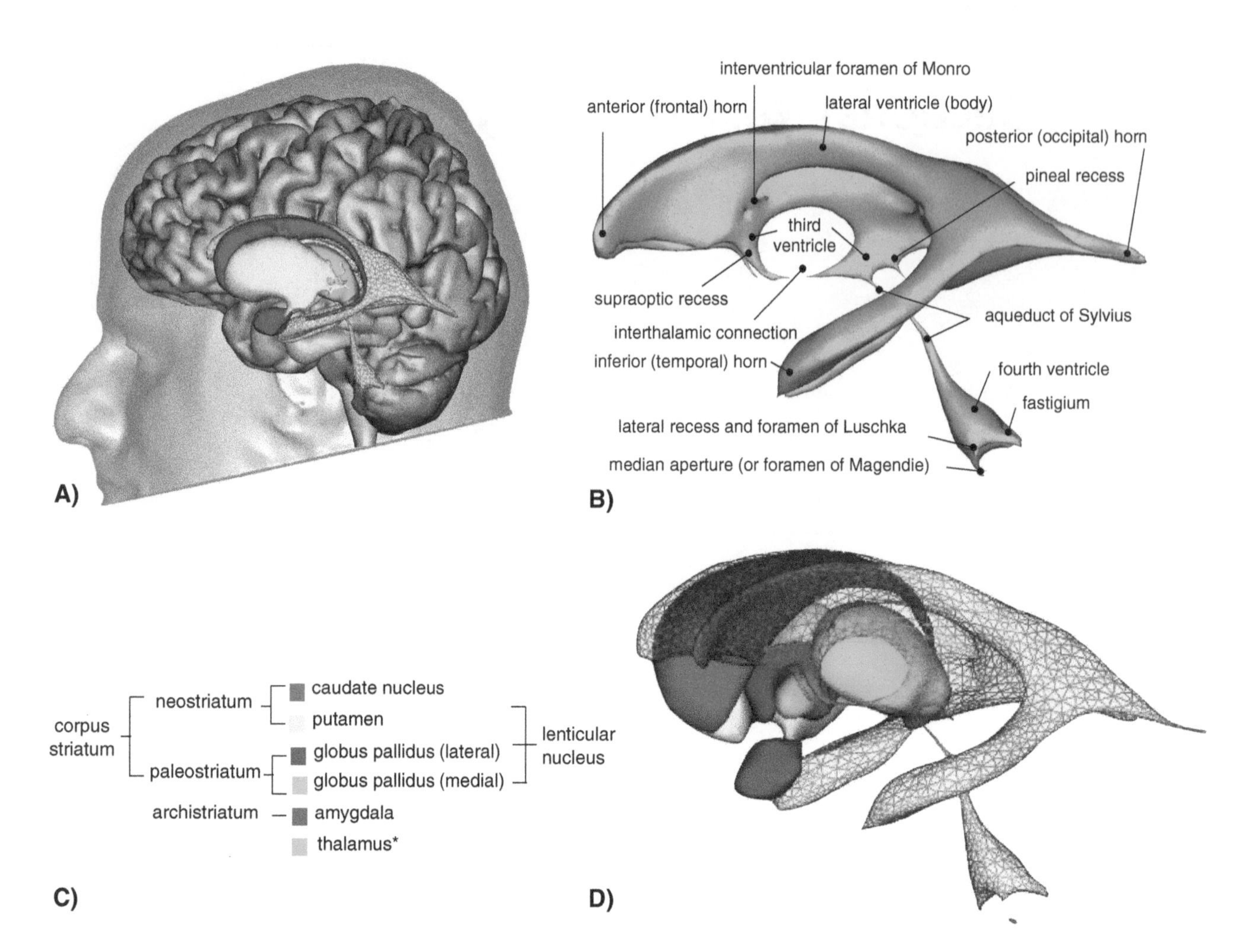

Figure 3.1 The anatomy of the ventricular system and deep grey nuclei reconstructed with MRI. A) Left lateral view of the ventricular system and deep grey nuclei. B) Left lateral view of the ventricular system. Note that the lateral ventricles, the interventricular foramen of Monro, the lateral recess, and the foramen of Luschka are bilateral structures. C) Nomenclature for the nuclei of the basal ganglia and thalamus (colours correspond to the image in A and D). *The thalamus is not part of the basal ganglia. D) Left antero-medial view of the thalamus and basal ganglia of the right hemisphere. The ventricular system is also represented in light blue.

Introduction

The method of sectioning the brain at different levels along the three orthogonal planes reveals anatomical structures that are otherwise hidden from external inspection. These include the ventricular system, the deep grey nuclei, and the white and grey matter. The ventricular system consists of several cavities filled with cerebrospinal fluid (CSF), communicating through small ducts. The lateral ventricles represent the largest cavities of the cerebral hemispheres and surround a group of grey matter agglomerates that form the subcortical telencephalic and diencephalic nuclei. These nuclei are separated and surrounded by many fibres that form the white matter of the cerebral hemispheres. The outer aspect of the cerebral white matter is delimitated by the cerebral cortex. Each of these structures will be described in some detail below.

Ventricular system

The cerebral ventricles consist of a series of interconnected spaces located at the centre of the brain (Figures 3.1 and 3.2). It is possible to distinguish two lateral ventricles within each cerebral hemisphere, a third ventricle between the two thalami, and a fourth ventricle posterior to the brainstem. The lateral ventricles divide into an anterior (frontal) horn, a body, a posterior (occipital) horn, and an inferior (temporal) horn (Figure 3.1B). The two lateral ventricles are separated by a series of midline structures, the genu of the corpus callosum anteriorly, then behind that by the septum pellucidum, and the splenium of the corpus callosum most posteriorly (Figure 3.2, axial slice number 7). The body of the corpus callosum forms the roof of the central part of each lateral ventricle. The caudate nucleus and the white matter of the corona radiata form the lateral walls of the lateral ventricles, whereas the hippocampus and the amygdala demarcate the medial wall of the inferior horn. The lateral ventricles communicate with the third ventricle via the two interventricular foramina of Monro (Figure 3.1B).

The third ventricle is a slit-like space located along the midline. It is bounded laterally by the thalami and inferiorly by the subthalamus and hypothalamus. Anteriorly are the lamina terminalis, fornix, and anterior commissure. The small supraoptic recess extends antero-ventrally just above the optic chiasm. Postero-dorsal is the pineal recess at the base of the pineal body. In a proportion of

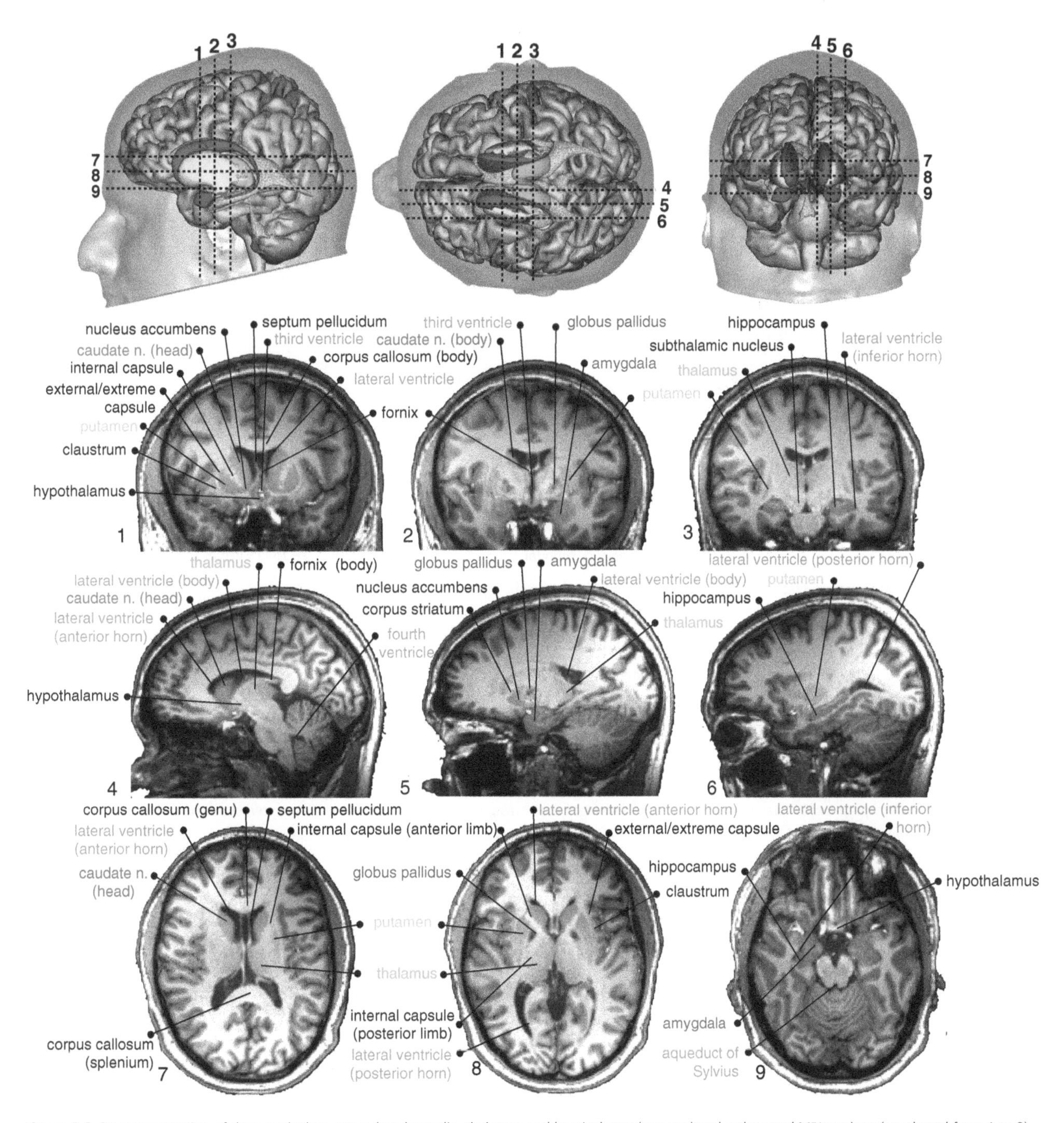

Figure 3.2 3D reconstruction of the ventricular system, basal ganglia, thalamus and hemispheres (top row) and orthogonal MRI sections (numbered from 1 to 9). The colours of the labels correspond to the colours of the 3D structures on the top row. The position of the coronal (images 1-3), sagittal (images 4-6) and axial (images 7-9) slices is indicated by the dashed lines in the top panel.

brains, the third ventricle is crossed by a small mass of neural tissue between the two thalami (the interthalamic connection).

The third ventricle continues into the fourth ventricle through the cerebral aqueduct of Sylvius. The fourth ventricle is a rhomboid-shaped cavity delimited by the brainstem anteriorly and the cerebellum posteriorly. The fourth ventricle narrows into the sagittal fastigium posteriorly and into the lateral recesses laterally. Caudally it continues into the central canal of the spinal cord. The fourth ventricle communicates with the subarachnoid space surrounding the cerebellum through a median aperture (foramen of Magendie) and the two lateral foramina of Luschka.

Subcortical telencephalic and diencephalic nuclei

Above the brainstem, the prosencephalon (or forebrain) is customarily divided into a caudal part, the diencephalon, and a rostral part, the telencephalon (i.e. cerebral hemispheres). Both diencephalon and telencephalon contain nuclei of grey matter located deeply within the brain.

The diencephalon

The most voluminous nuclei of the diencephalon are the thalami, ovoid nuclear structures located on both sides of the third ventricle

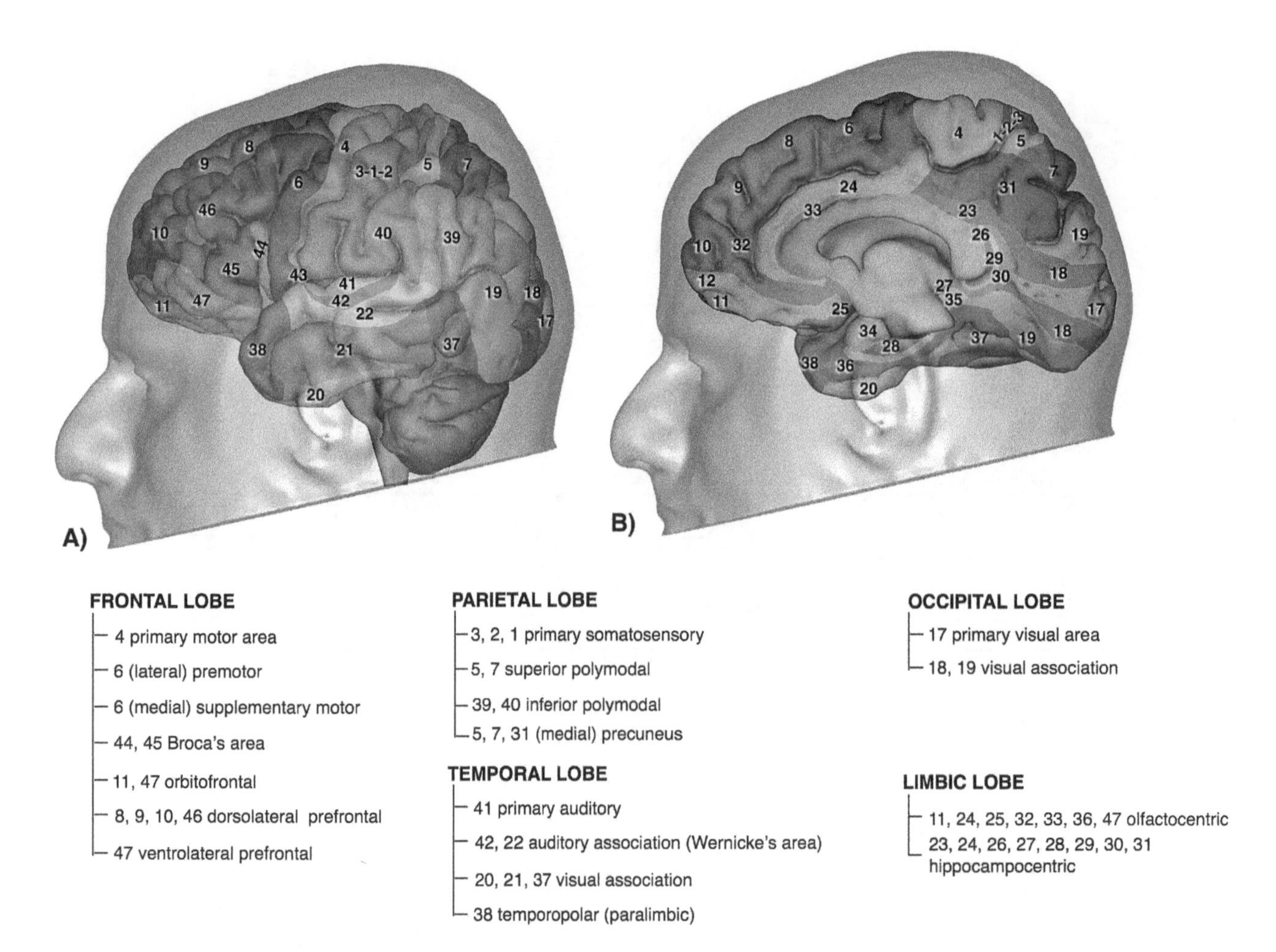

Figure 3.3 Brodmann's maps of the (A) lateral and (B) medial brain surface based on a cytoarchitectonic parcellation of the cerebral cortex. Areas 13–16 and 48–51 are missing in the original work of Brodmann (1909). Area 52 is located on the floor of the lateral sulcus (parainsular region) and is not visible on these maps.

(Figures 3.1 and 3.2). The thalami are divided into several nuclei (see Chapter 10), of which the anterior, medio-dorsal, ventral and lateral groups represent the major subdivisions. Beneath the thalamus lies the subthalamic nucleus of Luys and the hypothalamus, which are also part of the diencephalon.

Basal ganglia

In the telencephalon the basal ganglia are deeply located in the white matter of the cerebral hemispheres anterior to the thalamus and between the anterior and inferior horns of the lateral ventricles. They consist of two major functional divisions, the striatum and globus pallidus. The striatum is composed of two interconnected masses, the putamen and caudate nucleus. The globus pallidus is divided into: globus pallidus externus (or lateralis) and globus pallidus internus (or medialis). The putamen and globus pallidus together occupy a lens-shaped region (lenticular nucleus) lateral to the internal capsule, with the globus pallidus the more medial of the two. The caudate nucleus forms an arching bulge in the lateral wall of the lateral ventricle anteriorly, with its main mass located medial to the internal capsule. It has a large anterior head, a narrower body, and a long thin tail. The body curves over the thalamus and the narrow tail continues downwards and forwards in the temporal lobe along the wall of the lateral ventricle. Anteriorly and ventrally the putamen and caudate nuclei are continuous around the inferior edge of the internal capsule. The ventral fusion of the caudate and putamen is referred to as nucleus accumbens. From the caudate nucleus a series of small linear sheaths of grey matter radiate into the putamen. These confer a striped aspect to this region, which early anatomists indicated with the term striatum. The caudate and putamen are also designated with the term neostriatum due to their relatively recent appearance in the phylogeny scale. The amygdala is not part of the basal ganglia but it is anatomically and functionally related to them. Comparative anatomists use the terms paleostriatum and archistriatum to indicate the globus pallidus and the amygdala, respectively (Figure 3.1C).

The cerebral cortex

The cerebral cortex is a thin strip of grey matter covering the outermost surface of the cerebral hemispheres. After cutting the cerebral cortex along a plane perpendicular to its surface, it is possible to observe its multi-layered architecture. The layering pattern of the cerebral cortex changes across the brain (most of the cortex is six-layered) and according to this regional variation it can be parcellated into distinct areas (or fields) (Figure 3.3). Variations in

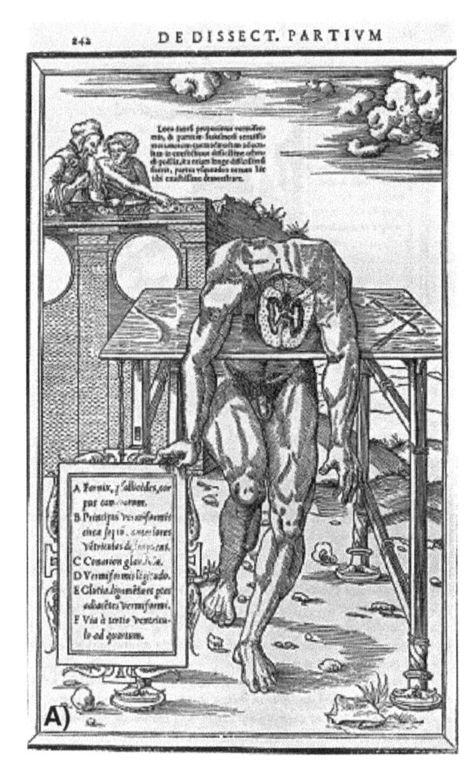

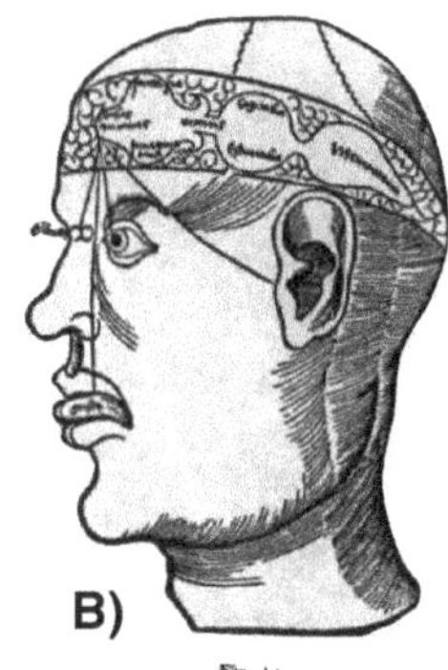

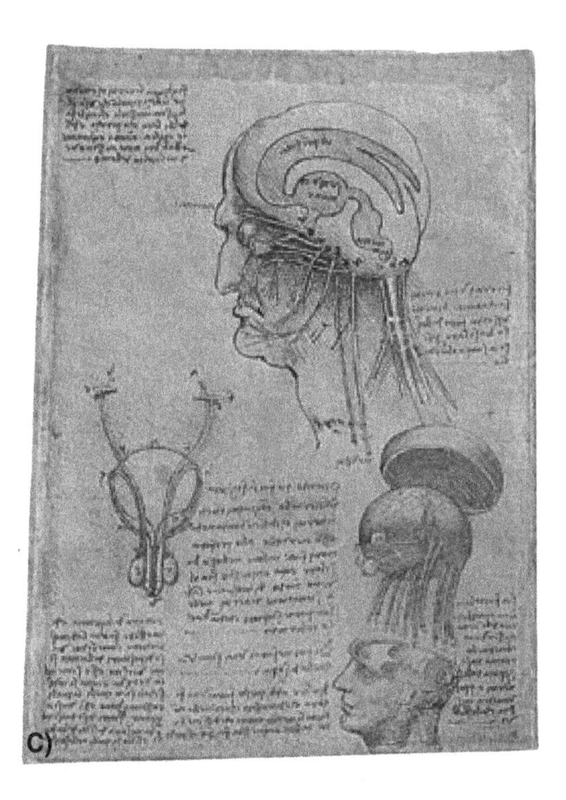

Figure 3.4 Ventricles revealed through sectional anatomy. A) The method for dissecting the brain *in situ* (without removing it from the skull) was used from the time of Herophilus (4th century BC) until the Renaissance. (Image from Estienne, 1545, reproduced with permission from the Wellcome Library, London.) B) In this drawing from Gregor Reisch's book (1503) the organs for vision, taste, smell, and hearing are connected to the anterior ventricle. Galen was the most fervent promotor of the ventricular theory. C) In the Renaissance, Leonardo da Vinci, as also many of his contemporaries, performed dissections of the human brain. Leonardo was also the first to create a wax cast of the ventricles. (Reproduced with kind permission from the Klassik Stiftung Weimar.)

the cell size, shape, and density across the layers have been used to divide the cortex into cytoarchitectonic areas. According to the work of different authors the number of these areas varies from a minimum of 17 (Campbell, 1905) to 107 (Economo and Koskinas, 1925). Brodmann's maps (1909), the most widely used in functional neuroimaging, are reproduced in Figure 3.3.

The frontal lobe is divided into ten fields, which are grouped into five main regions: i) Area 4 corresponds to the primary motor cortex and contains somatotopically organized neuronal bodies of the cortico-spinal projection fibres; ii) Area 6 occupies the premotor region and is divided into a lateral premotor cortex (PMC) and a medial supplementary motor area (SMA) and pre-supplementary motor area (pre-SMA); iii) Areas 44 and 45 correspond to Broca's area; iv) Dorsolateral (8, 9, 10, 46) and ventrolateral (47) prefrontal areas; v) Areas 11 and 47 are the main divisions of the orbitofrontal cortex (see also Chapters 5, 7, 10 and 11).

The parietal lobe is divided into nine fields, which are grouped into four main regions: i) Areas 3, 1, 2 correspond to the somatosensory cortex with a somatotopic organization similar to the primary motor cortex; ii) lateral Areas 5 and 7 form the superior polymodal parietal cortex; iii) Areas 39 and 40 correspond to the Geschwind's territory and occupy the inferior polymodal parietal cortex; iv) Area 31 and the medial Areas 5 and 7 form the precuneus. Area 43 is a transition region of the fronto-parietal operculum (see also Chapters 5, 7 and 8).

The temporal lobe is divided into seven fields, which are grouped into four main regions: i) Area 41 is the primary auditory cortex; ii) Areas 42 and 22 correspond to the auditory association cortex, part of which forms Wernicke's area; iii) Areas 20, 21, and 37 correspond to the temporal visual association cortex; iv) Area 38 occupies the temporopolar cortex, which is commonly considered one of the paralimbic areas (see Chapters 5, 7 and 8).

The occipital lobe is divided into Area 17, which corresponds to the primary visual cortex and Areas 18 and 19, which form the visual association cortex (see also Chapters 5 and 8).

The limbic lobe includes 15 fields divided into an olfactocentric group (Areas 11, 24, 25, 32, 33, 36, 47) and a hippocampocentric group (Areas 23, 24, 26, 27, 28, 29, 30, 31) (see also Chapters 5 and 11).

In addition to cell bodies the cerebral cortex also contains fibres that show two main orientations, tangential and radial. According to the pattern of distribution of the tangential and radial fibres, more than 200 myeloarchitectonic areas have been identified in the human cerebral cortex (Vogt and Vogt 1926).

From sectional neuroanatomy to function

Direct inspection of the brain surface does not reveal those hidden deep structures that in antiquity were considered the ultimate 'seat of the mind'. To gain access to those well-concealed cerebral regions, anatomists developed a new method of studying the brain, sectional anatomy. This method was originally undertaken by removing the upper part of the skull and dissecting the brain *in situ* by means of a series of axial slices beginning at the uppermost part

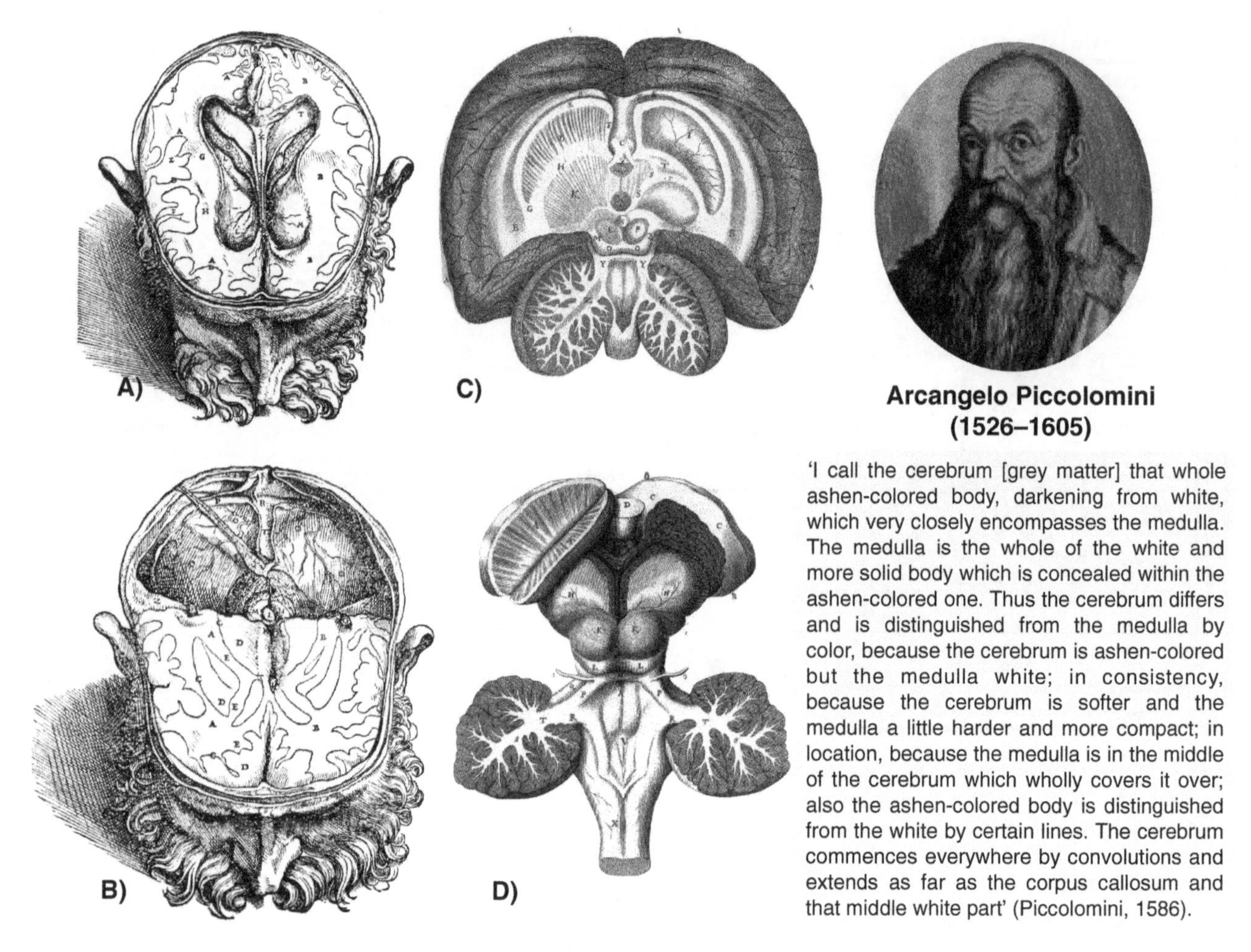

Figure 3.5 The study of sectional anatomy in the Renaissance. Images of the subcortical structures from Vesalius (1543) (A, B) and Willis (1664) (C, D). Note the difference in the method used to perform sections of the brain, *in situ* by Vesalius (classic method) and after removing the brain from the skull by Willis (Variolo's method). On the right is Piccolomini's portrait and his description of the differences between grey and white matter. A) Axial section through the corpus callosum and body of fornix. B) Vesalius never wrote about the separation between white and grey matter, but he certainly observed it in post-mortem dissections as suggested by this illustration taken from his book. C) Coronal section through the cerebral hemispheres showing the 'body with medullary streaks' (i.e. corpus striatum). D) Section through the left corpus striatum after removing the cerebral hemispheres. (C, D and the image of Arcangelo Piccolomini from the Wellcome Library, London)

of the cerebral hemispheres (Figure 3.4). Sectional anatomy originated around the end of the 4th century BC, probably with the work of Herophilus of Alexandria who paid particular attention to the ventricles and promoted the ventricular theory of brain function. Herophilus noticed that the ventricles are filled with fluid (i.e. liquor), which he considered the origin of the 'animal spirit' or 'soul' that flows through the nerves and moves the muscles. His work is entirely lost, but some of his ideas are reported in the writings of Galen:

Since all the nerves below the head spring either from the hindbrain [cerebellum] or the spinal cord, the [fourth] ventricle must be of considerable size, and receives the animal spirits previously compounded in the anterior ventricles; so that there must necessarily be a passage from them to it. (Translated in Dobson, 1925)

Galen himself performed brain dissections and after him the ventricular theory became the central dogma in the philosophy of mind (Figure 3.4). Galen is also credited for describing for the first time a number of other subcortical structures including the corpus callosum, the fornix, the thalamus, the pineal gland, and many others. His writings were so voluminous and his work so respected that after his death anatomists ceased to perform dissections for centuries and took his observations for granted. Furthermore, human dissections were considered not necessary by medieval philosophers, who were satisfied to solve anatomical problems by disputation alone. This resulted in the ventricular theory dominating over other alternative explanations throughout the Middle Ages and the Renaissance (Manzoni, 1998).

Human dissection was reintroduced in the 13th century. Among the anatomists of that time, Mondino de' Luzzi (c.1270–1326) is credited for restoring anatomy to its right place in medicine (Clarke and O'Malley 1996). Mondino made minor anatomical contributions and his ventricular theory was no different from his Greek predecessors; but his vividly figurative descriptions made his book the most widely used anatomical text for more than 250 years:

Before you proceed to the middle [third] ventricle, give consideration to that which intervenes between anterior and middle ventricles; there are three things, that is, anchae [thalami], which are kind of base of this anterior ventricle on the right and left [floor of the lateral ventricles], and they are of the substance of the brain in the shape of buttocks. At the side of each ancha, between the aforesaid ventricles, there is a blood red substance formed like a long worm such as an earthworm [choroid plexus], attached by ligaments and nervules on each side; by lengthening it constricts and closes the anchae and the way or passage from the anterior to the middle [ventricle], and the reverse. When a man wishes to cease cogitation, and again in consideration, he raises the walls and dilates the anchae so that the spirit may pass from one ventricle to the others. (Translated in Clarke and O'Malley, 1996)

Despite a certain moral resistance and religious bans, the practice of human dissection became gradually accepted in Europe. Unfortunately dissectors would often discourage students from describing what they really saw at the dissecting table and limited the use of human bodies to providing confirmation that Galen's observations were correct.

The first to use human dissection to confute Galen's doctrine was Andreas Vesalius (1514–1564) (Figure 3.5A, B). He urged his contemporaries to re-examine the structure and function of the brain derived from Galen and to reconsider the theory of ventricular localization on the basis of the evidence from comparative sectional observations:

All our contemporaries deny the particular powers of the principal soul to apes, dogs, horses, sheep, oxen, and other animals and, in brief, attribute only to man the faculty of reason, and that, insofar as I am able to understand, equally to all men. Nevertheless in dissection we see that men do not differ from animals by the possession of any special ventricle; not only is the number of ventricles the same, but all other things are very much alike in man and animal except in respect to the mass of the brain and a temperamental urge towards upright conduct. (Vesalius, 1543)

With Vesalius, anatomists began to doubt the ventricular theory and returned to the practice of dissection, this time as an experimental method to understand novel aspects of the human brain. In particular, the introduction of a revolutionary method for dissection by Costanzo Variolo (1543–1575) allowed direct access to previously undescribed features of the brain. Variolo examined the brain after removing it from the skull and then proceeded slicing the brain from the base upwards. He discovered in this way the pons, later named after him pons Variolii (Variolo, 1573). Other important anatomical descriptions followed. In 1587 Giulio Cesare Aranzi (1530–1589) described the hippocampus and its relationship to the temporal ventricles (Aranzi, 1587). In 1664 Thomas Willis (1621–1675) not only described the 'corpus striatum' (Figure 3.5C, D)—'bodies with medullary streaks'—but observed its degeneration in a patient who had suffered severe paralyses:

...for as often as I have opened the bodies of those who dyed of a long Palsie, and most grevious resolution of the Nerves, I have always found these bodies less firm than others in the Brain, discoloured like filth or dirt, and many chamferings obliterated. (Willis, 1664)

Following Willis' clinico-pathological observation, anatomists localized volition (i.e. voluntary movement) in the corpus striatum, an erroneous interpretation that was held true until the second half of the 19th century. Nevertheless, these anatomical discoveries were important in moving away from the ventricular theory and paving the way to a new revolutionary idea, the localization of brain function in the cortex. In 1586, Arcangelo Piccolomini described for the first time the anatomical features that distinguish the 'medulla' (i.e. white matter) from the 'cerebrum' (i.e. grey matter or cortex) (Figure 3.5). Piccolomini's observations marked the beginning of the anatomical study of the cortex, a field that expanded with the invention of the microscope. Marcello Malpighi and Antony van Leeuwenhoek (Figure 3.6) were the first to use a microscope to describe sections of the cortex:

I have discovered that the spread-out cortex is formed from a mass of minute glands... From the inner side extends a white 'nervous' fibre with resemblance to a vessel, and which can be observed owing to the brilliant whiteness of these bodies; hence the white medullary substance of the cerebrum arises from connected bunches of many fibrils. (Malpighi, 1666)

It is difficult to understand whether Malpighi's cortical 'glands'—Leeuwenhoek's gave a similar description of what he identified as cerebral 'globules'—were equivalent to neuronal cells or rather descriptions of optical artefacts such as air bubbles. Similarly, Malpighi's white 'fibres or fibrils' should not be interpreted as axonal fibres but probably as larger bundles (Clarke and O'Malley, 1996). Nevertheless, early microscopists had been able to show that, whereas the white matter was made of fibres, the grey matter consisted of corpuscular elements. They had not, however, become aware of any special arrangement of the latter, an observation made for the first time by Francesco Gennari (1752–1797) in 1776. He reported of a boundary line between the cortex and underlying white matter (later dubbed as the stripe of Gennari) (Figure 3.6B):

...especially when the brain has been dissected horizontally in layers, if one inspects the cortex where it is attached to the medullary substance [white matter] a whitish substance, like a line, appears... It varies not only in different brains but also in different parts of the same brain... In the medial part of the posterior lobes of the brain, not far from that part which rests on the tentorium, is the place (and this should be especially noted) in which I have observed this substance of which I have speak, collected into a small, very white line [Lineola albidior]. (Gennari, 1782)

Gennari refrained from any functional interpretation of his anatomical findings and humbly admitted that *'just as the use of so many other things is as yet concealed from us, so I do not know the purpose for which this substance was created.'* (Gennari, 1782). The ambition of correlating morphological variations of the cortex

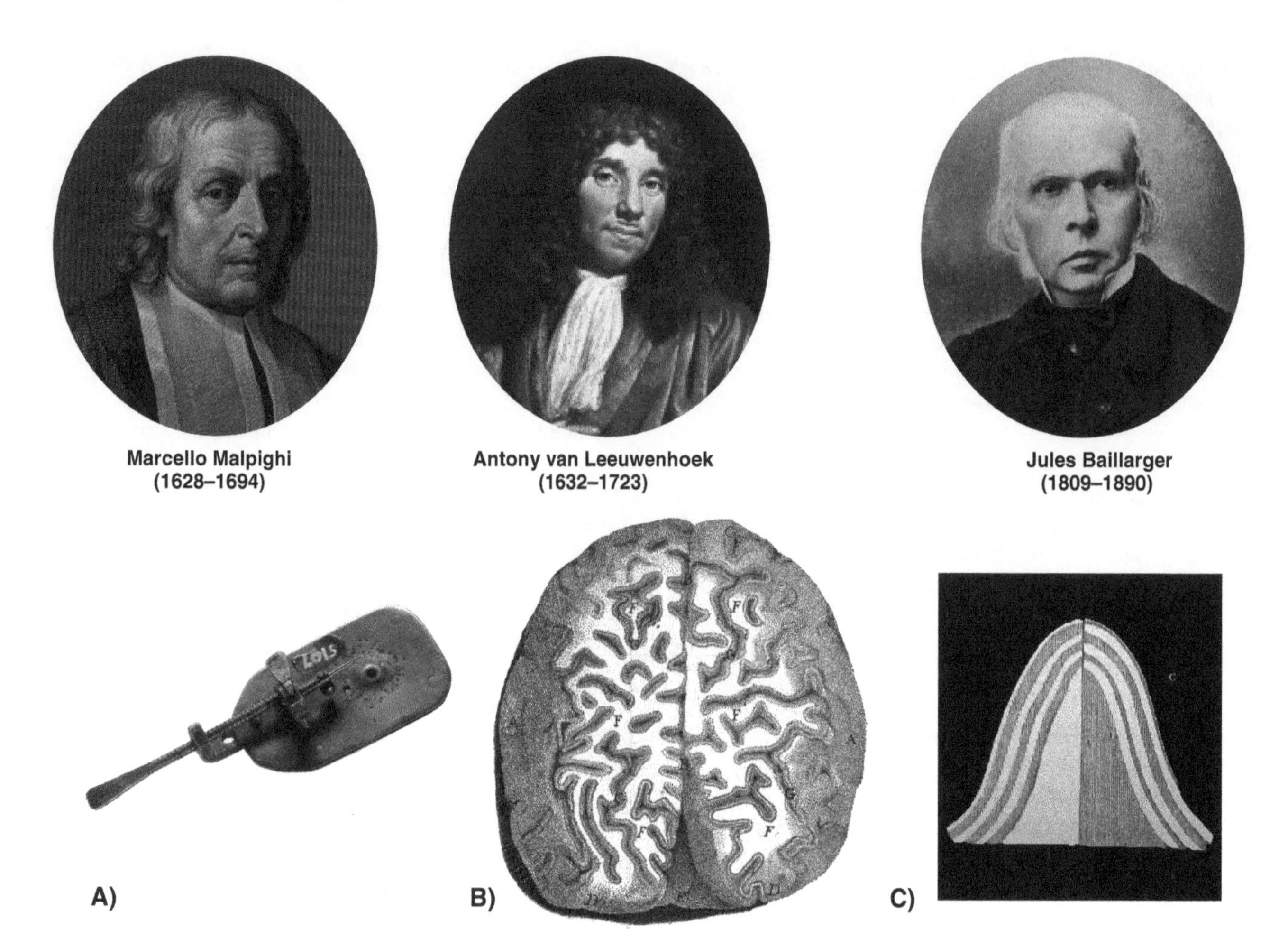

Figure 3.6 The study of the brain under the microscope and the discovery of the cortical multi-layered structure. A) One of the earliest microscopes, similar to those used by Malpighi and Leeuwenhoek in their investigations of the cerebral cortex. (Image from the Wellcome Library, London.) B) Image from Gennari (1782) where he describes for the first time the 'lineola albidior', a distinct feature of the calcarine cortex. In the original image the Gennari's line is visible also in other extra-calcarine posterior convolutions as a double rim of the cortex. (Image from the Wellcome Library, London.) C) The laminar structure of the cerebral cortex described by Baillarger in 1840 using a direct lighting reflected by the sample (left) or a light behind the sample, shining through the tissue (right) (from Testut, 1897). (Portraits from the Wellcome Library, London)

with intelligence and mental disorders was a challenge that French alienists were prepared to take on. In 1840, Jules Baillarge reported for the first time that the cortex is made up of layers (Figure 3.6C):

The researches which I have undertaken have made me recognize that the grey cortical substance of the convolutions of the brain is formed of six layers, arranged thus. The first, from within outwards, is grey, the second white, the third grey, the fourth white, the fifth grey, and the sixth whitish. These six layers, alternately grey and white, which remind one of the structure of a galvanic pile, can be seen with naked eye in most place. (Baillarger, 1840)

Baillarger had probably intended to go beyond the simple morphological similitude when he mentioned the galvanic pile, to imply possibly the notion of the cortex as a generator of 'animal electricity'. The method that Baillarger developed to reveal the cortical layers was particularly ingenious, a forerunner of contemporary polarized imaging (Axer et al., 1999) methods:

By a vertical cut I remove a very thin layer of grey cortical substance; I place it between two glass plates which I unite with wax to prevent all movement; I then expose it to the light of a lamp and examine it by that transmitted light. If the layer of grey substance studied in this way is homogenous and simple, it will allow light to pass through completely; if there are one or several white layers in its thickness, they will be recognized by their opacity. (Baillarger, 1840)

With this method he also recognized: '*...the existence of a great number of fibres penetrating from the central white matter into the cortical substance*', and concluded that '*the white matter at the summit of the convolutions, is entirely united to the grey matter by a large number of fibres. A simple juxtaposition of these two substances is thus inadmissible*' (Baillarger, 1840).

Microscopists soon confirmed these findings and complemented them with new data on the cellular composition of the cortical layers resulting from the technological advancements of that time.

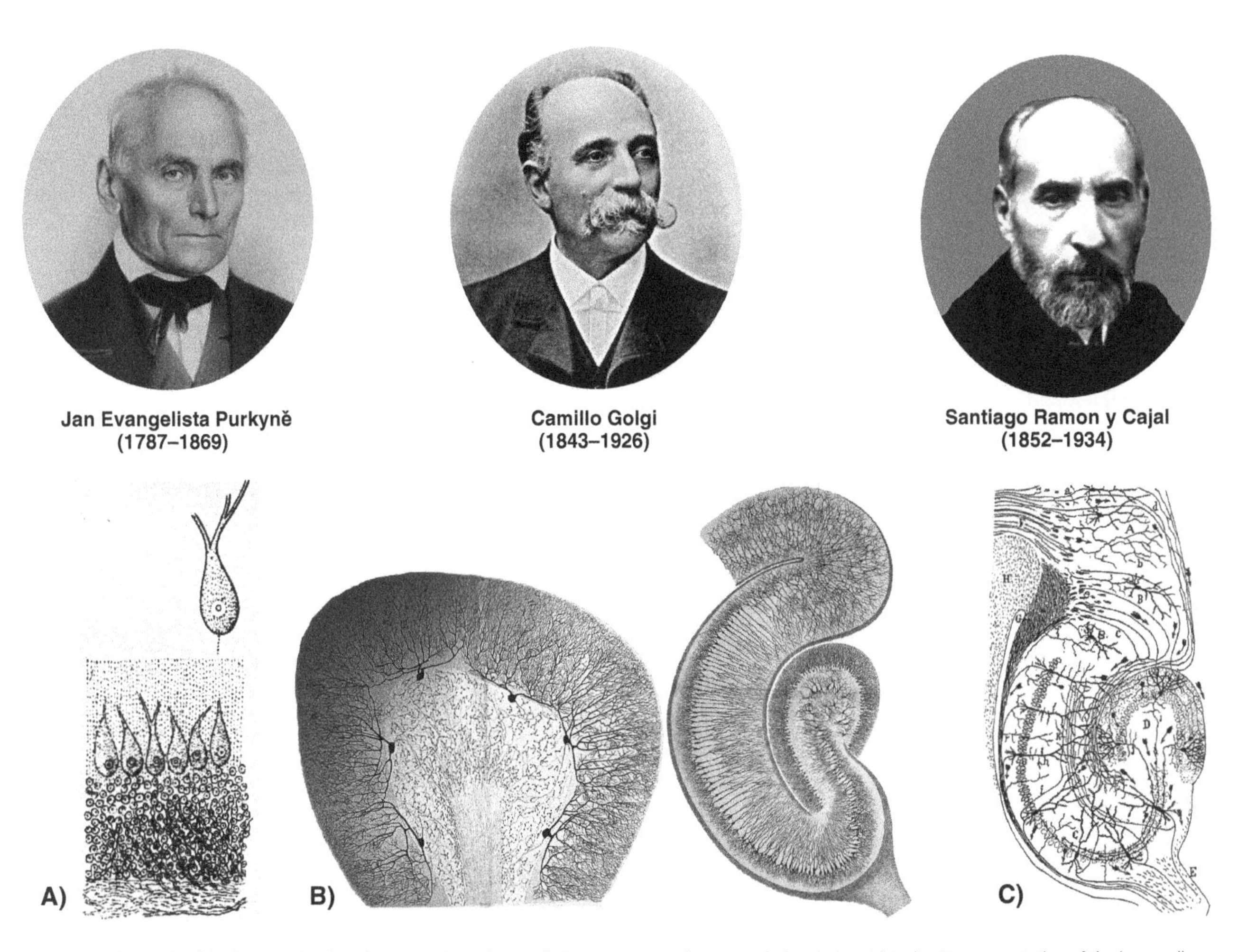

Figure 3.7 The study of the brain under the microscope from single cell discovery to complex network description. A) Purkyně's representation of the large cells in the cerebellar cortex (later named after him) (Purkyně, 1838). B) Golgi's representation of the anatomy of a cerebellar gyrus (Purkyně cells are the large neurons) and hippocampus (right) (Golgi, 1903–1929). C) Cajal's representation of the anatomy of the hippocampus where he uses arrows to indicate the direction of the electric impulses (Cajal, 1911). (Image of Camillo Golgi from the Wellcome Library, London)

Christian Gottfried Ehrenberg (1795–1876) and Gabriel Gustav Valentin (1813–1883) were among the first to use the achromatic compound microscope, which reduced the distortions from aberration. They identified cortical 'granules' (i.e. cells) of different size and shape in the cerebellar and cerebral cortex (Ehrenberg, 1833; Valentin, 1836). Jan Evangelista Purkyně combined histological staining with tissue fixation and discovered the large cells located in the cerebellar cortex that today bear his name (Figure 3.7A) (Purkyně, 1938). Vladimir Alexewitsch Betz (1834–1894) published in 1874 a paper on the cells of the primary motor and sensory cortex that he defined as the 'giant pyramids' of the fourth layer (today known as Betz cells). He also postulated that these *'cells have all the attributes of the so called motor-cells and very definitely continue as cerebral nerve fibres'* (Betz, 1874). Knowledge of the microscopic characteristics of the nervous fibres was also growing rapidly. Robert Remak (1815–1865) gave a remarkable account of the complex architecture of those fibres approaching and penetrating the cortex and separating *radiating* from *crossing* fibres:

The primary fibres which radiate from the central white matter towards the surface of the gyri are crossed by primary fibres in their course through the layers of the grey cortex, and pass through it in a direction parallel to the surface of the gyri… (Remak, 1841)

These microscopic studies of normal sectional anatomy culminated in the work of two giants in the history of neurohistology: Camillo Golgi, who discovered the method of silver impregnation ('reazione nera', i.e. black reaction) and supported the theory that a network connected all neurons ('syncytium') (Golgi, 1903–1929) and Ramon y Cajal, who vigorously promoted the neuron theory (i.e. the neuron as a single cell) (Cajal, 1911) (Figure 3.7C).

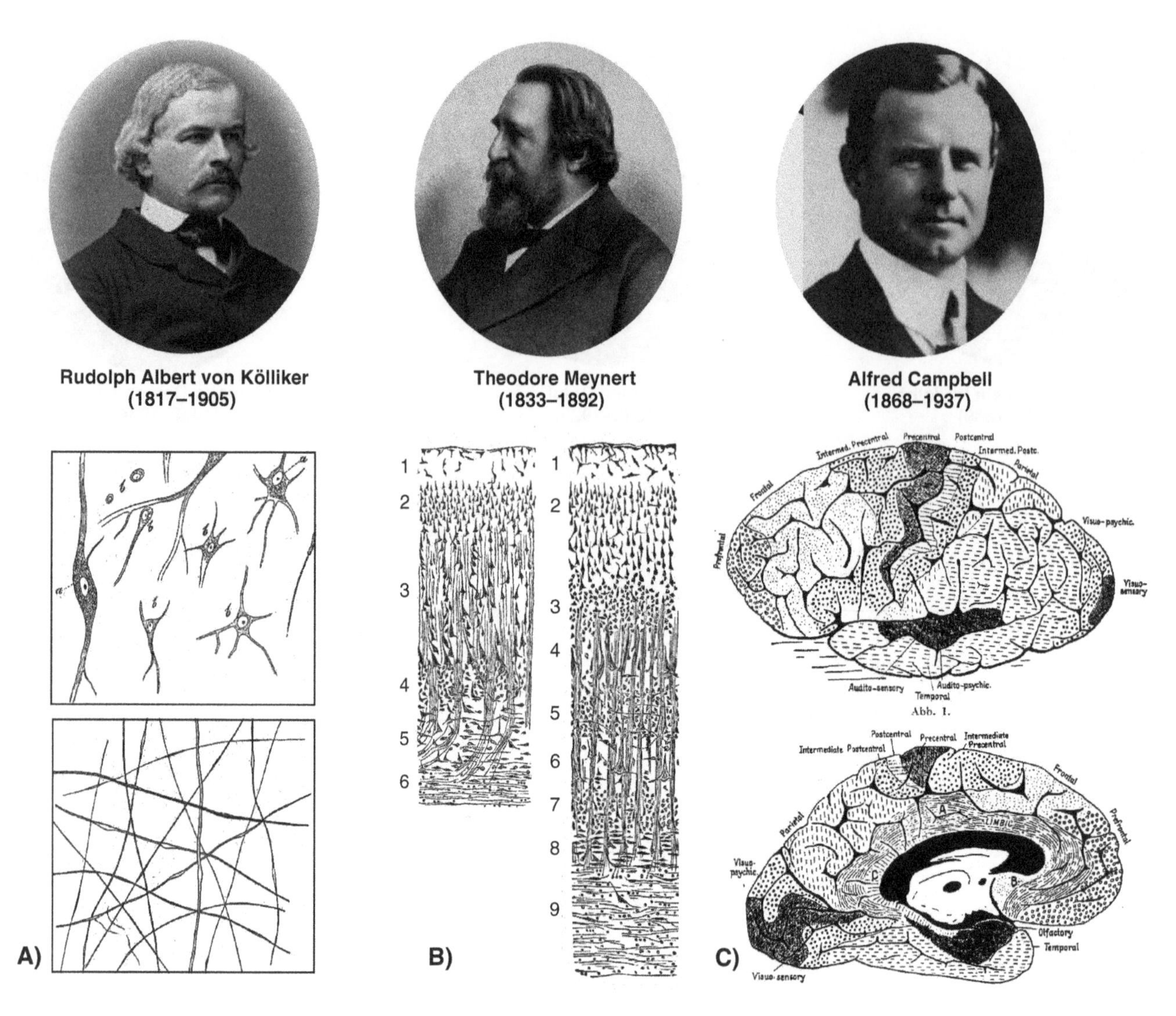

Figure 3.8 From cortical microscopy to cerebral cartography. (Image of Rudolph Albert von Kölliker from the Museum für Naturkunde Berlin; image of Theodore Meynert from the Wellcome Library, London.) A) Kölliker's (1852) microscopic description of the cells and fibres in the cerebral cortex. B) Meynert recognized six layers, from the surface they are: 1. molecular, 2. outer granular, 3. pyramidal, 4. inner granular, 5. Spindle cell layer, 6. medullary (multiform). Most of the cortex is six-layered, for example in the frontal lobe (left), but there are also cortical areas with a higher number of layers, as in the occipital cortex (right figure) (Meynert, 1885). C) Campbell's division of the cortex in 17 fields according to the interregional differences in the cortical cyto- and myeloarchitecture (Campbell, 1905)

The rate of discoveries increased significantly, and according to the great Swiss neurohistologist Rudolf Albert von Kölliker (1817–1905) the finer structure of the grey substance of the convolutions became fairly clear (Figure 3.8A):

The grey substance contains in its entire thickness both nerve cells and nerve fibers, and besides these a great deal of granular stroma... In the superficial white layer the cells are sparse and small... The middle or pure grey layer is the richest in cells which are gathered one against the other... Their shape is pyriform or fusiform, with three or many angles... Finally, in the innermost yellowish-red layer the cells are again rather sparse, though still very abundant... (Kölliker, 1849)

Kölliker also gave a detailed description of the cortical myeloarchitecture and proved conclusively that myelinated fibres arose from cortical cells. Thus, in the second half of the 19th century two related lines of investigation—the distribution and morphology of cortical cells (i.e. cytoarchitectonic) and the fibre pattern (i.e. myeloarchitectonic)—converged, and some of the most prestigious laboratories embarked on ambitious research programmes aimed at mapping the entire cortical mantle of the human brain by combining information from cyto- and myeloarchitectonics. Theodore Meynert, professor of nervous diseases in Vienna, pioneered this approach (Figure 3.8). Meynert studied the distribution of the neurons and their connections systematically across the cortical layers and different areas of the entire brain. He then tried to associate his anatomical findings with the specialization of functions by comparing differences among the cerebral lobes (e.g. frontal and occipital) (Figure 3.8B). His ultimate goal was to establish a scientific basis for mental illness and produce an anatomical classification of psychiatric disorders. For example, he observed that

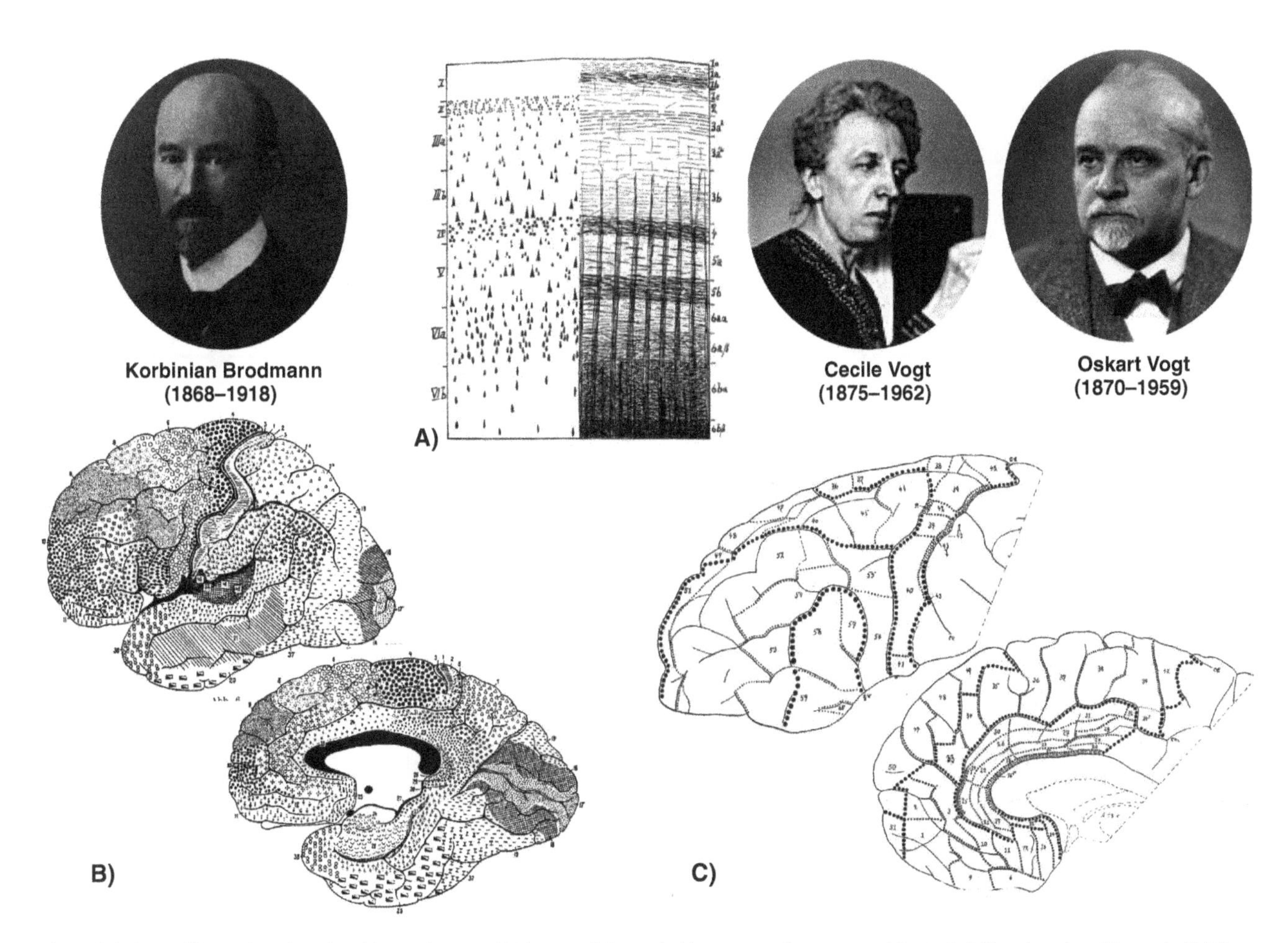

Figure 3.9 Cytoarchitectonic and myeloarchitectonic maps. A) Diagram of the cortical layers according to cytoarchitectonic (left) and myeloarchitectonic (right) anatomy (Vogt, 1910). B) The cytoarchitectonic map of the human cerebral cortex according to Brodmann (top left, lateral view; bottom right, medial view) (Brodmann, 1909). C) The myeloarchitectonic map of the human frontal lobe (top left, lateral view; bottom right, medial view) (Vogt, 1910). (Image of Oskart Vogt from the Wellcome Library, London)

'one cortical area (the uncinate convolution) consists of pyramids (cells) only and an extensive motor disturbance (epilepsy and epileptic-like convolutions) is followed by a constant disorder of it' (Meynert, 1867–68). He also divided the brain into two parts according to the structural differences of the cerebral cortex—the neopallium with grey outer surface and the archipallium with white outer surface.

Meynert's approach had a great influence on his contemporaries and inspired the work of future generations. In England, Alfred Campbell set out to complete Meynert's project. By combining the cyto- and myeloarchitectonic variations in the cortex, he was able to produce a map of distinctive cortical fields in *Homo sapiens* and other primate species and, within humans, of changes in the fields related to pathology (Figure 3.8C). Campbell succeeded in his endeavour in 1905 by publishing the monograph *Histological studies on the localisation of cerebral function* (Campbell, 1905). Although his map of cortical fields has been eclipsed by Brodmann's later contribution, Campbell's monograph was a monumental achievement for several reasons, the most important being the emphasis given to function. Campbell's project went beyond cytoarchitectonic cartography, attempting to integrate clinical, anatomical, and physiological evidence to provide a guide to function (ffytche and Catani, 2005). As is implicit in his title, the relationship between brain anatomy and brain function was central to Campbell's monograph, each of the chapters being devoted to a different brain region in terms of its histological anatomy and, together with clinicopathological, comparative, and physiological evidence, the functional insights that could be derived from such anatomical observations. Indeed, Campbell's cortical map is labelled not by numbers but by function. Thus, we find descriptions of 'visuo-sensory', 'audito-psychic', 'olfactory', 'pre-central motor', and 'post-central somatosensory' fields among the 17 he defined. Unfortunately, Campbell's path was destined to cross that of the Vogts and one of their most talented research assistants, Korbinian Brodmann (Figure 3.9).

Brodmann worked in Berlin's prestigious neurobiological laboratory on the cytoarchitectonic maps alone, the laboratory's myeloarchitectonic studies being carried out by Oskar and Cecile Vogt, a husband and wife team (Vogt and Vogt, 1926). Brodmann distinguished 44 sharply bordered areas in the left human brain, each of which he labelled with a number (Figure 3.9B). Brodmann admitted that although his ambitions extended beyond anatomy, his book ended up as a primarily anatomical text:

Although my studies of localisation are based on purely anatomical considerations and were initially conceived to resolve only

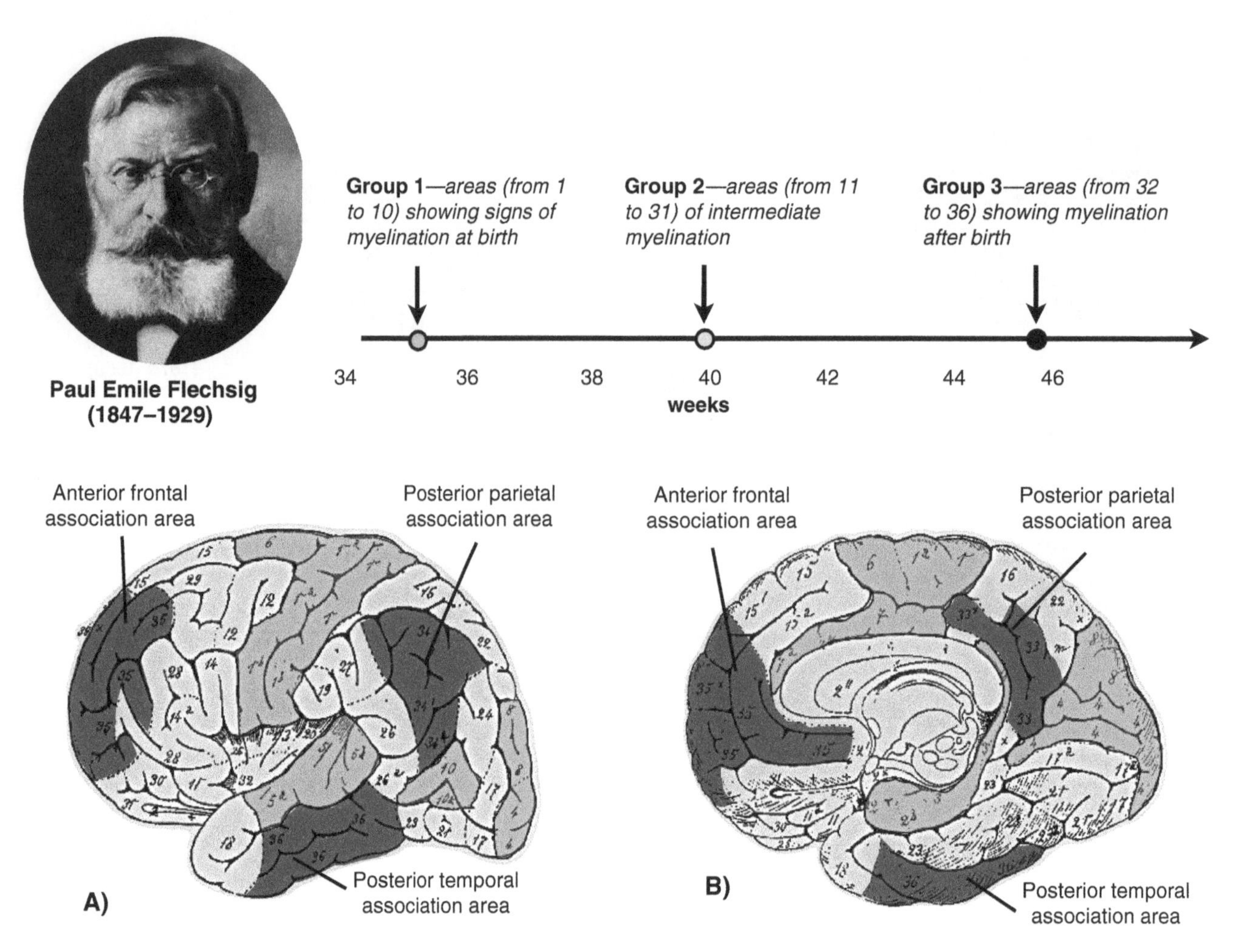

Figure 3.10 Flechsig's myelogenetic maps (1901) derived from the study of the myelination of the cortical areas of human fetuses. Primordial areas (group 1, orange-coloured) are myelinated at birth and have some correspondence with primary sensory and motor areas. Areas of group 2 (beige-coloured) and 3 (grey-coloured) are not fully myelinated at birth and correspond to secondary and tertiary associative areas.

anatomical problems, from the outset my ultimate goal was the advancement of a theory of function and its pathological deviations. (Brodmann, 1909)

Surprisingly when Brodmann wrote about function of the brain he was clearly disdainful of any localization in cortical areas:

Just as untenable as the idea of a 'concept cell' or an 'association layer' is the assumption of specific 'higher-order psychic centres'. Indeed recently theories have abounded which, like phrenology, attempt to localise complex mental activity such as memory, will, fantasy, intelligence, or spatial qualities such as appreciation of shape and position to circumscribed cortical zones. Older authors such as Goltz, Rieger, Wundt and recently, particularly outspokenly, Semon, have already quite rightly expressed their opposition to such a 'naive view' and pleaded simple psychological facts against it. (Brodmann, 1909)

Thus, despite his own cytoarchitectonic observations, Brodmann held higher functions to be the result of widespread brain activity:

In reality there is only one psychic centre: the brain as a whole with all its organs activated for every complex psychic event, either all together or most at the same time, and so widespread over the different parts of the cortical surface that one can never justify any separate specially differentiated 'psychic' centres within this whole. (Brodmann, 1909)

Paradoxically, in contemporary cognitive neurosciences, Brodmann's maps have become the *lingua franca* of cortical localization—numbers used universally as shorthand for a precise cortical locus and, in some cortical fields, a precise functional role.

On the other front of cerebral cartography, the Vogts set out an even harder task for themselves. By studying the variation of cortical myeloarchitectonic, Cecile and Oskar Vogt identified more than 200 areas, many of which represent subdivisions of the cytoarchitectonic areas of Brodmann (Figure 3.9C). Despite working four-handedly on the project and relying on the help of other assistants, they never finished their monumental project and their anatomical endeavour has remained incomplete for most of the temporal and occipital cortex.

Cortical maps of the human brain and other animals flourished in the first half of the 20th century, as much effort was dedicated to brain mapping. Beyond the obvious differences in number of fields or delineation of borders between areas, the work of brain cartographers differed in the ultimate functional goals of their

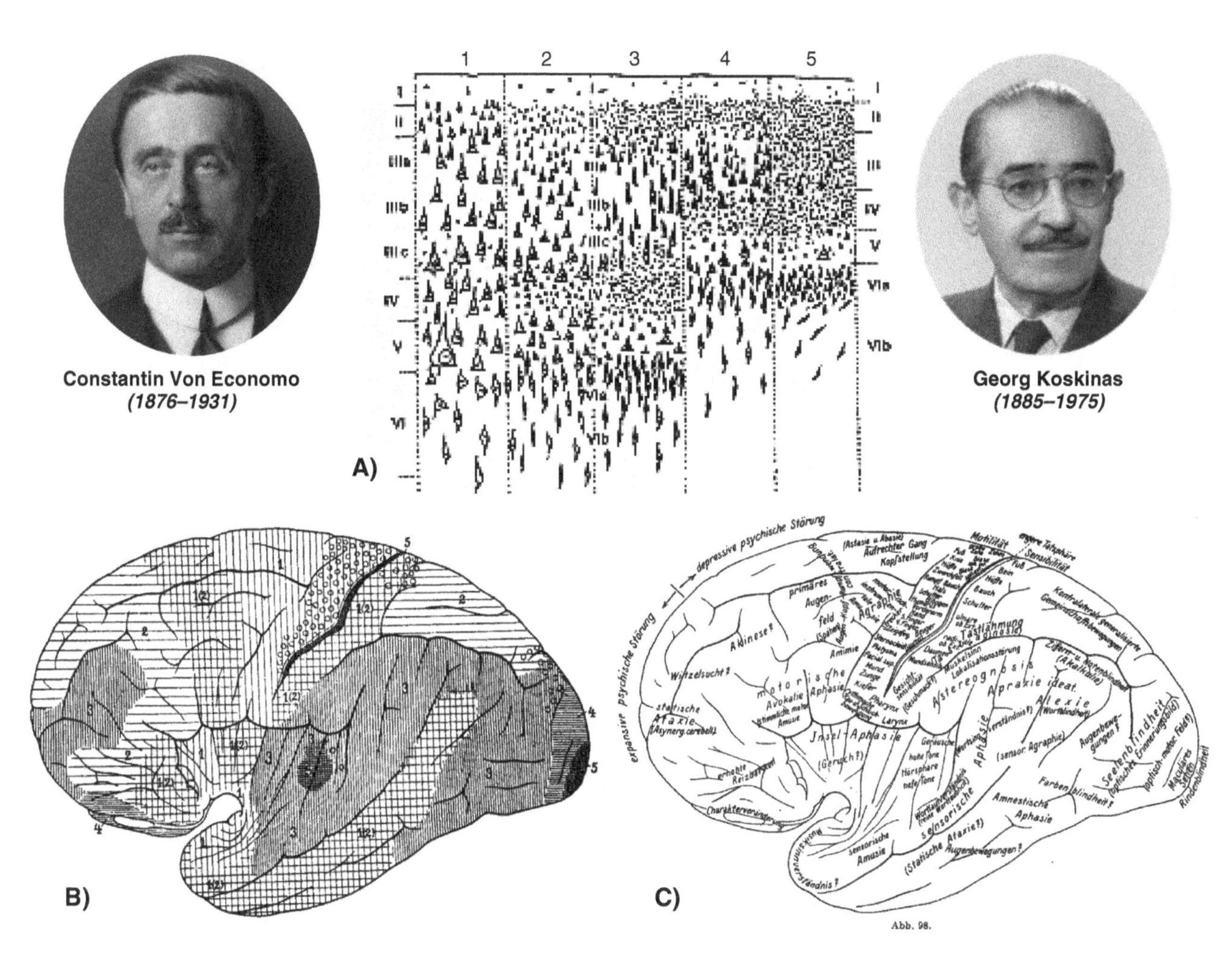

Figure 3.11 Cortical divisions of the human brain according to von Economo and Koskinas (1925). A) Diagram of the five principal types of neocortex: 1. agranular, 2. frontal, 3. parietal, 4. granular, 5. polar. B) Distribution of the five principal types of neocortex in lateral surface of the human brain. C) Corresponding functions of the cytoarchitectonic cortical areas. (von Economo and Koskinas, 1925)

research. Among those who have tried to gain functional insight from their anatomical maps, Paul Emile Flechsig and Constanin von Economo deserve a special mention. Flechsig's myelogenetic map contains 45 areas numbered according to the order of myelination during perinatal development (Figure 3.10). Primordial areas are already myelinated before birth and correspond to regions of origin or termination of motor and sensory projection fibres. Other areas become progressively myelinated postnatally and are considered to represent important relay stations between primary sensory areas. For this reason the postnatally myelinated areas are indicated with the term association areas. Flechsig was able to demonstrate that the human brain develops through stages, an early prenatal myelination of sensory-motor projection areas being a necessary prerequisite for later postnatal development of association areas subserving higher cognitive functions (Flechsig, 1901). The myelogenetic maps were very influential not only in the field of cortical mapping (Flechsig's maps were published even before the work of Campbell) but also for network theories of brain function that followed (see Chapter 4).

Constantin von Economo, together with his young assistant Georg Koskinas, published in 1925 *Die Cytoarchitektonik der Hirnrinde des erwachsenen Menschen,* an atlas with 112 microphotographs of histological sections and 810 pages of text (Figure 3.11). The work is based on the cytoarchitectonic analysis of 107 cortical areas, for each of which quantitative measurements were recorded about variations in cortical thickness and volume, form, size, and number of the cells, their density, grouping in blobs, stripes, and layers. The atlas is a monumental work, which, despite being considered by many the definitive text on cortical cartography, never met the favour of the scientific community. This is probably in part due to its encyclopaedic proportions and possibly to the general feelings against the 'crazy paving' school of cortical research, which peaked around that time (Le Gros Clark, 1952). However, one wonders why a similar destiny has not been reserved for other works on cortical mapping. Von Bonin and Bailey, for example, published *The neocortex of* Macaca mulatta in 1947. They adopted von Economo's method for designating cytoarchitectural areas and their work has had a long-lasting use ever since. The reason for a different fortune might simply lie in the methodological development that followed in the second half of the 20th century:

Cortical architecture can only be given functional meaning when correlated with data of a functional character derived using complementary techniques, preferably from the same brain. This is best

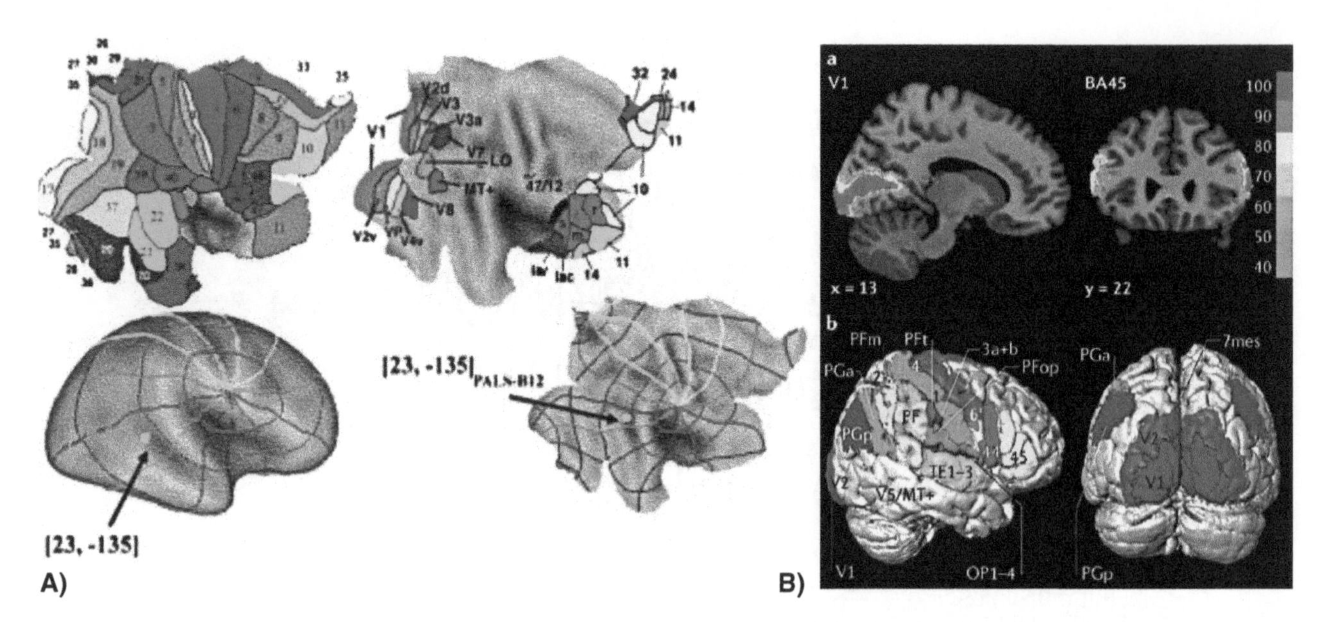

Figure 3.12 A) Flat maps of Brodmann's human fields according to Essen (reprinted from Van Essen et al., 2005 with permission from Elsevier) and B) probabilistic cytoarchitectonic maps according to Amunts et al. (Reprinted from Amunts et al., 2000 with permission from Elsevier.)

seen in microelectrode studies in animals in which the border of a sensory or motor representation, as defined by single- or multi-unit recording or by microstimulation and marked by a microlesion or injection of a dye, is matched to the cytoarchitecture of sections later cut from the same brain. It was this approach, as exemplified in the work of Mountcastle and Powell (1959) on the somatosensory cortex of the monkey and of Hubel and Wiesel (1962) on the visual cortex of the cat, that helped pull the study of cortical cytoarchitectonics out of the doldrums into which it had drifted, and which is now a widely used approach in animals. (Jones, 2008)

Thus, the answer seems to be that von Bonin and Baily happened to work on the best animal model for cortical mapping, the 'complementary methods' not being applicable to humans. Von Bonin and Baily were also dismissive of Brodmann's work, which they said to be 'full of errors of numbers'. This led future generations of physiologists to have no alternative to their maps, which also had the advantage of being accompanied by an English text and having a human equivalent map derived from the dissections of a single subject (Bailey and von Bonin, 1951).

In the last two decades functional neuroimaging has attempted to localize activities associated with the execution of specific tasks onto discrete cortical areas. Whilst the human maps of von Economo could have been a viable candidate for functional mapping, their position in the history was again deferred, this time by Brodmann's maps. This was mainly due to the arbitrary choice of Talairach and Tournoux to adopt Brodmann's nomenclature for their atlas. This gave shaky foundations to the field, as Talairach and Tournoux never performed a histological cytoarchitectonic study on the brain they used for their atlas (Talairach and Tournoux, 1988). This aspect may not be relevant to contemporary neuroimagers as they only desire a common nomenclature for sharing results across studies irrespective of their anatomical validity. There are, however, some problems such as the lack of data on the cytoarchitectonic divisions of the cortex buried within the sulci and the observation of significant interindividual variability that have convinced some laboratories to embark on the daunting task of producing cytoarchitectonic maps of the human brain that may partially overcome these limitations. The flat maps of Van Essen are a first attempt to map the cortical surface hidden in the sulci (Figure 3.12A) (Van Essen, 2005). But these maps are not derived from post-mortem histology and do not provide information on interindividual variability, a task that is currently undertaken by others (Figure 3.12B) (Amunts et al., 2000). The hope is that by perfecting early maps with modern objective quantitative methods and portraying interindividual variability one could obtain a better matching of *in vivo* functional with anatomical cortical parcellation for humans.

Other MRI methods for whole-brain analysis have been developed using a computerized sectional anatomy approach. Measurements of cortical thickness (MacDonald et al., 2000) and Voxel-based morphometry (VBM) (Ashburner and Friston, 2000) are examples of very powerful methods for exploring structural changes during normal (Paus et al. 1999; Good et al., 2001; Giorgio et al., 2010) and pathological (Honea et al., 2008; Schumann et al., 2010) development and correlation with symptoms profile in neurological (Gorno-Tempini et al., 2004) and psychiatric disorders (Ecker et al., 2010). However, the exact correspondence between these in vivo measurements and post-mortem histology remains to be confirmed.

In conclusion, throughout the centuries sectional anatomy has produced two of the most popular theories of brain function. While the ventricular theory seems to be completely abandoned, cortical localization of complex behaviour and cognitive functions is still a work in progress.

References

Amunts, K., Malikovic, A., Mohlberg, H., Schormann, T., and Zilles, K. (2000). Brodmann's areas 17 and 18 brought into stereotaxic space – where and how variable? *NeuroImage*, **11**(1), 66–84.

Aranzi, G.C. (1587). *De humano foetu liber tertio editus, ac recognitus, ejusdem anatomicarum observationum liber, ac De tumoribus secundum locos affectos liber nunc primum editi*. Venice: Venetiis: apud Jacobun Brechtanum. [Translated in Clarke, E. and O'Malley, C.D. (Eds.) (1996). *The Human Brain and Spinal Cord: A Historical Study Illustrated by Writings from Antiquity to the Twentieth Century*, second edition. San Francisco, CA: Norman Publishing.]

Ashburner, J. and Friston, K.J. (2000). Voxel-based morphometry – the methods. *NeuroImage*, **11**(6), 805–21.

Axer, H., Lippitz, B.E., and Keyserlingk, von D.G. (1999). Morphological asymmetry in anterior limb of human internal capsule revealed by confocal laser and polarized light microscopy. *Psychiat Res*, **91**(3), 141–54.

Bailey, P. and Bonin, G. von (1951). *The Isocortex of Man*. Urbana, IL: The University of Illinois Press.

Baillarger, J. (1840). Recherches sur la structure de la couche corticale des circonvolutions du cerveau. *Mém Acad R Méd*, **8**, 149–183. [Translated in Clarke, E. and O'Malley, C.D. (Eds.) (1996). *The Human Brain and Spinal Cord: A Historical Study Illustrated by Writings from Antiquity to the Twentieth Century*, second edition. San Francisco, CA: Norman Publishing.]

Betz, V.A. (1874). Anatomischer Nachweis zweier Gehirncentra. Centralbl Med Wissensch, **12**, 578–80. [Translated in Clarke, E., and O'Malley, C.D. (Eds.) (1996). *The Human Brain and Spinal Cord: A Historical Study Illustrated by Writings from Antiquity to the Twentieth Century*, second edition. San Francisco, CA: Norman Publishing.]

Bonin, G. von and Bailey, P. (1947). *The neocortex of* Macaca Mulatta. Urbana, IL: The University of Illinois Press.

Brodmann, K. (1909). *Vergleichende Lokalisationslehre der Großhirnrinde: in ihren Prinzipien dargestellt auf grund des zellebaues*. Leipzig: Barth.

Cajal, S.R. (1911). *Histologie du Systeme Nerveux de l'Homme et des Vertebretes*. Paris: A. Maloine.

Campbell, A.W. (1905). *Histological studies on the localisation of cerebral function*. Cambridge: Cambridge University Press.

Clarke, E. and O'Malley, C.D. (Eds.) (1996). *The Human Brain and Spinal Cord: A Historical Study Illustrated by Writings from Antiquity to the Twentieth Century*, second edition. San Francisco, CA: Norman Publishing.

Dobson, J. (1925). Herophilus of Alexandria. *Proc R Soc Med*, **18**, 19–32.

Ecker, C., Rocha-Rego, V., Johnston, P., et al. (2010). Investigating the predictive value of whole-brain structural MR scans in autism: a pattern classification approach. *NeuroImage*, **49**(1), 44–56.

Economo, C. von and Koskinas, G.N. (1925). *Die Cytoarchitektonik der Hirnrinde des erwachsenen Menschen*. Berlin: Springer-Verlag.

Ehrenberg, C.G. (1833). Notwendigkeit einer feineren mechanischen Zerlegung des. Gehirns und der Nerven vor der chemischen, dargestellt aus Beobachtungen von C.G. Ehrenberg. *Annln Phys*, **28**, 449–73.

Estienne, C. (1545). *De Dissectione Partium Corporis Humani Libri Tres*. Paris: Simon de Collines.

ffytche, D.H. and Catani M. (2005). Beyond localization: from hodology to function. *Phil Trans R Soc B*, **360**(1456), 767–79.

Flechsig, P.E. (1901). Developmental (myelogenetic) localisation of the cerebral cortex in the human subject. *Lancet*, **2**, 1027–9.

Gennari, F. (1782) De peculiari structura cerebri nonnulisque ejus morbis. Parma, Ex regio typographeo. [Translated in Clarke, E., and O'Malley, C.D. (Eds.) (1996). *The Human Brain and Spinal Cord: A Historical Study Illustrated by Writings from Antiquity to the Twentieth Century*, second edition. San Francisco, CA: Norman Publishing.]

Golgi, C. (1903–1929). *Opera Omnia*. Milan: Hoepli.

Good, C.D., Johnsrude, I.S., Ashburner, J., et al. (2001). Cerebral asymmetry and the effects of sex and handedness on brain structure: a voxel-based morphometric analysis of 465 normal adult human brains. *NeuroImage*, **14**(3), 685–700.

Giorgio, A., Watkins, K.E., Chadwick, M., et al. (2010). Longitudinal changes in grey and white matter during adolescence. *NeuroImage*, **49**(1), 94–103.

Gorno-Tempini, M.L., Dronkers, N.F., and Rankinet K.F., et al. (2004). Cognition and anatomy in three variants of primary progressive aphasia. *Ann Neurol*, **55**(3), 335–46.

Hubel, D. and Wiesel, T. (1962). Receptive fields, binocular interaction and functional architecture in the cat's visual cortex. *J Physiol (Lond)*, **160**, 106–54.

Honea, R.A., Meyer-Lindenberg, A., Hobbs, K.B., et al. (2008). Is gray matter volume an intermediate phenotype for schizophrenia? A voxel-based morphometry study of patients with schizophrenia and their healthy siblings. *Biol Psychiatry*, **63**(5), 465–74.

Kölliker, R.A. (1849). Neurologische Bemerkungen. *Zeit Wiss Zoo*, **1**, 135–63.

Kölliker, R.A. (1852). *Handbuch der Gewebelehre des Menschen*. Leipzig: Engelmann.

Le Gros Clarke, W.E. (1952). A note on cortical cyto-architectonics. *Brain*, **78**(1), 96–104.

Jones, E.G. (2008). Cortical maps and modern phrenology. *Brain*, **131**(8), 2227–33.

MacDonald, D., Kabani, N., Avis, D., and Evans, A.C. (2000) Automated 3-D extraction of inner and outer surfaces of cerebral cortex from MRI. *NeuroImage*, **12**(3), 340–56

Malpighi, M. (1666). De cerebri cortice. Bologna: Montius.

Manzoni, T. (1998). The cerebral ventricles, the animal spirits and the dawn of brain localization of function. *Arch Ital Biol*, **136**, 103–52.

Meynert, T. (1867–68). Der Bau der Grosshirnrinde und seine ortlichen Verschiedenheiten, nebst einem pathologisch-anatomischen Corollarium. *Vierteljahresschrift für Psychiatrie*, **1**, 77–93; 198–217, **2**, 88–113.

Meynert, T. (1885). *A Clinical Treatise on Diseases of the Fore-brain Based Upon a Study of Its Structure, Functions, and Nutrition* (Trans. B. Sachs). New York: G.P. Putnam's Sons.

Moutcastle, V. and Powell, T.P. (1959). Neural mechanisms subserving cutaneous sensibility, with special reference to the role of afferent inhibition in sensory perception and discrimination. *Bull Johns Hopkins Hosp*, **105**, 201–32.

Paus, T., Zijdenbos, A., Worsley, K., et al. (1999). Structural maturation of neural pathways in children and adolescents: in vivo study. *Science*, **283**(5409), 1908–11.

Piccolomini, A. (1586). Anatomicae praelectiones. Rome. [Translated in Clarke, E. and O'Malley, C.D. (Eds.) (1996). *The Human Brain and Spinal Cord: A Historical Study Illustrated by Writings from Antiquity to the Twentieth Century*, second edition. San Francisco, CA: Norman Publishing.]

Purkyně, J.E. (1838). Neueste Untersuchungen aus der Nerven und Hirnanatomie. *Ber Versamml Dtsch Nawrf Aertze Prag*, **15**, 177–80.

Reisch, G. (1503). *Margarita Philosophica*. Freiburg: Johann. Schott.

Remak, R. (1841). Anatomische Beobachtungen über das Gehirn, Rückenmark und Nervensystem. *Arch Anat Physiol* 506–22.

Schumann, C.M., Bloss, C.S., Barnes, C.C., et al. (2010). Longitudinal magnetic resonance imaging study of cortical development through early childhood in autism. *J Neurosci*, **30**(12), 4419–27.

Talairach, J. and Tournoux, P. (1988). *Co-planar Stereotaxic Atlas of the Human Brain: 3-Dimensional Proportional System – an Approach to Cerebral Imaging*. New York: Thieme Medical Publishers.

Testut, L. (1897). *Traite d'anatomie humaine*. Pars: O.Doin.

Valentin, G.G. (1836). Über den Verlauf und die letzten Enden der Nerven. *Nova Acta Phys-Med Acad Leopoldina (Breslau)*, **18**, 51–240.

van Essen, D.C. (2005). A population-average, landmark- and surface-based (PALS) atlas of human cerebral cortex. *NeuroImage*, **28**(3), 635–62.

Varolio, C. (1573). De nervis opticis nonnullisque aliis, praeter communem opinionem in humano capite observatis, epistolae. Padua: Meitti. [Translated in Clarke, E. and O'Malley, C.D. (Eds.) (1996). *The Human Brain and*

Spinal Cord: A Historical Study Illustrated by Writings from Antiquity to the Twentieth Century, second edition. San Francisco, CA: Norman Publishing.]

Vesalius (1543). *De humani corporis fabrica libri septem*. Basel: Oporinus. [Translated in Clarke, E. and O'Malley, C.D. (Eds.) (1996). *The Human Brain and Spinal Cord: A Historical Study Illustrated by Writings from Antiquity to the Twentieth Century*, second edition. San Francisco, CA: Norman Publishing.]

Vogt, C. and Vogt, O. (1926). Die vergleichend-architektonische und vergleichend-reizphysiologische Felderung der Grosshirnrinde unter besonderer Berücksichtigung der menschlichen. *Naturwissenschaften*, **14**, 1190–94.

Vogt, O. (1910). Die myeloarchitektonische Felderung des menschlichens Stirnhirns. *J Psychol Neurol*, **15**, 221–32.

Willis, T. (1664). *Cerebri anatome*. London: Martyn & Allestry.

CHAPTER 4

Connectional Neuroanatomy

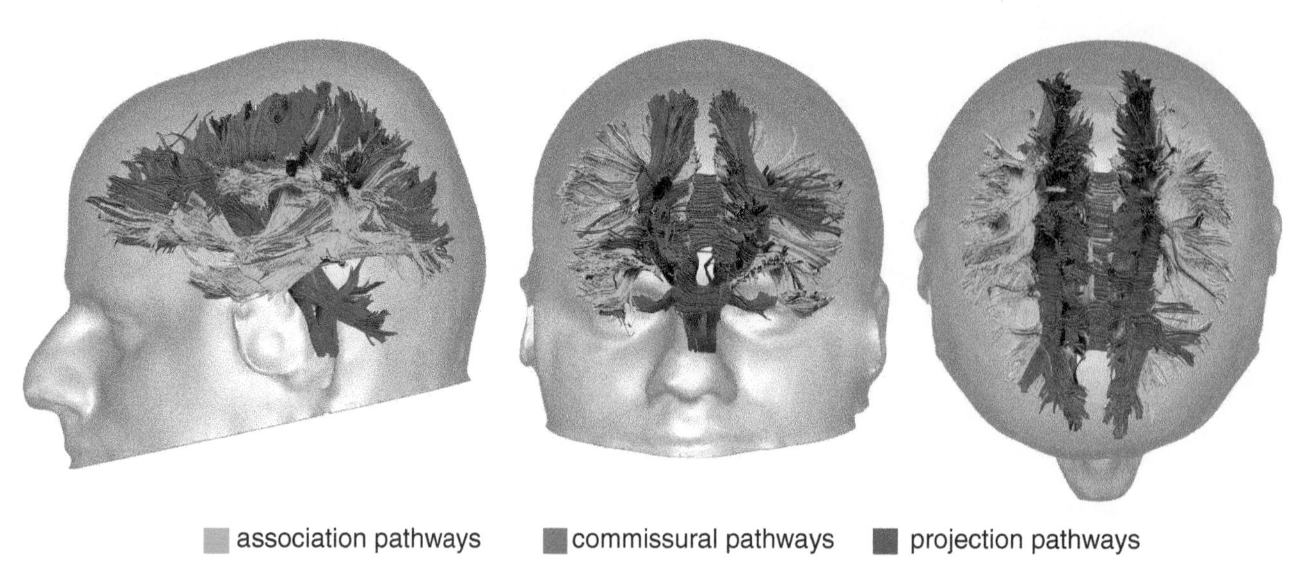

Figure 4.1 Diffusion tensor tractography reconstruction of the association, commissual, and projection pathways of the human brain.

Introduction

Between the cortex, the lateral ventricles, and the deep nuclei lies a mass of fibres connecting distant regions of the brain. Most of the cerebral fibres have both their origin and termination within the cortex, either in the same hemisphere (association fibres) or opposite hemispheres (commissural fibres) (Figure 4.1). Other fibres connect the cortex to subcortical regions (projection fibres). There are some fibres that, although being located in the cerebral hemispheres do not connect the cortex, but only subcortical structures (e.g. fibres of the anterior commissure connecting the two amygdalae). Fibres group together to form bundles of different diameter and several bundles form larger pathways called fasciculi (or tracts). White matter fasciculi are classified into association, commissural, and projection tracts, although some tracts may contain more than one type of fibre (e.g. the uncinate fasciculus contains both association and projection fibres).

The association pathways connect cortical regions within the same hemisphere and have an anterior–posterior (or posterior–anterior) direction. The major association tracts are the arcuate fasciculus, the cingulum, the uncinate, the inferior longitudinal fasciculus, and the inferior fronto-occipital fasciculus (Figure 4.2). The terminology used to indicate these tracts refers to either their shape (e.g. the uncinate for 'hook-like', cingulum for 'girdle' or 'belt'), their origin and termination (e.g. inferior fronto-occipital fasciculus), or their course and location (e.g. inferior longitudinal fasciculus). The long association tracts connect distant regions between lobes, while short U-shaped fibres connect neighbouring gyri within the same lobe (intralobar fibres) or different lobes (interlobar fibres). The association tracts are involved in higher cognitive functions, such as language, praxis, visuo-spatial processing, memory, and emotion (see also Chapters 7, 8, 11).

Commissural pathways are composed of fibres connecting the two halves of the brain. The major telencephalic commissures of the human brain include the corpus callosum, the anterior commissure, and the hippocampal commissure. The commissural pathways allow the transfer of inputs between the two halves of the brain and play a significant role in the functional integration of motor, perceptual, and cognitive functions (see also Chapter 9).

Projection pathways connect the cortex to subcortical structures, such as deep cerebral nuclei, brainstem nuclei, and spinal cord. Within the cerebral hemisphere the ascending projection fibres originate from subcortical nuclei (mainly thalamus) and terminate in the cortex, while descending projection fibres have the opposite orientation. Most of the projection fibres course through the corona radiata, internal capsule, cerebral peduncles, and brainstem. The fornix is also considered as a projection tract (see also Chapters 10 and 11).

Association pathways

Five major association tracts have been described in the human brain (Figure 4.2). The arcuate fasciculus is a lateral tract composed of long and short fibres connecting the perisylvian cortex of the frontal, parietal, and temporal lobes. The short fibres lie more lateral than the long fibres. The arcuate fasciculus of the left hemisphere is involved in language, praxis, and verbal working memory.

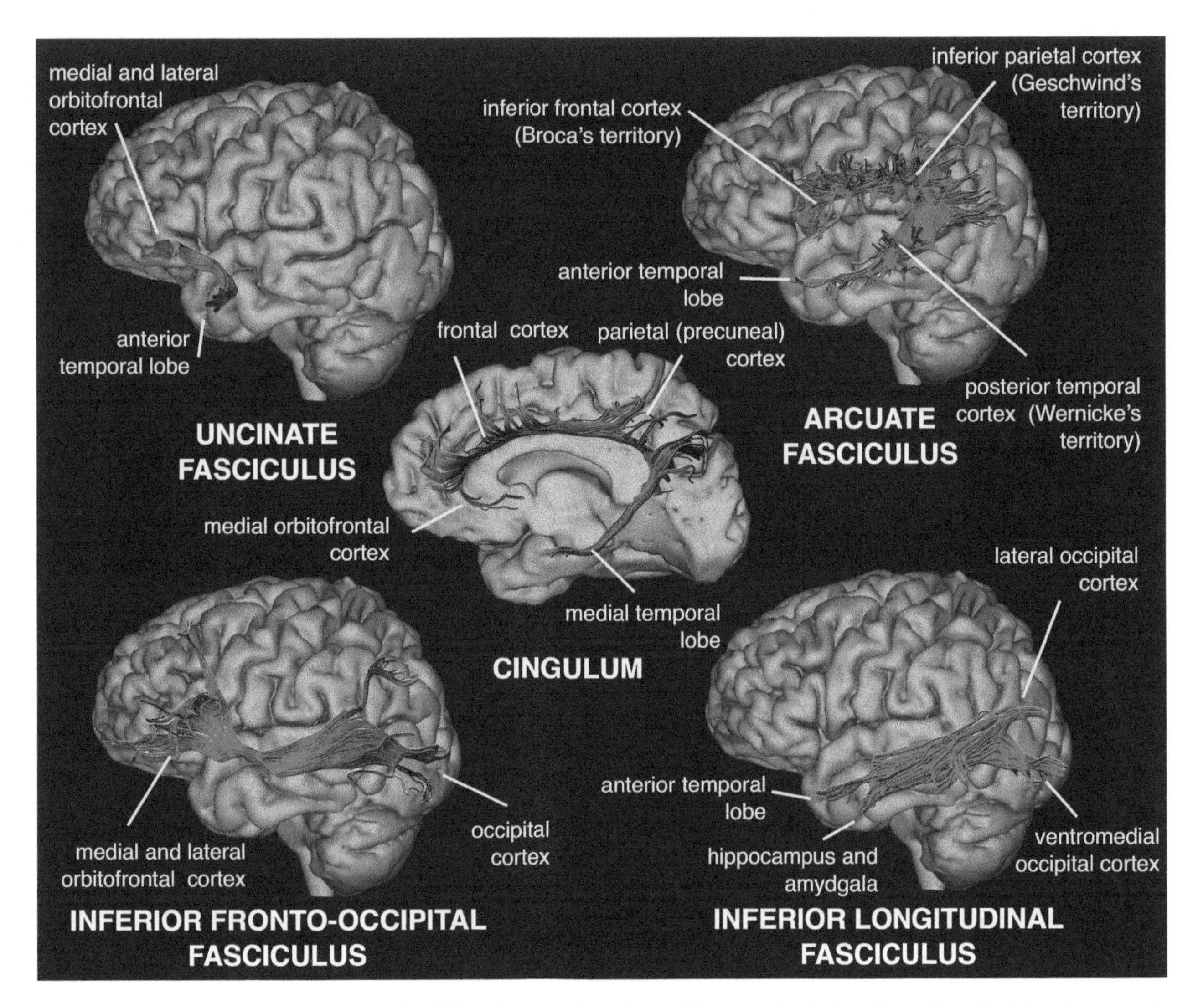

Figure 4.2 Diffusion tensor tractography reconstruction of the major association pathways of the cerebral hemispheres. Lateral view of the left uncinate, arcuate, inferior fronto-occipital and inferior longitudinal fasciculus. Medial view of the right cingulum.

The arcuate fasciculus of the right hemisphere is involved in visuospatial processing and some aspects of language such as prosody and semantics (see also Chapter 7).

The uncinate fasciculus is a ventral tract that connects the anterior temporal lobe with the medial and lateral orbitofrontal cortex. This fasciculus is part of the extended limbic system and is involved in memory, emotions, and language (see also Chapter 11).

The cingulum is a medial tract that runs within the cingulate gyrus around the corpus callosum. It contains fibres of different length, the longest of which run from the anterior temporal lobe to the orbitofrontal cortex. Short U-shaped fibres connect the medial frontal, parietal, occipital, and temporal lobes and different portions of the cingulate cortex. The cingulum is part of the limbic system and is involved in attention, memory, and emotion (see also Chapter 11).

The inferior longitudinal fasciculus is a ventral tract with long and short fibres connecting the occipital and temporal lobes. Its long fibres, which are medial to its short fibres, connect visual areas to the temporopolar cortex, amygdala and hippocampus. The inferior longitudinal fasciculus is involved in object and face perception, reading, visual memory, and in other functions related to language (see also Chapter 8).

The inferior fronto-occipital fasciculus is a ventral tract that connects the inferior and medial occipital lobe to the orbitofrontal cortex. In its occipital course the inferior fronto-occipital fasciculus runs parallel and medial to the inferior longitudinal fasciculus. On approaching the anterior temporal lobe, the fibres of the inferior fronto-occipital fasciculus gather together and enter the external capsule dorsal to the fibres of the uncinate fasciculus. The functions of the inferior fronto-occipital fasciculus are poorly understood, although it is possible that it participates in reading, attention, and visual processing. The existence of an inferior fronto-occipital fasciculus in the monkey brain is still debated (Schmahmann and Pandya, 2006; Catani, 2007) (see also Chapter 8).

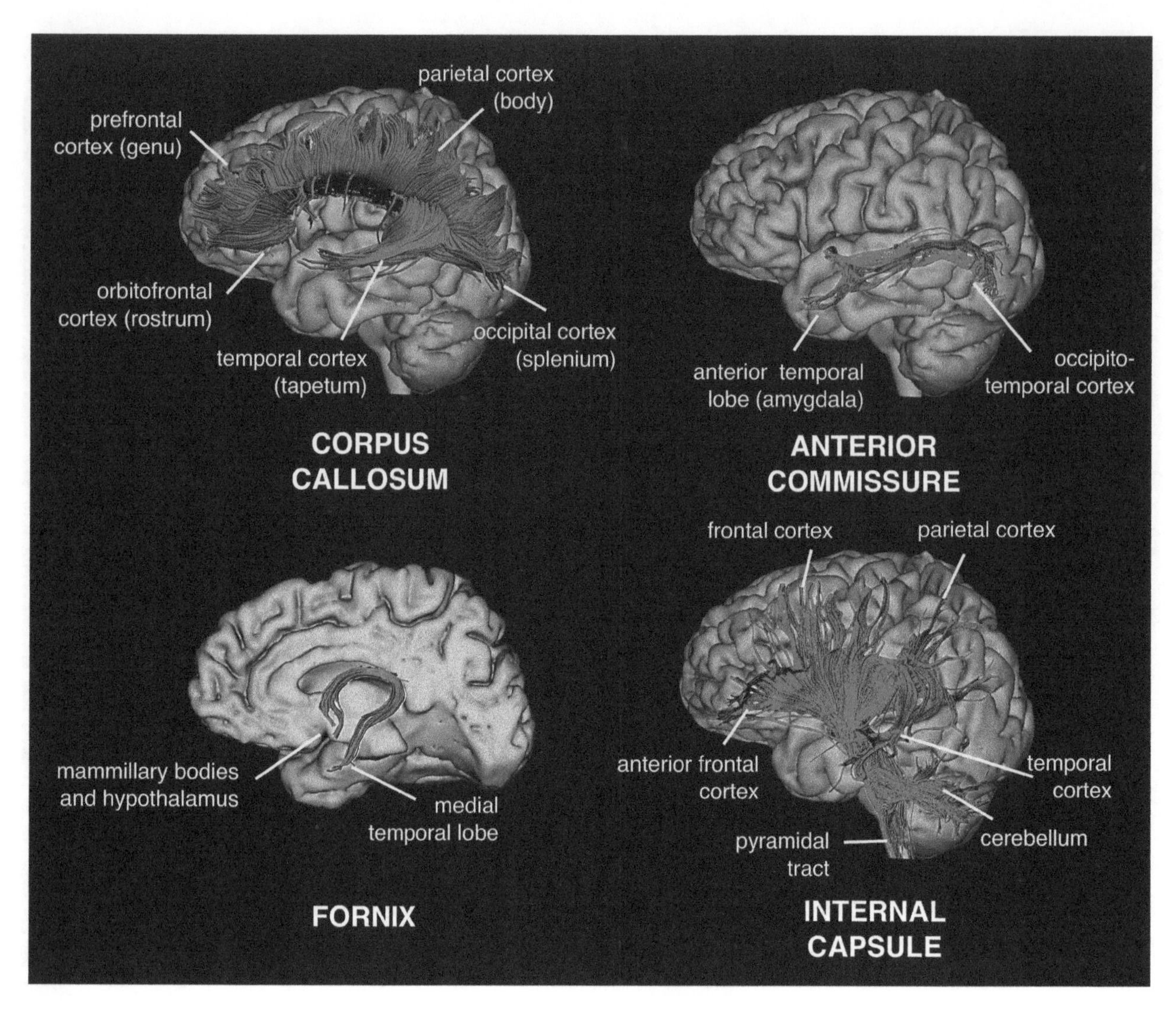

Figure 4.3 Diffusion tensor tractography reconstruction of the major commissural and projection tracts of the human brain. Left lateral view of the corpus callosum, anterior commissure and internal capsule. Left medial view of the fornix.

Commissural pathways

The two most important commissures of the cerebrum are the corpus callosum and the anterior commissure (Figure 4.3). The corpus callosum is the largest tract of the human brain and connects left and right cerebral hemispheres. It is conventionally divided into an anterior portion (genu) connecting the prefrontal and orbitofrontal regions, a central part (body) connecting precentral frontal regions and parietal lobes, and a posterior portion connecting the occipital lobes (splenium). The passage from the posterior body to the splenium is called the isthmus of the corpus callosum. The term tapetum is used for the lateral-inferior extensions of the splenial fibres connecting the posterior temporal lobes. The rostrum is a small part located just below the genu connecting the most medial regions of the orbitofrontal cortex. The fibres of the genu and rostrum arch anteriorly away from the midline and together they form the anterior forceps (or forceps minor). The fibres of the splenium arch posteriorly and form the posterior forceps (or forceps major). The corpus callosum allows transferring of inputs from one hemisphere to the other and is involved in several motor, perceptual and cognitive functions (see also Chapter 9).

The anterior commissure connects bilaterally the anterior and ventral temporal lobes (including the amygdalae) of the two hemispheres, and also the olfactory bulbs. The exact anatomy of its posterior connections (temporo-occipital) and its function in humans are not fully understood.

Projection pathways

The fornix is a projection tract that connects the medial temporal lobe (i.e. the hippocampus) to the septal nuclei, mammillary bodies, and hypothalamus. The fornix belongs to the limbic system and is involved in memory functions (Figure 4.3) (see also Chapter 11).

The internal capsule and corona radiata contain ascending fibres mainly from the thalamus and descending fibres from the fronto-parietal cortex to subcortical nuclei, including the basal ganglia, the brainstem nuclei, and the spinal cord. This complex projection system conveys sensorial information to the cortex and controls movement (see also Chapter 10).

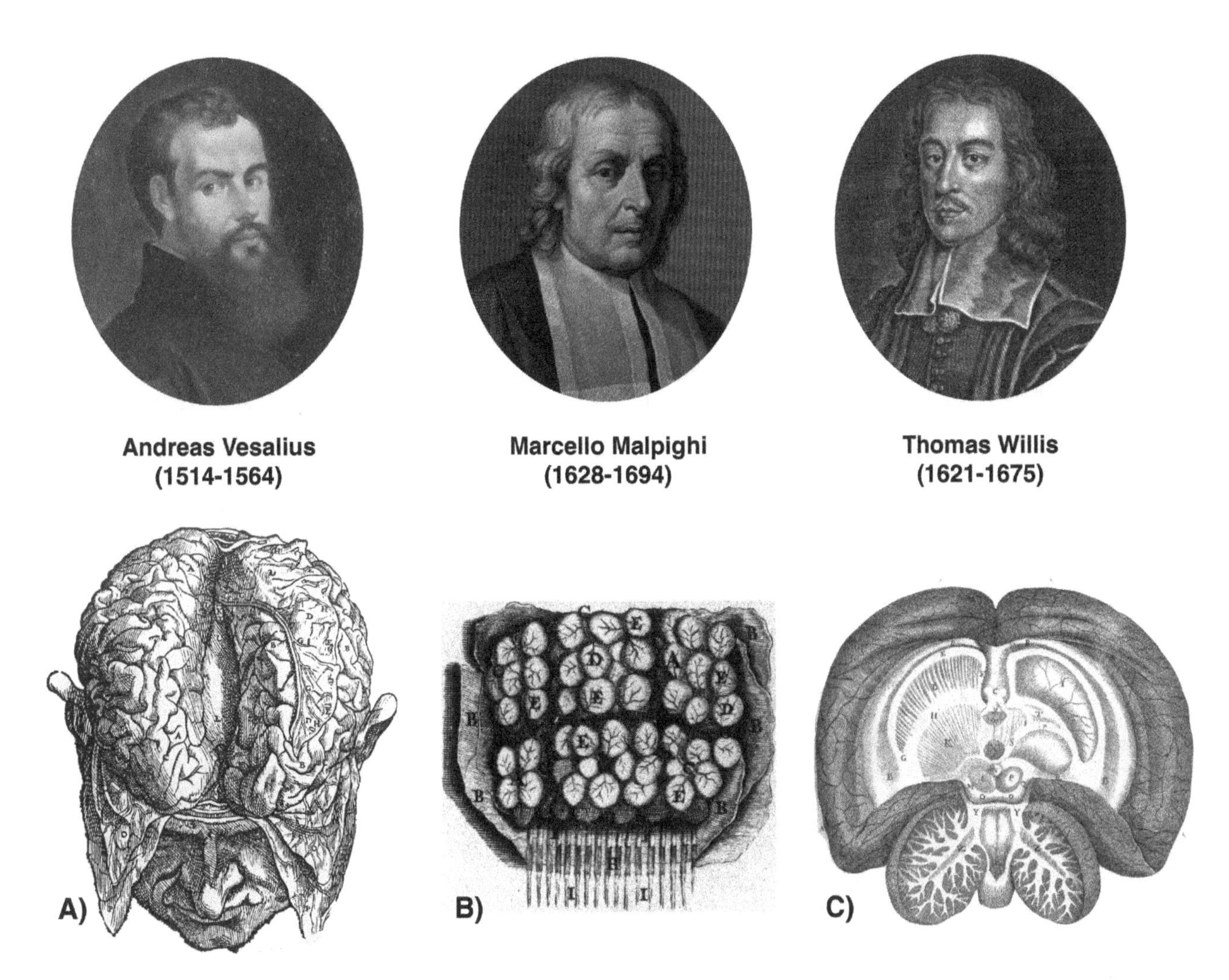

Figure 4.4 The study of white matter anatomy during the Renaissance before the development of advanced methods for fibre dissections. A) Vesalius (1543) described the corpus callosum following separation of the cerebral hemispheres along the longitudinal fissure (portrait adapted with permission of Musée des Beaux-Arts, Orleans); B) Malpighi (1666) used the microscope to describe the filamentous structure of the white matter fibres originating from the cortex; C) Willis (1664) observed 'streaks' in the corpus striatum and speculated on their ascending and descending course, and their possible sensory and motor functions. (Portraits of Malpighi and Willis from the Wellcome Library, London)

From connectional neuroanatomy to function

A differentiation between grey and white matter was never made in antiquity, however some of the tracts visible to external inspection were described. Galen, for example, mentions the corpus callosum and the fornix on several occasions, without referring to any possible function. In the Renaissance, Andreas Vesalius gives a more detailed description of the corpus callosum (Figure 4.4A) and suggests possible mechanical functions:

It relates the right side of the cerebrum to the left; finally, through that septum it supports and props the body formed like a tortoise [fornix] so that it may not collapse and, to the great detriment of all the functions of the cerebrum, crush the cavity common to the two [lateral] ventricles of the cerebrum. (Vesalius, 1543)

The use of a rudimental microscope allowed Marcello Malpighi (1666) to discover that the white matter of the brain is composed of fibres (Figure 4.4B) with a complex anatomy:

Regarding the marrow of the brain [white matter] or the corpus callosum... I was able to observe clearly [...] that the whole of this white part of the brain is divided into roundish fibres... All the fibres dispersed through the brain and cerebellum seem to have origin from the trunk of the spinal marrow [cord]... These fibres of the brain extend lengthwise not, however, parallel but often meeting and apparently gathered into a bundle; soon, however, they again divide and are carried above the other lateral fibres so that they resemble a loose net. (Malpighi, 1685)

In the years that Malpighi was reporting his findings Thomas Willis published his *Cerebri Anatome* (1664), in which he used sections of the brain to differentiate speculatively between ascending and descending tracts in the corpus striatum and motor and sensory pathways in the brainstem (Figure 4.4C):

If these [striatal] bodies are sectioned longitudinally through the middle, they appear marked with medullary streaks like rays; these streaks have a double purpose or tendency, that is, some descend

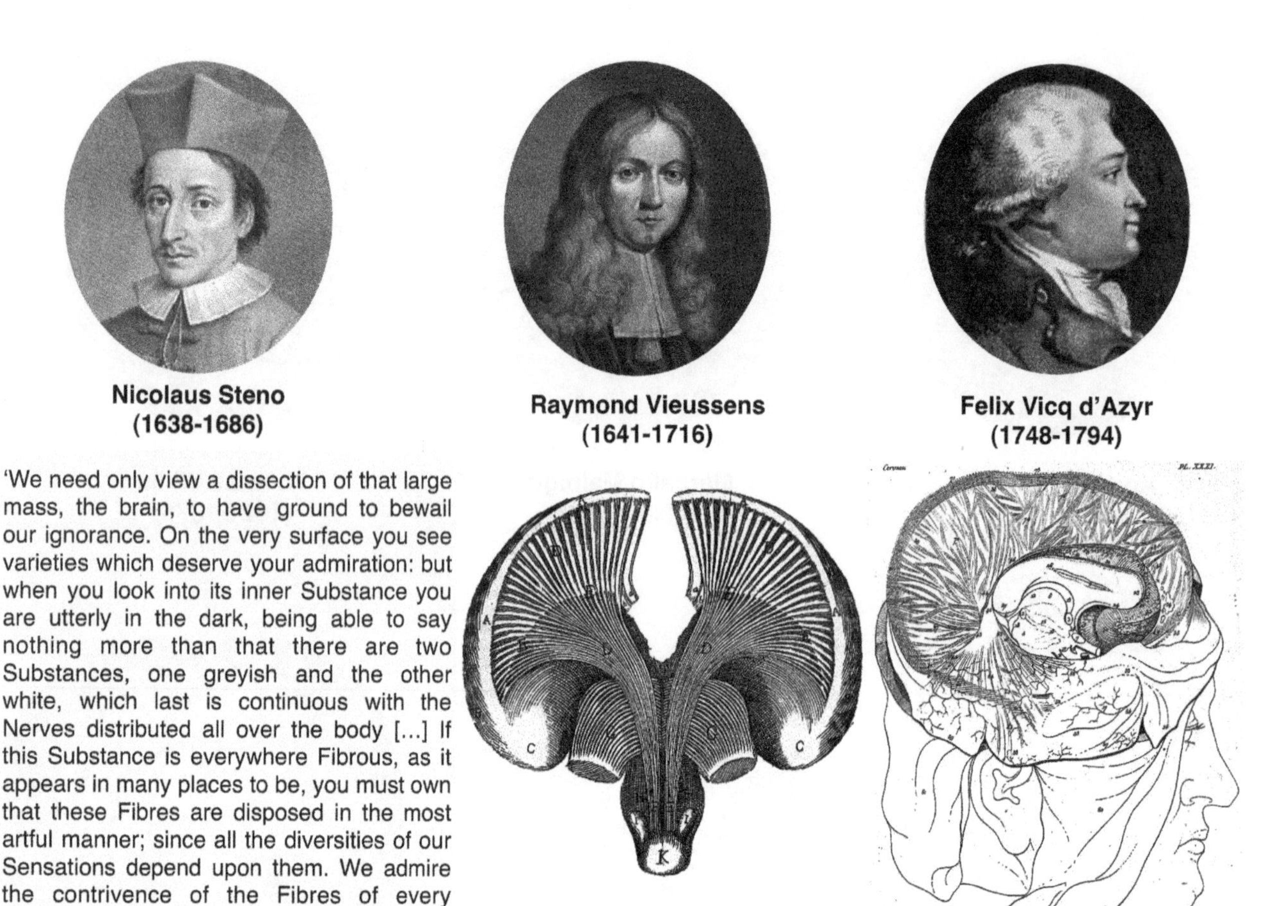

Figure 4.5 The beginning of the modern study of white matter connections based on methods for fibre dissection. A) Vieussens (1684) used post-mortem dissections to identify white matter tracts and was the first to separate the centrum ovale, which is composed of projection and association fibres from the commissural fibres of the corpus callosum (removed in the figure). B) Vicq d'Azyr (1786) used blunt dissection to separate commissural from association fibres.

from the top of this as if they were tracts from the cerebrum into the medulla oblungata, and others ascend from the lower part and meet up with the former... The medulla oblongata seems a broad, almost a royal, highway into which the animal spirits... are carried into all the nervous parts of the body; when the spirits are disposed in order in this common passage or, so to speak diatasso in regular series, they serve two purposes, that is, either they may be directed outwards towards the nerves, at which time they exert the locomotive faculty; or flow inwards towards their sources when the acts of sensation, or rather perceptions of sensible things, are performed. (Willis, 1664)

For Nicolaus Steno these speculations were beyond the available data and called for new methods to study the human brain connections (Figure 4.5). He was the first to suggest that one way of studying the white matter was *'to follow the nerve threads through the substance of the brain to find out where they go and where they end'* (Steno, 1669). To achieve this, neuroanatomists began to develop methods for hardening the soft brain tissue. Raymond Vieussens, for example, who boiled the brain in oil before dissecting it, distinguished the callosal fibres of the corpus callosum from those of the 'centrum ovale' (i.e. white matter of the cerebral hemispheres) (Figure 4.5A) (Vieussens, 1684). Despite these early anatomical achievements novel ideas about the functional correlates of brain connections were not forthcoming, and models of brain function remained tied to the pineal gland and spirits flowing from the ventricles into the hollow nerves.

Arguably the modern era of the study of white matter tracts begins with the work of Felix Vicq d'Azyr (Figure 4.5B), who identified other two groups of fibres in addition to the projection tracts and laid down the fundaments for a network-based theory of brain function:

It seems to me that the commissures are intended to establish sympathetic communications between the different parts of the brain... These communications can in general be divided into two classes: the first run from one hemisphere to the other, the second between different regions of the same hemisphere... When dealing

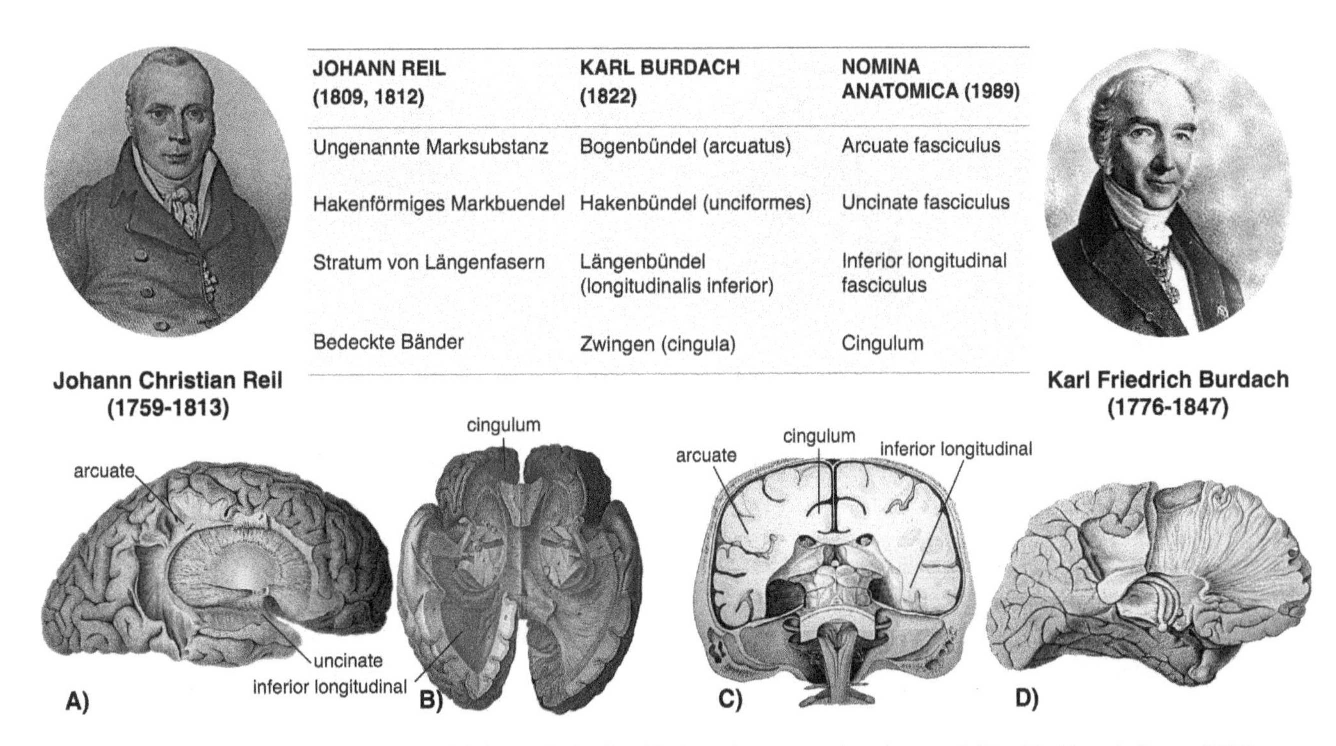

JOHANN REIL (1809, 1812)	KARL BURDACH (1822)	NOMINA ANATOMICA (1989)
Ungenannte Marksubstanz	Bogenbündel (arcuatus)	Arcuate fasciculus
Hakenförmiges Markbuendel	Hakenbündel (unciformes)	Uncinate fasciculus
Stratum von Längenfasern	Längenbündel (longitudinalis inferior)	Inferior longitudinal fasciculus
Bedeckte Bänder	Zwingen (cingula)	Cingulum

Figure 4.6 The discovery of the association tracts of the human brain. The table shows the correspondence between Reils's original terms in German (1809b, 1812a, 1812b), Burdach's German and Latin terminology (1822) and the names used in contemporary international nomenclature (Committe IAN, 1989). A, B) The development of a method based on the soaking of the brain in alcohol that made it more suitable for dissection led Reil to discover the major association white matter tracts of the human brain. C, D) Burdach confirmed the findings of Reil and introduced a novel nomenclature for the main association tracts that remains almost unchanged in the current international nomenclature.

with the origin of each nerve I shall show that the fibres which form them are very numerous and the majority proceed to very different regions, and that whether with reference to the nerves or with reference to the structure of the brain, everything is arranged in the system to multiply the connections of different parts of the brain so that inconveniences which would result from difficulty occasioned in any part of the brain, are prevented. (Vicq d'Azyr, 1786)

In the 19th century the continuous development of methods for fibre dissection led to a series of anatomical discoveries culminating in the identification of most of the association tracts of the human brain. The method of Johann Christian Reil, based on the soaking of the brain in alcohol (Reil, 1809a), allowed him to reveal for the first time the course of most of the association bundles running beneath the cerebral convolutions (Figure 4.6A, B). Among the tracts he identified are the medial curved fibres within the cingulate gyrus (i.e. Bedeckte Bänder) (Reil, 1812a) and the lateral arching connections coursing beneath the perisylvian fronto-parieto-temporal gyri that he described as the unnamed white matter substance (i.e. Ungenannte Marksubstanz) (Reil, 1812b). Other tracts that he identified on the most ventral part of the brain are the hooked-shaped fibres behind the insula (i.e. Hakenförmiges Markbuendel) (Reil, 1809b) and a longitudinal bundle between the occipital and the temporal lobes (i.e. Stratum von Längenfasern) (Reil, 1812b). Reil's findings were confirmed a decade later by Karl Friedrich Burdach in his *Vom Baue und Leben*, a three-volume textbook containing also the description of some previously unidentified tracts (Figure 4.6C, D) (Burdach, 1819–1826). For example, he comments on the existence of some fibres of the occipital-temporal connections that, instead of ending in the anterior temporal lobe, continue forward through the external capsule to project in the ventral part of the frontal lobe. This component was later identified as a separate tract, named the inferior fronto-occipital fasciculus (Curran, 1909). Burdach used Latin names for the major tracts, which became widely adopted and still remain almost unchanged in the current international anatomical nomenclature (Figure 4.6) (Committee IAN, 1989). Unfortunately Reil and Burdach's functional interpretation of the role of white matter fibres was either dismissive of the role of the association tracts or based on erroneous physiological speculations inspired by romantic philosophy.

One had to await the second half of the 19th century for the emergence of the hodological theme following the spread of 'associationist' models of cognitive functions from the realm of psychology to that of neurology and psychiatry (Meynert, 1885; James, 1890; Wundt, 1904). According to such models, the 'formation of concepts', the 'recall of memories', the 'naming of objects', and even the 'spontaneous and voluntary initiation of movement' required the associative convergence (or integration) of information from multiple sources. The association tracts seemed to be the ideal anatomical substrate for this integrative role. The idea of association has roots in Aristotle's writing and has been passed down the centuries from Epicurus through Hobbes to Hartley (Glassman and Buckingham, 2007). However, the credit for the formulation of

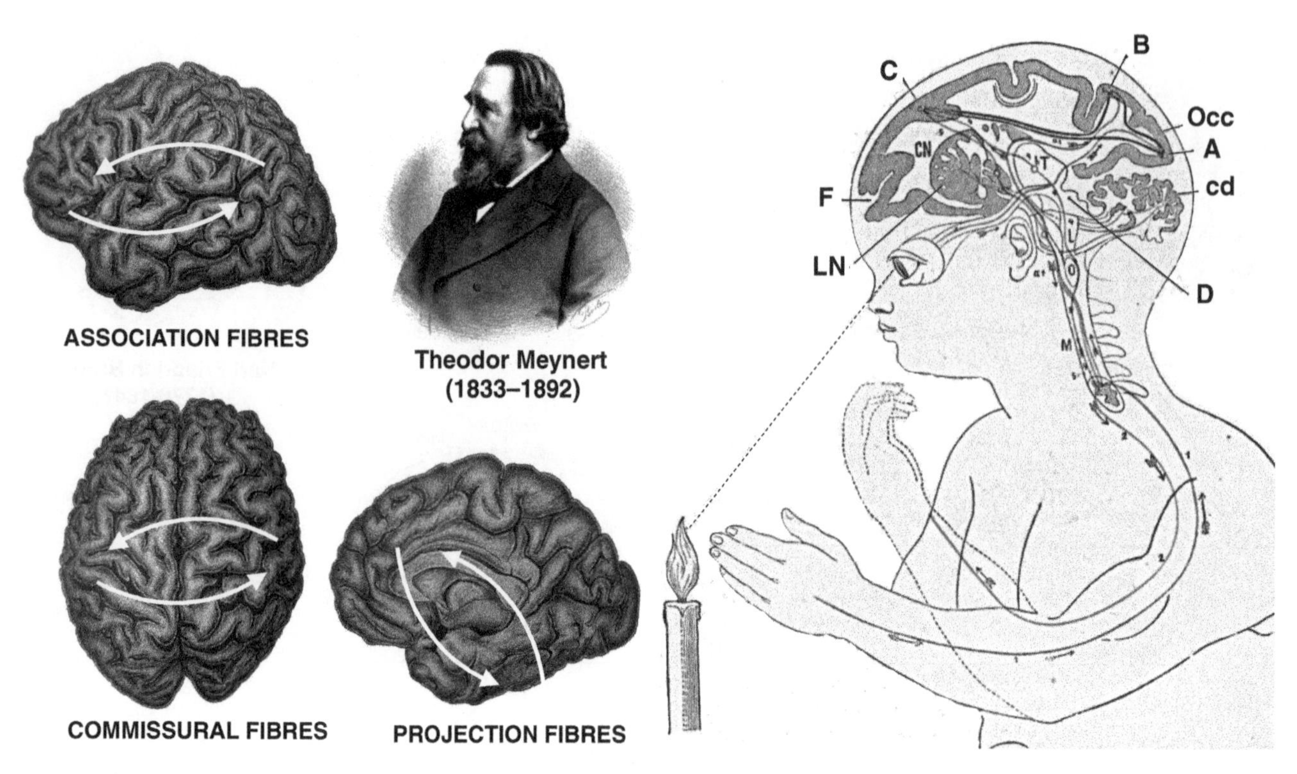

Figure 4.7 Meynert's classification of white matter tracts and the original diagram of the child touching a flame that he used as an example to explain his associationist theory applied to the conscious movement of the arm. Meynert (1885) wrote: *'Let us suppose a flame to have injured the hand of a child, and that the latter withdrew the hand from the flame. This movement will be effected, without the intervention of conscious impulses'* [through a spinal reflex. The visual image of the flame, the painful sensation and the sensation of the reflex movement will also be conveyed to cortical centres (A and B).] *'Since the centre C is connected with [the other sensory centres] the child need not actually burn his hand again before guarding against the flame; but the memory of the flame and of its effect (through association with the centre in which the painful sensation has been stored), will suffice, through the one or the other of these associations, to initiate a movement which will put the arm beyond the reach of the flame.'* Note that for Meynert the centre C in the frontal lobe is an associative centre where the convergence of multiple sensory inputs are translated into motor planning carried out by subcortical effectors (e.g. striatum). A, part of the visual cortex; B, part of the cortical centre for 'cutaneous sensation'; C, a centre for 'sensations of innervation'; cd, cerebellum; CN, caudate nucleus; D, mesencephalon; F, frontal cortex; L, pons Variolii; LN, nucleus lenticularis; M, medulla spinalis, terminating with a cross-section of the cervical spinal cord; O, medulla oblungata; Occ, occipital cortex; hT, thalamus opticus. Black connections correspond to association fibres, blue, centripetal projection fibres and red centrifugal projection fibres.

an associationist theory grounded on anatomy falls to a psychiatrist known by his contemporaries as the great brain-anatomist, Professor Theodor Meynert of Vienna. Meynert's ambitious clinical research programme aimed to establish the anatomical bases not only of mental disorders but also of specific symptoms; the success of its realization depending entirely on a deep anatomical knowledge of the human brain. He first classified white matter connections into projection, commissural, and association fibres (Figure 4.7). The last of these was further subdivided into two groups, the *U-shaped archiform fibres* and the *long association bundles*, according to their cortical projections and the length of their subcortical course:

The U-shaped bundles of the cortex do not necessarily extend simply from one convolution to one next adjoining, but they may skip one, two, three, or an entire series of convolutions, and may just join convolutions which are united among themselves to a convolution lying at some distance from these. The shortest fibrae propriae [or archiform fibres] lie nearest the cortex; the longest at the greatest depth, and are separated from other intervening fibrae propriae, the length of which increases gradatim from the surface inward... In examining the structure of the hemispheres, and remembering that different, distinctly limited and functionally separated portions of the cortex receive impressions from the various senses, we may naturally infer that the association-bundles, the fibrae propriae of the cortex, which form anatomical connections between the different cortical regions, effect the physiological associations of the images which are stored in these various parts. (Meynert, 1885)

Thus according to Meynert's model, the cortex, through its projection and association connection system, becomes not only a place for sensation and motor response, but also for higher cognitive functions and complex behaviour, such as 'logical functions', 'recollection', 'learning', and 'initiation of conscious movement' (Figure 4.7) (Meynert, 1885). Meynert was an outstanding anatomist of international repute who attracted young doctors eager to learn anatomy from all over Europe and the North America. Among them Carl Wernicke, Sergei Korsakoff, Auguste-Henri Forel, Paul Flechsig, Bernard Sachs, and Sigmund Freud. Meynert used his neuroanatomical findings to develop a theory of psychological function, but it was one of his most talented students, Carl Wernicke, who brought the associationist model to the clinic.

Carl Wernicke graduated in medicine in Breslau (now Wroclaw in Poland), where he performed most of his studies except for

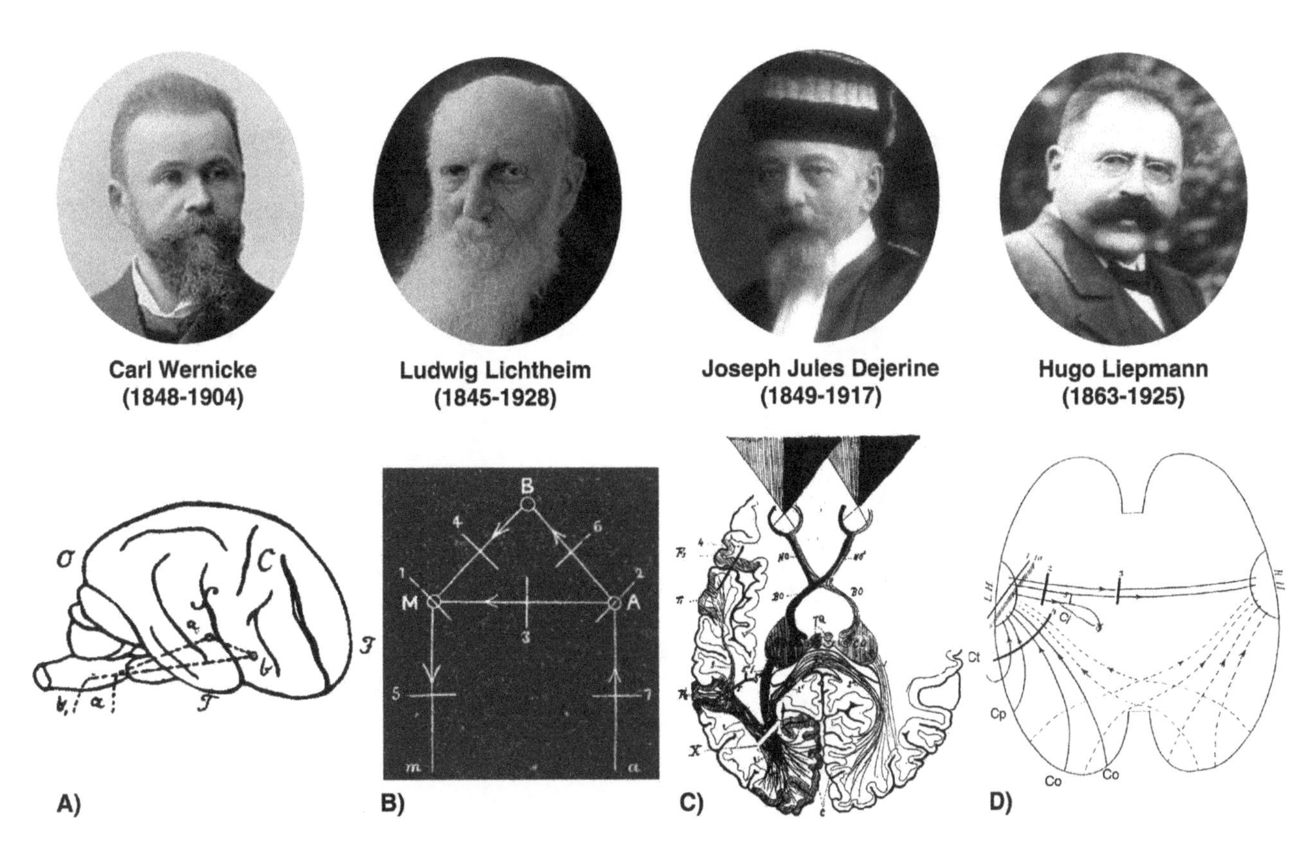

Figure 4.8 Classical disconnection syndromes described by 19th-century clinical neuroanatomists. A) Wernicke's (1874) diagram of the language network encompassing a frontal speech centre (Broca's centre) and a temporal comprehension centre (later termed the Wernicke's centre). He predicted that lesions to the connections between frontal and temporal language centres manifest with conduction aphasia. B) Lichtheim's (1885) diagram for a disconnectionist interpretation of aphasic disorders (1, Broca's aphasia, 2, Wernicke's aphasia; 3, conduction aphasia; 4, transcortical motor aphasia; 5, subcortical motor aphasia; 6, transcortical sensory aphasia; 7, subcortical sensory aphasia). (Reproduced from Lichtheim, © 1885 with permission from Oxford University Press.) C) Dejerine (1892) described the anatomy of the lesion of a patient presenting with acquired reading deficits. The autopsy revealed a lesion (X) to the connections between the visual verbal centre (angular gyrus) and the visual areas in both hemispheres. D) Liepmann's (1925) diagram of the networks underlying praxis consisting of direct connections between primary motor area (L.H.) and primary sensory areas in the temporal (Ct), occipital (Co) and parietal cortex (Cp). Apraxia may result from disconnection at different levels of the network, which he postulated to be asymmetrical (as indicated by the continuous and dashed lines connecting to L.H.). (Reproduced from Liepmann, 1925 with permission from Springer Science+Business Media) (Image of Hugo Liepmann reproduced with kind permission from Professor Bernd Holdorff)

a six-month period in Vienna with Meynert. Wernicke was greatly influenced by his teacher's associationist theory and in his MD thesis 'Der Aphasische Symptomencomplex' postulated that if higher functions arise through associative connections, disorders of higher function can derive from their breakdown. His greatest merit was to describe the first disconnection syndrome that was to become the prototype for all others—conduction aphasia (Leitungsaphasie) (Figure 4.8A) (Wernicke, 1874). Wernicke returned to Breslau between 1890 and 1904, where he established one of the most important clinics for neurology and psychiatry inspired by the associationist manifesto (Catani and ffytche, 2005) (see also Chapter 5). Here the disconnection paradigm was applied not only to conduction aphasia but also to other neurological syndromes (e.g. associative agnosia and apraxia) (Lissauer, 1890; Liepmann, 1900) and psychiatric disorders, such as schizophrenia (Wernicke, 1906). However, it soon became evident that the disconnectionist paradigm per se was not sufficient to explain lateralization of certain syndromes (e.g. aphasia and apraxia). Wernicke had to compromise with narrow localizationist theories to explain the lateralization of language disorders and he postulated the existence of specialized language centres in the left hemisphere. Similarly, Jules Dejerine described a disconnection syndrome characterized by the inability to read with intact writing, namely pure alexia, and localized the centre specialized for reading in the left angular gyrus (Figure 4.8C) (Dejerine, 1892).

Although cortical specialization became a widely adopted explanation for most of the lateralized neurological syndromes, it was one of Wernicke's students, Hugo Liepmann, who put forward an alternative explanation for the hemispheric lateralization: the anatomical lateralization of connections. Hugo Liepmann joined Wernicke's clinic as an assistant in 1895 and, when he left four years later, carried the Breslau associationist doctrine to Berlin. Here he developed an interest in the motor system, which led him to propose a disconnectionist account of goal-directed movement disorders—the apraxias. Liepmann's theory of apraxia, first published in 1900, was based on his case study of a 48-year-old imperial councillor (*Regierungsrat*), admitted to the Berlin psychiatric service with a diagnosis of mixed aphasia and dementia (Liepmann, 1900). Liepmann formulated a network model for praxis and hypothesized a disconnection of visual, auditory, and somatosensory areas from motor areas to explain the symptoms displayed by the councillor (Figure 4.8D). He then speculated on the leftward asymmetry of the praxis networks to explain the higher prevalence of left hemisphere lesions in these patients:

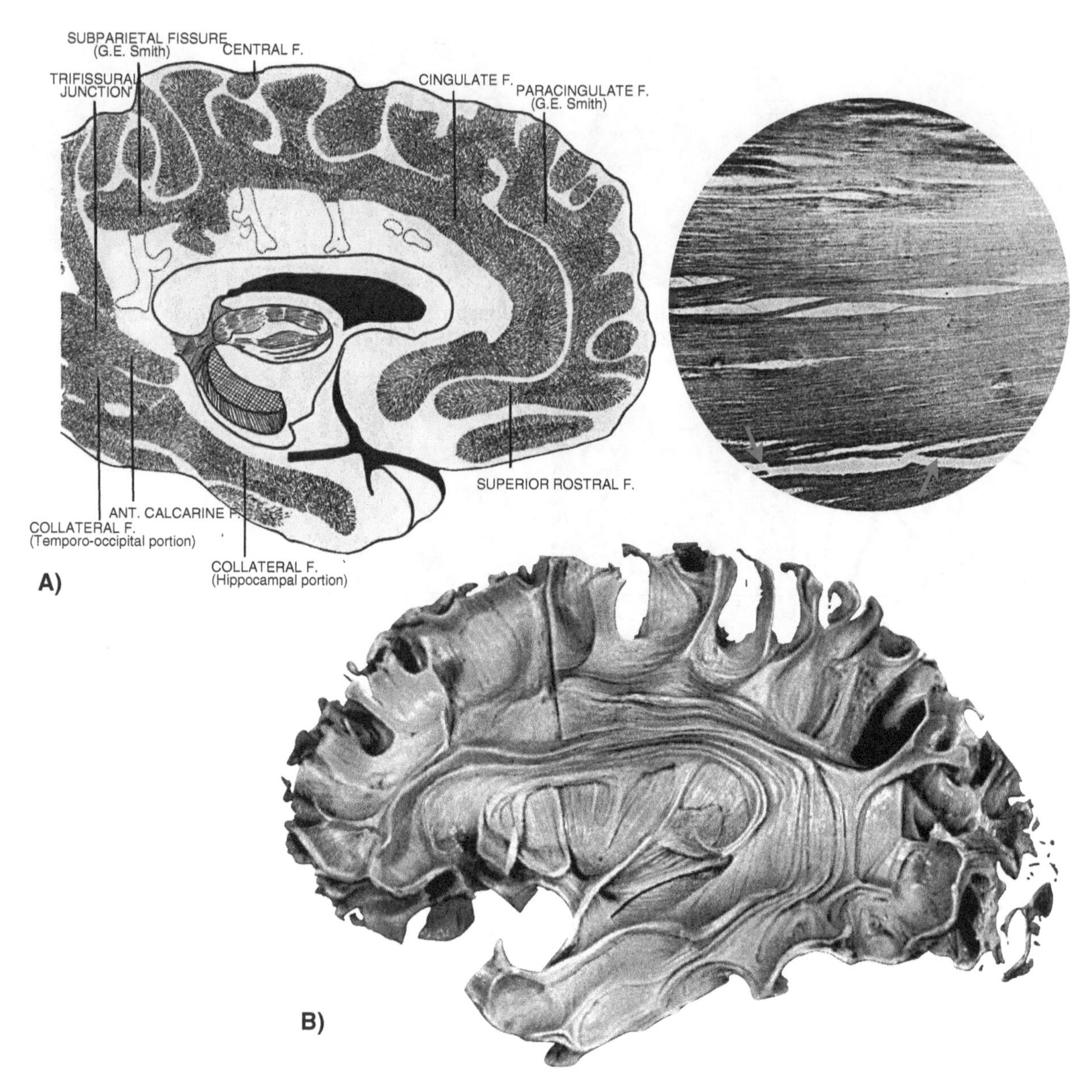

Figure 4.9 White matter anatomy delineated using post-mortem preparations and blunt dissections. A) Sagittal view of the medial connections of the left hemisphere from Rosett (1933). The method of using CO_2 to generate micro-explosions between fibres does not always separate fibres along a natural line of cleavage (red arrows indicate transected fibres). (Reproduced here by courtesy of Columbia University Press.) B) The method of Klingler (1935) uses freeze-thawing of water instead of CO_2. The preparation makes the brain suitable for blunt dissections of long tracts such as the arcuate fasciculus shown in the image (reproduced here by courtesy of Orell Füssli Verlag AG, Zürich, Switzerland).

Eupraxic movement results from the collaboration of many brain regions with the hand area. Lesions to the cortical regions and especially their connections with the hand centre at different points can impair praxis.[...] Especially the left hemisphere hand centre, including its connections to the rest of the brain and in particular in the same hemisphere, is irreplaceable, therefore, lesions to the left hemisphere are disastrous for praxis of all extremities. (Liepmann, 1908.)

Thus, unlike his predecessors Liepmann took an original position to explain the neurobiological underpinnings of left hemisphere dominance for praxis. His model does not imply the existence of a left dominant cortical area for complex movement control. Instead left dominance for praxis is considered as the result of an asymmetrical distribution of the sensory-motor pathways (Figure 4.8D). Liepmann's explanation was original but highly speculative in the absence of experimental evidence (Goldenberg, 2003). If the associationist school and the disconnectionist paradigm were to replace cortical localizationism in the neurology clinics, anatomical support was urgently needed.

Unfortunately most of the techniques for fibre dissections used by early neuroanatomists were inadequate for quantitative studies of the white matter tracts and did not advance significantly since the time of Reil. Some attempts were later made to ameliorate previous methods. Rosett (1933), for example, developed a technique based on the micro-explosions of the white matter. His method

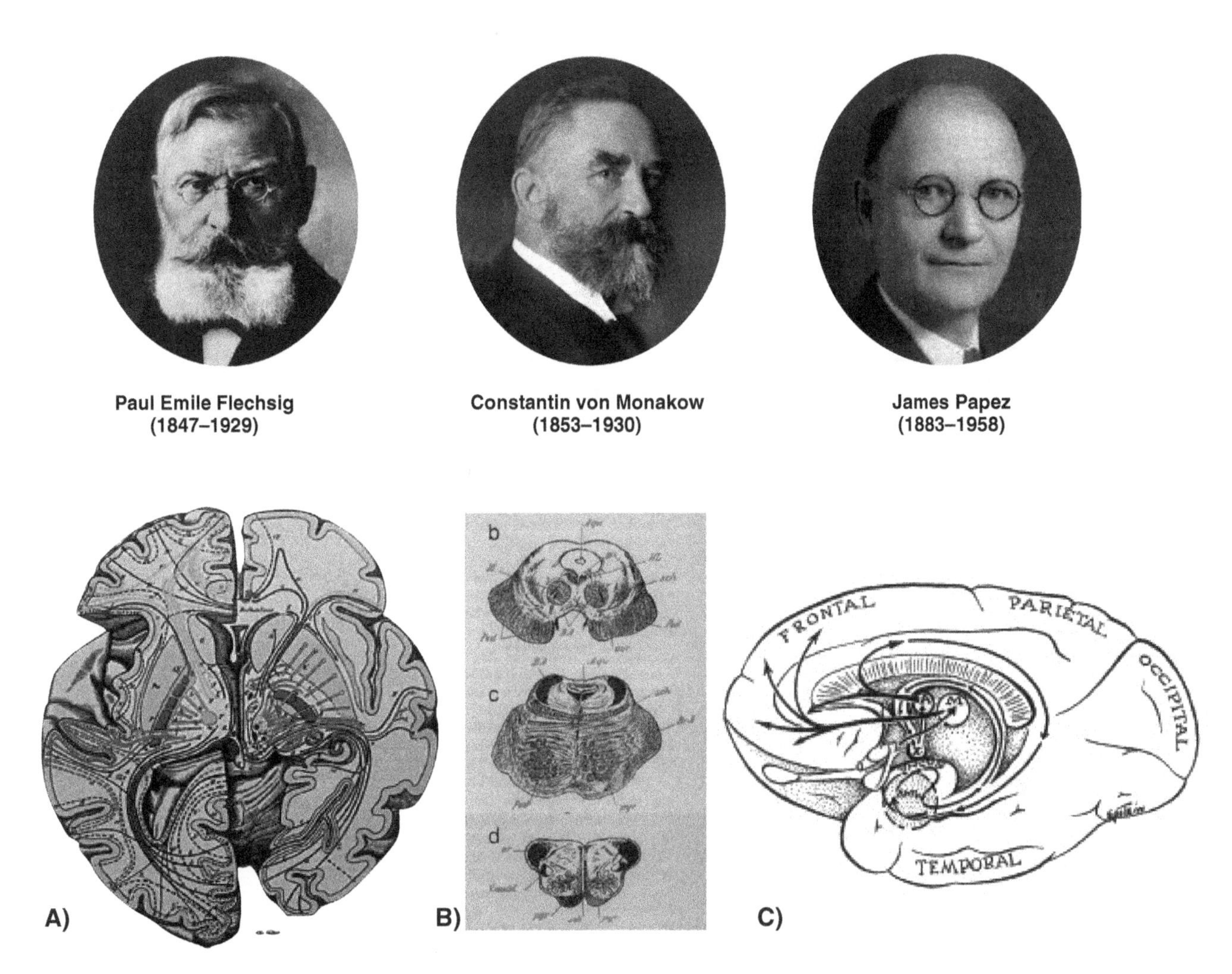

Figure 4.10 The study of the projection tracts flourished at the turn of the 20th century and culminated in the work of Papez that delineated a limbic circuit for behaviour and emotions. A) Drawing from Flechsig's 1896 book where the main projections from the thalamus and basal ganglia are described in detail using different colours (image of Paul Flechsig © Quelle: Archiv für Leipziger Psychiatriegeschichte, Universität Leipzig.) B) Monakow (1897) contributed to the study of the anatomy of several projection connections and developed the concept of diaschisis. In the figure the course of the degenerated pyramidal tract (red) is revealed using Marchi's staining method. C) The limbic system, as originally proposed by Papez (1937), is a circuit underlying emotions and social behaviour and is composed mainly of projection fibres. (From Osrow, 1955 with kind permission from the Psychoanalytic Quarterly Inc.) (Image of James Papez courtesy of the Archives, California Institute of Technology.)

consisted in the immersion of a previously fixed brain in a gas-compressed tank containing liquid carbon dioxide (CO_2). After quickly opening the valve of the tank the sudden reduction of pressure transforms the liquid CO_2 into a gas. The micro-explosions of the cerebral tissues cause a mechanical separation of the fibres along natural lines of cleavage (Figure 4.9A). Similarly, Klingler's preparation (1935) consisted of washing and freezing the brain for several weeks. This resulted in the mechanical separation of the fibres due to expansion of water passing from the liquid to the solid phase. These methods were able to show the 3D course of the major fibre tracts in the human brain (Figure 4.9B). However, dissection needed to be performed very carefully, and neuroanatomical knowledge, experience, and patience were necessary to obtain reliable results. Furthermore, uncertainty about the reliability of the methods to separate anatomically meaningful fibre bundles was a source of major concern. Except for the description of the inferior fronto-occipital fasciculus (Curran, 1909), little advancement derived from gross dissections.

At the same time, methods for neurohistology advanced at a faster pace. The introduction, for example, of staining for fibre degeneration, such as the method or Weigert-Pal or Marchi, and the study of serial sections of the specimens allowed visualization of small fibres in the normal brain or in pathological brains with cortico-subcortical lesions. Bernhard von Gudden, for example, perfected an experimental method in animals whereby he could produce secondary degeneration and atrophy of the nerve nuclei and their connections by removing the peripheral sense organs, such as the eyes, ears, or various cranial nerves (Gudden, 1870). Paul Emile Flechsig (Figure 4.10A) studied the developing pathways with his myelogenetic method consisting of staining myelin in brains of foetuses or newborns and mapping a chronological sequence of the white matter pathways of the brain. Considering that the projections tracts are among the first to myelinate, he was able to follow the origin and course of the corticospinal tract and describe for the first time asymmetry in its crossing (i.e. decussation) at the level of the medulla (Flechsig, 1876). Constantin von Monakow

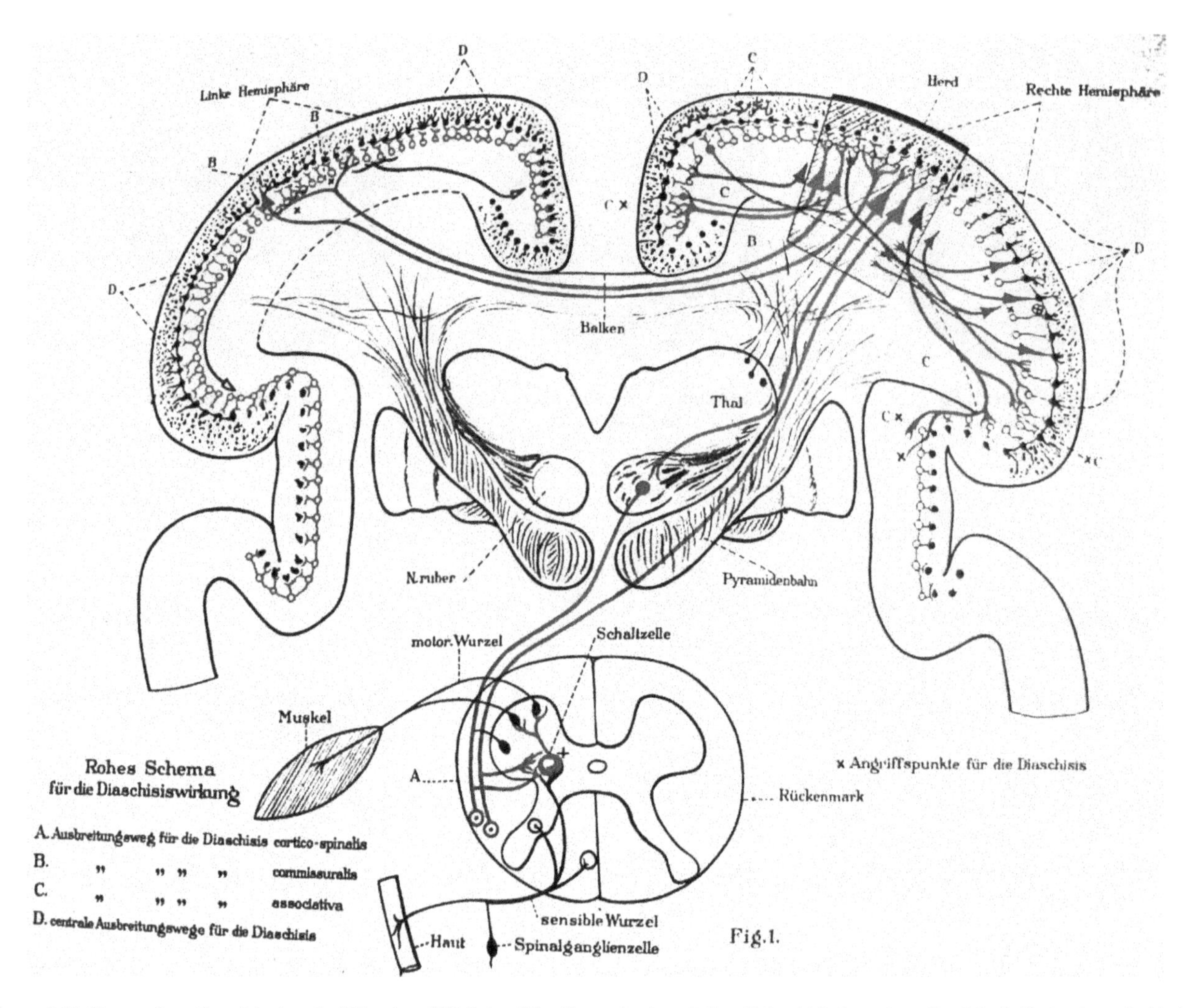

Figure 4.11 Diagram from the original work of Monakow (1914) describing the mechanism of disaschisis, which depends on the distant effects of a cortical lesion (in this case a lesion of the motor area indicated by a striped box).

(Figure 4.10B), an assistant of Gudden, continued the work of his mentor on the thalamic and motor projection fibres. Following 'extirpation' of circumscribed areas of the cortex he was able to follow retrogradely the degenerating tract to the thalamic nuclei and other subcortical nuclei. Monakow was also the first to draw attention to the effects of a lesion on other distant regions of the brain, a mechanism that he called *diaschisis* (Figure 4.11) (Monakow, 1914; Finger et al., 2004) (see also Chapter 5). The work of these authors together with the converging interest of neurophysiologists in the brainstem and thalamus shifted the attention from the associative to the projection fibres. This had two profound effects on the course taken by experimental clinical neuroscience. First, anatomists and physiologists concentrated their efforts in reconstructing cortico-subcortical projection circuits underlying previously unexplored areas of behaviour such as emotions and social behaviour. Between the two world wars many laboratories in North America began to report important anatomical models, including the limbic circuit of James Papez (Figure 4.10C) (Papez, 1937).

Second, neurosurgeons took an interest in this newly proposed models as they provided the anatomical rational for possible treatments of mental disorders. Carlyle Jacobsen and John Fulton, for example, observed reduction in aggressive behaviour without apparent loss of other mental functions in chimpanzees who had the frontal cortex removed by means of lobe cutting (i.e. lobotomy). They were also unable to provoke a form of experimental neurosis in these animals after the operation, supporting the idea that projecting connections from the thalamus to the frontal lobe mediate aggressive behaviour (Jacobsen et al., 1935; Fulton, 1952). These experimental results were presented in 1935 at the 2nd International Neurological Congress in London, attended by many clinical neuroscientists of that time. Among them was Antonio Egas Moniz, a neuropsychiatrist from the University of Lisbon, who on returning to his clinic had the idea of applying the fronto-thalamic disconnection approach to patients with severe psychoses. Together with the neurosurgeon Almeida Lima, they performed the first human leucotomy (i.e. 'white matter cutting') in 1935 (Figure 4.12A) (Moniz and Lima, 1936). They found a general improvement of

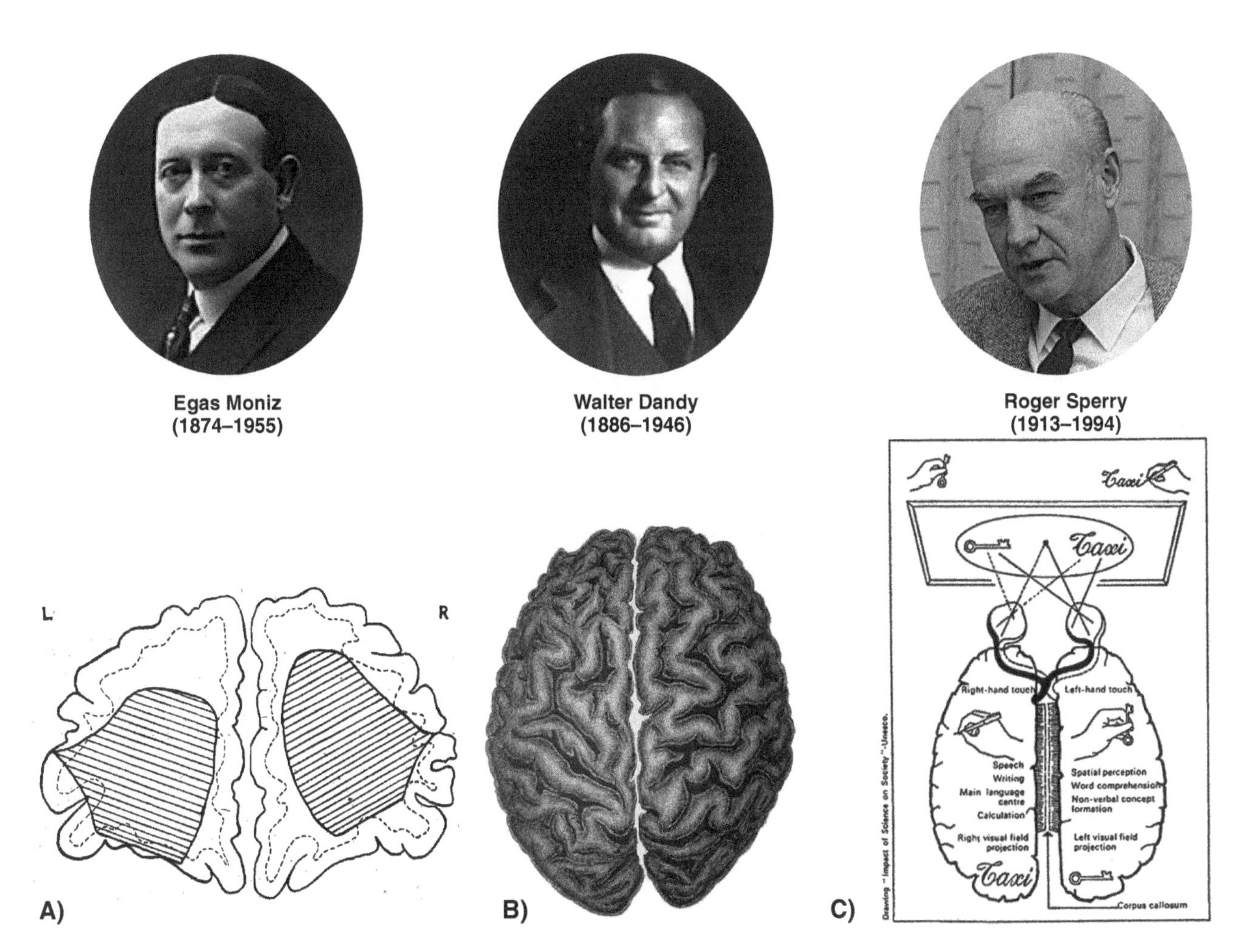

Figure 4.12 The disconnectionist approach to chronic mental illness and epilepsy adopted in neurosurgery in the 1930s and 1940s. A) Diagram illustrating the leucotomy procedure consisting of severing the connection between the thalamus and the frontal lobes (Image from Meyer and McLardy, 1948). B) Callosotomy is a surgical procedure that consists of cutting the callosal connections, thus preventing the spreading of the epileptic discharge from one hemisphere to the other. The work of Dandy was conducive to the development of the method. C) Callosotomized subjects were assessed by Roger Sperry (1974) using adapted neuropsychological tests that demonstrate the lack of transfer from one hemisphere to the other. Many complex functions were thus shown to be localized in only one hemisphere (illustration from the Nobel Committee for Physiology or Medicine, based on *Impact of Science on Society* published by UNESCO. Copyright the Nobel Committee) (Image of Roger Sperry courtesy of the Archives, California Institute of Technology.).

symptoms in severely agitated depressive patients and a modest effect on chronic schizophrenic patients (Moniz, 1937). Despite the high rate of side effects, the method was introduced in North America by an ambitious neurologist, Walter Freeman, who together with the neurosurgeon James Watts operated on the first American patient in 1936 (Freeman and Watts, 1945). Freeman simplified the neurosurgical procedure by adapting an ice pick for a trans-orbital approach to lobotomy through the medial upper corner of the eye cavity. The 'ice pick' lobotomy was easier to perform and accessible to the chronic psychiatric patients, most of them living in asylums. In a span of two decades Freeman performed more than 3000 frontal lobotomies (Freeman, 1957), but in the 1960s the introduction of chlorpromazine, an antipsychotic for rapid tranquillization, and the emergence of data on the severe side effects of lobotomy put an end to the era of radical surgery for severe emotional disorders (Finger, 1994).

Neurosurgery became also a viable approach for the treatment of refractory epilepsy after Walter Dandy's (1936) report of a patient who underwent a cut of the posterior half of the corpus callosum in order to remove a tumour of the pineal gland (Figure 4.12B). The apparent absence of symptoms following the callosal transection gave grounds for the use of callosotomy in severe epileptic patients in an attempt to limit the spread of epilepsy from one hemisphere to the other (Wagenen and Herren, 1940). The patients with a complete disconnection of the two hemispheres (i.e. 'split brain') initially studied by Akelaitis (1941) were reported to have very few negative effects on cognitive functions. Another case series produced by Bogen and Vogel (1963) and studied by Roger Sperry (1974) with more sophisticated psychological testing (Glickstein and Berlucchi, 2008) revealed not only that callosotomy had significant effects on cognition but provided unequivocal evidence for hemispheric lateralization of brain functions (Figure 4.12C). This work on

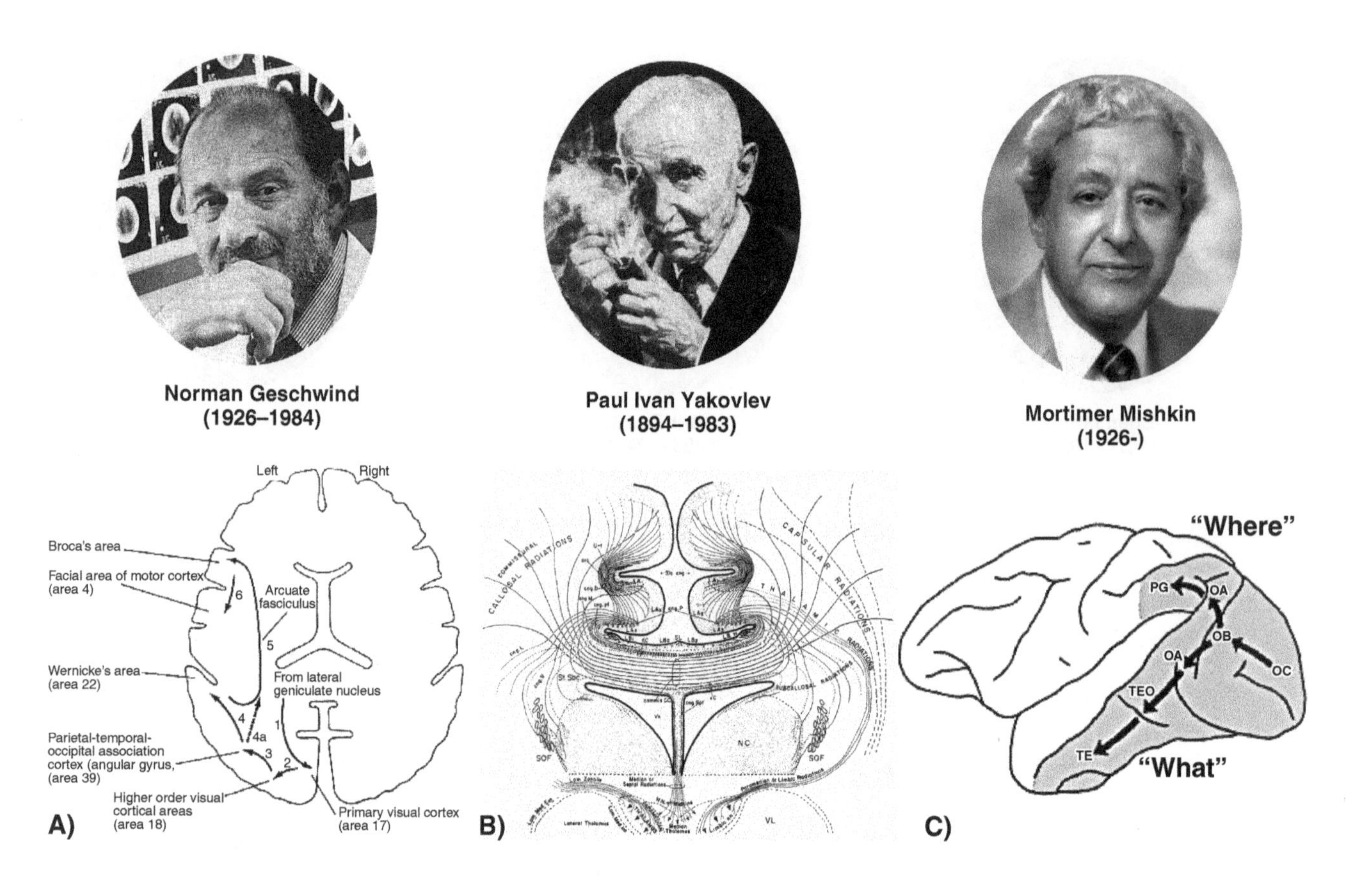

Figure 4.13 Geschwind's neo-associationist approach and the advancement of connectional anatomy derived from animal studies in the second half of the 20th century. A) The Geschwind-Wernicke model for reading described as a series of connections whose existence was not entirely demonstrated in the human brain. Note the serial and unidirectional flow of the visual inputs (i.e. bottom up processing) (Mayeux and Kandel, 1991). B) Yaklovev's (1961) representation of the connections of the cingulate cortex on a coronal slice derived from animal studies, using most of the methods developed by early neuroanatomists such as Gudden and Flechsig. C) The dorsal and ventral streams for visual processing described by Ungerleider and Mishkin (1982) derived from animal electrophysiology, axonal tracing and lesion studies. (Image from Ungerleider and Mishkin, 1982 with permission of The MIT Press.)

split brain patients reawakened the interest in clinico-anatomical correlation studies and began the shift away from holistic ideas that culminated in Geschwind's revived neo-associationism.

By his own admission, the work on split brain patients led Geschwind to *'re-examine the older clinical literature and to re-assess our patients with disturbances of the higher functions'* (Figure 4.13A) (Geschwind, 1965a). The outcome was his 1965 publication, which was to become a manifesto for the neo-associationist school (Geschwind, 1965a, 1965b). In a sense, Geschwind became for the 20th century what Wernicke was for the 19th century, a generation of neurologists being influenced by his approach. But he was not simply Wernicke's, Liepmann's and Dejerine's translator across centuries and languages, his own contribution being to enrich the classical associationist view with the idea of association cortex as an obligatory relay, and the first description of a case of tactile anomia after callosal damage (Catani and ffytche, 2005). He also benefited from the work of many anatomists that were re-exploring the work of early German authors on the phylogeny and ontogeny of connections (Figure 4.13B) (Yakovlev and Locke, 1961).

The disconnection syndromes, as reintroduced to post Second World War neurology by Geschwind, played a pivotal role in linking the insights of the 19th-century clinicians to contemporary neuroscience. Geschwind's annotations and reformulations, articulated in the language of neurons and axons, triggered fertile explorations of cortico-cortical connectivity and helped to interpret the results of behavioural and physiological experiments on monkeys from the vantage point of hodology (Catani and Mesulam, 2008). Mishkin's work on visual and memory pathways and their relation to sensory-limbic disconnection, for example, offered elegant confirmation of novel insights that Geschwind introduced to domains of cognition beyond aphasia (Figure 4.13C) (Mishkin, 1966). The enthusiasm generated by Geschwind helped to attract new talent to behavioural neurology. These clinician-investigators helped to transform the stylized diagrams of the 19th-century disconnection syndromes into the complex architecture of large-scale neural networks where

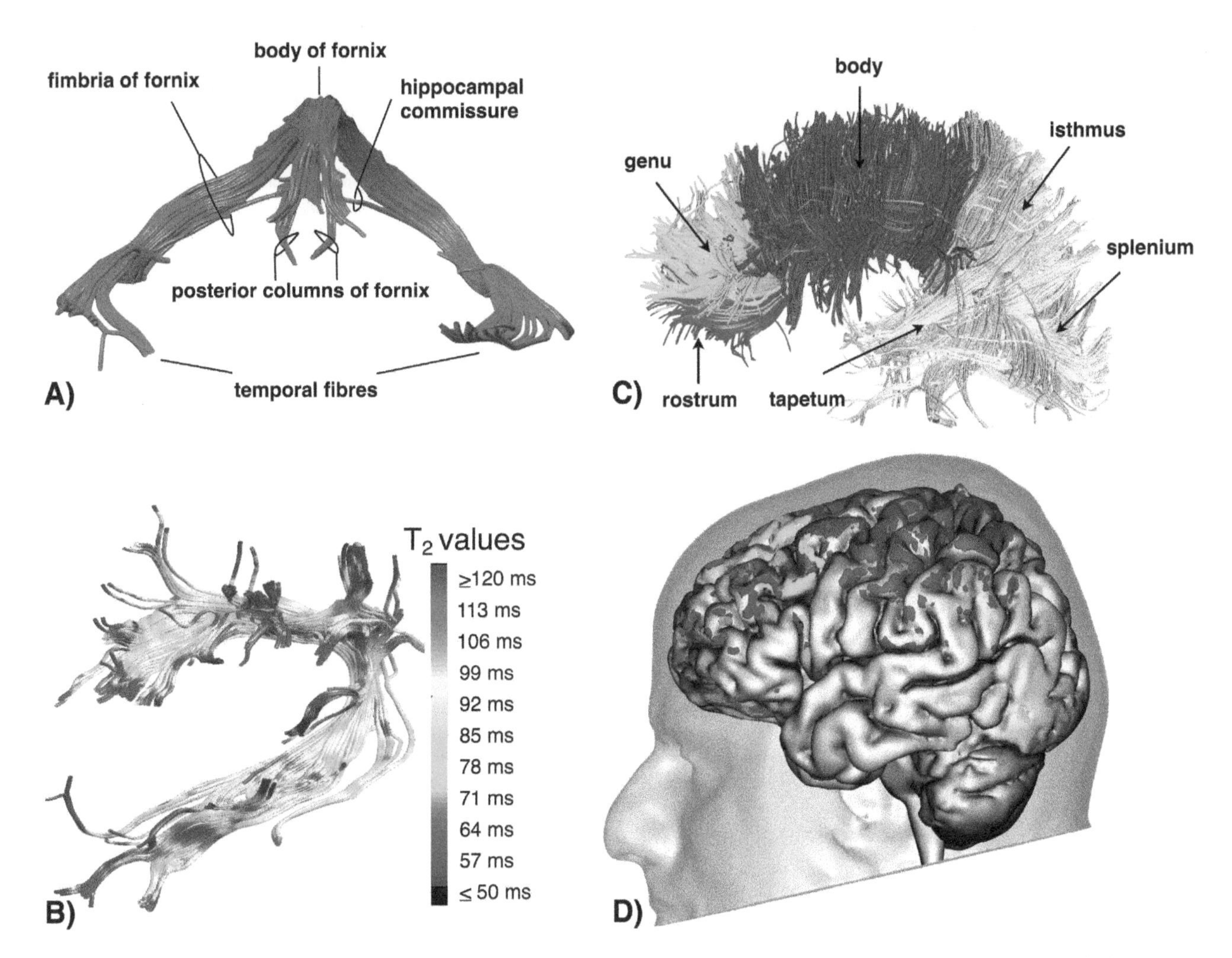

Figure 4.14 MRI methods based on diffusion tractography for the study of the *in vivo* connectional anatomy of the human brain. A) In this example, the tractography reconstruction of the fornix allows the study of its anatomy in detail (anterior view). B) Quantitative mapping of T2 values along the streamlines of the arcuate fasciculus. C, D) Tractography derived reconstructions of the anatomical subdivisions and corresponding cortical projections of the corpus callosum.

critical network epicentres constituted not only relays or convergence centres, but also hubs, nexuses, sluices for convergence, divergence, feedback loops, feed-forward connections, and transition points from serial to parallel processing (Damasio, 1989; Mesulam, 1990).

Although Geschwind never returned to the grand perspective of his 1965 theory, over the next 20 years he continued to work on clinicopathological correlations, extending the remit of his 1965 model beyond classical neurology into neurodevelopmental disorders, such as dyslexia. Together with his 1965 paper, this corpus of work founded behavioural neurology, strongly influenced the development of related disciplines, such as neuropsychiatry and neuropsychology, and prepared the ground for neural network and connectionist theories in the cognitive sciences (Catani and Mesulam, 2008) (see also Chapter 5).

But the dilemma that the behavioural neurologists of the neoassociationist school had to face stemmed principally from the lack of sufficient information on human neuroanatomy. An important consequence of this limitation was that it allowed for hypothetical *post hoc* disconnection explanations of deficits. For a given lesion, the absence of a predicted cognitive deficit (or for a given deficit, the absence of a predicted lesion) is amenable to explanation by any number of alternative pathways making it difficult to test or falsify the disconnectionist account. In contrast to the giant strides made in unravelling the connectivity of the monkey brain, the details of connection pathways in the human brain remained stuck in the methodology of the 19th century. In a scientific commentary in *Nature* in 1993 Crick and Jones voiced these concerns to the scientific community: *'to interpret the activity of the living human brains, their anatomy must be known in detail'* (Crick and Jones, 1993). They urged the *'development of new techniques since most of the methods used in the monkeys cannot be used on humans'*. A year later, in 1994, Basser, Mattiello, and LeBihan published their seminal paper where they describe diffusion tensor imaging (DTI) for the first time. Tractography based on DTI has introduced realistic prospects for visualizing association pathways and performing quantitative studies in the living human brain (Figure 4.14) (Basser et al., 2000; Catani et al., 2002;

Jones, 2008). At least in the short term, the future of the disconnection theme in neurology and psychiatry is likely to be shaped by the spectacular advances that are currently unfolding in the field of imaging. The combination of DTI with functional magnetic resonance imaging (fMRI), and other methods for studying functional links between regions, raises the possibility of identifying function-specific pathways in ways that have never been possible or conceivable in humans. But mapping functions onto single tracts is subject to the same criticisms directed to narrow cortical localizationism derived from other disciplines of anatomy (see also Chapter 5). Tractography is inevitably destined to proceed through trials of errors and successes along a path already traced by our predecessors; their words are a caveat to bear in mind in the journey to come:

... to follow the fascicles, their origin, trajectory and termination, this is the goal of those who, after Vicq d'Azyr, have studied this branch of the anatomy [hodology]. . . despite the improvement of the techniques available to us nowadays, there is still ... more than one obscure point to clarify. (Dejerine, 1895)

References

Akelaitis, A. (1941). Studies on the corpus callosum II. The higher visual functions in each homonymous field following complete section of the corpus callosum. *Arch Neurol Psychiat*, **45**(5), 788–96.

Basser, P.J., Mattiello, J., and LeBihan, D. (1994). MR diffusion tensor spectroscopy and imaging. *Biophys J*, **66**(1), 259–67.

Basser, P.J., Pajevic, S., Pierpaoli, C., Duda, J., and Aldroubi, A. (2000). In vivo fiber tractography using DT-MRI data. *Magn Reson Med*, **44**(4), 625–32.

Bogen, J. and Vogel, P. (1963). Treatment of generalized seizure by cerebral commissurotomy. *Surg Forum*, **14**, 431–3.

Burdach, K. (1819–1822–1826). *Vom Baue und Leben des Gehirns*. Leipzig: In der Dyk'schen Buchhandlung.

Catani, M. and ffytche, D.H. (2005). The rises and falls of disconnection syndromes. *Brain*, **128**(10), 2224–39.

Catani, M. (2007) From hodology to function. *Brain*, **130**(3), 602–5.

Catani, M., Howard, R.J., Pajevic, S., Jones, D.K. (2002). Virtual in vivo interactive dissection of white matter fasciculi in the human brain. *Neuroimage*, **17**(1):77-94.

Catani, M. and Mesulam, M. (2008) What is a disconnection syndrome? *Cortex*, **44**(8), 911–13.

Catani, M., Jones, D.K., and ffytche, D.H. (2005) Perisylvian language networks of the human brain. *Ann Neurol*, **57**(1), 8–16.

Clarke, E. and O'Malley, C.D. (Eds.) (1996). *The Human Brain and Spinal Cord: A Historical Study Illustrated by Writings from Antiquity to the Twentieth Century*, second edition. San Francisco: Norman Publishing.

Commitee, I.A.N. (1989). *Nomina anatomica*, sixth edition. Edinburgh; New York: Churchill Livingstone.

Crick, F. and Jones, E. (1993). Backwardness of human neuroanatomy. *Nature*, **361**(6408), 109–10.

Curran, E.J. (1909). A new association fiber tract in the cerebum. With remarks on the fiber dissection method of studying the brain. *J Comp Neurol*, **19**(6), 645–57.

Damasio, A.R. (1989). Time-locked multiregional retroactivation: A systems-level proposal for the neural substrates of recall and recognition. *Cognition*, **33**(1–2), 25–62.

Dandy, W.E. (1936). Operative experience in cases of pineal tumors. *Arch Surg*, **33**(1), 19–46.

Déjérine, J. (1892). Contibution a l'étude anatomo-pathologique et clinique des differentes variétés de cécité-verbale. *Mém Soc Biol*, **4**, 61–90.

Déjérine, J. (1895). *Anatomie des centres nerveux*. Paris: Rueff et Cie.

Finger S., Koehler, J.P., and Jagella, C. (2004). The Monakow concept of diaschisis. *Arch Neurol*, **61**(2), 283–8.

Finger, S. (1994). *Origins of Neuroscience*. Oxford: Oxford University Press.

Flechsig, P.E. (1876). *Die Leitungsbahnen im Gehirn und Rückenmark des Menschen auf Grund entwicklungsgeschichtlicher Untersuchungen*. Leipzig: Engelmann.

Flechsig, P.E. (1896). *Gehirn und Seele*. Leipzig: Verlag von Veit.

Freeman, W. & Watts, J.W. (1945). Prefrontal lobotomy. The problem of schizophrenia. *Am J Psychiat*, **101**, 739–48

Freeman, W. (1957). Frontal lobotomy 1936-56. A follow-up study of 3000 patients from one to twenty years. *Am J Psychiat*, **113**(10), 877–86.

Fulton J.F. (1952). *The frontal lobes and human behaviour*. Springfield IL: Charles C. Thomas.

Geschwind, N. (1965a). Disconnexion syndromes in animals and man. I. *Brain*, **88**(2), 237–94.

Geschwind, N. (1965b). Disconnexion syndromes in animals and man. II. *Brain*, **88**(3), 585–644.

Glassman, R.B. and Buckingham, H.W. (2007). David Hartley's neurophysiology of association. In Whitaker, H., Smith, C.U.M., and Finger, S. (Eds), *Brain, Mind and Medicine: Essays in Eighteenth-Century Neuroscience*. New York: Springer.

Glickstein, M. and Berlucchi, G. (2008). Classical disconnection studies of the corpus callosum. *Cortex*, **44**(8), 928–35.

Goldenberg, G. (2003). Apraxia and beyond: life and work of Hugo Liepmann. *Cortex*, **39**(3), 509–24.

Gudden, B.A. von (1870). Experimentaluntersuchungen bei das peripherischer und centrale Nervensystem. *Arch Psychiatr Nervenkr*, **2**, 693–723.

Jacobsen, C.F., Wolf, J.B., and Jackson, T.A. (1935). An experimental analysis of the functions of the frontal association areas in primates. *J Nerv Mental Dis*, **82**, 1–14.

James, W. (1890). *The Principles of Psychology, Vol. 1–2*. New York: Henry Holt.

Jones, D.K. (2008). Studying connections in the living human brain with diffusion MRI. *Cortex*, **44**(8), 936–52.

Kemper, T. (1984). Paul Ivan Yakovlev 1894–1983. *Arch Neurol*, **41**(5), 536–40.

Klingler, J. (1935). Erleichterung der makroskopischen Präparation des Gehirn durch den Gefrierprozess. *Schweiz Arch Neurol Psychiat*, **36**, 247–56.

Lichtheim, L. (1885). On aphasia. *Brain*, **7**, 433–84.

Liepmann, H. (1900). Das Krankheitsbild der Apraxie (motorische Asymbolie) auf Grund eines Falles von einseitiger Apraxie. *Monatssch Psychiat Neurol*, **8**, 15–44, 102–32, 182–97.

Liepmann, H. (1908). *Drei Aufsatze aus dem Apraxiegebiet*. Berlin: Karger.

Liepmann, H. (1925). Apraktische Störungen. In Kramer, I. (Ed.) *Lehrbuch der Nervenkrankheiten* Berlin: Springer.

Lissauer, H. (1890). Ein Fall Von Seelenblindheit Nebst Einim Beitrage Zur Theori Derselben. *Arch Psychiat Nervenkrankh*, **21**, 222–70.

Malpighi, M. (1666). *De cerebri cortice*. Bologna: Montius.

Malpighi, M. (1685). De Externo Tactus Organo Anatomica Observatio. Naples: Aegidium Londum. [Translated in Clarke, E. and O'Malley, C. D. (Eds.) (1996). *The Human Brain and Spinal Cord: A Historical Study Illustrated by Writings from Antiquity to the Twentieth Century*. San Francisco: Norman Publishing.]

Mayeux, R., and Kandel, E.R. (1991). Disorders of language: the aphasias. In Kandel, E.R., Schwartz J., and Jessel, T. *Principles of Neural Science*, third edition. New York: Elsevier.

Mesulam M.M. (1990). Large-scale neurocognitive networks and distributed processing for attention, language, and memory. *Ann Neurol*, **28**(5), 597–613.

Meyer, A., and McLardy, T. (1948). Posterior cuts in prefrontal leucotomy: a clinico-pathological study. *J Ment Sci*, **94**(396), 555–64.

Meynert, T. (1885). *A Clinical Treatise on Diseases of the Fore-brain Based Upon a Study of Its Structure, Functions, and Nutrition* (Sachs, B. Trans.). New York: G.P. Putnam's Sons.

Mishkin M. (1966). *Visual mechanisms beyond the striate cortex. In Russel R. (Ed.) Frontiers in physiological psychology*. New York: Academic Press.

Monakow, C. von (1897). *Gehirnpathologie*. Vienna: A. Hölder.

Monakow C. von (1914). *Die Lokalisation im Grosshirn und der Abbau der Funktion durch kortikale Herde*. Wiesbaden: JF Bergmann.

Moniz, E. (1937). Prefrontal leucotomy in the treatment of mental disorders. *Am J Psychiat*, **93**(6), 1379–85.

Moniz, E. and Lima, A. (1936). Premiers essais de psychochirurgie – Technique et résultats. *Lisboa Médica*, **13**, 152.

Ostow, M. (1955) A psychoanalytic contribution to the study of brain function. *Psychoanalytic Quarterly*, **24**, 383–423.

Papez, J.W. (1937). A proposed mechanism of emotion. *Arch Neurol Psychiat*, **38**(4), 725–43.

Reil, J.C. (1809a). Untersuchungen über den Bau des grossen Gehirns im Menschen. *Arch Physiol*, **9**, 136–46.

Reil, J.C. (1809b). Die Sylvische Grube oder das Thal, das gestreifte große Hirnganglium, dessen Kapsel und die Seitentheile des großen Gehirns. *Arch Physiol*, **9**, 195–208.

Reil, J.C. (1812a). Nachträge zur anatomie des großen und kleinen Gehirns. *Arch Physiol*, **11**, 345–76.

Reil, J.C. (1812b). Die vördere Commissur im großen Gehirn *Arch Physiol*, **11**, 89–100.

Rosett, J. (1933). *Intercortical Systems of the Human Cerebrum*. New York: Columbia University Press.

Schmahmann, J.D. and Pandya, D.N. (2006). *Fibre Pathways of the Brain*. Oxford: Oxford University Press.

Sperry, R.W. (1974), *Lateral Specialization in the Surgically Separated Hemispheres* (The Neurosciences. 3rd Study Program; New York: Rockefeller University Press).

Steno, N. (1669). Discours de Monsieur Stenon sur l'anatomie du cerveau. Paris, Ninville. [Translated in Clarke, E. and O'Malley, C. D. (Eds.) (1996). *The Human Brain and Spinal Cord: A Historical Study Illustrated by Writings from Antiquity to the Twentieth Century*, second edition. San Francisco, CA: Norman Publishing.]

Ungerleider, L. G. and Mishkin, M. (1982). *Two cortical visual systems*. In Ingle, D.J., Goodale, M.A., and Mansfield, R.J.W. (Eds) *Analysis of Visual Behavior*, pp. 549–86. Cambridge: MA: MIT Press.

Vesalius, A. (1543). *De humani corporis fabrica libri septem*. Basel: Oporinus. [Translated in Clarke, E. and O'Malley, C.D. (Eds.) (1996). *The Human Brain and Spinal Cord: A Historical Study Illustrated by Writings from Antiquity to the Twentieth Century*, second edition. San Francisco, CA: Norman Publishing.]

Vicq d'Azyr, F. (1786). *Traité d'anatomie et de physiologie*. Paris: Didot l'Aine. [Translated in Clarke, E. and O'Malley, C. D. (Eds.) (1996). *The Human Brain and Spinal Cord: A Historical Study Illustrated by Writings from Antiquity to the Twentieth Century*, second edition. San Francisco, CA: Norman Publishing.]

Vieussens, R. de (1684). *Neurographia universalis*. Lyons: Certe.

Wagenen, W. van, and Herren, R. (1940). Surgical division of commissural pathways in the corpus callosum. Relation to spread of an epileptic attack. *Arch Neurol Psychiat*, **44**(4), 740–59.

Wernicke, C. (1874). *Der Aphasische Symptomencomplex. Ein psychologische Studie auf anatomischer Basis* (Eggert, G. Trans.). Breslau: Cohn and Weigert.

Wernicke, C. (1906). *Grundrisse der Psychiatrie*. Leipzig: Thieme.

Willis, T. (1664). *Cerebri anatome*. London: Martyn and Allestry.

Wundt, W. (1904). *Principles of Physiological Psychology. Translated from the Fifth German Edition (1902) by Edward Bradford Titchener*. London: Swan Sonnenschein.

Yakovlev, P.I. and Locke S. (1961). Limbic nuclei of thalamus and connections of limbic cortex. III. Corticocortical connections of the anterior cingulate gyrus, the cingulum, and the subcallosal bundle in monkey. *Arch Neurol*, **5**, 364–400.

CHAPTER 5

The Clinico-Anatomical Correlation Method

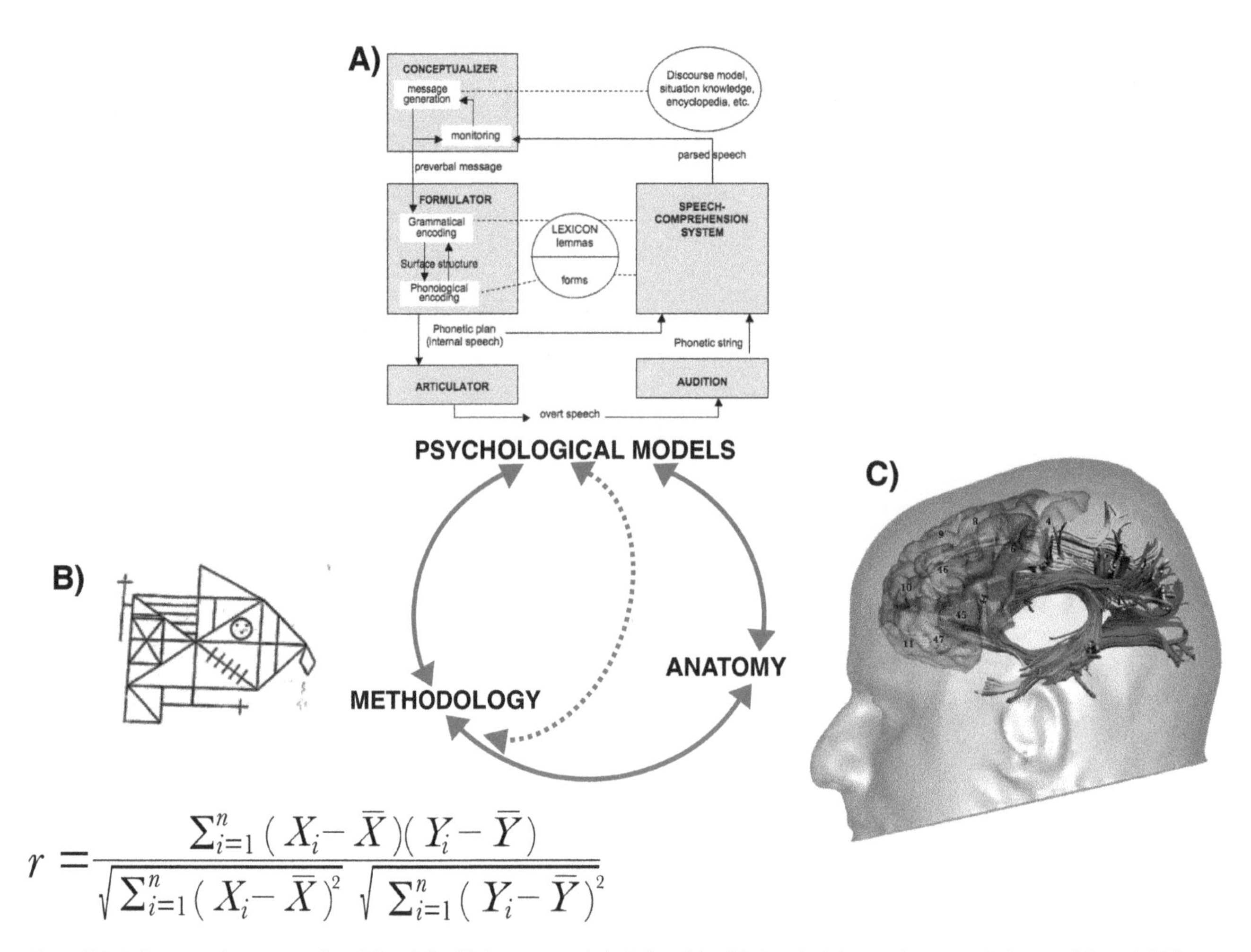

Figure 5.1 A diagrammatic representation of the relationship between psychological models, clinical methodology, and anatomy. At the top of the cycle (A) is an example of a psychological model derived from experimental psychology studies (Levelt and Willem, 1989). B) The figure of Rey and the equation of correlation are intended to illustrate the advances in the assessment methodology for the clinical assessment, such as standardized testing and group statistics. C) The cortical fields of Brodmann and the connections of the frontal lobe are illustrated using novel imaging MRI methods that we envisage will impact on the future development of anatomical models. In different periods the contribution of anatomy to the iterative development of psychological models has varied, as shown by the dotted lines (a loop where anatomy is excluded) and continuous lines (where anatomy is included) in the figure. Current advances in brain imaging promise to reintegrate anatomy into the loop (Catani and ffytche, 2010).

Introduction

The clinico-anatomical correlation method is the circular process that allows brain function to be inferred by studying the correspondence between clinical manifestations and lesion location (Figure 5.1). The validity of this approach depends on: i) the level of sophistication of the methodology used for the patient's clinical characterization (e.g. neuropsychological tests, clinical scales, etc.); ii) the spatial resolution of the anatomical investigations; and iii) the theoretical constructs and hypotheses being tested (e.g. psychological models) (Damasio and Damasio, 1989).

Throughout the last two centuries these three aspects have advanced unevenly, thus impeding clinical neuroscientists from taking full advantage of the correlative method (Figure 5.1). In the 1970s to 1980s, for example, advancements in the methodology for clinical assessment (i.e. wider range of neuropsychological batteries, normative values, use of statistical methods) were not paralleled by the availability of neuropathological investigations (De Renzi, 2001). In the last decade the extraordinary flourishing of techniques for structural and functional imaging promises new ways to further advance our ability to map lesions onto cortical regions and underlying connections (Catani and ffytche, 2010). This chapter looks into the correlative method in detail, mainly from an anatomical angle, and introduces a general framework for clinico-anatomical correlation in contemporary neuroscience (i.e. the hodotopic framework) (Catani and ffytche, 2005).

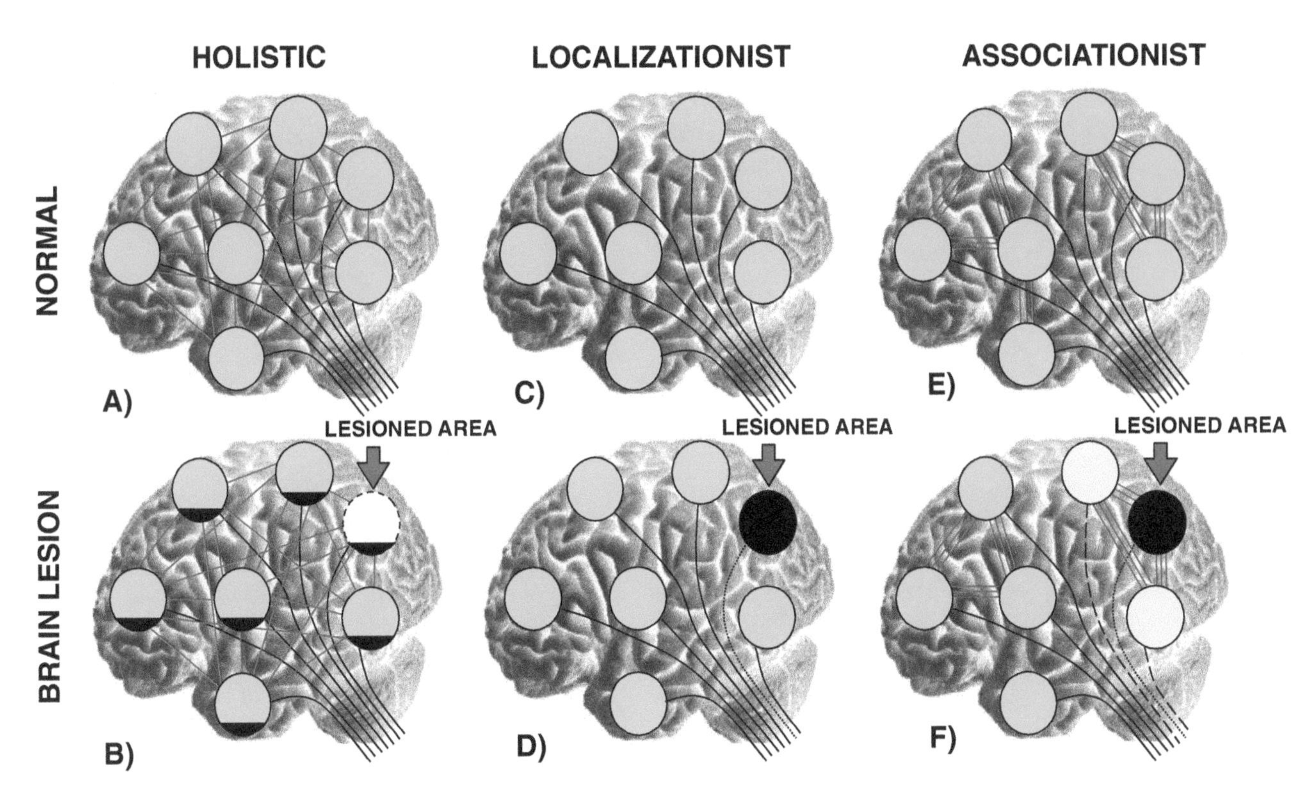

Figure 5.2 Theories of brain function and consequences of a discrete cortical lesion. A, B) Holistic models consider all regions as mutually interconnected through a network of homogeneously distributed association fibres. A) Cognitive functions are the result of the simultaneous activity of all regions acting as a whole through the association pathways. B) In case of damage to one area (white), the network allows the redistribution of the lost function to undamaged brain regions. It follows that for holistic approaches function cannot be localized, the symptoms resulting from the loss of a 'quantity' of cerebral cortex rather than a localized area. C, D) Localizationist models give little importance to interregional connections as each function is carried out by discrete independent regions, whose damage (black) results in the complete functional loss of the lesioned area; hence, localization of symptoms corresponds to localization of functions. E, F) Associationist models consider the brain as organized in parallel distributed networks around cortical epicentres. Primary sensory and motor functions are localized but higher cognitive functions are distributed within large-scale networks. A cortical lesion causes functional loss of the damaged area (black) and partial dysfunction of those regions connected to the lesioned area (yellow).

General considerations

Correlating symptoms with lesion location is not an easy task, in part due to factors that are poorly understood and often beyond the control of the clinician. According to John Hughlings Jackson (1894), for example, the clinical manifestations of a brain disorder depend on the 'four factors of the insanity': i) the depth of the dissolution; ii) the onset and rapidity of the process; iii) the kind of brain in which the dissolution occurs; and iv) the influence of external and internal circumstances upon the patient. Similarly, Macdonald Critchley in his 1953 monograph on the parietal lobe argued:

Patients are apt to differ widely; one showing but few abnormal signs, and these mild in nature; another displaying an intricate clinical picture with elaborate psychomotor and psychosensory features [...] The pathological nature of the disease process is significant – that is to say, whether it be neoplastic, ischaemic, atrophic or inflammatory; and with this factor the degree of abruptness of onset, the question of progression or regression of symptoms, and the length of the clinical history are, of course, important. [...] The age of the patient at the time he develops his parietal affection may be momentous, especially if it be a question of disease of an immature as opposed to a mature nervous system. Lastly, we believe that the type of personality of the patient before the onset of parietal disease may influence considerably the eventual clinical picture. (Critchley, 1953)

In other words Jackson and Critchley highlighted that the clinico-anatomical method is far from being an exact science, because of the many variables intervening in the correlative process. Yet despite these important criticisms, the common observation that clinical patterns do occur in frequent association with lesions localized in specific regions of the brain suggests that the method is probably valid and worth pursuing. Whether we can rely on it to localize brain function is a matter that has been debated for many years and not yet been settled.

Localization of symptoms or functions?

John Hughlings Jackson was among the first to warn that localization of symptoms does not necessarily imply localization of function. He argued that it is entirely possible that some symptoms can be explained by a secondary effect of the damage on other regions, such as, for example, some positive symptoms resulting from a

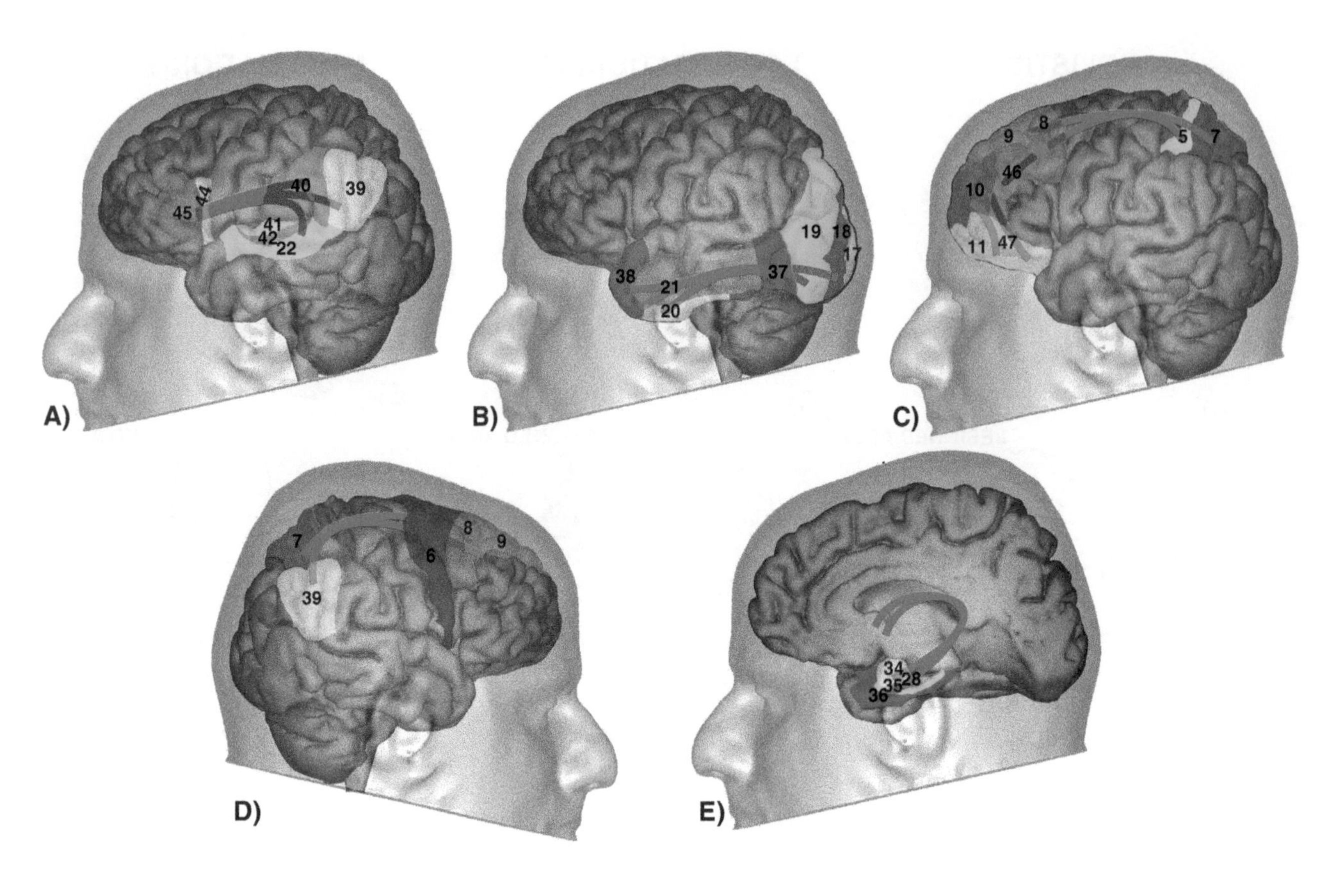

Figure 5.3 Mesulam's large-scale networks for cognition and behaviour. A) Left hemisphere-dominant language network with epicentres in Wernicke's and Broca's areas. B) Face-object identification network with epicentres in occipito-temporal and temporopolar cortex. C) Executive function-comportment network with epicentres in lateral prefrontal cortex, orbitofrontal cortex, and posterior parietal cortex. D) Right hemisphere-dominant spatial attention network with epicentres in the dorsal posterior parietal cortex, frontal eye fields, and cingulate gyrus. E) Memory-emotion network with epicentres in the hippocampal-entorhinal regions and the amygdaloid complex (Mesulam, 1990; 2000).

'release' mechanism (Jackson, 1881, 1894). Jackson's writings had little impact on his contemporaries, but were used some decades later as the ensign of the resurgent holistic movement. Crucial to holism was the assumption that all areas are mutually interconnected through short- and long-range fibres (Figure 5.2A). According to the holistic theory, this architectural property of the brain explains the ability of other parts of the cortex to take over functions within the competence of the damaged area, a property designated by Karl Lashley as 'equipotentiality' of the cortex. For Lashley, for example, the occipital region was important but not critical for vision (Lashley, 1950). Hence, localization of cortical damage is irrelevant, as lesions are more useful for inferring what the unaffected regions of the brain do without the lesioned area than what the lesioned area itself does when it is part of the intact brain (Figure 5.2B) (Finger, 1994). Anatomical and functional evidence in animals and in humans accumulated in the last 50 years do not support the principle of cortical equipotentiality, as it has been shown that many areas are not reciprocally connected and are unable to compensate for a loss of function elsewhere.

Opposite to the holistic view is the cortical localizationist approach, which considers the clinico-anatomical correlation method as key to understanding brain function (Figure 5.2C, D). Karl Kleist, for example, took advantage of the abundance of wounded soldiers with head injury to pinpoint the location of many neurological deficits, creating a cortical map whose close resemblance to phrenological cartography was unnerving for many clinicians (Kleist, 1934). Again the observation that many patients with similar symptoms had lesions in different brain regions and the exceeding amount of associative connections in the human brain, which allows intense inter-regional communication and functional integration, speak against a pure cortical localizationist model.

A third theory originates from the observation that association tracts seem to connect nodal areas forming the epicentres of specialized networks (Figure 5.2E, F). The associationist theory originally elaborated by Meynert (1885) and Wernicke (1874), was re-formulated by Geschwind in the 1960s and his neoassociationist school in more recent years (Geschwind, 1965a, 1965b; Damasio and Damasio, 1989; Mesulam, 1990; Ross, 2010). According to Mesulam (2000), for example, there are at least five large-scale networks in the human brain dedicated to language, face-and-object recognition, executive function-comportment, spatial attention, and memory-emotion (Figure 5.3). The nodes of these networks can be divided into critical versus participating areas, where lesions to critical areas cause an irreversible cognitive impairment. The extension and location of the critical and participating areas are still to be determined for most of the networks.

Cortical or subcortical localization of functions?

To many, the question may appear rather trivial and the answer quite obvious in view of the fact that synapses, which are considered the computational unit, are localized in the grey matter. The cortical versus subcortical (here intended as white matter) dichotomy can be reformulated into the clinical domain as cortical versus disconnection syndromes. In the left hemisphere, for example, language deficits have been classically categorized as either cortical (e.g. Broca's or Wernicke's aphasia) or subcortical (e.g. conduction aphasia) syndromes. For some disorders of the right hemisphere, such as neglect, the debate is still very alive. Thus, for some authors, neglect is a disconnection syndrome of the superior longitudinal fasciculus (Bartolomeo et al., 2007, Doricchi et al., 2008), while for others it is a pure cortical syndrome (Karnath et al., 2009). We consider the problem as ill-posed for two main reasons. The first reason is quite empirical and refers to the fact that pure cortical lesions may not exist. This was already highlighted by Critchley in 1953:

The conception of a purely cortical defect is open to question. Even if such a type of pathology can be said ever to exist, it must admittedly be very rare. True, some obscure types of laminar brain atrophy implicate only the most superficial cell-layers of the cortex, but even here, the axons of the affected cells cannot but share in the pathology, and the lesion thus becomes extended deeply into the white matter, or into contiguous parts of the cortex which are not obviously atrophic themselves. The existence of complex corticocortical ramifications, and of corticosubcortical 'feed-back' processes illustrates the fallacy of looking upon small cortical areas of disease as isolated defects.

Furthermore, while in cortical lesions the damage always extends to the underlying white matter, in pure white matter lesions the cortex is not affected. This suggests that a network dysfunction is the common denominator for all brain disorders, and a tract-based nomenclature should be preferred to a cortical localizationism, which is certainly inappropriate for at least pure subcortical lesions.

The second reason refers to the theoretical construct underlying our current understanding of higher cognitive functions. We have seen before how complex functions result from the activity of extended large-scale networks, whose epicentres act as converging or diverging nodes within a specialized network. These epicentres act as transmodal areas for binding modality-specific fragments of information into coherent experiences, memories, and thoughts (Mesulam, 2000). Wernicke's area, for example, by virtue of its strategic location between primary auditory cortex and other sensory association areas, acts as a gateway for cross-modal integration, by associating a heard name with a seen or felt object. Transmodal areas are not 'warehouses for storing memories' but rather 'signposts' for converging or diverging inputs at a crossroad. It follows that if one accepts *'cognition as emergent properties of large-scale neural networks, function is instantiated in both the grey and white matter'* (Ross, 2010). Hence, any attempt to oppose cortical to subcortical function is as futile as separating the axon from the cell body. However, different mechanisms act for cortical and subcortical lesions, which may influence severity of symptoms and recovery.

What is a disconnection syndrome?

In a brain composed of localized but connected areas, disconnection leads to dysfunction. This simple formulation underlies a range of 19th-century neurological disorders, referred to collectively as disconnection syndromes. The term disconnection is generally used to indicate classical syndromes following lesions to white matter connections (Catani and ffytche, 2005). The term became popular in the second half of the 19th century following Wernicke's description of the disconnection syndrome that was to become the prototype for all others—conduction aphasia (Wernicke, 1874). He argued that the *'production of spontaneous movement, that is, the consciously formulated word, would be brought about by the rearousal of the motor image through the associated memory image of the sound.'* Spontaneous speech, in his opinion, resulted from the interaction of distant cortical areas. Consequently, he interpreted the characteristic paraphasic speech of patients with conduction aphasia as the expression of the inability of temporal regions to monitor Broca's area speech output through subinsular connections. Wernicke's model was the forerunner of current parallel distributed network models of cognition (Figure 5.4A) (Ross, 2010). His greatest merit was to anchor his ideas into the clinical-anatomical correlation method, where he coupled a careful description of the behavioural disturbances of his patients to the anatomical findings from post-mortem dissections.

Shortly after Wernicke's description of conduction aphasia, Lichtheim extended the disconnection paradigm to give a comprehensive account of different aphasic syndromes (Lichtheim, 1885). Lichtheim translated Wernicke's ideas into simple and intuitive diagrams that became standard references for clinicians. However, Lichtheim also introduced hypothetical centres and connections backed by little supportive evidence. His diagrams served the purpose of fitting a theoretical framework that best explained clinical empirical observations without necessarily an anatomical correspondence. These diagrams promoted a mechanical view of brain function where connections represented 'transferring devices' between stores of specialized information localized in individual cortical areas (Figure 5.4B) (Catani and Mesulam, 2008).

A concept closely related to the disconnection mechanism is Monakow's diaschisis (Monakow, 1914). It refers to the distant effect that follows damage to a cortical area (Figure 5.4C). Diaschisis was introduced to highlight the reversible changes that may occur after a brain lesion and the potential for recovery (Finger et al., 2004). Unfortunately, Monakow's ideas were used to show that any attempt to localize brain function by means of clinico-anatomical correlation was meaningless and the disconnection theme fell out of favour between the two world wars.

In the 1960s, Norman Geschwind brought new credibility to the localizationist approach by re-interpreting the functional role of connections and specialized cortical areas according to evidence arising from experimental physiology in monkey. He also extended the disconnection paradigm beyond white matter lesions to lesions of the association cortex (Figure 5.4D). In Geschwind's 1965 model, even a lesion confined to the association cortex could cause a disconnection syndrome, little distinction being made between such lesions and those restricted to white matter tracts. He argued that *'lesions of association cortex, if extensive enough, act to disconnect primary receptive or motor areas from other regions of the cortex in the same or in the opposite hemisphere'* (Geschwind, 1965a). Based on this broader view, Geschwind reappraised disorders of higher cognitive functions as disconnection syndromes resulting either from a lesion of the white matter or the cortex acting as a

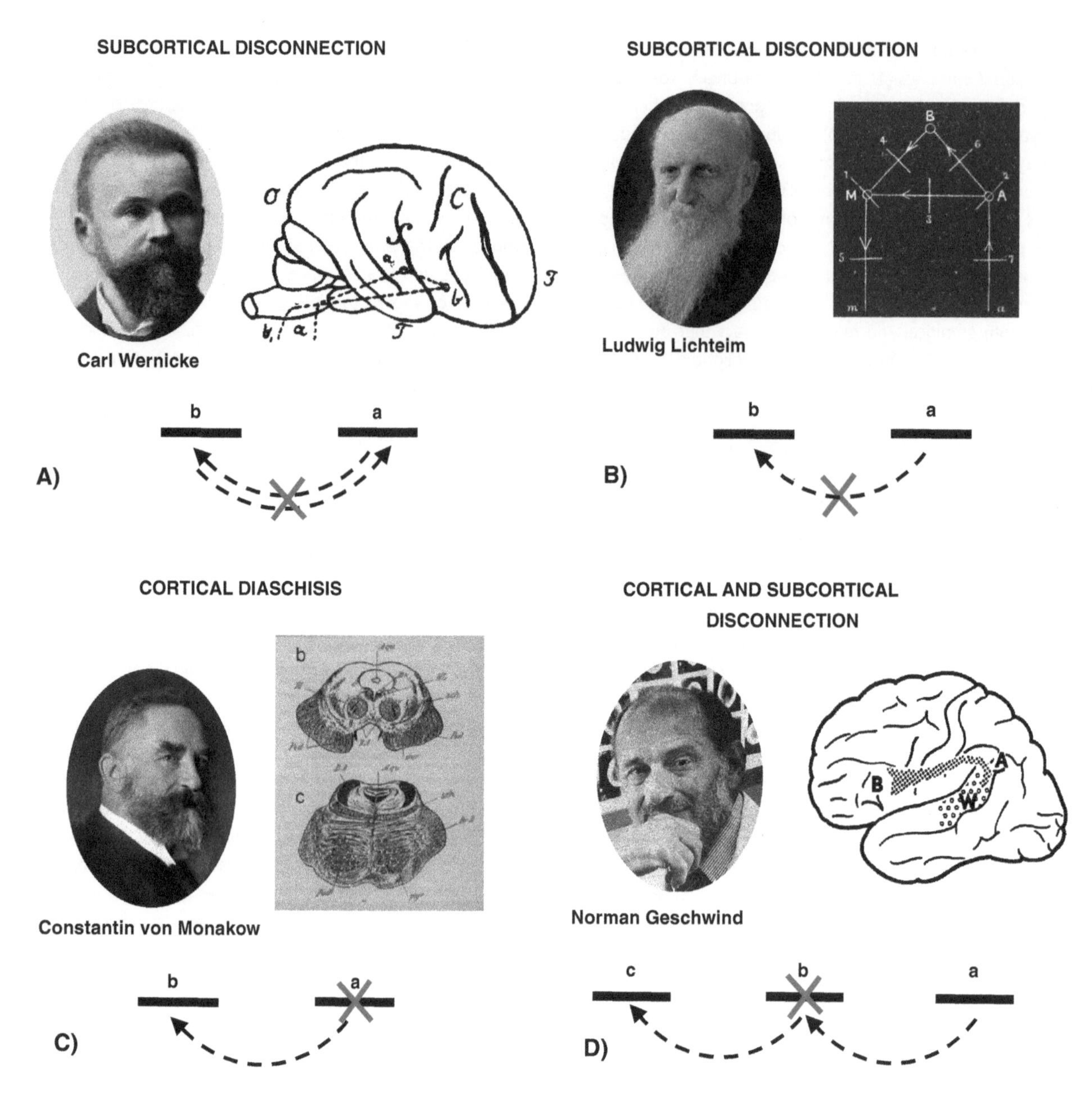

Figure 5.4 Different interpretations of the disconnection paradigm and its anatomical correlates. The figures on the right are the historical originals. The diagrams beneath visualize the authors' interpretation of the disconnection mechanisms. A) For Wernicke, connections, by means of their 'associative function', represent the anatomical substrate of higher cognitive functions allowing coordinated interaction between distant cortical regions. According to this interpretation verbal fluency, for example, requires bidirectional communication between temporal and frontal areas and constant monitoring of the posterior auditory area (a) over Broca's area (b). For Wernicke, the paraphasic speech of patients with conduction aphasia was evidence for the loss of the activity of area (a) over area (b) secondary to a lesion of the underlying white matter fibres (indicated by the red X). B) According to Lichtheim the language network is composed of cortical 'centres' serially connected by long tracts. The flow of information is unidirectional (e.g. from a to b) as also indicated by the arrows in the original drawing (from A to M). Lesions to white matter pathways impair the transfer (or conduction) of information from one centre to the other. The function of cortical centres remains relatively intact for other activities that do not require the involvement of the severed connections. C) For Monakow, the lesion to a cortical area has distant effects on areas connected to the damaged area. His concept of 'diaschisis' broadens the disconnection paradigm beyond white matter to include cortical lesions. D) For Geschwind, disconnection syndromes went also beyond white matter to imply lesions of the association cortex itself, the latter acting as relay station between primary and sensory cortical areas. In his 1965 model, even a pure lesion of the association cortex could cause a disconnection syndrome, little distinction being made between such lesion and those restricted to white matter tracts. According to this model, for example, he reappraised Wernicke's aphasia as a cortical disconnection syndrome.

relay station (Figure 5.5). But Geschwind admitted that his intuitions, pending experimental anatomical evidence, were to be regarded as 'speculative'.

Jerry Lettvin, whose work on the visual pathways of the frog anticipated the discoveries of Mountcastle, Wiesel and Hubel, was a close childhood friend of Geschwind. Upon reading Geschwind's 1965 'Disconnexion Syndromes in Animals and Man', Lettvin is quoted as having asked, mostly in jest, *'So, Norman, you discovered that neurons have axons. What's new?'* (Catani and Mesulam, 2008). This friendly jab can become a valid criticism if the concept

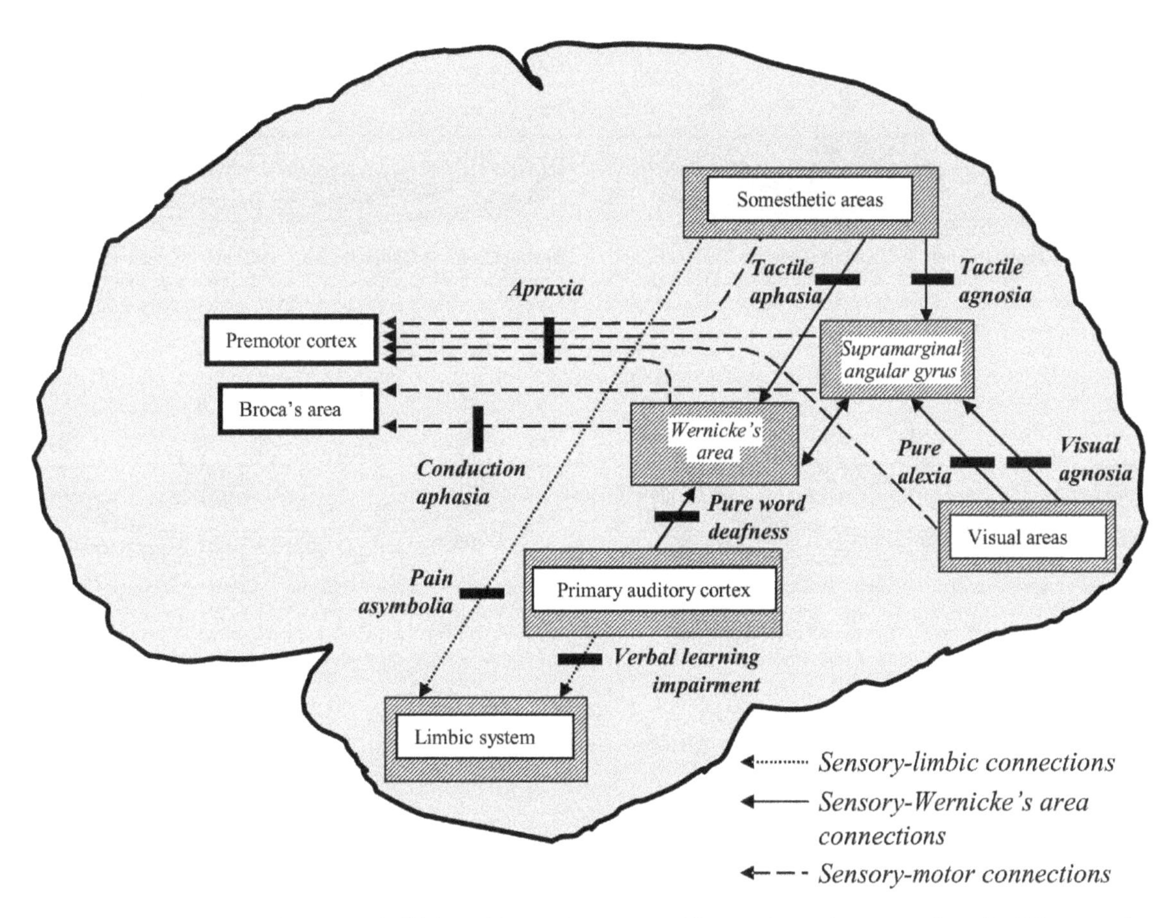

Figure 5.5 Diagram showing Geschwind's re-appraisal of high cognitive disorders in terms of disconnection syndromes according to his 1965 papers. The connections and the corresponding syndromes are classified into three types: sensory–limbic disconnection syndromes (dotted lines), sensory–motor disconnection syndromes (dashed lines); sensory–Wernicke's area disconnection syndromes (solid lines) (Catani and ffytche, 2005).

of disconnection is allowed to become too broad. If area A projects to B, damaging A would obviously disconnect it from B, but attributing the resultant dysfunction of B to a 'disconnection syndrome', rather than to the destruction of A, would add little insight and might trivialize the usage of the 'disconnection' terminology. Furthermore in the absence of detailed anatomical maps of human brain connections, the disconnection paradigm allows for hypothetical post-hoc disconnection explanations of higher function deficits. For a given lesion, the absence of a predicted deficit (or for a given deficit, the absence of a predicted lesion) is amenable to explanation by any number of alternative pathways making it difficult to test or falsify the disconnectionist account.

The advent of new techniques is beginning to redress the imbalance between our knowledge of the cortex and that of its connections. Tractography, although lacking the anatomical precision of post-mortem tracing, is helping advance our understanding of human white matter anatomy and may have an important contribution to make in testing the disconnection paradigm. Other techniques exploring mathematical connectivity between areas in electrophysiological and neuroimaging data provide complementary evidence on the functional roles of such connections (e.g. resting state analysis). Together these techniques allow us to broaden the remit of contemporary clinico-anatomical correlation frameworks beyond disorders of disconnection and cortical deficit to include disorders of hyperconnection and cortical hyperfunction within a hodotopic framework (Catani and ffytche, 2005).

A hodotopic framework for clinico-anatomical correlation

The hodotopic framework as originally proposed by Catani and ffytche (2005) is summarized in Figures 5.6 and 5.7. In the figures a simplified large-scale network of cortical territories and white matter connections is represented. A series of specialized areas are connected by U-shaped association fibres to form a functional territory. Territories are connected by longer white matter association tracts. What are the functional consequences of pathology in this model? The framework uses a new terminology to allow us to extend the model beyond classical neurological disorders of white matter disconnections and cortical deficits to include those involving white matter hyperconnectivity and cortical hyperfunction (e.g. epilepsy, schizophrenia). According to the new terminology, a topological mechanism (from the Greek *topos* = place) refers generally to a dysfunction of the cortex irrespective of whether the dysfunction is one of deficit, hyperfunction, or a combination

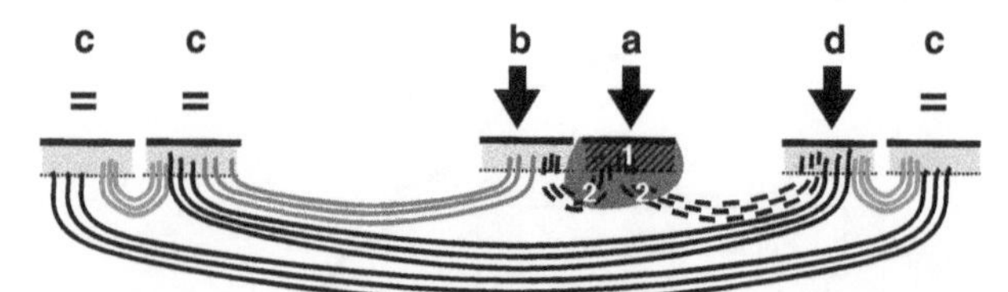

A) **Topological (hypofunctional) + diaschisis (hypofunctional):** a circumscribed cortico-subcortical lesion (grey area) causes hypofunctioning of areas a, b, d through a combined topological (1) and diaschisis (2) mechanism.

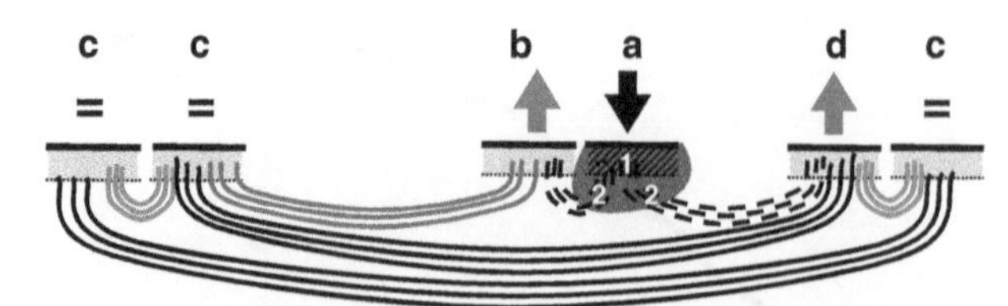

B) **Topological (hypofunctional) + diaschisis (hyperfunctional):** if area a exerts an inhibitory effect on b and d, its damage (grey area) causes hyperfunctioning of areas b and d through a diaschisis (2) mechanism.

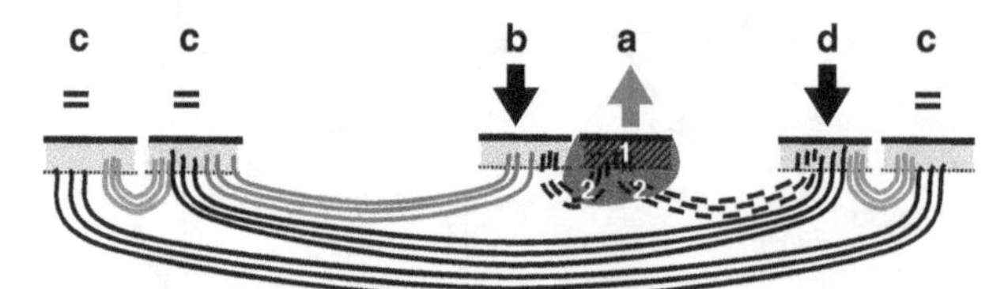

C) **Topological (hyperfunctional) + diaschisis (hypofunctional):** a circumscribed cortico-subcortical pathology (grey area) causes hyperfunctioning of area a through a topological mechanism and hypofunctioning of areas b and d through a diaschisis (2) mechanism.

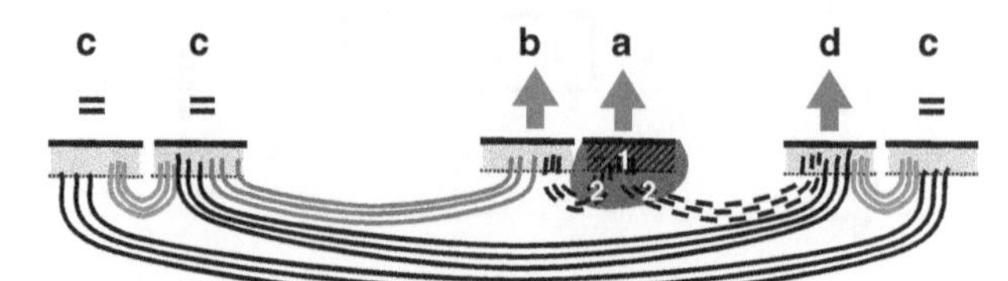

D) **Topological (hyperfunctional) + diaschisis (hyperfunctional):** a circumscribed cortico-subcortical pathology (grey area) causes hyperfunctioning of area a through a topological mechanism and hyperfunctioning of areas b and d through a diaschisis mechanism.

Figure 5.6 A hodotopic framework for clinico-anatomical correlation. Here the effects of a superficial cortical-subcortical lesion are described in terms of a combined topological effect and diaschisis. Red arrows indicate hyperfunctioning areas, black arrows hypofunctioning areas.

of the two. In the visual domain, for example, topologically-related dysfunction ranges from deficits related to the loss of a specialized cortical region (e.g. prosopagnosia from lesions of the face-specialized cortex in the fusiform gyrus) (Sergent et al., 1992), to specific positive symptoms (e.g. face hallucinations related to the hyperexcitibility and spontaneous activation of the face-specialized cortex) (ffytche et al., 1998). A hodological mechanism (from the Greek *hodos* = road or path) refers generally to dysfunction related to connecting pathways, irrespective of whether the dysfunction is one of disconnection, hyperconnection, or a combination of the two. Hodologically-related dysfunctions range from classical disconnection syndromes such as the conduction aphasia of pure subcortical lesions (Naeser et al., 1982), to the combination of local hyperconnectivity and distant hypoconnectivity in autism (Courchesne and Pierce, 2005).

Cortical pathology

Figures 5.6 and 5.7 illustrate the different combinations of topological and hodological mechanisms. Figure 5.6 relates to the pathology of a cortical area and its underlying U-shaped fibres, such as might be caused by localized vascular, neoplastic, epileptic, or neurosurgical lesions. Cortical pathologies may cause two kinds of topological dysfunction, either hypoactivity or hyperactivity of the damaged area. The dysfunction may go beyond the affected cortical site to include cortical regions connected to it. These effects arise through a hodological mechanism (i.e. diaschisis). As for pure hodologically-based dysfunction, these remote effects will demonstrate either a metabolic diaschisis (e.g. reduced cerebellar metabolism for controlateral frontal lesions) or a dynamic diaschisis, only being apparent for tasks normally requiring both interconnected regions (De Renzi and Vignolo, 1962; Kempler et al., 1988; Price et al., 2001).

Subcortical pathology

The second mechanism relates to pure subcortical white matter lesions leading to cortical dysfunction (hyper- or hypofunctioning) through a hodological effect (i.e. disconnection) (Figure 5.7A, B). For some tasks this dysfunction may simply reflect the failure or excess of transfer of outputs from one area to another (Lichtheim's subcortical disconduction); however, for tasks requiring the simultaneous cooperation of cortical regions (e.g. synchronous bimanual coordination), one can consider the function itself to be distributed (a functional loop), the lesion disrupting or enhancing the function as a whole (Wernicke's subcortical disconnection). Whether for a serial or distributed task, the dysfunction of connected regions may only be apparent for those tasks requiring both connected areas, the function of each area individually being normal when they are part of a different network. For some 'irritative' lesions of the white matter the hodological effect could result in the hyperactivity of the entire network or a combined hyper- and hypofunctional effect.

Cortico-subcortical pathology

The third type of dysfunction involves both hodological and topological mechanisms (Figure 5.7C, D). This is the pattern most likely to be encountered clinically and would typically be caused by a large stroke, tumour, or traumatic brain injury. Here, the lesion involves both cortical and subcortical structures with superficial and deep white matter affected. In this case, combined topological and hodological effects produce widespread cortical dysfunction. Auditory hallucinations in schizophrenia provide an example of combined hodological and topological hyperfunction, with increased activation of Broca's, Wernicke's, and Geschwind's territory (Lennox et al., 2000; Shergill et al., 2000) and indirect, diffusion tensor tractography evidence of decreased anatomical

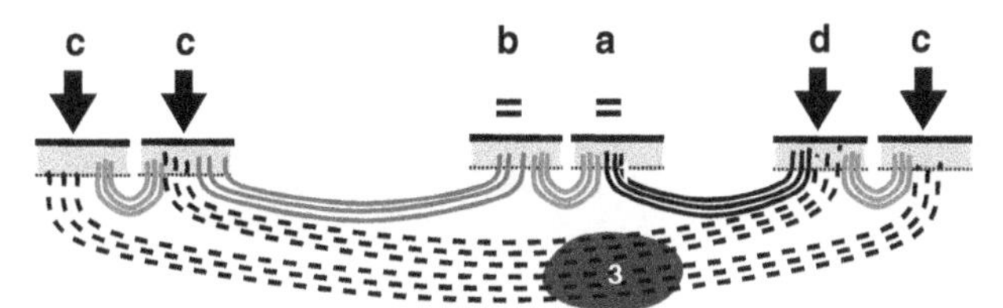

A) Disconnection (hypofunctional): a subcortical lesion (grey area) severes deep tracts (dashed lines) causing hypofunctioning of areas d and c through a disconnection (3) mechanism.

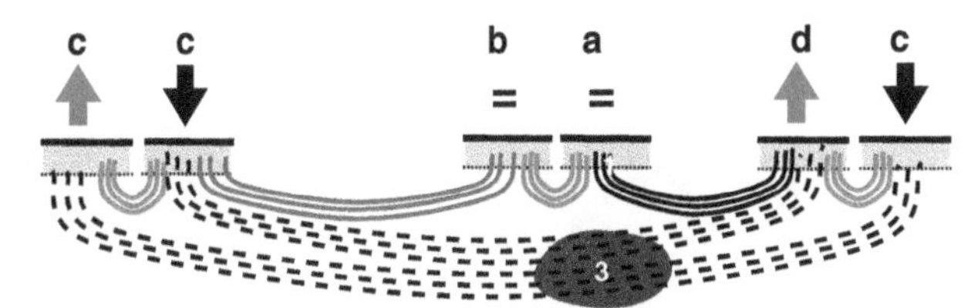

B) Disconnection (mix hypo- and hyperfunctional): a subcortical lesion (grey area) severes deep tracts (dashed lines) causing a combined hypo- and hyperfunctional effect on areas d and c through a disconnection (3) mechanism.

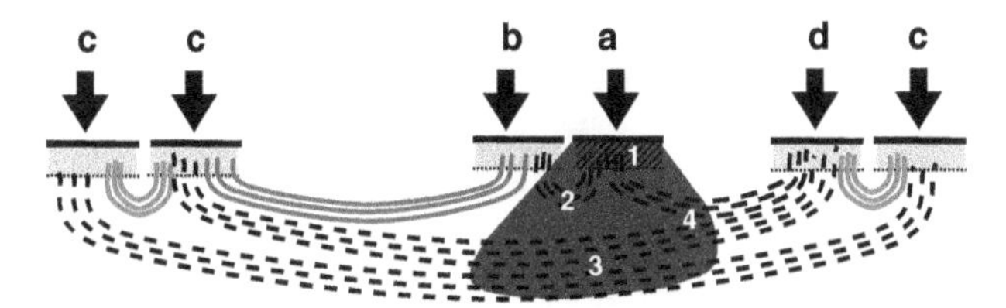

C) Hodotopic effects (hypofunctional): A cortico-subcortical lesion (grey area) causes hypofunctioning of all areas through a combined topological (1), diaschitic (2), and disconnection (3) mechanism. Note that other combinations are possible, like the combined diaschitic and disconnection effect (4) on area d.

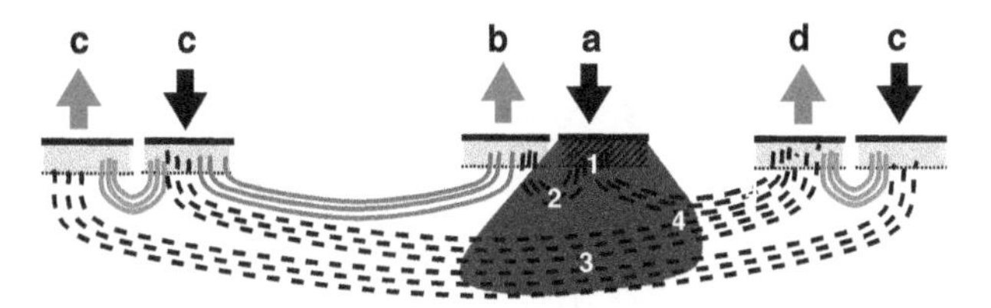

D) Hodotopic effects (mix hypo- and hyperfunctional): A cortico-subcortical lesion (grey area) causes a combined hypo- and hyperfunctional effecton all areas through a combined topological (1), diaschitic (2), and disconnection (3) mechanism. 4) Combined diaschitic and disconnection effect on area d.

Figure 5.7 A hodotopic framework for clinico-anatomical correlation. A-B) The effects of a deep lesion on long tracts connecting distant areas. C-D) The effects of a cortico-subcortical lesion result in combined topological and hodological (diaschisis, disconnection) dysfunction. Here the effects of a deep lesion are described in terms of a combined topological effect, disconnections and diaschisis. Red arrows indicate hyperfunctioning areas, black arrows hypofunctioning areas.

connectivity between these regions (Hubl et al., 2004; Catani et al., 2011).

In general, the hodotopic framework adds clinically useful features to existing models, in particular, extending them beyond classical neurological deficits and disconnections to encompass a broader range of disorders. Of course, its main clinical usefulness will come not from the generalizations outlined above, but from its application to specific functional domains. Figure 5.8 gives three clinical examples within the visual domain (ffytche et al., 2010). Topological visual syndromes occur following discrete lesions of the specialized occipital areas. Typical examples of topological hypofunctional disorders of vision include visual perception deficits for colour (achromatopsia), motion (akinetopsia), and faces (prosopagnosia). Increased perceptual activity within cortex specialized for a given visual attribute underlies both hallucinations and illusions of that attribute. Typical examples of topological hyperfunctional disorders are hallucinations linked to calcarine area and surrounding cortex (phosphenes). Similarly, colour, object, face hallucinations are each linked to hyperactivity of their respective region of cortical specialization. Considering that specialized cortical areas are composed of several hundreds of thousand of interconnected neurons, often through intracortical connections or short U-shaped fibres, topological disorders can also be classified as local small network syndromes.

Hodological visual syndromes can be separated into tract-specific (e.g. single tract syndrome) and extended network syndromes. Hemianopsia following damage along the origin course and termination of the optic radiations is a typical example of a hypofunctional tract-specific syndrome. Visual hypoemotionality is also a classical visual-limbic disconnection syndrome affecting a subset of the inferior longitudinal fasciculus fibres (Bauer, 1982). Visual hallucinations in frontal lobe pathologies, or some forms of synaesthesia are typical cases of hyperfunctional tract-specific syndromes (Schneider, 1961). Some complex syndromes seem to result from a dysfunction of an extended network of cortical and subcortical areas connected by several tracts. Unilateral neglect, for example, has been described in association with lesions of the inferior parietal cortex, inferior frontal lobe, superior temporal cortex, and occipital lobe, or their connecting tracts (i.e. superior longitudinal fasciculus, inferior longitudinal fasciculus, inferior fronto-occipital fasciculus, etc.) (Doricchi et al., 2008).

At least in the short term, the future of the clinico-anatomical correlation method in neurology and psychiatry is likely to be shaped by advances that are currently unfolding in the field of imaging. In the past, lesions could only be referred to cytoarchitectonic atlases, which provided little information on white matter tracts (e.g. Talairach and Tournoux, 1988). As a consequence, the clinico-anatomical correlation method could only link specific deficits with a discrete cortical area. Today, lesions can be mapped onto both cytoarchitectonic and white matter atlases, the latter derived for example from diffusion MRI (Mori et al., 2005) or post-mortem dissections (Bürgel et al., 2006). This adds a new layer of complexity to the correlation method by linking deficits to inferred dysfunction within distributed networks (Rudrauf et al., 2008). Furthermore, we are now able to track pathways between distal cortical areas and obtain measurements of the microstructural integrity of white

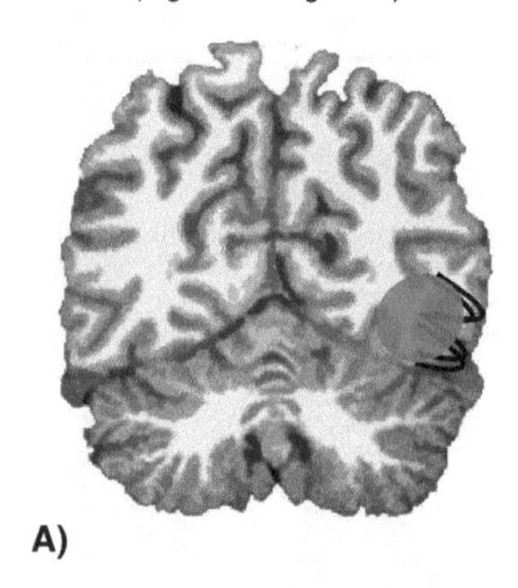

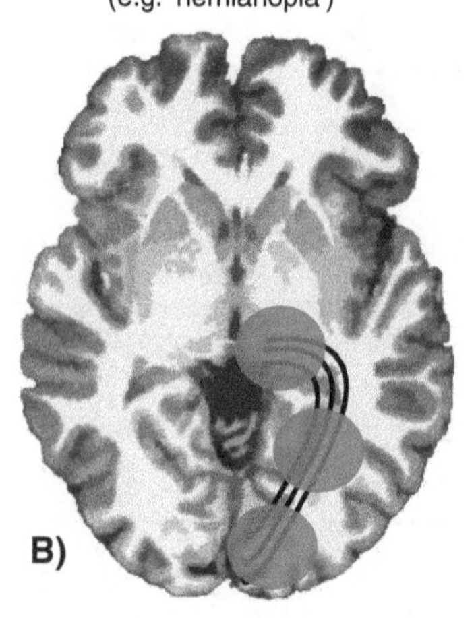

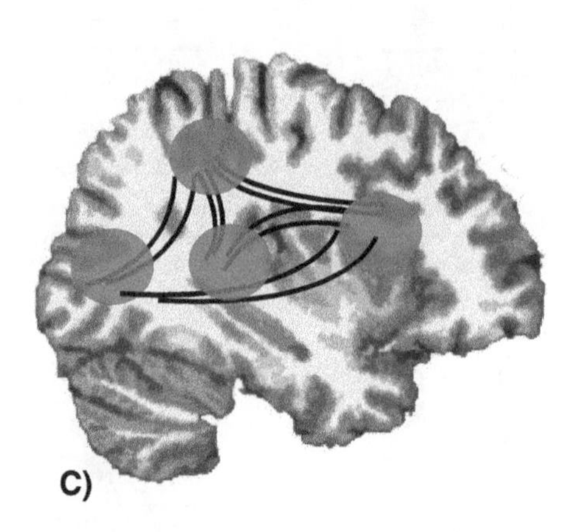

Figure 5.8 An example of the application of the hodotopic framework to the clinico-anatomical correlation of three distinct visual disorders. A) A lesion to a circumscribed cortical region has an effect on the local connectivity. B) Different lesions along the course of the optic radiation cause homonymous hemianopsia. C) Lesions affecting different long association tracts can manifest with similar neglect symptoms.

matter tracts in the living human brain. This facilitates the application of the clinico-anatomical correlation method outside the neurology clinic within the practice of functional psychiatry where a 'lesion' remains inferred, rather than demonstrable.

By combining diffusion MRI with other functional imaging methods we will be able to go beyond a one-to-one correspondence of lesion site and deficit and understand the precise mechanisms (i.e. hypofunctioning or hyperfunctioning areas) underlying a wide range of symptoms. Both topological and hodological explanations will be used to test specific hypotheses and generate unbiased interpretations. This is as true today as at its beginning, when, for example, one-sided interpretation of the anatomical correlates of speech arrest deficit brought into existence Broca's area instead of Broca's fasciculus (Figure 5.9).

The final part of this chapter is a synopsis of the most common syndromes associated with lesions localized within a cerebral lobe. Our hope is that for many 'lobe syndromes' evidence will come to clarify possible underlying topological and hodological mechanisms.

Synopsis of cerebral lobe syndromes

Frontal lobe

The principal subdivisions of the frontal lobe (Figure 5.10) are: i) precentral cortex (BA 4 and inferior 6); ii) premotor cortex (BA 6, 8, 44, 45); iii) prefrontal cortex (BA 9, 10, 46, 47); and iv) orbitofrontal cortex (BA 11, 47). The clinical manifestations commonly associated with frontal lobe lesions can be grouped into four syndromes.

Motor syndrome. The primary motor cortex located in the posterior part of the precentral gyrus gives origin to almost 50% of the fibres of the cortico-spinal tract. Lesions to the primary motor cortex and its connections manifest with contralateral motor deficits of the limbs or face (e.g. hemiparesis or complete hemiplegia) (Newton et al., 2006). In patients with left hemisphere damage, the extension of the lesion to the anterior segment of the arcuate fasciculus may impair the ability to execute or carry out learned purposeful movements (i.e. apraxia) of the left arm and face in addition to hemiparesis of the right arm (see also disorders of motility in the parietal syndromes) (Heilman and Watson, 2008; Ramayya et al., 2010). Anterior to the primary motor cortex, at the junction between the precentral gyrus and the superior frontal sulcus, is the frontal eye field. The frontal eye field is connected to the inferior parietal lobule through the second branch of the superior longitudinal fasciculus (SLF II) (Anderson et al., 2011; Thiebaut de Schotten et al., 2011). Lesions to the frontal eyefield and its connections manifest with gaze abnormalities. Finally, the anarchic hand syndrome (one hand acting autonomously as if having its own will) is a rare condition associated with lesions of the medial frontal lobe and its callosal or intralobar connections (Marchetti, 1998). Abnormal neuronal activity within the precentral region can manifest with motor seizures. The 'Jacksonian march' is a partial seizure characterised by a succession of involuntary movements of the fingers and arm due to spreading of the epileptic activity along the motor homunculus.

Cognitive syndrome. The dorsolateral prefrontal cortex is connected to the parietal cortex, temporal lobe, and basal ganglia by the cingulum, superior longitudinal fasciculus, arcuate fasciculus, and internal capsule. Patients with dorsolateral lesions show cognitive impairment characterized by memory deficit, altered serial motor sequencing, poor response inhibition, impaired cognitive estimation, and impaired abstract thinking. Executive functions for goal-directed behaviour (planning, rule learning, focusing, hierarchical organization, switching, monitoring) are also affected (Stuss and Knight, 2002). These patients are often easily distractible and show impaired mental flexibility together with motor perseveration. Some syndromes are lateralized. For example, impaired ability

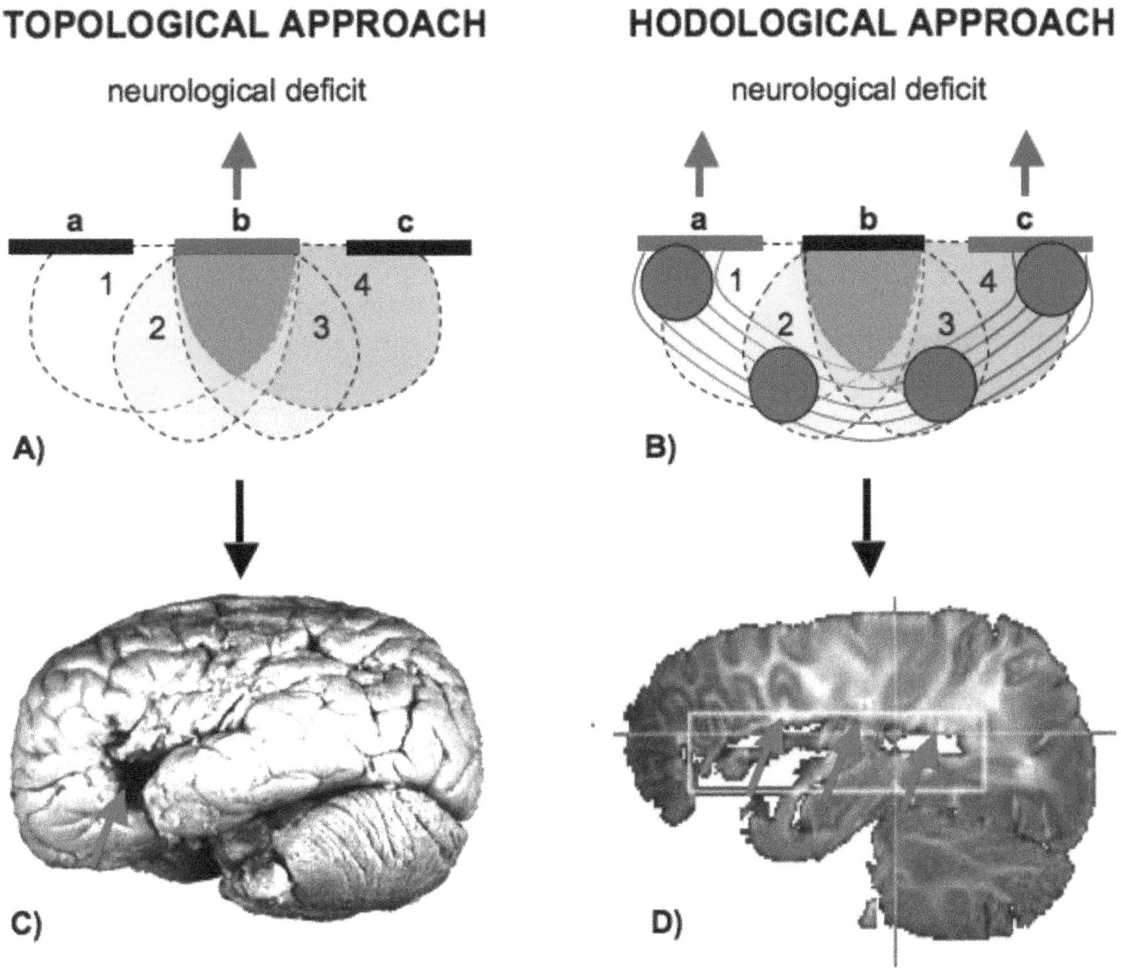

Figure 5.9 The interpretation of a correlative study varies according to the general framework adopted. A) A strict topological approach considers brain functions as localized in specific cortical regions. Within this framework the critical area for the same neurological deficit manifested by a group of stroke patients (four in the example, where each area, from 1 to 4, represents the extension of the lesion in each patient) is located at the cortical region of maximum lesion overlap (region b in the example). B) The hodological approach to brain–behaviour correlation includes a consideration of brain pathways that pass through the damaged area. Within this framework, the neurological deficit could also be attributed to a disconnection between a and c because all lesions affect the same a to c pathway at different levels (red circles). Note that A and B represent the same experiment (i.e. same patients and image analysis); however, the conclusions drawn from it are opposite due to the different theoretical approach. The same considerations apply to single case studies. C) Image of the brain of Broca's aphasic patient (Leborgne) showing a lesion to the inferior frontal cortex. Broca, who worked within a topological framework, considered that his patient's speech deficit was the consequence of the cortical lesion in the inferior frontal lobe (indicated by the red arrow). D) Sagittal MRI image of the same brain shown in (C). Clearly the lesion extends into the white matter of the arcuate fasciculus (red arrows) of the left hemisphere (Dronkers et al., 2007). If Broca had worked within a hodological framework and performed dissections of his patient's brain, arguably he would have attributed the speech deficit to a lesion of the arcuate fasciculus, possibly resulting in the formulation of a Broca's tract rather than a Broca's area (Catani and Mesulam, 2008).

to convey meaning and emotions through the modulation of speech intonation, stress, rhythm, and gestural expression (anterior affective-aprosodia) (Ross, 1981) or unilateral neglect (Husain et al., 2000) often occur with lesions of the right inferior frontal gyrus, while apraxia (Heilman and Watson, 2008) and Broca's aphasia (Catani and Mesulam, 2008) are more frequent in left inferior frontal lesions.

Abulic syndrome. The medial prefrontal cortex is connected to the medial parietal, occipital, and temporal lobe by the cingulum and the first branch of the SLF (SLF I) (Thiebaut de Schotten et al., 2011). Lesions to the medial frontal lobe manifest with apathy, whose central features are blunt affect, loss of motivation, and reduced goal-directed movements (Mesulam, 2000). Apathy has affective, emotional, cognitive, and motor components. Affective and emotional aspects of apathy manifest with blunt affect and absence of interests. The cognitive aspects of apathy include decreased engagement with usual activities, lack of curiosity and interest in learning, impoverished generative thinking, and reduced ability to sustain effort. These affective and cognitive symptoms are associated with anterior medial frontal lesions (extending into the anterior cingulate cortex) (Rankin et al., 2006).

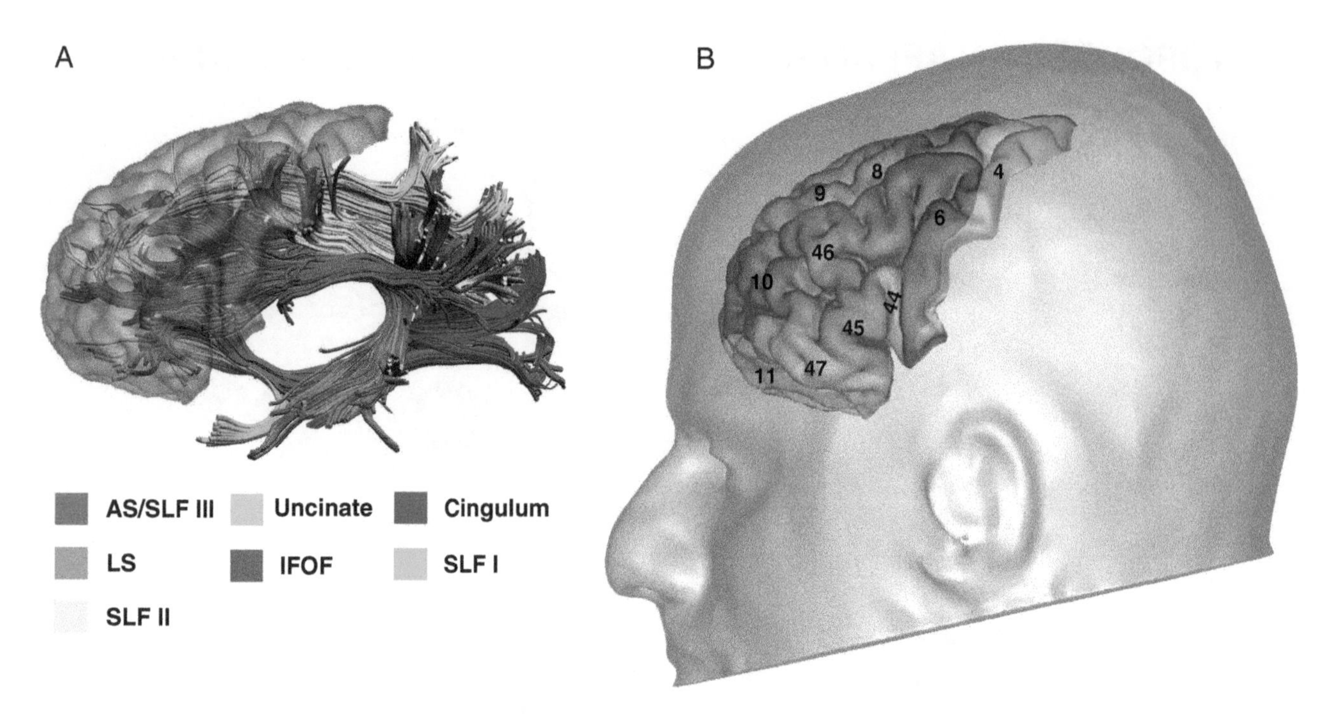

Figure 5.10 A) Lateral view of the cortical anatomy of the frontal lobe and its main associative connections (B). Numbers indicate cytoarchitectonic areas according to Brodmann's nomenclature. The major association pathways of the frontal lobes are the cingulum for the medial surface, the uncinate and inferior fronto-occipital fasciculus (IFOF) for the ventral surface, the superior longitudinal fasciculus (SLF), and the anterior (AS) and long segments (LS) of the arcuate fasciculus for the dorsolateral surface.

Finally, motor symptoms of apathy manifest with marked reduction of spontaneous movements (i.e. limb akinesia and mutism) and are usually associated with more posterior lesions of the medial region of the superior frontal lobe (e.g. supplementary and pre-supplementary motor areas) (Mega et al., 1997).

Behavioural syndrome. The orbitofrontal cortex is connected to the anterior temporal and ventral temporo-occipital cortices through the uncinate and inferior fronto-occipital fasciculi, respectively. The most medial portion of the orbitofrontal cortex is also connected to the medial dorsolateral prefrontal region through the cingulum. Patients with orbitofrontal lesions present with personality changes characterized by disinhibition, social inappropriateness, and sexual preoccupation. Other frontal lobe signs that frequently accompany the behavioural syndrome are related to automatic motor behaviour. Patients, for example, automatically imitate the examiner's movements without being told to do so (imitation behaviour), grip objects (grasping), or use (utilization behaviour) tools presented to them (Mesulam, 2000; Salloway et al., 2001; Rosen et al., 2005). Neuropsychiatric manifestations are also frequently described in these patients. These include reduced empathy, impulsivity, distractibility, emotional lability, depression, and more rarely hypomania or mania. Cognitive deficits are not frequent in patients with frontal behavioural symptoms (Mega et al., 1997).

Symptoms that characterize the cognitive, abulic and behavioural syndromes are frequently observed also in developmental and neuropsychiatric conditions. These include attention deficit and hyperactivity disorders, autism, schizophrenia, fronto-temporal dementias, affective and anxiety disorders, and obsessive compulsive disorder (For reviews of the clinical anatomy of the frontal lobe syndromes see Fuster, 1997; Mesulam, 2000; Salloway et al., 2001; Stuss and Knight, 2002; Cummings and Mega, 2003; Darby and Walsh, 2005; Nieuwenhuys et al., 2008.) (See also Chapters 7, 9, 10, 11).

Parietal lobe

The parietal lobe includes: i) post-central gyrus (BA 3, 1, 2); ii) superior parietal lobule (BA 5, 7); iii) inferior parietal lobule (BA 39, 40); iv) precuneus (medial BA 5, 7); and v) posterior parietal gyrus (BA 19) (Figure 5.11). The parietal syndromes can be divided into five groups.

Disorders of somatosensory and tactile function. The post-central gyrus is connected to neighbouring regions by U-shaped fibres and receives dense projections from the thalamus. Lesions to the post-central gyrus and its connections can manifest with either isolated or combined impaired sensation for pain, temperature, touch, and vibration. Sensation can be absent (anaesthesia), reduced (hypoaesthesia), or increased (hyperaesthesia). If more than one modality is altered (either reduced or increased), the patient reports multiple symptoms such as tingling, burning, numbness, feeling of pins and needles (dysaesthesia) (Bassetti et al., 1993). Proprioception (the ability to detect joint motion and limb position) can also be impaired in isolation or in combination with

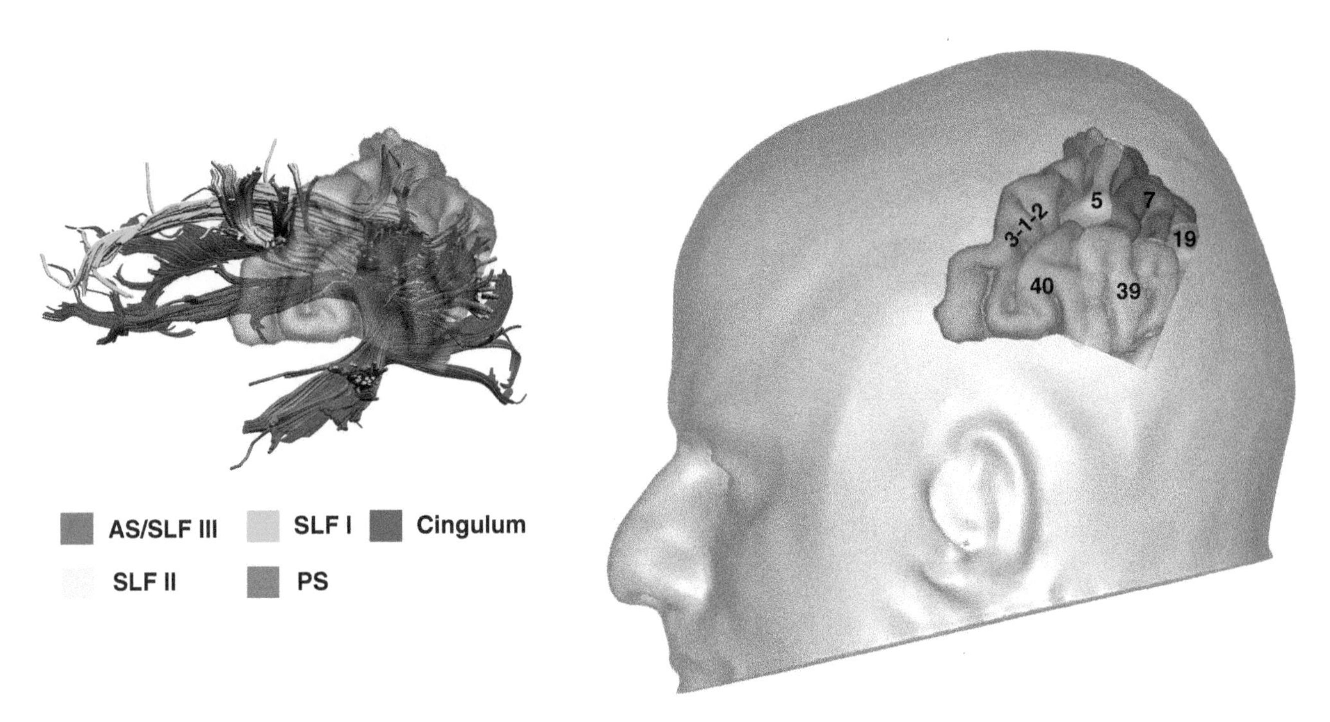

Figure 5.11 Lateral view of the cortical anatomy of the parietal lobe and its main associative connections. Numbers indicate cytoarchitectonic areas according to Brodmann's nomenclature. The major association pathways of the parietal lobes are the cingulum for the medial surface, the superior longitudinal fasiculus (SLF), and the anterior (AS) and posterior (PS) segments of the arcuate fasciculus for the dorsolateral surface.

other somatosensory modalities. Spontaneous tactile sensation can present in the form of tactile perseveration (persistence of touch sensation after cessation of the stimulus), polyaesthesia (a single tactile stimulus is reported as several), and hallucination of touch (tactile sensation in the absence of any stimulus) (ffytche et al., 2010). Other forms of sensation require higher integration and are associated with more extensive lesions of the parietal lobe. These include impaired tactile recognition of three-dimensional objects (astereognosis) or letters (agraphaesthesia), inability to localize sensory stimuli (atopaesthesia), and failure to detect the stimulus on one side of the body when touched simultaneously on both sides (extinction) (Critchley, 1953).

Disorders of motility. The parietal regions posterior to the postcentral gyrus connect extensively with the occipital lobe through U-shaped fibres and project to the frontal lobes through the superior longitudinal and arcuate fasciculi. Patients with lesions of the precuneus and superior parietal lobe may present uncoordinated voluntary movements that lack speed, smoothness, and appropriate direction (ataxia). Ataxia is not related to disorders of motor, sensory, and visual function. Some patients are impaired solely in reaching and grasping for objects (optic ataxia). In these cases the lesion can be located more posterior in the precuneus and posterior parietal gyrus (Karnath and Perenin, 2005). More frequently patients with parietal lesions present with the inability to carry out complex movements despite intact coordination and preserved sensory, visual, and motor functions (apraxia). Apraxic patients seem to have lost the ability to initiate and perform previously learned skilled movements involving the use of tools (limb kinetic apraxia). Other patients have difficulties in putting together one-dimensional units so as to form two-dimensional figures or patterns (constructional apraxia) (Heilman and Watson, 2008). More rarely, there is a dissociation between intact spontaneous execution of a motor act and the inability to imiitate or pantomime the act (ideomotor apraxia) (Alexander et al., 1992). Apraxias occur with left-sided lesions of the inferior parietal lobule (BA 39, 40) and its connections to the occipital and frontal lobes through the superior longitudinal and arcuate system and U-shaped fibres (Ramayya et al., 2010). Oculomotor apraxia is a disorder of gaze in which subjects are unable to disengage fixation to move their gaze from one object to another and is associated with dorsal parieto-occipital lesions.

Disorders of spatially-directed attention. Lesions of the right inferior parietal lobule can present with a range of spatial disorders. Patients can manifest reduced awareness of parts of their body (motor or sensory neglect), the space close to their body (peripersonal neglect), or the visual stimuli from one hemi-field (visuospatial neglect). Neglect can affect motor, somatosensory, and visual modalities together or separately.

Pure motor neglect is characterized by spontaneous underutilization of one side, with normal utilization of the same side on command. These patients show no defects of strength, reflexes, or sensibility of the neglected side (Laplane and Degos, 1983).

Pure somatosensory neglect is a rare condition in patients without sensorial defects who fail to detect somatosensory stimulations on one side of the body (Vallar et al., 1993). Visual neglect has several degrees, from reduced space-directed attention to one side for simultaneous bilateral stimulation (extinction) to complete

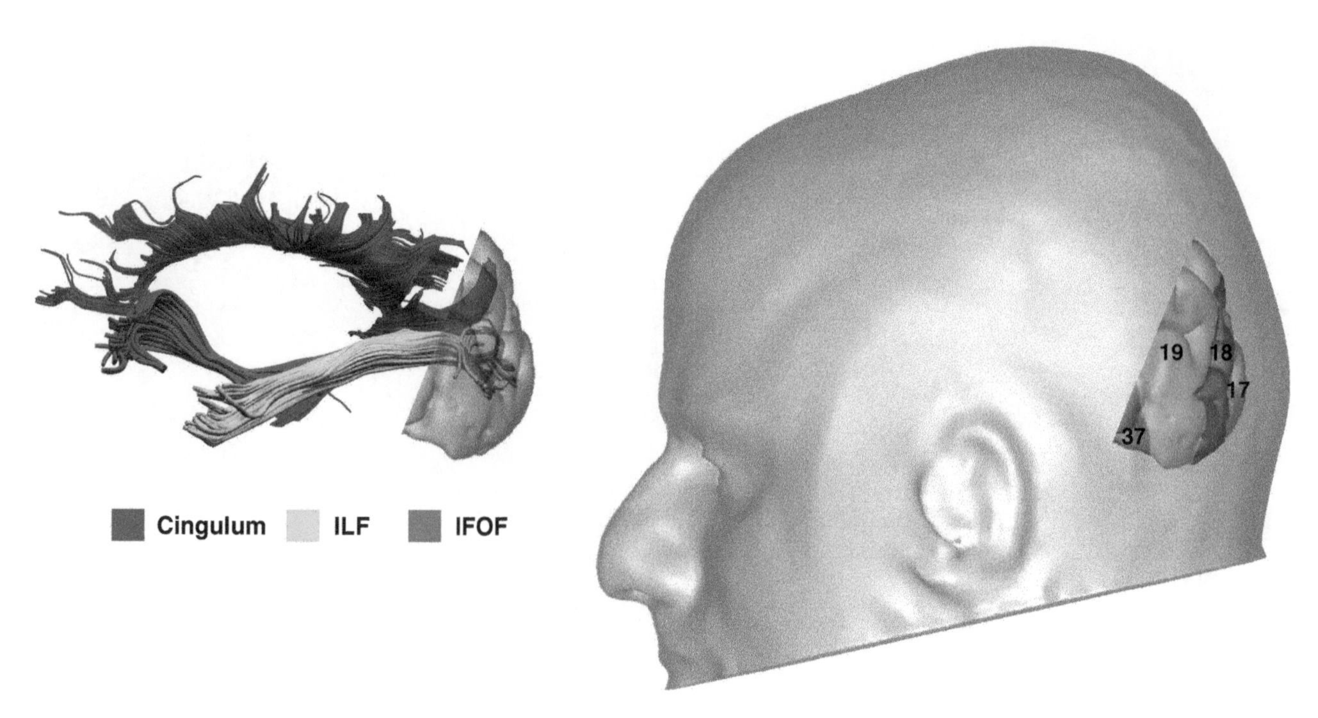

Figure 5.12 Lateral view of the cortical anatomy of the occipital lobe and its main associative connections. Numbers indicate cytoarchitectonic areas according to Brodmann's nomenclature. The major association pathways of the occipital lobes are the cingulum for the medial surface, the inferior longitudinal fasciculus (ILF), and the inferior fronto-occipital fasciculus (IFOF).

unawareness of one half of the space. Neglect is frequently associated with lesions of the right inferior parietal lobe and its connections to the frontal and temporal lobes through the arcuate fasciculus and superior longitudinal fasciculus (Husain et al., 2000; Thiebaut de Schotten et al., 2005; Doricchi et al., 2008).

Disorders of symbolic thought and memory. Disorders such as impaired manipulation of numbers in mathematical operations (acalculia), impaired reading (alexia) or writing (agraphia) are frequently associated with lesions of the left inferior parietal lobule and its connections through the arcuate fasciculus (BA 39, 40). Other disorders observed in patients with inferior parietal lobule lesions include impaired repetition of words (conduction aphasia), reduced auditory comprehension (Wernicke's aphasia), deficits in verbal working memory, and more rarely inability to name (anomia) colours, objects, faces (Critchley, 1953) or recognize mirror stimuli (mirror agnosia) (Priftis et al., 2003).

Complex visual defects. The inferior parietal lobe and posterior parietal gyrus are connected to visual areas of the temporal and occipital lobes through U-shaped fibres. Damage to these cross-modal transition zones can impair the ability to integrate memory, vision, and proprioception. Patients may report an inability to revisualize objects or scenes (disorders of mental imagery), the feeling that experiences, typically visual, seem strange or unreal (derealization), and out-of-body experiences (ffytche et al., 2010). Other complex syndromes have been attributed to lesions of the parietal lobe. Balint syndrome is characterized by an inability to perceive simultaneous stimuli (simultagnosia), optic ataxia, and oculomotor apraxia due to bilateral degeneration of the parieto-occipital cortex. The pseudothalamic syndrome manifests with somatosensory loss, astereognosis, tactile extinction, hemiplegia, and focal sensory epilepsy. Finally, Gerstmann's syndrome is associated with a left parietal lesion manifesting with finger agnosia, left–right disorientation, acalculia, and pure agraphia (Vallar, 2007; Rusconi et al., 2010). A neurodevelopmental Gerstmann's syndrome has also been described in children (Benson, 1970). (For reviews of the clinical anatomy of the parietal lobe syndromes see Head and Holmes, 1911; Dejerine, 1914; Critchley, 1953; Bassetti et al., 1993; Nieuwenhuys et al., 2008.) (See also Chapters 7, 8, 9, 10).

Occipital lobe

The occipital lobe is divided into a primary striate visual cortex (BA 17), and a much more extended extrastriate cortex, which corresponds to the occipital visual association areas (BA 18, 19) (Figure 5.12). Disorders associated with occipital lobe lesions can be grouped into three categories.

Disorders of simple visual perception. The primary visual cortex receives inputs from the lateral geniculate nucleus and connects via U-shaped fibres directly with the neighbouring extrastriate cortex. Patients with lesions to the primary visual area or the optic radiation may present with an area of altered vision (scotoma). The loss of vision can extend to a quadrant (quadranopsia), an entire hemifield (homonymous hemianopsia) or to both hemifields if the lesion is bilateral (complete cerebral blindness). Hallucinations linked to primary visual cortex pathology are of simple featureless forms and colours (phosphenes) (ffytche et al., 2010).

Disorders of the dorsal stream. The associative occipital cortex is composed of several regions linked through multiple, parallel, cortico-cortical connections divided into dorsal and ventral visual streams (Ungerleider and Mishkin, 1982). The dorsal stream is related to spatial aspects of vision (the 'where' stream). The function of the dorsal stream has been recently revised in light of new evidence for its role in the visual control of skilled actions. Accordingly, the dorsal pathway has been renamed the 'how' stream (i.e. vision for action) (Milner and Goodale, 2008; Kravitz et al., 2011). Patients with dorsal stream lesions present with selective loss of motion vision (akinetopsia) or motion hallucinations. In visual alloaesthesia the world is perceived in an incorrect orientation, for example, inverted, tilted, or right–left reversed. Optic ataxia is a disturbance of limb guidance such that subjects are unable to reach for objects in an otherwise intact visual field. Optic ataxia can also be found in lesions of the superior occipital lobe and its connections. Increased activity in the dorsal occipital regions can cause motion hallucinations.

Disorders of the ventral stream. The ventral stream is composed of associative interconnected cortical areas specialized for colour, face, object, and letter vision (the 'what' stream) (Ungerleider and Mishkin, 1982). Patients with lesions of the ventral cortex and its connections present with selective loss of colour vision (achromatopsia) (Zeki, 1990), face perception (prosopagnosia) (Fox et al., 2008), object perception (visual object agnosia) (Catani et al., 2003), and impaired reading (alexia) (Cohen et al., 2000). Similarly colour, object, and text or letter-string hallucinations are each linked to pathology of their respective region of cortical specialization. Face hallucinations and illusions characterized by distorted facial features (prosopometamorphopsia) are likely to relate to a region specialized for face features on the lateral convexity of the occipital lobe (the occipital face area) while hallucinations of normal faces or facial intermetamorphosis (a change in the visually perceived identity of a face) are likely to relate to activity within an area specialized for faces on the ventral occipito-temporal surface (i.e. fusiform face area) (ffytche and Howard, 1999). Change in perception of objects (micropsia), unilateral neglect, and selective loss of visual imagery have also been described in patients with ventral occipital lesions.

Complex visual syndromes. In some patients the involvement of long-range fibres connecting the occipital lobe to more anterior regions manifests with complex visual syndromes. Patients with cortical blindness due to extensive occipital lesions can sometime deny their visual impairment (Anton's syndrome) (Anton, 1889). Patients with Anton's syndrome may also report visual experiences, which traditionally have been interpreted as confabulations (a false memory or false report of visual perceptual experience) or as spontaneous visual imagery.

Reduplicative phenomena are thought to relate to the disconnection of visual and affective or memory regions due to lesions of the inferior longitudinal fasciculus and inferior fronto-occipital fasciculus (ffytche et al., 2010). In these disorders, familiar people, places, and objects are perceived as duplicates that have replaced the real person, place, or object. Similar disconnection accounts are given of disorders characterized by derealization, reduced emotional tone to visual experience (visual hypoemotionality) (Bauer, 1982) and impaired registering of visual experiences in the short-term memory (visual amnesia) (Ross, 2008).

The Fregoli syndrome in which unfamiliar people are perceived as familiar (typically as a person in disguise with malevolent intent) can be interpreted as hyperconnection between visual, emotional, and memory networks. Similarly, the strong affective and imagery components of flashbacks in post-traumatic stress disorder suggest hyperconnection between visual, emotional, and memory regions. (For reviews of the clinical anatomy of the occipital lobe syndromes see Milner and Goodale, 1993; ffytche and Howard, 1999; Darby and Walsh, 2005; Milner and Goodale, 2008; ffytche et al., 2010 Kravitz et al., 2011) (See also Chapters 8, 9, 10).

Temporal lobe

The main divisions of the temporal lobe include: i) primary auditory cortex (BA 41); ii) auditory association cortex (BA 22, 42); iii) visual association cortex (BA 20, 21, 37); and iv) temporopolar cortex (BA 38) (Figure 5.13). Lesions to each of these regions cause four distinct temporal lobe syndromes.

Cortical deafness for sounds and words. The primary auditory cortex (BA 41) receives acoustic radiations from both medial geniculate nuclei and connects to adjacent areas through U-shaped fibres. Lesions to the left primary auditory area impair the ability to recognize words ('word deafness' or sensory aphasia), while right-sided lesions impair recognition of non-verbal sounds ('sound deafness' or acoustic agnosia).

Disorders of language. The auditory association cortex extends along the superior and part of the middle temporal gyrus and connects to the inferior parietal lobe and the posterior frontal lobe through the posterior and long segments of the arcuate fasciculus, respectively (Catani et al., 2005). The most frequent syndrome associated with lesions to the posterior part of this region in the left hemisphere is Wernicke's aphasia, characterized by impaired auditory comprehension and repetition with normal verbal fluency (Hillis et al., 2002). Equivalent lesions in the right hemisphere may cause impairment in understanding emotional aspects of language (receptive affective-aprosodia) (Ross, 1981). The auditory association cortex is also connected to more posterior visual occipital regions and more anterior temporal areas through the inferior longitudinal fasciculus and U-shaped fibres. Patients with lesions to the above connections may show difficulty in using words to name objects (nominal aphasia), or inability to repeat a series of words with normal repetition of single words (auditory amnesic aphasia). Auditory hallucinations are also frequently observed with lesions or stimulation (e.g. intraoperative cortical stimulation) of the auditory association cortex.

Disorders of multimodal visual processing. The visual association cortex is located in the middle and inferior temporal gyri (BA 20, 21, 37) and connects to the occipital lobe through the inferior longitudinal fasciculus and to the frontal lobe through the long segment of the arcuate fasciculus and the uncinate fasciculus. The most posterior part is connected to the inferior parietal lobe through the posterior segment of the arcuate fasciculus. Objects appearing larger (macropsia), smaller (micropsia), nearer (pelopsia), or further away (teleopsia) have been attributed to a dysfunction of

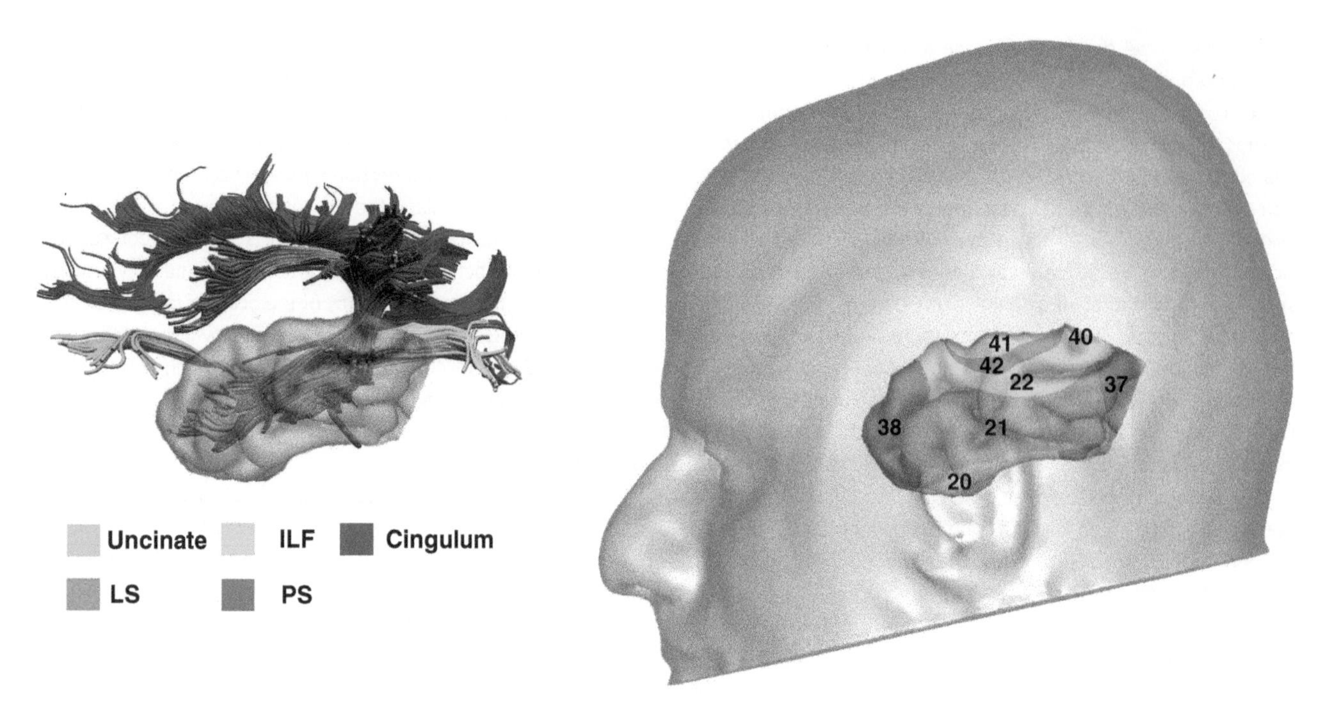

Figure 5.13 Lateral view of the cortical anatomy of the temporal lobe and its main associative connections. Numbers indicate cytoarchitectonic areas according to Brodmann's nomenclature. The major association pathways of the occipital lobes are the cingulum for the medial surface, the uncinate for the anterior temporal region, the inferior longitudinal fasciculus (ILF) for the ventral and lateral surface, the long (LS) and posterior segments (PS) of the arcuate fasciculus for the posterior temporal regions. The inferior fronto-occipital fasciculus (IFOF), which courses through the temporal lobe without sending projections to the temporal cortex, is often damaged in lesions extending deep into the white matter of the temporal lobe.

object constancy (i.e. the integration of object distance with visual angle at the retina to estimate size) associated with lesions to the posterior regions of the temporal lobe. Autoscopic phenomena describe a range of experiences in which the self is duplicated in the external space. In autoscopy, visual perspective remains in the physical body (the duplicate self is seen in the external world). In out-of-body experience, the physical body is seen from the perspective of the external self (Blanke et al., 2002). In autoscopy, perspective changes between the external self and the physical self in rapid alternation. These phenomena are thought to relate to the disintegration of visual, proprioceptive, tactile, and vestibular modalities and have been linked to transient dysfunction in the region of the temporo-parietal junction (ffytche et al., 2010). Lesions to the ventral aspect of the fusiform gyrus and inferior occipito-temporal gyrus cause prosopagnosia (Fox et al., 2008), mainly for lesions of the right hemisphere, and alexia in the left hemisphere (Epelbaum et al., 2008).

Disorders of memory and behaviour. The temporopolar region morphologically belongs to the temporal lobe but functionally is considered as part of the limbic system. It is connected to more posterior temporal and occipital regions through the inferior longitudinal fasciculus and U-shaped fibres and to the frontal lobe via the uncinate fasciculus. Semantic dementia is a memory disorder characterized by a progressive inability to associate meaning with sensory perception. These patients show marked atrophy of the anterior temporal lobe and its main connections (Agosta et al., 2010; Galantucci et al., 2011). Lesions to the uncinate fasciculus may also present with reduced verbal fluency and naming deficits, especially for famous faces (Papagno et al., 2011). Other complex syndromes associated with anterior temporal dysfunction are discussed in the section on limbic disorders. (For reviews of the clinical anatomy of the temporal lobe syndromes see Gloor, 1997; Mesulam, 2000; Darby and Walsh, 2005.) (see also Chapters 8, 9 and 11).

Limbic lobe

The limbic system includes a central cortico-subcortical network (centred around the hippocampus–fornix–thalamus) and a group of paralimbic cortical areas connected through the cingulum and uncinate fasciculus (Figure 5.14). The limbic lobe syndromes are divided into three distinct groups.

Disorders of the hippocampal–hypothalamic division. Lesions to the hippocampal–centred division of the limbic system cause severe amnesia for events before (retrograde) and after (anterograde) the onset of the lesion (Markowitsch, 2000). Retrograde amnesia is more severe for recent than remote events, while anterograde amnesia affects explicit learning for new events but leaves implicit-procedural memory for motor and perceptual tasks intact. Some patients, in addition to memory deficits, show difficulties in spatial orientation due to the inability to derive directional information from landmark cues in familiar and new environments (Aguirre and D'Esposito, 1999). This form of memory dysfunction is associated with lesions to the posterior parahippocampal gyrus, retrosplenial cingulate cortex, and posterior precuneus (Vann et al., 2009, Valenstein et al., 1987). Degeneration of the medial temporal lobe structures is a feature of Alzheimer's disease. Korsakoff syndrome is

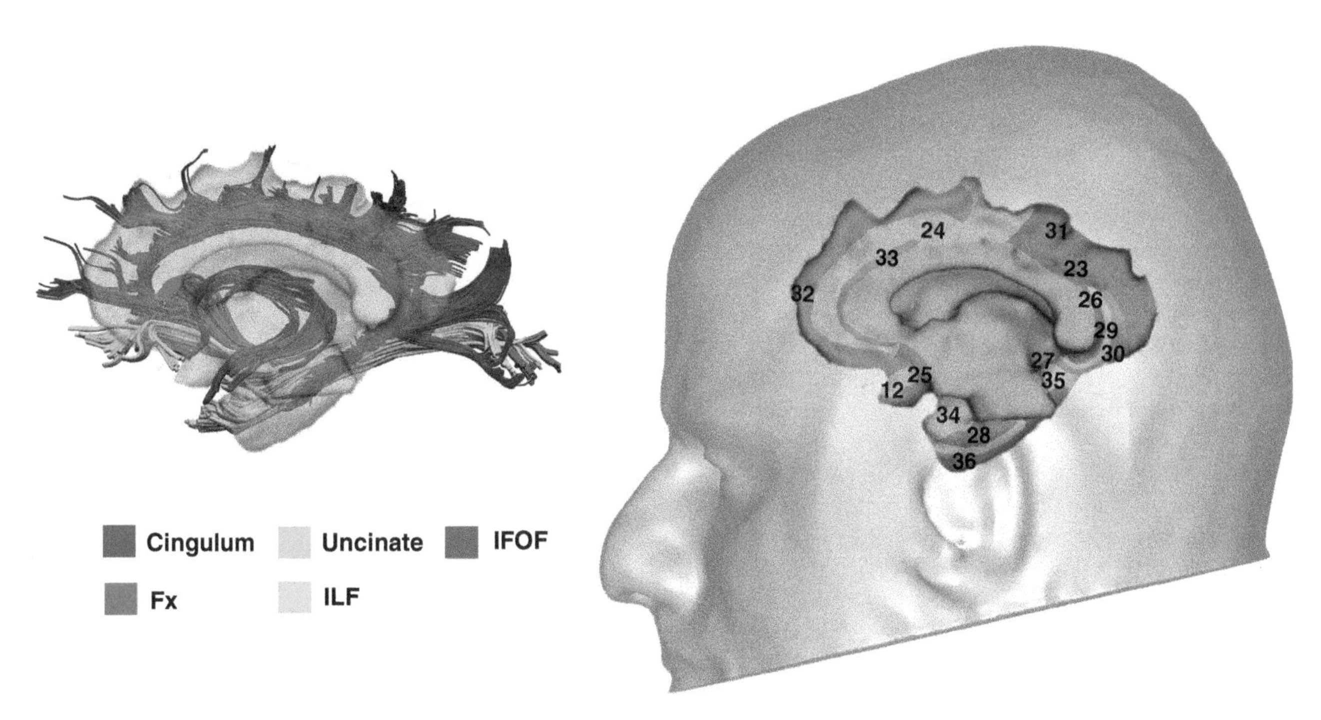

Figure 5.14 Medial view of the cortical anatomy of the limbic lobe and its main associative connections. Numbers indicate cytoarchitectonic areas according to Brodmann's nomenclature. The major association pathways of the limbic lobes are the cingulum, the uncinate, the inferior longitudinal fasciculus (ILF), and the fornix.

a severe amnesia with confabulation due to alcoholic degeneration of the mammillo-thalamic tract synaptically linked to the hippocampus via the fornix.

Disorders of the orbitofrontal–amygdala division. Patients with lesions of the anterior temporal lobe, including the amygdala and its connections to olfactory and orbitofrontal cortex, present with symptoms reminiscent of the classical Klüver–Bucy syndrome described in monkeys (Klüver and Bucy, 1939). These include abnormal emotional placidity (absence of reaction), hyperorality (strong oral tendency in examining objects), hypermetamorphosis (tendency to attend and react to visual stimuli), altered dietary preferences, increased sexual activity, and visual agnosia (Terzian and Ore, 1955). In patients with amygdala lesion, the autonomic responses normally associated with learning and recall are abolished. Lesions confined to the amygdalae are extremely rare in humans and result in mildly impaired modulation of social behaviour and abnormalities in the visual examination of faces (Adolphs et al., 1994). In patients with temporolimbic epilepsy a wide range of acute and chronic psychiatric symptoms have been observed, from panic attacks to aggressive outbursts, depression, and psychosis (Waxman and Geschwind, 1974; Flugel et al., 2006). Personality changes are often associated with abnormalities in the orbitofrontal cortex, medial anterior cingulate, amygdala, and their reciprocal connections (i.e. the uncinate fasciculus).

Disorders of the paralimbic areas. The paralimbic areas include the temporopolar region, insula, cingulate gyrus, orbitofrontal cortex, and parahippocampal gyrus. Lesions to paralimbic regions may present with emotional indifference to pain (pain asymbolia), altered olfaction, impaired ability to express emotions, reduced attention and motivation. Severe apathy and personality changes are frequently associated with orbitofrontal and anterior cingulate pathology. Degeneration of the anterior temporal and posterior orbitofrontal cortex presents with semantic dementia, alteration of personality traits and behaviour, and mood changes. The anterior temporal and orbitofrontal cortex are also commonly affected in traumatic brain injury. Stimulation of the cingulate and parahippocampal cortices causes memory flashbacks, dreamlike states, and mood alterations. Obsessive–compulsive disorder has been associated with increased activity in the cingulate cortex causing 'rigid hyperattentiveness' to thoughts and emotions.

Reduced olfactory sensation is observed in patients with Alzheimer's disease and Parkinson's disease. Olfactory hallucinations are commonly reported in epilepsy. Anomia for olfactory stimuli presented to the right nostril (right olfactory anomia) (Gordon and Sperry, 1969) and memory deficits (Zaidel and Sperry, 1974) have been described in split-brain patients with complete commissurotomy (complete severing of the corpus callosum and anterior and posterior commissure) but not in those with intact anterior commissure (Ledoux et al., 1977; Risse et al., 1978). The right olfactory anomia in these patients is due to a disconnection between the right olfactory cortex and the left fronto-temporal language areas. (For reviews of the clinical anatomy of the limbic lobe syndromes see Mesulam, 2000; Cummings and Mega, 2003; and Nieuwenhuys et al., 2008) (See also Chapter 11).

Conclusions

The clinical-anatomical correlation method has produced a spectacular amount of data about the organization of the human brain. Several decades of case reports and case–control studies have generated anatomical insights that stratified in textbooks and were passed on to the next generation in an almost crystallized form. We have, however, highlighted that clinical neuroanatomy is still a work in progress, where the development of new methodologies can help to overcome the impasse of current cortical localizationist theories. Atlases have always played an important part in the advancement of our anatomical understanding of neurological and psychiatric disorders. Previous atlases were the offspring of the excitement surrounding the renaissance of clinical anatomy, and we hope that the *Atlas of Human Brain Connections* will, in a small way, capture the excitement and potential of 21st-century anatomy. At the same time we are well aware that, at present, the hodotopic approach to clinico-anatomical correlation is more aspirational than proven, and stands or falls on empirical contributions yet to come. For this reason we concluded this chapter with a brief synopsis of the most frequent symptoms and syndromes classically attributed to individual cerebral lobes. We have not detailed the anatomy of each single disorder beyond broadly defined regions but we have shown side-to-side the cortical and connectional anatomy of each lobe. Our goal is to stimulate clinicians to develop a critical approach to correlative studies and broaden their view of clinical anatomy beyond the cortical surface to encompass the hodology of the lesion.

References

Adolphs, R., Tranel, D., Damasio, H., and Damasio, A. (1994). Impaired recognition of emotion in facial expressions following bilateral damage to the human amygdala. *Nature*, **372**(6507), 669–72.

Agosta, F., Henry, R.G., Migliaccio, R., et al. (2010). Language networks in semantic dementia. *Brain*, **133**(1), 286–99.

Aguirre, G.K., and D'Esposito, M. (1999). Topographical disorientation: a synthesis and taxonomy. *Brain*, **122**(9), 1613–28.

Alexander, M.P., Baker, E., Naeser, M.A., et al. (1992). Neuropsychological and neuroanatomical dimensions of ideomotor apraxia. *Brain*, **115**(1), 87–107.

Anderson, E.J., Jones, D.K., O'Gorman, R.L., Leemans, A., Catani, M., Husain, M. (2011). Cortical network for gaze control in humans revealed using multimodal MRI. *Cereb Cortex*, (in press).

Anton, G. (1889). Ueber die Selbstwahrnehmung der Herderkrankungen des Gehirns durch den Kranken bein Rindenblindheit und Rindentaubheit. *Archiv für Psychiatrie*, **32**, 86–127.

Bartolomeo, P., Thiebaut de Schotten, M., and Doricchi, F. (2007). Left unilateral neglect as a disconnection syndrome. *Cereb Cortex*, **17**(11), 2479–90.

Bassetti, C., Bogousslavsky, J., and Regli, F. (1993). Sensory syndromes in parietal stroke. *Neurology*, **43**(10), 1942–9.

Bauer, R.M. (1982). Visual hypoemotionality as a symptom of visual-limbic disconnection in man. *Arch Neurol*, **39**(11), 702–8.

Benson, D.F. and Geschwind, N. (1970) Developmental Gerstmann syndrome. *Neurology*, **20**(3), 293–8.

Blanke, O., Ortigue, S., Landis, T., et al. (2002). Stimulating illusory own-body perceptions. *Nature*, **419**(6904), 269–70.

Bürgel, U., Amunts, K., Hoemke, L., Mohlberg, H., Gilsbach, J. M., and Zilles, K. (2006). White matter fiber tracts of the human brain: three-dimensional mapping at microscopic resolution, topography and intersubject variability. *NeuroImage*, **29**(4), 1092–105.

Catani, M. and ffytche, D.H. (2005). The rises and falls of disconnection syndromes. *Brain*, **128**(10), 2224–39.

Catani, M. and ffytche, D.H. (2010). On 'the study of the nervous system and behaviour. *Cortex*, **46**(1), 106–9.

Catani, M. and Mesulam, M. (2008). What is a disconnection syndrome? *Cortex*, **44**(8), 911–13.

Catani, M., Jones, D.K., Donato, R., et al. (2003). Occipito-temporal connections in the human brain. *Brain*, **126**(9), 2093–107.

Catani, M., Jones, D.K., and ffytche, D.H. (2005). Perisylvian language networks of the human brain. *Ann Neurol*, **57**(1), 8–16.

Catani, M., Craig, M.C., Forkel, S.J., et al. (2011). Altered Integrity of Perisylvian Language Pathways in Schizophrenia: Relationship to Auditory Hallucination. *Biol Psychiatry*, DOI 10.1016/j.biopsych.2011.06.013

Cohen, L., Dehaene, S., Naccache, L., et al. (2000). The visual word form area: spatial and temporal characterization of an initial stage of reading in normal subjects and posterior split-brain patients. *Brain*, **123**(2), 291–307.

Courchesne, E. and Pierce K. (2005). Why the frontal cortex in autism might be talking only to itself: local over-connectivity but long-distance disconnection. *Curr Opin Neurobiol*, **15**(2), 225–30.

Critchley, M. (1953). *The parietal lobe*. New York: Hafner Publishing Company.

Cummings, J.L. and Mega, M.S. (2003). *Neuropsychiatry and Behavioral Neuroscience*. New York: Oxford University Press.

Damasio, H. and Damasio, A. (1989). *Lesion Analysis in Neuropsychology*. New York: Oxford University Press.

Darby, D. and Walsh, K. (2005). *Walsh's Neuropsychology*, fifth edition. Amsterdam: Elsevier.

De Renzi, E. (2001). A farewell. *Cortex*, **37**(1), 5–8.

De Renzi, E. and Vignolo, L. A. (1962). The token test: a sensitive test to detect receptive disturbances in aphasics. *Brain*, **85**(4), 665–78.

Dejerine, J. (1914). *Sémiologie des affections du système nerveux*. Paris: Masson.

Doricchi, F., Thiebaut de Schotten, M., Tomaiuolo, F., and Bartolomeo, P. (2008). White matter (dis)connections and gray matter (dys)functions in visual neglect: gaining insights into the brain networks of spatial awareness. *Cortex*, **44**(8), 983–95.

Dronkers, N.F., Plaisant, O., Iba-Zizen, M.T., and Cabanis, E.A. (2007). Paul Broca's historic cases: high resolution MR imagimg of the brains of Leborgne and Lelong. *Brain*, **130**(5), 1432–41.

Epelbaum, S., Pinel, P., Gaillard, R., et al. (2008). Pure alexia as a disconnection syndrome: New diffusion imaging evidence for an old concept. *Cortex*, **44**(8), 962–74.

ffytche, D.H., and Howard, R.J. (1999). The perceptual consequences of visual loss: positive pathologies of vision. *Brain*, **122**(7), 1247–60.

ffytche, D.H., Howard, R.J., Brammer, M.J., David, A., Woodruff, P., and Williams, S. (1998). The anatomy of conscious vision: an fMRI study of visual hallucinations. *Nat Neurosci*, **1**(8), 738–42.

ffytche D.H., Blom, J.D., and Catani, M. (2010). Disorders of visual perception. *J Neurol Neurosurg Psychiat*, **81**(11), 1280–7.

Finger, S. (1994). *Origins of Neuroscience*. New York: Oxford University Press.

Finger, S., Koehler, J.P., and Jagella, C. (2004). The Monakow concept of diaschisis. *Arch Neurol*, **61**(2), 283–8.

Flugel, D., Cercignani, M., Symms, M.R., et al. (2006). Diffusion tensor imaging findings and their correlation with neuropsychological deficits in patients with temporal lobe epilepsy and interictal psychosis. *Epilepsia* **47**(5), 941–4.

Fox, C., Iaria, G., and Barton, J. (2008). Disconnection in prosopagnosia and face processing. *Cortex*, **44**(8), 996–1009.

Fuster, J.M. (1997). *The Prefrontal Cortex: Anatomy, Physiology, and Neuropsychology of the Frontal Lobe*. New York: Raven Press.

Galantucci, S., Tartaglia, M.C., Wilson, S.M., Henry, M.L., et al. (2011). White matter damage in primary progressive aphasias: a diffusion tensor tractography study. *Brain* **134**(10), 3011–29.

Geschwind, N. (1965a). Disconnexion syndromes in animals and man. I. *Brain*, **88**(2), 237–94.

Geschwind, N. (1965b). Disconnexion syndromes in animals and man. II. *Brain*, **88**(3), 585–644.

Gordon, H.W. and Sperry, R.W. (1969). Lateralization of olfactory perception in the surgically separated hemispheres of man. *Neuropsychologia*, **7**(2), 111–20.

Gloor, P. (1997). *The Temporal Lobe and the Limbic system*. New York: Oxford University Press.

Head, H. and Holmes, G. (1911). Sensory disturbances from cerebral lesions. *Brain*, **34**(2–3), 102–254.

Heilman, K.M. and Watson, R.T. (2008). The disconnection apraxias. *Cortex*, **44** (8), 975–82.

Hillis, A.E., Wityk, R.J., Barker, P.B., et al. (2002). Subcortical aphasia and neglect in acute stroke: the role of cortical hypoperfusion. *Brain*, **125**(5), 1094–104.

Hubl, D., Koenig, T., Strik, W., et al. (2004). Pathways that make voices: white matter changes in auditory hallucinations. *Arch Gen Psychiatry*, **61**(7), 658–68.

Husain, M., Mattingley, J.B., Rorden, C., et al. (2000). Distinguishing sensory and motor biases in parietal and frontal neglect. *Brain*, **123**(8), 1643–59.

Jackson, J.H. (1881). Remarks on dissolution of the nervous system as exemplified by certain post-epileptic conditions. *Med Press Circ*, 329–32.

Jackson, J.H. (1894). The factors of insanities. *Med Press Circ*, **108**, 615–19.

Karnath, H.O. and Perenin, M.T. (2005). Cortical control of visually guided reaching: evidence from patients with optic ataxia. *Cereb Cortex*, **15**(10), 1561–9.

Karnath, H.O., Rorden, C., and Ticini, L.F. (2009). Damage to white matter fiber tracts in acute spatial neglect. *Cereb Cortex*, **19**(10), 2331–7.

Kempler, D., Metter, E.J., Jackson, C.A., et al. (1988). Disconnection and cerebral metabolism. The case of conduction aphasia. *Arch Neurol*, **45**(3), 275–9.

Kleist, K. (1934). *Gehirnpathologie*. Leipzig: J.A. Barth.

Klüver, H., and Bucy, P.C. (1939). Preliminary analysis of functions of the temporal lobes in monkeys. *Arch Neurol Psychiatry*, **42**(6), 979–1000.

Kravitz, D.J., Saleem, K.S., Baker, C.I., and Mishkin, M. (2011). A new neural framework for visuospatial processing. *Nat Rev Neurosci*, **12**(4), 217–30.

Laplane, D. and Degos, J.D. (1983). Motor neglect. *J Neurol Neurosurg Psychiat*, **46**(2), 152–8.

Lashley, K.S. (1950). In search of the engram. *Symp Soc Exp Biol*, **4**, 454–82.

Ledoux, J.E., Risse, G.L., Springer, S.P., et al. (1977). Cognition and commissurotomy. *Brain*, **100(1**), 87–104.

Lennox, B.R., Park, S.B.G., Medley, I., et al. (2000). The functional anatomy of auditory hallucinations in schizophrenia. *Psychiatr Res*, **100**(1), 13–20.

Levelt, W.J.M. (1989). *Speaking: from intention to articulation*. Cambridge: MIT Press.

Lichtheim, L. (1885). On aphasia. *Brain*, **7**, 433–84.

Marchetti, C. and Della Sala, S. (1998). Disentangling the alien and anarchic hand. *Cognit Neuropsychiat*, **3**(3), 191–207.

Markowitsch, H.J. (2000). Memory and amnesia. In Mesulam, M. (Ed.) *Principles of Behavioural and Cognitive Neurology*, second edition, pp. 257–93. New York: Oxford University Press.

Mega, M.S., Cummings, J.L., Salloway, S., et al. (1997). The limbic system: an anatomic, phylogenetic, and clinical perspective. *J Neuropsychiatry Clin Neurosci*, **9**(3), 315–30.

Mesulam, M.M. (1990). Large-scale neurocognitive networks and distributed processing for attention, language, and memory. *Ann Neurol*, **28**(5), 597–613.

Mesulam, M. (2000). Behavioural neuroanatomy: large-scale networks, association cortex, frontal syndromes, the limbic system, and the hemispheric specializations. In Mesulam, M. (Ed.) *Principles of Behavioural and Cognitive Neurology*, pp. 1–120. New York: Oxford University Press.

Meynert, T. (1885). *A clinical treatise on diseases of the forebrain based upon a study of its structure, functions, and nutrition*. New York: G.P. Putnam's Sons.

Milner, A.D. and Goodale, M.A. (1993). *The Visual Brain in Action*. Oxford: Oxford University Press.

Milner, A.D., and Goodale, M.A. (2008). Two visual systems re-viewed. *Neuropsychologia*, **46**(3), 774–85.

Monakow, C. von. (1914). *Die Lokalisation im Grosshirn und der Abbau der Funktion durch kortikale Herde*. Wiesbaden: JF Bergmann.

Mori, S., Wakana, S., Nagae-Poetscher, L.M., and van Zijl, P.C.M. (2005). *MRI atlas of human white matter*. Amsterdam: Elsevier.

Naeser, M.A., Alexander, M.P., Helm-Estabrooks, N., Levine, H.L., Laughlin, S.A., and Geschwind, N. (1982). Aphasia with predominantly subcortical lesion sites: description of three capsular/putaminal aphasia syndromes. *Arch Neurol*, **39**(1), 2–14.

Newton, J.M., Ward, N.S., Parker, G.J., et al. (2006). Non-invasive mapping of corticofugal fibres from multiple motor areas-relevance to stroke recovery. *Brain*, **129**(7), 1844–58.

Nieuwenhuys, R., Voogd, J., and Huijzen, C. (2008). *The human central nervous system*, fourth edition. Berlin: Springer-Verlag.

Papagno, C., Miracapillo, C., Casarotti, A., et al. (2011). What is the role of the uncinate fasciculus? Surgical removal and proper name retrieval. *Brain*, **134**(2), 405–14.

Price, C.J., Warburton, E.A., Moore, C.J., Frackowiak, R.S., and Friston, K.J. (2001). Dynamic diaschisis: anatomically remote and context-sensitive human brain lesions. *J Cogn Neurosci*, **13**(4), 419–29.

Priftis, K., Rusconi, E., Umilta, C., and Zorzi M. (2003). Pure agnosia for mirror stimuli after right inferior parietal lesion. *Brain*, **126**(4), 908–19.

Ramayya, A.G., Glasser, M.F., and Rilling, J.K. (2010). A DTI investigation of neural substrates supporting tool use. *Cereb Cortex*, **20**(3), 507–16.

Rankin, K.P., Gorno-Tempini, M.L., Allison, S.C., et al. (2006). Structural anatomy of empathy in neurodegenerative disease. *Brain*, **129**(11), 2945–56.

Risse, G.L., Ledoux, J.E., Springer, S.P., et al. (1978). The anterior commissure in man: functional variation in a multisensory system. *Neuropsychologia*, **16**(1), 23–31.

Rosen, H.J., Allison, S.C., Schauer, G.F., et al. (2005). Neuroanatomical correlates of behavioural disorders in dementia. *Brain*, **128**(11), 2612–25.

Ross, E.D. (1981). The aprosodias. Functional-anatomic organization of the affective components of language in the right hemisphere. *Arch Neurol*, **38**(9), 561–9.

Ross, E.D. (2008). Sensory-specific amnesia and hypoemotionality in humans and monkeys: Gateway for developing a hodology of memory. *Cortex*, **44**(8), 1010–22.

Ross, E.D. (2010). Cerebral localization of functions and the neurology of language: fact versus fiction or is it something else? *Neuroscientist*, **16**(3), 222–43.

Rudrauf, D., Mehta, S., and Grabowski, T. (2008). Disconnection's renaissance take shape: formal incorporation in group-level lesion studies. *Cortex*, **44**(8), 1084–96.

Rusconi, E., Pinel, P., Dehaene, S., and Kleinschmidt A. (2010). The enigma of Gerstmann's syndrome revisited: a telling tale of the vicissitudes of neuropsychology. *Brain*, **133**(2), 320–32.

Salloway, S., Malloy, P., and Duffy, J. (Eds.) (2001). *The frontal lobes and neuropsychiatric illness*. Washington, DC: American Psychiatric Press.

Schneider, R.C., Crosby, E.C., Bagchi, B.K., and Calhoun, H.D. (1961). Temporal or occipital lobe hallucinations triggered from frontal lobe lesions. *Neurology*, **11**, 172–9.

Sergent, J., Ohta, S., and Macdonald, B. (1992). Functional neuroanatomy of face and object processing. Brain, **115**(1), 15–36.

Shergill, S.S., Brammer, M.J., Williams, S.C.R., Murray, R.M., and McGuire, P. K. (2000). Mapping auditory hallucinations in schizophrenia using functional magnetic resonance imaging. *Arch Gen Psychiatry*, **57**(11), 1033–8.

Stuss, D. T. and Knight, R. T. (Eds.) (2002). *Principles of Frontal Lobe Function*. Oxford: Oxford University Press.

Talairach, J., and Tournoux, P. (1988). *Co-planar Stereotaxic Atlas of the Human Brain: 3-Dimensional Proportional System – An Approach to Cerebral Imaging*. New York: Thieme Medical Publishers.

Terzian, H., and Ore, G.D. (1955). Syndrome of Klüver and Bucy; reproduced in man by bilateral removal of the temporal lobes. *Neurology* **5**(6), 373–80.

Thiebaut de Schotten, M., Dell'Acqua, F., Forkel, S., et al. (2011). A lateralized brain network for visuo-spatial attention. *Nat Neurosci*, **14**(10), 1245–6.

Thiebaut de Schotten, M., Urbanski, M., Duffau, H., et al. (2005). Direct evidence for a parietal-frontal pathway subserving spatial awareness in humans. *Science*, **309**(5744), 2226–8.

Ungerleider, L.G., and Mishkin, M. (1982). Two cortical visual systems. In Ingle, D.J., Goodale, M.A., and Mansfield, R.J.W. (Eds.) *Analysis of visual behaviour*, pp. 549–86. Cambridge, MA: MIT Press.

Valenstein, E., Bowers, D., Verfaellie, M., Heilman, K.M., Day, A., Watson, R.T. (1987). Retrosplenial amnesia. *Brain*, **110**(6), 1631–46.

Vallar, G., Bottini, G., Rusconi, M.L., and Sterzi, R. (1993). Exploring somatosensory hemineglect by vestibular stimulation. *Brain*, **116**(1), 71–86.

Vallar, G. (2007). Spatial neglect, Balint-Homes' and Gerstmann's syndrome, and other spatial disorders. *CNS Spectr*, **12**(7), 527–36.

Vann, S.D., Aggleton, J.P., and Maguire, E.A. (2009). What does the retrosplenial cortex do? *Nat Rev Neurosci*, **10**(11), 792–802.

Waxman, S.G., and Geschwind, N. (1974). Hypergraphia in temporal lobe epilepsy. *Neurology*, **24**(7), 629–36.

Wernicke, C. (1874). *Der Aphasische Symptomencomplex. Ein psychologische Studie auf anatomischer Basis* (Eggert, G. Trans). Breslau: Cohn & Weigert.

Zaidel, D. and Sperry, R.W. (1974). Memory impairment after commissurotomy in man. *Brain*, **97**(2), 263–72.

Zeki, S. (1990). A century of cerebral achromatopsia. *Brain*, **113**(6), 1721–77.

CHAPTER 6

Atlas of Human Brain Connections (all tracts)

Axial maps

Axial slice –50

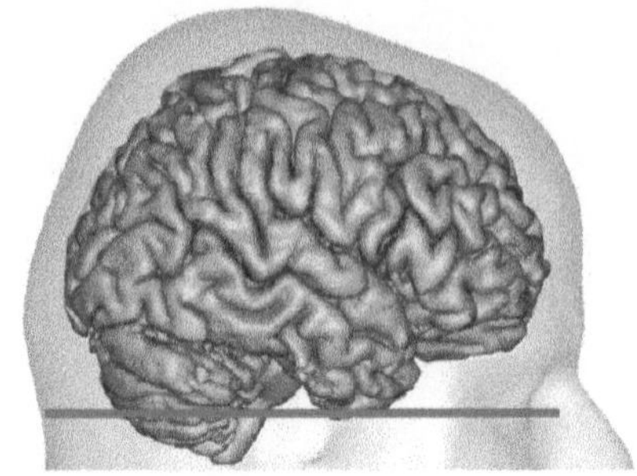

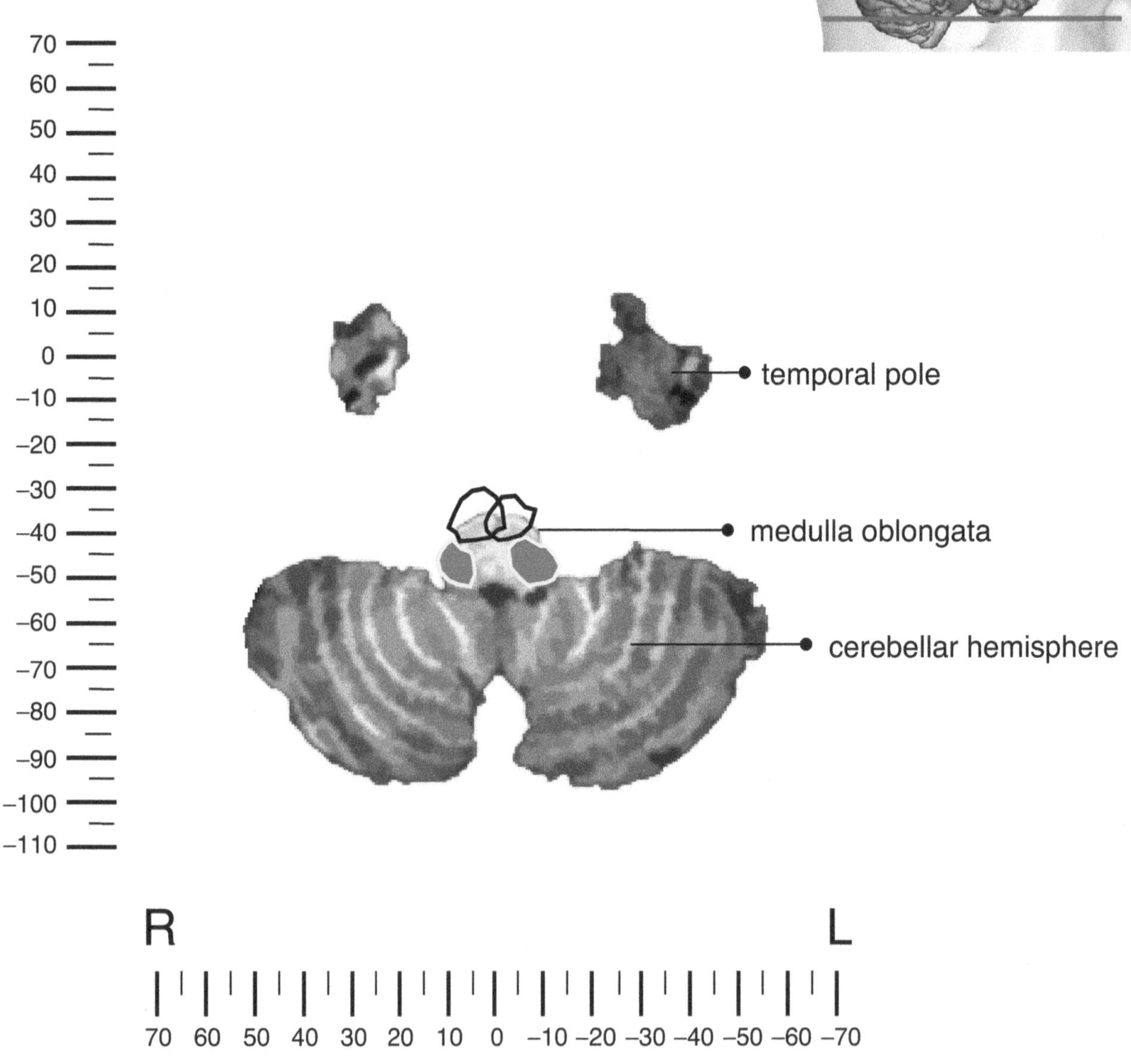

TRACT LEGEND

CST	Cortico-Spinal Tract / IC
ICP	Inf. Cerebellar Peduncle

Axial slice –48

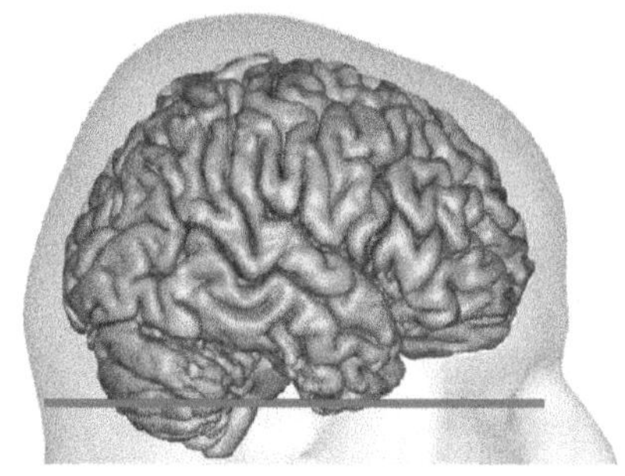

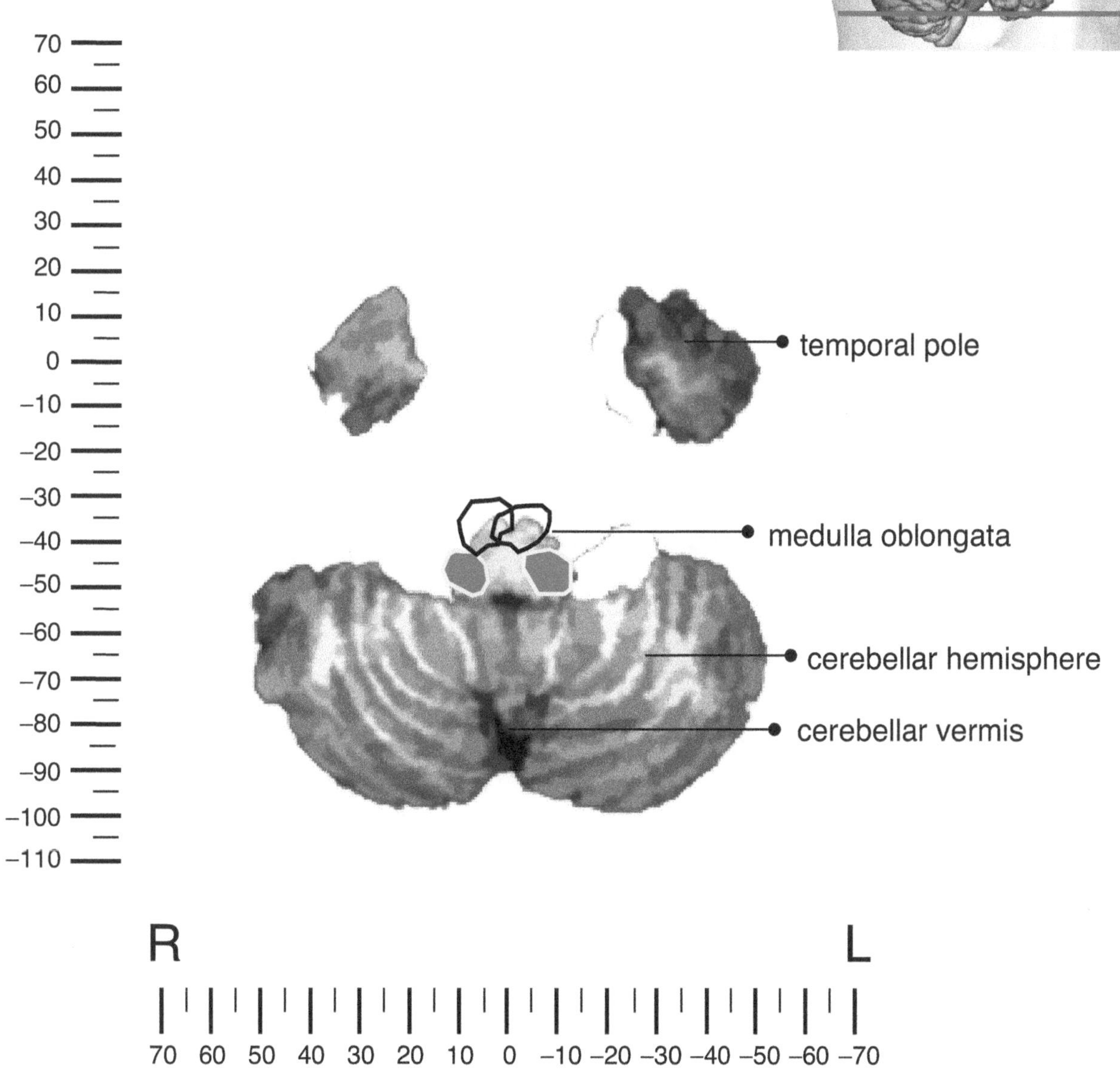

TRACT LEGEND

CST	Cortico-Spinal Tract / IC
ICP	Inf. Cerebellar Peduncle

Axial slice –46

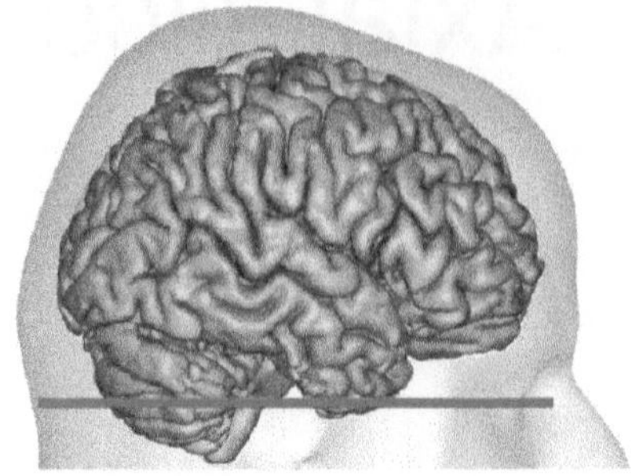

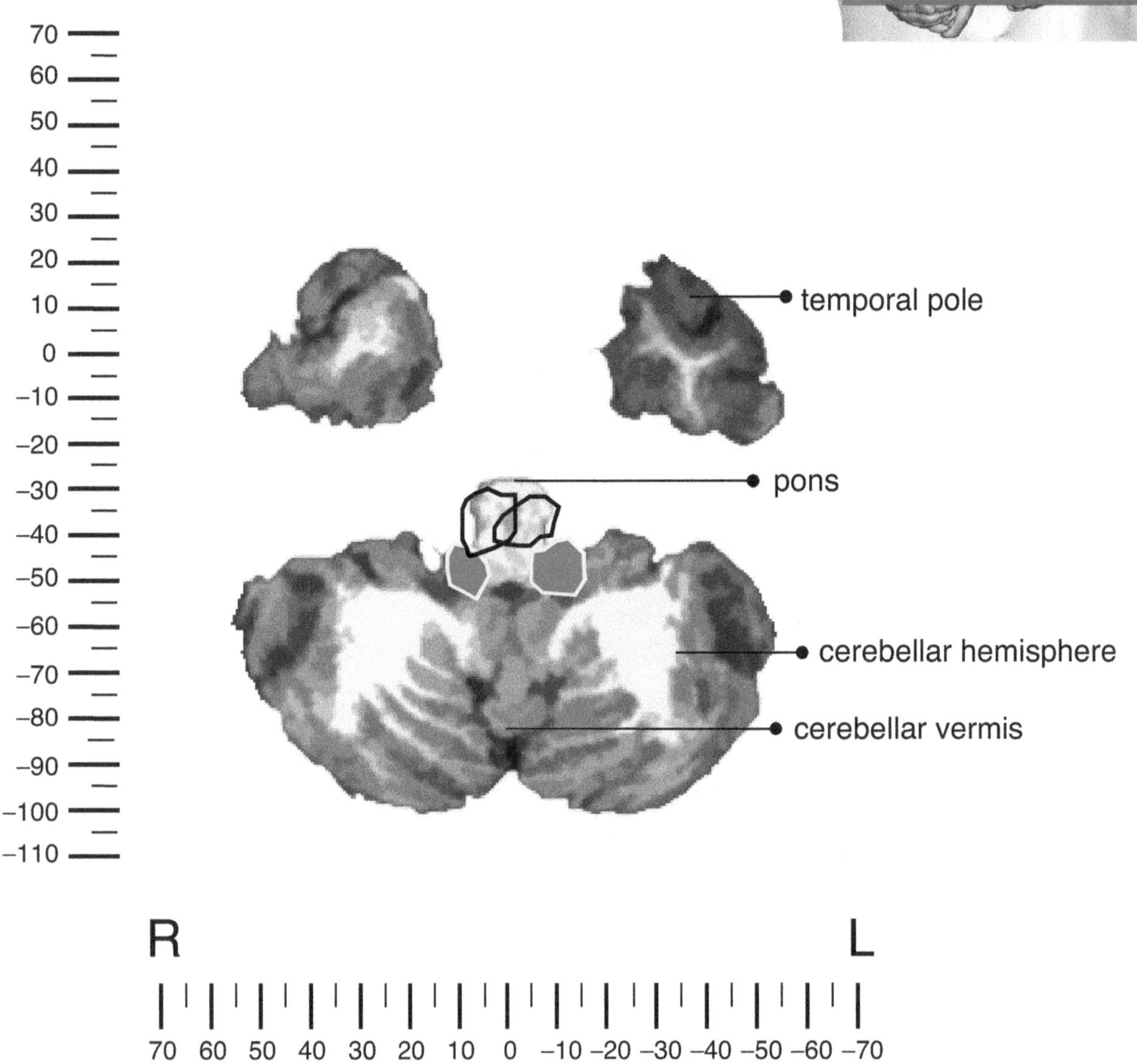

TRACT LEGEND

CST	Cortico-Spinal Tract / IC
ICP	Inf. Cerebellar Peduncle

Axial slice –44

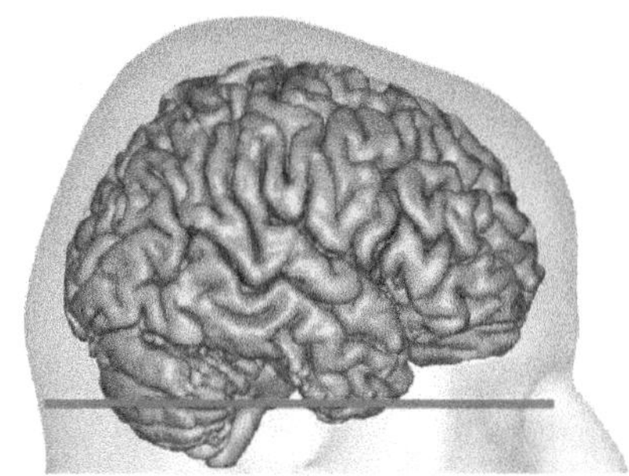

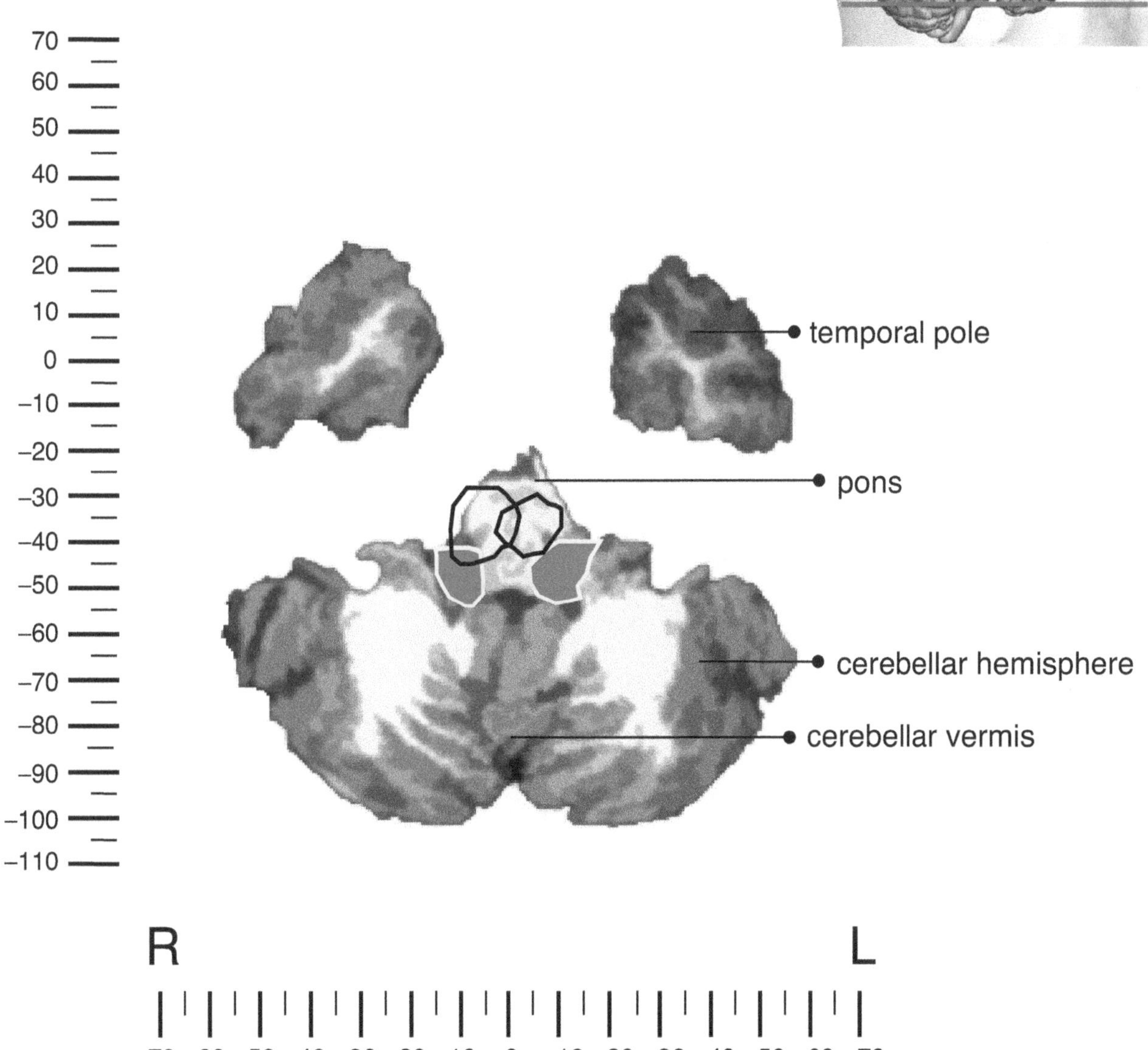

TRACT LEGEND

CST	Cortico-Spinal Tract / IC
ICP	Inf. Cerebellar Peduncle

Axial slice –42

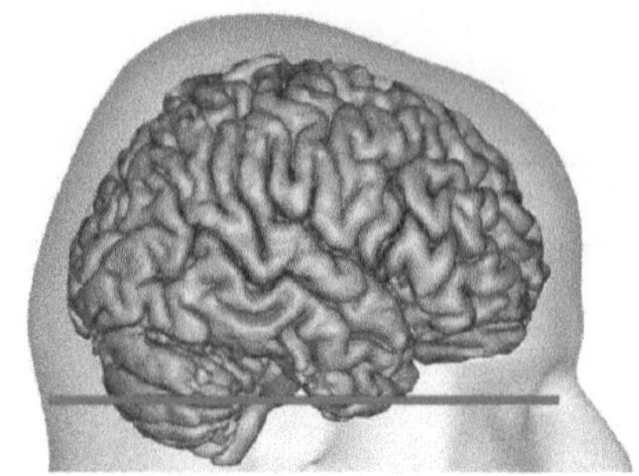

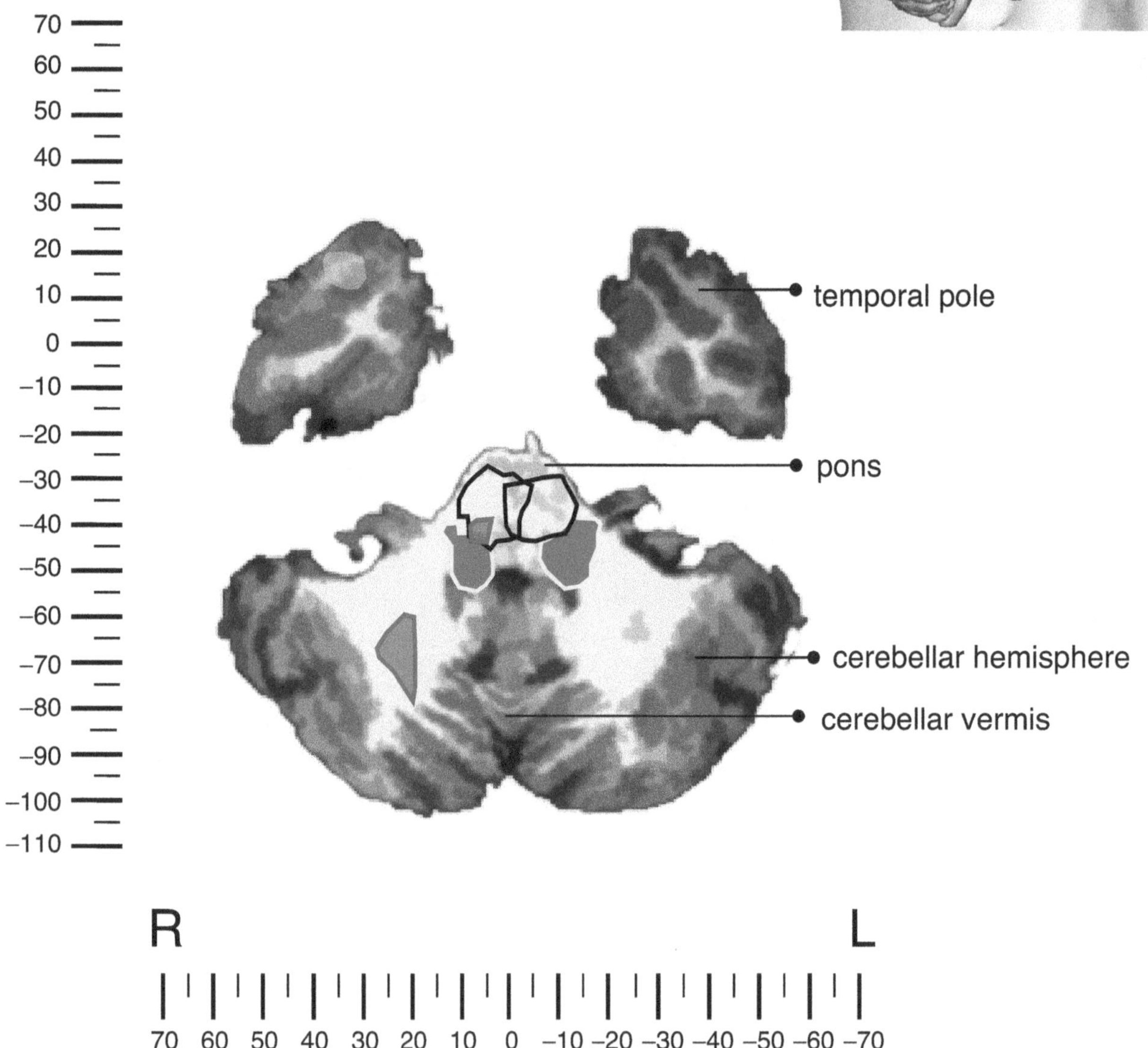

TRACT LEGEND

CST	Cortico-Spinal Tract / IC
CPC	Cortico-Ponto-Cerebellar Tract
ICP	Inf. Cerebellar Peduncle
AC	Anterior Commissure

Axial slice –40

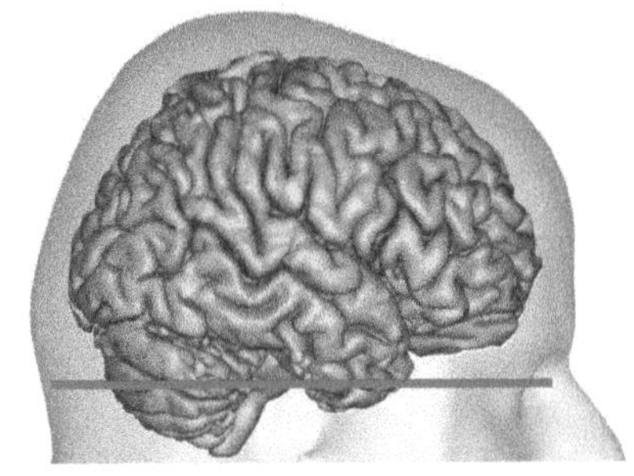

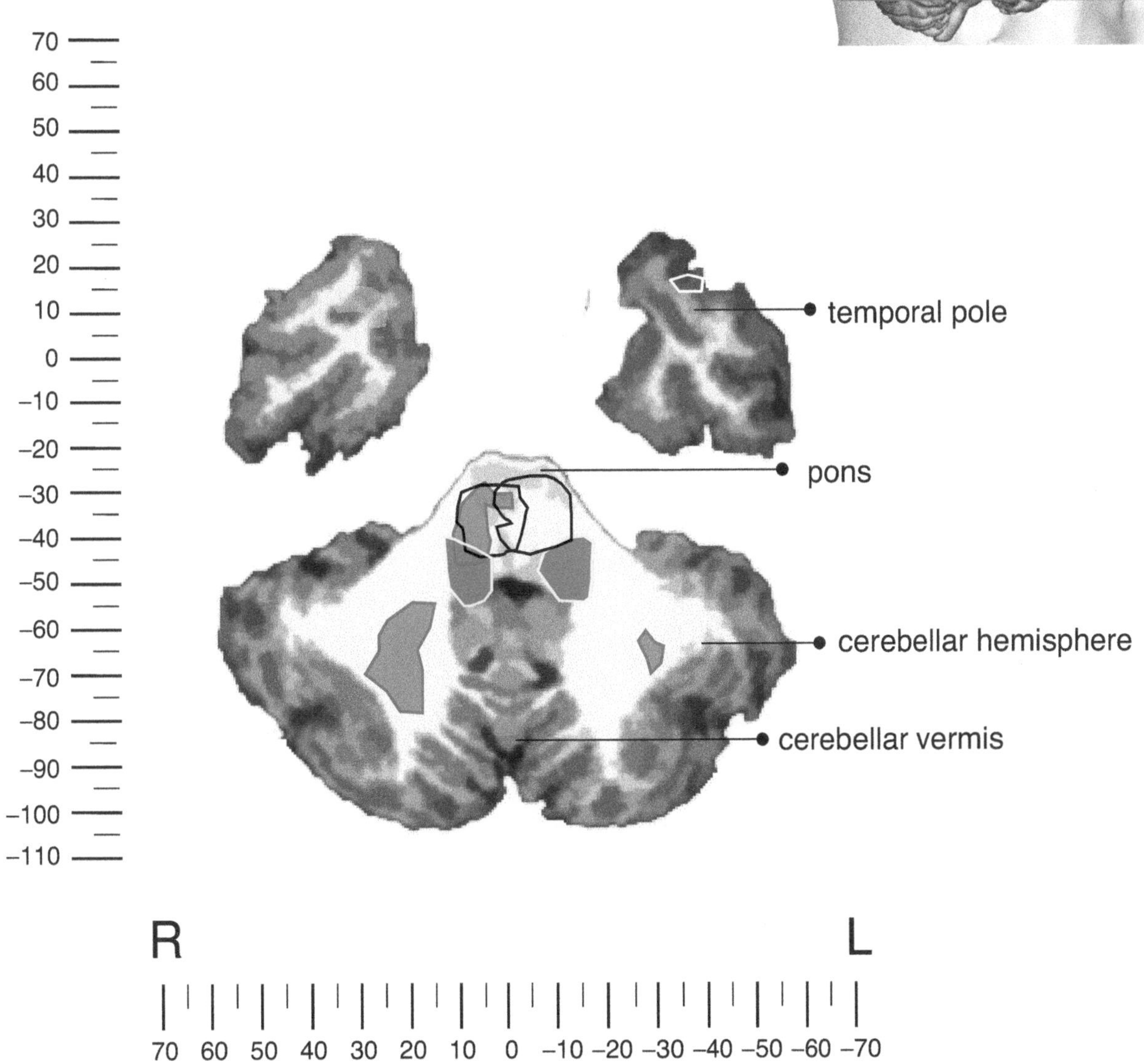

R L

70 60 50 40 30 20 10 0 –10 –20 –30 –40 –50 –60 –70

TRACT LEGEND

CST	Cortico-Spinal Tract / IC	CPC	Cortico-Ponto-Cerebellar Tract
ICP	Inf. Cerebellar Peduncle	AC	Anterior Commissure
		Fx	Fornix

Axial slice –38

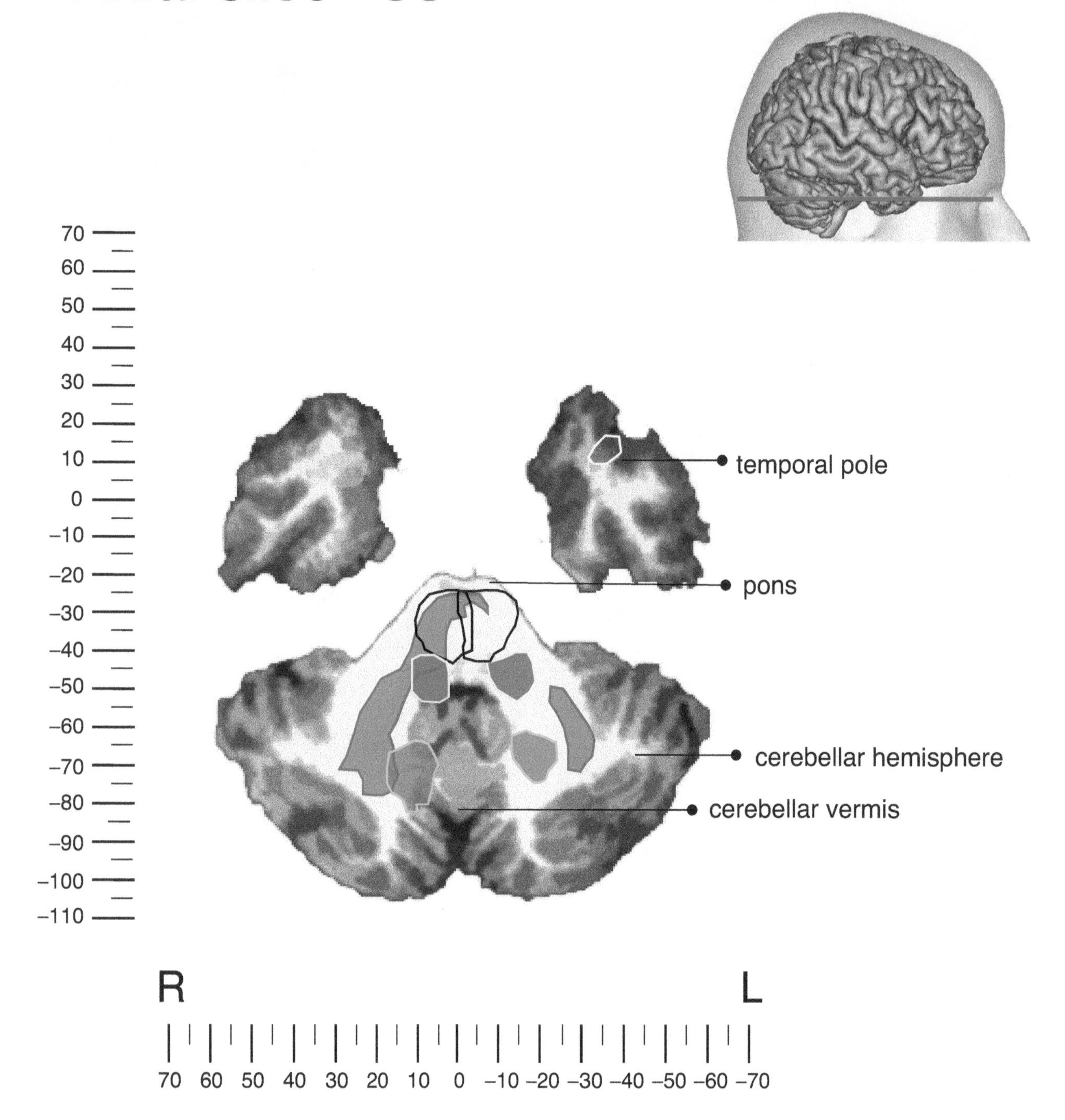

TRACT LEGEND

CST	Cortico-Spinal Tract / IC	CPC	Cortico-Ponto-Cerebellar Tract
ICP	Inf. Cerebellar Peduncle	AC	Anterior Commissure
SCP	Sup. Cerebellar Peduncle	Fx	Fornix

Axial slice –36

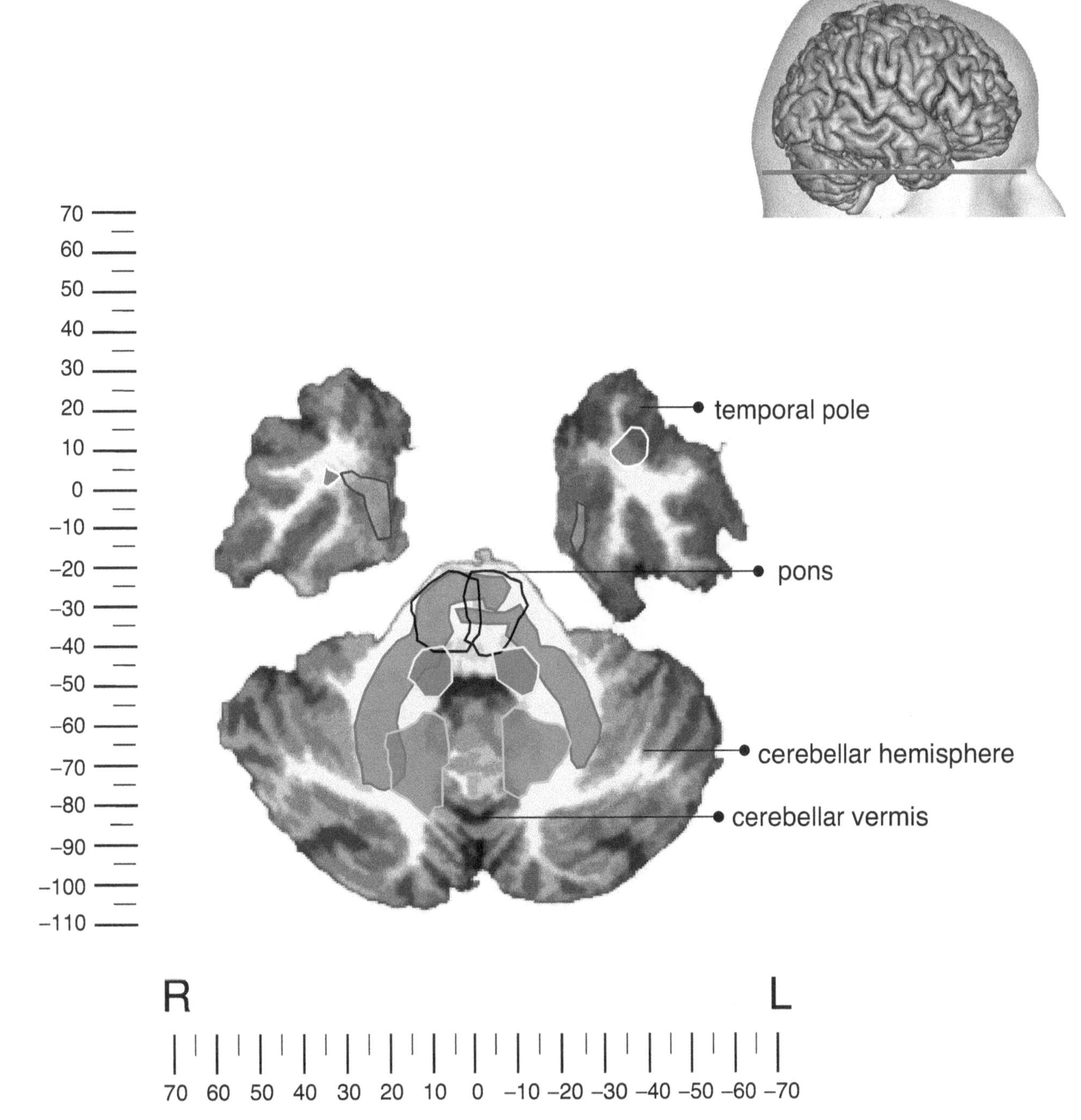

TRACT LEGEND

CST	Cortico-Spinal Tract / IC	CPC	Cortico-Ponto-Cerebellar Tract	Cing	Cingulum
ICP	Inf. Cerebellar Peduncle	AC	Anterior Commissure		
SCP	Sup. Cerebellar Peduncle	Fx	Fornix		

Axial slice –34

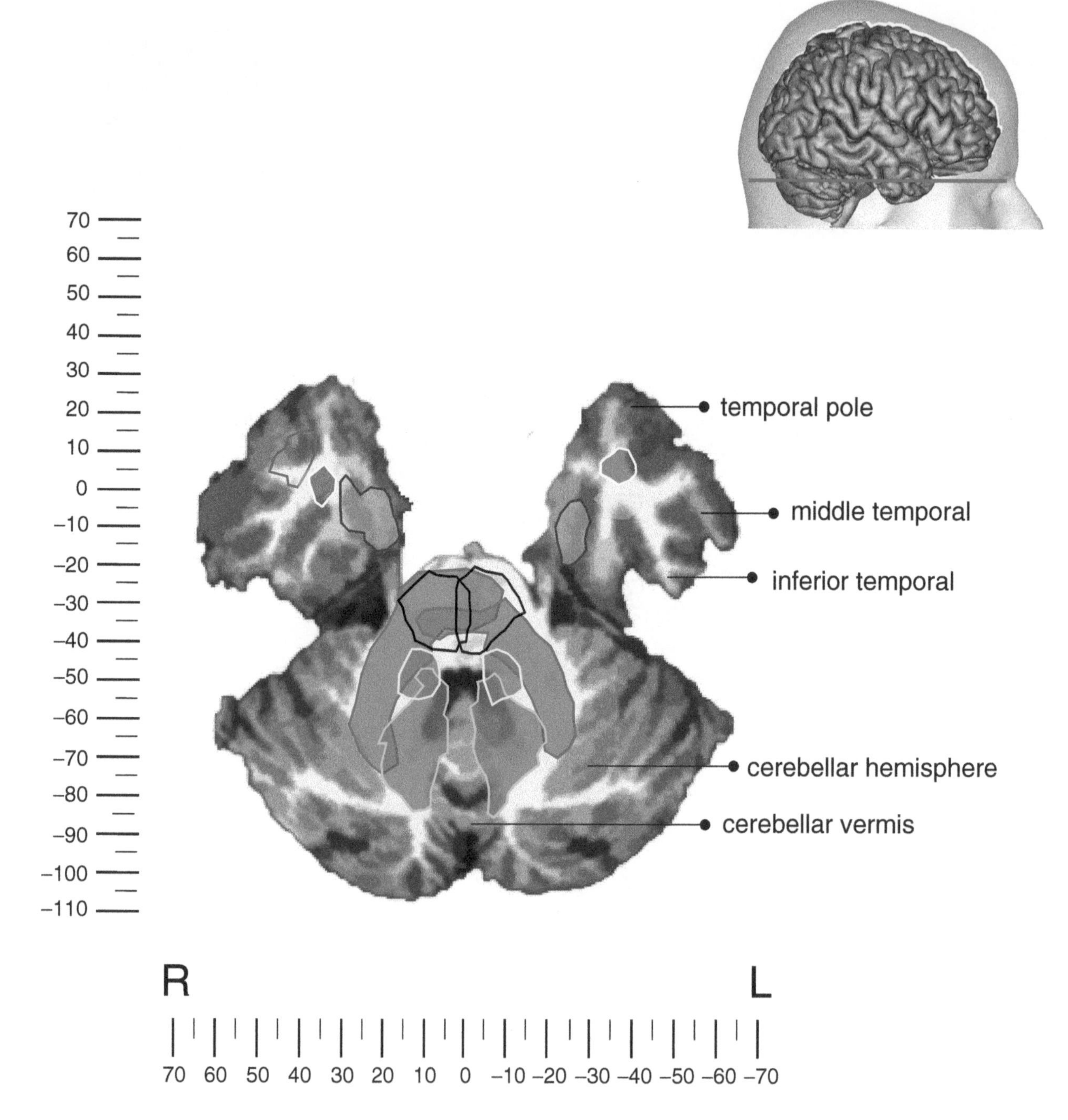

TRACT LEGEND

CST	Cortico-Spinal Tract / IC	CPC	Cortico-Ponto-Cerebellar Tract	Cing	Cingulum
ICP	Inf. Cerebellar Peduncle	AC	Anterior Commissure		
SCP	Sup. Cerebellar Peduncle	Fx	Fornix		

Axial slice –32

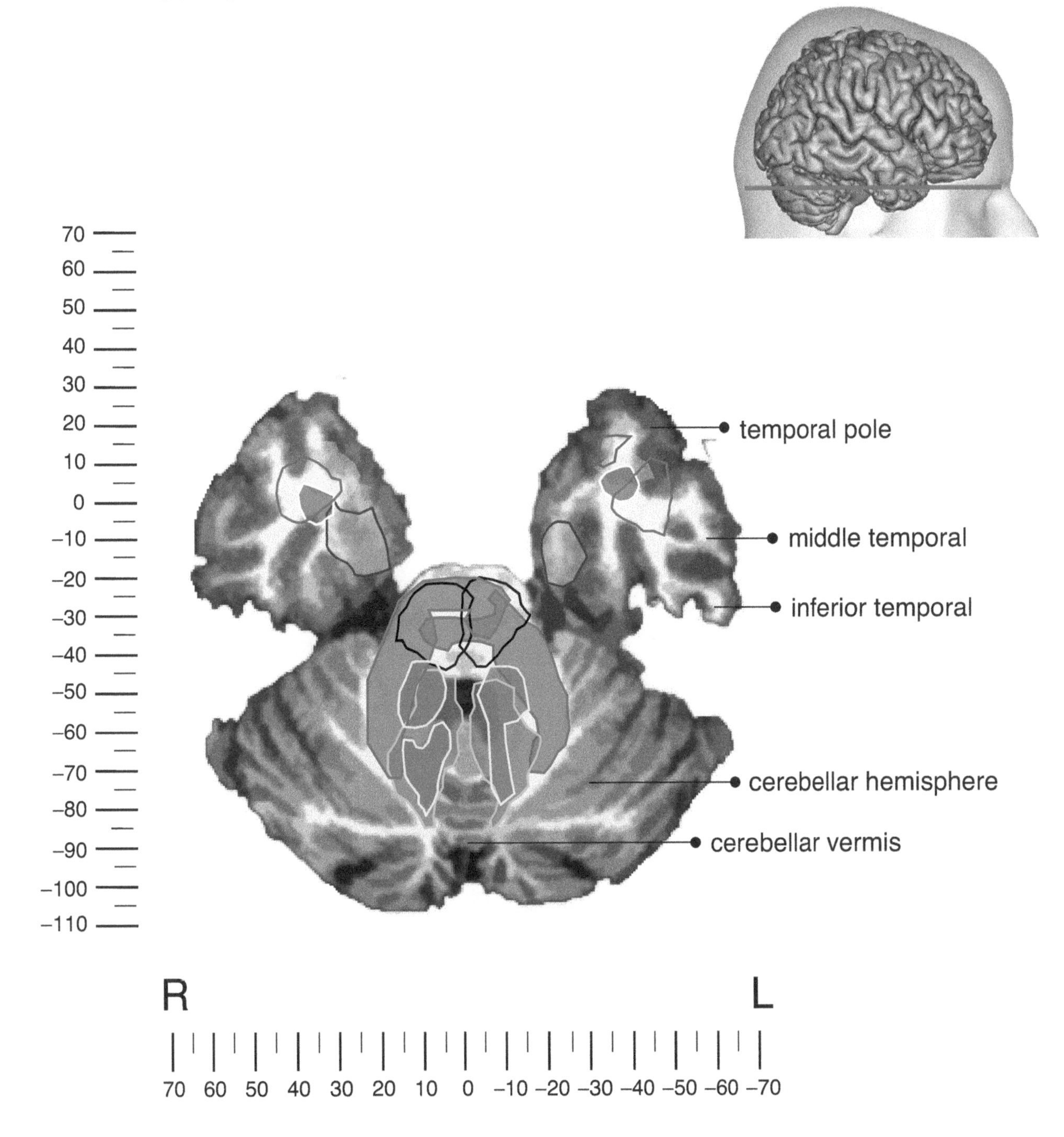

TRACT LEGEND

CST	Cortico-Spinal Tract / IC	CPC	Cortico-Ponto-Cerebellar Tract	Cing	Cingulum
ICP	Inf. Cerebellar Peduncle	AC	Anterior Commissure	Unc	Uncinate
SCP	Sup. Cerebellar Peduncle	ILF	Inf. Longitudinal Fasc.	Fx	Fornix

Axial slice –30

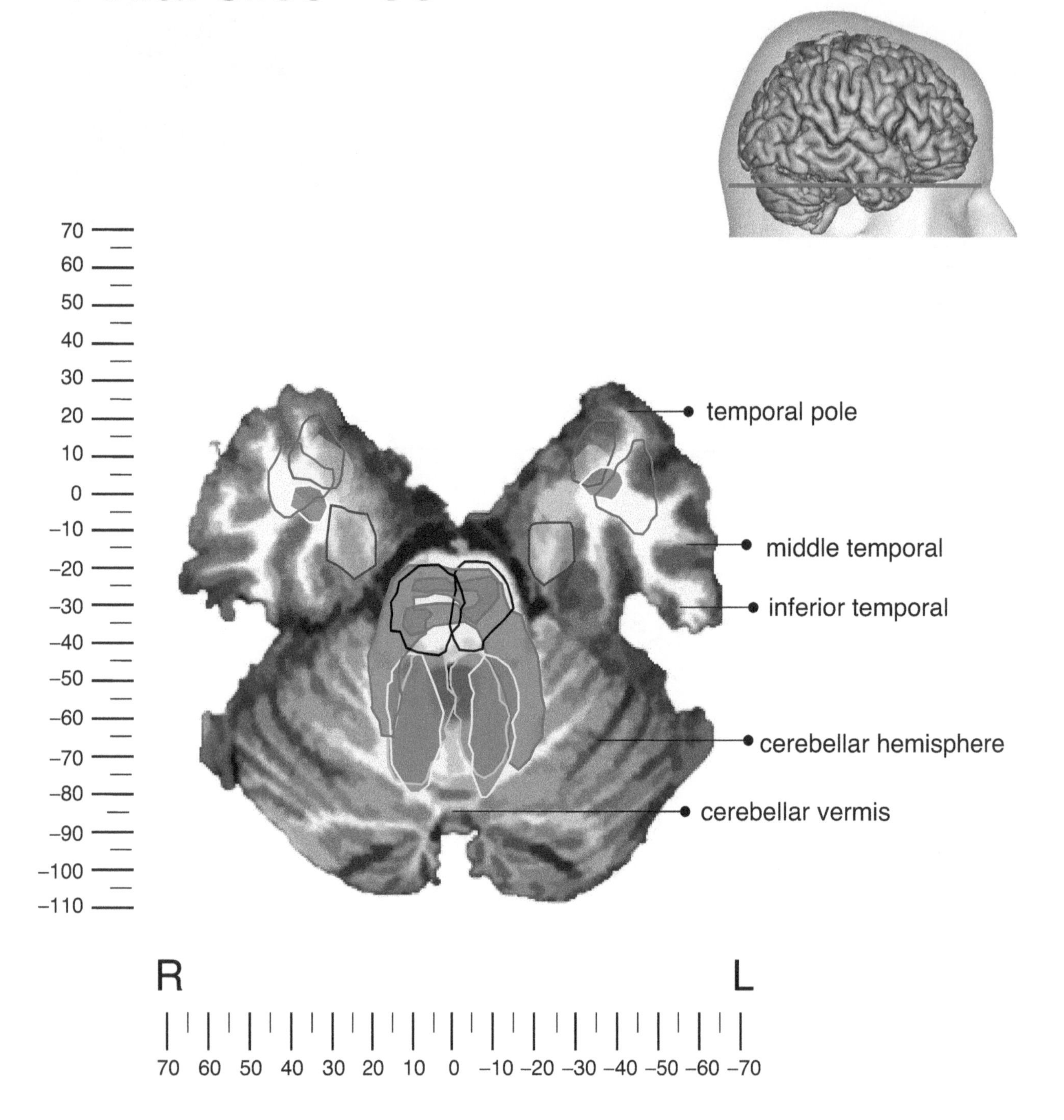

TRACT LEGEND

Abbreviation	Tract	Abbreviation	Tract	Abbreviation	Tract
CST	Cortico-Spinal Tract / IC	CPC	Cortico-Ponto-Cerebellar Tract	Cing	Cingulum
ICP	Inf. Cerebellar Peduncle	AC	Anterior Commissure	Unc	Uncinate
SCP	Sup. Cerebellar Peduncle	ILF	Inf. Longitudinal Fasc.	Fx	Fornix

Axial slice –28

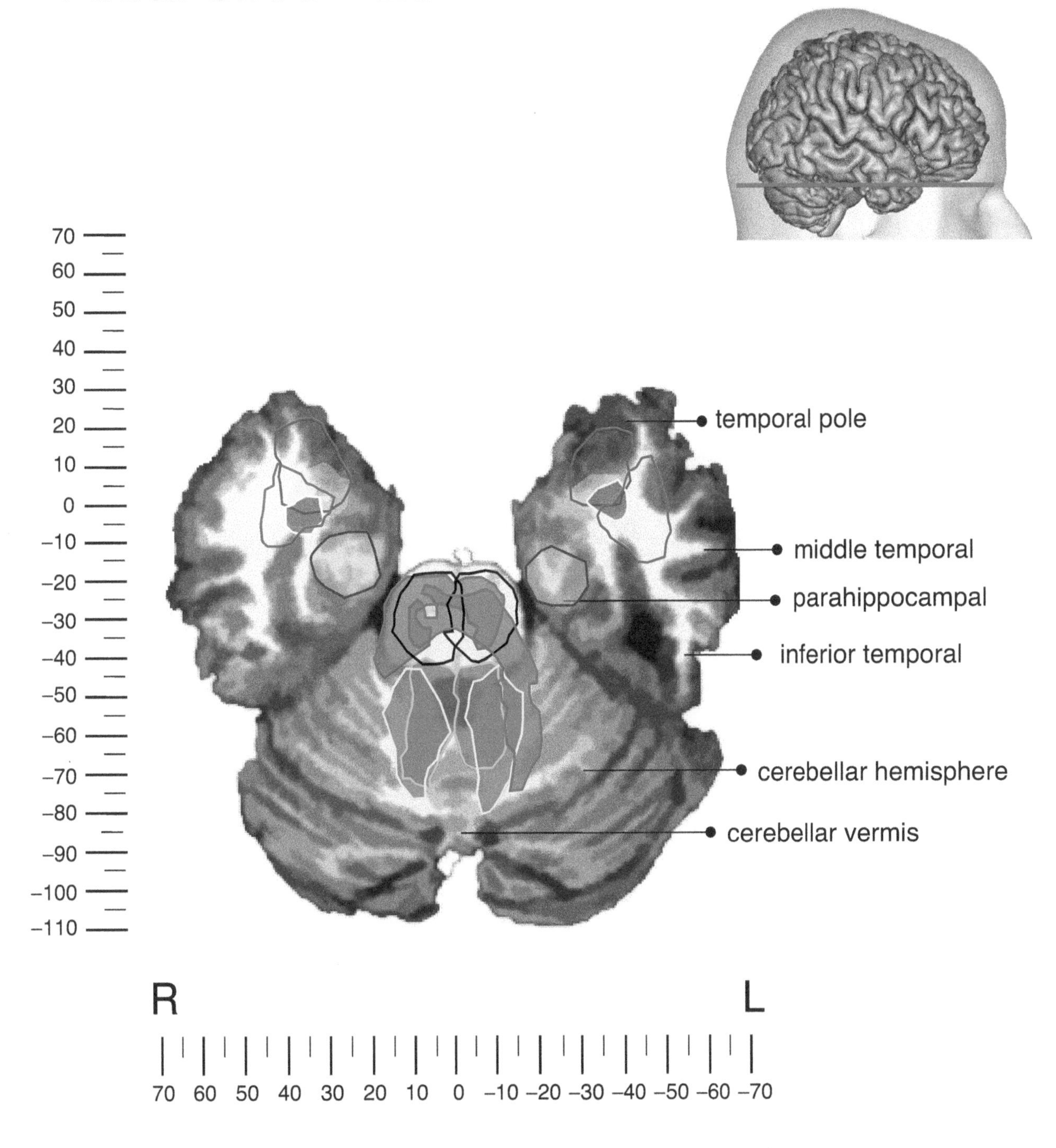

TRACT LEGEND

CST	Cortico-Spinal Tract / IC	CPC	Cortico-Ponto-Cerebellar Tract	Cing	Cingulum
ICP	Inf. Cerebellar Peduncle	AC	Anterior Commissure	Unc	Uncinate
SCP	Sup. Cerebellar Peduncle	ILF	Inf. Longitudinal Fasc.	Fx	Fornix

Axial slice −26

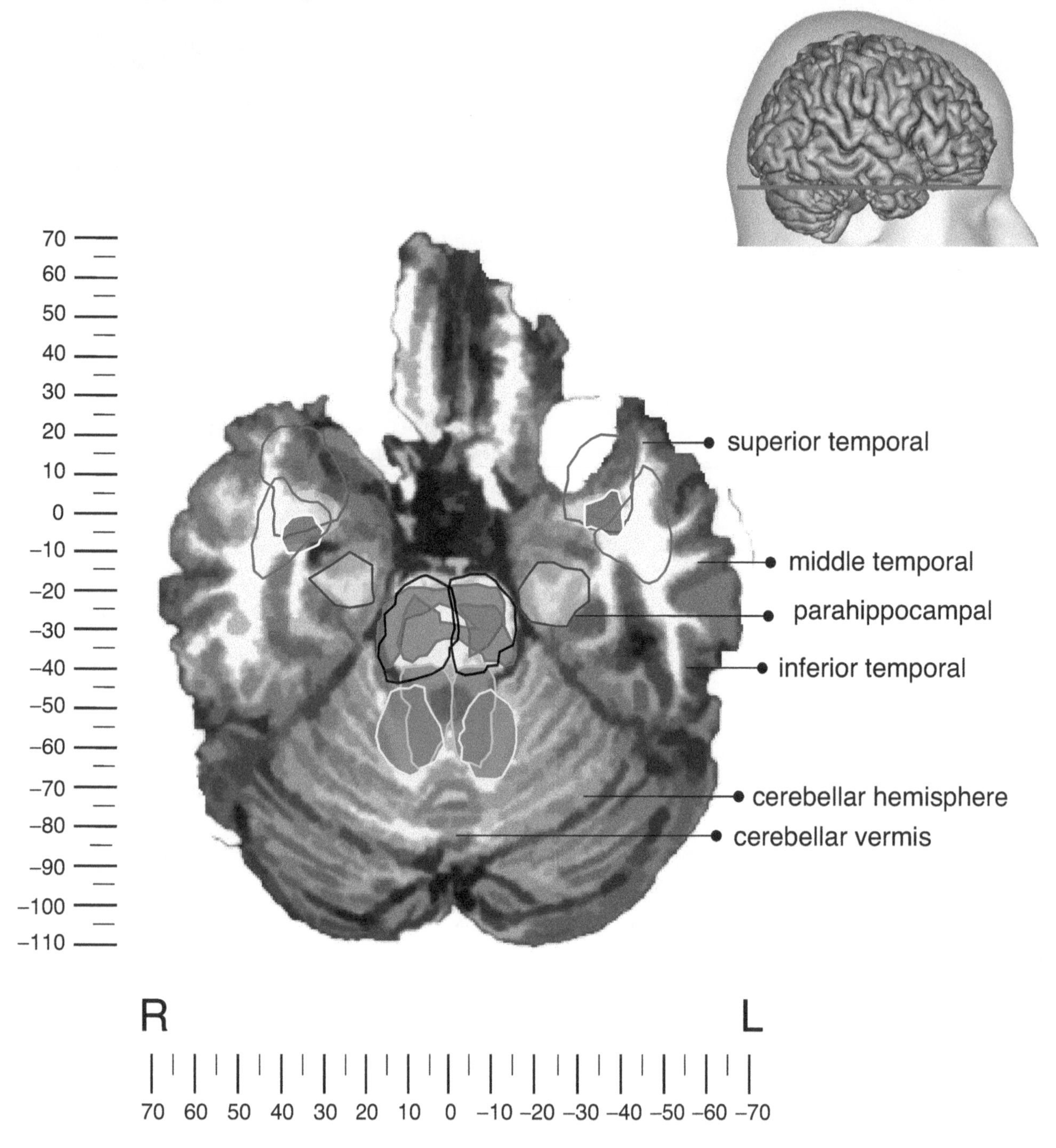

TRACT LEGEND

CST	Cortico-Spinal Tract / IC	CPC	Cortico-Ponto-Cerebellar Tract	Cing	Cingulum
ICP	Inf. Cerebellar Peduncle	AC	Anterior Commissure	Unc	Uncinate
SCP	Sup. Cerebellar Peduncle	ILF	Inf. Longitudinal Fasc.	Fx	Fornix

Axial slice –24

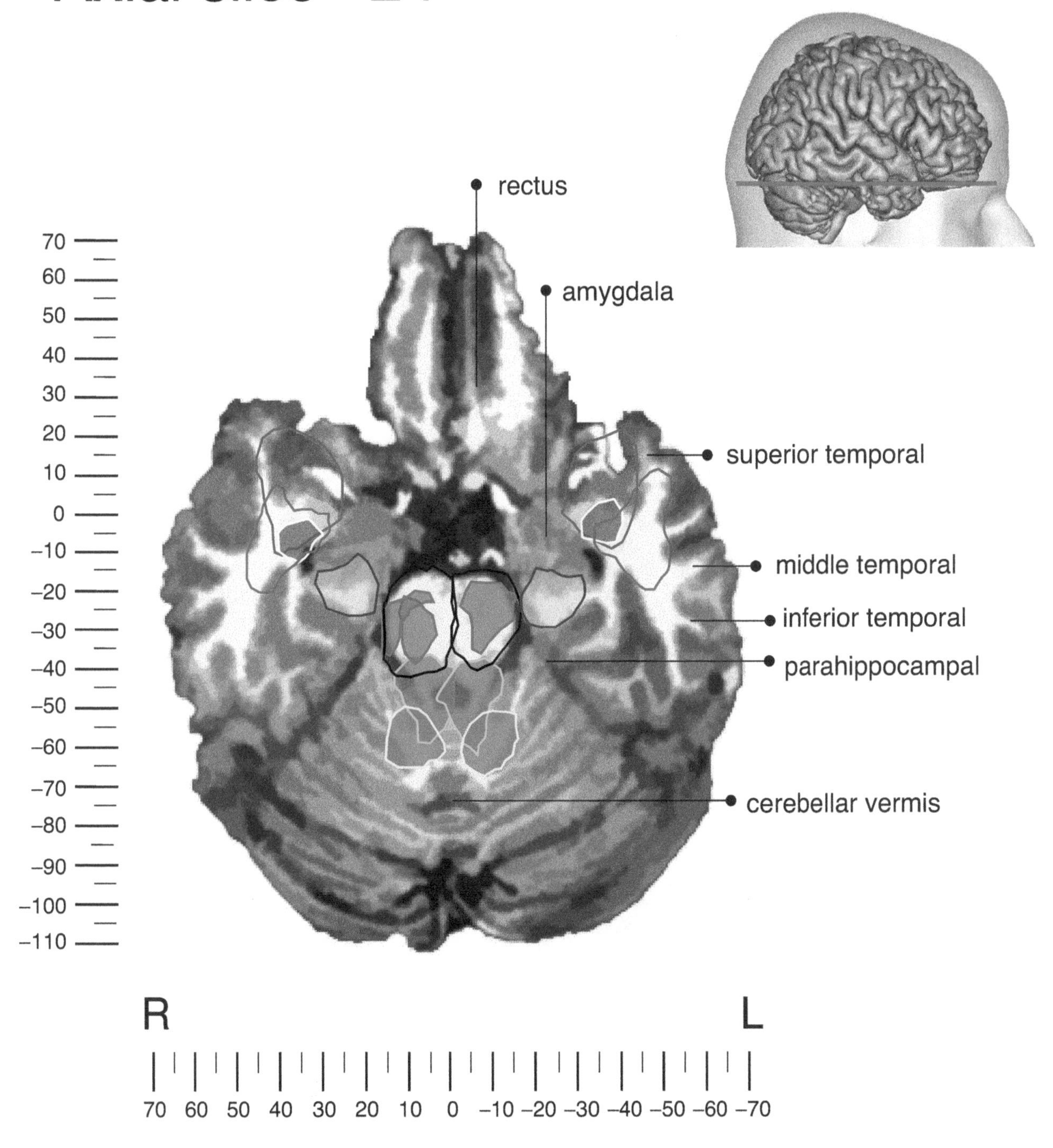

TRACT LEGEND

CST	Cortico-Spinal Tract / IC	CPC	Cortico-Ponto-Cerebellar Tract	Cing	Cingulum
ICP	Inf. Cerebellar Peduncle	AC	Anterior Commissure	Unc	Uncinate
SCP	Sup. Cerebellar Peduncle	ILF	Inf. Longitudinal Fasc.	Fx	Fornix

Axial slice –22

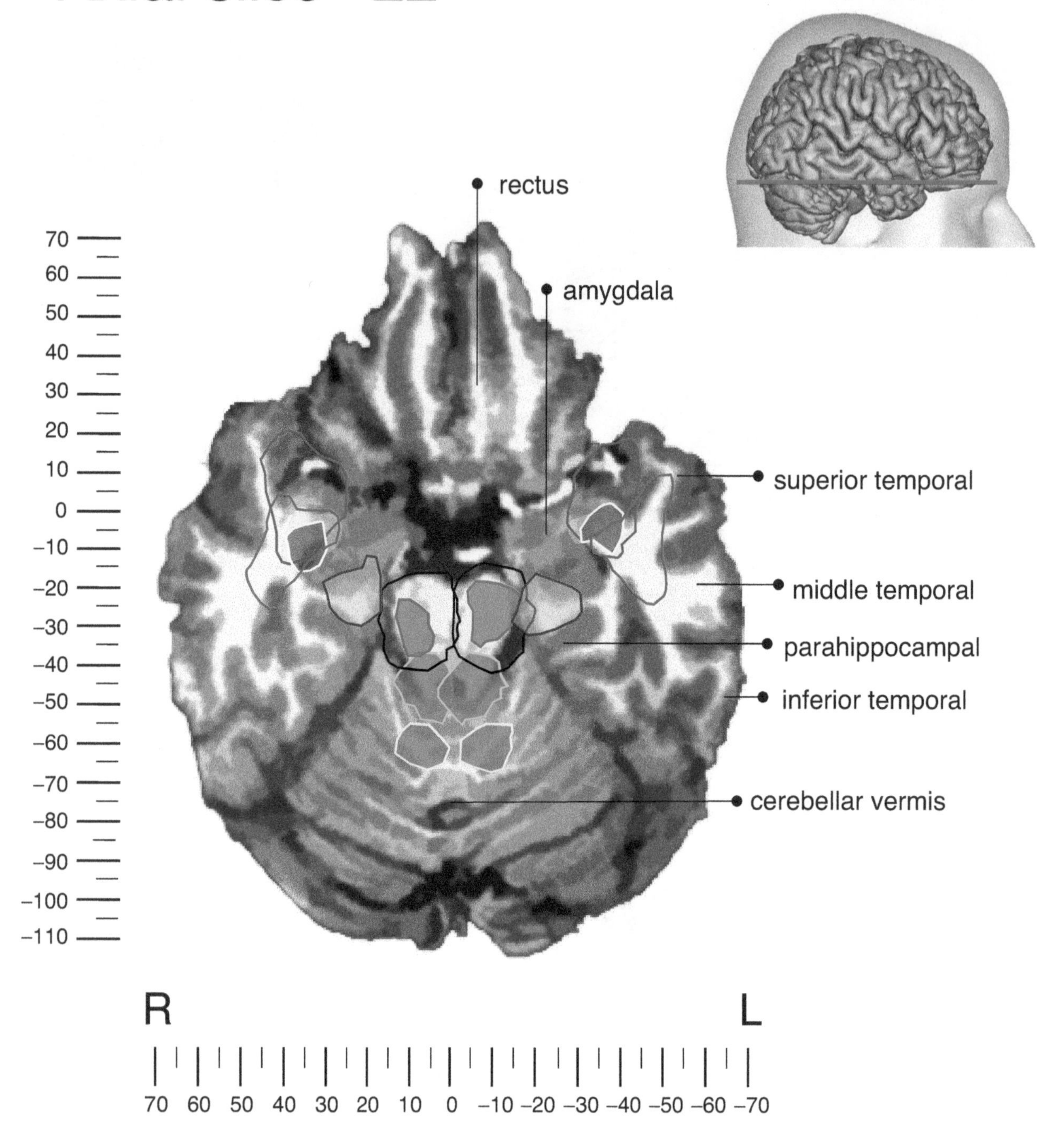

TRACT LEGEND

CST	Cortico-Spinal Tract / IC	CPC	Cortico-Ponto-Cerebellar Tract	Cing	Cingulum
ICP	Inf. Cerebellar Peduncle	AC	Anterior Commissure	Unc	Uncinate
SCP	Sup. Cerebellar Peduncle	ILF	Inf. Longitudinal Fasc.	Fx	Fornix

Axial slice –20

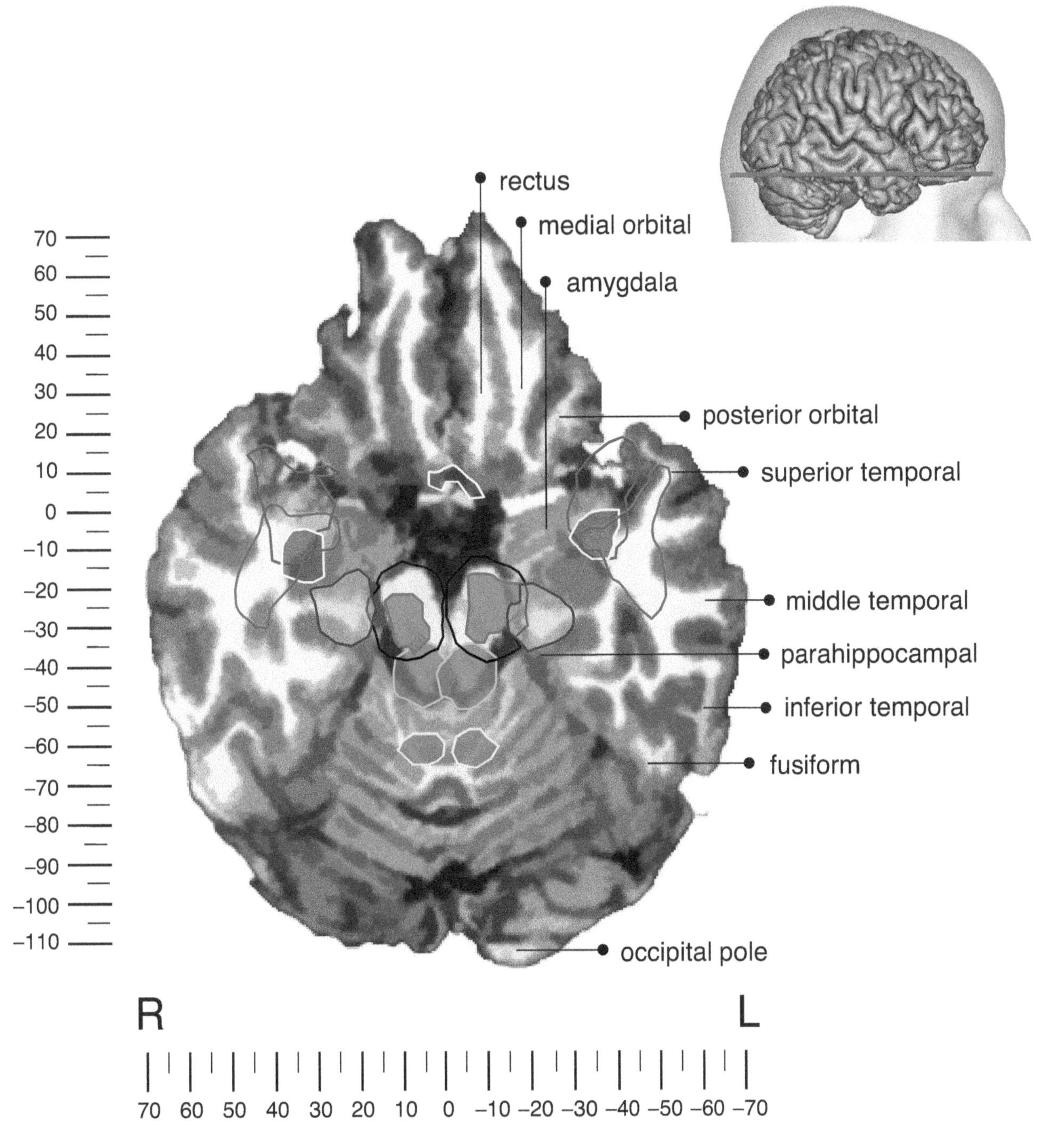

TRACT LEGEND

CST	Cortico-Spinal Tract / IC	CPC	Cortico-Ponto-Cerebellar Tract	Cing	Cingulum
ICP	Inf. Cerebellar Peduncle	AC	Anterior Commissure	Unc	Uncinate
SCP	Sup. Cerebellar Peduncle	ILF	Inf. Longitudinal Fasc.	Fx	Fornix

Axial slice –18

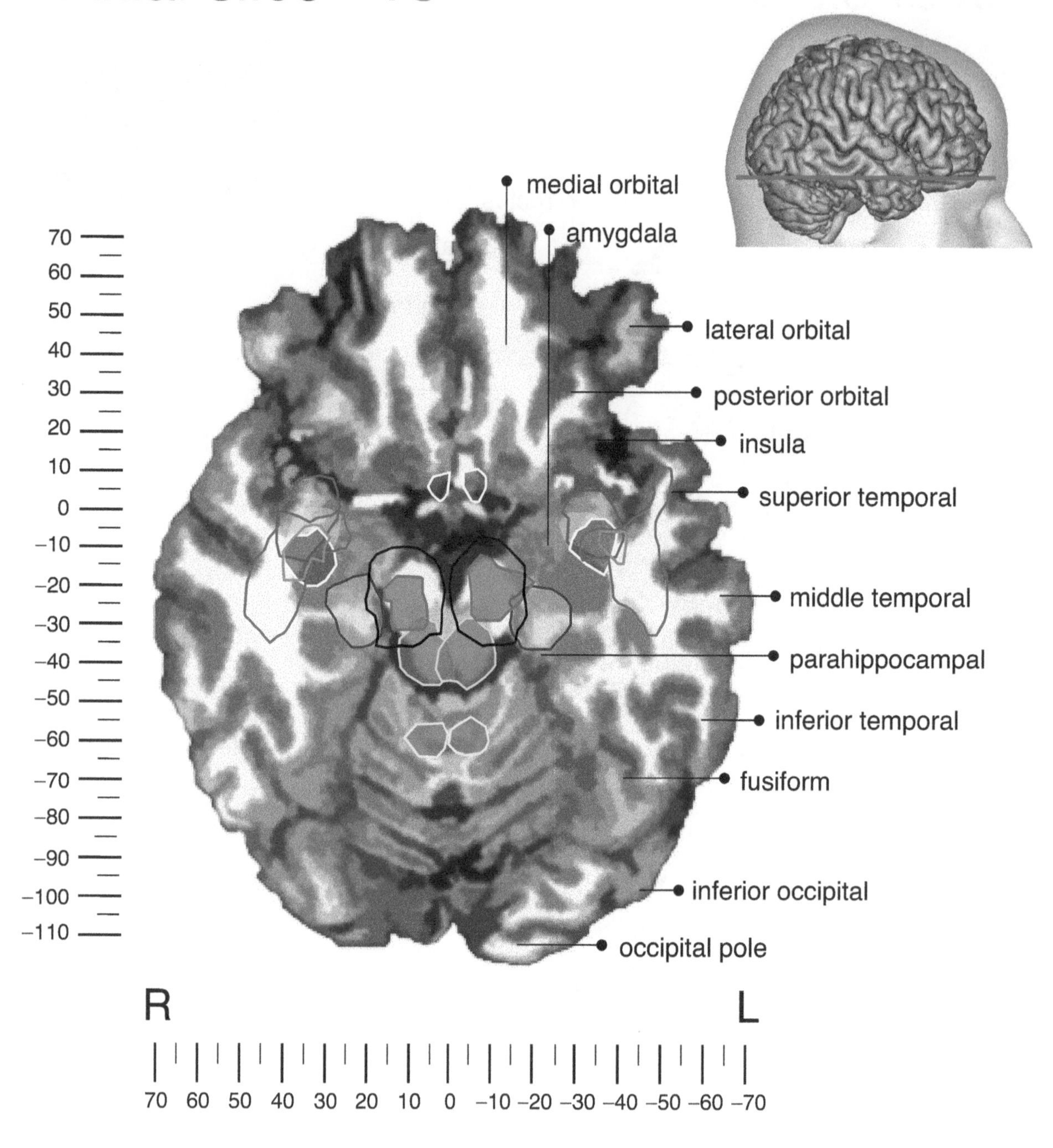

TRACT LEGEND

CST	Cortico-Spinal Tract / IC	CPC	Cortico-Ponto-Cerebellar Tract	Cing	Cingulum
ICP	Inf. Cerebellar Peduncle	AC	Anterior Commissure	Unc	Uncinate
SCP	Sup. Cerebellar Peduncle	ILF	Inf. Longitudinal Fasc.	Fx	Fornix
IFOF	Inf. Fronto-Occipital Fasc.				

Axial slice –16

medial orbital
anterior orbital
lateral orbital
posterior orbital
insula
superior temporal
middle temporal
inferior temporal
fusiform
inferior occipital
occipital pole

70 60 50 40 30 20 10 0 –10 –20 –30 –40 –50 –60 –70 –80 –90 –100 –110

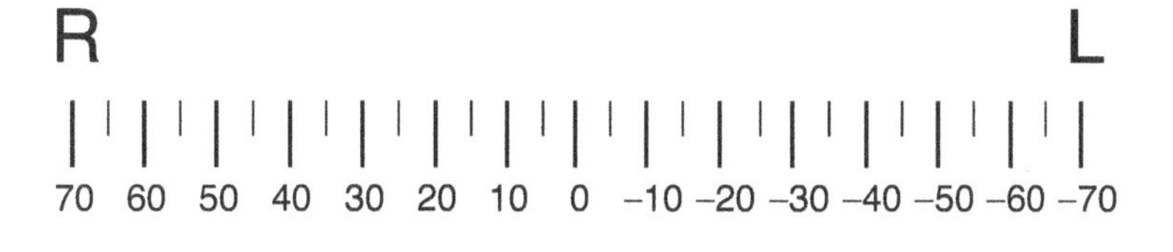

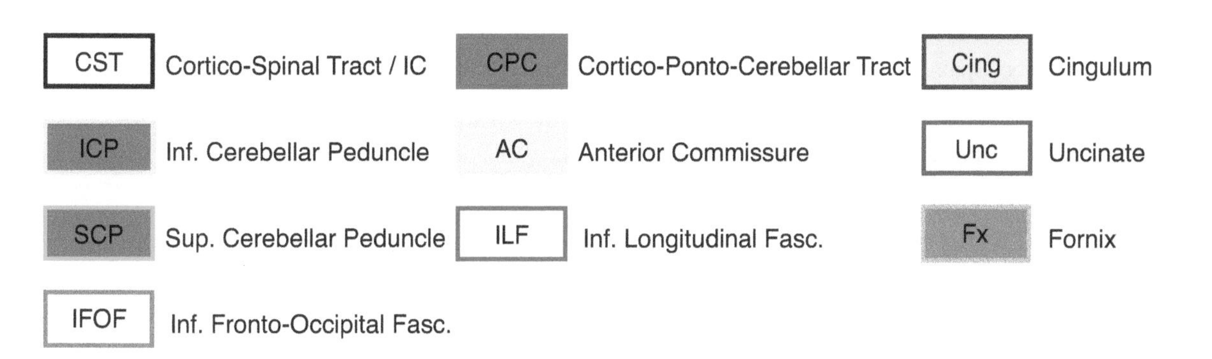

Axial slice –14

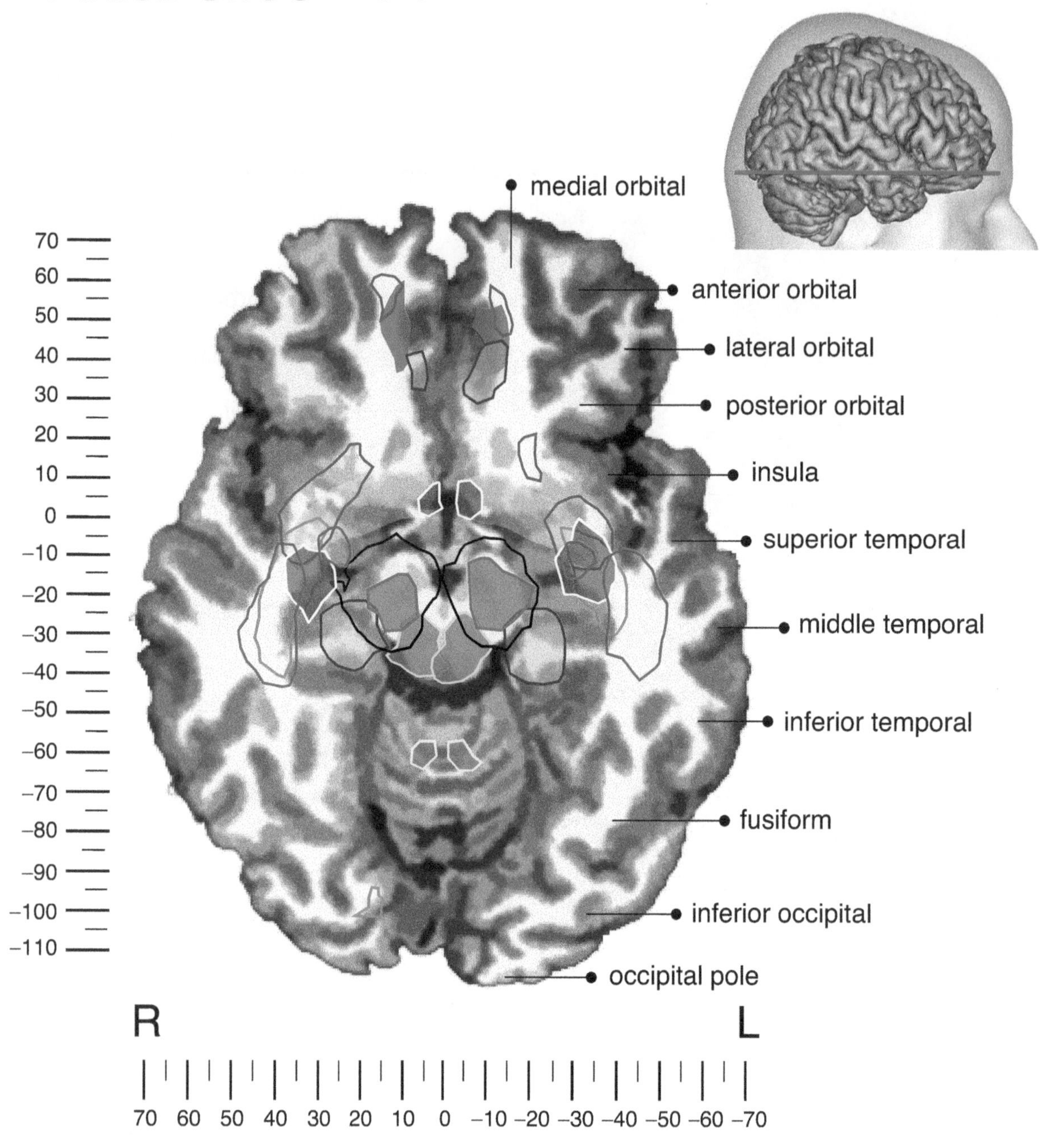

TRACT LEGEND

Abbr.	Tract	Abbr.	Tract	Abbr.	Tract
CST	Cortico-Spinal Tract / IC	CPC	Cortico-Ponto-Cerebellar Tract	Cing	Cingulum
ICP	Inf. Cerebellar Peduncle	AC	Anterior Commissure	Unc	Uncinate
SCP	Sup. Cerebellar Peduncle	ILF	Inf. Longitudinal Fasc.	Fx	Fornix
IFOF	Inf. Fronto-Occipital Fasc.	CC	Corpus Callosum		

Axial slice −12

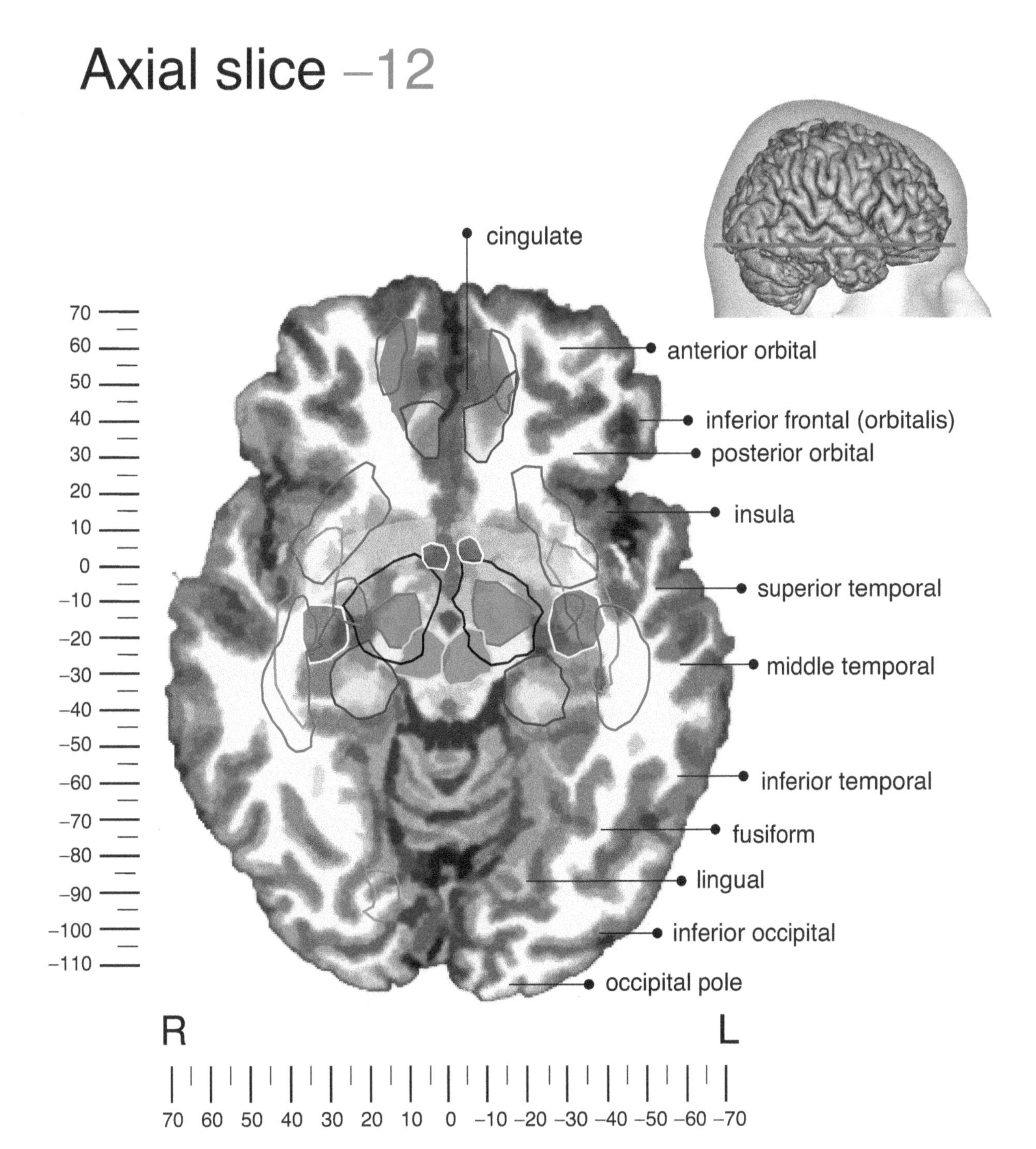

TRACT LEGEND

CST	Cortico-Spinal Tract / IC	CPC	Cortico-Ponto-Cerebellar Tract	Cing	Cingulum
CC	Corpus Callosum	AC	Anterior Commissure	Unc	Uncinate
SCP	Sup. Cerebellar Peduncle	ILF	Inf. Longitudinal Fasc.	Fx	Fornix
IFOF	Inf. Fronto-Occipital Fasc.				

Axial slice –10

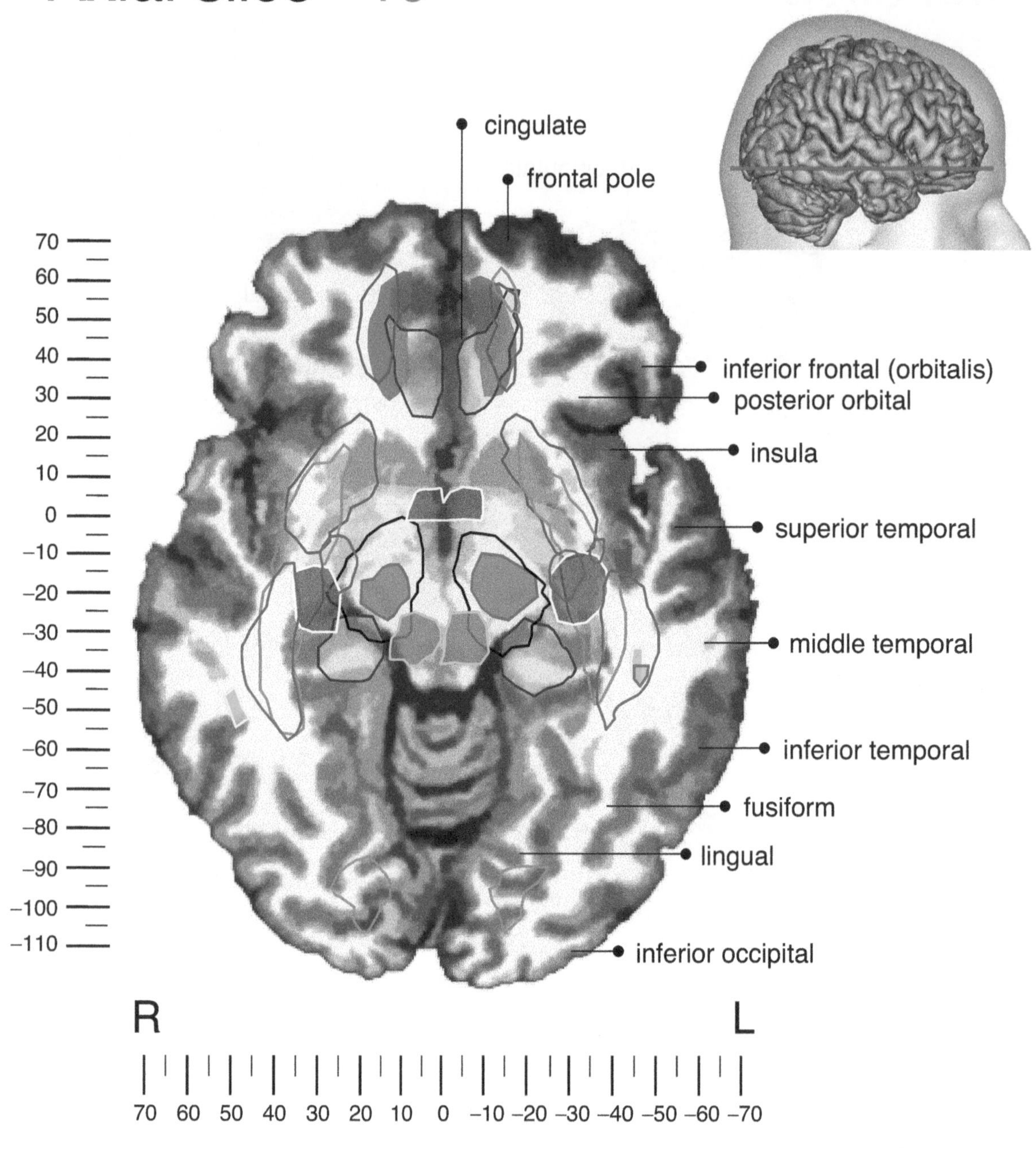

TRACT LEGEND

CST	Cortico-Spinal Tract / IC	CPC	Cortico-Ponto-Cerebellar Tract	Cing	Cingulum
CC	Corpus Callosum	AC	Anterior Commissure	Unc	Uncinate
SCP	Sup. Cerebellar Peduncle	ILF	Inf. Longitudinal Fasc.	Fx	Fornix
IFOF	Inf. Fronto-Occipital Fasc.	AF-L	Arcuate (long segment)	AF-P	Arcuate (post. segment)

Axial slice –8

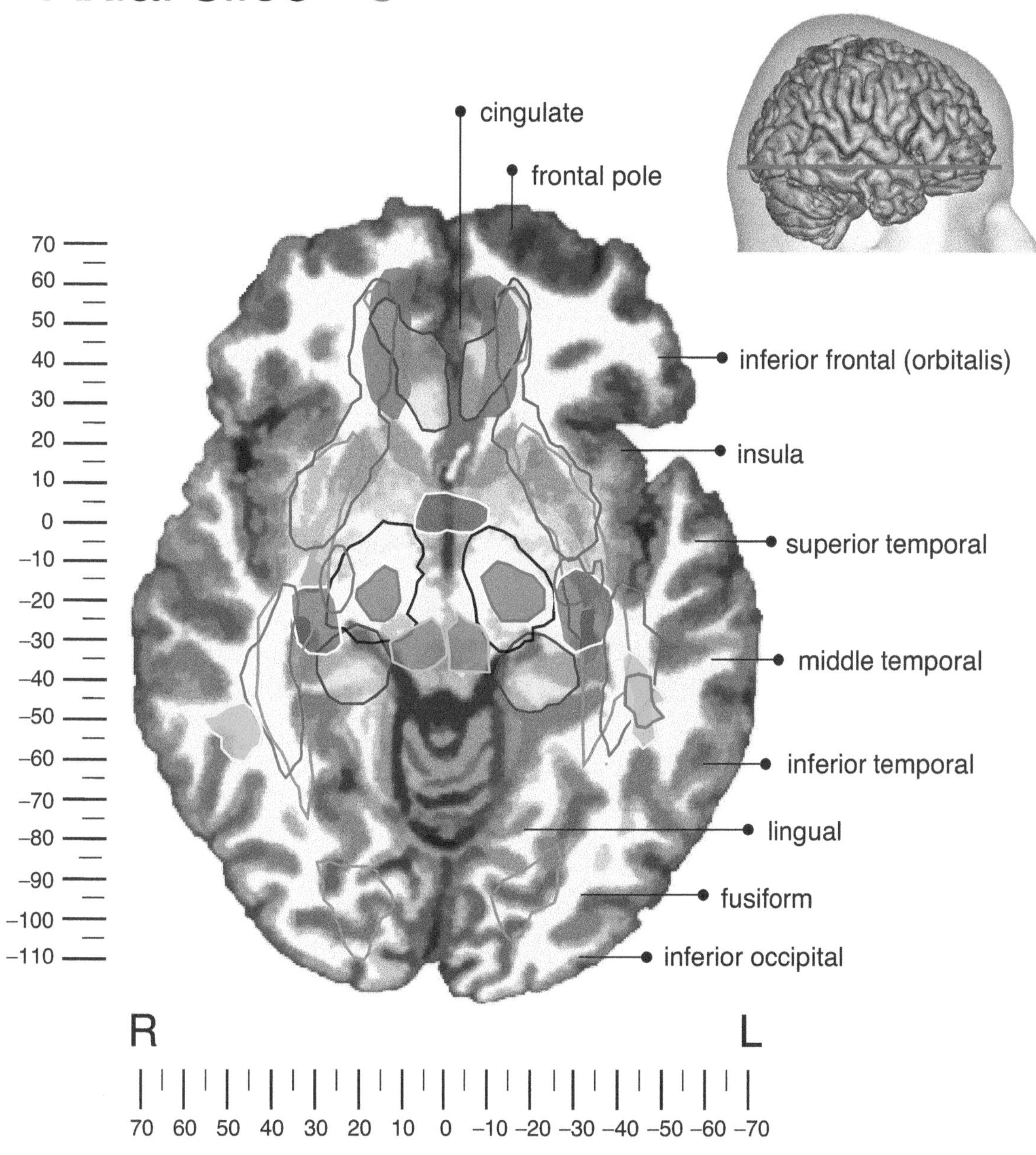

TRACT LEGEND

CST	Cortico-Spinal Tract / IC	CPC	Cortico-Ponto-Cerebellar Tract	Cing	Cingulum
CC	Corpus Callosum	AC	Anterior Commissure	Unc	Uncinate
SCP	Sup. Cerebellar Peduncle	ILF	Inf. Longitudinal Fasc.	Fx	Fornix
IFOF	Inf. Fronto-Occipital Fasc.	AF-L	Arcuate (long segment)	AF-P	Arcuate (post. segment)
OR	Optic Radiations				

Axial slice –6

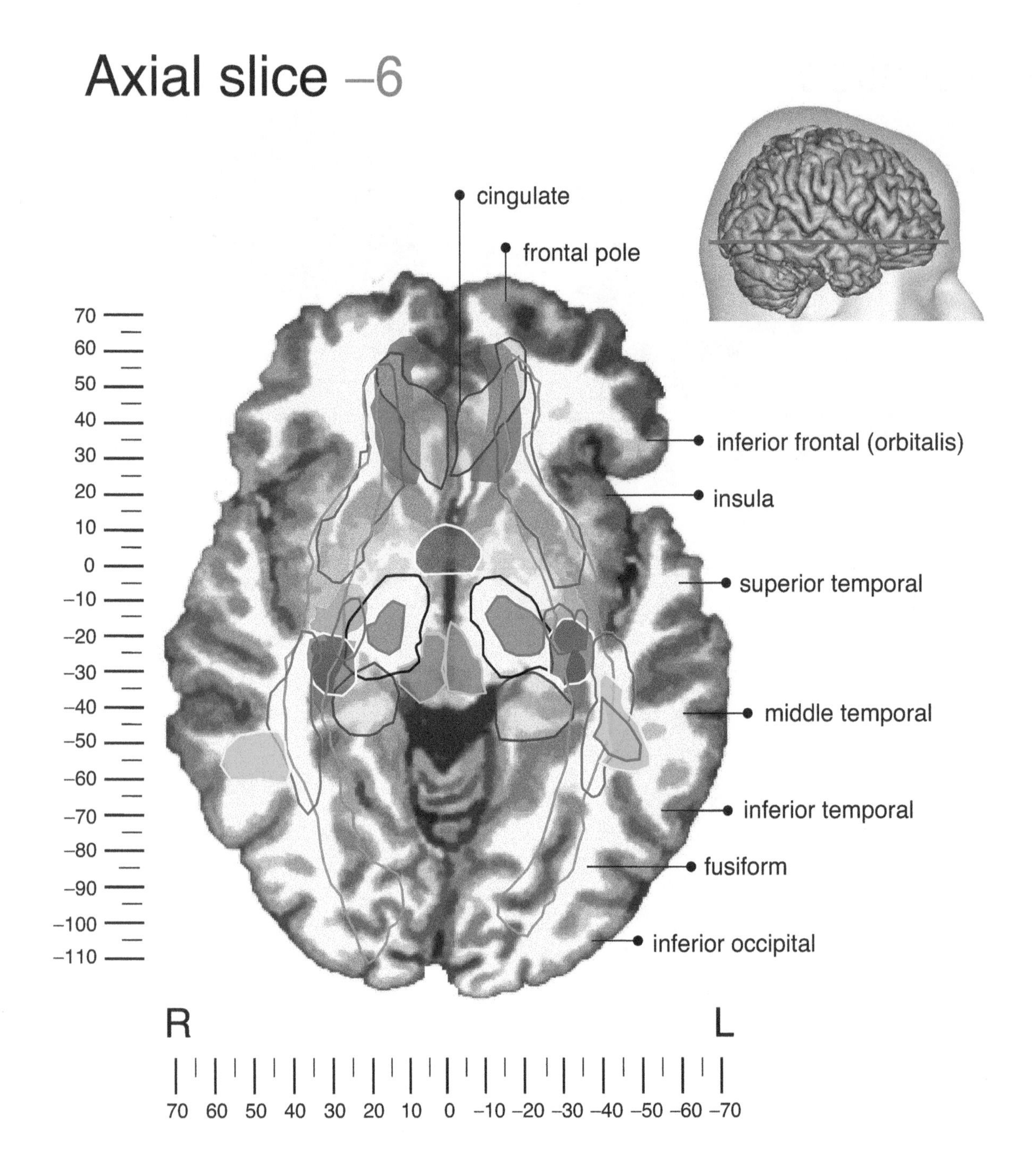

TRACT LEGEND

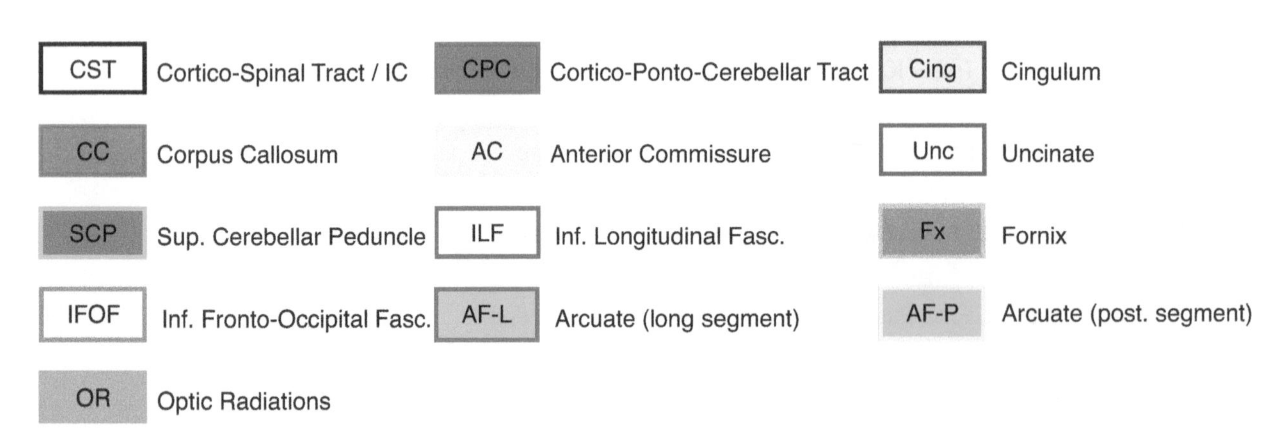

Axial slice –4

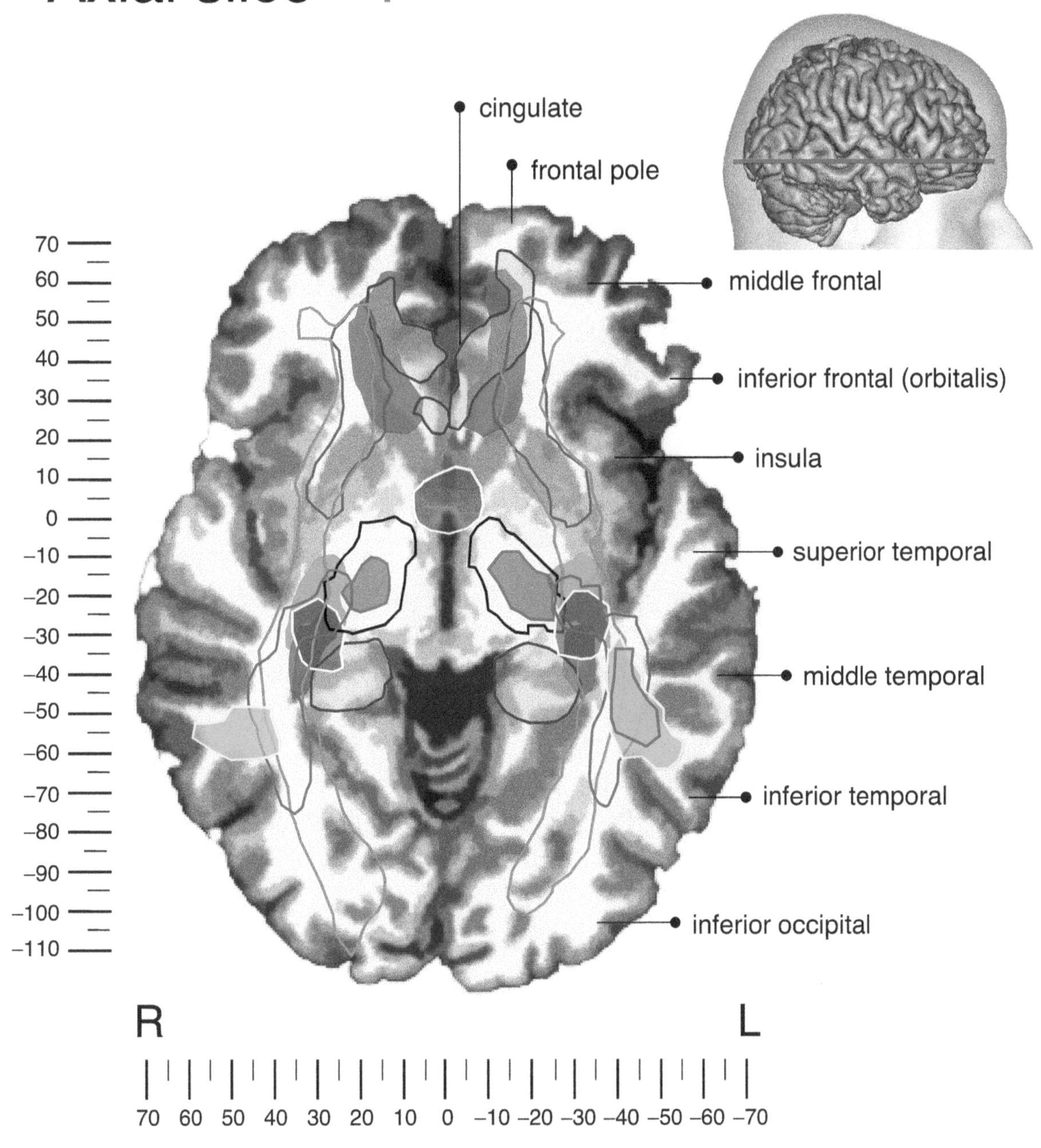

TRACT LEGEND

CST	Cortico-Spinal Tract / IC	CPC	Cortico-Ponto-Cerebellar Tract	Cing	Cingulum
CC	Corpus Callosum	AC	Anterior Commissure	Unc	Uncinate
IFOF	Inf. Fronto-Occipital Fasc.	ILF	Inf. Longitudinal Fasc.	Fx	Fornix
OR	Optic Radiations	AF-L	Arcuate (long segment)	AF-P	Arcuate (post. segment)

Axial slice –2

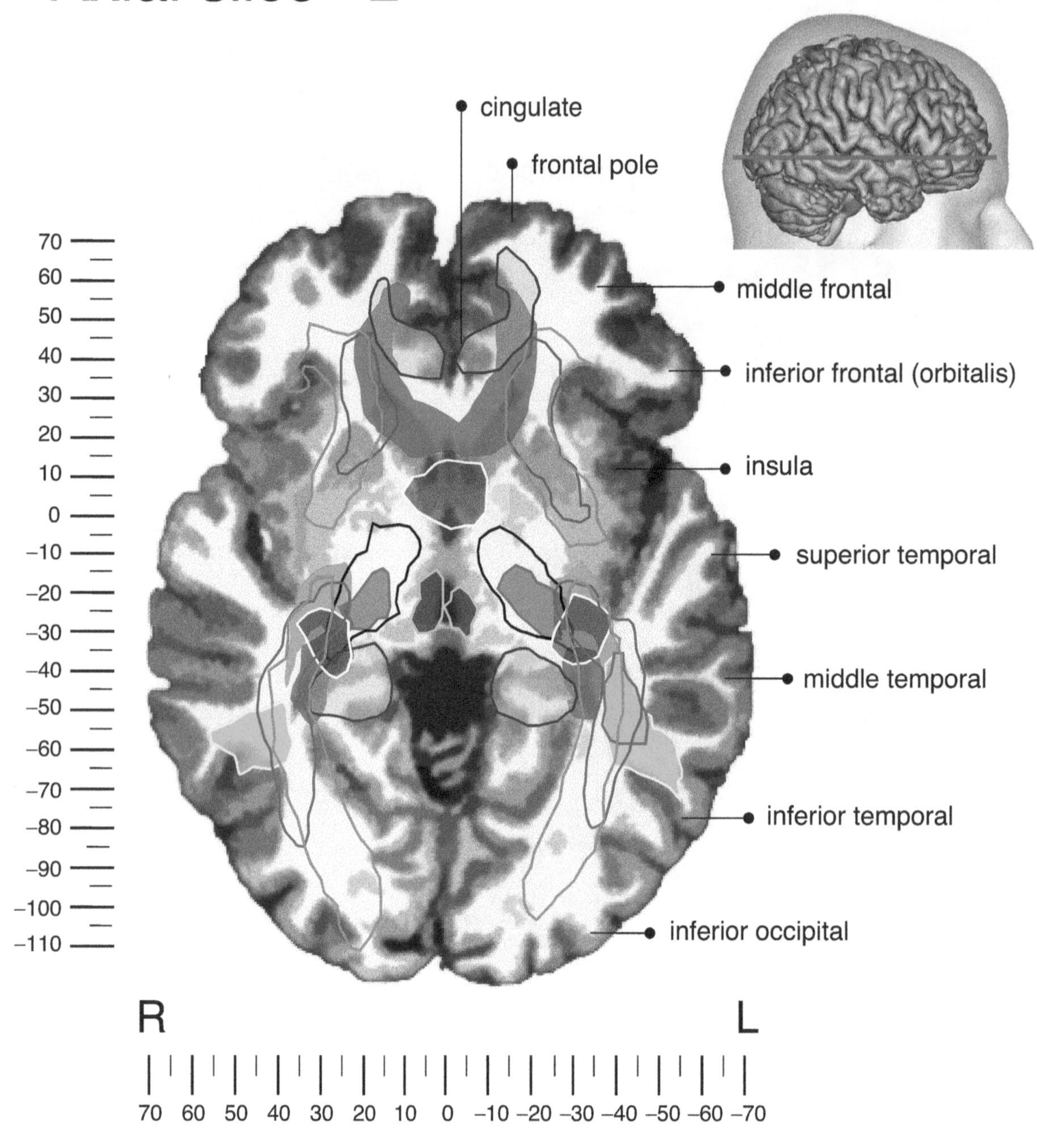

TRACT LEGEND

CST	Cortico-Spinal Tract / IC	CPC	Cortico-Ponto-Cerebellar Tract	Cing	Cingulum
CC	Corpus Callosum	AC	Anterior Commissure	Unc	Uncinate
IFOF	Inf. Fronto-Occipital Fasc.	ILF	Inf. Longitudinal Fasc.	Fx	Fornix
OR	Optic Radiations	AF-L	Arcuate (long segment)	AF-P	Arcuate (post. segment)

Axial slice 0

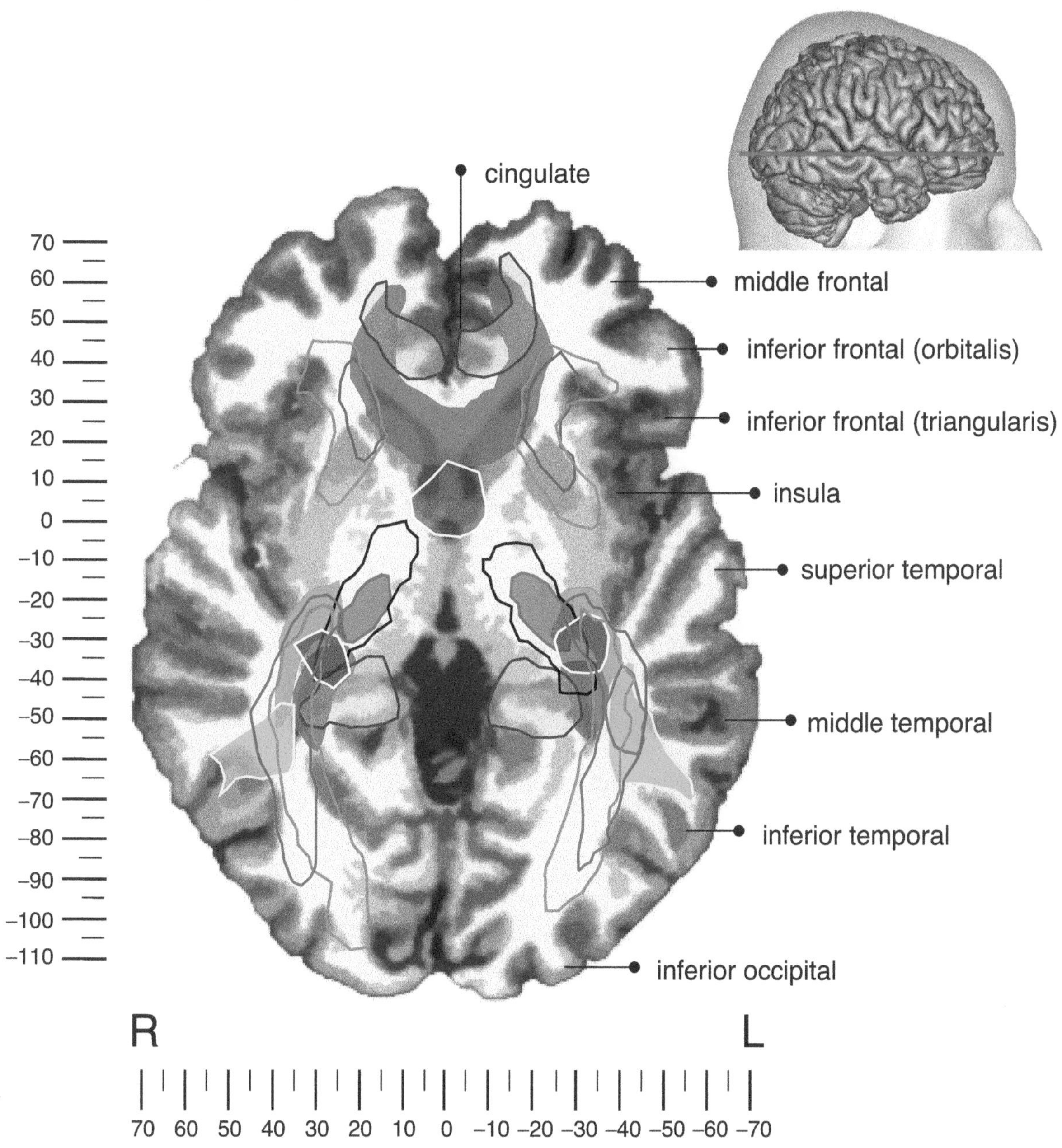

TRACT LEGEND

CST	Cortico-Spinal Tract / IC	CPC	Cortico-Ponto-Cerebellar Tract	Cing	Cingulum
CC	Corpus Callosum	AC	Anterior Commissure	Unc	Uncinate
IFOF	Inf. Fronto-Occipital Fasc.	ILF	Inf. Longitudinal Fasc.	Fx	Fornix
OR	Optic Radiations	AF-L	Arcuate (long segment)	AF-P	Arcuate (post. segment)

Axial slice 2

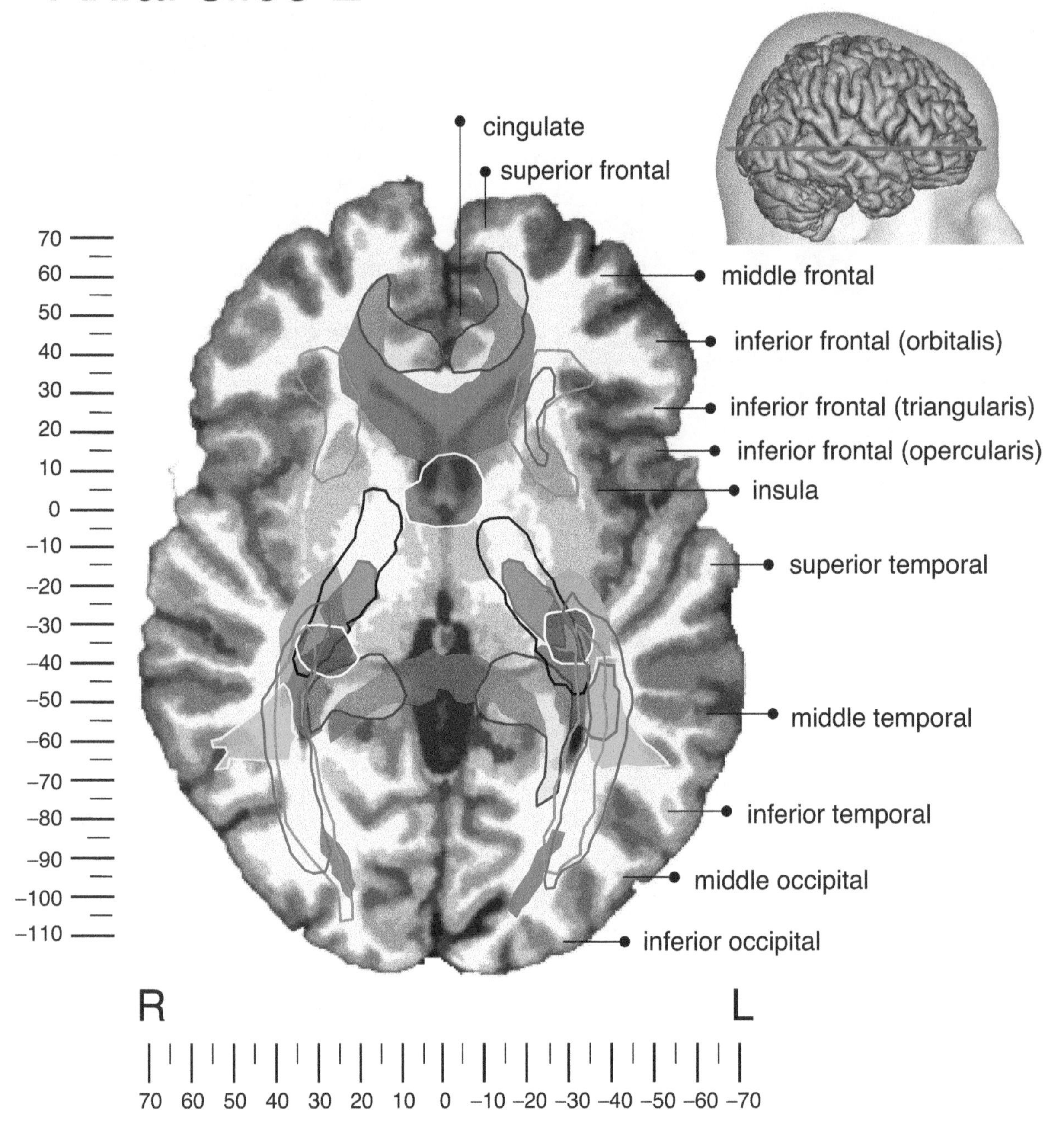

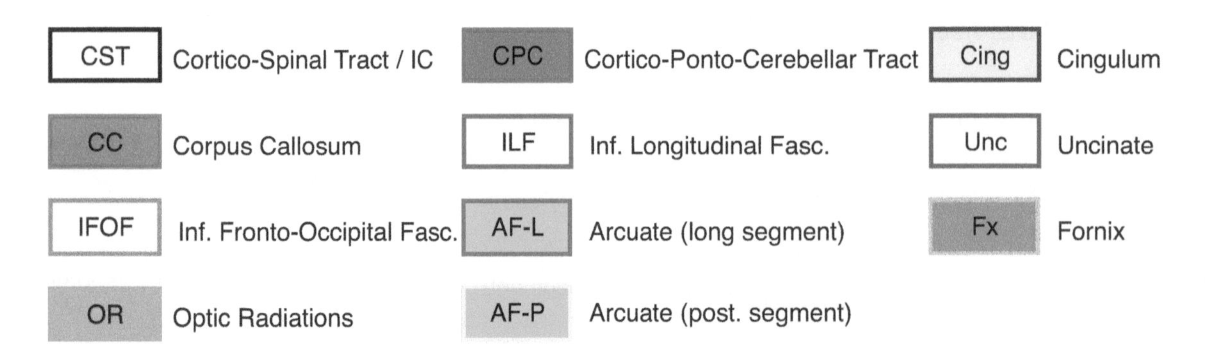

Axial slice 4

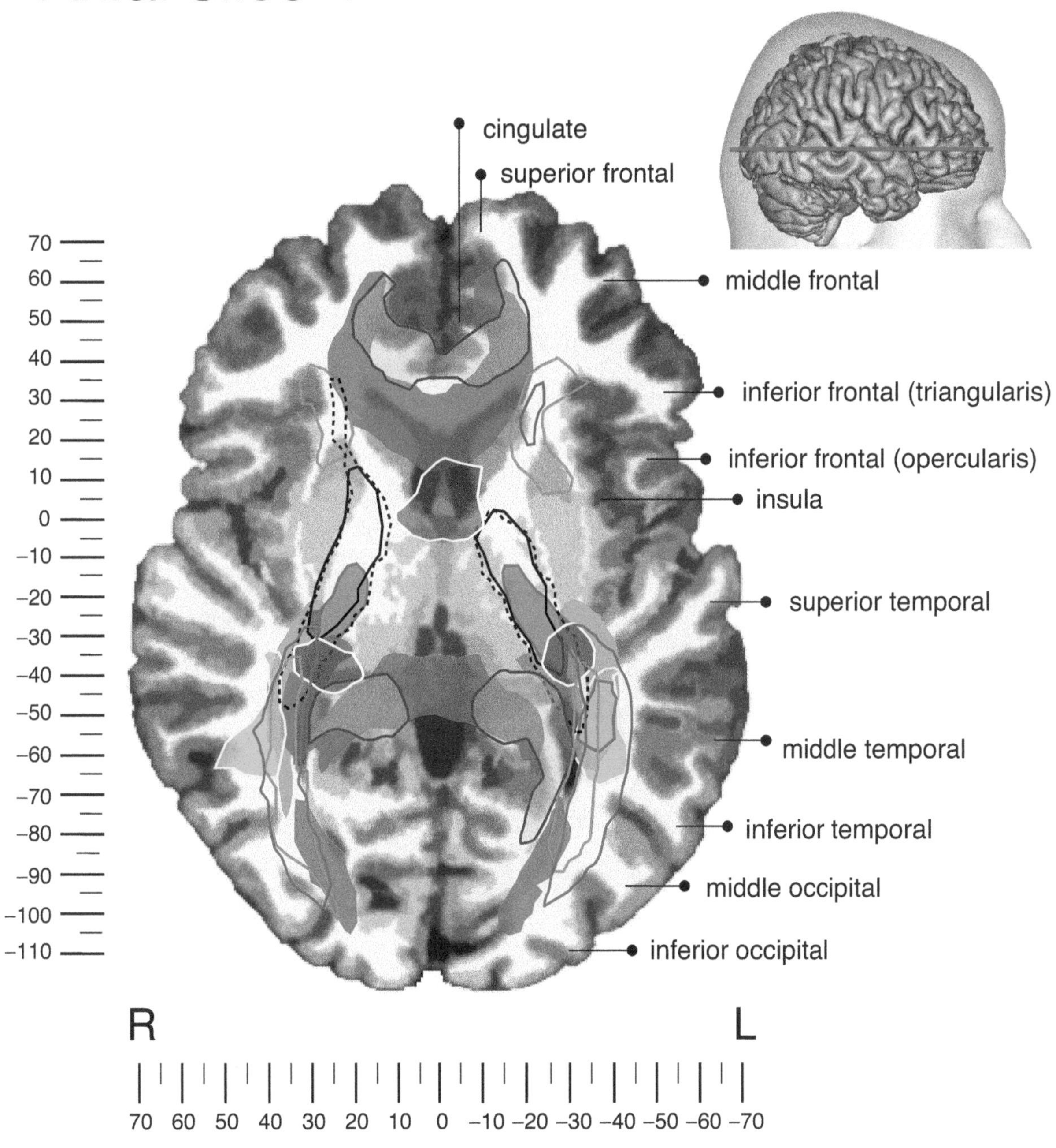

TRACT LEGEND

CST	Cortico-Spinal Tract	CPC	Cortico-Ponto-Cerebellar Tract	Cing	Cingulum
CC	Corpus Callosum	ILF	Inf. Longitudinal Fasc.	Unc	Uncinate
IFOF	Inf. Fronto-Occipital Fasc.	AF-L	Arcuate (long segment)	Fx	Fornix
OR	Optic Radiations	AF-P	Arcuate (post. segment)	IC	Internal Capsule

Axial slice 6

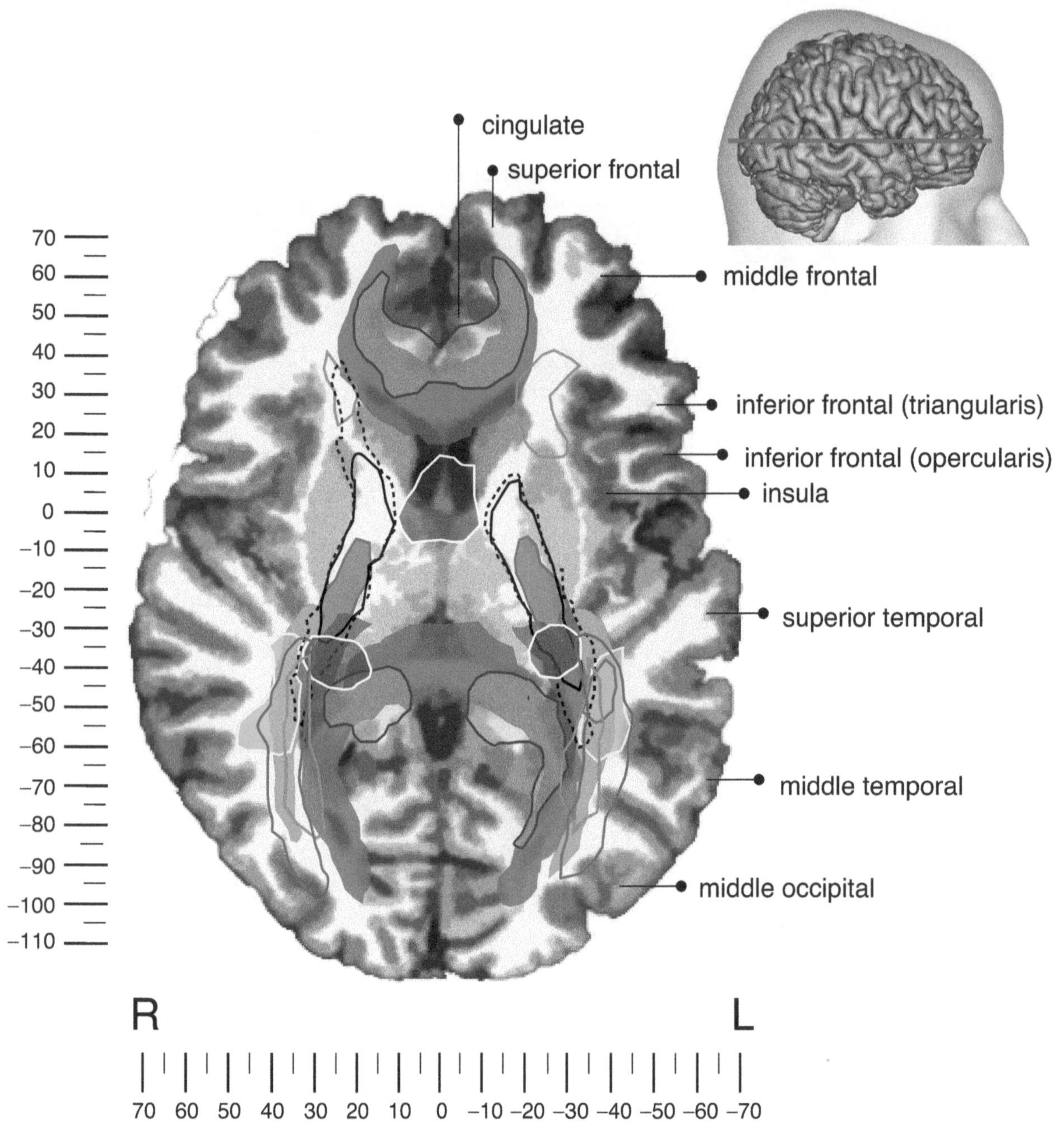

TRACT LEGEND

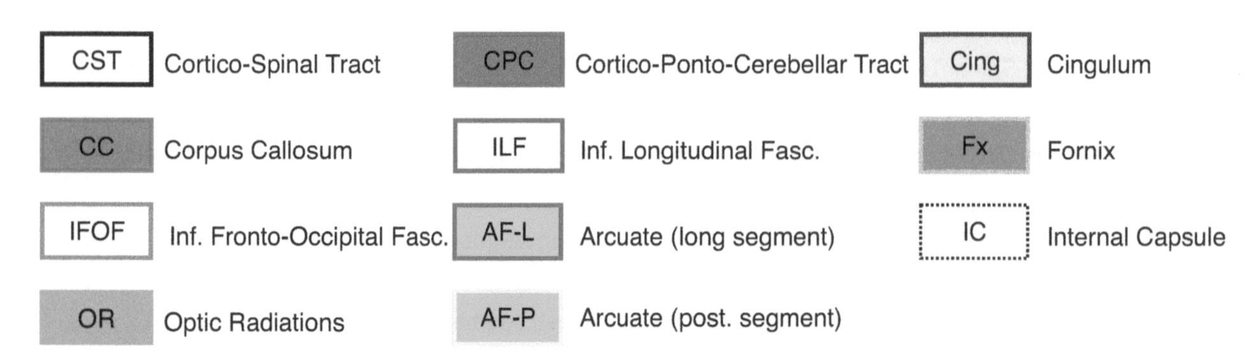

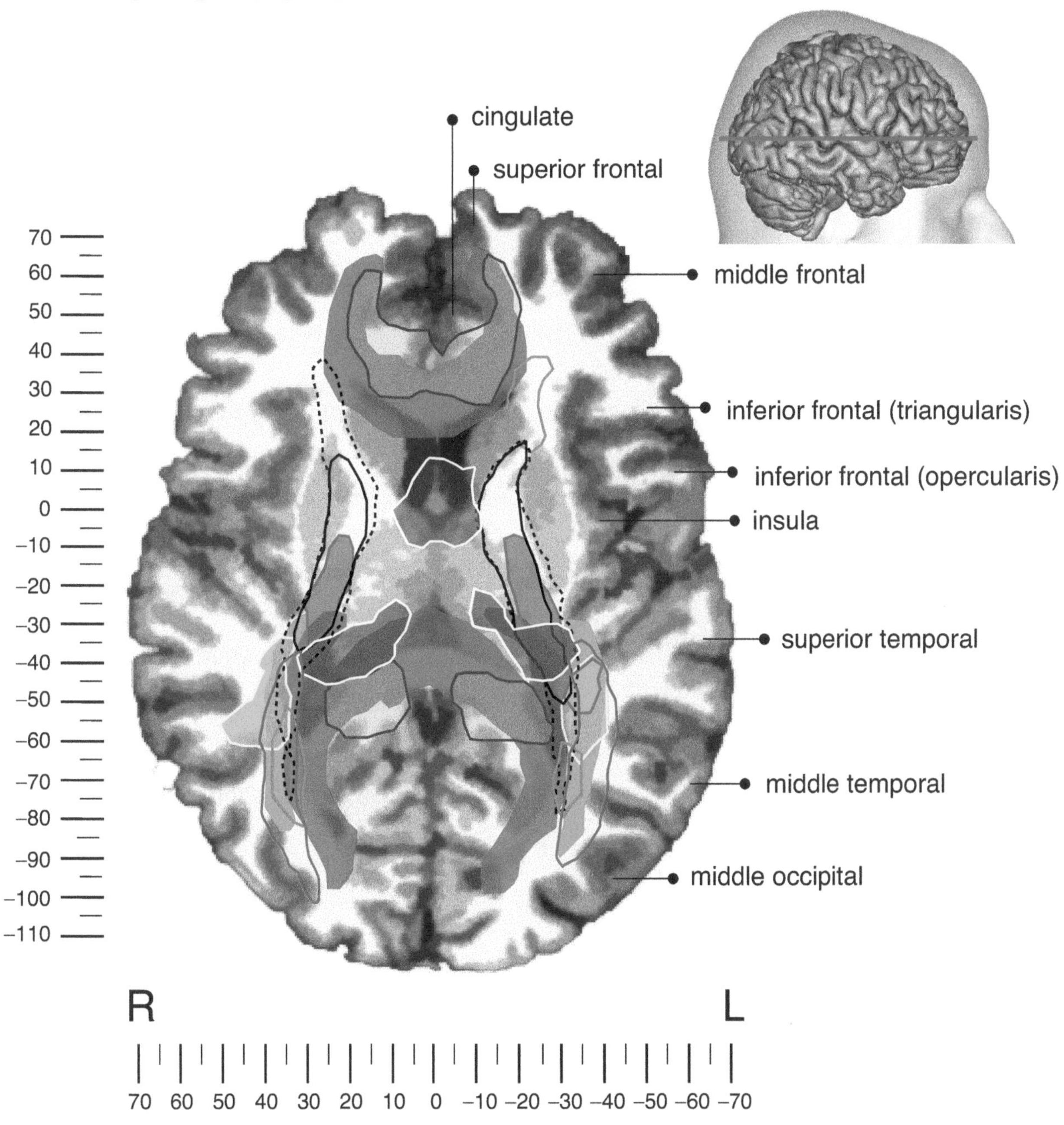

TRACT LEGEND

CST	Cortico-Spinal Tract	CPC	Cortico-Ponto-Cerebellar Tract	Cing	Cingulum
CC	Corpus Callosum	ILF	Inf. Longitudinal Fasc.	Fx	Fornix
IFOF	Inf. Fronto-Occipital Fasc.	AF-L	Arcuate (long segment)	IC	Internal Capsule
OR	Optic Radiations	AF-P	Arcuate (post. segment)		

Axial slice 10

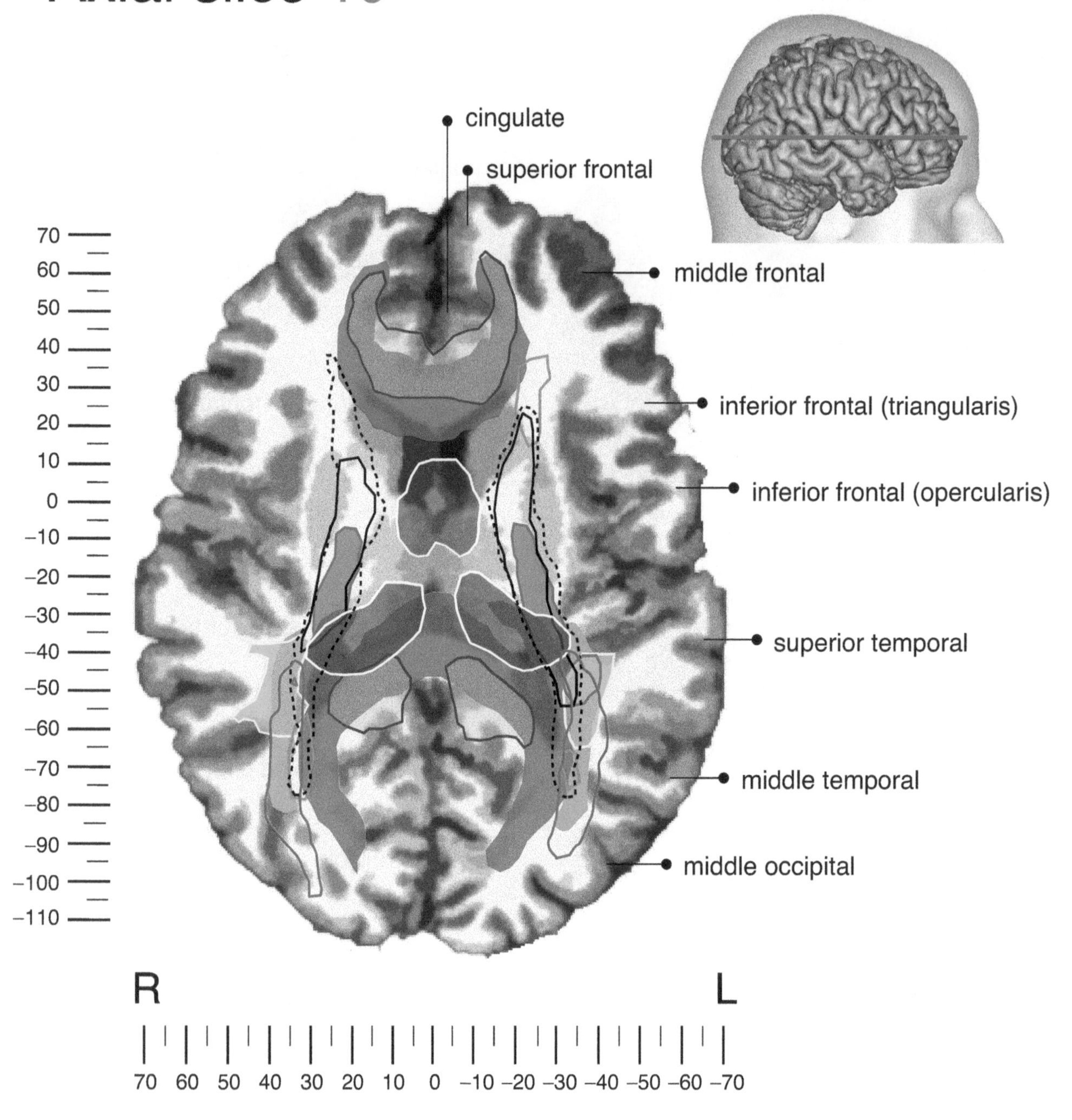

TRACT LEGEND

CST	Cortico-Spinal Tract	CPC	Cortico-Ponto-Cerebellar Tract	Cing	Cingulum
CC	Corpus Callosum	ILF	Inf. Longitudinal Fasc.	Fx	Fornix
IFOF	Inf. Fronto-Occipital Fasc.	AF-L	Arcuate (long segment)	IC	Internal Capsule
OR	Optic Radiations	AF-P	Arcuate (post. segment)		

Axial slice 12

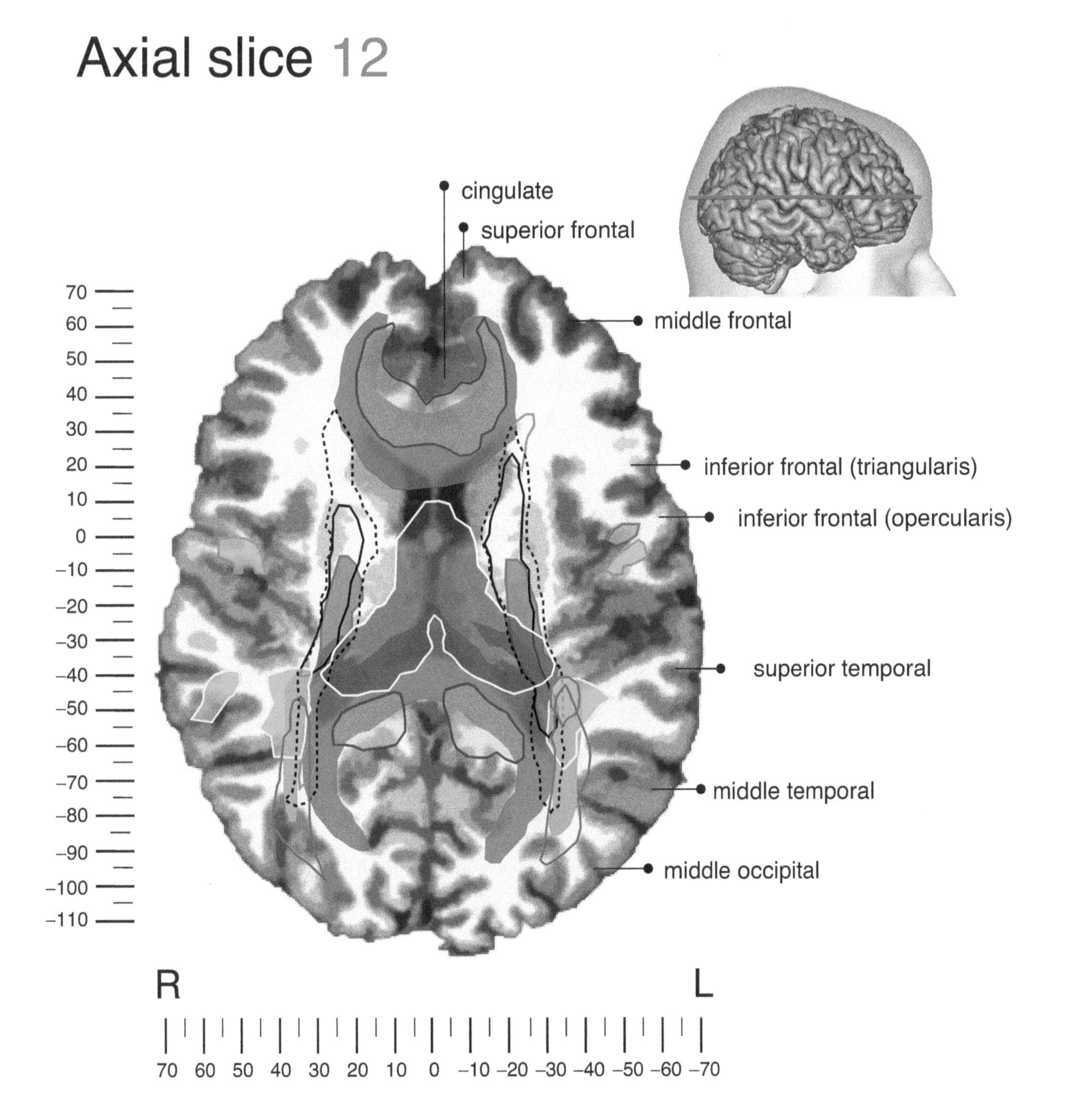

TRACT LEGEND

CST	Cortico-Spinal Tract	CPC	Cortico-Ponto-Cerebellar Tract	Cing	Cingulum
CC	Corpus Callosum	ILF	Inf. Longitudinal Fasc.	Fx	Fornix
IFOF	Inf. Fronto-Occipital Fasc.	AF-L	Arcuate (long segment)	IC	Internal Capsule
OR	Optic Radiations	AF-P	Arcuate (post. segment)	AF-A	Arcuate (ant. segment)

Axial slice 14

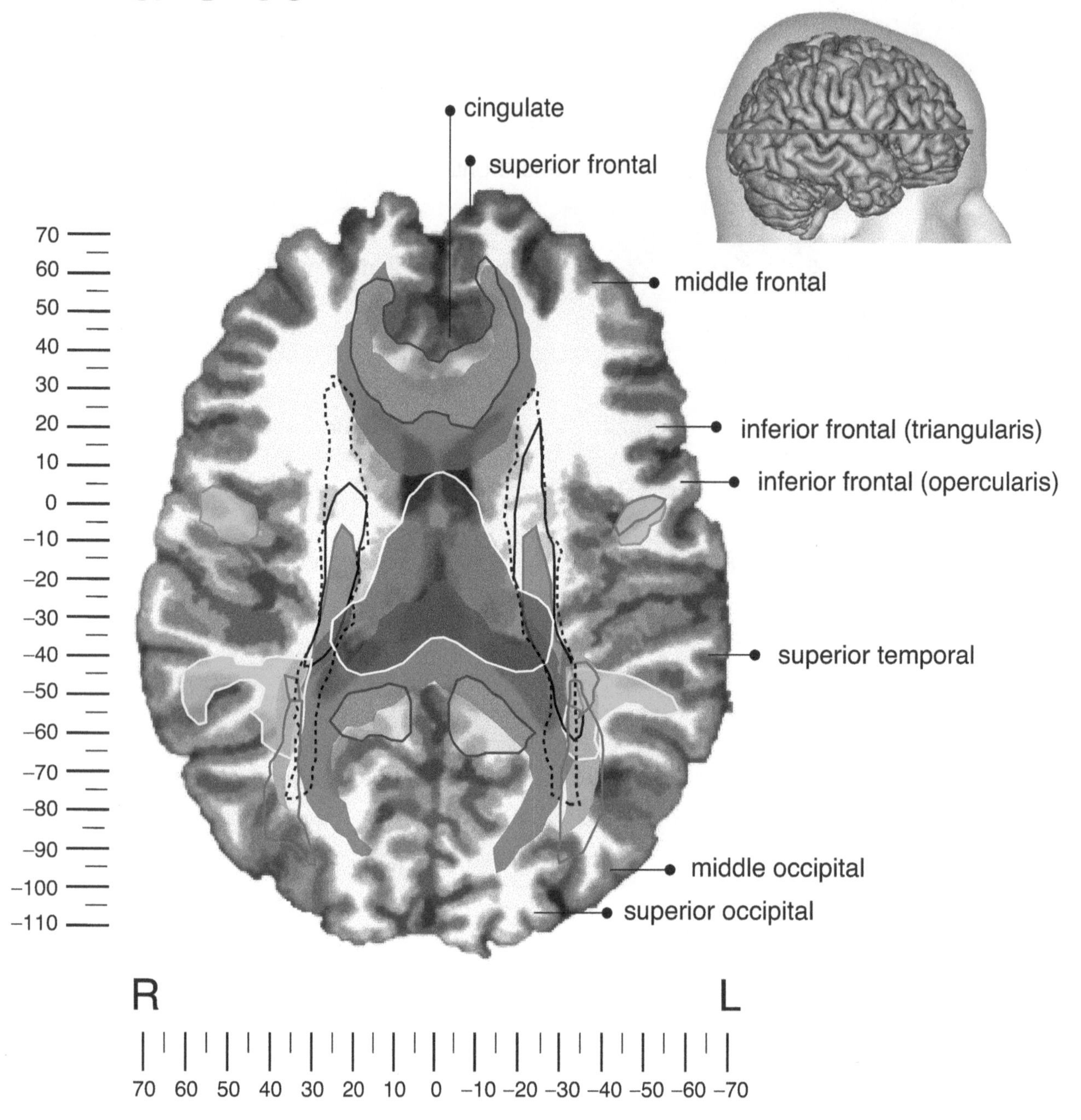

TRACT LEGEND

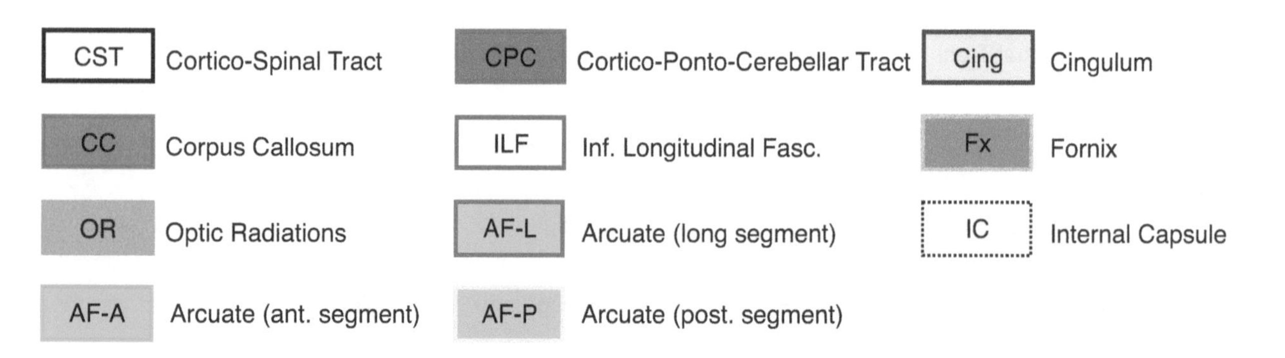

Axial slice 16

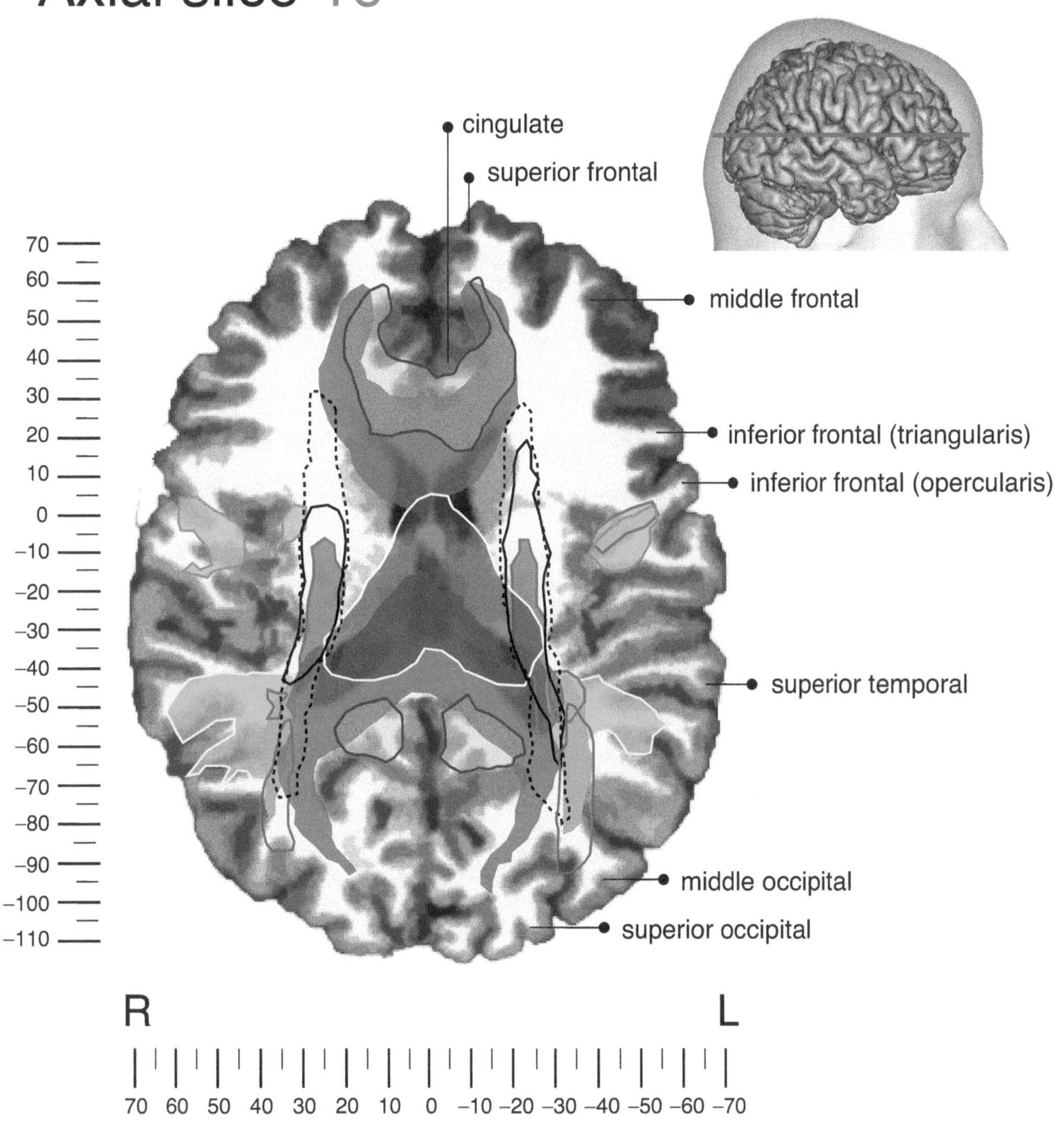

TRACT LEGEND

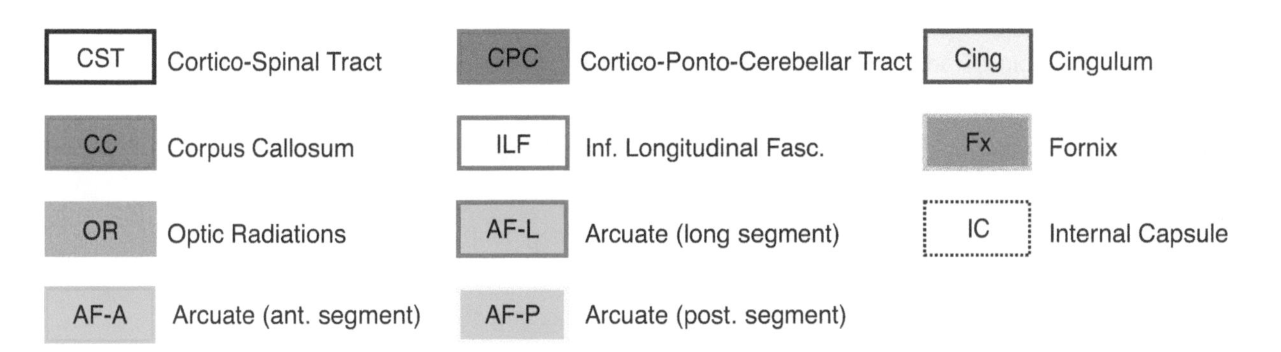

Axial slice 18

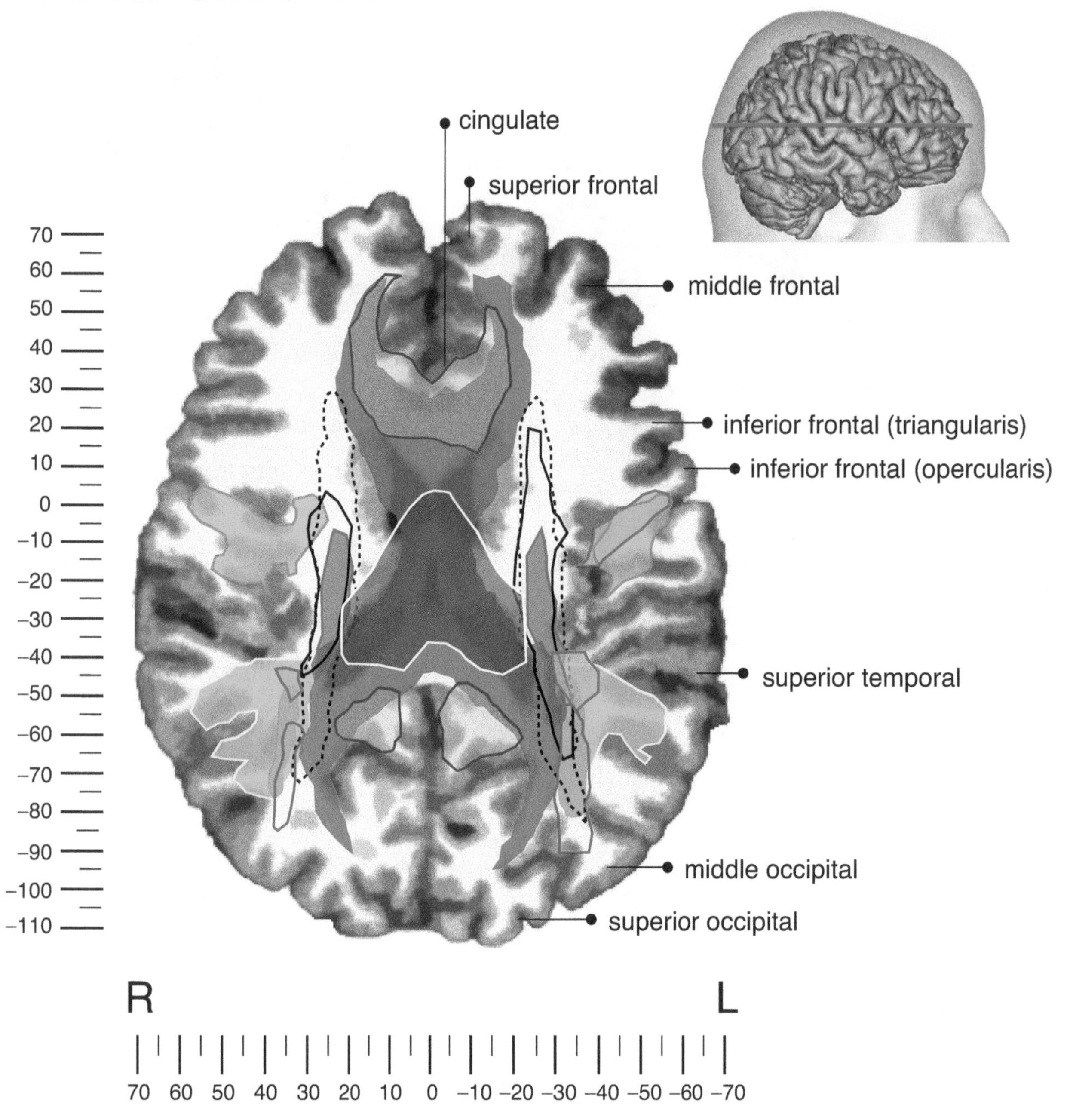

TRACT LEGEND

CST	Cortico-Spinal Tract	CPC	Cortico-Ponto-Cerebellar Tract	Cing	Cingulum
CC	Corpus Callosum	ILF	Inf. Longitudinal Fasc.	Fx	Fornix
OR	Optic Radiations	AF-L	Arcuate (long segment)	IC	Internal Capsule
AF-A	Arcuate (ant. segment)	AF-P	Arcuate (post. segment)		

Axial slice 20

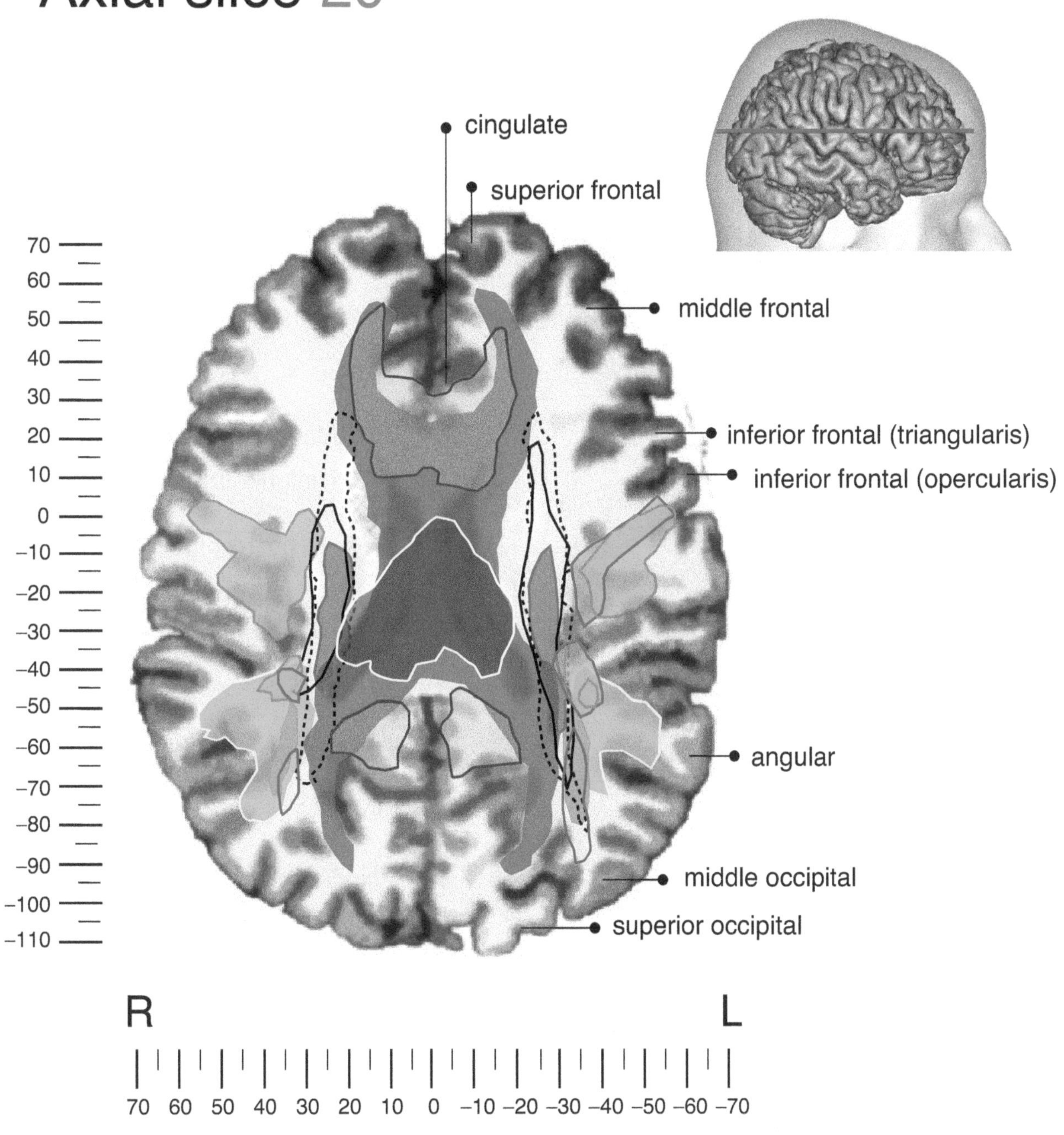

TRACT LEGEND

CST	Cortico-Spinal Tract	CPC	Cortico-Ponto-Cerebellar Tract	Cing	Cingulum
CC	Corpus Callosum	ILF	Inf. Longitudinal Fasc.	Fx	Fornix
OR	Optic Radiations	AF-L	Arcuate (long segment)	IC	Internal Capsule
AF-A	Arcuate (ant. segment)	AF-P	Arcuate (post. segment)		

Axial slice 22

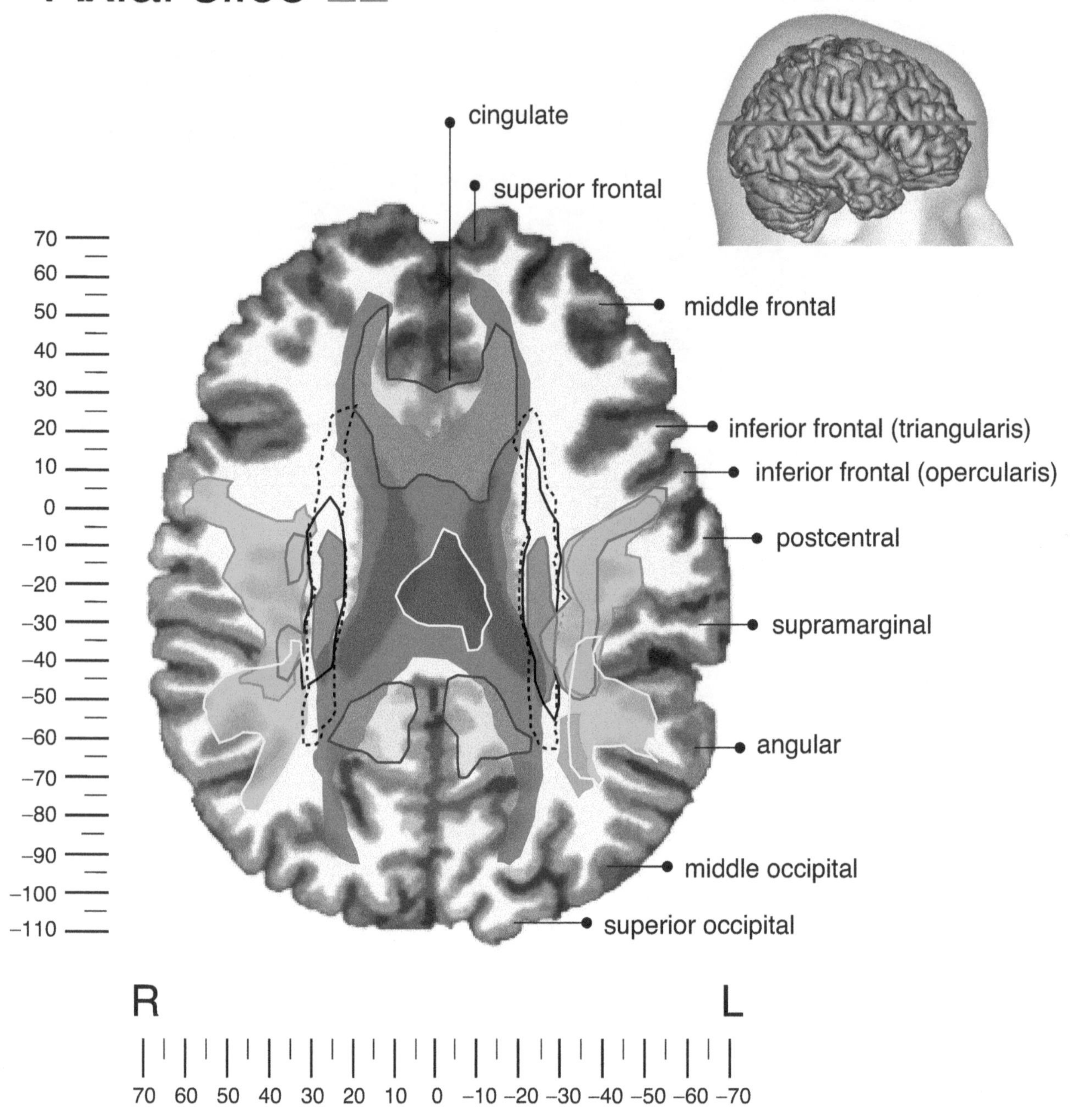

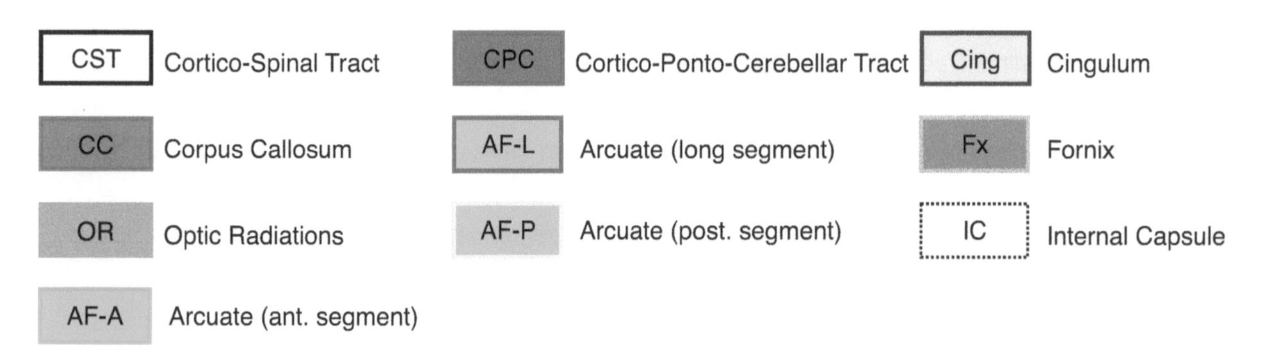

Axial slice 24

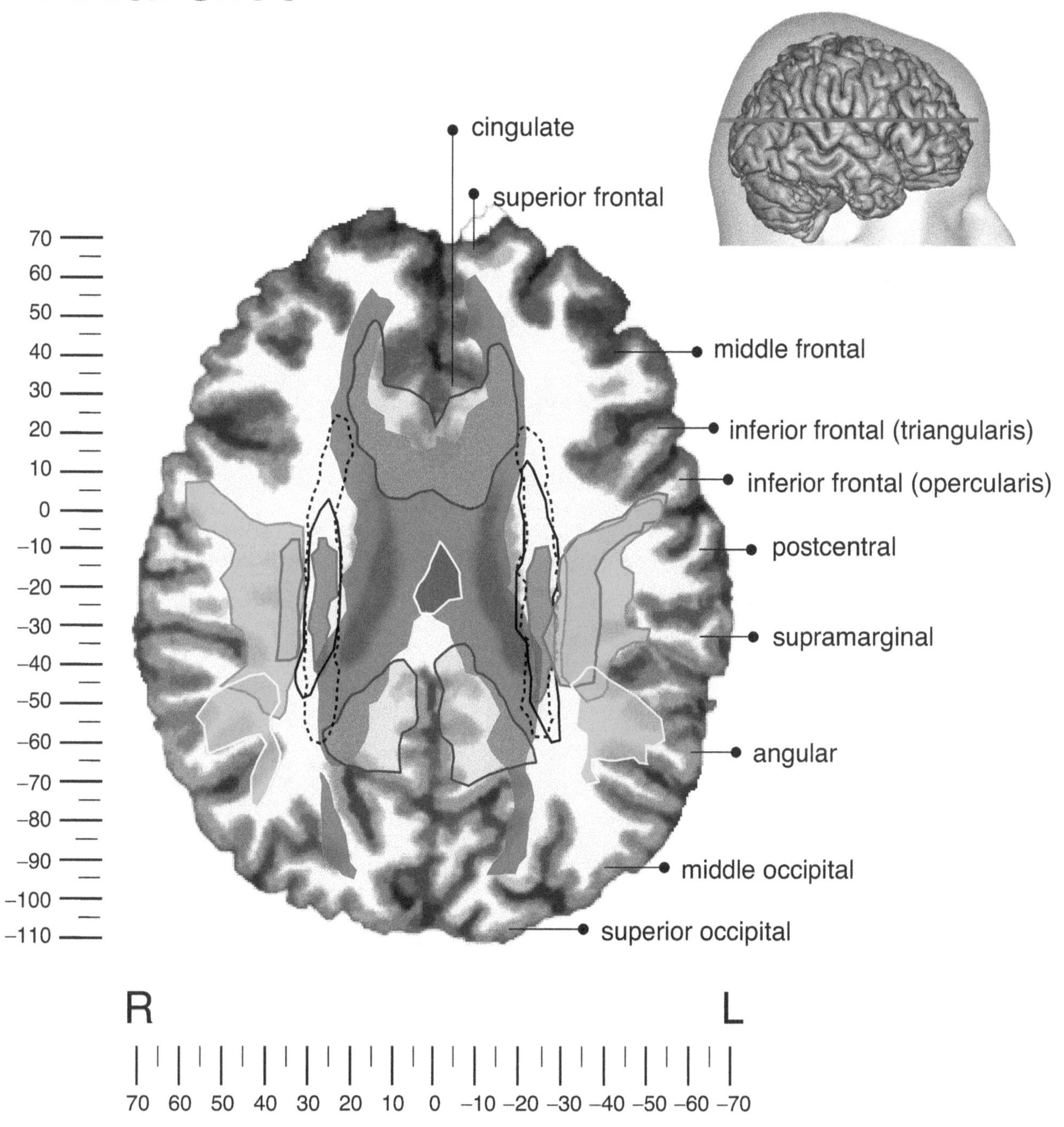

TRACT LEGEND

Abbreviation	Tract
CST	Cortico-Spinal Tract
CPC	Cortico-Ponto-Cerebellar Tract
Cing	Cingulum
CC	Corpus Callosum
AF-L	Arcuate (long segment)
Fx	Fornix
AF-A	Arcuate (ant. segment)
AF-P	Arcuate (post. segment)
IC	Internal Capsule

Axial slice 26

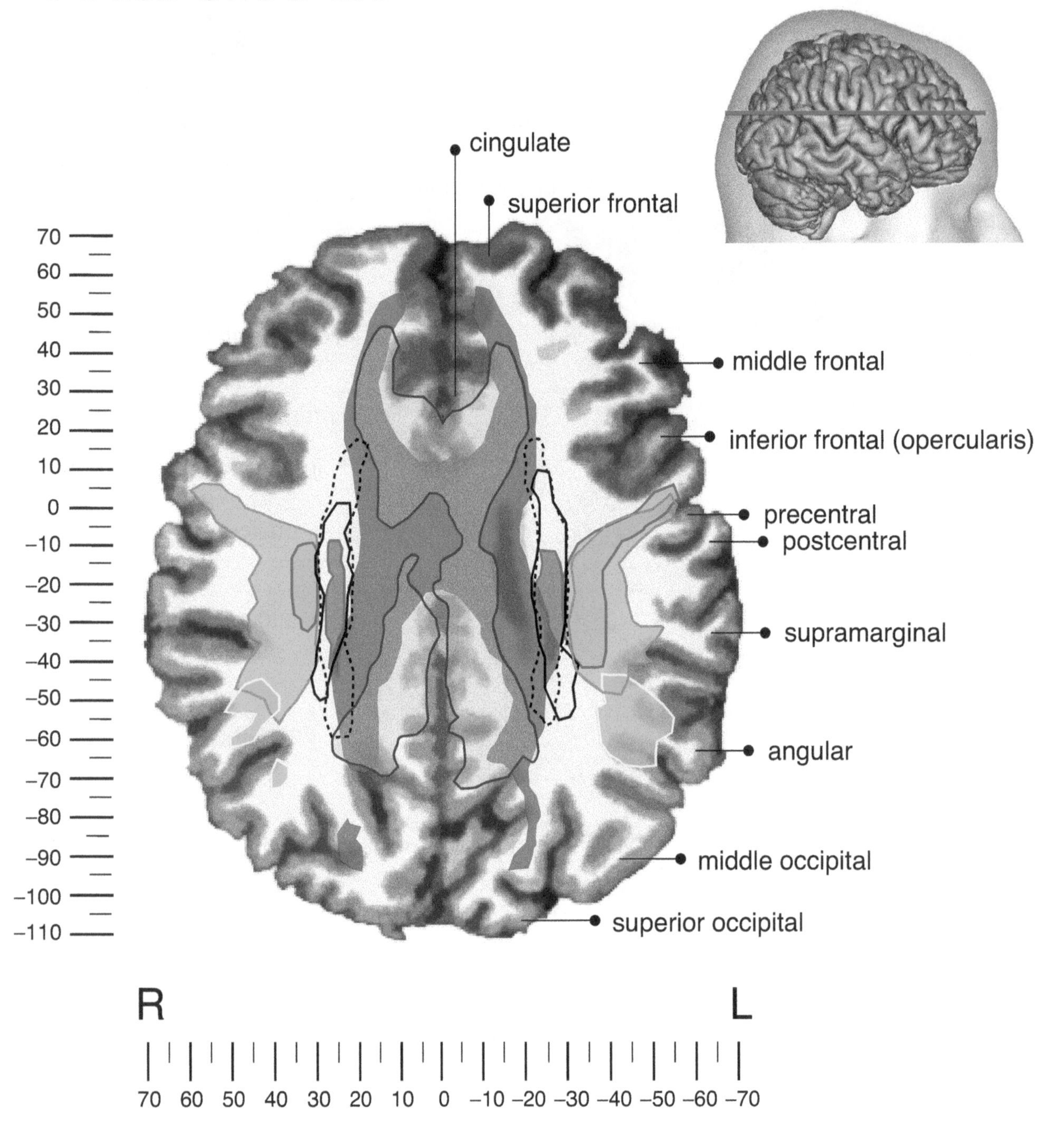

TRACT LEGEND

CST	Cortico-Spinal Tract	CPC	Cortico-Ponto-Cerebellar Tract	Cing	Cingulum
CC	Corpus Callosum	AF-L	Arcuate (long segment)	IC	Internal Capsule
AF-A	Arcuate (ant. segment)	AF-P	Arcuate (post. segment)		

Axial slice 28

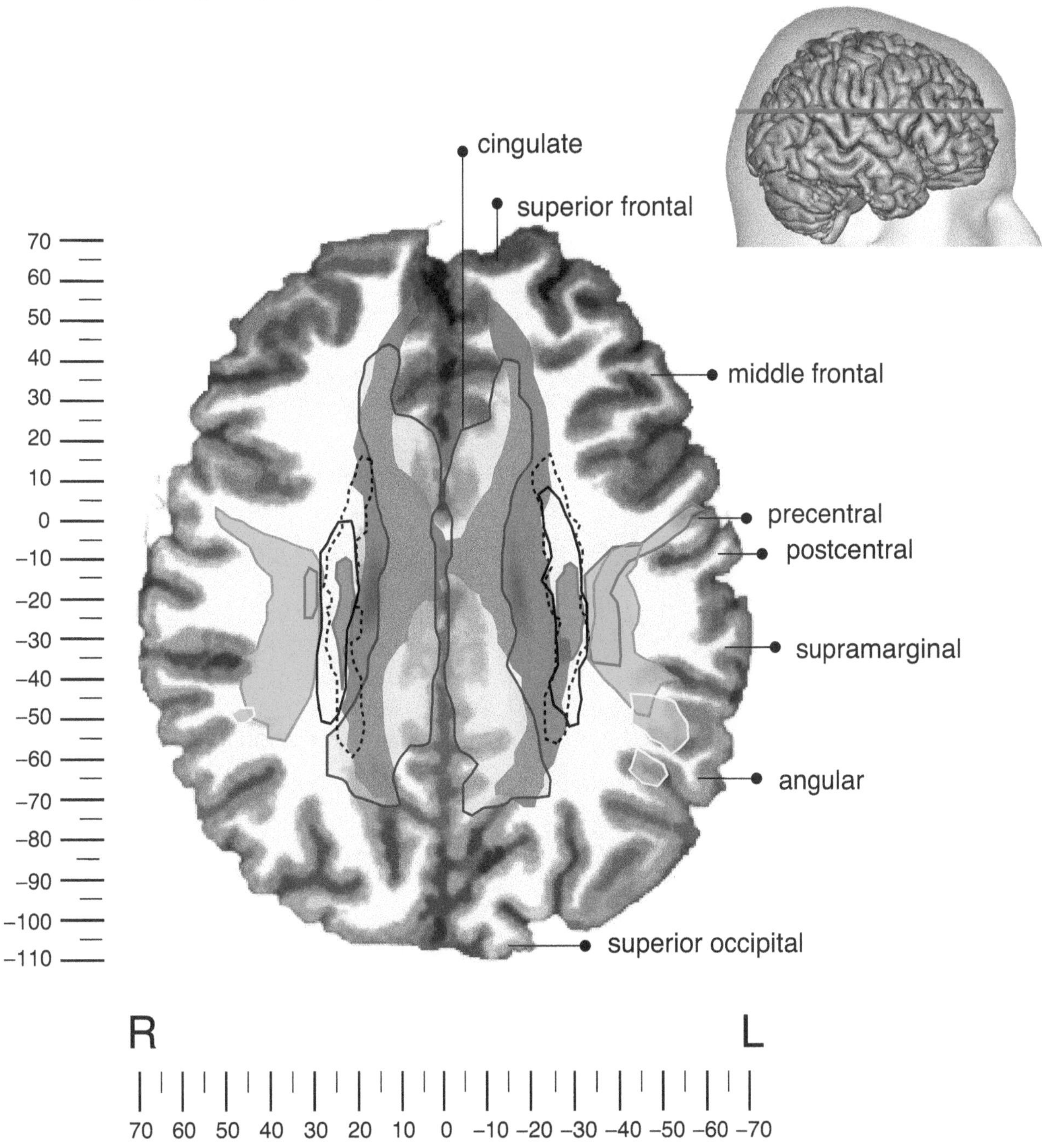

TRACT LEGEND

CST	Cortico-Spinal Tract	CPC	Cortico-Ponto-Cerebellar Tract	Cing	Cingulum
CC	Corpus Callosum	AF-L	Arcuate (long segment)	IC	Internal Capsule
AF-A	Arcuate (ant. segment)	AF-P	Arcuate (post. segment)		

Axial slice 30

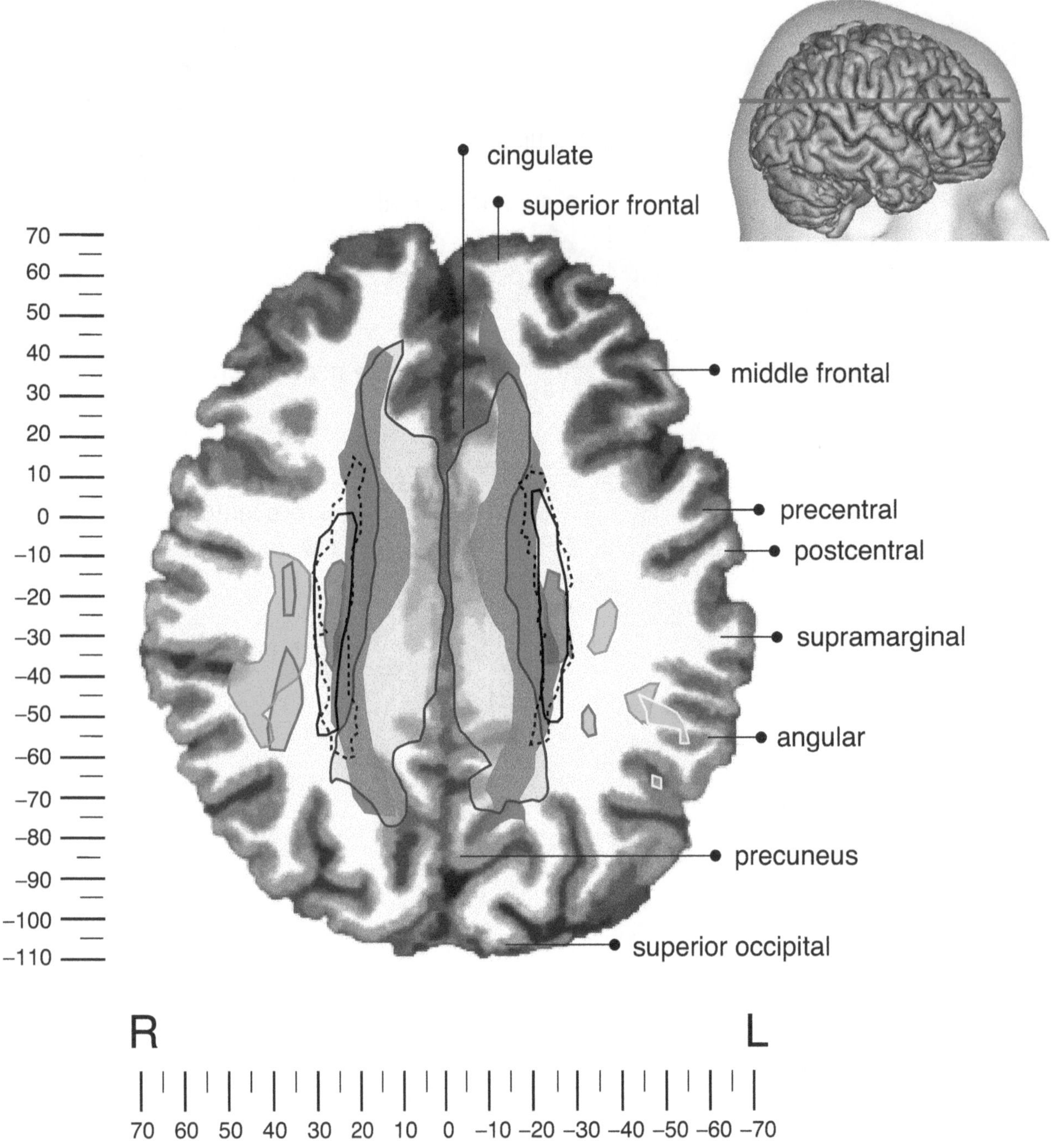

TRACT LEGEND

CST	Cortico-Spinal Tract	CPC	Cortico-Ponto-Cerebellar Tract	Cing	Cingulum
CC	Corpus Callosum	SLF II	Sup. Longitudinal Fasc. II	IC	Internal Capsule
AF-A	Arcuate (ant. segment)	AF-P	Arcuate (post. segment)		

Axial slice 32

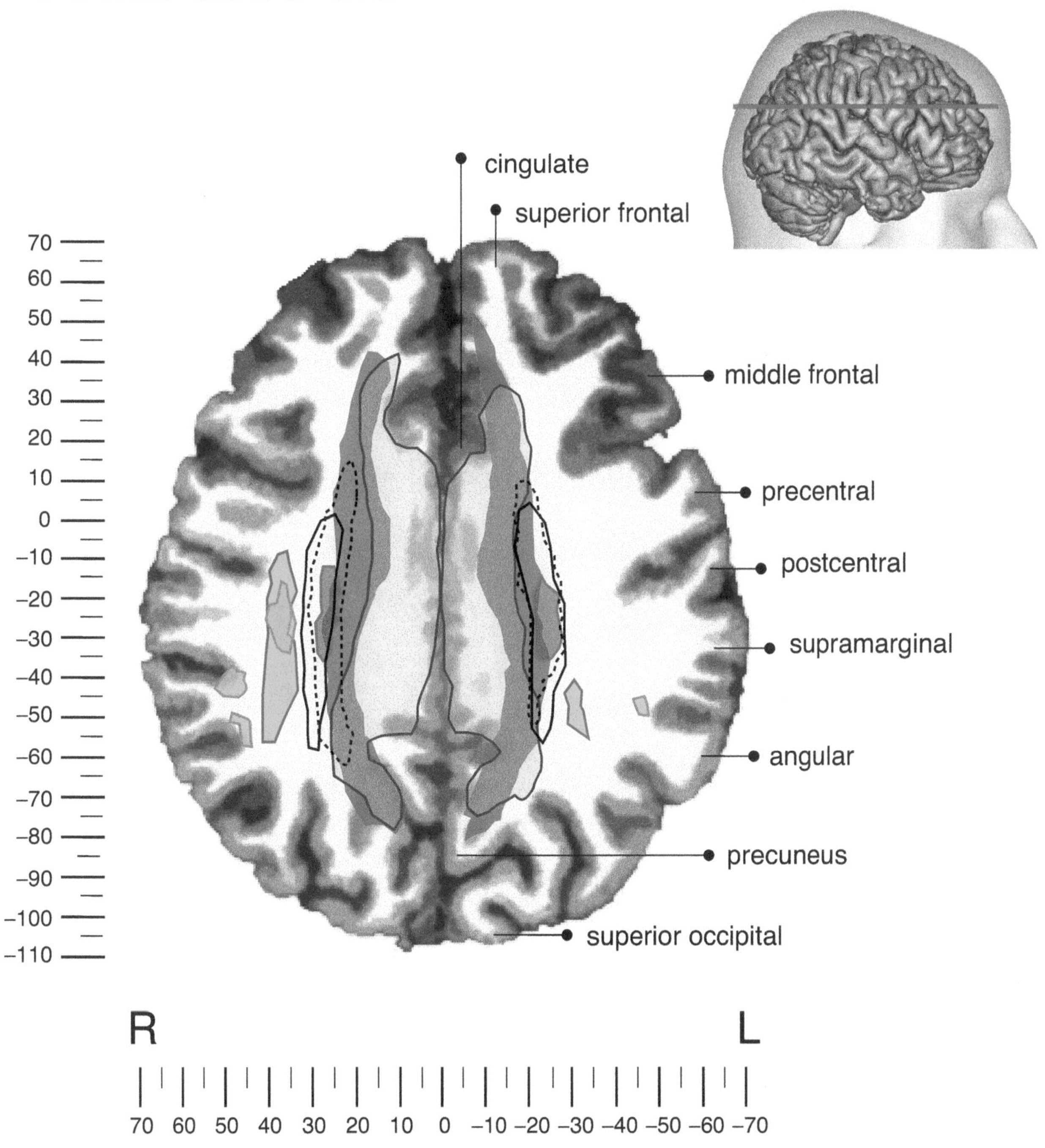

TRACT LEGEND

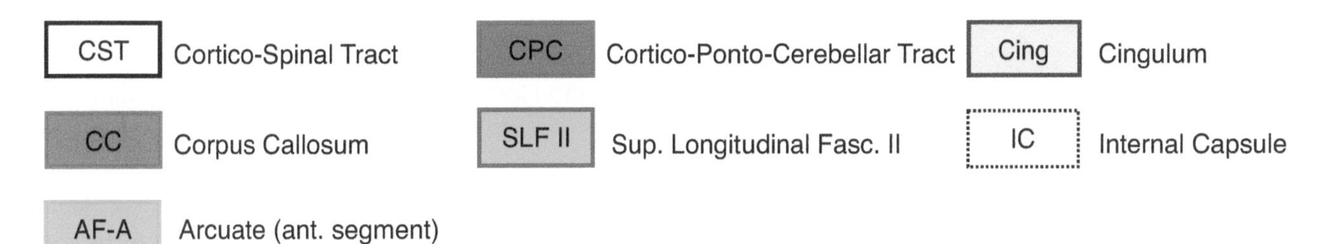

Axial slice 34

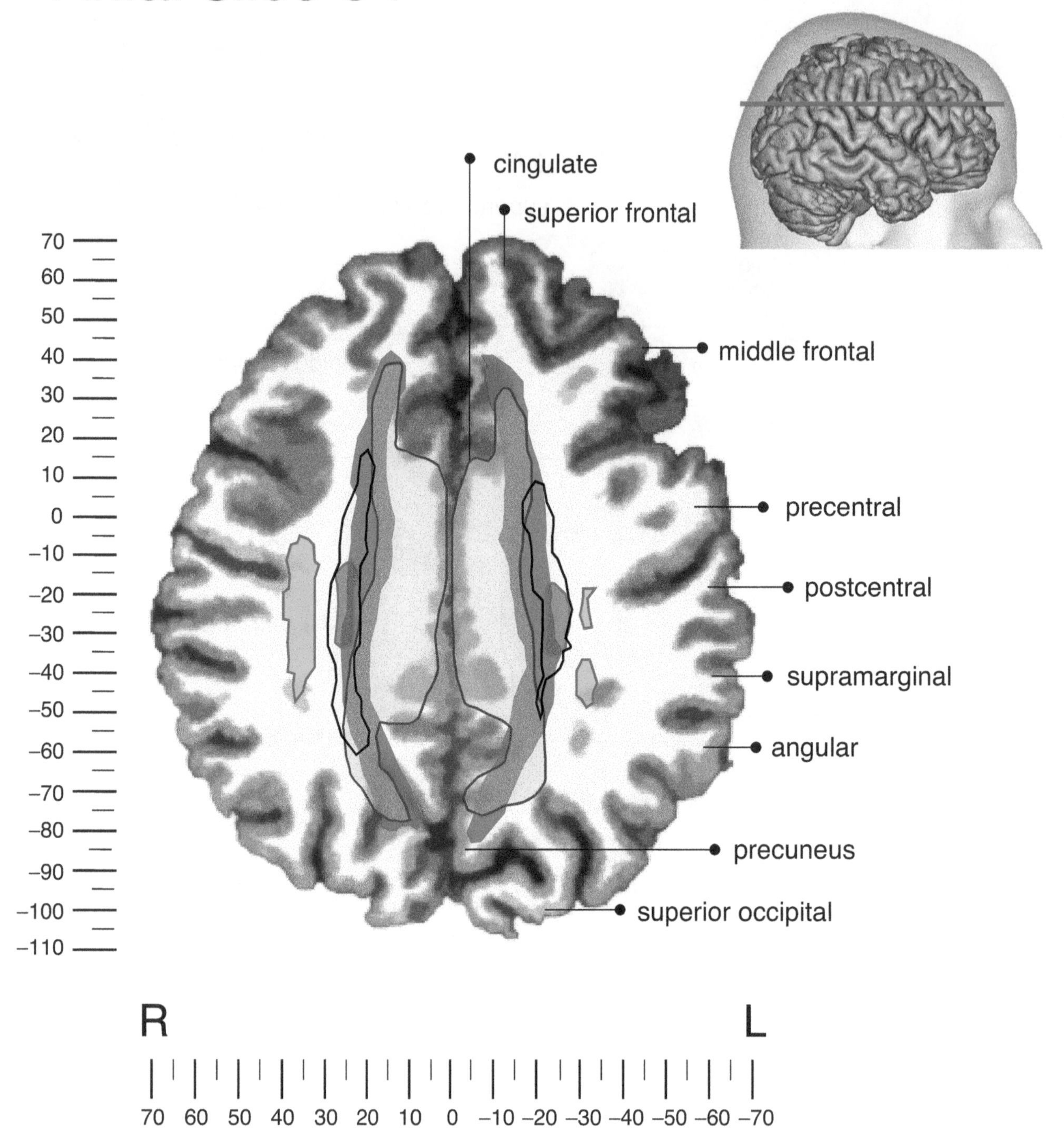

TRACT LEGEND

CST	Cortico-Spinal Tract / IC	CPC	Cortico-Ponto-Cerebellar Tract	Cing	Cingulum
CC	Corpus Callosum	SLF II	Sup. Longitudinal Fasc. II		

Axial slice 36

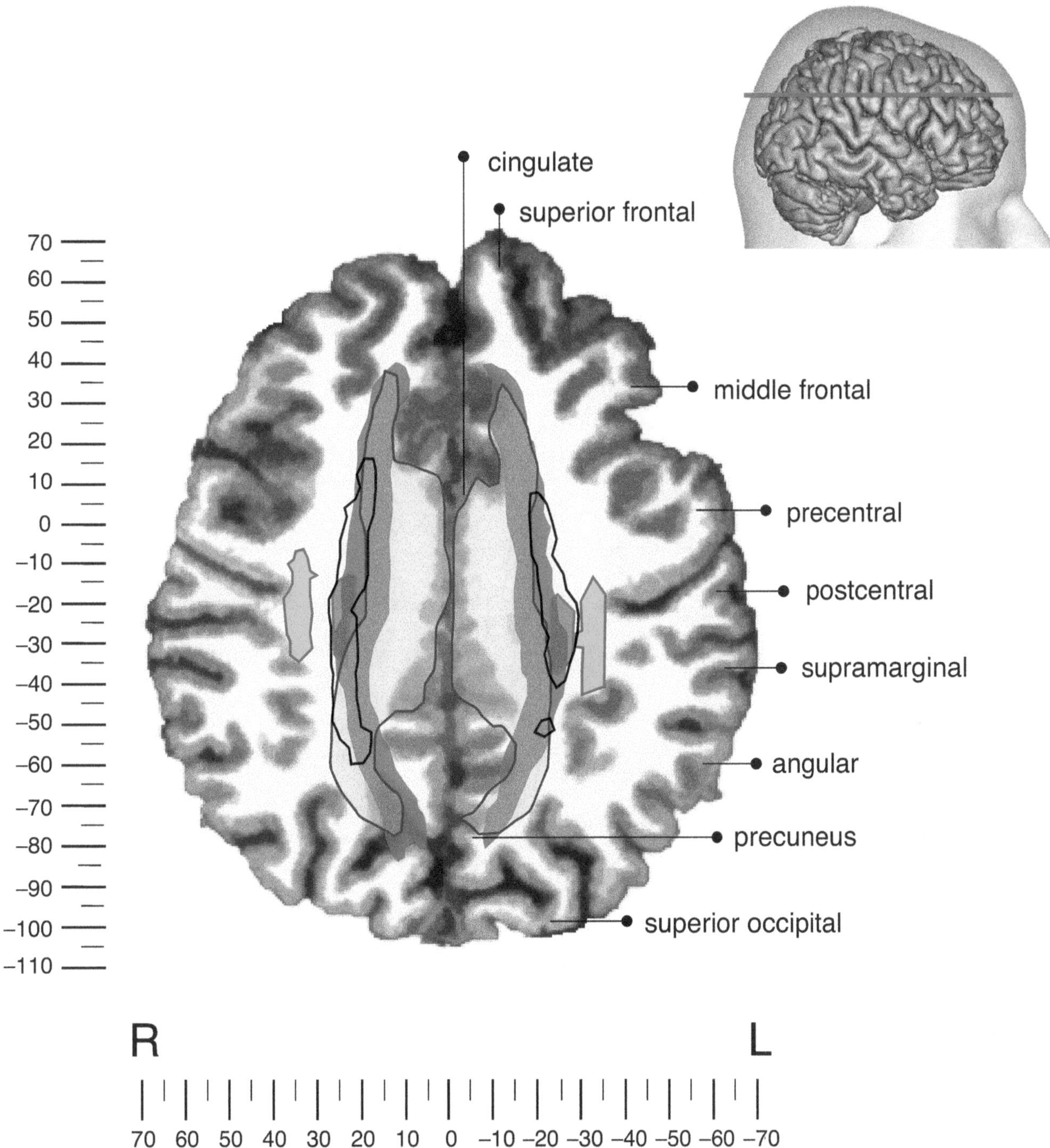

TRACT LEGEND

CST	Cortico-Spinal Tract / IC	CPC	Cortico-Ponto-Cerebellar Tract	Cing	Cingulum
CC	Corpus Callosum	SLF II	Sup. Longitudinal Fasc. II		

Axial slice 38

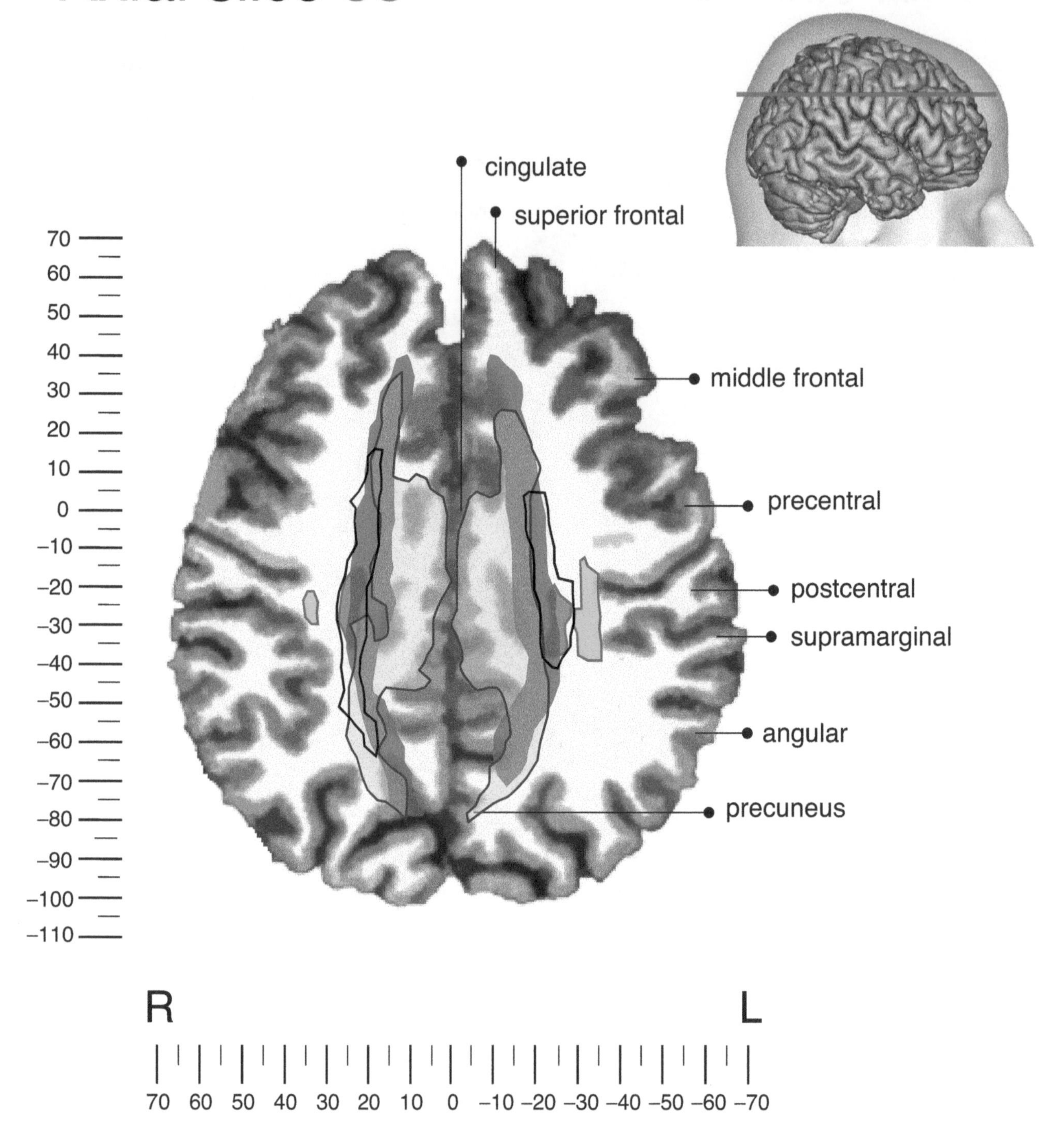

TRACT LEGEND

CST Cortico-Spinal Tract

CPC Cortico-Ponto-Cerebellar Tract

Cing Cingulum

CC Corpus Callosum

SLF II Sup. Longitudinal Fasc. II

Axial slice 40

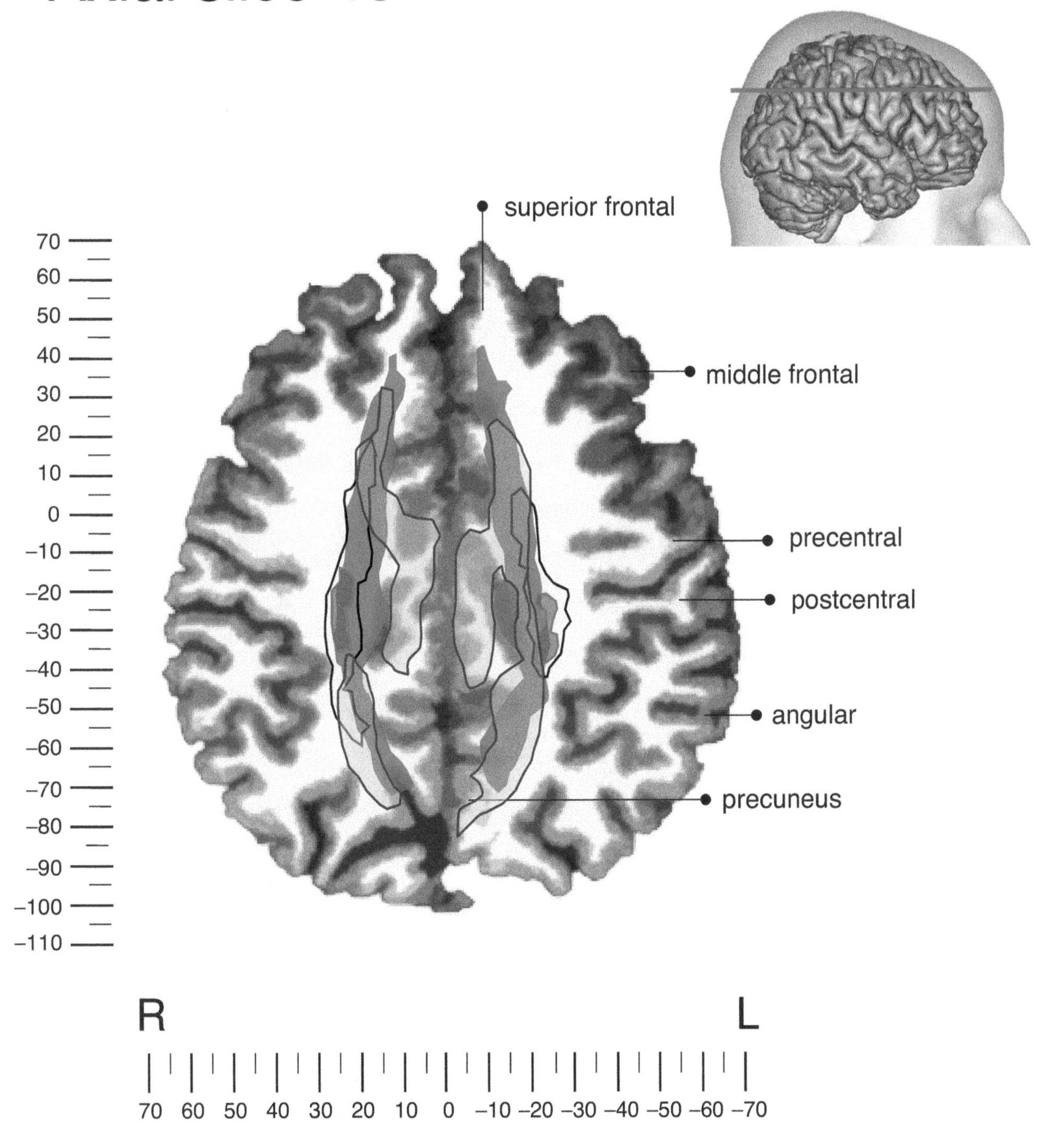

TRACT LEGEND

Axial slice 42

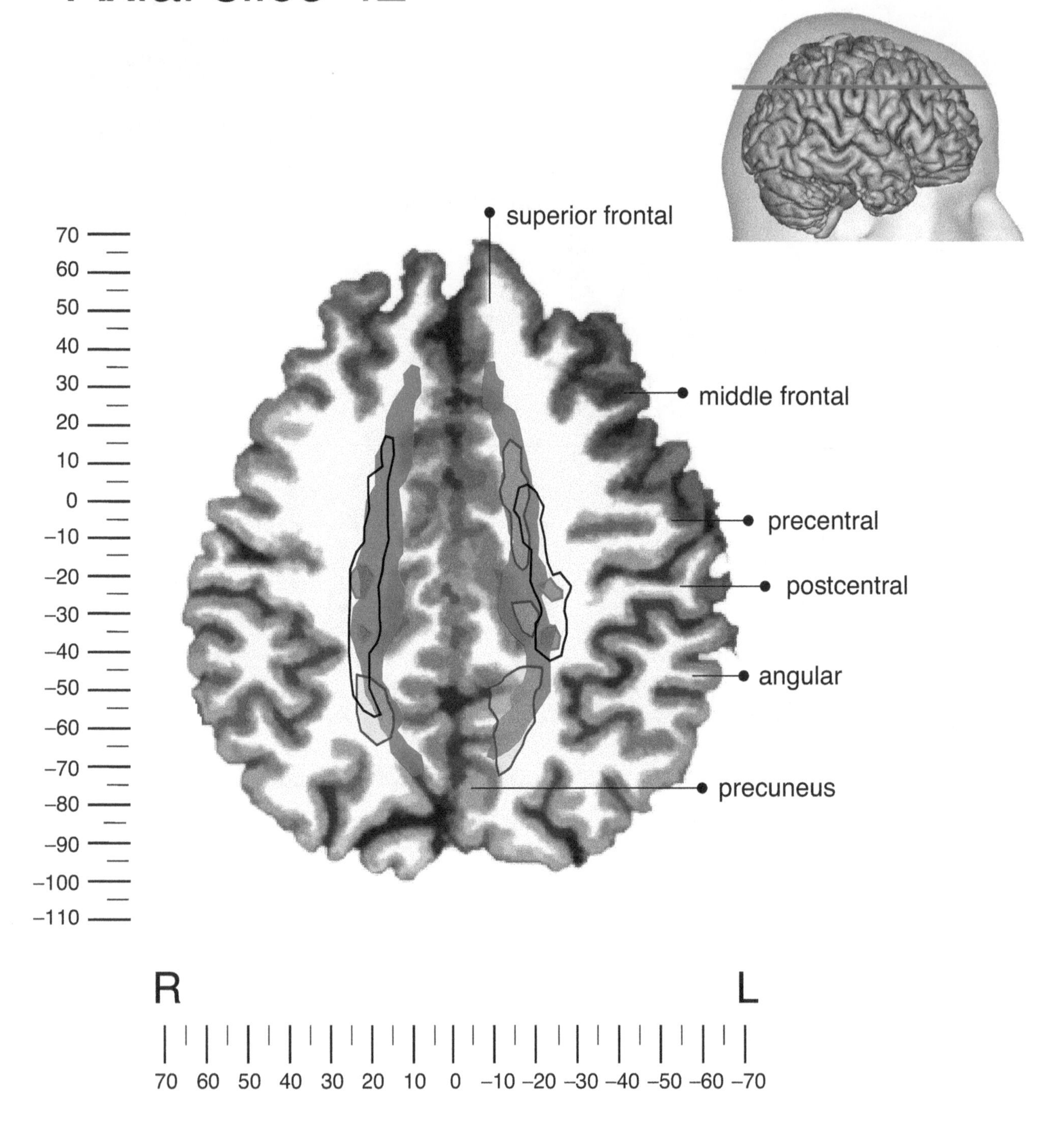

TRACT LEGEND

CST Cortico-Spinal Tract / IC

CPC Cortico-Ponto-Cerebellar Tract

Cing Cingulum

CC Corpus Callosum

Axial slice 44

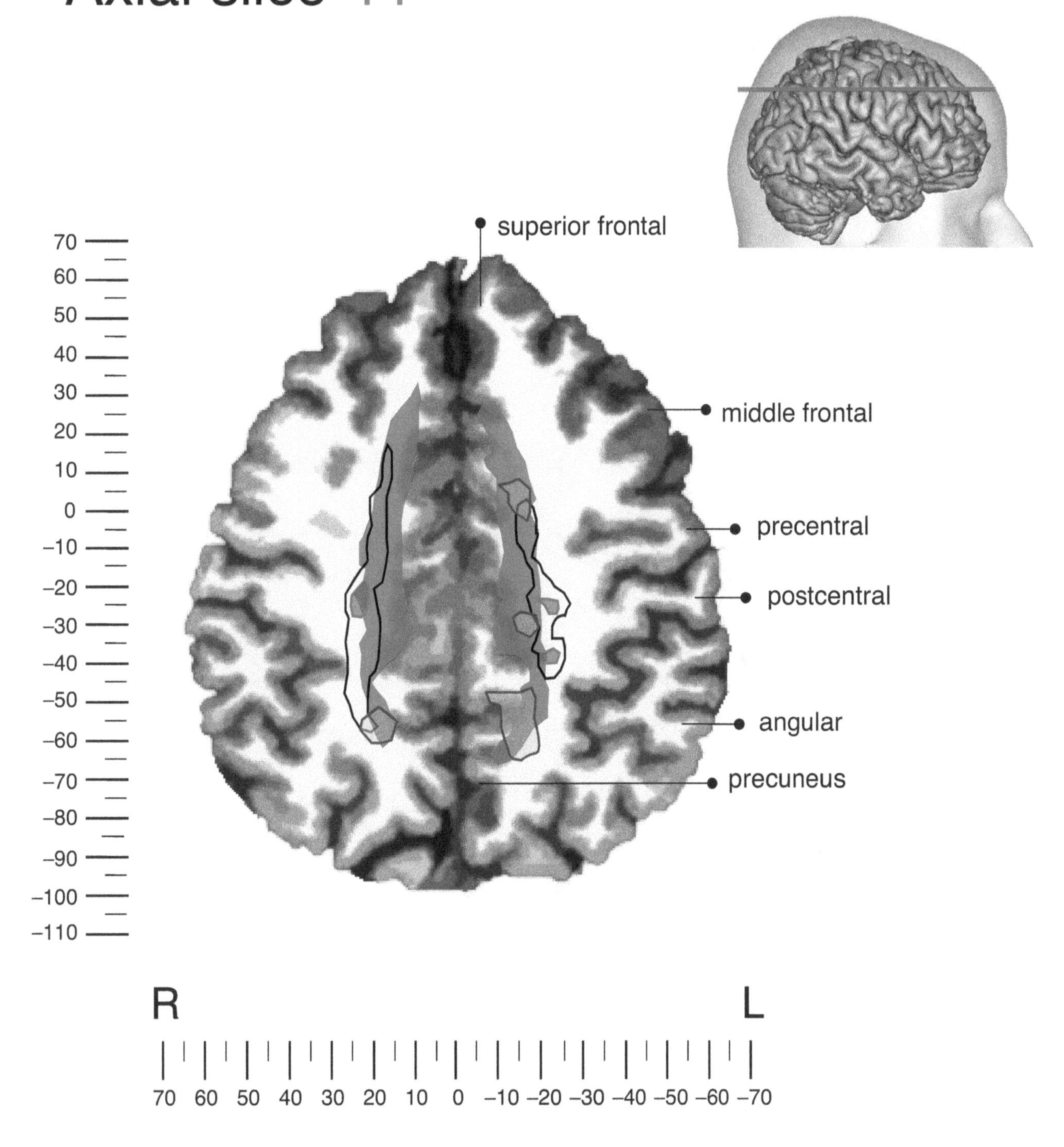

TRACT LEGEND

CST	Cortico-Spinal Tract / IC
CPC	Cortico-Ponto-Cerebellar Tract
Cing	Cingulum
CC	Corpus Callosum

Coronal maps

Coronal slice –74

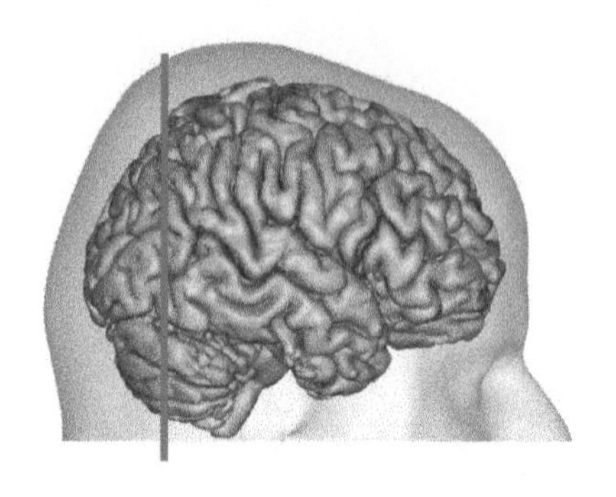

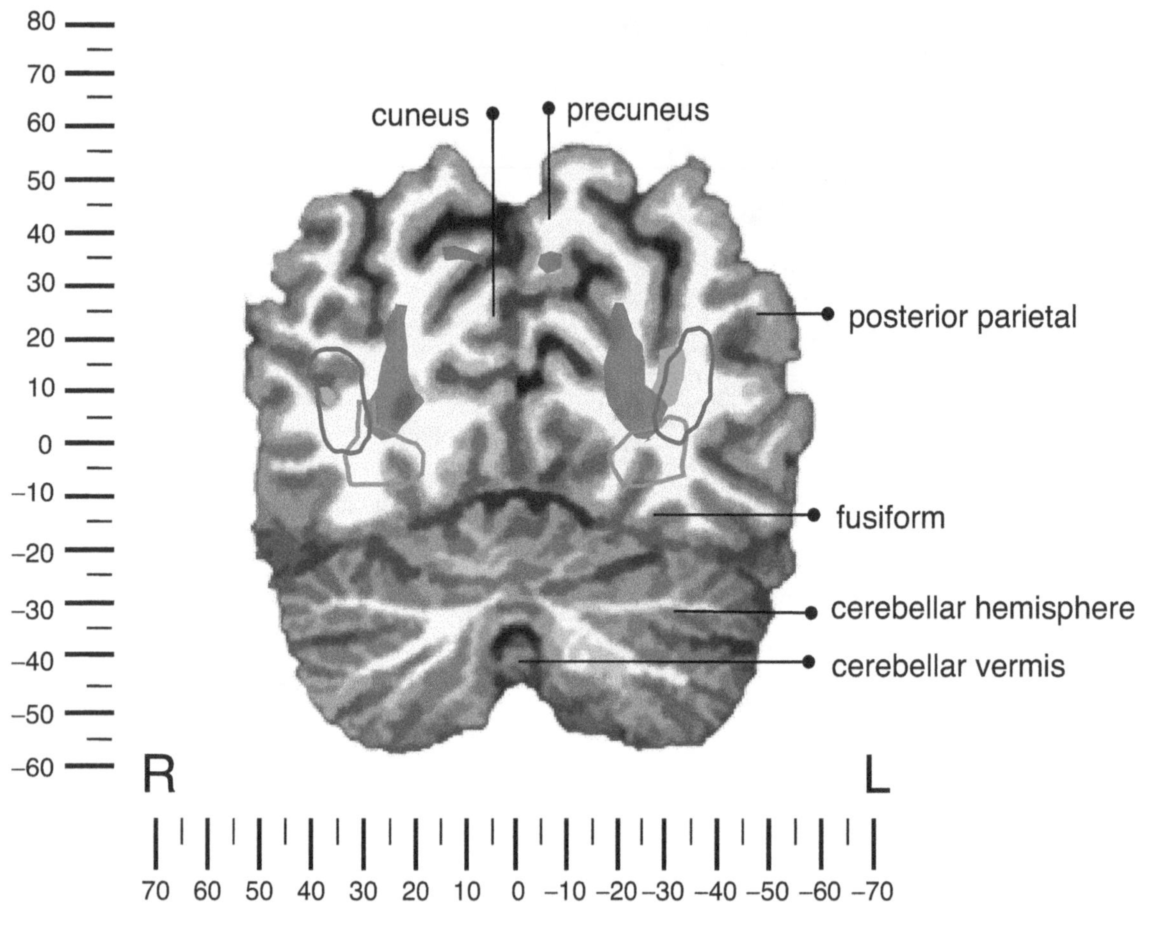

TRACT LEGEND

CC Corpus Callosum

IFOF Inf. Fronto-Occipital Fasc.

OR Optic Radiations

ILF Inf. Longitudinal Fasc.

Coronal slice –72

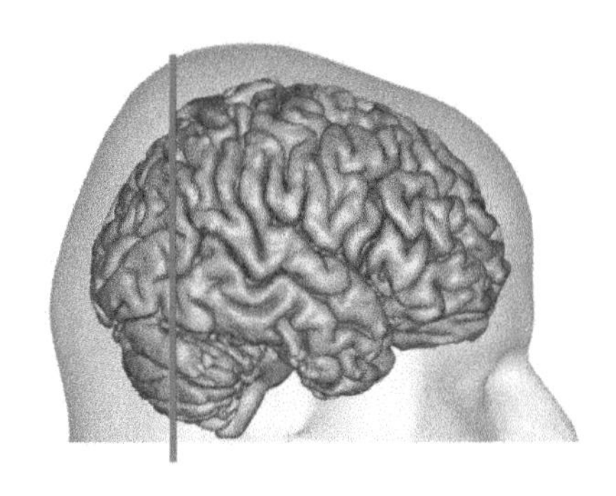

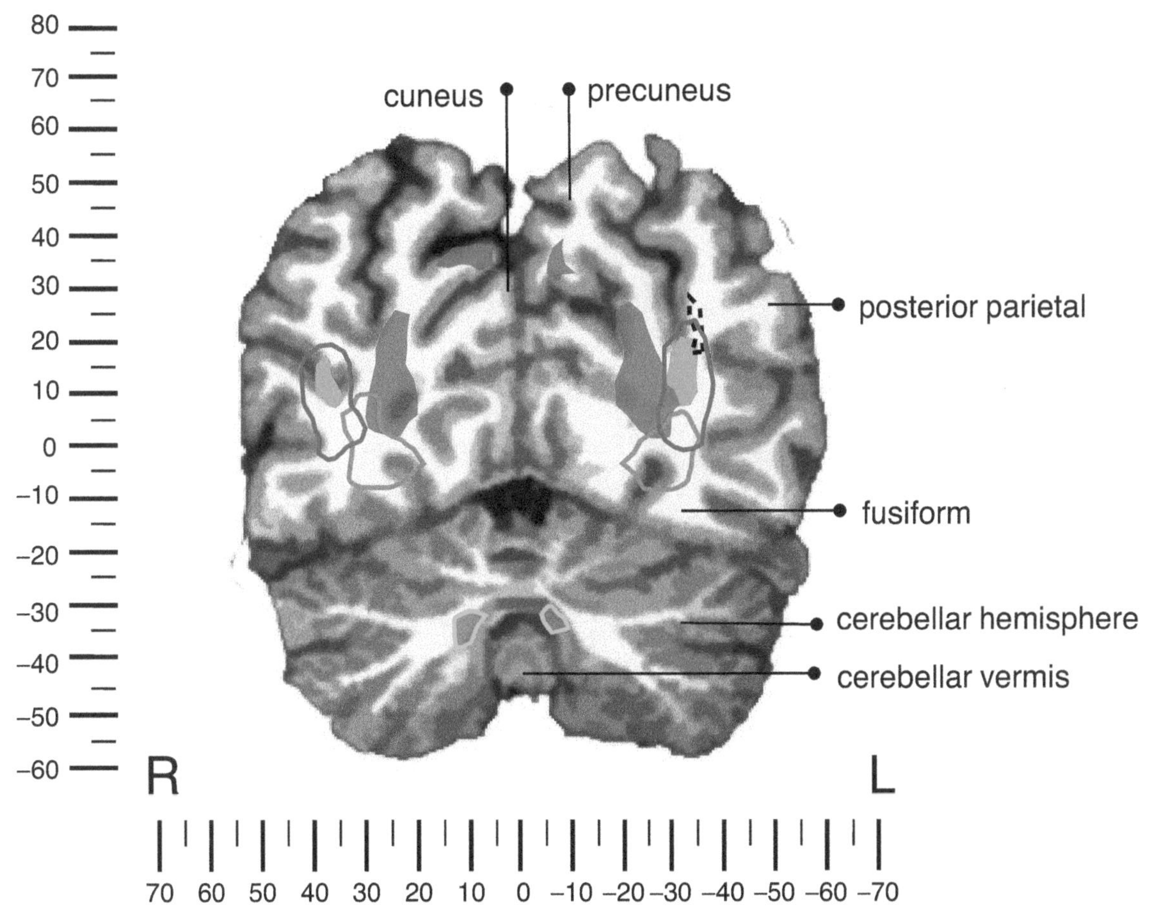

TRACT LEGEND

CC	Corpus Callosum	IFOF	Inf. Fronto-Occipital Fasc.	OR	Optic Radiations
ILF	Inf. Longitudinal Fasc.	SCP	Sup. Cerebellar Peduncle	IC	Internal Capsule

Coronal slice –70

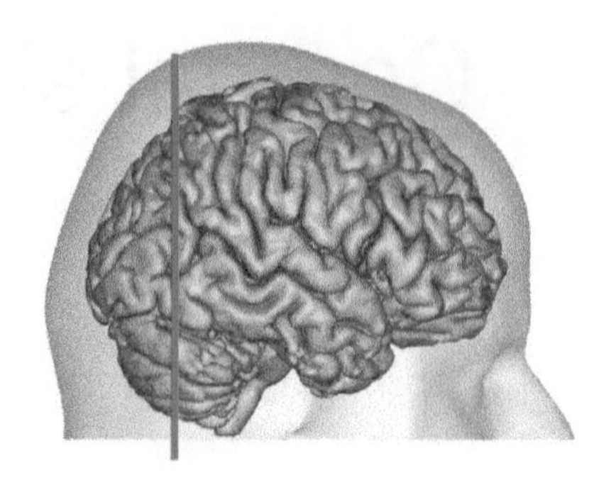

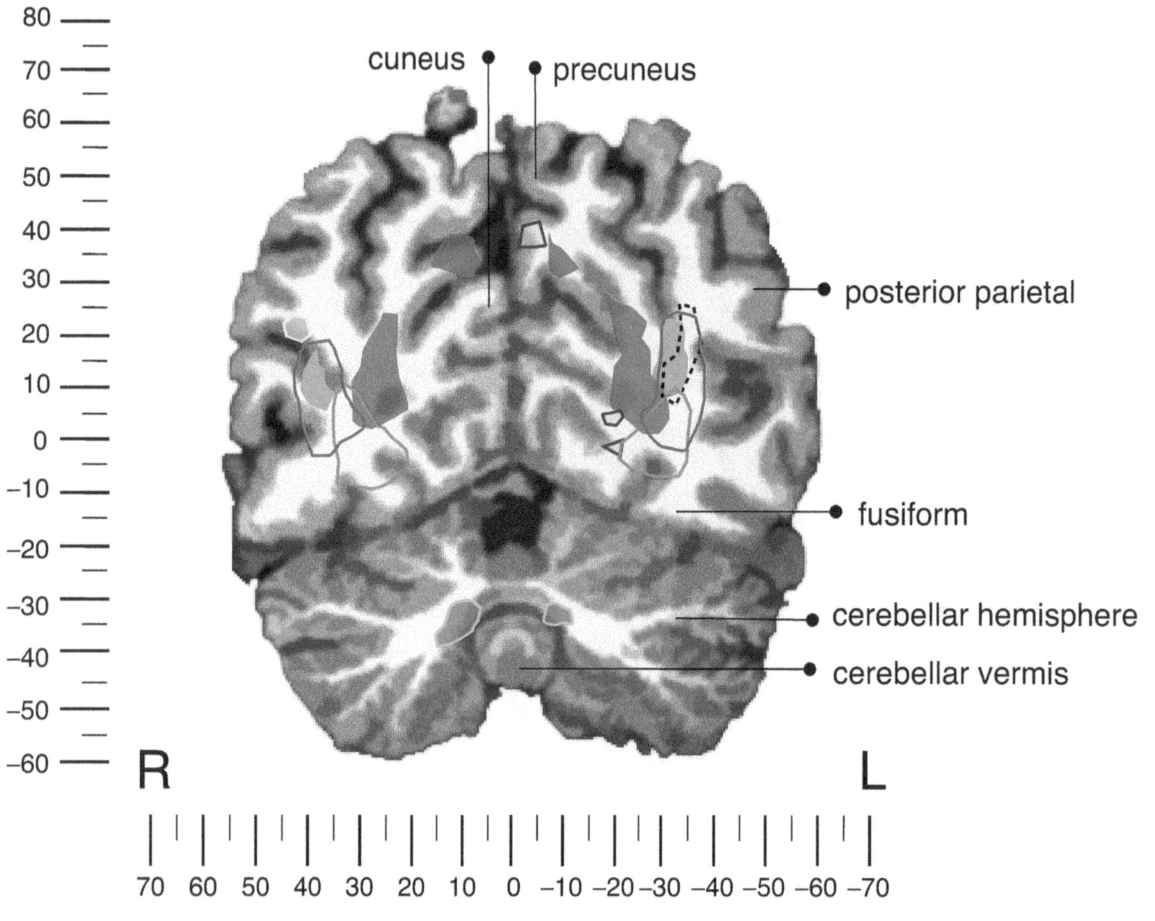

TRACT LEGEND

CC	Corpus Callosum	IFOF	Inf. Fronto-Occipital Fasc.	OR	Optic Radiations
ILF	Inf. Longitudinal Fasc.	SCP	Sup. Cerebellar Peduncle	IC	Internal Capsule
AF-P	Arcuate (post. segment)	Cing	Cingulum		

Coronal slice –68

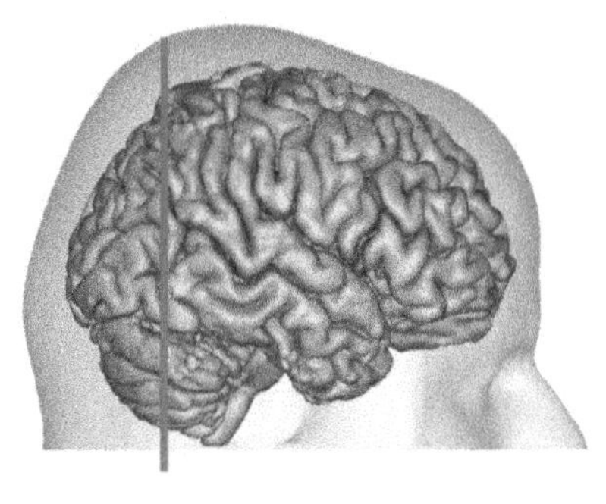

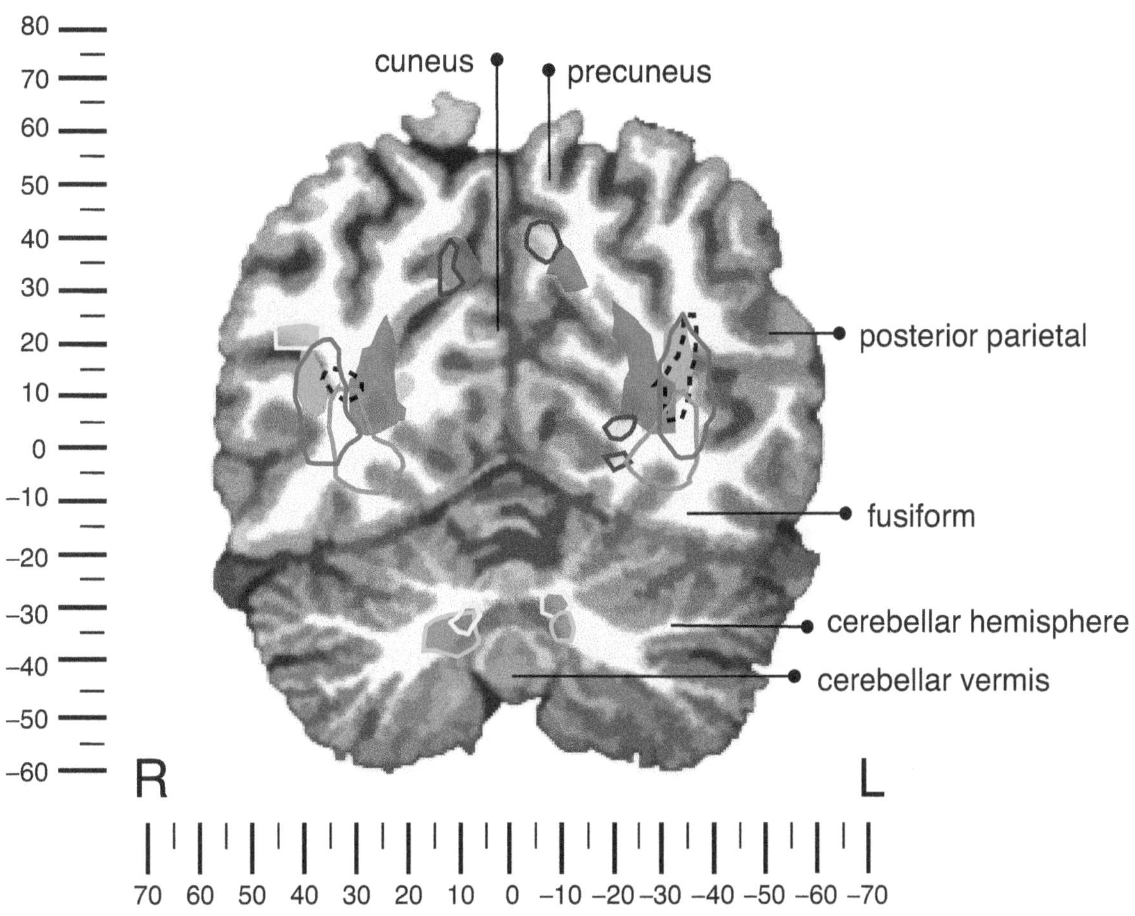

TRACT LEGEND

CC	Corpus Callosum	IFOF	Inf. Fronto-Occipital Fasc.	OR	Optic Radiations
ILF	Inf. Longitudinal Fasc.	SCP	Sup. Cerebellar Peduncle	IC	Internal Capsule
AF-P	Arcuate (post. segment)	ICP	Inf. Cerebellar Peduncle	Cing	Cingulum

Coronal slice –66

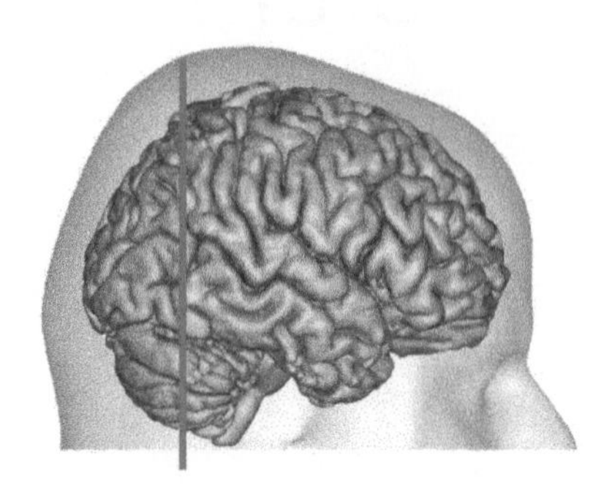

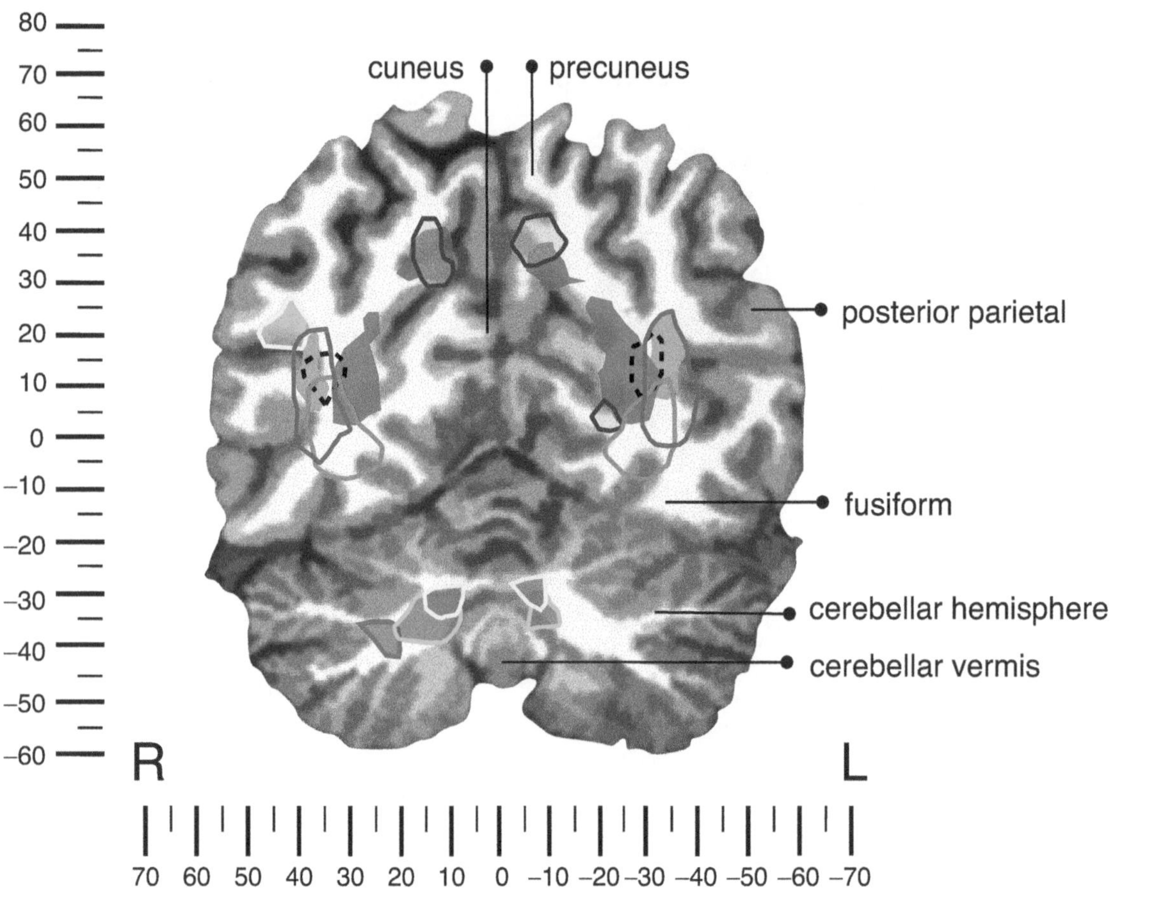

TRACT LEGEND

CC	Corpus Callosum
ILF	Inf. Longitudinal Fasc.
AF-P	Arcuate (post. segment)
IFOF	Inf. Fronto-Occipital Fasc.
SCP	Sup. Cerebellar Peduncle
ICP	Inf. Cerebellar Peduncle
CPC	Cortico-Ponto-Cerebellar Tract
OR	Optic Radiations
IC	Internal Capsule
Cing	Cingulum

Coronal slice –64

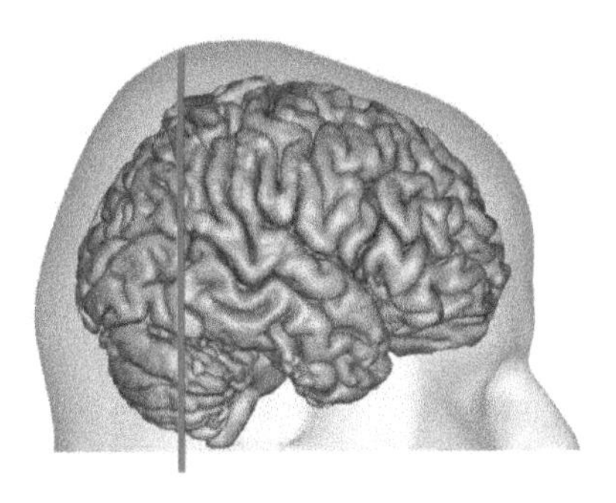

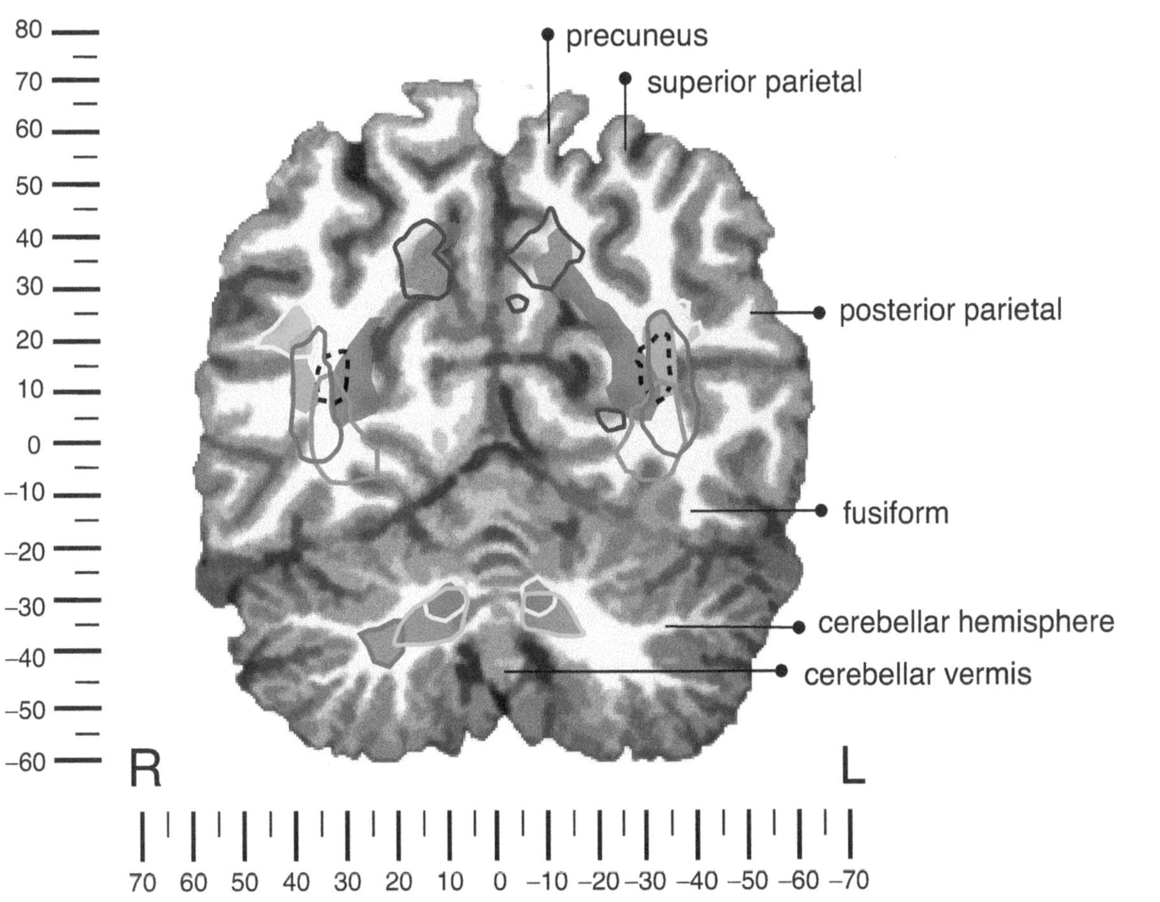

TRACT LEGEND

CC	Corpus Callosum	SCP	Sup. Cerebellar Peduncle	OR	Optic Radiations
ILF	Inf. Longitudinal Fasc.	ICP	Inf. Cerebellar Peduncle	IC	Internal Capsule
AF-P	Arcuate (post. segment)	CPC	Cortico-Ponto-Cerebellar Tract	Cing	Cingulum
IFOF	Inf. Fronto-Occipital Fasc.				

Coronal slice –62

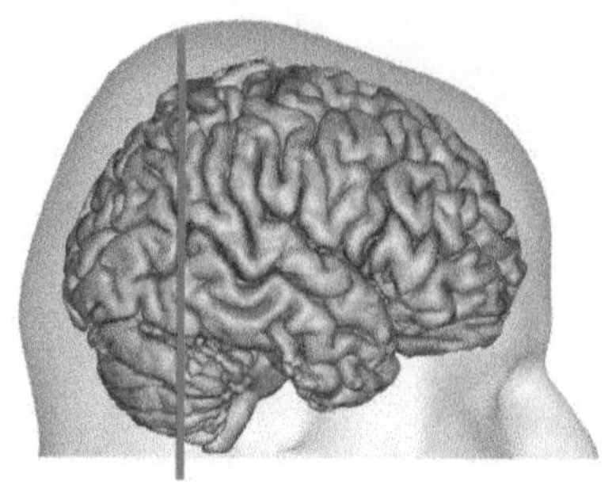

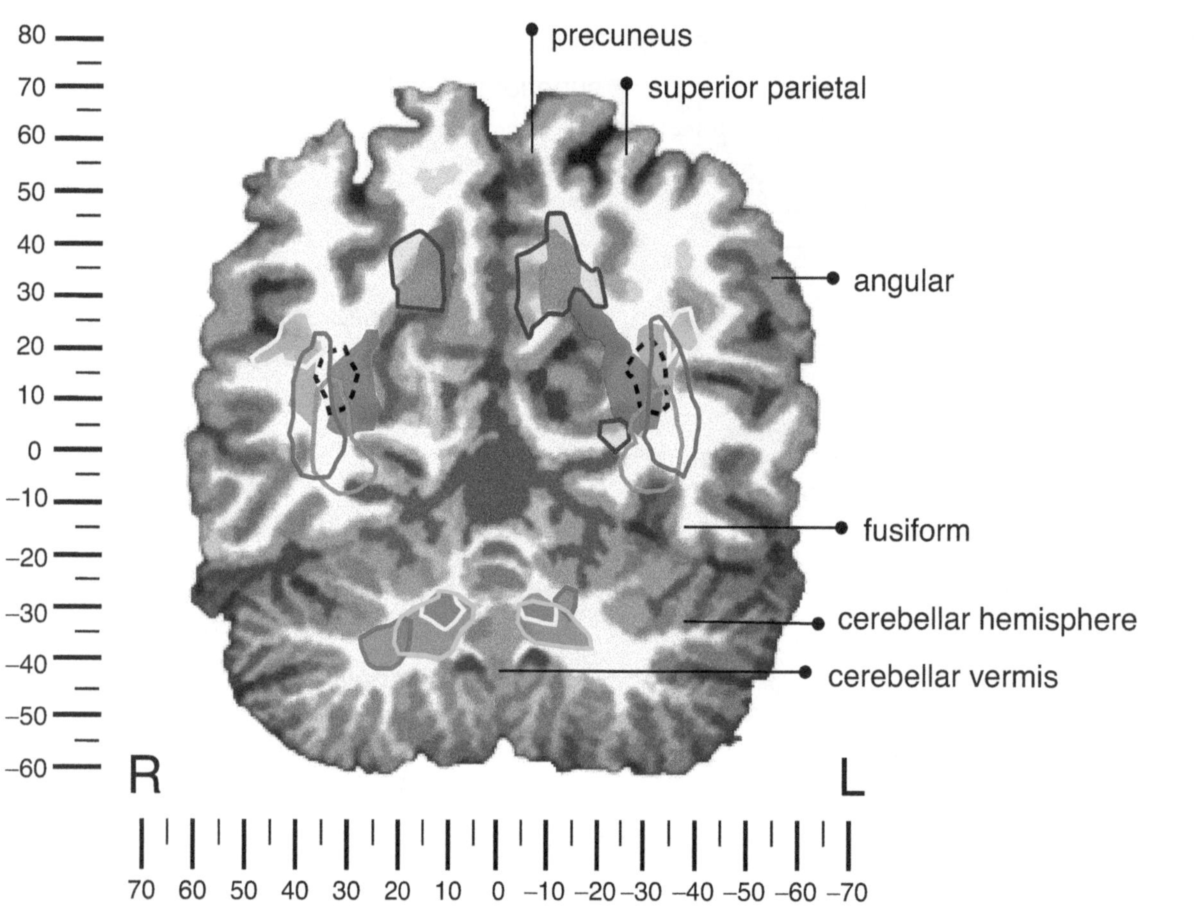

TRACT LEGEND

CC	Corpus Callosum	SCP	Sup. Cerebellar Peduncle	OR	Optic Radiations
ILF	Inf. Longitudinal Fasc.	ICP	Inf. Cerebellar Peduncle	IC	Internal Capsule
AF-P	Arcuate (post. segment)	CPC	Cortico-Ponto-Cerebellar Tract	Cing	Cingulum
IFOF	Inf. Fronto-Occipital Fasc.				

Coronal slice –60

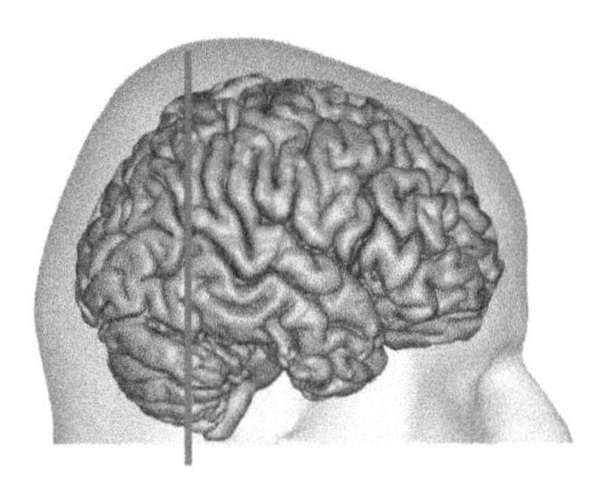

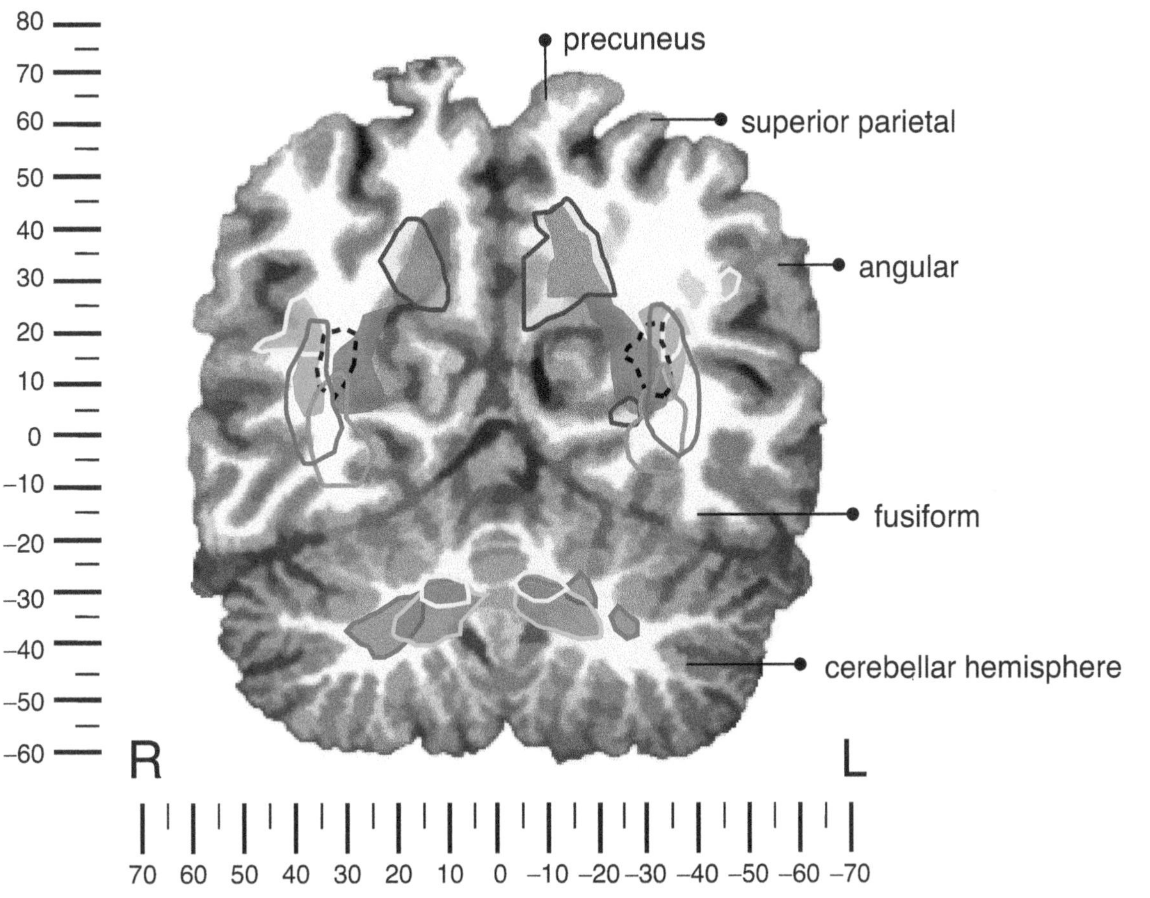

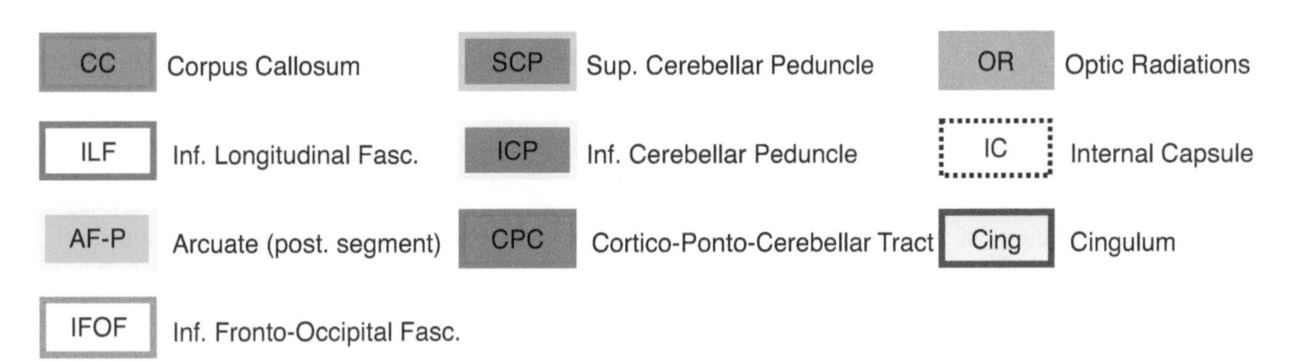

Coronal slice –58

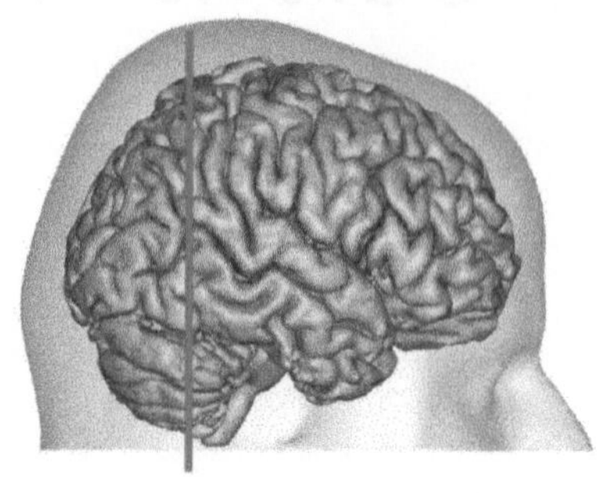

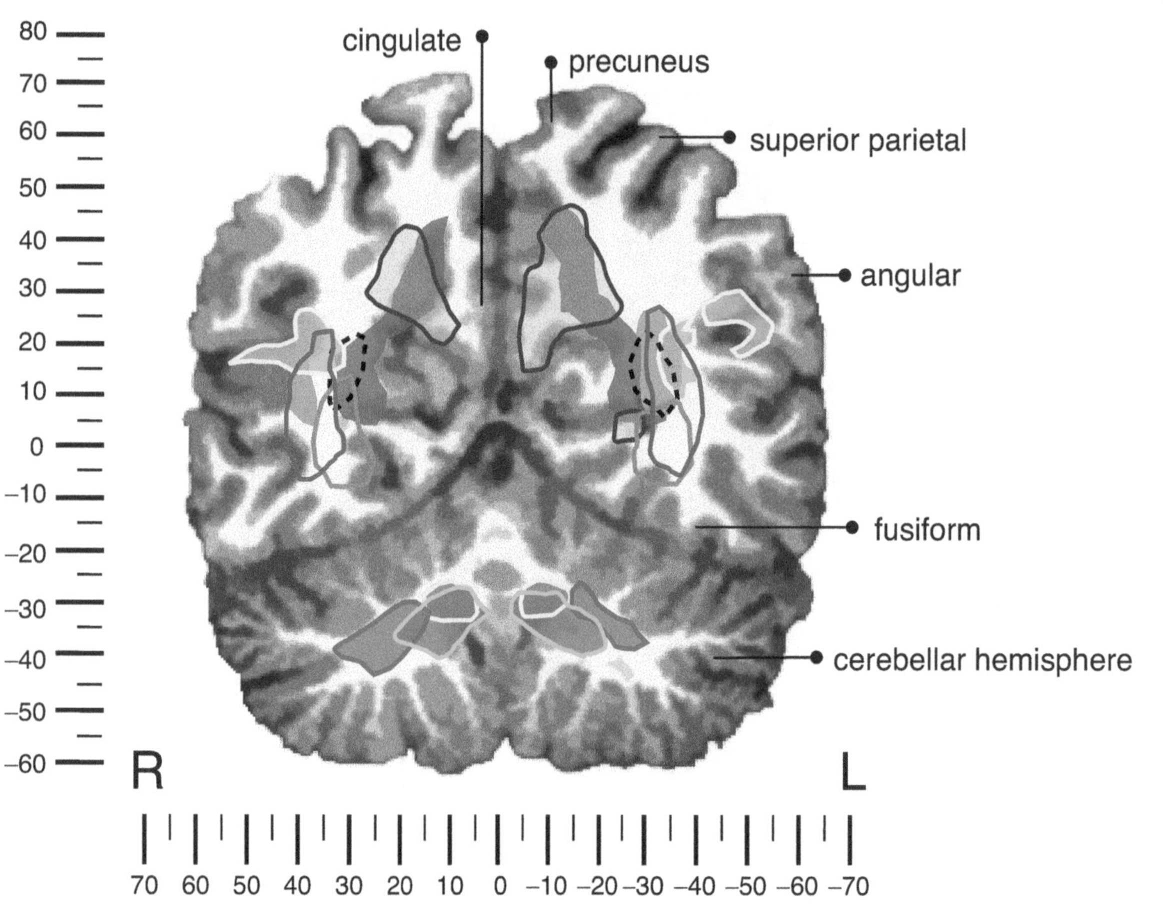

TRACT LEGEND

CC	Corpus Callosum	SCP	Sup. Cerebellar Peduncle	OR	Optic Radiations
ILF	Inf. Longitudinal Fasc.	ICP	Inf. Cerebellar Peduncle	IC	Internal Capsule
AF-P	Arcuate (post. segment)	CPC	Cortico-Ponto-Cerebellar Tract	Cing	Cingulum
IFOF	Inf. Fronto-Occipital Fasc.				

Coronal slice –56

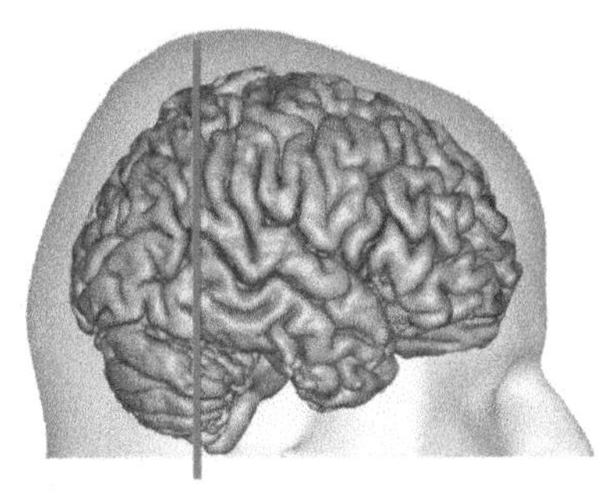

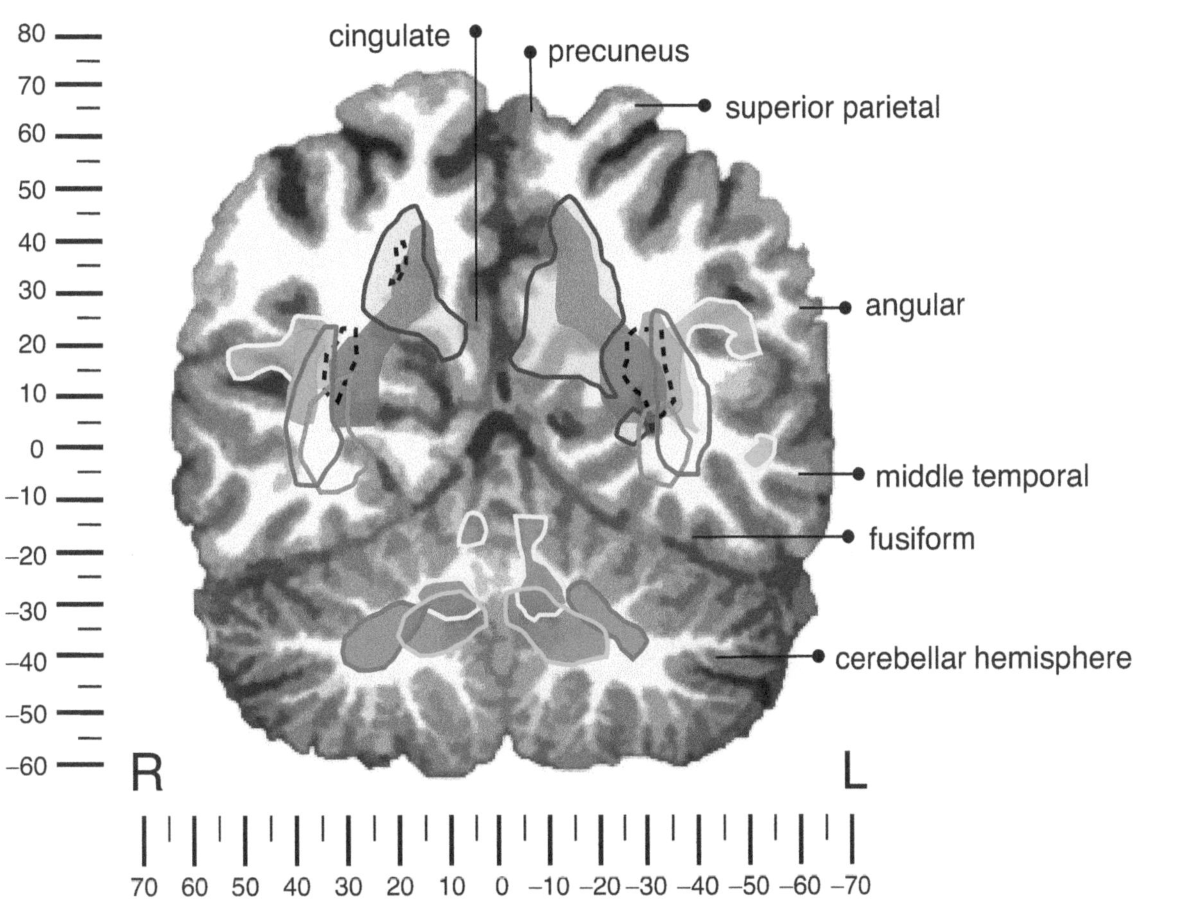

TRACT LEGEND

CC	Corpus Callosum	SCP	Sup. Cerebellar Peduncle	OR	Optic Radiations
ILF	Inf. Longitudinal Fasc.	ICP	Inf. Cerebellar Peduncle	IC	Internal Capsule
AF-P	Arcuate (post. segment)	CPC	Cortico-Ponto-Cerebellar Tract	Cing	Cingulum
IFOF	Inf. Fronto-Occipital Fasc.				

Coronal slice –54

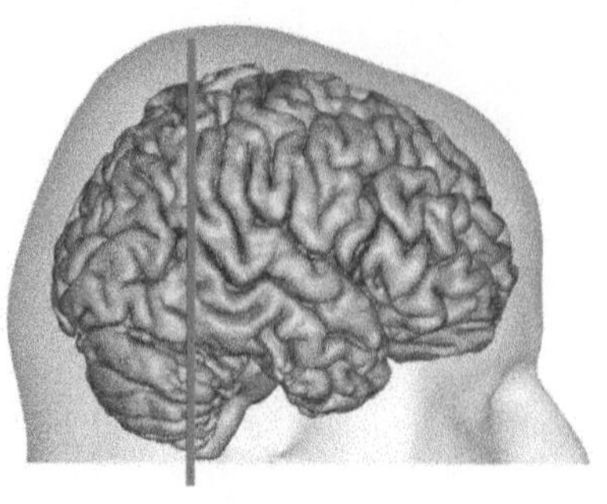

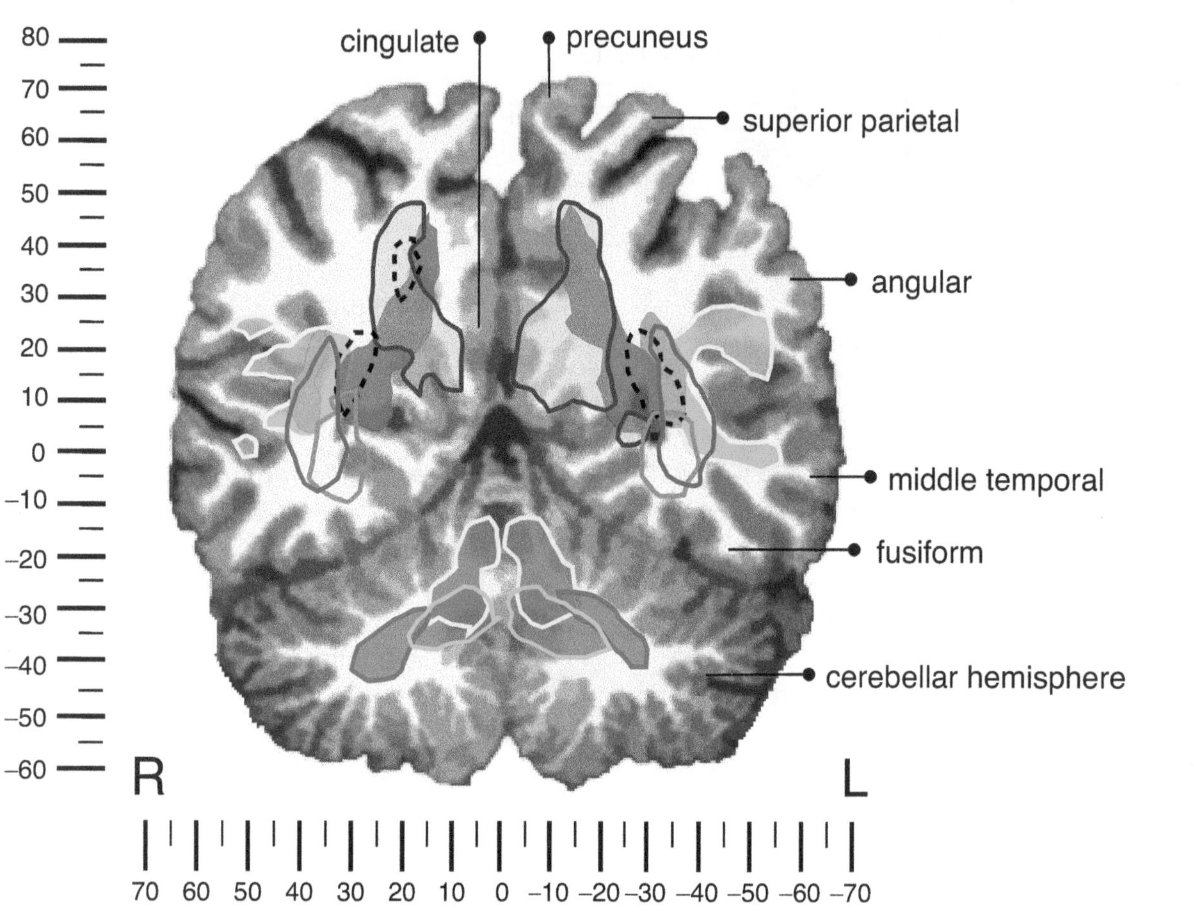

TRACT LEGEND

CC	Corpus Callosum	SCP	Sup. Cerebellar Peduncle	OR	Optic Radiations
ILF	Inf. Longitudinal Fasc.	ICP	Inf. Cerebellar Peduncle	IC	Internal Capsule
AF-P	Arcuate (post. segment)	CPC	Cortico-Ponto-Cerebellar Tract	Cing	Cingulum
IFOF	Inf. Fronto-Occipital Fasc.				

Coronal slice –52

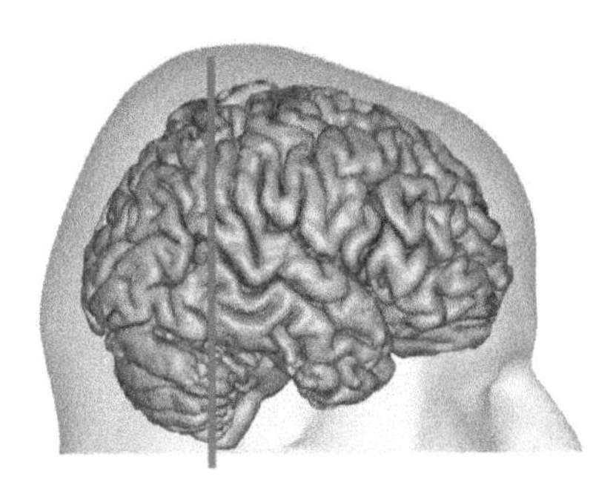

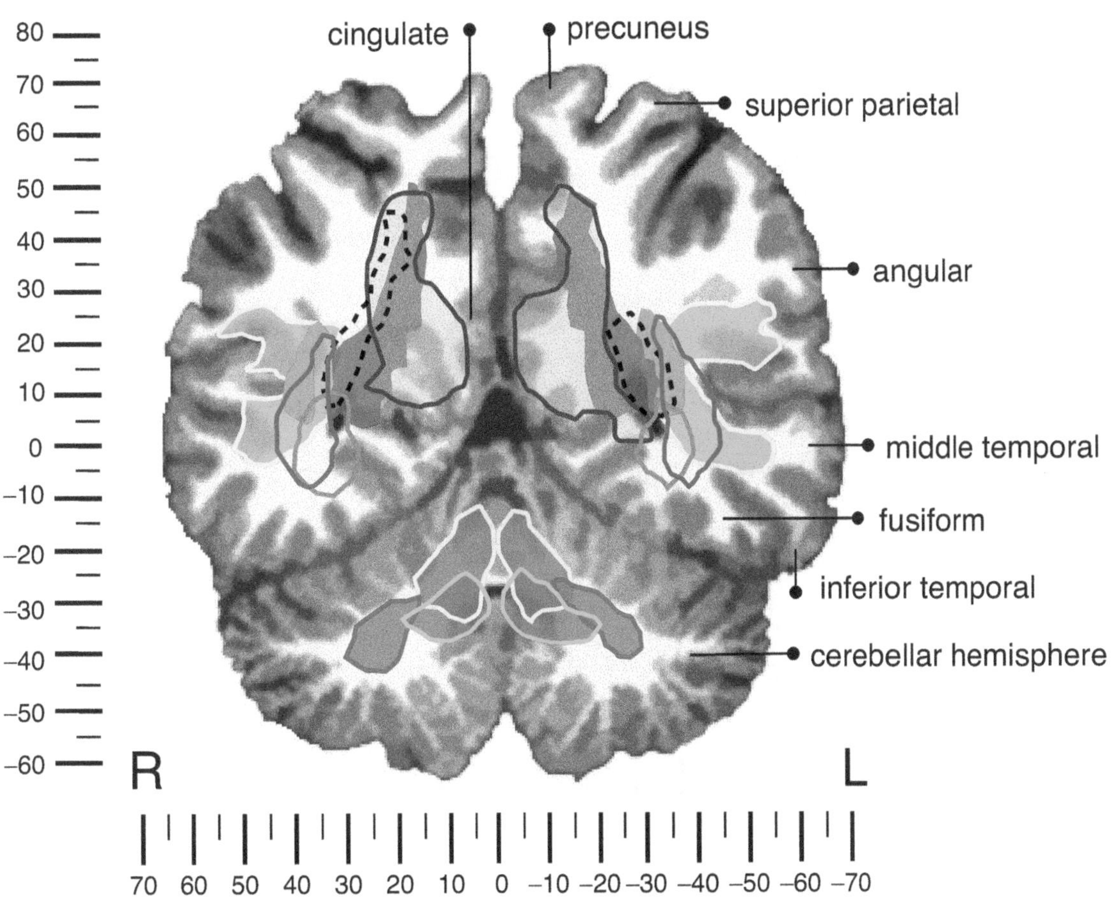

TRACT LEGEND

CC	Corpus Callosum	SCP	Sup. Cerebellar Peduncle	OR	Optic Radiations
ILF	Inf. Longitudinal Fasc.	ICP	Inf. Cerebellar Peduncle	IC	Internal Capsule
AF-P	Arcuate (post. segment)	CPC	Cortico-Ponto-Cerebellar Tract	Cing	Cingulum
IFOF	Inf. Fronto-Occipital Fasc.				

Coronal slice –50

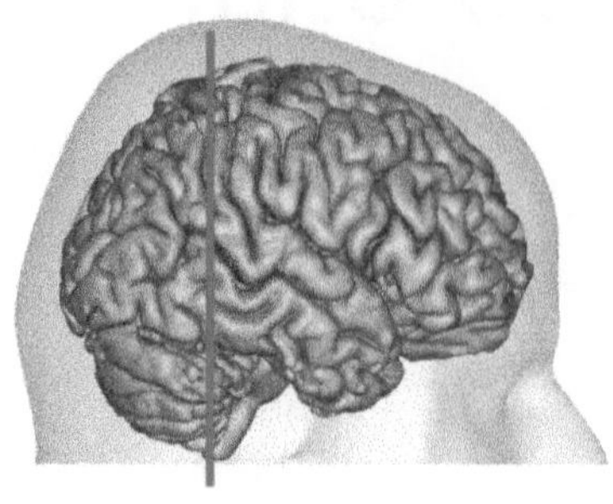

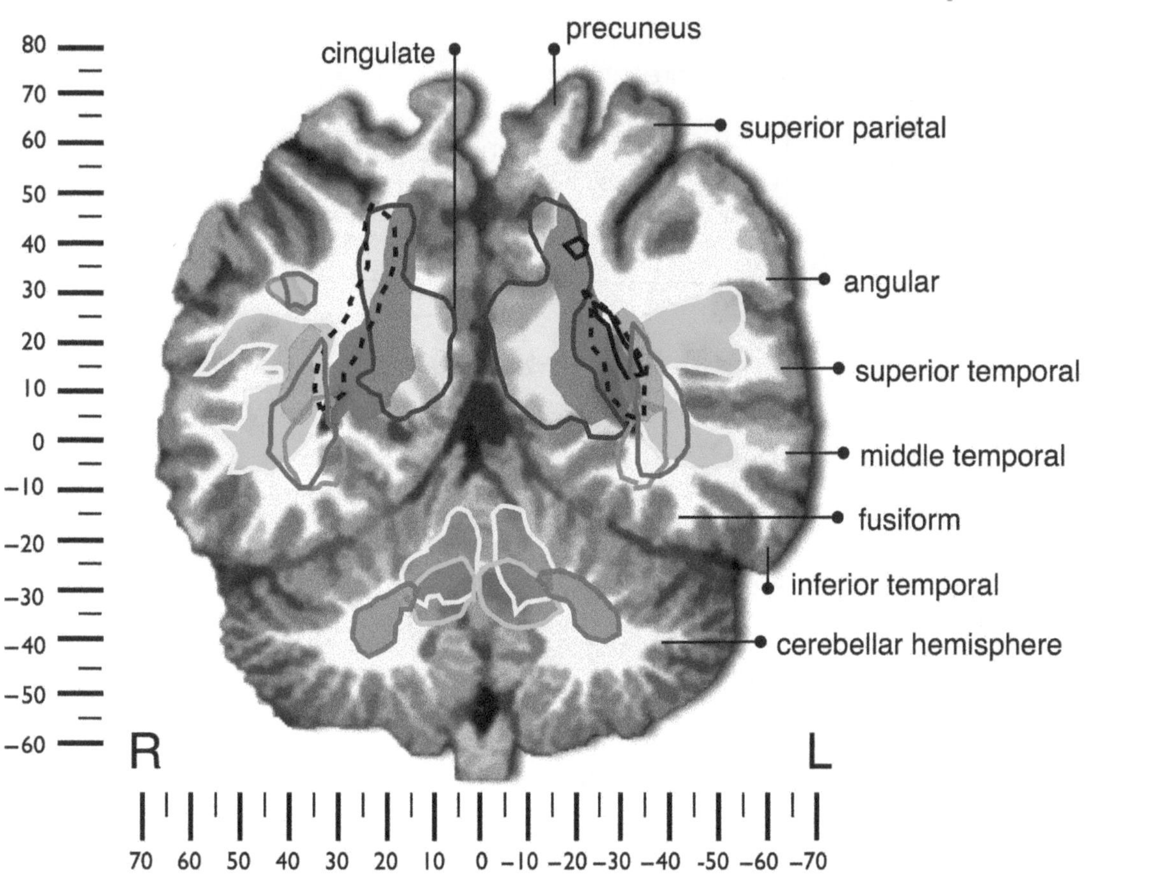

TRACT LEGEND

CC	Corpus Callosum	SCP	Sup. Cerebellar Peduncle	OR	Optic Radiations
ILF	Inf. Longitudinal Fasc.	ICP	Inf. Cerebellar Peduncle	IC	Internal Capsule
AF-P	Arcuate (post. segment)	CPC	Cortico-Ponto-Cerebellar Tract	Cing	Cingulum
IFOF	Inf. Fronto-Occipital Fasc.	SLF II	Sup. Longitudinal Fasc. II	CST	Cortico-Spinal Tract
AF-A	Arcuate (ant. segment)				

Coronal slice –48

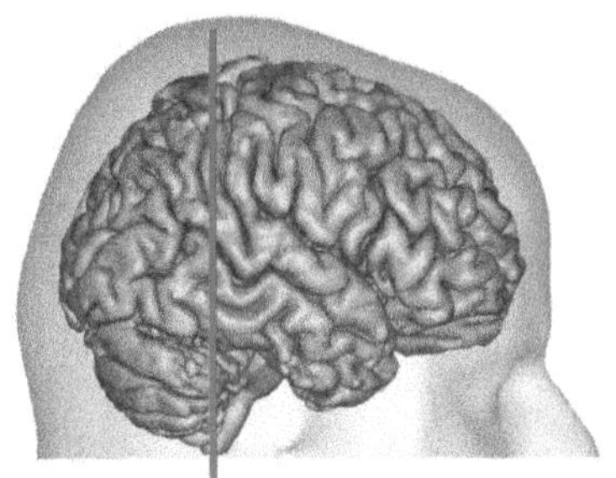

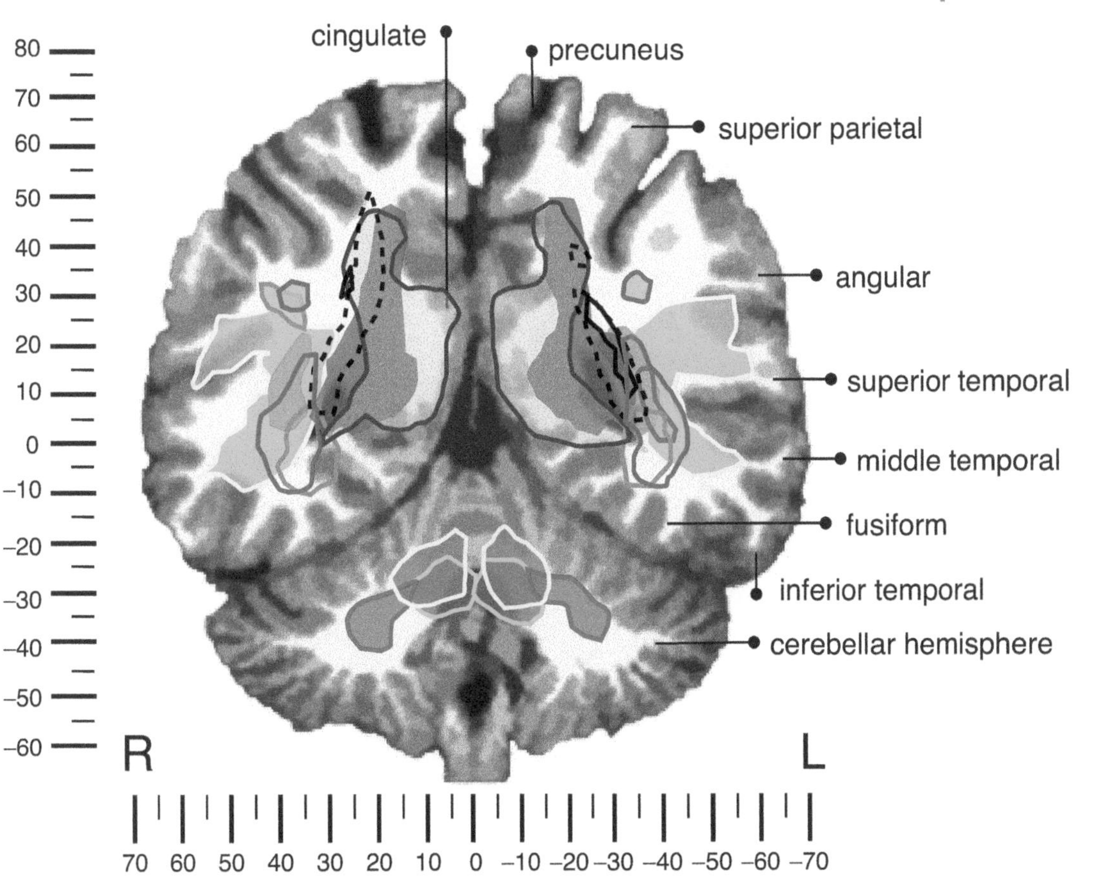

TRACT LEGEND

CC	Corpus Callosum	SCP	Sup. Cerebellar Peduncle	OR	Optic Radiations
ILF	Inf. Longitudinal Fasc.	ICP	Inf. Cerebellar Peduncle	IC	Internal Capsule
AF-P	Arcuate (post. segment)	CPC	Cortico-Ponto-Cerebellar Tract	Cing	Cingulum
AF-L	Arcuate (long segment)	SLF II	Sup. Longitudinal Fasc. II	CST	Cortico-Spinal Tract
AF-A	Arcuate (ant. segment)	IFOF	Inf. Fronto-Occipital Fasc.		

Coronal slice –46

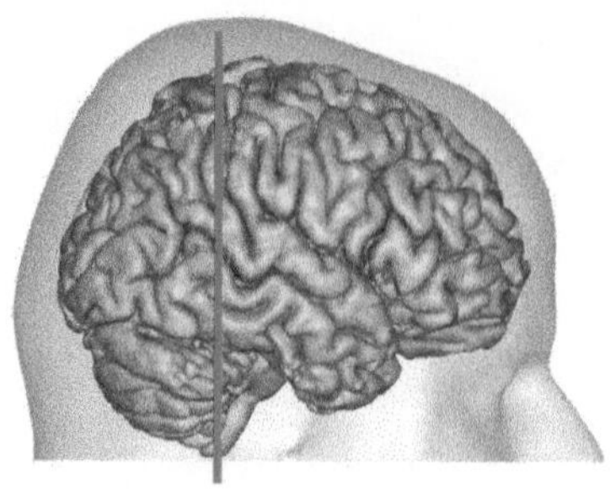

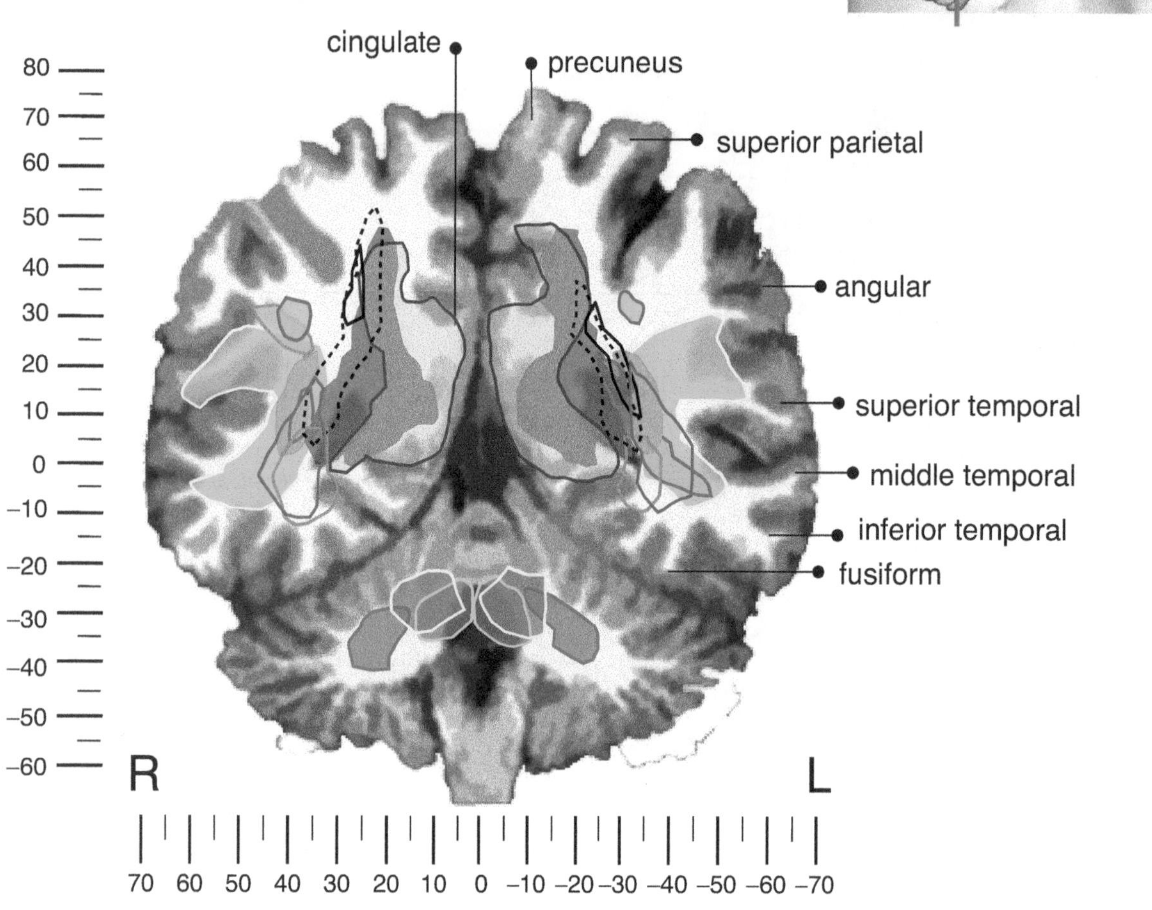

TRACT LEGEND

CC	Corpus Callosum	SCP	Sup. Cerebellar Peduncle	OR	Optic Radiations
ILF	Inf. Longitudinal Fasc.	ICP	Inf. Cerebellar Peduncle	IC	Internal Capsule
AF-P	Arcuate (post. segment)	CPC	Cortico-Ponto-Cerebellar Tract	Cing	Cingulum
AF-L	Arcuate (long segment)	SLF II	Sup. Longitudinal Fasc. II	CST	Cortico-Spinal Tract
AF-A	Arcuate (ant. segment)	IFOF	Inf. Fronto-Occipital Fasc.		

Coronal slice –44

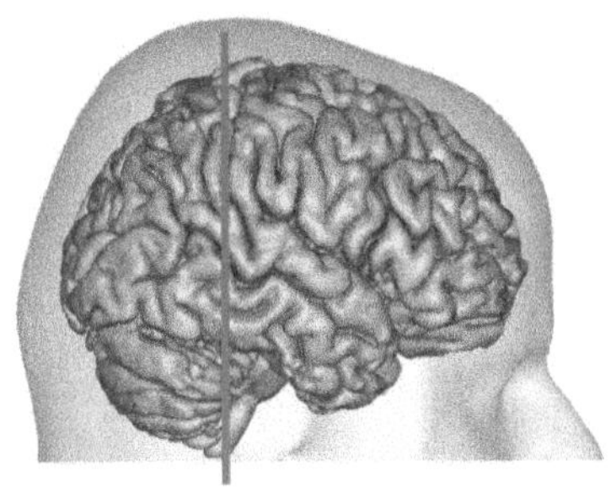

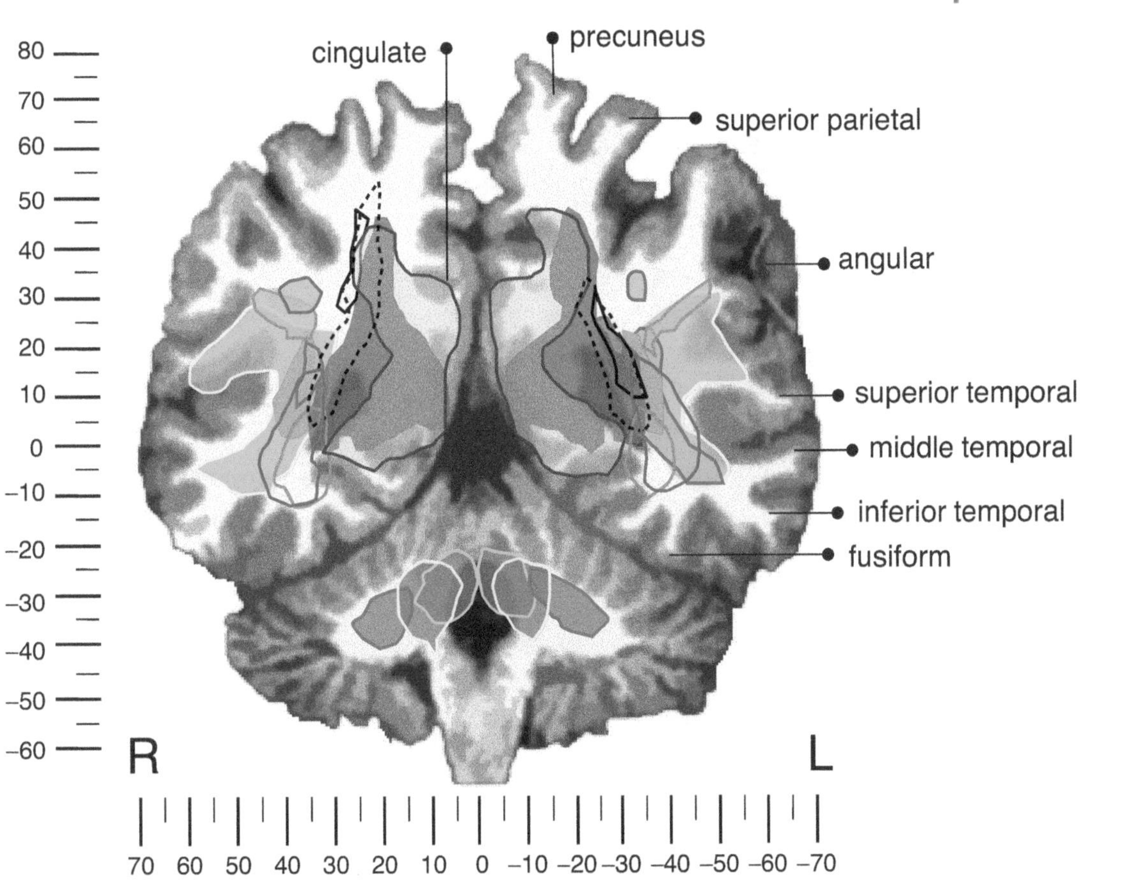

TRACT LEGEND

CC	Corpus Callosum	SCP	Sup. Cerebellar Peduncle	OR	Optic Radiations
ILF	Inf. Longitudinal Fasc.	ICP	Inf. Cerebellar Peduncle	IC	Internal Capsule
AF-P	Arcuate (post. segment)	CPC	Cortico-Ponto-Cerebellar Tract	Cing	Cingulum
AF-L	Arcuate (long segment)	SLF II	Sup. Longitudinal Fasc. II	CST	Cortico-Spinal Tract
AF-A	Arcuate (ant. segment)	IFOF	Inf. Fronto-Occipital Fasc.		

Coronal slice –42

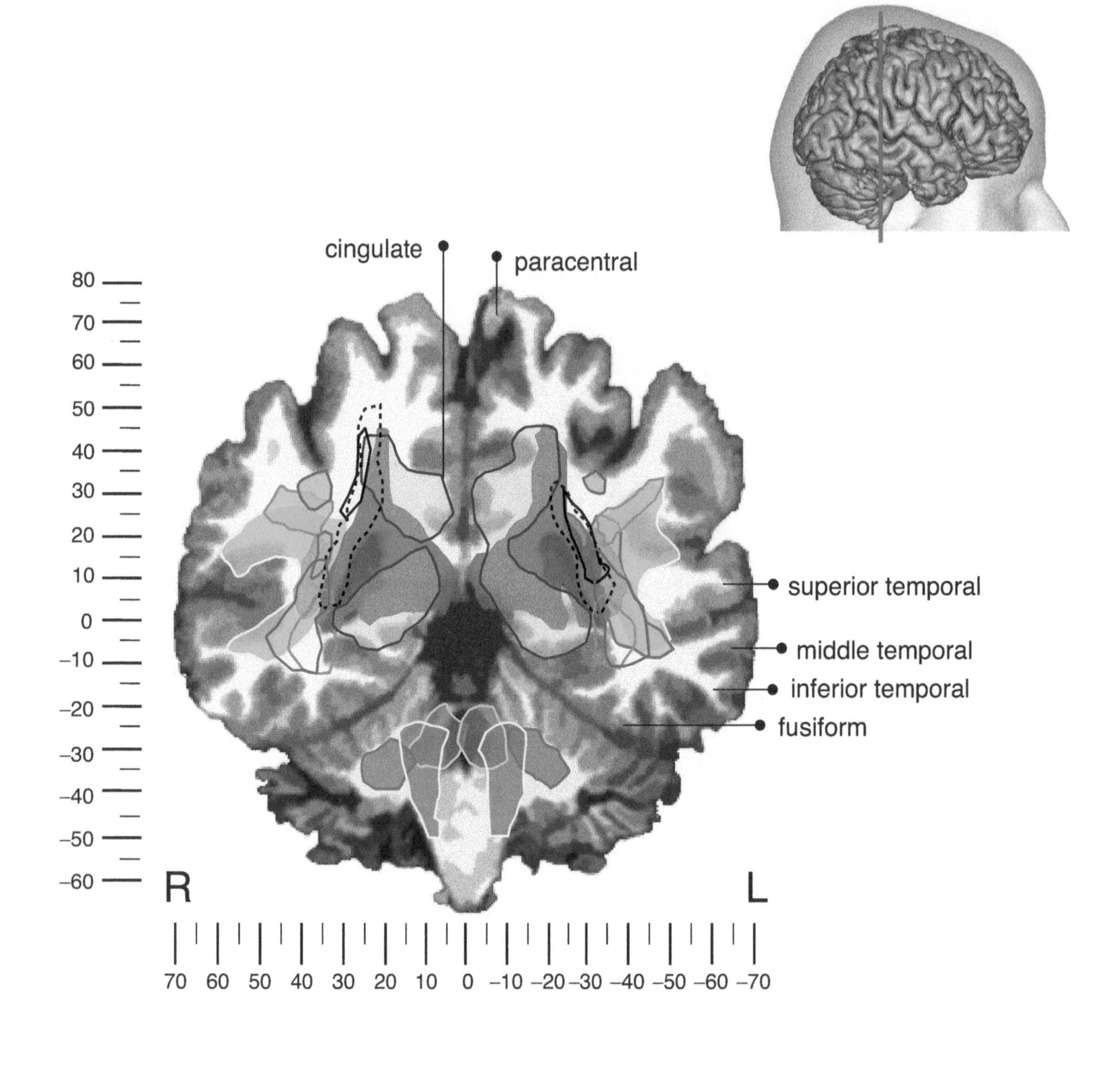

TRACT LEGEND

CC	Corpus Callosum	SCP	Sup. Cerebellar Peduncle	OR	Optic Radiations
ILF	Inf. Longitudinal Fasc.	ICP	Inf. Cerebellar Peduncle	IC	Internal Capsule
AF-P	Arcuate (post. segment)	CPC	Cortico-Ponto-Cerebellar Tract	Cing	Cingulum
AF-L	Arcuate (long segment)	SLF II	Sup. Longitudinal Fasc. II	CST	Cortico-Spinal Tract
AF-A	Arcuate (ant. segment)	IFOF	Inf. Fronto-Occipital Fasc.		

Coronal slice –40

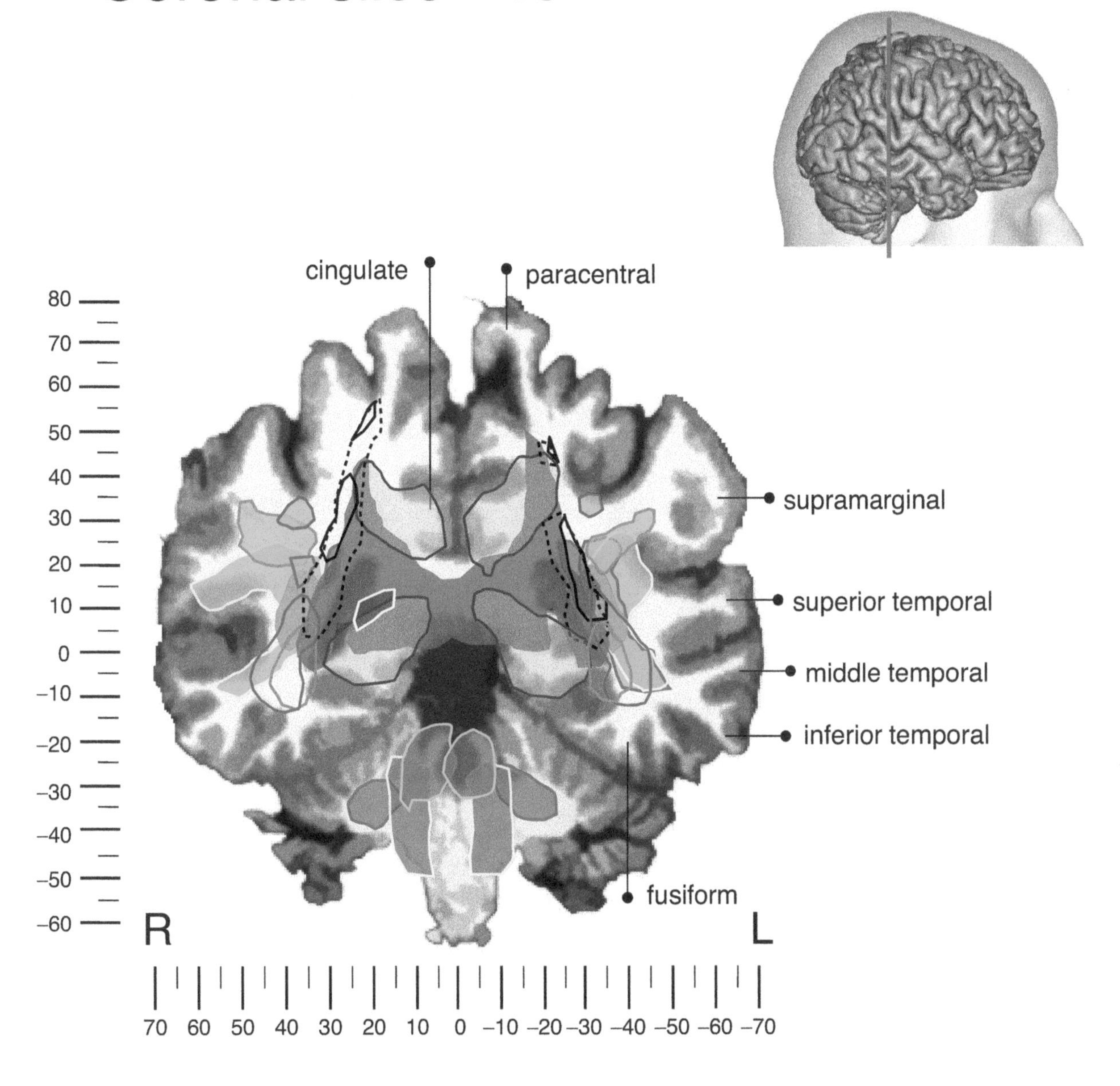

TRACT LEGEND

CC	Corpus Callosum	SCP	Sup. Cerebellar Peduncle	OR	Optic Radiations
ILF	Inf. Longitudinal Fasc.	ICP	Inf. Cerebellar Peduncle	IC	Internal Capsule
AF-P	Arcuate (post. segment)	CPC	Cortico-Ponto-Cerebellar Tract	Cing	Cingulum
AF-L	Arcuate (long segment)	SLF II	Sup. Longitudinal Fasc. II	CST	Cortico-Spinal Tract
AF-A	Arcuate (ant. segment)	IFOF	Inf. Fronto-Occipital Fasc.	Fx	Fornix

Coronal slice –38

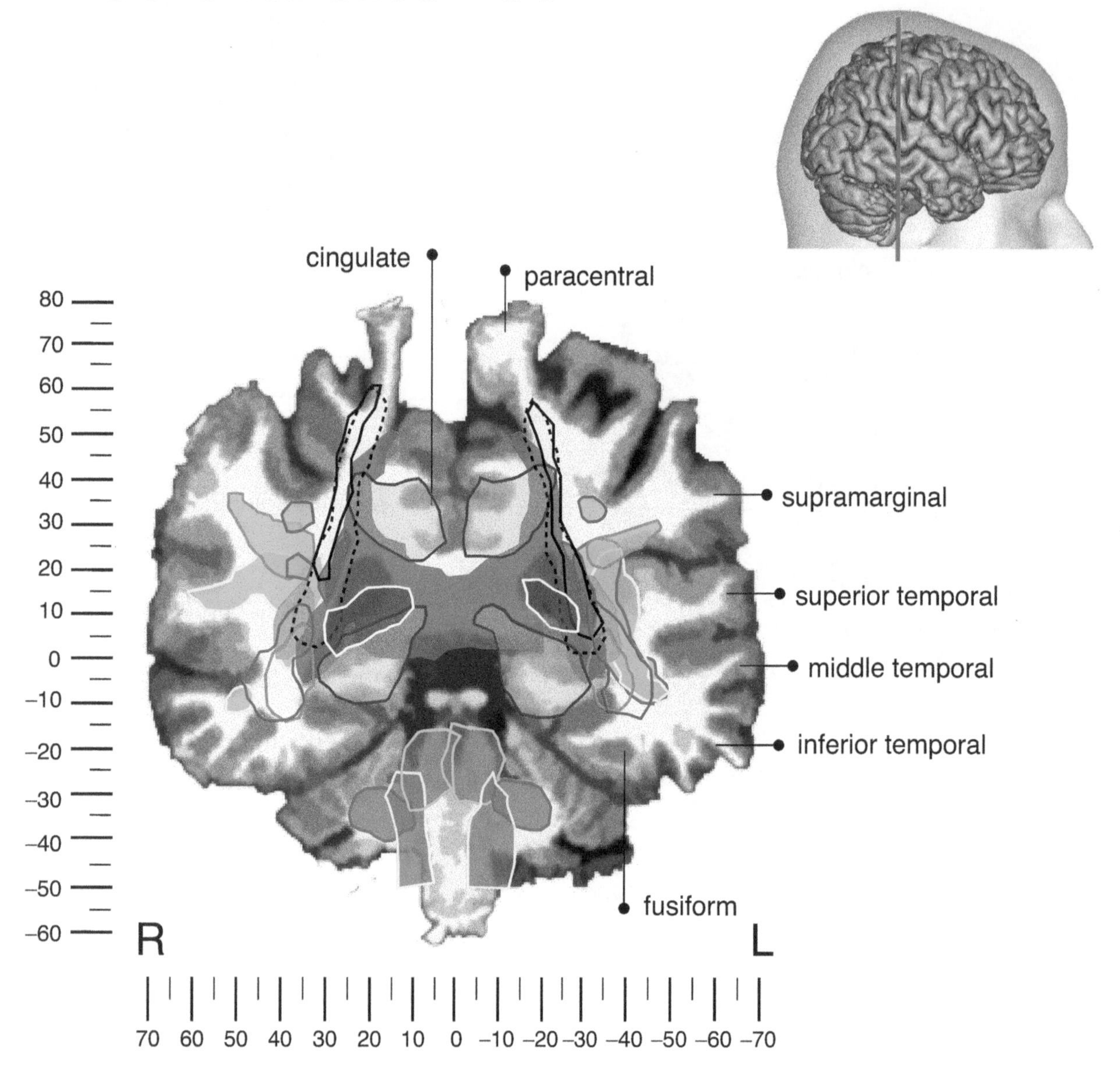

TRACT LEGEND

CC	Corpus Callosum	SCP	Sup. Cerebellar Peduncle	OR	Optic Radiations
ILF	Inf. Longitudinal Fasc.	ICP	Inf. Cerebellar Peduncle	IC	Internal Capsule
AF-P	Arcuate (post. segment)	CPC	Cortico-Ponto-Cerebellar Tract	Cing	Cingulum
AF-L	Arcuate (long segment)	SLF II	Sup. Longitudinal Fasc. II	CST	Cortico-Spinal Tract
AF-A	Arcuate (ant. segment)	IFOF	Inf. Fronto-Occipital Fasc.	Fx	Fornix

Coronal slice –36

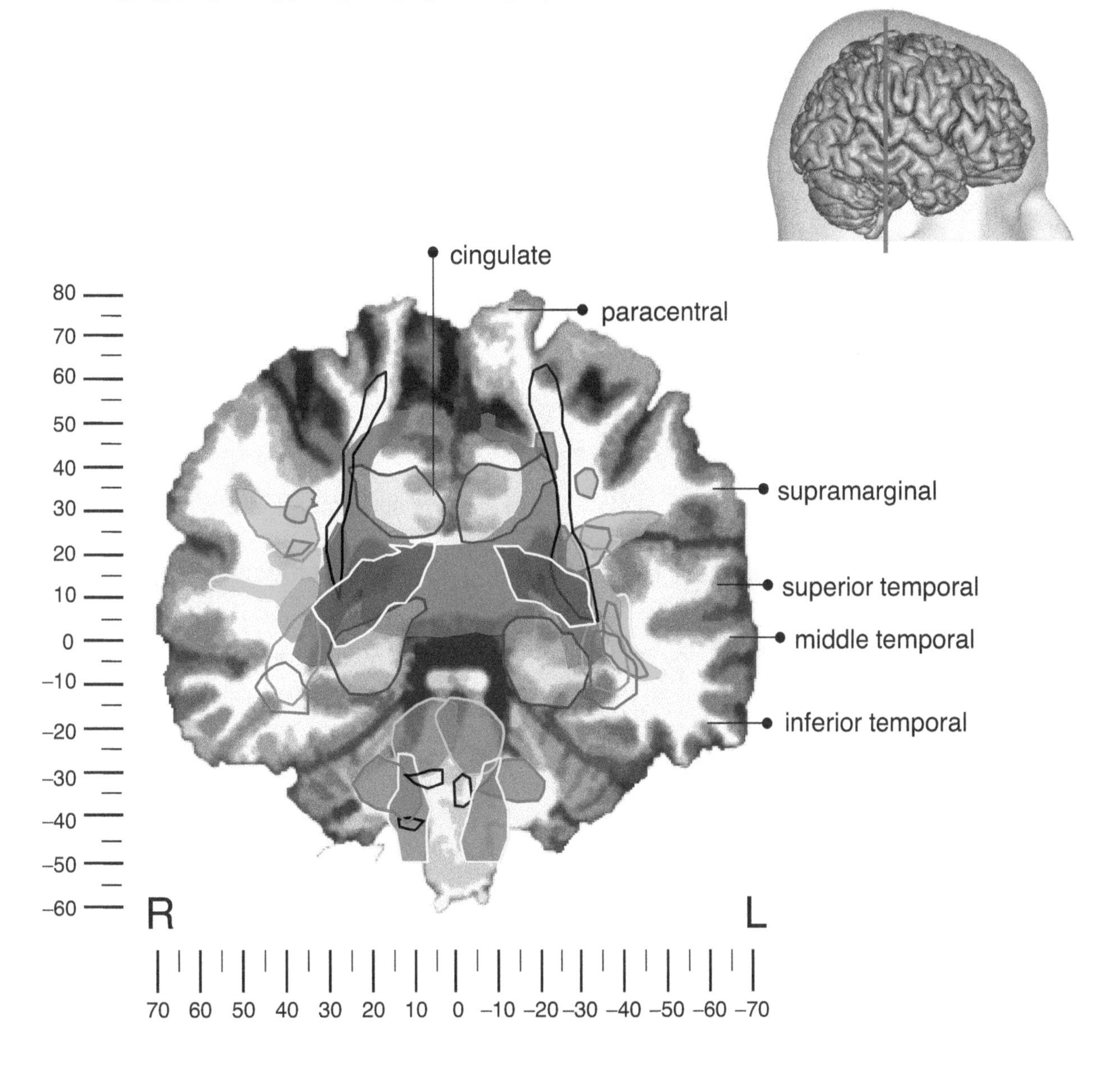

TRACT LEGEND

CC	Corpus Callosum	SCP	Sup. Cerebellar Peduncle	OR	Optic Radiations
ILF	Inf. Longitudinal Fasc.	ICP	Inf. Cerebellar Peduncle	Cing	Cingulum
AF-P	Arcuate (post. segment)	CPC	Cortico-Ponto-Cerebellar Tract	CST	Cortico-Spinal Tract / IC
AF-L	Arcuate (long segment)	SLF II	Sup. Longitudinal Fasc. II	Fx	Fornix
AF-A	Arcuate (ant. segment)	IFOF	Inf. Fronto-Occipital Fasc.		

Coronal slice –34

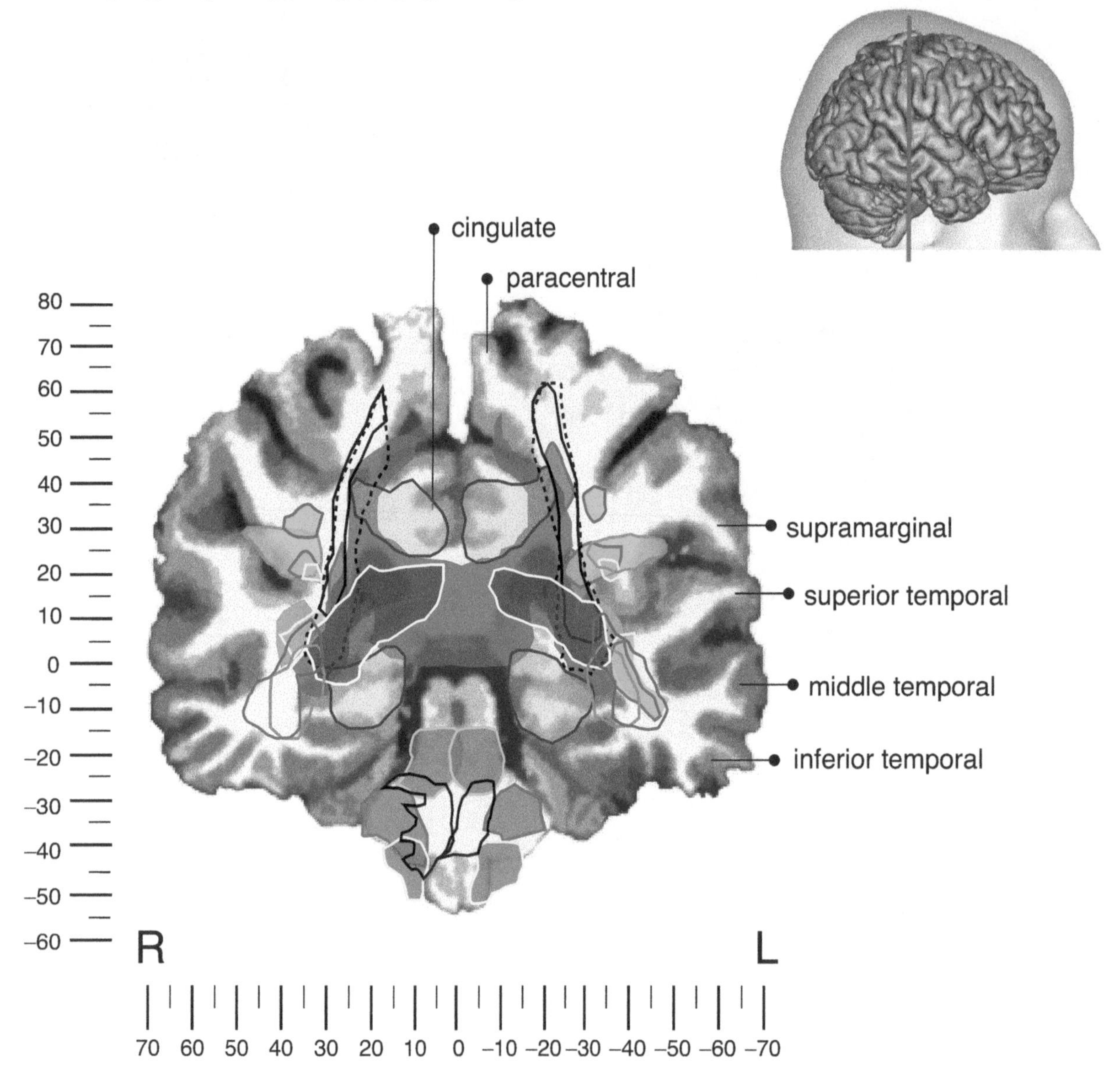

TRACT LEGEND

CC	Corpus Callosum	SCP	Sup. Cerebellar Peduncle	OR	Optic Radiations
ILF	Inf. Longitudinal Fasc.	ICP	Inf. Cerebellar Peduncle	IC	Internal Capsule
AF-P	Arcuate (post. segment)	CPC	Cortico-Ponto-Cerebellar Tract	Cing	Cingulum
AF-L	Arcuate (long segment)	SLF II	Sup. Longitudinal Fasc. II	CST	Cortico-Spinal Tract
AF-A	Arcuate (ant. segment)	IFOF	Inf. Fronto-Occipital Fasc.	Fx	Fornix

Coronal slice –32

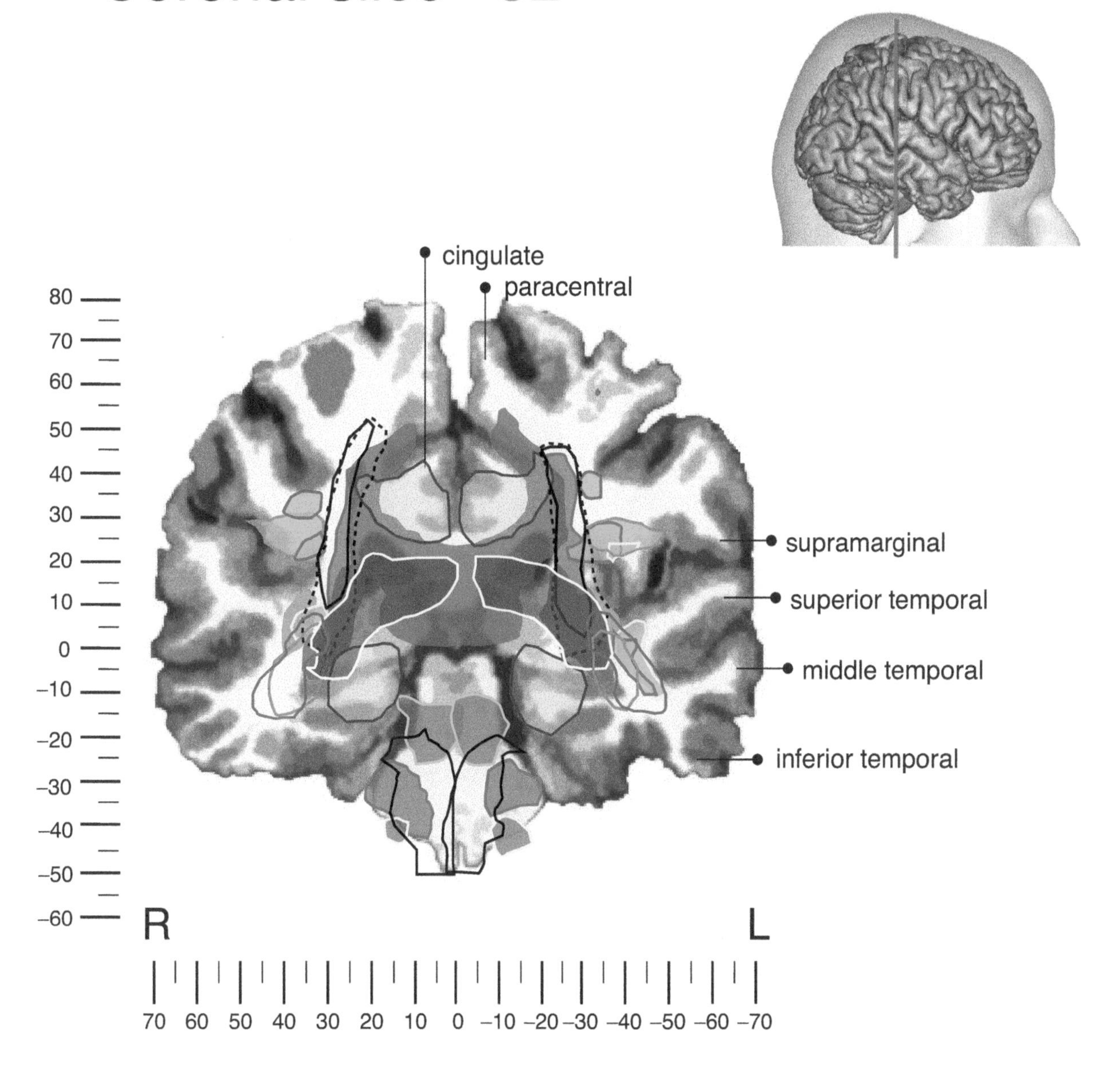

TRACT LEGEND

CC	Corpus Callosum	SCP	Sup. Cerebellar Peduncle	OR	Optic Radiations
ILF	Inf. Longitudinal Fasc.	ICP	Inf. Cerebellar Peduncle	IC	Internal Capsule
AF-P	Arcuate (post. segment)	CPC	Cortico-Ponto-Cerebellar Tract	Cing	Cingulum
AF-L	Arcuate (long segment)	SLF II	Sup. Longitudinal Fasc. II	CST	Cortico-Spinal Tract
AF-A	Arcuate (ant. segment)	IFOF	Inf. Fronto-Occipital Fasc.	Fx	Fornix

Coronal slice –30

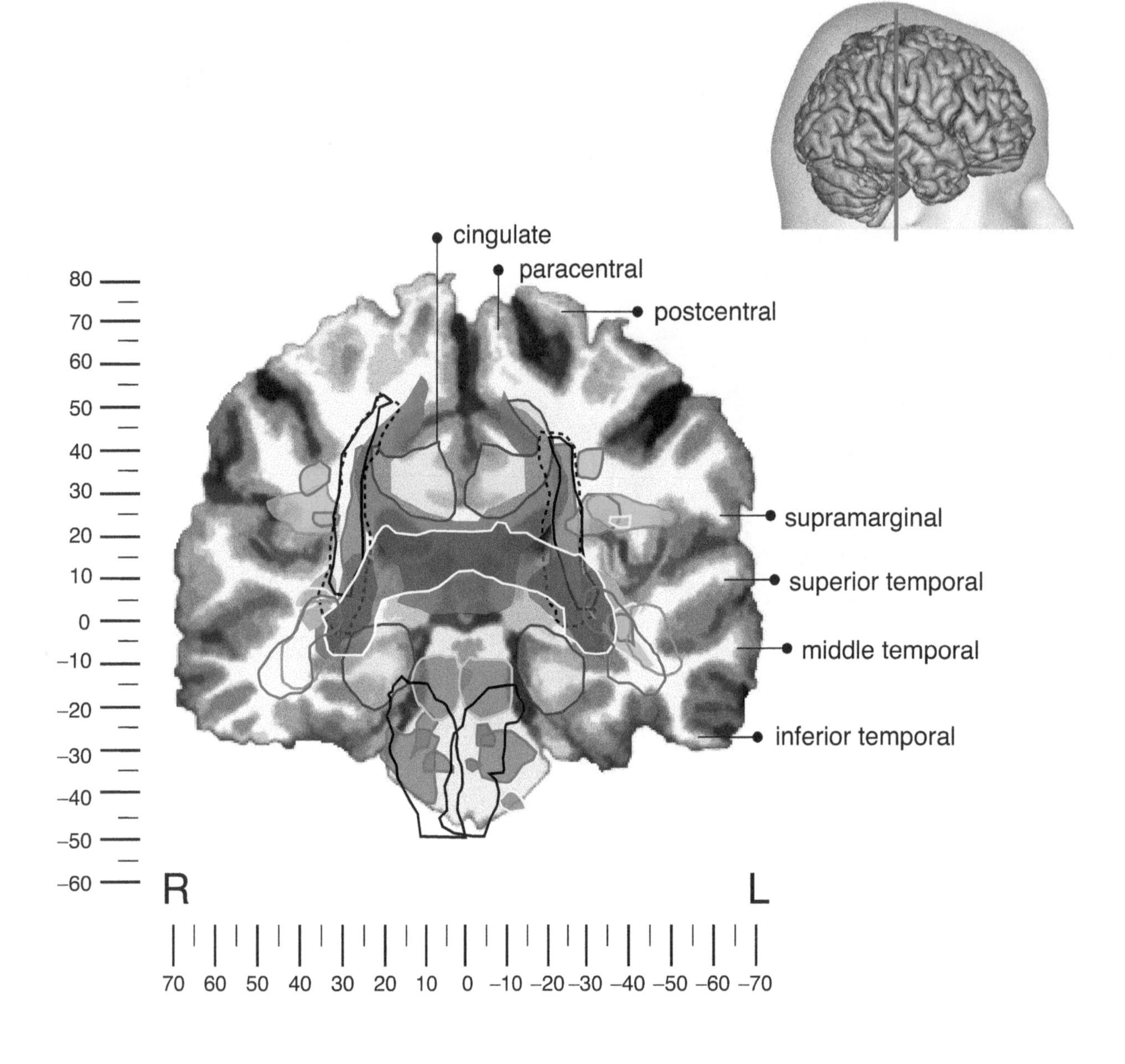

TRACT LEGEND

CC	Corpus Callosum	SCP	Sup. Cerebellar Peduncle	OR	Optic Radiations
ILF	Inf. Longitudinal Fasc.	ICP	Inf. Cerebellar Peduncle	IC	Internal Capsule
AF-P	Arcuate (post. segment)	CPC	Cortico-Ponto-Cerebellar Tract	Cing	Cingulum
AF-L	Arcuate (long segment)	SLF II	Sup. Longitudinal Fasc. II	CST	Cortico-Spinal Tract
AF-A	Arcuate (ant. segment)	IFOF	Inf. Fronto-Occipital Fasc.	Fx	Fornix

Coronal slice –28

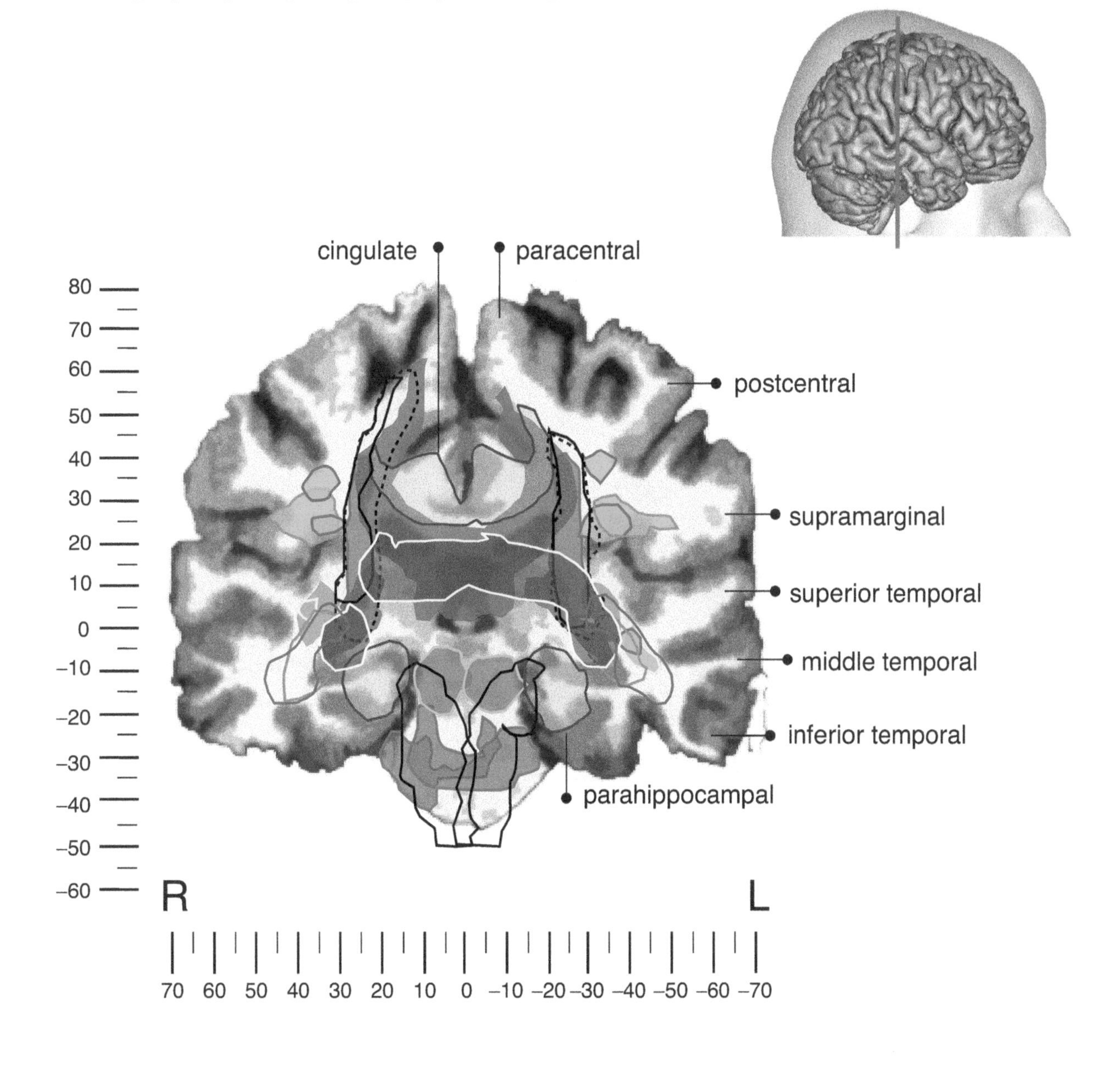

TRACT LEGEND

CC	Corpus Callosum	SCP	Sup. Cerebellar Peduncle	OR	Optic Radiations
ILF	Inf. Longitudinal Fasc.	CST	Cortico-Spinal Tract	IC	Internal Capsule
AF-P	Arcuate (post. segment)	CPC	Cortico-Ponto-Cerebellar Tract	Cing	Cingulum
AF-L	Arcuate (long segment)	SLF II	Sup. Longitudinal Fasc. II	Fx	Fornix
AF-A	Arcuate (ant. segment)	IFOF	Inf. Fronto-Occipital Fasc.		

Coronal slice –26

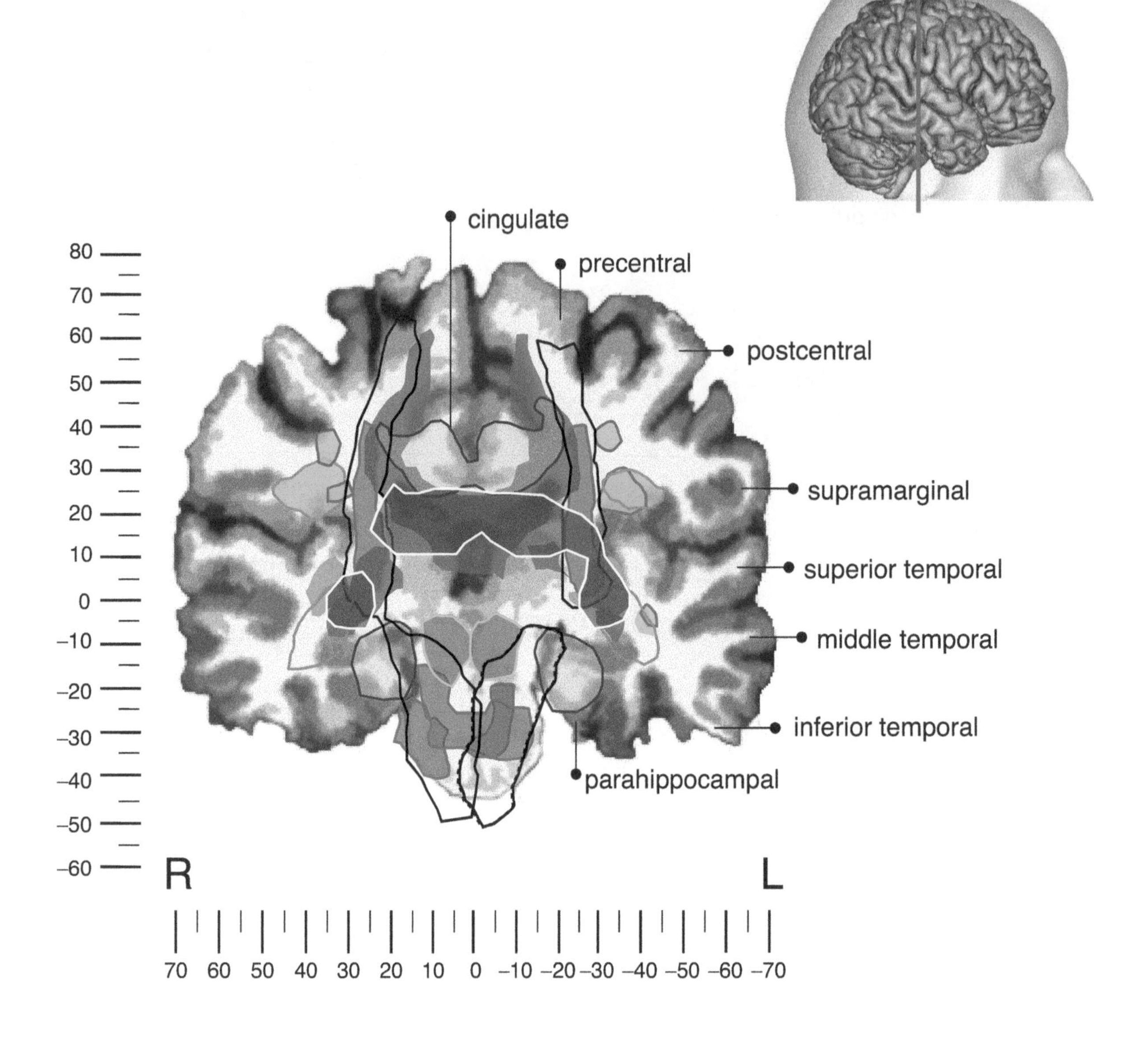

TRACT LEGEND

CC	Corpus Callosum	SCP	Sup. Cerebellar Peduncle	OR	Optic Radiations
ILF	Inf. Longitudinal Fasc.	CST	Cortico-Spinal Tract / IC	Cing	Cingulum
AF-P	Arcuate (post. segment)	CPC	Cortico-Ponto-Cerebellar Tract	Fx	Fornix
AF-L	Arcuate (long segment)	SLF II	Sup. Longitudinal Fasc. II		
AF-A	Arcuate (ant. segment)	IFOF	Inf. Fronto-Occipital Fasc.		

Coronal slice –24

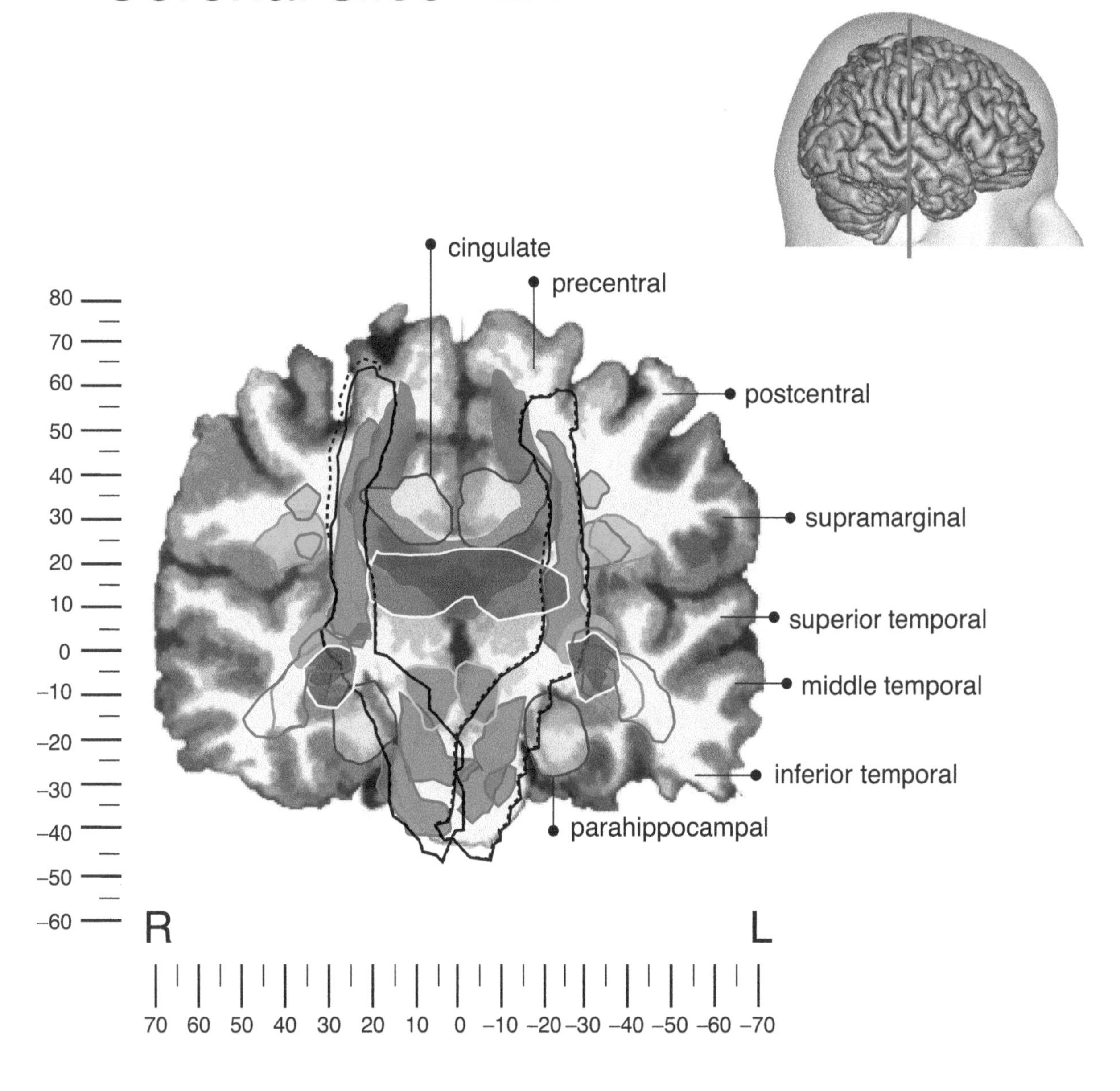

TRACT LEGEND

CC	Corpus Callosum	SCP	Sup. Cerebellar Peduncle	OR	Optic Radiations
ILF	Inf. Longitudinal Fasc.	CST	Cortico-Spinal Tract	IC	Internal Capsule
IFOF	Inf. Fronto-Occipital Fasc.	CPC	Cortico-Ponto-Cerebellar Tract	Cing	Cingulum
AF-L	Arcuate (long segment)	SLF II	Sup. Longitudinal Fasc. II	Fx	Fornix
AF-A	Arcuate (ant. segment)				

Coronal slice –22

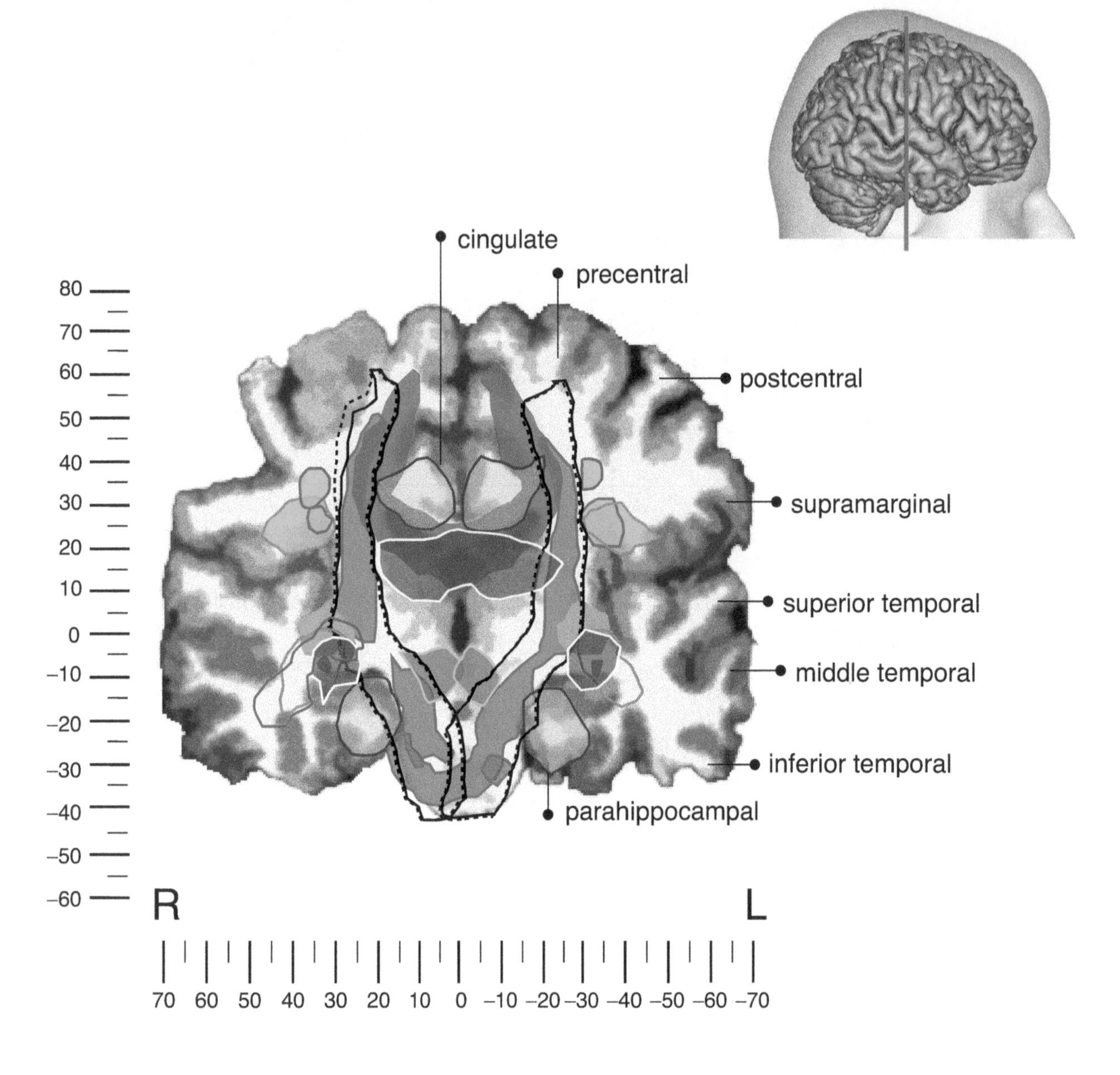

TRACT LEGEND

CC	Corpus Callosum	SCP	Sup. Cerebellar Peduncle	OR	Optic Radiations
ILF	Inf. Longitudinal Fasc.	CST	Cortico-Spinal Tract	IC	Internal Capsule
IFOF	Inf. Fronto-Occipital Fasc.	CPC	Cortico-Ponto-Cerebellar Tract	Cing	Cingulum
AF-L	Arcuate (long segment)	SLF II	Sup. Longitudinal Fasc. II	Fx	Fornix
AF-A	Arcuate (ant. segment)				

Coronal slice –20

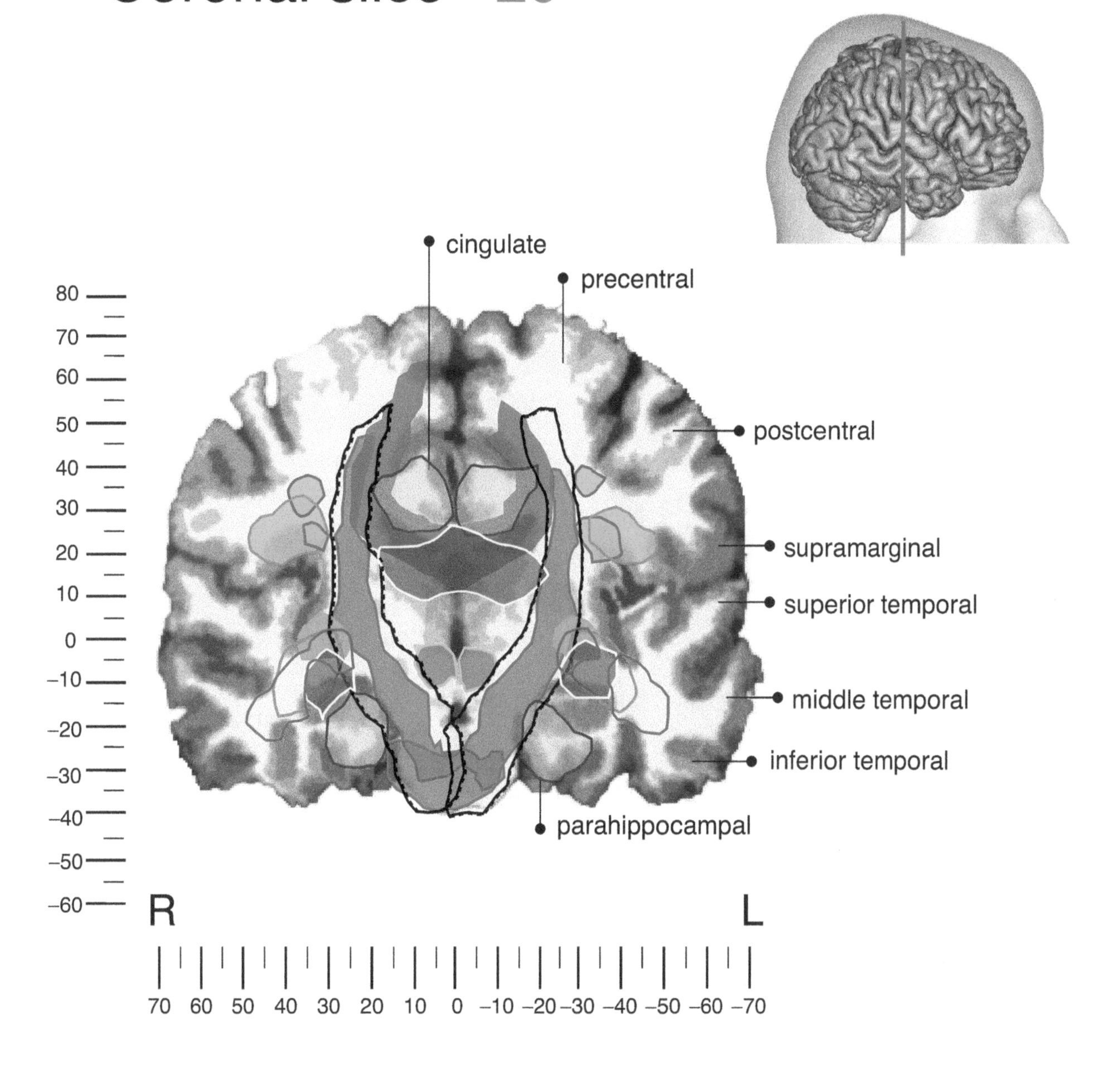

TRACT LEGEND

CC	Corpus Callosum	SCP	Sup. Cerebellar Peduncle	OR	Optic Radiations
ILF	Inf. Longitudinal Fasc.	CST	Cortico-Spinal Tract	IC	Internal Capsule
IFOF	Inf. Fronto-Occipital Fasc.	CPC	Cortico-Ponto-Cerebellar Tract	Cing	Cingulum
AF-L	Arcuate (long segment)	SLF II	Sup. Longitudinal Fasc. II	Fx	Fornix
AF-A	Arcuate (ant. segment)				

Coronal slice −18

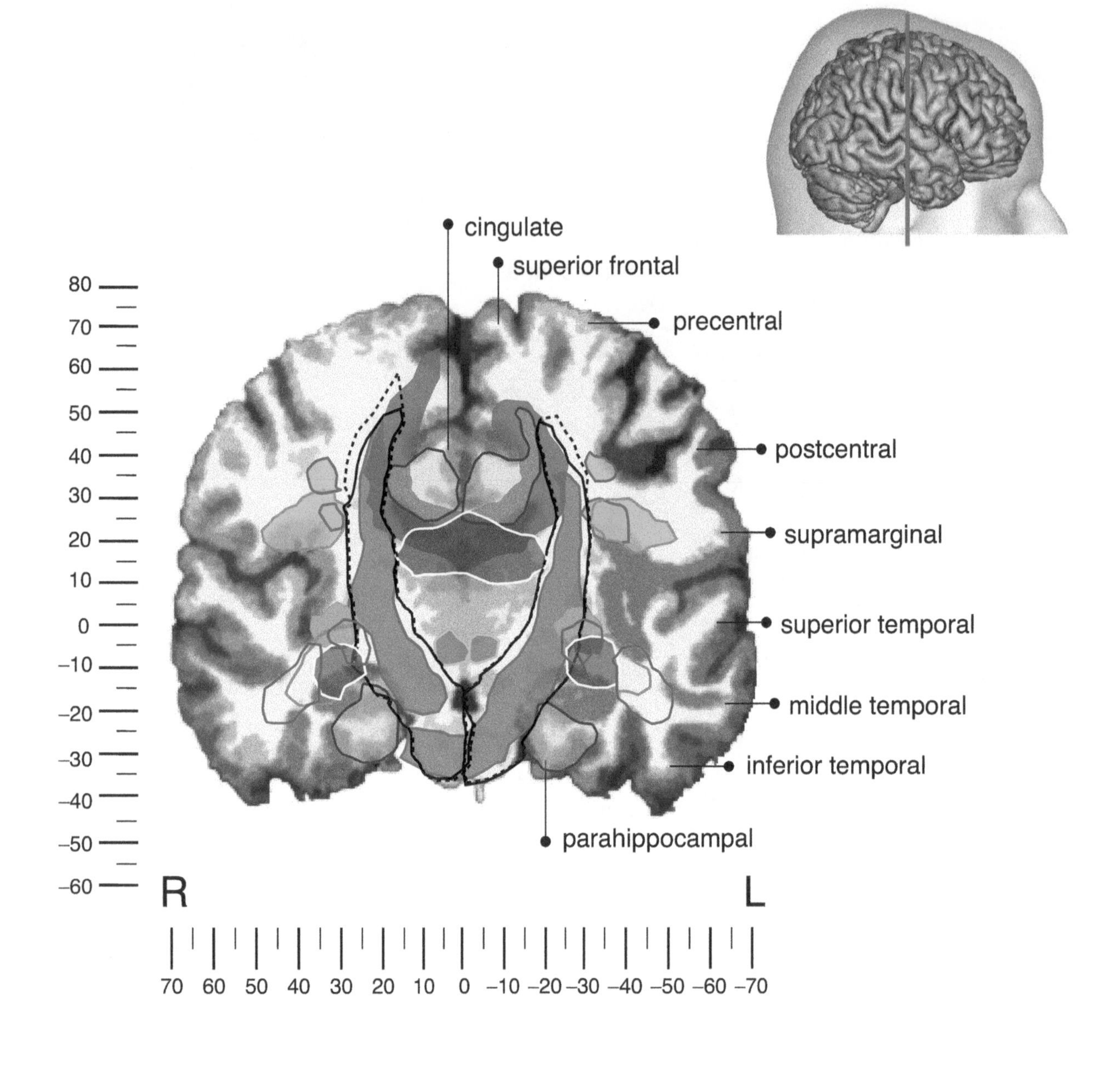

TRACT LEGEND

CC	Corpus Callosum	SCP	Sup. Cerebellar Peduncle	OR	Optic Radiations
ILF	Inf. Longitudinal Fasc.	CST	Cortico-Spinal Tract	IC	Internal Capsule
IFOF	Inf. Fronto-Occipital Fasc.	CPC	Cortico-Ponto-Cerebellar Tract	Cing	Cingulum
AF-L	Arcuate (long segment)	SLF II	Sup. Longitudinal Fasc. II	Fx	Fornix
AF-A	Arcuate (ant. segment)				

Coronal slice –16

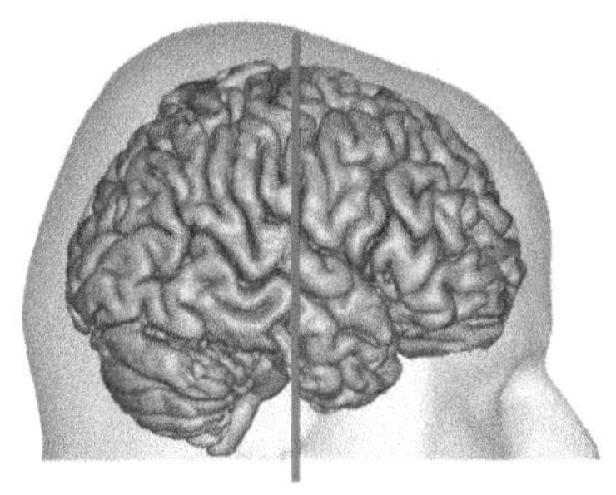

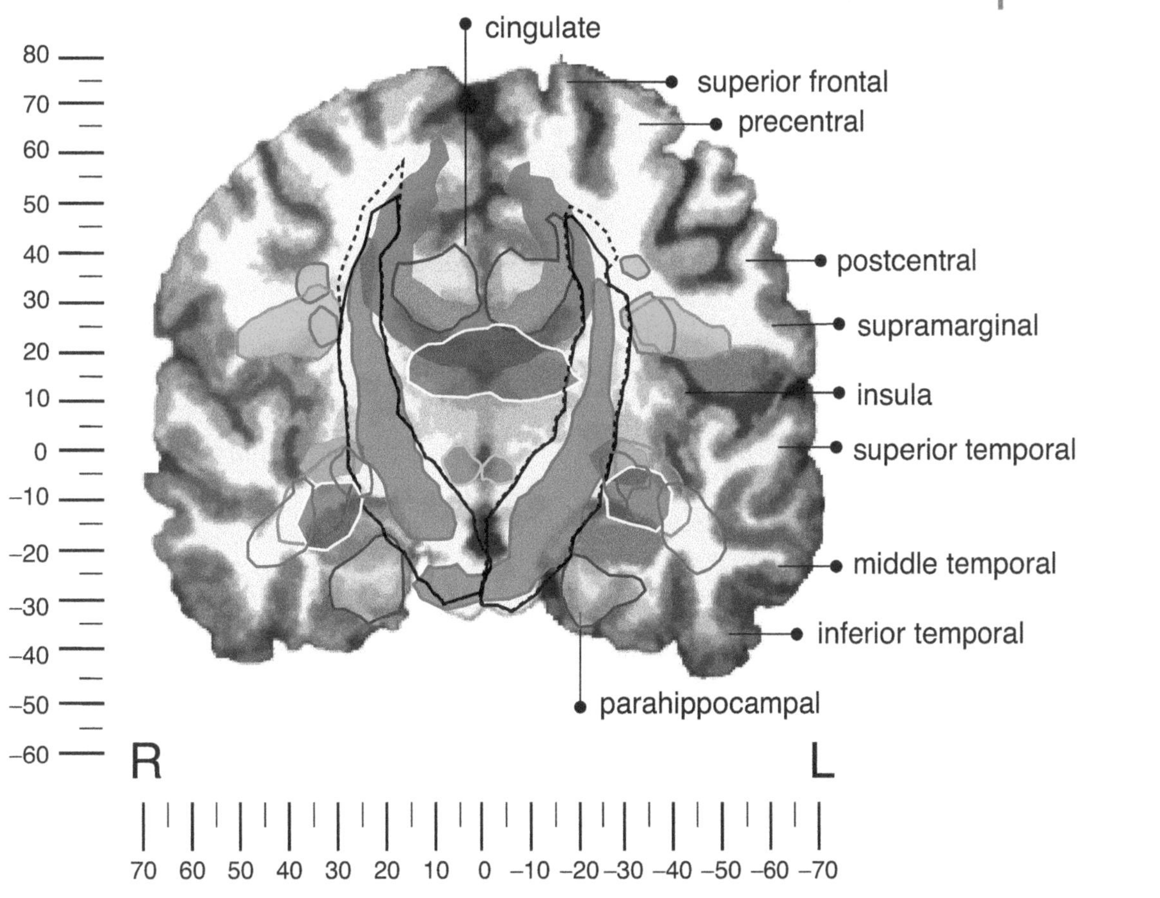

TRACT LEGEND

CC	Corpus Callosum	SCP	Sup. Cerebellar Peduncle	OR	Optic Radiations
ILF	Inf. Longitudinal Fasc.	CST	Cortico-Spinal Tract	IC	Internal Capsule
IFOF	Inf. Fronto-Occipital Fasc.	CPC	Cortico-Ponto-Cerebellar Tract	Cing	Cingulum
AF-L	Arcuate (long segment)	SLF II	Sup. Longitudinal Fasc. II	Fx	Fornix
AF-A	Arcuate (ant. segment)				

Coronal slice –14

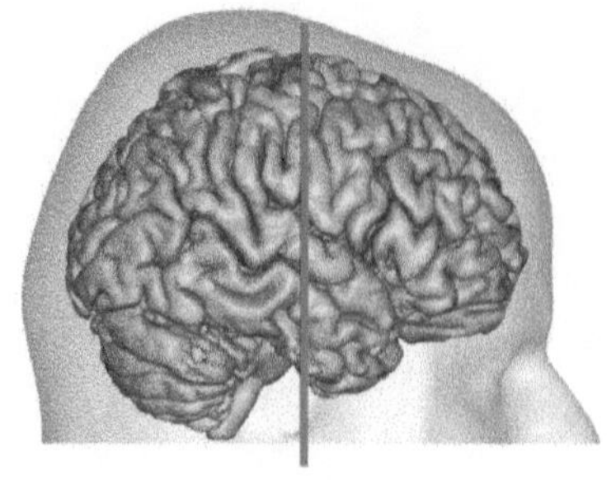

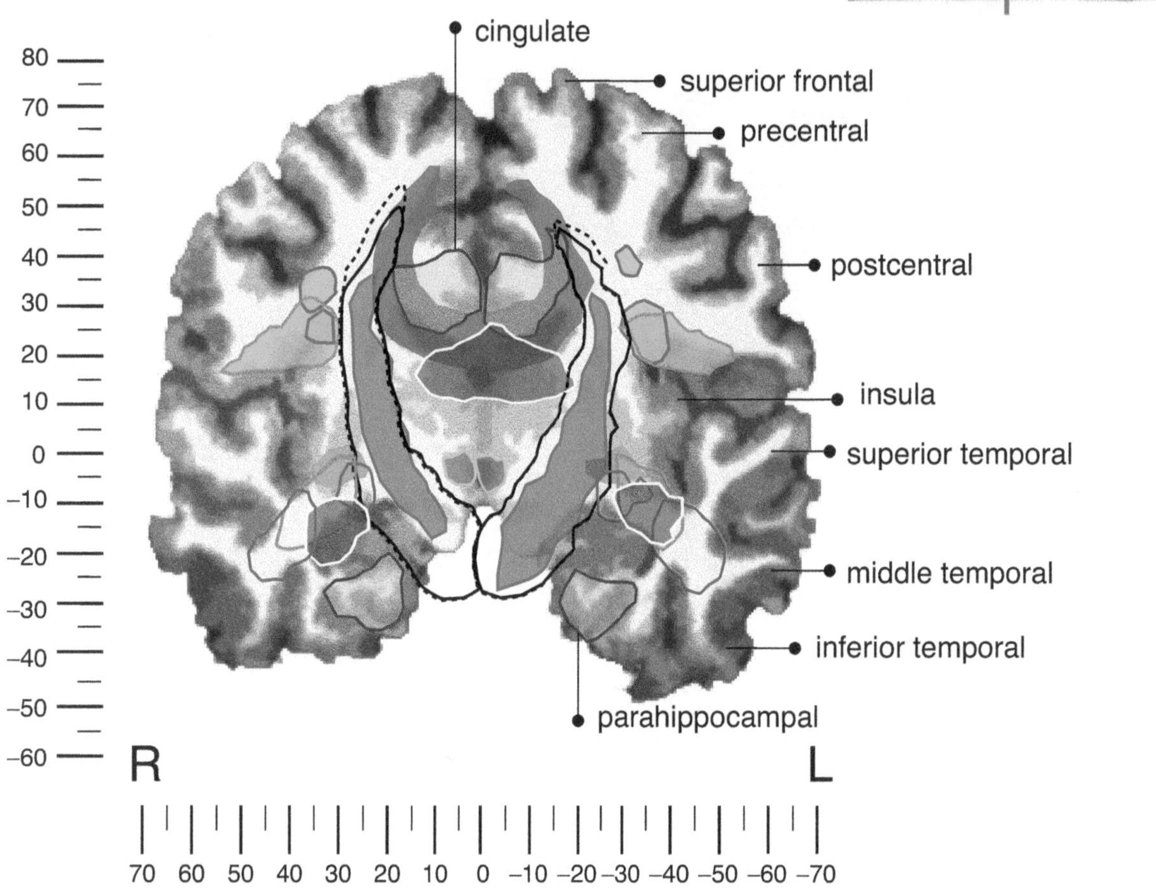

TRACT LEGEND

CC	Corpus Callosum	SCP	Sup. Cerebellar Peduncle	OR	Optic Radiations
ILF	Inf. Longitudinal Fasc.	CST	Cortico-Spinal Tract	IC	Internal Capsule
IFOF	Inf. Fronto-Occipital Fasc.	CPC	Cortico-Ponto-Cerebellar Tract	Cing	Cingulum
AF-L	Arcuate (long segment)	SLF II	Sup. Longitudinal Fasc. II	Fx	Fornix
AF-A	Arcuate (ant. segment)	AC	Anterior Commissure		

Coronal slice −12

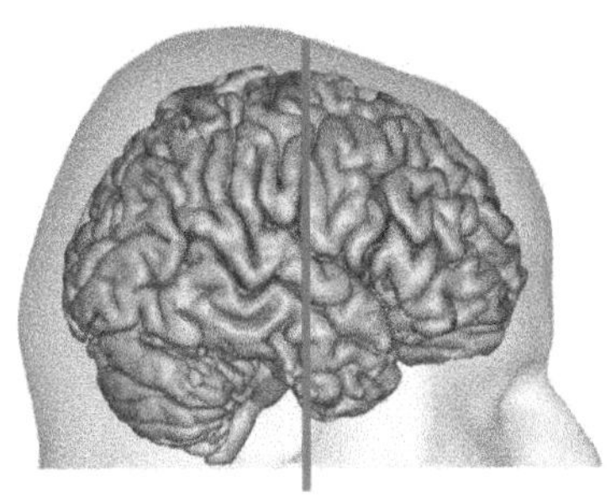

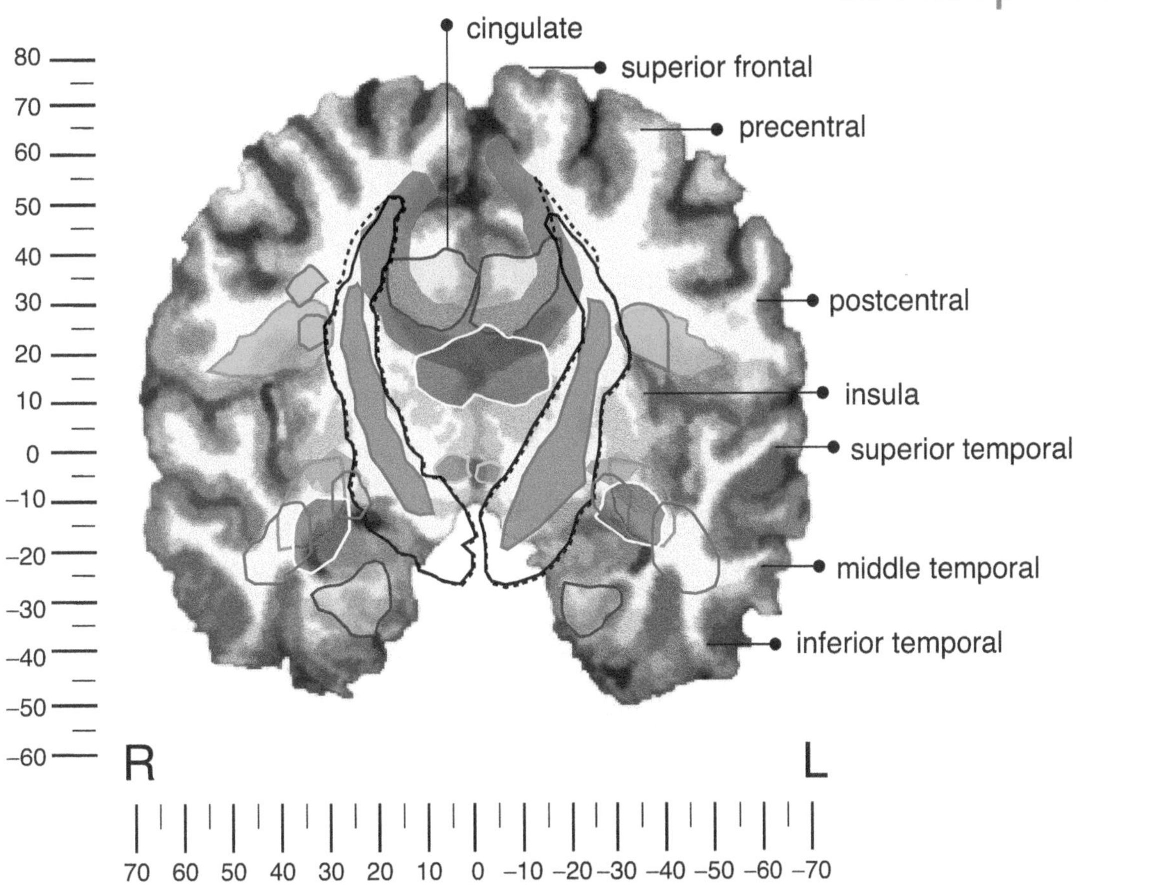

TRACT LEGEND

CC	Corpus Callosum	SCP	Sup. Cerebellar Peduncle	OR	Optic Radiations
ILF	Inf. Longitudinal Fasc.	CST	Cortico-Spinal Tract	IC	Internal Capsule
IFOF	Inf. Fronto-Occipital Fasc.	CPC	Cortico-Ponto-Cerebellar Tract	Cing	Cingulum
AF-L	Arcuate (long segment)	SLF II	Sup. Longitudinal Fasc. II	Fx	Fornix
AF-A	Arcuate (ant. segment)	AC	Anterior Commissure		

Coronal slice –10

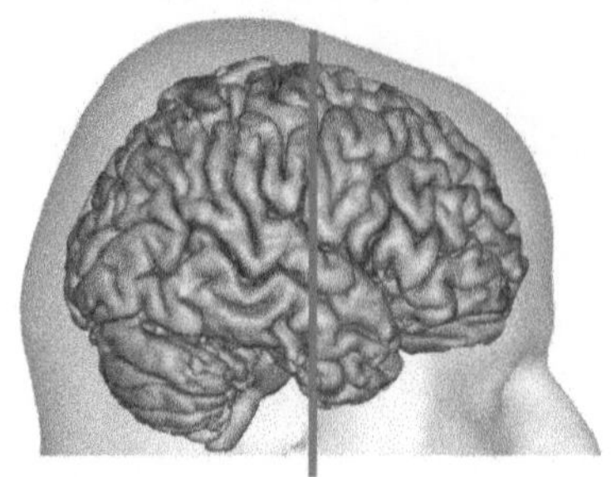

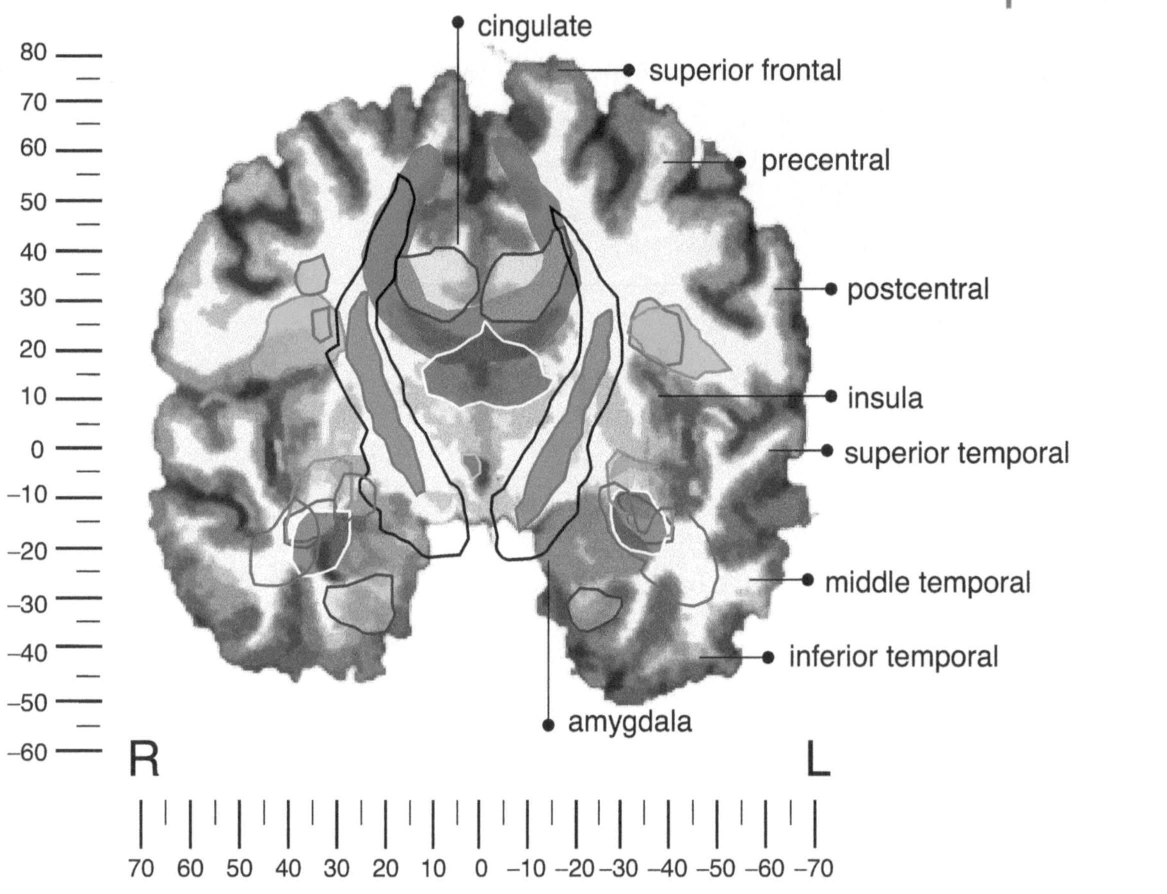

TRACT LEGEND

CC	Corpus Callosum	SCP	Sup. Cerebellar Peduncle	OR	Optic Radiations
ILF	Inf. Longitudinal Fasc.	CST	Cortico-Spinal Tract / IC	Cing	Cingulum
IFOF	Inf. Fronto-Occipital Fasc.	CPC	Cortico-Ponto-Cerebellar Tract	Fx	Fornix
AF-L	Arcuate (long segment)	SLF II	Sup. Longitudinal Fasc. II	Unc	Uncinate
AF-A	Arcuate (ant. segment)	AC	Anterior Commissure		

Coronal slice –8

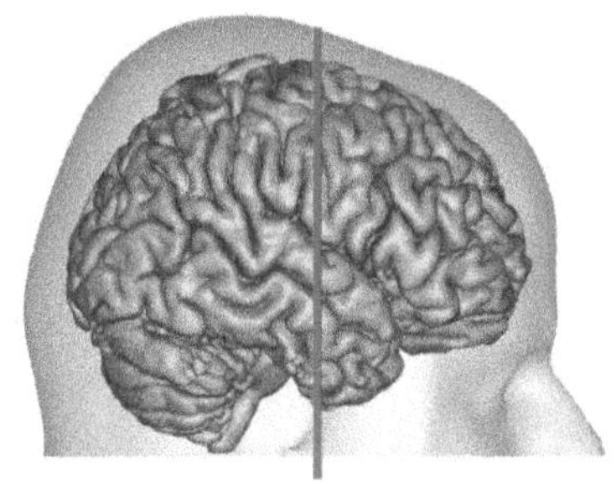

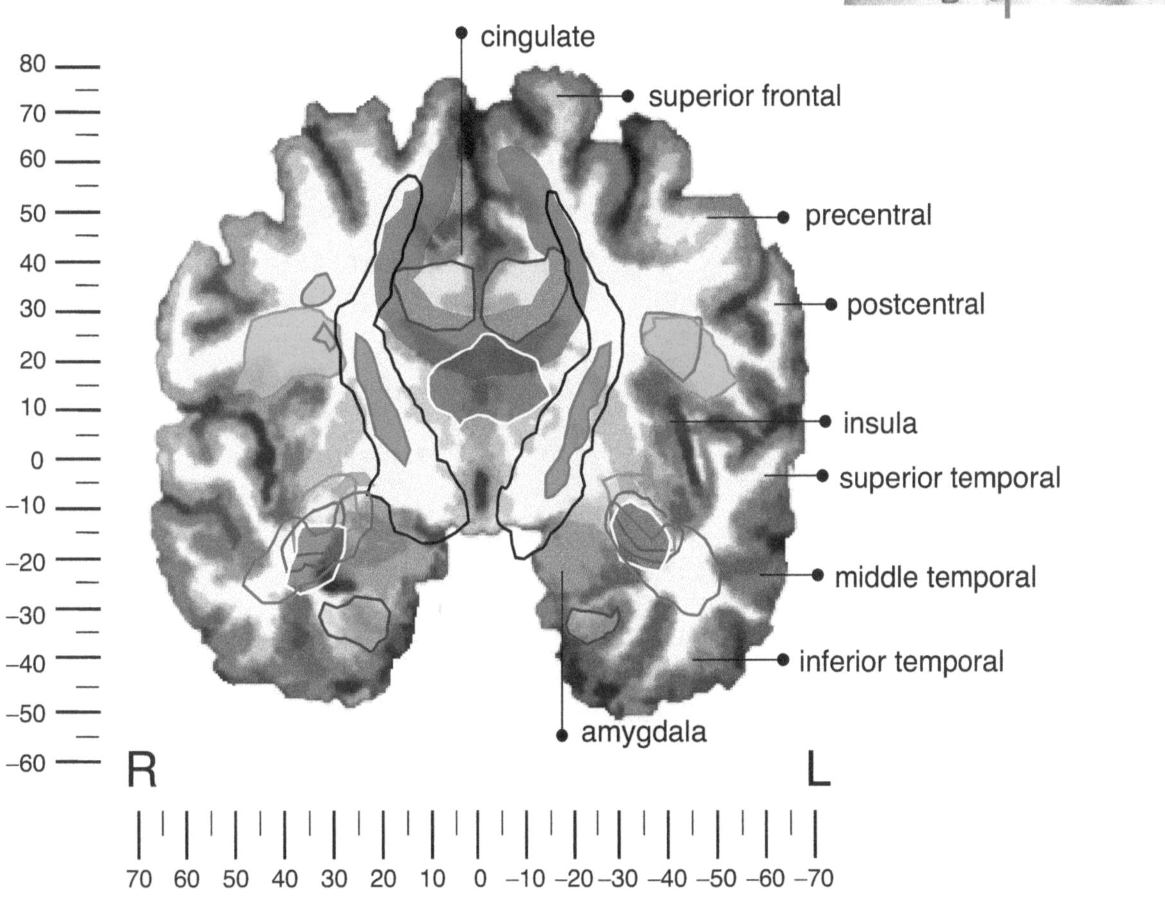

TRACT LEGEND

CC	Corpus Callosum	OR	Optic Radiations	AC	Anterior Commissure
ILF	Inf. Longitudinal Fasc.	CST	Cortico-Spinal Tract / IC	Cing	Cingulum
IFOF	Inf. Fronto-Occipital Fasc.	CPC	Cortico-Ponto-Cerebellar Tract	Fx	Fornix
AF-L	Arcuate (long segment)	SLF II	Sup. Longitudinal Fasc. II	Unc	Uncinate
AF-A	Arcuate (ant. segment)				

Coronal slice –6

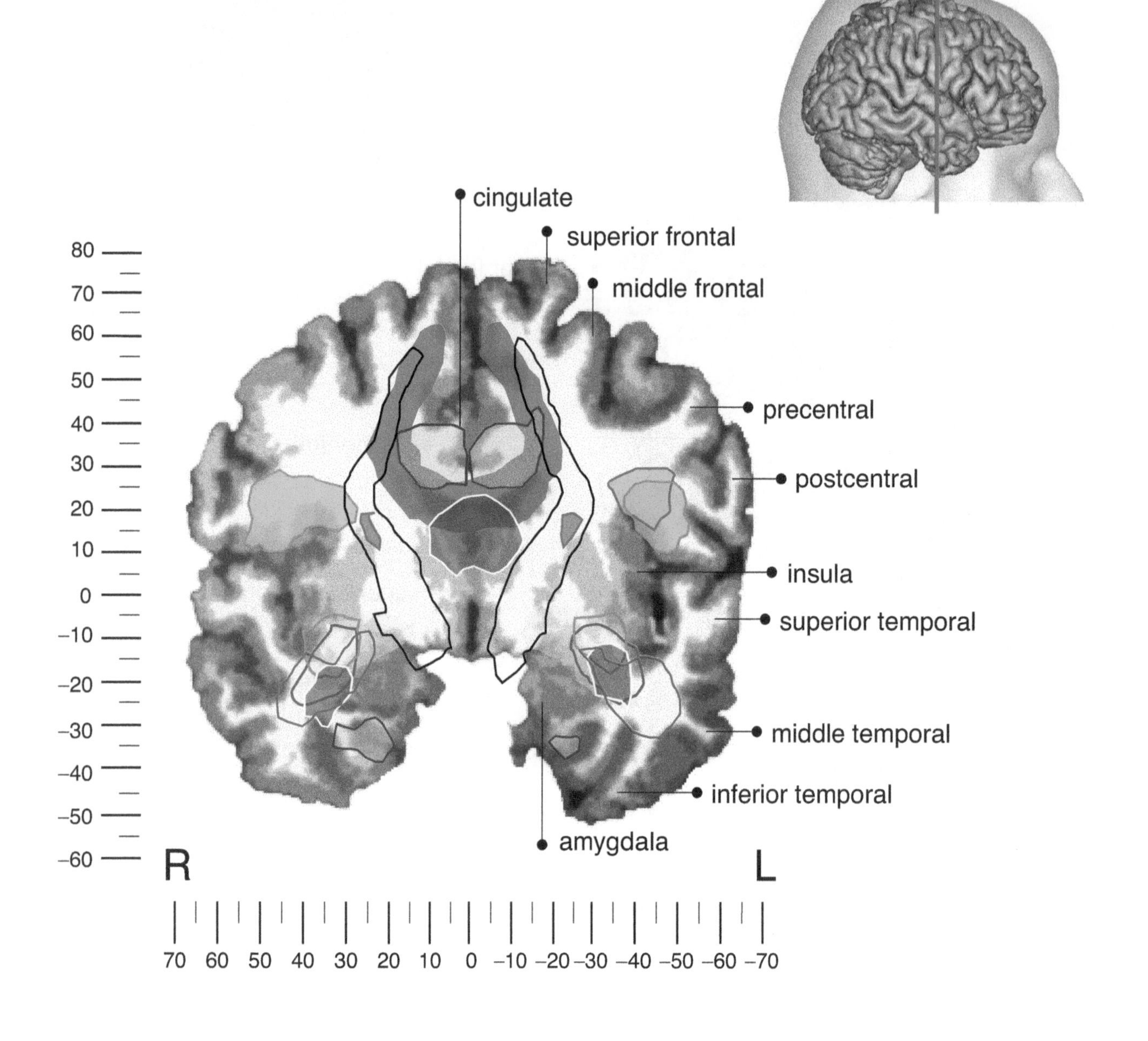

TRACT LEGEND

CC	Corpus Callosum	CST	Cortico-Spinal Tract / IC	AC	Anterior Commissure
ILF	Inf. Longitudinal Fasc.	CPC	Cortico-Ponto-Cerebellar Tract	Cing	Cingulum
IFOF	Inf. Fronto-Occipital Fasc.	AF-A	Arcuate (ant. segment)	Fx	Fornix
AF-L	Arcuate (long segment)	Unc	Uncinate		

Coronal slice –4

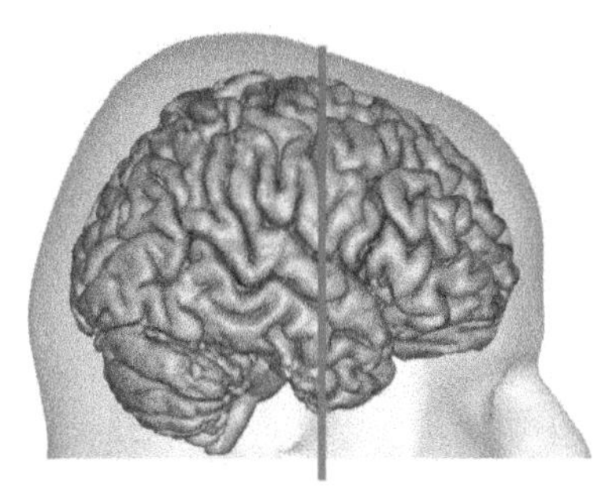

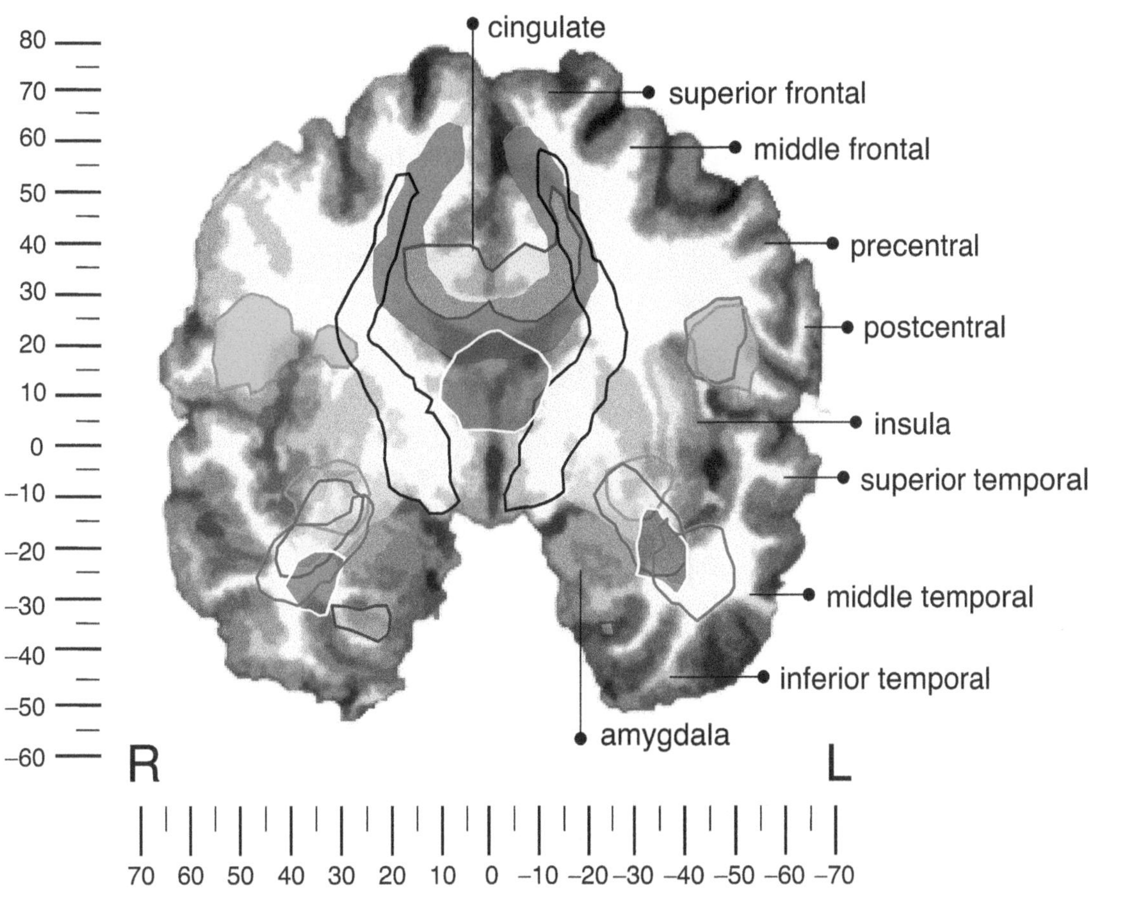

TRACT LEGEND

CC	Corpus Callosum	CST	Cortico-Spinal Tract / IC	AC	Anterior Commissure
ILF	Inf. Longitudinal Fasc.	Unc	Uncinate	Cing	Cingulum
IFOF	Inf. Fronto-Occipital Fasc.	AF-A	Arcuate (ant. segment)	Fx	Fornix
AF-L	Arcuate (long segment)				

Coronal slice –2

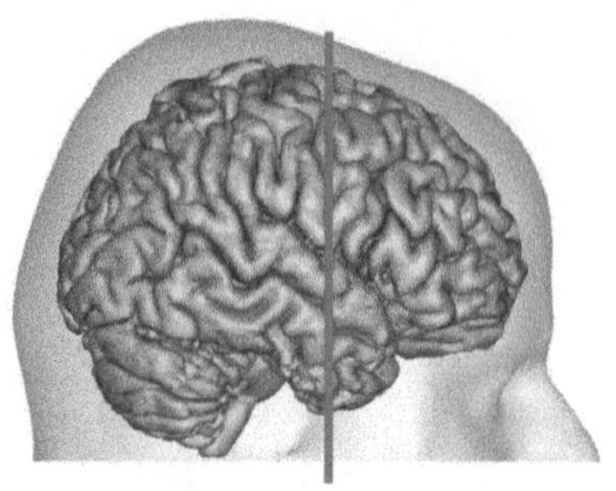

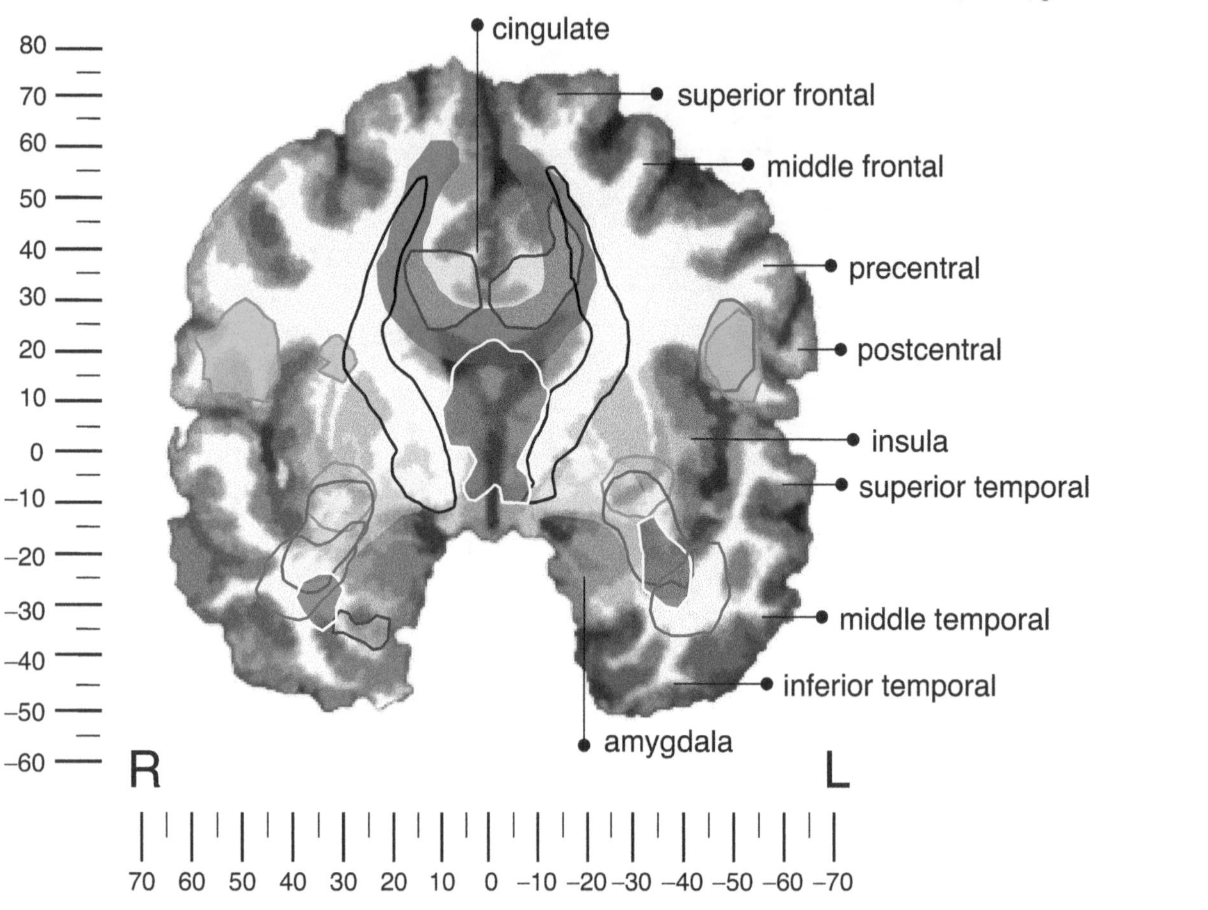

TRACT LEGEND

CC	Corpus Callosum	CST	Cortico-Spinal Tract / IC	AC	Anterior Commissure
ILF	Inf. Longitudinal Fasc.	Unc	Uncinate	Cing	Cingulum
IFOF	Inf. Fronto-Occipital Fasc.	AF-A	Arcuate (ant. segment)	Fx	Fornix
AF-L	Arcuate (long segment)				

Coronal slice 0

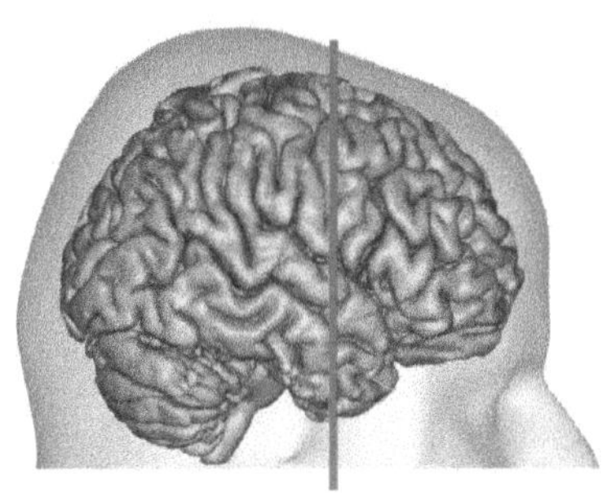

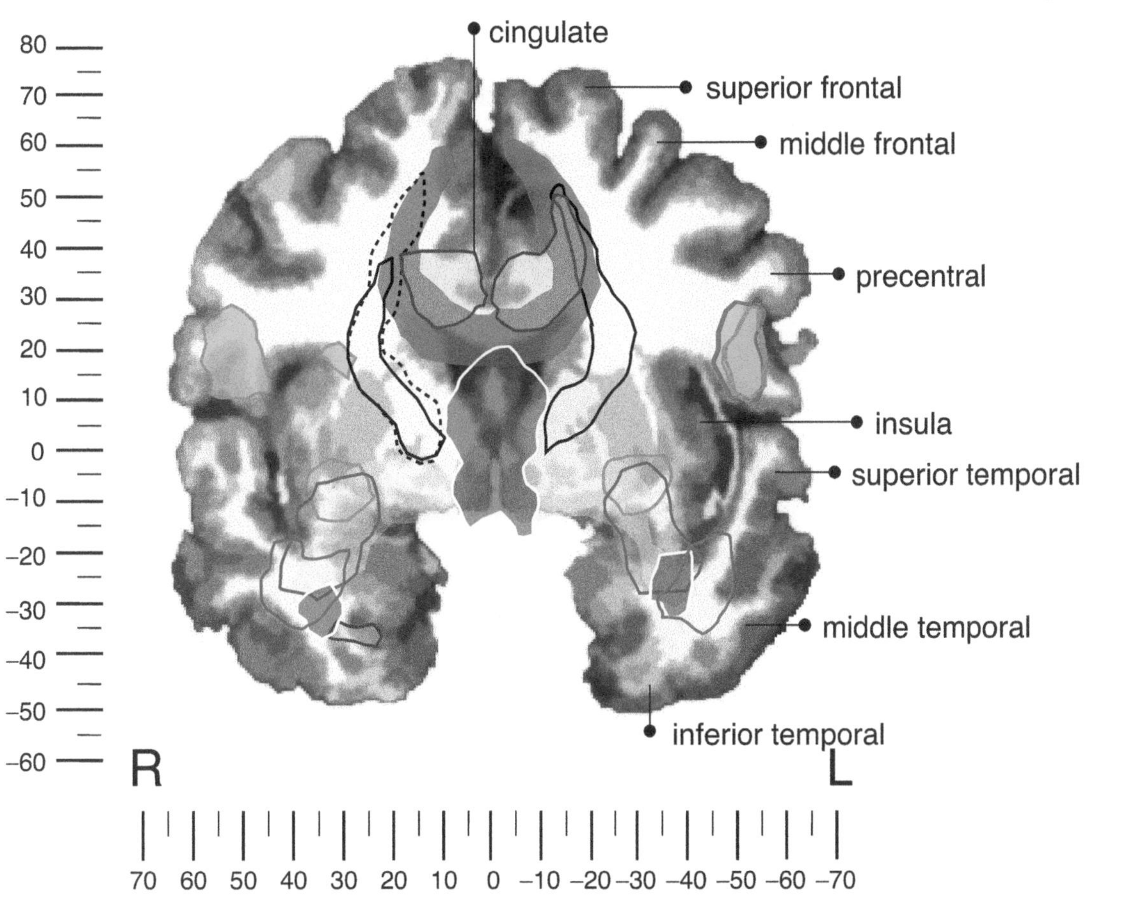

TRACT LEGEND

CC	Corpus Callosum	IC	Internal Capsule	AC	Anterior Commissure
ILF	Inf. Longitudinal Fasc.	CST	Cortico-Spinal Tract	Cing	Cingulum
IFOF	Inf. Fronto-Occipital Fasc.	Unc	Uncinate	Fx	Fornix
AF-L	Arcuate (long segment)	AF-A	Arcuate (ant. segment)		

Coronal slice 2

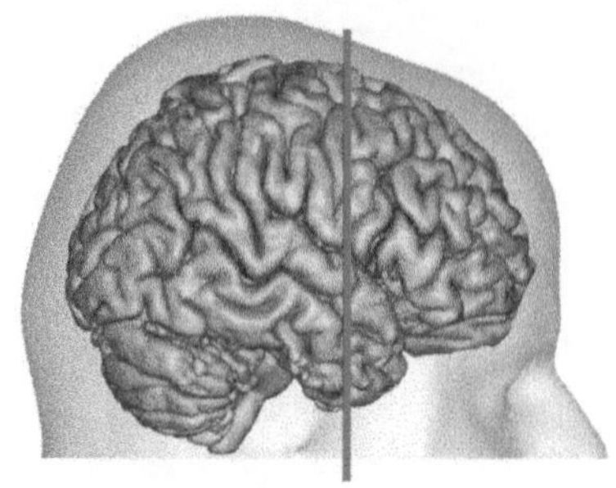

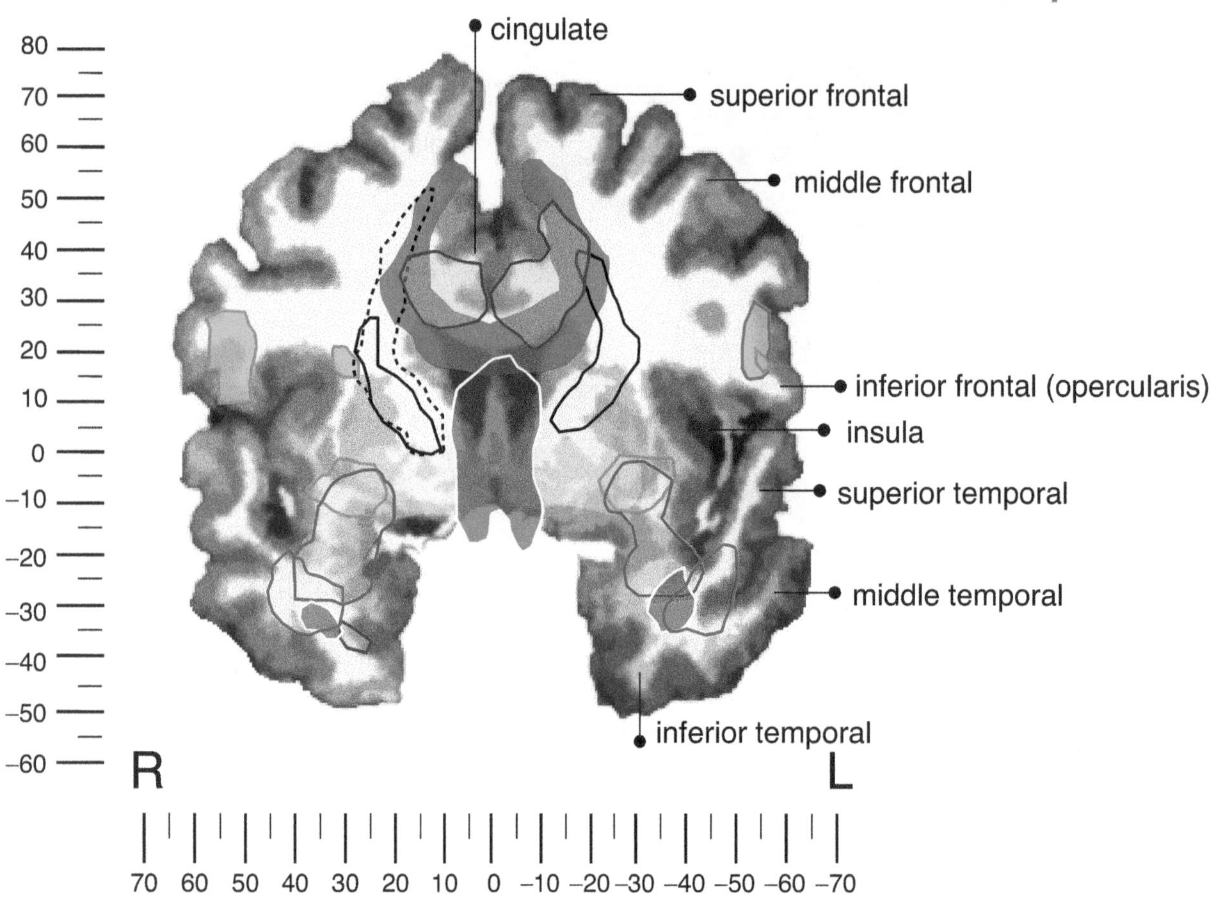

TRACT LEGEND

CC Corpus Callosum
IC Internal Capsule
AC Anterior Commissure
ILF Inf. Longitudinal Fasc.
CST Cortico-Spinal Tract
Cing Cingulum
IFOF Inf. Fronto-Occipital Fasc.
Unc Uncinate
Fx Fornix
AF-L Arcuate (long segment)
AF-A Arcuate (ant. segment)

Coronal slice 4

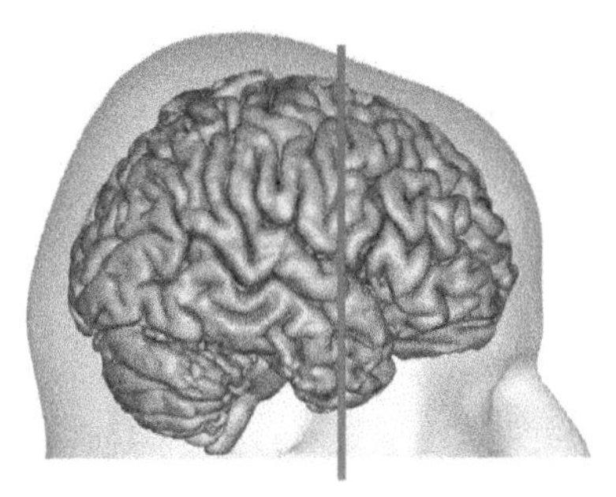

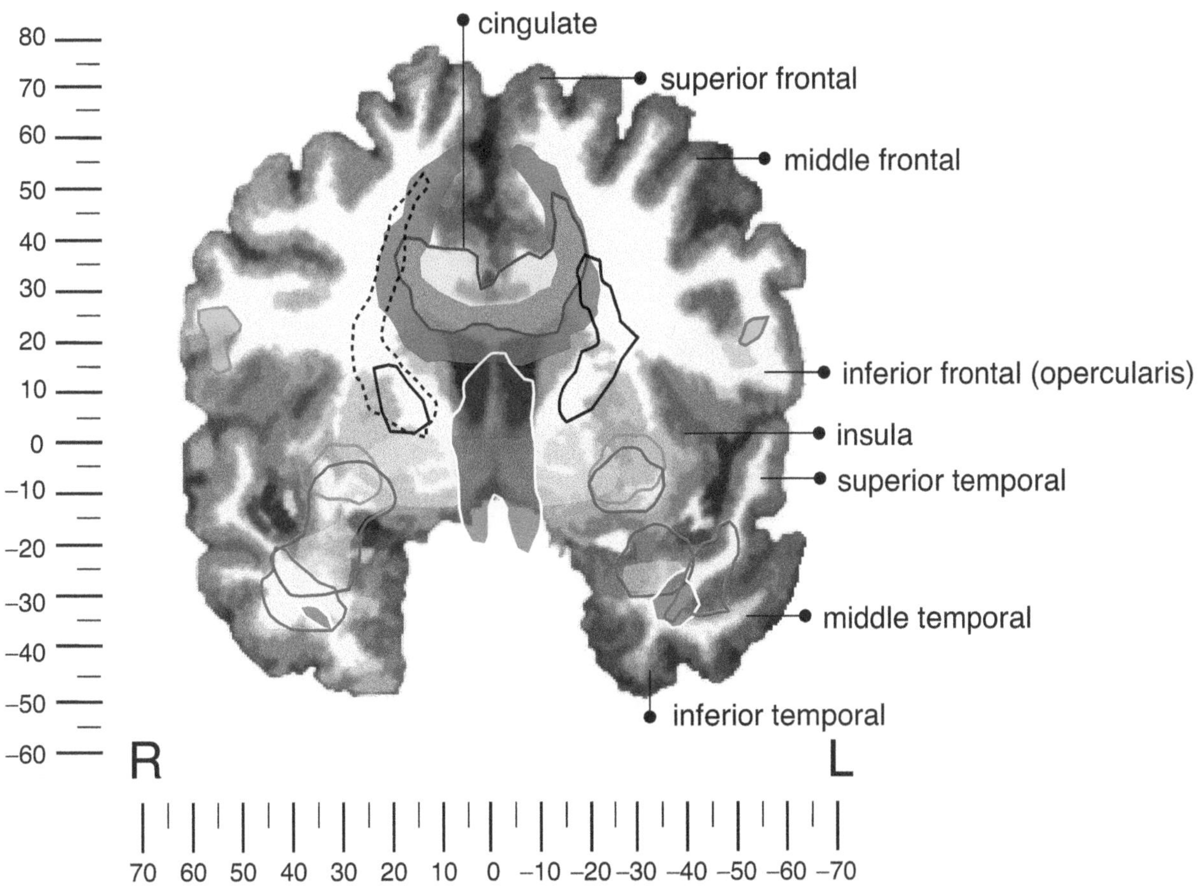

TRACT LEGEND

CC	Corpus Callosum	IC	Internal Capsule	AC	Anterior Commissure
ILF	Inf. Longitudinal Fasc.	CST	Cortico-Spinal Tract	Cing	Cingulum
IFOF	Inf. Fronto-Occipital Fasc.	Unc	Uncinate	Fx	Fornix
AF-L	Arcuate (long segment)	AF-A	Arcuate (ant. segment)		

Coronal slice 6

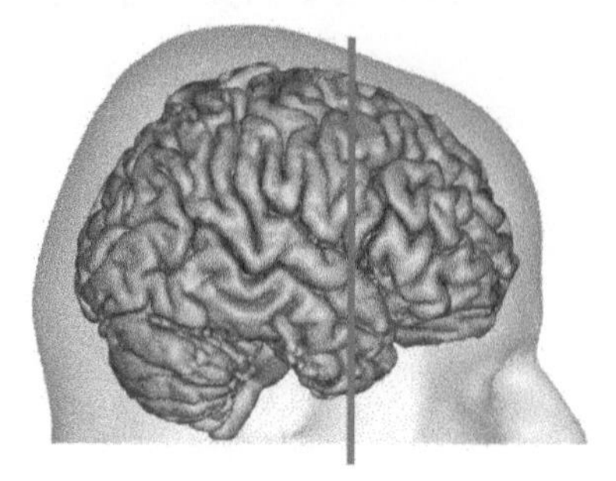

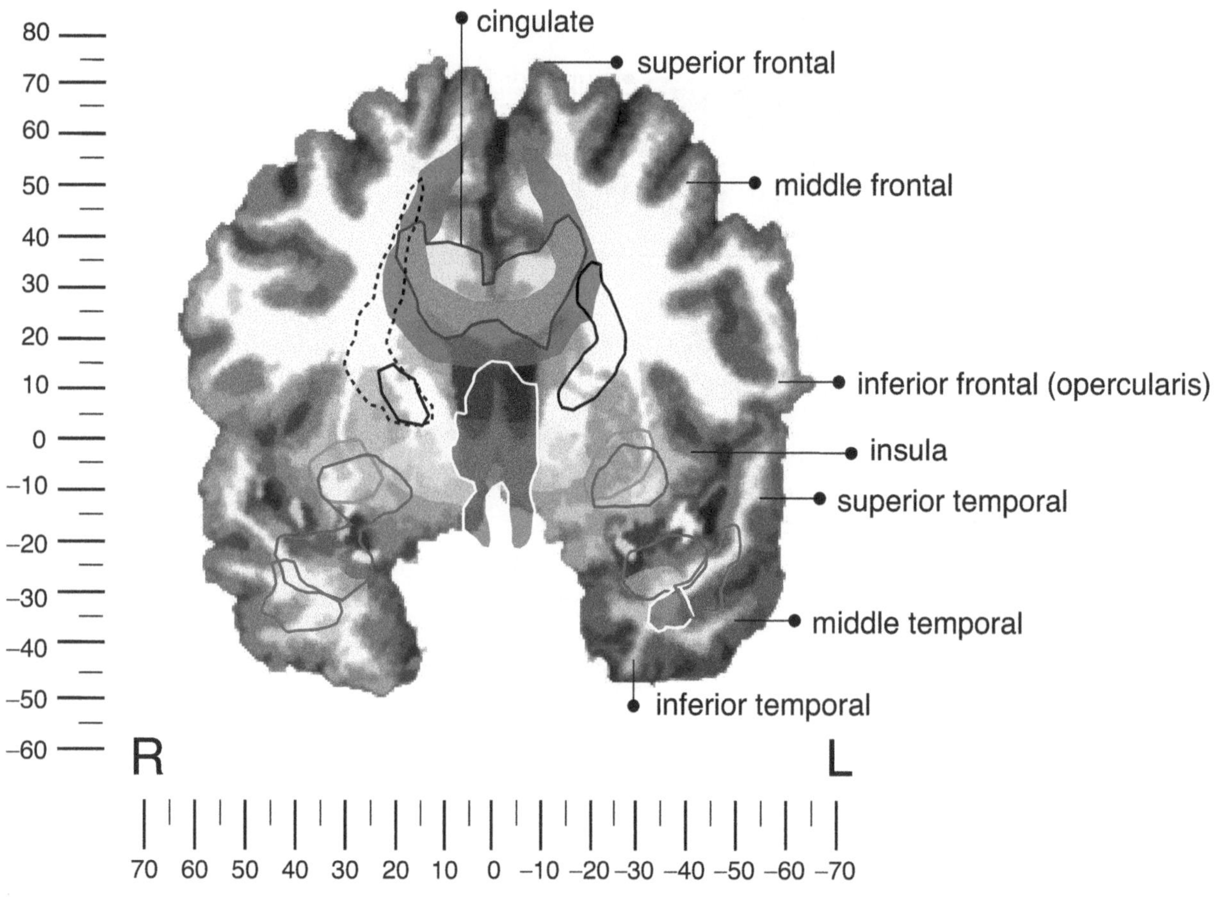

TRACT LEGEND

CC	Corpus Callosum	IC	Internal Capsule	AC	Anterior Commissure
ILF	Inf. Longitudinal Fasc.	CST	Cortico-Spinal Tract	Cing	Cingulum
IFOF	Inf. Fronto-Occipital Fasc.	Unc	Uncinate	Fx	Fornix

Coronal slice 8

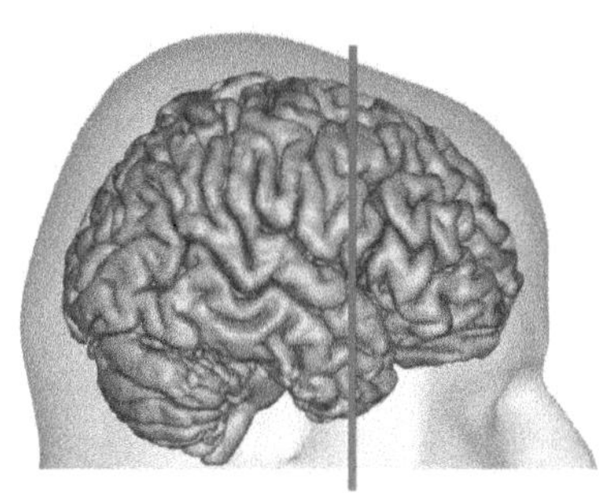

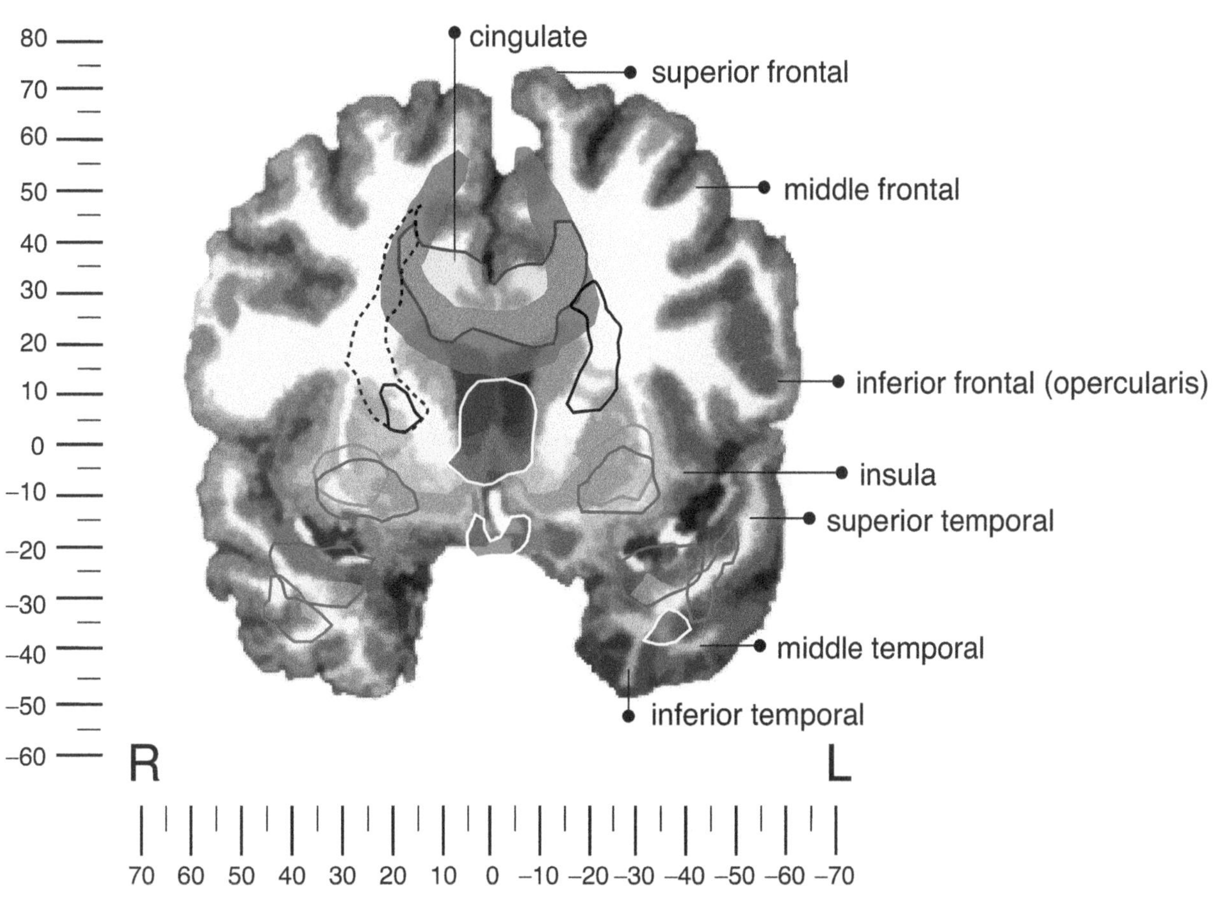

TRACT LEGEND

CC	Corpus Callosum	IC	Internal Capsule	AC	Anterior Commissure
ILF	Inf. Longitudinal Fasc.	CST	Cortico-Spinal Tract	Cing	Cingulum
IFOF	Inf. Fronto-Occipital Fasc.	Unc	Uncinate	Fx	Fornix

Coronal slice 10

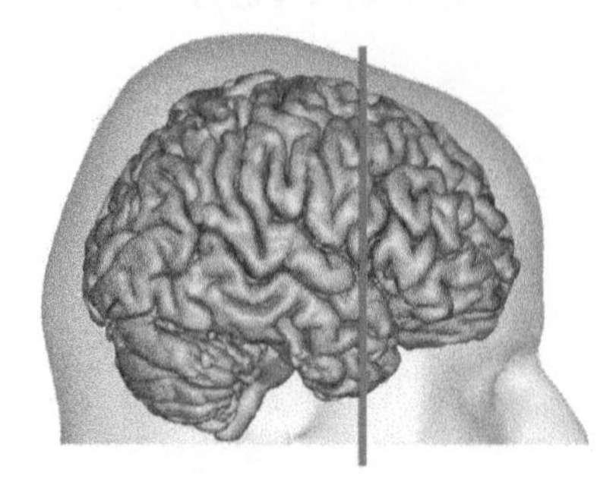

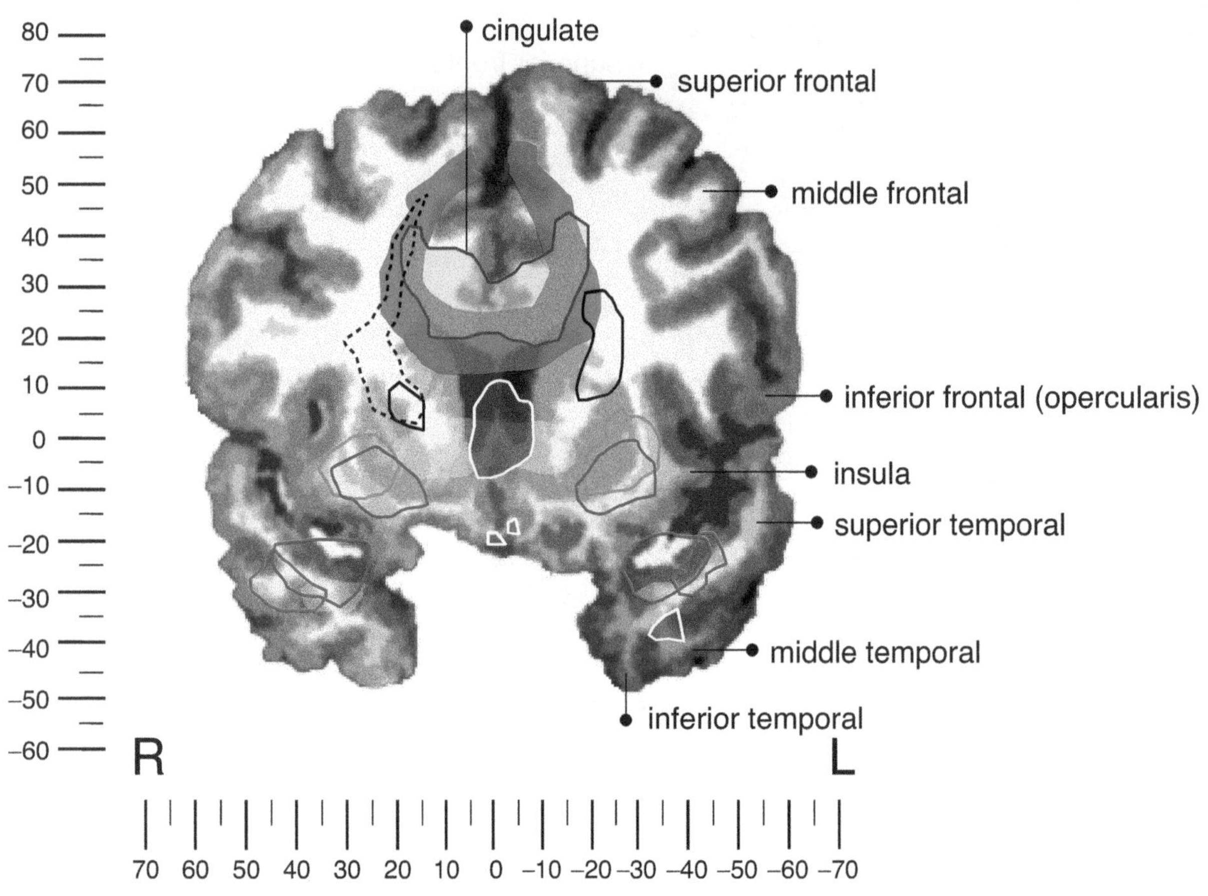

TRACT LEGEND

CC	Corpus Callosum	IC	Internal Capsule	AC	Anterior Commissure
ILF	Inf. Longitudinal Fasc.	CST	Cortico-Spinal Tract	Cing	Cingulum
IFOF	Inf. Fronto-Occipital Fasc.	Unc	Uncinate	Fx	Fornix

Coronal slice 12

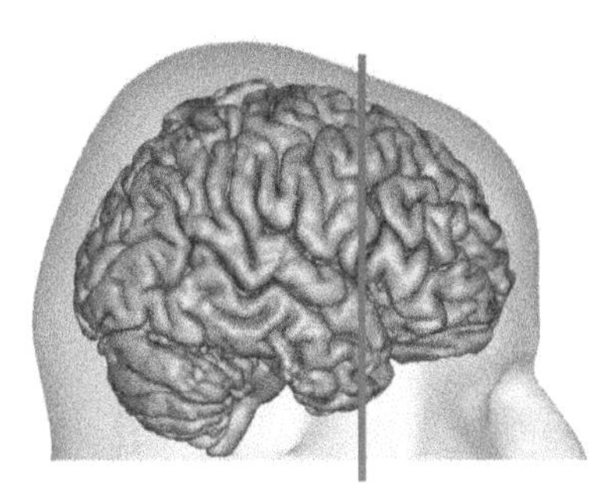

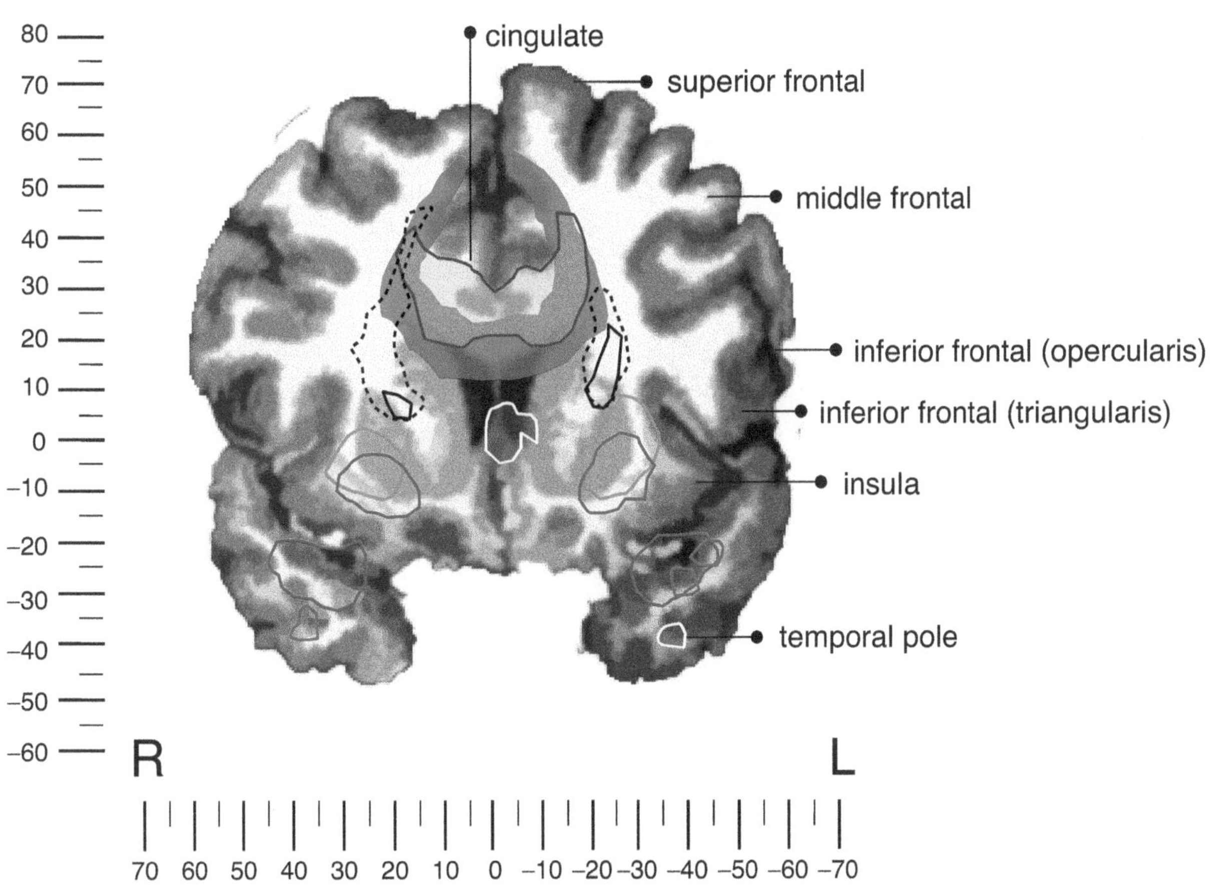

TRACT LEGEND

CC	Corpus Callosum	IC	Internal Capsule	AC	Anterior Commissure
ILF	Inf. Longitudinal Fasc.	CST	Cortico-Spinal Tract	Cing	Cingulum
IFOF	Inf. Fronto-Occipital Fasc.	Unc	Uncinate	Fx	Fornix

Coronal slice 14

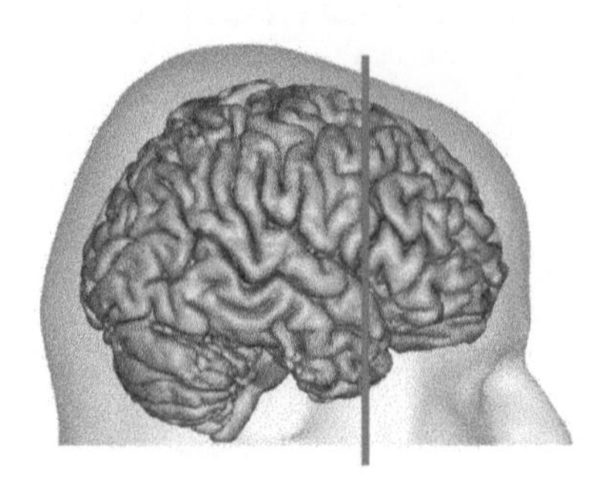

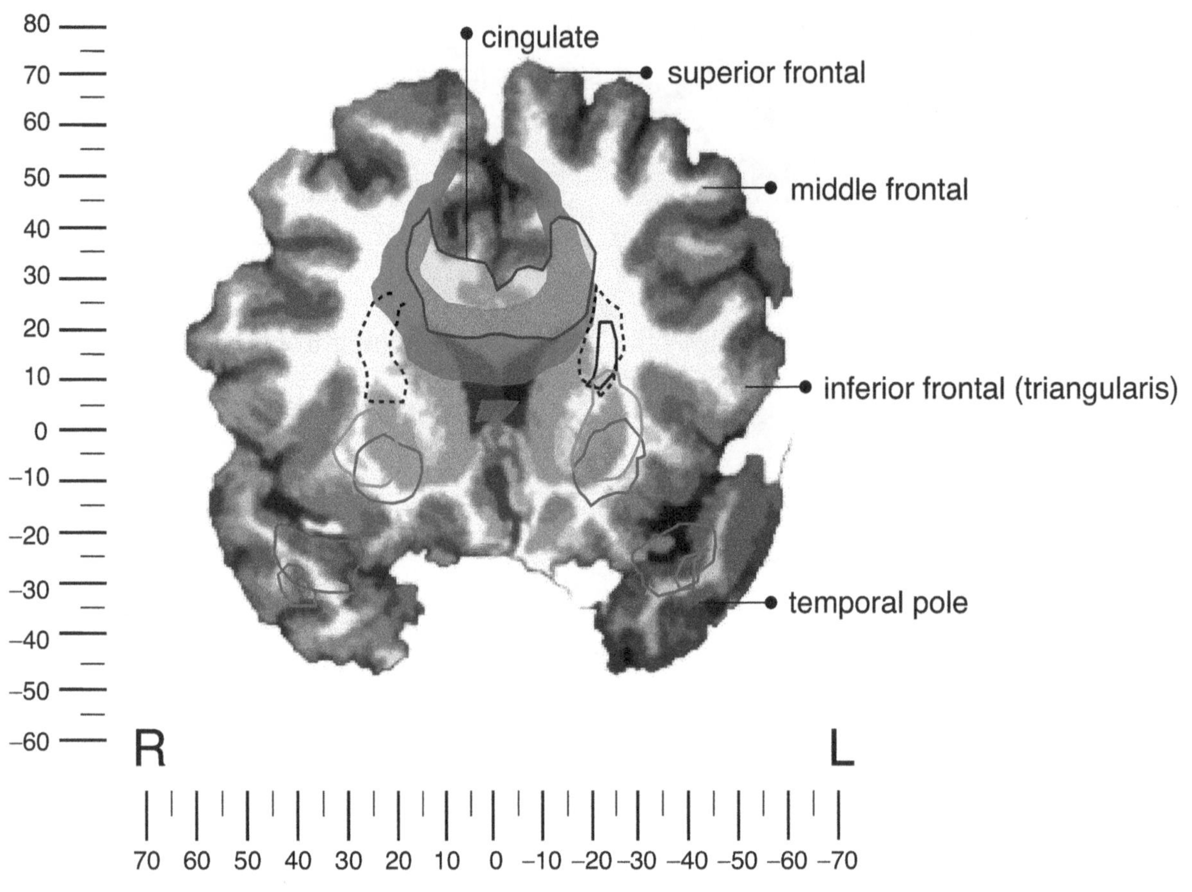

TRACT LEGEND

CC	Corpus Callosum	IC	Internal Capsule	AC	Anterior Commissure
ILF	Inf. Longitudinal Fasc.	CST	Cortico-Spinal Tract	Cing	Cingulum
IFOF	Inf. Fronto-Occipital Fasc.	Unc	Uncinate		

Coronal slice 16

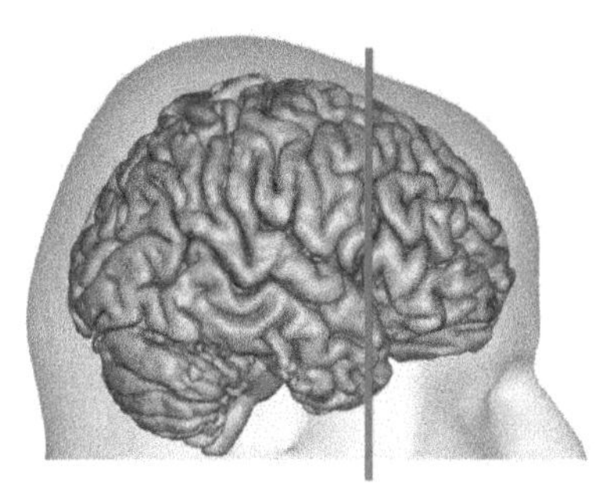

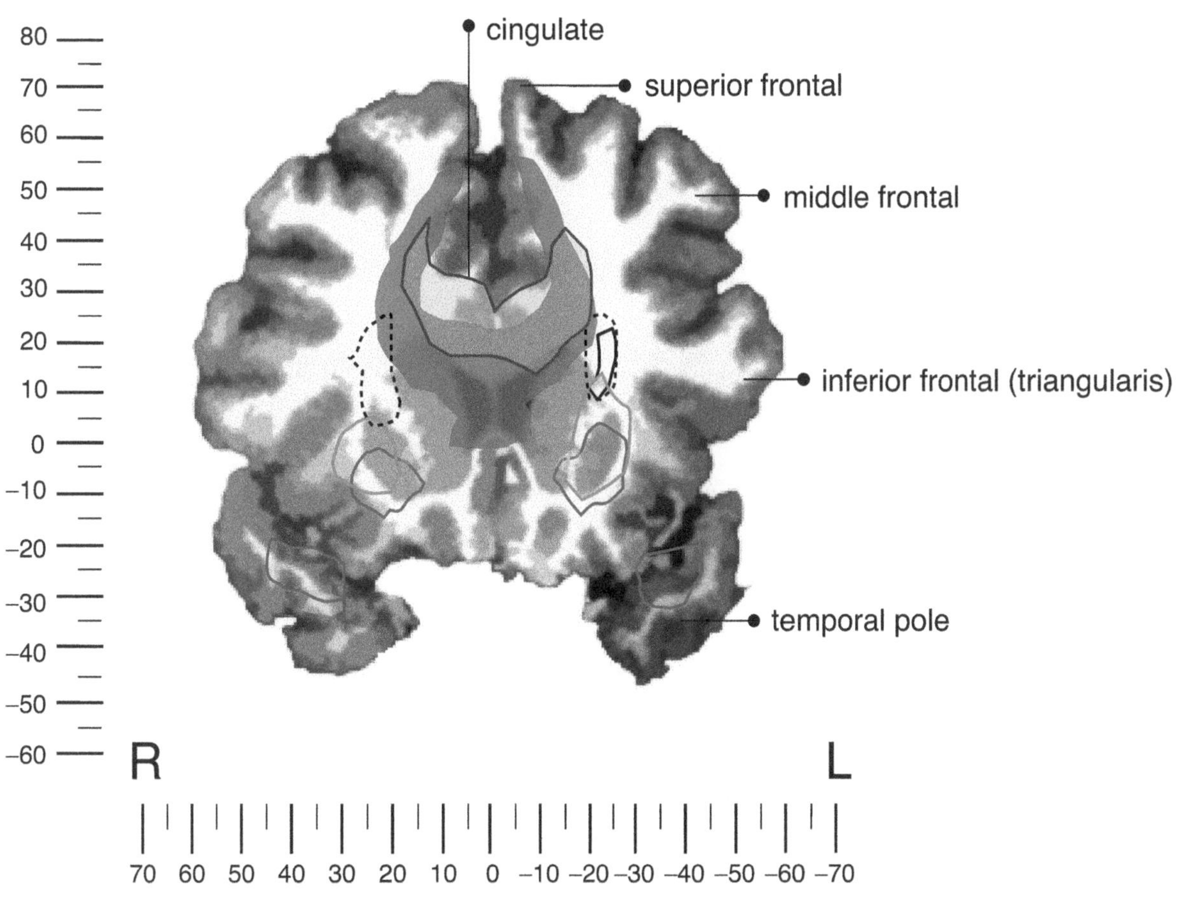

TRACT LEGEND

CC	Corpus Callosum	IC	Internal Capsule	AC	Anterior Commissure
Unc	Uncinate	CST	Cortico-Spinal Tract	Cing	Cingulum
IFOF	Inf. Fronto-Occipital Fasc.				

Coronal slice 18

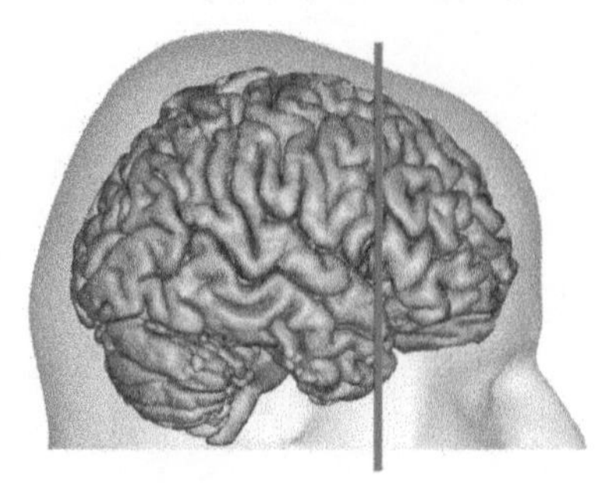

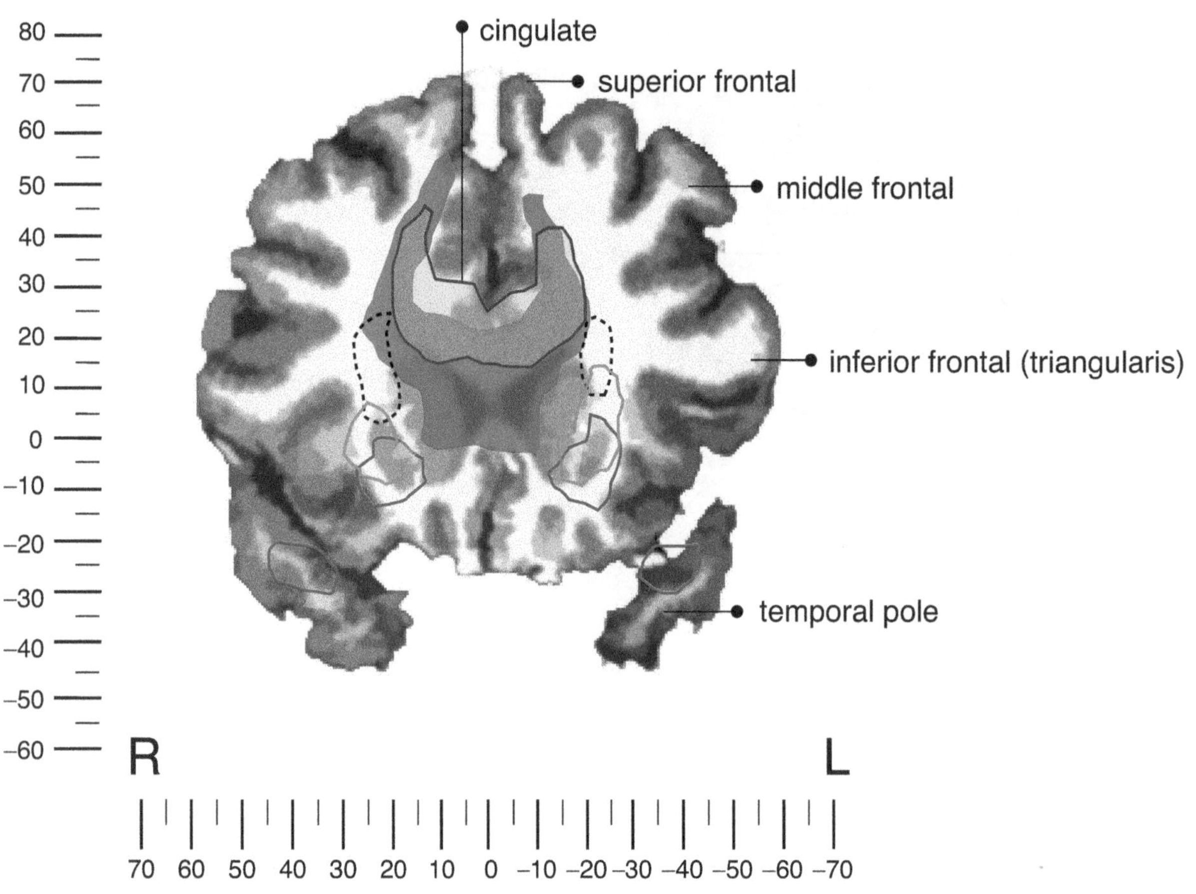

TRACT LEGEND

CC	Corpus Callosum	IC	Internal Capsule	AC	Anterior Commissure
Unc	Uncinate	IFOF	Inf. Fronto-Occipital Fasc.	Cing	Cingulum

Coronal slice 20

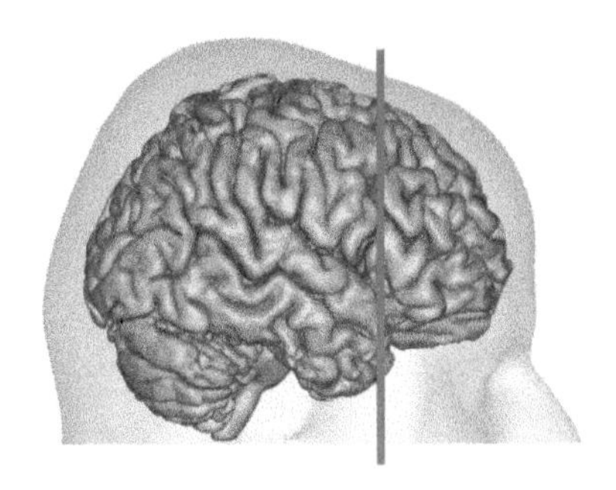

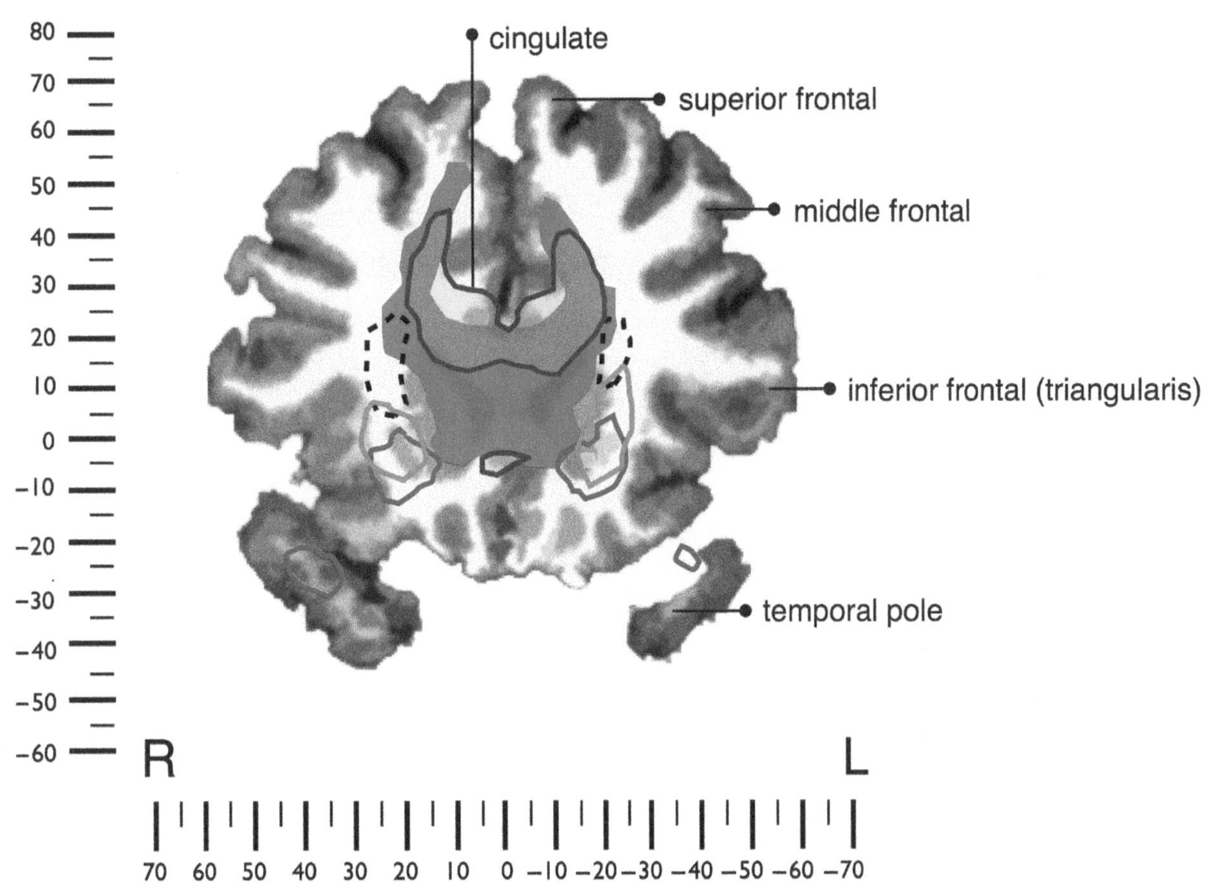

TRACT LEGEND

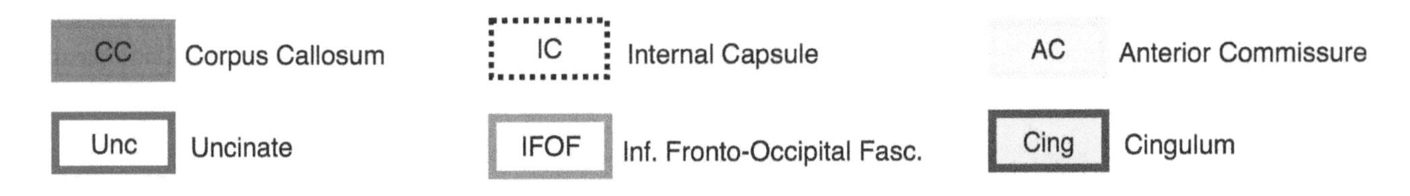

Coronal slice 22

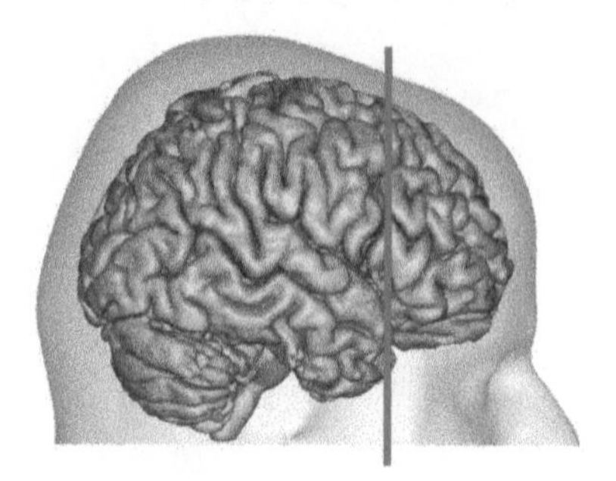

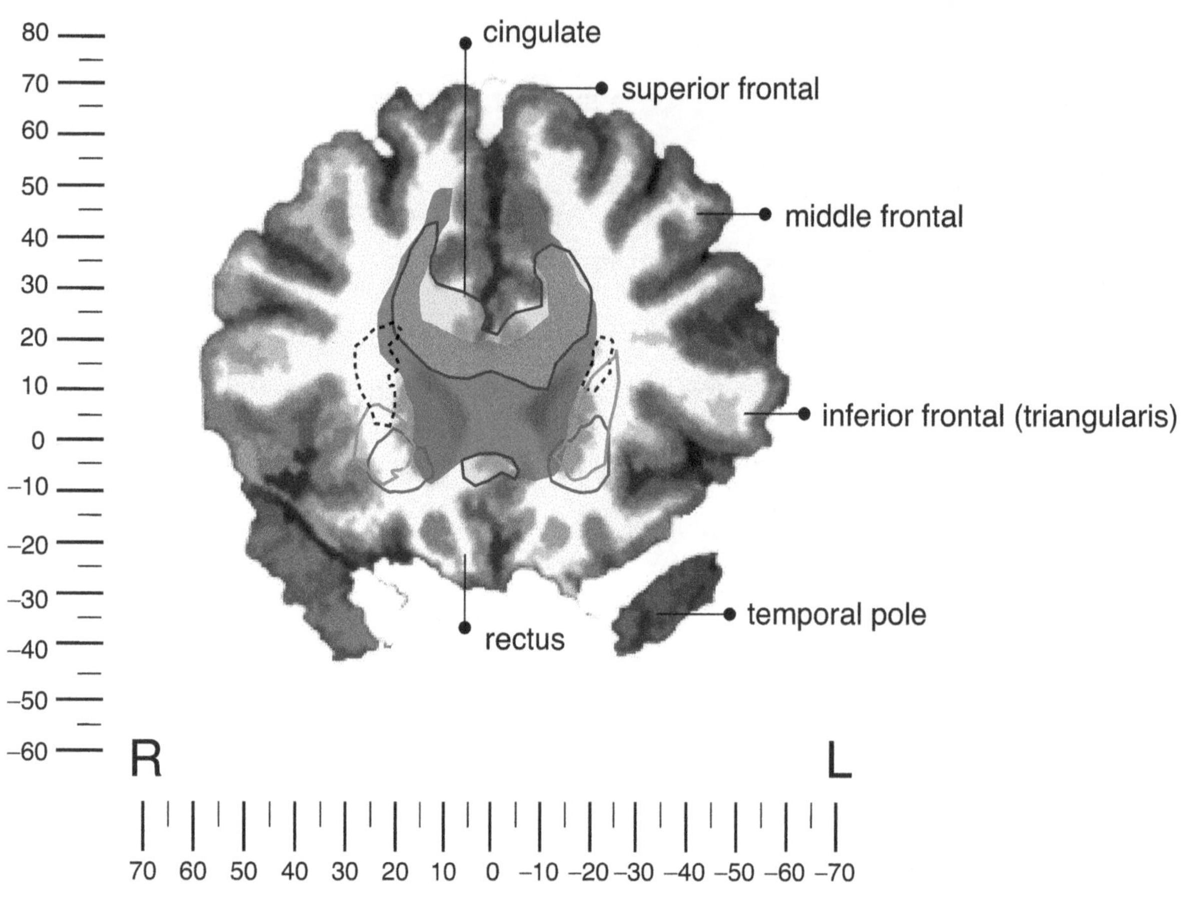

TRACT LEGEND

CC	Corpus Callosum	IC	Internal Capsule	AC	Anterior Commissure
Unc	Uncinate	IFOF	Inf. Fronto-Occipital Fasc.	Cing	Cingulum

Coronal slice 24

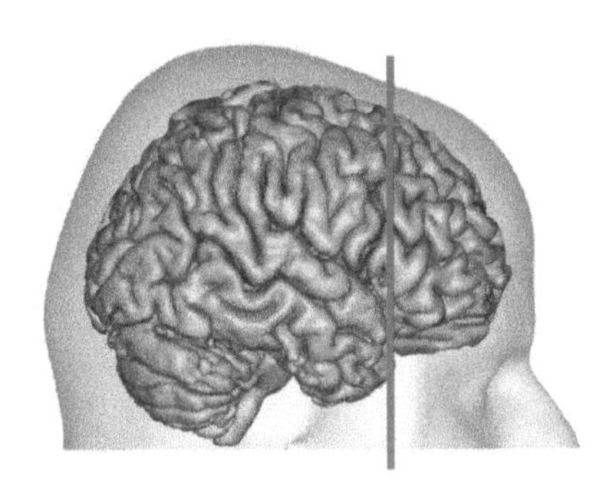

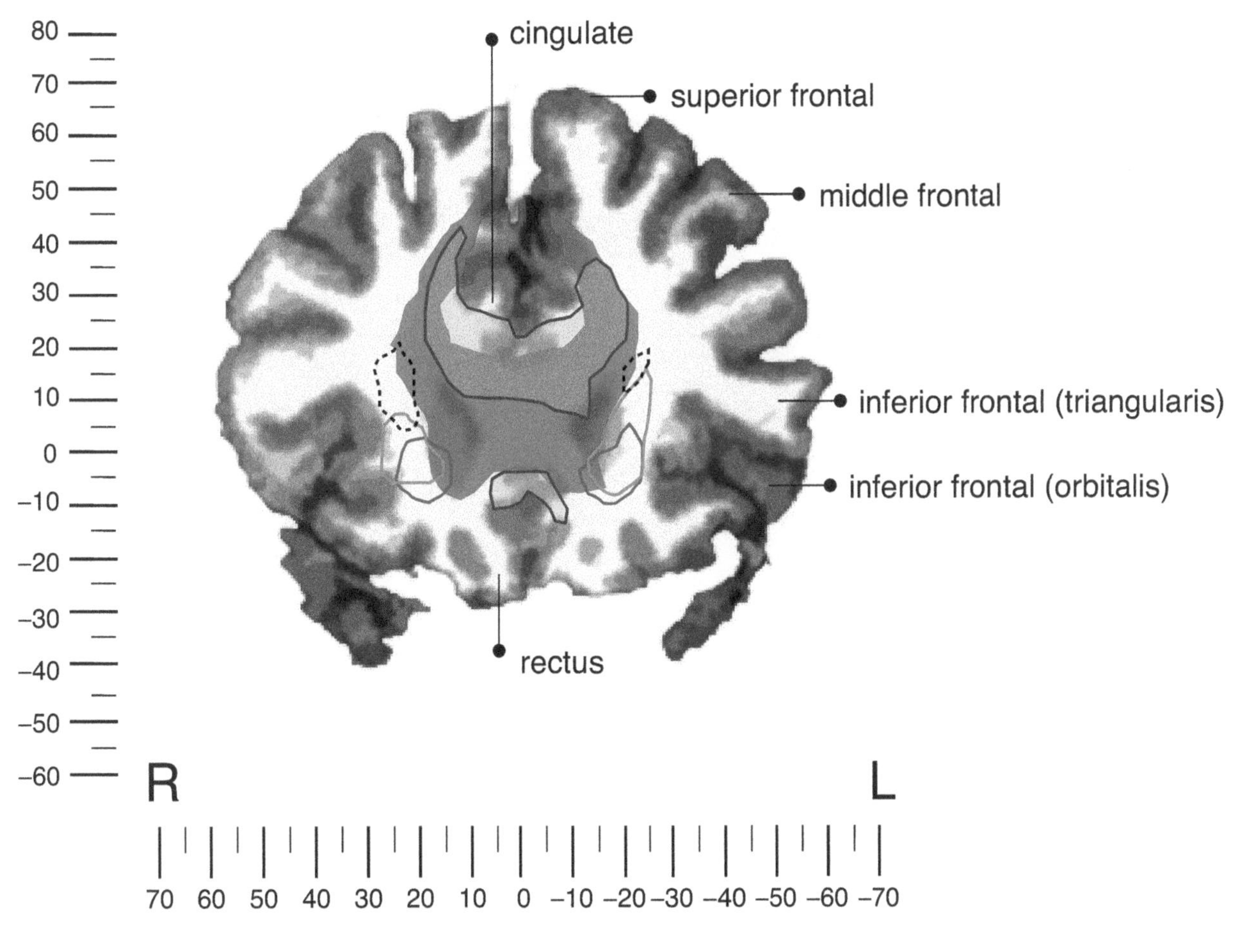

TRACT LEGEND

CC	Corpus Callosum	IC	Internal Capsule	Cing	Cingulum
Unc	Uncinate	IFOF	Inf. Fronto-Occipital Fasc.		

Coronal slice 26

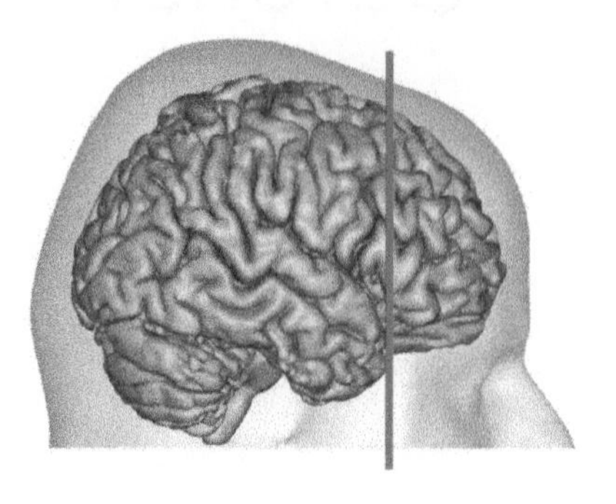

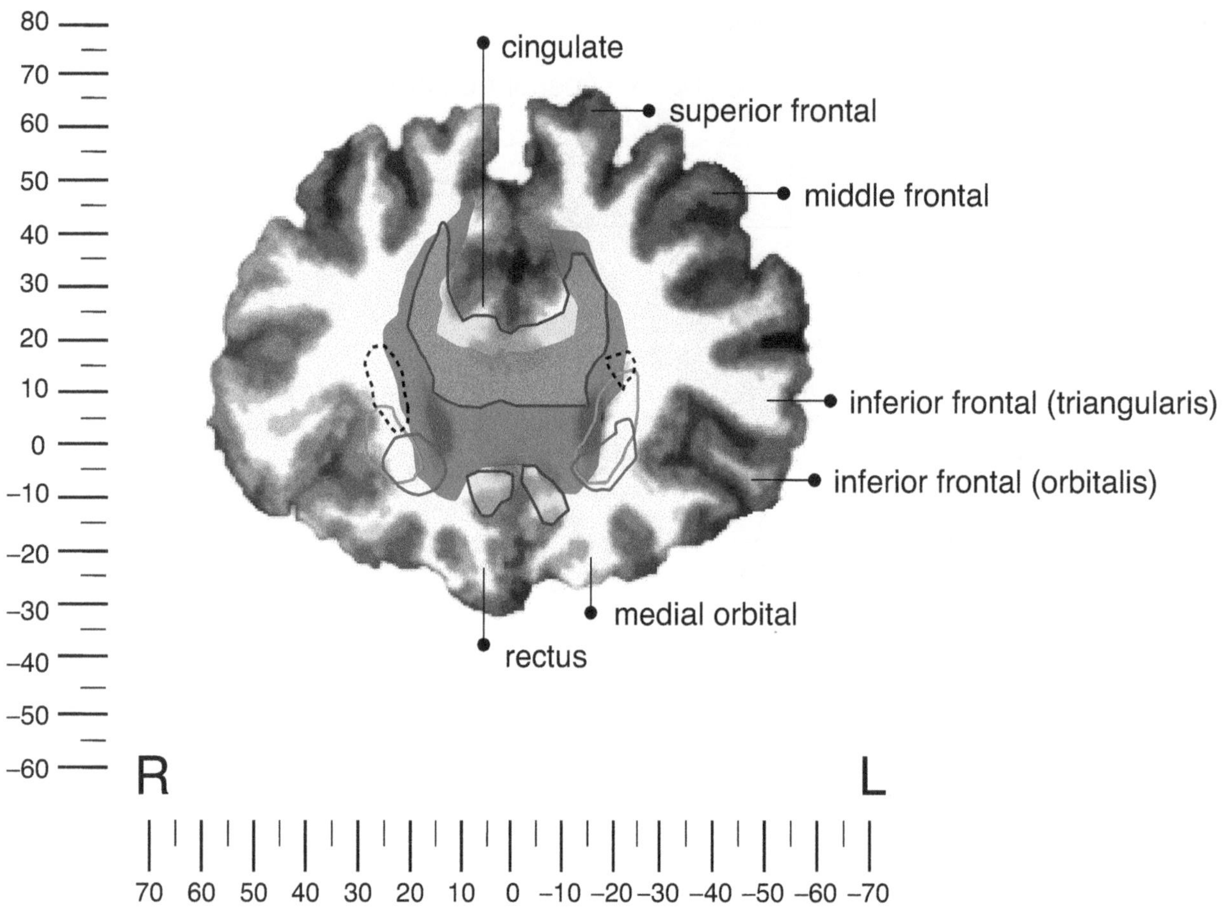

TRACT LEGEND

CC	Corpus Callosum	IC	Internal Capsule	Cing	Cingulum
Unc	Uncinate	IFOF	Inf. Fronto-Occipital Fasc.		

Coronal slice 28

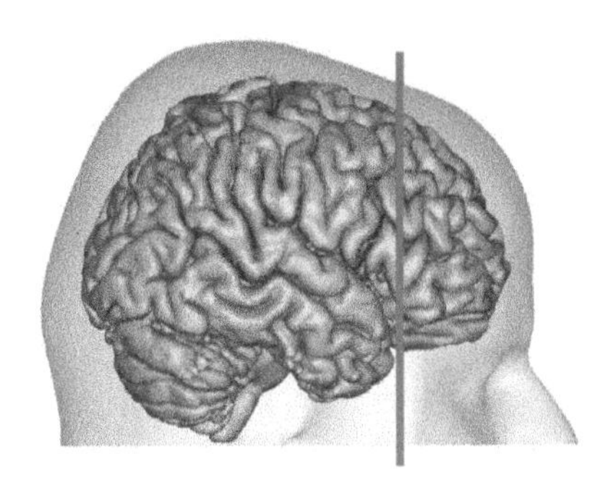

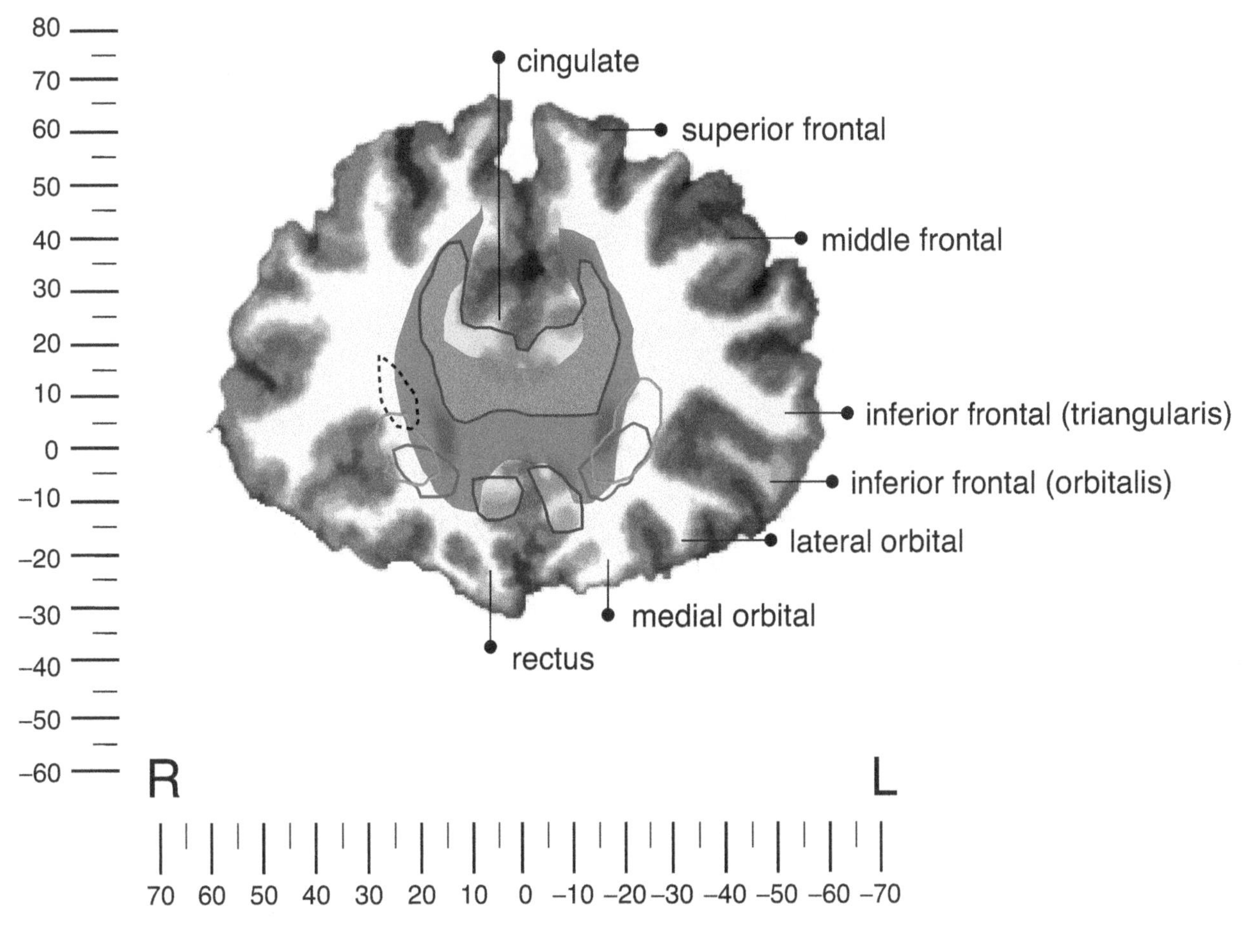

TRACT LEGEND

CC	Corpus Callosum	IC	Internal Capsule	Cing	Cingulum
Unc	Uncinate	IFOF	Inf. Fronto-Occipital Fasc.		

Coronal slice 30

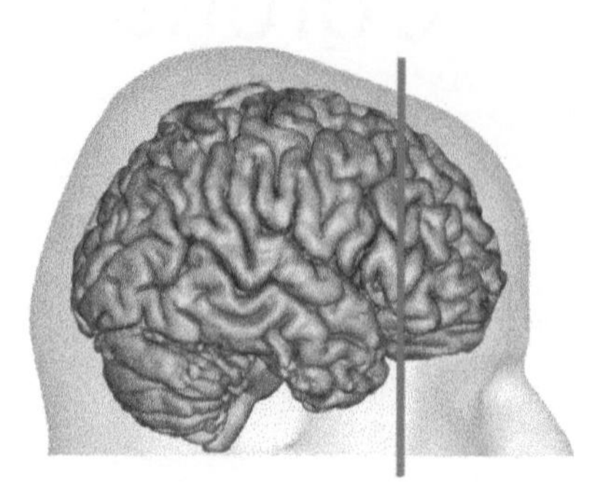

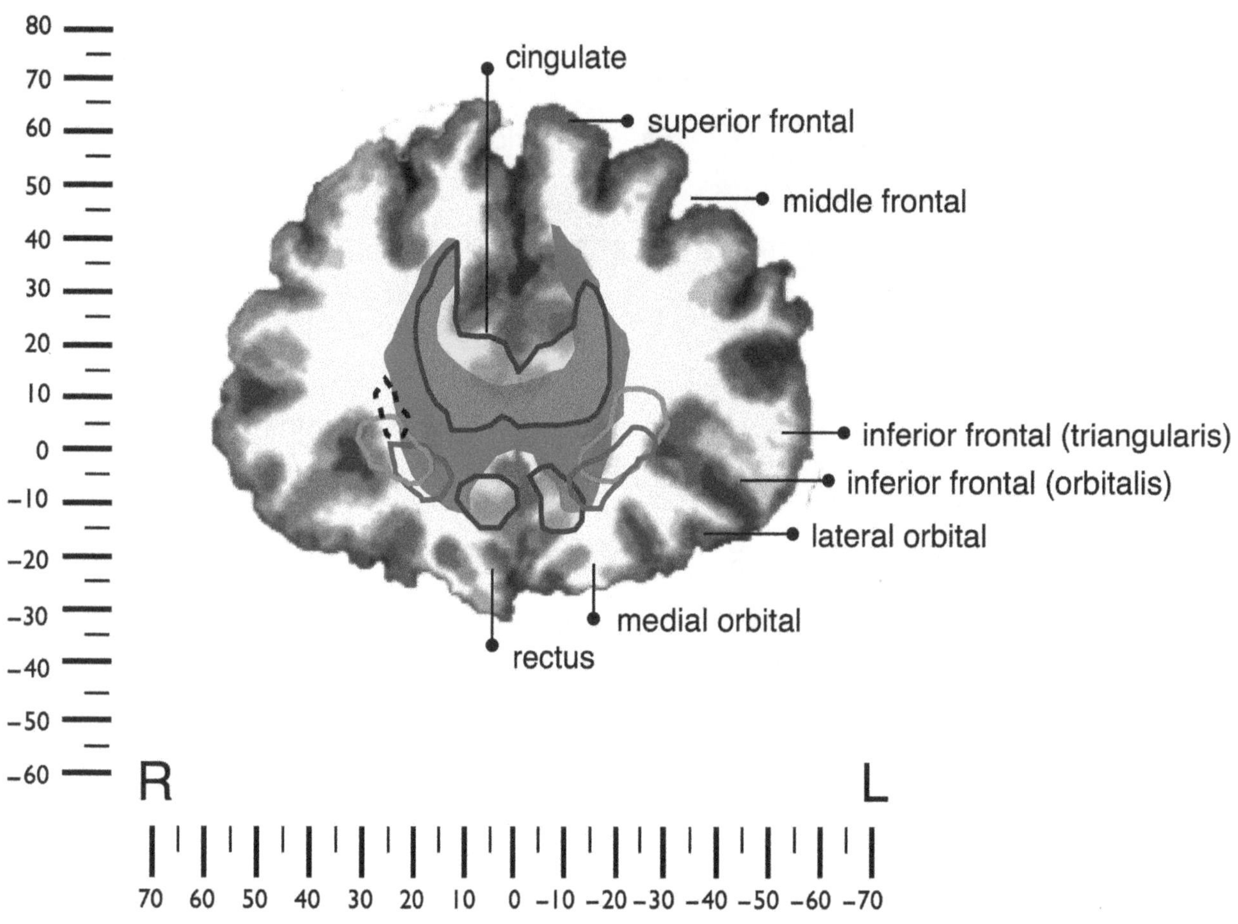

TRACT LEGEND

Symbol	Tract	Symbol	Tract	Symbol	Tract
CC	Corpus Callosum	IC	Internal Capsule	Cing	Cingulum
Unc	Uncinate	IFOF	Inf. Fronto-Occipital Fasc.		

Coronal slice 32

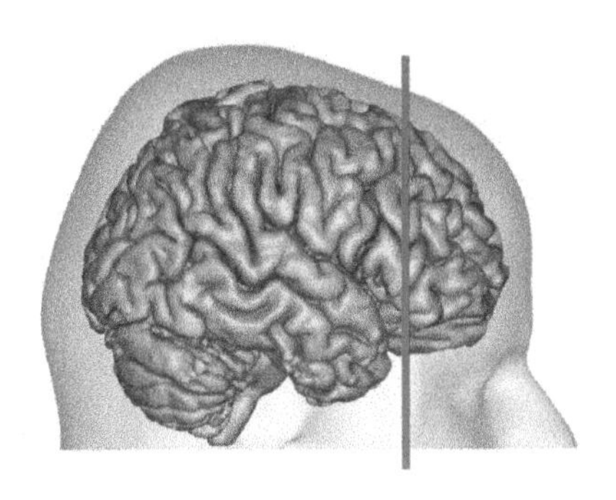

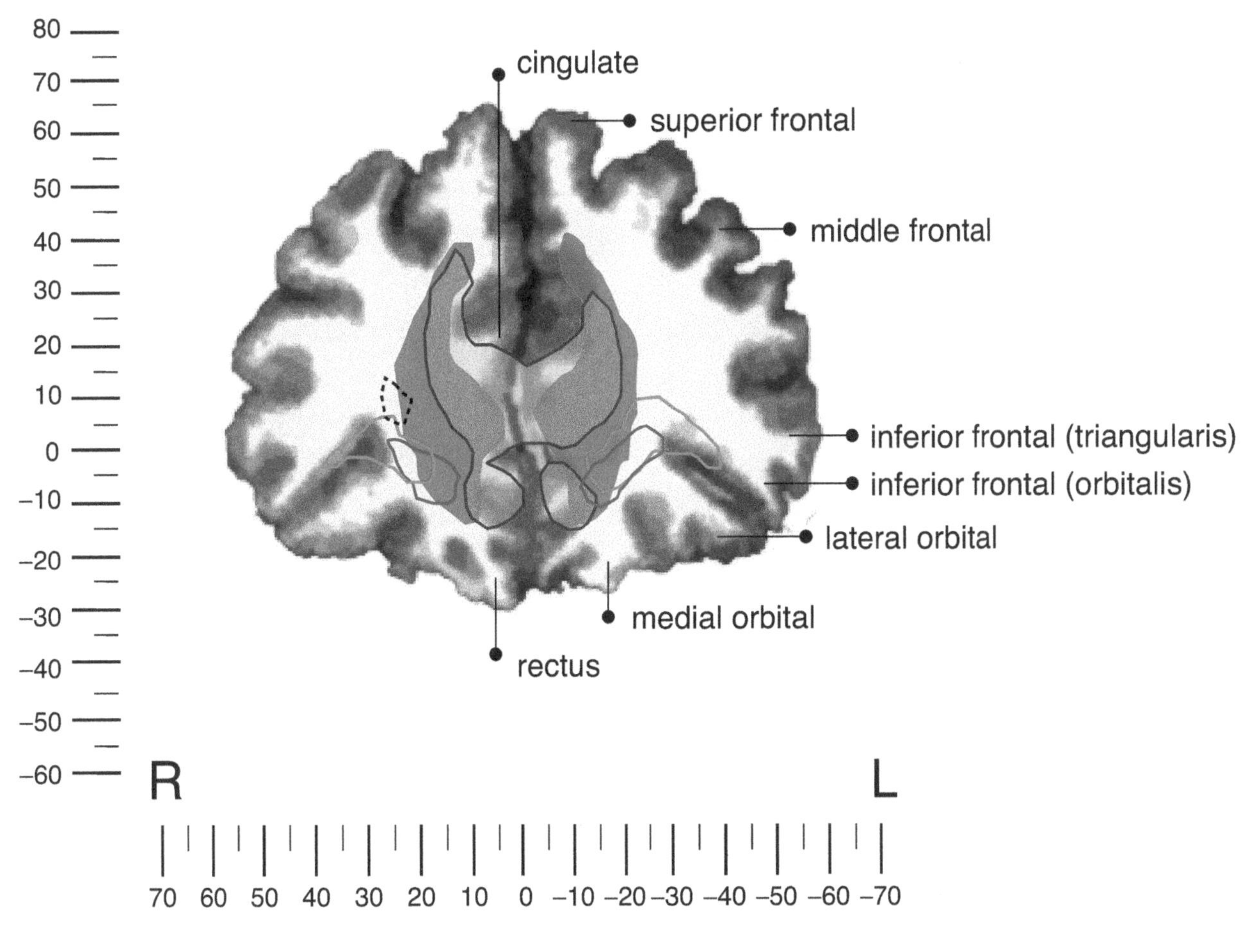

TRACT LEGEND

CC Corpus Callosum

IC Internal Capsule

Cing Cingulum

Unc Uncinate

IFOF Inf. Fronto-Occipital Fasc.

Coronal slice 34

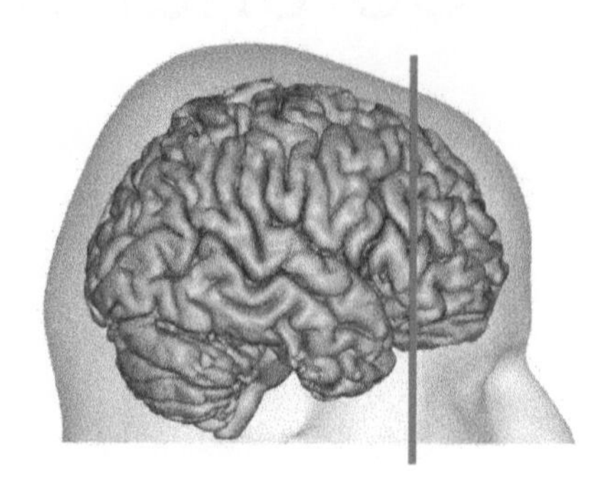

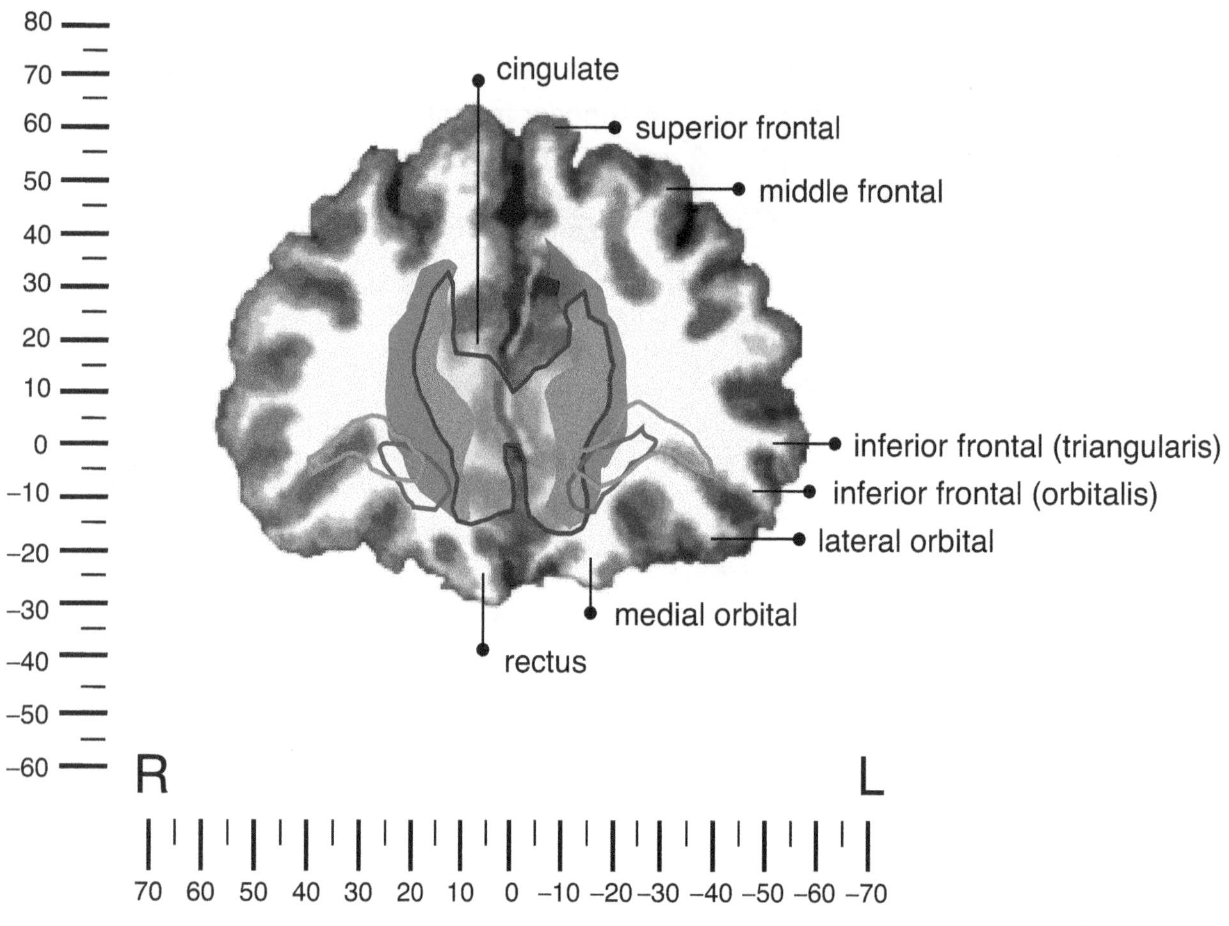

TRACT LEGEND

CC	Corpus Callosum
IFOF	Inf. Fronto-Occipital Fasc.
Cing	Cingulum
Unc	Uncinate

Coronal slice 36

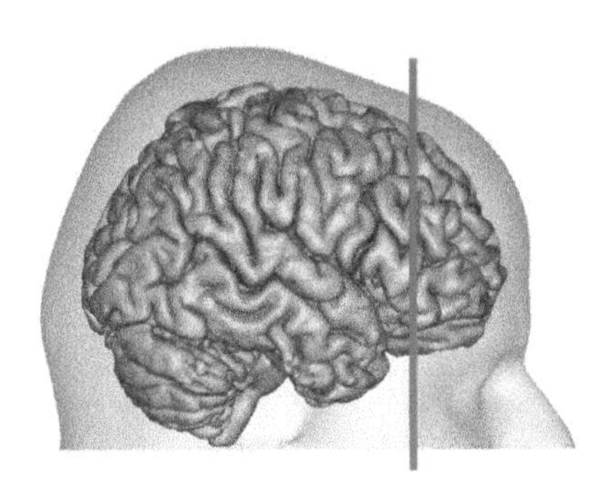

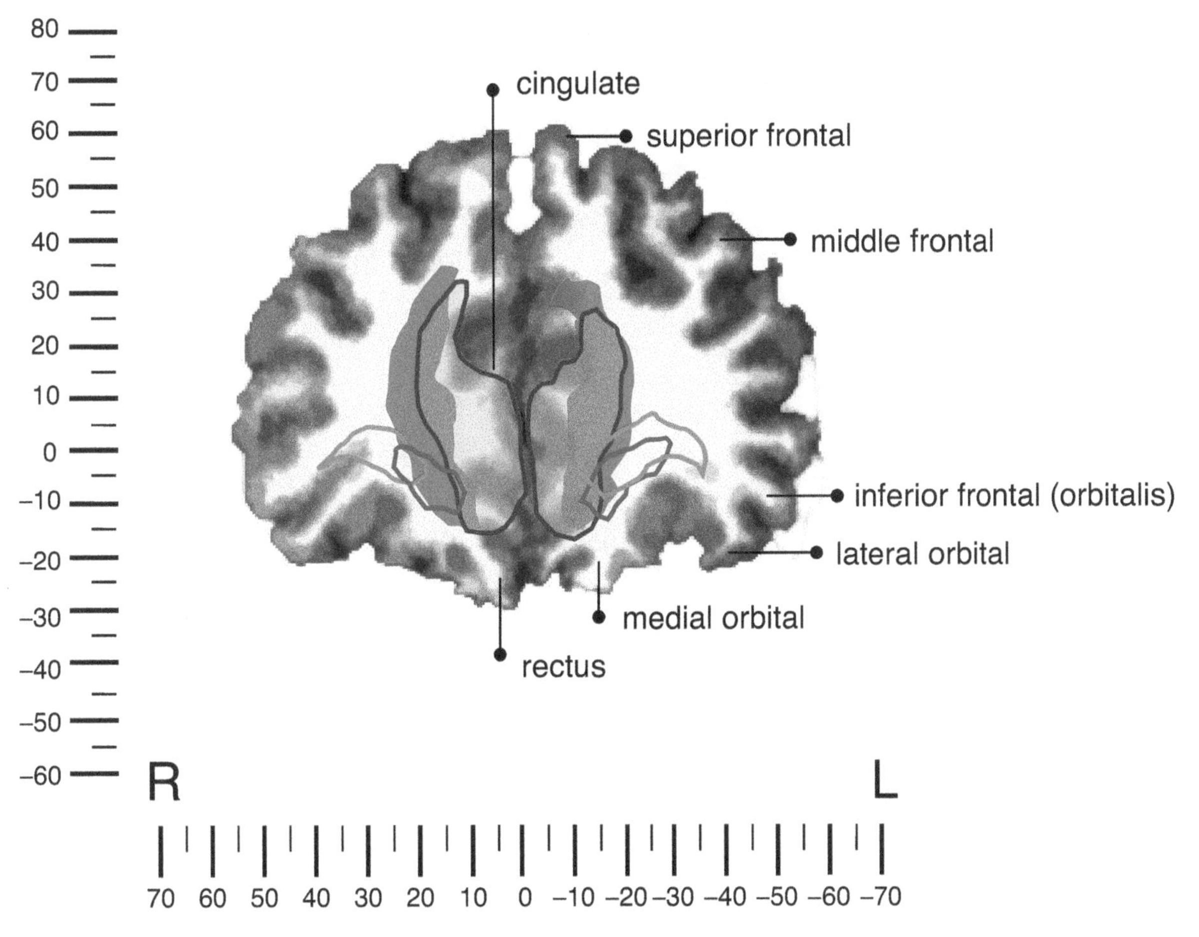

TRACT LEGEND

CC	Corpus Callosum
IFOF	Inf. Fronto-Occipital Fasc.
Cing	Cingulum
Unc	Uncinate

Coronal slice 38

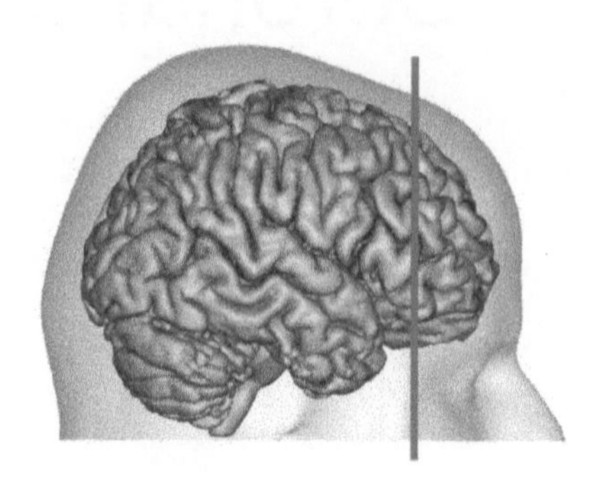

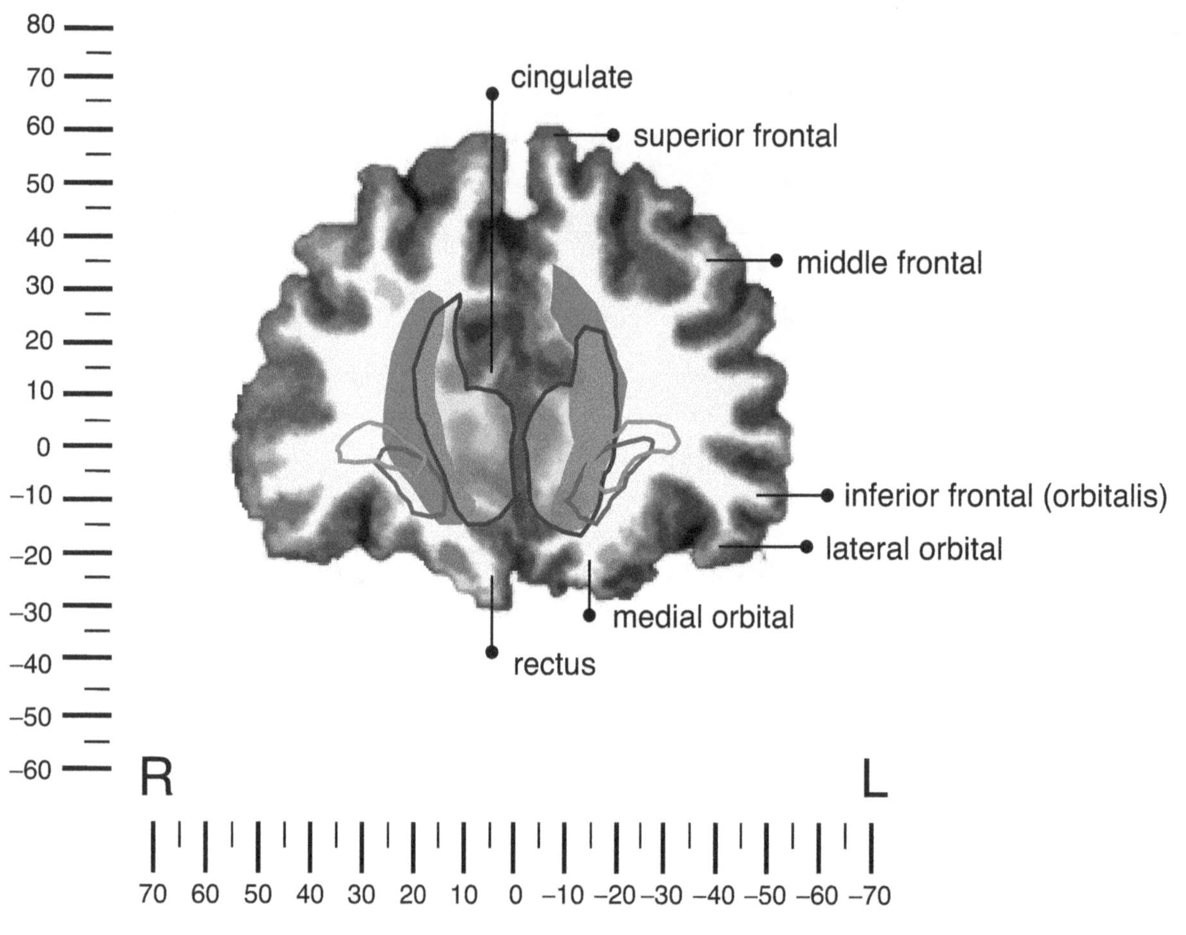

TRACT LEGEND

CC	Corpus Callosum
IFOF	Inf. Fronto-Occipital Fasc.
Cing	Cingulum
Unc	Uncinate

Coronal slice 40

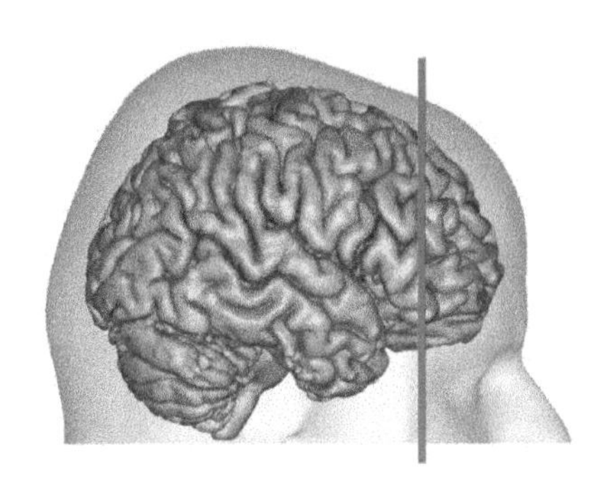

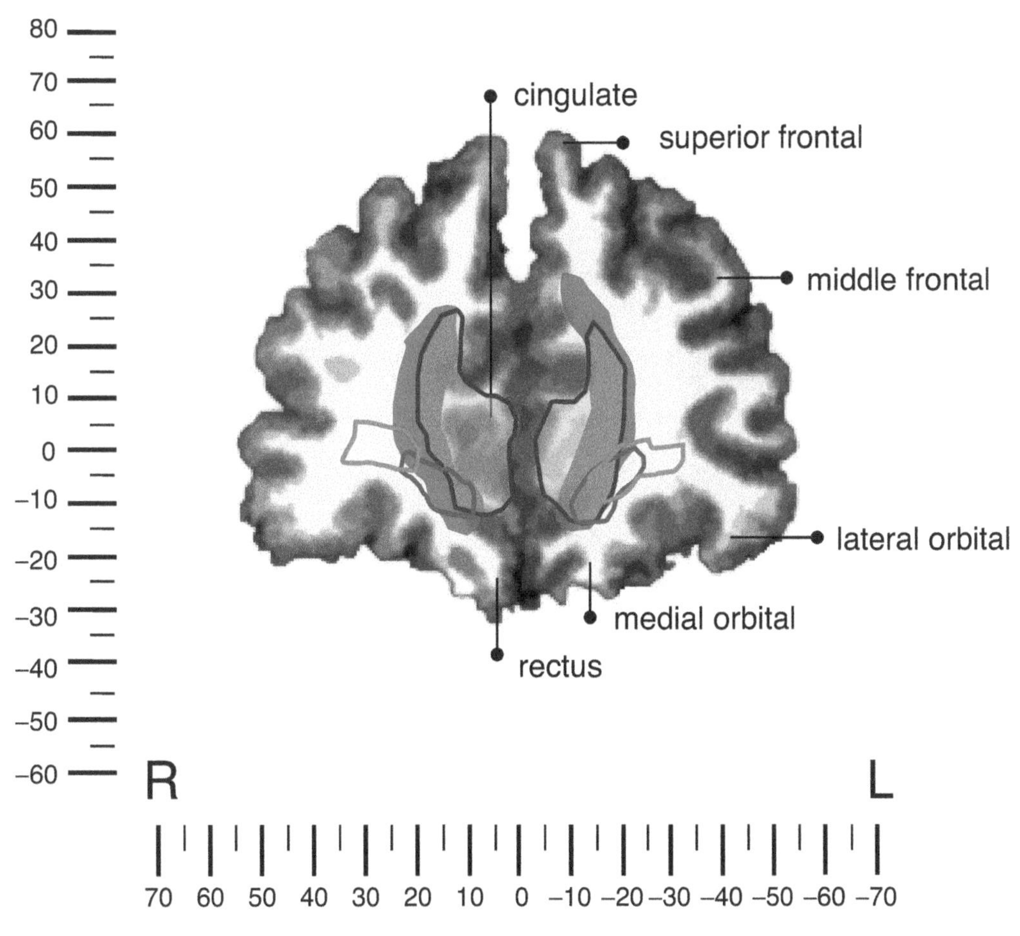

TRACT LEGEND

CC Corpus Callosum

IFOF Inf. Fronto-Occipital Fasc.

Cing Cingulum

Unc Uncinate

Coronal slice 42

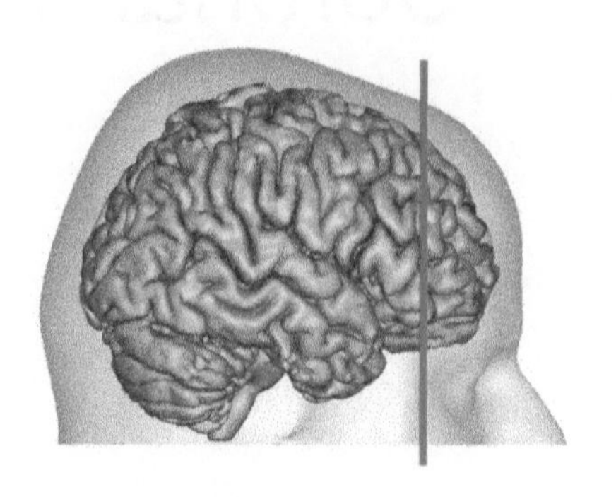

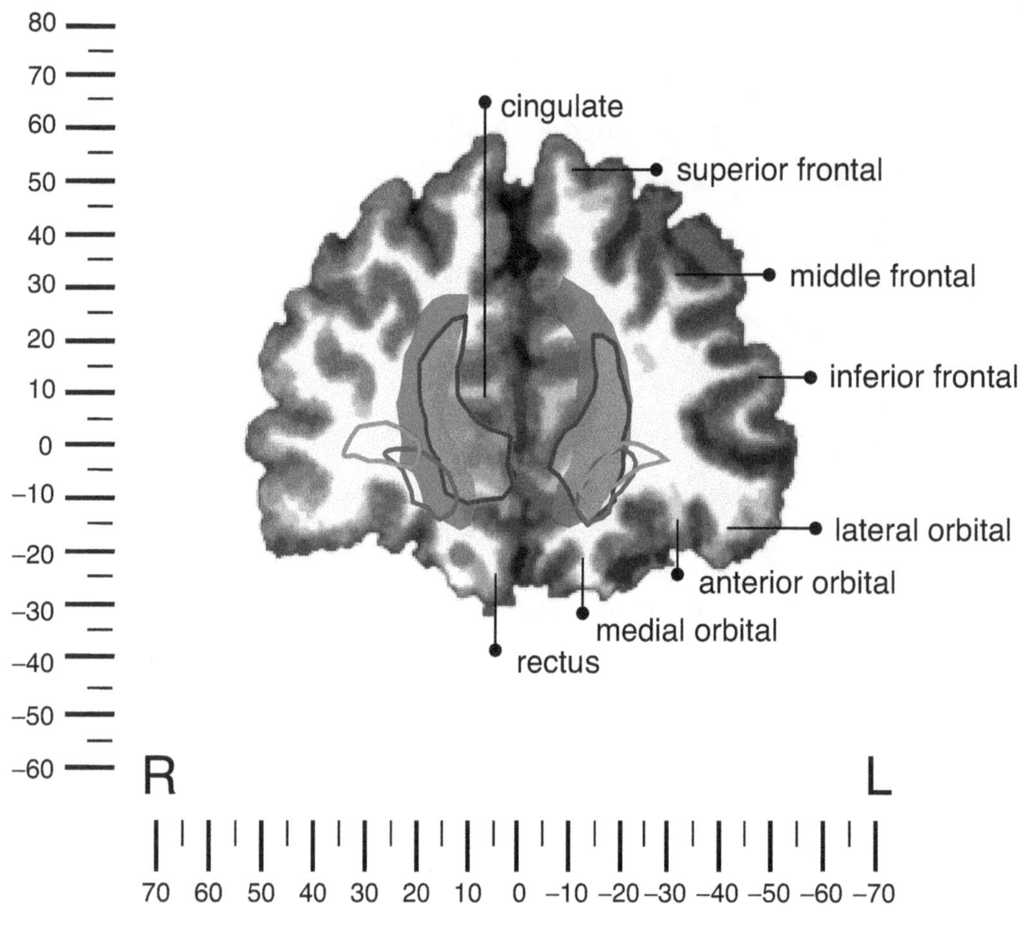

TRACT LEGEND

CC Corpus Callosum

IFOF Inf. Fronto-Occipital Fasc.

Cing Cingulum

Unc Uncinate

Coronal slice 44

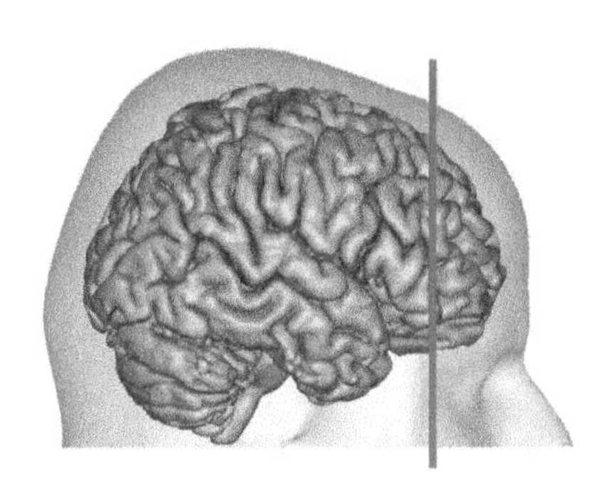

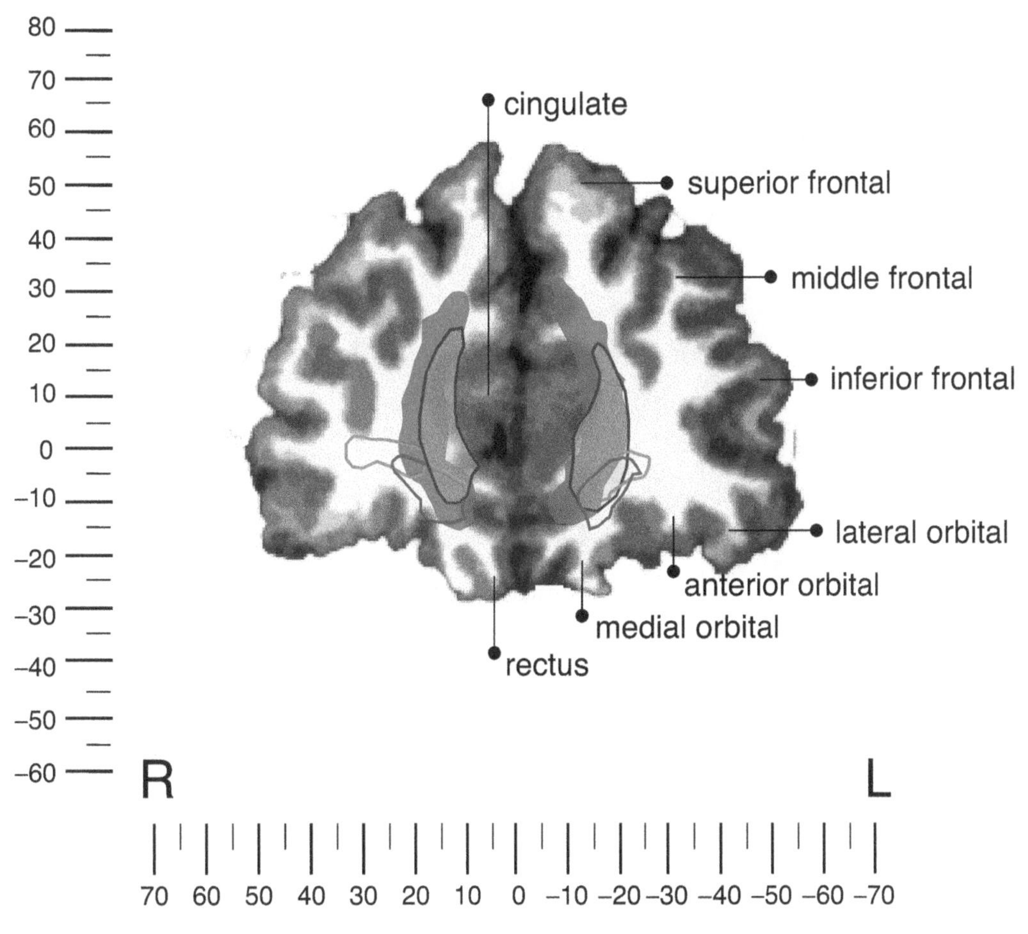

TRACT LEGEND

CC Corpus Callosum

IFOF Inf. Fronto-Occipital Fasc.

Cing Cingulum

Unc Uncinate

Sagittal maps

Sagittal slice 54

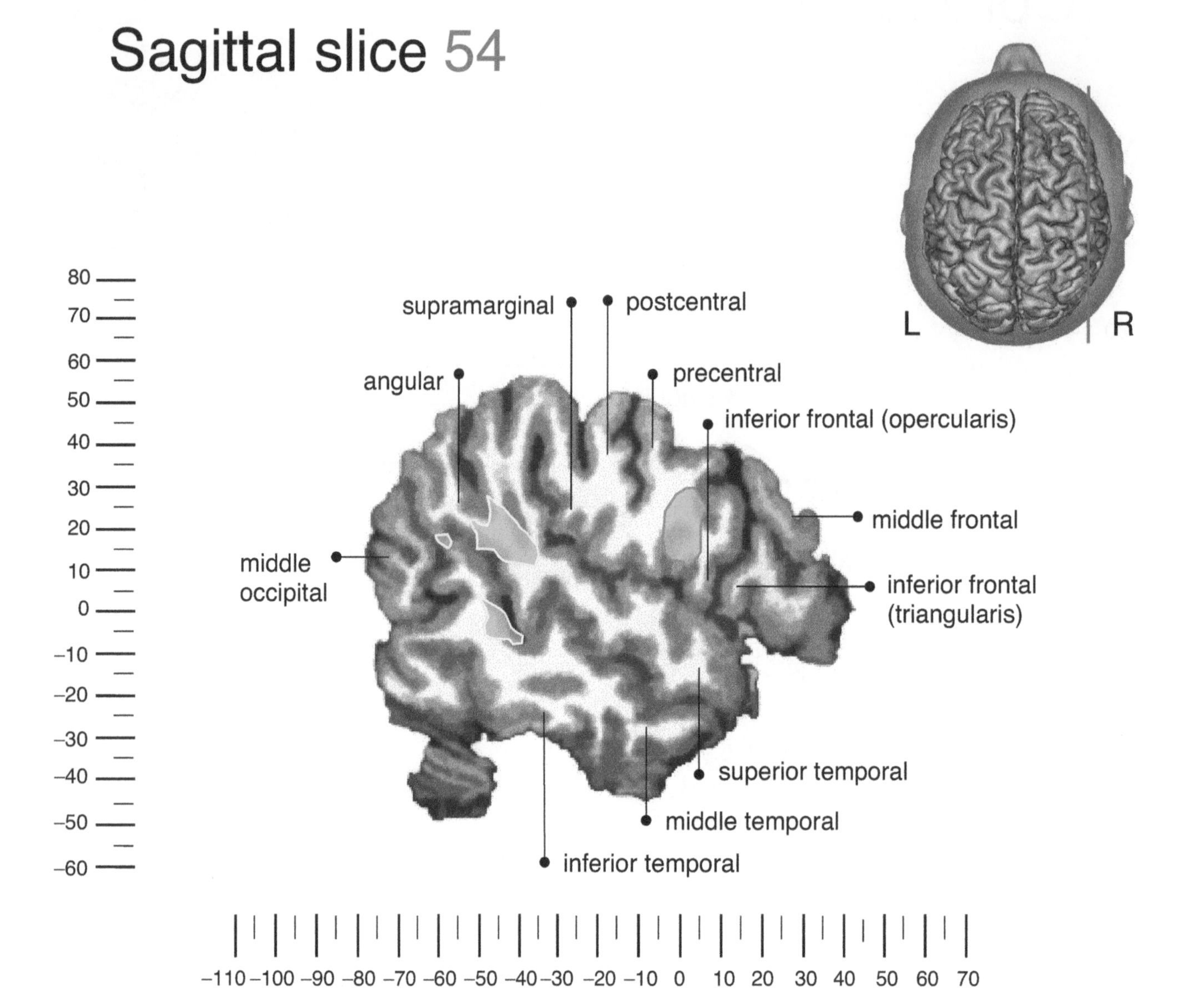

TRACT LEGEND

AF-P	Arcuate (post. segment)	AF-A	Arcuate (ant. segment)

Sagittal slice 52

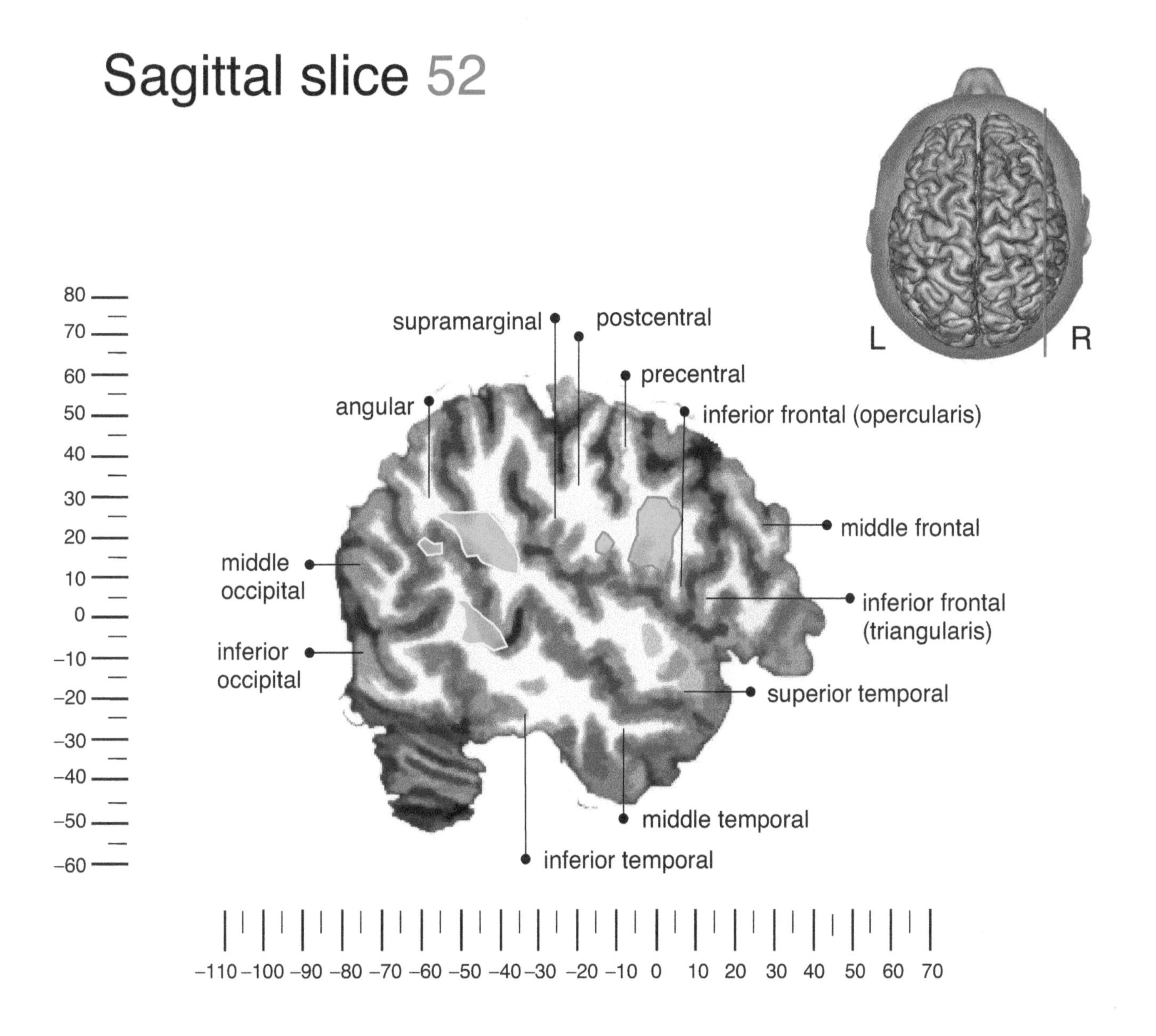

TRACT LEGEND

AF-P	Arcuate (post. segment)	AF-A	Arcuate (ant. segment)

Sagittal slice 50

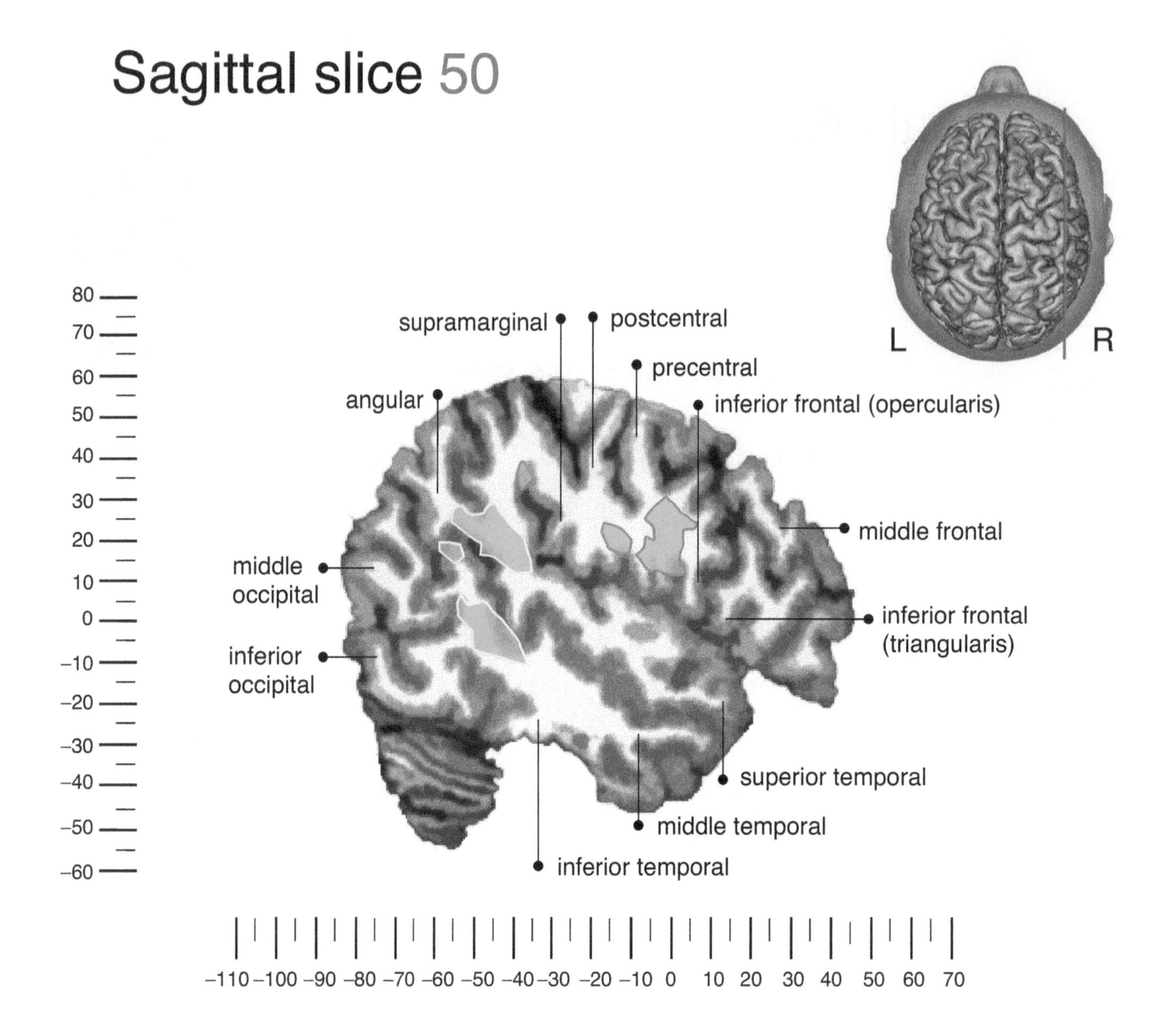

TRACT LEGEND

AF-P	Arcuate (post. segment)	AF-A	Arcuate (ant. segment)

Sagittal slice 48

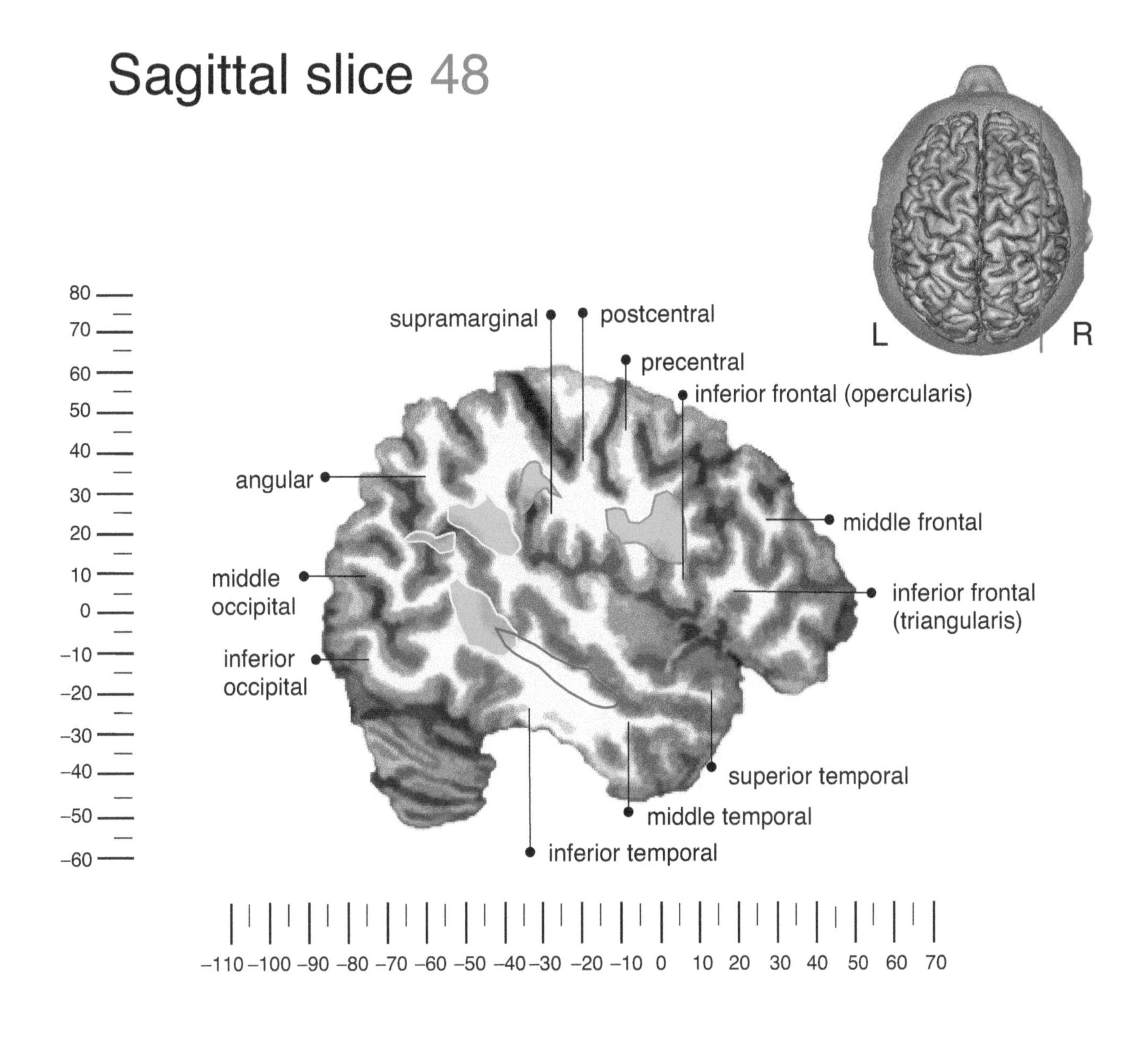

TRACT LEGEND

AF-P	Arcuate (post. segment)	AF-A	Arcuate (ant. segment)	ILF	Inf. Longitudinal Fasc.

Sagittal slice 46

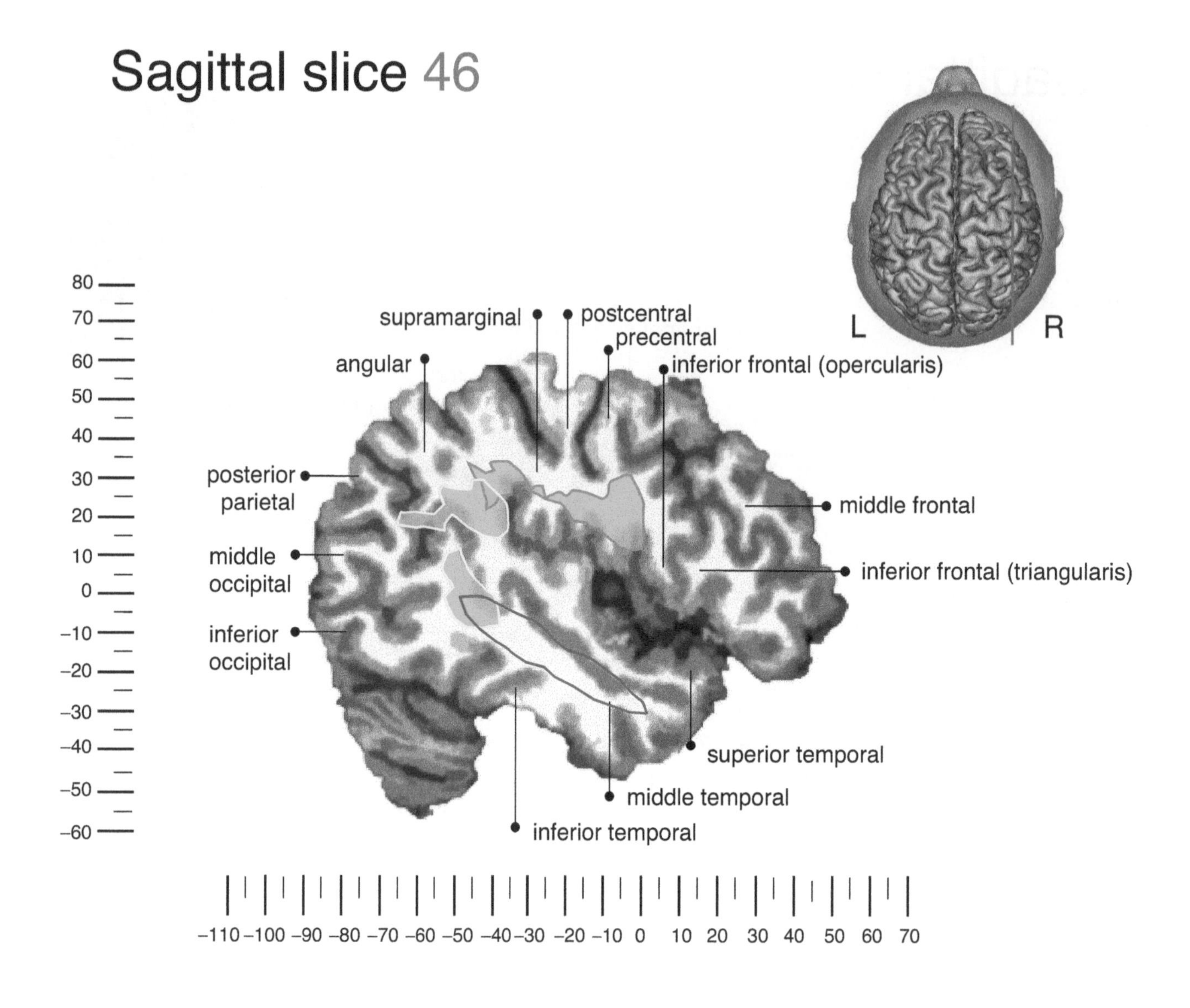

TRACT LEGEND

AF-P	Arcuate (post. segment)	AF-A	Arcuate (ant. segment)	ILF	Inf. Longitudinal Fasc.

Sagittal slice 44

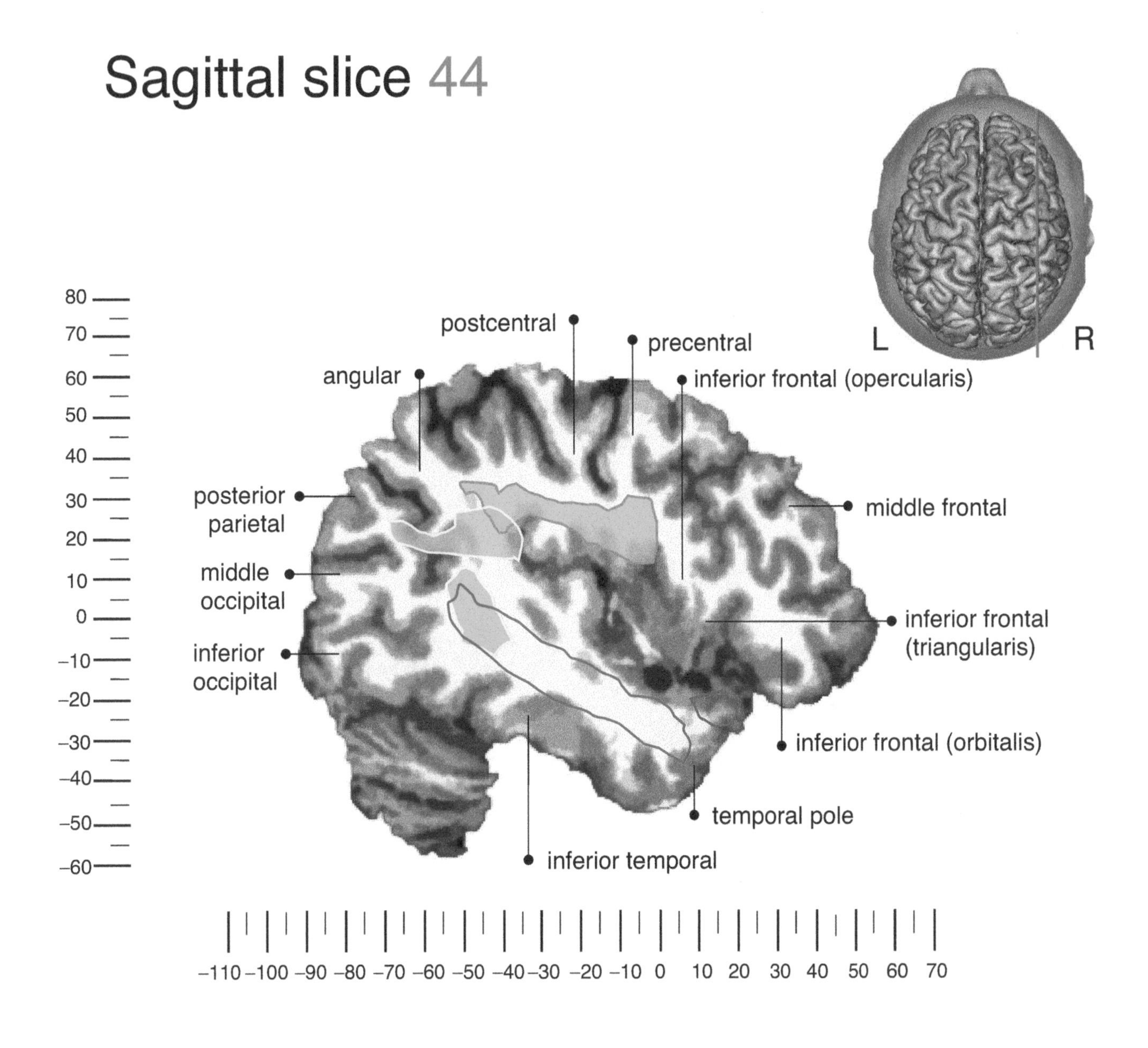

TRACT LEGEND

AF-P	Arcuate (post. segment)
AF-A	Arcuate (ant. segment)
ILF	Inf. Longitudinal Fasc.
Unc	Uncinate

Sagittal slice 42

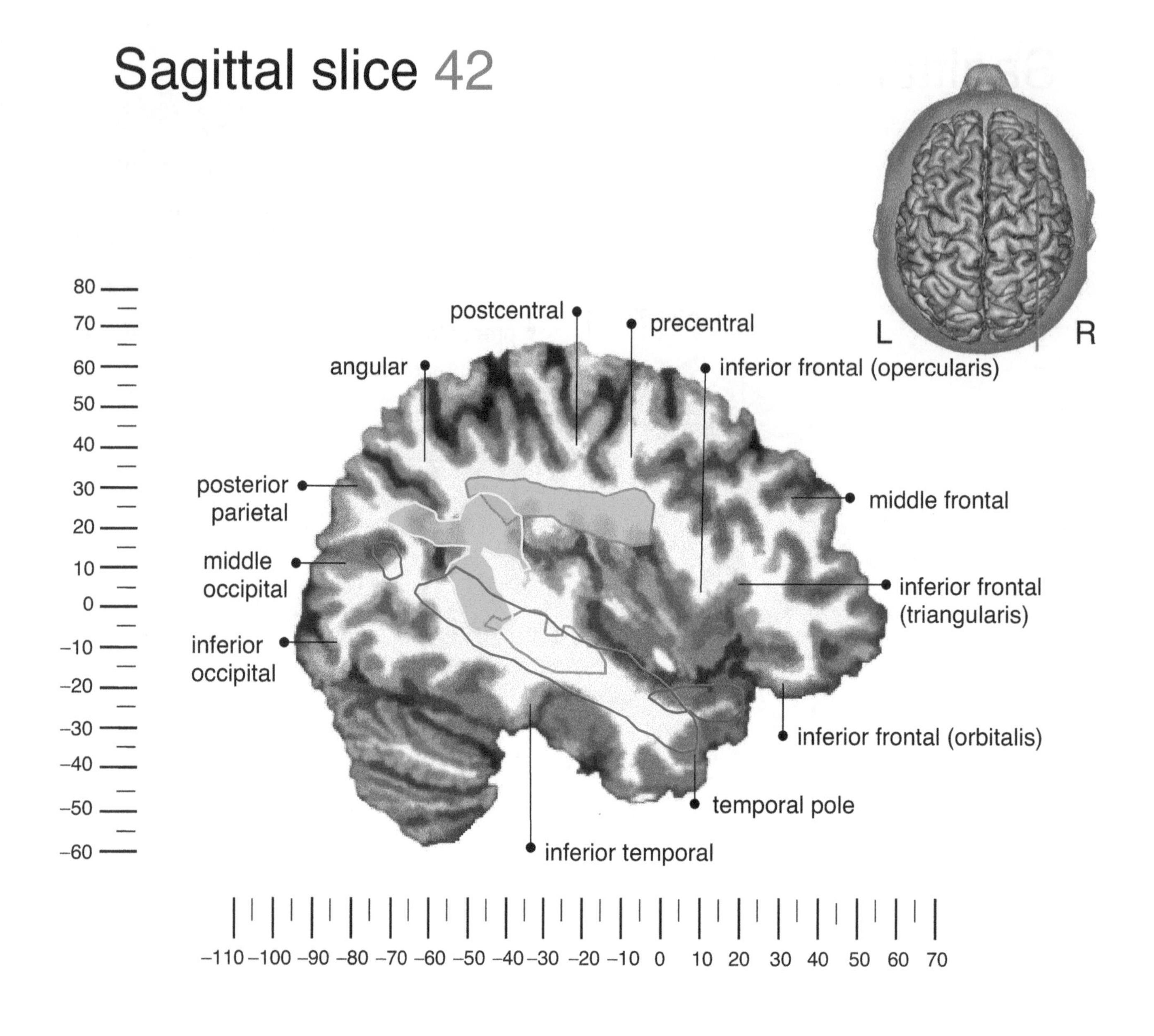

TRACT LEGEND

AF-P	Arcuate (post. segment)	AF-A	Arcuate (ant. segment)	ILF	Inf. Longitudinal Fasc.
Unc	Uncinate	IFOF	Inf. Fronto-Occipital Fasc.		

Sagittal slice 40

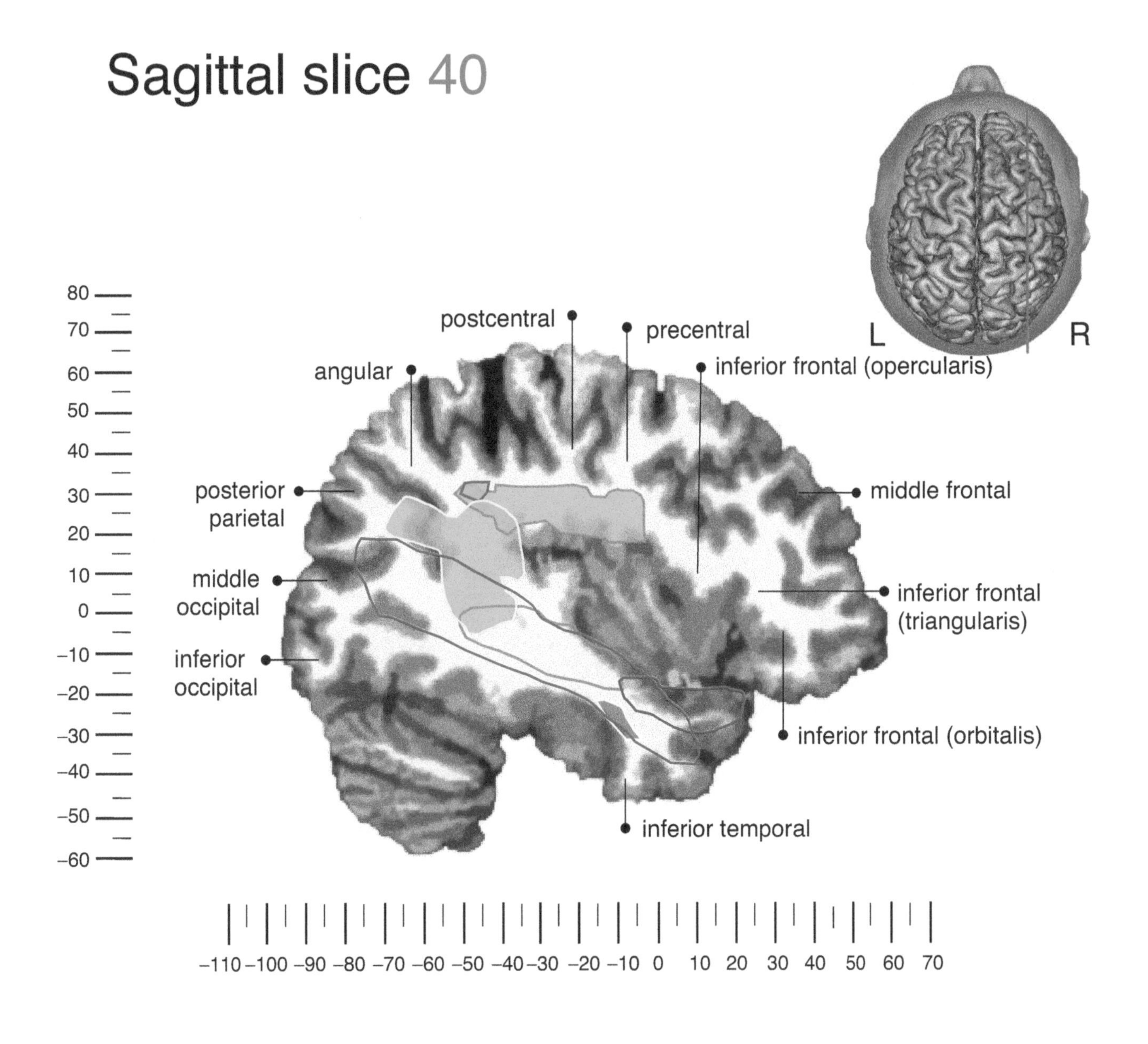

TRACT LEGEND

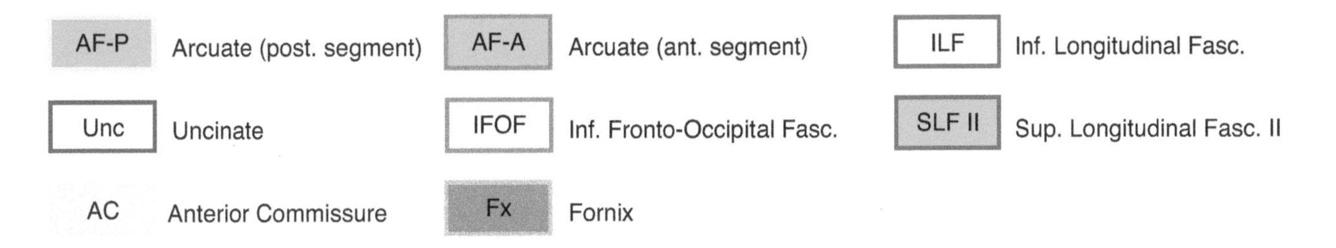

Sagittal slice 38

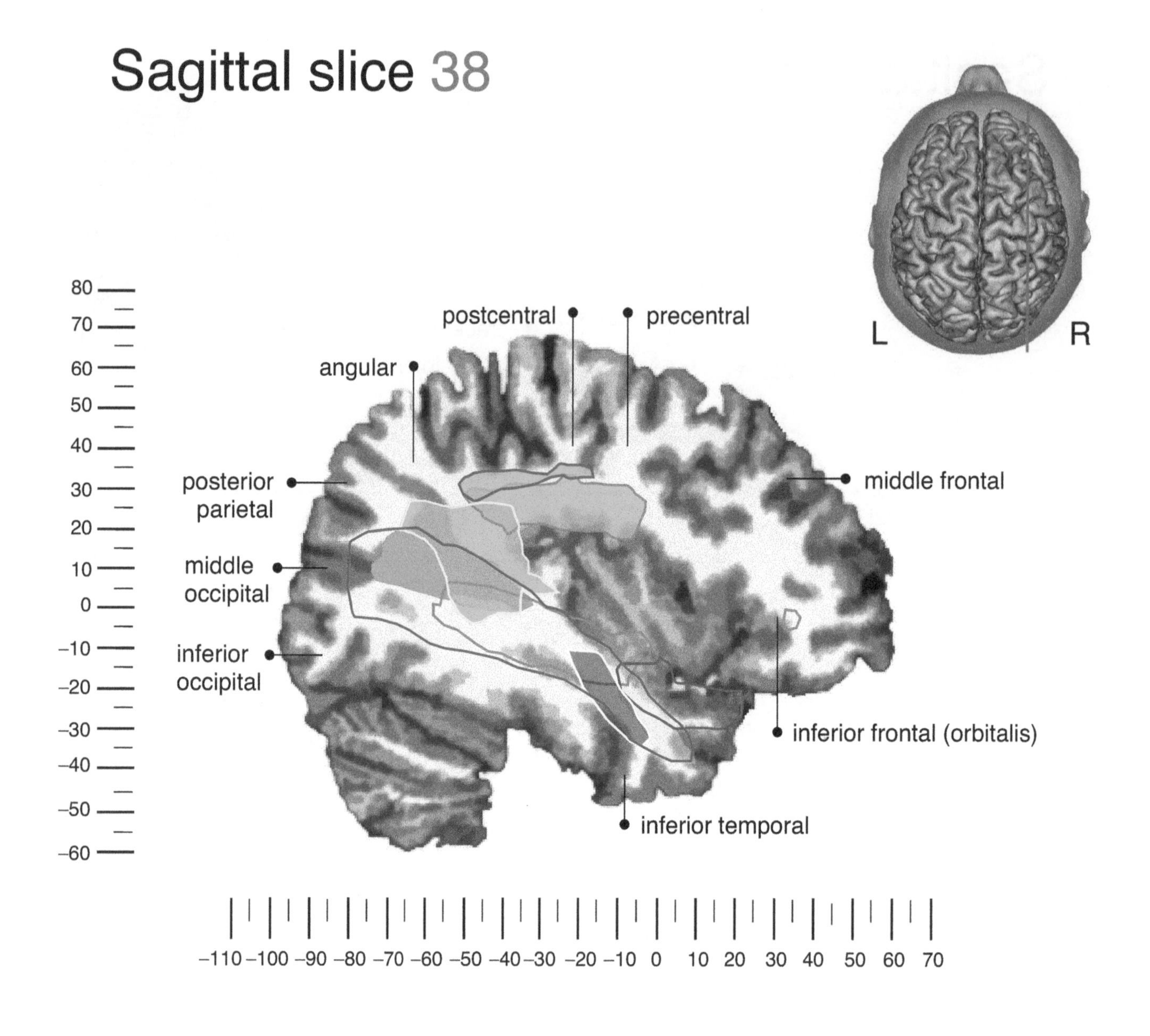

TRACT LEGEND

AF-P	Arcuate (post. segment)	AF-A	Arcuate (ant. segment)	ILF	Inf. Longitudinal Fasc.
Unc	Uncinate	IFOF	Inf. Fronto-Occipital Fasc.	SLF II	Sup. Longitudinal Fasc. II
AC	Anterior Commissure	Fx	Fornix		

Sagittal slice 36

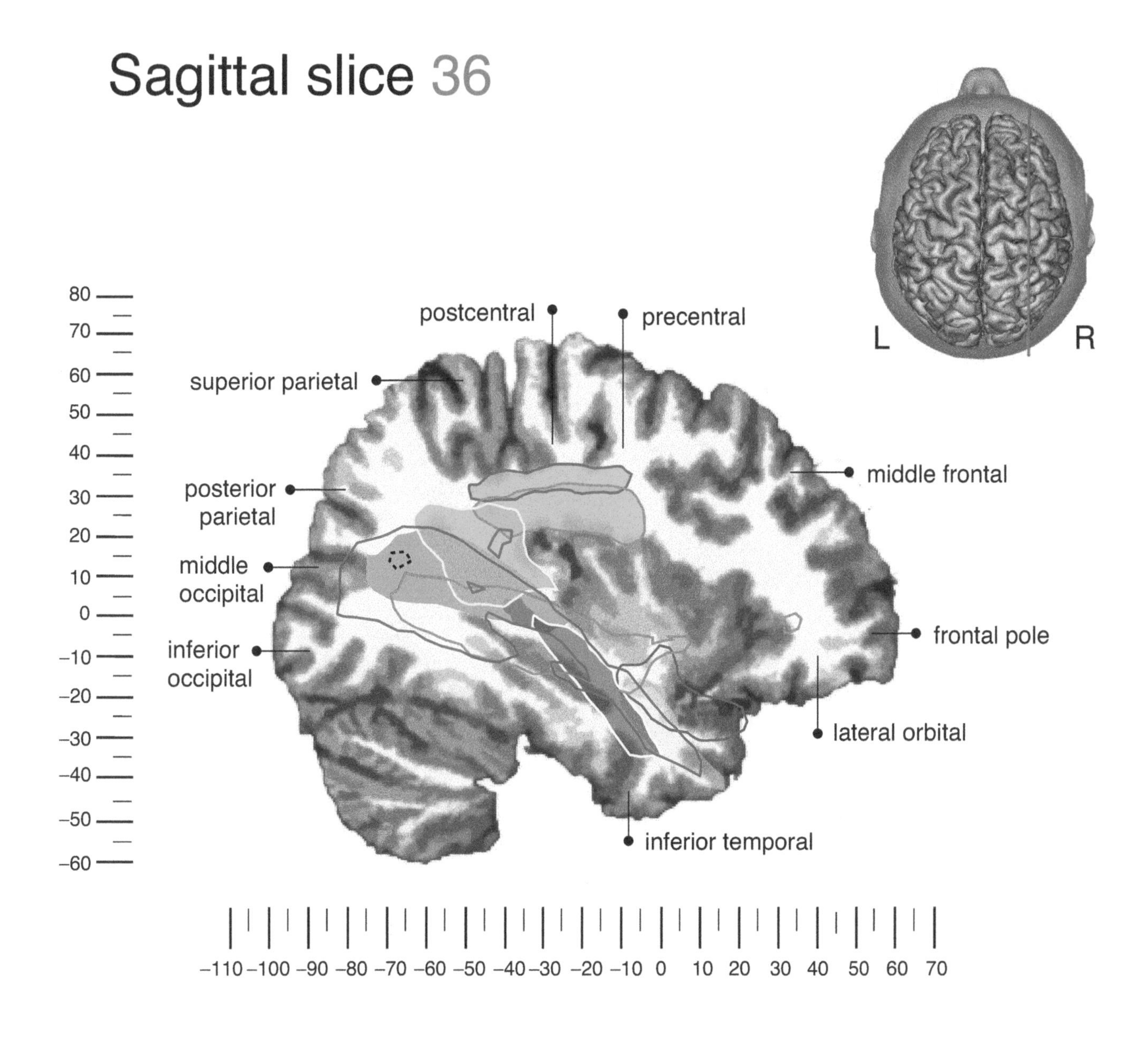

TRACT LEGEND

AF-P	Arcuate (post. segment)	AF-A	Arcuate (ant. segment)	ILF	Inf. Longitudinal Fasc.
Unc	Uncinate	IFOF	Inf. Fronto-Occipital Fasc.	SLF II	Sup. Longitudinal Fasc. II
AC	Anterior Commissure	Fx	Fornix	CC	Corpus Callosum
OR	Optic Radiations	IC	Internal Capsule	AF-L	Arcuate (long segment)

Sagittal slice 34

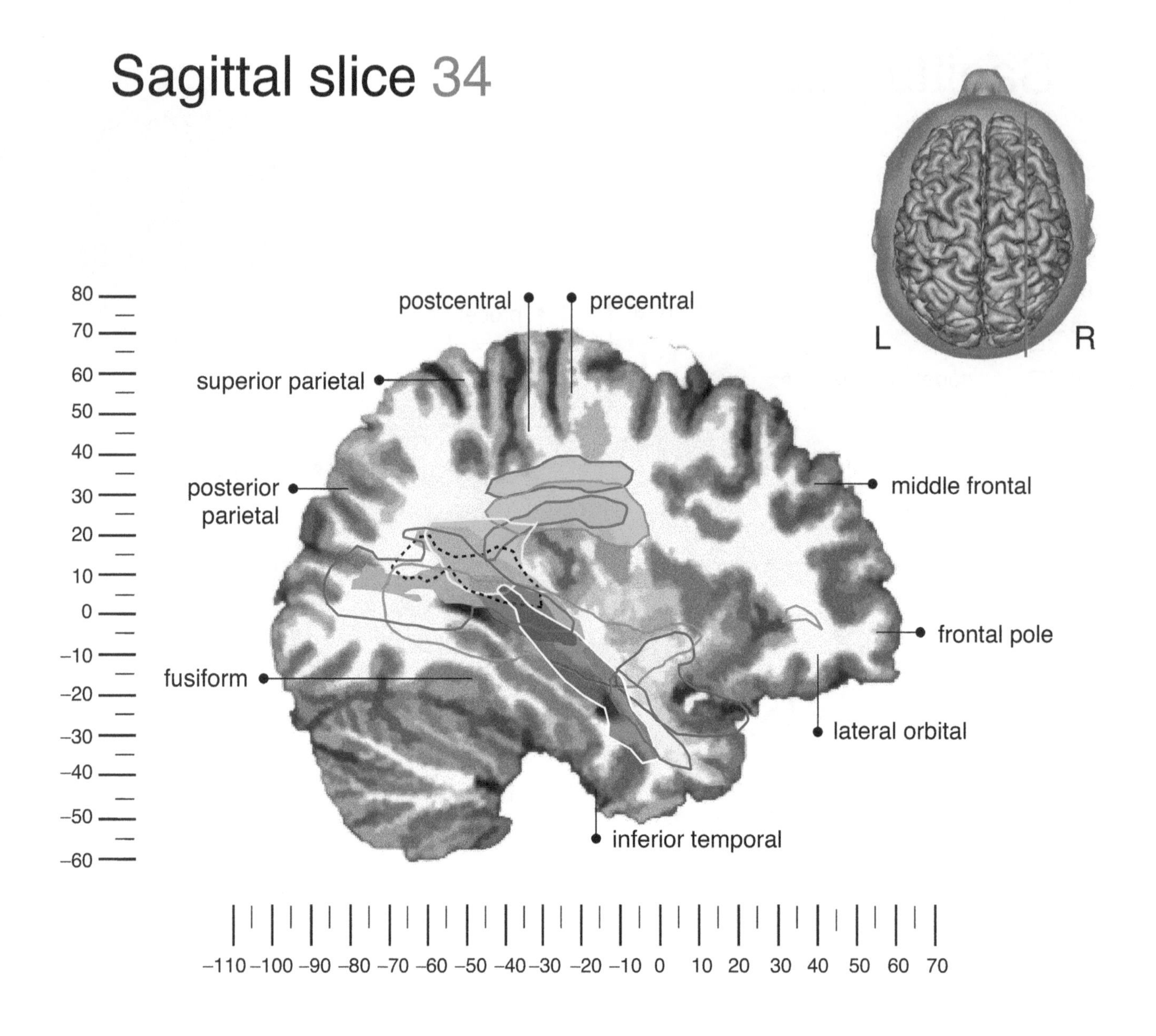

TRACT LEGEND

AF-P	Arcuate (post. segment)	AF-A	Arcuate (ant. segment)	ILF	Inf. Longitudinal Fasc.
Unc	Uncinate	IFOF	Inf. Fronto-Occipital Fasc.	SLF II	Sup. Longitudinal Fasc. II
AC	Anterior Commissure	Fx	Fornix	CC	Corpus Callosum
OR	Optic Radiations	IC	Internal Capsule	AF-L	Arcuate (long segment)

Sagittal slice 32

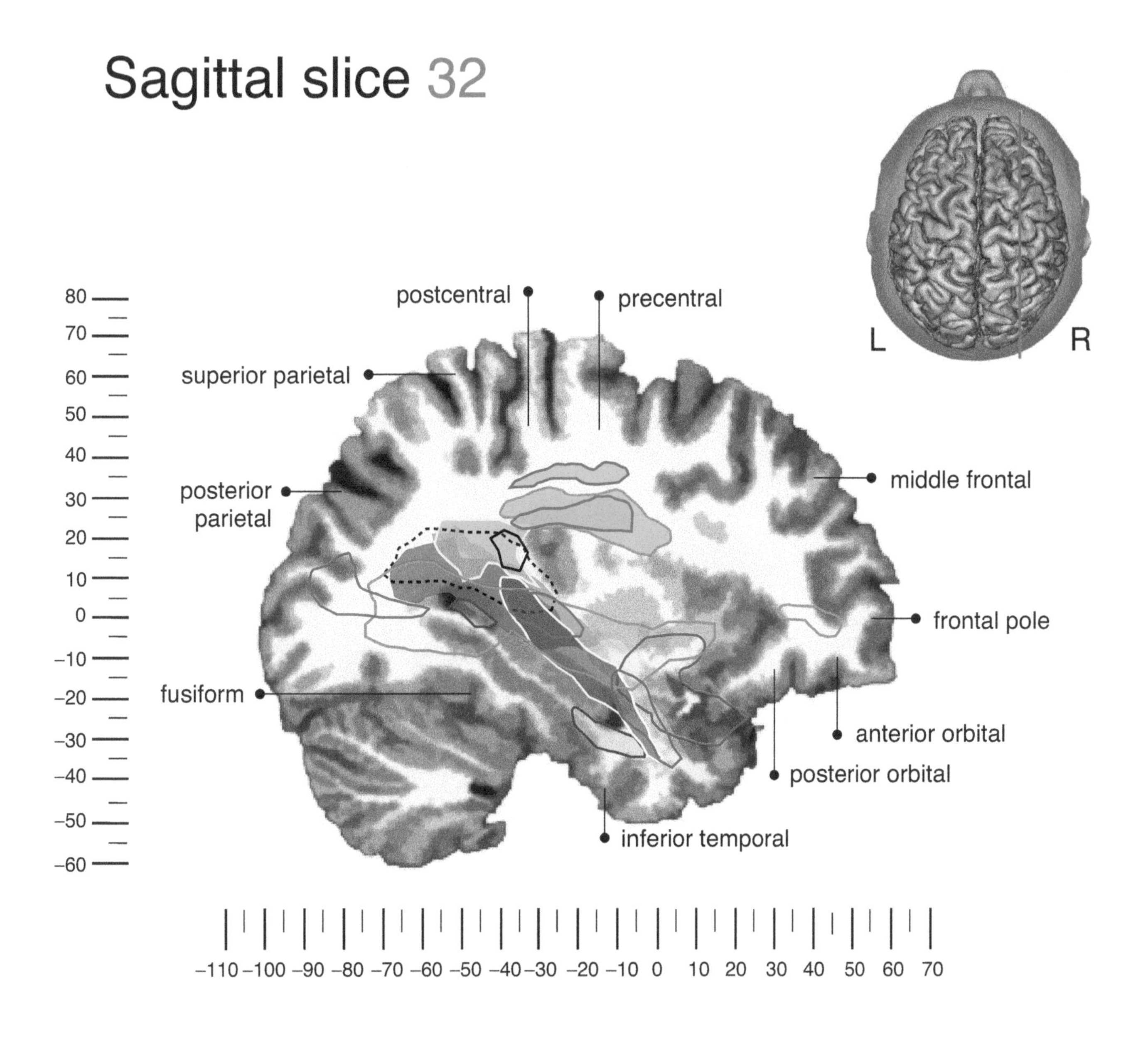

TRACT LEGEND

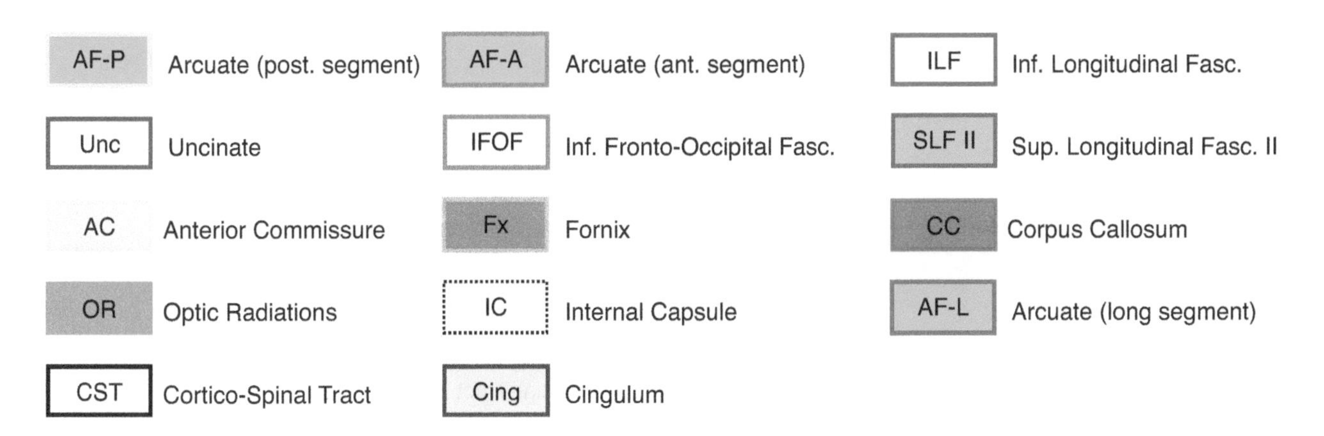

Sagittal slice 30

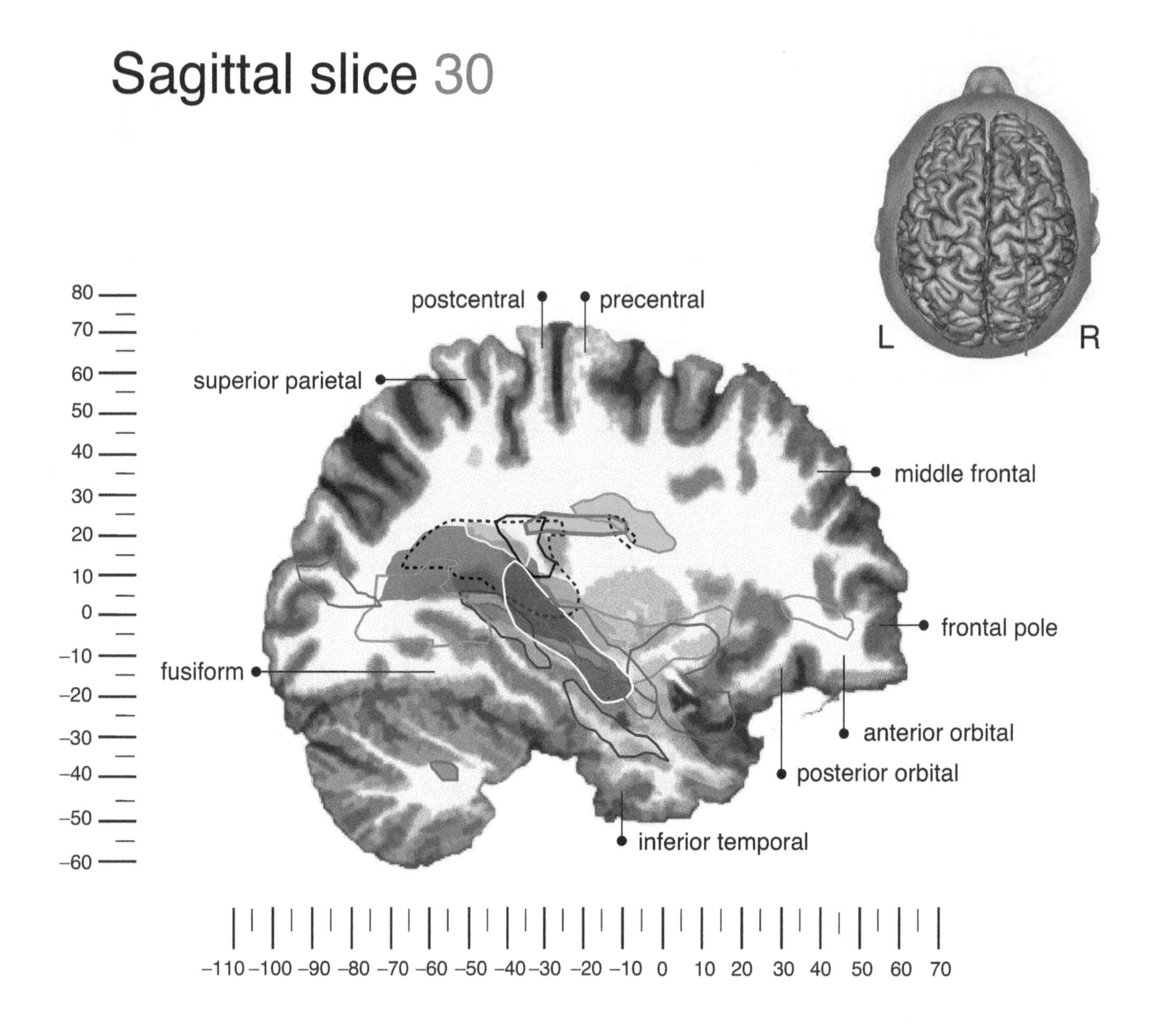

TRACT LEGEND

AF-P	Arcuate (post. segment)	AF-A	Arcuate (ant. segment)	ILF	Inf. Longitudinal Fasc.
Unc	Uncinate	IFOF	Inf. Fronto-Occipital Fasc.	Cing	Cingulum
AC	Anterior Commissure	Fx	Fornix	CC	Corpus Callosum
OR	Optic Radiations	IC	Internal Capsule	AF-L	Arcuate (long segment)
CST	Cortico-Spinal Tract	CPC	Cortico-Ponto-Cerebellar Tract		

Sagittal slice 28

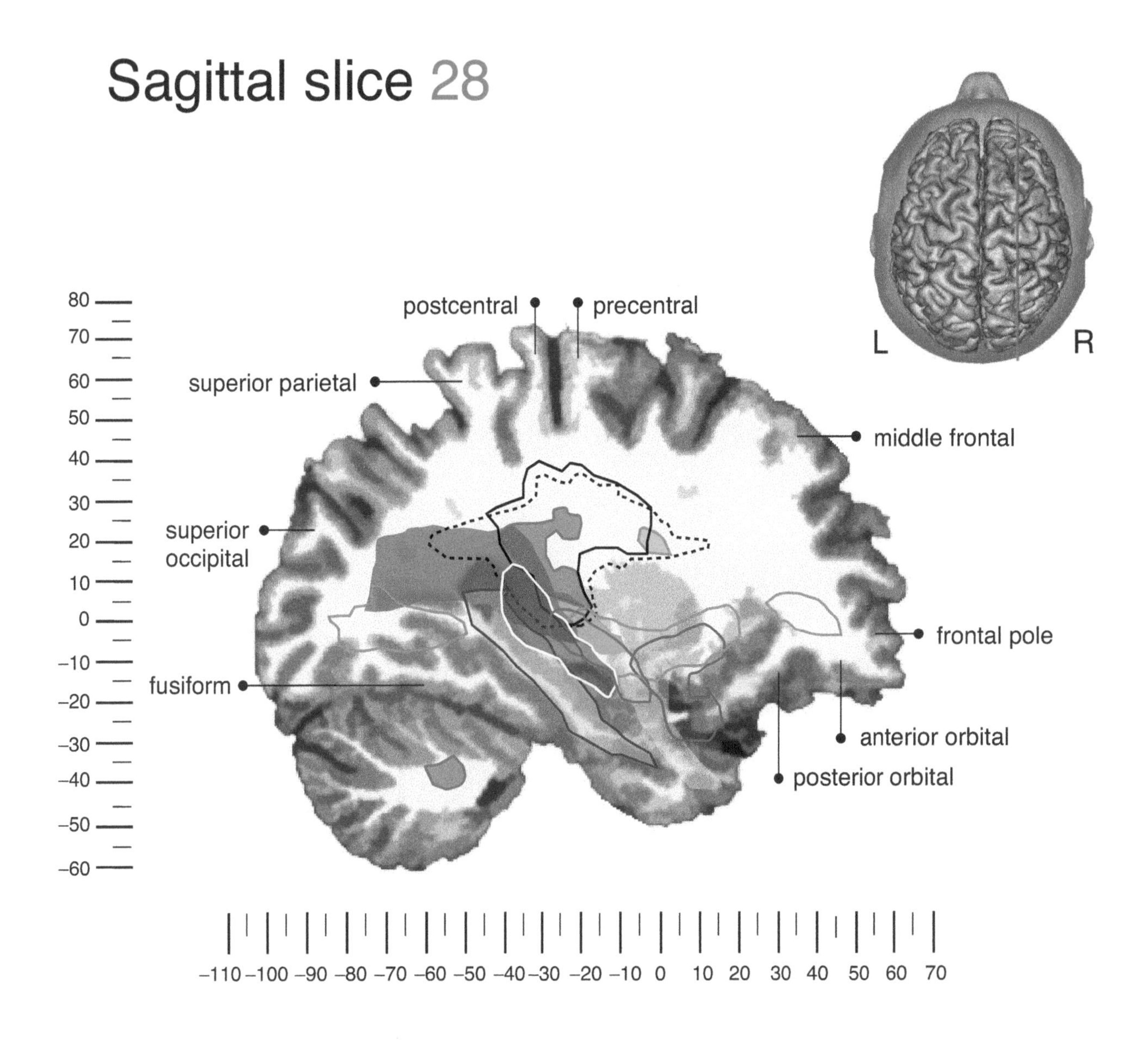

TRACT LEGEND

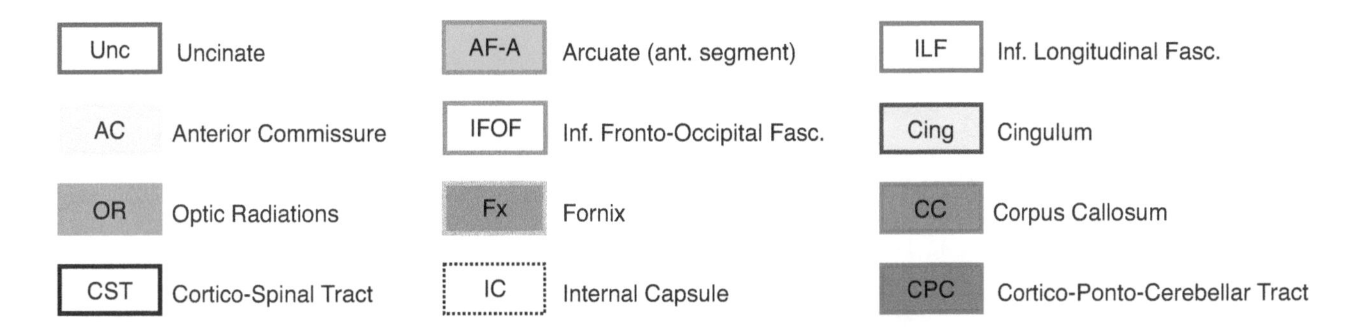

Sagittal slice 26

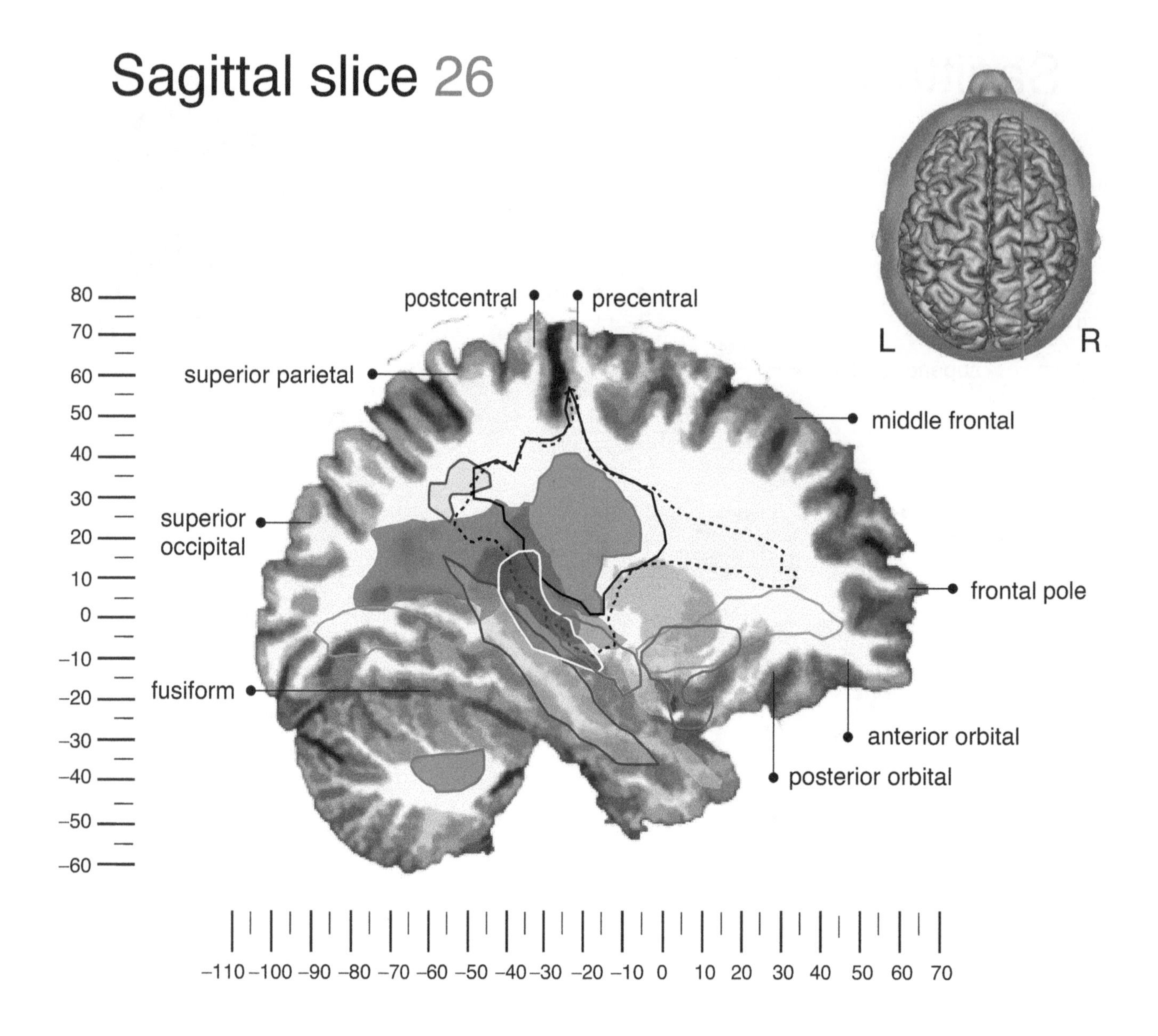

TRACT LEGEND

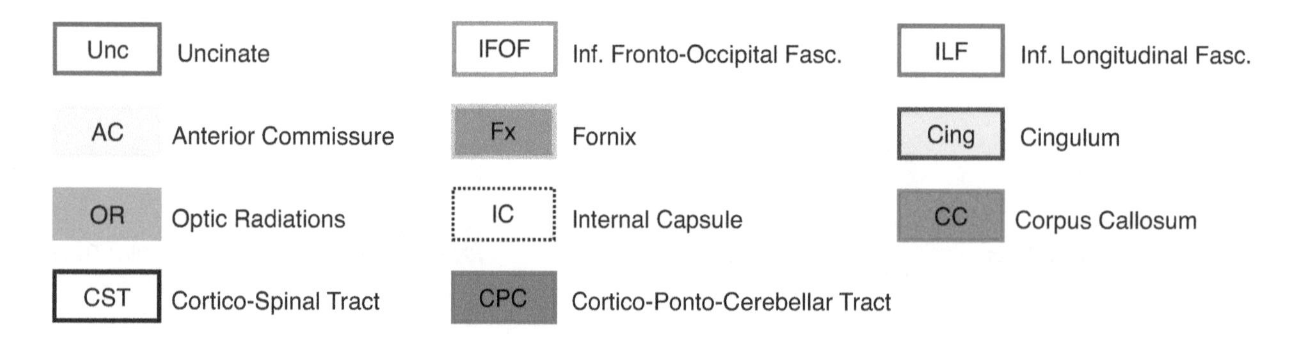

Sagittal slice 24

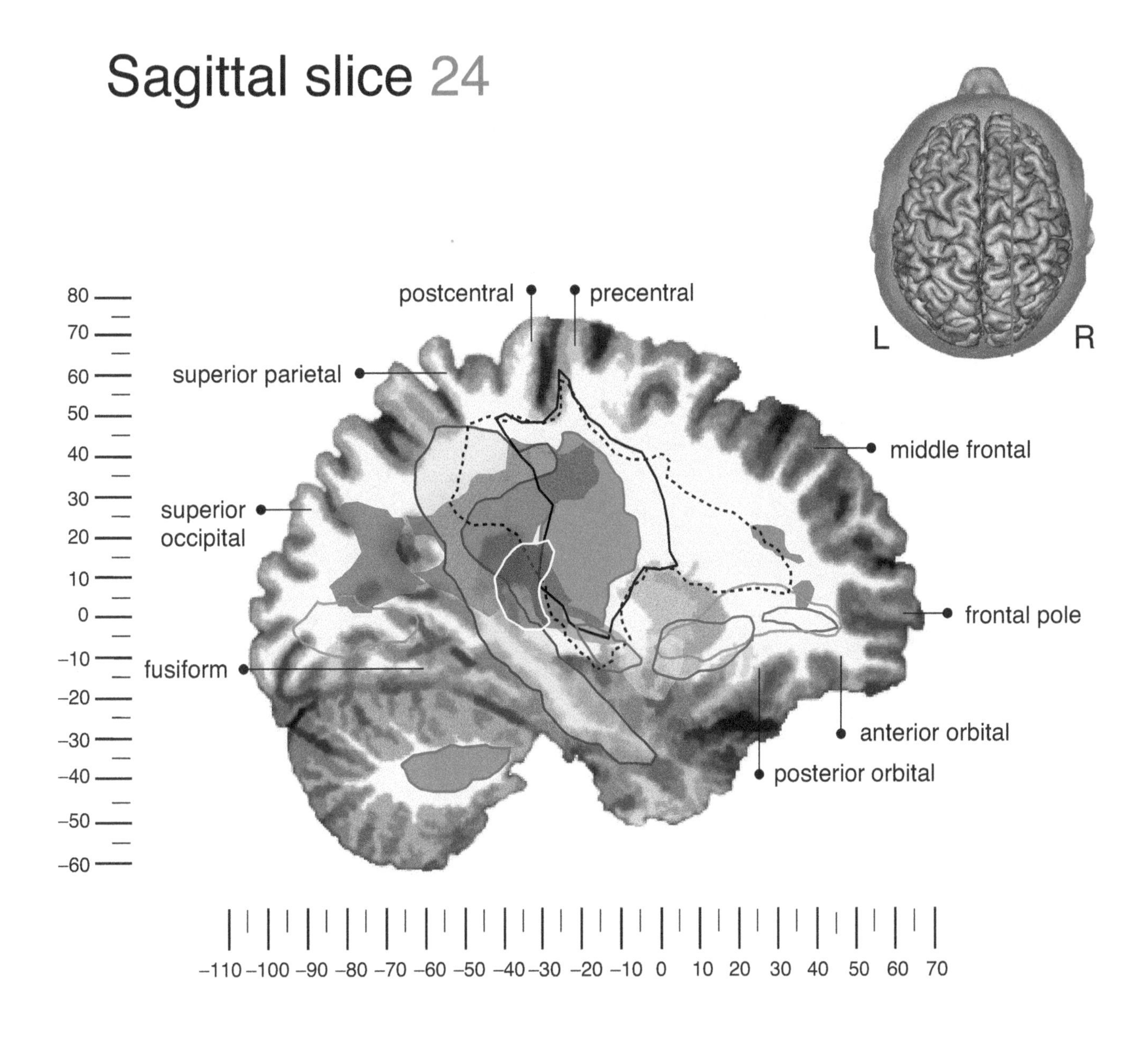

TRACT LEGEND

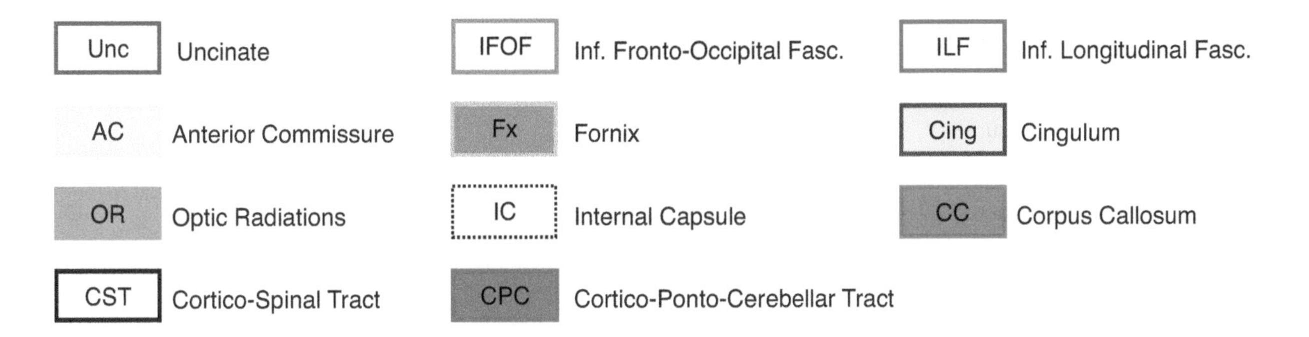

Sagittal slice 22

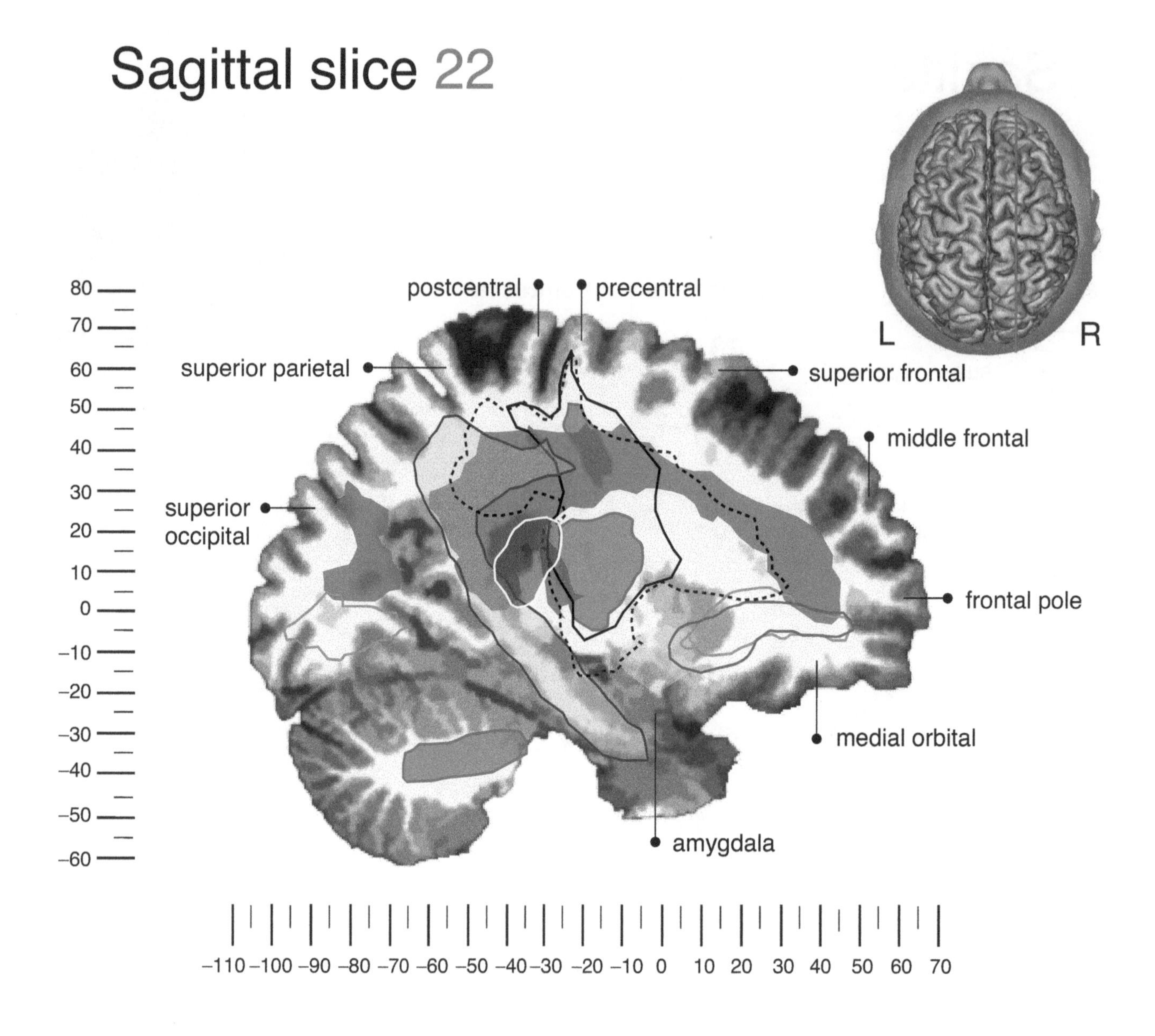

TRACT LEGEND

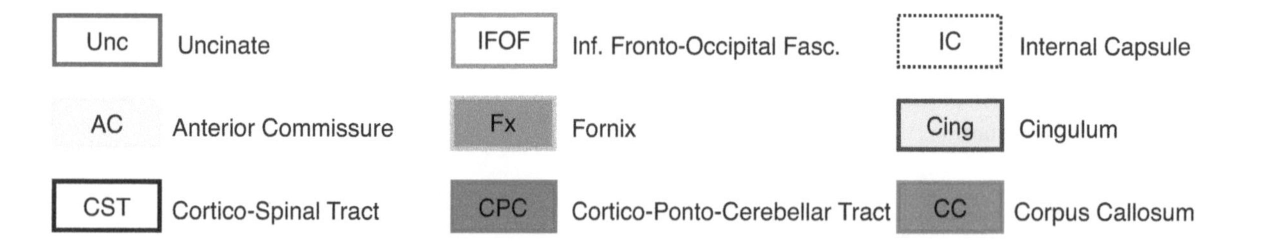

Sagittal slice 20

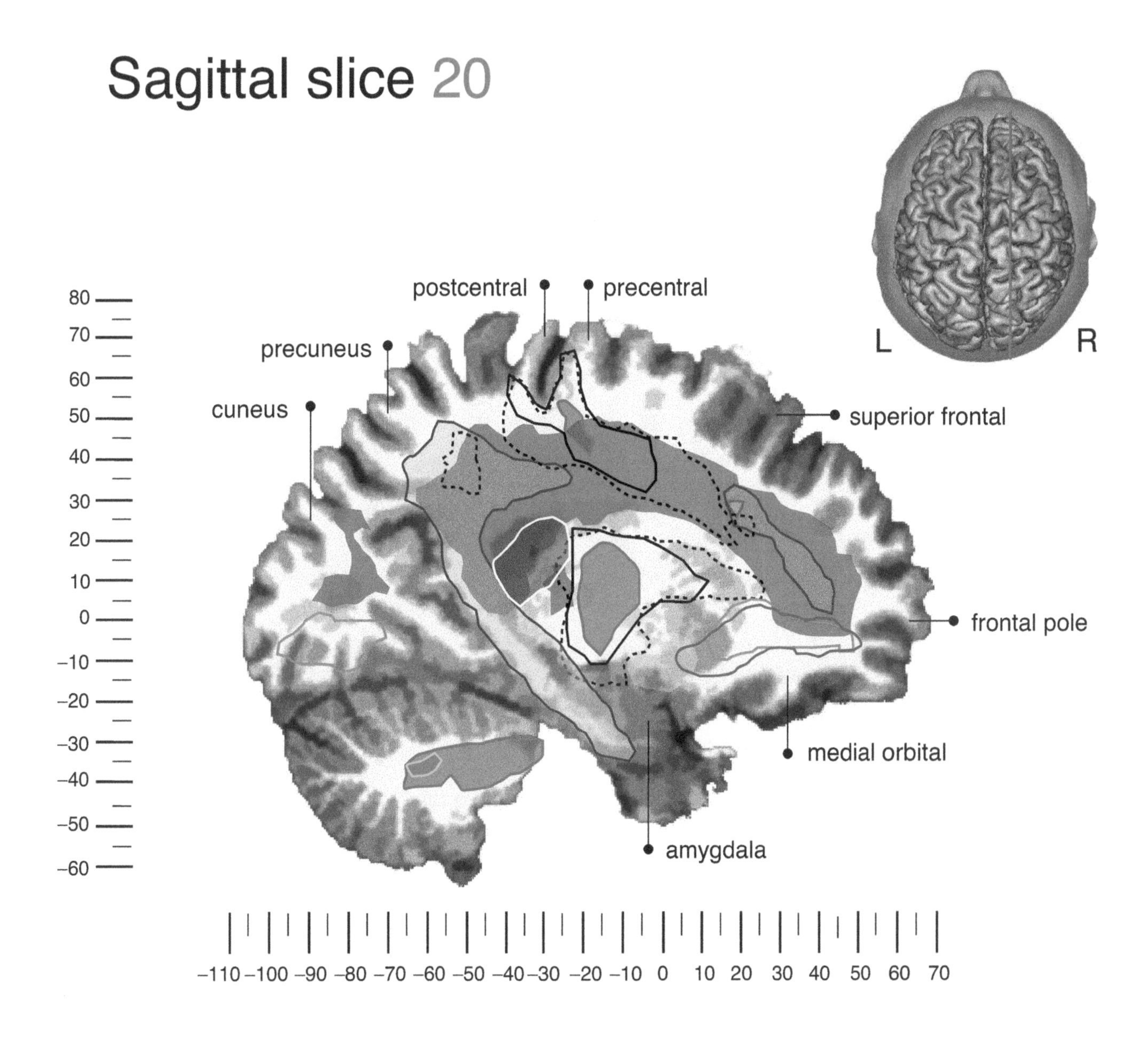

TRACT LEGEND

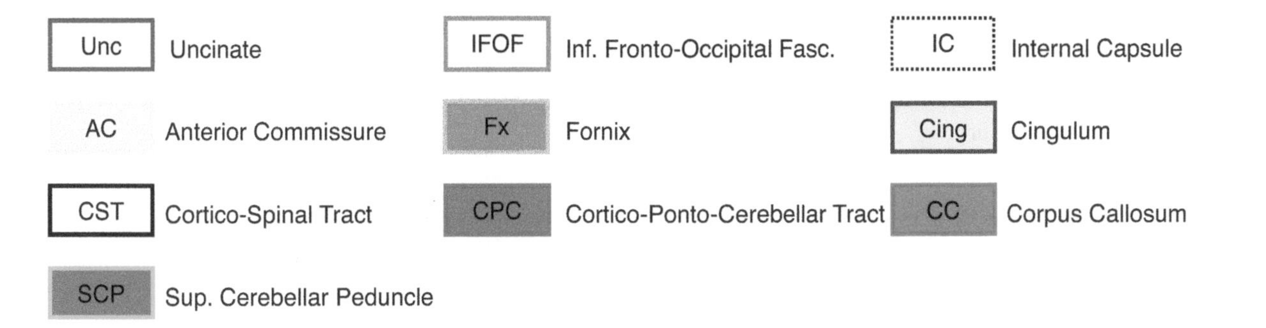

Sagittal slice 18

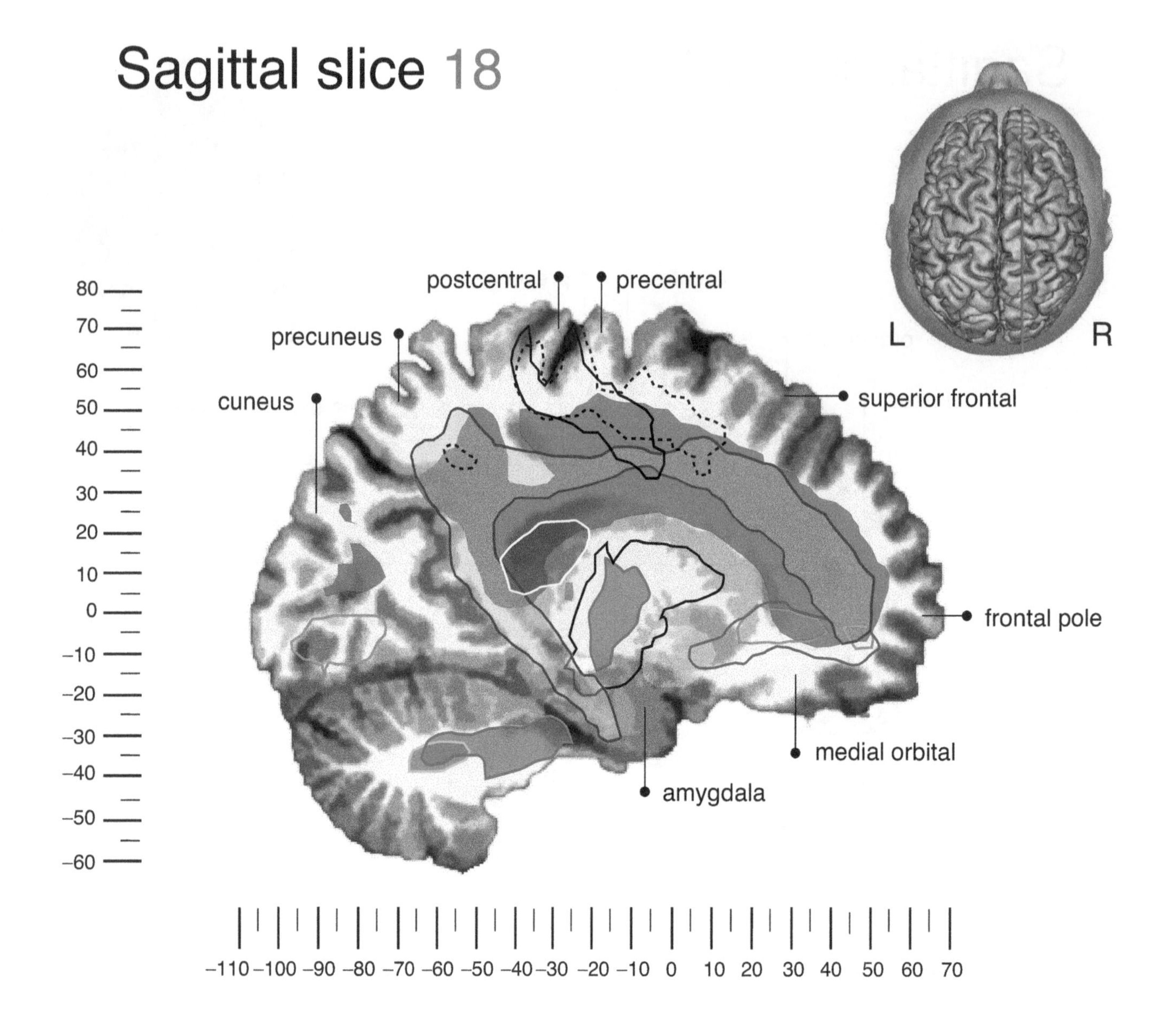

TRACT LEGEND

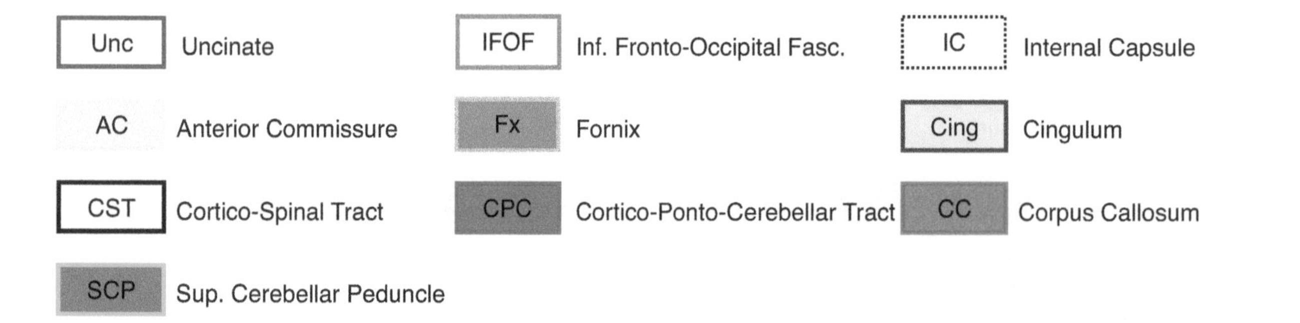

Sagittal slice 16

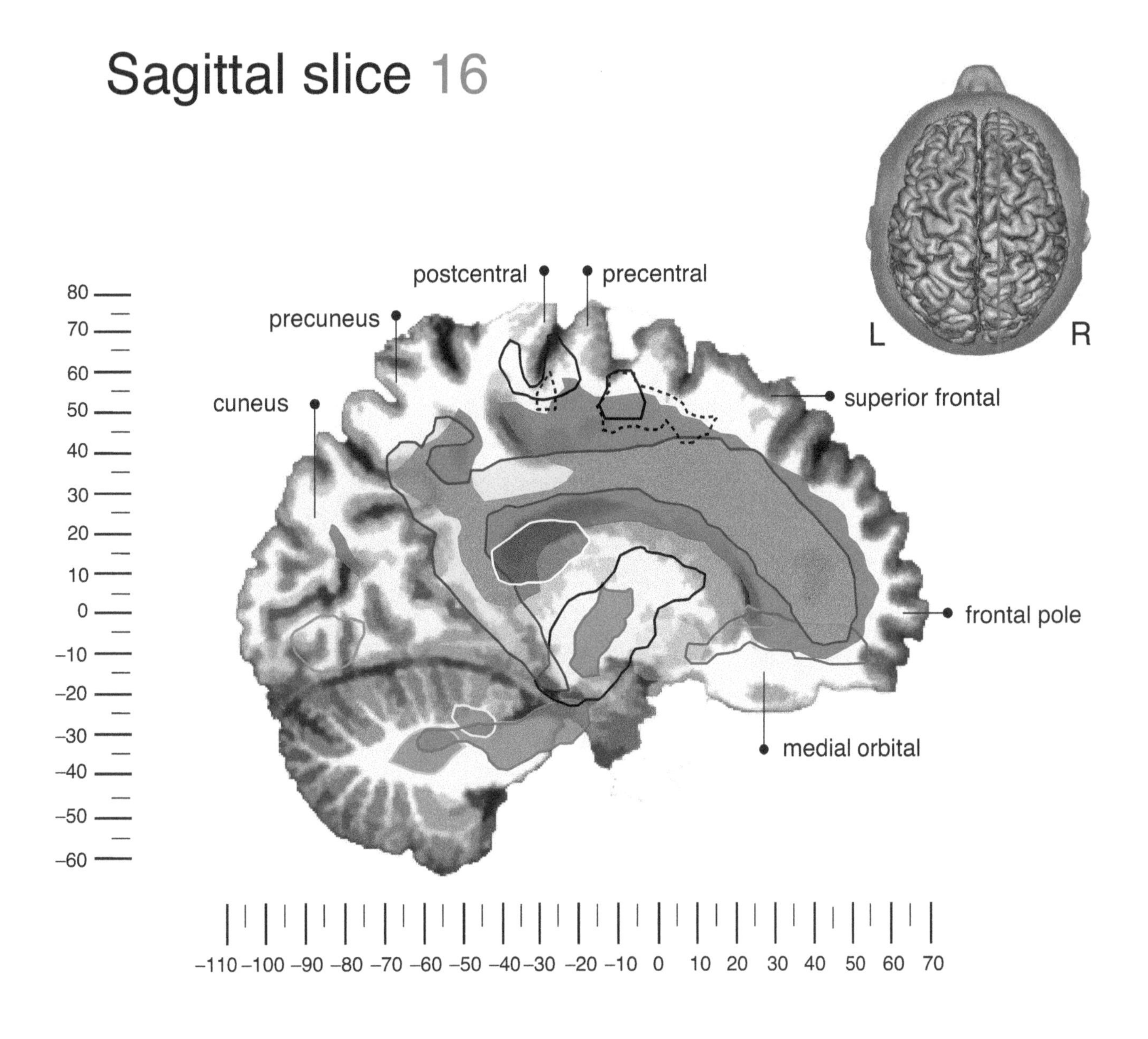

TRACT LEGEND

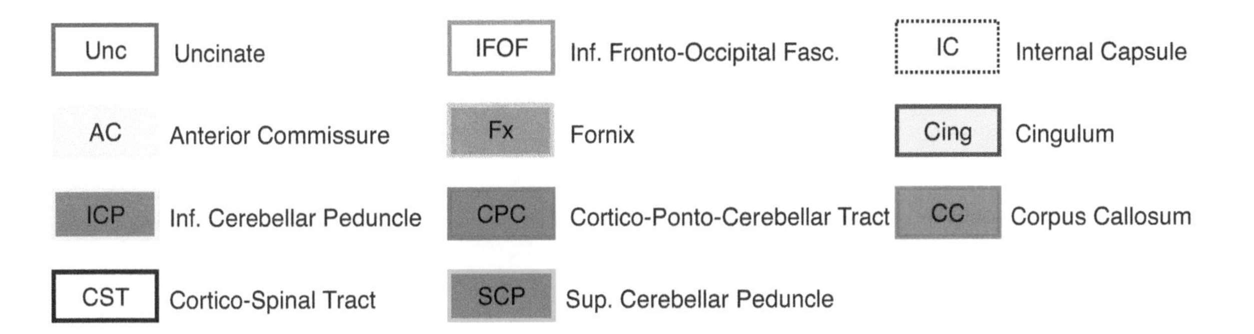

Sagittal slice 14

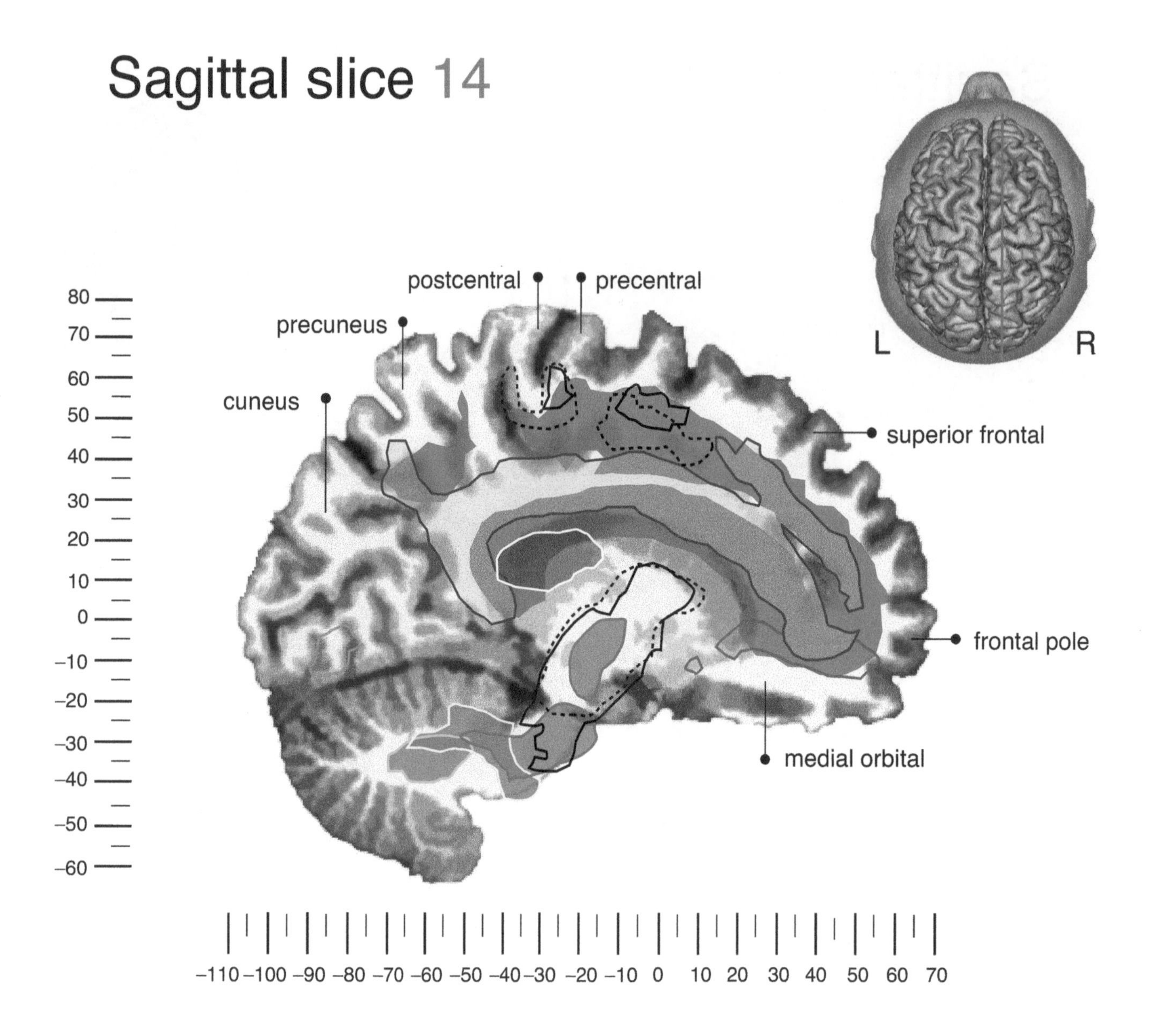

TRACT LEGEND

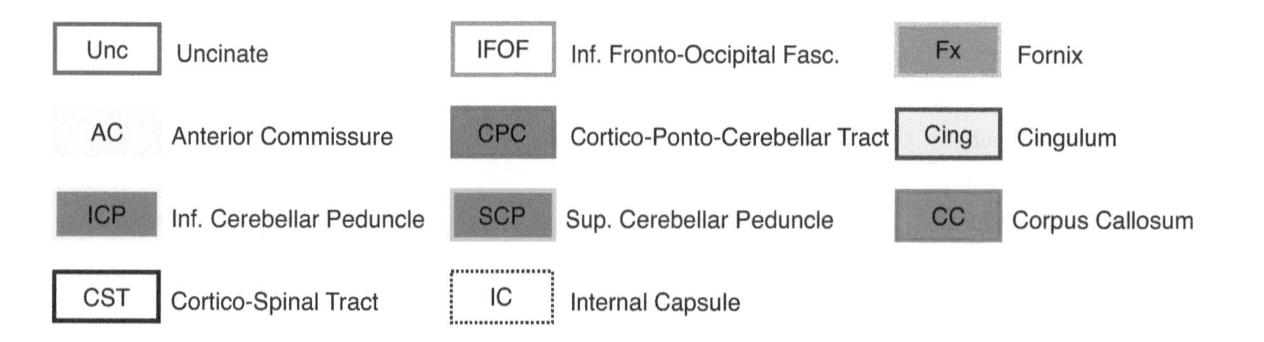

Sagittal slice 12

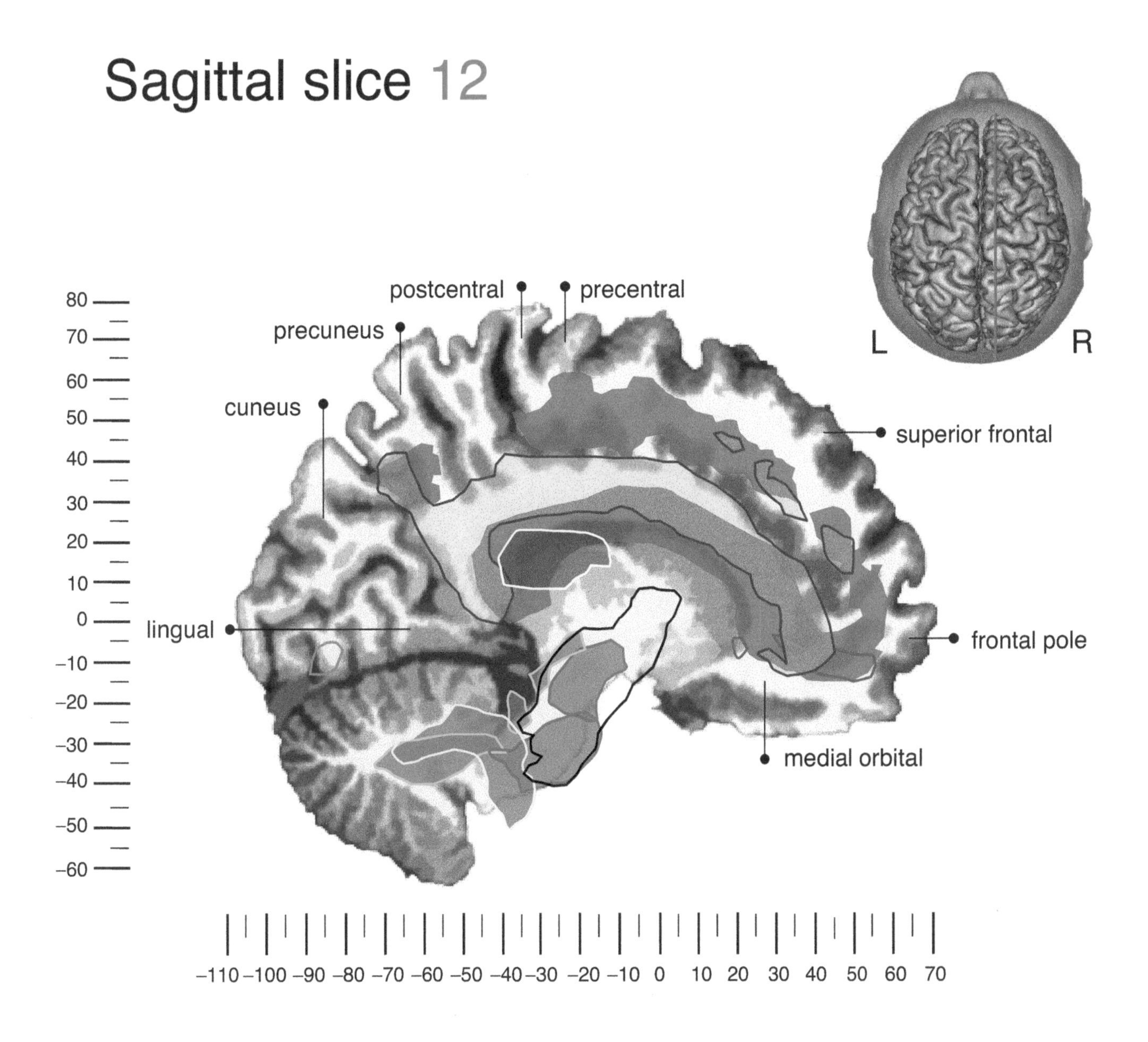

TRACT LEGEND

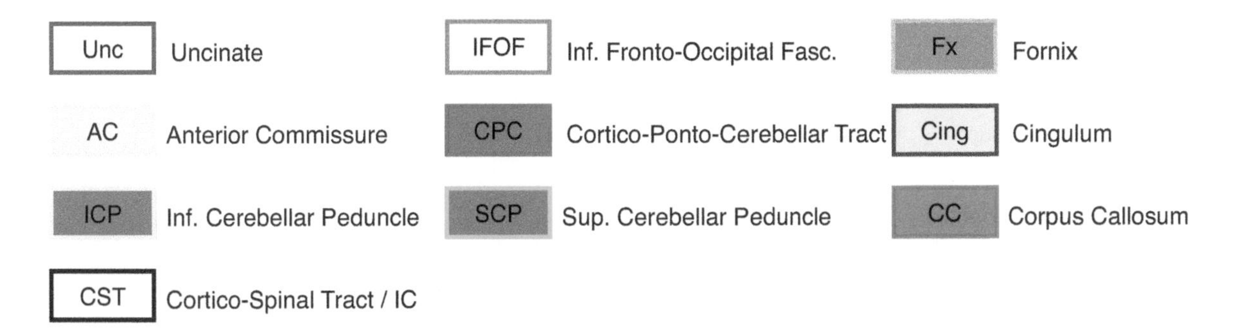

Sagittal slice 10

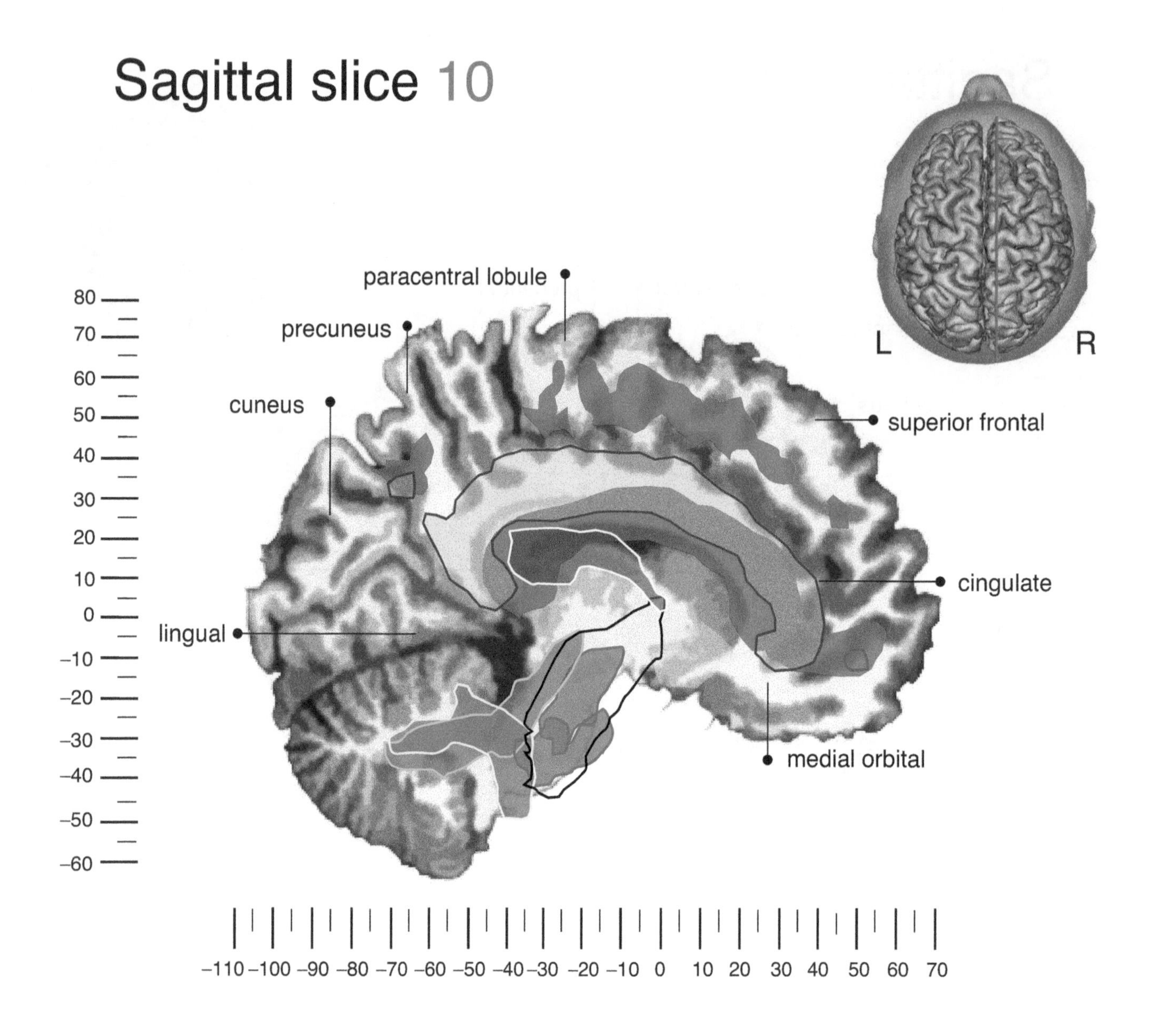

TRACT LEGEND

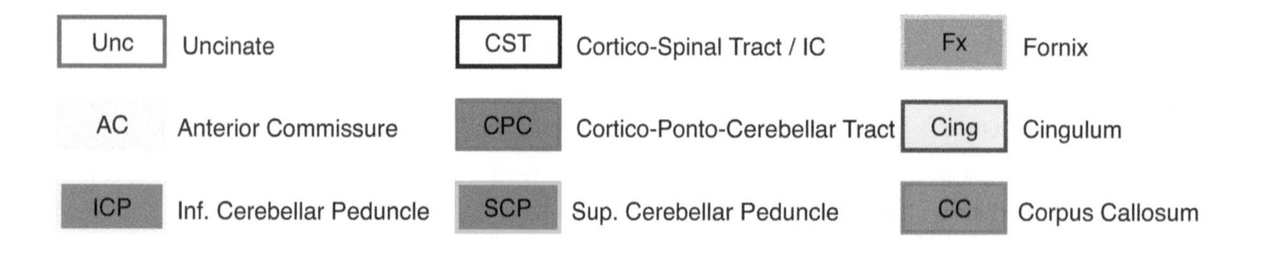

Sagittal slice 8

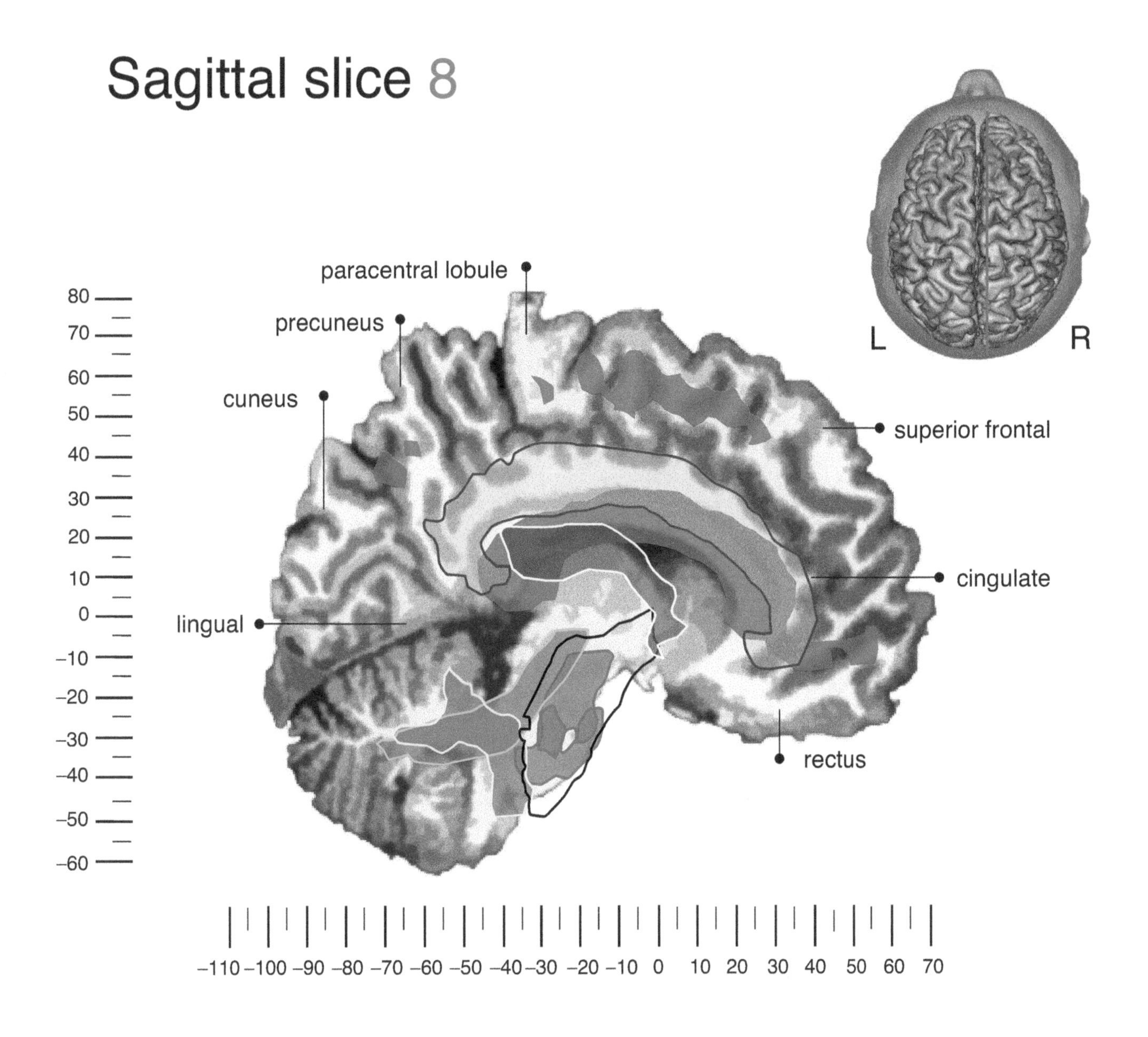

TRACT LEGEND

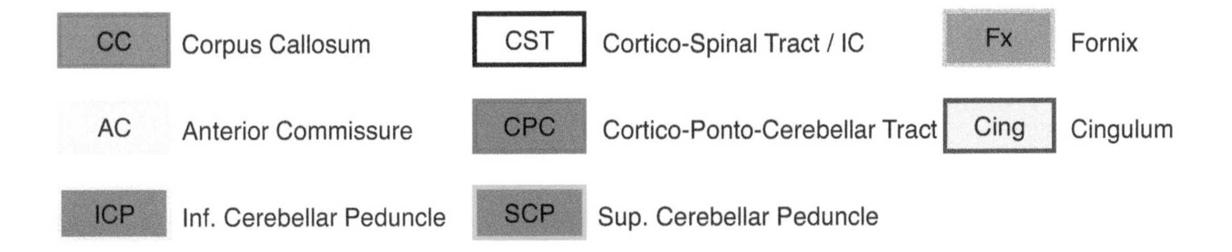

Sagittal slice 6

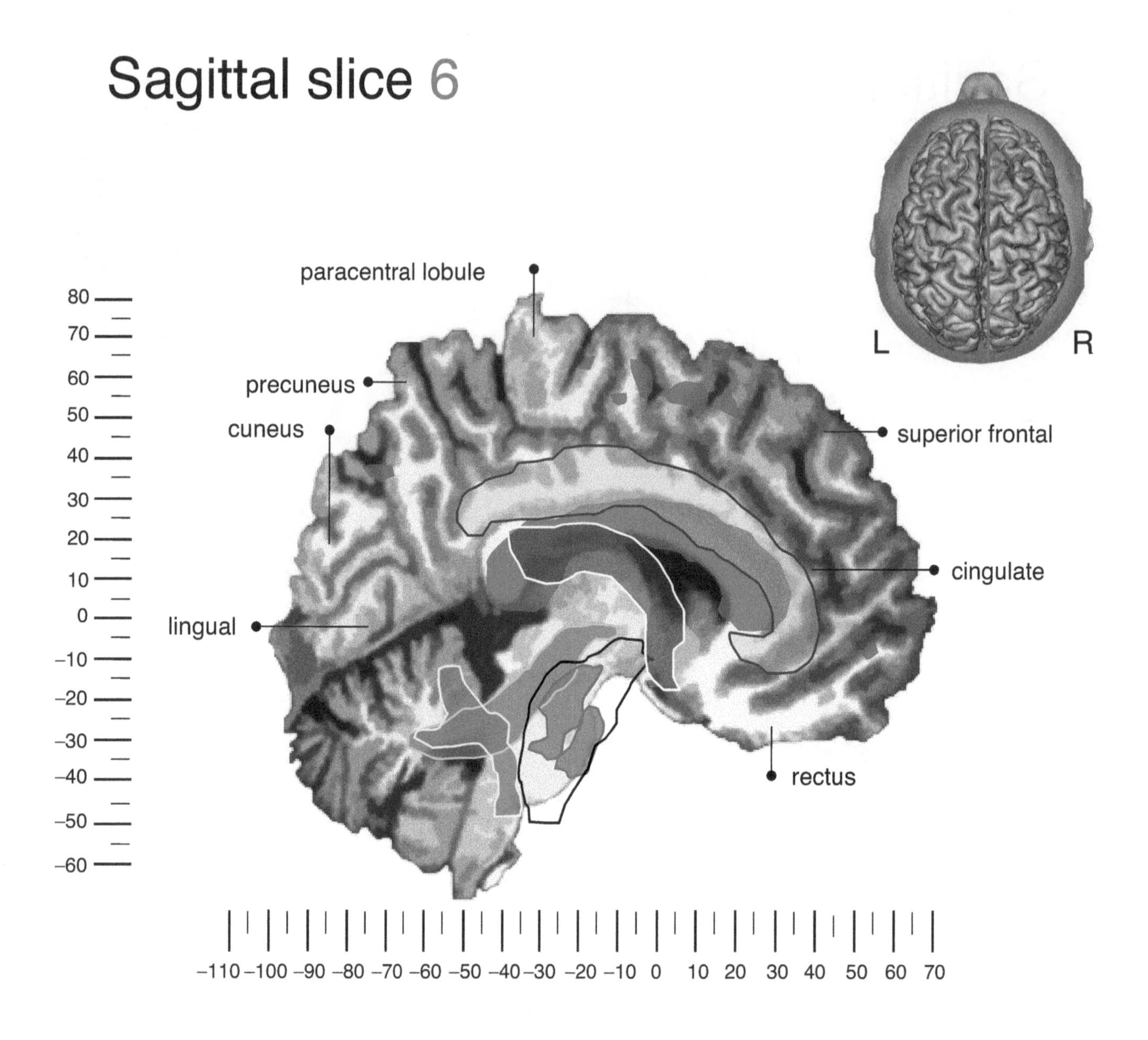

TRACT LEGEND

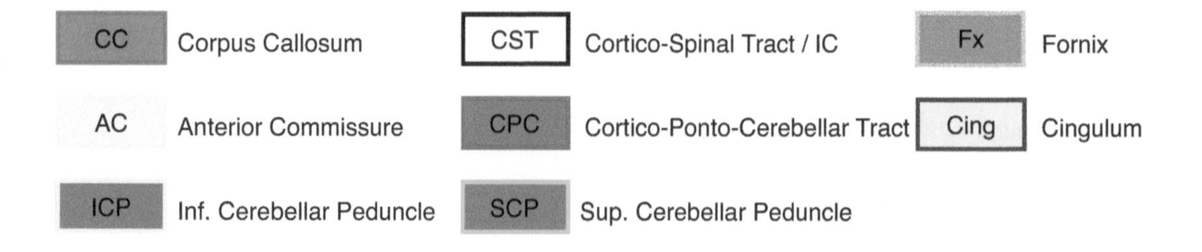

Sagittal slice 4

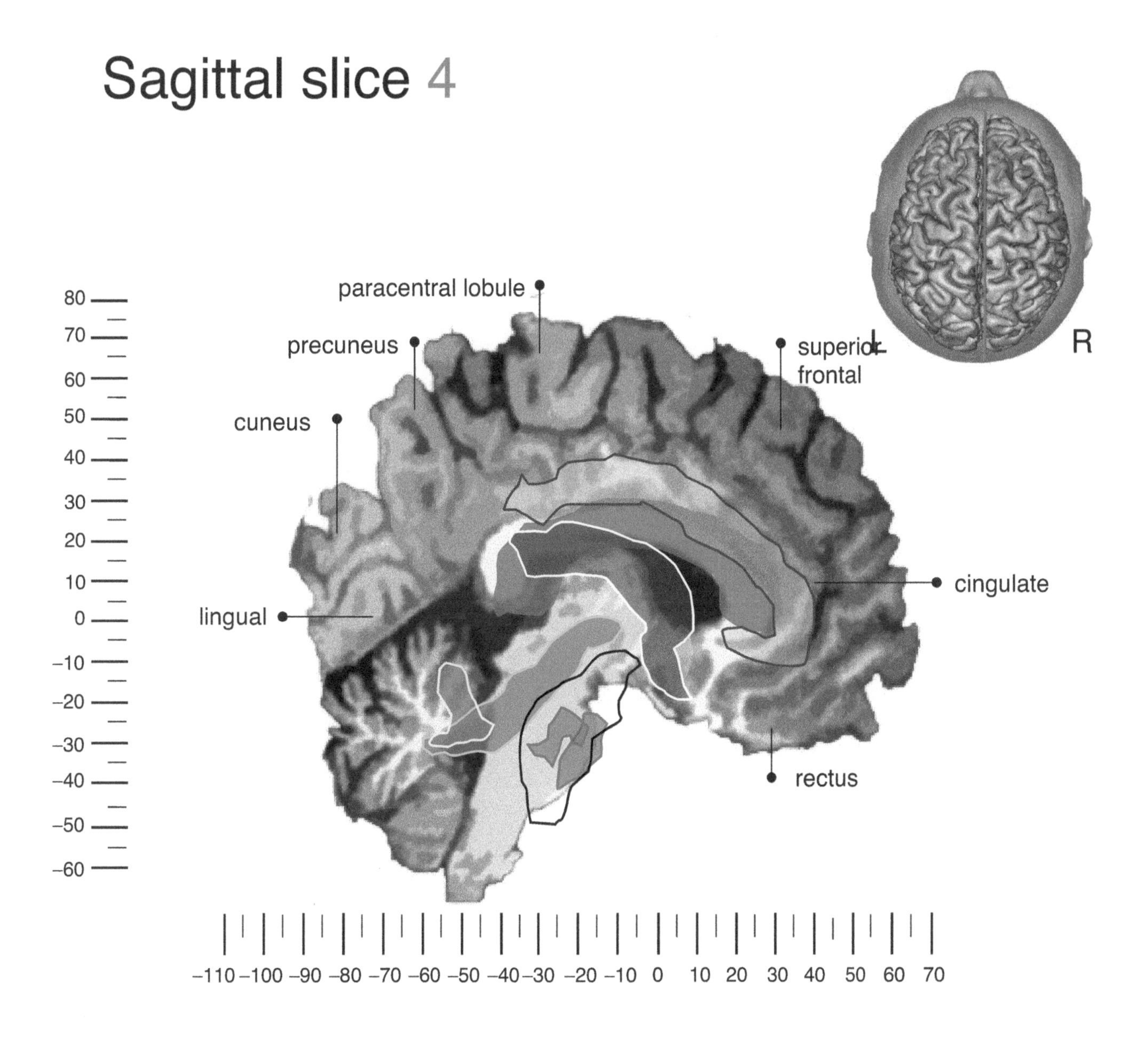

TRACT LEGEND

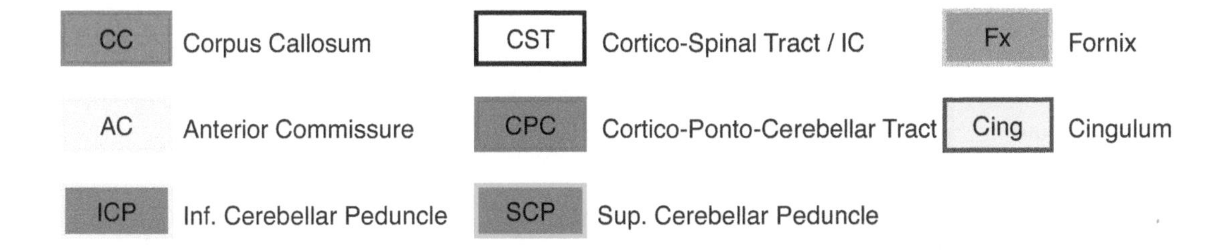

Sagittal slice 2

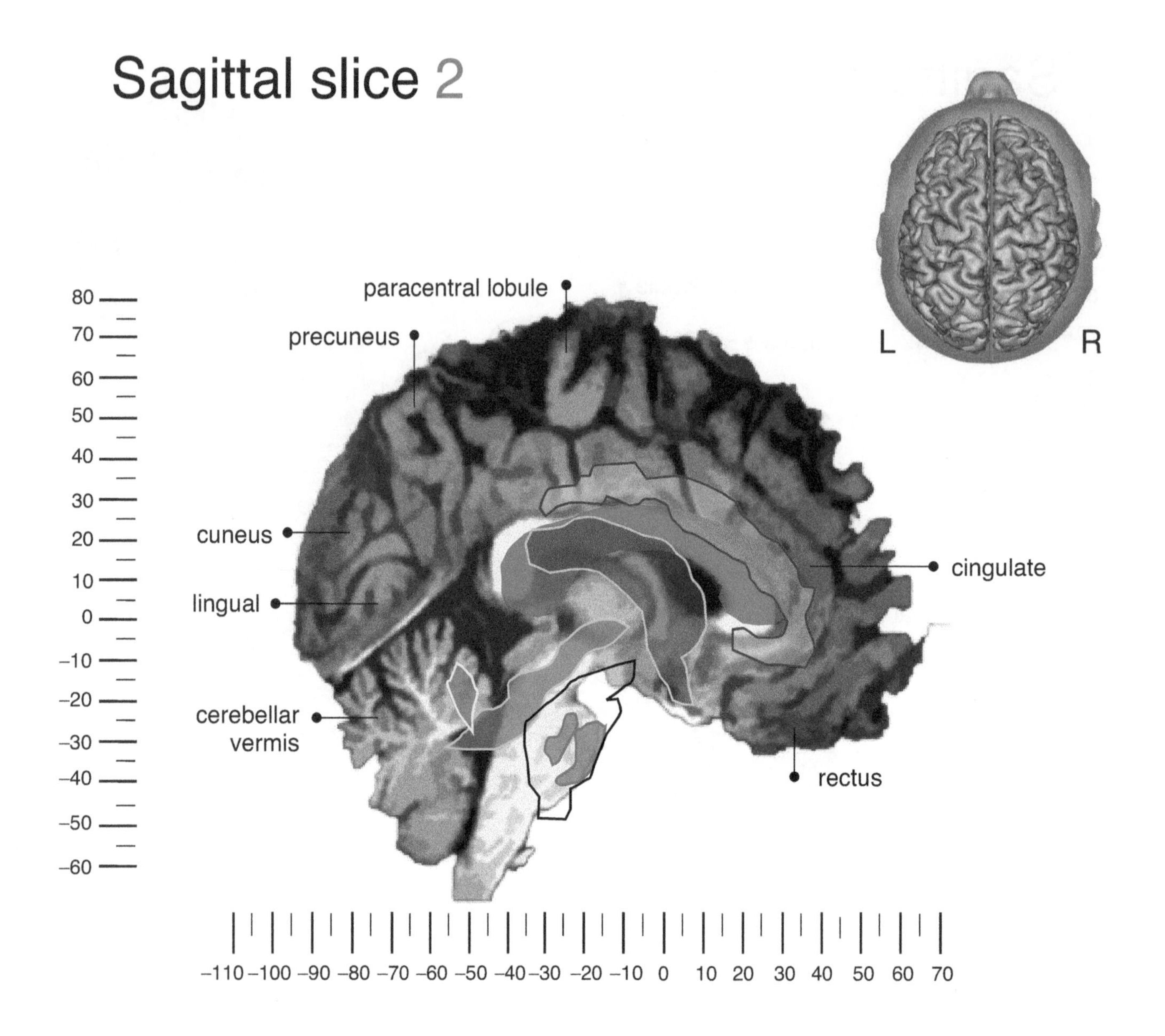

TRACT LEGEND

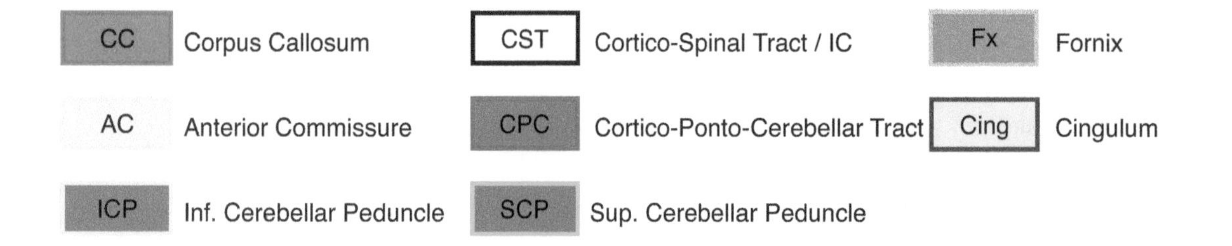

Sagittal slice 0

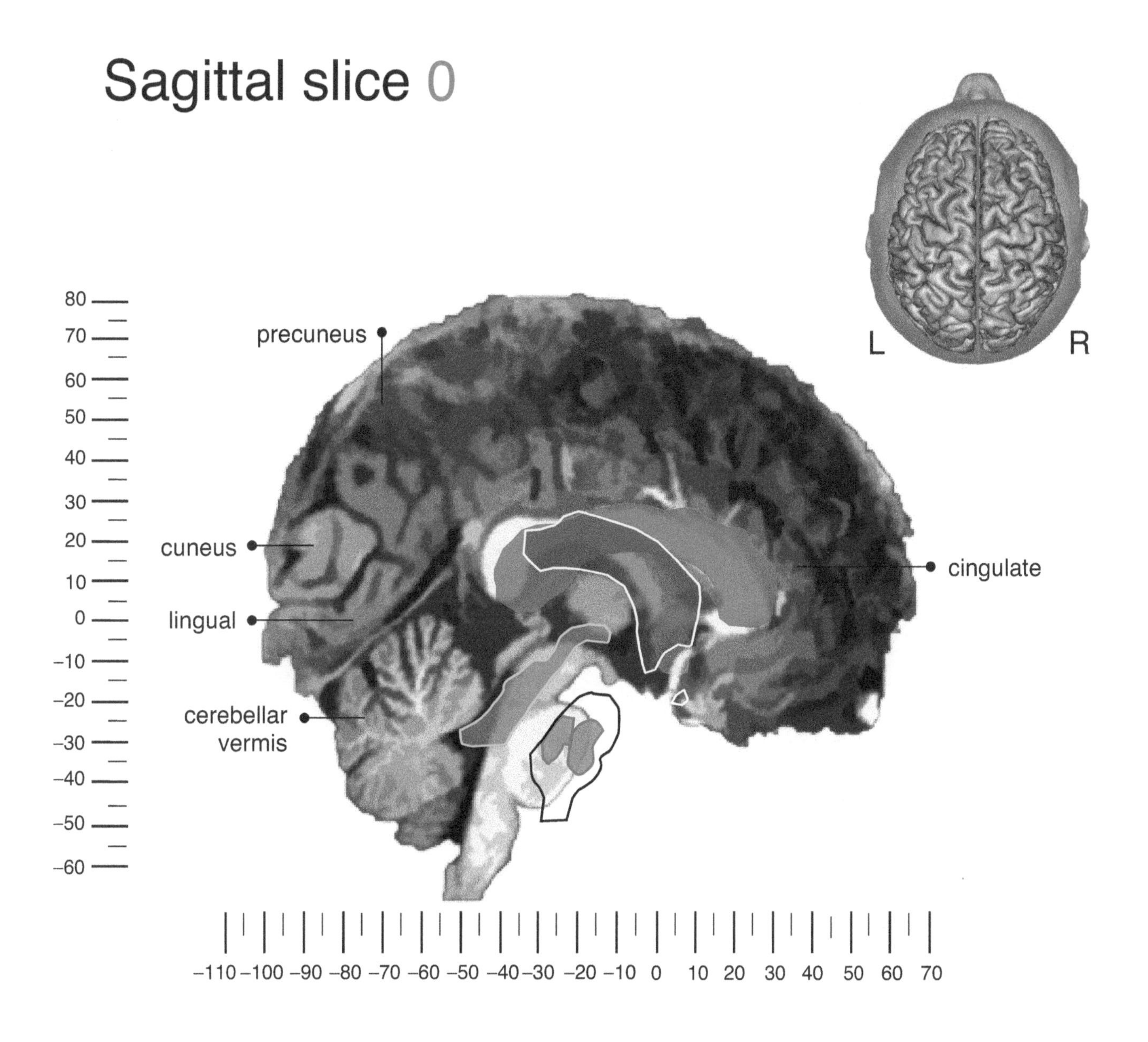

TRACT LEGEND

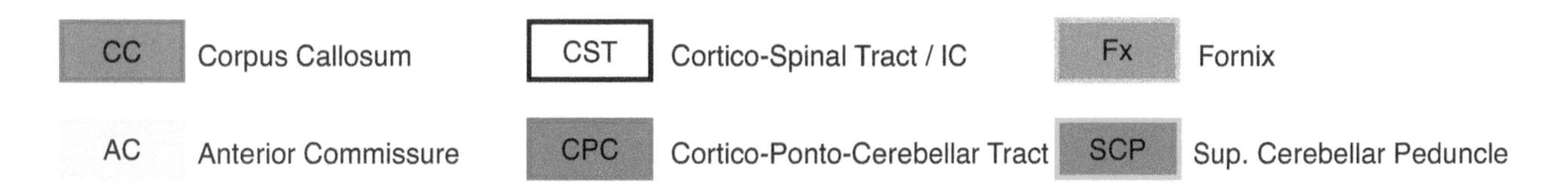

Sagittal slice –2

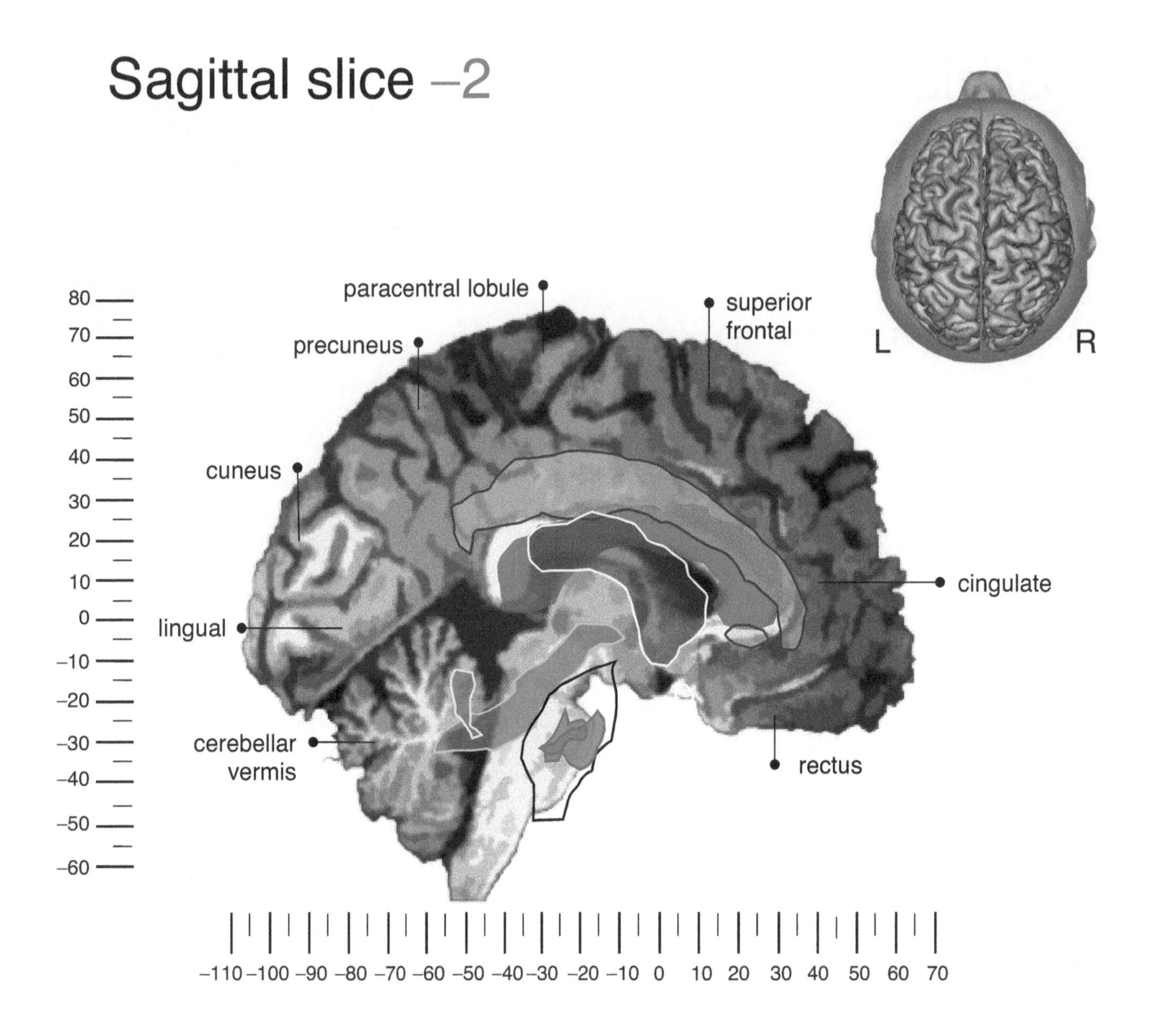

TRACT LEGEND

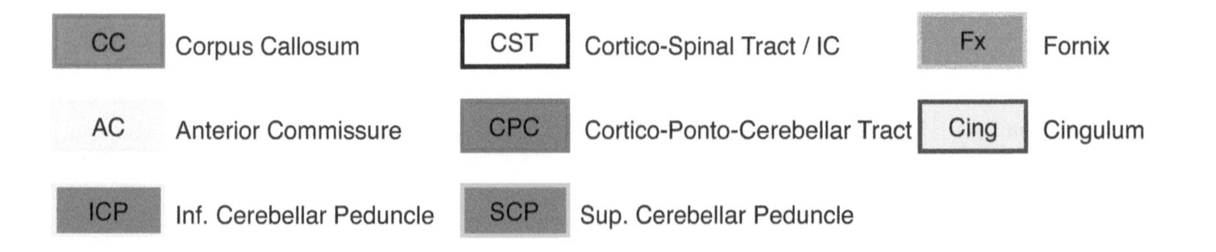

Sagittal slice –4

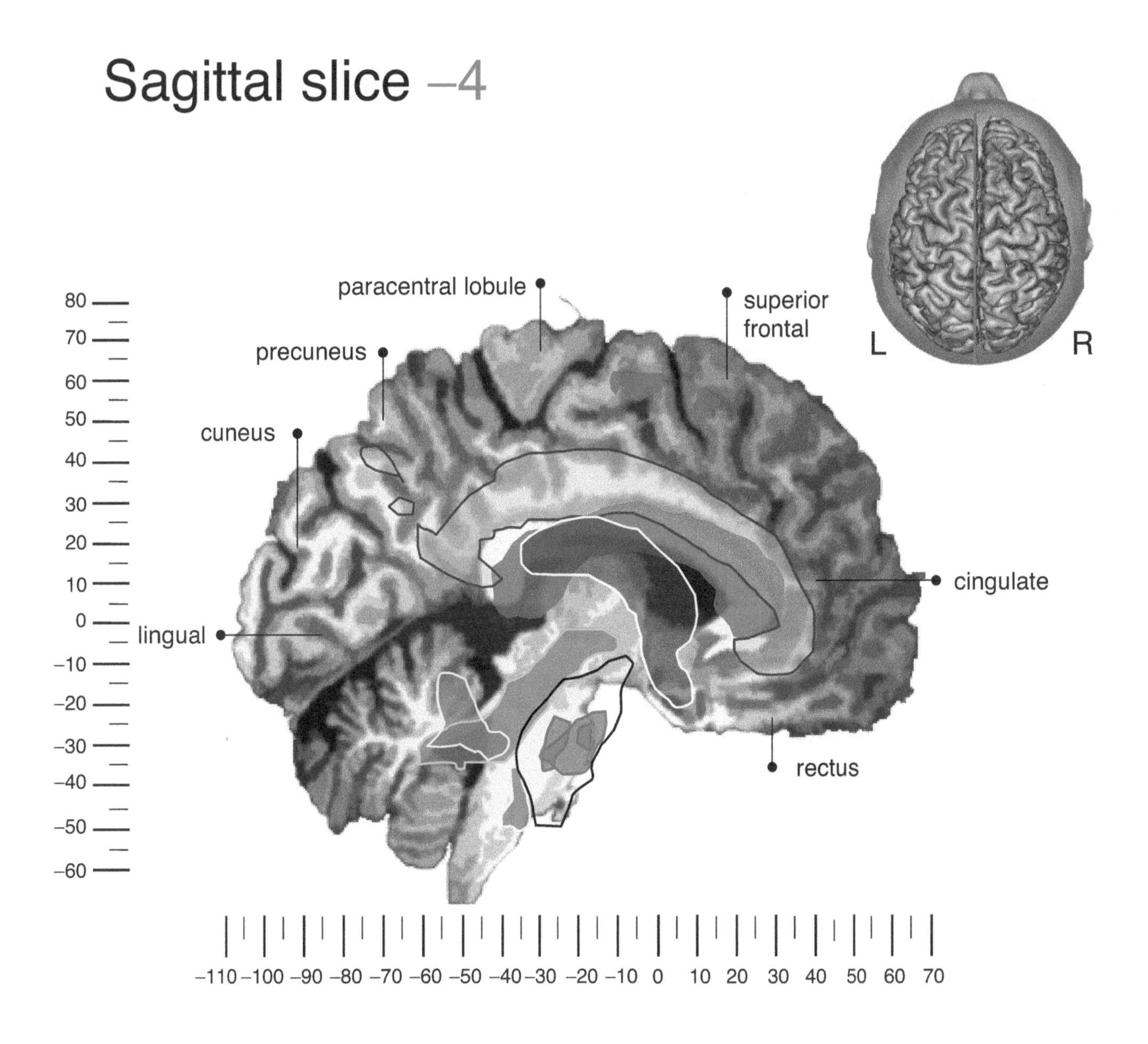

TRACT LEGEND

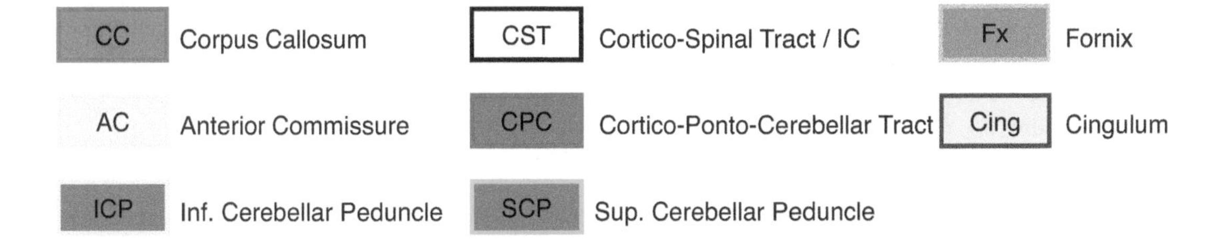

Sagittal slice –6

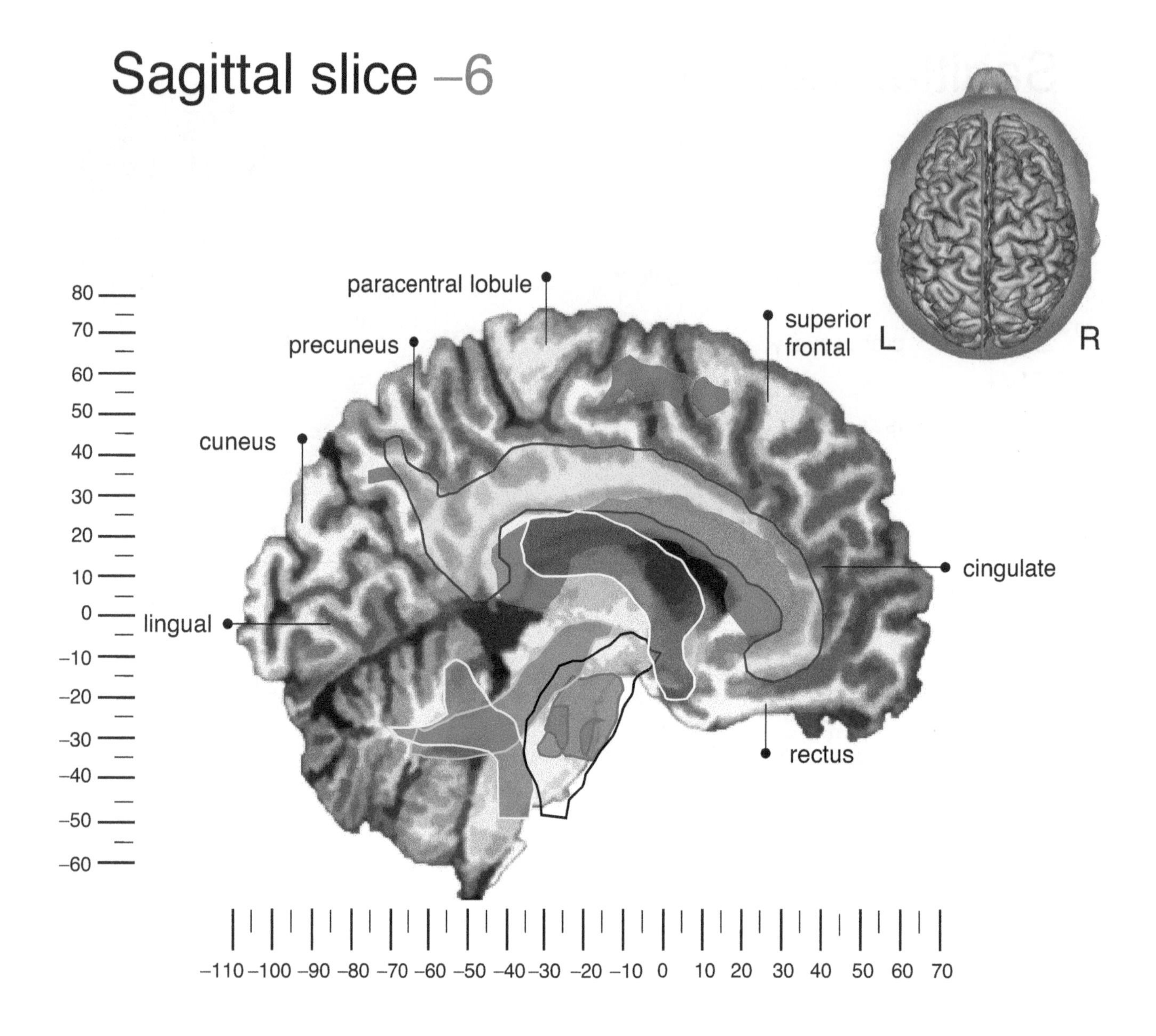

TRACT LEGEND

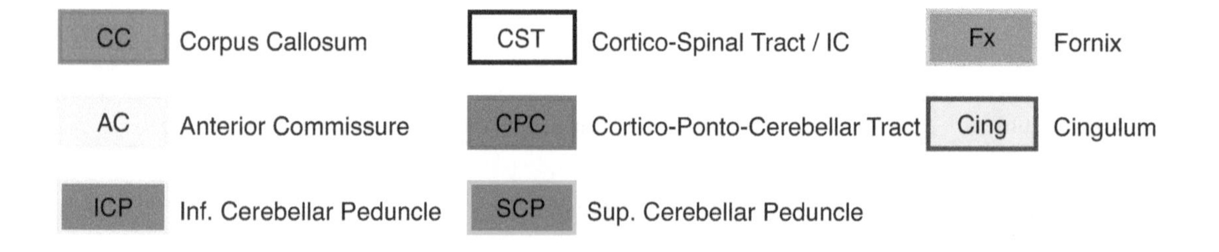

Sagittal slice –8

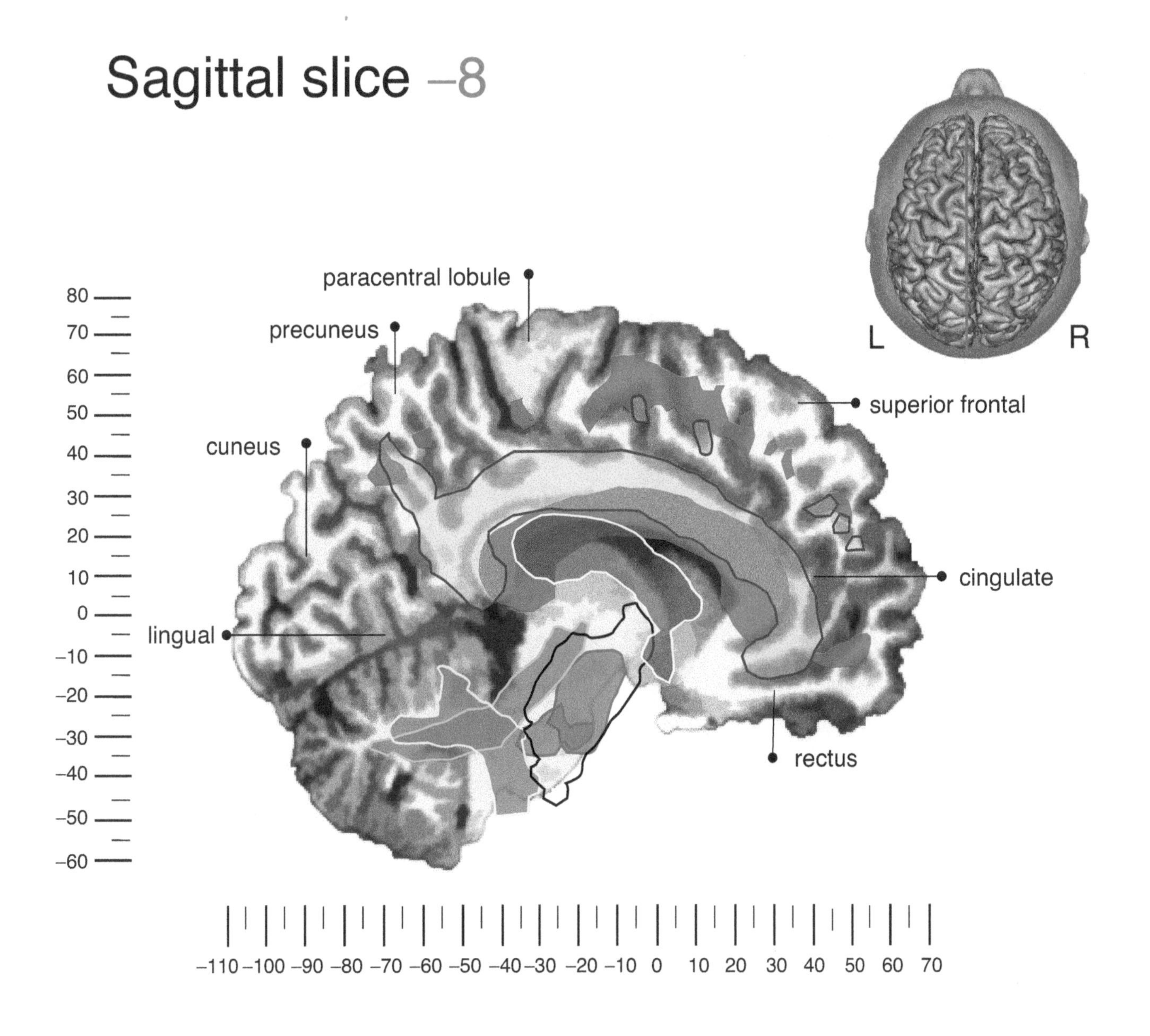

TRACT LEGEND

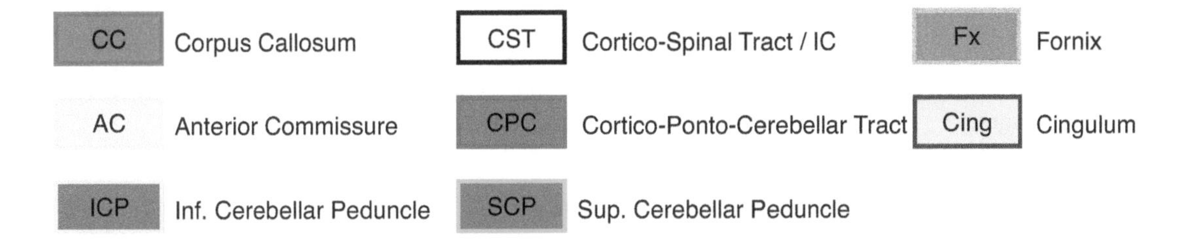

Sagittal slice –10

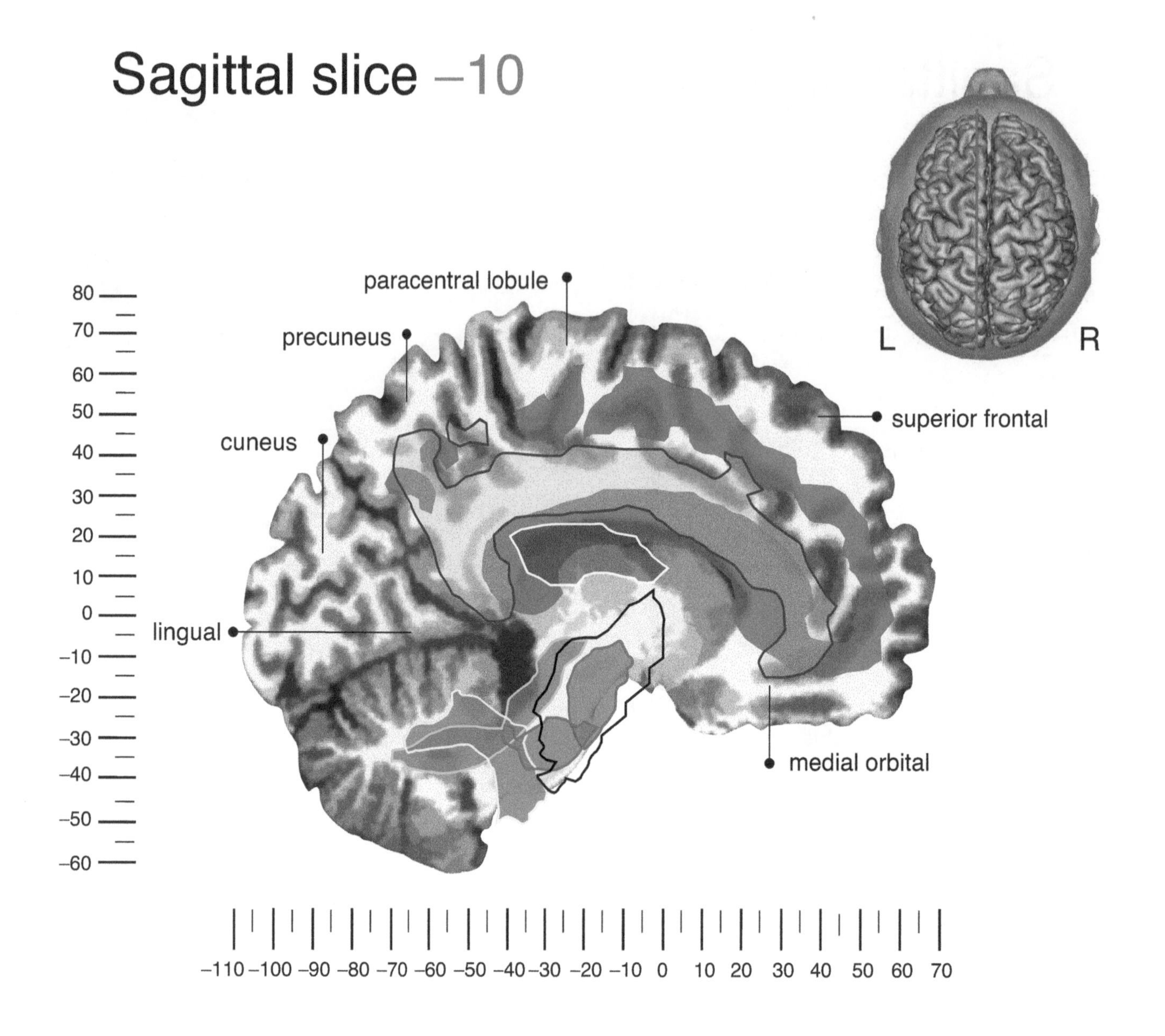

TRACT LEGEND

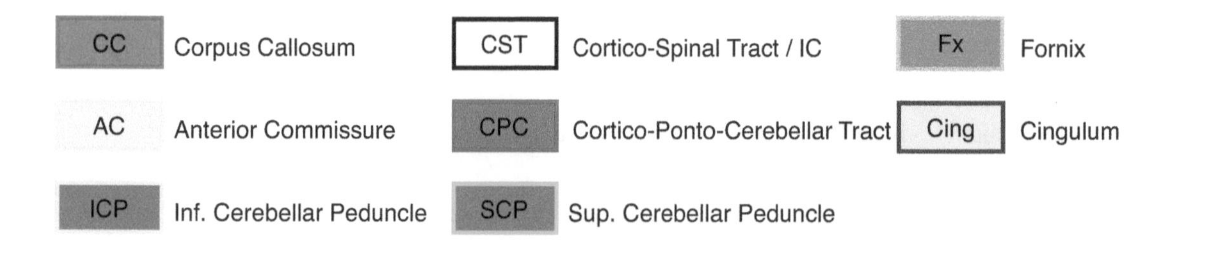

Sagittal slice –12

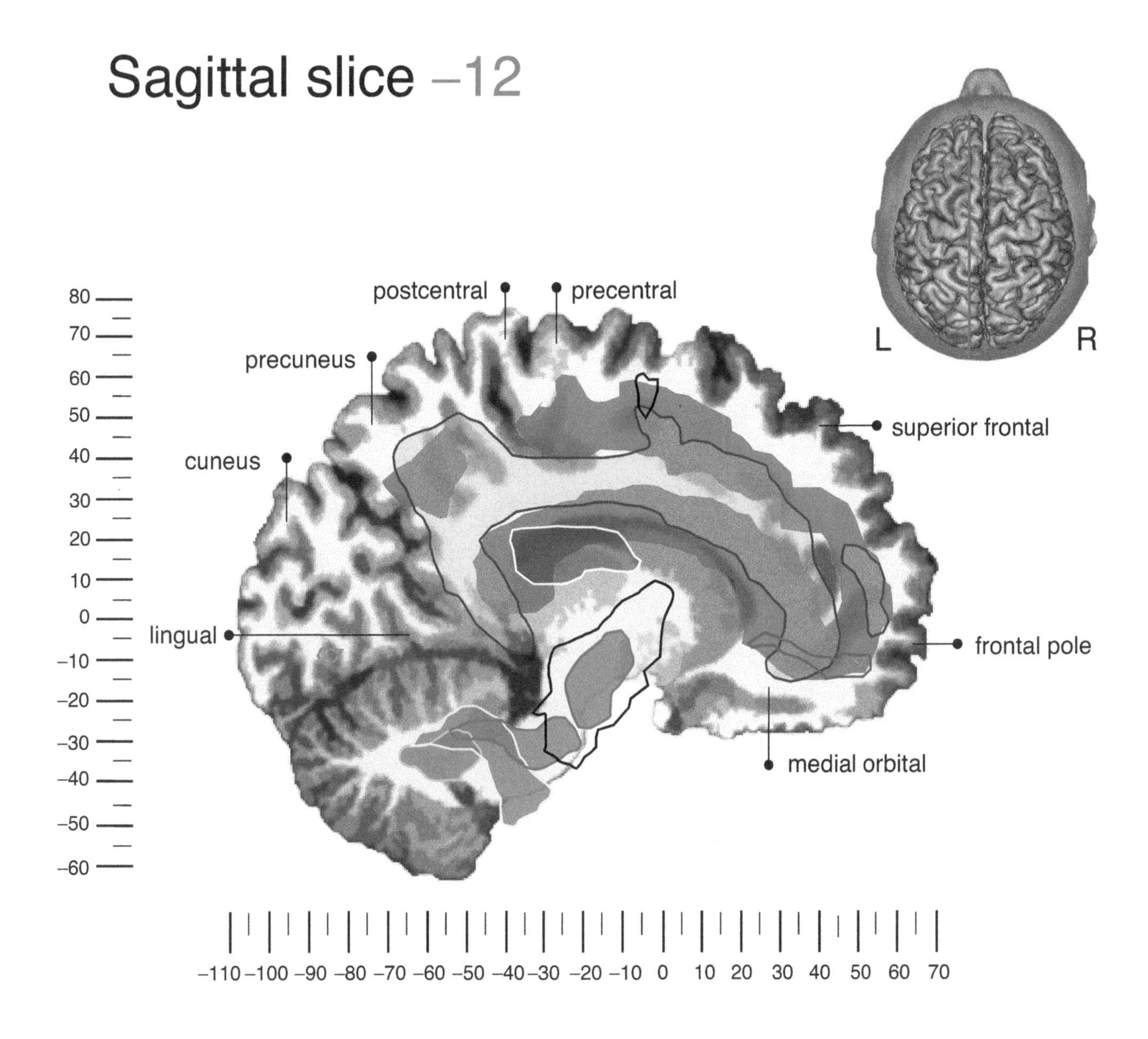

TRACT LEGEND

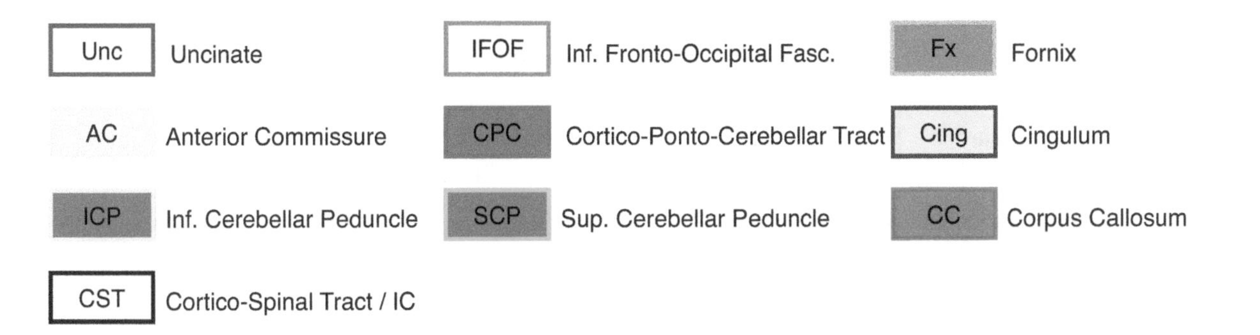

Sagittal slice –14

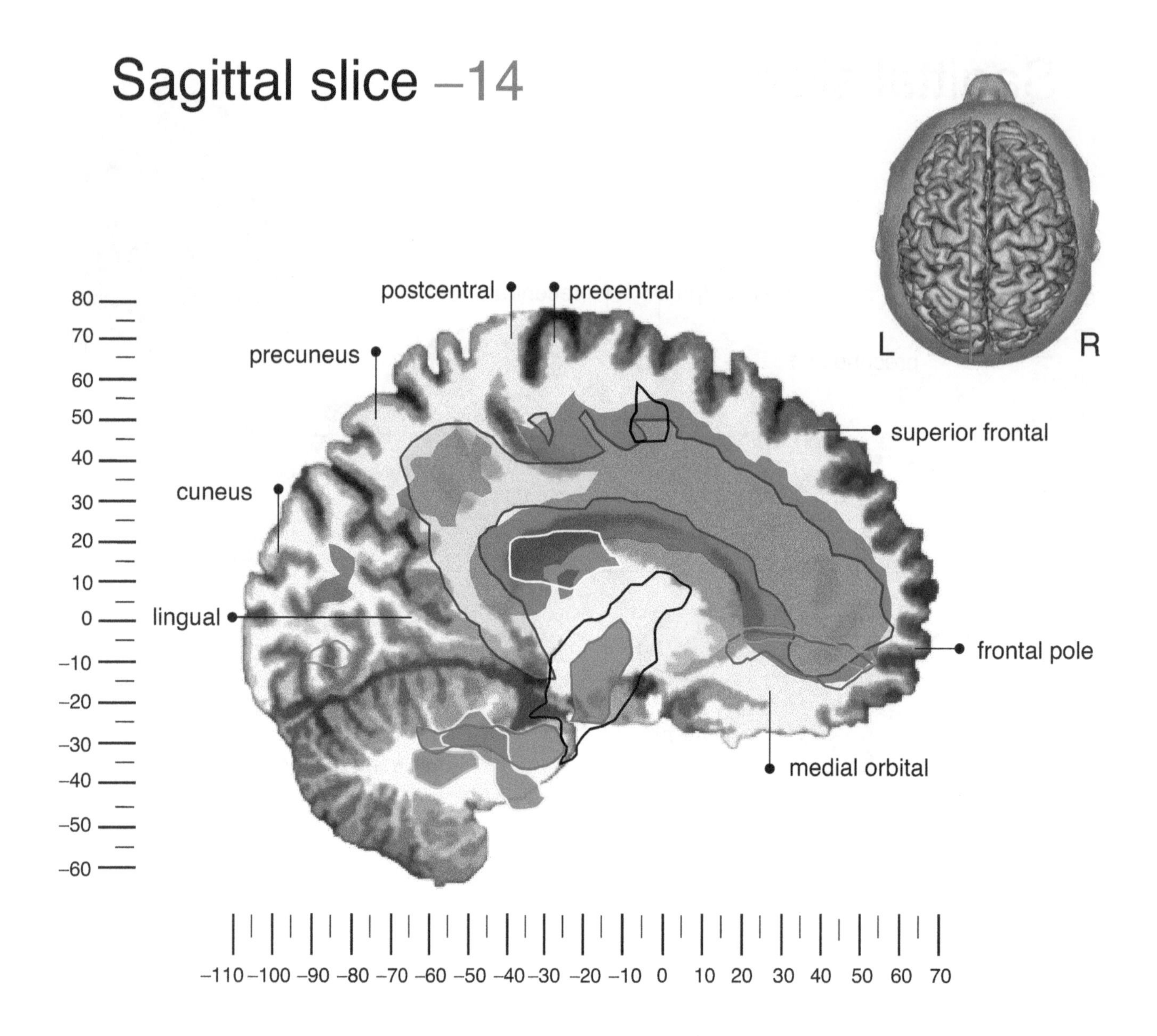

TRACT LEGEND

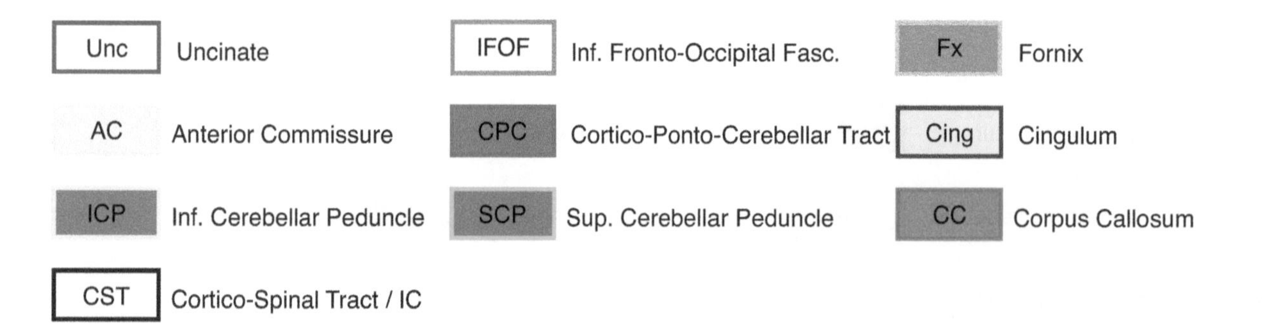

Sagittal slice –16

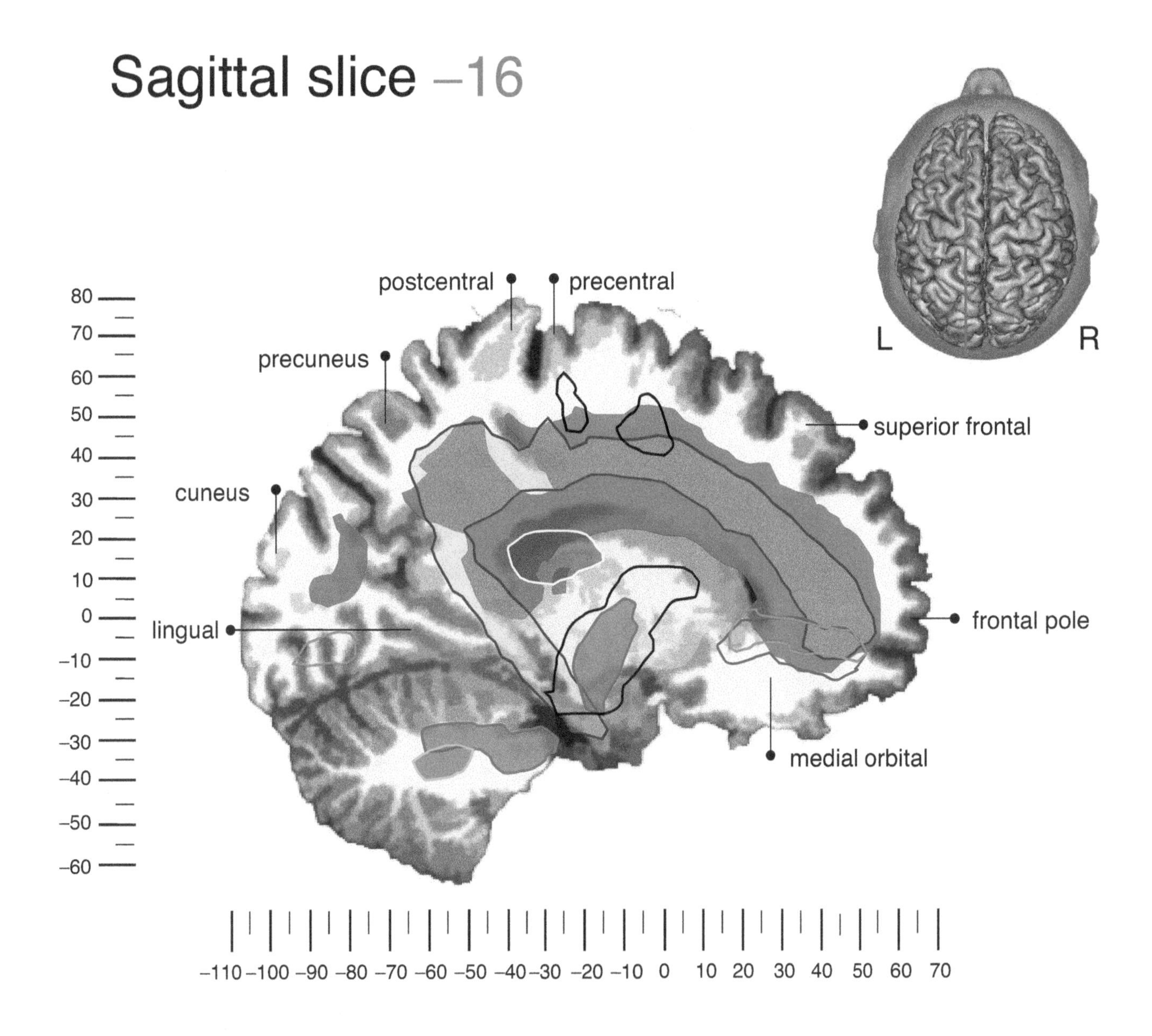

TRACT LEGEND

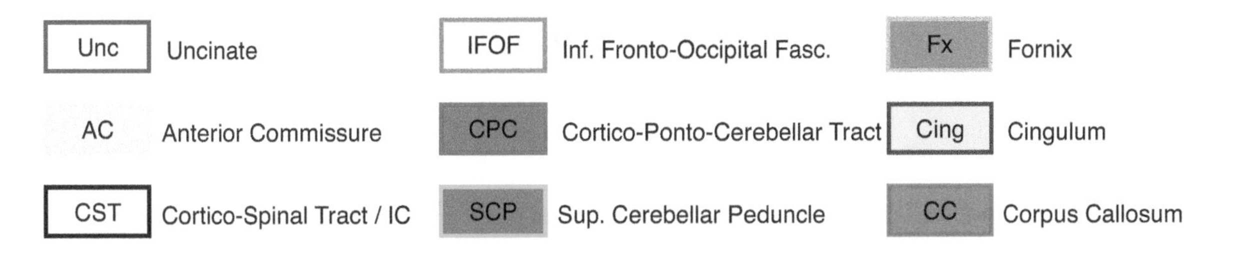

Sagittal slice −18

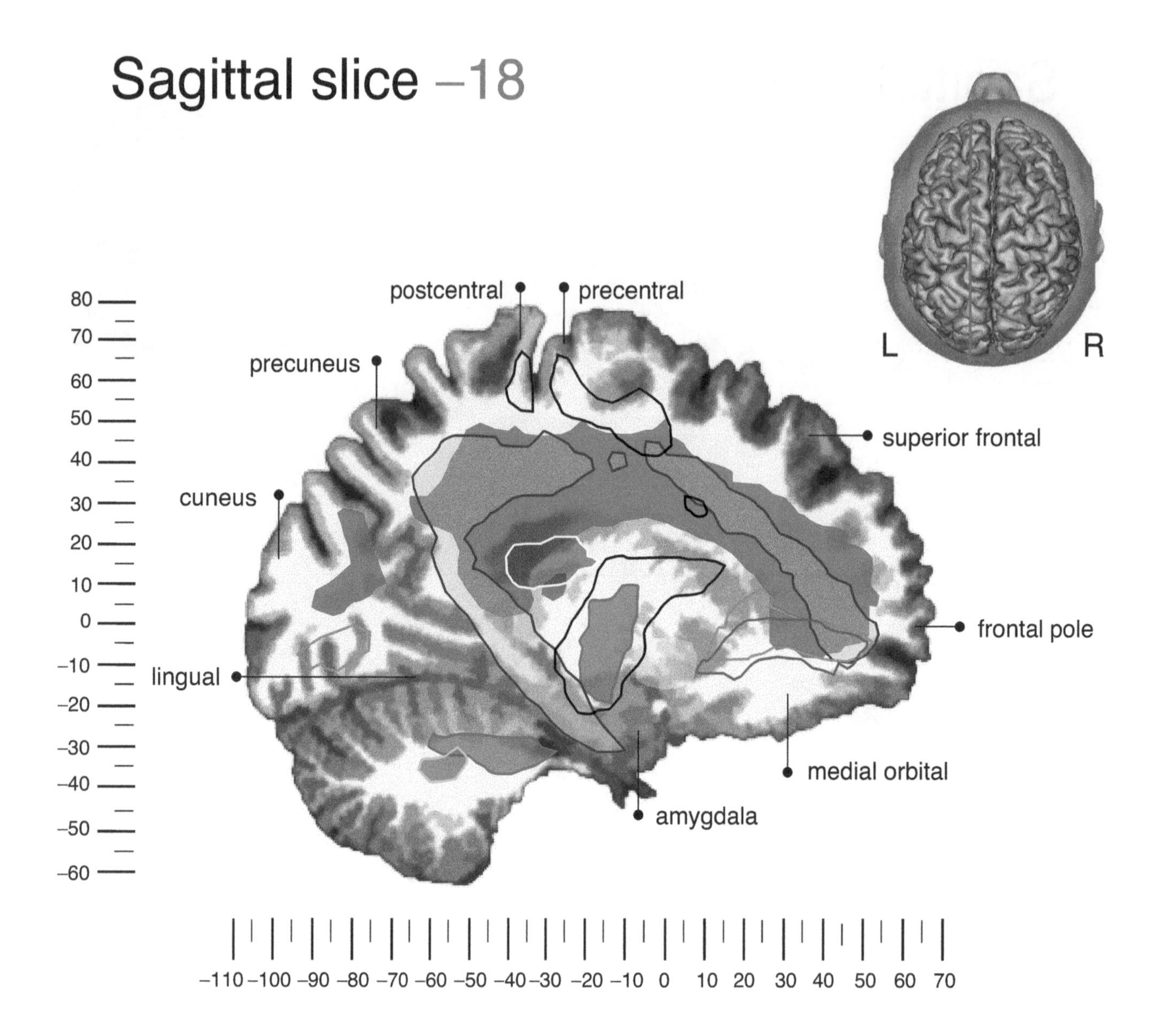

TRACT LEGEND

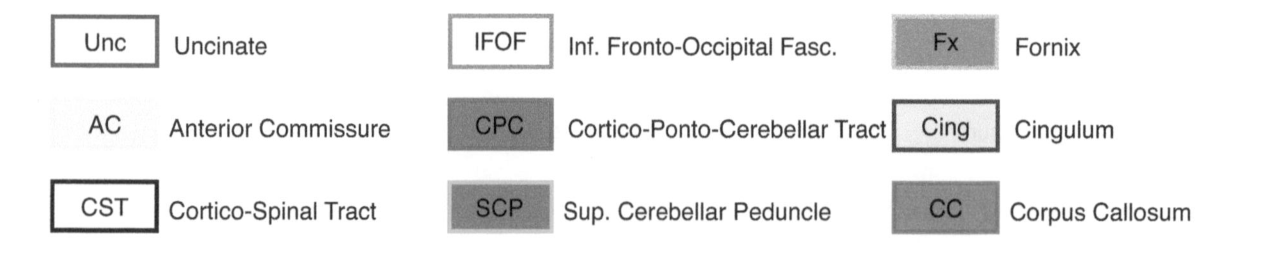

Sagittal slice –20

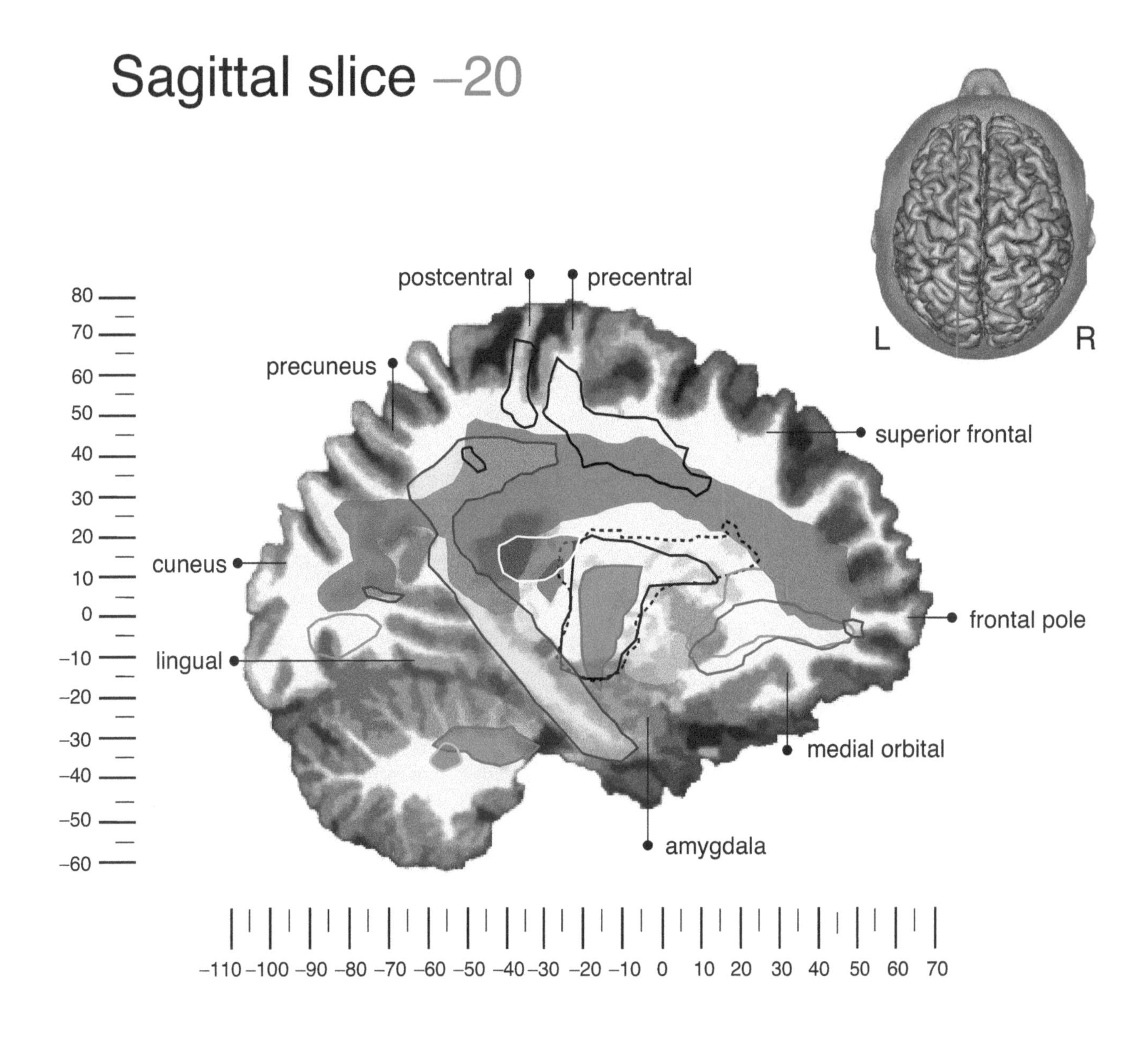

TRACT LEGEND

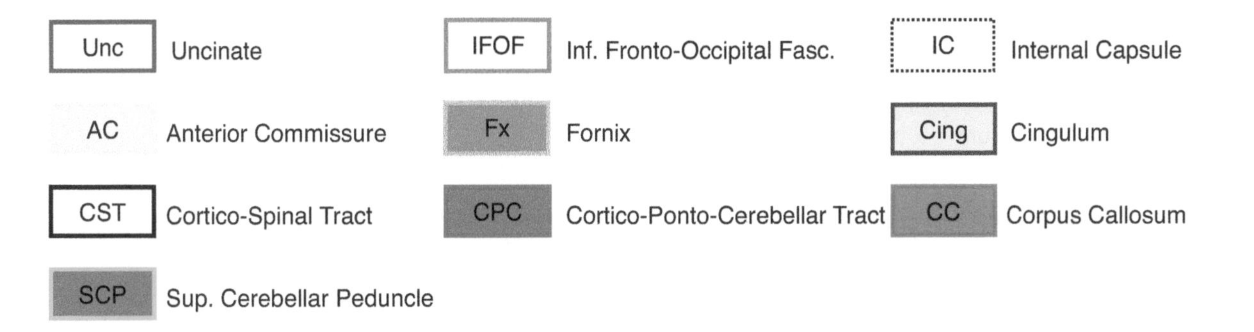

Sagittal slice –22

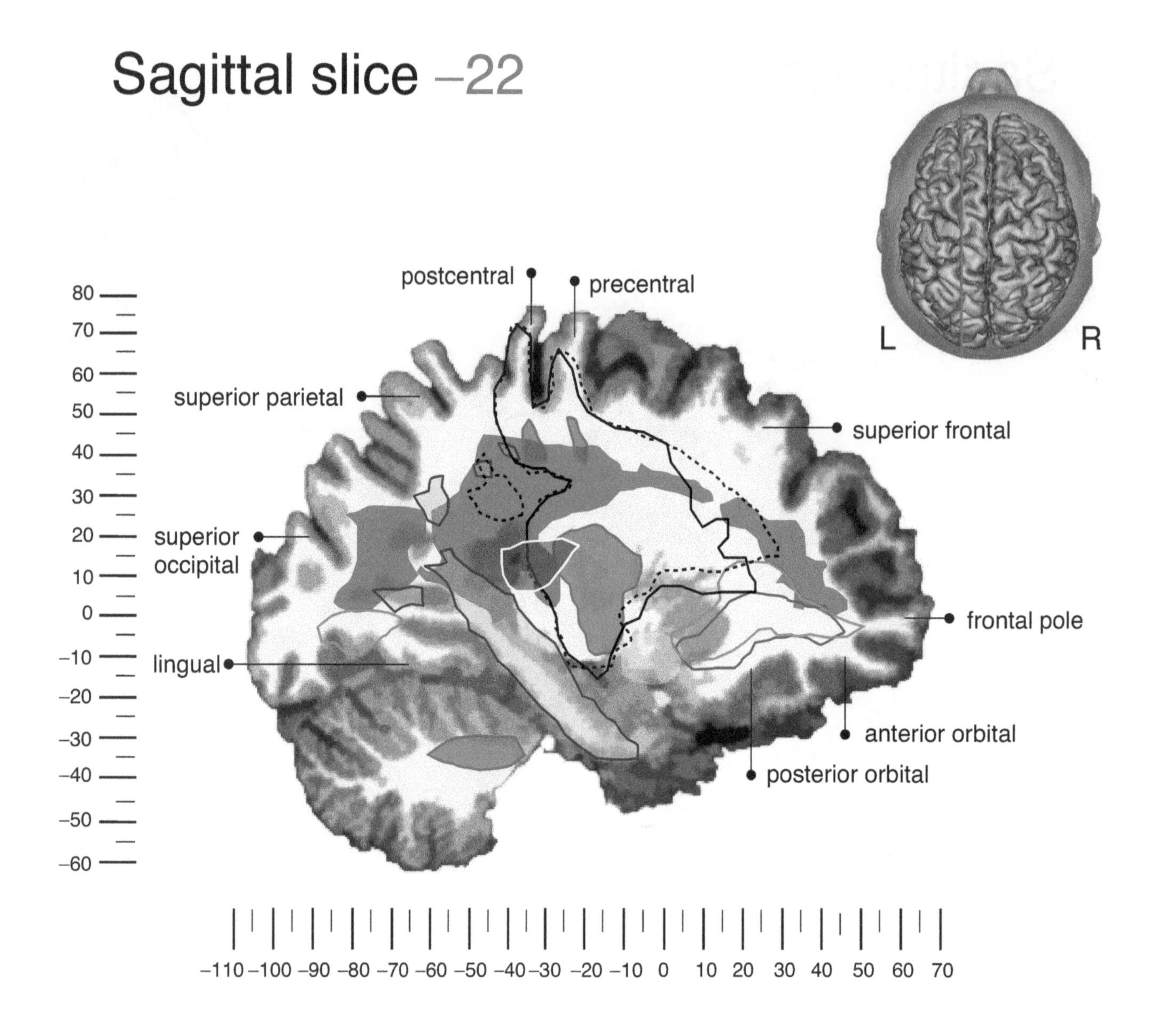

TRACT LEGEND

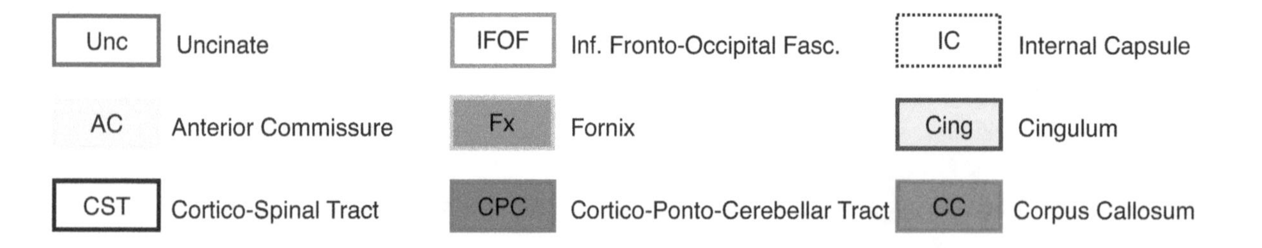

Sagittal slice –24

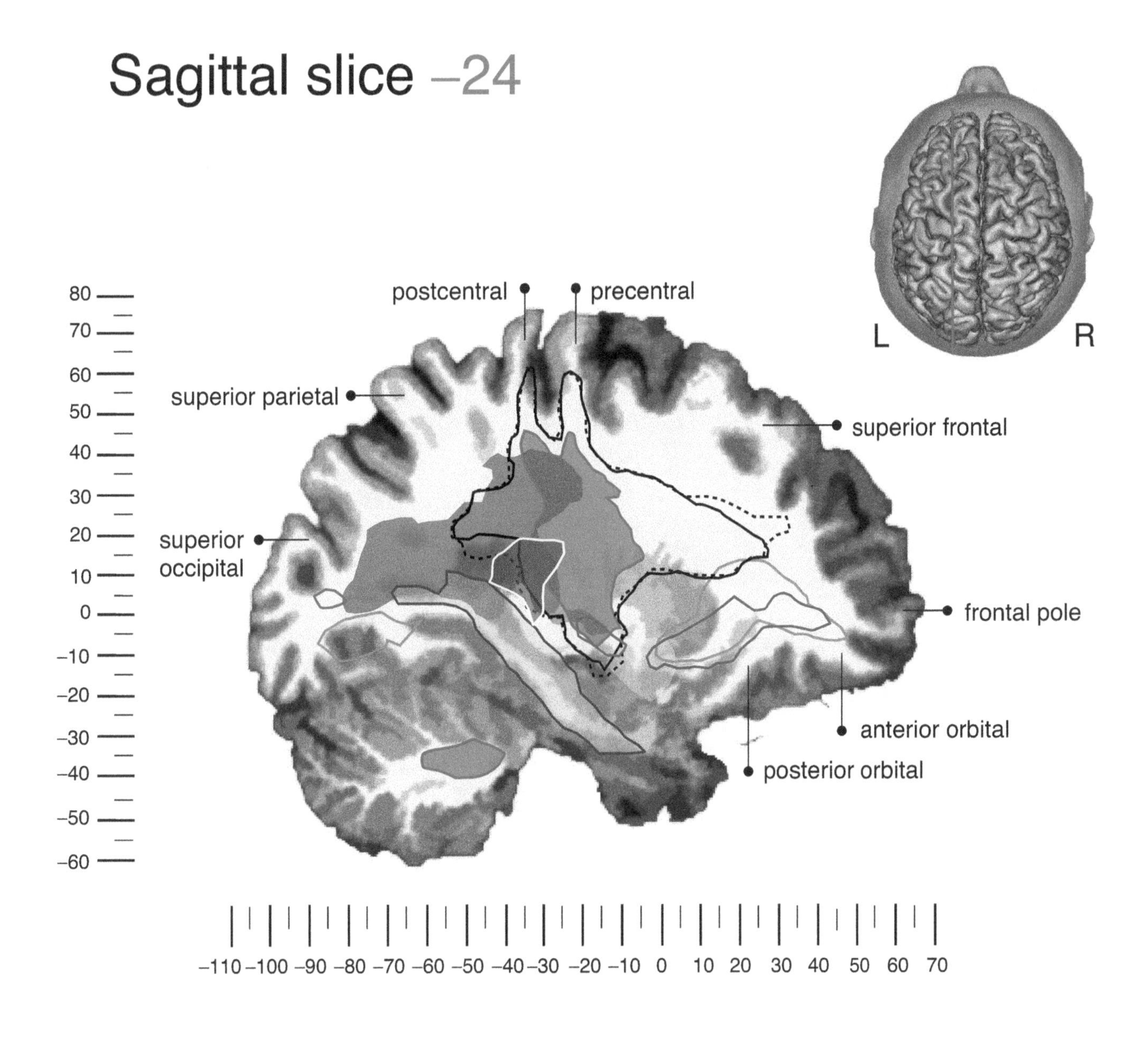

TRACT LEGEND

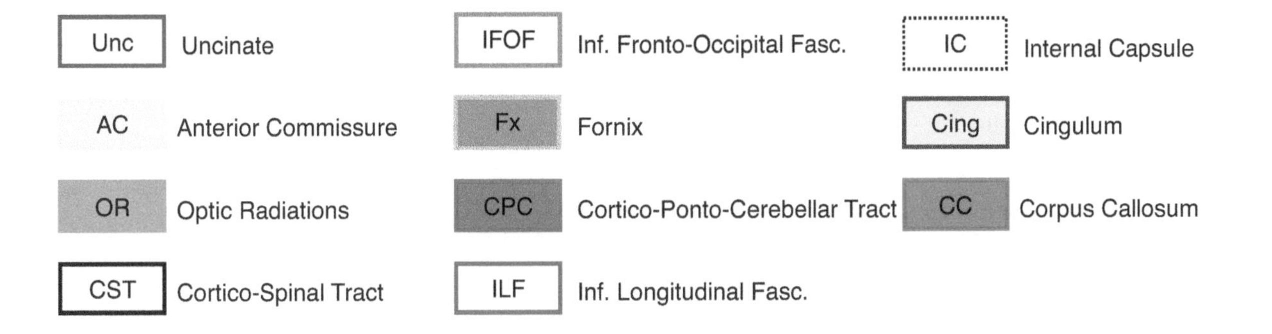

Sagittal slice –26

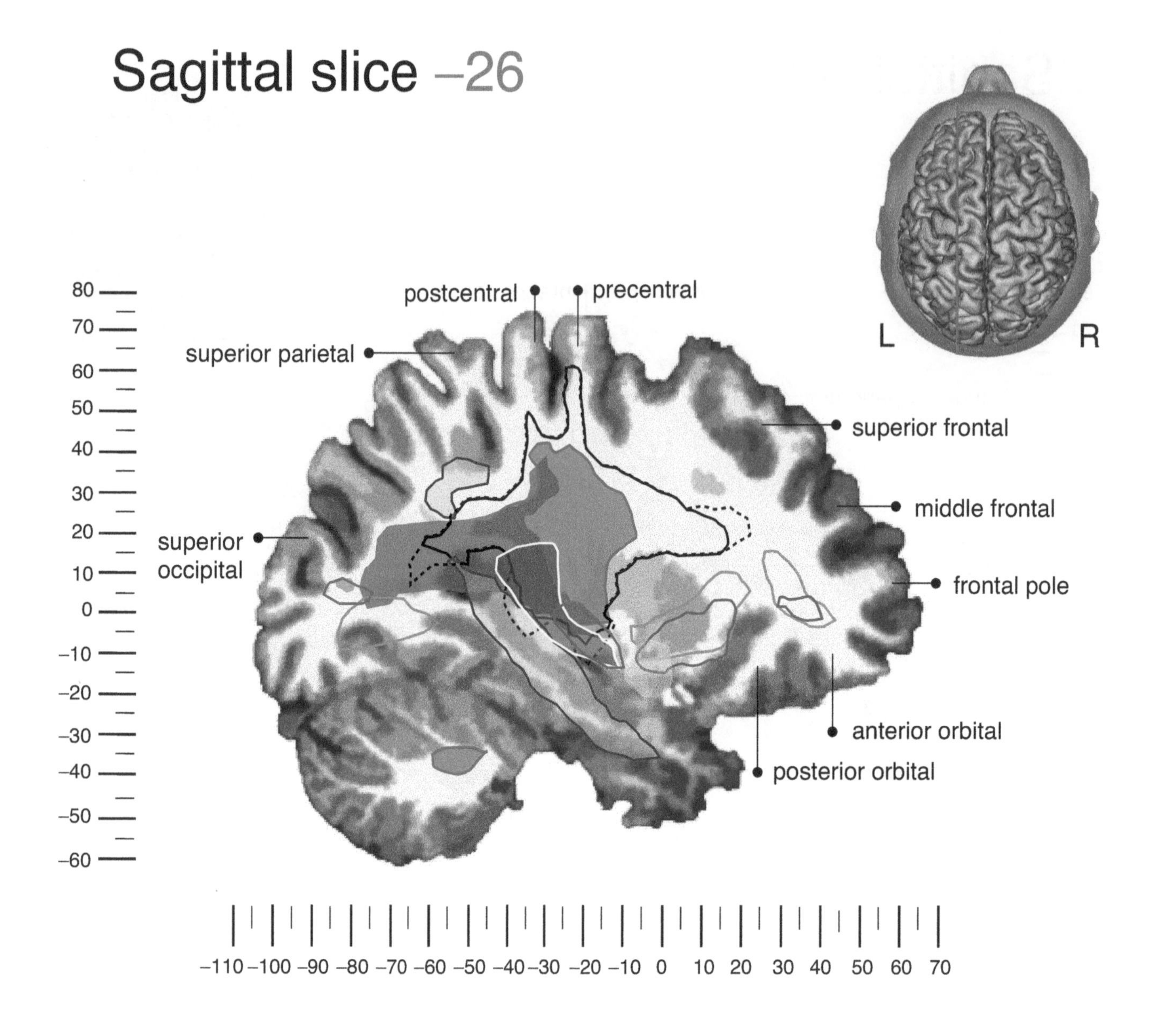

TRACT LEGEND

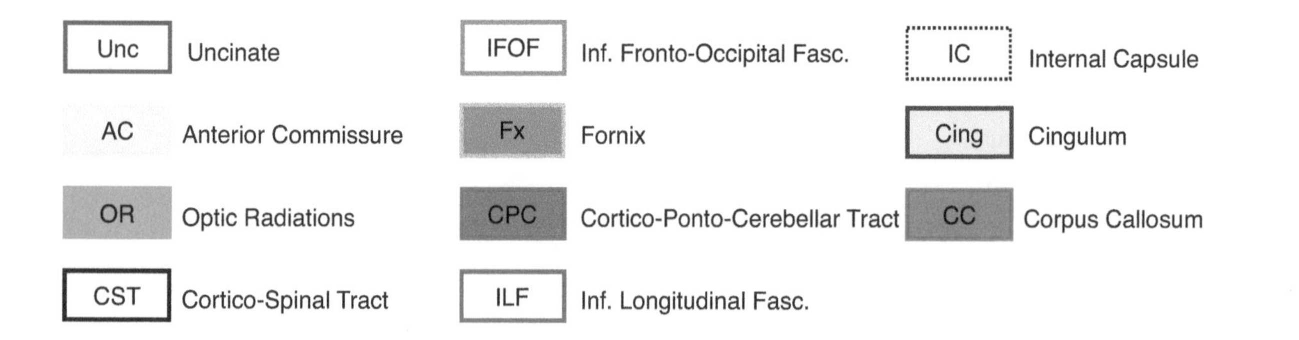

Sagittal slice −28

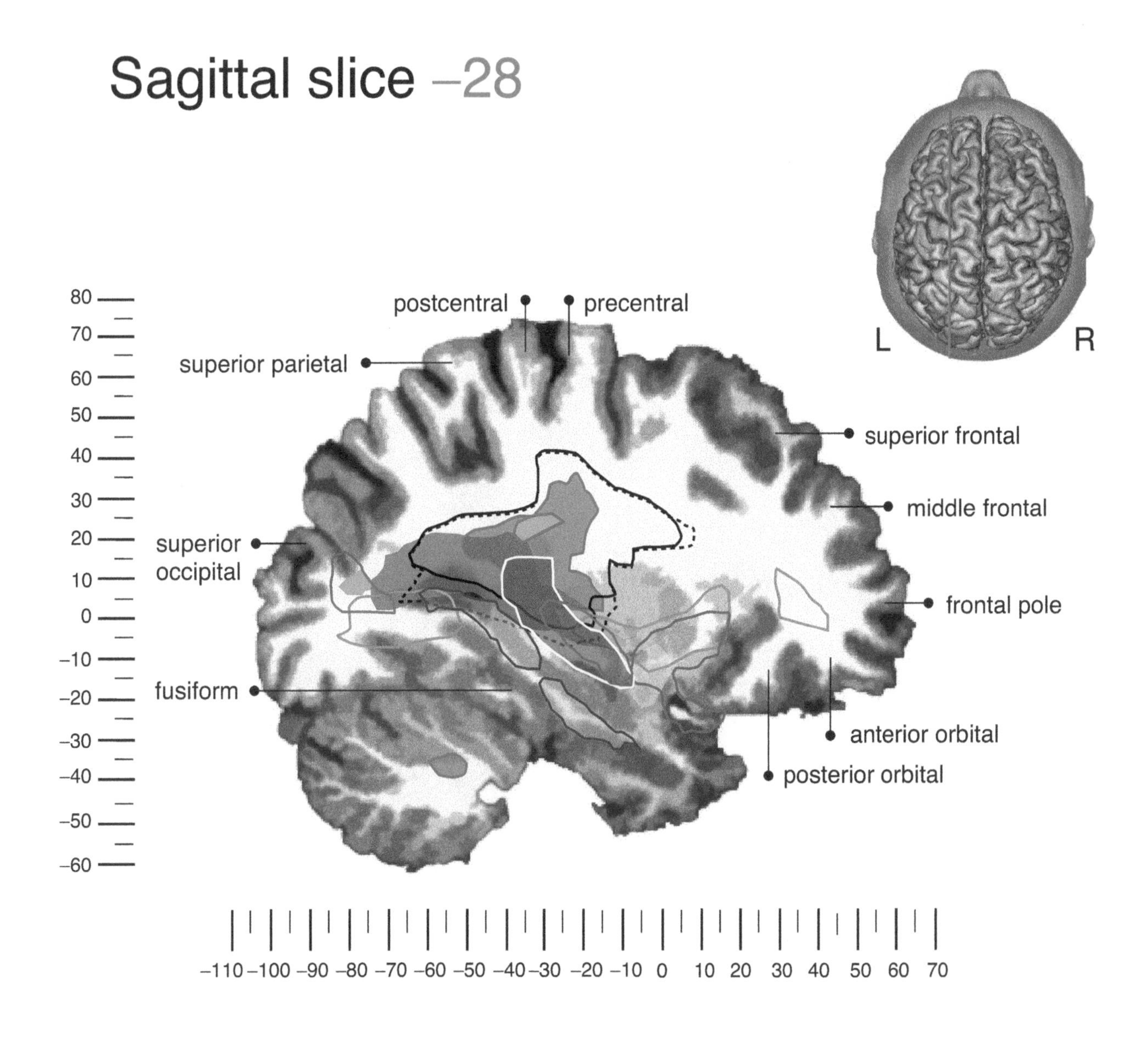

TRACT LEGEND

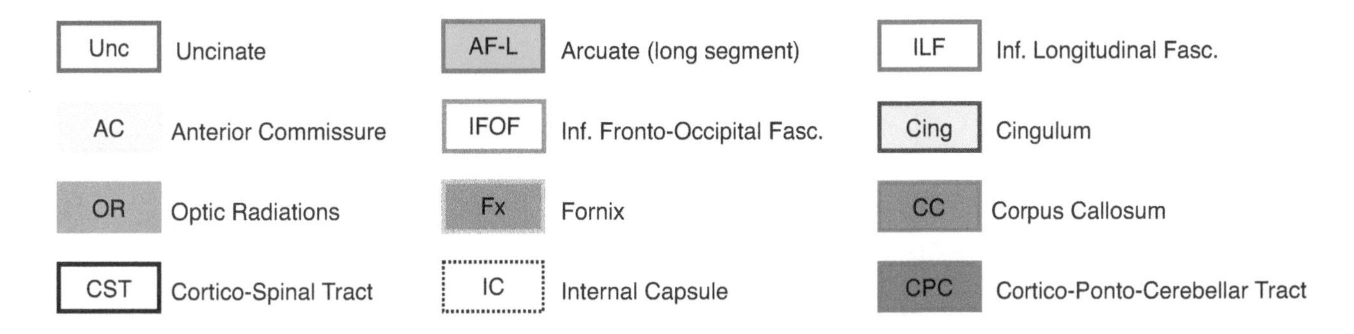

Sagittal slice –30

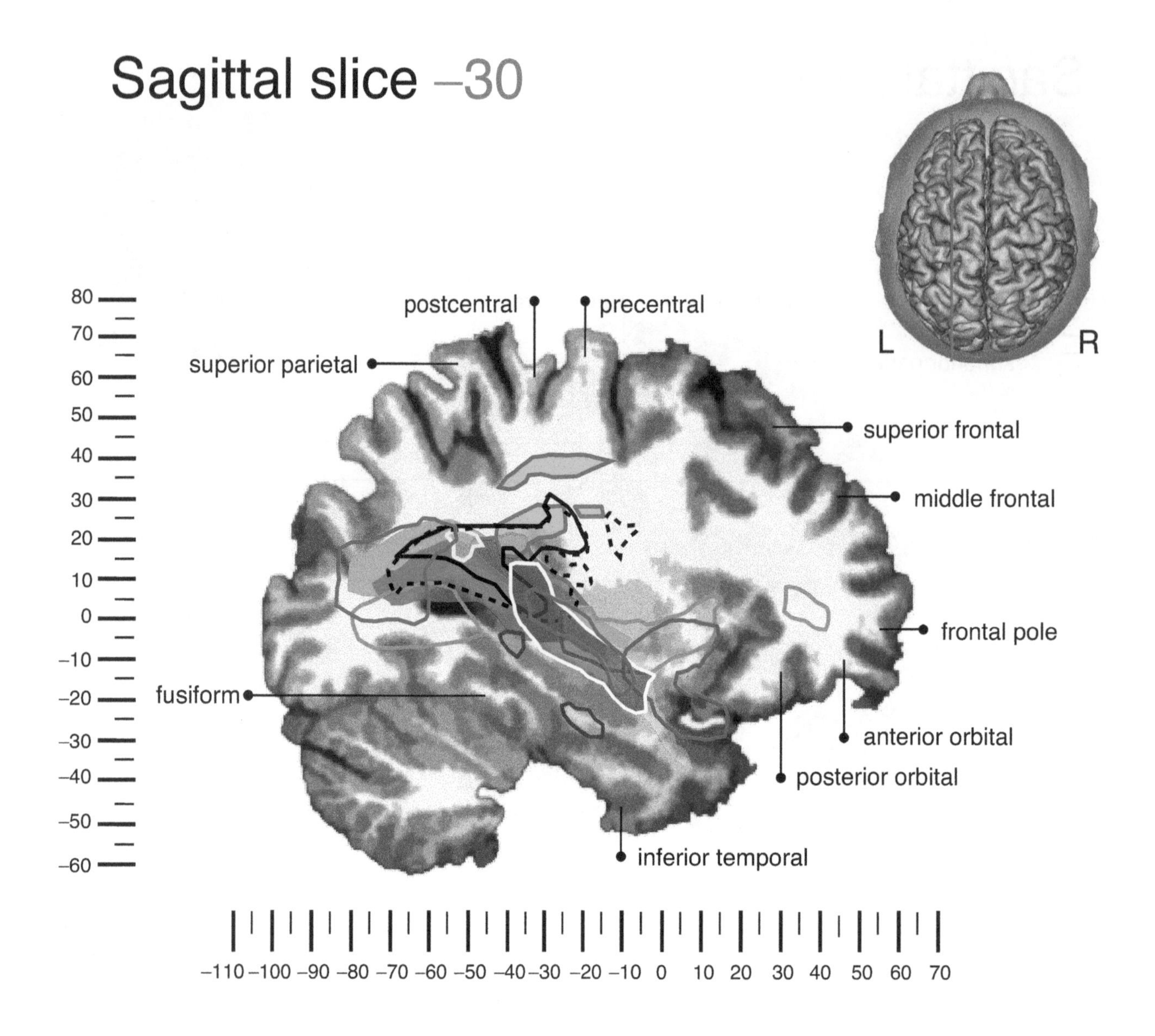

TRACT LEGEND

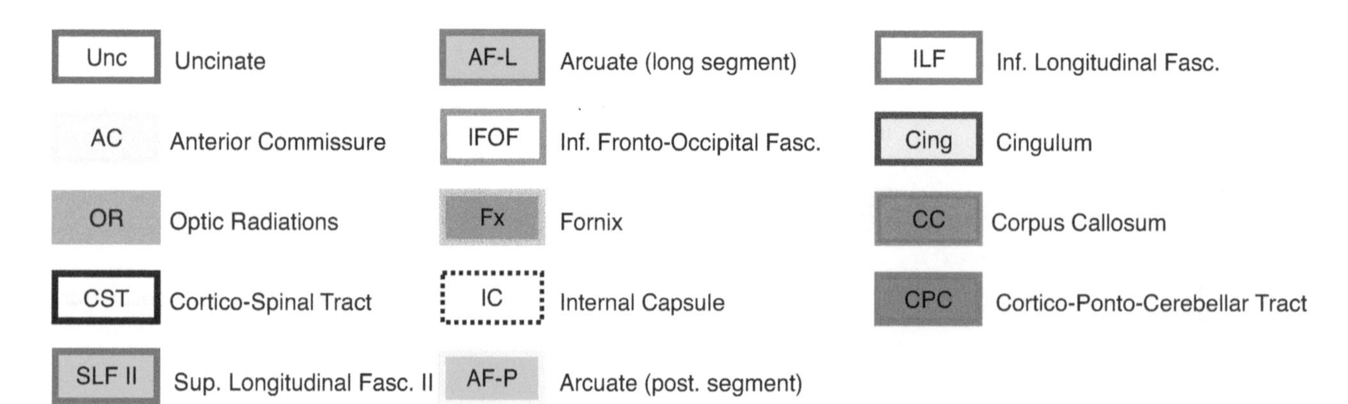

Sagittal slice −32

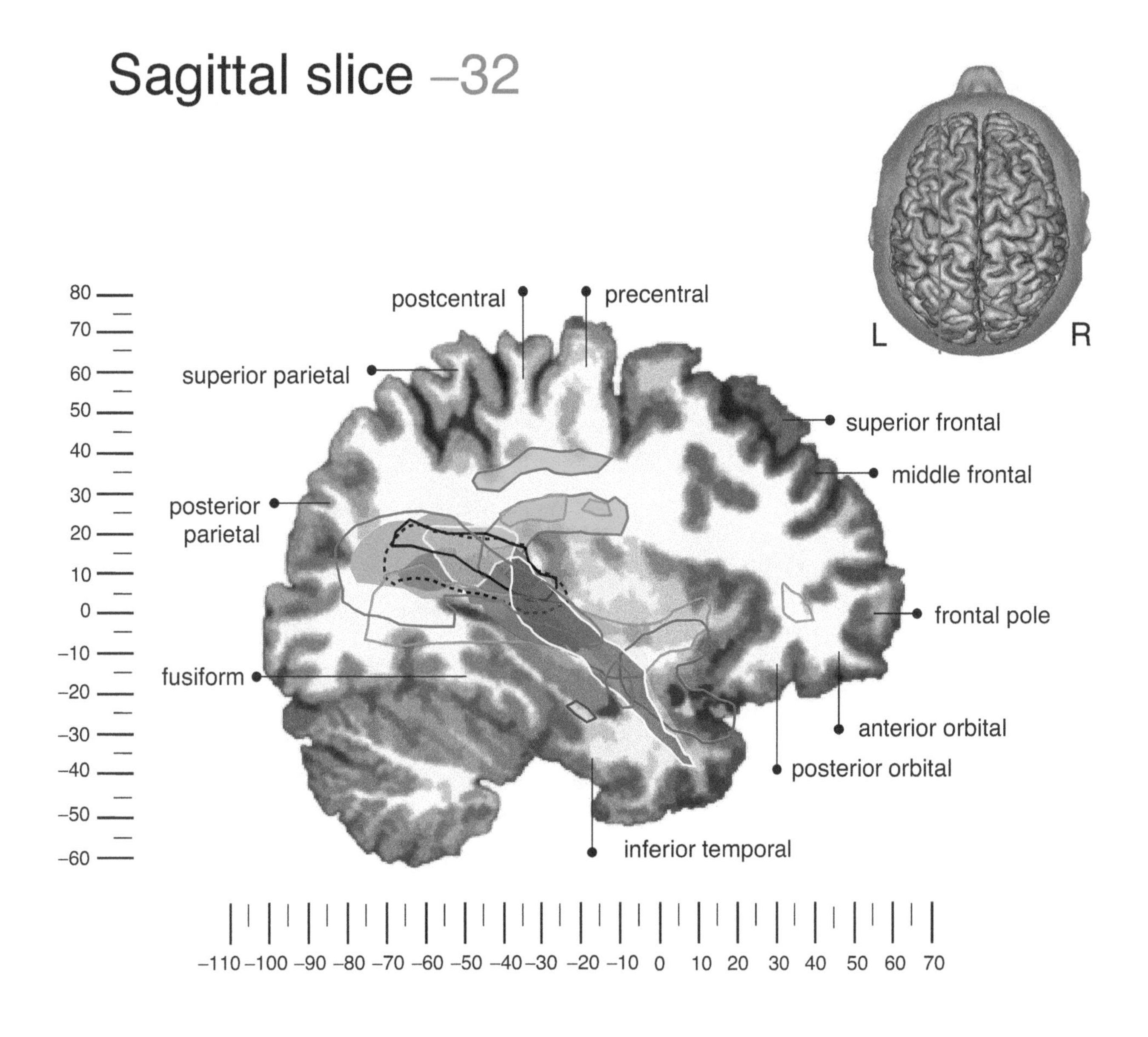

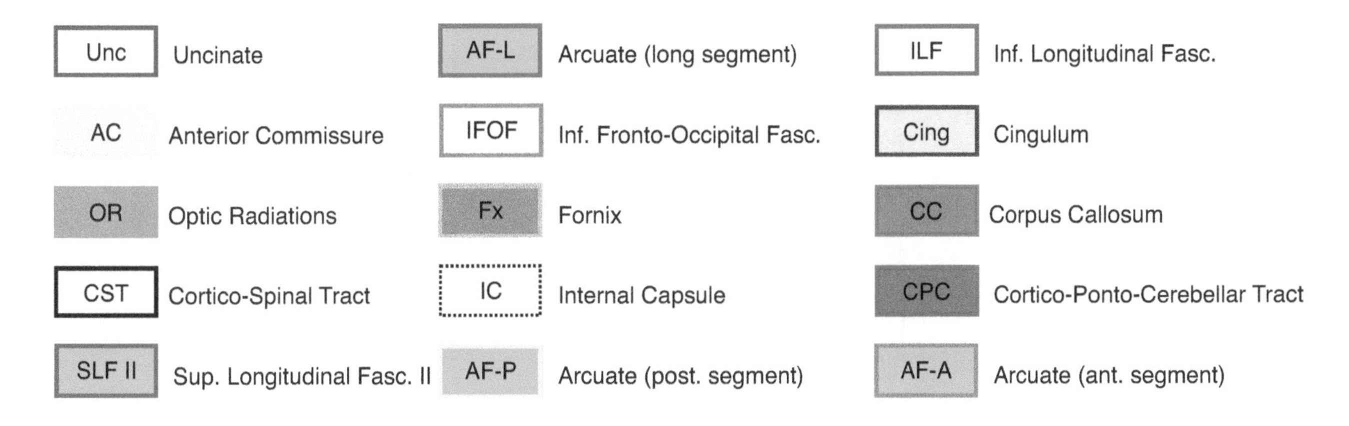

Sagittal slice –34

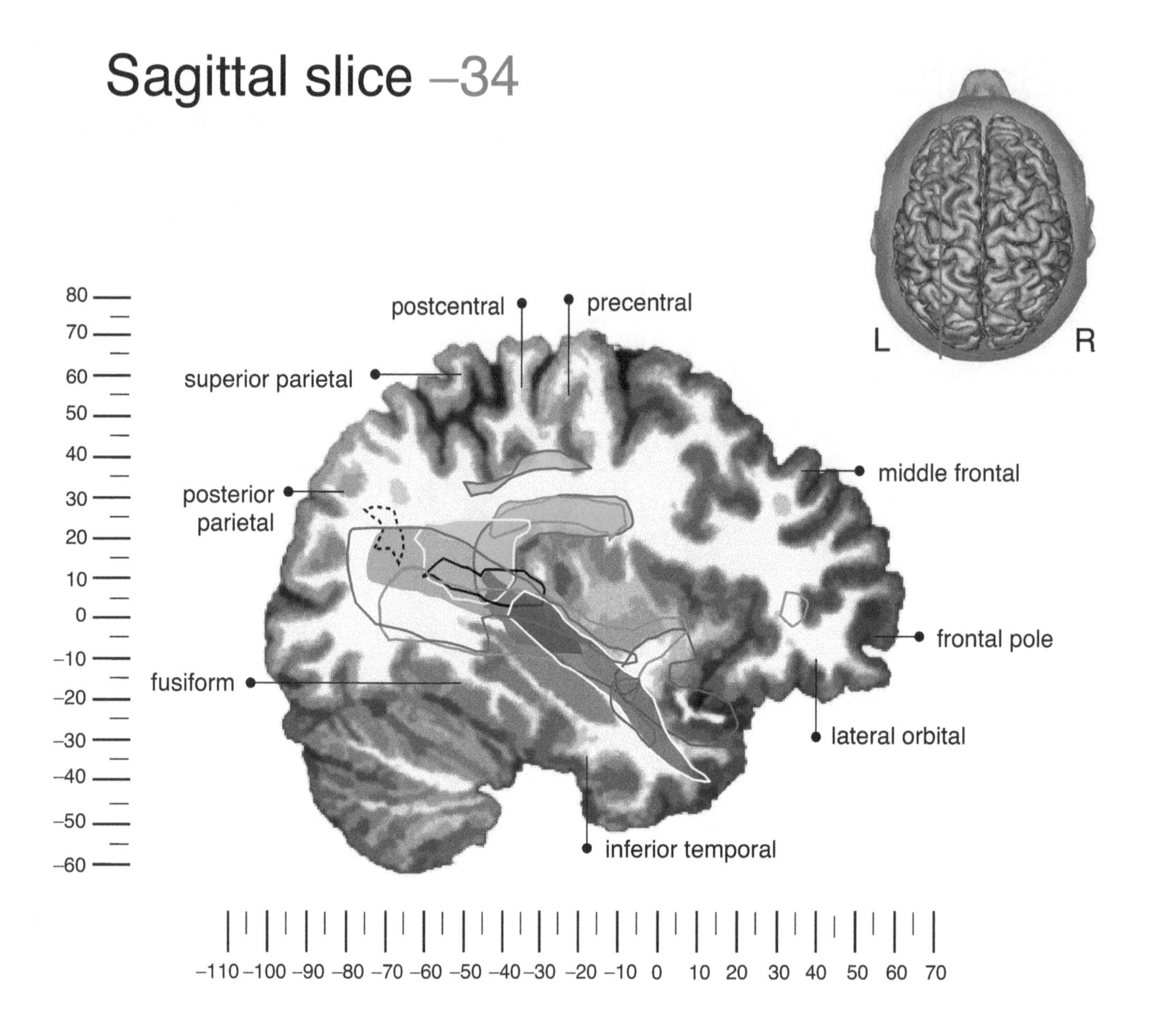

TRACT LEGEND

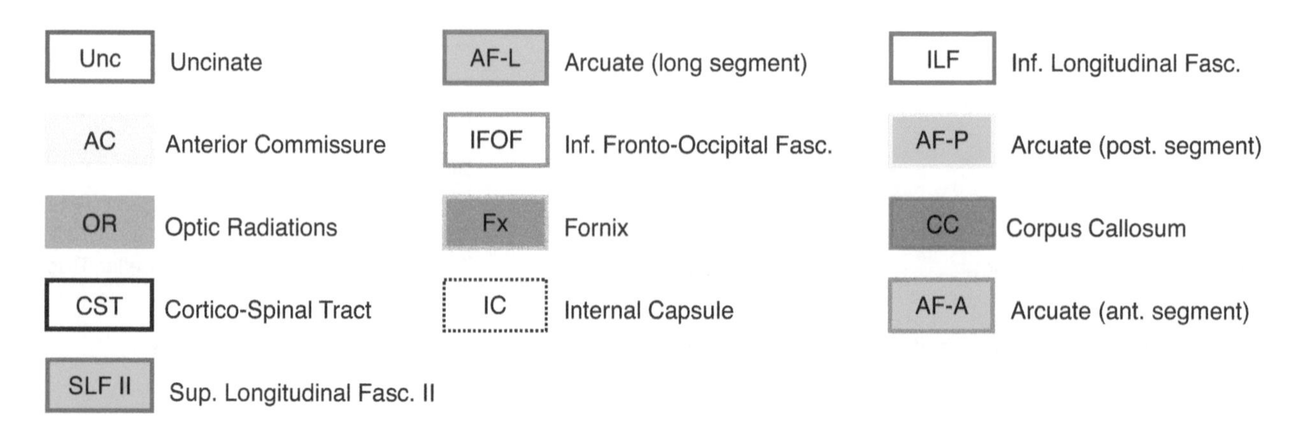

Sagittal slice –36

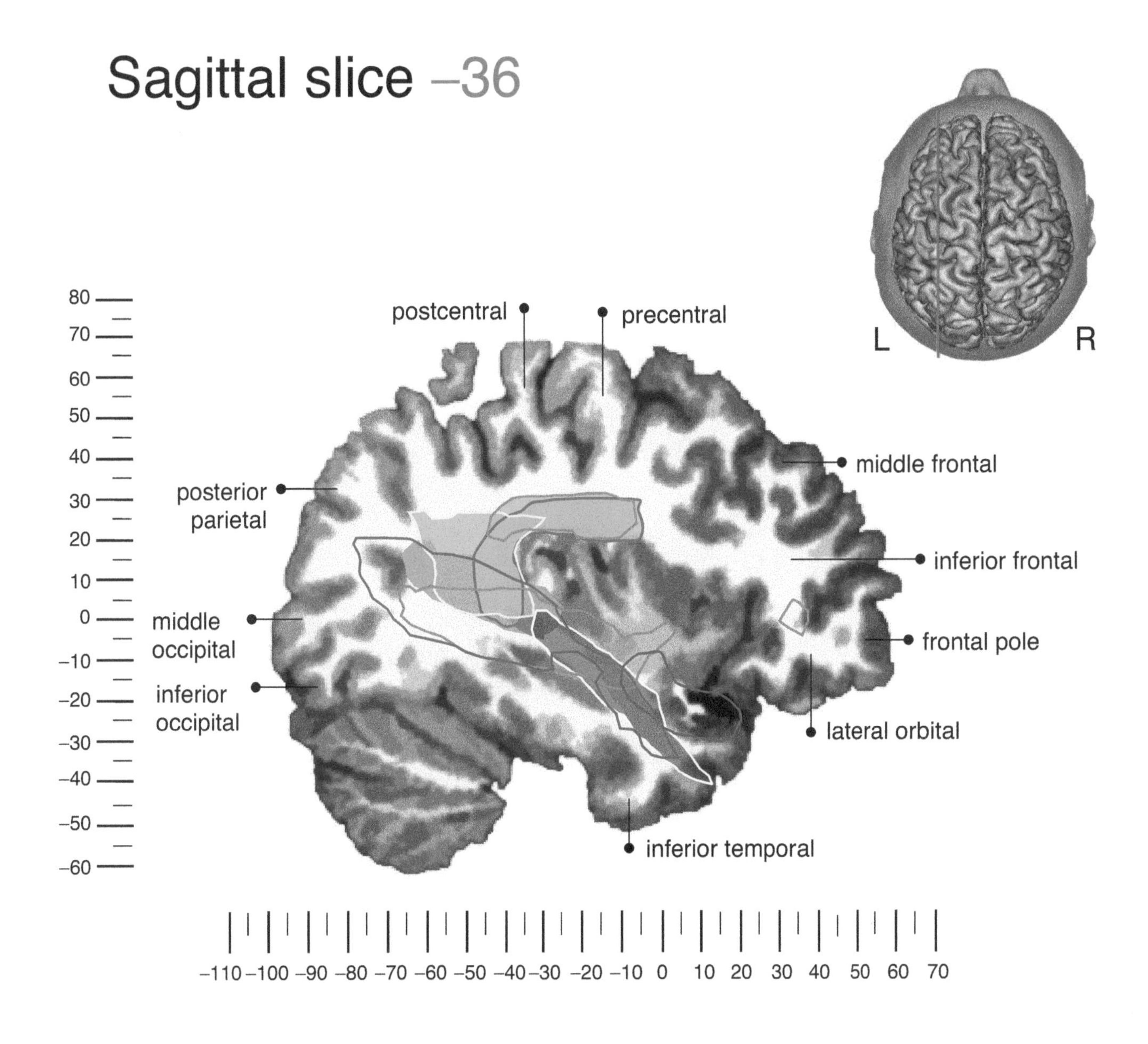

TRACT LEGEND

Unc	Uncinate	AF-L	Arcuate (long segment)	ILF	Inf. Longitudinal Fasc.
AC	Anterior Commissure	IFOF	Inf. Fronto-Occipital Fasc.	AF-P	Arcuate (post. segment)
OR	Optic Radiations	Fx	Fornix	CC	Corpus Callosum
AF-A	Arcuate (ant. segment)				

Sagittal slice –38

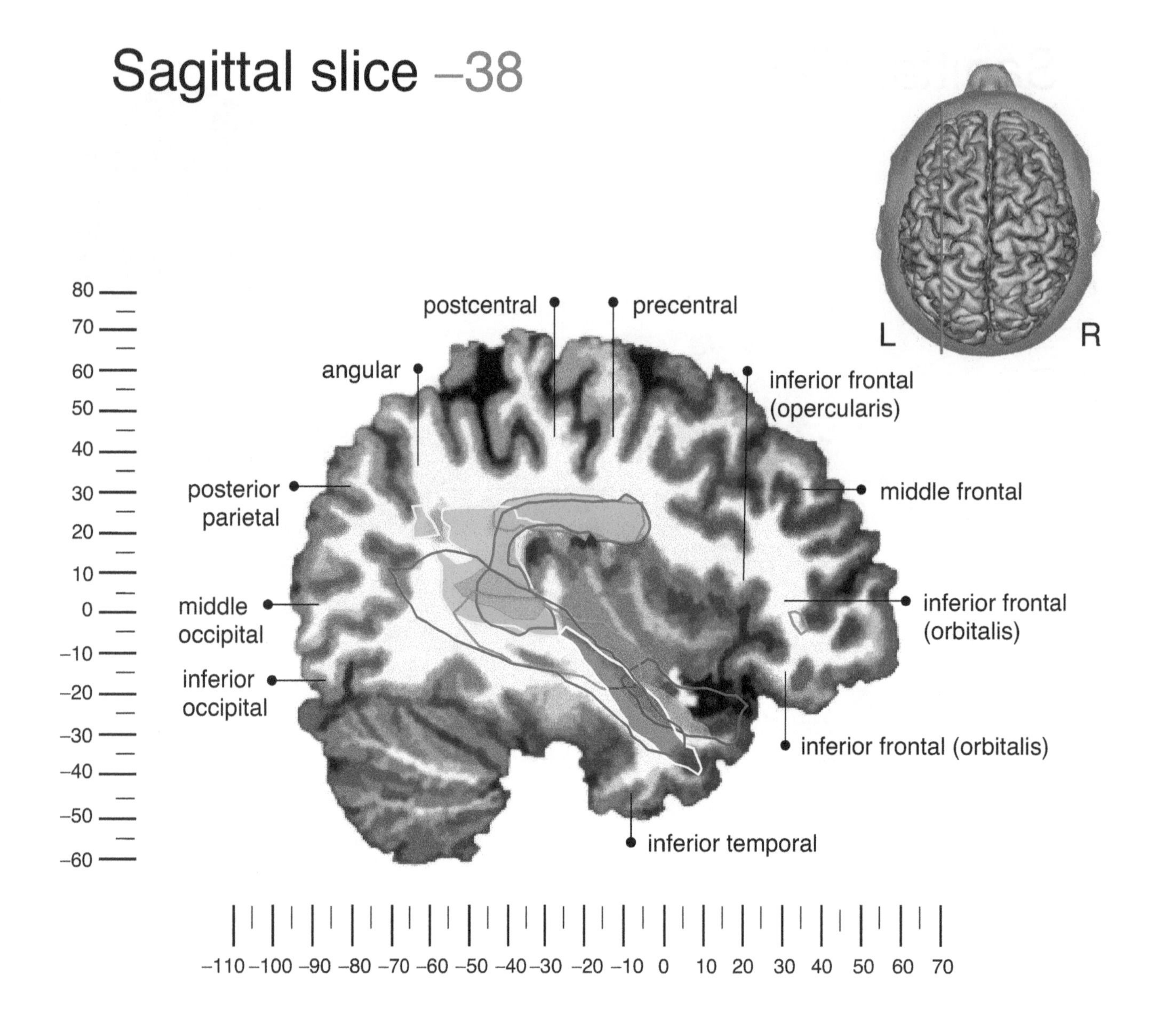

TRACT LEGEND

Unc	Uncinate	AF-L	Arcuate (long segment)	ILF	Inf. Longitudinal Fasc.
AC	Anterior Commissure	IFOF	Inf. Fronto-Occipital Fasc.	AF-P	Arcuate (post. segment)
OR	Optic Radiations	Fx	Fornix	AF-A	Arcuate (ant. segment)

Sagittal slice –40

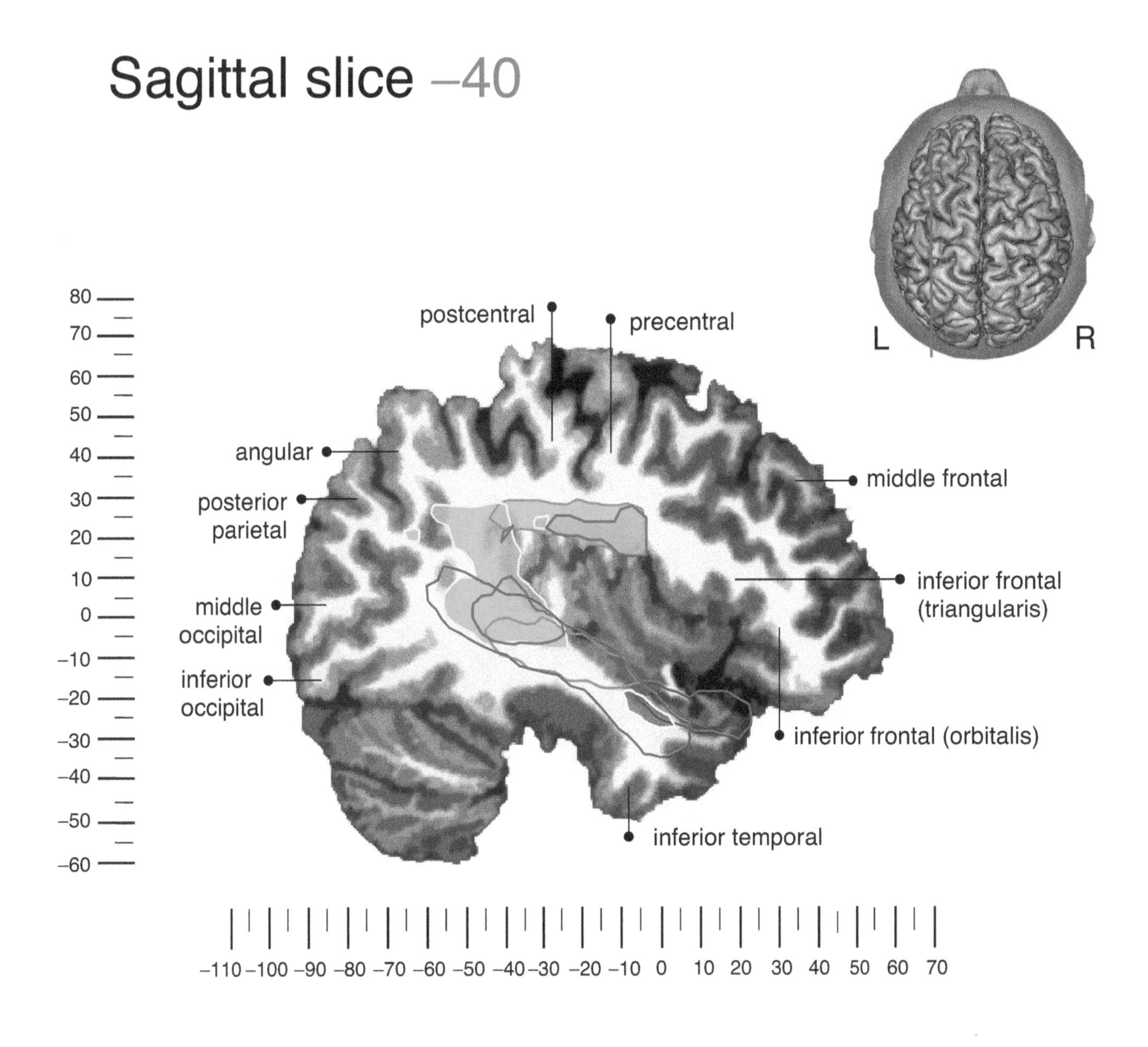

TRACT LEGEND

Unc	Uncinate	AF-L	Arcuate (long segment)	ILF	Inf. Longitudinal Fasc.
AC	Anterior Commissure	IFOF	Inf. Fronto-Occipital Fasc.	AF-P	Arcuate (post. segment)
OR	Optic Radiations	Fx	Fornix	AF-A	Arcuate (ant. segment)

Sagittal slice –42

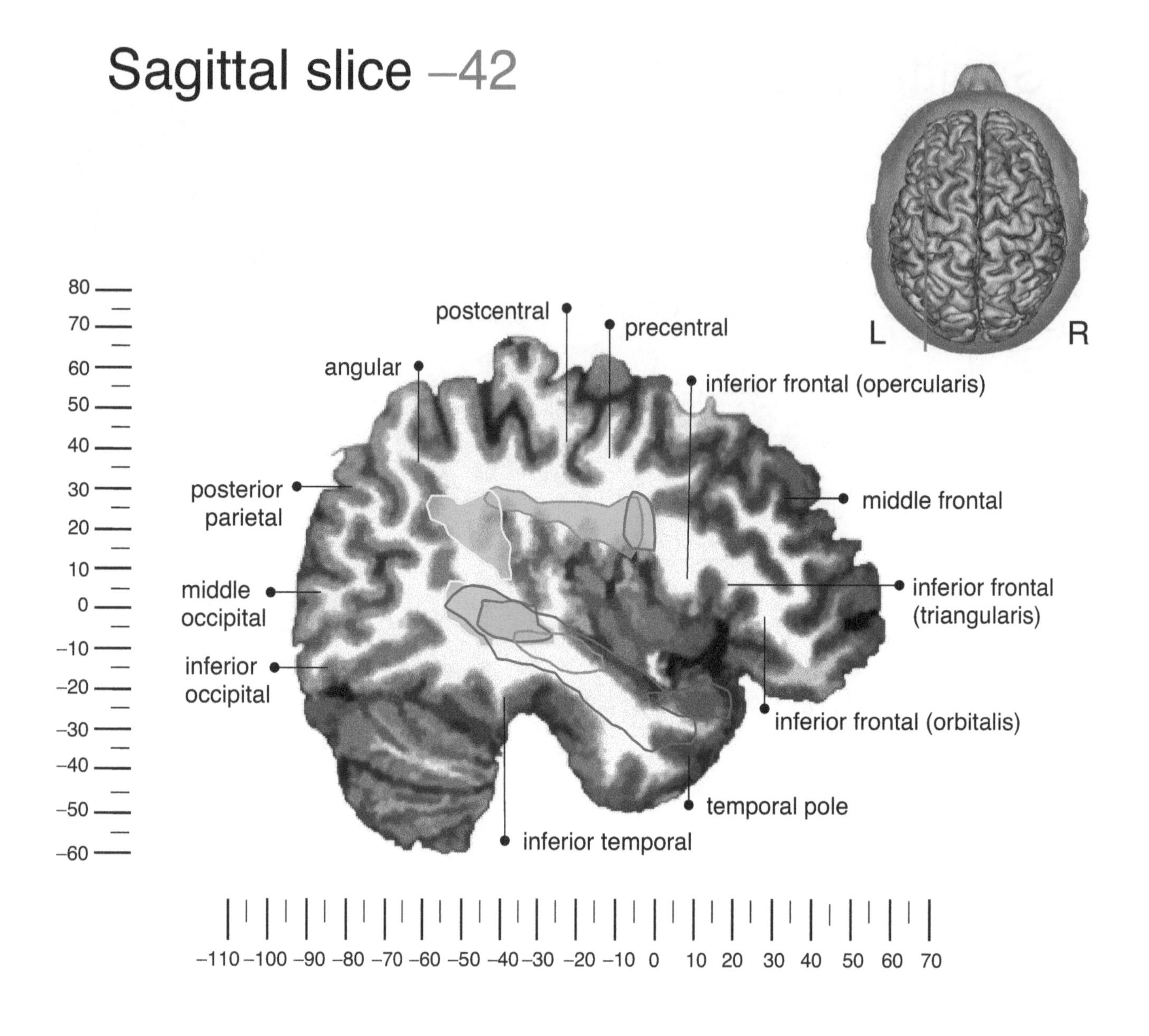

TRACT LEGEND

Unc	Uncinate	AF-L	Arcuate (long segment)	ILF	Inf. Longitudinal Fasc.
AF-A	Arcuate (ant. segment)	IFOF	Inf. Fronto-Occipital Fasc.	AF-P	Arcuate (post. segment)

Sagittal slice –44

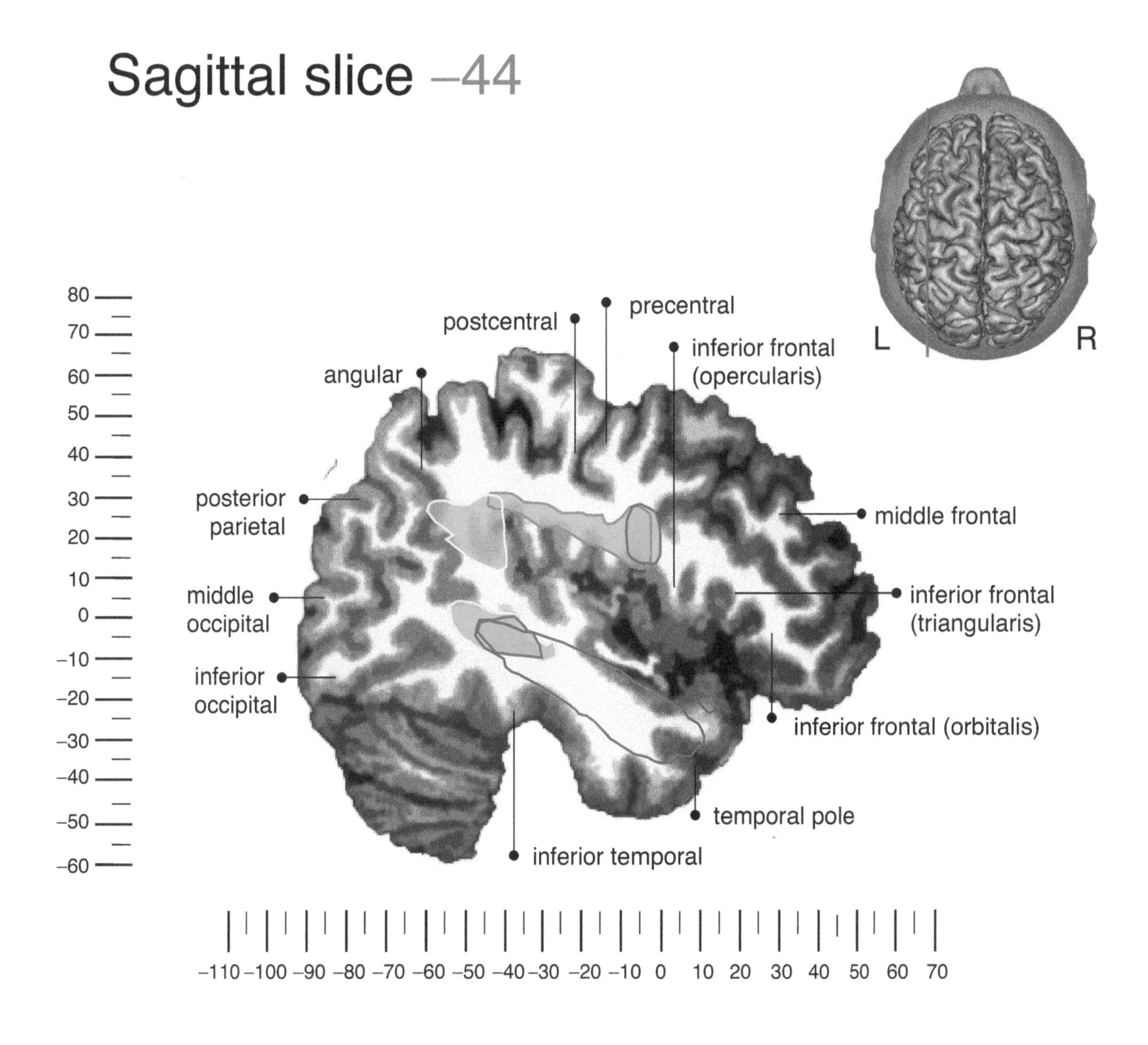

TRACT LEGEND

Unc	Uncinate	AF-L	Arcuate (long segment)	ILF	Inf. Longitudinal Fasc.
AF-A	Arcuate (ant. segment)	AF-P	Arcuate (post. segment)		

Sagittal slice –46

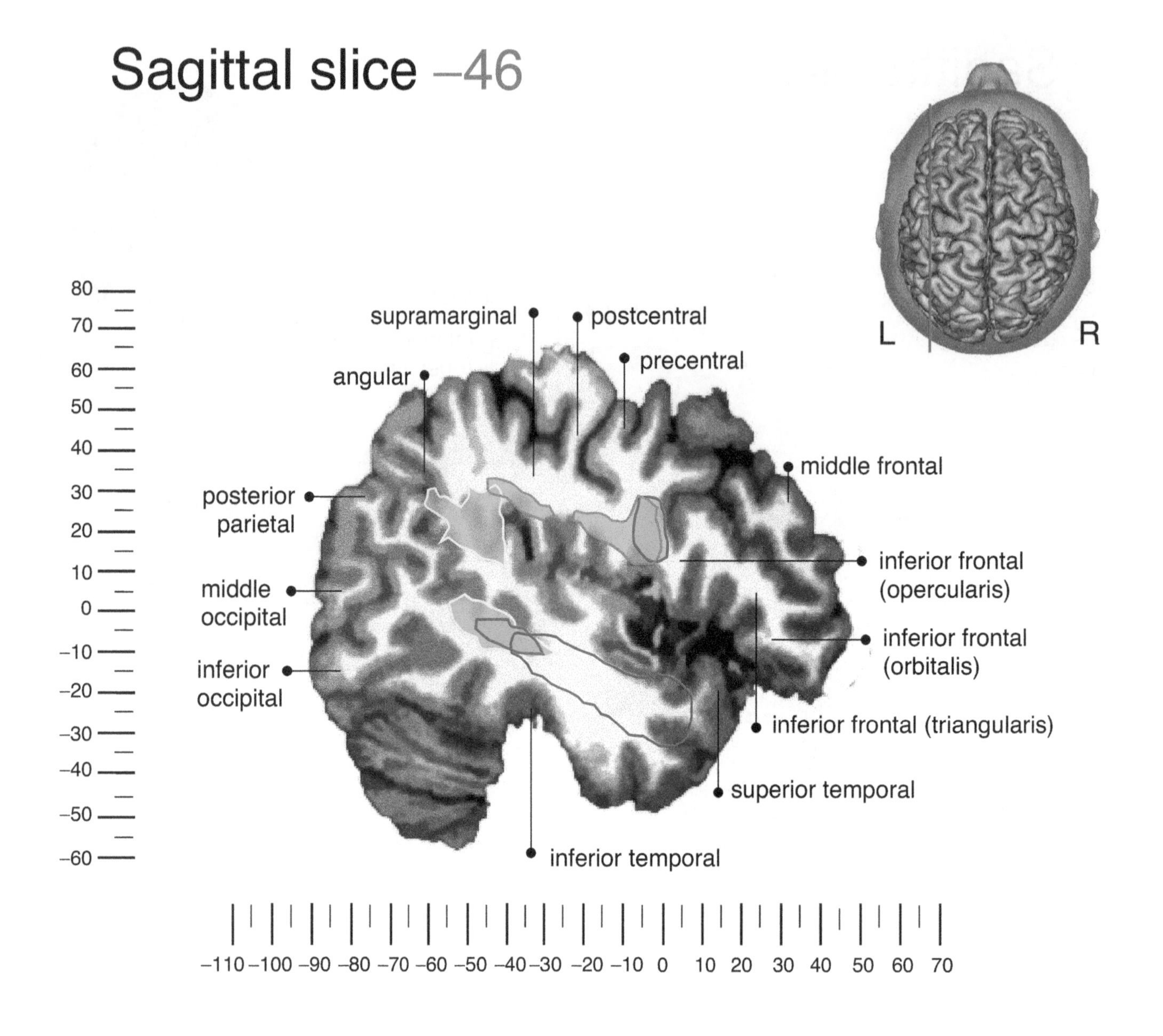

TRACT LEGEND

AF-A	Arcuate (ant. segment)	AF-L	Arcuate (long segment)	AF-P	Arcuate (post. segment)
ILF	Inf. Longitudinal Fasc.				

Sagittal slice –48

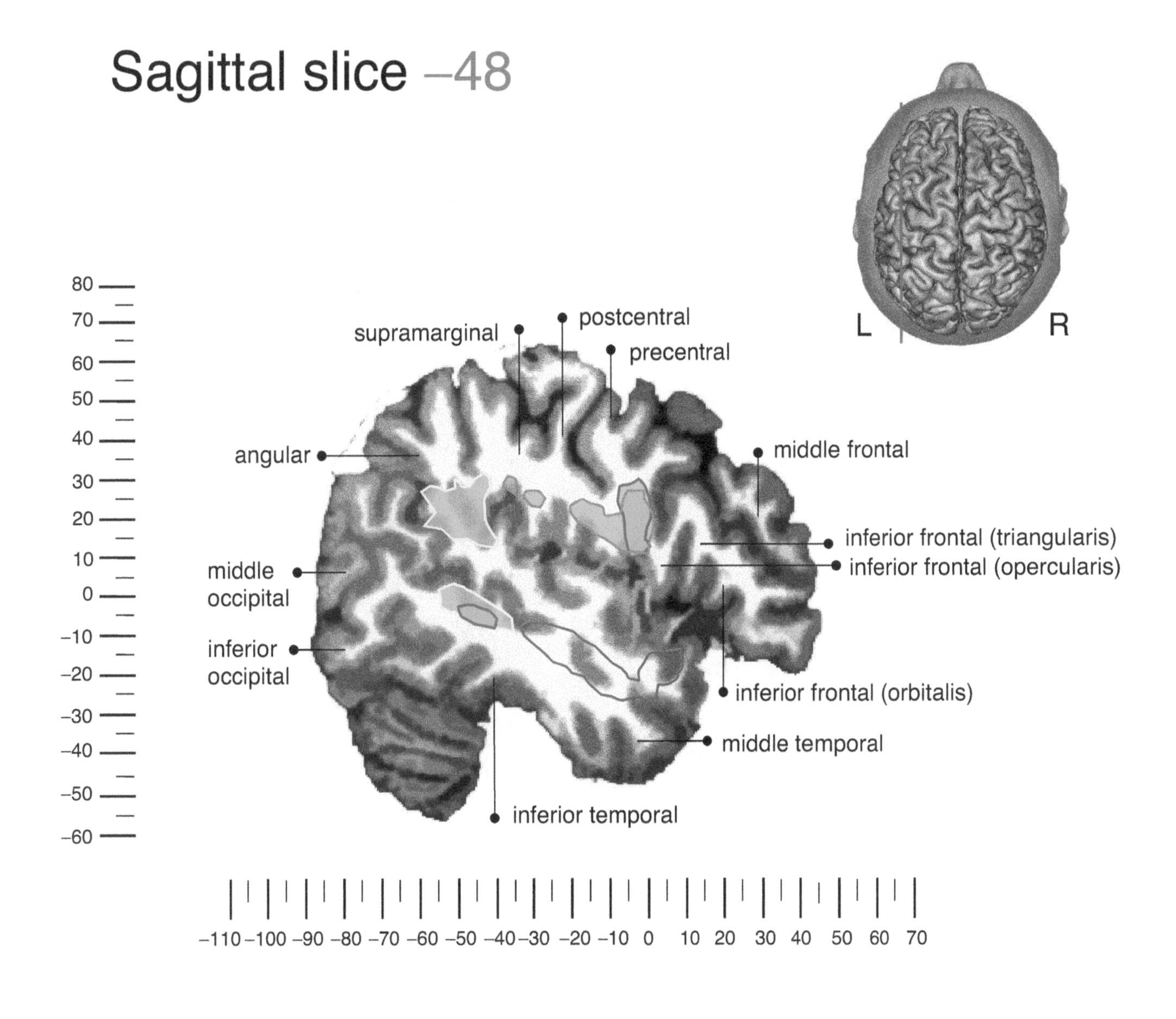

TRACT LEGEND

AF-A	Arcuate (ant. segment)
AF-L	Arcuate (long segment)
AF-P	Arcuate (post. segment)
ILF	Inf. Longitudinal Fasc.

Sagittal slice –50

L R

80 70 60 50 40 30 20 10 0 –10 –20 –30 –40 –50 –60

postcentral
supramarginal
precentral
inferior frontal (opercularis)
angular
middle frontal
middle occipital
inferior frontal (triangularis)
inferior occipital
superior temporal
middle temporal
inferior temporal

TRACT LEGEND

AF-A	Arcuate (ant. segment)
AF-L	Arcuate (long segment)
AF-P	Arcuate (post. segment)
ILF	Inf. Longitudinal Fasc.

Sagittal slice –52

L
R

80
70
60
50
40
30
20
10
0
–10
–20
–30
–40
–50
–60

postcentral
supramarginal
precentral
inferior frontal (opercularis)
angular
middle frontal
middle occipital
inferior frontal (triangularis)
inferior occipital
superior temporal
middle temporal
inferior temporal

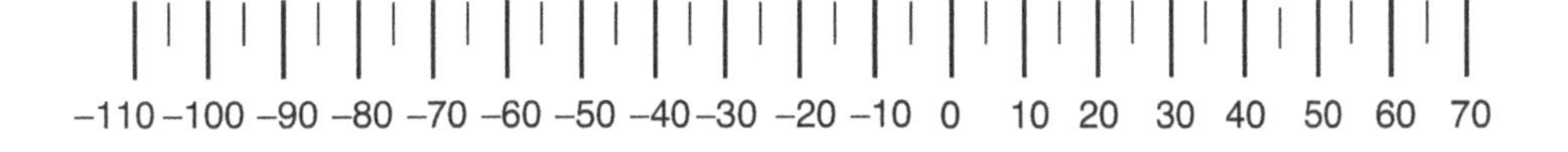

TRACT LEGEND

AF-A	Arcuate (ant. segment)	AF-L	Arcuate (long segment)	AF-P	Arcuate (post. segment)

Sagittal slice –54

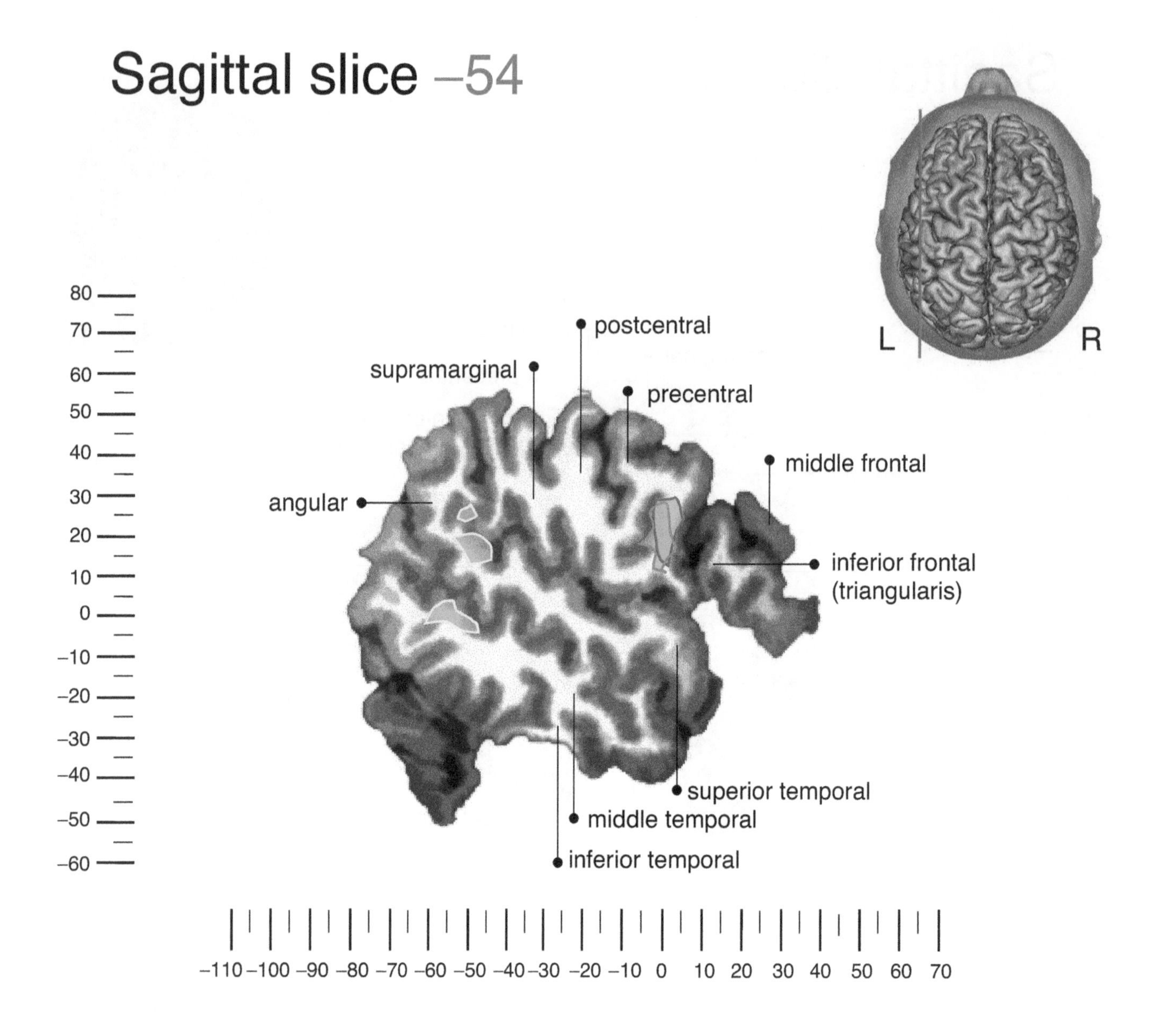

TRACT LEGEND

AF-A	Arcuate (ant. segment)	AF-L	Arcuate (long segment)	AF-P	Arcuate (post. segment)
ILF	Inf. Longitudinal Fasc.				

CHAPTER 7

Perisylvian Pathways

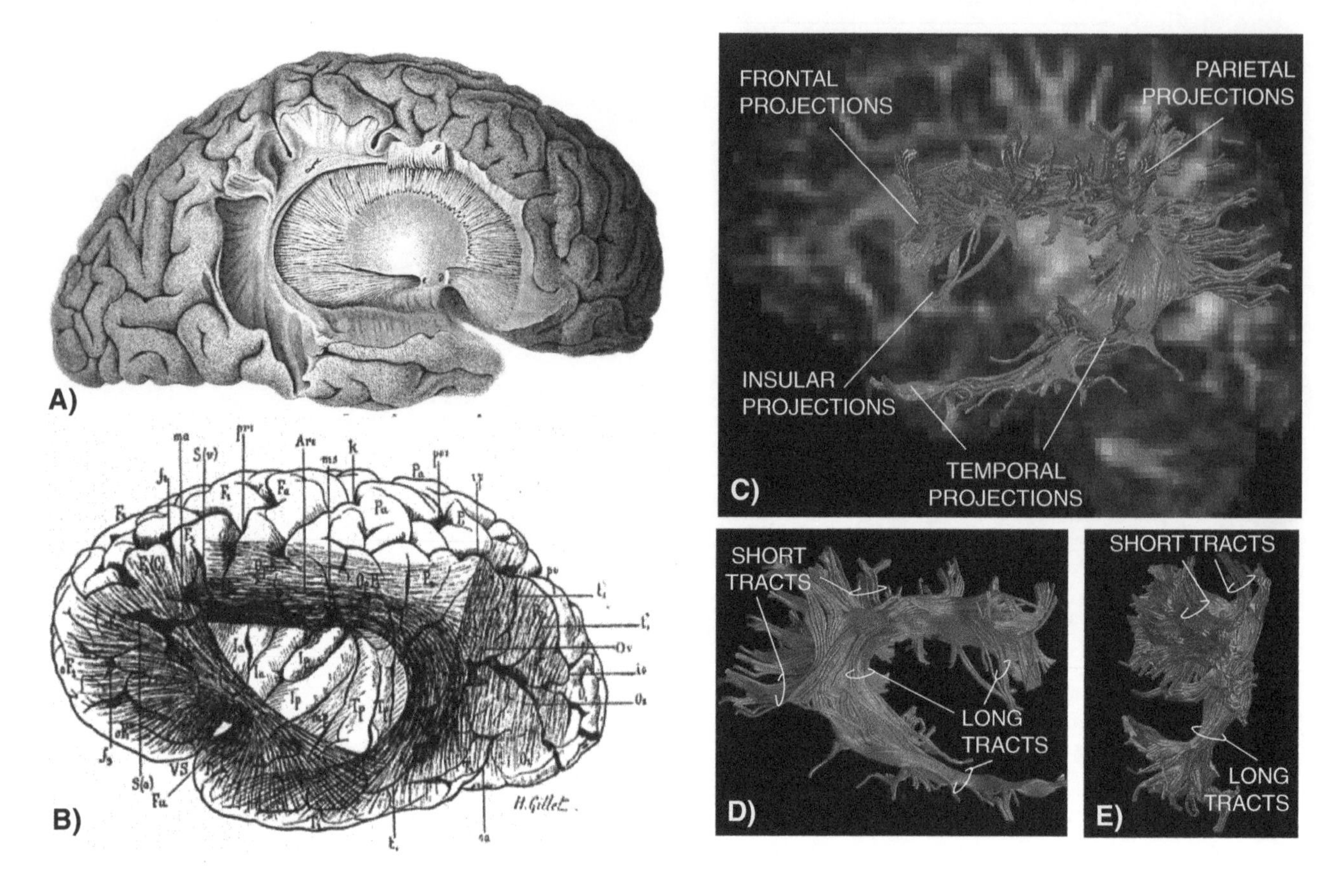

Figure 7.1 A) Reil's (1812) and B) Dejerine's (1895) post-mortem dissections of the arcuate fasciculus. C) Left lateral, D) right medial, and E) posterior view of the arcuate fasciculus reconstructed with diffusion tensor tractography (Catani et al., 2002).

Introduction

Johann Christian Reil was the first to identify a group of fibres running deeply in the white matter of the temporal, parietal, and frontal regions around the Sylvian fissure, which he described as the *Ungenannte Marksubstanz* (unnamed white matter) (Figure 7.1A) (Reil, 1812). In 1822 Karl Burdach confirmed Reil's finding and designated the perisylvian tracts collectively as the *Fasciculus Arcuatus*, because of the arching shape of its longest fibres (Burdach, 1822). The work of the German neuroanatomists was subsequently revisited by Dejerine (1895) who believed that the arcuate fasciculus was mainly composed of short U-shaped associative fibres connecting neighbouring perisylvian regions (Figure 7.1B). Dejerine was also the first to use the term superior longitudinal fasciculus and arcuate fasciculus interchangeably. In the meantime, post-mortem studies of patients with aphasia led Paul Broca (1861) and Carl Wernicke (1874) to discover the cortical 'centres' dedicated to speech production (Broca's area in the frontal lobe) and auditory comprehension (Wernicke's area in the temporal lobe). Wernicke was also the first to postulate that language relies on the integrity of a 'psychic reflex arc' between temporal and frontal regions. However, the arcuate fasciculus was not part of Wernicke's original anatomical model, as he thought that the temporal and frontal language areas were indirectly connected by fibres passing through the external capsule and relaying in the cortex of the insula (Wernicke, 1874). It was Monakow (1885), and later Dejerine (1895), to propose the arcuate fasciculus as the tract directly connecting Broca's and Wernicke's area. This anatomical model was subsequently revitalized by Norman Geschwind—with an additional emphasis on the importance of the connections to the angular gyrus—and became known as the Wernicke–Geschwind model (Geschwind, 1965, 1970; Catani and ffytche, 2005). Although the existence of the arcuate fasciculus has been confirmed in humans by many studies using blunt dissections and axonal staining of degenerating axons, these methods have not shed much light on its detailed anatomy. For this reason more powerful methods for tracing axonal pathways have been used in the monkey to identify a homologue of the arcuate fasciculus (Petrides and Pandya, 1988; Schmahmann and Pandya, 2006). But the absence of language in non-human primates raises doubts on the possibility of translating *tout court* these findings to man. The advent of diffusion tensor imaging tractography has therefore opened a new era in the field. The first tractography studies applied to the perisylvian pathways showed that the anatomy of the arcuate fasciculus is more complex than previously thought (Figures 7.1C,D,E and 7.2) (Catani et al., 2002, 2005a; Parker et al., 2005). In addition to the long segment directly connecting Wernicke's and Broca's territories (i.e. the arcuate fasciculus *sensu strictu*), there is an indirect pathway consisting of an anterior and a posterior segment linking the inferior parietal lobule (Geschwind's territory) to Broca's territory

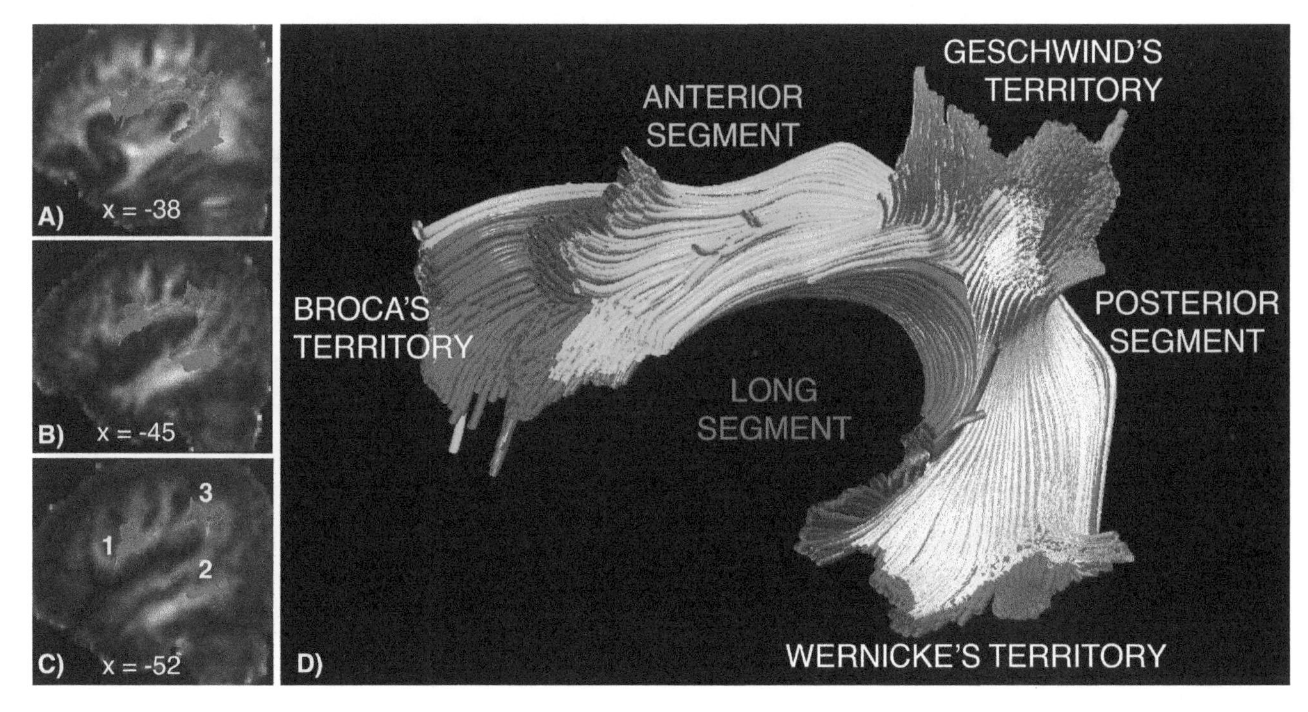

Figure 7.2 A–C) Sagittal sections of the arcuate fasciculus in the average dataset (10 subjects). The lateralmost section (Talairach x = −52) shows three cortical termination territories (numbered 1–3). The three cortical projection zones were used as starting point for the tracking of the three segments shown in D. D) Tractography reconstruction of the arcuate fasciculus using the two-regions of interest approach in the average brain. Broca's and Wernicke' territories are connected through direct and indirect pathways. The direct pathway (long segment shown in red) runs medially and corresponds to classical descriptions of the arcuate fasciculus. The indirect pathway runs laterally and is composed of an anterior segment (green) connecting the inferior parietal cortex (Geschwind's territory) and Broca's territory and a posterior segment (yellow) connecting Geschwind's and Wernicke's territories (Catani et al., 2005).

and Wernicke's territory, respectively. Additional support for the existence of the three perisylvian segments comes from human intraoperative electrocorticography (Matsumoto et al., 2004), functional connectivity (Schmithorst and Holland, 2007), and recent post-mortem dissections (Lawes et al., 2008).

Functional anatomy of the perisylvian cortical territories

Classical models of language distinguish a posterior temporal region (i.e. Wernicke's area corresponding to BA 22) (Wernicke, 1874) dedicated to auditory comprehension, and an anterior region in the posterior part of the inferior frontal gyrus (i.e. Broca's area corresponding to BA 44 and 45) dedicated to speech production (Broca, 1861). The distribution of the arcuate terminations found by tractography studies extends beyond the classical limits of language areas and, for this reason, the term territory rather than area is preferred. Broca's territory, for example, includes part of the posterior middle frontal gyrus and inferior precentral gyrus, whilst Wernicke's territory includes the posterior part of the middle temporal gyrus. This wider distribution is consistent with several lines of evidence suggesting that Broca's and Wernicke's territories are formed by smaller cortical regions specialized for different aspects of language. For example, more posterior regions of Broca's territory are engaged in phonological tasks (e.g. 'saying words beginning with the letter F'), whereas more anterior regions specialize in syntactic functions (i.e. combining words according to correct grammatical language rules) and semantic functions (e.g. naming animals or objects) (Bookheimer, 2002). A similar segregation has been suggested for Wernicke's territory, with superior-anterior specialization for phonological tasks and posterior-inferior specialization for semantic tasks (Cannestra et al., 2000; Castillo et al., 2001).

The parallel pathways model of the arcuate fasciculus highlights the importance of a third territory, the inferior parietal cortex, as a separate primary language region with dense connections to classical language areas through the indirect pathway. The inferior parietal lobe corresponds to BA 39 (angular) and BA 40 (supramarginal), and although its involvement in language has been recognized for some time, its exact role is still largely unknown. The development of this region, which is one of the latest to myelinate in the human brain (Flechsig, 1901), is thought to have coincided with the emergence of language during evolution (Geschwind, 1965). Recent functional neuroimaging studies have shown that Geschwind's territory is part of an extended network activated during comprehension of global coherence of narratives (Martin-Loeches et al., 2008), processing of concrete concepts (Sabsevitz et al., 2005), episodic memory retrieval of words (Vilberg and Rugg, 2008), and verbal working memory (Jacquemot and Scott, 2006). Also, thanks to its

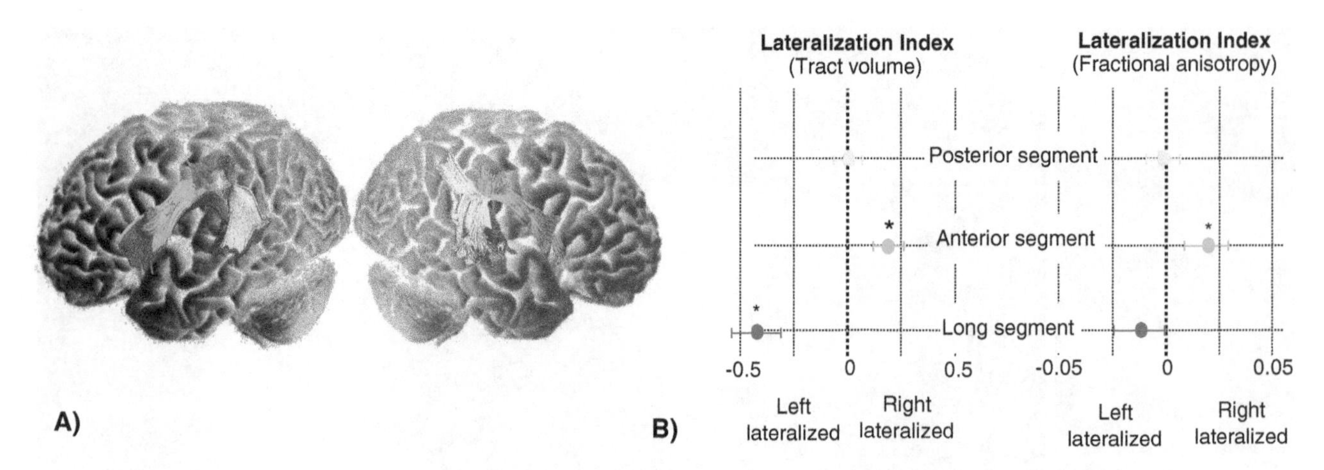

Figure 7.3 Left–right hemispheric differences in the perisylvian pathways. A) In an average data set derived from 10 healthy subjects, direct connections (red) between temporal and frontal regions in the left hemisphere are absent from the right (Catani et al., 2007). B) Tract-specific measurements of the volume and fractional anisotropy (an index that reflects the microstructural organization of fibres) of the three segments of the arcuate in a group of 40 young adults. The volume and fractional anisotropy of the anterior segment show a significant right lateralization. The volume of the long segment is left lateralized with a statistically non-significant left lateralization also of the fractional anisotropy. The posterior segment is bilateral and symmetrical (Thiebaut de Schotten et al., 2011a).

anatomical position, Geschwind's territory is a zone of convergence and integration of sensory and motor information. It is therefore well suited to play a key role in the self-awareness of speech and actions in general (Jardri et al., 2007). It is possible that like Broca's and Wernicke's territories, Geschwind's territory is a heterogeneous region with functionally segregated subdivisions. For example, with respect to working memory, the posterior portion of Geschwind's territory (i.e. angular gyrus) together with Wernicke's territory seem to be involved in the phonological storage of words (input buffer), while the anterior supramarginal gyrus interacts with Broca's territory for the active rehearsal of stored words (output buffer) (Vallar et al., 1997; Jacquemot and Scott, 2006). A similar segregation within Geschwind's territory can be made with regard to other functions, such as praxis and reading.

Lateralization of perisylvian pathways

Hemispheric asymmetry is a key feature of the human brain. Left–right differences in perisylvian anatomy have been shown using microscopic examination of post-mortem specimens (Galuske et al., 2000), structural T_1-weighted MRI (Paus et al., 1999) and DTI (Buchel, et al., 2004; Nucifora et al., 2005; Hagmann et al., 2006; Powell et al., 2006; Catani et al., 2007; Vernooij et al., 2007; Thiebaut de Schotten et al., 2011a). In young adults, tractography shows that of the three segments the posterior indirect is bilateral and symmetrical, while the anterior indirect and the long segments of the arcuate fasciculus are lateralized to the right and the left, respectively (Figure 7.3) (Catani et al., 2007; Thiebaut de Schotten et al., 2011a). However, the lateralization of these segments is quite heterogeneous among the general population. Volumetric measurements of the long segment, for example, show that more than half of the adult population (~60%) has an extreme degree of leftward lateralization (i.e. a complete absence or minimal presence in the right hemisphere) (Figure 7.4A) (Catani et al., 2007). The remaining approximately 40% shows either a moderate leftward lateralization (~20%) or a bilaterally symmetric pattern (~20%). Similar results are reported for left-handed subjects (Vernooij et al., 2007) and younger groups of children and adolescents (aged 6–17 years) (Eluvathingal et al., 2007). Preliminary reports also suggest a

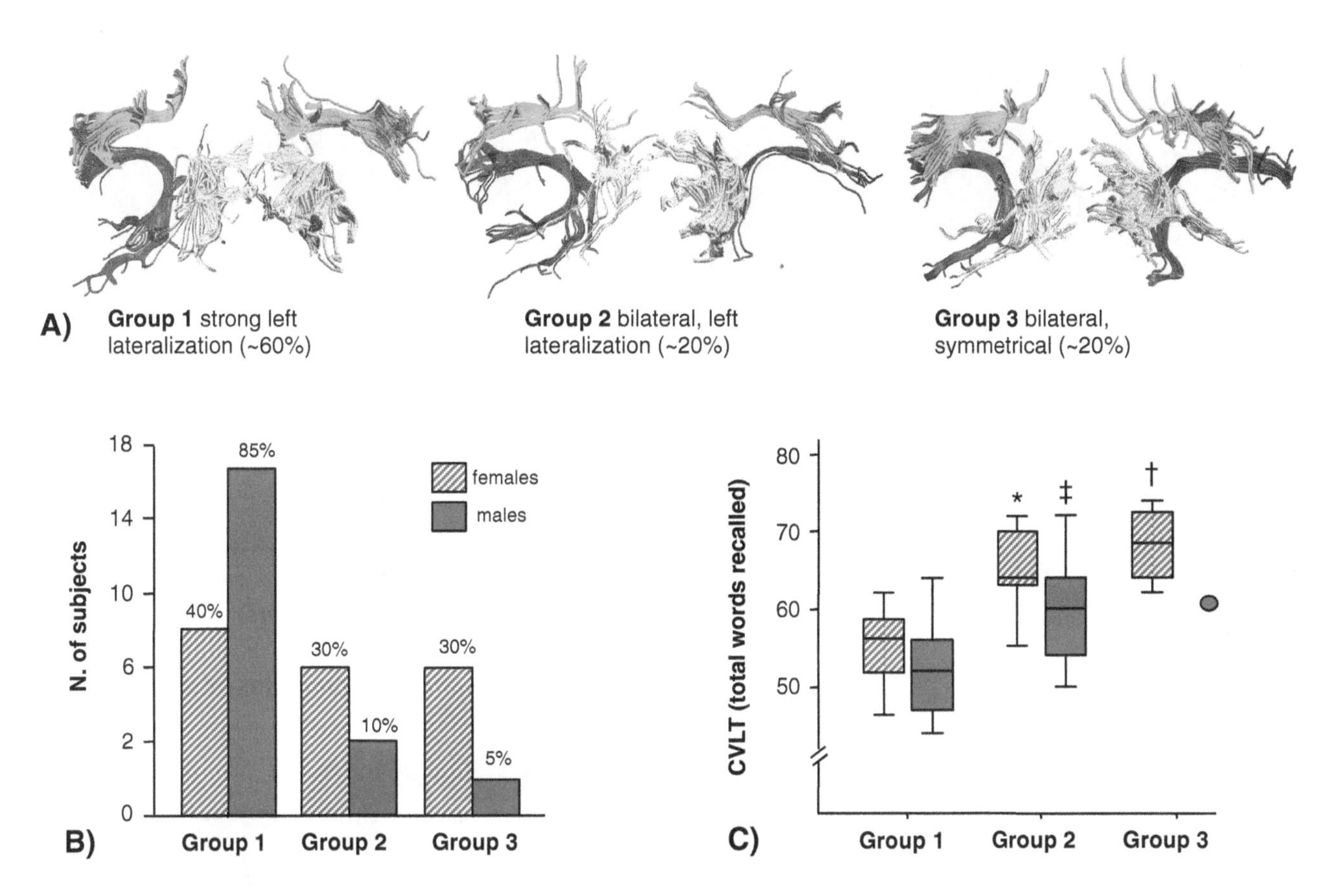

Figure 7.4 Hemispheric lateralization, gender dimorphism, and behavioural correlates of the long segment. A) Distribution of lateralization patterns of the direct long segment in a group of 40 young adults. B) Distribution of the lateralization groups between genders (20 males and 20 females). C) Performances in the California Verbal Learning Test (CVLT) according to the lateralization pattern and gender (striped colours are females, the red circle represents the only male belonging to group 3) (*$p < 0.05$ vs Group 1; †$p < 0.01$ vs. Group 1; ‡$p < 0.05$ vs. Group 1) (Catani et al., 2007).

gender dimorphism with regard to the lateralization of the long segment. Hagmann et al. (2006) reported that compared to right-handed subjects, left-handed males (but not left-handed females) show a more bilateral distribution of the long segment. An opposite trend has been found in the right-handed population with females being more likely to have a bilateral pattern compared to males (Figure 7.4B) (Catani et al., 2007).

The lateralization of the perisylvian pathways seems to be a dynamic process that begins very early in life and continues throughout adolescence and early adulthood. Voxel-based and tractography analysis of diffusion tensor datasets acquired in infants aged 1–4 months already shows some degree of left lateralization in the fractional anisotropy (i.e. an indirect index of axonal organization or myeline maturation) of the temporal portion of the arcuate fasciculus. This asymmetry does not seem to be specific to the language pathways as it also occurs in the sensory-motor pathways of the cortico-spinal tract (Dubois et al., 2008). Lebel et al. (2008) have recently shown that the fractional anisotropy of the arcuate fasciculus increases by 25% from the age of 5 years to 30 years, although the majority of these changes occur before the age of 20 years. Our recent tractography analysis of the parallel pathways model reveals that the three segments differ in their developmental trajectories. By the age of 10 years the long segment is larger in the left hemisphere compared to the right, while the anterior segment is larger in the right compared to the left. Both segments remain lateralized throughout adolescence and early adulthood. The interhemispheric asymmetry of the posterior segment shows a more dynamic pattern, with a greater right volume before

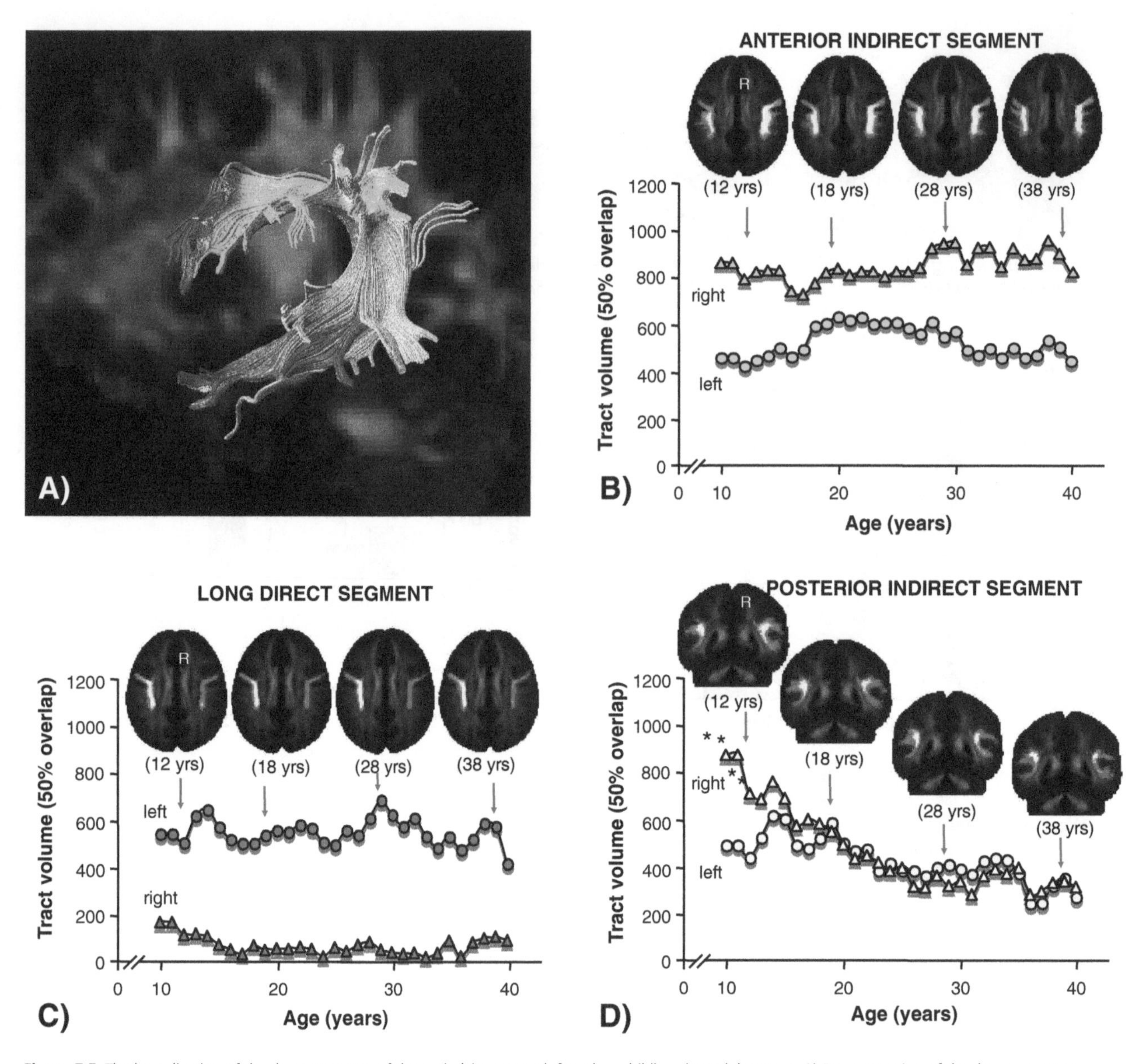

Figure 7.5 The lateralization of the three segments of the perisylvian network from late childhood to adolescence. A) Reconstruction of the three segments of the left arcuate fasciculus in a 20-year-old man. B) The graph shows the changes in volume of the right and left anterior segment between the age of 10 and 40 years. The volume of the anterior segment is right lateralized in 12-year-old subjects and remains right lateralized throughout adolescence and early adulthood. C) An opposite pattern of lateralization is observed for the long direct segment. D) The posterior segment shows a rightwards lateralization before adolescence and becomes progressively more bilateral and smaller throughout adolescence and adulthood (Catani et al., 2010).

adolescence, and a progressive bilateral distribution throughout adolescence and early adulthood (Figure 7.5) (Pugliese et al., 2008). These modifications in volume are likely to reflect biological changes in white matter that accompany cortical pruning during adolescence.

To what extent do these structural asymmetries correlate with functional lateralization and language performances? Preliminary studies combining tractography and fMRI reported no correlation between the lateralization of the arcuate fasciculus volume and the degree of functional lateralization as determined by fMRI during tasks of verbal fluency, verb generation, and reading comprehension (Powell et al., 2006). The functional lateralization seems to correlate better with the lateralization of the microstructural organization of fibres as measured by fractional anisotropy (Powell et al., 2006; Vernooij et al., 2007). Other studies have correlated the anatomical lateralization of perisylvian segments with behavioural performances in verbal memory and visuospatial attention tasks (Catani et al., 2007; Thiebaut de Schotten et al., 2011b). The results are so far somehow surprising. For example, the extreme left lateralization of the direct long segment is associated with worse performances in a complex verbal memory task that relies on semantic clustering for retrieval (i.e. California Verbal Learning Test, CVLT) (Figure 7.4C). Furthermore, the right lateralization of the anterior segment is not correlated with visuospatial attention

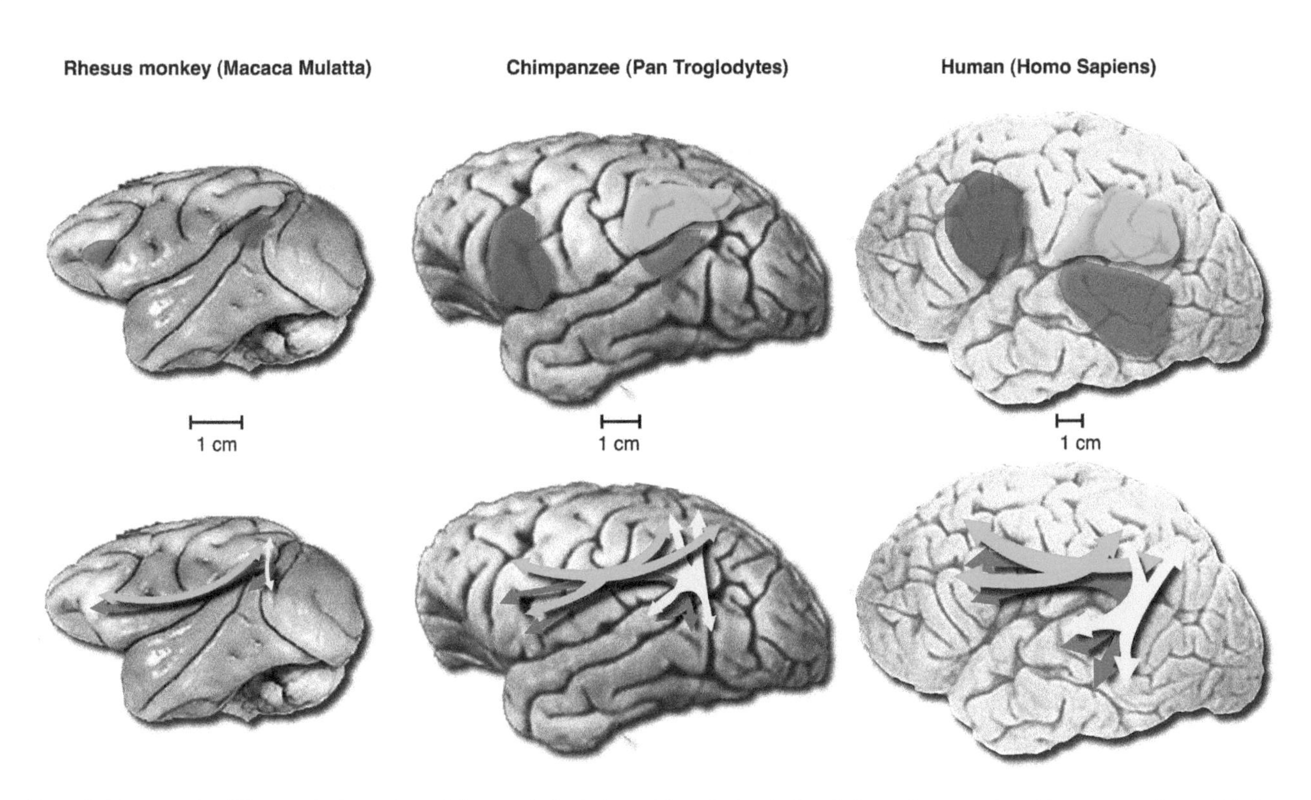

Figure 7.6 Comparative differences in the anatomy of the arcuate fasciculus and its cortical projections between rhesus monkey, chimpanzee, and man. Note the progressive increase from monkey to man of the cortical areas corresponding to Broca's (blue), Wernicke's (purple) and Geschwind's (cyan) territory. This increase is paralleled by a progressive expansion of the long (red) and posterior (yellow) segments along the phylogeny scale.

performances on line bisection and visual detection tests. Conversely a significant correlation was found between visuo-spatial attentional performances and lateralization of a more dorsal fronto-parietal tract that usually shows a degree of rightward asymmetry inferior to that of the anterior segment (Thiebaut de Schotten et al., 2011b). Overall these findings support the notion that lateralization of perisylvian pathways is an important aspect of human brain organization, but the correlation with functional activation and behaviour is not straightforward. The most asymmetric tracts show little correlation with interhemispheric dominant functions, and paradoxically, it can be argued that a bilateral representation might ultimately be advantageous for certain cognitive functions (Catani et al., 2007).

Comparative anatomy of the perisylvian networks

Differences of neuronal connectivity in animal brains account for most of the behavioural differences between species (Striedter, 2005). Hence, by comparing human and simian connectional anatomy we may unveil the architectural backbone of human cognition and identify, for example, the evolutionary changes underlying the development of language. One theory of the evolution of language from early primates to human is that it involved a change in the strengths of perisylvian connections (Aboitiz and Garcia, 1997a, 1997b). These authors argue that two evolutionary tendencies occurred (Figure 7.6). First, posterior temporal and inferior parietal regions became increasingly connected, linking the auditory system and a pre-existing parietal–premotor loop involved in the generation of complex vocalizations. Second, the development of connections between posterior superior temporal and inferior frontal regions linked auditory to orofacial premotor regions. One may speculate that these two tendencies correspond to the evolution of the posterior and long segments of the arcuate fasciculus, respectively, the anterior segment being, phylogenetically, the oldest component of the perisylvian network. This hypothesis can now be tested using tractography and comparing the parallel pathways model described in humans with the findings from both axonal tracing and tractography studies in monkeys. Perisylvian connections in the monkey brain have been studied extensively using axonal tracing techniques; however, their significance with respect to language remains controversial because homologies between cortical areas in monkeys and humans are unclear (Deacon, 1992;

Table 7.1 Neurological manifestations of perisylvian disorders

Symptoms	Clinical presentation	Tracts involved
Broca's aphasia	Non-fluent speech, impaired repetition, intact comprehension	Anterior and long segment
Wernicke's aphasia	Impaired auditory comprehension and repetition, fluent speech	Posterior and long segment
Conduction aphasia	Repetition deficits, paraphasias, relatively intact comprehension and fluency	Long segment
Transcortical motor aphasia	Impaired fluency but intact repetition and comprehension	Anterior segment
Transcortical sensory aphasia	Impaired comprehension but intact fluency and repetition	Posterior segment
Global aphasia	Severe deficit in language comprehension, production, and repetition	Anterior, posterior, and long segments
Pure alexia	Acquired reading impairment often associated with colour agnosia. Spontaneous writing and writing to dictation are preserved	Posterior segment
Dyslexia	Developmental reading difficulties	Anterior segment, posterior segment
Dysgraphia	Deficits in spontaneous writing and on dictation	Anterior segment
Working memory deficits	Impaired digit span	Anterior, posterior, and long segments
Apraxia	Loss of the ability to perform voluntary skilled movements without sensorimotor deficits	Anterior and posterior segments
		Right hemisphere
Unilateral spatial neglect	Failure to respond or reorient to novel or meaningful stimuli presented to the left hemispace	Anteior, posterior, and long segments, SLF II
Expressive amusia	Inability to sing or play a previously learned instrument. Loss of the ability to write musical annotations	Anterior segment, long segment
Receptive amusia	Difficulties to recognize out-of-tune notes, familiar melodies, or read musical annotations	Posterior segment, long segment
Anterior affective-aprosodia	Flatness of speech and loss of gestural abilities involving the face and limbs	Anterior segment, long segment
Posterior affective-aprosodia	Inability to comprehend or repeat emotional gestures and discern speech intonation	Posterior segment, long segment

Schmahmann and Pandya, 2006). Furthermore the findings on the segments of the arcuate fasciculus are contradictory, as some studies conclude that there is no equivalent of the arcuate fasciculus in the monkey (Petrides and Pandya, 1988; Schmahmann and Pandya, 2006; Thiebaut de Schotten et al., 2012), whereas other studies report a pattern of connectivity between perisylvian areas similar to the arcuate segments described in humans (Deacon, 1992; Petrides and Pandya, 2002). A recent tractography study suggests that a rudiment of the arcuate fasciculus may exist in monkeys with an increase of its complexity in chimpanzees and men (Rilling et al., 2008). Overall, both axonal tracing and more recent tractography studies support the theory that evolution of language from monkey to human involved a change in the pattern of perisylvian connections (Catani et al., 2005).

Perisylvian disorders

Patients with lesions to the perisylvian pathways often manifest disorders of language (Mesulam, 1990; Catani and Mesulam, 2008;

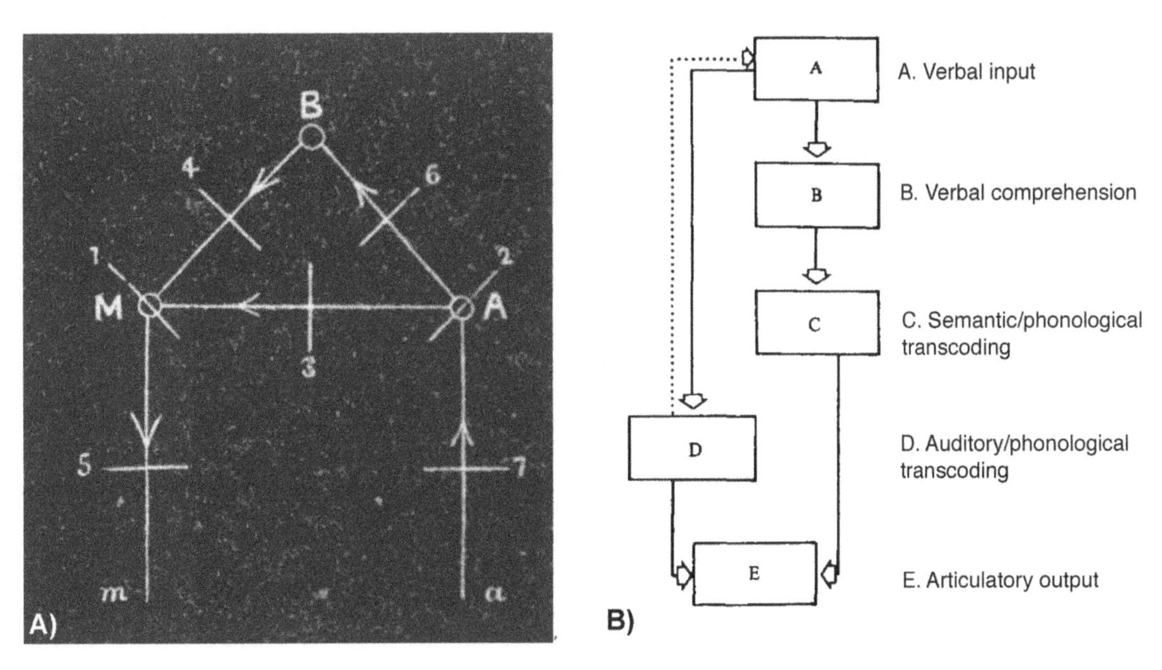

Figure 7.7 A) Lichtheim's original model of aphasia disorders was intended as 'aide-memoire' and according to his intentions it had little anatomical correspondence, except for centre 'M' that correspond to Broca's area, and centre 'A' to Wernicke's area. Centre 'B' is the concept centre. Numbers correspond to different forms of aphasia (1, Broca's aphasia; 2, Wernicke's aphasia; 3, conduction aphasia; 4, transcortical motor aphasia; 6, transcortical sensory aphasia; lesions to 5 and 7 do not manifest with aphasia) (Lichtheim, 1885). B) McCarthy and Warrington's model of repetition. The letters do not correspond to the letter in A). For an explanation see text (McCarthy and Warrington, 1984).

Ross, 2010), praxis (Heilmann and Watson, 2008), prosody (Ross et al., 1981; Ross, 2010), and visuospatial attention (Mesulam, 1990; Vallar, 1998; Doricchi et al., 2008). In addition, the perisylvian cortex is frequently involved in disorders of working memory (Baddeley, 2007), social cognition (Frith and Frith, 2007), and calculation (Rusconi et al., 2010). For some conditions (e.g. classical neurological disorders of higher cognitive functions) the anatomical correlates are well known, whilst for others (e.g. most of the psychiatric disorders) the underlying anatomical changes are still largely unknown. The major syndromes associated with perisylvian disorders and their anatomical correlates are listed in Table 7.1.

Aphasias

Since the work of Wernicke and other pioneers in aphasiology, the functions traditionally associated with the arcuate fasciculus are repetition of words and speech output monitoring. It follows that lesions to the direct connections between Wernicke's and Broca's territories manifest with conduction aphasia, characterized by the inability to repeat words and paraphasic speech (i.e. the use of incorrect words or phonemes while speaking). In most patients with conduction aphasia, Wernicke's and Broca's areas are not affected and therefore comprehension and verbal fluency are intact (Figure 7.7A). Conduction aphasia was central to the 19th-century models of aphasia where emphasis was placed on 'centres' and 'connections'. A seminal contribution was published by Lichtheim in 1885. He introduced a hypothetical third 'concept' centre (i.e. 'the part where concepts are elaborated') and described two additional forms of aphasia deriving from its disconnection from Broca's and Wernicke's areas (Figure 7.7A). In transcortical motor aphasia, spontaneous speech is reduced, but comprehension and repetition are intact. In transcortical sensory aphasia comprehension is reduced but fluency and repetition are spared. Transcortical motor aphasia is associated with frontal lesions affecting the anterior segment of the arcuate fasciculus (Schiff et al., 1983). The location of the lesion in transcortical sensory aphasia is entirely consistent with damage to the posterior segment between the posterior temporal and parietal lobes (Damasio and Geschwind 1984; Boatman et al., 2000) (Figure 7.8A). Although there are some striking similarities between Lichtheim's model and the three segments of the arcuate fasciculus, this does not necessarily imply that Lichtheim's 'concept' centre is localized in Geschwind's territory. One could argue, however, that the inferior parietal may represent the gateway for converging multimodal sensory inputs into aspects of language (e.g. semantic knowledge) (Geschwind, 1965).

Although Lichtheim's model has represented an invaluable aide-memoire for clinicians, its validity has been questioned in light of more sophisticated psychological testing tools developed in the second half of the 20th century. In 1984, McCarthy and Warrington, for example, proposed a more complex model for repetition (Figure 7.7B), based on the report of three aphasic patients. Two patients had a temporo-parietal lesion that resulted in impaired repetition tasks but relatively preserved spontaneous speech and comprehension (classical conduction aphasia). In contrast, the third patient had a superficial parietal lesion resulting in impaired spontaneous speech but relatively intact repetition (transcortical motor aphasia). The patients with a deep lesion were facilitated in repetition tasks that required active semantic processing (e.g. repeating a word in the context of a sentence), whereas the patient with a superficial lesion remained impaired. The opposite dissociation was observed in tasks that required passive repetition (e.g. repeating syllables or pseudo-words). The patients with a deep lesion were impaired, whereas the patient with a superficial lesion

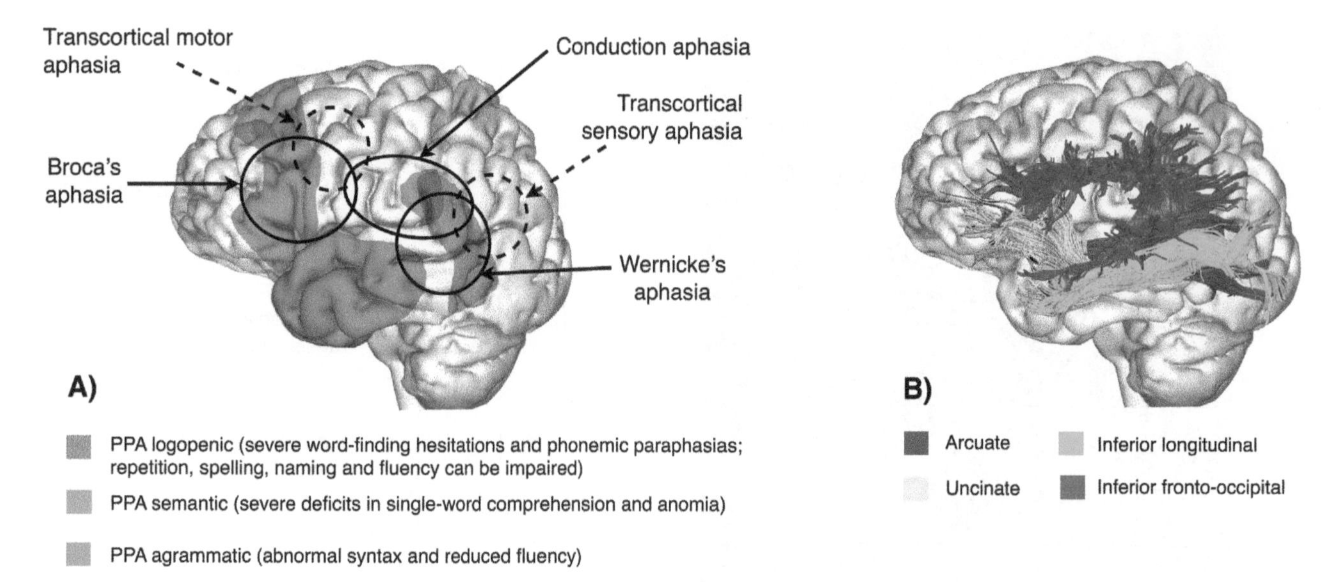

Figure 7.8 A) Classical lesion studies (circles) in stroke patients (Based on Damasio and Geschwind, 1984) identify the arcuate fasciculus as the tract underlying language function. Recent cortical morphometry studies (modified from Mesulam et al., 2009) in patients with different forms of primary progressive aphasia (PPA) suggest that the networks underlying language functions extend beyond the arcuate fasciculus to include more ventral networks. B) Tractography reconstruction of the dorsal (arcuate fasciculus) and ventral (inferior longitudinal, inferior fronto-occipital, and uncinate tract) pathways affected in aphasia.

performed better. The authors proposed a two-route model of speech production: a direct pathway between Wernicke's and Broca's areas acting in fast, automatic word repetition, and an indirect pathway where a stage of verbal comprehension and semantic/phonological transcoding intervenes between verbal input and articulatory output (Figure 7.7B). It is possible to hypothesize that the deep intracerebral lesions in the first two patients would have affected predominantly the direct long segment pathway that lies medial to the indirect pathway. In contrast, the superficial cortical lesion of the third patient would have affected predominantly the indirect pathway or its parietal cortical relay station. Hence, according to this model both direct and indirect pathways of the arcuate fasciculus are involved in repetition, the direct engaging in tasks involving auditory/phonological transcoding (e.g. repeating pseudowords), while the posterior and anterior indirect pathways cooperate in tasks involving verbal comprehension and semantic/phonological transcoding.

In the 1980s, with the advent of neuroimaging and standardized neuropsychological testing, clinicians soon realized that a great clinical and anatomical variability existed among aphasic patients. It was observed, for example, that conduction aphasias fall into two distinct groups: the Broca-like syndrome in which the deficit in repetition is accompanied by a relative impairment in fluency and the Wernicke-like syndrome in which the deficit in repetition is accompanied by a relative impairment in comprehension (Naeser et al., 1982; Kempler et al., 1988). One obvious explanation for this dichotomy is that more anterior lesions encroach on Broca's territory, whereas more posterior lesions encroach on Wernicke's territory; the two resulting aphasias being a mixture of cortical deficits (Broca's or Wernicke's) and subcortical conduction deficits. However, the observation of cases of Broca-like or Wernicke-like conduction aphasias in patients with pure subcortical lesions (Naeser et al., 1982) suggests this explanation is incorrect. Thus, an alternative explanation is that Broca-like and Wernicke-like conduction aphasias result from lesions involving the direct and the indirect pathways of the arcuate at different points along their course. A lesion located anteriorly and involving the long and anterior segments would lead to Broca-like conduction aphasia, whereas a lesion located posteriorly involving the long and posterior segments would lead to a Wernicke-like conduction aphasia (Catani et al., 2005). This explanation is consistent with clinical evidence that Broca-like conduction aphasias occur with subcortical frontal lesions, whereas Wernicke-like conduction aphasias occur with subcortical temporoparietal lesions (Naeser et al., 1982).

Other studies have challenged the classical division of aphasias into anterior (expressive) and posterior (receptive) forms. Basso et al. (1985), for example, found that 15% of patients presenting with Broca's aphasia show no involvement of Broca's area. These cases that on appearance do not fit the classical anatomical models can be explained using the hodological approach. Figure 7.9A–B shows one of these 'atypical' cases where the patient presented impaired repetition and reduced fluency (i.e. classical Broca's aphasia) following surgical resection of a brain tumour in the superior aspect of the left frontal gyrus. Broca's area was spared but the tumour and the perilesional oedema encroached on the anterior and long segments of the arcuate fasciculus, thus explaining the deficits in repetition and fluency.

In the last decade some authors have questioned the role of the arcuate fasciculus in language (Schmahmann and Pandya, 2006; Hickok and Poeppel, 2007), in particular in relation to repetition (Bernal and Ardilla, 2009) and new anatomical models have been put forward. Most of these models are derived mainly from animal work (Schmahmann and Pandya, 2006; Hickok and Poeppel, 2007) and are difficult to reconcile with classical lesion studies. Nonetheless, a view that language should not be confined to the cortical territories of the arcuate fasciculus, is recently supported

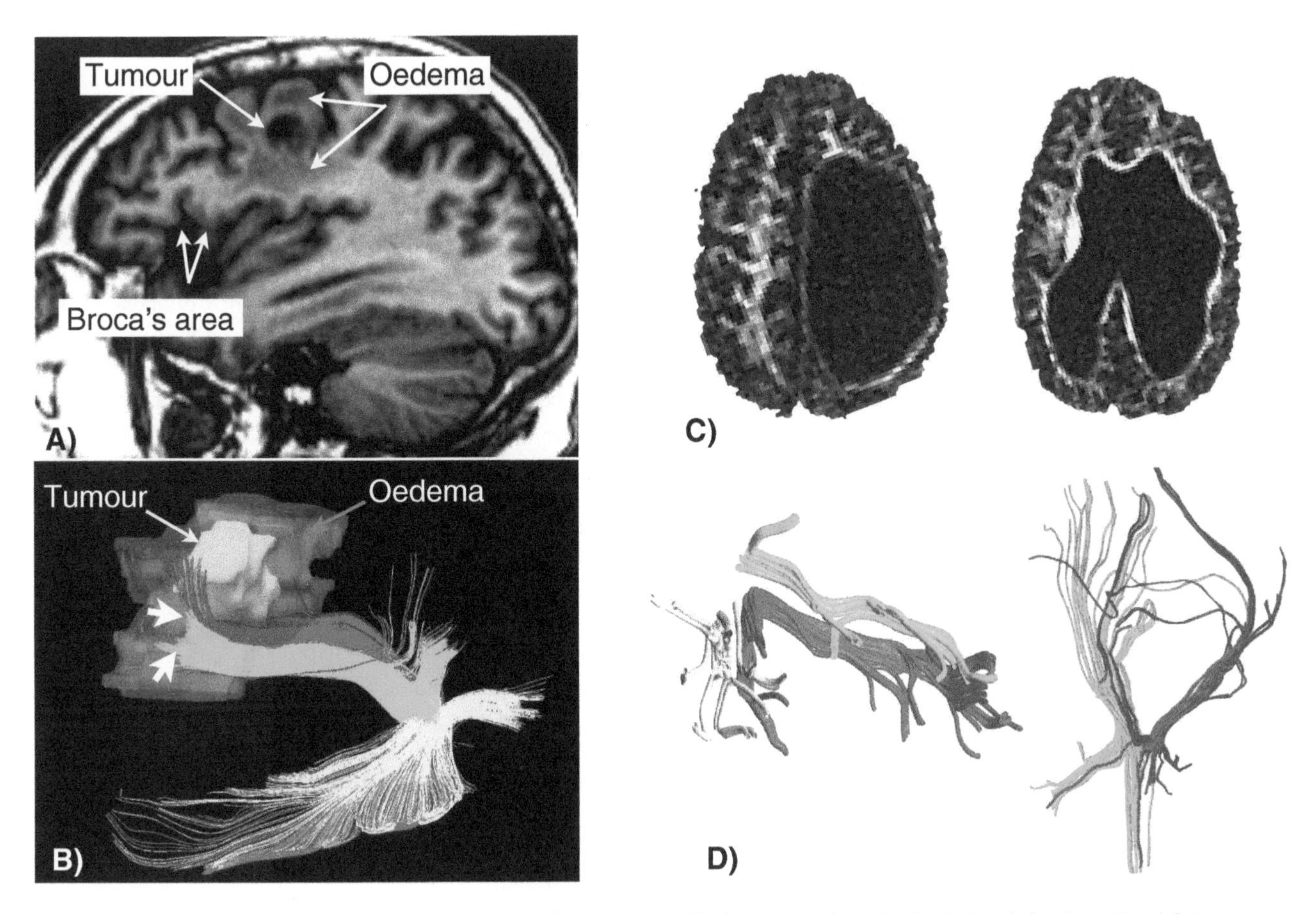

Figure 7.9 A, B) Tractography reconstruction of the arcuate fasciculus in a patient with a brain tumour in the left hemisphere. Isolated repetition deficits were present before surgery. Impaired fluency deficits appeared after surgery despite Broca's area being spared at surgery. In this patient the most likely explanation for persistency of repetition deficits and impaired fluency (Broca's type of aphasia) is a damage to the anterior and long segment of the arcuate fasciculus (indicated by white arrows in figure B). (Dataset courtesy of Dr Alberto Bizzi, Istituto Neurologico 'Carlo Besta', Milan). C, D) Diffusion tensor imaging of an 18-year-old male born very preterm at 26 weeks of gestation. C) The axial fractional anisotropy images show marked dilatation of the lateral ventricles, particularly in the left hemisphere. D) Tractography reconstruction of the perisylvian language pathways was only possible for the right hemisphere (left image). The asymmetry of the cortico-spinal tract (right image) is also evident with a more severe damage to the left side (red tract). Considering the normal IQ of this subject and the left-handedness, it is possible that the tractography findings show a compensatory reorganization of the language and motor tracts that may have followed the white matter damage at birth.

also by findings in patients with disorders of language associated with focal cortical degeneration (i.e. primary progressive aphasias) (Mesulam et al., 2009). In these patients the atrophy of the anterior temporal lobe and its connections correlates with naming deficits, suggesting an involvement of the uncinate and inferior longitudinal fasciculus (Agosta et al., 2010). Other language functions such as syntax and repetition remain however related to the arcuate fasicuclus. A recent study, for example, found that in patients with left hemisphere stroke the deficits in repetition correlate with severity of damage to the arcuate fasciculus independently from involvement of the cortical language areas (Breier et al., 2008). This suggests that conduction aphasia is a disconnection syndrome of the arcuate fasciculus and confirms the role of the arcuate fasciculus in repetition. It still remains to be explained why conduction aphasia is usually a transitory deficit, the majority of patients recovering almost completely despite permanent damage to the arcuate fasciculus. This observation raises the question of possible compensatory mechanisms, either within the same hemisphere or between the two hemispheres, which could operate during recovery not only in conduction aphasia but in all forms of aphasia in general. A possible alternative pathway is represented by the ventral route connecting language areas through the inferior longitudinal fasciculus, the uncinate fasciculus, and some fibres of the external capsule running just behind the insula (Figure 7.8B) (Anwander et al., 2007; Barrick, et al. 2007; Saur et al., 2008). In patients with arcuate damage, this ventral route could take over some functions as suggested by a diffusion imaging study which found that recovery in chronic aphasic patients was critically dependent on the integrity of the anterior temporal white matter fibres (Meinzer et al., 2010). Similar compensatory mechanisms in the contralateral hemisphere could operate. Figure 7.9C-D shows, as example of extreme contralateral compensation, the diffusion images of a man born very preterm (at 26 weeks) who developed normal language despite the complete absence of the arcuate fasciculus fibres in the left hemisphere due to leucomalacia at birth.

Disorders of writing and reading

Other forms of language impairment (e.g. written) often present with perisylvian lesions. Agraphia or dysgraphia (i.e. impaired writing) usually co-occurs with aphasia but occasionally presents as an isolated deficit. This observation led to a proposal for a distinct centre for writing (i.e. Exner's centre) in the posterior part of the second frontal gyrus (Exner, 1881). In reviewing the literature for disorders of writing, Nielsen (1947) concluded that agraphia without associated neuropsychological signs is rare, but when it occurs it is associated with a lesion of either the middle frontal gyrus or the angular gyrus. Nielsen argued that Exner's area works in association with the angular gyrus and the motor area to produce writing. He also proposed that information from the angular gyrus is carried to Exner's area by fibres that pass close to Broca's area. According to this view, the proximity between Broca's area and the fibres underlying writing functions can explain the frequent association of agraphia and aphasia. We argue that Nielsen's fronto-parietal network corresponds in part to the anterior segment of the arcuate fasciculus.

Children with dyslexia show microstructural changes in the anterior segment of the arcuate fasciculus that correlate with the severity of the reading difficulties (Rimrodt et al., 2010). These findings support the classical models of alexia (a reading impairment in patients with acquired brain lesions) which consider the angular gyrus as the 'reading centre' (Dejerine, 1892). This view has been challenged in the last decade and a more ventral occipito-temporal area has been proposed, the so-called visual-word-form area (Cohen et al., 2000). A recent diffusion tensor imaging study in stroke patients seems to reconcile the above positions, by showing that lesions affecting the angular gyrus and the basal temporal area (or the connections between those areas) are responsible for reading and spelling deficits (Philipose et al., 2007). Furthermore, functional imaging studies suggest a distinct role in reading for the anterior and posterior parts of Geschwind's territory (Graves et al., 2010). Tasks requiring mapping orthography to phonology (a written word to its sound) are associated with increased activity in the supramarginal gyrus, whereas tasks requiring reading words with high semantic representation activate the angular gyrus. Overall, imaging studies confirm the involvement of the indirect pathway in reading, with different functions for the posterior and anterior segments of the arcuate fasciculus.

Memory disorders

Perisylvian networks are invovled in verbal working memory and long-term memory functions. Working memory is a form of short-term memory that describes a limited capacity system dedicated to temporarily maintain and simultaneous manipulate information (Baddeley, 2003). The multi-component model proposed by Baddeley and Hitch (1974), comprises a control system of limited attentional capacity—the central executive—assisted by two subsidiary storage systems, the phonological loop, which is based on sound and language, and the visuospatial sketchpad. The phonological loop comprises a phonological store, which can hold memory traces for a few seconds before they fade, and an articulatory rehearsal process that allows the retrieval of memory traces (analogous to subvocal speech). This model of working memory provides an interface between perception, long-term memory, and action, with many important implications for language development. Baddeley (2007) has recently proposed that the three segments of the arcuate fasciculus participate in different functions in working memory tasks (Figure 7.10A). This view is supported by a number of functional and structural studies (Paulesu et al., 1993; Gruber, 2001; Vallar and Papagno, 2002; Schulze et al., 2011). In tasks involving the temporary storage of verbal input, activity in the left Wernicke's and Geschwind's territories has been observed. Furthermore, patients with phonological short-term memory deficits in the absence of more general language impairment typically have lesions in the left temporo-parietal area (Vallar and Papagno, 2002). Hence, activity mediated by the posterior segment is associated with functions of the phonological input buffer (Baddeley, 2007; Jacquemot and Scott, 2008). On the other hand, the subvocal rehearsal of stored memory traces depends on the activation of the left Broca's territory and its connections to Wernicke's and Geschwind's territories. This suggests that the anterior and long segments of the arcuate fasciculus mediate the rehearsal of memory traces from the short-term memory storage system.

Long-term memory deficits are typical of patients with Alzheimer's disease. These patients show concomitant cortical atrophy, amyloid deposition, and reduced metabolism in the regions interconnected by the posterior segment of the arcuate fasciculus (together with the posterior cingulate region, see also Chapter 11). Interestingly, these regions overlap with the cortical areas that deactivate while subjects engage in cognitive tasks—the so-called default network (Figure 7.10B).

Apraxias and discalculia

The execution of purposeful actions or movements requiring utilization of objects and utensils (i.e. praxis) rely on the integrity of the perisylvian system (Liepmann, 1900; Goldenberg, 2003). Patients with conceptual apraxia might have no understanding of the mechanical knowledge of a tool (e.g. they may use a screw-driver to hammer a nail) as if they have lost the tool–action semantic representations. These patients often have Alzheimer's disease or semantic dementia that involves the angular gyrus and its temporal connections (i.e. posterior segment of the arcuate). Other apraxic patients may be able to recognize a tool and its function but fail to assemble the spatial and temporal representation of a gestural plan necessary for a correct execution of goal directed actions (e.g. failing to imitate or execute on command the action of hammering). This form is called ideomotor apraxia and it is often associated with lesions involving more anterior regions of Geschwind's territory (i.e. the supramarginal gyrus) and its connections (i.e. anterior and posterior segments of the arcuate fasciculus) (Alexander et al., 1992). In another condition, patients have a normal conceptual representation of tools and movement planning but fail to execute actions with fine dexterity. This form is called limb-kinetic apraxia and is often described in patients with frontal lesions affecting the primary motor cortex and its callosal and projection connections (Heilman et al., 1982; Buxbaum 2001; Frey, 2007; Heilman and Watson, 2008; Ramayya et al., 2010).

Finally, patients with Gerstmann's syndrome—a combination of discalculia (deficits in calculation), left and right disorientation, finger agnosia, and dysgraphia/alexia—often have extended inferior parietal lesions (Rusconi et al., 2010). It is possible that the association of symptoms in Gerstmann's syndrome results from the

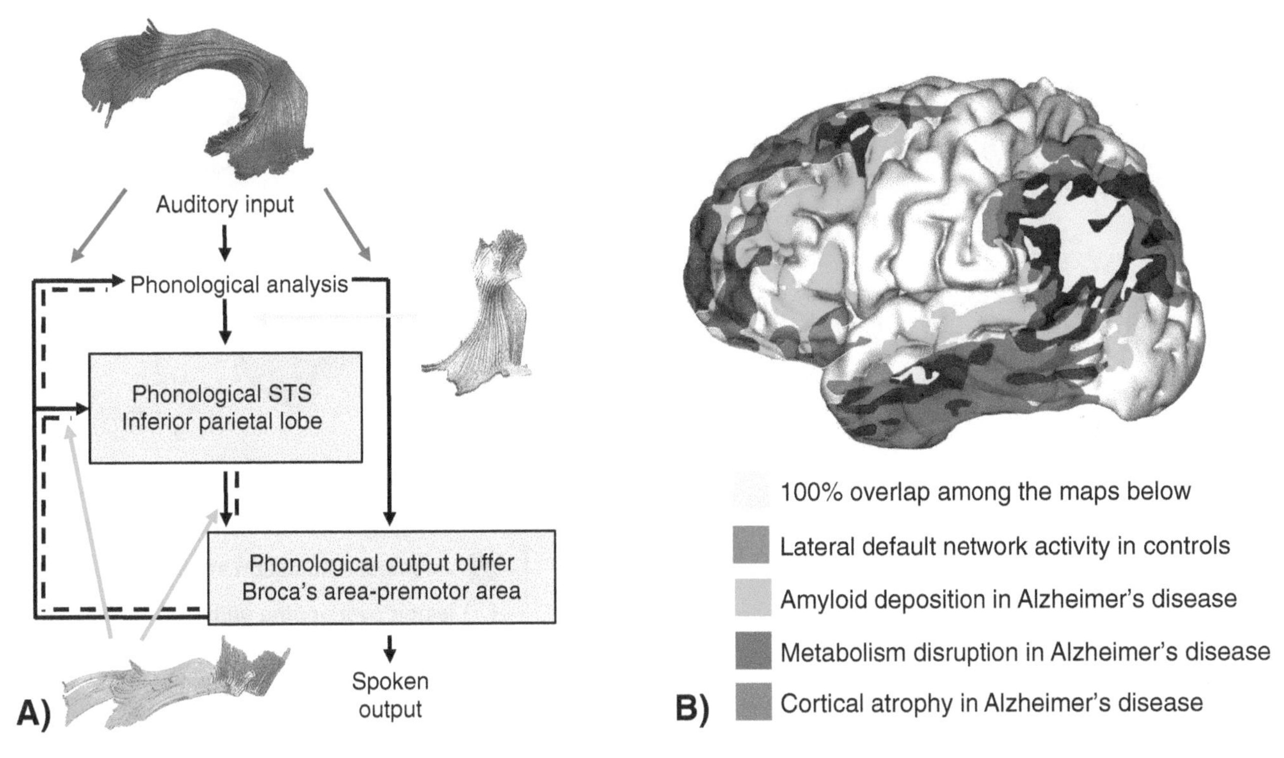

Figure 7.10 A) Baddeley's model of verbal working memory and possible anatomical substrates of its different components (from Baddeley 2007). The model is composed of an auditory cortex for phonological analysis, a system for short-term storage (STS) in the inferior parietal lobe, and a frontal region for programming speech output. Auditory information gains direct access to a phonological store, after which it passes to an output buffer for recall or recycling through rehearsal (dashed circuit). B) Overlap (yellow) between the cortical areas of the default network (red) and the regions that in Alzheimer's disease show amyloid deposition (green), reduced metabolism (blue) and cortical atrophy (purple) (Original data from Buckner et al., 2005).

disruption of an extended network of fibres converging to the inferior parietal lobe, such as commissural pathways (left–right disorientation), inferior-parietal and postcentral gyrus connections (finger agnosia), anterior segment fibres from the supramarginal gyrus (discalculia), and angular gyrus connections to the frontal lobe and temporo-occipital regions (dysgraphia/alexia). A partial disconnection syndrome could explain the presentation of isolated deficits or combination of some but not all four symptoms. This interpretation is supported by a tractography study that found a correlation between the microstructural properties of the anterior segment of the arcuate fasciculus and arithmetic approximation skills (Tsang et al., 2009).

Right hemisphere syndromes

The most common disorders associated with right hemisphere perisylvian lesions are those affecting the ability to process visuospatial information from the left hemifield, which usually manifests with neglect (Mesulam, 1990; Vallar, 1998). In most of the patients the deficits are transitory and become apparent only with clinical manoeuvres or tests that elicit the hemispatial neglect. The most common are pen and pencil bedside tests where the patient is asked to bisect lines or mark randomly distributed letters or figures (e.g. bells) (Mesulam, 2000). In these tests the patient with neglect ignores the visual information from the left side of the sheet. In milder forms of neglect the deficit becomes apparent only when the examiner asks the patient to detect bilateral visual stimuli presented simultaneously (i.e. the extinction phenomenon). The neglect can sometimes be so severe that it makes it difficult for the person to ambulate independently even in a familiar environment. These patients often bump onto tables and chairs or fail to walk through a door. Findings from functional imaging studies suggest the existence of two parallel networks dedicated to visuospatial attention (Figure 7.11A) (Corbetta and Shulman, 2002). A dorsal fronto-parietal network engages during tasks requiring strategic and voluntary orientation of attention towards visual targets. A ventral network engages in tasks where the attention is automatically grabbed by unexpected visual targets. The activity of the two networks and their interaction depends on the three branches of the superior longitudinal fasciculus (Thiebaut de Schotten et al., 2011b).

A dilemma that currently faces the field of spatial attention is that lesion overlap methods show hemispatial neglect to be caused by damage located more ventrally, and only in the inferior network (Figure 7.11B), and sometimes even outside the perisylvian regions (e.g. along the inferior longitudinal or inferior fronto-occipital

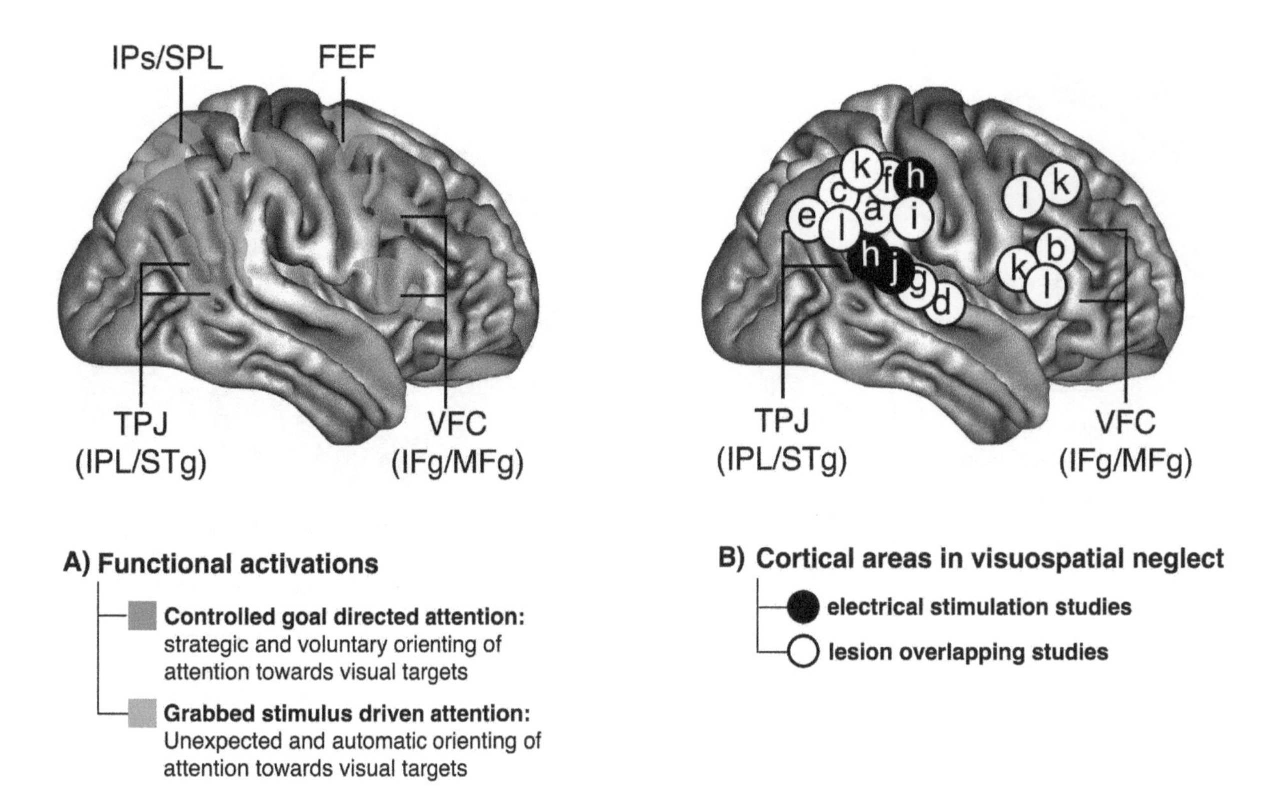

Figure 7.11 A) The functional anatomy of the fronto-parietal networks underlying visuo-spatial attention. The data are derived from several functional magnetic resonance imaging studies and show the existence of a dorsal fronto-parietal network activated in tasks requiring controlled goal directed attention and a ventral network dedicated to automatic reorienting of spatial attention to unexpected stimuli (Corbetta and Shulman, 2002). B) Disorders of spatial attention are frequently associated with either cortical or subcortical lesions of the right perisylvian regions connected by the three segments of the arcuate fasciculus and second branch of the superior longitudinal fasciculus. (Data derived from: a, Vallar and Perani, 1986; b, Husain and Kennard, 1996; c, Leibovitch et al., 1998; d, Karnath et al., 2001; e, Mort et al., 2003; f, Doricchi and Tomaiuolo, 2003; g, Karnath et al., 2004; h, Thiebaut de Schotten et al., 2005; i, Corbetta et al., 2005; j, Gharabaghi et al., 2006; k, Committeri et al., 2007; l, Verdon et al., 2010). FEF, frontal eye field; IFg, inferior frontal gyrus; IPL, inferior parietal lobule; IPs, intraparietal sulcus; MFg, middle frontal gyrus; SPL, superior parietal lobule; STg, superior temporal gyrus; TPJ, temporo-parietal junction; VFC, ventral frontal cortex (Thiebaut de Schotten et al., 2011b).

fasciculus) (Bird et al., 2006; Park et al., 2006; Urbanski et al., 2008; Tomaiuolo et al., 2010). It is possible to imagine how in the future the combination of lesion overlap methods with tractography may help to resolve this dilemma by showing that the effective outcome of the ventral lesions is to disrupt underlying pathways interconnecting both dorsal and ventral networks.

Normal musical competence depends on the integrity of a right temporo-frontal network (Hyde et al., 2011). In subjects with tone-deafness (i.e. congenital amusia) a reduced fractional anisotropy and volume of the right long segment of the arcuate fasciculus has been reported using tractography (Loui et al., 2009). Amusia is often encountered also in patients with acquired brain injury. Patients with receptive amusia have difficulties in recognizing out-of-tune notes and familiar melodies. If they were musicians, reading musical annotations can be difficult or impossible for them. Patients with expressive amusia lose the ability to sing or to play a previously learned instrument. Writing musical annotations can also be impaired in expressive amusia. Receptive amusia is associated with posterior temporal lesions, whereas expressive amusia is often seen in patients with more anterior frontal lesions (Sarkamo et al., 2009).

Disorders of language can occasionally manifest in patients with right perisylvian lesions. They may complain of difficulties in understanding communicative intent conveyed by the structure of paragraphs and chapters, or difficulties in processing metaphors or idiomatic (i.e. non-literal) types of expressions (Ross, 2010). More often patients with right perisylvian lesions present with impairment in the ability to convey meaning and emotions through the modulation of speech intonation, stress, and rhythm (i.e. prosody), or facial, limb, and gestural expression (Ross, 1981). These disorders are generally classified according to the lesion location into anterior (frontal) affective-aprosodias, characterized by flatness of speech and loss of gestural abilities involving the face and limbs, and posterior (temporal) affective-aprosodias, characterized by the loss of the ability to comprehend affective prosody. Conduction affective-aprosodia has also been reported (Ross, 2010).

Neurodevelopmental disorders

The continuous formation and remodelling of long-range white matter connections occurring very early in life is a necessary prerequisite for the development of normal cognitive functions. Alterations of this process could be responsible for the delay in normal cognitive development or the emergence in childhood and adolescence of those symptoms which are typical of neurodevelopmental disorders such as autism and schizophrenia.

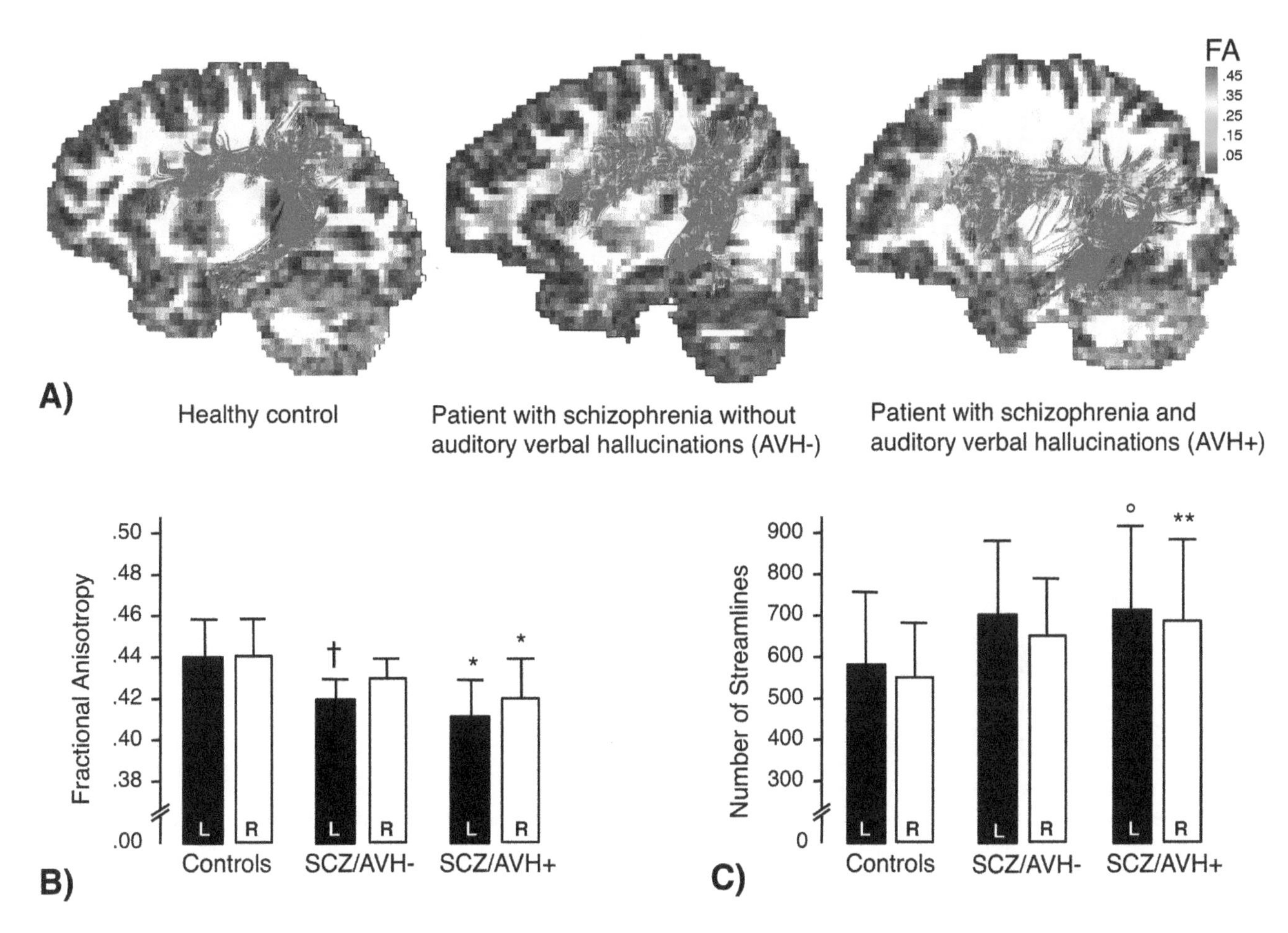

Figure 7.12 A) Tractography dissections of the arcuate fasciculus in schizophrenia. B, C) Fractional anisotropy values mapped onto the tractography reconstruction of the arcuate fasciculus in a healthy control, a patient with schizophrenia without auditory hallucinations (AVH–), and a patient with schizophrenia and auditory hallucinations (AVH+). *Differences are significant at $p < 0.001$ vs. healthy controls. **Differences are significant at $p=0.029$ vs. healthy controls. †Differences are significant at $p=0.042$ vs. healthy controls. °Differences are significant at $p=0.006$ vs. healthy controls (Catani et al., 2011).

Children with global developmental delay fail to achieve most, if not all, developmental milestones in speech and language, motor, cognitive, social, and emotional skills. Sundaram et al. (2008), performed tractography dissections of the arcuate fasciculus in a group of children with global developmental delay and in children with typical neurodevelopment. The authors were not able to dissect the long segment of the arcuate fasciculus bilaterally in half of the children with developmental delay, while a normal pattern was found in all typically developing children. The other tracts of the temporal lobe (i.e. inferior longitudinal fasciculus, uncinate, cingulum) were grossly intact, suggesting a specific developmental abnormality in the arcuate fasciculus.

Autism spectrum disorder is a relatively common neurodevelopmental disorder characterized by a triad of repetitive and stereotypic behaviour, impaired language communication, and deficits in social reciprocity. Structural MRI studies show white matter increase in the regions containing the fibres of the arcuate fasciculus (Herbert et al., 2004; Radua et al., 2011). Kumar et al. (2010) used diffusion tensor tractography to show that in young children with autism spectrum disorder the average length of the long segment of the arcuate fasciculus is greater in the right compared to the left. This pattern of asymmetry is different compared to those of children with neurotypical development who show a more symmetrical distribution. In adolescents with high-functioning autism, the long segment of the arcuate fasciculus shows an increase of radial diffusivity, suggesting that anatomical abnormalities of the axonal membrane and/or myelin persist later in life (Fletcher et al., 2010).

Schizophrenia is a neurodevelopmental disorder that usually manifests in late adolescence or early adulthood with auditory hallucinations, delusions, thought disorder, and abnormalities of the emotional sphere (e.g. lack of motivation, reduced social interactions, etc.). The hypothesis that schizophrenia symptoms may result from a breakdown in communication between cortical areas was originally formulated by Wernicke (1906) and Kraepelin (1913). More recently emission tomography (McGuire et al., 1995), functional MRI (Lawrie et al., 2002; Stephan et al., 2006) and electroencephalography (Ford et al., 2002) studies have reported that schizophrenia is associated with abnormal interactions between frontal, parietal, and temporal brain regions during tasks that engage language processing. These abnormalities are especially

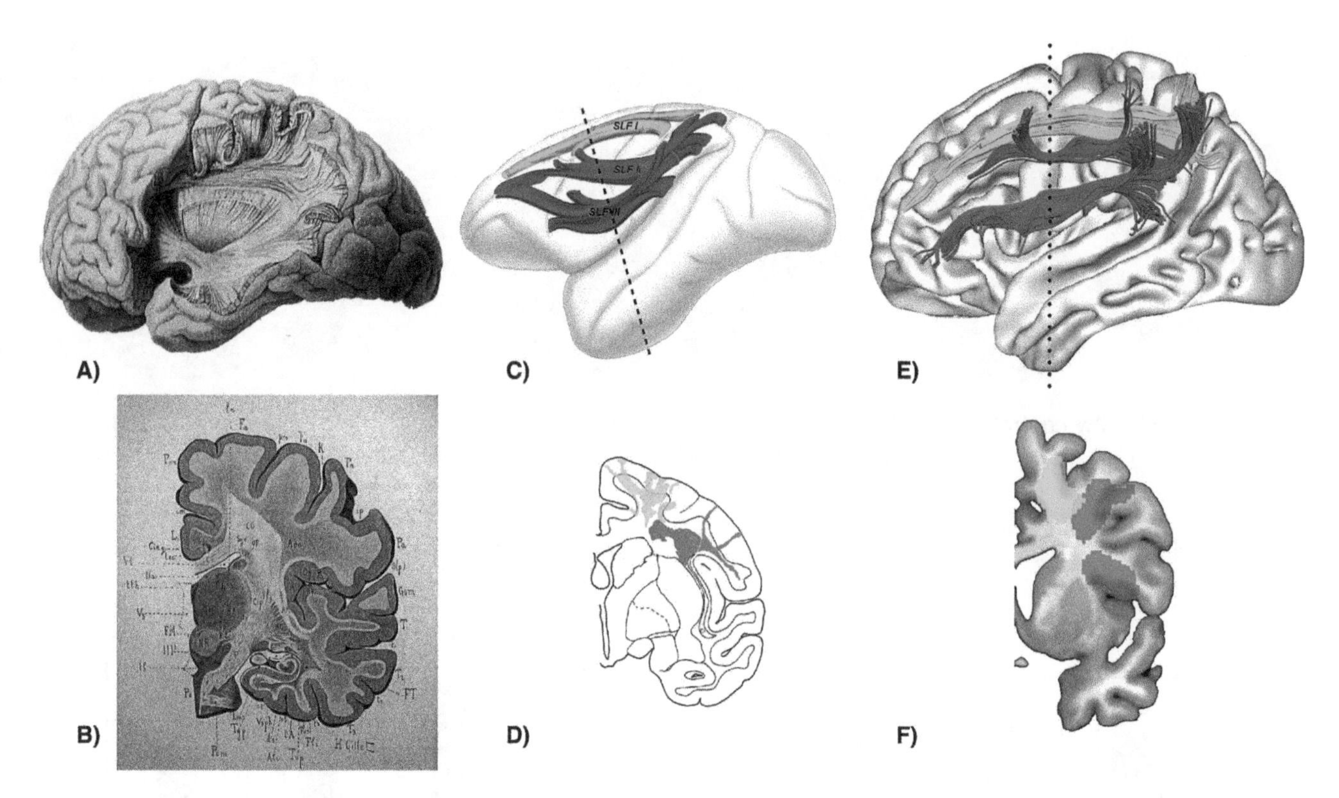

Figure 7.13 The anatomy of the superior longitudinal fasciculus. A, B) Post-mortem dissection of the superior longitudinal fasciculus in the human brain (A, from Mayo, 1827; B, from Dejerine, 1895). C, D) The three branches of the superior longitudinal fasciculus in the macaque brain (Schmahmann and Pandya 2006). E, F) Spherical deconvolution tractography (Dell'Acqua et al., 2010) of the three branches of the superior longitudinal fasciculus in the human brain (Thiebaut de Schotten al. 2011b).

marked in patients with a vulnerability towards experiencing auditory verbal hallucinations; particularly when these patients are performing cognitive tasks that require the generation and/or appraisal of inner or overt speech (Allen et al., 2007). The perisylvian areas implicated in these studies have also been found to be activated when patients are actually experiencing auditory verbal hallucinations (McGuire et al., 1993; Lennox et al., 2000; Shergill et al., 2000). Diffusion tensor tractography studies show structural changes of the fibres of the arcuate fasciculus (i.e. reduced fractional anisotropy) in young patients (Jones et al., 2006; Phillips et al., 2009; Catani et al., 2011) but not in patients with late-onset schizophrenia-like psychosis (Jones et al., 2005; Voineskos et al., 2010). These changes are particularly evident in the posterior and long segments of schizophrenic patients (Skudlarski et al., 2010; Catani et al., 2011). This reduction is bilateral, in patients with auditory hallucinations, but limited to the left hemisphere in patients without auditory hallucinations (Figure 7.12). These findings support the hypothesis that there may be selective vulnerability of specific anatomical connections in schizophrenia and that extensive bilateral damage may render schizophrenic patients more vulnerable to auditory hallucinations (Catani et al., 2011).

Superior longitudinal fasciculus system

At the beginning of this chapter we mentioned that Dejerine (1895) used the terms arcuate fasciculus and superior longitudinal fasciculus as synonyms (Figure 7.13A, B). This equivalence of terms is, however, anatomically incorrect and we feel the need for some clarification. In the monkey brain, a group of longitudinal fibres connecting the dorsolateral fronto-parietal cortex has been described using axonal tracing (Pandya and Kuypers, 1969) and cortical electrophysiological recording (Rizzolatti et al., 1998). Petrides and Pandya (1984) used the term superior longitudinal fasciculus (SLF) to define this system of fronto-parietal connections and identified three separate branches (Figure 7.13C, D). The most dorsal component of the superior longitudinal fasciculus (SLF I) projects to the parietal precuneus and the supplementary motor area at the medial and superior surface of the superior frontal gyrus. The second branch of the superior longitudinal fasciculus (SLF II) projects to the posterior region of the inferior parietal lobule (including the intraparietal sulcus) and the lateral aspect of the superior (i.e. frontal eye field) and middle frontal gyrus. The most ventral branch of the superior longitudinal fasciculus (SLF III) projects to the inferior parietal lobule (supramarginal and anterior angular gyrus) and the posterior region of the inferior frontal gyrus (Schmahmann and Pandya, 2006). Classical human post-mortem studies have dissected only the ventral component of the superior longitudinal fasciculus, whereas recent tractography studies suggest that simian-human similarities extend to all three branches (Makris et al., 2005; Thiebaut de Schotten et al., 2008, 2011b) (Figure 7.13).

The SLF I is often confused with the fibres of the cingulum bundle as both tracts project to the same frontal and parietal regions. The two tracts are, however, separated by the most medial projections of the corpus callosum and the cingulate sulcus, the SLF I being more lateral and dorsal compared to the cingulum. The SLF I processes the spatial coordinates of trunk and inferior limbs and contributes to the preparatory stages of movement planning (e.g. anticipation), oculomotor coordination, visual reaching, and possibly voluntary orientation of attention (Duffy and Burchfiel, 1971; Johnson, et al., 1996; Leiguarda and Marsden, 2000; Corbetta and Schulman 2002).

The SLF II is often confused with the superior fronto-occipital fasciculus. In the monkey both tracts project to the same regions but are separated by the fibres of the internal capsule, SLF II being more lateral compared to the superior fronto-occipital fasciculus. In humans the superior fronto-occipital fasciculus may not exist (see also Chapter 8). In the left hemisphere the SLF II is involved in processing the spatial coordinates of the upper limbs, and in other functions similar to the SLF I (Leiguarda and Marsden, 2000; Goldenberg and Karnath, 2006). In the right hemisphere the SLF II participates in attention, visuospatial processing, and spatial working memory (Lynch et al., 1977; Goldberg and Segraves, 1987; Alivisatos and Milner, 1989; Blatt et al. 1990; Corbetta et al., 1993; Courtney et al., 1998; Koski et al., 1998; Levy and Goldman-Rakic, 2000; Thiebaut de Schotten, et al., 2005).

Finally the SLF III corresponds to the anterior segment of the arcuate fasciculus and the two terms are currently used interchangeably. Future studies will be necessary to establish whether, according to its functional role, the SLF III/anterior segment should be considered as part of the sensory-motor SLF system or the arcuate language network.

References

Aboitiz, F. and Garcia, R.V. (1997a). The evolutionary origin of the language areas in the human brain. A neuroanatomical perspective. *Brain Res Rev*, **25**, 381–96.

Aboitiz, F. and Garcia, R.V. (1997b). The anatomy of language revisited. *Biol Res*, **30**(4), 171–83.

Agosta, F., Henry, R.G., Migliaccio, R., et al. (2010). Language networks in semantic dementia. *Brain*, **133**(1), 286–99.

Alexander, M.P., Baker, E., Naeser, M.A., Kaplan, E., and Palumbo, C. (1992). Neuropsychological and neuroanatomical dimensions of ideomotor apraxia. *Brain*, **115**(1), 87–107.

Alivisatos, B. and Milner, B. (1989). Effects of frontal or temporal lobectomy on the use of advance information in a choice reaction time task. *Neuropsychologia*, **27**(4), 495–503.

Allen, P., Aleman, A., and McGuire, P.K. (2007). Inner speech models of auditory verbal hallucinations: evidence from behavioural and neuroimaging studies. *Int Rev Psychiatry*, **19**(4), 407–15.

Anwander, A., Tittgemeyer, M., von Cramon, D.Y., Friederici, A.D., and Knosche, T.R. (2007). Connectivity-based parcellation of Broca's area. *Cereb Cortex*, **17**(4), 816–25.

Baddeley, A.D. (2003). Working memory: looking back and looking forward. *Nat Rev Neurosci*, **4**(10), 829–39.

Baddeley, A.D. (2007). *Working memory, thought and action*. Oxford: Oxford University Press.

Baddeley, A.D. and Hitch, G. (1974). Working memory. In Bower, G.H. (Ed.) *The psychology of learning and motivation: Advances in research and theory*, Vol. 8, pp. 47–89. New York: Academic Press.

Barrick, T.R., Lawes, I.N., Mackay, C.E., and Clark, C.A. (2007). White matter pathway asymmetry underlies functional lateralization. *Cereb Cortex*, **17**(3), 591–8.

Basso, A., Lecours, A.R., Moraschini, S., and Vanier, M. (1985). Anatomoclinical correlations of the aphasias as defined through computerized tomography: exceptions. *Brain Lang*, **26**(2), 201–29.

Bernal, B. and Ardila, A. (2009). The role of the arcuate fasciculus in conduction aphasia. *Brain*, **132**(9), 2309–16.

Bird, C.M., Malhotra, P., Parton, A., Coulthard, E., Rushworth, M.F., and Husain, M. (2006). Visual neglect after right posterior cerebral artery infarction. *J Neurol Neurosurg Psychiatry*, **77**(9), 1008–12.

Blatt, G.J., Andersen, R.A., and Stoner, G.R. (1990). Visual receptive field organization and cortico-cortical connections of the lateral intraparietal area (area LIP) in the macaque. *J Comp Neurol*, **299**(4), 421–45.

Boatman, D., Gordon, B., Hart, J., et al. (2000). Transcortical sensory aphasia: revisited and revised. *Brain*, **123**(8), 1634–42.

Bookheimer, S.Y. (2002). Functional MRI of language: new approaches to understanding the cortical organization of semantic processing. *Annu Rev Neurosci*, **25**, 151–88.

Breier, J.I., Hasan, K.M., Zhang, W., Men, D., and Papanicolaou, A.C. (2008). Language dysfunction after stroke and damage to white matter tracts evaluated using diffusion tensor imaging. *Am J Neuroradiol*, **29**(3), 483–7.

Broca, P. (1861). Sur le volume et la forme du cerveau suivant les individus et suivant les races. *Bull Soc Anthropol*, **2**, 139–207.

Buchel, C., Raedler, T., Sommer, M., et al. (2004). White matter asymmetry in the human brain: a diffusion tensor MRI study. *Cereb Cortex*, **14**(9), 945–51.

Buckner, R.L., Snyder, A.Z., Shannon, B.J., et al. (2005). Molecular, structural, and functional characterization of Alzheimer's disease: evidence for a relationship between default activity, amyloid, and memory. *J Neurosci*, **25**(34), 7709–17.

Burdach, K. (1822). *Vom Bau und Leben des Gehirns und Rückenmarks, Vol. 2.* Leipzig: In der Dyk'schen Buchhandlung.

Buxbaum, L.J. (2001). Ideomotor apraxia: a call to action. *Neurocase*, **7**(6), 445–58.

Cannestra, A.F., Bookheimer S.Y., Pouratian, N., et al. (2000). Temporal and topographical characterization of language cortices using intraoperative optical intrinsic signals. *Neuroimage*, **12**(1), 41–54.

Castillo, E.M., Simos, P.G., Davis, R.N., et al. (2001). Levels of word processing and incidental memory: dissociable mechanisms in the temporal lobe. *Neuroreport*, **12**(16), 3561–6.

Catani, M. and ffytche, D.H. (2005). The rises and falls of disconnection syndromes. *Brain*, **128**(10), 2224–39.

Catani, M. and Mesulam, M. (2008). The arcuate fasciculus and the disconnection theme in language and aphasia: history and current state. *Cortex*, **44**(8), 953–61.

Catani, M., Howard, R.J., Pajevic, S., and Jones, D.K. (2002). Virtual in vivo interactive dissection of white matter fasciculi in the human brain. *NeuroImage*, **17**(1), 77–94.

Catani, M., Jones, D.K., and ffytche, D.H. (2005). Perisylvian language networks of the human brain. *Ann Neurol*, **57**(1), 8–16.

Catani, M., Allin, M.P., Husain, M., et al. (2007). Symmetries in human brain language pathways correlate with verbal recall. *Proc Natl Acad Sci USA*, **104**(43), 17163–8.

Catani, M., Forkel, S., and Thiebaut de Schotten, M. (2010). Asymmetry of human brain pathways. In Hugdahl, K. and Westernhausen, R. (Eds.). *The two halves of the brain: Information processing in the cerebral hemispheres*, pp. 177–210. Cambridge:, MA: MIT Press.

Catani, M., Craig, M.C., Forkel, S.J., et al. (2011). Altered Integrity of Perisylvian Language Pathways in Schizophrenia: Relationship to Auditory Hallucination. *Biol Psychiatry*, **70**(12), 1143-50.

Cohen, L., Dehaene, S., Naccache, L., et al. (2000). The visual word form area: spatial and temporal characterization of an initial stage of reading in normal subjects and posterior split-brain patients. *Brain*, **123**(2), 291–307.

Committeri, G., Pitzalis, S., Galati, G., et al. (2007). Neural bases of personal and extrapersonal neglect in humans. *Brain*, **130**(2), 431–41.

Corbetta, M. and Shulman, G.L. (2002). Control of goal-directed and stimulus-driven attention in the brain. *Nat Rev Neurosci*, **3**(3), 201–15.

Corbetta, M., Miezin, F.M., Shulman, G.L., and Petersen, S.E. (1993). A PET study of visuospatial attention. *J Neurosci*, **13**(3), 1202–26.

Corbetta, M., Kincade, M., Lewis, C. et al. (2005). Neural basis and recovery of spatial attention deficits in spatial neglect. *Nat Neurosci*, **8**(11), 1603–10.

Courtney, S.M., Petit, L., Maisog, J.M., Ungerleider, L.G., and Haxby, J.V. (1998). An area specialized for spatial working memory in human frontal cortex. *Science*, **279**(5355), 1347–51.

Damasio, A.R. and Geschwind, N. (1984). The neural basis of language. *Ann Rev Neurosci*, **7**, 127–47.

Deacon, T.W. (1992). Cortical connections of the inferior arcuate sulcus cortex in the macaque brain. *Brain Res*, **573**(1), 8–26.

Dejerine, J. (1892). Contribution a l'étude anatomo-pathologique et clinique des differentes variétés de cécité-verbale. *Mém Soc Biol*, **4**, 61–90.

Dejerine, J. (1895). *Anatomie des Centres Nerveux*, Vol. 1. Paris: Rueff et Cie.

Dell'Acqua, F., Scifo, P., Rizzo, G., et al. (2010). A modified damped Richardson-Lucy algorithm to reduce isotropic background effects in spherical deconvolution. *NeuroImage*, **49**(2), 1446–58.

Doricchi, F. and Tomaiuolo, F. (2003). The anatomy of neglect without hemianopia: a key role for parietal-frontal disconnection? *Neuroreport*, **14**(17), 2239–43.

Doricchi, F., Thiebaut de Schotten, M., Tomaiuolo, F., and Bartolomeo, P. (2008). White matter (dis)connections and gray matter (dys)functions in visual neglect: gaining insights into the brain networks of spatial awareness. *Cortex*, **44**(8), 983–95.

Dubois, J., Dehaene-Lambertz, G., Perrin, M., et al. (2008). Asynchrony of the early maturation of white matter bundles in healthy infants: quantitative landmarks revealed noninvasively by diffusion tensor imaging. *Hum Brain Mapp*, **29**(1), 14–27.

Duffy, F.H. and Burchfiel, J.L. (1971). Somatosensory system: organizational hierarchy from single units in monkey area 5. *Science*, **172**(980), 273–5.

Eluvathingal, T.J., Hasan, K.M., Kramer, L., Fletcher, J.M., and Ewing-Cobbs, L. (2007). Quantitative diffusion tensor tractography of association and projection fibers in normally developing children and adolescents. *Cereb Cortex*, **17**(12), 2760–8.

Exner, S. (1881). *Untersuchungen über die Localization der Functionen in der Grosshirnrinde des Menschen*. Wien: Wilhelm Braumüller.

Flechsig, P.E. (1901). Developmental (myelogenetic) localisation of the cerebral cortex in the human subject. *Lancet*, **2**, 1027–9.

Fletcher, P.T., Whitaker, R.T., Tao, R., et al. (2010). Microstructural connectivity of the arcuate fasciculus in adolescents with high-functioning autism. *NeuroImage*. **51**(3), 1117–25.

Ford, J.M., Mathalon, D.H., Whitfield, S., Faustman, W.O., and Roth, W.T. (2002). Reduced communication between frontal and temporal lobes during talking in schizophrenia. *Biol Psychiat*, **51**(6), 485–92.

Frey, S.H. (2007). What puts the how in where? Tool use and the divided visual streams hypothesis. *Cortex*, **43**(3), 368–75.

Frith, C.D. and Frith, U. (2007). Social cognition in humans. *Curr Biol*, **17**(16), 724–32.

Galuske, R.A., Schlote, W., Bratzke, H., and Singer, W. (2000). Interhemispheric asymmetries of the modular structure in human temporal cortex. *Science*, **289**(5486), 1946–9.

Geschwind, N. (1965). Disconnexion syndromes in animals and man. I. *Brain*, **88**(2), 237–294.

Geschwind, N. (1970). The organization of language and the brain. *Science*, **170**(961), 940–4.

Gharabaghi, A., Fruhmann Berger, M., Tatagiba, M. and Karnath, H.O. (2006). The role of the right superior temporal gyrus in visual search-insights from intraoperative electrical stimulation. *Neuropsychologia* **44**(12), 2578–81.

Goldberg, M.E. and Segraves, M.A. (1987). Visuospatial and motor attention in the monkey. *Neuropsychologia*, **25**(1A), 107–18.

Goldenberg, G. (2003). Apraxia and beyond: life and work of Hugo Liepmann. *Cortex*, **39**(3), 509–24.

Goldenberg, G. and Karnath, H.O. (2006). The neural basis of imitation is body part specific. *J Neurosci*, **26**(23), 6282–7.

Graves, W.W., Desai, R., Humphries, C., Seidenberg, M.S., and Binder, J.R. (2010). Neural systems for reading aloud: a multiparametric approach. *Cereb Cortex*, **20**(8), 1799–815.

Gruber, O. (2001). Effects of domain-specific interference on brain activation associated with verbal working memory task performance. *Cereb Cortex*, **11**(11), 1047–55.

Hagmann, P., Cammoun, L., Martuzzi, R., et al. (2006). Hand preference and sex shape the architecture of language networks. *Hum Brain Mapp*, **27**(10), 828–35.

Heilman, K.M., Rothi, L.J., and Valenstien, E. (1982). Two forms of ideomotor apraxia. *Neurology*, **32**(4), 342–6.

Heilman, K.M. and Watson, R.T. (2008). The disconnection apraxias. *Cortex*, **44**(8), 975–82.

Herbert, M.R., Ziegler, D.A., Deutsch, C.K., et al. (2005). Brain asymmetries in autism and developmental language disorder: a nested whole-brain analysis. *Brain*, **128**(1), 213–26.

Hickok, G. and Poeppel, D. (2007). The cortical organization of speech processing. *Nat Rev Neurosci*, **8**(5), 393–402.

Husain, M. and Kennard, C. (1996). Visual neglect associated with frontal lobe infarction. *J Neurol*, **243**(9), 652–7.

Hyde, K.L., Zatorre, R.J., and Peretzet, I. (2011). Functional MRI evidence of an abnormal neural network for pitch processing in congenital amusia. *Cereb Cortex*, **21**(2), 292–9.

Jacquemot, C. and Scott, S.K. (2006). What is the relationship between phonological short-term memory and speech processing? *Trends Cogn Sci*, **10**(11), 480–6.

Jardri, R., Pins, D., Bubrovszky, M., et al. (2007). Self awareness and speech processing: an fMRI study. *NeuroImage*, **35**(4), 1645–53.

Johnson, P.B., Ferraina, S., Bianchi, L., and Caminiti, R. (1996). Cortical networks for visual reaching: physiological and anatomical organization of frontal and parietal lobe arm regions. *Cereb Cortex*, **6**(2), 102–19.

Jones, D.K., Catani, M., Pierpaoli, C., et al. (2005). A diffusion tensor magnetic resonance imaging study of frontal cortex connections in very-late-onset schizophrenia-like psychosis. *Am J Geriatr Psychiatry*, **13**(12), 1092–9.

Jones, D.K., Catani, M., Pierpaoli, C., et al. (2006). Age effects on diffusion tensor magnetic resonance imaging tractography measures of frontal cortex connections in schizophrenia. *Hum Brain Mapp*, **27**(3), 230–8.

Karnath, H.O., Ferber, S., and Himmelbach, M. (2001). Spatial awareness is a function of the temporal not the posterior parietal lobe. *Nature*, **411**(6840), 950–3.

Karnath, H.O., Berger, M.F., Küker, W., et al. (2004). The anatomy of spatial neglect based on voxelwise statistical analysis: a study of 140 patients. *Cereb Cortex*, **14**(10), 1164–72.

Kempler, D., Metter, E.J., Jackson, C.A., et al. (1988). Disconnection and cerebral metabolism: the case of conduction aphasia. *Arch Neurol*, **45**(3), 275–9.

Koski, L.M., Paus, T., and Petrides, M. (1998). Directed attention after unilateral frontal excisions in humans. *Neuropsychologia*, **36**(12), 1363–71.

Kraepelin, E. (1913). *Psychiatrie. Ein Lehrbuch fur Studirende und Aertzte. Achte,vollständig umgearbeitete Aufl age, III. Band, II. Teil Klin psychiat.* Leipzig: Barth Verlag.

Kumar, A., Sundaram, S.K., Sivaswamy, L., et al. (2010). Alterations in frontal lobe tracts and corpus callosum in young children with autism spectrum disorder. *Cereb Cortex*, **20**(9), 2103–13.

Lawes, I.N., Barrick, T.R., Murugam, V., et al. (2008). Atlas-based segmentation of white matter tracts of the human brain using diffusion tensor tractography and comparison with classical dissection. *NeuroImage*, **39**(1), 62–79.

Lawrie, S.M., Buechel, C., Whalley, H.C., Frith, C.D., Friston, K.J., and Johnstone, E.C. (2002). Reduced frontotemporal functional connectivity in schizophrenia associated with auditory hallucinations. *Biol Psychiat*, **51**(12), 1008–11.

Lebel, C., Walker, L., Leemans, A., Philips, L., and Beaulieu, C. (2008). Microstructural maturation of the human brain from childhood to adulthood. *NeuroImage*, **40**(3), 1044–55.

Leiguarda, R.C. and Marsden, C. D. (2000). Limb apraxias: higher-order disorders of sensorimotor integration. *Brain*, **123**(5), 860–79.

Leibovitch, F.S., Black, S.E., Caldwell, C.B., et al. (1998). Brain-behavior correlations in hemispatial neglect using CT and SPECT: the Sunnybrook Stroke Study. *Neurology*, **50**(4), 901–8.

Lennox, B.R., Park, S.B., Medley, I., Morris, P.G., and Jones, P.B. (2000). The functional anatomy of auditory hallucinations in schizophrenia. *Psychiat Res*, **100**(1), 13–20.

Levy, R. and Goldman-Rakic, P.S. (2000). Segregation of working memory functions within the dorsolateral prefrontal cortex. *Exp Brain Res*, **133**(1), 23–32.

Lichtheim, L. (1885). On aphasia. *Brain*, **7**, 433–84.

Liepmann, H. (1900). Das Krankheitsbild der Apraxie (motorische Asymbolie) auf Grund eines Falles von einseitiger Apraxie. *Monatssch Psychia Neurol*, **8**, 15–44, 102–32, 182–97.

Loui, P., Alsop D., and Schlaug G. (2009). Tone deafness: a new disconnection syndrome? *J Neurosci*, **29**(33), 10215–20.

Lynch, J.C., Mountcastle, V.B., Talbot, W.H., and Yin, T.C. (1977). Parietal lobe mechanisms for directed visual attention. *J Neurophysiol*, **40**(2), 362–89.

Makris, N., Kennedy, D.N., McInerney, S., et al. (2005). Segmentation of subcomponents within the superior longitudinal fascicle in humans: a quantitative, in vivo, DT-MRI study. *Cereb Cortex*, **15**(6), 854–69.

Martin-Loeches, M., Casado, P., Hernandez-Tamames, J.A., and Alvarez-Linera, J. (2008). Brain activation in discourse comprehension: a 3T fMRI study. *NeuroImage*, **41**(2), 614–22.

Matsumoto, R., Nair, D.R., LaPresto, E., et al. (2004). Functional connectivity in the human language system: a cortico-cortical evoked potential study. *Brain*, **127**(10), 2316–30.

Mayo, H. (1827). *A series of engravings intended to illustrate the structure of the brain and spinal chord in man*. London: Burgess and Hill.

McCarthy, R. and Warrington, E. (1984). A two-route model of speech production. *Brain*, **107**(2), 463–85.

McGuire, P.K., Shah, G.M., and Murray, R.M. (1993). Increased blood flow in Broca's area during auditory hallucinations in schizophrenia. *Lancet*, **342**(8873), 703–6.

McGuire, P.K., Silbersweig, D.A., Wright, I., et al. (1995). Abnormal monitoring of inner speech: a physiological basis for auditory hallucinations. *Lancet*, **346**(8975), 596–600.

Meinzer, M., Mohammadi, S., Flöel, A., et al. (2010). Integrity of the hippocampus and surrounding white matter is correlated with language training success in aphasia. *NeuroImage*, **53**(1), 283–90.

Mesulam, M.M., (1990). Large-scale neurocognitive networks and distributed processing for attention, language, and memory. *Ann Neurol*, **28**(5), 597–613.

Mesulam, M.M., Wieneke, C., Rogalski, E., Cobia, D., Thompson, C., and Weintraub, S. (2009). Quantitative template for subtyping primary progressive aphasia. *Arch Neurol*, **66**(12), 1545–51.

Monakow, C. (1885). Neue experimentelle Beiträge zur Anatomie der Schleife: vorläufige Mitteilung. *Neurologisches Centralblatt*, **12**, 265–8.

Mort, D.J., Malhotra P., Mannan, S.K., et al. (2003). The anatomy of visual neglect. *Brain*, **126**(9), 1986–97.

Naeser, M.A., Alexander, M.P., Helm-Estabrooks, N., et al. (1982). Aphasia with predominantly subcortical lesion sites: description of three capsular/putaminal aphasia syndromes. *Arch Neurol*, **39**(1), 2–14.

Nielsen, J.M. (1947). *Agnosia, apraxia, aphasia: their value in cerebral localization*, second edition. New York: Paul B. Hoeber, Inc., Medical Book Department of Harper & Bros.

Nucifora, P.G., Verma, R., Melhem, E.R., Gur, R.E., and Gur, R.C. (2005). Leftward asymmetry in relative fiber density of the arcuate fasciculus. *Neuroreport*, **16**(8), 791–4.

Pandya, D.N. and Kuypers, H.G. (1969). Cortico-cortical connections in the rhesus monkey. *Brain Res*, **13**(1), 13–36.

Park, K.C., Lee, B.H., Kim, E.J., et al. (2006) Deafferentation-disconnection neglect induced by posterior cerebral artery infarction. *Neurology*, **66**(1), 56–61.

Parker, G.J., Luzzi, S., Alexander, D.C., et al. (2005). Lateralization of ventral and dorsal auditory-language pathways in the human brain. *NeuroImage*, **24**(3), 656–66.

Paulesu, E., Frith, C.D., and Frackowiak, R.S.J. (1993). The neural correlates of the verbal component of working memory. *Nature*, **362**(6418), 342–5.

Paus, T., Zijdenbos, A., Worsley, K., et al. (1999). Structural maturation of neural pathways in children and adolescents: in vivo study. *Science*, **283**(5409), 1908–11.

Petrides, M. and Pandya, D.N. (1984). Projections to the frontal cortex from the posterior parietal region in the rhesus monkey. *J Comp Neurol*, **228**(1), 105–16.

Petrides, M. and Pandya, D.N. (1988). Association fiber pathways to the frontal cortex from the superior temporal region in the rhesus monkey. *J Comp Neurol*, **273**(1), 52–66.

Petrides, M. and Pandya, D.N. (2002). Comparative cytoarchitectonic analysis of the human and the macaque ventrolateral prefrontal cortex and corticocortical connection patterns in the monkey. *Eur J Neurosci*, **16**(2), 291–310.

Philipose, L.E., Gottesman, R.F., Newhart, M., et al. (2007). Neural regions essential for reading and spelling of words and pseudowords. *Ann Neurol*, **62**(5), 481–92.

Phillips, O.R., Nuechterlein, K.H., Clark, K.A., et al. (2009). Fiber tractography reveals disruption of temporal lobe white matter tracts in schizophrenia. *Schizophr Res*, **107**(1), 30–8.

Powell, H.W., Parker, G.J., Alexander, D.C., et al. (2006). Hemispheric asymmetries in language-related pathways: a combined functional MRI and tractography study. *NeuroImage*, **32**(1), 388–99.

Radua, J., Via, E., Catani, M., and Mataix-Cols, D. (2011). Voxel-based meta-analysis of white matter abnormalities in autism spectrum disorders. *Psychol Med*, **41**(7), 1539–50.

Ramayya, A.G., Glasser, M.F., and Rilling, J.K. (2010). A DTI investigation of neural substrates supporting tool use. *Cereb Cortex*, **20**(3), 507–16.

Reil, J. (1812). Die vördere commissur im groben gehirn. *Arch Physiol*, **11**, 89–100.

Rilling, J. K., Glasser, M.F., Preuss, T.M., et al. (2008). The evolution of the arcuate fasciculus revealed with comparative DTI. *Nat Neurosci*, **11**(4), 426–8.

Rimrodt, S.L., Peterson, D.J., Denckla, M.B., Kaufmann, W.E., and Cutting, L.E. (2010). White matter microstructural differences linked to left perisylvian language network in children with dyslexia. *Cortex*, **46**(6), 739–49.

Rizzolatti, G., Luppino, G., and Matelli, M. (1998). The organization of the cortical motor system: new concepts. *Electroencephalogr Clin Neurophysiol*, **106**(4), 283–96.

Ross, E.D. (1981). The aprosodias: functional-anatomic organization of the affective components of language in the right hemisphere. *Arch Neurol*, **38**(9), 561–9.

Ross, E.D. (2010). Cerebral localization of functions and the neurology of language: fact versus fiction or is it something else? *Neuroscientist*, **16**(3), 222–43.

Rusconi, E., Pinel, P., Dehaene, S., and Kleinschmidt, A. (2010). The enigma of Gerstmann's syndrome revisited: a telling tale of the vicissitudes of neuropsychology. *Brain*, **133**(2), 320–32.

Sabsevitz, D.S., Medler, D.A., Seidenberg, M., and Binder, J.R. (2005). Modulation of the semantic system by word imageability. *NeuroImage*, **27**(1), 188–200.

Sarkamo, T., Tervaniemi, M., Soinila, S., et al. (2009). Cognitive deficits associated with acquired amusia after stroke: A neuropsychological follow-up study. *Neuropsychologia*, **47**(12), 2642–51.

Saur, D., Kreher, B.W., Schnell, S., et al. (2008). Ventral and dorsal pathways for language. *Proc Natl Acad Sci USA*, **105**(46), 18035–40.

Schiff, H.B., Alexander, M.P., Naeser, M.A., et al. (1983). Aphemia: clinical-anatomical correlations. *Arch Neurol*, **40**(12), 720–7.

Schmahmann, J.D., and Pandya, D.N. (2006). *Fiber Pathways of the Brain*. New York: Oxford University Press.

Schmithorst, V.J. and Holland, S.K. (2007). Sex differences in the development of neuroanatomical functional connectivity underlying intelligence found using Bayesian connectivity analysis. *NeuroImage*, **35**(1), 406–19.

Schulze, K., Zysset, S., Mueller, K., Friederici, A.D., and Koelsch, S. (2011). Neuroarchitecture of verbal and tonal working memory in nonmusicians and musicians. *Hum Brain Mapp*, **32**(5), 771–83.

Shergill, S.S., Brammer, M.J., Williams, S.C., Murray, R.M., and McGuire, P.K. (2000). Mapping auditory hallucinations in schizophrenia using functional magnetic resonance imaging. *Arch Gen Psychiatry*, **57**(11), 1033–8.

Skudlarski, P., Jagannathan, K., Anderson, K., et al. (2010). Brain connectivity is not only lower but different in schizophrenia: a combined anatomical and functional approach. *Biol Psychiatry*, **68**(1), 61–9.

Stephan, K.E., Baldeweg, T., and Friston, K.J. (2006). Synaptic plasticity and dysconnection in schizophrenia. *Biol Psychiat*, **59**(10), 929–39.

Striedter, G. (2005). *Principles of Brain Evolution*. Sunderland: Sinauer.

Sundaram, S.K., Sivaswamy, L., Makki, M.I., Behen, M.E., and Chugani, H.T. (2008). Absence of arcuate fasciculus in children with global developmental delay of unknown etiology: a diffusion tensor imaging study. *J Pediat*, **152**(2), 250–5.

Thiebaut de Schotten, M., Urbanski, M., Duffau, H., et al. (2005). Direct evidence for a parietal-frontal pathway subserving spatial awareness in humans. *Science*, **309**(5744), 2226–8.

Thiebaut de Schotten, M., ffytche, D., Bizzi, A., et al. (2011a). Atlasing location, asymmetry and inter-subject variability of white matter tracts in the human brain with MR diffusion tractography. *NeuroImage*, **54**(1), 49–59.

Thiebaut de Schotten, M., Dell'Acqua, F., Forkel, S., et al. (2011b). A lateralized brain network for visuo-spatial attention. *Nat Neurosci*, **14**(10), 1245–46.

Thiebaut de Schotten, M., Dell'acqua, F., Valabregue, R., et al. (2012). Monkey to human comparative anatomy of the frontal lobe association tracts. *Cortex*, **48**(1), 82–96.

Tomaiuolo, F., Voci, L., Bresci, M., et al. (2010). Selective visual neglect in right brain damaged patients with splenial interhemispheric disconnection. *Exp Brain Res*, **206**(2) 209–17.

Tsang, J.M., Dougherty, R.F., Deutsch, G.K., Wandell, B.A., and Ben-Shachar, M. (2009). Frontoparietal white matter diffusion properties predict mental arithmetic skills in children. *Proc Natl Acad Sci USA*, **106**(52), 22546–51.

Urbanski, M., Thiebaut de Schotten, M., Rodrigo, S., et al. (2008). Brain networks of spatial awareness: evidence from diffusion tensor imaging tractography. *J Neurol Neurosurg Psychiatry*, **79**(5), 598–601.

Vallar, G., Di Betta, A.M., and Silveri, M.C. (1997). The phonological short-term store-rehearsal system: patterns of impairment and neural correlates. *Neuropsychologia*, **35**(6), 795–812.

Vallar, G. (1998). Spatial hemineglect in humans. *Trends Cogn Sci*, **2**(3), 87–97.

Vallar, G. and Perani, D. (1986). The anatomy of unilateral neglect after right-hemisphere stroke lesions. A clinical/CT-scan correlation study in man. *Neuropsychologia*, **24**(5), 609–22.

Vallar, G. and Papagno, C. (2002). Neuropsychological impairments of verbal short- term memory. In Baddeley, A.D., Kopelman, M.D., and Wilson, B.A. (Eds.) *The Handbook of Memory Disorders*, second edition, pp. 249–70. Chichester: Wiley.

Verdon, V., Schwartz, S., Lovblad, K.O., Hauert, C.A., and Vuilleumier, P. (2010). Neuroanatomy of hemispatial neglect and its functional components: a study using voxel-based lesion-symptom mapping. *Brain*, **133**(3), 880–94.

Vernooij, M.W., Smits, M., Wielopolski, P.A., Houston, G.C., Krestin, G.P., and van der Lugt, A. (2007). Fiber density asymmetry of the arcuate fasciculus in relation to functional hemispheric language lateralization in both right- and left-handed healthy subjects: a combined fMRI and DTI study. *NeuroImage*, **35**(3), 1064–76.

Vilberg, K.L. and Rugg, M.D. (2008). Memory retrieval and the parietal cortex: a review of evidence from a dual-process perspective. *Neuropsychologia*, **46**(7), 1787–99.

Voineskos, A.N., Lobaugh, N.J., Bouix, S., et al. (2010). Diffusion tensor tractography findings in schizophrenia across the adult lifespan. *Brain*, **133**(5), 1494–504.

Wernicke, C. (1874). *Der Aphasische Symptomencomplex. Ein psychologische Studie auf anatomischer Basis*. Breslau: Cohn & Weigert.

Wernicke, C. (1906). *Grundrisse der Psychiatrie*. Leipzig: Thieme.

Arcuate fasciculus (atlas)

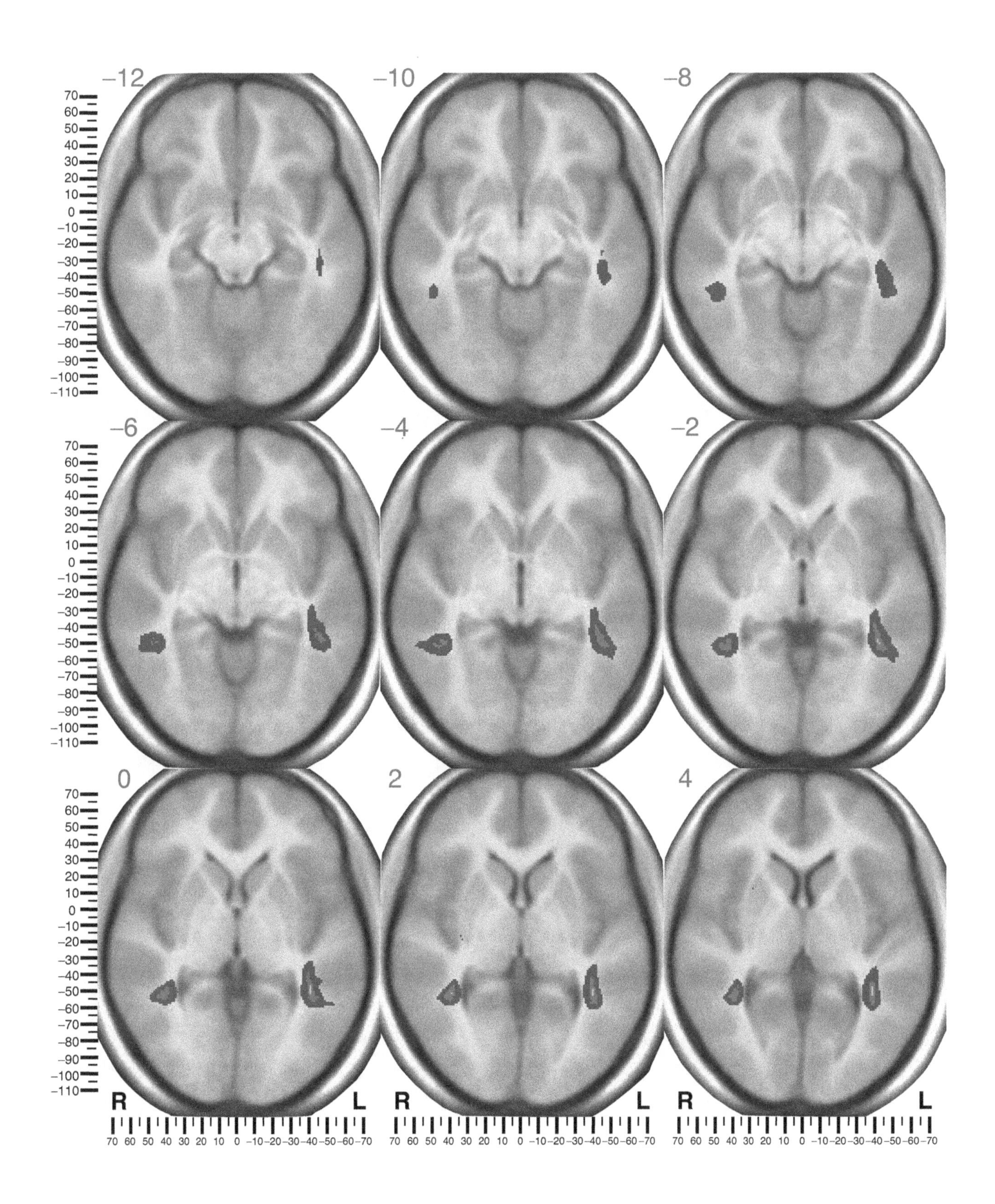

Arcuate: inferior axial view

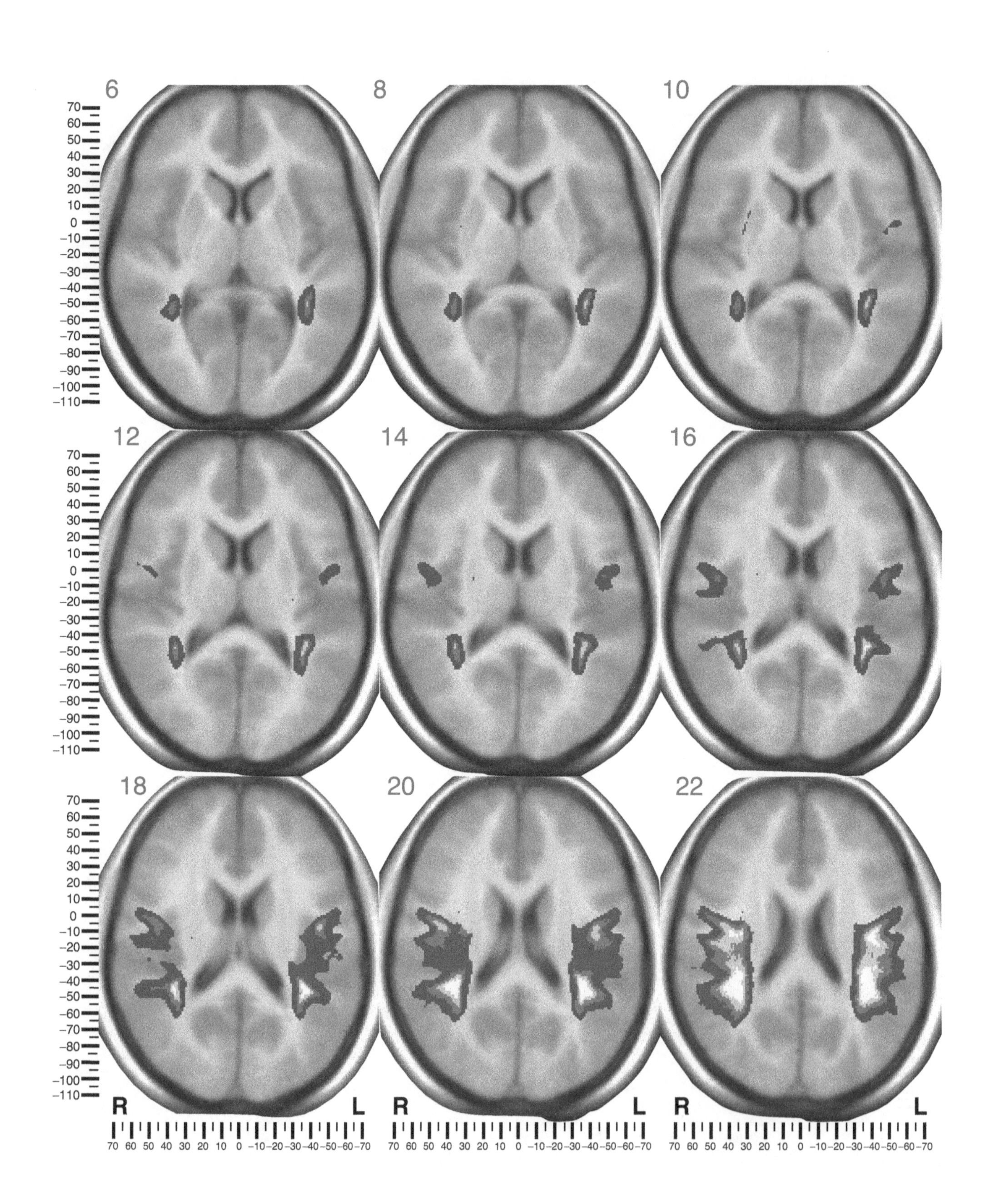

Arcuate: inferior axial view

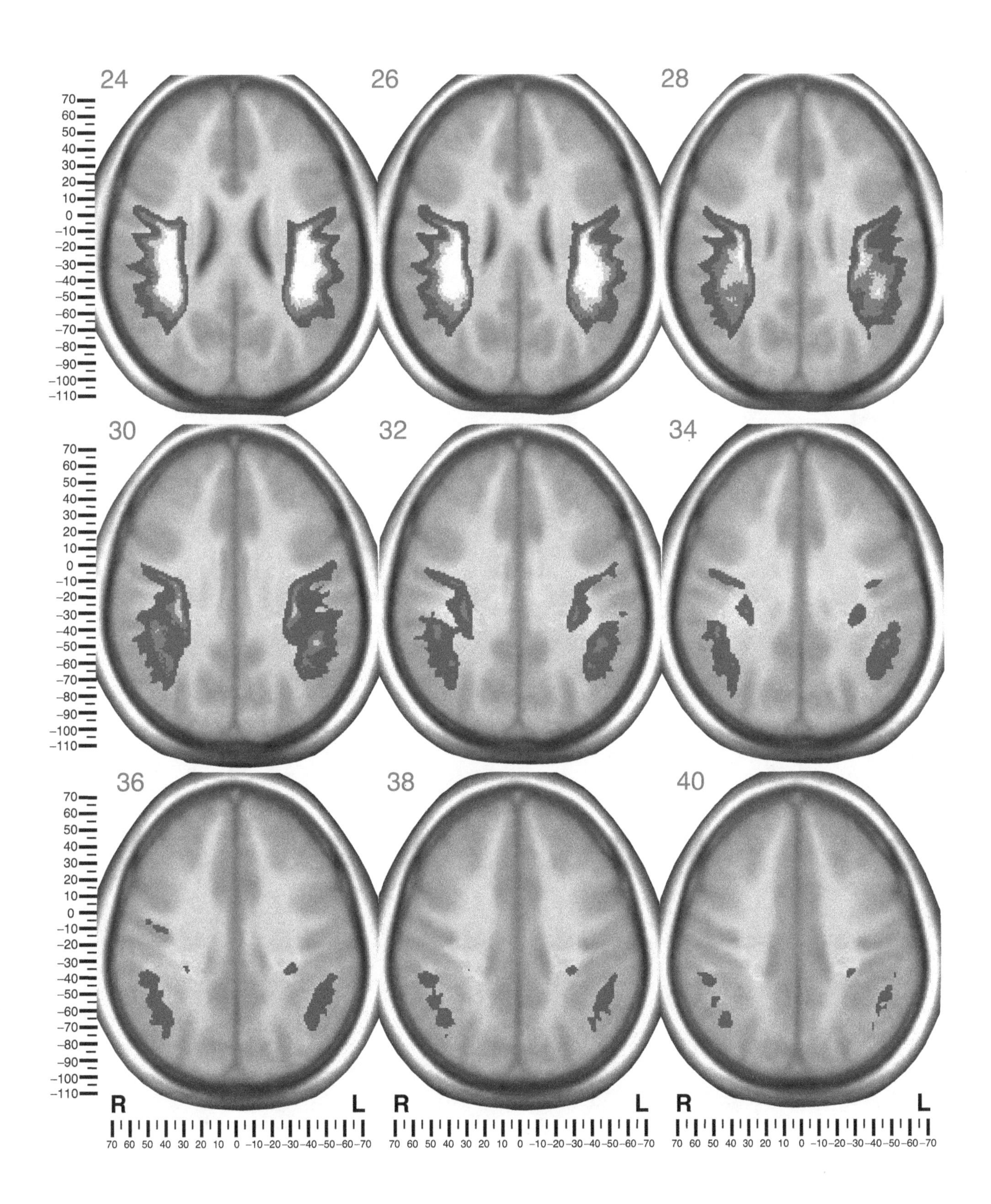

Arcuate: anterior coronal view

p < .05 | > 50% | > 75% | > 90%

Arcuate: anterior coronal view

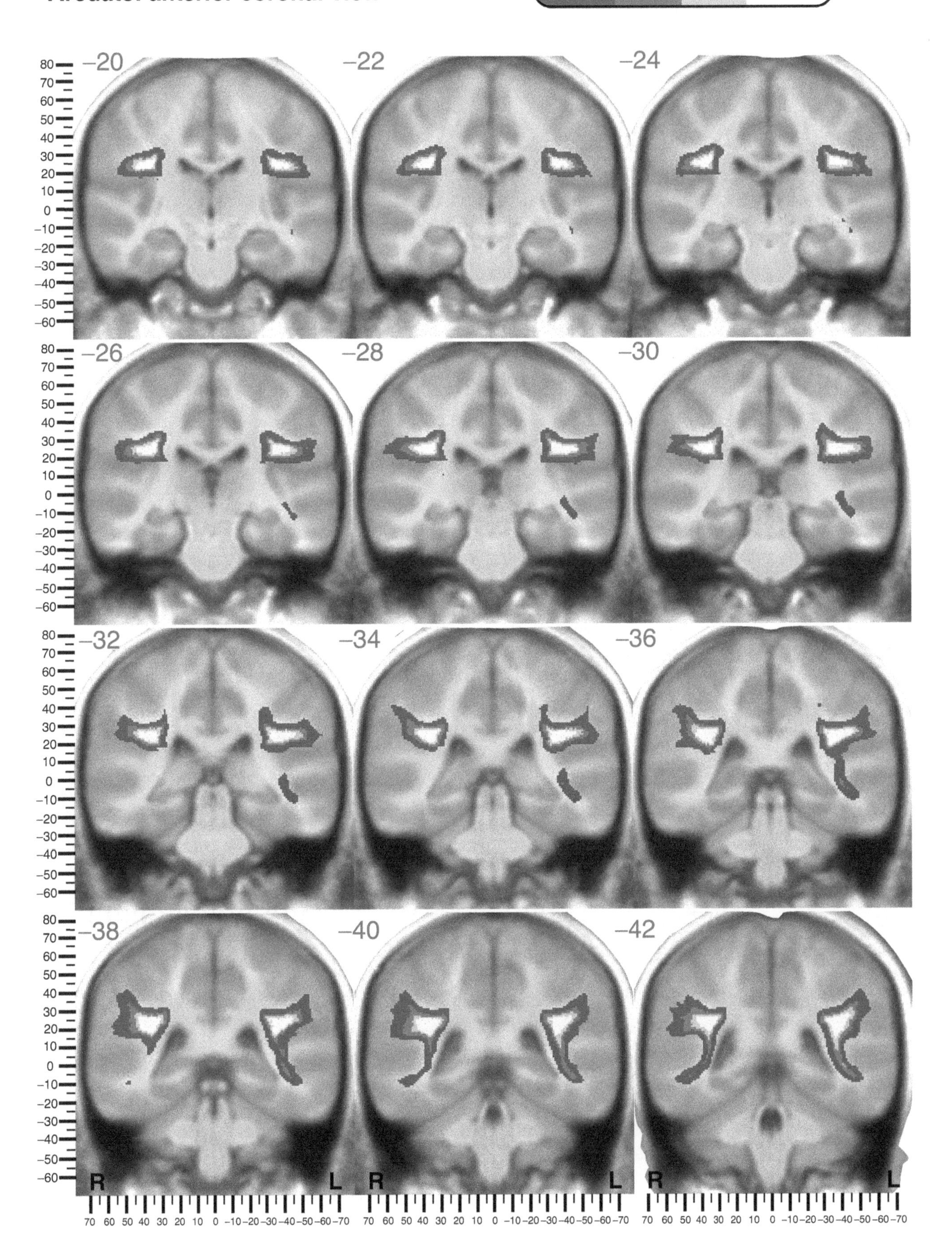

Arcuate: anterior coronal view

p < .05 | > 50% | > 75% | > 90%

−44 −46 −48 −50 −52 −54 −56 −58 −60 −62 −64

R L

80 70 60 50 40 30 20 10 0 −10 −20 −30 −40 −50 −60

70 60 50 40 30 20 10 0 −10 −20 −30 −40 −50 −60 −70

Arcuate (right): lateral sagittal view

p < .05 | > 50% | > 75% | > 90%

56 54 52 50 48 46 44 42

Arcuate (right): lateral sagittal view

p < .05 | > 50% | > 75% | > 90%

40 38 36 34 32 30 28 26

Arcuate (left): medial sagittal view

p < .05 | > 50% | > 75% | > 90%

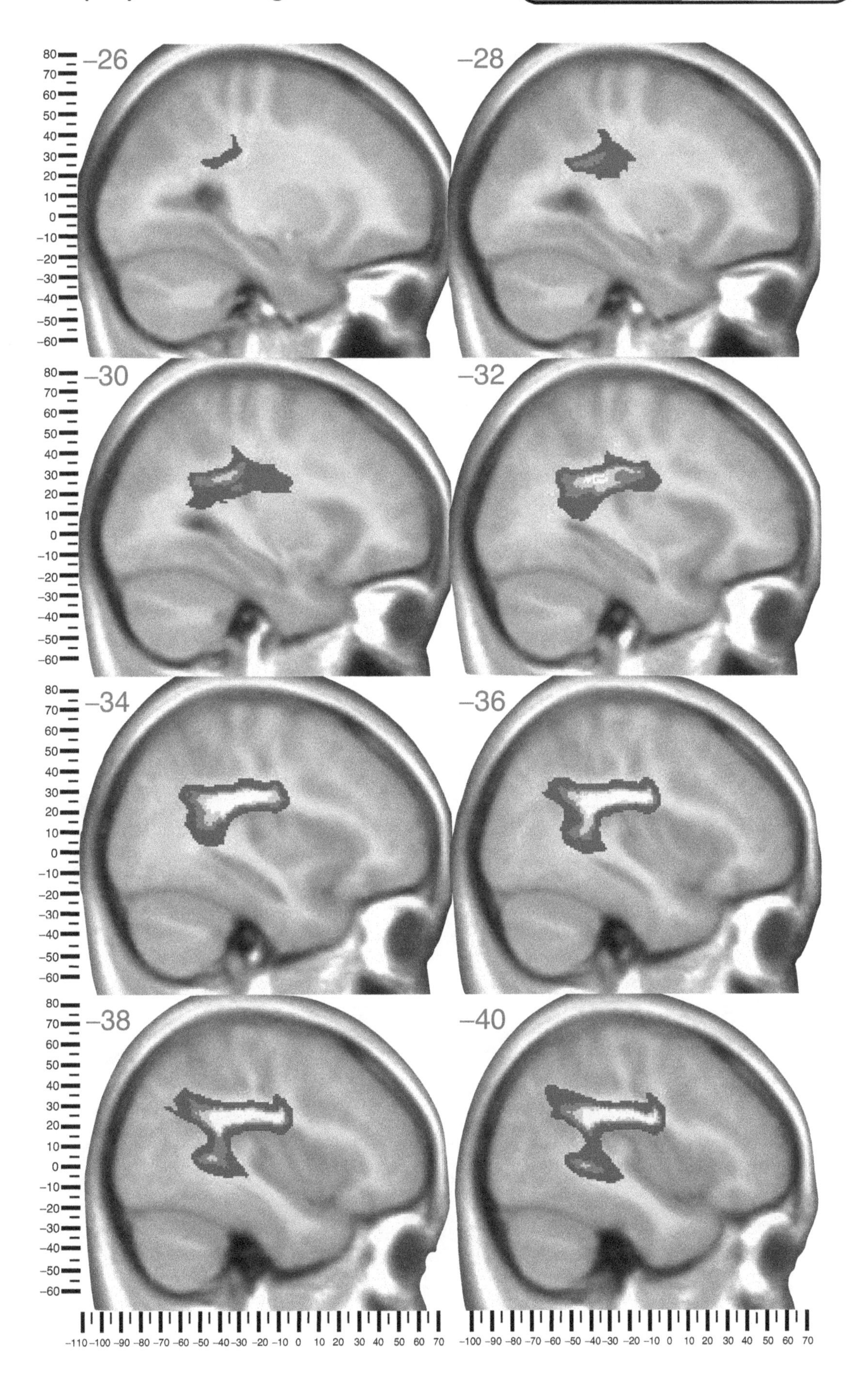

Arcuate (left): medial sagittal view

p < .05 | > 50% | > 75% | > 90%

−42 −44 −46 −48 −50 −52 −54 −56

Long segment (atlas)

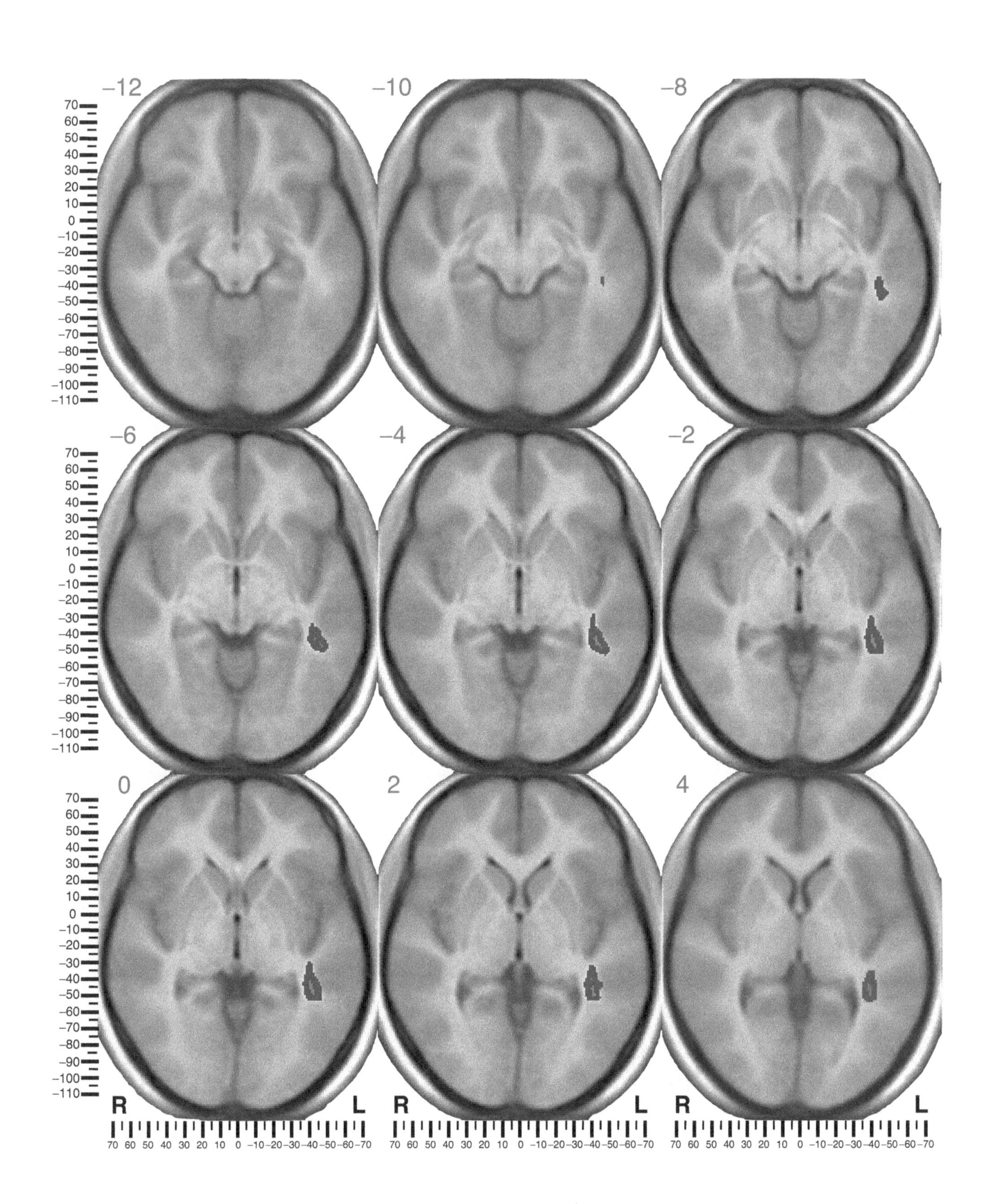

Long segment: inferior axial view

p < .05 | > 50% | > 75% | > 90%

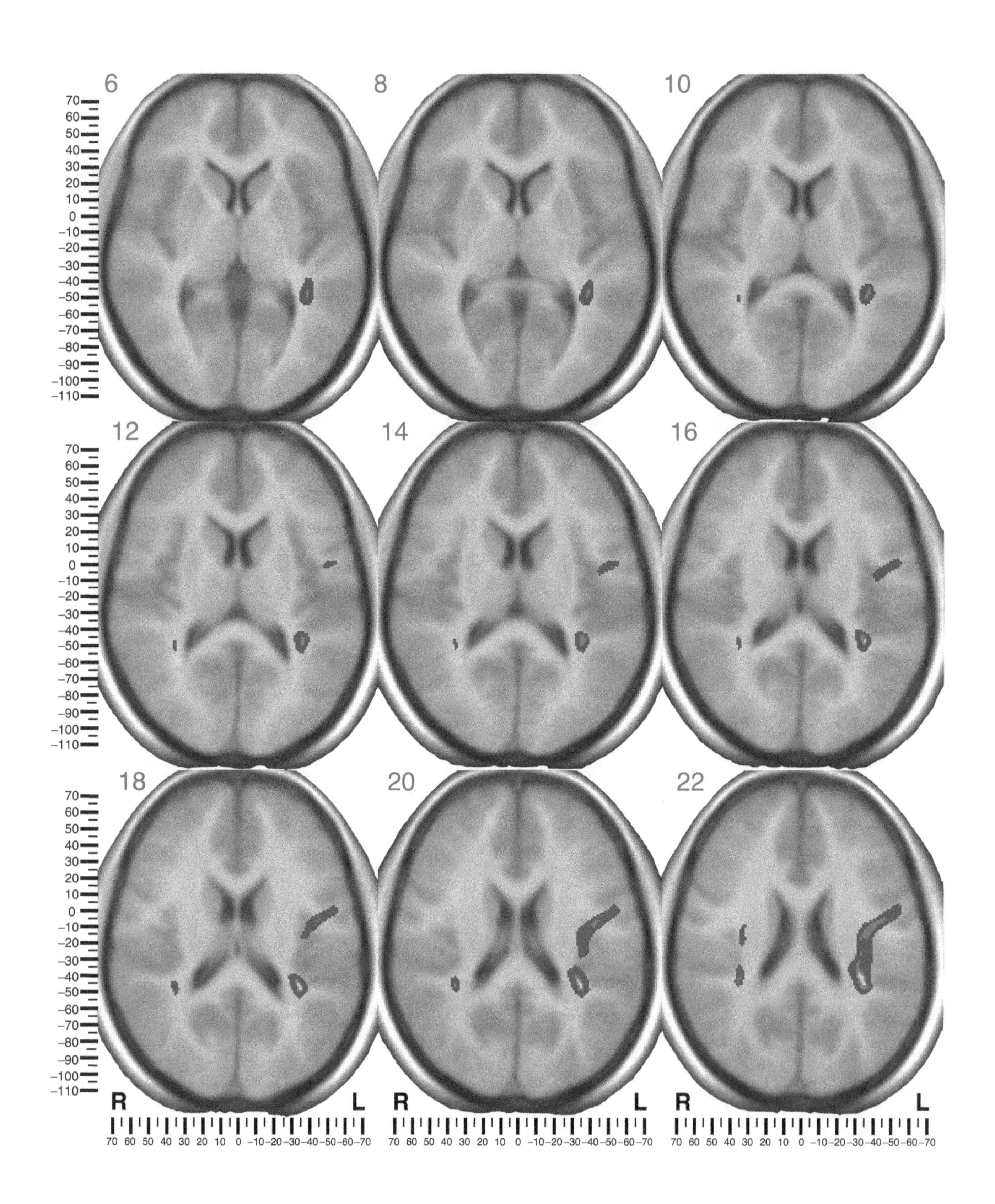

Long segment: inferior axial view

p < .05 | > 50% | > 75% | > 90%

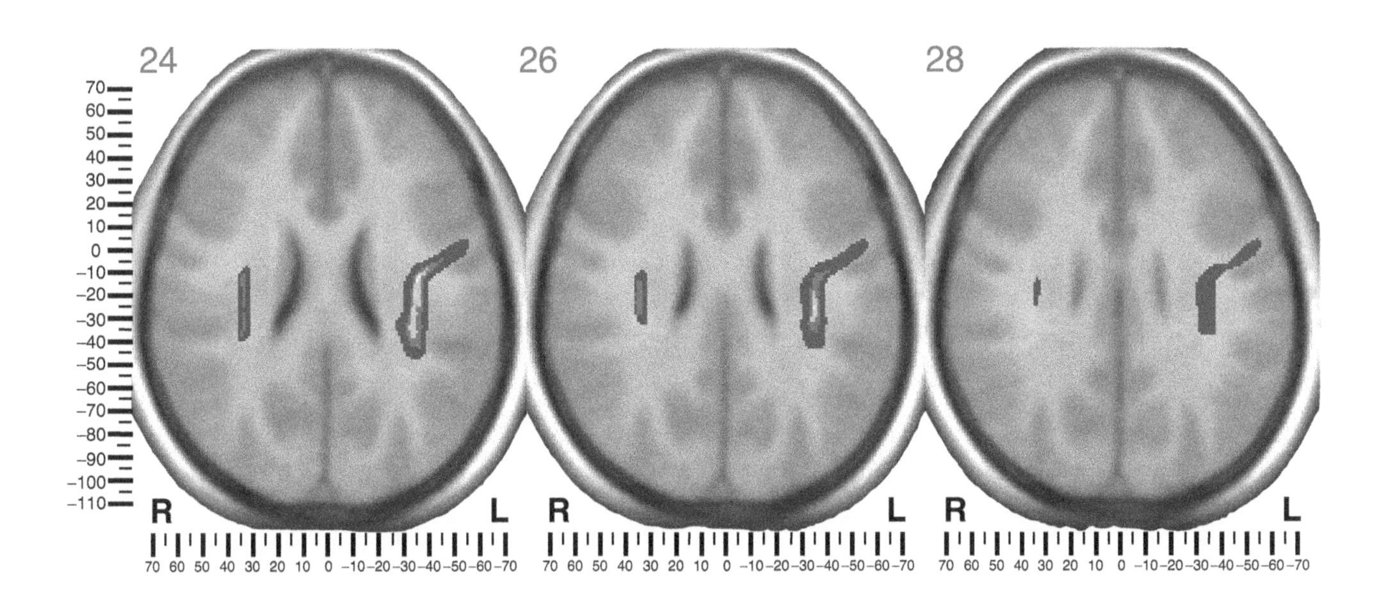

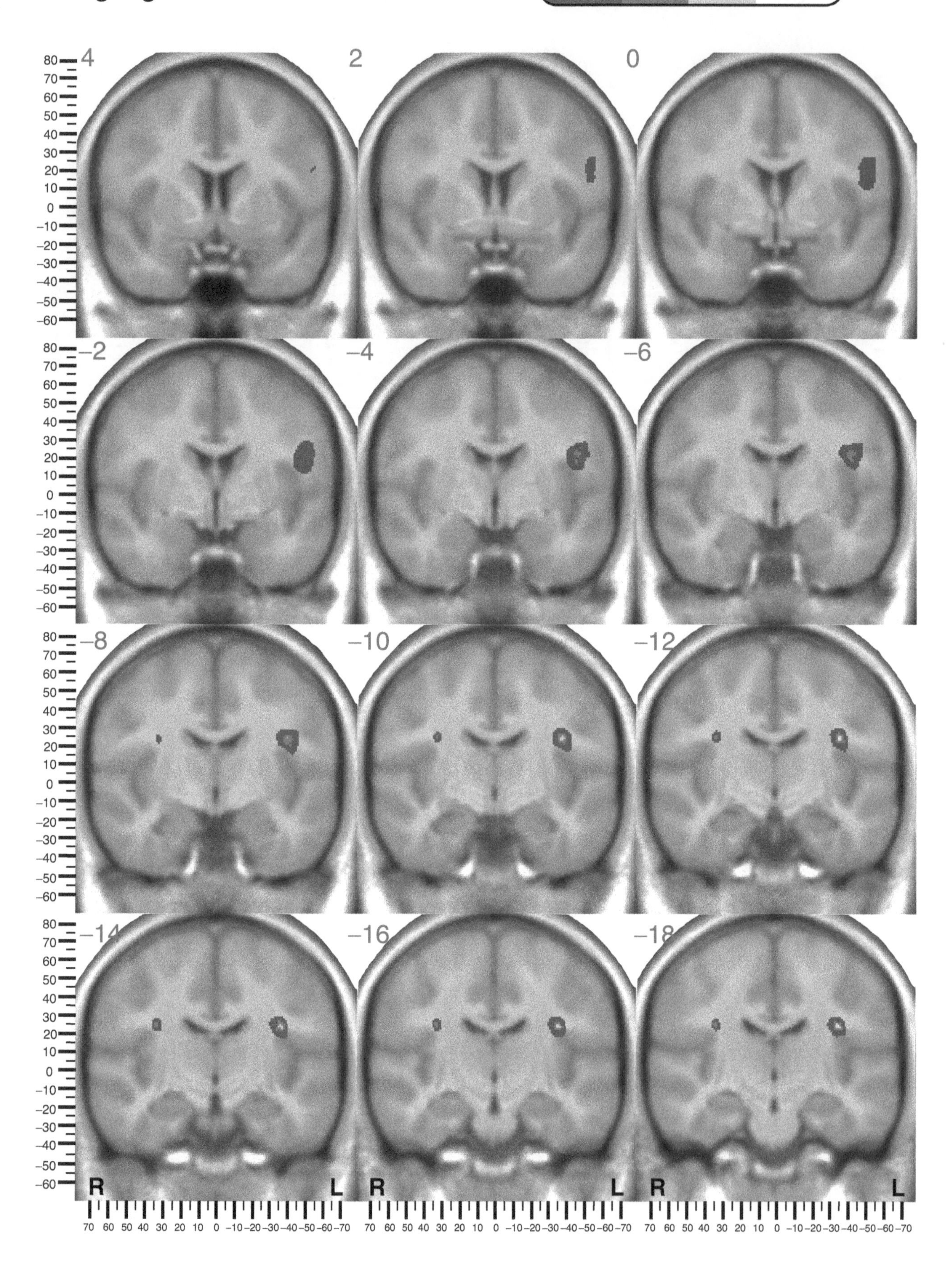

Long segment: anterior coronal view
p < .05
> 50%
> 75%
> 90%
4
2
0
–2
–4
–6
–8
–10
–12
–14
–16
–18
R
L
80 70 60 50 40 30 20 10 0 –10 –20 –30 –40 –50 –60
70 60 50 40 30 20 10 0 –10 –20 –30 –40 –50 –60 –70

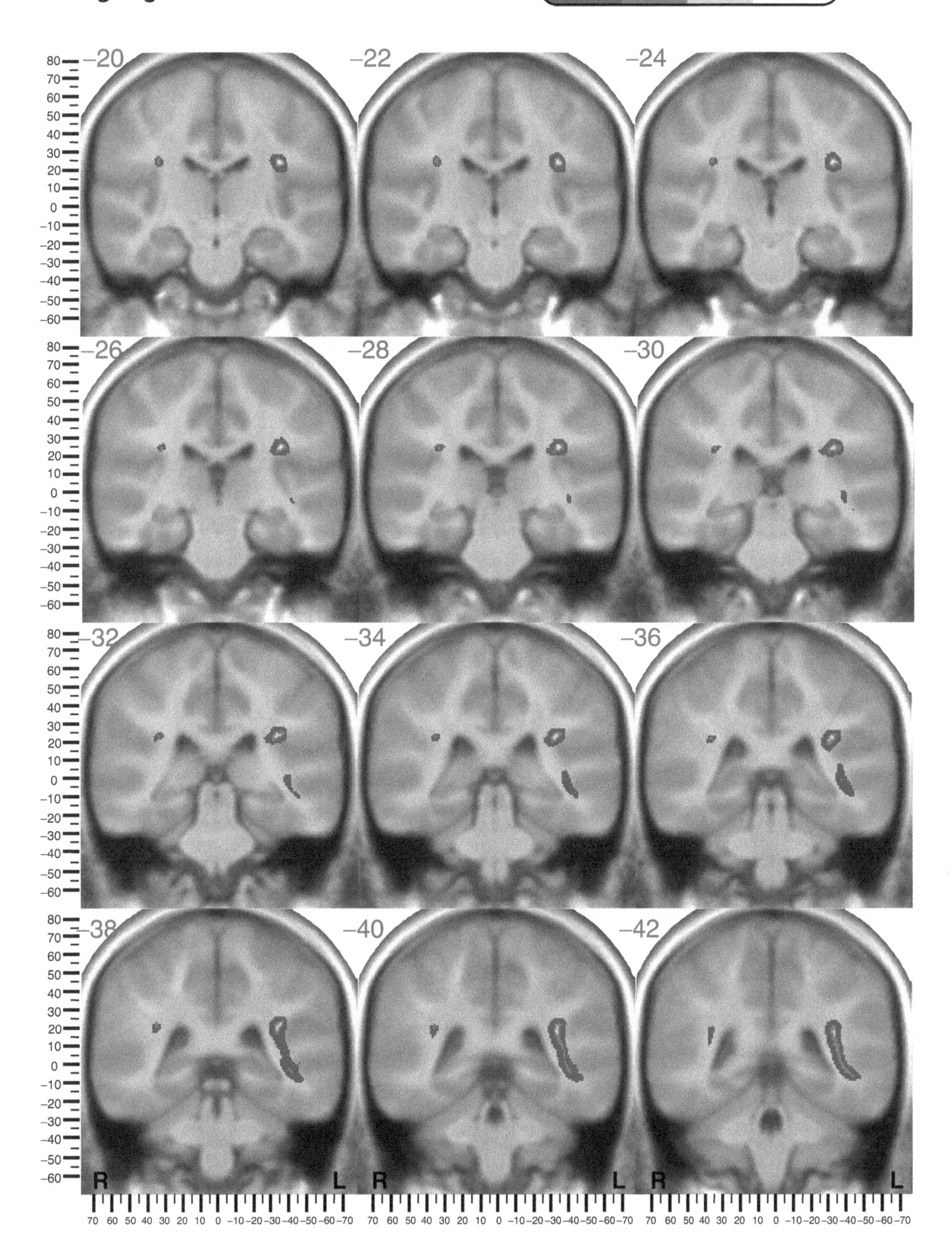
Long segment: anterior coronal view
p < .05
> 50%
> 75%
> 90%
-20
-22
-24
-26
-28
-30
-32
-34
-36
-38
-40
-42
80
70
60
50
40
30
20
10
0
-10
-20
-30
-40
-50
-60
R
L
70 60 50 40 30 20 10 0 -10 -20 -30 -40 -50 -60 -70

Long segment: anterior coronal view

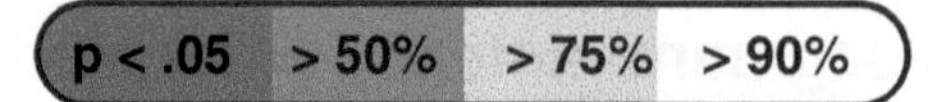

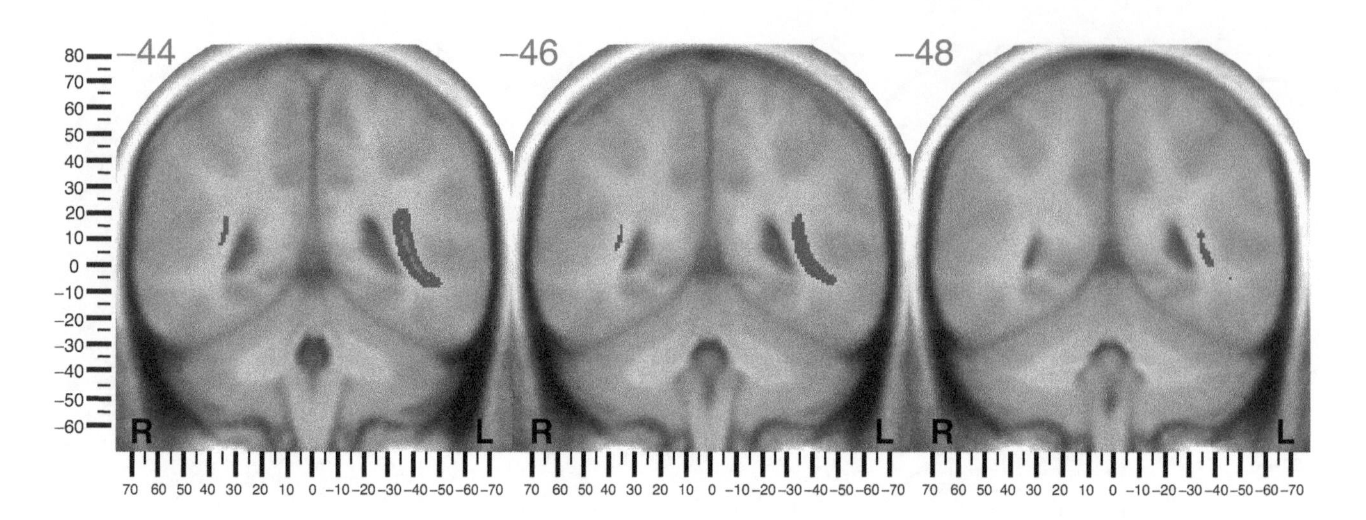

Long segment (right): lateral sagittal view

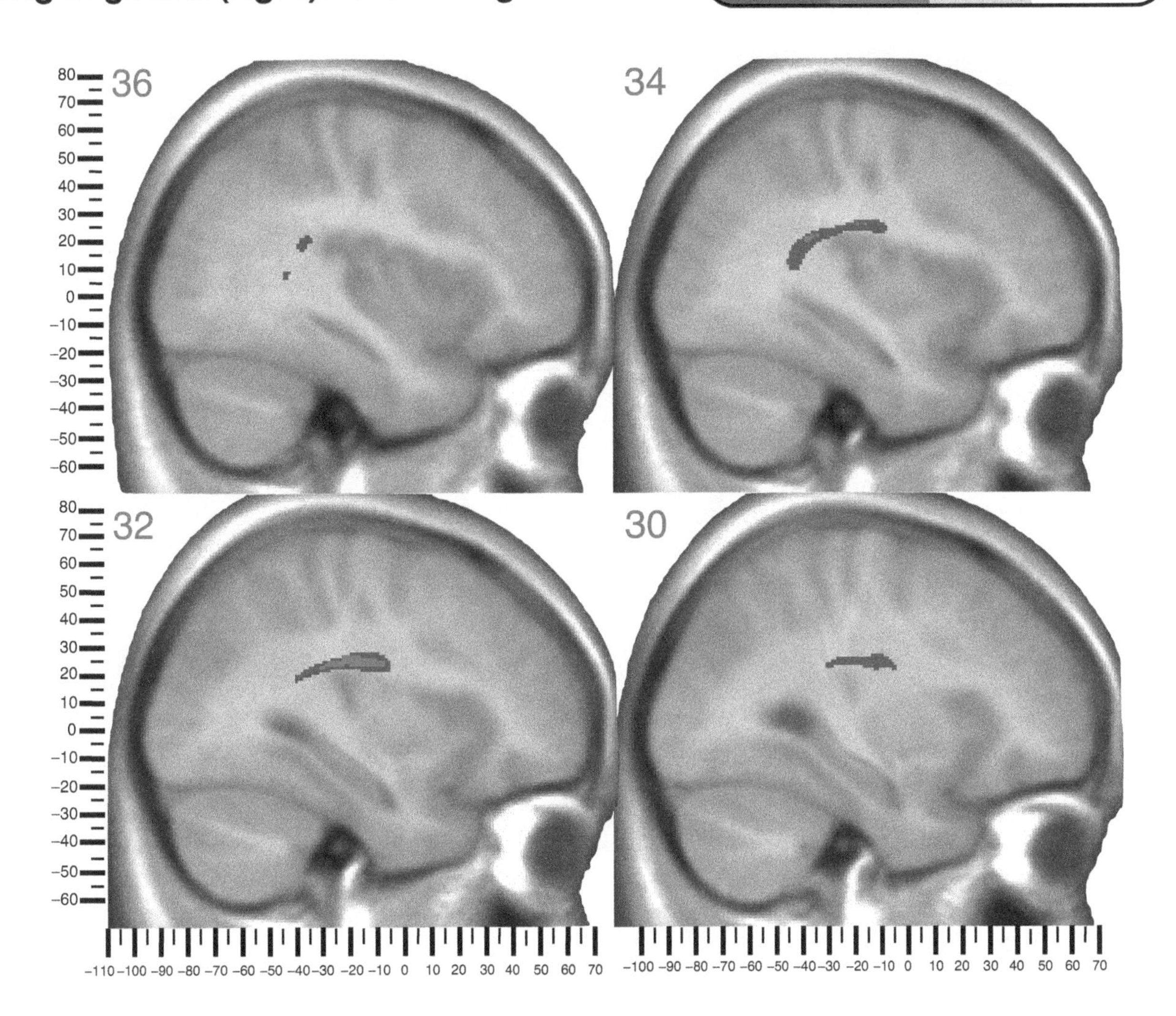

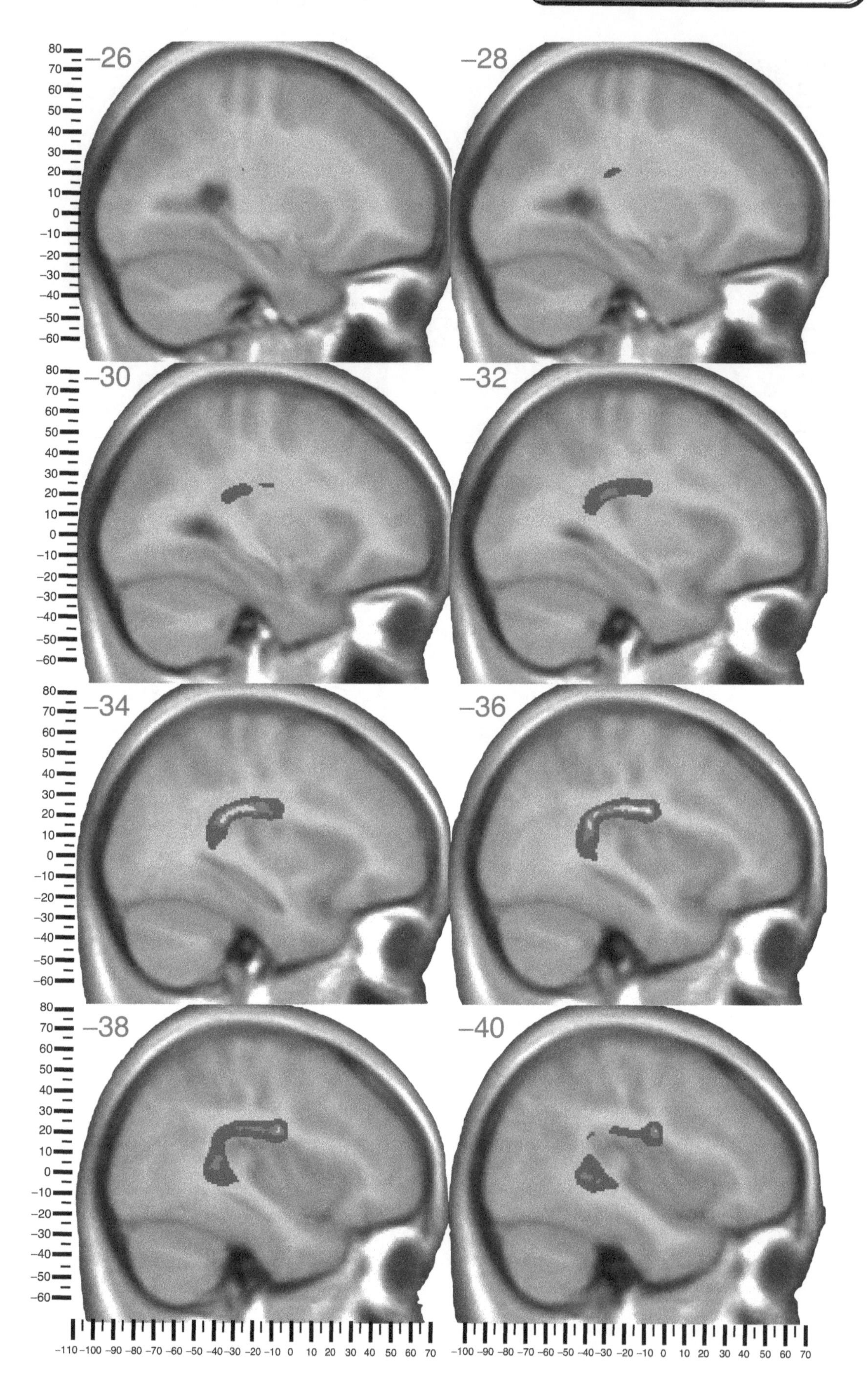

Long segment (left): medial sagittal view
p < .05
> 50%
> 75%
> 90%
-26
-28
-30
-32
-34
-36
-38
-40

Long segment (left): medial sagittal view

p < .05 | > 50% | > 75% | > 90%

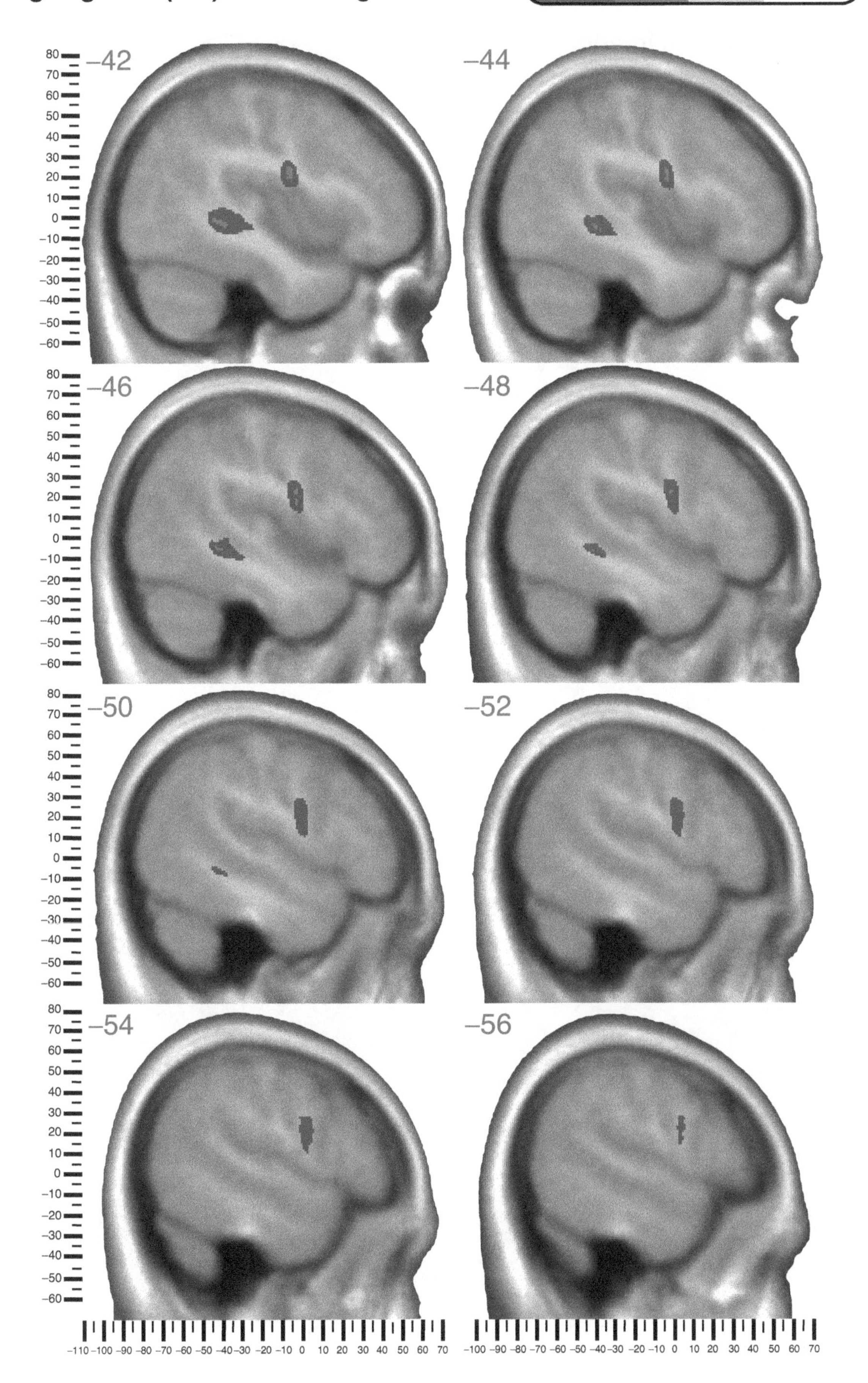

Anterior segment (atlas)

Anterior segment: inferior axial view

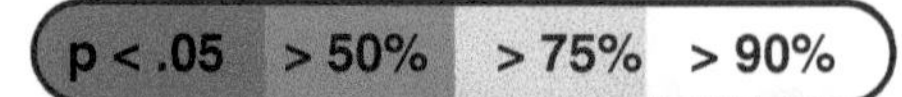

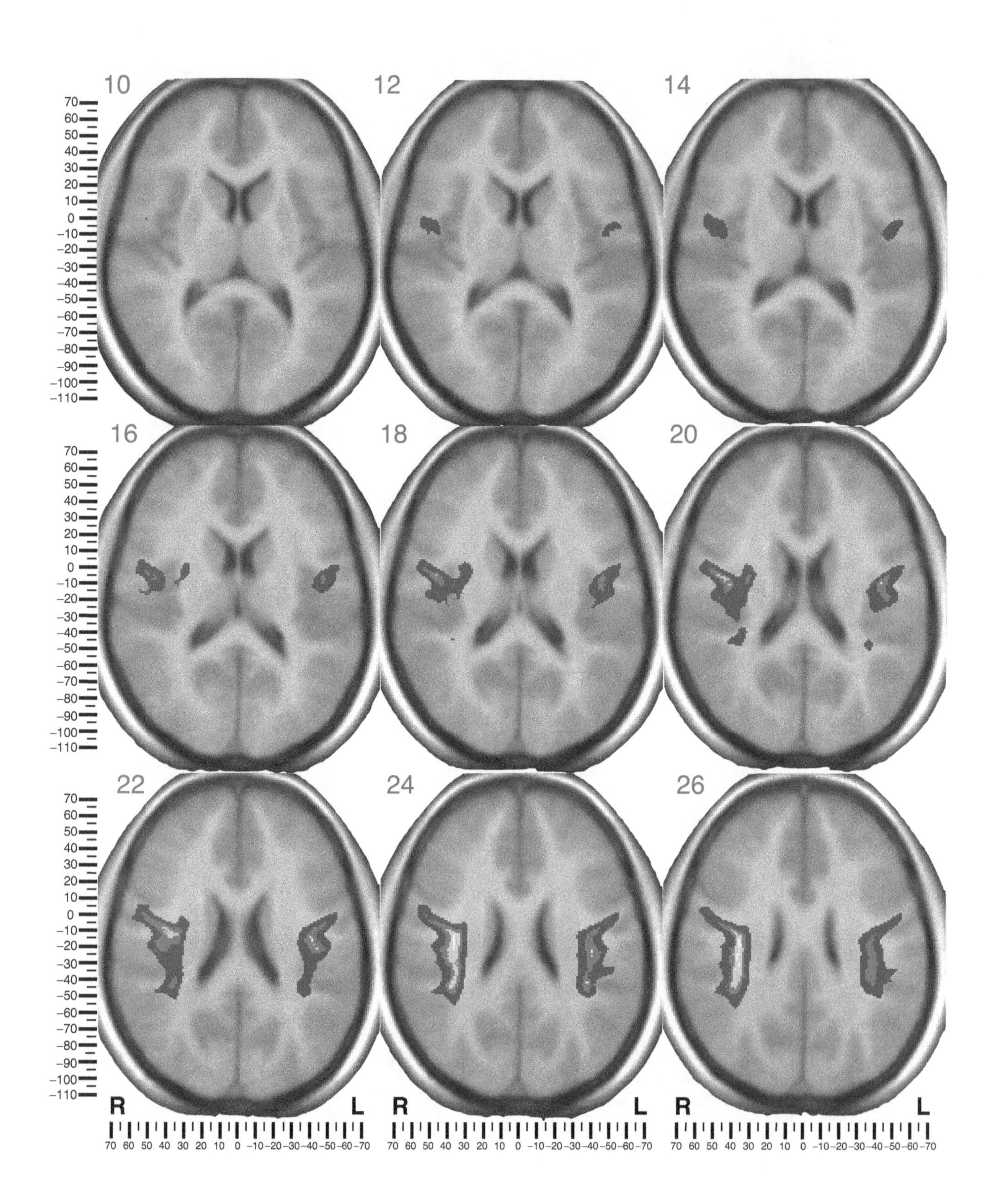

Anterior segment: inferior axial view

p < .05 | > 50% | > 75% | > 90%

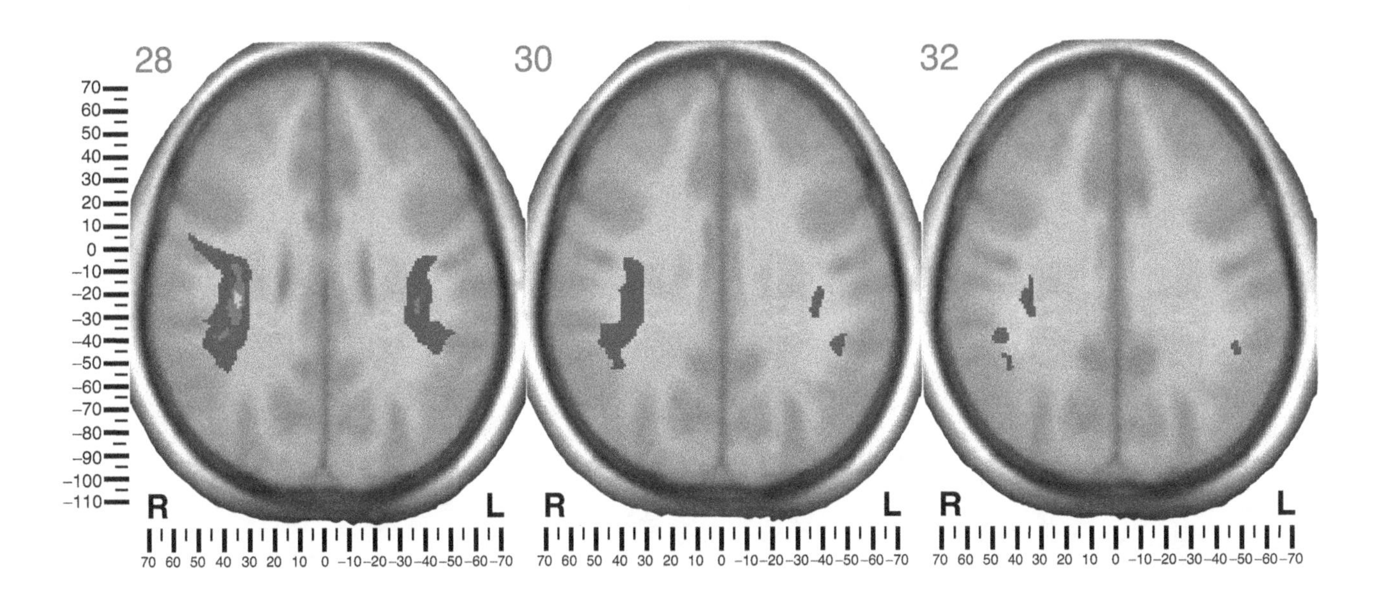

Anterior segment: anterior coronal view

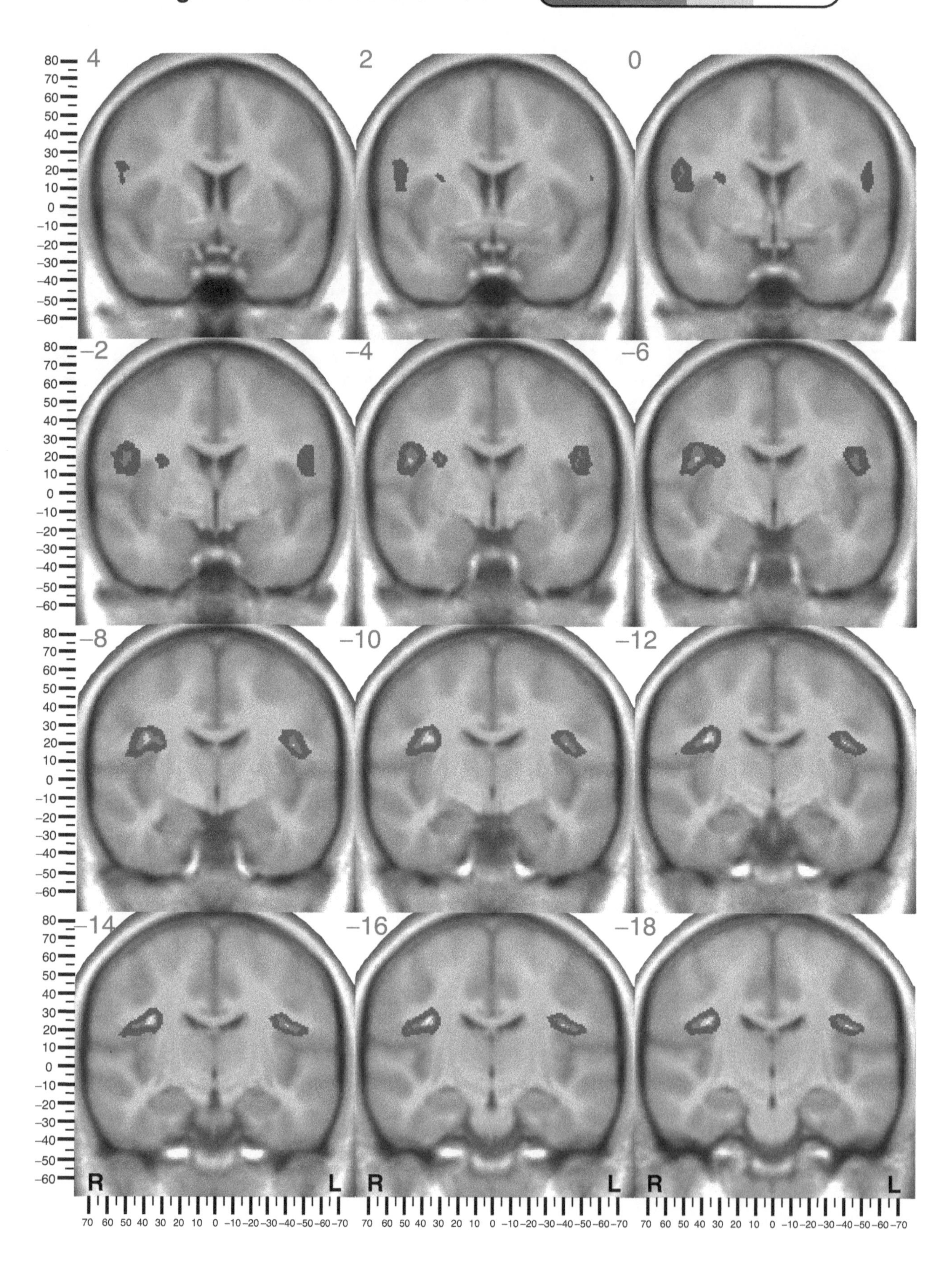

Anterior segment: anterior coronal view

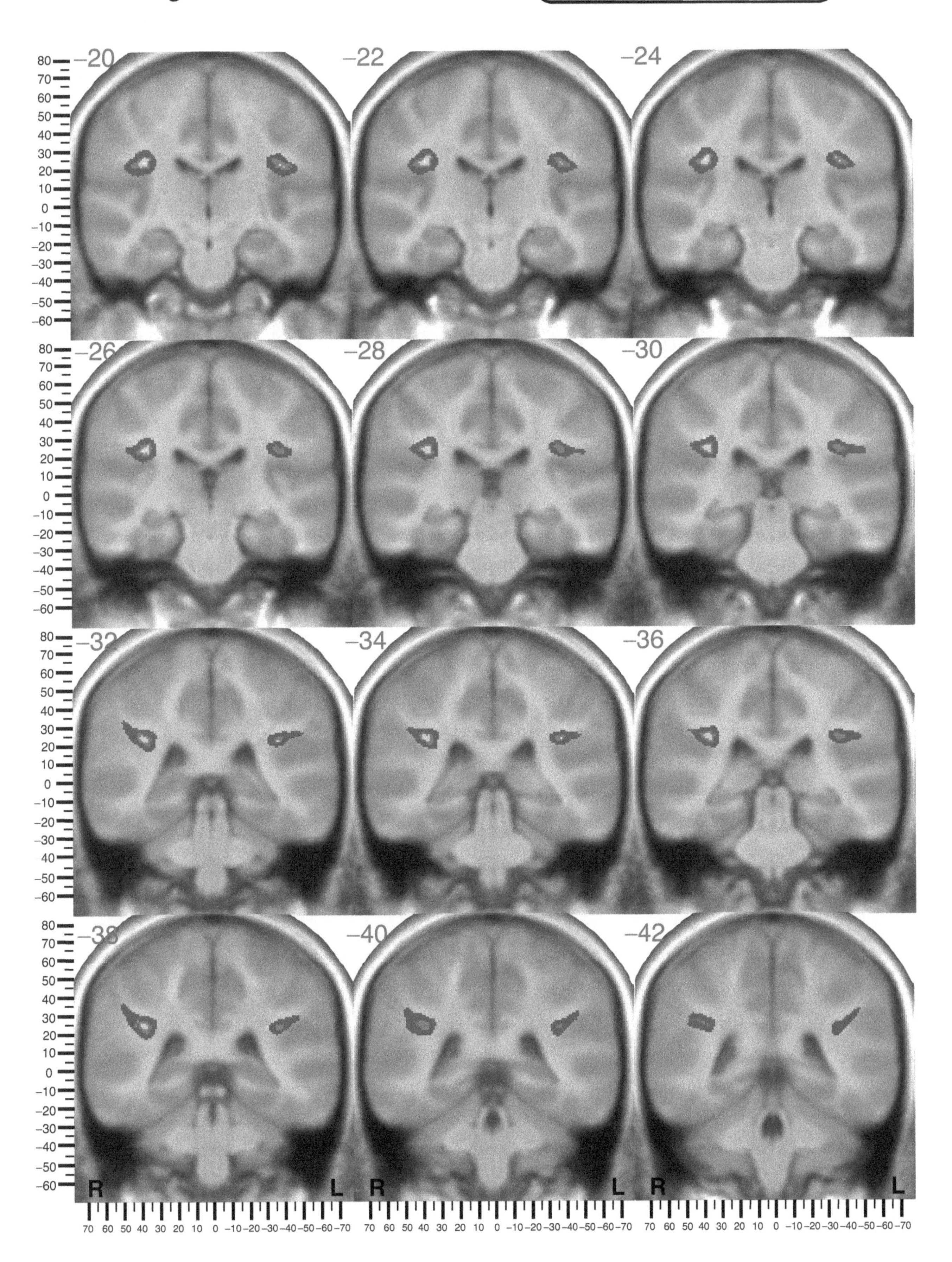

Anterior segment: anterior coronal view

p < .05 | > 50% | > 75% | > 90%

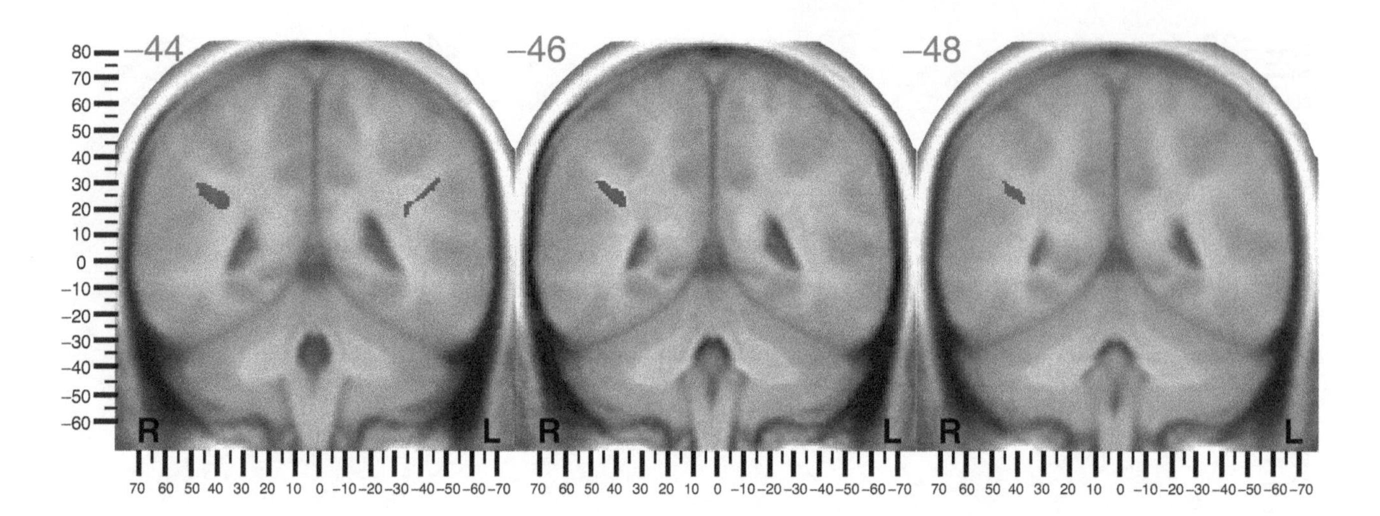

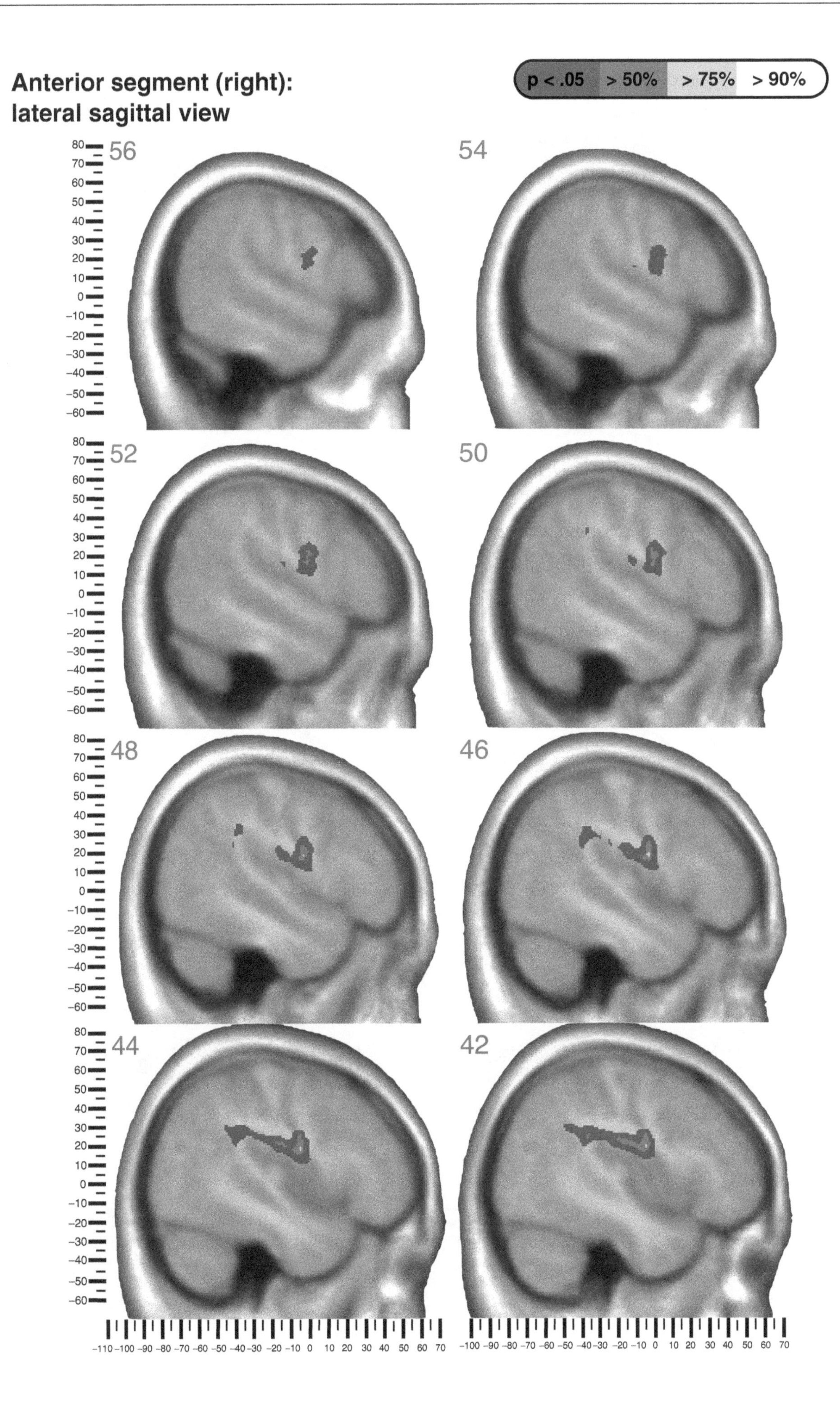
Anterior segment (right):
lateral sagittal view
p < .05
> 50%
> 75%
> 90%
56
54
52
50
48
46
44
42

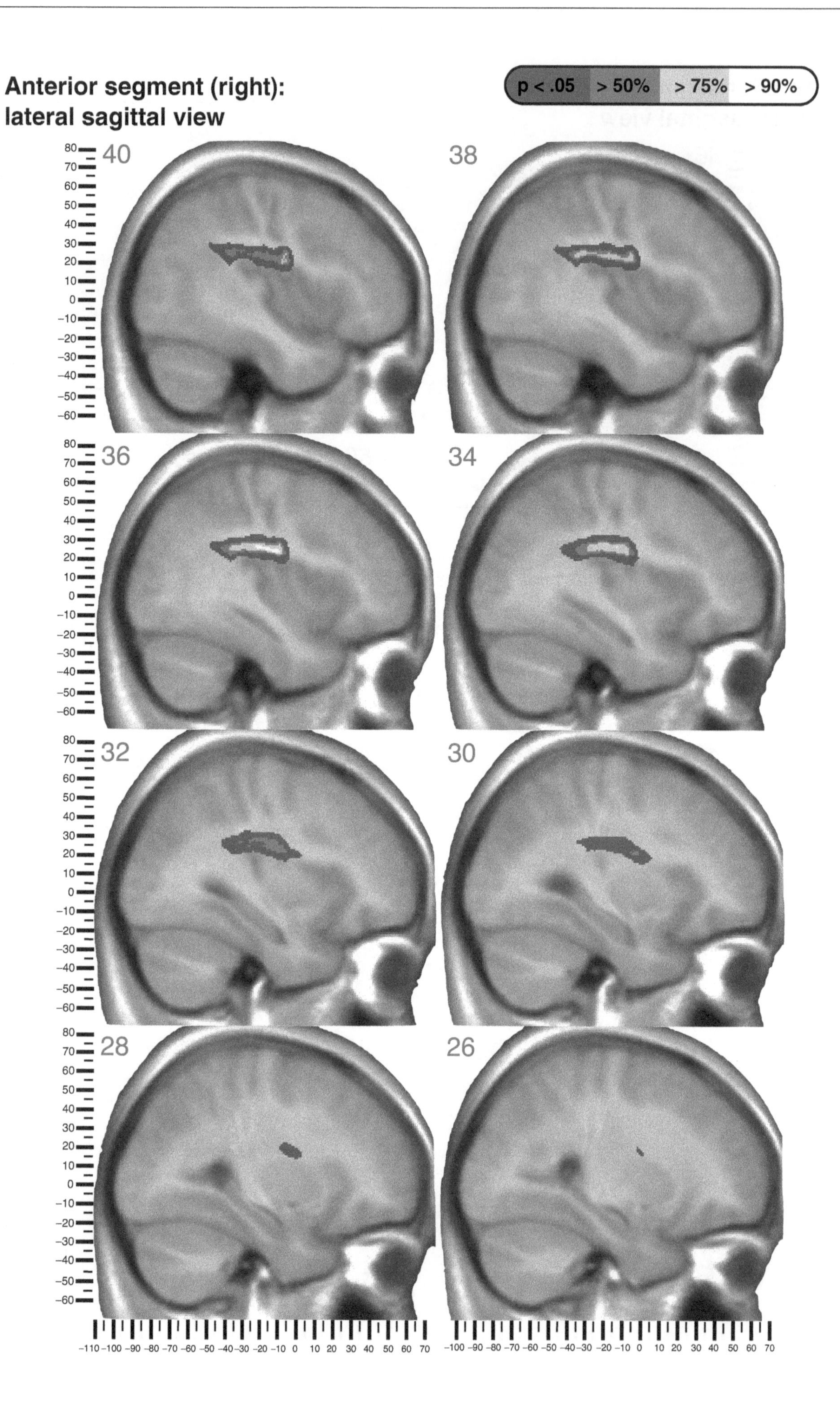

Anterior segment (right):
lateral sagittal view
p < .05
> 50%
> 75%
> 90%
40
38
36
34
32
30
28
26
80
70
60
50
40
30
20
10
0
-10
-20
-30
-40
-50
-60
-110 -100 -90 -80 -70 -60 -50 -40 -30 -20 -10 0 10 20 30 40 50 60 70
-100 -90 -80 -70 -60 -50 -40 -30 -20 -10 0 10 20 30 40 50 60 70

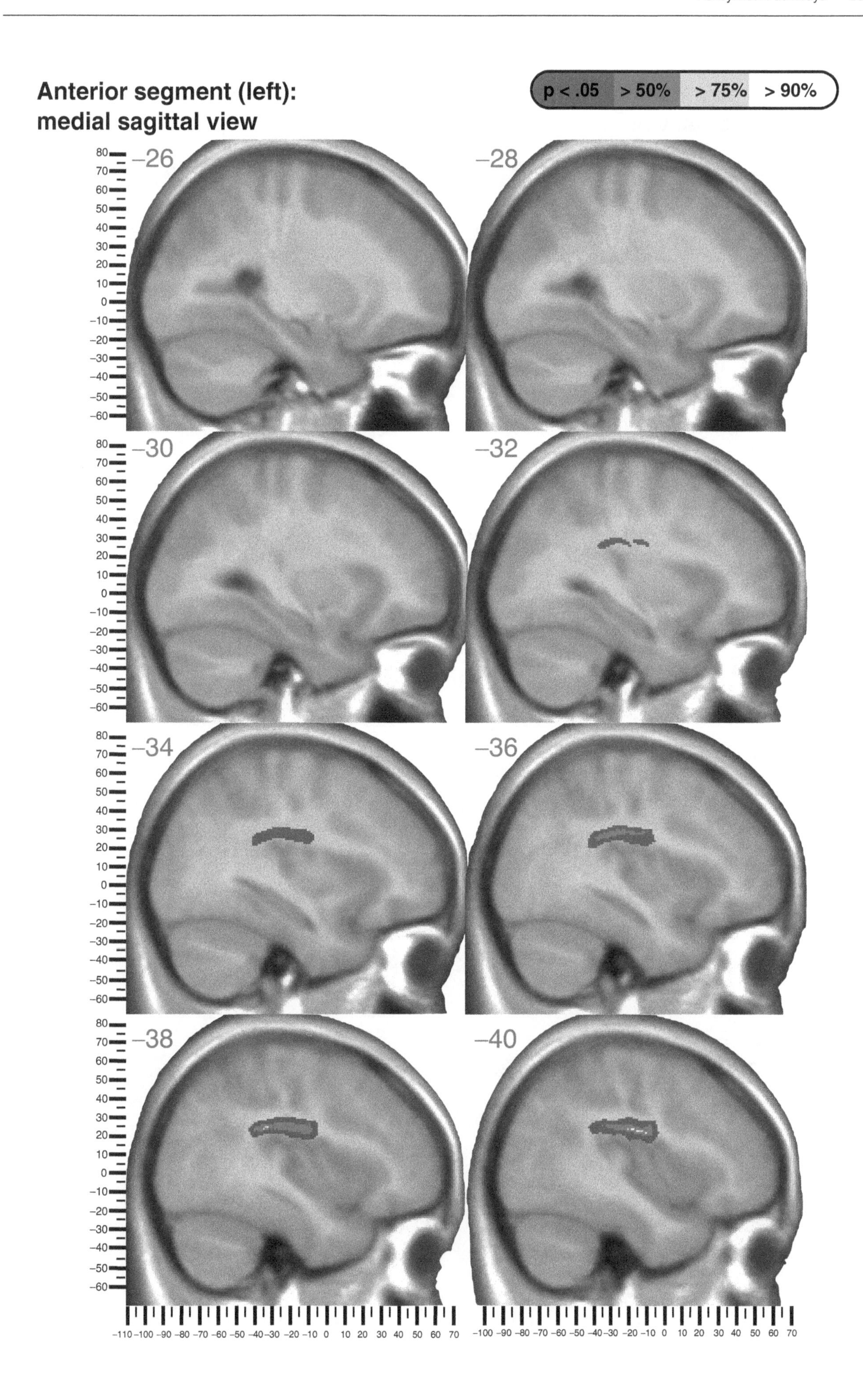

Anterior segment (left): medial sagittal view
p < .05
> 50%
> 75%
> 90%
-26
-28
-30
-32
-34
-36
-38
-40
80
70
60
50
40
30
20
10
0
-10
-20
-30
-40
-50
-60
-110 -100 -90 -80 -70 -60 -50 -40 -30 -20 -10 0 10 20 30 40 50 60 70
-100 -90 -80 -70 -60 -50 -40 -30 -20 -10 0 10 20 30 40 50 60 70

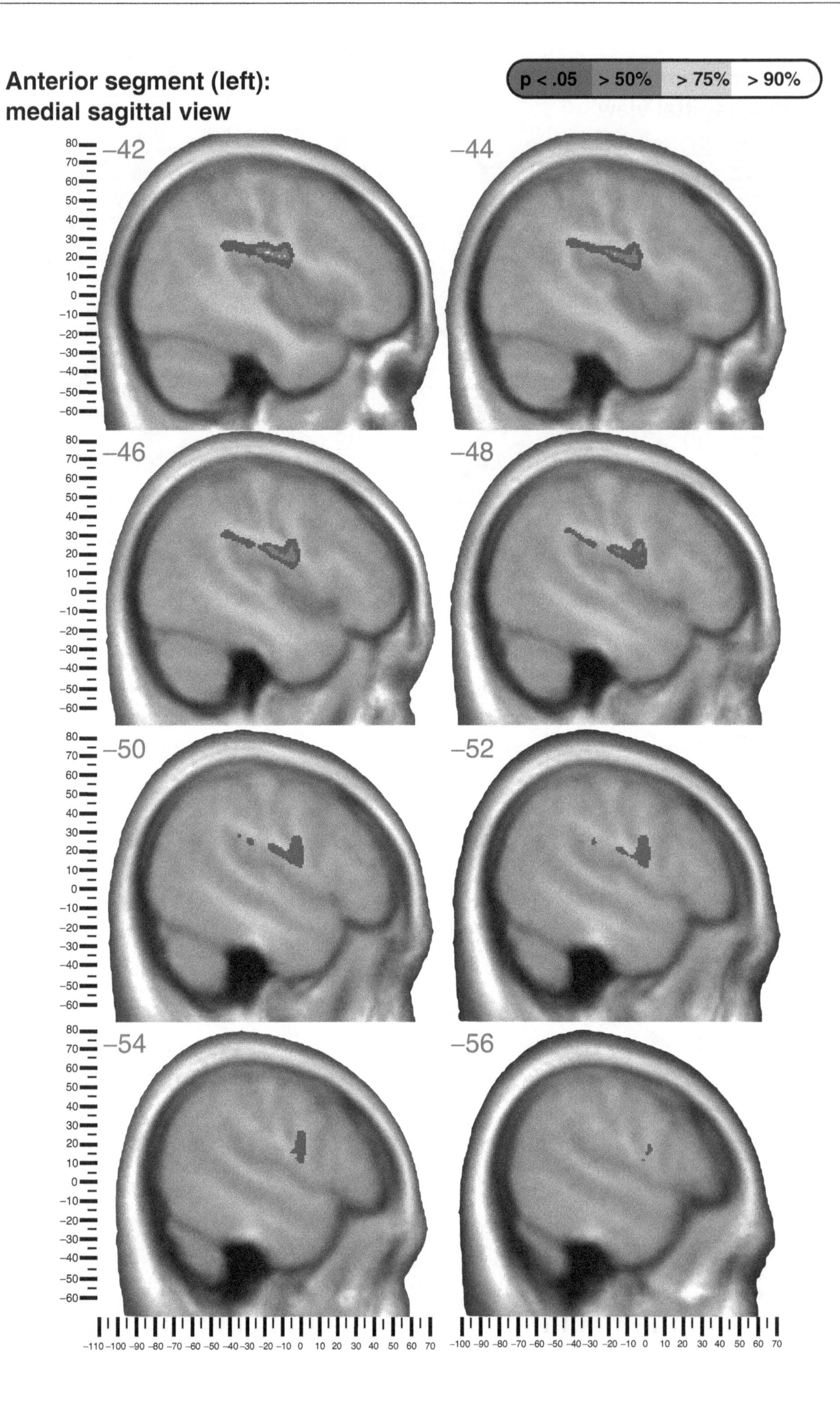
Anterior segment (left):
medial sagittal view
p < .05
> 50%
> 75%
> 90%
-42
-44
-46
-48
-50
-52
-54
-56
80
70
60
50
40
30
20
10
0
-10
-20
-30
-40
-50
-60
-110 -100 -90 -80 -70 -60 -50 -40 -30 -20 -10 0 10 20 30 40 50 60 70
-100 -90 -80 -70 -60 -50 -40 -30 -20 -10 0 10 20 30 40 50 60 70

Posterior segment (atlas)

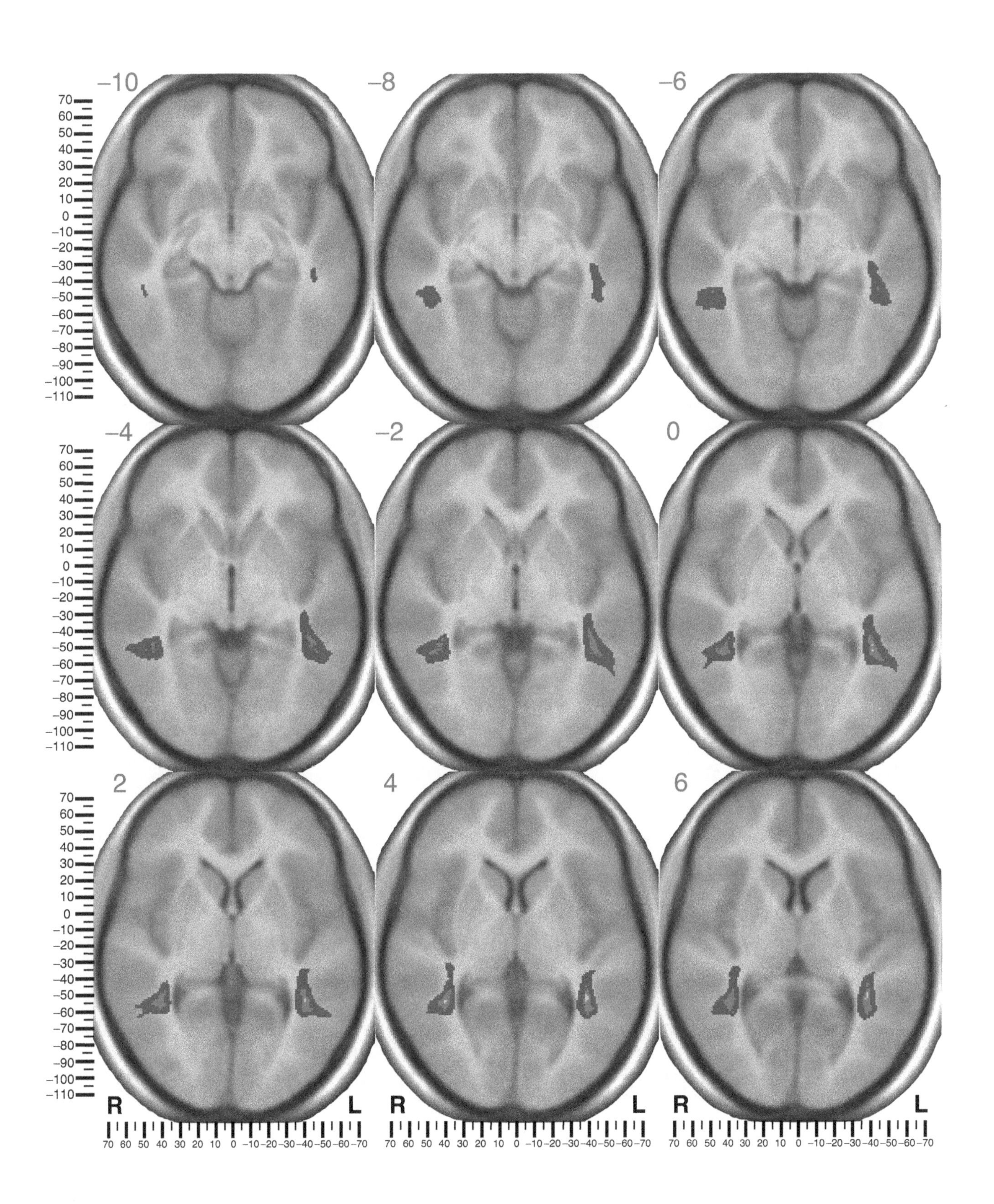

Posterior segment: inferior axial view

p < .05 | > 50% | > 75% | > 90%

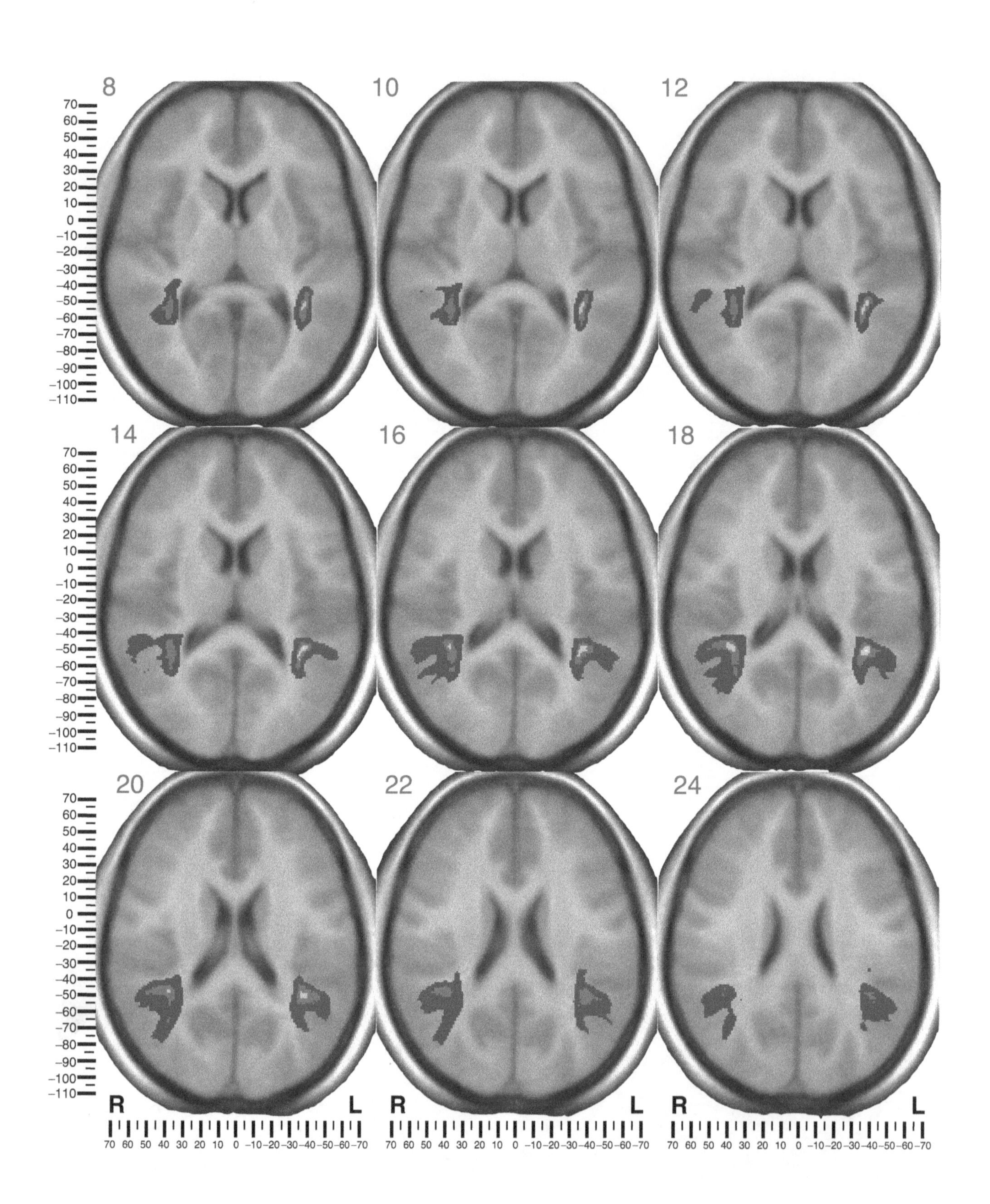

Posterior segment: inferior axial view

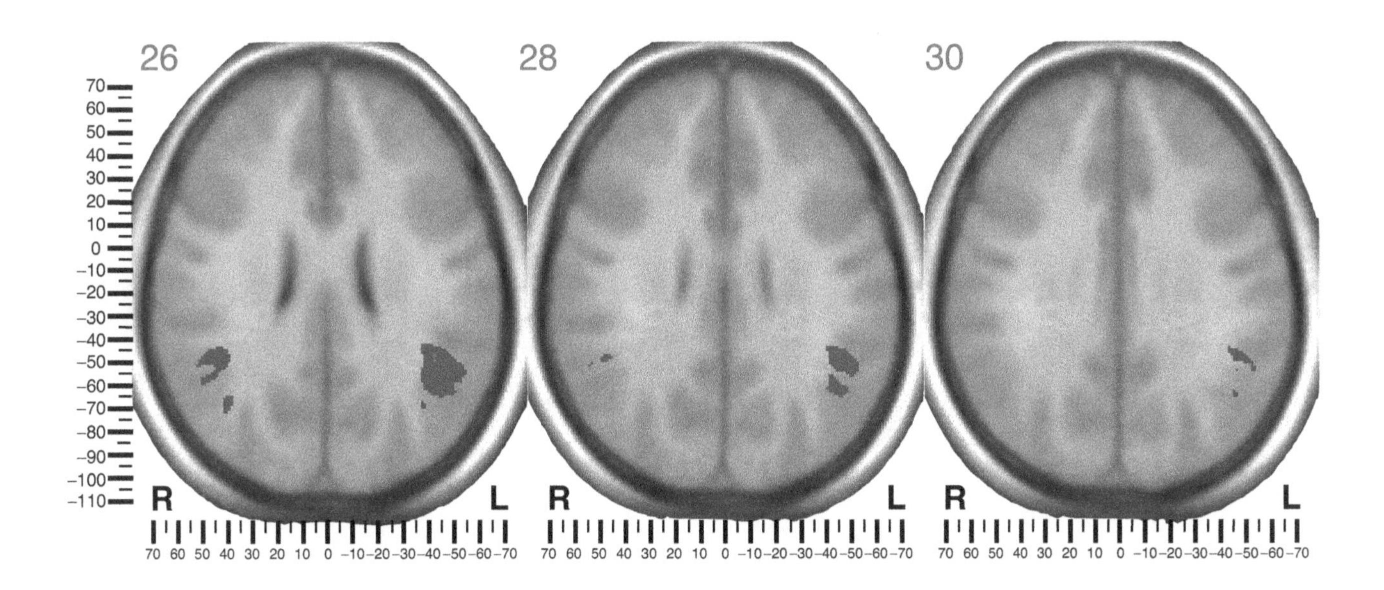

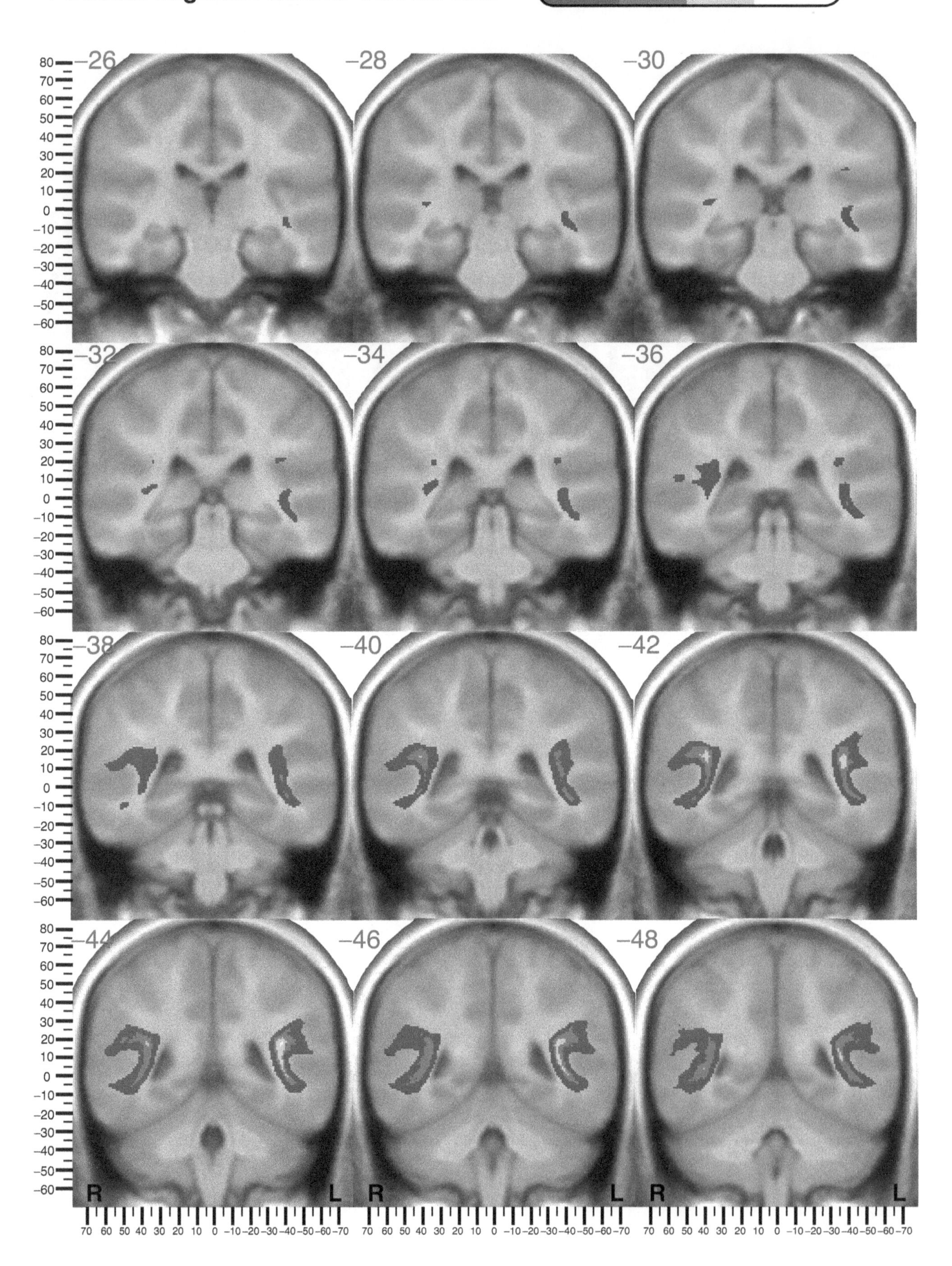

Posterior segment: anterior coronal view
p < .05
> 50%
> 75%
> 90%
-26
-28
-30
-32
-34
-36
-38
-40
-42
-44
-46
-48
R
L
80
70
60
50
40
30
20
10
0
-10
-20
-30
-40
-50
-60
70 60 50 40 30 20 10 0 -10 -20 -30 -40 -50 -60 -70

Posterior segment: anterior coronal view

p < .05 > 50% > 75% > 90%

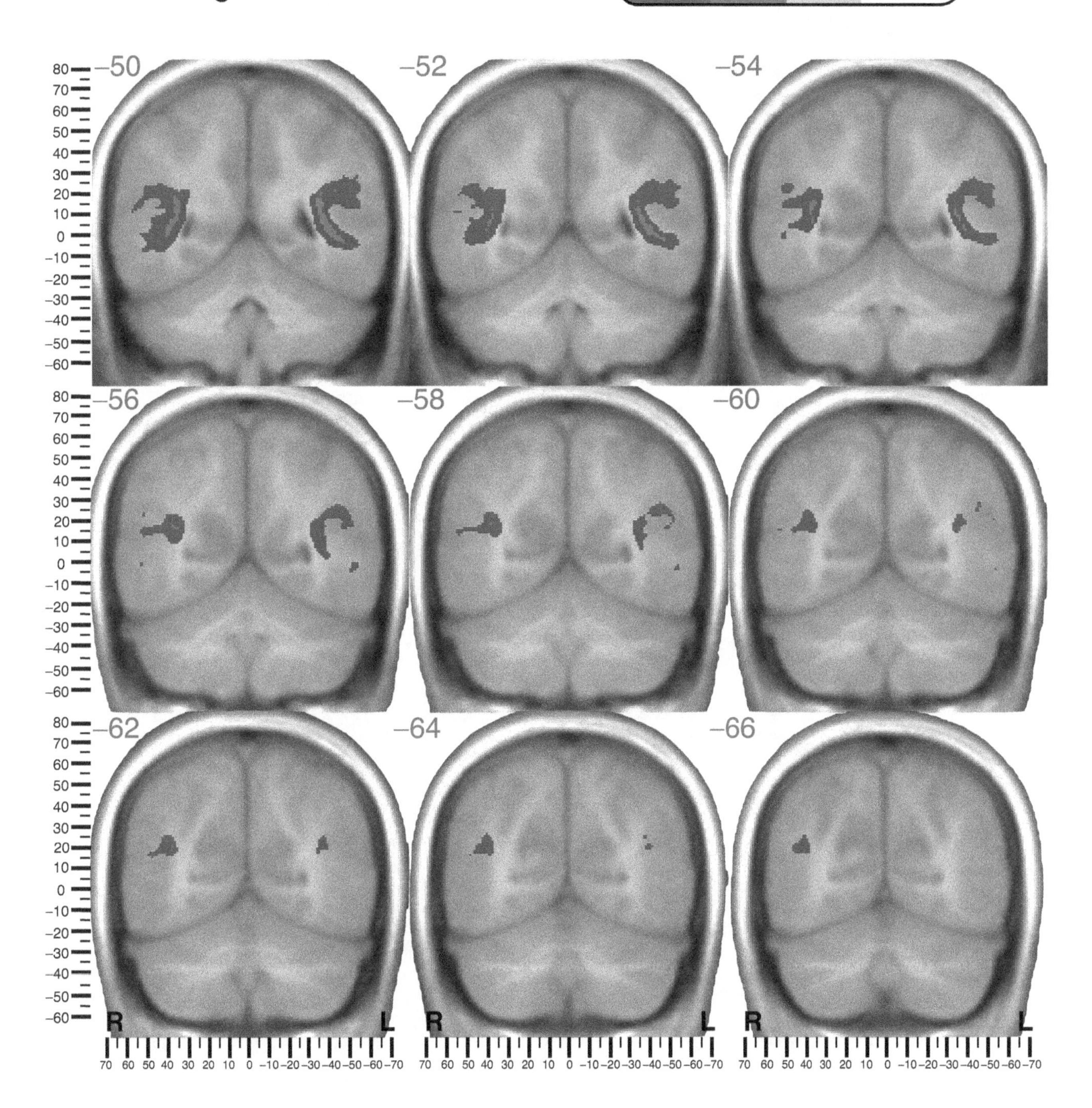

Posterior segment (right): lateral sagittal view

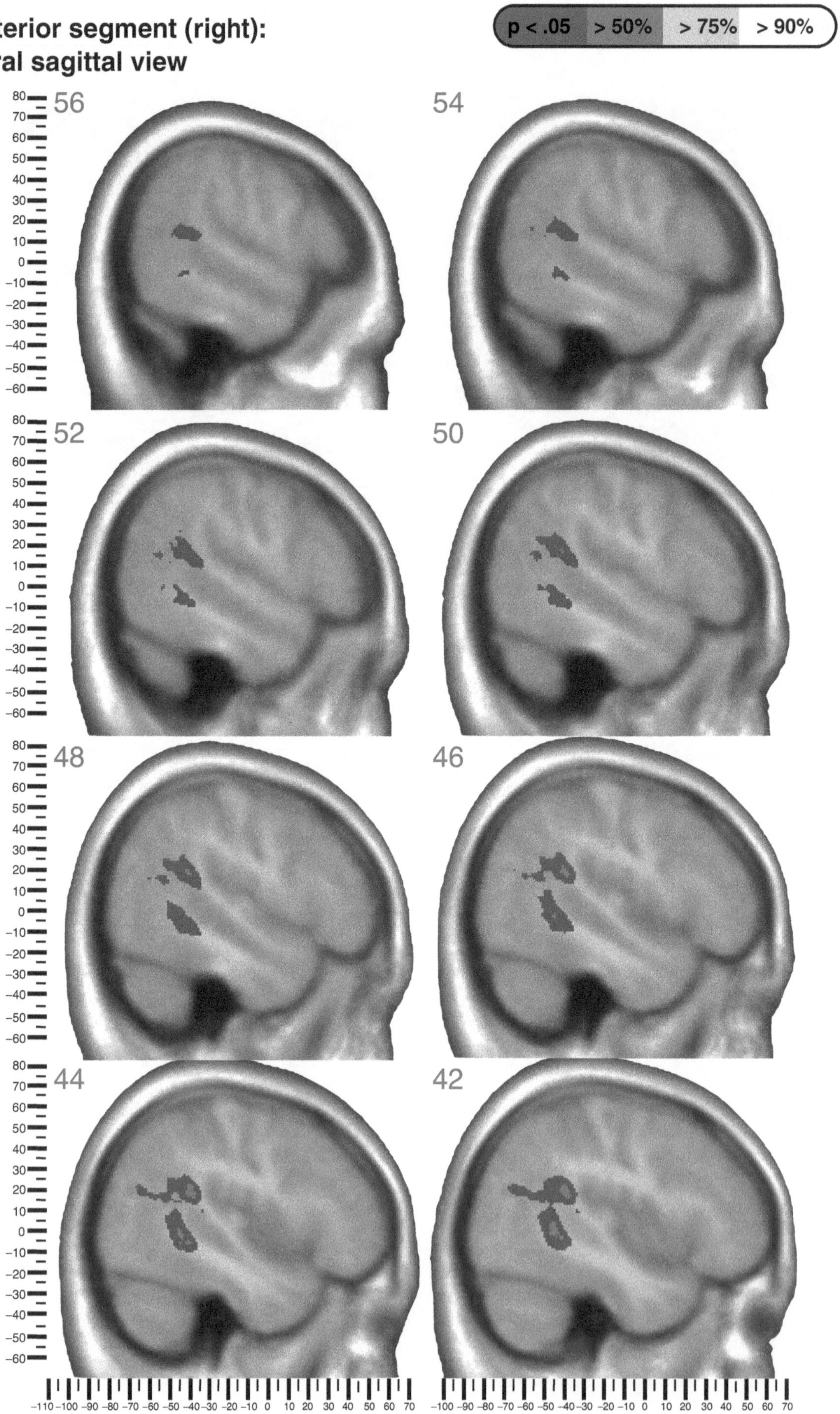

p < .05 > 50% > 75% > 90%

80 70 60 50 40 30 20 10 0 −10 −20 −30 −40 −50 −60

56 54

80 70 60 50 40 30 20 10 0 −10 −20 −30 −40 −50 −60

52 50

80 70 60 50 40 30 20 10 0 −10 −20 −30 −40 −50 −60

48 46

80 70 60 50 40 30 20 10 0 −10 −20 −30 −40 −50 −60

44 42

−110 −100 −90 −80 −70 −60 −50 −40 −30 −20 −10 0 10 20 30 40 50 60 70

−100 −90 −80 −70 −60 −50 −40 −30 −20 −10 0 10 20 30 40 50 60 70

Posterior segment (right): lateral sagittal view

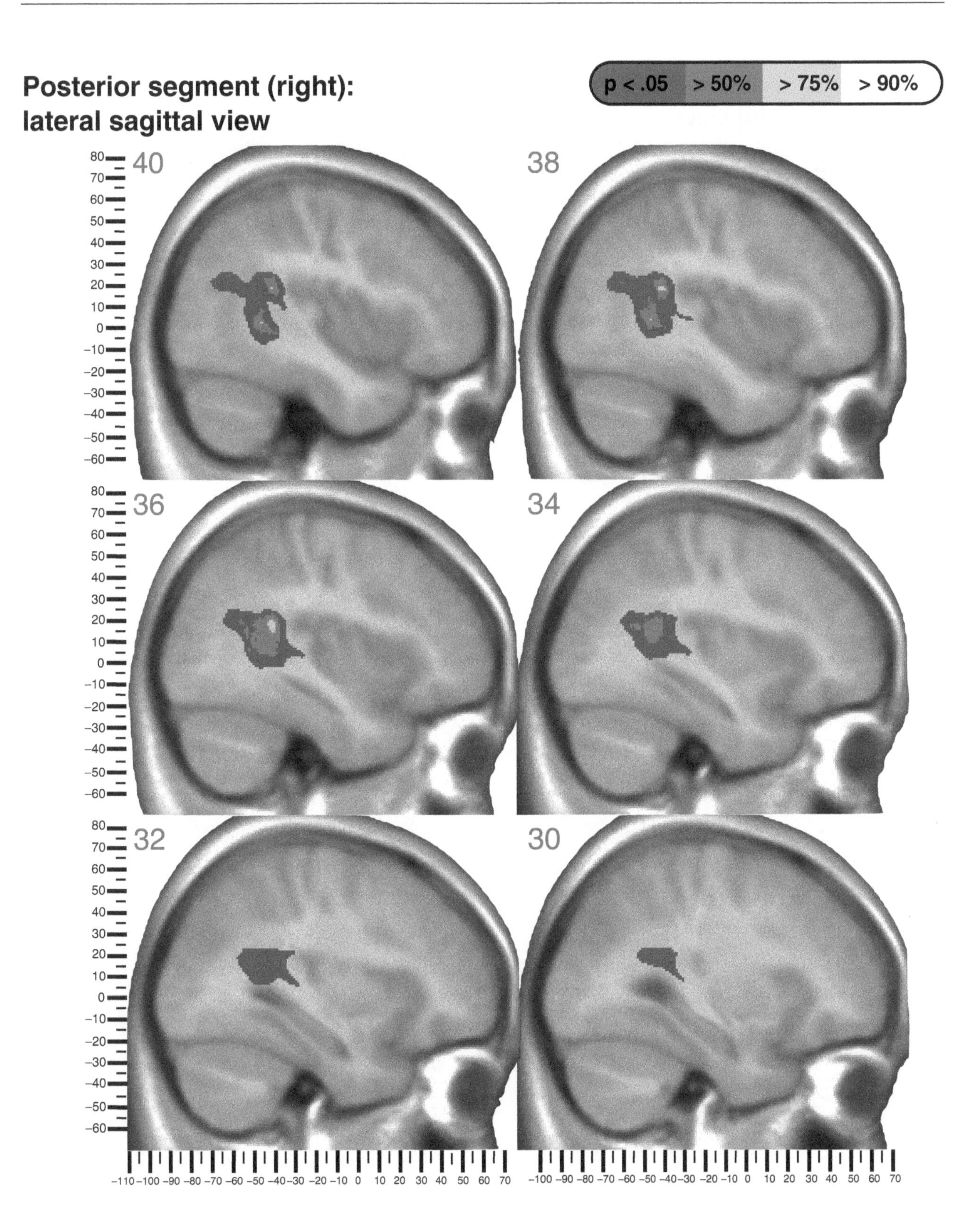

Posterior segment (left): medial sagittal view

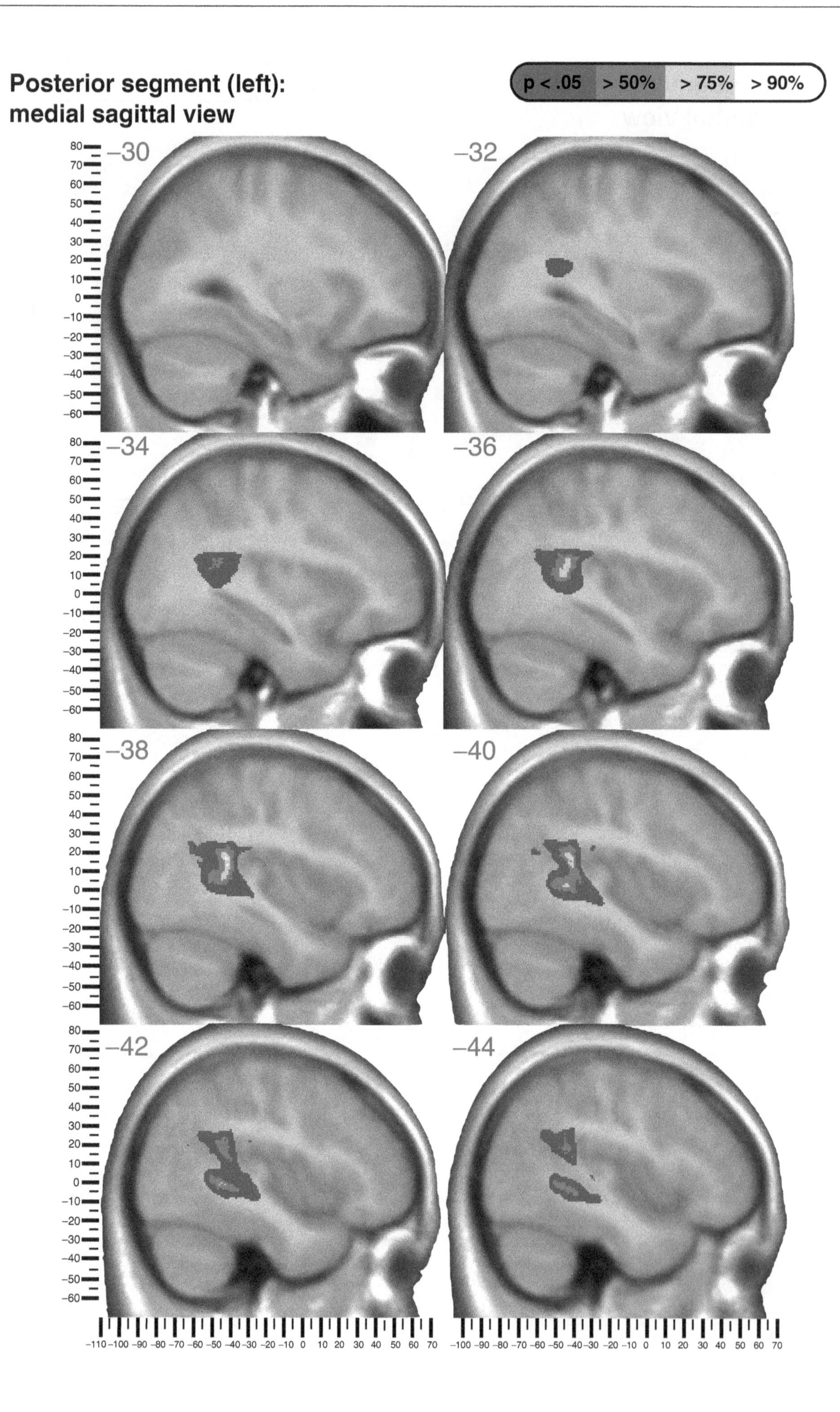

Posterior segment (left): medial sagittal view

p < .05 > 50% > 75% > 90%

–46 –48

–50 –52

–54 –56

CHAPTER 8

Occipital Visual Pathways

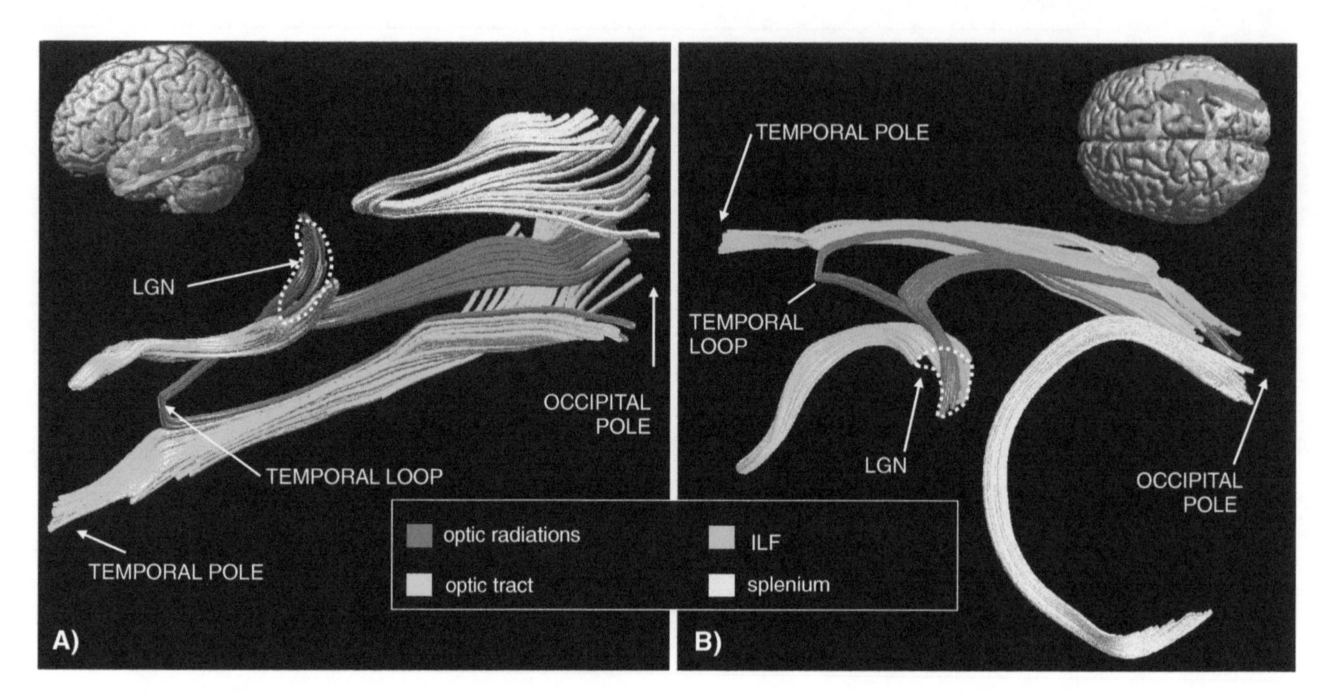

Figure 8.1 A) Medial sagittal and B) dorsal view of the right visual pathways reconstructed with tractography. The optic tract (only the retro-chiasmatic portion is shown in the figure) and optic radiations (main dorsal and ventral loop) convey visual inputs from the retina to the primary visual cortex. The inferior longitudinal fasciculus (ILF) connects occipital optic regions to the anterior temporal lobe. The splenium connects the occipital visual areas of both hemispheres. By following the distribution of terminating fibres it is also possible to define approximately the shape of the area corresponding to the lateral geniculate nucleus (LGN). This bean-shaped metathalamic nucleus is defined by the dotted white line (Catani et al., 2003).

Introduction

In the occipital lobe, a system of short and long tracts is dedicated to visual processing of faces, objects, letters, and space. Along this system it is possible to recognize three main classes of white matter connections: optic radiation, autochthonous fibres (short association or U-shaped fibres connecting adjacent occipital gyri), and long association fibres connecting occipital lobe with more anterior brain regions. A fourth group of commissural fibres connect the two occipital lobes (see Chapter 9).

The optic radiation

On anatomical dissection, a large swath of fibres connects the thalamus with the occipital lobe, namely the optic radiation of Gratiolet (1854). Work on embryological myelination, showed that the radiation, while appearing homogeneous, consists of two distinct pathways. A subset of fibres, myelinated at birth, connects the lateral geniculate nucleus (LGN) to the calcarine cortex, a pathway referred to as the 'narrowly defined' optic radiation. The remaining fibres of the radiation of Gratiolet are myelinated postnatally and connect the extra-calcarine cortex with the pulvinar of the thalamus (referred to as the pulvinar-occipital connections). Flechsig called attention to the peculiar course of the ventral fibres of the optic radiation, and named the sharp turn made by the ventral optic radiation the 'temporal knee' (Flechsig, 1896). This ventral fibres were subsequently studied by other authors (Archambault, 1906; Meyer, 1907) and its discovery attributed erroneously to Adolf Meyer. For this reason, it has been suggested that the term 'Meyer's loop' should be replaced with 'Flechsig–Meyer's loop' (for a historical review, see Polyak, 1957).

Figure 8.1 shows the tractography reconstruction of the optic radiation in its narrow sense, together with the splenial fibres of the corpus callosum and the inferior longitudinal fasciculus. As fibres from the optic nerve leave the chiasm and enter the optic tract they describe an 'S' shaped course around the midbrain and terminate in the antero-ventral portion of the lateral geniculate nucleus. The fibres of the optic tract enter the lateral geniculate nucleus antero-ventrally whilst the fibres of the optic radiation leave the lateral geniculate nucleus from its posterior dorso-lateral surface. The optic radiation then divides into two bundles: a smaller ventral temporal bundle (the Flechsig–Meyer's loop) and a larger dorsal bundle. The temporal bundle first projects forward and laterally towards the temporal pole, then after a short run, describes a sharp arc around the temporal horn of the lateral ventricle and continues backward and medially towards the occipital pole, where it terminates in the lower calcarine lip. The fibres of the dorsal bundle leave the lateral geniculate nucleus and assemble into a thick, compact lamina. After a short lateral course, its fibres bend posteriorly and medially towards the occipital pole and terminate in the upper calcarine lip. In addition to these major pathways, some visual inputs pass directly from the lateral geniculate nucleus or pulvinar to the motion-specialized cortex on the lateral occipital surface (Zeki and ffytche, 1998) (see Chapter 9).

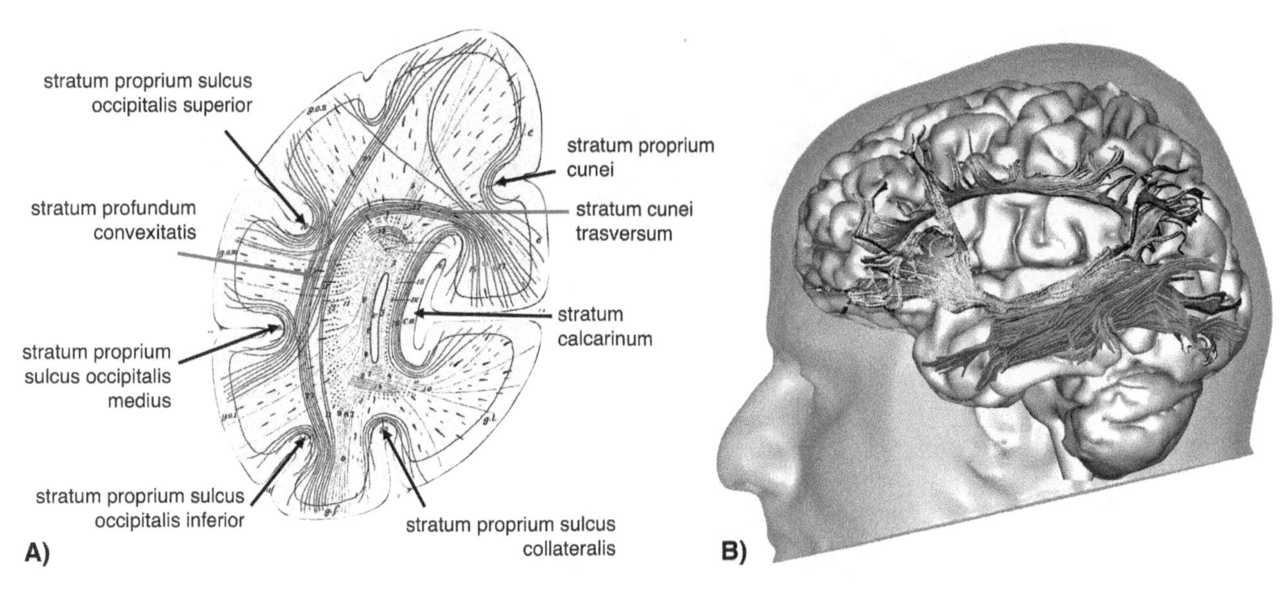

Figure 8.2 Short and long association tracts of the occipital lobe. A) Sachs' original diagram (1892) of the major autochthonous intralobar fibres of the occipital lobe. Black arrows indicate short U-shaped fibres connecting adjacent gyri. Red arrows indicate intralobar fibres connecting distant occipital gyri. B) Long association tracts of the occipital lobe reconstructed with diffusion tensor tractography. (Catani et al., 2002)

Autochthonous connections

Two classes of U-shaped fibres connecting regions within the same lobe can be identified in the white matter of the occipital lobe (Figure 8.2A) (Sachs, 1892; Déjérine, 1895). A group of vertical, horizontal, and longitudinal short fibres connects neighbouring occipital gyri. Another group of fibres runs deeper and connects relatively distant gyri within the occipital lobe. The specific function of these groups of intralobar occipital fibres is largely unknown. In general they constitute a diverging system of connections allowing the hierarchical propagation of visual inputs from primary visual areas to higher order association areas (Van Essen and Maunsell, 1983; Felleman and Van Essen, 1991; Mesulam, 2000). They also provide communication between specialized occipital areas (e.g. colours, motion, faces, etc.) and mediate the convergence of visual inputs to higher order areas (Zeki and Shipp, 1988; ffytche and Catani, 2005). Some of these fibres form a series of longitudinal projection systems that can be traced from occipital unimodal visual association areas to heteromodal association areas of the lateral temporal lobe (Figure 8.3) (Tusa and Ungerleider, 1985; Catani et al., 2003).

Long association fibres

The major long association tracts of the occipital lobe are the inferior longitudinal fasciculus, the cingulum, and the inferior fronto-occipital fasciculus (Figure 8.2B) (ffytche and Catani, 2005).

The inferior longitudinal fasciculus was originally described by Reil (1812) who erroneously considered it as a cortico-subcortical projection tract. It was later identified correctly by Burdach (1822) as an association tract connecting occipital with temporal regions (Figures 8.2B and 8.3). The inferior longitudinal fasciculus does not constitute a single pathway, but contains fibres of different length. The occipital branches of the inferior longitudinal fasciculus connect with a number of occipital regions, including the extrastriate areas on the dorso-lateral occipital cortex (e.g. descending occipital gyrus), the ventral surface of the posterior lingual and fusiform gyri, and the medial regions of the cuneus (Catani et al., 2003). These branches run anteriorly parallel and lateral to the fibres of the splenium and optic radiation and, at the level of the posterior horn of the lateral ventricle, gather into a single bundle. In the temporal lobe, the inferior longitudinal fasciculus continues anteriorly and projects to the middle and inferior temporal gyri, temporal pole, parahippocampal gyrus, hippocampus, and amygdala. An observation originally emphasized by Campbell (1905), and consistent with axonal tracing (Schmahmann and Pandya, 2006) and tractography findings (Catani et al., 2003), is that long associative fibres, such as those of the inferior longitudinal fasciculus, arise from the extrastriate cortex but not the calcarine striate cortex.

A second set of long association fibres forms part of the posterior cingulum and connects the most medial regions of the dorsal cuneus with the precuneus and medial temporal lobe (Figure 8.2B). The anatomy of the cingulum is described in greater detail in Chapter 11.

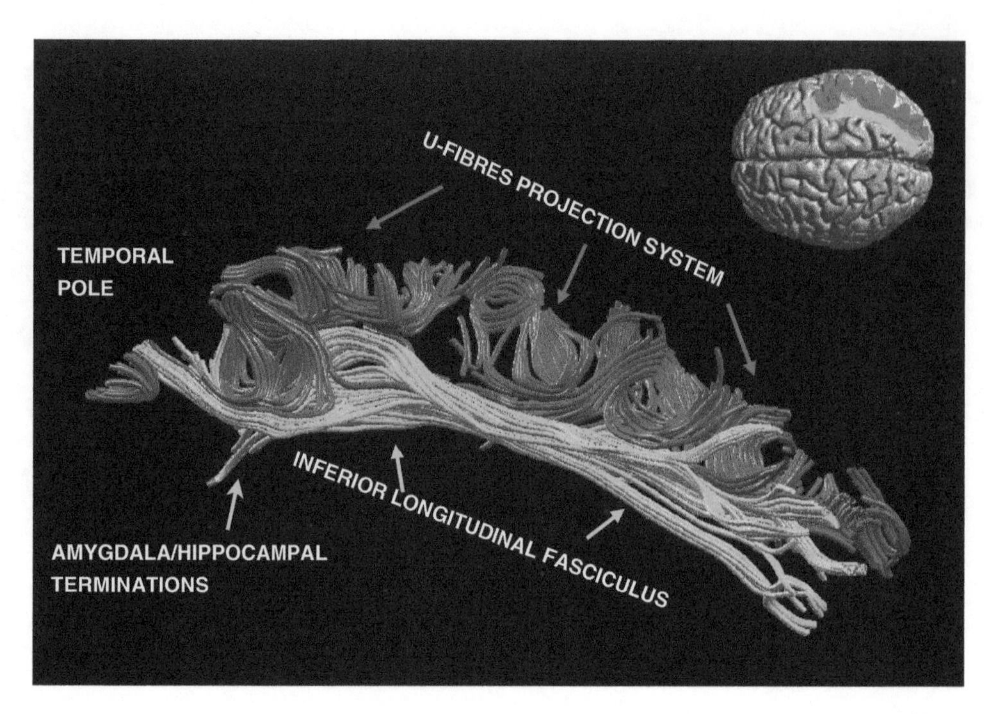

Figure 8.3 Diffusion tensor tractography reconstruction of the right human direct and indirect occipito-temporal connection system viewed from above. Occipital and temporal autochthonous U-shaped fibres (red) form the lateral indirect projection system. The long fibres of the inferior longitudinal fasciculus (green) form the medial direct pathway connecting extrastriate occipital cortex with anterior temporal cortex, amygdala, and hippocampus (Catani et al., 2003).

Finally, a third group of fibres running medial to the inferior longitudinal fasciculus and lateral to the optic radiation forms the inferior fronto-occipital fasciculus (Figure 8.2B). The existence of a direct connection between occipital and frontal lobes was originally suggested by Burdach (1822) and was confirmed by several authors, including Trolard (1906) and later Curran (1909) who proposed calling it the 'inferior' fronto-occipital fasciculus to distinguish it from the more dorsal fibres of the superior fronto-occipital fasciculus (see later) (Figure 8.4). The fronto-occipital fasciculus originates from the inferior and medial surface of the occipital lobe, with a minor contribution probably from the medial parietal lobe (Catani et al., 2002; Martino et al., 2010). As it leaves the temporal lobe, the inferior fronto-occipital fasciculus narrows in section and its fibres gather together at the level of the external capsule just above the uncinate fasciculus.

In the frontal lobe its fibres spread to form a thin sheet curving dorsolaterally to terminate mainly in the inferior frontal gyrus. The most ventral fibres continue anteriorly and terminate in the medial fronto-orbital region and frontal pole (Catani et al., 2002).

While the anatomy of the inferior fronto-occipital fasciculus in humans is well documented, the existence of a superior fronto-occipital fasciculus has been questioned. The superior fronto-occipital fasciculus was described for the first time in patients with agenesis of the corpus callosum (Onufrowicz, 1887) where it was thought to represent fibres initially programmed to form the corpus callosum but unable to transverse the hemispheric midline (see also Chapter 9) (Paul et al., 2007). A few years later, Muratoff (1893) used the technique of ablation degeneration in dogs and found a bundle of poorly myelinated fibres running above the head of the caudate nucleus and beneath the corpus callosum, which he termed the 'subcallosal fasciculus'. Following the description of the superior fronto-occipital fasciculus in acallosal brains and the subcallosal fasciculus in dogs, neuroanatomists began to wonder whether a similar tract exists in the normal human brain. The question has remained unanswered, although there are two competing views in the literature. Dejerine and the French school (including Talairach and Tournoux, 1988) supported the theory that the subcallosal fasciculus exists in the normal brain and represents an association pathway between occipital and frontal lobes. This view is opposed to the German school, which considered the subcallosal tract as a system of projection fibres connecting to the caudate nucleus (i.e. the 'corona radiata' of the caudate nucleus) (Meynert, 1885). More recently Türe et al. (1997) proposed that the subcallosal fasciculus includes ascending thalamic projections directed anteriorly towards the frontal lobe.

The fibres of the 'subcallosal fasciculus' can be easily visualized in coronal and axial diffusion images and dissected *in vivo* using diffusion tractography. Figure 8.4 shows the tractographic reconstruction of the tracts passing through the 'subcallosal region'. Three sets of fibres can be identified. The first connects the lateral prefrontal cortex to the superior parietal gyrus (associative fronto-parietal fibres). A second group converges towards the anterior horn of the lateral ventricles and connects the frontal cortex with the caudate nucleus (fronto-striatal fibres). Finally a third group originates from the medial thalamus and projects to the frontal lobe (thalamo-frontal fibres).

These findings do not support the existence of 'subcallosal' association fibres connecting occipital and frontal lobe regions as originally proposed by Dejerine (1895). For this reason we agree with Schmahmann and Pandya (2006) who proposed the eponym 'Muratoff bundle' for the subcallosal projection fibres originating from the caudate nucleus. These authors have, however, used the

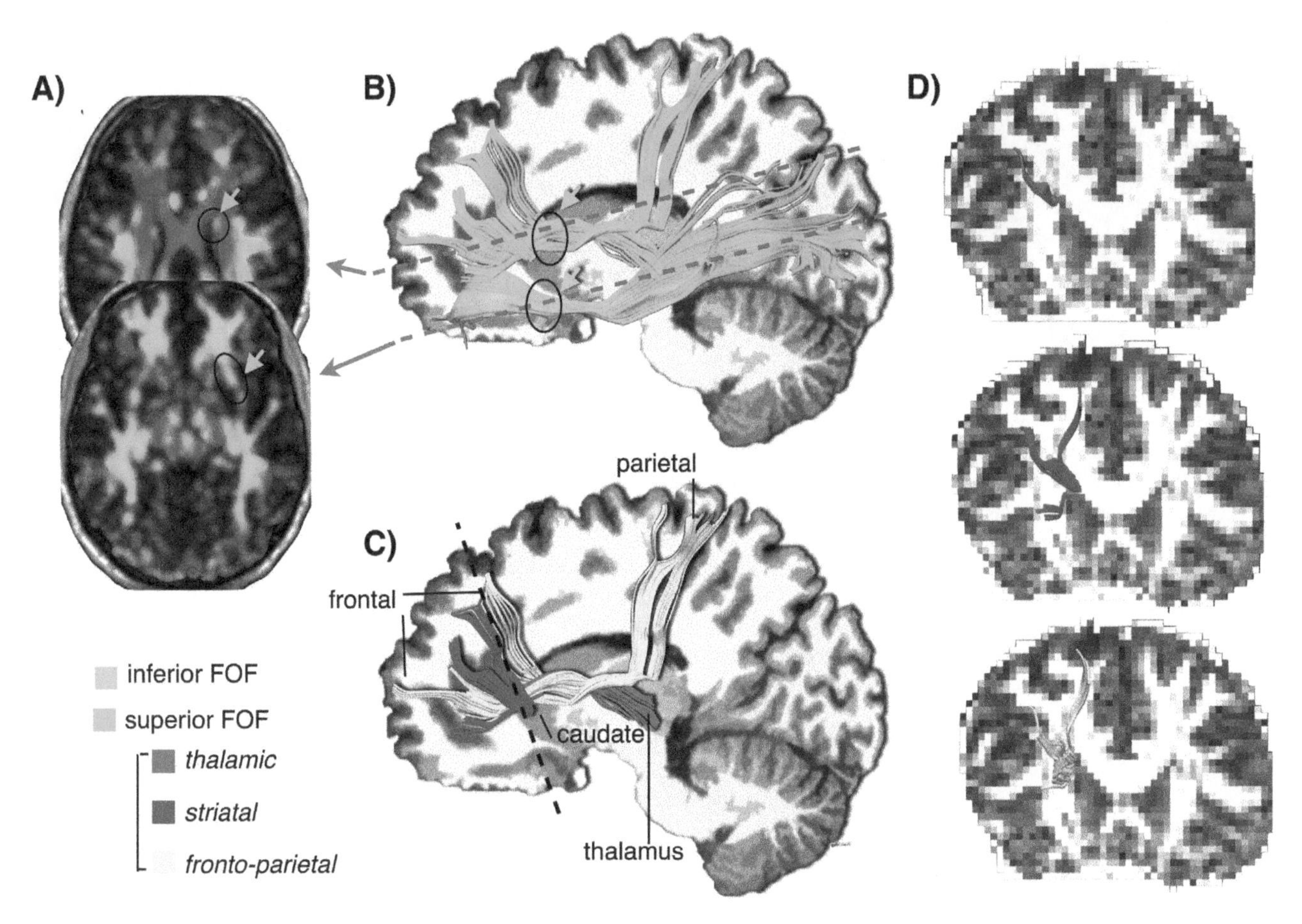

Figure 8.4 Tractography reconstruction of the superior and inferior fronto-occipital fibre system. A–B) The inferior fronto-occipital fasciculus (inferior FOF, cyan) originates in the occipital lobe and projects to the ventral regions of the frontal lobe (figure B). Its central fibres pass through the anterior floor of the external capsule (circle indicated with cyan arrow in the left colour-coded maps in figure A). B–D) The streamlines of the superior fronto-occipital fasciculus (superior FOF, green) originate from the anterior and dorsal frontal regions and group together in the proximity of the anterior horn of the lateral ventricle. Here they assume a longitudinal course above the caudate nucleus, medially to the corona radiata, and beneath the corpus callosum (figure B). In humans the superior FOF has three posterior terminations (figure C–D): the thalamic fibres (red) originate from the thalamus, the fronto-striatal fibres (blue) terminate to the head of the caudate nucleus, the association fibres (yellow) terminate in the parietal cortex. While the existence of the first two subgroups of fibres has been demonstrated in man using post-mortem dissections, the anatomy of the fronto-parietal association fibres remains to be confirmed.

term 'superior fronto-occipital fasciculus' to indicate a group of fibres running close to the Muratoff bundle and connecting frontal and occipital lobes in the monkey brain. The evidence for the existence of a similar bundle in the human brain is lacking, although a human equivalent of the superior fronto-occipital fasciculus could be the fronto-parietal tract passing through the subcallosal region. This human–simian difference (fronto-parietal vs fronto-occipital) could be related to in the relatively minor extension of the occipital visual cortex in humans. This interpretation awaits experimental confirmation.

Functional and comparative anatomy of cortical visual areas

The idea of a point-to-point correspondence between the retinal image and its visual representation in the brain is called retinotopy. The origin of the retinotopic concept can be traced back to the prescient Arab scholar Al-Hasan ibn al-Haythamand and found iconographically reproduced in the works of René Descartes and Isaac Newton (Gross, 1998). The idea, however received experimental support only in the second half of the 19th century with cortical ablation studies conducted in animals by Panizza, Henschen, Munk, and Minkowski (Finger, 1994; Clarke and O'Malley, 1996). Henschen was the first to understand that the representation of the lower visual field is in the superior border of the calcarine fissure and that of the upper field is in the inferior border. He also correctly aligned the horizontal meridian of the visual field with the base of the calcarine fissure but incorrectly placed the peripheral representation at the occipital pole and the foveal vision (centre of the visual field) in the anterior part of the calcarine fissure (Henschen, 1893).

The resurgence of military conflicts at the dawn of the 20th century provided unique case series of wounded soldiers with localized occipital lesions and partial visual deficits. This permitted Tatsuji Inouye (1909) in Japan and Gordon Holmes (1918) in England to produce more accurate retinotopic maps of the calcarine cortex and correctly localize the foveal representation in the occipital pole

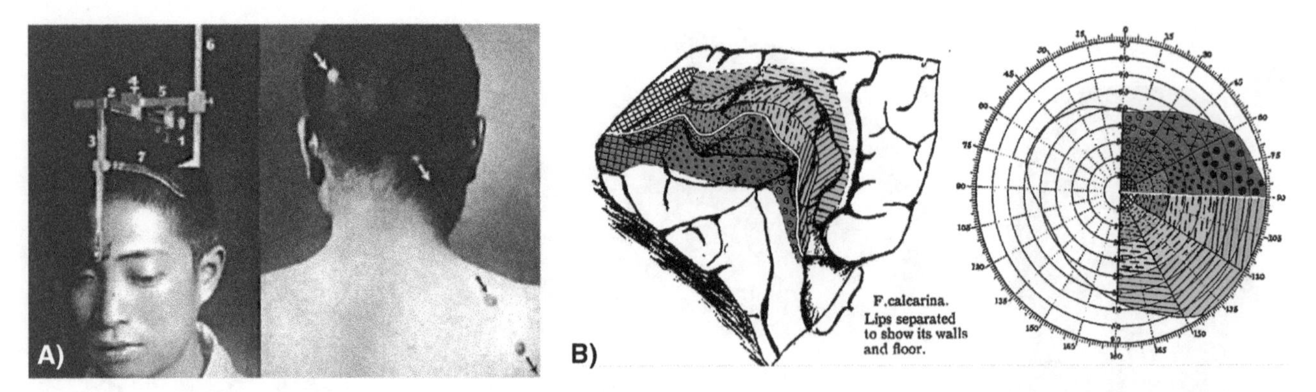

Figure 8.5 A) The cranio-coordinometer (left) used by Inouye (1909) to plot the trajectory of bullets through the head and the calcarine cortex. Inouye was a Japanese ophthalmologist whose job was to objectively quantify for pension purposes the visual impairment of soldiers wounded during the Russo-Japanese war. In the 28 cases examined he made careful plots of visual field loss by calculating a straight trajectory from the sites of entrance. Some of Inouye's casualties were in a prone position when they were hit and the bullet entered the skull at one point, exited from a second, and then re-entered the body. These cases were particularly useful for demonstrating the precise trajectory of the bullet and the exact location of the lesion in the calcarine cortex. Inouye produced the first detailed retinotopic map of the calcarine cortex and recognized that the occipital lobe contains neighbouring visual areas with less precise topographic maps. He published his results in German but his work remained largely unknown. B) A coloured version of the original diagram from Gordon Holmes (1918) showing the different areas of the right visual hemifield projected onto the calcarine cortex. Note that the horizontal meridian of the visual hemifield (indicated with a white line) corresponds to the base of the calcarine sulcus, while the vertical meridian is represented along the external borders of the calcarine cortex. Also the foveal region (centre of the hemifield in the right figure) projects onto a wider region (i.e. tip of the occipital pole in the left figure) compared to peripheral retina.

(Glickstein and Whitteridge, 1987; Fishman, 1997). Another property of the calcarine cortex that emerged from their studies is that more cortex is devoted proportionately to the representation of the foveal retina than to the periphery (Figure 8.5B) (Inouye, 1909; Holmes, 1918).

The field advanced greatly with the use of more sophisticated electrophysiological methods in the animal brain that became available in the second half of the 20th century, especially single-neuron recording. Hubel and Wiesel, in a series of experiments, demonstrated the existence of neurons responding selectively to orientation, movement, and speed of movement of the visual stimulus and to stimuli presented to either the left or the right eye (ocular dominance) (Hubel and Wiesel, 1959, 1962, 1965). They also showed that these neurons, responding selectively to particular orientations or to ocular dominance, are arranged in vertical columns and parallel stripes (approximately 0.5 mm wide in humans). The study of the visual system extended beyond the calcarine cortex, and other occipital retinotopic maps were described in the animal and labelled with the progressive numbering V1, V2, V3, etc (Talbot and Marshall, 1940; Hubel and Wiesel, 1965). Other groups recorded neuronal activity in response to visual stimuli in areas in the temporal and parietal lobe. Most of these areas were found to have different features compared with those described by Hubel and Wiesel. For example, neurons within area IT respond to shapes, colours, faces, and to the effects of attention but not to specific orientation of the stimulus. Other areas in the temporal lobe respond to both auditory and visual stimuli (polimodal area STP, for example) (Gross et al., 1993; Gross, 1998).

Current functional neuroimaging methods for mapping retinotopic cortex and functionally specialized areas suggest the existence of more than eight retinotopic areas (and many subdivisions) (Figure 8.6), with homologies but also dissimilarities between animal and human brains (Zeki and Shipp, 1988; Tootell et al., 1998; Sereno and Tootell, 2005; Wandell and Winawer, 2011). The visual system as we know it today consists of multiple map-like areas, each specialized for different visual attributes, connected by feed-forward and feedback pathways (Zeki and Shipp, 1988; Fellman and Van Essen, 1991).

The primary visual cortex (termed V1, area 17, or striate calcarine cortex in different classificatory schemes) is situated in monkeys and humans at the occipital pole extending along the medial surface of the cuneus. In the monkey brain V1 extends also to the lateral occipital surface and occupies a relatively larger surface (Brindley, 1972).

Beyond V1, further consistencies between the monkey and human visual system have been found in the specialization of cortical visual areas for different visual attributes, specialization in this context referring to a selectivity of response to some classes of

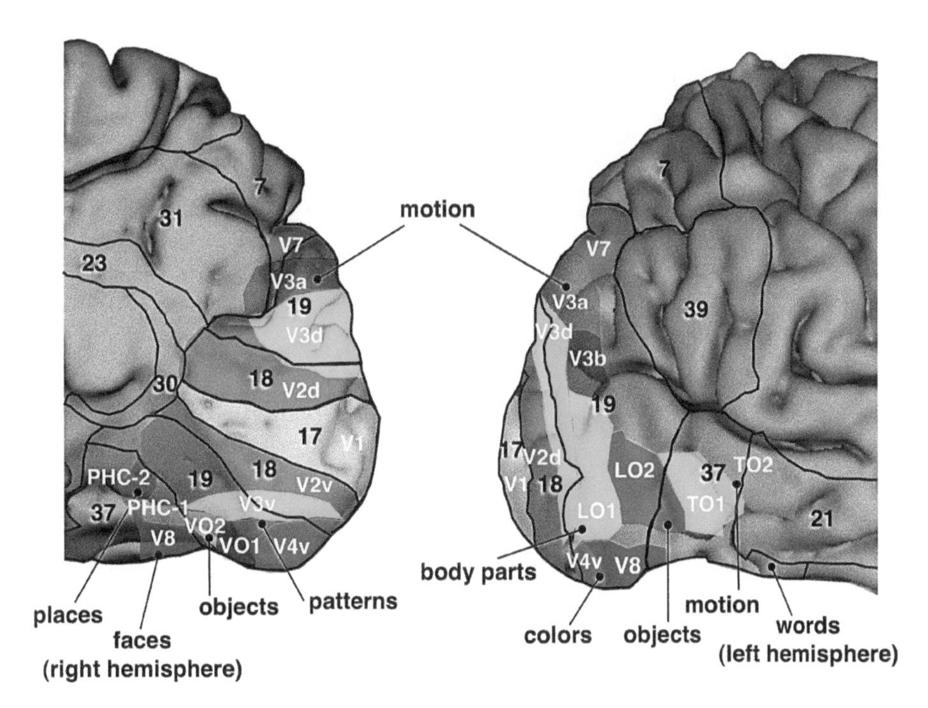

Figure 8.6 Cytoarchitectonic, retinotopic, and functionally specialized areas of the medial and lateral surface of the human occipital cortex. Brodmann cytoarchitectonic fields are demarcated by black lines and indicated with numbers (e.g. 17, 18, 19, etc.) (Brodmann, 1909). Coloured retinotopic areas are derived from functional imaging studies and indicated with combined white letters and numbers (e.g. V1, V2d, V2v, etc.) (Tootell et al., 1998; Sereno and Tootell, 2005; Wandell and Winawer, 2011). A direct correspondence between cytoarchitectonic and retinotopic maps is evident only for BA 17 and V1. The approximate location of the areas specialized for patterns, objects, places, colours, motion, words, body parts, and faces is also indicated. The functionally specialized areas do not overlap with the boundaries of the cytoarchitectonic and retinotopic maps. Furthermore for many specialized areas there is not a general consensus on their exact location in the human brain. The colour area, for example, is localized in V4v according to some authors (Zeki et al., 1998) and in V8 according to others (Sereno et al., 1995).

visual stimuli compared to others. For example, different areas on the ventral occipito-temporal surface respond preferentially to patterns, colours, familiar faces, places, and landscapes, objects, text, and letter strings (Figure 8.6) (Zeki et al., 1991; Sereno and Tootell, 2005; Wandell and Winawer, 2011). The lateral occipito-parietal and occipito-temporal cortices contain specializations for visual motion, body parts, face features, intermediate-level object processing, and eye and mouth movements. Such specialized regions are typically bilateral, although the specialization for text and letter strings is predominantly left-hemispheric while that for faces is predominantly right-hemispheric.

Despite decades of cortical mapping, a precise and detailed location of retinotopic and functionally specialized areas is far from complete. While some agreement for interspecies homologies has been reached for V1, V2, and MT/V5 (Sereno and Tootell, 2005), the function, location, and comparative anatomy of the other areas remain to be established.

Functional and comparative anatomy of visual pathways

In 1982, Leslie Ungerleider and Mortimer Mishkin proposed that specialized occipital, temporal, and parietal cortices are linked through two parallel, cortico-cortical visual systems for spatial and object vision. In their model the visual information is processed along a dorsal occipito-parietal stream for spatial vision (the 'where' stream) and a ventral occipito-temporal stream for object vision, colours, faces, and visual language (the 'what' stream). The two-streams model was largely based on anatomical, electrophysiological, and behavioural findings derived from studies in the animal brain (Mishkin et al., 1983; Van Essen and Maunsell, 1983; Felleman and Van Essen, 1991). The model has been recently modified to take into account new evidence for a role of the dorsal pathway in the visual control of skilled actions. Accordingly, the dorsal

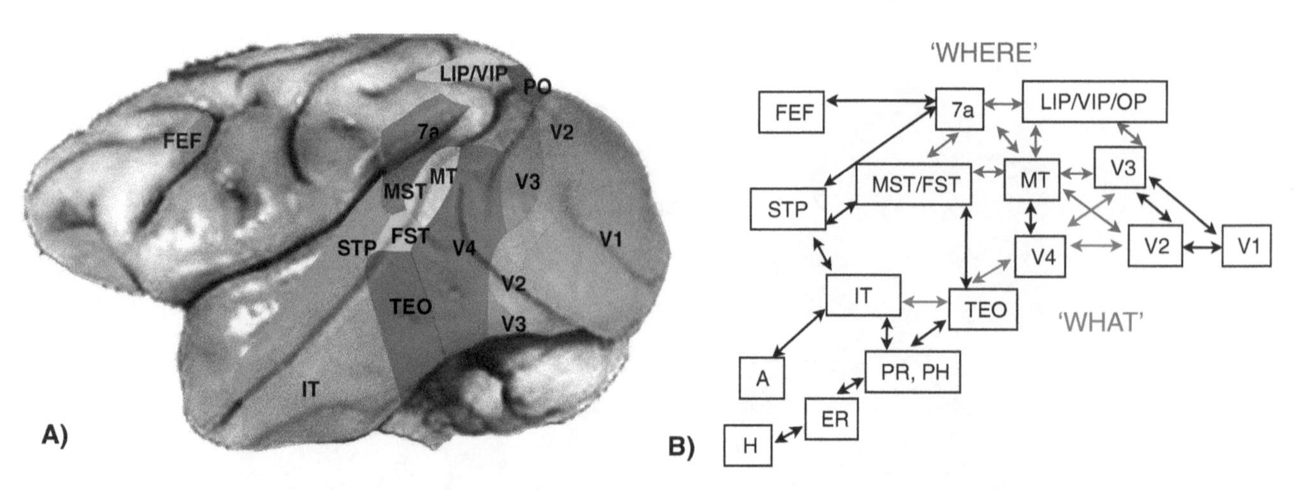

Figure 8.7 A) The major retinotopic maps of the macaque brain extend beyond the occipital lobe to more anterior parietal and temporal cortices (Gross et al., 1993). The precision of the retinotopic maps progressively reduces away from the occipital pole and more anterior areas (e.g. STP, TE, TEO) lack a precise organization. Copyright © 1992 Society for Industrial and Applied Mathematics. Reprinted with permission. All rights reserved. B) Diagram of the connections between the major retinotopic maps showing the dorsal and ventral visual streams for spatial and object vision (Ungerleider and Mishkin, 1982; Gross et al., 1993). The information from the lateral geniculate nucleus reaches V1, where the basic visual features (orientation, speed of movement, etc.) are encoded. From here visual information is processed along two diverging parallel pathways. The occipital temporal connections mediate object vision along the so-called ventral 'what' stream, while the dorsal and lateral occipito-parieto-temporal connections process spatial vision along the dorsal 'where' stream (Ungerleider and Mishkin, 1982) (see text for a recent revision of the role of the dorsal stream). Note that both streams share many intermediate relay stations and are interconnected through U-shaped fibres. Abbreviations: A, amygdala; ER, entorhinal cortex; FEF, frontal eye field; FST, fundus of the superior temporal sulcus; H, hippocampus; IT, inferior temporal area; LIP, lateral intraparietal area; MST, medial superior temporal area; MT, middle temporal area; PH parahippocampal cortex; PO, parieto-occipital area; PR, perirhinal cortex; TEO, temporo-occipital area; V1, V2, V3, V4, first, second, third, fourth visual areas; VIP, ventral intraparietal area. From David, M. Goodale and R. Mansfield, *Analysis of Visual Behavior*, figure from pages 549-586, © Massachusetts Institute of Technology, by permission of the MIT Press.

pathway has been renamed the 'how' stream (i.e. vision for action) (Milner and Goodale, 2008; Kravitz et al., 2011).

The communication along the two streams is likely to be mediated by short U-shaped connections that can be visualized in the human brain with tractography. The function of the U-shaped fibres is to segregate and integrate visual processing within and between the two visual streams (Fellemann and Van Essen, 1991; Zeki and Shipp, 1988). All U-shaped fibres are bidirectional feedforward and feedback connections and mediate bottom-up processing and top-down control (Gross et al., 1993; Zeki and Shipp, 1988).

Tractography studies suggest that these inter-species consistencies extend to long-range association white matter tracts. Striking similarities have been found between occipito-temporal connections identified by diffusion tensor tractography in man (Catani et al., 2003) and those described in the monkey (Kuypers et al., 1965; Tusa and Ungerleider, 1985; Schmahmann et al., 2007). Tractography studies found no evidence for direct connections between the striate cortex (V1) and the anterior temporal lobe (Catani et al., 2003). The main projection to the ventro-lateral surface of the temporal lobe is from V4, an area that also projects to the parahippocampal gyrus (Martin-Elkins and Horel, 1992). Furthermore, the amygdala sends back-projections to V4 and V2, but not to V1 (Amaral and Price, 1984; Iwai and Yukie, 1987).

Evidence for the existence of direct and indirect connections between extrastriate occipital and anterior temporal lobe in the human brain also derives from electrophysiological studies. Wilson (1983) found that after visual stimulation the onset latency of some parahippocampal gyrus cells differed from those in the occipital cortex by 2 ms—a difference too small to be caused by multisynaptic transmission through the U-shaped fibre occipito-temporal projection system. In fact, other cells were found in the same region with latencies of 200 ms, consistent with multi-stage hierarchical transmission. The implication of the latency findings is that there are two pathways by which signals reach the parahippocampal gyrus—an indirect long latency pathway and a direct short latency pathway.

The ventral fusiform branch of the inferior longitudinal fasciculus falls within the coordinates of human V4 (McKeefry and Zeki, 1997) and connects to lateral temporal and medial temporal parahippocampal regions as does V4 in the monkey. It would seem likely therefore that, in addition to the lateral temporal branches identified anteriorly with tractography, further medial and ventral branches arise along the entire extent of the inferior longintudinal fasciculus. In man, V2 is located in the lingual gyrus, cuneus, and lateral surface of the occipital lobe (Horton and Hoyt, 1991; Sereno et al., 1995), showing a perfect correspondence with the locations of the tractography-defined occipital branches of the inferior

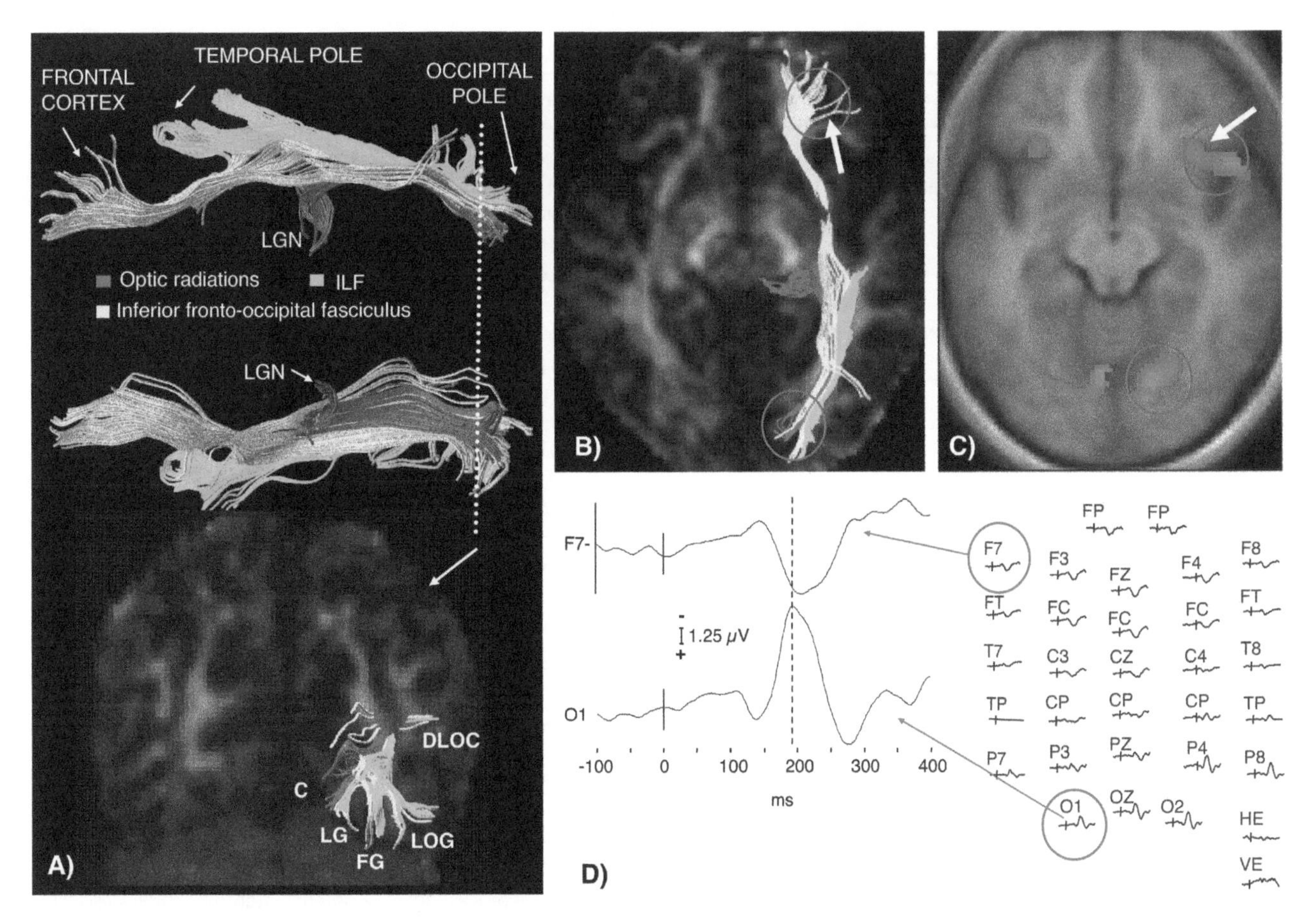

Figure 8.8 Multimodal imaging of the inferior fronto-occipital pathways suggests that the involvement of the lateral orbitofrontal cortex in conscious perceptual tasks may be mediated by this tract. A–B) Tractography reconstruction of the inferior fronto-occipital fasciculus connecting frontal and occipital lobes (C, cuneus; LG, lingual gyrus; FG, fusiform gyrus; LOG, lateral occipital gyrus; DLOC, dorsolateral occipital cortex). C) As revealed by fMRI, irrespective of whether subjects perceive the stimulus or not, activation is found bilaterally in the posterior inferior frontal lobe (indicated by the white arrow), with more activity here when the stimulus is perceived (subjects report actually seeing the stimulus) compared to when it is not (subject report not seeing the stimulus) (Pins and ffytche, 2003). D) From the temporal perspective of the evoked potential method, activity in this lateral orbitofrontal region (F7) is linked in some way to visual function as it exhibits a prominent positive wave, peaking some 10 ms after the late negative wave in the occipital lobe (O1) (194 ms occipital negativity, 205 ms frontal positivity) (Pins and ffytche, 2003; ffytche and Catani, 2005).

longitudinal fasciculus. These findings indicate that, in addition to the serial, hierarchical pathway from occipital to anterior temporal regions through U-shaped fibre connections, signals pass directly to the parahippocampal gyrus, amygdala, polar temporal, and frontal cortex and return directly and indirectly from these anterior temporal and limbic regions to early visual areas (Catani et al., 2003). What functions might be served by direct occipito-temporal (and temporo-occipital) pathways and how do they differ from those of the indirect pathway formed by the U-shaped fibres? The question is not easy to answer as case reports of occipito-temporal disconnection syndromes typically involve lesions of white matter and occipito-temporal cortex, transecting both the direct and indirect occipito-temporal connections. One possible difference is that the indirect system relates to visual perceptual qualities (shape, colour, patterns, etc.) and the direct with emotional qualities and visual memory. Perhaps one function of the direct feed-forward pathway is to prime medial temporal structures to facilitate the consolidation of visual memories. Whatever the function of the feed-forward direct projection, the direct feedback projection is easier to interpret. Once the emotional valence of a visual stimulus has been identified, it would seem reasonable to assume that signals will be fed back directly to early visual areas, enhancing the visual processing of emotionally significant stimuli. In support of this view, recent imaging studies have found neuromodulatory effects of the amygdala on extrastriate visual cortex (Morris et al., 1998; Pessoa et al., 2002) and psychophysical studies have identified an equivalent modulatory effect of emotional content on visual perceptual processing (Anderson and Phelps, 2001). Furthermore, in carrying visual information from occipital areas to the temporal lobe, the inferior longitudinal fasciculus is likely to play an important role in linking object representations to their lexical labels (Mummery et al., 1999).

Table 8.1 Early visual pathways syndromes

Lesion location	Disorder	Disorder description
Optic radiations, calcarine cortex	Visual field defects (scotoma, quadrantanopsia, hemianopsia)	A partial loss of vision in a small region of the visual field (scotoma), a single quadrant (quadrantonopsia), or an entire hemifield (hemianopsia). The visual deficit is always opposite to the lesion (e.g. left hemifield due to a right calcarine lesion). A left or right upper quadrantonopsia is indicative of a lesion to the Flechsig–Meyer loop.
	Cortical blindness	Complete visual loss due to extensive bilateral occipital lesions. Some patients maintain the ability to consciously perceive and discriminate visual motion in their otherwise blind visual field when stimulated with fast motion (Riddoch phenomenon). Rarely the blind patient denies his visual deficit and gives confabulatory responses (Anton syndrome).
	Aperceptive agnosia	Inability to recognize, discriminate, or copy objects in a patient with otherwise normal visual ability. Lesions are always close to the primary visual area or its connections to neighbouring gyri.
	Simple hallucinations (or phosphenes)	Spontaneous increase of activity in the primary visual cortex due to local irritative pathology (e.g. epilepsy), or a distant hodological effect (e.g. de-afferentation caused by eye disease in Charles Bonnet syndrome).

The similarities between monkey and human occipito-temporal connections do not extend to the fronto-occipital fasciculus. Early human post-mortem and recent tractography findings of an inferior fronto-occipital fasciculus in the normal brain have not been replicated in the macaque brain using axonal tracing methods (Petrides and Pandya, 2007; Schmahmann et al., 2007). This has led to the hypothesis that the inferior fronto-occipital fasciculus may be a tract unique to the human brain (Catani, 2007). The functions of the inferior fronto-occipital fasciculus remain largely unknown. The occipital areas connected by the inferior fronto-occipital fasciculus include the lingual gyrus and cuneus, which engage in tasks requiring retrieving information from visual memory, spatial attention, and recognition of colours, letters, and faces. The anterior projections of the inferior fronto-occipital fasciculus terminate in the medial and lateral orbitofrontal cortex, which activates in tasks requiring top-down modulation (either activation or inhibition) of visual perception and short-term memory. Also these frontal regions seem especially important in guiding decision-making and social behaviour in relation to the emotional value of the external sensory stimuli (Rudrauf et al., 2008). Kawasaki et al. (2001) used single-neuron cortical recording in an epileptic patient to demonstrate a modulation of the rate of spiking of individual neurons in the medial orbitofrontal cortex as early as 120–170 ms (rapid response) during the presentation of faces expressing negative emotions or aversive visual scenes. Interestingly, the neurons of the right lateral orbitofrontal cortex responded only to aversive scenes and not to fearful faces suggesting a segregation of function within the lateral and medial frontal projections of the inferior fronto-occipital fasciculus. In addition to the rapid response they found a delayed neuronal activity consisting of prolonged increase in firing rate. This suggests that the prefrontal cortex may guide the processing of emotional stimuli at multiple temporal stages encompassing both early unconscious emotional processes and delayed processing, which together contribute to a conscious awareness of the emotion. Although these studies focus on emotional visual processing, the inferior fronto-occipital fasciculus seems to play a role in the rapid top-down modulation of visual processing in general (i.e., not restricted to processing emotional value of visual percepts) (Figure 8.8) (Pins and ffytche, 2003; Bar et al., 2006), including top-down amplification of visual percepts characteristic of conscious visual processing (Dehaene et al., 2006) and executive control of voluntary visual recall (Tomita et al., 1999).

A different pattern of functional connectivity links the occipital and frontal areas in states of reduced consciousness. For example, Braun et al. (1998) demonstrated that REM sleep is associated with activation of the extrastriate visual cortices and concomitant reduction of activity in lateral orbital and prefrontal regions. The role of the inferior fronto-occipital fasciculus in sleep is confirmed by patients with acquired inability to dream (anoneria) following lesions of the lateral orbitofrontal cortex (Solms 1997). The exact role of the inferior fronto-occipital fasciculus in functions such as mental rotation (Schendan et al., 2008), space-directed attention (Urbanski et al., 2008), reading (Shaywitz et al., 2002; Mechelli et al., 2004), and semantic processing (Duffau et al., 2005), remains to be clarified.

The functions of the cingulum and possible clinical correlates are discussed in Chapter 11.

Disorders of visual perception

If visual perception is the day-to-day conscious visual experience of the world when we open our eyes, a visual perceptual disorder might be considered a deviation from this experience (ffytche et al., 2010). The disorders of visual perception can be divided into three groups according to whether the lesion affects afferent visual projections to

Table 8.2 Ventral and dorsal visual pathways syndromes

Lesion location	Disorder	Disorder description
Ventral visual stream (specialized for objects, patterns, colours, faces, body parts, and words)	Object agnosia, achromatopsia, prosopagnosia, alexia	Inability to recognize objects, colours, faces, and words due to de-afferentation or direct lesion of the specialized areas.
	Object, colour, face, text illusion/hallucination, visual functional hallucinations	Hyperactivity of specialized cortex can cause visual illusions of distorted objects (metamorphopsias), faces (prosometamorphopsias), or text. Objects can appear larger (macropsia), smaller (micropsia), nearer (pelopsia) or further away (teleopsia). Seeing objects, colours, faces, or letters in the absence of a real visual stimulus can present spontaneously or sometimes can be triggered by a normal visual percept (functional hallucinations).
Dorsal visual stream (specialized for motion, spatial coordinates, and depth)	Akinetopsia	Loss of motion vision following lesions of lateral occipital motion-specialized cortex. Lesions disconnecting motion-specialized cortex and spatial coordinate frames in the parietal lobe may present with 'cinematographic vision', where moving objects appear as a series of static freeze frames at different spatial locations.
	Illusions and hallucinations of object position and movement	Increased perceptual activity within dorsal and lateral occipital and parietal cortex can manifest with distorted perception of multiple copies of the same object (polyopia), objects remaining fixed in retinal coordinates with eye movements (visual perseveration), or a pattern spreading from an object to its surrounds (illusory visual spread). Hallucinations typically involve moving objects (motion hallucination) and persons or animals that pass sideways out of the visual field (passage hallucination).

primary visual cortex (Table 8.1), the dorsal and ventral visual streams (Table 8.2), or the extended visual networks connecting the occipital lobe with the limbic and language regions (Table 8.3).

Cortical blindness is the total loss of vision that follows extensive bilateral lesions of the occipital lobes affecting either the calcarine cortex and its underlying white matter, or the optic radiations alone (Babinski, 1914; Critchley, 1953). Some patients with cortical blindness not only deny their visual deficits but confabulate and behave as if they were sighted. This condition is also known as Anton syndrome (Anton, 1899). The Riddoch phenomenon is a variant of cortical blindness in which motion-specialized cortex remains intact (Riddoch, 1917). The patients maintain the ability to consciously perceive visual motion in their otherwise blind visual field (Zeki and ffytche, 1998). Unilateral lesions of the optic radiations or the calcarine cortex usually manifest with partial visual defects limited to one quadrant of the visual field (quadrantopsia) or the entire hemifield (hemianopsia).

The visual agnosias are a heterogeneous group of disorders of object recognition, usually associated with discrete occipital lesions sparing the calcarine cortex. Lissauer (1890) distinguished two broad subtypes. In apperceptive agnosia the patient is unable to form an accurate perceptual representation of the structure of an object. This is the most severe form of visual agnosia where the patient is unable to recognize objects in any form due to lesions very close to primary visual areas. In this case visual inputs reach the primary visual cortex without progressing further along the occipital stream.

In the second subtype, the associative visual agnosia, the formation of the object percept is intact, but this information cannot be matched to object memories in order to recognize that the subject has encountered the object before. These forms are associated with lesions, which spare early visual cortex but involve its white matter outputs, thus resulting in visual sensory images being disconnected from other brain areas. In patients with associative visual agnosia, the ability to visually perceive an object is largely preserved but the visual percept fails to elicit the wider associations required for recognition. Similarly, damage to other specialized occipital areas can result in isolated forms of agnosias, such as achromatopsia (the selective loss of colour vision following lesions of colour specialized cortex) (Steffan, 1881; Zeki 1990), prosopagnosia (face perception deficit) (Wigan, 1844; Bodamer, 1947; Fox et al., 2008), and alexia (the inability to read words) (Dejerine, 1892; Epelbaum et al., 2008). Similar to object agnosias, disorders of colour, face, and letter perception have a variety of subtypes likely to reflect the direct involvement of the specialized cortical areas or their connections. The patient described by Epelbaum et al. (2008) is a classical example of disconnection alexia, in this case following the surgical excision of a small epileptogenic cortical region in the left ventral occipital lobe. Pre- and post-operative diffusion tensor tractography showed a reduction of the number of short U-shaped fibres connecting the area in the fusiform gyrus specialized for words (i.e. visual word form area) with surrounding occipital areas. The fMRI of the patient during reading tasks showed activation of the visual word form area during letter-by-letter reading but not for fast word presentation. This suggests that the visual word form area was anatomical intact and could be activated using an alternative route. In patients with congenital prosopagnosia, where conventional structural imaging is usually normal, the fractional anisotropy of the inferior longitudinal fasciculus and inferior fronto-occipital fasciculus is often reduced and this reduction correlates significantly with impaired performance in face processing tasks (Thomas et al., 2009). Akinetopsia is the selective loss of motion vision following lesions of lateral occipital motion-specialized cortex; the motion

Table 8.3 Extended visual pathways syndromes

Lesion location	Disorder	Disorder description
Perisylvian language and motor regions, limbic areas, frontal cortex (arcuate fasciculus, cingulum inferior fronto-occipital fasciculus, inferior longitudinal fasciculus, superior longitudinal fasciculus)	Colour and object anomia	Disconnection of visual and language areas results in specific deficits of colour or object naming.
	Visual amnesia	Memory deficit for visually presented material due to a lesion of the inferior longitudinal fasciculus disconnecting visual areas and hippocampus.
	Visual hypoemotionality, visual de-realization, Capgras syndrome, fear recognition deficits	These disconnection syndromes follow damage to the fibres of the inferior longitudinal and inferior fronto-occipital fasciculus connecting visual areas and anterior limbic structures (e.g. amygdala and orbitofrontal cortex). The patients report lack of emotional response to images (visual hypo-emotionality) or threatening situations (fear recognition deficits). Others complain of visual experiences being unreal (de-realization) or faces of known persons having been replaced by impostors (Capgras syndrome).
	Visual imagery deficits, visual anoneria	Inability to retrieve visual mental representations either during the day or while dreaming.
	Hemispatial visual neglect	Lesions of the inferior longitudinal fasciculus or inferior fronto-occipital fasciculus disconnecting visual areas and right fronto-parietal visuospatial network manifesting with a failure to respond or reorient to novel or meaningful stimuli presented to the left hemispace.
	Imitative and oculomotor apraxia, utilization behaviour, optic apraxia	Disconnection between motor and visual areas can result in the inability to imitate actions (imitative apraxia) or in the automatic use of objects placed within the field of view and reach (utilization behaviour). Balint syndrome is a degenerative disorder characterized by simultagnosia, optic ataxia, and oculomotor apraxia.
	Synaesthesia across sensory modalities, reflex hallucinations	A percept in one sensory modality evoking a parallel percept usually in a different sensory modality (e.g. seeing coloured music) due to abnormal hyperconnectivity between distant sensory regions. Sometimes the percept triggers reflex hallucinations in a different modality (e.g. seeing a butterfly whenever the thumb is touched).
	Flashbacks, memory hallucinations, phobias, Fregoli phenomenon	A wide range of disorders related to hyperconnectivity between visual and limbic areas resulting in an excessive emotional response to visual perception or memories.

deficit is not absolute with a relative sparing of the perception of slow movement.

Increased perceptual activity within occipital cortical areas or an extended cortico-cortical network underlies several syndromes characterized by positive symptoms such as hallucinations (perception of a false non-existing sensory input) and illusions (distorted perception). Hallucinations linked to the primary visual cortex are of simple featureless forms and colours, or take the form of television-like static (i.e. visual snow). These phenomena are termed phosphenes or photopsias by some authors although they are perhaps better referred to as simple hallucinations given that they differ from other types of visual hallucination only in terms of the location of cortex involved (ffytche, 2008). Colour and object hallucinations are each linked to their respective region of cortical specialization and the same is likely to be the case for motion hallucinations (moving objects) and passage hallucinations (visual hallucinations of persons or animals that pass sideways out of the visual field) (ffytche et al., 1998, ffytche and Howard, 1999). Size and object distortions are referred to collectively as metamorphopsias. Objects appearing larger (macropsia) or smaller (micropsia) have been attributed to a dysfunction of object constancy within the temporal lobe (the integration of object distance with retinal visual angle to estimate size). The same mechanism may be responsible for objects appearing nearer (pelopsia) or further away (teleopsia). Text or letter-string hallucinations that lack semantic content, such as jumbled letters, are linked to the cortical area in the fusiform gyrus specialized for word forms.

Occasionally text hallucinations present with a clear semantic content, such as the visual hallucinations of the patient reported by ffytche et al. (2004) where the sentences were disturbing visual command hallucinations (e.g. 'eat the fish', 'don't take your tablets', 'they are after your money'). Illusions of distorted faces (prosopometamorphopsia) relate to activity on the lateral convexity of the occipital lobe (Heo et al., 2004), a region specialized for face features (the occipital face area), and the same region is likely to account for hallucinations of distorted faces. Hallucinations of normal (undistorted) faces relate to activity within the ventral occipito-temporal cortex (fusiform face area) (ffytche et al., 1998), and the same region is likely to account for facial intermetamorphosis (a change in the visually perceived identity of a face).

Polyopia (perceiving multiple copies of the same object often arranged in rows and columns), visual perseveration (an object remaining fixed in retinal coordinates with eye movements), delayed palinopsia (an object returning to field of view), and illusory visual spread (the spread of a pattern from an object to its surrounds) have been hypothesized as linked to a range of coordinate systems within the dorsolateral occipital-temporo-parietal cortex (Ceriani et al., 1998). The same region is likely to underlie entomopia (a variety of polyopia in which the copies of an object are arranged like the compound eye of an insect), the trailing phenomenon (a variety of visual perseveration in which a series of discontinuous stationary images trail behind a moving object), positive afterimages (afterimages in which objects are seen in their true as opposed to complementary colours or black and white tones as in the negative afterimages).

The nature of the visual percept disorder is not always indicative of the location of the lesion, which may be distant and operate through a hodological mechanism. Disconnection of cortical visual areas from subcortical inputs as a result of visual pathway lesions or eye disease is associated with simple and complex visual hallucinations and illusions. In patients with Charles Bonnet syndrome, for example, the increased activity in the occipital visual cortex that accompanies the visual hallucinations is secondary to eye disease and deafferentation of the lateral geniculate nucleus (Madill and ffytche, 2005; ffytche, 2008). A similar mechanism may operate in brainstem lesions associated with peduncular hallucinations (Benke, 2006). In some patients the site of the lesion is in regions of the brain, which are classically not considered as part of the visual system. The first patient described by Schneider et al. (1961) experienced visual hallucinations of complex scenes that disappeared following the surgical resection of a well-encapsulated abscess in the white matter of the orbitofrontal region. A second patient with medial frontal lobe meningioma complained post-operatively of seeing sparks and stars of light (i.e. phosphenes). The visual hallucinations in these patients are likely to result from activity propagated from the frontal lobe to the occipital cortex along the inferior fronto-occipital fasciculus. This conclusion is consistent with evidence from a recent intra-operative stimulation study in which visual hallucinations of faces were evoked by stimulation of the inferior frontal gyrus, but not the anterior temporal or premotor cortex (Vignal et al., 2000).

The direct connection between frontal and occipital cortices mediated by the inferior fronto-occipital fasciculus seems to have a role also in the retrieval of visual images during the awake state (visual imagery) or sleep (visual dreaming). Visual imagery may be impaired with a relative preservation of visual perception (Brain, 1941; Guariglia et al., 1993). More often, however, visual imagery deficits are associated with visual agnosia (Basso et al., 1980; Farah, 1984). The loss of visual imagery is observed in patients with lesions along the course of the inferior fronto-occipital fasciculus (Basso et al., 1980; Guariglia et al., 1993). Deficits of visual imagery may also be accompanied by loss of visual dreaming (visual anoneria). The Charcot–Wilbrand syndrome is defined by the combination of anoneria and visual agnosia. Solms (1997) distinguishes global anoneria, a defective dreaming in all sensory modalities, from dreaming deficits restricted to the visual domain. Global anoneria is also frequently associated with lesions along the course of the inferior fronto-occipital fasciculus (Basso et al., 1980; Solms, 1997).

Many other visual disorders have been explained as disconnection syndromes resulting in the failure to transfer inputs between two interconnected areas, especially when the deficit occurs in association with different lesion locations within a more extended system involving other domains (e.g. language, emotions, visuospatial attention, etc.). Deficits in naming visually presented objects, colours, and faces may result from anterior lesions disconnecting specialized occipital cortex and language areas (e.g. uncinate fasciculus). Visual neglect has been described for lesions of the inferior longitudinal fasciculus (Bird et al., 2006) and the inferior fronto-occipital fasciculus (Urbanski et al., 2008) disconnecting occipital areas from the perisylvian network in the right hemisphere.

Similar disconnection accounts are given of disorders affecting the visual-limbic networks, such as de-realization (the feeling that experiences, typically visual, seem strange or unreal), visual hypoemotionality (a lack of emotional tone to visual experience) (Bauer, 1982; Habib 1986), and visual amnesia (a deficit of registering visual experiences in short-term memory) (Ross, 1980). These forms, especially when presenting as isolated deficits, are likely to result from partial disconnection of the inferior longitudinal fasciculus and/or the inferior fronto-occipital fasciculus. The patient described by Habib (1986), for example, was a woman who reported 'lack of emotive reaction elicited by visual stimuli' but exhibited normal emotional response to auditory stimuli. Bauer (1982), who studied skin conductance changes in a prosopagnosic patient, found increased autonomic responses to sexual narratives but not to viewing female nudes. Visual hypoemotionality has been interpreted as a visuo-limbic disconnection of the inferior longitudinal fasciculus fibres connecting occipital areas and amygdala. The patient with visual amnesia described by Ross (1980) was unable to learn novel, non-verbalizable visual stimuli due to a lesion apparently restricted to occipito-hippocampal connections. Nevertheless, in this case visual information was able to reach the medial temporal structures through the indirect U-fibre projection system and elicit old memories. A recent study showed that lesions to the right inferior fronto-occipital fasciculus significantly predict an overall facial emotion recognition impairment with specific impairments for sadness, anger, and fear (Philippi et al., 2009). Among the patients they studied, a subject presented with an isolated disconnection syndrome characterized by selective impairment of face expression recognition associated with a lesion to the right inferior fronto-occipital fasciculus. Finally patients with lesions to the orbitofrontal regions that receive projections from the inferior fronto-occipital fasciculus are found to be impaired in the ability to become aware of brief visual stimuli (Del Cul et al., 2009). These data have been used to support the hypothesis that the anterior frontal areas are involved in conscious visual processing (Crick and Koch, 1995; ffytche and Catamo, 2005; Dehaene et al., 2006).

In some patients the deficit involves their ability to retain full insight into their problem, suggesting an involvement of a more extended network. The Capgras delusion is a misidentification syndrome in which the subject claims that familiar people are impostors (Capgras and Reboul-Lachaux, 1923). The subject states that the 'impostors' look similar to the people they are replacing, often describing them as 'doubles', but does not accept them as the real person (Lucchelli and Spinnler, 2007). On occasion this can be specific to the visual domain, as some patients do not show a similar delusion for the voices of the same people they consider visually to be impostors (Hirstein and Ramachandran, 1997). One possible explanation is a dissociation between an intact conscious face recognition and an altered affective response to the face, so that the encounter does not feel quite 'real' (Fox et al., 2008). This is supported by studies using skin conductance where the Capgras patients, similar to patients with visual hypoemotionality, have reduced affective response to familiar faces (Ellis et al., 1997; Hirstein and Ramachandran, 1997; Brighetti et al., 2007). In addition, the Capgras patients develop a delusional explanation for their symptoms, in a similar way to confabulatory amnesics with medial orbitofrontal lesions (Benson et al., 1996; Schneider, 2008). It is possible that the Capgras delusion results from a combined hodological mechanism affecting the connections between the occipital face processing areas, the amygdala, and the orbitofrontal cortex.

Some other conditions have been interpreted as disorders derived from increased connectivity between sensory modalities or within modalities. In patients with synaesthesia, the normal, day-to-day perceptual experience in one sensory modality is accompanied by a parallel hallucinatory percept in a different modality. Examples in the visual domain include visual percepts of colour evoking gustatory hallucinations (coloured taste, or colour–taste synaesthesia) or auditory percepts of music evoking parallel hallucinatory visual percepts of colour (coloured music, or music-colour synaesthesia). Other parallel sensory experiences include reflex hallucinations, in which a sensory stimulus in one modality evokes a hallucination in another (e.g. a command hallucination ordering to 'cross' triggered by the red traffic light) and functional hallucinations, in which hallucinations are triggered in the same modality as the inducing stimulus (e.g. seeing a number that triggers the visual hallucination of a distorted face). The Frégoli syndrome, in which unfamiliar people are perceived as familiar (typically as a person in disguise with malevolent intent), can be conceived as a hyperconnection within visual emotional or memory networks. Similarly, the strong affective and imagery components of post-traumatic stress disorder (PTSD) flashbacks, flashbulb memories (detail-perfect memories of events that at the time of witnessing had evoked a high level of surprise and emotional arousal), and memory hallucinations (fantastic false memories conjured up retrospectively) suggest a hyperconnection between visual, emotional, and memory regions. Autoscopic phenomena describe a range of experiences in which the self is duplicated in external space. In autoscopy, visual perspective remains in the physical body (the duplicate self is seen in the external world). In out-of-body experience (OBE) the physical body is seen from the perspective of the external self. In 'heautoscopy' the hallucinated 'double' is identified as oneself, despite the lack of an exact physical resemblance. These phenomena are thought to relate to the disintegration of visual, proprioceptive, tactile, and vestibular modalities and have been linked to transient hyperfunction of an occipito-temporo-parietal network (ffytche et al., 2010).

Other manifestations could be interpreted as the result of a combined topological and hodological mechanism involving simultaneous hypo- and hyper-connectivity along different segments of the direct and indirect visual pathways. Patients prone to LSD-related visual flashbacks have EEG hypoconnectivity between the occipital lobe and other cortical regions but EEG hyperconnectivity within the occipital lobe when resting with closed eyes (a state conducive to hallucinations) (Abraham and Duffy, 2001).

References

Abraham, H.D. and Duffy, F.H. (2001). EEG coherence in post-LSD visual hallucinations. *Psychiat Res*, **107**(3), 151–63.

Amaral, D.G. and Price, J.L. (1984). Amygdalo-cortical projections in the monkey (*Macaca fascicularis*). *J Comp Neurol*, **230**(4), 465–96.

Anderson, A.K. and Phelps, E.A. (2001). Lesions of the human amygdala impair enhanced perception of emotionally salient events. *Nature*, **411**(6835), 305–9.

Anton, G. (1899). Über die Selbstwahrnehmung der Herderkrankungen des Gehirns durch den Kranken bei Rindenblindheit und Rindentaubheit. *Archiv für Psychiatrie und Nervenkrankheiten, Berlin*, **32**, 86–127.

Archambault, L. (1906). Le faisceau longitudinal inférieur et le faisceau optique central: quelques considérations sur les fibres d'association du cerveau. *Rev Neurol (Paris)*, **4**, 1206–22.

Babinski, J. (1914). Contribution a l'étude des troubles mentaux dans l'hémiplégie organique (anosognosie). *Revue Neurologique*, **27**, 845–8.

Basso, A., Bisiach, E., Luzzatti, C., et al. (1980). Loss of mental imagery: a case study. *Neuropsychologia*, **18**(4–5), 435–42.

Bauer, R.M. (1982). Visual hypoemotionality as a symptom of visual-limbic disconnection in man. *Arch Neurol*, **39**(11), 702–8.

Bar, M., Kassam, K.S., Ghuman, A.S., et al. (2006). Top-down facilitation of visual recognition. *Proc Natl Acad Sci USA*, **103**(2) 449–54.

Benke, T. (2006). Peduncular hallucinosis: a syndrome of impaired reality monitoring. *J Neurol*, **253**(12), 1561–71.

Benson, D.F., Djenderedjian, A., Miller, B.L., et al. (1996). Neural basis of confabulation. *Neurology*, **46**(5), 1239–43.

Bird, C.M., Malhotra, P., Parton, A., Coulthard, E., Rushworth, M.F., and Husain, M. (2006). Visual neglect after right posterior cerebral artery infarction. *J Neurol Neurosurg Psychiat*, **77**(9), 1008–12.

Bodamer, J. (1947). Die Prosop-Agnosie (die Agnosie des Physiognomieerkennens). *Archiv fur Psychiatrie und Nervenkrankheiten*, **179**, 6–53.

Brain, R.W. (1941). Visual disorientation with special reference to lesion of the right brain hemisphere. *Brain*, **64**(4), 244–72.

Braun, A.R., Balkin, T.J., Wesensten, N.J., et al., (1998). Dissociated pattern of activity in visual cortices and their projections during human rapid eye movement sleep. *Science*, **279**(5347), 91–5.

Brighetti, G., Bonifacci, P., Borlimi, R., and Ottaviani, C. (2007). 'Far from the heart far from the eye': evidence from the Capgras delusion. *Cognit Neuropsychiat*, **12**(3), 189–97.

Brindley, G.S. (1972). The variability of the human striate cortex. *J Physiol*, **225**(2), 1P–3P.

Burdach, K. (1822). *Vom Baue und Leben des Gehirns*. Leipzig: In der Dyk'schen Buchhandlung.

Brodmann K. (1909). *Vergleichende Lokalisationslehre der Großhirnrinde: in ihren Prinzipien dargestellt auf Grund des Zellenbaues*. Leipzig: Barth.

Campbell, A.W. (1905). *Histological studies on the localisation of cerebral function*. Cambridge: Cambridge University Press.

Capgras, J. and Reboul-Lachaux, J. (1923). L'illusion des 'sosies' dans un delire systematise chronique. *Bull Soc Clin Med Ment*, **2**, 6–16.

Catani, M. (2007). From hodology to function. *Brain*, **130**(3), 602–5.

Catani, M., Howard, R.J., Pajevic, S., and Jones, D.K. (2002). Virtual in vivo interactive dissection of white matter fasciculi in the human brain. *NeuroImage*, **17**(1), 77–94.

Catani, M., Jones, D.K., Donato, R., and ffytche, D.H. (2003). Occipito-temporal connections in the human brain. *Brain*, **126**(9), 2093–107.

Ceriani, F., Gentileschi, V., Muggia, S., et al. (1998). Seeing objects smaller than they are: micropsia following right temporo-parietal infarction. *Cortex*, **34**(1) 131–8.

Clarke, E. and O'Malley, C.D. (1996). *The Human Brain and Spinal Cord: A Historical Study Illustrated by Writings from Antiquity to the Twentieth Century*. San Francisco, CA: Norman Publishing.

Crick, F. and Koch, C. (1995). Are we aware of neural activity in primary visual cortex? *Nature*, **375**(6527), 121–3.

Critchley, M. (1953). *The parietal lobe*. New York: Hafner Publishing Company.

Curran, E.J. (1909). A new association fibre tract in the cerebrum. *J Comp Neurol Psychol*, **19**(6), 645–56.

Dehaene, S., Changeux, J.P., Naccache, L., Sackur, J., and Sergent, C. (2006). Conscious, preconscious, and subliminal processing: a testable taxonomy. *Trends Cogn Sci*, **10**(5), 204–11.

Dejerine, J. (1892). Contribution a l'étude anatomo-pathologique et clinique des differentes variétés de cécité-verbale. *Mém Soc Biol*, **4**, 61–90.

Dejerine, J. (1895). *Anatomie des Centres Nerveux, Vol. 1*. Paris. Rueff et Cie.

Del Cul, A., Dehaene, S., Reyes, P., Bravo, E., Slachevsky, A. (1999). Causal role of prefrontal cortex in the threshold for access to consciousness. *Brain*, **132**(9), 2531–40.

Duffau, H., Gatignol, P., Mandonnet, E., Peruzzi, P., Tzourio-Mazoyer, N., and Capelle, L. (2005). New insights into the anatomo-functional connectivity of the semantic system: a study using cortico-subcortical electrostimulations. *Brain*, **128**(4), 797–810.

Ellis, H.D., Young, A.W., Quayle, A.H., and De Pauw, K.W. (1997). Reduced autonomic responses to faces in Capgras delusion. *Proc Biol Sci*, **264**(1384), 1085–92.

Epelbaum, S., Pinel, P., Gaillard, R., et al. (2008). Pure alexia as a disconnection syndrome: New diffusion imaging evidence for an old concept. *Cortex*, **44**(8), 962–74.

Farah, M.J. (1984). The neurological basis of mental imagery: a componential analysis. *Cognition*, **18**(1–3), 245–72.

Felleman, D.J. and Van Essen, D.C. (1991). Distributed hierarchical processing in the primate cerebral cortex. *Cereb Cortex*, **1**(1), 1–47.

ffytche, D.H. (2008). The hodology of hallucinations. *Cortex*, **44**(8), 1067–83.

ffytche, D.H. and Catani, M. (2005). Beyond localization: from hodology to function. *Phil Trans R Soc B*, **360**(1456), 767–79.

ffytche, D.H., Howard, R.J., Brammer, M.J., et al. (1998). The anatomy of conscious vision: an fMRI study of visual hallucinations. *Nat Neurosci*, **1**(8), 738–42.

ffytche, D.H., and Howard, R.J. (1999). The perceptual consequences of visual loss: positive pathologies of vision. *Brain*, **122**(7), 1247–60.

ffytche, D.H., Lappin, J.M., and Philpot M. (2004). Visual command hallucinations in a patient with pure alexia. *J Neurol Neurosurg Psych*, **75**(1), 80–6.

ffytche, D.H., Blom, J.D., and Catani, M. (2010). Disorders of visual perception. *J Neurol Neurosurg Psychiat*, **81**(11), 1280–7.

Finger, S. (1994). *Origins of Neuroscience*. Oxford: Oxford University Press.

Fishman, R.S. (1997). Gordon Holmes, the cortical retina, and the wounds of war. The seventh Charles B. Snyder Lecture. *Doc Ophthalmol*, **93**(1–2), 9–28.

Flechsig, P.E. (1896). *Gehirn und Steele*. Leipzig: Veit.

Fox, C.J., Iaria, G., and Barton, J.J.S. (2008). Disconnection in prosopagnosia and face processing. *Cortex*, **44**(8), 996–1009.

Guariglia, C., Padovani, A., Pantano, P. et al. (1993). Unilateral neglect restricted to visual imagery. *Nature*, **364**(6434), 235–7.

Glickstein, M. and Whitteridge, D. (1987). Tatsuji Inouye and the mapping of the visual fields on the human cerebral cortex. *Trends Neurosci*, **10**(9), 350–3.

Gratiolet, P. (1854). Note sur les expansions des racines cérébrales du nerf optique et sur leur terminaison dans une région déterminée de l'écorce des hémisphères. *Comp Rend Hebdomad Séances Acad Scie Paris*, **29**, 274–78.

Gross, C.G. (1998). *Brain, vision, memory. Tales in the History of Neuroscience*. Cambridge MA: MIT Press.

Gross, C.G., Rodman, H.R., Gochin, P.M., and Colombo, M.W. (1993). Inferior temporal cortex as a pattern recognition device. In Baum, E. (Ed.) *Computational Learning and Cognition*, pp. 44–72. Philadelphia, PA: Society for Industrial and Applied Mathematics.

Habib, M. (1986). Visual hypo-emotionality and prosopagnosia associated with right temporal lobe isolation. *Neuropsychologia*, **24**(4), 577–82.

Henschen, S.E. (1893). On the visual path and centre. *Brain*, **16**, 170–80.

Heo K., Cho, Y.J., Lee, S.K. et al. (2004). Single-photon emission computed tomography in a patient with ictal metamorphopsia. *Seizure*, **13**(4), 250–3.

Hirstein, W. and Ramachandran, V.S. (1997). Capgras syndrome: a novel probe for understanding the neural representation of the identity and familiarity of persons. *Proc Biol Sci*, **264**(1380), 437–44.

Holmes, G.M. (1918). Disturbance of vision by cerebral lesions. *Br J Ophthalmol*, **2**, 353–84.

Horton, J.C. and Hoyt, W.F. (1991). The representation of the visual field in human striate cortex: a revision of the classic Holmes map. *Arch Ophthalmol*, **109**(6), 816–24.

Hubel, D.H. and Wiesel, T.N. (1959). Receptive fields of single neurones in the cat's striate cortex. *J Physiol*, **148**(3), 574–91.

Hubel D.H. and Wiesel, T.N. (1962). Receptive fields, binocular interaction and functional architecture in the cat's visual cortex. *J Physiol* **160**(1), 106–54.

Hubel, D.H. and Wiesel, T.N. (1965). Receptive fields and functional architecture in two nonstriate visual areas (18 and 19) of the cat. *J Neurophysiol*, **28**(2), 229–89.

Iwai, E. and Yukie, M. (1987). Amygdalofugal and amygdalopetal connections with modality-specific visual cortical areas in macaques (*Macaca fuscata, M. mulatta*, and *M. fascicularis*). *J Comp Neurol*, **261**(3), 362–87.

Kawasaki, H., Kaufman, O., Damasio H., et al. (2001). Single-neuron responses to emotional visual stimuli recorded in human ventral prefrontal cortex. *Nat Neurosci*, **4**(1), 15–6.

Kravitz, D.J., Saleem, K.S., Baker, C.I., and Mishkin, M. (2011). A new neural framework for visuospatial processing. *Nat Rev Neurosci*, **12**(4), 217–30.

Kuypers, H.G., Szwarcbart, M.K., Mishkin, M., and Rosvold, H.E. (1965). Occipitotemporal corticocortical connections in the Rhesus monkey. *Exp Neurol*, **11**(2), 245–62.

Inouye, T. (1909). *Die Sehstörungen bei Schussverletzungen der kortikalen Sehsphäre nach Beobachtungen an Versundeten der letzten Japanische Kriege*. W. Engelmann.

Lissauer, H. (1890). Ein Fall von Seelenblindheit nebst einem Beitrage zur Theorie derselben. *Arch Psychiat Nervenkr*, **21**, 222–70.

Lucchelli, F. and Spinnler, H. (2007). The case of lost Wilma: a clinical report of Capgras delusion. *Neurol Sci*, **28**(4), 188–95.

Madill, S.A. and ffytche, D.H. (2005). Charles Bonnet syndrome in patients with glaucoma and good acuity. *Br J Ophthalmol*, **89**(6), 785–6.

Martin-Elkins, C.L. and Horel, J.A. (1992). Cortical afferents to behaviorally defined regions of the inferior temporal and parahippocampal gyri as demonstrated by WGA-HRP. *J Comp Neurol*, **321**(2), 177–92.

Martino, J., Brogna, C., Robles, S. G., Vergani, F., and Duffau, H. (2010). Anatomic dissection of the inferior fronto-occipital fasciculus revisited in the lights of brain stimulation data. *Cortex*, **46**(5), 691–9.

McKeefry, D.J. and Zeki, S. (1997). The position and topography of the human colour centre as revealed by functional magnetic resonance imaging. *Brain*, **120**(12), 2229–42.

Mechelli, A., Price, C.J., Friston, K.J., and Ishai, A. (2004). Where bottom-up meets top-down: neuronal interactions during perception and imagery. *Cereb Cortex*, **14**(11), 1256–65.

Meyer, A. (1907). The connections of the occipital lobes and the present status of the cerebral visual affections. *Trans Ass Am Phys*, **22**, 7–16.

Meynert, T. (1885). *A clinical treatise on diseases of the fore-brain based upon a study of its structure, functions, and nutrition* (Sachs, B. Trans.). New York: G.P. Putnam's Sons.

Mesulam. M.M. (2000). *Principles of behavioral and cognitive neurology*. New York: Oxford University Press.

Milner, A.D. and Goodale, M.A. (2008). Two visual systems re-viewed. *Neuropsychologia*, **46**(3), 774–785.

Mishkin, M., Ungerleider, L.G., and Macko, K.A. (1983). Object vision and spatial vision: two cortical pathways. *Trends Neurosci*, **6**(10), 414–17.

Morris, J.S., Friston, K.J., Büchel, C., et al. (1998). A neuromodulatory role for the human amygdala in processing emotional facial expressions. *Brain*, **121**(1), 47–57.

Mummery, C.J., Patterson, K., Wise, R.J., Vandenberghe, R., Price, C.J., and Hodges, J.R. (1999). Disrupted temporal lobe connections in semantic dementia. *Brain*, **122**(1), 61–73.

Muratoff W. (1893). Secundäre Degenerationen nach Durchschneidung des Balkens. *Neurologisches Centralblatt*, **12**, 714– 729.

Onufrowicz, W. (1887). Das balkenlose Mikrocephalengehirn Hoffman. Ein Beitrag zur pathologischen und normalen Anatomie des menschlichen Gehirnes. *Archiv für Psychiatrie*, **18**, 305–28.

Paul, L.K., Brown, W.S., Adolphs, R., et al. (2007). Agenesis of the corpus callosum: genetic, developmental and functional aspects of connectivity. *Nat Rev Neurosci*, **8**(4), 287–99.

Pessoa, L., McKenna, M., Gutierrez, E., and Ungerleider, L.G. (2002). Neural processing of emotional faces requires attention. *Proc Natl Acad Sci USA*, **99**(17), 11458–63.

Petrides, M. and Pandya, D.N. (2007). Efferent association pathways from the rostral prefrontal cortex in the macaque monkey. *J Neurosci*, **27**(43), 11573–86.

Philippi, C.L., Mehta, S., Grabowski, T., Adolphs, R., and Rudrauf, D. (2009). Damage to association fiber tracts impairs recognition of the facial expression of emotion. *J Neurosci*, **29**(48), 15089–99.

Pins, D. and ffytche, D.H. (2003). The neural correlates of conscious vision. *Cereb Cortex*, **13**(5), 461–74.

Polyak, S. (1957). *The vertebrate visual system*. Chicago, IL: University of Chicago Press.

Reil, J.C. (1812). Nachträge zur anatomie des großen und kleinen Gehirns. *Arch Physiol*, **11**, 345–76.

Riddoch, G. (1917). Dissociation of visual perceptions due to occipital injuries, with especial reference to appreciation of movement. *Brain*, **40**(1), 15–57.

Ross, E.D. (1980). Sensory-specific and fractional disorders of recent memory in man. I. Isolated loss of visual recent memory. *Arch Neurol*, **37**(4), 193–200.

Rudrauf, D., David, O., Lachaux J.P., et al. (2008). Rapid interactions between the ventral visual stream and emotion-related structures rely on a two-pathway architecture. *J Neurosci*, **28**(11), 2793–803.

Sachs, H. (1892). *Das Hemisphärenmark des menschlichen grosshirns*. Leipzig: Verlag Von Georg Thieme.

Schendan, H.E. and Stern, C.E. (2008). Where vision meets memory: prefrontal-posterior networks for visual object constancy during categorization and recognition. *Cereb Cortex*, **18**(7), 1695–711.

Schmahmann, J.D. and Pandya, D.N. (2006). *Fiber Pathways of the Brain*. New York: Oxford University Press.

Schmahmann, J.D., Pandya, D.N., Wang, R., et al. (2007). Association fiber pathways of the brain: Parallel observations from diffusion spectrum imaging and autoradiography. *Brain*, **130**(3), 630–53.

Schneider, A. (2008). *The confabulating mind: how the brain creates reality*. Oxford: Oxford University Press.

Schneider, R.C., Crosby, E.C., Bagchi, B.K., and Calhoun, H.D. (1961). Temporal or occipital lobe hallucinations triggered from frontal lobe lesions. *Neurology*, **11**, 172–9.

Sereno, M.I., Dale, A.M., Reppas, J.B., et al. (1995). Borders of multiple visual areas in humans revealed by functional magnetic resonance imaging. *Science*, **268**(5212), 889–93.

Sereno, M.I. and Tootell, R.B. (2005). From monkeys to humans: what do we now know about brain homologies? *Curr Opin Neurobiol*, **15**(2), 135–44.

Shaywitz, B.A., Shaywitz, S.E., Pugh, K.R., et al. (2002). Disruption of posterior brain systems for reading in children with developmental dyslexia. *Biol Psychiat*, **52**(2), 101–10.

Solms, M. (1997). *The Neuropsychology of Dreams: a Clinico-Anatomical Study*. Mahwah: Lawrence Erlbaum Associates.

Steffan, P. (1881) Beitrag zur Pathologie des Farbensinnes. *Archiv fir Ophthalmologie*, **27**, 1–24.

Talairach, J. and Tournoux, P. (1988). *Co-planar stereotaxic atlas of the human brain: an approach to medical cerebral imaging*. Stuttgart; New York: Thieme Medical Publishers.

Talbot, S.A. and Marshall, W.H. (1941). Physiological studies on neural mechanisms of visual localization and discrimination. *Am J Ophthal* **24**, 1255–63.

Thomas, C., Avidan, G., Humphreys, K., Jung, K-J, Gao, F., and Behrmann, M. (2009). Reduced structural connectivity in ventral visual cortex in congenital prosopagnosia. *Nat Neurosci*, **12**(21), 29–31.

Tomita, H., Ohbayashi, M., Nakahara, K., Hasegawa, I., and Miyashita, Y. (1999). Top-down signal from prefrontal cortex in executive control of memory retrieval. *Nature*, **401**(6754), 699–703.

Tootell, R.B., Hadjikhani, N., Mendola, J.D., Marrett, S., and Dale, A.M. (1998). From retinotopy to recognition: fMRI in human visual cortex. *Trends Cog Sci*, **2**(5), 174–83.

Türe, U., Yaşargil, M.G., Pait, T.G. (1997). Is there a superior occipitofrontal fasciculus? A microsurgical anatomic study. *Neurosurgery*, **40**(6),1226–32.

Türe, U., Yaşargil, M.G., Friedman, A.H., et al. (2000). Fiber dissection technique: lateral aspect of the brain. *Neurosurgery*, **47**(2) 417–26.

Tusa, R.J. and Ungerleider, L.G. (1985). The inferior longitudinal fasciculus: a reexamination in humans and monkeys. *Ann Neurol*, **18**(5), 583–91.

Trolard, P. (1906). Le faisceau longitudinal inferieur du cerveau. *Revue Neurologique* **14**, 440–6.

Ungerleider, L.G., and Mishkin, M. (1982). Two cortical visual systems. In Ingle, D.J., Goodale, M.A., and Mansfield, R.J.W. (Eds.) *Analysis of visual behaviour*, pp. 549–86. Cambridge, MA: MIT Press.

Urbanski, M., Thiebaut de Schotten, M., Rodrigo, S., et al. (2008). Brain networks of spatial awareness: evidence from diffusion tensor imaging tractography. *J Neurol Neurosurg Psychiat*, **79**(5), 598–601.

Van Essen, D.C. and Maunsell, J.H.R. (1983). Hierarchical organization and functional streams in the visual cortex. *Trends Neurosci*, **6**, 370–5.

Vignal, J.P., Chauvel, P., and Halgren, E. (2000). Localised face processing by the human prefrontal cortex: stimulation-evoked hallucinations of faces. *Cognit Neuropsychol*, **17**(1), 281–91.

Wandell, B.A. and Winawer, J. (2011). Imaging retinotopic maps in the human brain. *Vision Res*, **51**(7), 718–37.

Wigan, A.L. (1844). *A New View of Insanity: The Duality of the Mind Proved by the Structure, Functions, and Diseases of the Brain, and by the Phenomena of Mental Derangement and Shown to Be Essential to Moral Responsibility*. London: Longman, Brown, Green and Longmans.

Wilson, C.L., Babb, T.L., Halgren, E., and Crandall, P. H. (1983). Visual receptive fields and response properties of neurons in human temporal lobe. *Brain*, **106**(2), 473–502.

Zeki, S. (1990). A century of cerebral achromatopsia. Brain, **113**(6), 1721–77.

Zeki, S. and ffytche, D.H. (1998). The Riddoch syndrome: insights into the neurobiology of conscious vision. *Brain*, **121**(1), 25–45.

Zeki, S. and Shipp, S. (1988). The functional logic of cortical connections. *Nature*, **335**(6188), 311–17.

Zeki, S., Watson, J.D.G., Lueck, C.J., Friston, K.J., Kennard, C., and Frackowiak, R.S.J. (1991). A direct demonstration of functional specialization in human visual cortex. *J Neurosci*, **11**(3), 641–9.

Zeki, S., McKeefry, D.J., Bartels, A., and Frackowiak, R.S. (1998). Has a new color area been discovered? *Nat Neurosci*, **1**(5), 335–6.

Optic radiations (atlas)

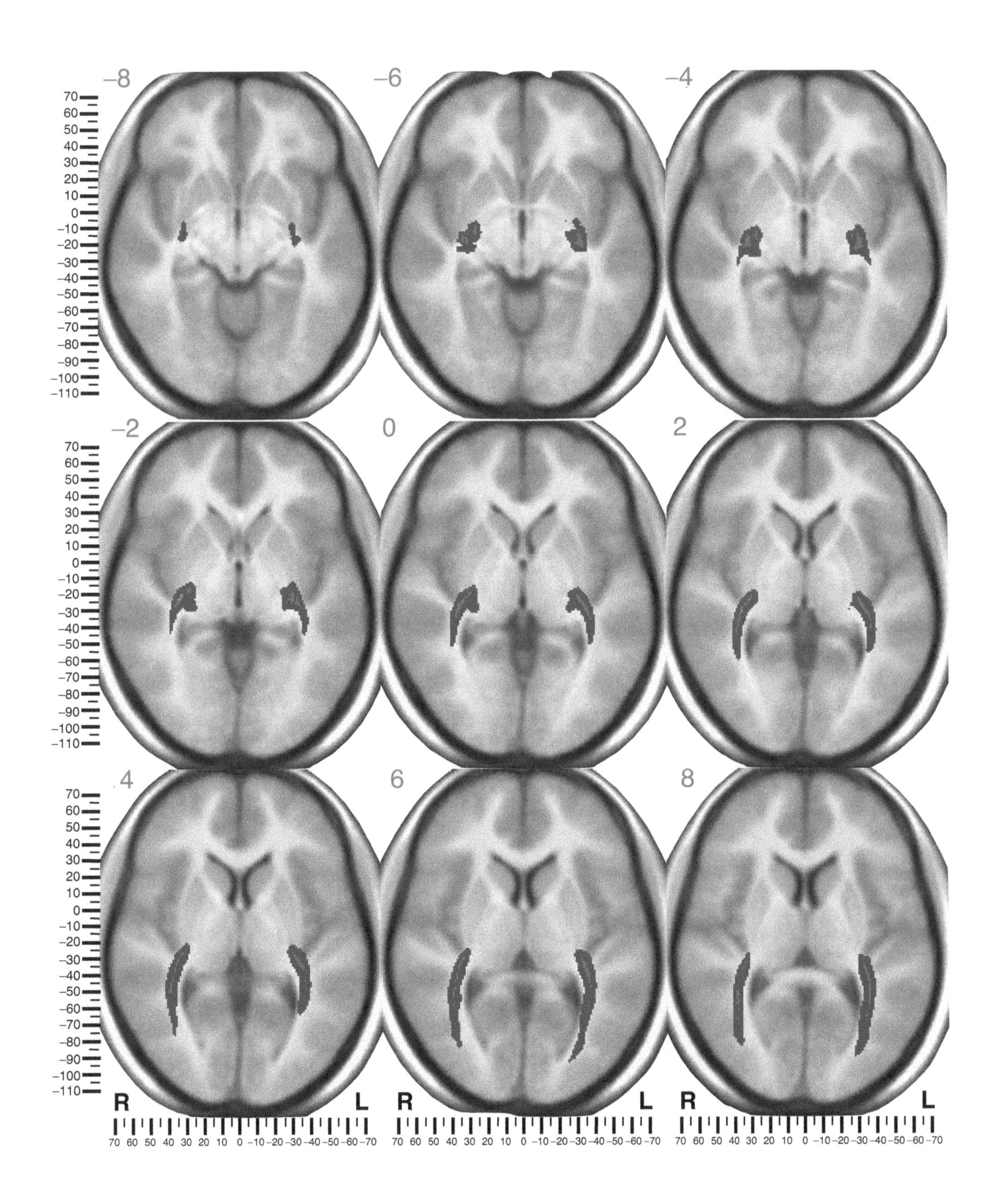

Optic radiations: inferior axial view

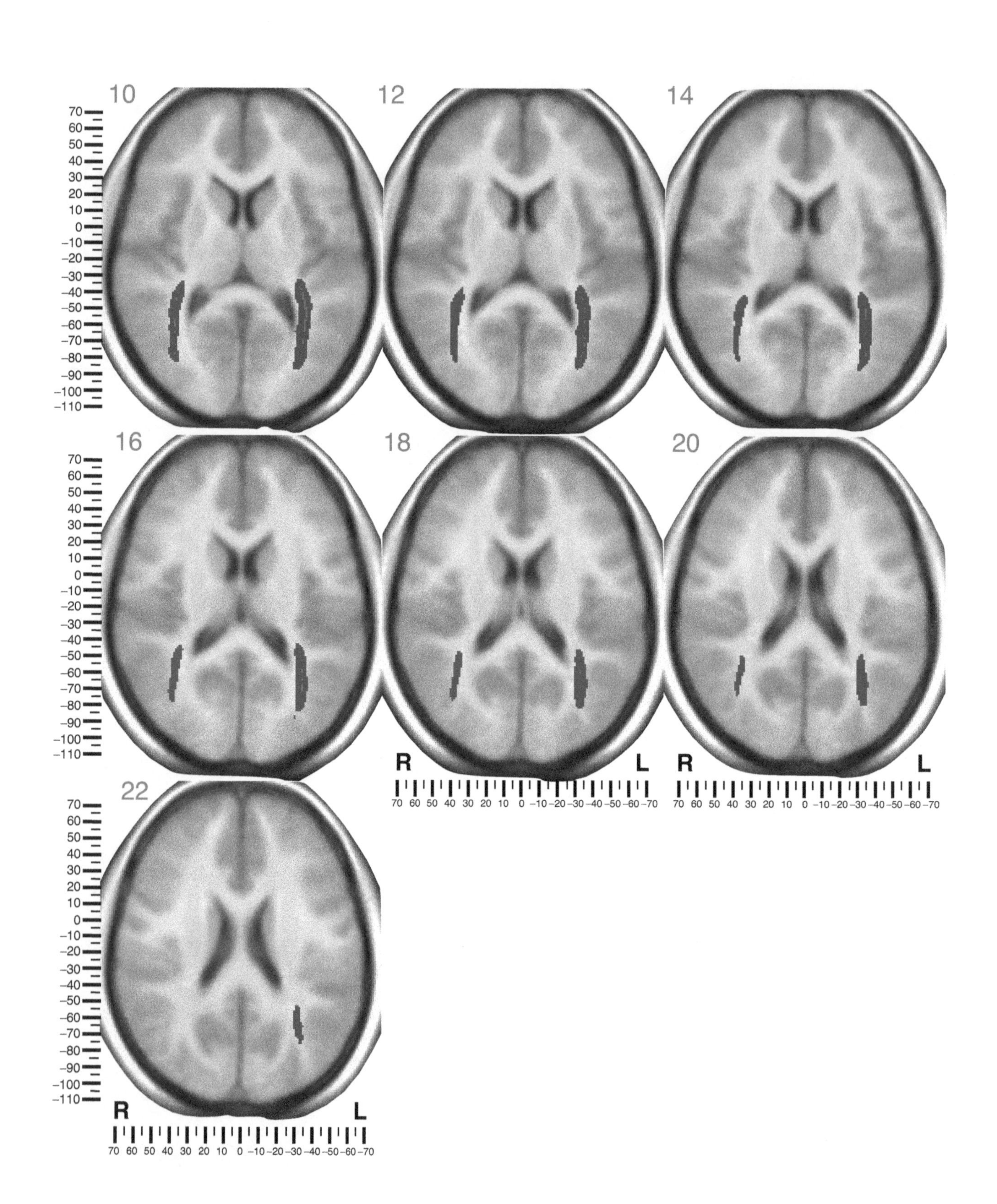

Optic radiations: anterior coronal view

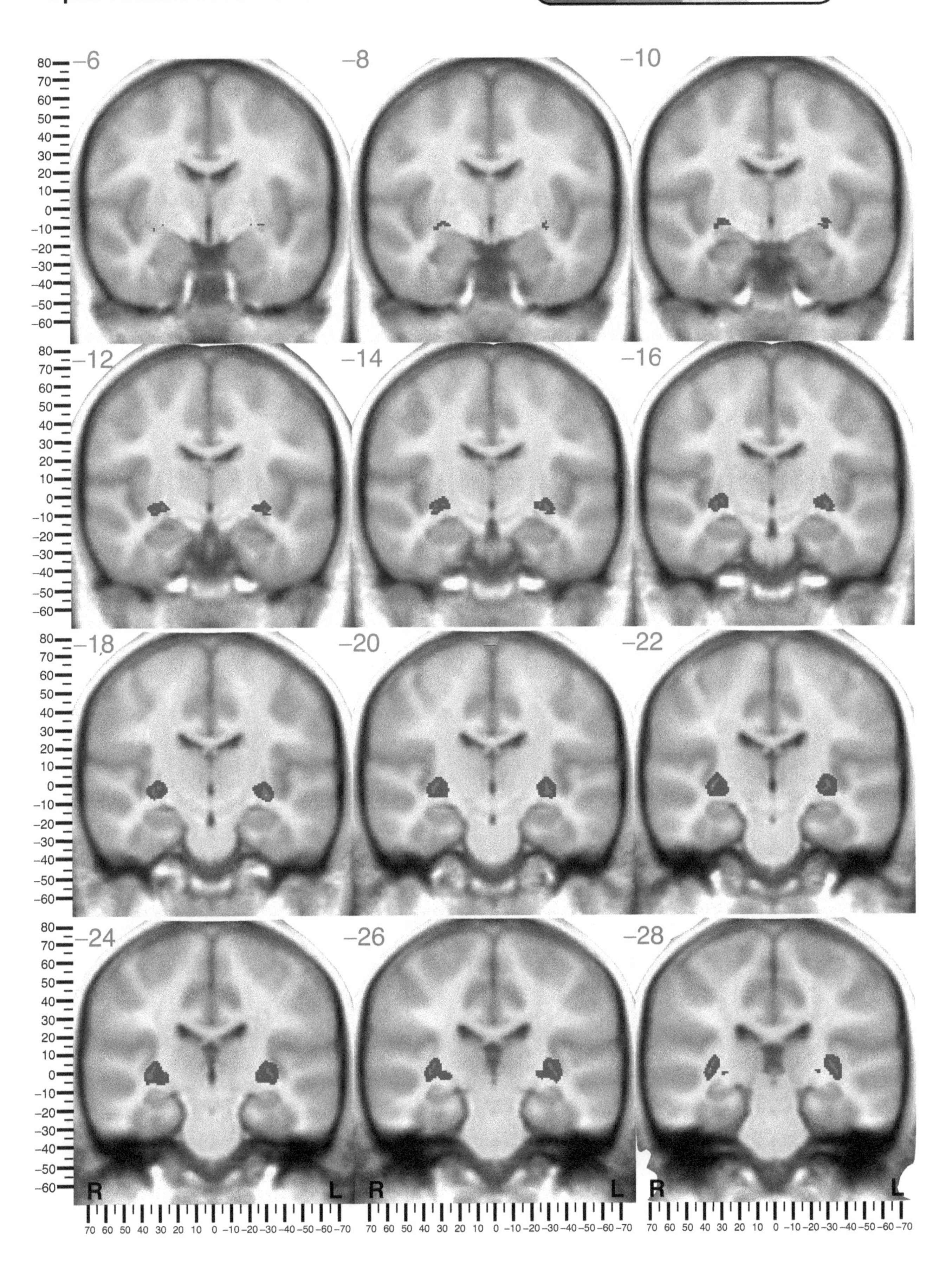

Optic radiations: anterior coronal view

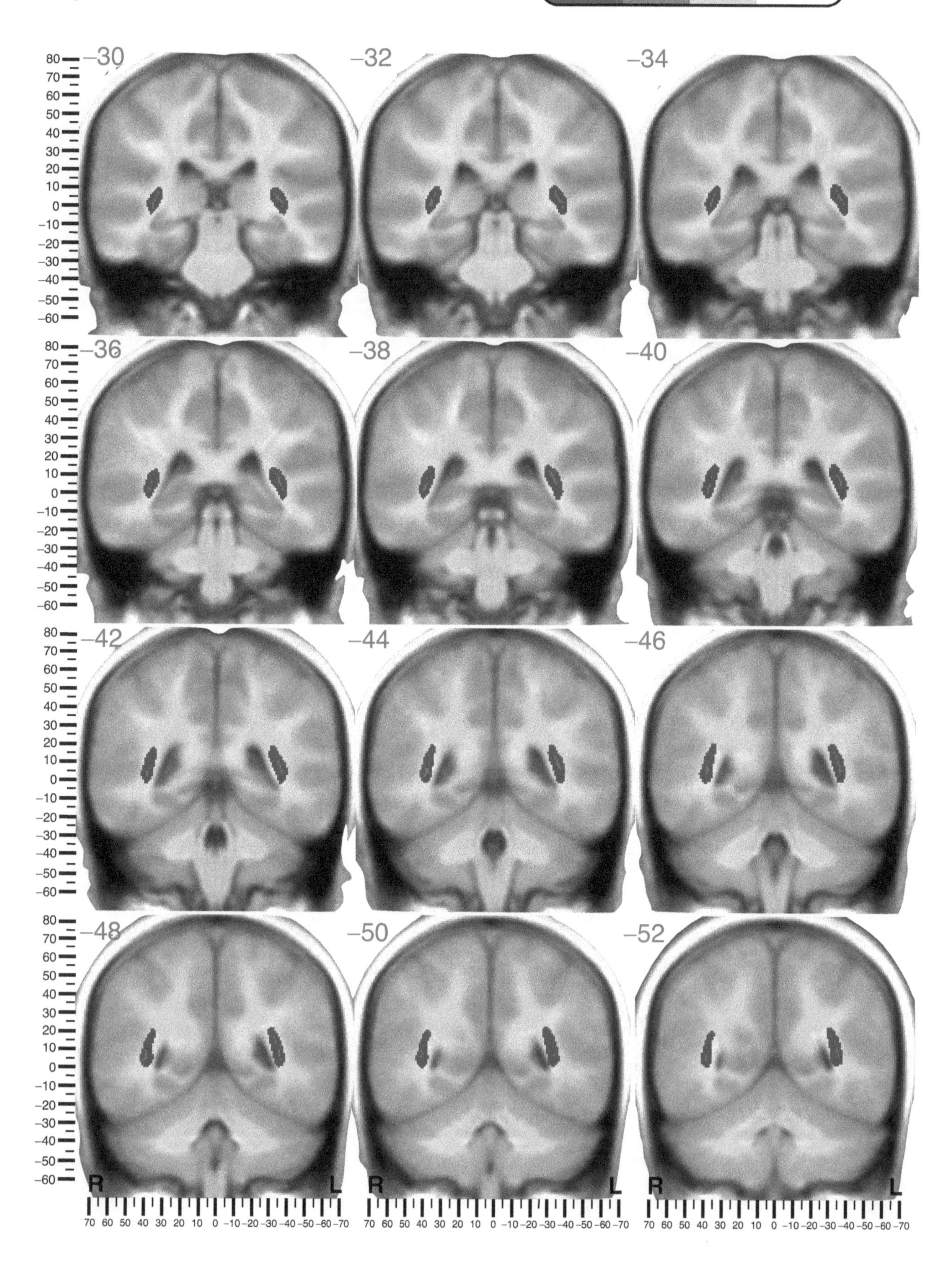

Optic radiations: anterior coronal view

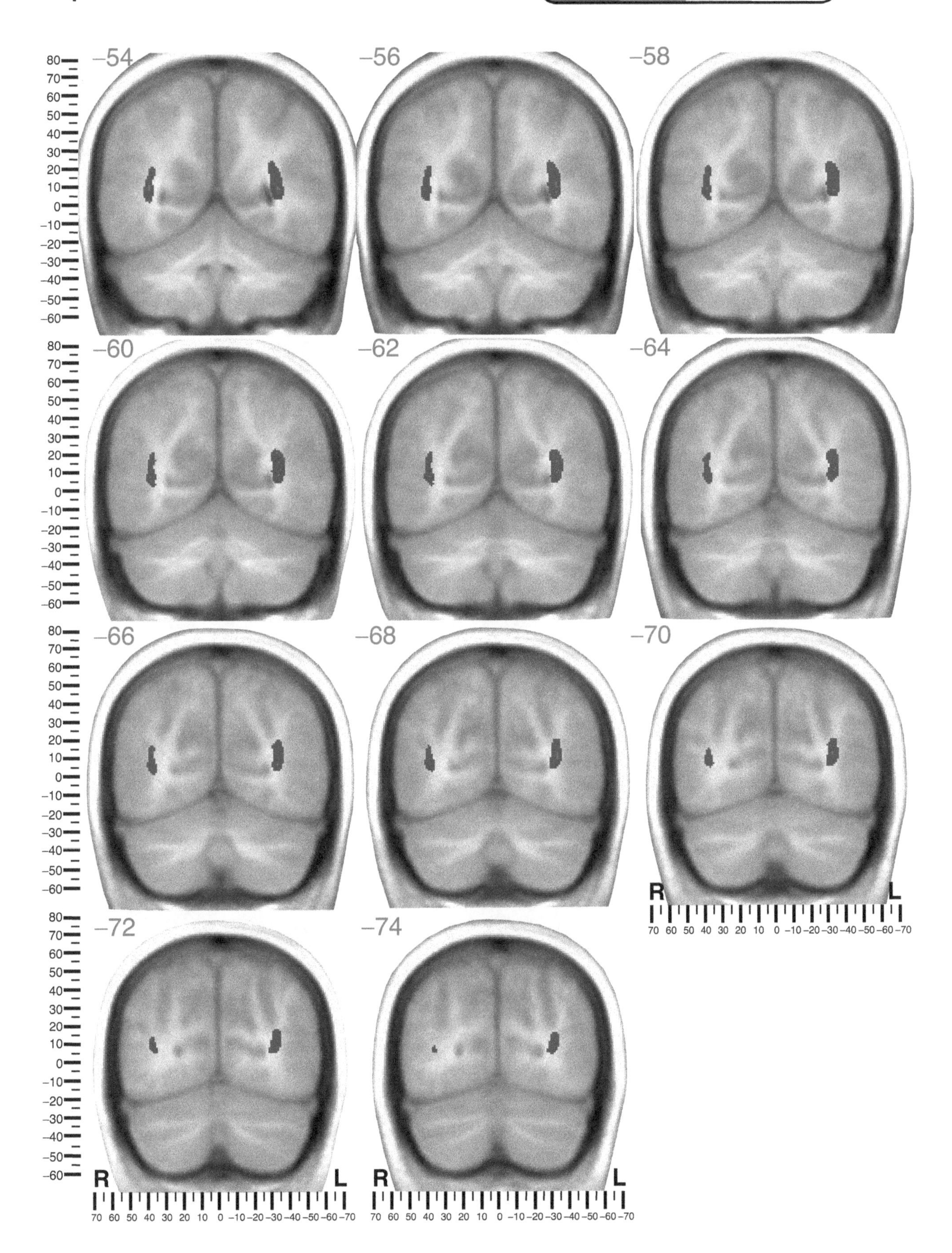

Optic radiations (right): lateral sagittal view

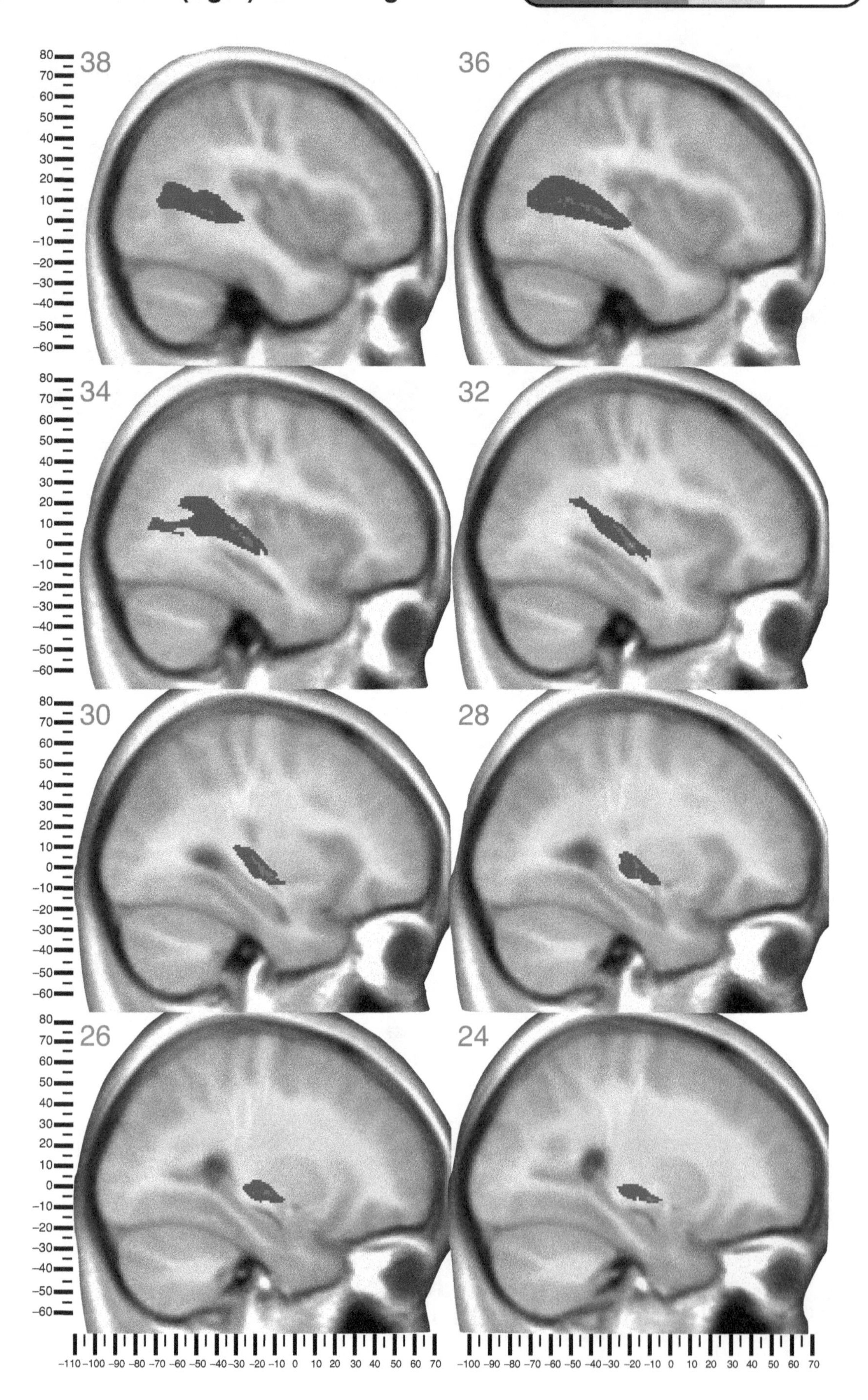

Optic radiations (left): medial sagittal view

p < .05 | > 50% | > 75% | > 90%

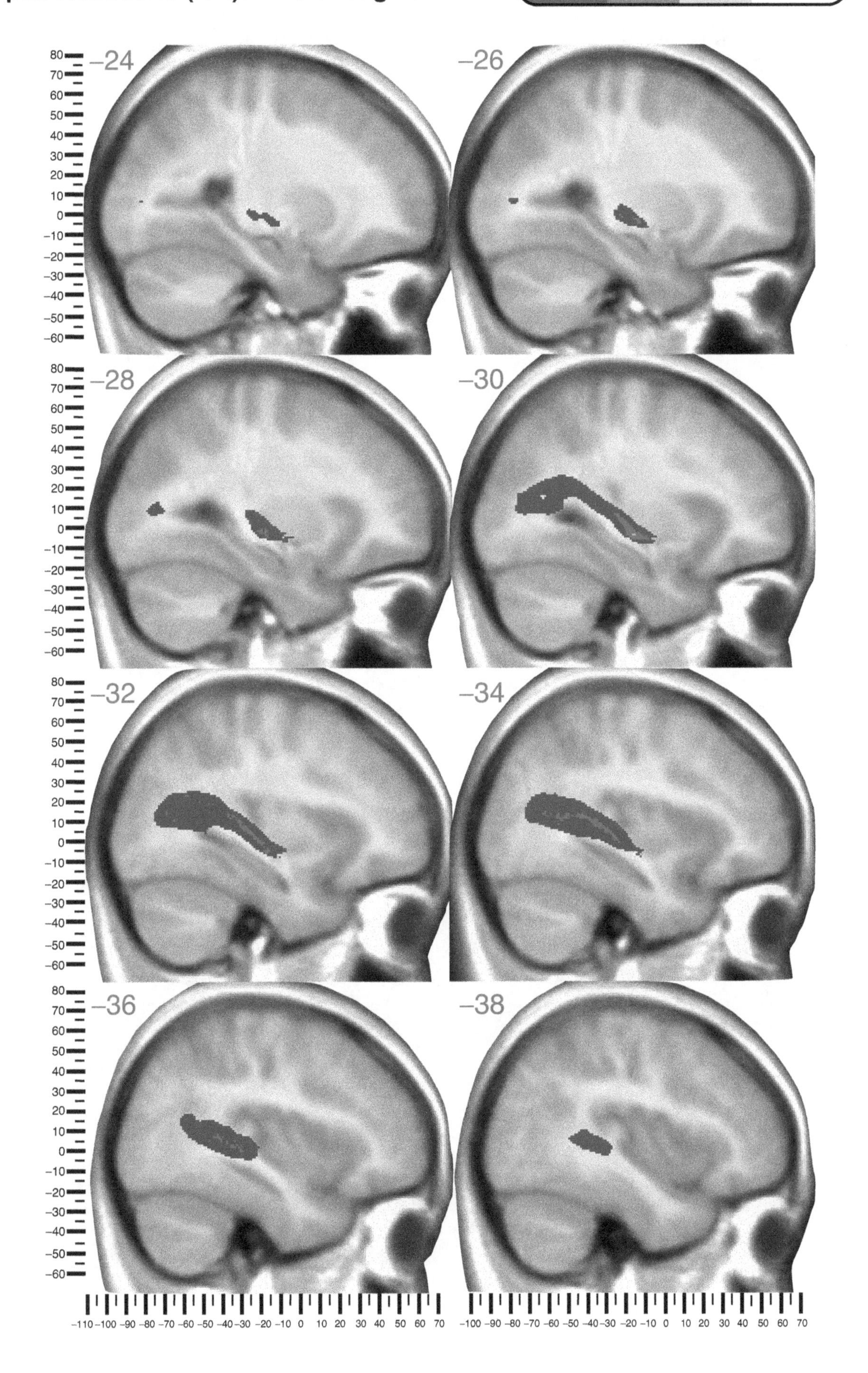

Inferior longitudinal fasciculus (atlas)

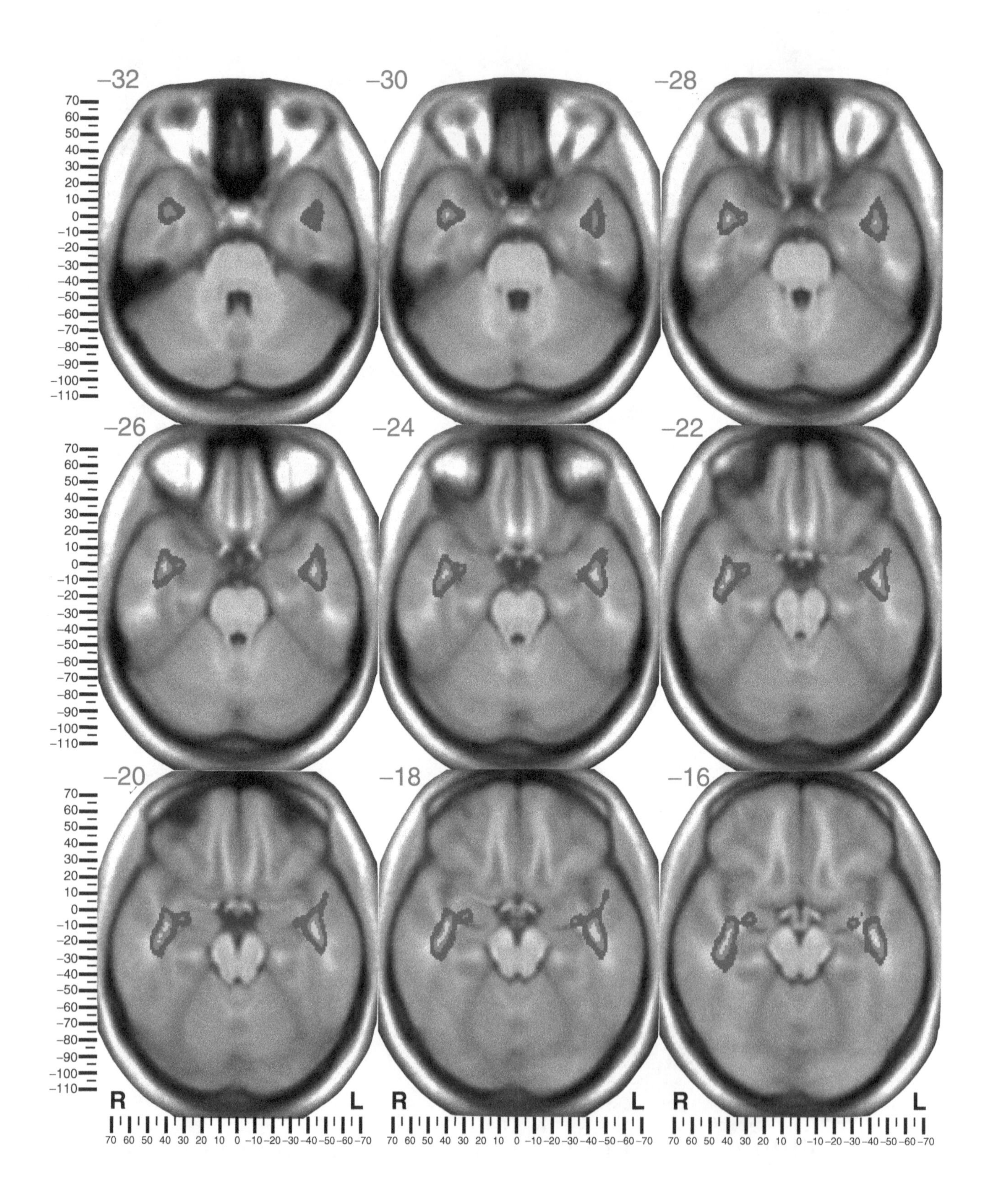

Inferior longitudinal: inferior axial view

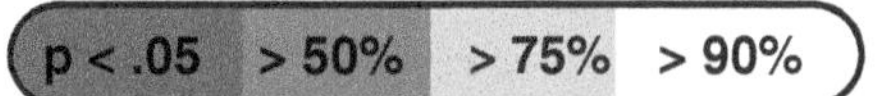

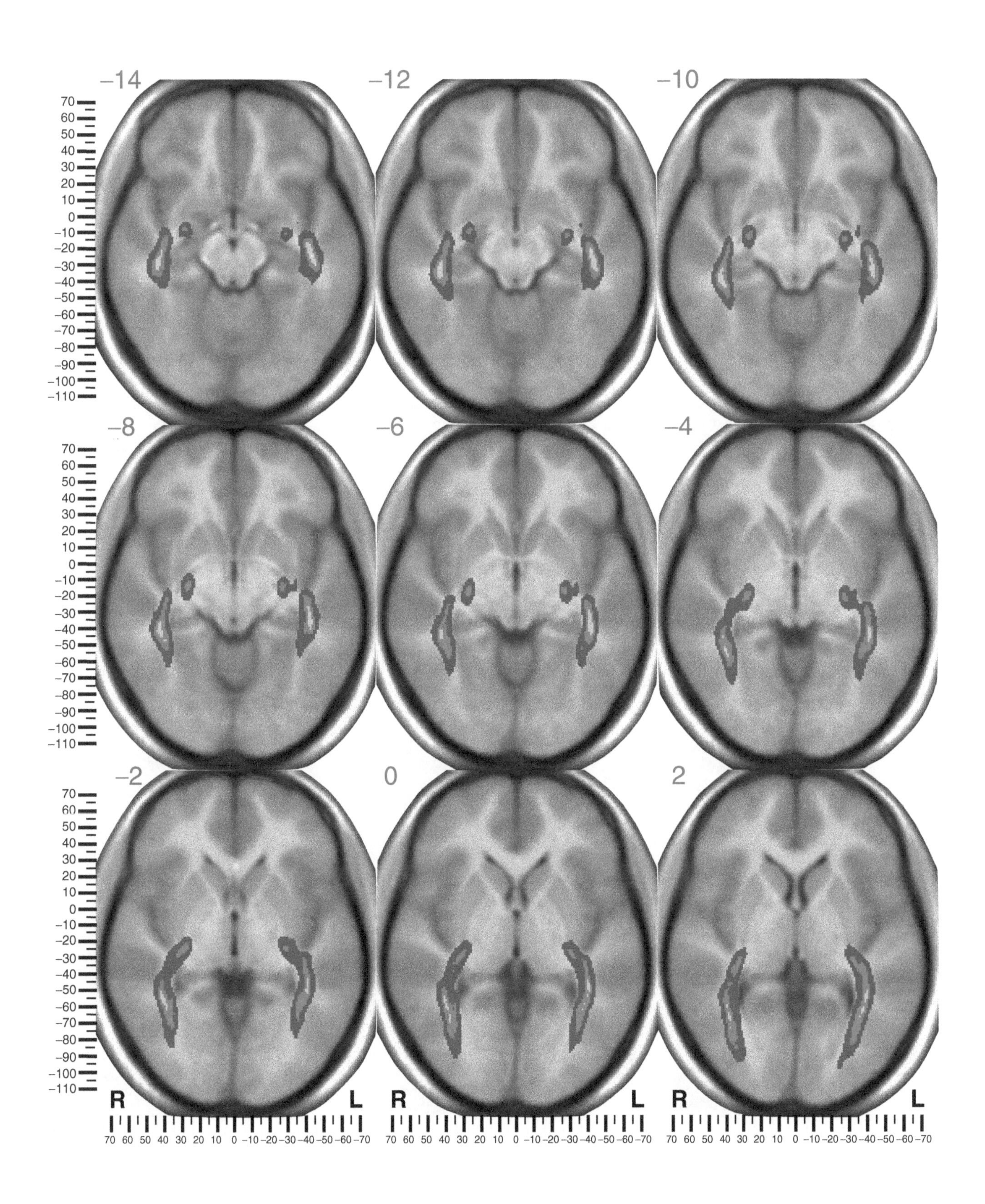

Inferior longitudinal: inferior axial view

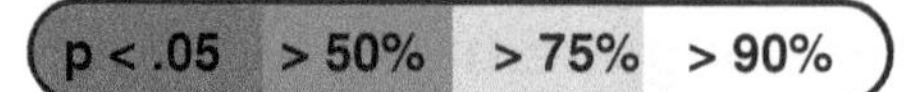

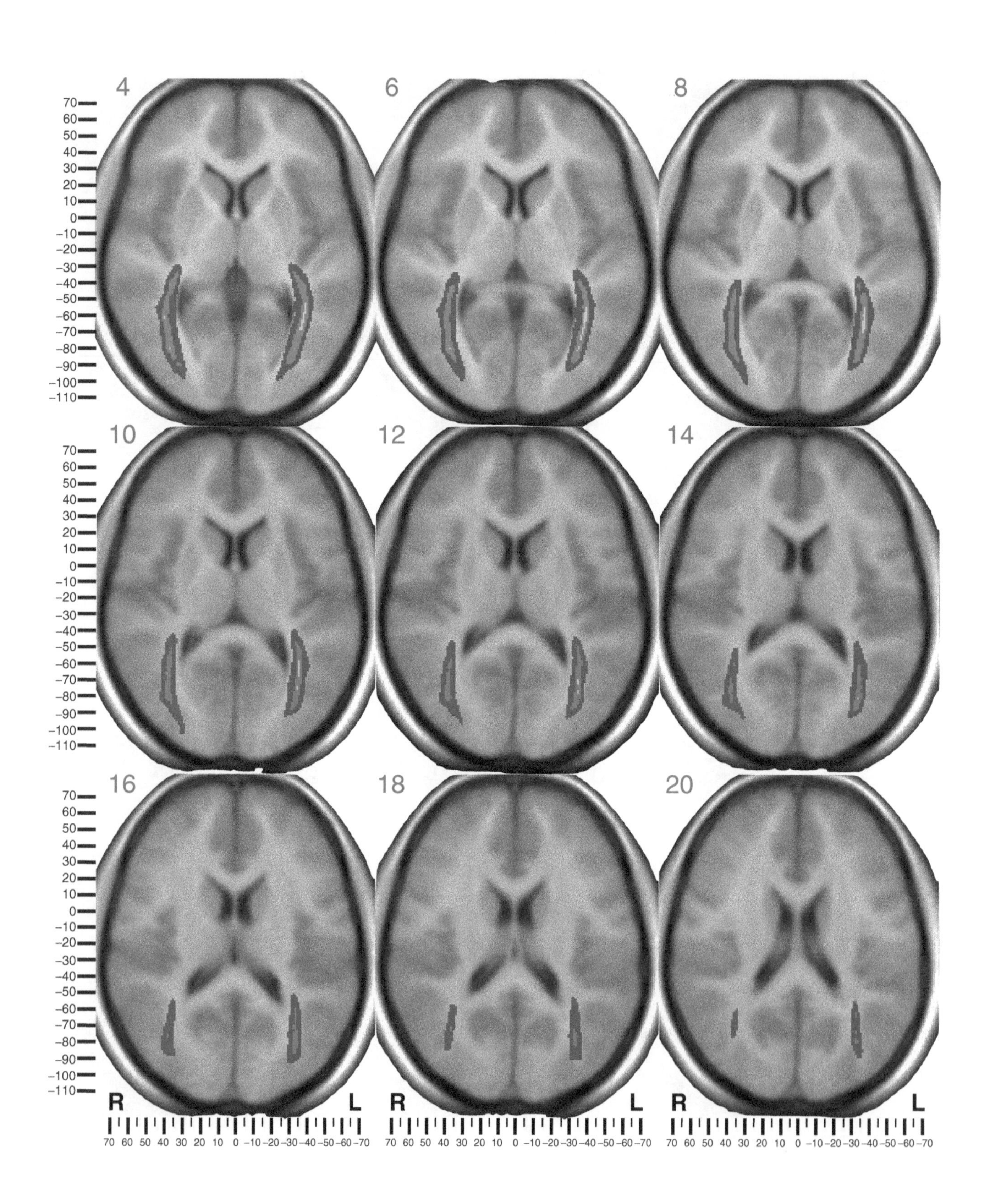

Inferior longitudinal: anterior coronal view

p < .05 | > 50% | > 75% | > 90%

Inferior longitudinal: anterior coronal view

p < .05 | > 50% | > 75% | > 90%

–12 –14 –16 –18 –20 –22 –24 –26 –28 –30 –32 –34

R L

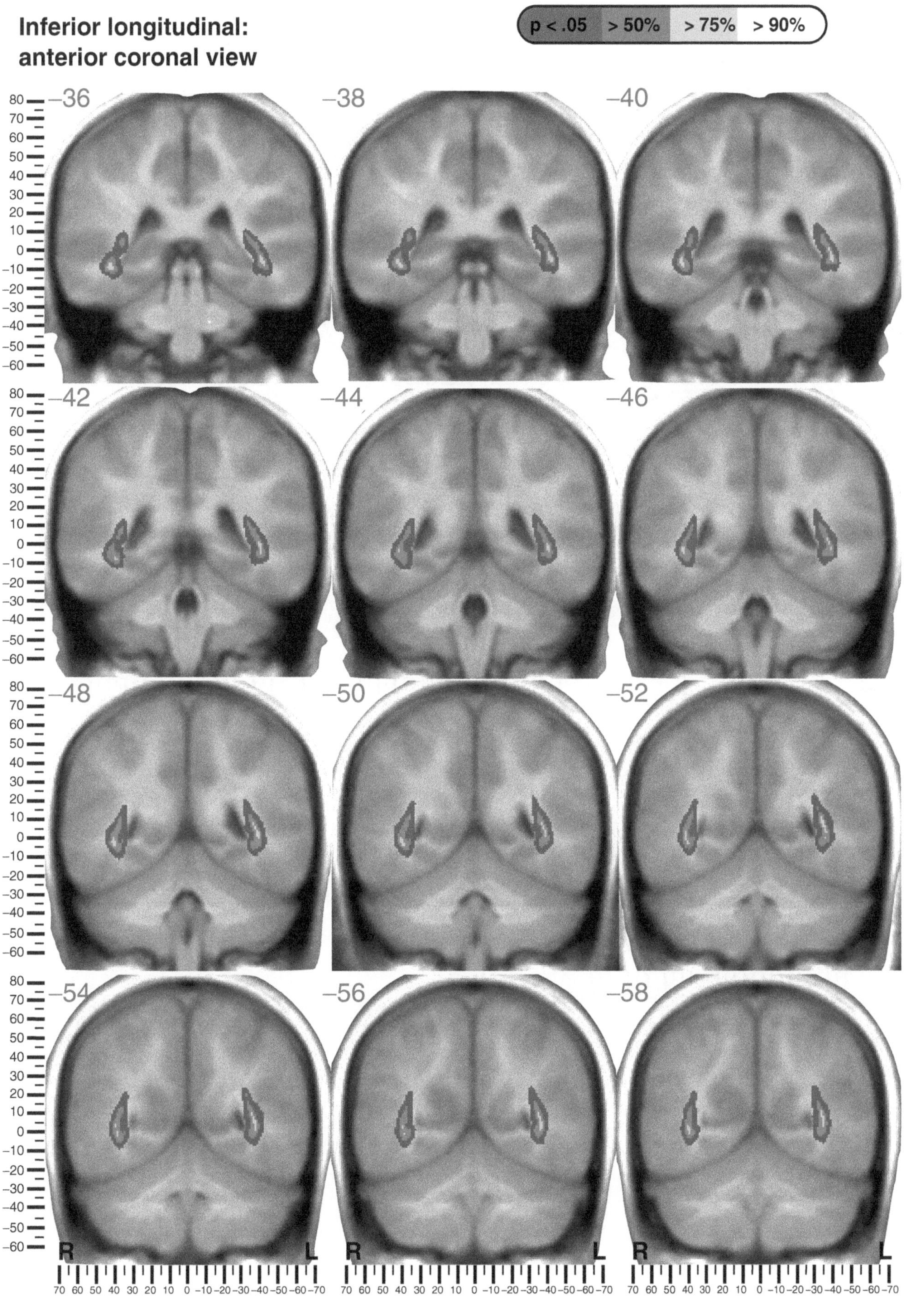

Inferior longitudinal:
anterior coronal view

p < .05 | > 50% | > 75% | > 90%

–36 –38 –40 –42 –44 –46 –48 –50 –52 –54 –56 –58

R L

Inferior longitudinal: anterior coronal view

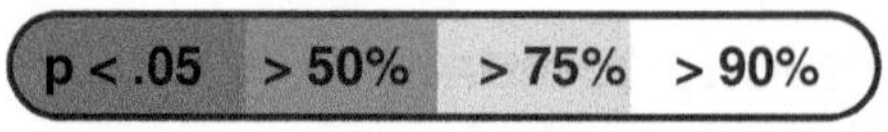

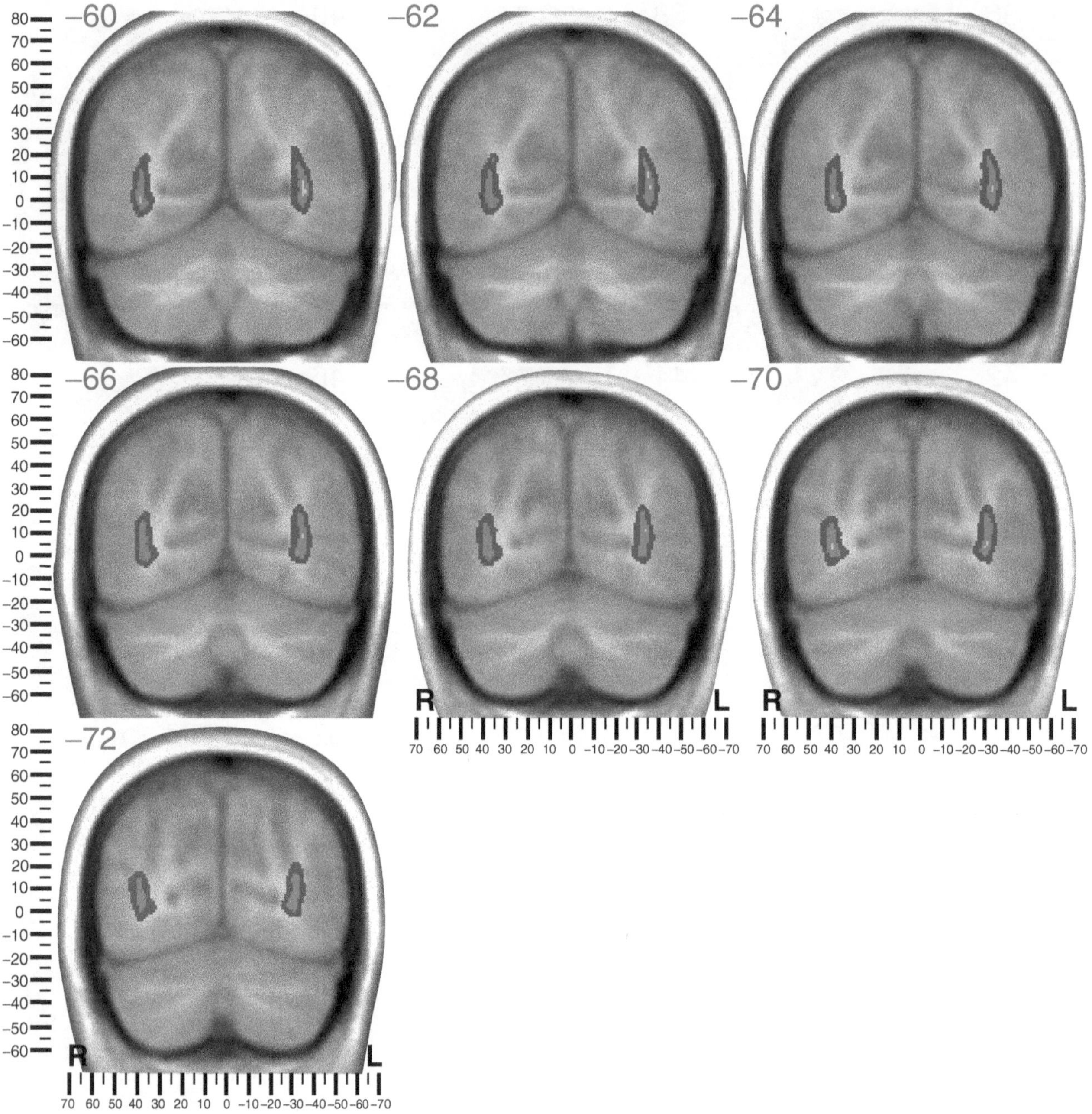

Inferior longitudinal (right): medial sagittal view

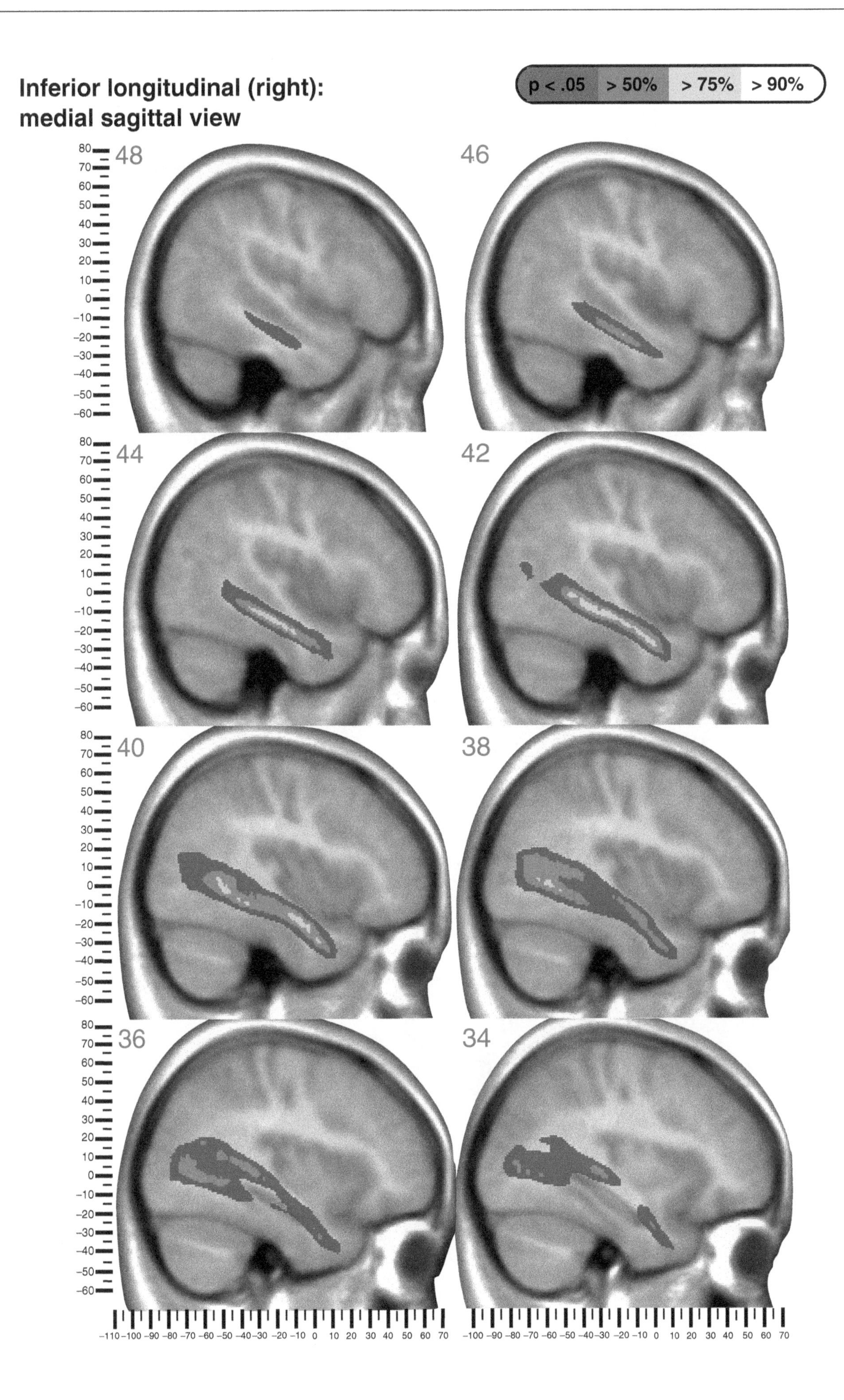

Inferior longitudinal (right): medial sagittal view

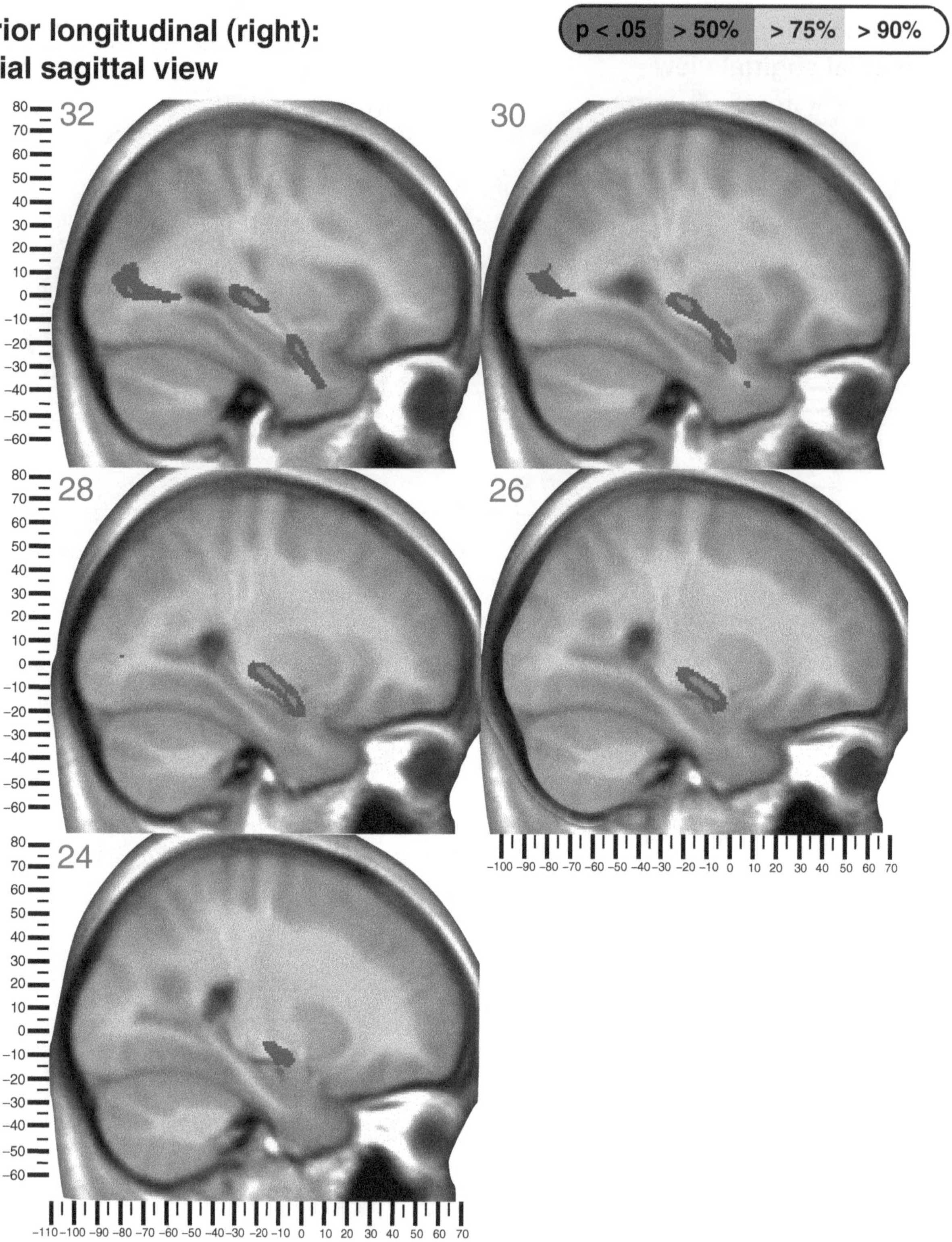

p < .05 | > 50% | > 75% | > 90%

32

30

28

26

24

Inferior longitudinal (left): lateral sagittal view

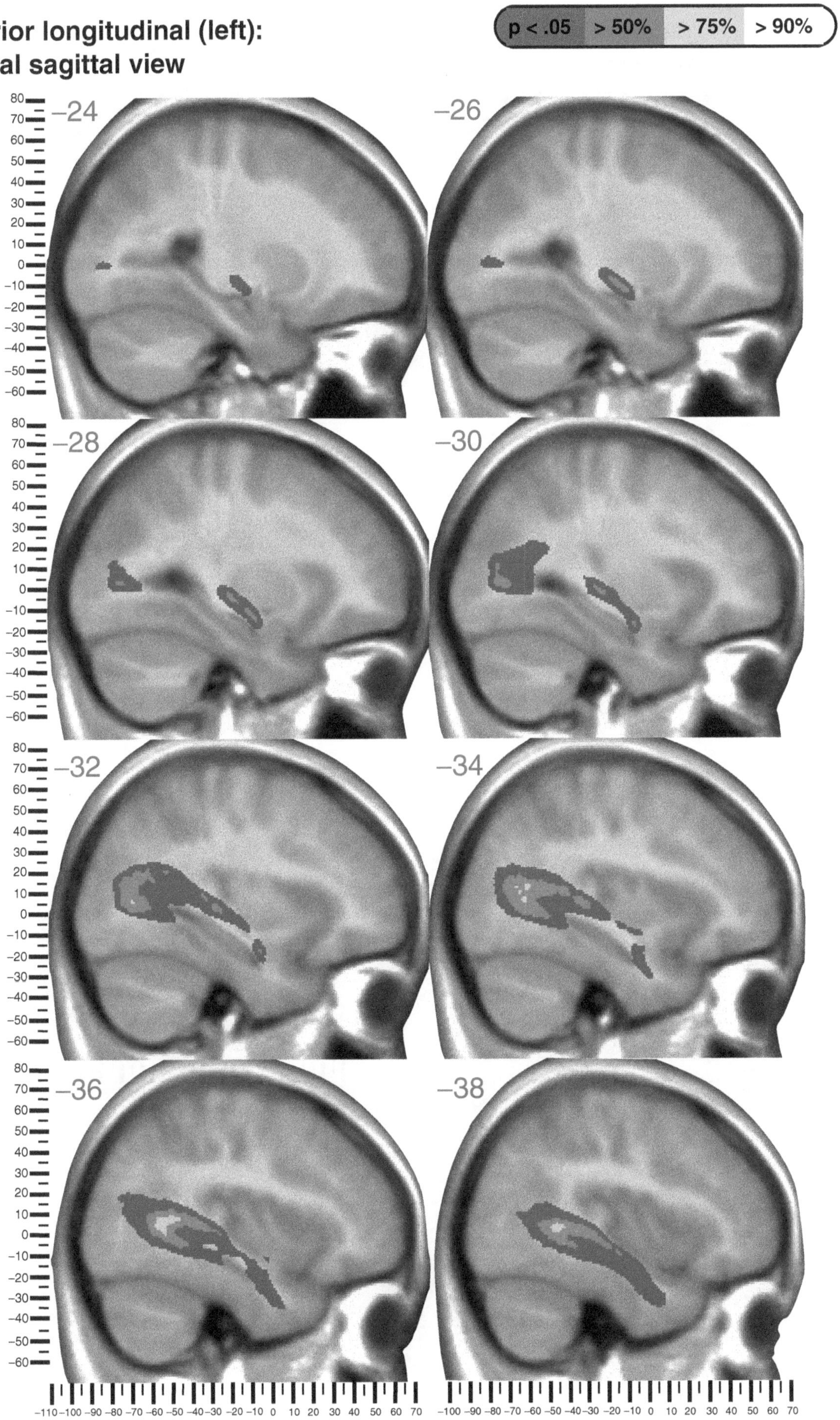

p < .05 | > 50% | > 75% | > 90%

-24 -26 -28 -30 -32 -34 -36 -38

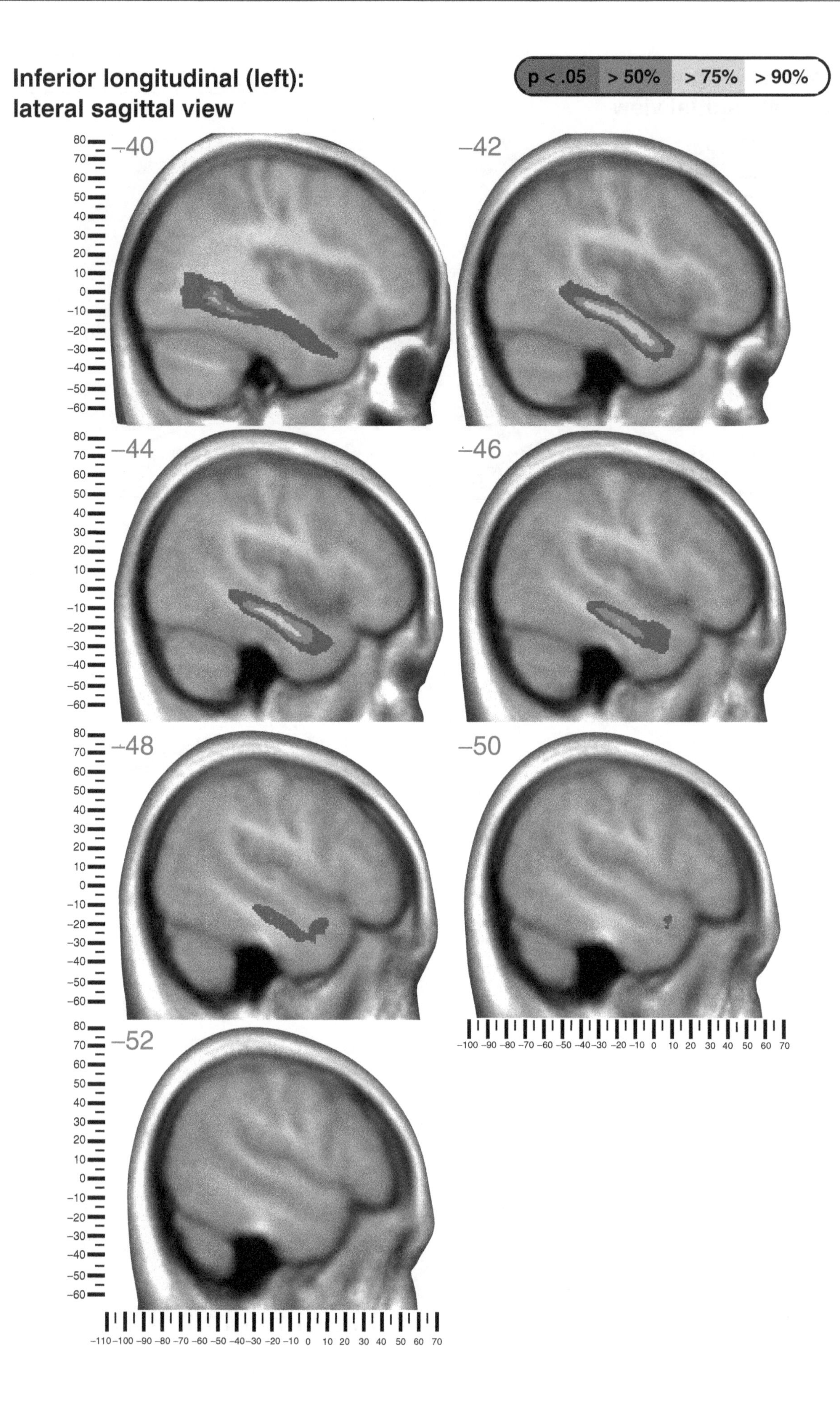
Inferior longitudinal (left):
lateral sagittal view
p < .05
> 50%
> 75%
> 90%
−40
−42
−44
−46
−48
−50
−52

Inferior fronto-occipital fasciculus (atlas)

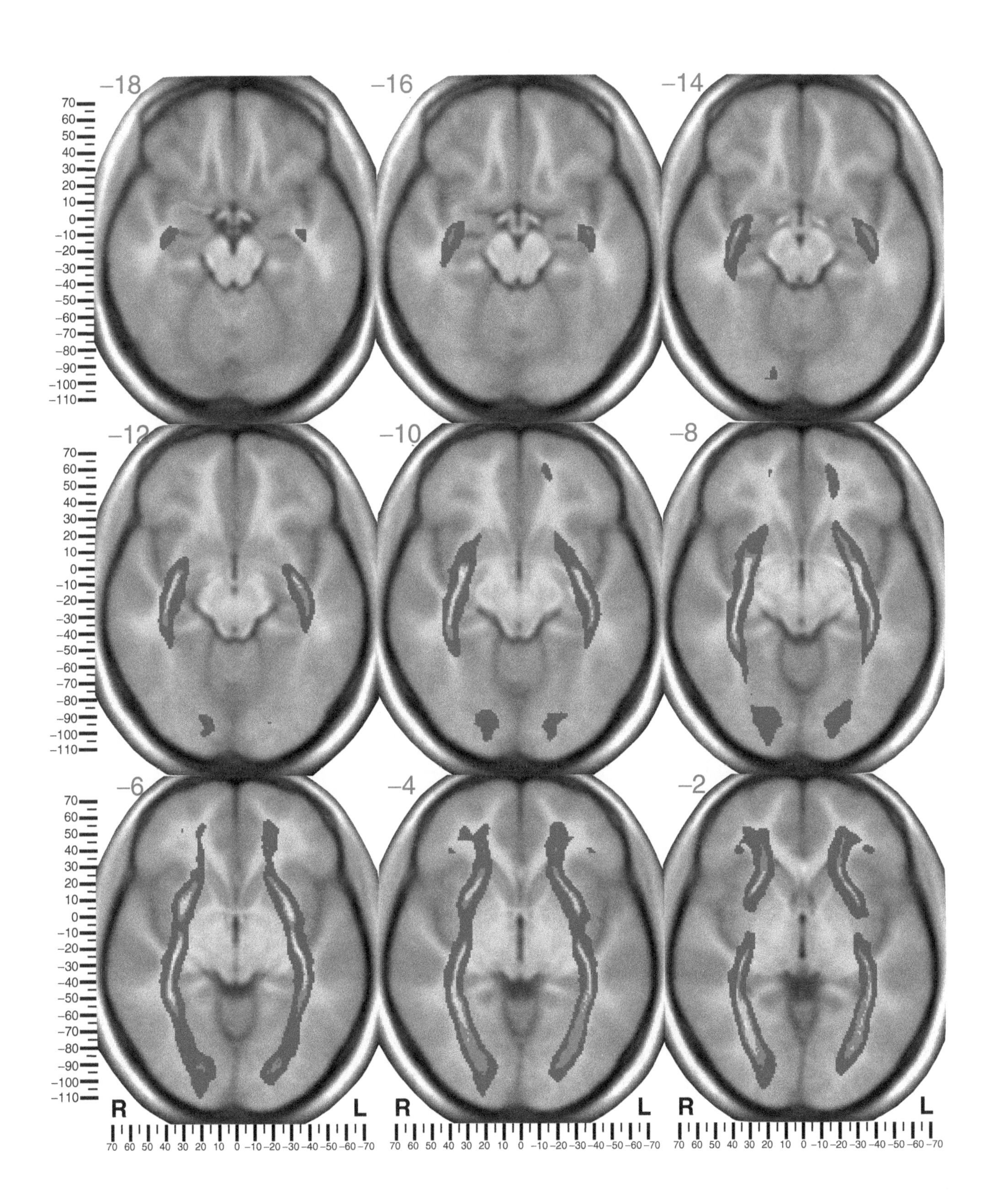

Inferior fronto-occipital: inferior axial view

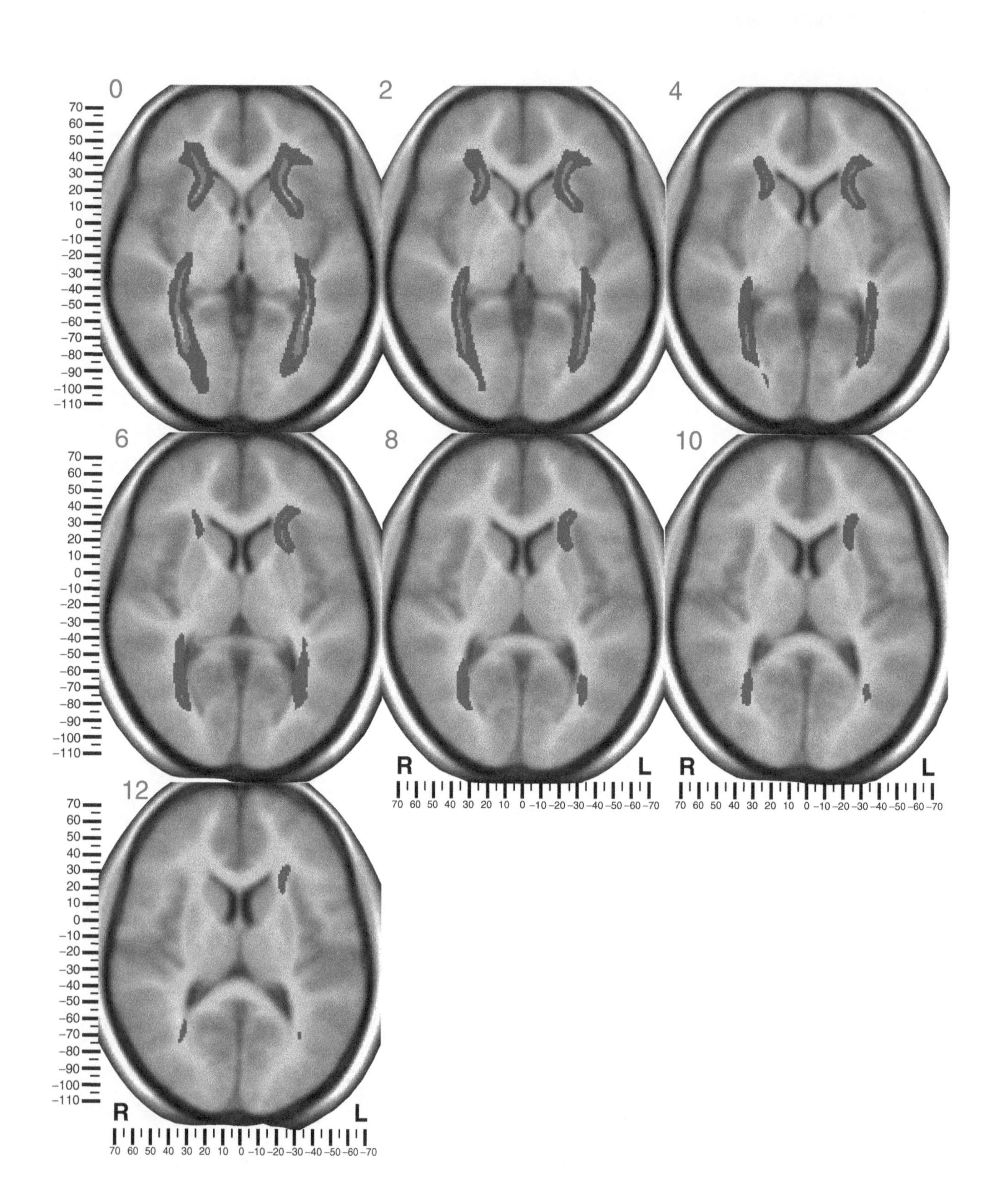

Inferior fronto-occipital: anterior coronal view

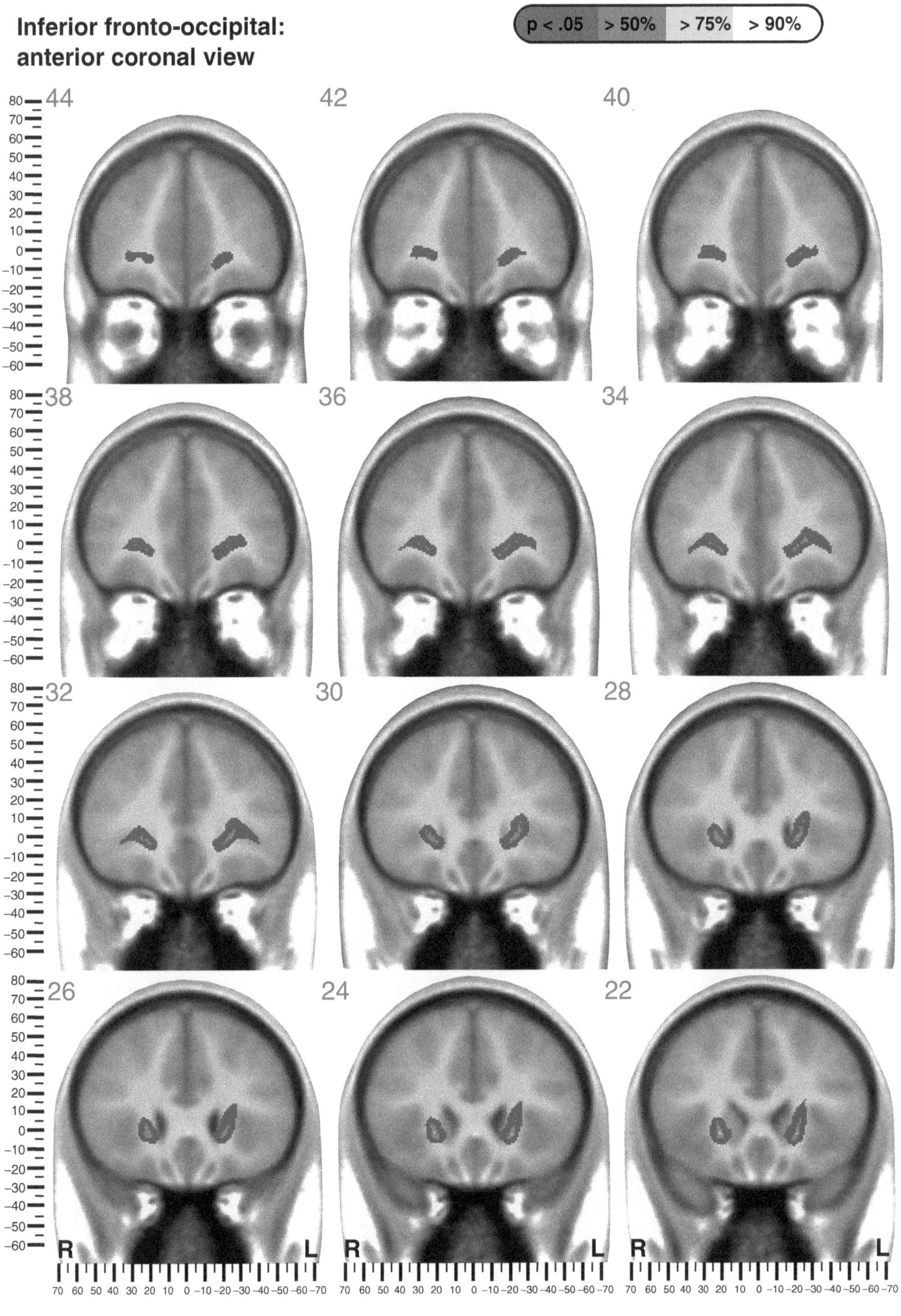

p < .05 > 50% > 75% > 90%

80 70 60 50 40 30 20 10 0 −10 −20 −30 −40 −50 −60

44 42 40

80 70 60 50 40 30 20 10 0 −10 −20 −30 −40 −50 −60

38 36 34

80 70 60 50 40 30 20 10 0 −10 −20 −30 −40 −50 −60

32 30 28

80 70 60 50 40 30 20 10 0 −10 −20 −30 −40 −50 −60

26 24 22

R L R L R L

70 60 50 40 30 20 10 0 −10 −20 −30 −40 −50 −60 −70

70 60 50 40 30 20 10 0 −10 −20 −30 −40 −50 −60 −70

70 60 50 40 30 20 10 0 −10 −20 −30 −40 −50 −60 −70

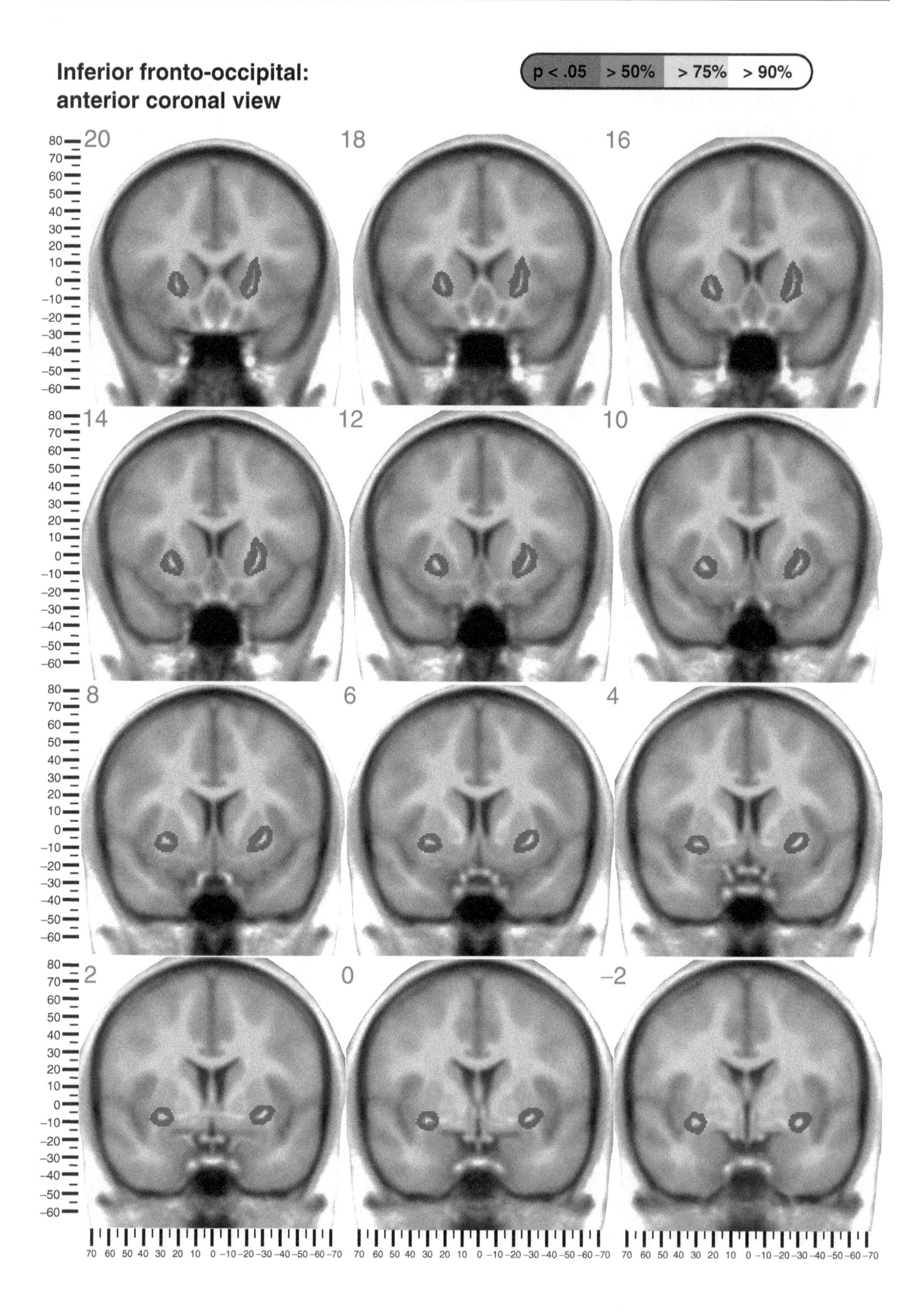
Inferior fronto-occipital:
anterior coronal view
p < .05
> 50%
> 75%
> 90%
20
18
16
14
12
10
8
6
4
2
0
–2

Inferior fronto-occipital: anterior coronal view

p < .05 | > 50% | > 75% | > 90%

-4 -6 -8 -10 -12 -14 -16 -18 -20 -22 -24 -26

80 70 60 50 40 30 20 10 0 -10 -20 -30 -40 -50 -60

R L

70 60 50 40 30 20 10 0 -10 -20 -30 -40 -50 -60 -70

Inferior fronto-occipital: anterior coronal view

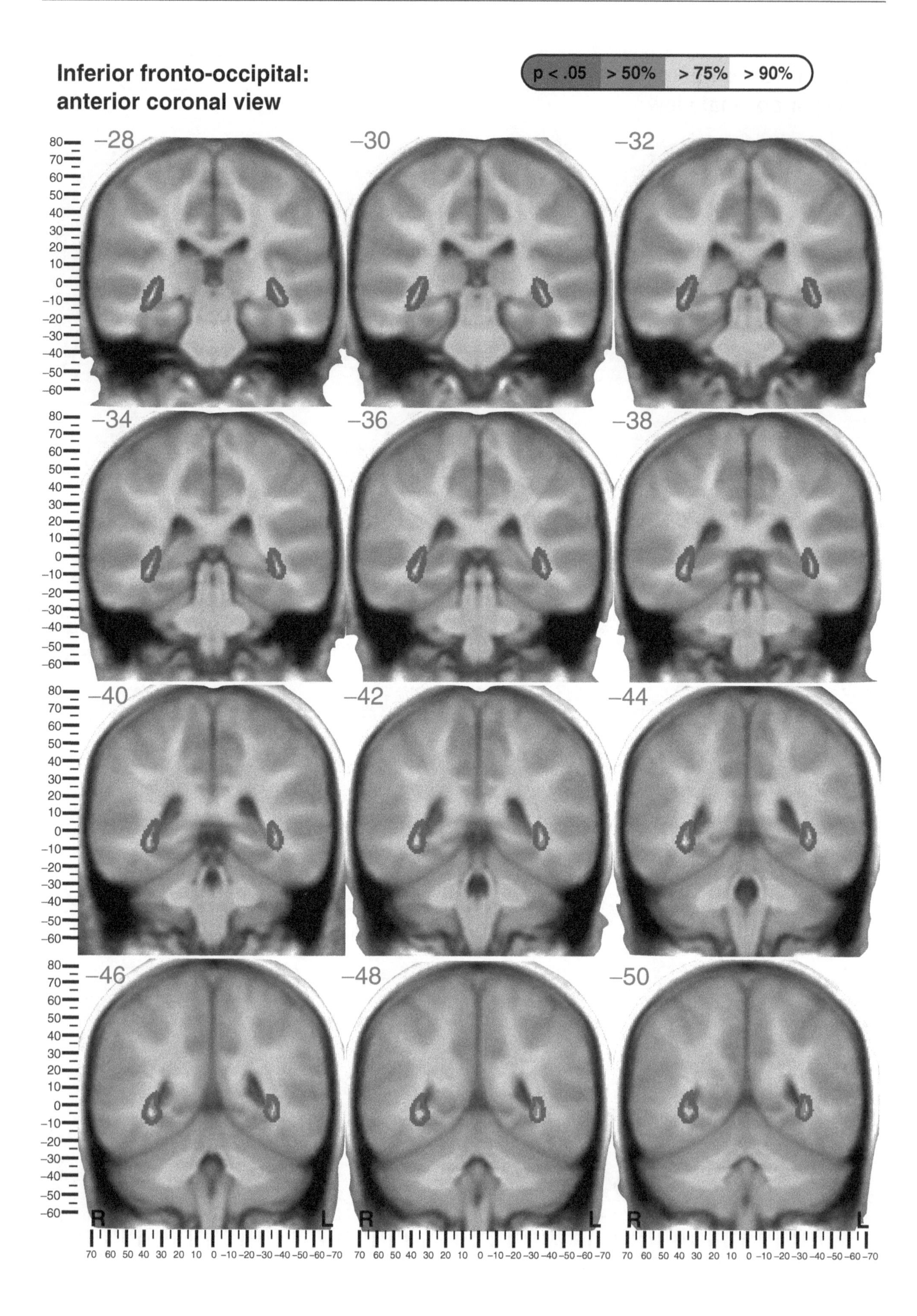

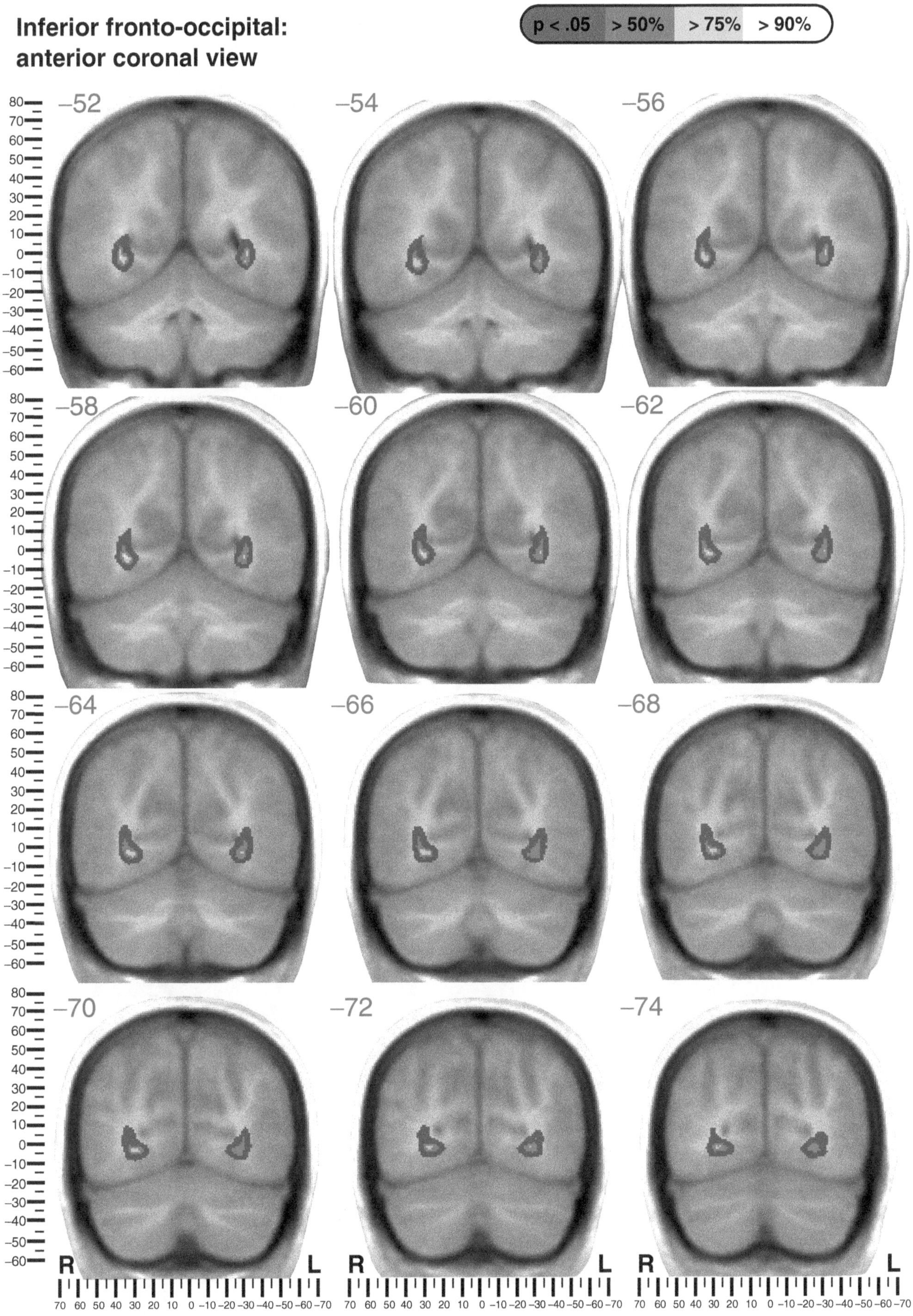

Inferior fronto-occipital: anterior coronal view

p < .05 > 50% > 75% > 90%

–52 –54 –56 –58 –60 –62 –64 –66 –68 –70 –72 –74

R L

70 60 50 40 30 20 10 0 –10 –20 –30 –40 –50 –60 –70

80 70 60 50 40 30 20 10 0 –10 –20 –30 –40 –50 –60

Inferior fronto-occipital (right): lateral sagittal view

p < .05 | > 50% | > 75% | > 90%

42 40

80 70 60 50 40 30 20 10 0 –10 –20 –30 –40 –50 –60

38 36

80 70 60 50 40 30 20 10 0 –10 –20 –30 –40 –50 –60

34 32

80 70 60 50 40 30 20 10 0 –10 –20 –30 –40 –50 –60

30 28

80 70 60 50 40 30 20 10 0 –10 –20 –30 –40 –50 –60

–110 –100 –90 –80 –70 –60 –50 –40 –30 –20 –10 0 10 20 30 40 50 60 70

–100 –90 –80 –70 –60 –50 –40 –30 –20 –10 0 10 20 30 40 50 60 70

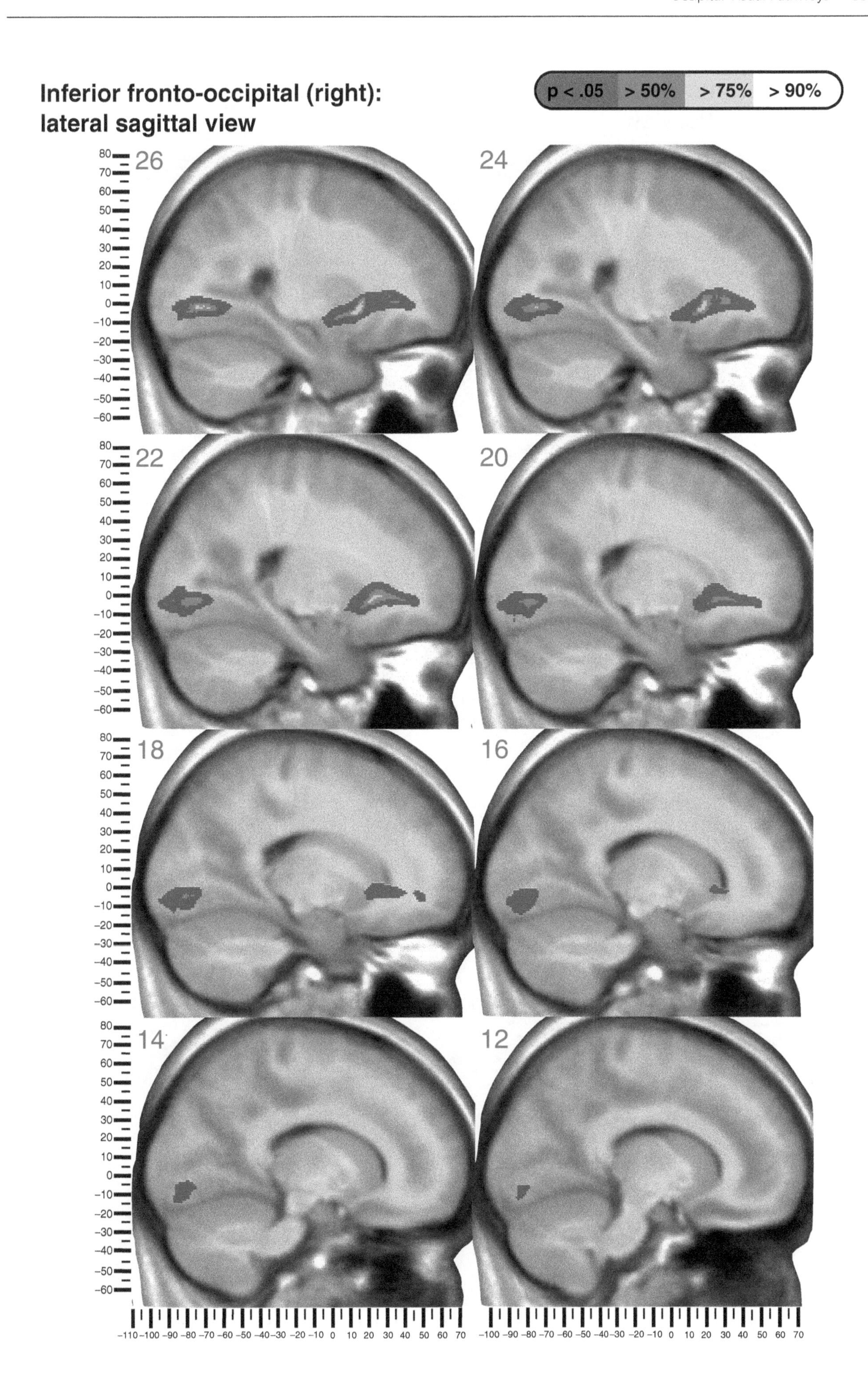
Inferior fronto-occipital (right):
lateral sagittal view
p < .05
> 50%
> 75%
> 90%
26
24
22
20
18
16
14
12

Inferior fronto-occipital (left): medial sagittal view

p < .05 | > 50% | > 75% | > 90%

-12 -14 -16 -18 -20 -22 -24 -26

Inferior fronto-occipital (left): medial sagittal view

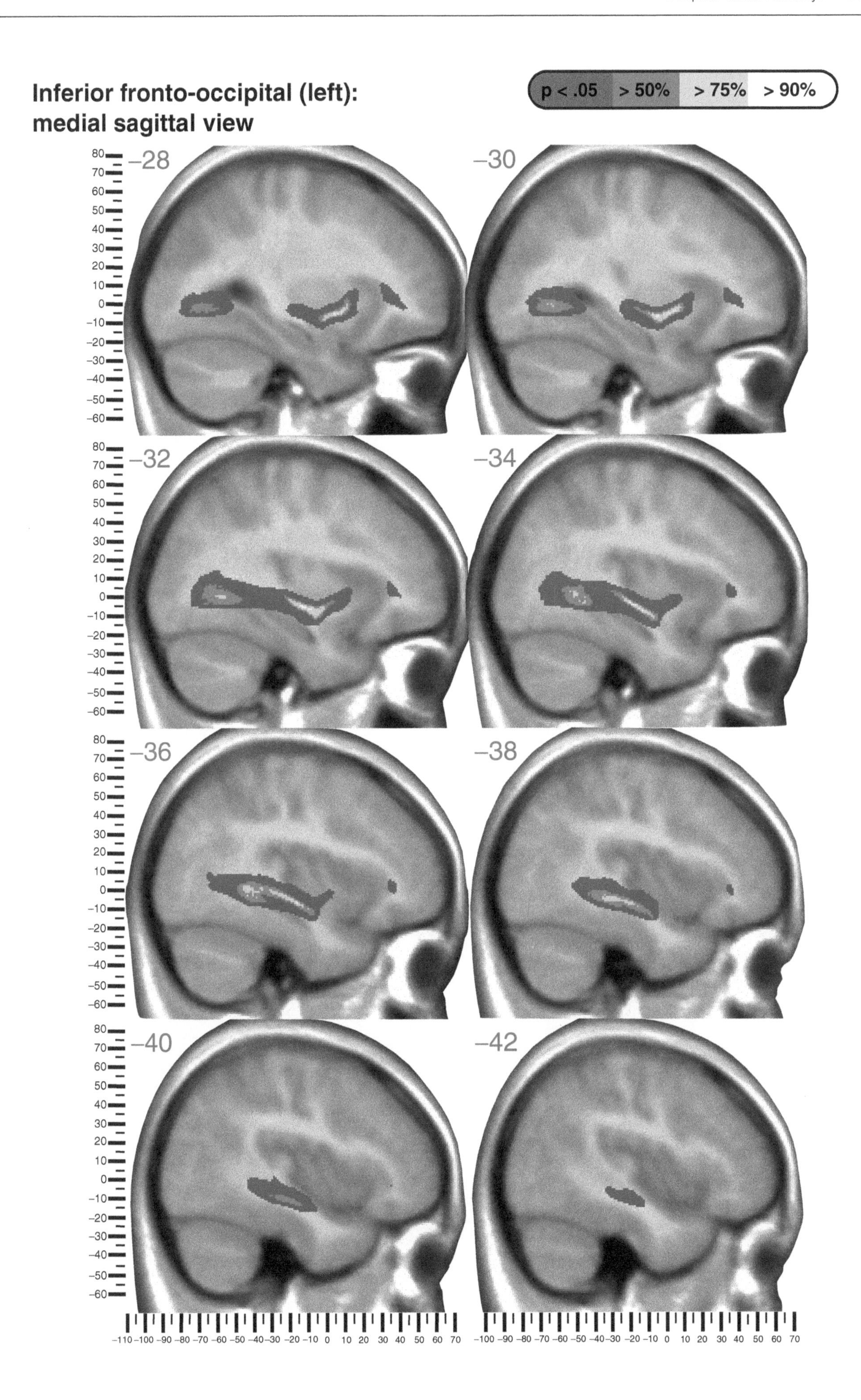

CHAPTER 9

Commissural Pathways

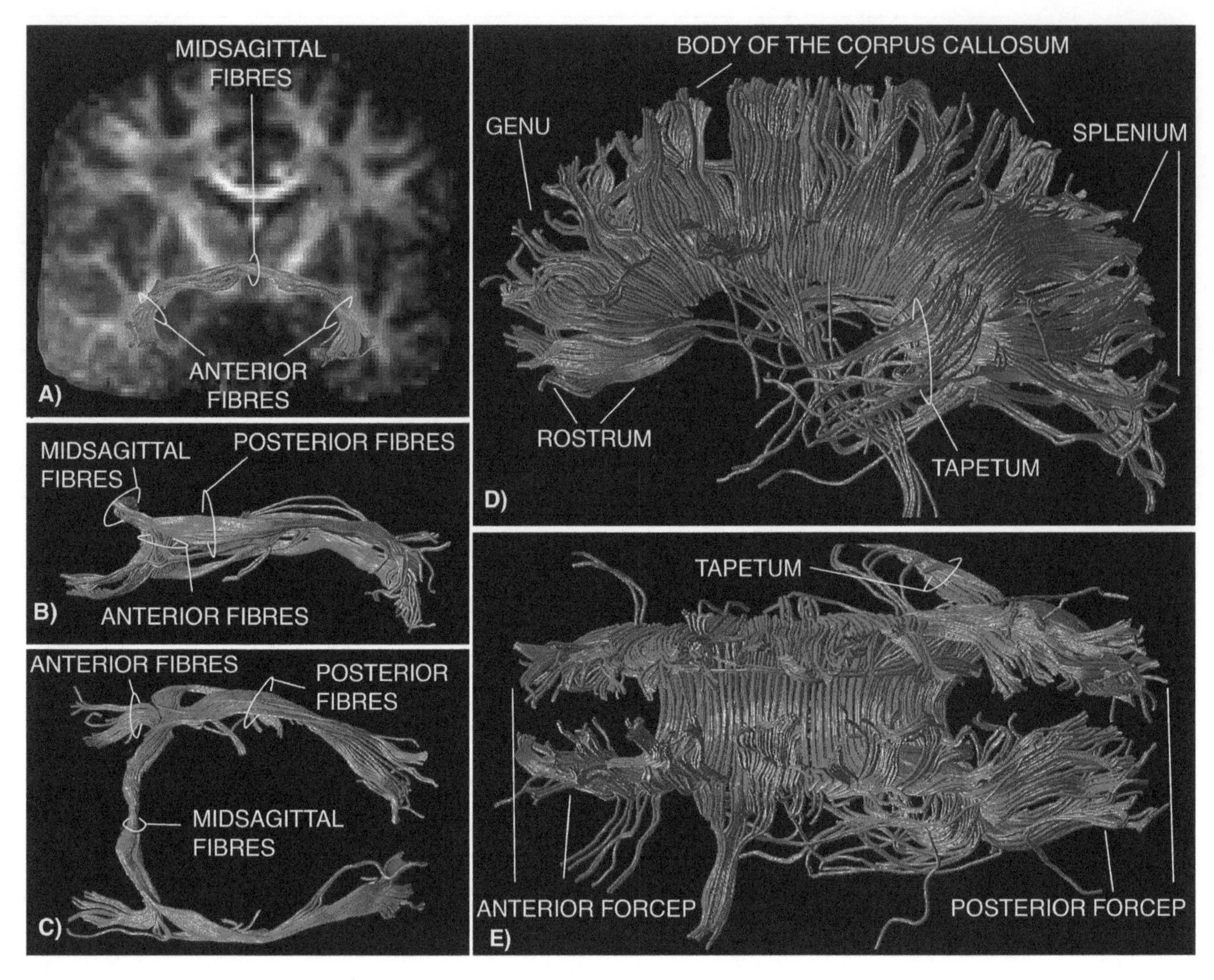

Figure 9.1 Diffusion tensor tractography reconstruction of the anterior commissure and the corpus callosum. A) anterior, B) left lateral, and C) superior view of the anterior commissure. D) Left lateral and E) superior view of the corpus callosum (Catani et al., 2002).

Introduction

Commissural pathways are composed of fibres connecting broadly similar regions in the two cerebral hemispheres. In the human brain they consist of the anterior commissure, corpus callosum, and the hippocampal commissure of the fornix. These three commissures can be easily visualized with current tractography methods (Figure 9.1; for the hippocampal commissure see Figure 11.2 in Chapter 11).

Other small commissures are composed of short fibres connecting adjacent paramedian structures of the thalamus (massa intermedia), the tegmentum (tectal commissure of Forel), the habenula (habenular commissure), and the brainstem (reticular commissure). A general assumption underlying the concept of commissural connections is that the information is transferred between homologous cortical or subcortical regions. There are, however, a significant number of heterotopic commissural fibres connecting non-homologous regions, at least in the corpus callosum (Clarke, 2003). Furthermore, the communication between the two sides can also occur through non-commissural tracts, such as the superior cerebellar peduncle fibres that connect the deep cerebellar nuclei of one side with the thalamic nuclei of the contralateral side. The commissural pathways play a significant role in the development of interhemispheric specialization and integration of several motor, perceptual, and cognitive functions (Gazzaniga, 2000). The anterior commissure and corpus callosum will be described in this chapter, while the hippocampal is dealt separately in Chapter 11.

Anterior commissure

The anterior commissure is a small bundle of fibres shaped like the handlebars of an old bicycle straddling the midline (Figure 9.1A–C). It is a familiar landmark in neuroradiology (e.g. distances in Talairach coordinates are measured from the anterior commissure as origin) (Talairach and Tournoux, 1988; Mai et al., 1997). It crosses the midline as a compact cylindrical bundle between anterior and posterior columns of the fornix and runs laterally, at first through the anterior perforated substance, and then between the globus pallidus and putamen before dividing into an anterior and posterior branch. The more anterior fibres connect the amygdalae (Turner et al., 1979), hippocampal gyri, and temporal poles (Demeter et al., 1990), while more posterior fibres connect the ventral temporal

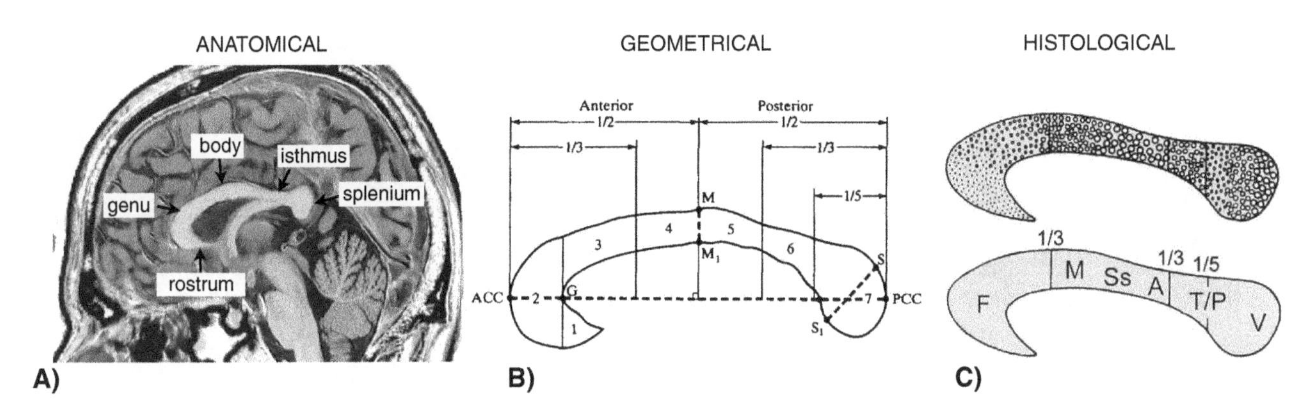

Figure 9.2 Subdivisions of the mid-sagittal callosal area according to classical descriptive (A) anatomy, (B) geometrical criteria, and (C) neurohistology. B) In the geometrical subdivision the horizontal dashed line passes through the anteriormost (ACC) and posteriormost (PCC) extremes of the corpus callosum. The line is then used as a baseline to divide the corpus callosum into seven regions. This method identify also three linear measures: M and M1 are the superior and inferior limits, respectively, of the callosum at its midpoint; S and S1 denote the superior and inferior limits of the splenium, chosen such that S-S1 is the length of the maximal perpendicular distance between two parallel lines drawn as tangents to the superior and inferior surfaces of the splenium; G is the anteriormost point on the inner convexity of the anterior callosum (Witelson, 1989). C) In the histological division the letters indicate: F, frontal; M motor; Ss somatosensory; A, auditory; T/P, temporo-parietal; V, visual (Aboitiz et al., 1992).

and occipital cortices. The anterior division follows the course of the uncinate fasciculus into the anterior temporal lobe, while the more posterior division runs parallel to the inferior longitudinal fasciculus, optic radiations, and inferior fronto-occipital fasciculus.

Compared to tractography, axonal tracing studies in animals show a far more complex anatomy (Crosby et al., 1962; Di Virgilio et al., 1999) with anterior fibres also projecting to olfactory regions (e.g. olfactory bulb, anterior perforated substance, etc.). These anterior olfactory-linked fibres are scarce in humans (Lauer, 1945; Kiernan, 1998). The exact cortical projections of the posterior fibres are not well defined in humans. Some authors have suggested that they reach the inferior occipital cortex in humans but not in the monkey brain (Meynert, 1885; Rockland and Pandya, 1986; Di Virgilio et al., 1999; Catani et al., 2002).

Experimental studies in monkeys show that the anterior commissure has an excitatory role and is responsible for transmitting visual information to the contralateral hemisphere (Rocha-Miranda et al., 1975; Gross et al., 1977; Seacord et al., 1979; Cook, 1986). The functions of the anterior commissure in humans are poorly understood. Behavioural testing following surgical division of the corpus callosum has shown that the anterior commissure is unable to transfer information related to visual tasks interhemispherically (Gazzaniga and Freedman, 1973). This is probably correlated with the higher ratio of anterior commissure cross-sectional area to total neocortical commissural area in monkeys compared to humans (Foxman et al., 1986). Anomia for olfactory stimuli presented to the right nostril (right olfactory anomia) (Gordon and Sperry, 1969) and memory deficits (Zaidel and Sperry, 1974) have been described in split-brain patients with complete commissurotomy (i.e. complete severing of the corpus callosum and anterior and hippocampal commissure) but not in those with an intact anterior commissure (Ledoux et al., 1977; Risse et al., 1978). The right olfactory anomia in these patients is due to a disconnection between the right olfactory cortex and the left fronto-temporal language areas.

Corpus callosum

The corpus callosum is the largest commissural tract in the human brain, consisting of 200–300 million axons of varying size and degree of myelination (Tomasch, 1954; Aboitiz et al., 1992). Figure 9.1 shows the anatomy of the corpus callosum as derived from diffusion tensor tractography dissections. The corpus callosum forms the roof of the lateral ventricles and its fibres are conventionally divided into an anterior forceps (or forceps minor), a middle portion (body), and a posterior forceps (or forceps major). Some splenial fibres stretch either side out into the temporal lobe and form the tapetum of the corpus callosum (Crosby et al., 1962). The term forceps (from Latin, 'pincers') derives from the horseshoe shape of its most anterior and posterior fibres viewed in axial projections (Figure 9.1E). The anterior forceps connects the prefrontal and orbitofrontal regions. The body of the corpus callosum contains fibres connecting the premotor and precentral frontal cortex and the parietal lobes. Fibres from the splenium make up the posterior forceps and connect the occipital lobes. The group of fibres known as the tapetum sweeps laterally over the inferior horn of the lateral ventricle, then inferiorly to link bilaterally those regions of the temporal lobe not connected by the anterior commissure (Déjérine, 1895; Crosby et al., 1962).

Other criteria have been adopted to subdivide the corpus callosum. The most common methods segment the corpus callosum as seen in mid-sagittal section. Thus, classical anatomical textbooks divide the corpus callosum (from anterior to posterior) into rostrum, genu, body, isthmus, and splenium (Figure 9.2A) (Crosby et al.,

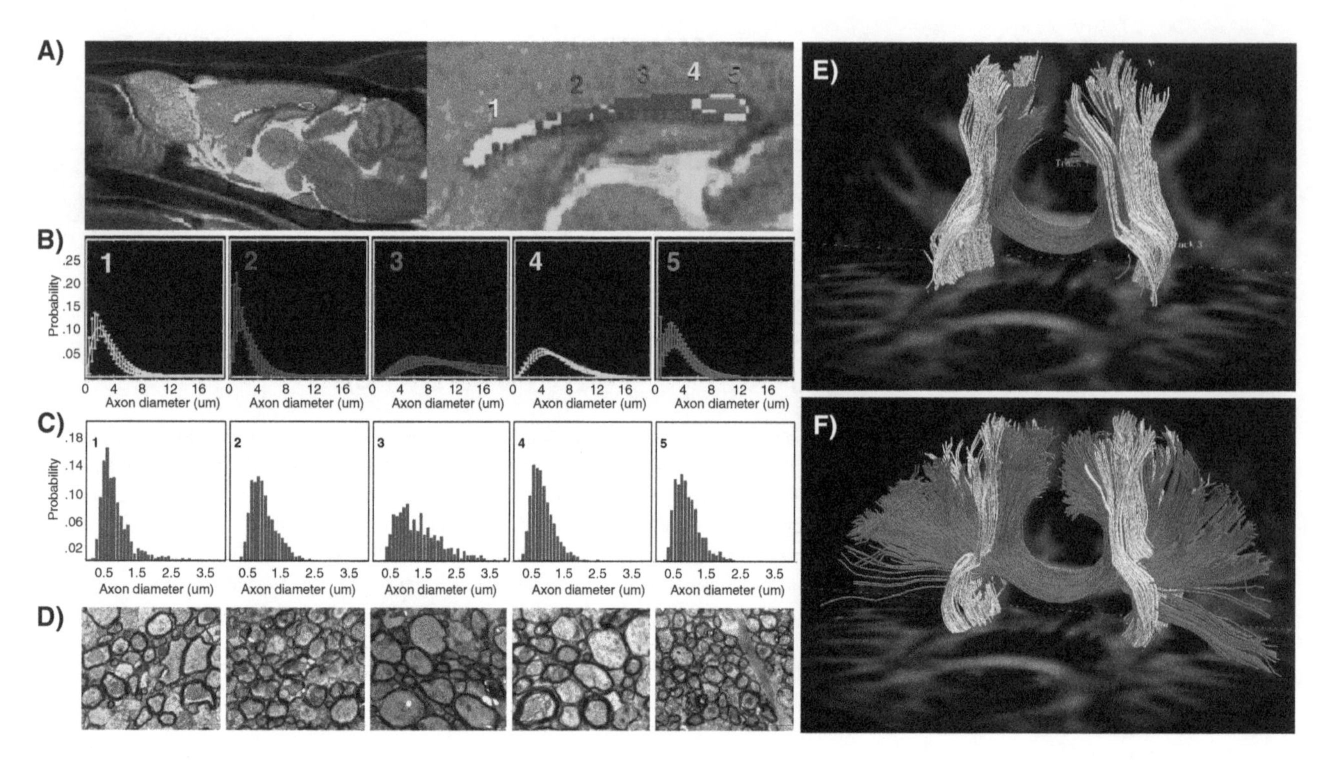

Figure 9.3 Diffusion MRI methods for studying the *in vivo* anatomy of the corpus callosum. A–D) Estimation of axon diameter distribution in the rat brain (Barazany et al., 2009). A) Mid-sagittal T2-weighted MRI (7 Tesla) of the rat brain and segmentation of the corpus callosum in five clusters (magnified in the right panel) according to the *in vivo* estimation of the averaged axon diameter distribution (shown in B). C, D) In vivo estimation was validated with histological analysis. Similar methods are also available for the in vivo estimation of the axon diameter in the human brain (Dell'Acqua et al., 2010a). E) In humans tractography algorithms based on diffusion tensor models fail to reconstruct the most lateral projections of the corpus callosum (red) due to the crossing with fibres of the corticospinal tract (yellow). F) Callosal fibres projecting to the most lateral regions are visualized using tractography methods based on spherical deconvolution (Dell'Acqua et al., 2010b; Catani and Dell'Acqua et al., 2011)

1962). The rostrum is a beak-shaped portion in the most ventral region of the anterior corpus callosum. The genu is the curved anterior limit of the corpus callosum dorsal to the rostrum. The body is the central and largest horizontal part. The isthmus is a narrow zone of transition between the body and the splenium. The splenium forms the posterior limit of the corpus callosum. Geometrical criteria for a precise division of the different parts of the corpus callosum have been proposed for research purposes (Witelson, 1989; Woodruff et al., 1993). Witelson (1989), for example, divides the mid-sagittal area of the corpus callosum into seven regions (Figure 9.2B), where region 1 corresponds to the rostrum, region 2 to the genu, regions 3, 4, and 5 to the body, region 6 to the isthmus, and region 7 to the splenium.

Histological studies showed that there is some correspondence between the seven regions of Witelson and the pattern of distribution of callosal axons according to their diameter (Figure 9.2C). Aboitiz et al. (1992) argue that the axons originating from sensorimotor areas, auditory temporal cortex, and occipital regions have a large diameter (larger than 2 µm), and are highly myelinated and fast conducting. Other regions in the rostrum, genu, anterior body, and anterior splenium contain smaller, slow-conducting fibres.

In vivo callosal anatomy

In vivo estimation of axon diameter distribution within the corpus callosum has been obtained using recent advanced methods in diffusion MRI (Figure 9.3) (Assaf et al., 2008; Barazany et al., 2009). Using a series of diffusion-weighted MRI brain scans at 7 Tesla, Barazany et al. (2009) were able to infer the *in vivo* distribution of axon diameters along the mid-sagittal area of the corpus callosum in a rat brain which had a broad correspondence to the distribution estimated with histology (Figure 9.3A–D). Similar estimations are also possible in the living human brain within clinically acceptable time (Dell'Acqua et al., 2010a).

Diffusion tensor tractography has been applied by a number of research groups to work out the correspondence between callosal mid-sagittal regions and cortical projections (Huang et al., 2005; Hofer and Frahm, 2006; Zarei et al., 2006; Park et al., 2008; Chao et al., 2009). These studies suggest that, for example, Witelson's division of the mid-sagittal area may not be indicative of the precise cortical ending of the callosal fibres passing through each of the seven regions (see also Figures 9.4 and 9.5). One limitation of these approaches based on diffusion tensor imaging, however, is that the tensor model assumes that fibres at each voxel are well described by a single orientation estimate. This means that when the fibres of corpus callosum cross those of the internal capsule tractography is unable to reconstruct them after this point, thus showing only the most medial tracts (Figure 9.3E). More recent tractographic developments seem to have partially overcome this problem, being based on high angular resolution diffusion imaging (HARDI) methods (Frank, 2001), such as spherical deconvolution and appropriate processing of multi-peak diffusion profiles (Tuch et al., 2003; Tournier et al., 2004; Alexander et al., 2005; Behrens et al., 2007;

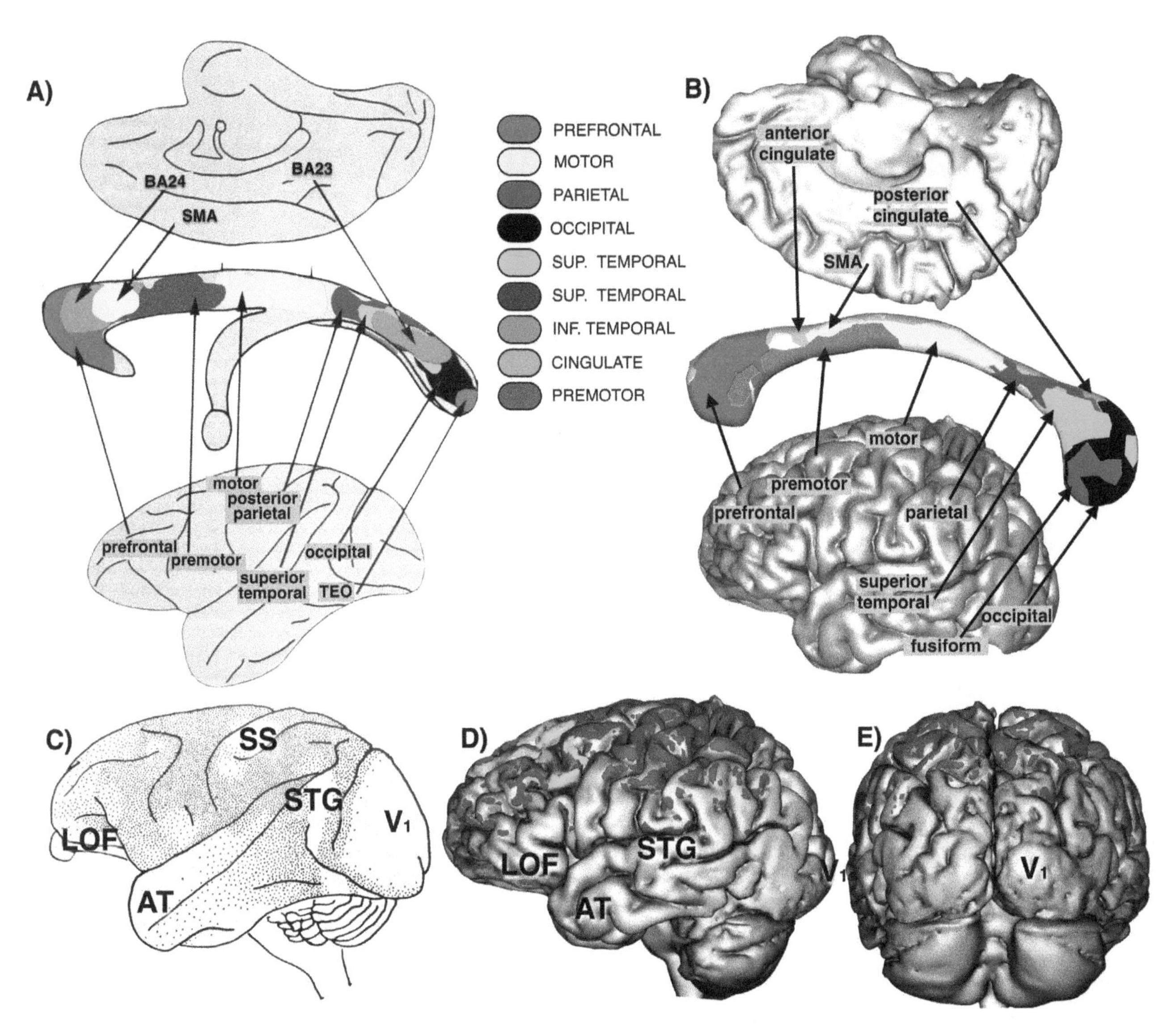

Figure 9.4 Comparative anatomy of the corpus callosum in simian and human brains. A, B) Correspondence between the post-mortem segmentation of the mid-sagittal area of the corpus callosum in the monkey (*Macaca mulatta*, left image) based on axonal tracing methods (Schmahmann and Pandya, 2006) and human brain using spherical deconvolution tractography (right image).). (Dell'Acqua et al., 2010). C–E) Comparison between the post-mortem histological maps of the callosal cortical projections in the monkey brain (C) (from Myers 1965) and the *in vivo* maps of the callosal projections in the human brain based on spherical deconvolution tractography (D, E). The labelled regions are not connected by callosal fibres and correspond to the lateral orbitofrontal region (LOF), anterior temporal (AT) areas, the superior temporal gyrus (STG), somatosenory cortex (SS), and the primary visual area (V1). Colours in D and E do not correspond to the colours in A and B but indicate the genu (blue and green), body (red), and splenium plus tapetum (yellow).

Dell'Acqua et al., 2010b). With these methods a complete reconstruction of the callosal fibres is obtained (Figure 9.3F) (Catani and Dell'Acqua, 2011).

Comparative anatomy

Figure 9.4 shows the topography of the mid-sagittal area of the corpus callosum and its cortical projections as derived from post-mortem studies in the rhesus monkey (*Macaca mulatta*) (Myers, 1965; Schmahmann and Pandya, 2006) and *in vivo* spherical deconvolution tractography in the human brain (Dell'Acqua et al., 2010b). This preliminary comparison suggests a similar topography of the callosal projections between the two species, though with some differences. The projection to the motor area, for example, seems to extend more posteriorly in humans compared to the monkey (yellow area in Figure 9.4A, B). This is probably due to the greater expansion of the cortical frontal areas, and therefore of prefrontal callosal connections, in humans (Hofer and Frahm, 2006).

Figure 9.4C shows the callosal projections of the monkey brain in detail. In addition to an absence of commissural fibres in the primary visual cortex, there is absence or scarcity of callosal projections to

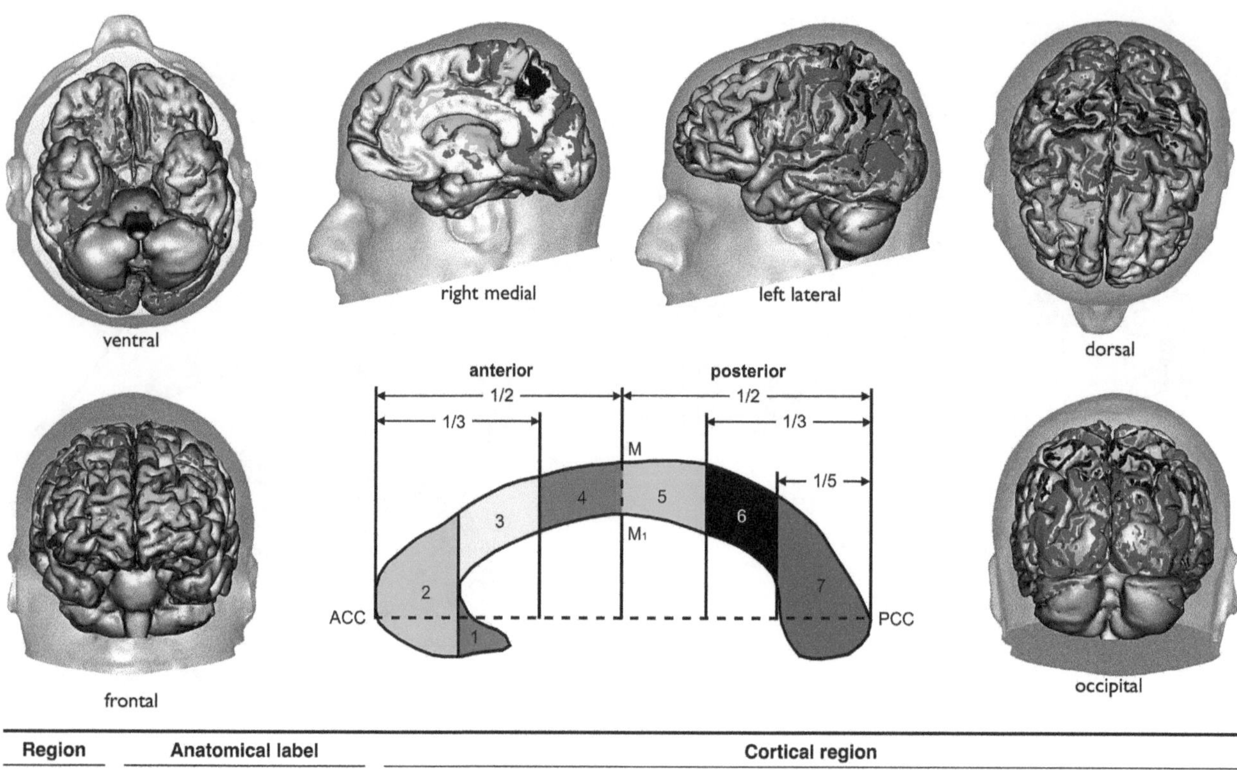

Region	Anatomical label	Cortical region	
		Monkey (axonal tracing)	*Human (tractography)*
1	Rostrum	Caudal/orbital prefrontal, inferior premotor	Orbital and polar frontal
2	Genu	Prefrontal	Prefrontal, premotor, polar, and orbital frontal
3	Rostral body	Premotor, supplementary motor	Motor and premotor
4	Anterior midbody	Motor	Motor, somatosensory, premotor
5	Posterior midbody	Somatosensory, posterior parietal	Somatosensory, parietal, motor
6	Isthmus	Superior temporal, posterior parietal	Parietal and someasthesic
7	Splenium	Occipital, inferior temporal	Occipital, posterior temporal and parietal

Figure 9.5 Subregional topography of the mid-sagittal area of the human corpus callosum according to Witelson's division (diagram) and corresponding cortical projections. The table compares the cortical projections of Witelson's regions as derived from axonal tracing studies in monkey and human spherical deconvolution tractography.

the anterior third of the temporal lobe and the auditory receptive area, at the dorsal surface of the superior temporal gyrus. There is also absence of projections to the hand region of the parietal lobe (BA 3, 1 and 2) and a near absence in the dorsal region of the anterior parietal lobe. In the frontal lobe the hand and foot regions lying hidden on the anterior bank of the central sulcus (area 4) and the orbitofrontal and dorsal prefrontal cortex receive sparse connections. With these exceptions, the cortical surface throughout seems to receive commissural fibres with great intensity (Myers, 1965). In humans a similar pattern of callosal projections can be replicated *in vivo* with spherical deconvolution tractography (Figure 9.4D, E) (Dell'Acqua et al., 2010b). A possible difference is the presence of callosal connections to the boundary region between area 17 and 18, which has been observed in the human but not the monkey brain (Clarke and Miklossy, 1990).

Callosal topography has been linked to theories explaining evolution of specialized brain functions. It has been proposed that evolutionary pressures guiding brain size and lateralization of specialized areas are accompanied by reduced interhemispheric connectivity and greater intrahemispheric connectivity (Aboitiz et al., 1992; Corballis et al., 2000; Gazzaniga, 2000). According to this hypothesis areas receiving less input from a homologous contralateral cortical area would become specialized by virtue of being deprived of transcallosal connections. Similarly, by being connected through smaller diameter, slower-conducting transcallosal fibres, specialized areas reduce unnecessary crosstalk and maintain separate parallel processing systems (Doron and Gazzaniga, 2008). The absence of direct callosal connections in certain areas, as shown in Figure 9.4C–E can therefore be explained by the fact that both lateral orbitofrontal cortex and ventral temporo-occipital areas show selectivity for highly specialized and lateralized functions such as face perception in the right hemisphere (Kanwisher et al., 1997; Gauthier et al., 1999), or speech and visual words in the left hemisphere (Epelbaum et al., 2008). These theories remain speculative, although future tractography studies combining both methodologies for callosal microstructure (e.g. axonal diameter) and cortical projection topography are likely to advance our understanding of human interhemispheric integration in relation to the evolution of the human brain.

Functional anatomy and tractography studies

The corpus callosum has also been divided into different subregions according to the pattern of its cortical projections. It is generally thought that this pattern matches the geometrical subdivision of the mid-sagittal area according to the Witelson's criteria and that each subregion contains fibres specialized for motor and sensory functions (Figure 9.5). Thus, for example, subregion 5 (posterior mid-body) of the mid-sagittal area contains fibres connecting the parietal lobes and is involved in transferring tactile information from the primary somatosensory area of one hemisphere to the other. The general organization proposed in the table contained in Figure 9.5 is, however, derived mainly from work on monkeys (Pandya et al., 1971; Schmahmann and Pandya, 2006) and we have previously seen that recent human tractography results suggest that it may be an oversimplification, with many subregions overlapping. Furthermore, the tractography reconstruction suggests that the pattern of distribution of the callosal fibres may be shifted posteriorly in the human as compared to the monkey brain (see also Figure 9.4). This could explain some of the differences between experimental results in callosotomized monkeys and split-brain human patients (e.g. preservation of somatosensory transfer in patients with resection of the posterior body and isthmus). Much of our knowledge about the functional specificity of the callosal subregions comes from experimental studies in animals (Glickstein and Berlucchi, 2008) and testing patients who have undergone a complete or partial severing of the corpus callosum (see Gazzaniga, 2000 for a review). After partial callosotomy, the function of disconnected callosal subregions can be assessed with neuropsychological testing. Patients with anterior callosotomy (rostrum, genu, and body) usually show no impairment in most of the cases (Gordon et al., 1971; Risse et al., 1989; Aglioti et al., 2000). Patients with only the posterior third of the splenium spared from callosotomy show deficits in all sensory modalities of interhemispheric transfer except for visual information (Risse et al., 1989). Finally, disconnection of the splenium abolishes interhemispheric transfer of visual modalities (Gazzaniga and Freedman, 1973). There are, however, many exceptions among callosotomized patients and many cases of patients with non-surgical damage to subregions of the corpus callosum that suggest a functional segregation also for more anterior subregions (e.g. genu and body) (Table 9.1).

The combination of diffusion MRI with fMRI, and behavioural measurements allows for a finer grained study of functional subregional topography. Dougherty et al. (2005) found that streamlines connecting extrastriate visual areas are topographically organized within the splenium, with representation from the fovea to the periphery proceeding in an anterior to posterior direction. In a study by Baird et al. (2005) participants were asked to recognize and name objects presented from unusual and canonical perspectives. This would require interhemispheric integration between the right parietal cortex, specialized for recognizing objects from unusual perspectives, and the left inferior frontal cortex, which is specialized for object naming. Their results indicate that shorter reaction times were associated with higher fractional anisotropy values in the splenium, whereas longer reaction times were associated with higher fractional anisotropy values in the genu. These results suggest that interhemispheric transfer is a dynamic process and that the performances in the same task depend on the anatomical features of the callosal connections mediating the communication between the two hemispheres. Johansen-Berg et al. (2007) found a direct correlation between performances in bimanual coordination and fractional anisotropy levels in the body of the corpus callosum. A recent probabilistic tractography study showed that the number of callosal streamlines reaching the midline is dependent on the side used to start tracking (Putnam et al., 2010). The authors found that a greater numbers of streamlines propagate from the right extrastriate visual areas through the splenium than from the left side. Although the authors have interpreted this result as reflecting an anatomical feature of the splenium that could explain classical findings of a faster right-to-left transfer of visual information, other methodological considerations could explain these findings (e.g. inter-hemispheric asymmetry of other tracts crossing the callosal fibres).

Callosal disorders and clinical syndromes

The clinical studies of the corpus callosum can be grouped in the following four broad categories: i) complete (agenesis) or partial (dysgenesis) absence of the corpus callosum; ii) partial or complete callosotomy in patients with treatment-resistant epilepsy (split-brain patients); iii) acquired callosal lesions (e.g. tumours, vascular lesions, demyelinating disorders, etc.); iv) neurodevelopmental psychiatric disorders, (e.g. autism and schizophrenia) and other metabolic or toxic disorders (e.g. alcoholism). Irrespective of the aetiology, callosal syndromes can present with similar clinical manifestations (Table 9.1).

Dysgenesis of the corpus callosum

Development of the corpus callosum begins at approximately 8 weeks of gestation and continues at different rates for its different parts. During fetal development the genu grows at a faster rate than the body and the splenium, but after birth, the splenium takes the lead in relative growth (Rakic and Yakovlev, 1968). Anomalies of development can result in a partial or complete absence of the corpus callosum, and several degrees of atrophy or dysmyelination. Callosal dysgenesis has been associated with a number of aetiological factors (genetic, vascular, infectious, metabolic, etc.) but the nature of the insult to the developing corpus callosum is far less important than the timing of the insult (Barkovich and Norman, 1988). Thus, insults occurring at around 8 weeks of gestation often result in complete callosal absence, whereas later insults cause partial dysgenesis or atrophy of a developing corpus callosum.

The clinical correlates of callosal dysgenesis are difficult to determine. The first case of callosal agenesis was described by Reil in 1812 and subsequently reported by many authors, most of the cases showing various degrees of cognitive and motor impairment. Hyndman and Penfield (1937) reported certain gross psychological observations in five patients with ventriculography evidence of callosal dysgenesis and concluded that manifestations in these patients were related to the degree of callosal agenesis. In most of these cases, however, other associated brain abnormalities are likely to contribute to the clinical manifestations. Disorders frequently associated with callosal dysgenesis include Chiari malformation (callosal abnormalities are found in 10% of subjects with Chiari I and in 80% of Chiari II), neuronal migration disorders (e.g. grey matter heterotopias, lissencephaly, pachygyria, schizencephaly), Dandy–Walker malformation, holoprosencephaly, and genetic syndromes (e.g. X-linked lissencephaly and ambiguous genitalia syndrome,

Table 9.1 Callosal disconnection syndromes

Callosal Syndromes	Lesion site	Description
Alien hand syndrome "Diagonistic dyspraxia"	Genu, anterior body	Involuntary intermanual conflict resulting in one hand (usually left) interfering with the correct execution of the movements of the other hand (Akelaitis 1945; Bogen, 1987).
Anarchic hand syndrome	Body	Complex involuntary goal-directed movements of one hand (i.e. the anarchic hand) that the patient perceives as his own hand but with a will of its own (Goldstein, 1908; Goldberg et al., 1981; Della Sala et al., 1991; Marchetti and Della Sala, 1998).
Left hand tactile anomia and agnosia	Posterior body	Inability to name objects presented to the left hand (Geschwind and Kaplan, 1962; Gaymard et al., 1993). The inability to recognize objects after tactile presentation (i.e. tactile agnosia) has also been described (Balsamo et al., 2008).
Left hand apraxia	Body-splenium	Impaired ability to execute with the left hand goal-directed movements on imitation (when presented to the right hemifield) or under verbal command (Liepmann, 1920; Zaidel and Sperry, 1977; Graff-Radford et al., 1987; Gaymard et al., 1993; Kazui and Sawada, 1993; Heilman and Watson, 2008).
Left hand agraphia	Body splenium	Failure to write correctly with the left hand (Geschwind and Kaplan, 1962; Gazzaniga et al., 1962; Sugishita et al., 1980; Gersh and Damasio, 1981).
Hemispatial Neglect	Body-splenium	Failure to respond or reorient to novel or meaningful stimuli presented to one hemifield (usually the left) (Kashiwagi et al., 1990; Heilman and Adams, 2003; Lausberg et al., 2003; Park et al., 2006; Tomaiuolo et al., 2010).
"Main étrangère"	Isthmus-splenium	A condition characterized by the patient's inability to recognize his left hand as his own when the hand is held in the other hand and is out of the visual field (Brion and Jedynak, 1972; Bogen, 1979).
Dichotic listening suppression	Isthmus-splenium	When auditory stimuli are simultaneously presented to both ears, patients find easier to recognize verbal linguistic stimuli (e.g. words, digits, etc.) presented to the right ear and tend to ignore those presented to the left ear (left ear suppression) (Milner et al., 1968; Sugishita et al., 1995; Pollmann et al., 2002). An opposite preference is observed when patients are asked to execute dichotic non-verbal tasks (right ear suppression) (Sparks and Geschwind, 1968).
Bilateral crossed optic ataxia	Posterior body-splenium	Specific visuomotor difficulty in reaching objects located contralaterally to the hand used (Gazzaniga et al., 1965; Gaymard et al., 1993).
Deficits in binocular stereopsis and vergence eye movements	Splenium	Reduced ability to perceive tridimensional objects and inability to produce appropriate convergent or divergent eye movements while fixating targets on the midsagittal plane (Berlucchi and Aglioti, 1999).
Optic aphasia	Splenium	Failure to name objects presented visually, with preserved ability to name the same objects to definition (Freund, 1889; Luzzatti et al., 1998).
Color anomia	Splenium	Inability to name visually presented colours (Geschwind and Fusillo, 1966).
Alexia without agraphia (pure alexia)	Splenium	Inability to read with preserved ability to write due to a disconnection between both occipital visual areas and the visual word form system (Dejerine, 1892; Geschwind and Fusillo, 1966; Epelbaum et al., 2008).
Left visual hemianomia and hemialexia	Splenium	Inability to name objects (or read words) presented tachiscopically in the left half visual field. The right hemisphere maintains a normal capacity for visual recognition but has lost its access to the contralateral language regions (Gazzaniga et al., 1965; Sperry et al., 1969; Berlucchi and Aglioti, 1999).

XLAG; Aicardi syndrome, and the corpus callosum hypoplasia, retardation, adducted thumbs, spasticity, and hydrocephalus—CRASH syndrome) (for a complete list see Paul et al., 2007).

With the advent of neuroimaging it has become clear that most people with isolated agenesis of the corpus callosum have no symptoms or minimal deficits. They often present to the attention of the physician with aspecific complaints, such as headache or dizziness, and a brain scan reveals the anatomical anomaly. Tractography has been used recently to study the course of callosal fibres in patients with incomplete agenesis of the corpus callosum (Wahl et al., 2009) and in patients with Chiari II malformation (Herweh et al., 2009). Figure 9.6 shows the MRI images of a subject

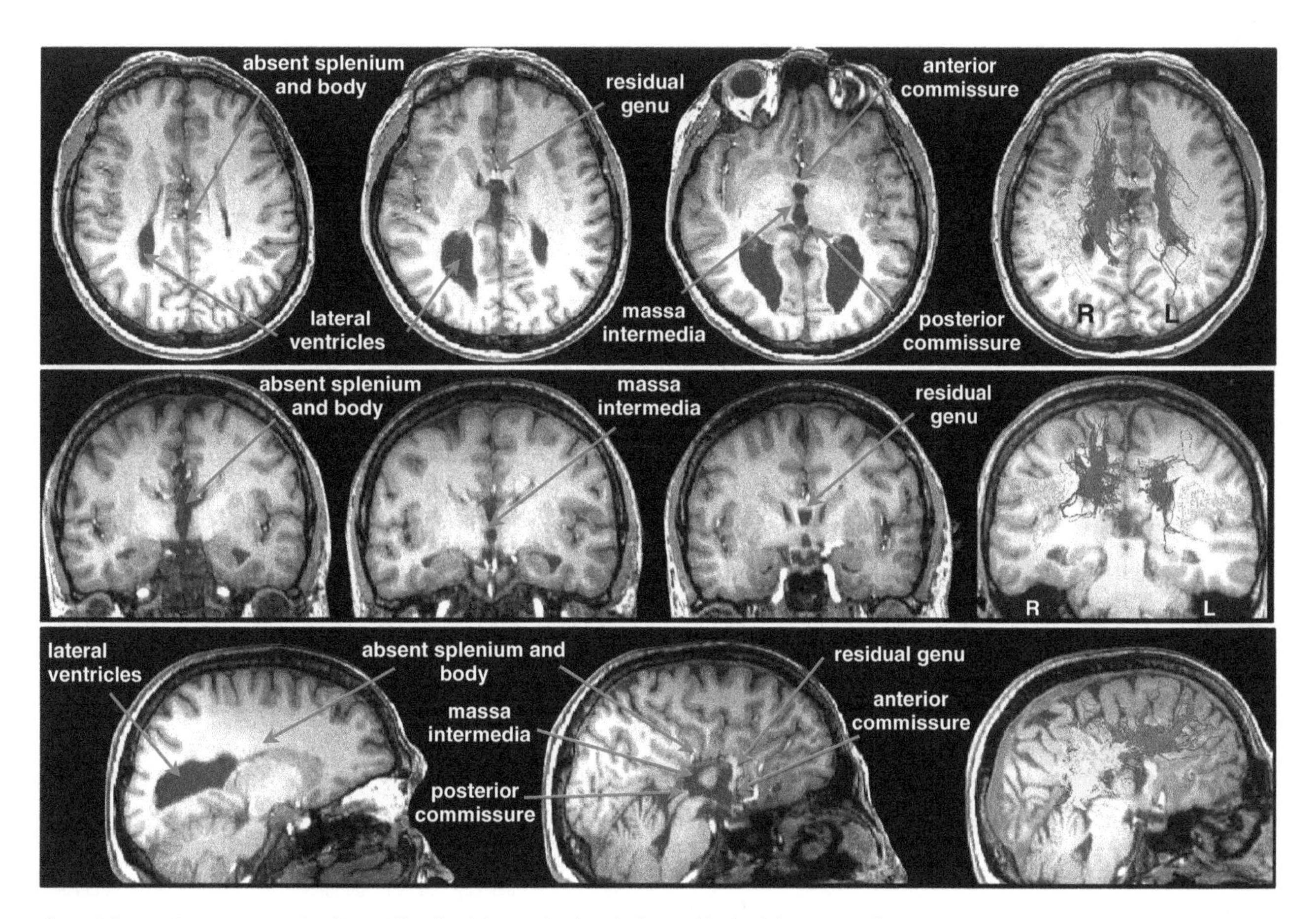

Figure 9.6 MRI of an asymptomatic subject with callosal dysgenesis. The splenium and body of the corpus callosum are not developed, while there are some residual portions of the genu. Other commissural structures are evident such as the anterior commissure, the massa intermedia, and the posterior (hippocampal) commissure. The lateral ventricles are enlarged due to the lack of callosal fibres. In the right column the fibres of the arcuate fasciculus (yellow) and superior fronto-occipital fasciculus (red) are reconstructed with DTI-tractography. The superior fronto-occipital fasciculus derives from the uncrossed fibres of the corpus callosum.

with almost complete agenesis of the corpus callosum; this patient had normal cognitive functions. The tractography reconstruction shows the trajectory of a longitudinal bundle (superior fronto-occipital fasciculus), which is commonly found in acallosal brains. The tract is thought to derive from the deviation of those callosal fibres unable to cross the midline during fetal development. Its existence in the healthy brain is still debated (see Chapter 8). The lack of significant deficits in acallosal subjects is thought to be related to the ability of other commissures (especially those in the brainstem) to compensate for the absence of callosal fibres (Jeeves, 1994; Milner, 1994).

Split-brain patients

The technique of severing the commissural fibres in epileptic patients was introduced in the 1930s with a view to control the spread of ictal discharges from one hemisphere to the other (Van Wagenen and Herren, 1940). It provided an opportunity for understanding the functions of the corpus callosum and the lateralization of cerebral functions by assessing in isolation each hemisphere. Split-brain patients can have complete commissurotomy (section of the corpus callosum, anterior commissure, hippocampal commissure, and, in some cases, the massa intermedia), complete callosotomy alone (only the corpus callosum), or partial section of the anterior corpus callosum (e.g. patients with anterior frontal seizures). The clinical effects of callosal section can last from weeks to a few months after the operation (acute syndrome) or persist in a stable condition for many years (chronic syndrome) (Bogen 1987). The acute syndrome often manifests with reduced spontaneous speech and sometimes complete mutism. Difficulty in coordinating bimanual movements is often observed, ranging from mild akinesia to intermanual conflict in which the hands act at cross-purposes. Patients sometimes complain that their left hand behaves in a 'foreign' or 'alien' manner and they often express surprise at apparently purposeful left-hand actions, which are not under their voluntary control (Bogen, 1993). Left-hand apraxia, motor perseveration, and grasping are also observed. Symptoms vary across patients and the exact pathological mechanism is not known, but oedema from retraction (of one hemisphere to allow surgical approach) in addition to the callosal disconnection has been suggested. After the acute phase, patients with complete callosotomy usually show minimal impairment in their daily living activities. Special testing methods, usually involving the presentation of the sensory input to one side (e.g. tachiscopic visual presentation, dichotic listening, etc.), can expose otherwise unreported deficits, such as the lack of interhemispheric transfer of lateralized functions (e.g. left-hand tactile anomia) or non-lateralized functions (e.g. the inability to retrieve with one hand an object palpated with the other). Split-brain subjects progressively acquire a variety of strategies for circumventing their interhemispheric transfer deficits. A common example is for the patient to speak out loud the name of an object that he holds in the right hand; because the right hemisphere can understand many individual words, the object can then be recognized when the patient is asked to palpate the object with the left hand and his eyes closed (Berlucchi and Aglioti, 1999).

Acquired neurological disorders

Classical isolated callosal disconnection syndromes have been described in patients with acquired callosal disorders (Table 9.1). These include left-hand apraxia (Liepmann, 1920) and left tactile anomia (Geschwind and Kaplan, 1962). In other syndromes the callosal damage is associated with lesions to other tracts or cortical regions (e.g. splenium and occipito-temporal connections in pure alexia) (Déjérine, 1892).

Psychiatric disorders

The corpus callosum is affected in a wide range of neuropsychiatric conditions, including attention deficit and hyperactivity disorder (ADHD), Gilles de la Tourette syndrome (Yazgan and Kinsbourne, 2003), autism (Alexander et al., 2007), schizophrenia (Cowell et al., 2003), and depression (Ballmaier et al., 2008). Toxic and metabolic disorders affecting the corpus callosum include fetal alcohol syndrome, preterm birth (Kontis et al., 2009), and alcoholism (e.g. the Marchiafava–Bignami disease is a progressive demyelination of the corpus callosum manifesting with fits, dementia and coma in patients with alcoholism or nutritional deficits).

References

Aboitiz, F., Scheibel, A.B., Fisher, R.S., and Zaidel, E. (1992). Fiber composition of the human corpus callosum. *Brain Res*, **598**(1–2), 143–53.

Aglioti, S.M., Tassinari, G., Fabri, M., et al. (2001). Taste laterality in the split brain. *Eur J Neurosci*, **13**(1), 195–200.

Akelaitis, A.J. (1945). Studies of the corpus callosum: IV. Diagonistic dyspraxia in epileptics following partial and complete section of the corpus callosum. *Am J Psychiat*, **101**, 594–9.

Alexander, A.L, Lee, J.E., Lazar, M., et al. (2007). Diffusion tensor imaging of the corpus callosum in autism. *NeuroImage*, **34**(1), 61–73.

Alexander, D.C. (2005). Multiple-fiber reconstruction algorithms for diffusion MRI. *Ann NY Acad Sci*, **1064**, 113–33.

Assaf, Y., Blumenfeld-Katzir, T., Yovel, Y., and Basser, P.J. (2008). AxCaliber: a method for measuring axon diameter distribution from diffusion MRI. *Magn Reson Med*, **59**(6), 1347–54.

Baird, A., Colvin, M., Vanhorn, J., Inati, S., and Gazzaniga, M. (2005). Functional connectivity: Integrating behavioral, diffusion tensor imaging, and functional magnetic resonance imaging data sets. *J Cogn Neurosci*, **17**(4), 687–93.

Ballmaier, M., Kumar, A., Elderkin-Thompson, V., et al. (2008). Mapping callosal morphology in early- and late-onset elderly depression: an index of distinct changes in cortical connectivity. *Neuropsychopharmacology*, **33**(7), 1528–36.

Balsamo, M., Trojano, L., Giamundo, A., and Grossi, D. (2008). Left hand tactile agnosia after posterior lesion. *Cortex* **44**(8), 1030–6.

Barazany, D., Basser, P.J., and Assaf Y. (2009). In vivo measurement of axon diameter distribution in the corpus callosum of rat brain. *Brain*, **132**(5), 1210–20.

Barkovich, A.J. and Norman, D. (1988). Anomalies of the corpus callosum: correlation with further anomalies of the brain. *Am J Roentgenol*, **151**(1), 171–9.

Behrens, T.E., Berg, H.J., Jbabdi, S., Rushworth, M.F., and Woolrich, M.W. (2007). Probabilistic diffusion tractography with multiple fibre orientations: What can we gain? *NeuroImage*, **34**(1), 144–55.

Berlucchi, G. and Aglioti (1999). Interhemispheric disconnection syndromes. In Denes, G. and Pizzamiglio, L. (Eds) *Handbook of Clinical and Experimental Neuropsychology*. Hove: Hove Psychology Press.

Bogen, J.E. (1979). The callosal syndrome. In Valenstein, E. and Heilman, K.M. (Eds) *Clinical neuropsychology*, pp. 308–59. Oxford: Oxford University Press.

Bogen, J.E. (1987). Physiological consequences of complete or partial commissural section. In Apuzzo M.L.J. (Ed.) *Surgery of the third ventricle*, pp. 175–94. Baltimore: Williams and Wilkins.

Bogen, J.E. (1993). The callosal syndromes. In Valenstein, E. and Heilman, K.M. (Eds) *Clinical neuropsychology*, third edition, pp. 337–407. Oxford: Oxford University Press.

Brion, S. and Jedynak, C.P. (1972). Trouble du transfert interhemispherique à propos de trois observations de tumeurs du corps calleux: le signe de la main étrangère. *Rev Neurol (Paris)*, **126**(4), 257–66.

Catani, M. and Dell'Acqua, F. (2011). Mapping white matter pathways with diffusion imaging tractography: focus on neurosurgical applications. in Hughes Duffau (ed.), *Brain mapping: from neural basis of cognition to surgical applications*. Wien: Springer.

Catani, M., Howard, R.J., Pajevic, S., Jones, D.K. (2002) Virtual in vivo interactive dissection of white matter fasciculi in the human brain. *NeuroImage*, **17**(1), 77–94.

Chao, Y.P, Cho, K.H., Yeh, C.H., Chou, K.H., Chen, J.H., and Lin, C.P. (2009). Probabilistic topography of human corpus callosum using cytoarchitectural parcellation and high angular resolution diffusion imaging tractography. *Hum Brain Mapp*, **30**(10), 3172–87.

Clarke, S. (2003) The role of homotopic and heterotopic callosal connections in humans. In Zaidel, E. and Iacoboni, M. (Eds.) *The parallel brain*, pp. 461–72. Cambridge, MA: MIT Press.

Clarke, S. and Miklossy, J. (1990). Occipital cortex in man: organization of callosal connections, related myelo- and cytoarchitecture, and putative boundaries of functional visual areas. *J Comp Neurol*, **298**(2), 188–214.

Cook, N. (1986). *The brain code: mechanisms of information transfer and the role of the corpus callosum*. London: Methuen.

Corballis, P., Funnell, M., and Gazzaniga, M. (2000). An evolutionary perspective on hemispheric asymmetries. *Brain Cogn*, **43**(1–3), 112–17.

Cowell, P.E., Denenberg, V., Boehm, G., Kertesz, A., and Nasrallah, H. (2003). Using the corpus callosum as an effective probe in the study of schizophrenia. In: Zaidel, E. and Iacoboni, M. (Eds.) *The parallel brain*, pp. 433–44. Cambridge: MIT Press.

Crosby, E.C., Humphrey, T., and Lauer, E.W. (1962). *Correlative Anatomy of the Nervous System*. New York: Macmillian Co.

Déjérine, J. (1892). Contribution a l'etude anatomo-pathologique et clinique des differentes varietes de cecite verbale. *Mém Soc Biol*, **4**, 61–90.

Déjérine, J. (1895). *Anatomie des Centres Nerveux, Vol. 1*. Paris: Rueff et Cie.

Dell'Acqua, F., Thiebaut de Schotten, M., Chick, J., et al. (2010a). Correspondence between in vivo axonal diameter and cortical connectivity of the corpus callosum, Proc. 16th Annual meeting of the Organization for Human Brain Mapping, Barcelona, Spain.

Dell'Acqua, F., Scifo, P., Rizzo, G., et al. (2010b). A modified damped Richardson-Lucy algorithm to reduce isotropic background effects in spherical deconvolution. *NeuroImage*, **49**(2), 1446–58.

Della Sala, S., Marchetti, C., and Spinnler, H. (1991). Right-sided anarchic (alien) hand: a longitudinal study. *Neuropsychologia*, **29**(11), 1113–27.

Demeter, S., Rosene, D.L., and van Hoesen, G.W. (1990). Fields of origin and pathways of the interhemispheric commissures in the temporal lobe of macaques. *J Comp Neurol*, **302**(1), 29–53.

Di Virgilio, G., Clarke, S., Pizzolato, G., and Schaffner, T. (1999). Cortical regions contributing to the anterior commissure in man. *Exp Brain Res*, **124**(1), 1–7.

Doron, K.W. and Gazzaniga, M.S. (2008). Neuroimaging techniques offer new perspectives on callosal transfer and interhemispheric communication. *Cortex*, **44**(8), 1023–9.

Dougherty, R., Ben-Shachar, M., Bammer, R., Brewer, A., and Wandell, B. (2005). Functional organization of human occipital-callosal fiber tracts. *Proc Natl Acad Sci USA*, **102**(20), 7350–5.

Epelbaum, S., Pinel, P., Gaillard, R., et al. (2008). Pure alexia as a disconnection syndrome: new diffusion imaging evidence for an old concept. *Cortex*, **44**(8), 962–74.

Foxman, B.T., Oppenheim, J., Petito, C.K., and Gazzaniga, M.S. (1986). Proportional anterior commissure area in humans and monkeys. *Neurology*, **36**(11), 1513–17.

Frank, L.R. (2001). Anisotropy in high angular resolution diffusion-weighted MRI. *Magn Reson Med*, **45**(6), 935–9.

Freund, C.S. (1889). Ueber optische aphasie und seelenblindheit. *Arch Psychiat Nervenkrankh* **20**, 276–97, 371–416. [English translation by Beaton, A., Davidoff, J., and Erstfeld, U. (1991). On optic aphasia and visual agnosia. *Cogn Neuropsychol*, **8**(1), 21–38.]

Gauthier, I., Tarr, M., Anderson, A., Skudlarski, P., and Gore, J. (1999). Activation of the middle fusiform 'face area' increases with expertise in recognizing novel objects. *Nat Neurosci*, **2**(6), 568–73.

Gaymard, B., Rivaud, S., Rigolet, M.H., et al. (1993). Bilateral crossed optic ataxia in a corpus callosum lesion. *J Neurol Neurosurg Psychiat*, **56**(3), 323–4.

Gazzaniga, M.S. (2000). Cerebral specialization and interhemispheric communication: Does the corpus callosum enable the human condition? *Brain*, **123**(7), 1293–326.

Gazzaniga, M.S. and Freedman, H. (1973). Observations on visual processes after posterior callosal section. *Neurology*, **23**(10), 1126–30.

Gazzaniga, M.S., Bogen, J.E., and Sperry, R.W. (1962). Some functional effects of severing the cerebral commissures in man. *Proc Nat Acad Sci*, **48**(10), 1765–69.

Gazzaniga, M.S., Bogen, J.E., and Sperry, R.W. (1965). Observations on visual perception after disconnexion of the cerebral hemispheres in man. *Brain*, **88**(2), 221–36.

Gersh, F. and Damasio, A. (1981). Praxis and writing of the left hand may be served by different callosal pathways. *Arch Neurol*, **38**(10), 634–6.

Geschwind, N. and Fusillo, M. (1966). Color-naming defects in association with alexia. *Arch Neurol*, **15**(2), 137–46.

Geschwind, N. and Kaplan, E. (1962). A human cerebral disconnection syndrome. A preliminary report. *Neurology*, **12**, 675–85.

Glickstein, M. and Berlucchi, G. (2008). Classical disconnection studies of the corpus callosum. *Cortex*, **44**(8), 914–27.

Goldberg, G., Mayer, N.H., and Toglia, J.U. (1981). Medial frontal cortex infarction and the alien hand sign. *Arch Neurol*, **38**(11), 683–6.

Goldstein, K. (1908). Zur Lehre von der motorisschen Apraxie. *J Psychol Neurol*, **XI**, 169–87.

Gordon, H.W., and Sperry, R.W. (1969). Lateralization of olfactory perception in the surgically separated hemispheres of man. *Neuropsychologia*, **7**(2), 111–20.

Gordon, H.W., Bogen, J.E., and Sperry, R.W. (1971). Absence of deconnexion syndrome in two patients with partial section of the neocommissures. *Brain*, **94**(2), 327–36.

Graff-Radford, N.R., Welsh, K., and Godersky, J. (1987). Callosal apraxia. *Neurology*, **37**(1), 100–5.

Gross, C.G., Bender, D.B., and Mishkin, M. (1977). Contributions of the corpus callosum and the anterior commissure to visual activation of inferior temporal neurons. *Brain Res*, **131**(2), 227–39.

Heilman, K.M., and Adams, D.J. (2003). Callosal neglect. *Arch Neurol*, **60**(2), 276–9.

Heilman, K.M., and Watson, R.T. (2008). The disconnection apraxias. *Cortex*, **44**(8), 975–82.

Herweh, C., Akbar, M., Wengenroth, M., et al. (2009). DTI of commissural fibers in patients with Chiari II-malformation. *NeuroImage*, **44**(2), 306–11.

Hofer, S. and Frahm, J. (2006). Topography of the human corpus callosum revisited-comprehensive fiber tractography using diffusion tensor magnetic resonance imaging. *NeuroImage*, **32**(3), 989–94.

Huang, H., Zhang, J., Jiang, H., et al. (2005). DTI tractography based parcellation of white matter: application to the mid-sagittal morphology of corpus callosum. *NeuroImage*, **26**(1), 195–205.

Hyndman, O.R. and Penfield, W. (1937). Agenesis of the corpus callosum; its recognition by ventriculography. *Arch Neurol Psychiat*, **37**(6), 1251–7.

Jeeves, M.A. (1994). Callosal agenesis-a natural split brain: overview. In Lassonde, M. and Jeeves, M.A. (Eds.) *Callosal agenesis: a natural split brain?*, pp. 235–46. New York: Plenum Press.

Johansen-Berg, H., Della-Maggiore, V., Behrens, T.E., Smith, S.M., and Paus, T. (2007). Integrity of white matter in the corpus callosum correlates with bimanual co-ordination skills. *NeuroImage*, **36**(Suppl 2), T16–21.

Kanwisher, N., Mcdermott, J., and Chun, M. (1997). The fusiform face area: a module in human extrastriate cortex specialized for face perception. *J Neurosci*, **17**(11), 4302–11.

Kashiwagi, A., Kashiwagi, T., Nishikawa, T., Tanabe, H., and Okuda, J. (1990). Hemispatial neglect in a patient with callosal infarction. *Brain*, **113**(4), 1005–23.

Kazui, S. and Sawada, T. (1993). Callosal apraxia without agraphia. *Ann Neurol*, **33**(4), 401–13.

Kiernan, J.A. (1998). *Barr's Human Nervous System: An Anatomical View Point.* Philadelphia, PA: Lippincot–Raven.

Kontis, D., Catani, M., Cuddy, M., et al. (2009). Diffusion tensor MRI of the corpus callosum and cognitive function in adults born preterm. *Neuroreport*, **20**(4), 424–8.

Lauer, E.W. (1945). The nuclear pattern and fiber connections of certain basal telencephalic centers in the macaque. *J Comp Neurol*, **82**(3), 215–54.

Lausberg, H., Kita, S., Zaidel, E., and Ptito, A. (2003). Split-brain patients neglect left personal space during right-handed gestures. *Neuropsychologia*, **41**(10), 1317–29.

Ledoux, J.E., Risse, G.L., Springer, S.P., Wilson, D.H., and Gazzaniga, M.S. (1977). Cognition and commissurotomy. *Brain*, **100**(1), 87–104.

Liepmann, H. (1920). Apraxie. *Ergebn Ges Med*, **1**, 516–43.

Luzzatti, C., Rumiati, R., and Ghirardi, G. (1998). A functional model of visuo-verbal disconnection and the neuroanatomical constraints of optic aphasia. *Neurocase*, **4**(1), 71–87.

Mai, J.K., Assheuer, J., and Paxinos, G. (1997). *Atlas of the human Brain.* San Diego, CA: Academic Press.

Marchetti, C. and Della Sala, S. (1998). Disentangling the alien and anarchic hand. *Cognit Neuropsychiat*, **3**(3), 191–207.

Meynert, T. (1885). *A Clinical Treatise on Diseases of the Fore-brain Based Upon a Study of Its Structure, Functions, and Nutrition* (Sachs, B. Trans.). New York: G.P. Putnam's Sons.

Milner, A.D. (1994). Visual integration in callosal agenesis. In Lassonde, M. and Jeeves, M.A. (Eds.) *Callosal agenesis: a natural split brain?*, pp. 171–83. New York: Plenum Press.

Milner, B., Taylor, L., and Sperry, R. W. (1968). Lateralized suppression of dichotically presented digits after commissural section in man. *Science*, **161**(837), 184–6.

Myers, R.E. (1965). Organization of forebrain commissures. In Ettlinger, E.G. (Ed.) *Functions of the corpus callosum. CIBA Foundation Study Group 20*, pp. 133–43. London: J&A Churchill.

Pandya, D., Karol, E., and Heilbronn, D. (1971). The topographical distribution of interhemispheric projections in the corpus callosum of the rhesus monkey. *Brain Res*, **32**(1), 31–43.

Park, K.C., Lee, B.H., Kim, E.J., et al. (2006). Deafferentation-disconnection neglect induced by posterior cerebral artery infarction. *Neurology*, **66**(1), 56–61.

Park, H., Kim, J., Lee, S., et al. (2008). Corpus callosal connection mapping using cortical gray matter parcellation and DT-MRI. *Hum Brain Mapp*, **29**(5), 503–16.

Paul, L.K., Brown, W.S., Adolphs, R., et al. (2007). Agenesis of the corpus callosum: genetic, developmental and functional aspects of connectivity. *Nat Rev Neurosci*, **8**(4), 287–99.

Pollmann, S., Maertens, M., von Cramon, D.Y., Lepsien, J., and Hugdahl, K. (2002). Dichotic listening in patients with splenial and nonsplenial callosal lesions. *Neuropsychology* **16**(1), 56–64.

Putnam, M.C., Steven, M.S., Doron, K.W., Riggall, A.C., and Gazzaniga, M.S. (2010). Cortical projection topography of the human splenium: hemispheric asymmetry and individual differences. *J Cogn Neurosci*, **22**(8), 1662–9.

Rakic, P. and Yakovlev, P.I. (1968). Development of the corpus callosum and cavum septi in man. *J Comp Neurol*, **132**(1), 45–72.

Reil, J.C. (1812). Mangel des mittleren und freyenn Theils des Balkens im Menschengehirn. *Arch Physiol*, **II**, 341–4.

Risse, G.L., Ledoux, J.E., Springer, S.P., Wilson, D.H., and Gazzaniga, M.S. (1978). The anterior commissure in man: functional variation in a multisensory system. *Neuropsychologia*, **16**(1), 23–31.

Risse, G., Gates, J., Lund, G., Maxwell, R., and Rubens, A. (1989). Interhemispheric transfer in patients with incomplete section of the corpus callosum. Anatomic verification with magnetic resonance imaging. *Arch Neurol*, **46**(4), 437–43.

Rocha-Miranda, C.E., Bender, D.B., Gross, C.G., and Mishkin, M. (1975). Visual activation of neurons in inferotemporal cortex depends on striate cortex and forebrain commissures. *J Neurophysiol*, **38**(3), 475–91.

Rockland, K.S. and Pandya, D.N. (1986). Topography of occipital lobe commissural connections in the rhesus monkey. *Brain Res*, **365**(1), 174–8.

Schmahmann, J.D. and Pandya, D.N. (2006). *Fiber Pathways of the Brain*. New York: Oxford University Press.

Seacord, L., Gross, C.G., and Mishkin, M. (1979). Role of inferior temporal cortex in interhemispheric transfer. *Brain Res* **167**(2), 259–72.

Sparks, R. and Geschwind, N. (1968). Dichotic listening in man after section of neocortical commissures. *Cortex*, **4**, 3–16.

Sperry, R.W., Gazzaniga, M.S., and Bogen, J.E. (1969). Interhemispheric relationships: the neocortical commissures; syndromes of hemispheric disconnection. In Vinken, P.J. and Bruyn, G.W. (Eds.) *Handbook of Clinical Neurology Vol 4; Disorders of speech, perception and symbolic behaviour*, pp. 237–90. Amsterdam: Elsevier.

Sugishita, M., Toyokura, Y., Yoshioka, M., and Yamada, R. (1980). Unilateral agraphia after section of the posterior half of the truncus of the corpus callosum. *Brain Lang*, **9**(2), 215–25.

Sugishita, M., Otomo, K., Yamazaki, K., Shimizu, H., Yoshioka, M., and Shinohara, A. (1995). Dichotic listening in patients with partial section of the corpus callosum. *Brain*, **118**(2), 417–27.

Talairach, J. and Tournoux, P. (1988). *Co-planar Stereotaxic Atlas of the Human Brain: 3-Dimensional Proportional System – an Approach to Cerebral Imaging*. New York: Thieme Medical Publishers.

Tomaiuolo, F., Voci, L., Bresci M., et al. (2010). Selective visual neglect in right brain damaged patients with splenial interhemispheric disconnection. *Exp Brain Res*. **206**(2), 209–17.

Tomasch, J. (1954). Size, distribution, and number of fibres in the human corpus callosum. *Anat Rec*, **119**(1), 119–35.

Tournier, J.D., Calamante, F., Gadian, D.G., and Connelly, A. (2004). Direct estimation of the fiber orientation density function from diffusion-weighted MRI data using spherical deconvolution. *NeuroImage*, **23**(3), 1176–85.

Tuch, D.S., Reese, T.G., Wiegell, M.R., and Wedeen, V.J. (2003). Diffusion MRI of complex neural architecture. *Neuron*, **40**(5), 885–95.

Turner, B.H., Mishkin, M., and Knapp, M.E. (1979). Distribution of the anterior commissure to the amygdaloid complex in the monkey. *Brain Res*, **162**(2), 331–7.

Van Wagenen, W. and Herren, R. (1940). Surgical division of commissural pathways in the corpus callosum. Relation to spread of an epileptic attack. *Arch Neurol Psychiat*, **44**(4), 740–59.

Wahl, M., Strominger, Z., Jeremy, R.J., et al. (2009). Variability of homotopic and heterotopic callosal connectivity in partial agenesis of the corpus callosum: a 3T diffusion tensor imaging and Q-ball tractography study. AJNR *Am J Neuroradiol*, **30**(2), 282–9.

Witelson, S.F. (1989). Hand and sex differences in the isthmus and genu of the corpus callosum. A postmortem morphological study. *Brain*, **112**(3), 799–835.

Woodruff, P.W.R., Pearlson, G.D., Geer, M.J., Barta, P.E., and Chilcoat, H.D. (1993). A computerized magnetic resonance imaging study of corpus callosum morphology in schizophrenia. *Psycholog Med*, **23**(1), 45–56.

Yazgan, M.Y. and Kinsbourne, M. (2003). Functional consequences of changes in callosal area in Tourette's syndrome and attention deficit/hyperactivity disorder. In Zaidel, E. and Iacoboni, M. (Eds.) *The parallel brain*, pp. 423–32. Cambridge: MIT Press.

Zaidel, D. and Sperry, R.W. (1977). Some long-term motor effects of commissurotomy in man. *Neuropsychologia*, **15**(2), 193–204.

Zaidel, D. and Sperry, R.W. (1974). Memory impairment after commissurotomy in man. *Brain*, **97**(2), 263–72.

Zarei, M., Johansen-Berg, H., Smith, S., et al. (2006). Functional anatomy of interhemispheric cortical connections in the human brain. *J Anat*, **209**(3), 311–20.

Anterior commissure (atlas)

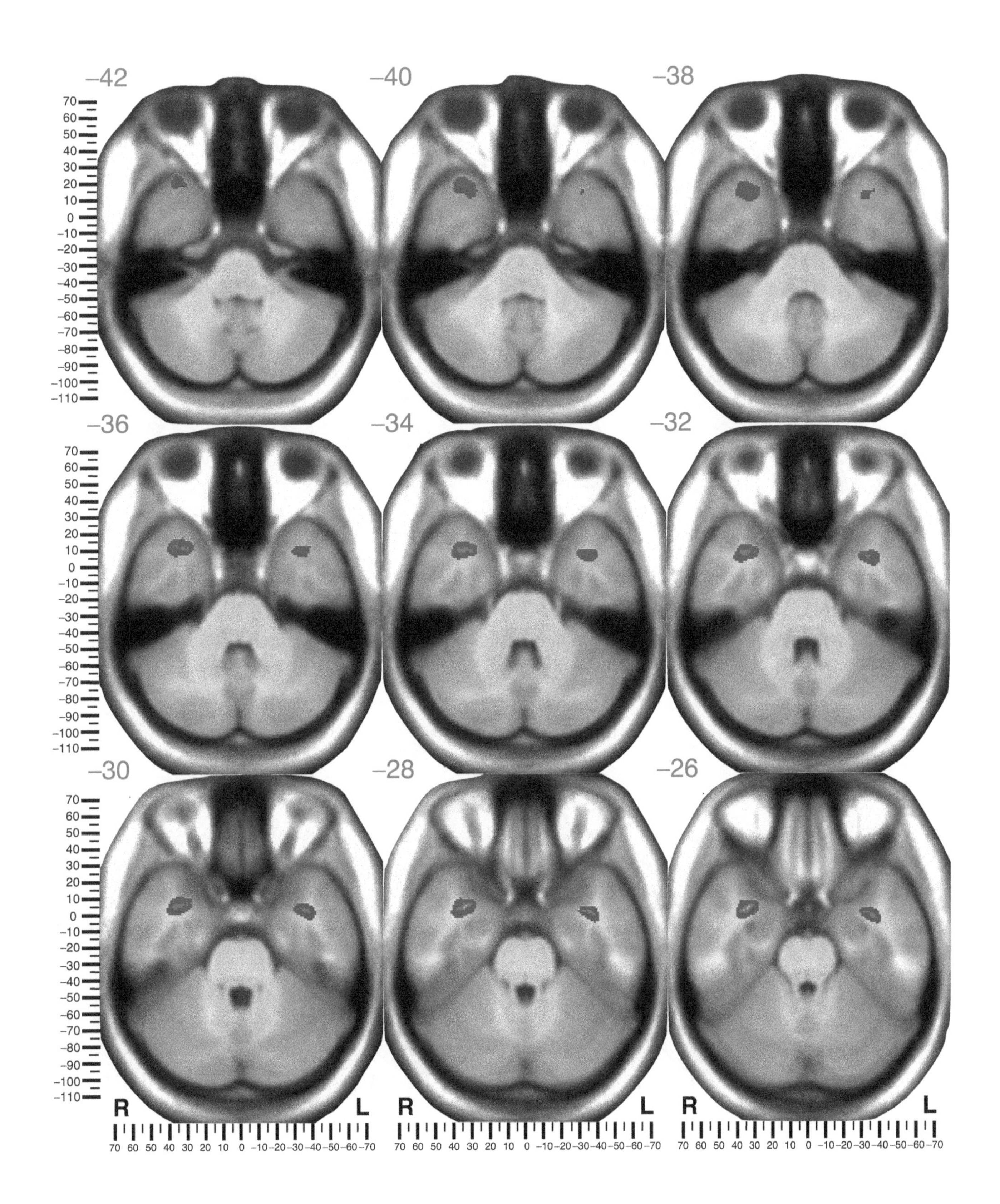

Anterior commissure: inferior axial view

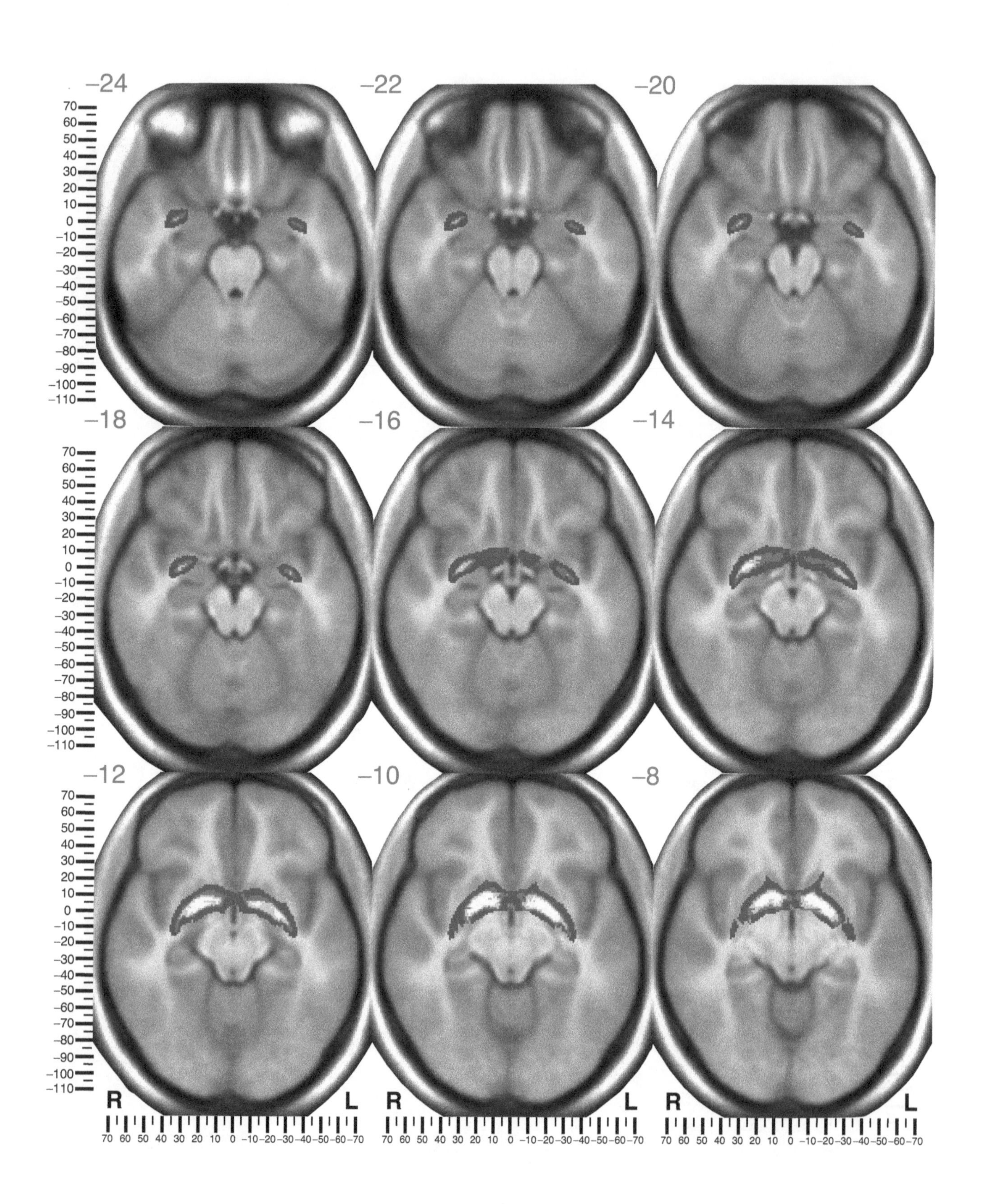

Anterior commissure: inferior axial view

p < .05 | > 50% | > 75% | > 90%

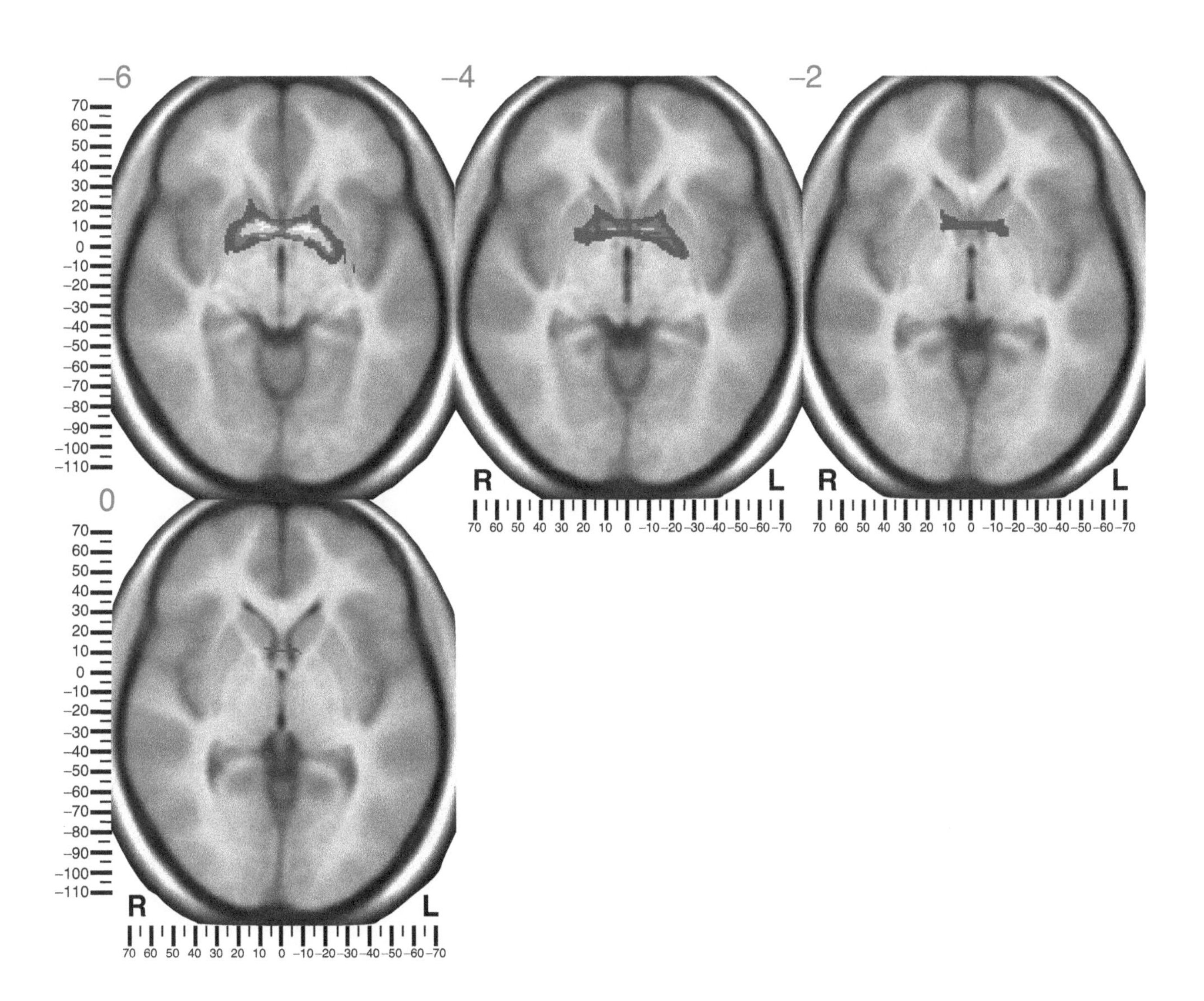

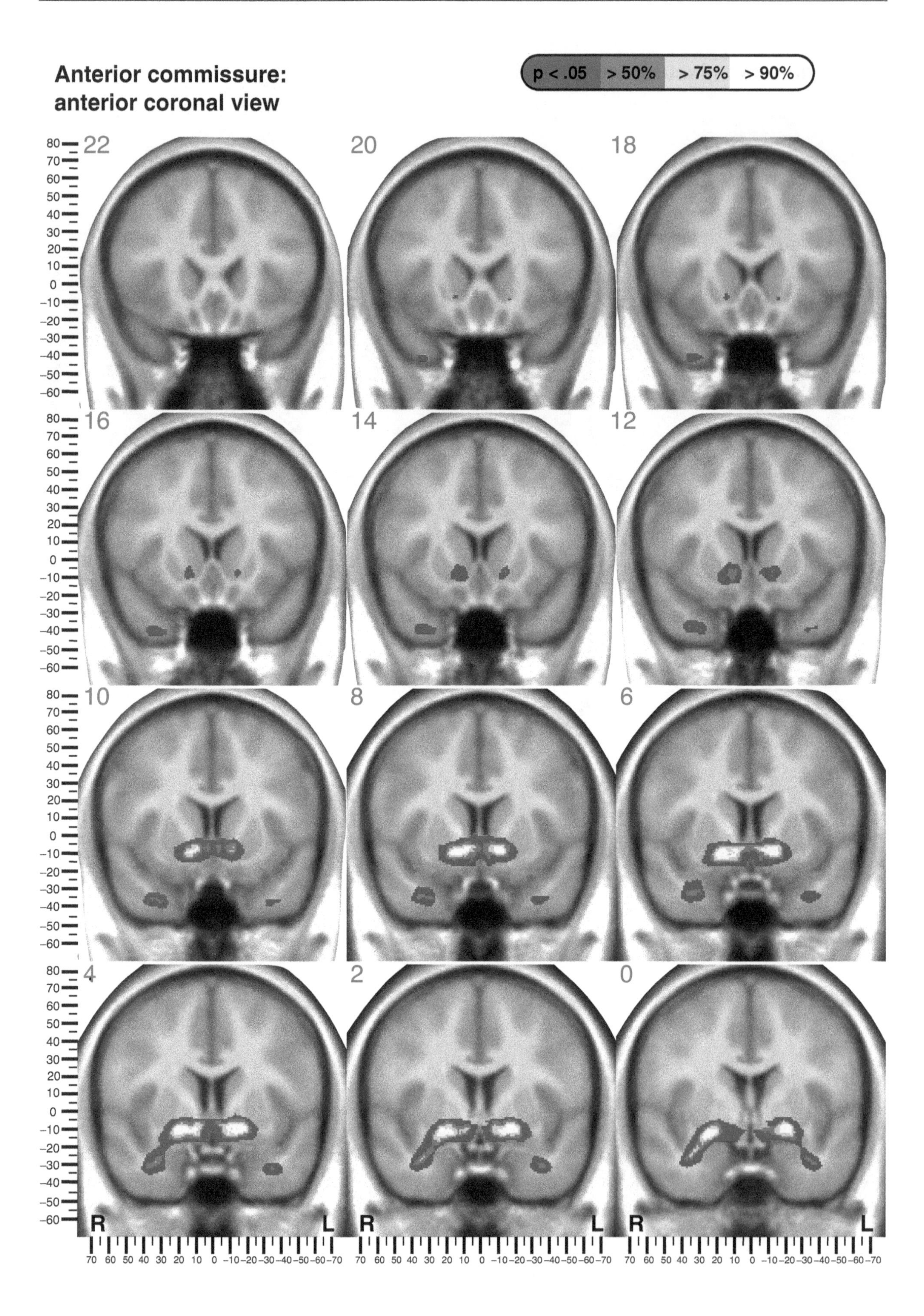
Anterior commissure:
anterior coronal view
p < .05
> 50%
> 75%
> 90%
22
20
18
16
14
12
10
8
6
4
2
0
80
70
60
50
40
30
20
10
0
−10
−20
−30
−40
−50
−60
R
L
70 60 50 40 30 20 10 0 −10 −20 −30 −40 −50 −60 −70

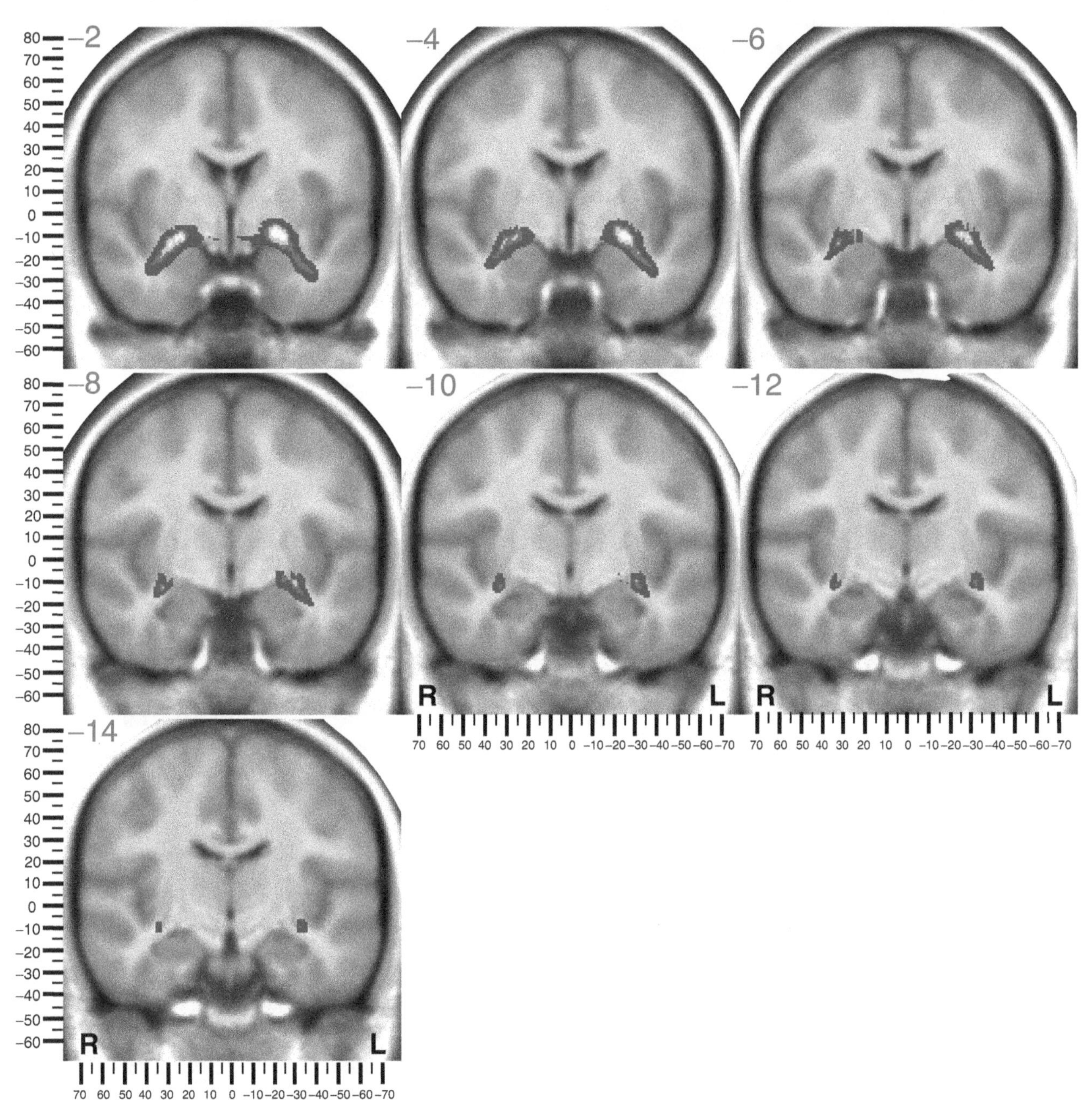
Anterior commissure:
anterior coronal view
p < .05
> 50%
> 75%
> 90%
-2
-4
-6
-8
-10
-12
-14
R
L
80
70
60
50
40
30
20
10
0
-10
-20
-30
-40
-50
-60
70 60 50 40 30 20 10 0 -10 -20 -30 -40 -50 -60 -70

Anterior commissure: lateral sagittal view

p < .05 | > 50% | > 75% | > 90%

38 36 34 32 30 28 26 24

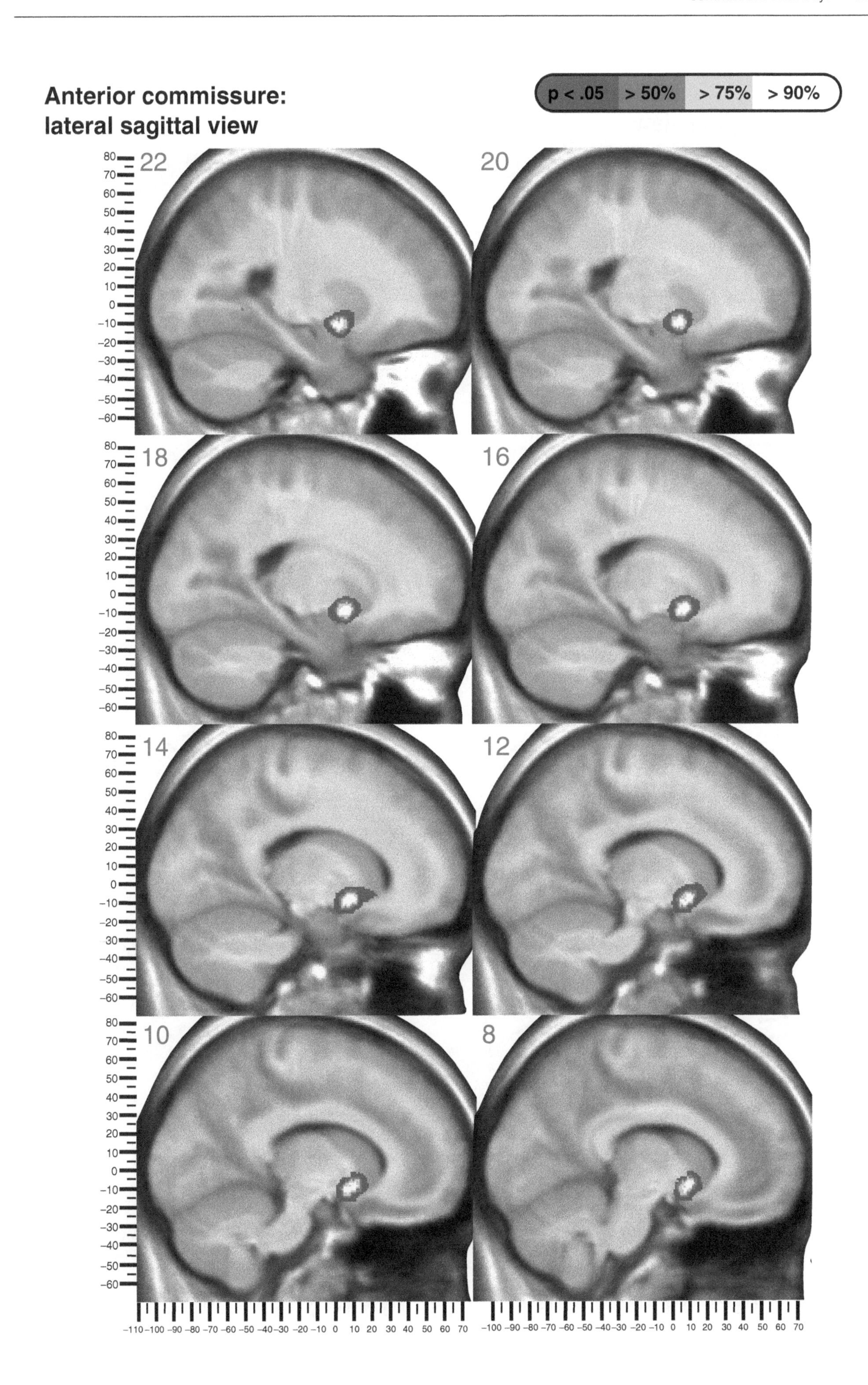
Anterior commissure: lateral sagittal view
p < .05
> 50%
> 75%
> 90%
22
20
18
16
14
12
10
8

Anterior commissure: lateral sagittal view

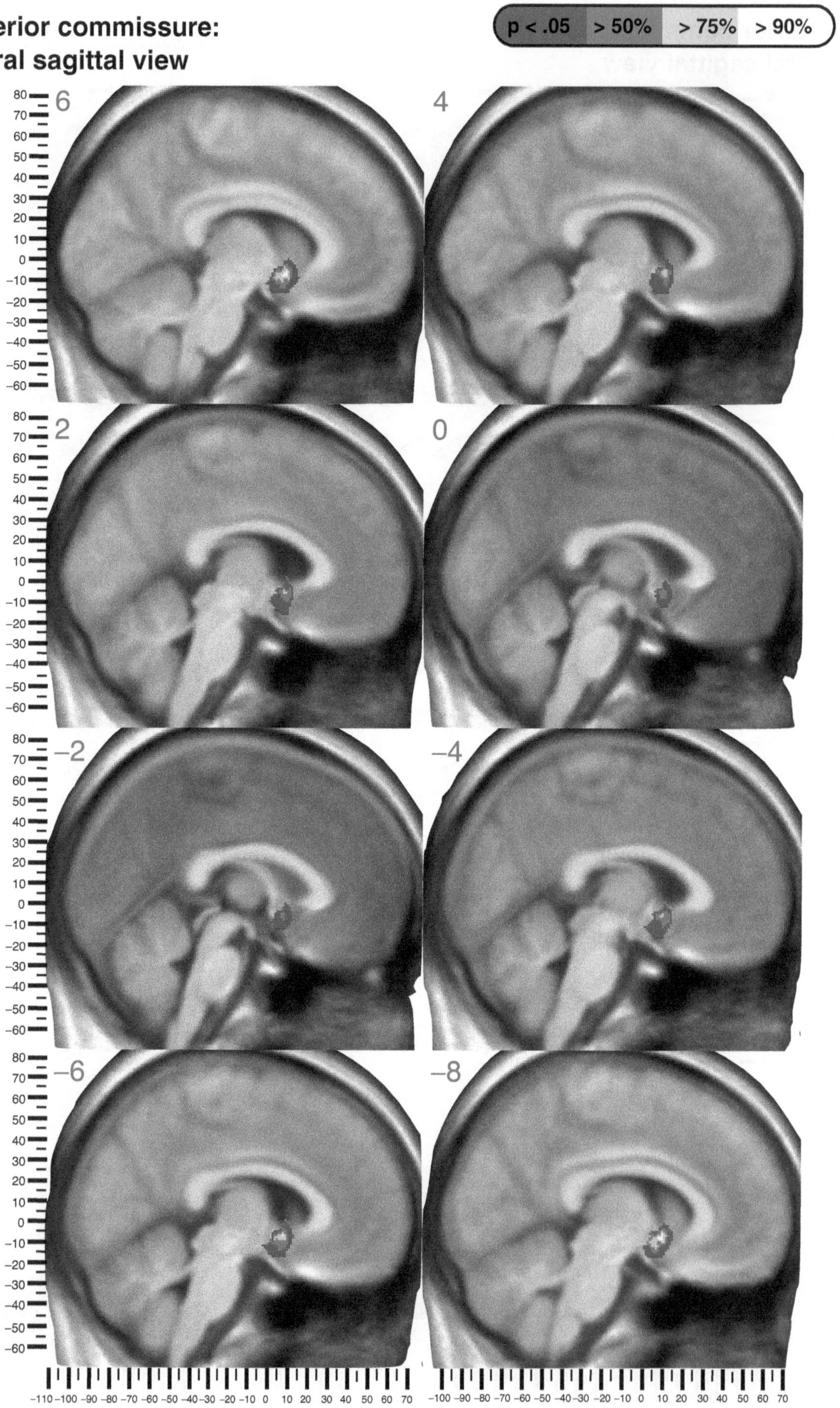

p < .05 | > 50% | > 75% | > 90%

6 4

2 0

–2 –4

–6 –8

80 70 60 50 40 30 20 10 0 –10 –20 –30 –40 –50 –60

–110 –100 –90 –80 –70 –60 –50 –40 –30 –20 –10 0 10 20 30 40 50 60 70

–100 –90 –80 –70 –60 –50 –40 –30 –20 –10 0 10 20 30 40 50 60 70

Anterior commissure: medial sagittal view

p < .05 | > 50% | > 75% | > 90%

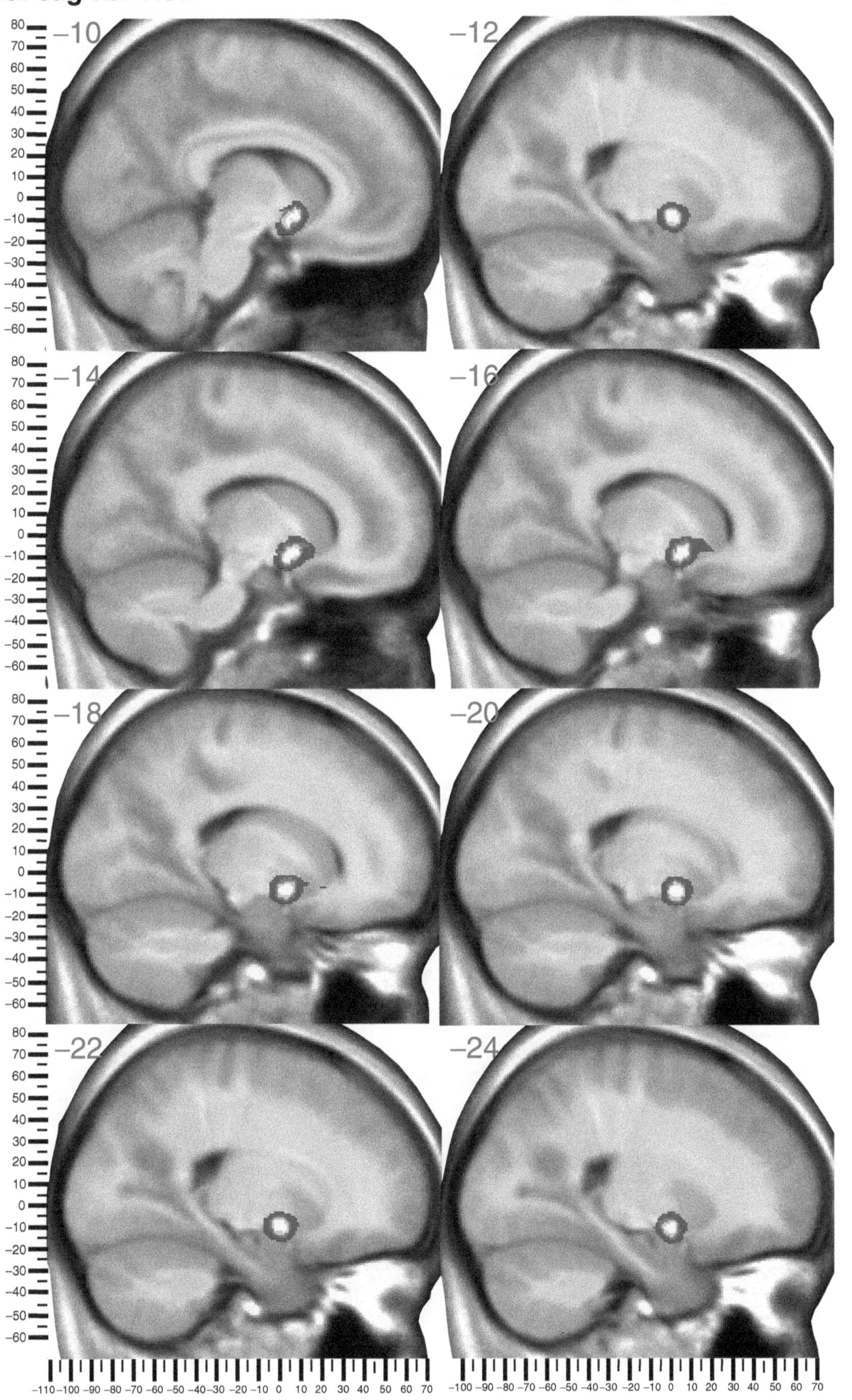

Anterior commissure: medial sagittal view

p < .05 | > 50% | > 75% | > 90%

−26 −28 −30 −32 −34 −36 −38 −40

Corpus callosum (atlas)

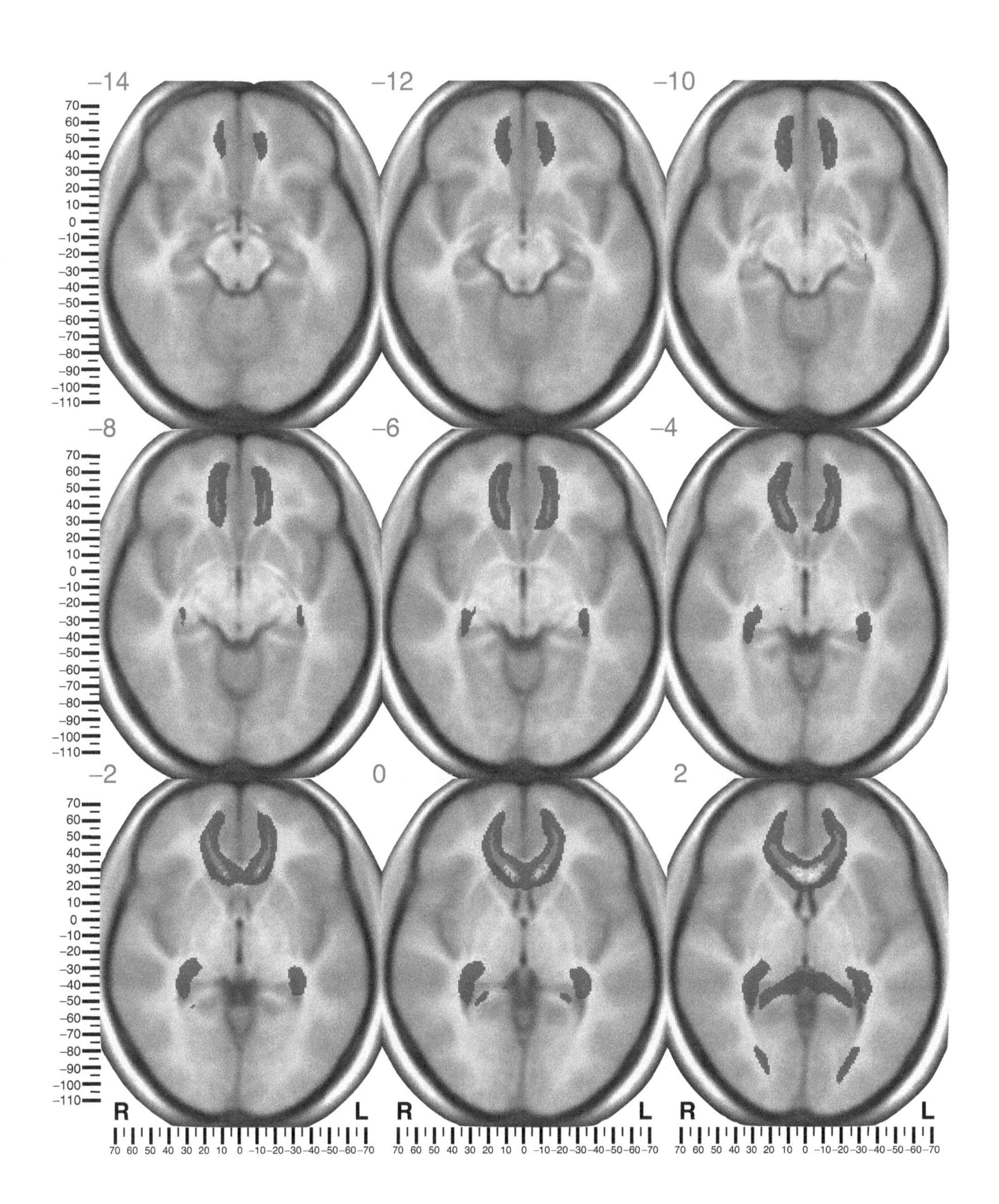

Corpus callosum: inferior axial view

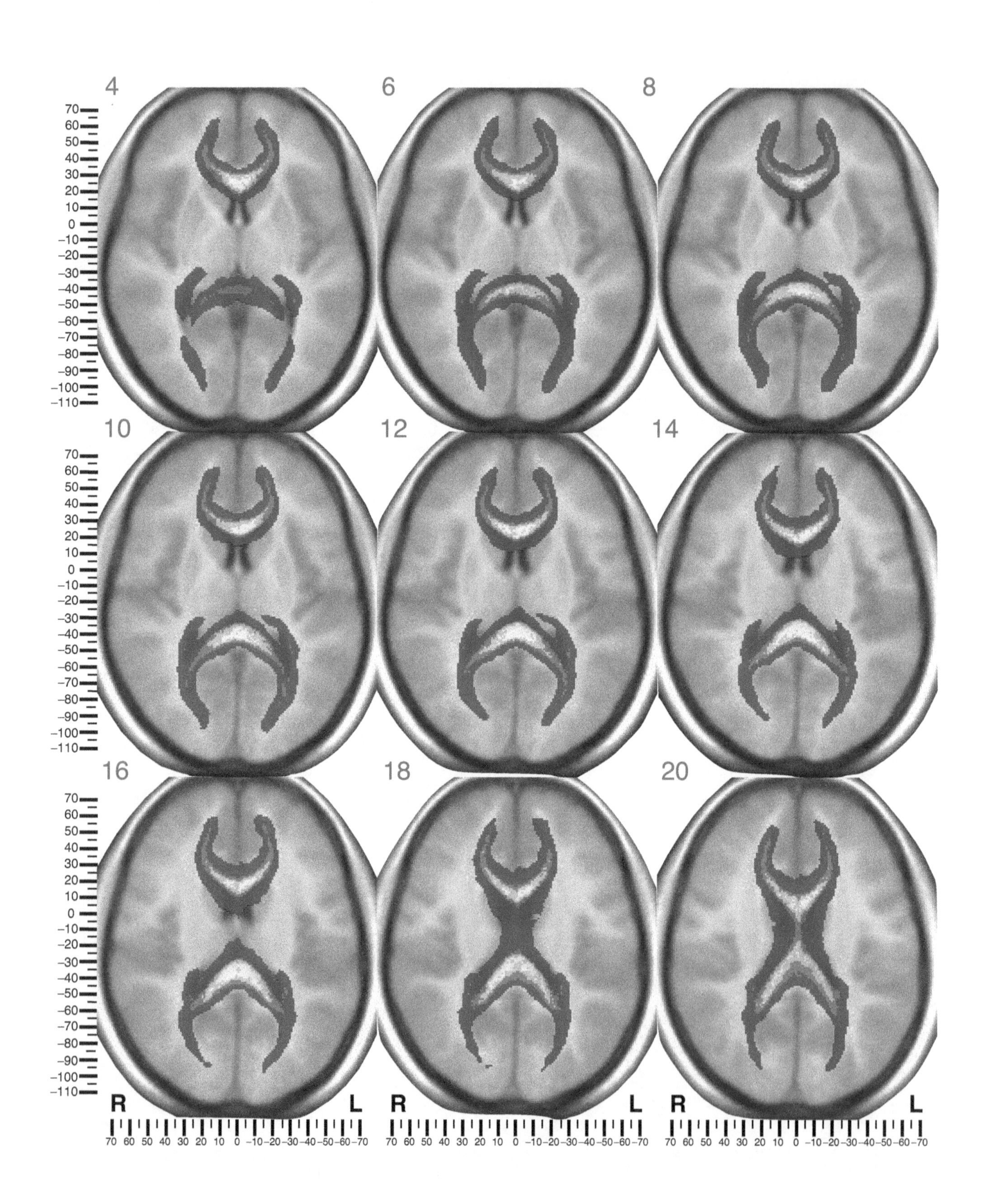

Corpus callosum: inferior axial view

p < .05 | > 50% | > 75% | > 90%

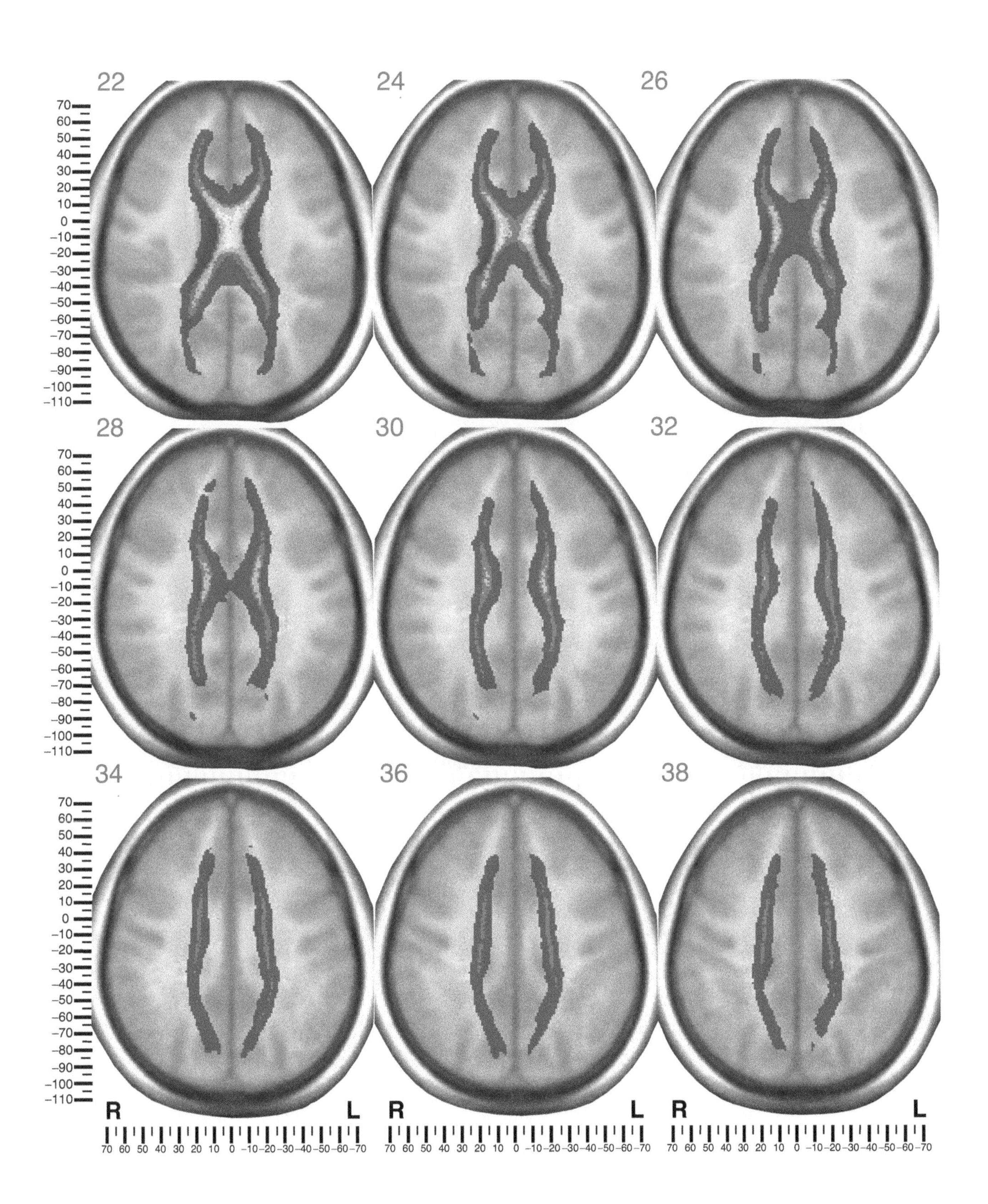

Corpus callosum: inferior axial view

p < .05 | > 50% | > 75% | > 90%

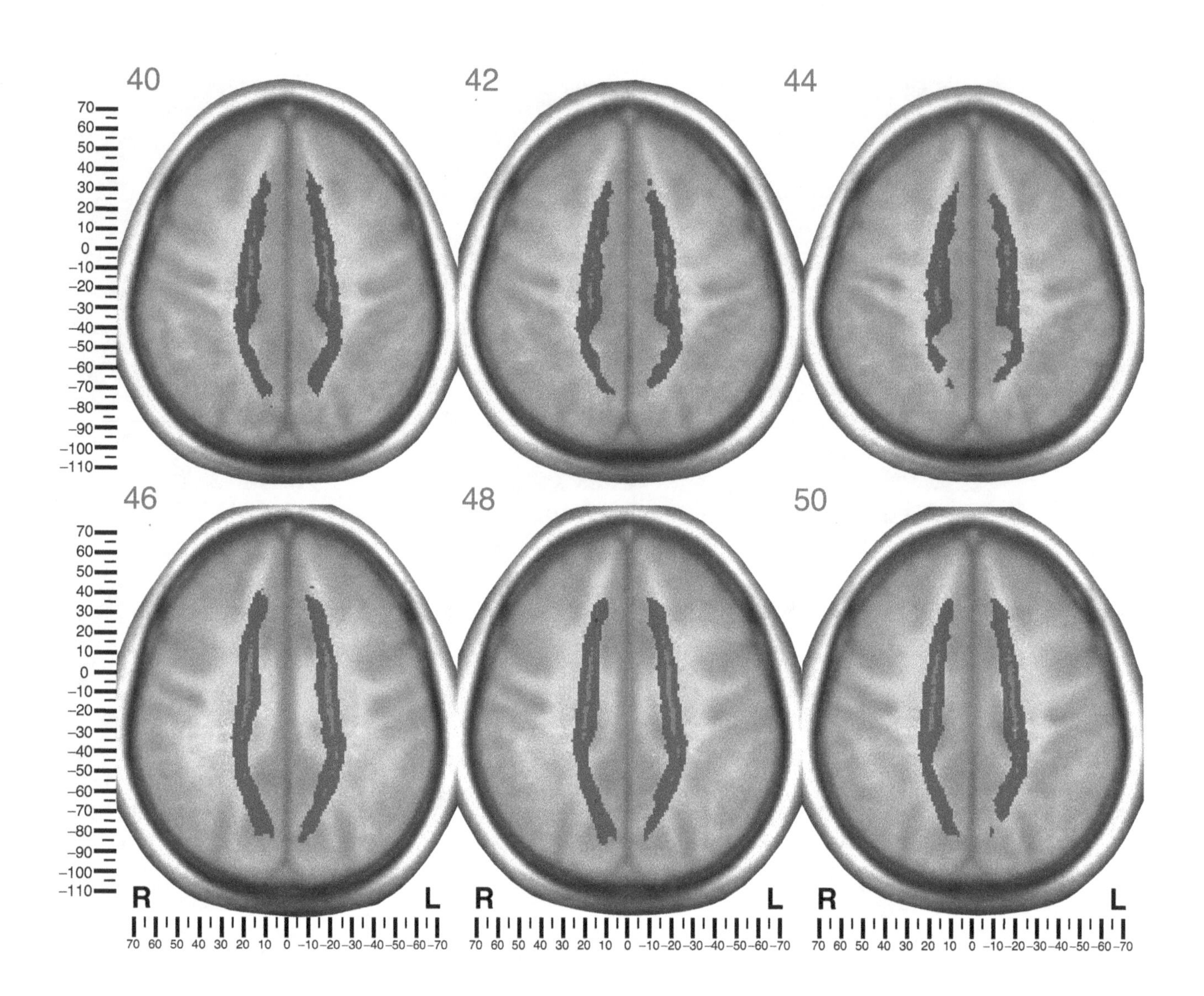

Corpus callosum: anterior coronal view

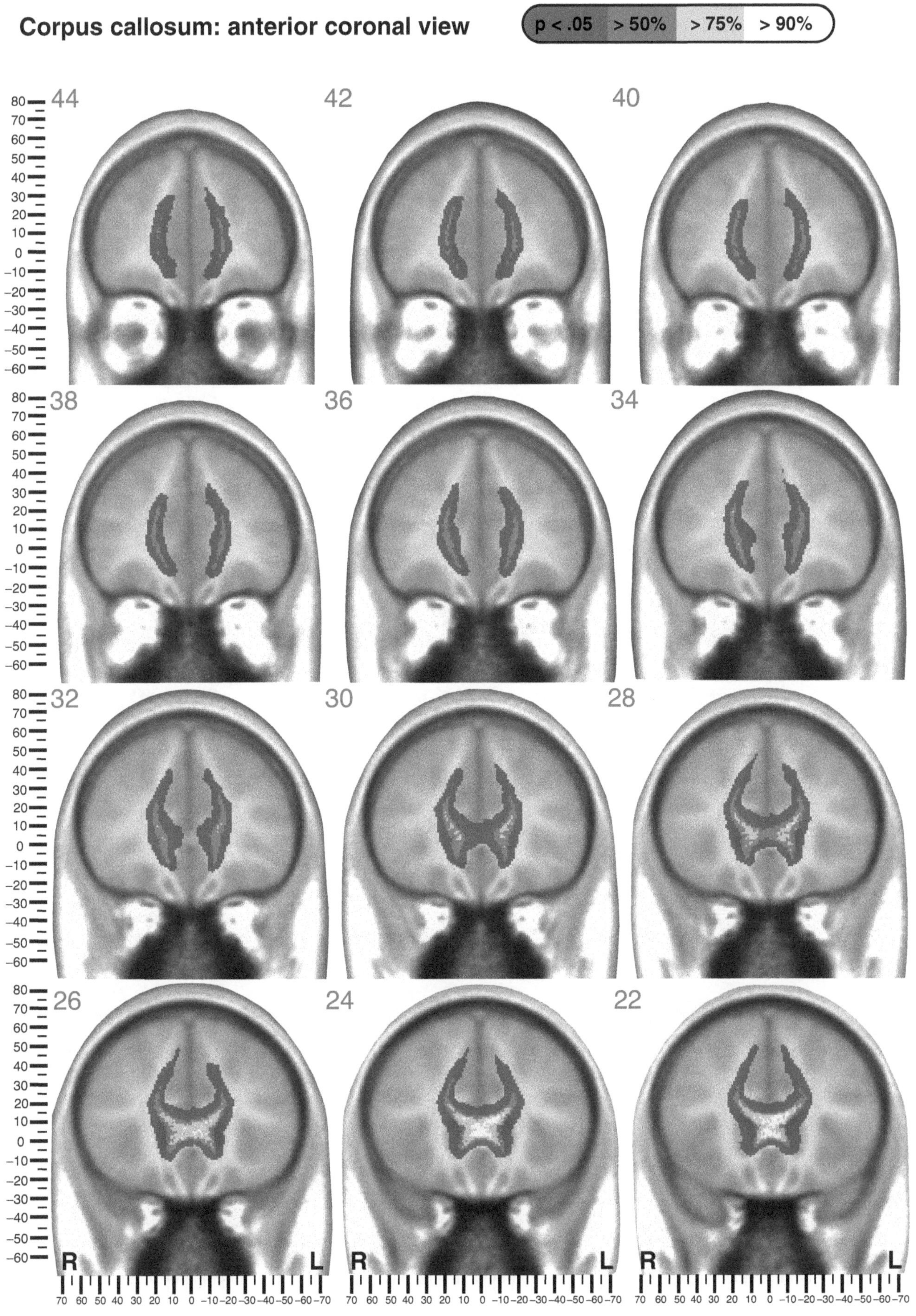

p < .05 | > 50% | > 75% | > 90%

Corpus callosum: anterior coronal view

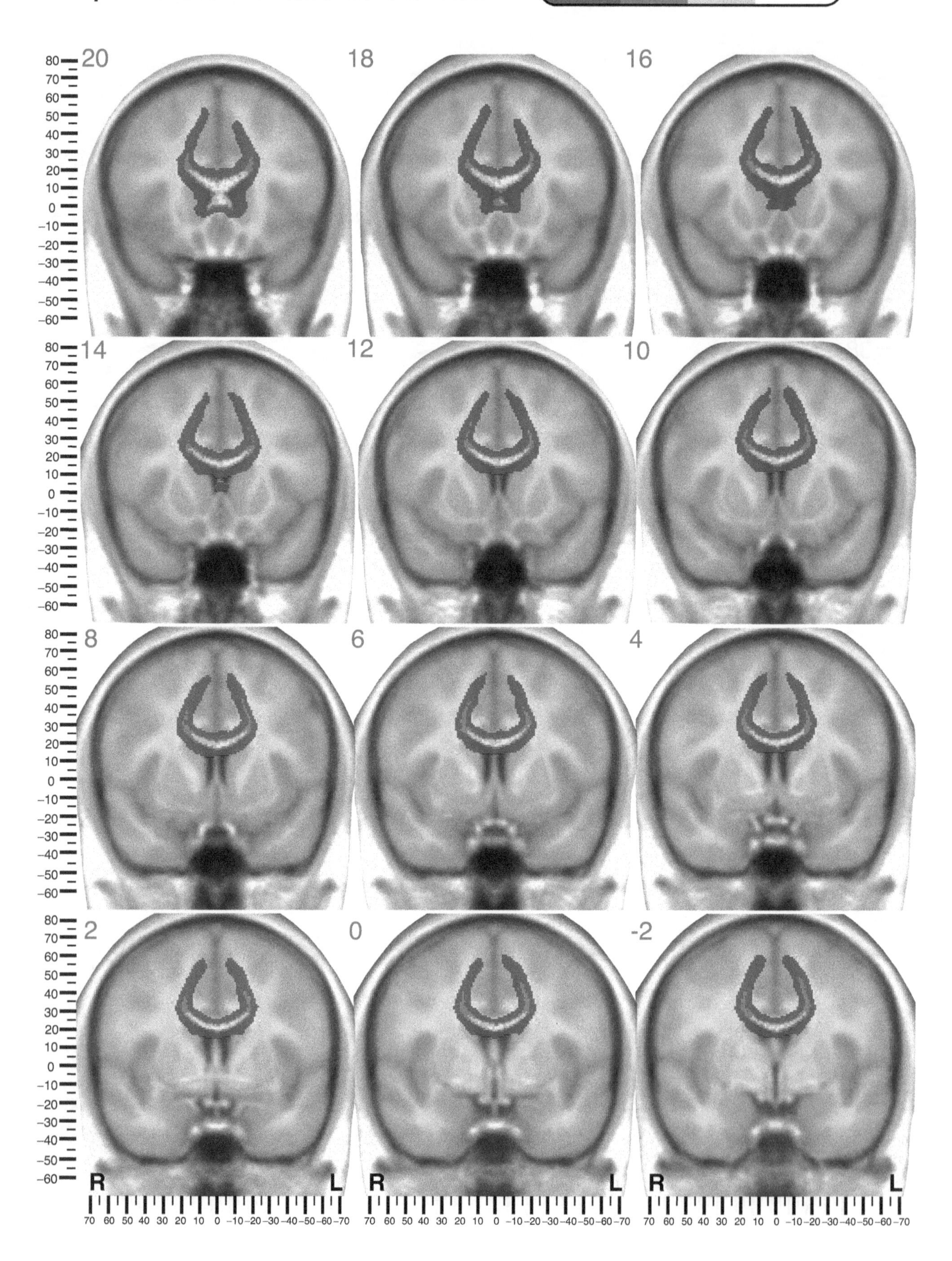

Corpus callosum: anterior coronal view

p < .05 | > 50% | > 75% | > 90%

–4 –6 –8

–10 –12 –14

–16 –18 –20

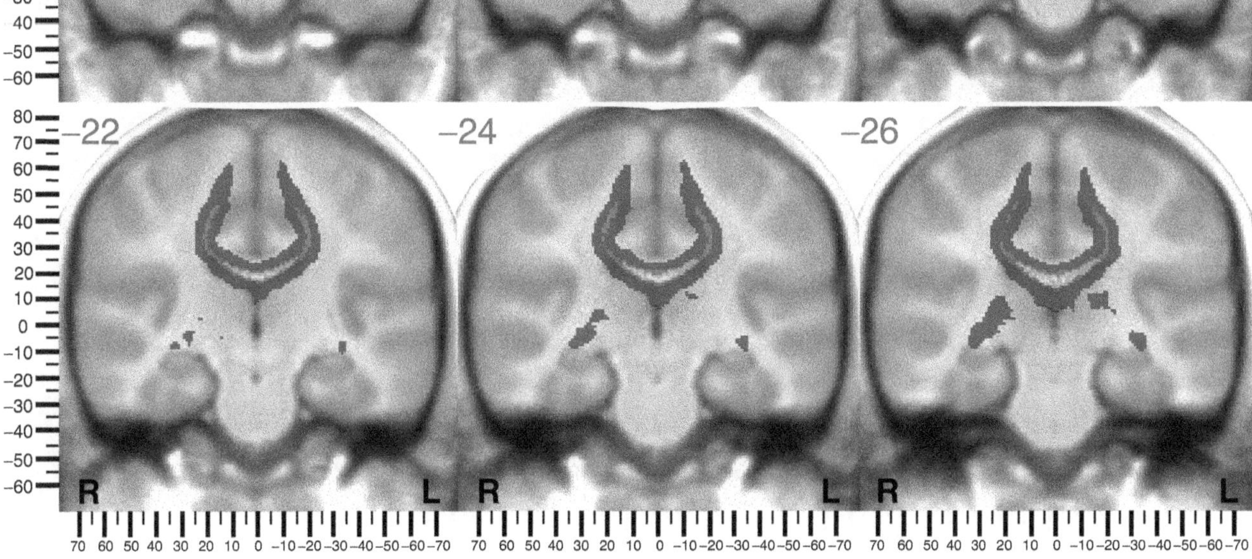

Corpus callosum: anterior coronal view

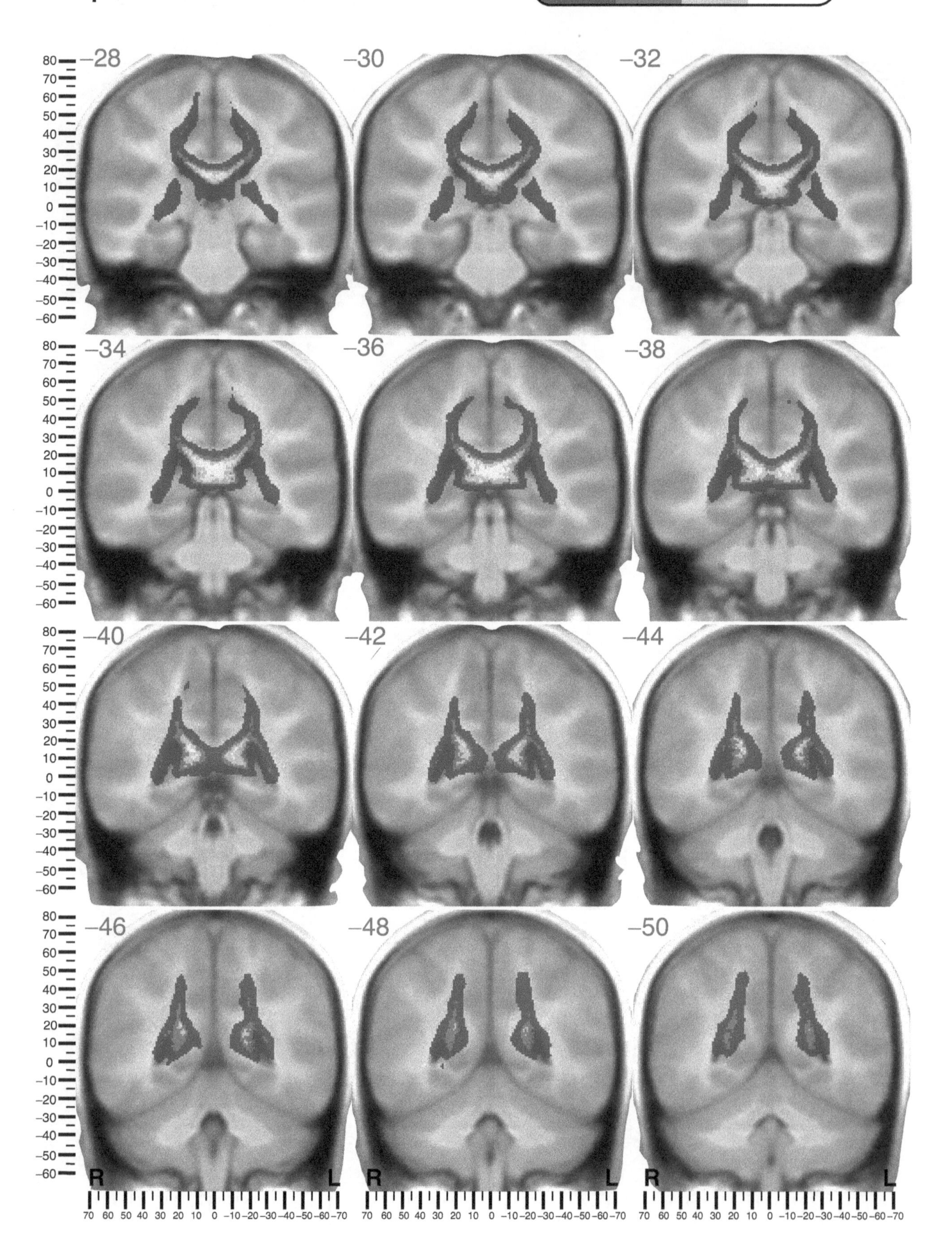

Corpus callosum: anterior coronal view

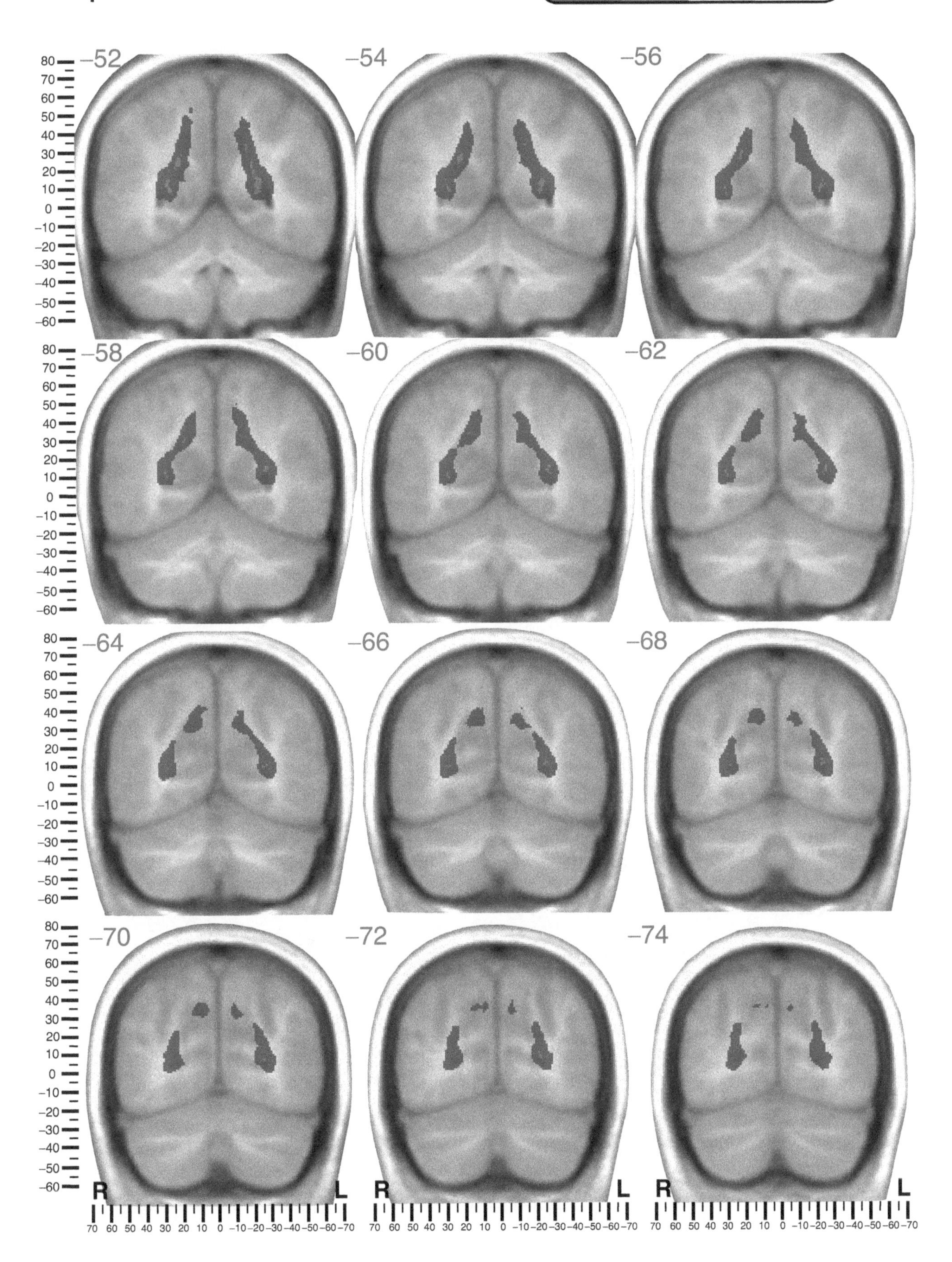

Corpus callosum: lateral sagittal view

p < .05 | > 50% | > 75% | > 90%

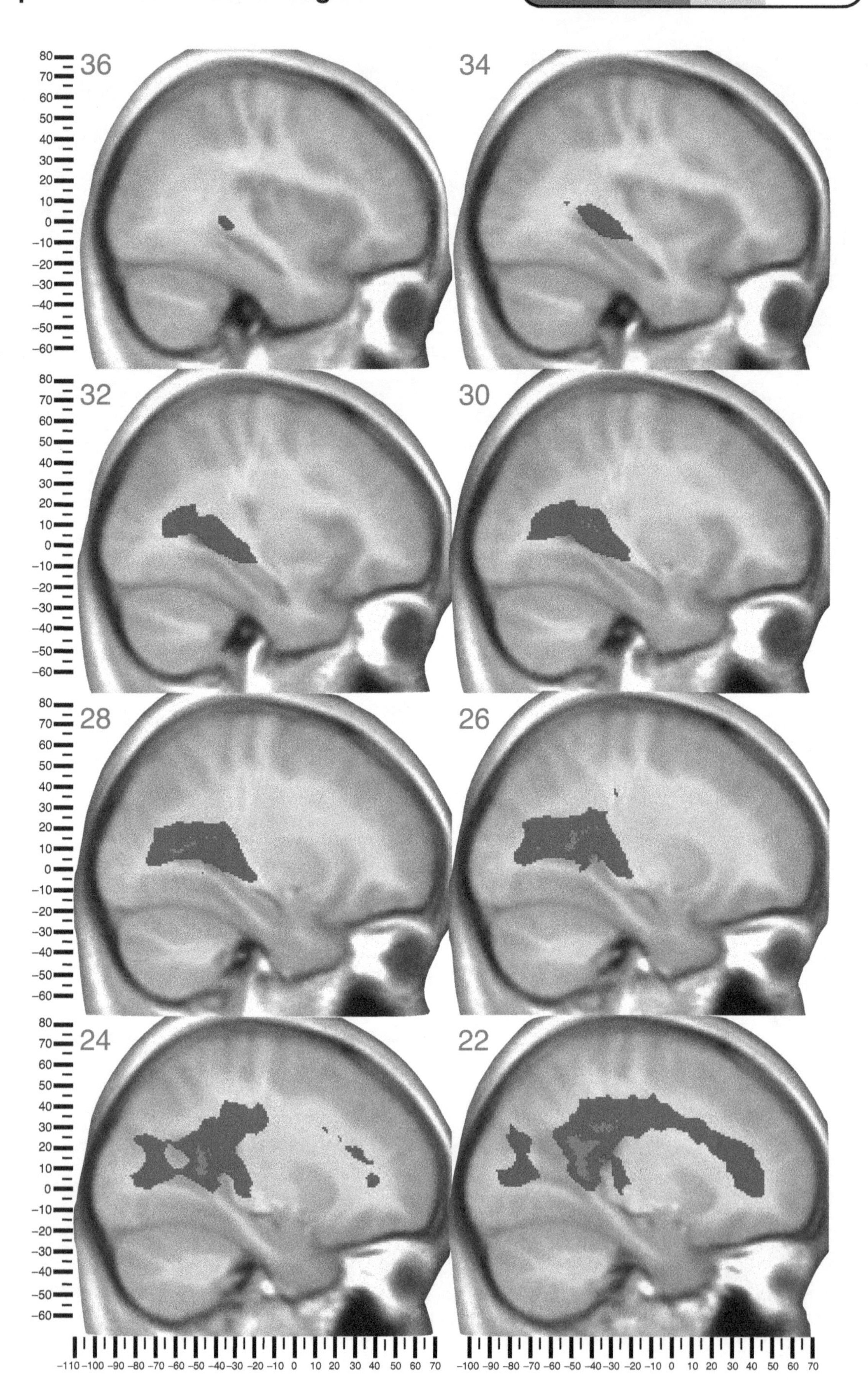

Corpus callosum: lateral sagittal view

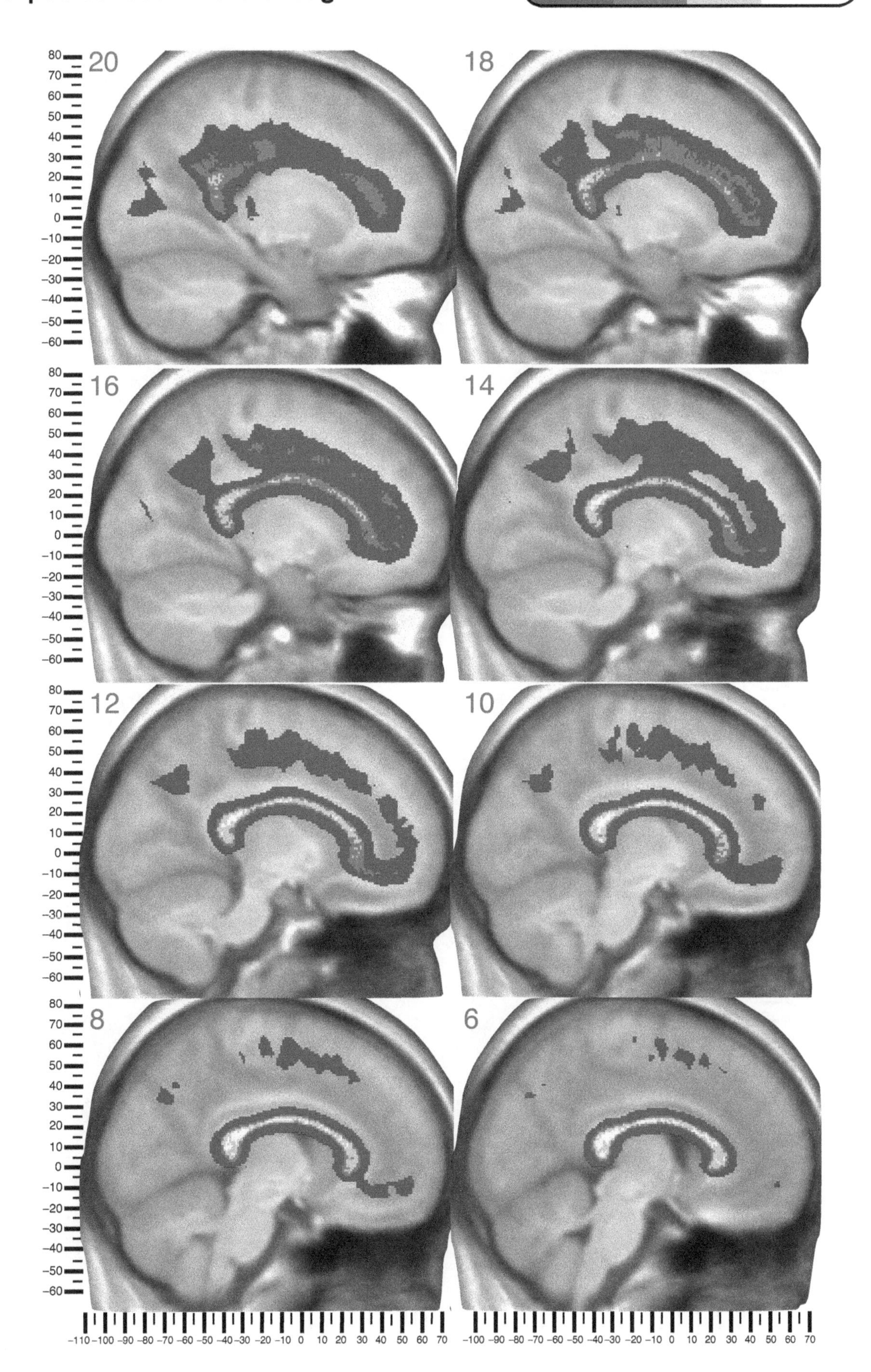

Corpus callosum: medial sagittal view

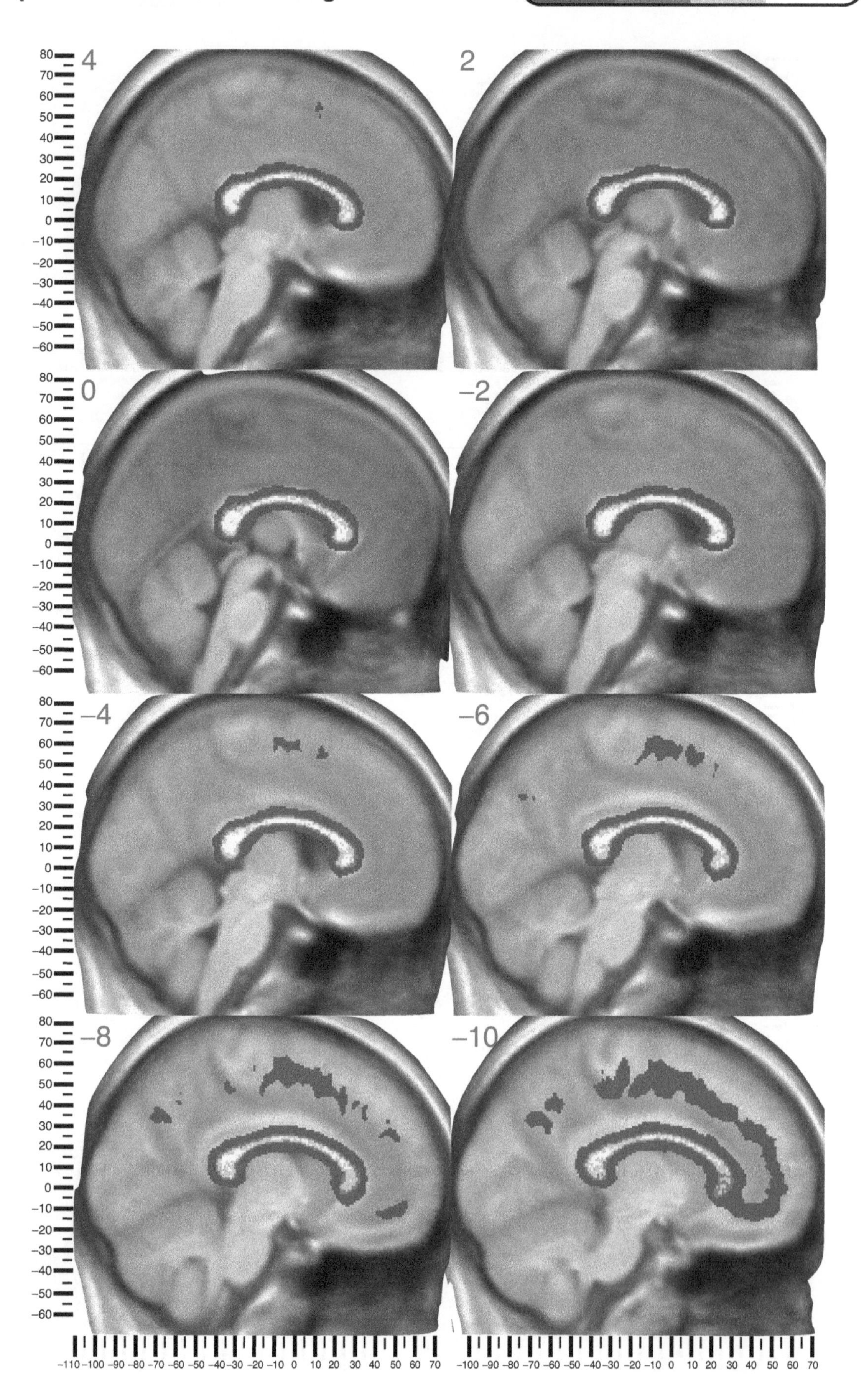

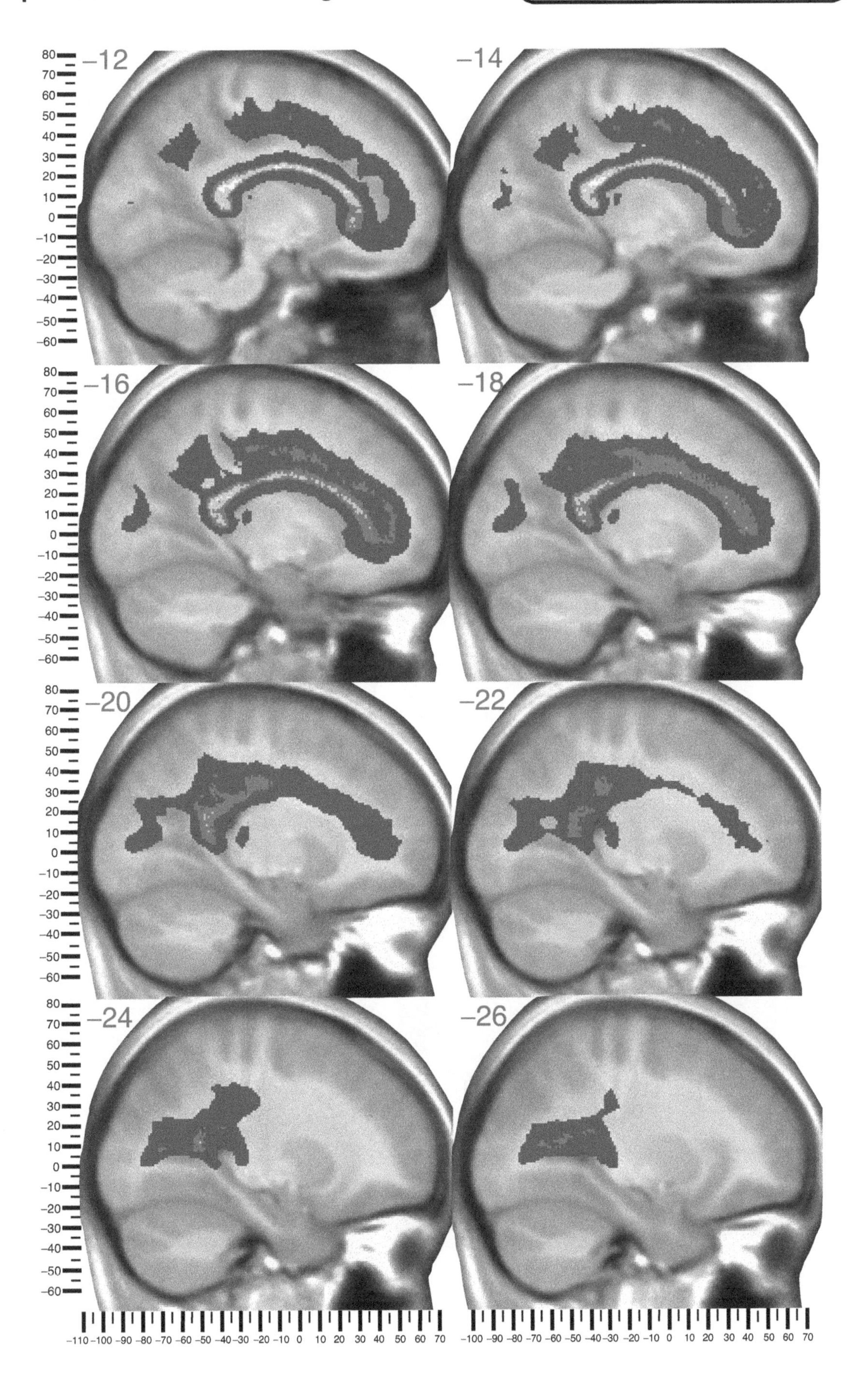
Corpus callosum: medial sagittal view
p < .05
> 50%
> 75%
> 90%
–12
–14
–16
–18
–20
–22
–24
–26
80
70
60
50
40
30
20
10
0
–10
–20
–30
–40
–50
–60
–110 –100 –90 –80 –70 –60 –50 –40 –30 –20 –10 0 10 20 30 40 50 60 70
–100 –90 –80 –70 –60 –50 –40 –30 –20 –10 0 10 20 30 40 50 60 70

Corpus callosum: medial sagittal view

p < .05 | > 50% | > 75% | > 90%

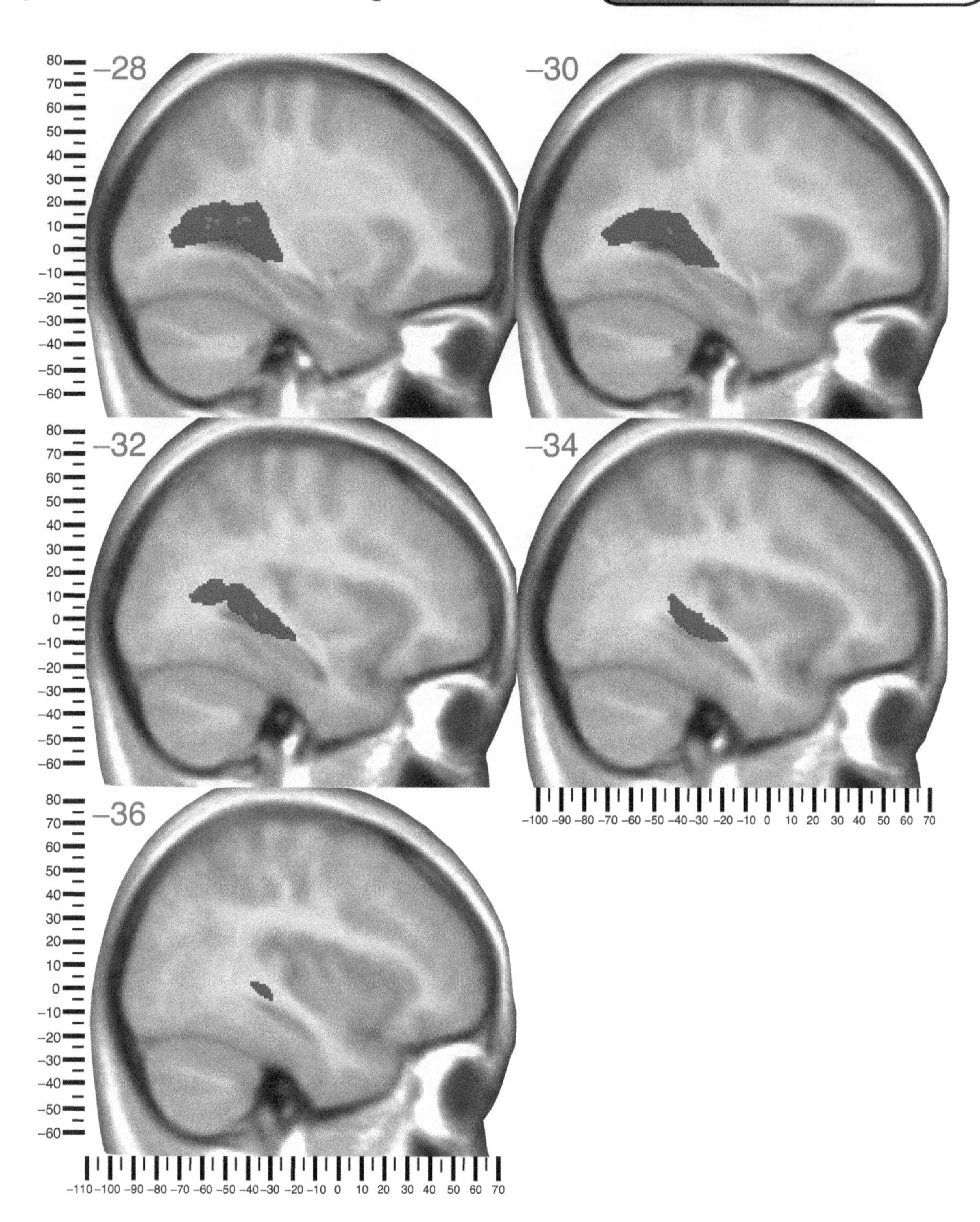

CHAPTER 10

Projection Systems

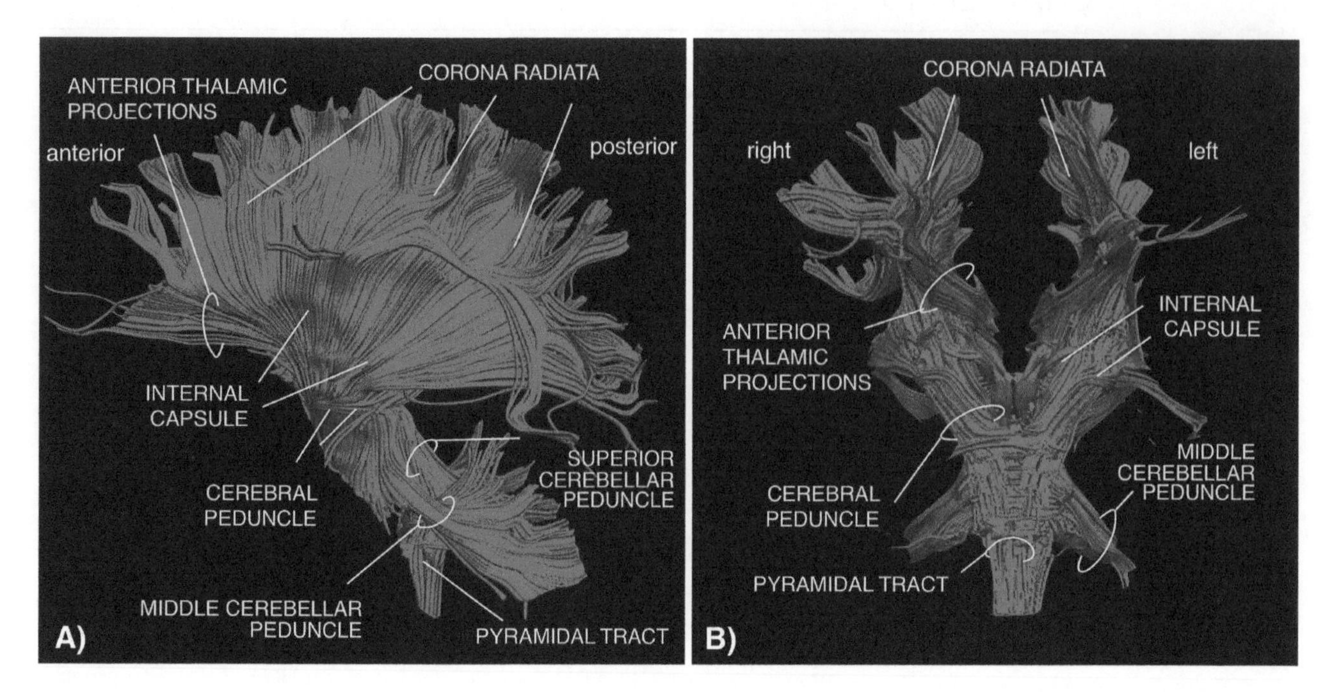

Figure 10.1 A) Left lateral and B) anterior view of the ascending (sensory) and descending (motor) projection pathways reconstructed with diffusion tensor tractography. The projections pathways connect the cerebral cortex, to the thalamus, basal ganglia, brainstem, cerebellum, and spinal cord. The corona radiata is a fan-like array of fibres containing all projection tracts originating from or projecting to the cerebral cortex. The internal capsule is continuous with the corona radiata and corresponds to the segment of fibres between the thalamus and the basal ganglia. Within the internal capsule the descending motor pathways are intertwined with the ascending thalamic fibres, as seen in the most dorsal sections, and with the afferent thalamic fibres from the brainstem and spinal cord in the most ventral sections. In the midbrain the descending motor tracts are enclosed within the cerebral peduncles, while the ascending spinal and cerebellar fibres are located more medially and posteriorly. The corticospinal tract is also indicated with the term pyramidal tract in its course between the pons and the spinal cord.

Introduction

Sensory and motor information travels through a complex system of ascending and descending projection fibres that functionally connect the peripheral and central nervous systems. Ascending sensory pathways are comprised of two major somatosensory routes: the spinothalamic tract runs within the dorsal column of the spinal cord and the medial lemniscus of the brain stem and conveys somatosensory information from the body; the trigeminal-thalamic pathway carries somatosensory information mainly from the head (face, oral and nasal cavities, etc.). Other incoming sensory projections are formed by the auditory, visual and olfactory pathways. The three major descending cortico-subcortical projection systems are concerned mainly with motor functions, and include the corticospinal and corticobulbar tracts, and the cortical efferents to the basal ganglia, and the projections to the cerebellum via the pontine nuclei. The projections to the basal ganglia and cerebellum are indirectly reciprocal, so that the cortex also receives projections from these centres via the thalamus, to create complex cortico-basal ganglion and cortico-cerebellar circuits. These projection systems are only partially segregated anatomically, as they share several subcortical relay stations (e.g. the thalamus). Figure 10.1 illustrates the trajectories of the major projection pathways reconstructed with diffusion tensor tractography.

The anatomical nomenclature of these projection pathways changes along their course through the cerebral hemispheres and brainstem, and names derived from descriptive anatomy have often been adopted. For example, the fan-like array of fibres between the cortical surface, basal ganglia, and thalamus is called the corona radiata because of its crown-like shape, formed partly by the ascending thalamic fibres radiating out to various cortical destinations, and partly by descending motor fibres converging towards the base of the cerebrum.

More inferiorly, the fibres of the corona radiata enter the internal capsule, a space confined between the thalamus and caudate nucleus medially and the lentiform nucleus laterally. In horizontal section, the internal capsule has the appearance of a shallow letter V, divided into an anterior limb, a middle genu, and posterior limb. Within the internal capsule the fibres have an orderly arrangement from front to back according to their destination or origin, and in the case of the somatosensory and motor fibres, the parts of the body they serve (Figures 10.1 and 10.2) (Beevor and Horsely, 1890; Dejerine 1901; Crosby et al., 1962; Behrens et al., 2003; Newton et al., 2006). The anterior limb of the internal capsule carries ascending thalamic projections concerned with cognitive, limbic, basal ganglionic, cerebellar, and other complex functions of the frontal lobe. The anterior limb also contains descending fronto-striatal fibres to the basal ganglia, and fronto-pontine fibres to the pontine nuclei, which in turn project to the cerebellum. More posteriorly, the genu contains ascending somatosensory thalamic projections from the trigeminal pathway to the postcentral gyrus, and descending corticobulbar motor fibres serving cranial nerve somatomotor nuclei. The posterior limb contains ascending thalamic fibres to the parietal, occipital, and temporal lobe, and

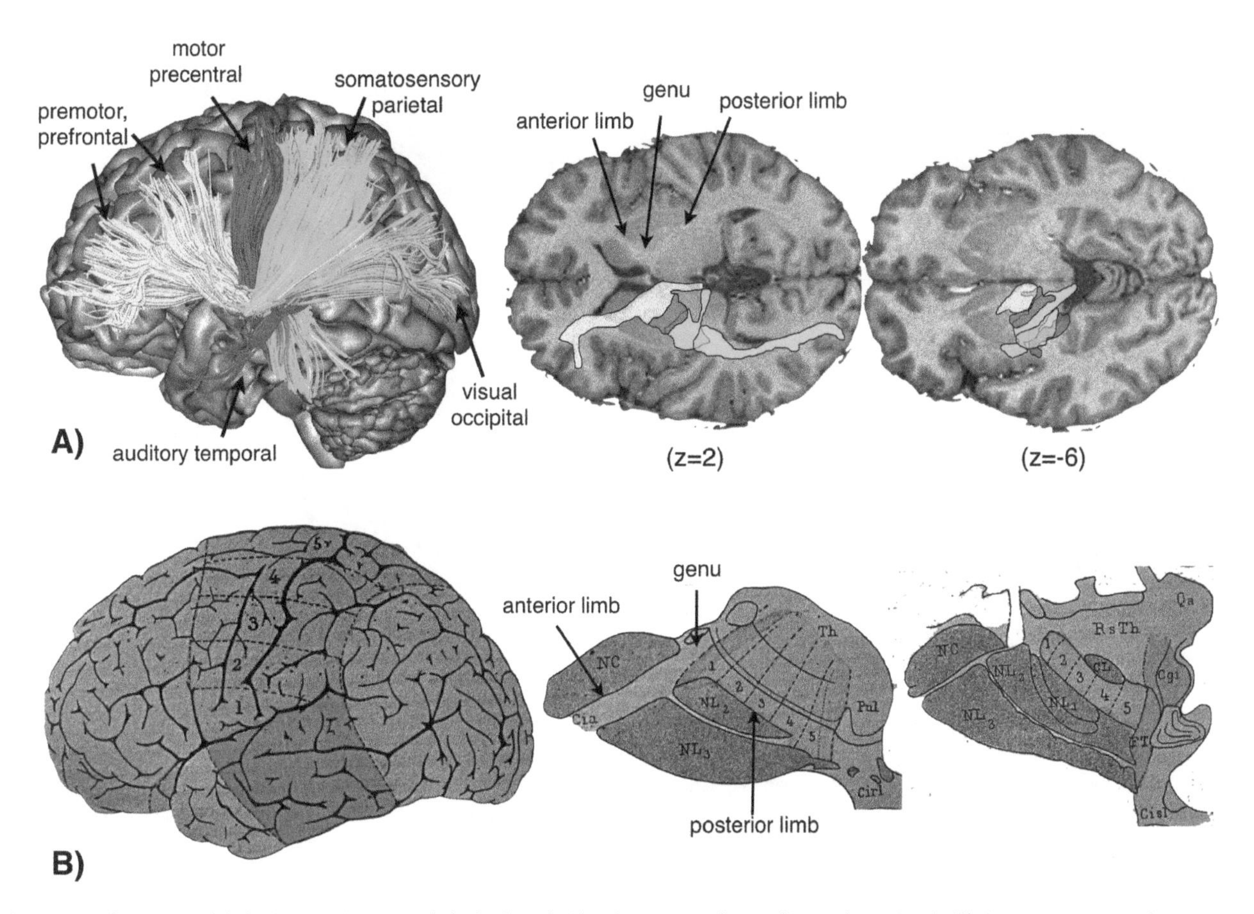

Figure 10.2 The motor and thalamic projections are orderly distributed within the corona radiata and internal capsule. A) Diffusion tensor tractography reconstruction shows that the motor and thalamic pathways follow a fronto-parieto-occipital-temporal distribution along the anterior and posterior limbs of the internal capsule. The coloured tracts indicate, from front to back, the anterior (yellow), superior (red and green), posterior (blue), and inferior (purple) thalamic peduncles (see also Figure 10.5). B) Dejerine's (1901) original diagram showing a more detailed (numbered) distribution of the corticospinal tract in the posterior limb of the left internal capsule sectioned horizontally at two levels, left being the more superior. Note that Dejerine's numbered zones correspond to the fronto-parietal areas indicated on the dorsolateral aspect of the cerebral hemisphere. Abbreviations: NC, caudate nucleus, NL, nucleus lentiform; Th, thalamus; Pul, pulvinar.

descending corticopontine and corticospinal motor fibres (Figure 10.2). At the most posterior end of the internal capsule, visual projection fibres from the lateral geniculate nucleus of the thalamus run towards the primary visual cortex in the occipital lobe, via the capsule's retrolenticular part, whilst auditory projection fibres transit its sublenticular region en route to the primary auditory cortex in the superior temporal gyrus.

In the midbrain, the descending motor (corticospinal, corticobulbar and corticopontine) fibres collect inferiorly and posteriorly at the base of the internal capsule into compact bundles within the most anterior region of the cerebral peduncles. Ascending in the midbrain, posterior to the cerebral peduncles, are three major sets of somatosensory afferents to the thalamus. Dorsolaterally, the fibres of the spinothalamic (anterolateral) tract convey pain, temperature, and coarse touch information from the contralateral side of the body. More ventromedially, the medial lemniscus carries conscious proprioceptive, fine discriminative touch, and vibration sensory information from the contralateral side of the body. Finally, the trigeminal lemniscus is formed by trigeminothalamic fibres that carry somatosensory information from the opposite trigeminal sensory nucleus to the thalamus.

Corticospinal and corticobulbar projections

The corticospinal tract is the main descending projection system between the motor cortex and the spinal cord and thereby subserves motor control functions for all voluntary muscles of the arms, trunk, and legs. A smaller division of the descending motor pathways, the corticobulbar tract, originates mainly from the ventro-lateral region of the precentral gyrus' motor strip and projects to the motor nuclei of cranial nerves that control facial, masticatory, tongue, pharyngeal, and laryngeal striated muscles. The topographical distribution of the cortical motor neurons is also defined pictorially by the term 'motor homunculus', which imagines the body to be arched over the cortical surface and distorted to reflect the relative areas of the cortex devoted to different body regions (Penfield and Boldrey, 1937). Approximatly half of the corticospinal fibres originate from the precentral gyrus (primary motor cortex), whereas axons of the other half arise from various regions, including the premotor and somatosensory cortices, anterior cingulate gyrus, and other parts of the parietal cortex (Figure 10.2A). In the internal capsule the fibres of the corticospinal tract maintain a precise order correlated with the body parts they serve (somatotopic order) (Figure 10.2B)

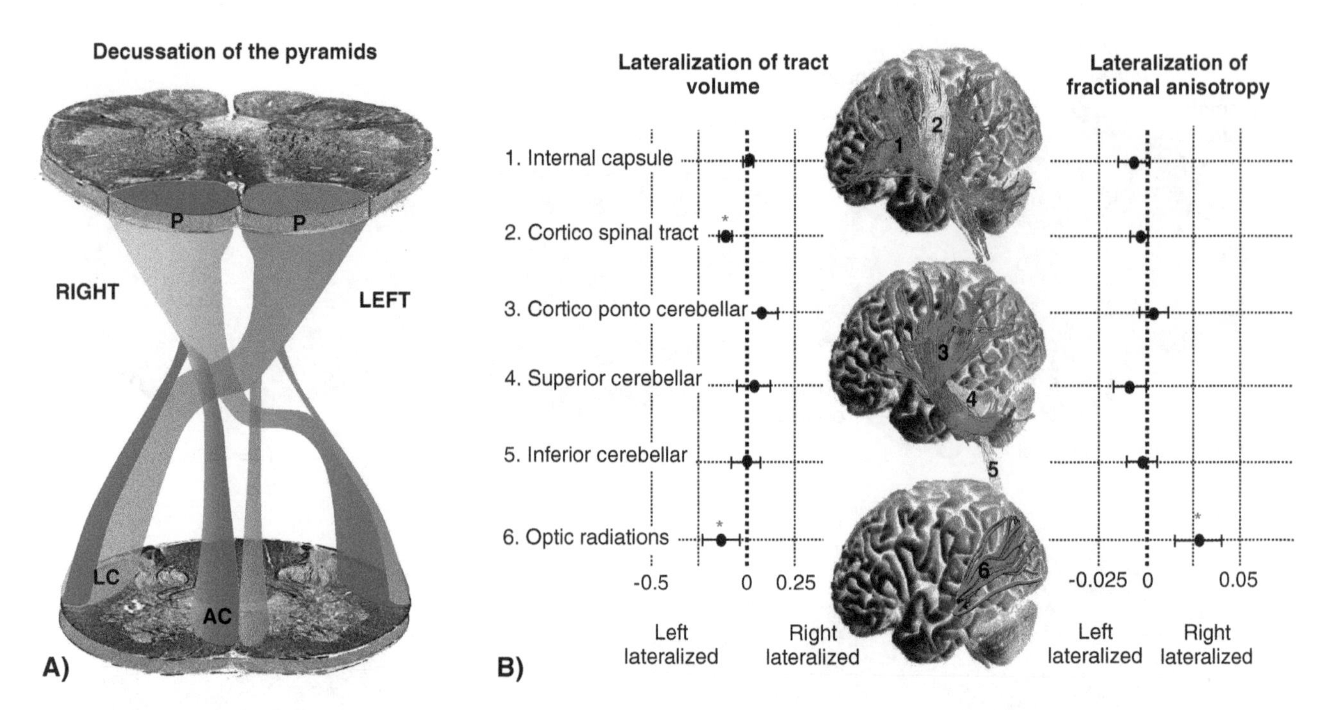

Figure 10.3 A) A diagram of the decussation of the corticospinal tract between a section of the medulla oblongata at the level of the pyramids (P) and inferiorly, a section of the spinal cord. The great majority of the fibres of the cortico-spinal tract cross the midline into the lateral column (LC) of the contralateral spinal cord, to form the lateral corticospinal tract. In 40% of subjects the crossing pattern is asymmetrical with a larger number of fibres crossing from left to right. A small group of ipsilateral fibres continues into the lateral column or into the anterior column (AC) as the anterior (ventral) corticospinal tract. Also, note that the left pyramidal tract decussates higher compared to the right pyramidal tract in 90% of subjects. B) Diffusion tensor tractography of the major projection tracts shows a leftwards lateralization of the volume of the cortico-spinal tract and optic radiations (values are means with 95% confidence interval; * $p < 0.05$ after correction for multiple comparisons). The optic radiations show also a rightwards lateralization of the fractional anisotropy. (Thiebaut de Schotten et al., 2011).

(Dejerine, 1901; Newton et al., 2006). Thus, the fibres from the ventro-lateral cortex (corresponding to the muscles close to the mouth) are anterior to the fibres from the dorso-lateral (hand) and dorso-medial (leg) cortex, etc.

The corticospinal tract enters the brainstem through the cerebral peduncles and passes through the pons between the transverse pontine fibres. In the medulla oblongata the corticospinal fibre bundle forms a ridge anteriorly (the so-called pyramid) (Figure 10.3), then just before entering the spinal cord, most of the corticospinal fibres cross from one side to the other (decussation of the pyramids) and enter the lateral column of the spinal cord. This decussation of the corticospinal tract was first described by Domenico Mistichelli (1709) and François Pourfour du Petit (1710), but it was only when Paul Emile Flechsig (1876) applied his myelogenetic method to brain sections of newborns that neurohistological evidence for corticospinal crossing and tract asymmetry emerged. Flechsig found that most of the cortico-spinal fibres cross the midline and form the lateral column of the contralateral spinal cord. In up to 40% of the specimens the crossing pattern is asymmetrical with a larger portion of fibres crossing from left to right (Flechsig, 1876). The majority of the uncrossed ipsilateral fibres form the anterior column of the spinal cord, but a smaller proportion of uncrossed fibres contributes to the lateral column of the same side. Yakovlev and Rakic (1966) used Flechsig's method in a larger sample, and they not only confirmed his findings but found that the decussating bundle of the left corticospinal tract crosses the midline at a more dorsal level in almost 90% of the cases. Considering that many of these dorsal fibres are likely to originate from the hand region of the left precentral gyrus that controls the right hand, these observations were thought to be of relevance to handedness. This hypothesis was not confirmed by Kertesz and Geschwind (1971) who studied the variation of the crossing pattern with respect to manual preference in adults and found no correlation with handedness. An asymmetrical pattern has also been reported for the number of pyramidal fibres of the spinal cord, with 75% of cases showing a greater number on the right compared to the left (Nathan et al., 1990). A more recent post-mortem study measured the volume of the corticospinal tract within the corona radiata and internal capsule, and found a greater volume on the left compared to the right in 70% of the subjects (Rademacher et al., 2001). The discrepancy between the percentage reported by Flechsig (1876) and that by Rademacher et al. (2001) may be related to the different ages of the samples used in the two studies, fetuses in the first and old adults in the latter.

A magnetic resonance study (T_1-weighted images) on a large cohort of children and adolescents has shown a greater age-related increase in the white matter density of the left internal capsule compared to the right (Paus et al., 1999). In adults the larger white matter volume of the left internal capsule is independent of handedness (Good et al., 2001; Hervé et al., 2006). Eluvathingal et al. (2007) used tractography to show higher fractional anisotropy on the left corticospinal tract compared to the right in 70% of healthy children and adolescents. In young adults the overall volume of the corticospinal tract has a moderate, but statistically significant, leftward asymmetry in the portion running between the cortex and the bulbar pyramids (Thiebaut de Schotten et al., 2011). This asymmetrical pattern is present in 70% of the subjects, while 12.5% show a rightward asymmetry (i.e. right larger than left). The remaining 17.5% show a bilateral and almost symmetrical distribution.

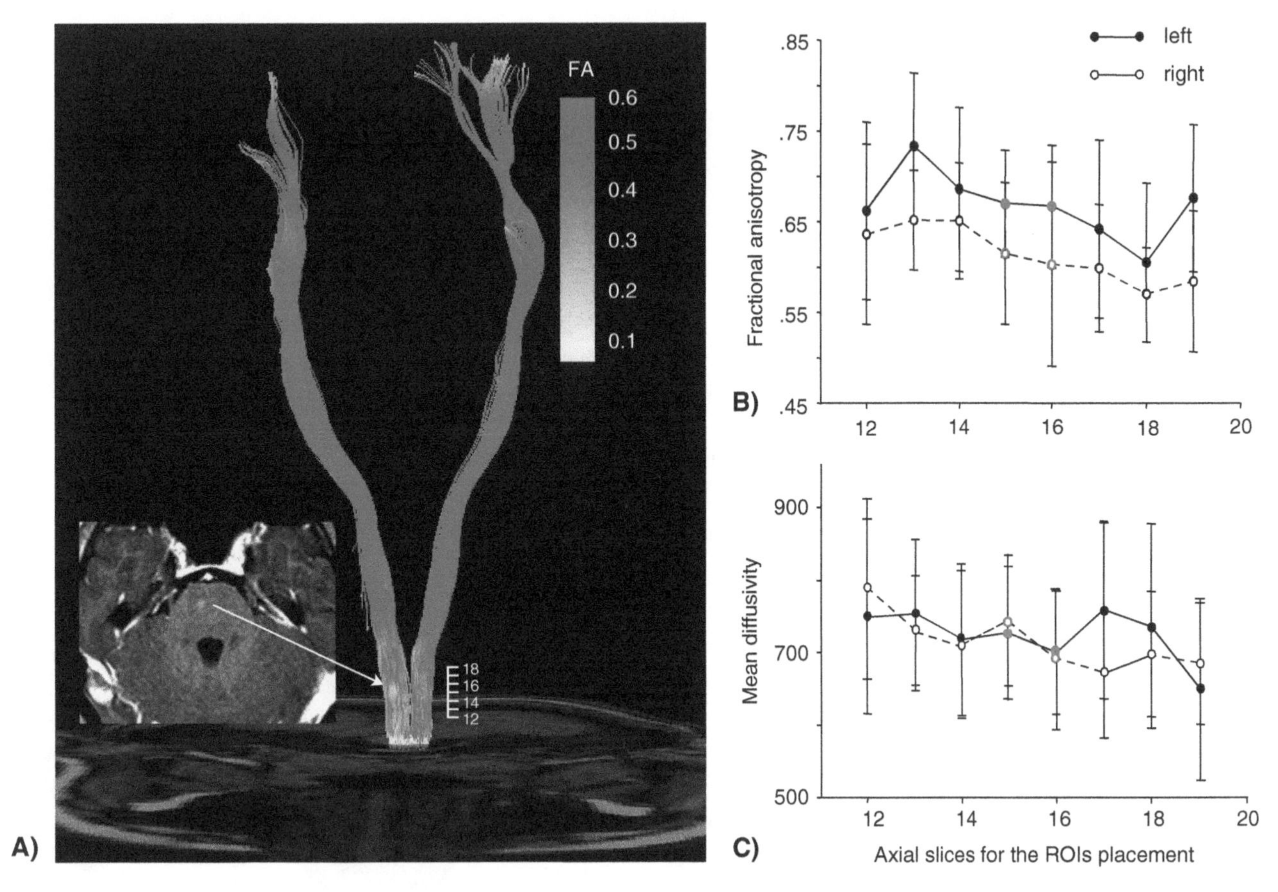

Figure 10.4 Frontal view of the corticospinal tracts reconstructed from a 22-year-old woman who presented with acute left-sided hemiplegia. A) Tractography shows a small rounded region of reduced fractional anisotropy in the right cortico-spinal tract (indicated by the white arrow). On axial T1-weighted spin-echo images the region shows enhancement after gadolinium suggesting a brain capillary telangiectasia. B, C) Measurements of fractional anisotropy and mean diffusivity (MD) along the pyramidal tract were made using eight regions of interest (ROI) selected on axial slices on both side. The ROIs around the lesion are indicated in red and show no significant reduction of fractional anisotropy compared to the contralateral side, suggesting that the fibres surrounding the lesion are compressed but intact. The patient made a complete recovery after 6 months of physiotherapy. This case exemplifies the clinical utility of combined tractography and ROI quantitative measurements.

The distribution above is similar in both genders. These results are very similar to the findings derived from previous post-mortem dissections (Nathan et al., 1990; Rademacher et al., 2001).

Lesions of the corticospinal tract manifest with a pyramidal syndrome whose characteristics vary according to the level of the lesion, the extension of the damage, and the time of onset. Unilateral lesions to the segment of the corticospinal tract between the motor cortex and the midbrain cause reduced motor strength on the contralateral side of the body. Motor deficits can range from weakness (hemiparesis) to complete absence of movement (hemiplegia). The patients have difficulty in holding objects in their hands, raising their arm or leg, walking or standing. If the corticobulbar tract is affected, facial weakness becomes apparent (e.g. asymmetry of the mouth, deviation of the tongue to the affected side, etc.) with difficulties in swallowing (dysphagia) and articulating of words (dysarthria). The pseudobulbar syndrome is a selective disconnection syndrome of the cortico-bulbar tract and other limbic connections in which bulbar motor deficits are accompanied by emotional lability (i.e. patients typically report episodes of laughter or crying without an apparent motivation or sometimes in response to stimuli that normally elicit a different emotional reaction). The pyramidal syndrome changes over time. Muscle tone and reflexes are usually reduced in the acute stage but become progressively increased with time and the limbs assume an anti-gravity posture characteristic of chronic spasticity (e.g. upper arm flexion and leg extension). These modifications are the consequence of the lack of corticospinal inhibition on the reflex arc at the spinal level. Disorders associated with spasticity are anoxic lesions occurring at birth (e.g. cerebral palsy), head injury, multiple sclerosis, upper motor neuron disease, tumours, and stroke. When the lesion is in the brainstem the damage to the corticospinal tract can extend to other structures, and the patient manifests with alternating syndromes characterized by a combination of contralateral pyramidal symptoms and ipsilateral deficits due to damage to nuclei of the cranial nerves. The locked-in syndrome is a rare condition where the patient is conscious but completely paralysed except for eye movements due to a bilateral lesion to the cerebral peduncles.

Diffusion tensor tractography can facilitate the localization of small lesions along the cortico-spinal tract. Figure 10.4 shows the diffusion tractography reconstruction of a 22-year-old woman who presented with sudden weakness of the left arm and leg. The patient's neurological deficit was initially considered of a psychological nature (i.e. a conversion disorder) due to the absence of an evident pathology on structural magnetic resonance images obtained without contrast agent. The presence of an area of reduced fractional anisotropy in the right cortico-spinal tract prompted a

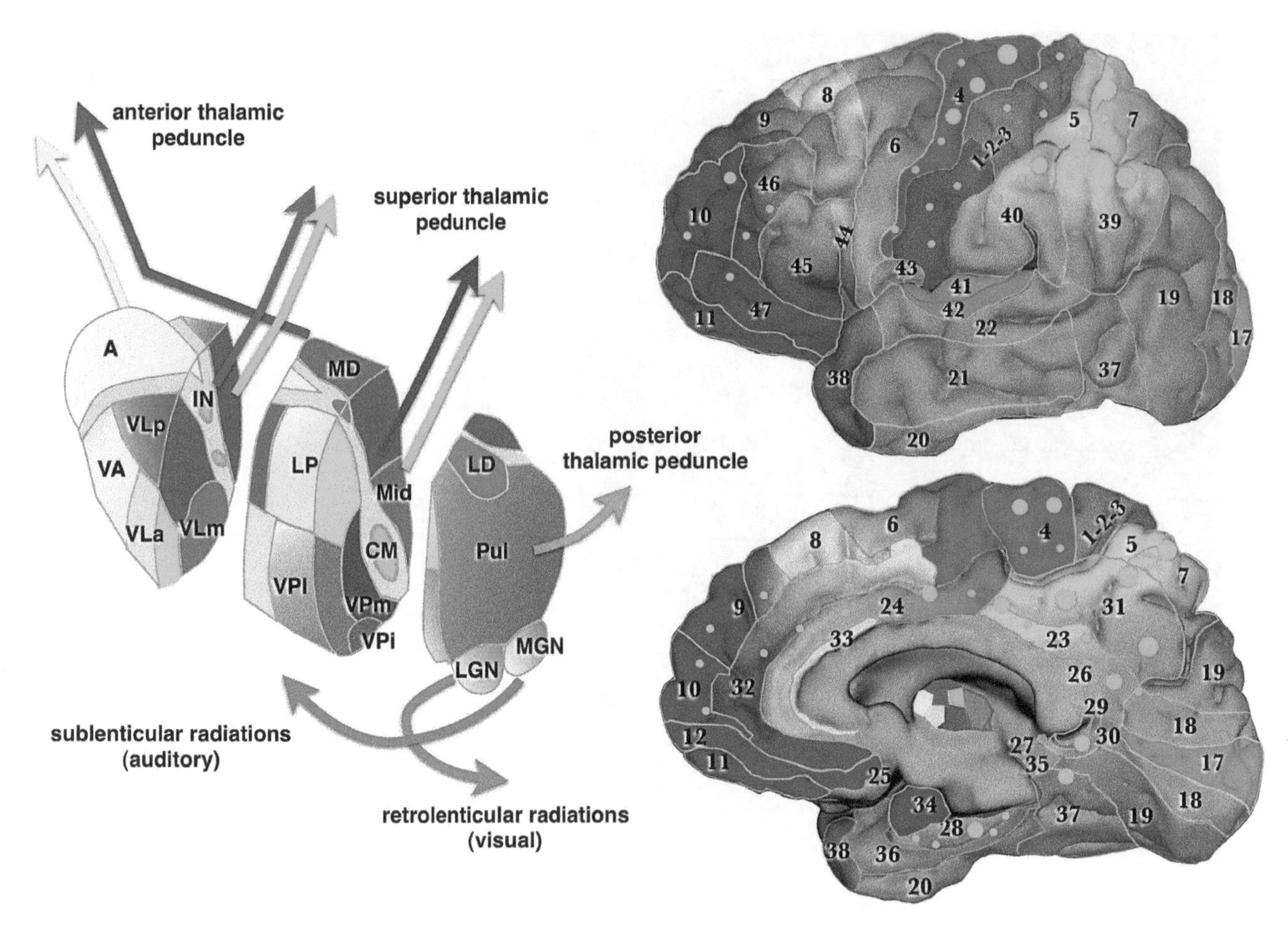

Figure 10.5 Diagram of the major thalamic nuclei and representation of the thalamic cortical projections. The ascending projection fibres of the internal capsule are conventionally divided into: anterior thalamic peduncle, superior thalamic peduncle, posterior thalamic peduncle, retrolenticular and sublenticular radiations. Green dots on the cortical surface indicate the connection from the intralaminar nuclei (IN) and the centromedian nucleus (CM). Note that the size of the green dots indicates the density of projections and the numbers correspond to Brodmann areas. The ventral medial nucleus (VM) is not visible on the dorsolateral view of the thalamus. The internal medullary lamina is coloured in light grey. A, anterior group; LGN, lateral geniculate nucleus; LD, laterodorsal; LP, lateroposterior; MD, mediodorsal; MGN, medial geniculate nucleus; Mid, midline; Pul, pulvinar; VA, ventral anterior; VLa, ventral lateral anterior; VLm, ventral lateral medial; VLp, ventral lateral posterior; VPi, ventral posterior inferior; VPl, ventral posterior lateral; VPm, ventral posterior medial. (Nomenclature is based on Morel et al., 1997).

second magnetic resonance investigation with contrast agent that showed a brain capillary telangiectasia at the pons level.

Ascending thalamic projections

The thalamic radiations originate from the thalamic nuclei and after a short course within the internal capsule enter the corona radiata to terminate in the cortex of the ipsilateral hemisphere. Diffusion tensor imaging tractography has been used to visualize the thalamic projections and segment the thalamic nuclei according to their cortical projection zones (Behrens et al., 2003). Similar to the fibres of the cortico-spinal tract, the different thalamic projections are aligned in an orderly manner along the anterior and posterior limbs of the internal capsule (Figure 10.2). The thalamus receives its afferent connections from the basal ganglia (mainly globus pallidus, substantia nigra, and nucleus subthalamicus), the cerebellum, brainstem nuclei, the spinal cord, and the limbic system. The efferent thalamic projections to the cerebral cortex radiate anteriorly to the frontal cortex (anterior thalamic peduncle), superiorly to the precentral frontal regions and parietal cortex (superior thalamic peduncle), posteriorly to the occipito-temporal cortex (posterior thalamic peduncle), and infero-anteriorly to the temporal cortex and amygdala (inferior thalamic peduncle) (Figure 10.5).

The pattern of connectivity of the thalamic nuclei determines their functional role (Figure 10.5) (Table 10.1). The thalamus contains a sheet of white matter, the internal medullary lamina, that separates the mediodorsal thalamic nucleus (MD) from the ventral and lateral thalamic nuclei. The internal medullary lamina envelops anteriorly the anterior thalamic nuclei and posteriorly a group of small nuclei, the intralaminar nuclei, of which the centromedian (CM) is the largest. The anterior thalamic nuclei receive information from the hippocampus via the fornix and hypothalamic mammillary bodies via the mammillo-thalamic tract; in turn the anterior thalamic nuclei project mainly to the anterior cingulate cortex and medial superior frontal gyrus (e.g. supplementary and pre-supplementary motor areas) (Aggleton, 2008). The anterior thalamic nuclei are considered part of the limbic system and damage to them has been

Table 10.1 Nomenclature of the principal thalamic groups and associated functional and clinical correlates

Thalamic groups	Principal nuclei	Function	Symptoms
Anterior thalamic group	Anterodorsal (AD) Anteroventral (AV) Anteromedial (AM)	Limbic functions such as memory and learning, emotion, drive and motivation	Amnesia, language difficulties (reduced spontaneous speech, anomia)
Intralaminar nuclei	Centromedian (CM)	Regulation of sleep, wakefulness, attention, multimodal sensory processing, affective component of pain perception	Hemispatial and motor neglect
Medio-dorsal nucleus (MD) and Midline (Mid) group		Drive, motivation, emotion, executive functions, working memory, attention, autonomic and sleep-wake cycle regulation.	Apathy, abulia, disinhibition, working memory deficits, sleep dysregulation
Ventral group	Anterior (VA)	Complex behaviour, motor programming	Dystonia, language impairment (reduced fluency, perseveration, stuttering), behavioural problems
	Lateral – anterior (VLa) – medial (VLm) – posterior (VLp)	Motor, language, and memory	Ataxia, mild motor weakness, language, memory difficulties
	Posterior – lateral (VPl) – medial (VPm) – inferior (VPi)	 Somatosensory (body and limb) Somatosensory (head and neck), gustatory Vestibular	Dejerine-Roussy disease (thalamic pain syndrome); contralateral hemianesthesia (typically for all sensory modalities) of body and limbs (VPl) or head and neck (VPm)
	Medial (VM)	Nociception and thermoception	Hypoesthesia or hypersthesia for pain and temperature
Lateral group	Dorsal (LD) Posterior (LP) Pulvinar (Pul)	Visual-sensory-motor integration and visual salience (discriminating relevant from irrelevant visual stimuli)	Impaired visual discrimination, hemispatial neglect, language deficits, psychosis
Metathalamus	Lateral geniculate nucleus (LGN)	Visual perception	Contralateral homonymous hemianopia (loss of vision in the same visual field on both eyes)
	Medial geniculate nucleus (MGN)	Auditory perception	Central deafness

associated with profound amnesia in stroke patients (see also Chapter 11). A recent diffusion tensor imaging study in patients with Alzheimer's disease showed reduced fractional anisotropy of the anterior thalamic nuclei and surrounding white matter of the internal medullary lamina (Zarei et al., 2010). These data seem to support earlier post-mortem studies that found severe neuronal loss and tangle formation in the anterior thalamic nuclei of patients with Alzheimer's disease (Xuereb et al., 1991).

The pattern of connectivity of the intralaminar nuclei is less specific compared to the other thalamic groups. The intralaminar nuclei receive extensive projections from the brainstem (reticular formation, vestibular nuclei, superior colliculi), deep cerebellar nuclei, basal ganglia, and spinal/trigeminal grey matter. The cortical projections of the intralaminar nuclei lack topographical organization and are diffuse to parietal, frontal, and limbic cortices. The effects of direct lesions or disconnection of the intralaminar nuclei have been extensively studied in animals (Van der Werf et al., 2002). Currently, the evidence shows that this group of nuclei is involved in the regulation of sleep, wakefulness, attention, arousal, pain perception, and multimodal sensory processing. The centromedian nucleus is the target for deep brain stimulation used for the treatment of chronic pain (Weigel and Krauss, 2004), epilepsy (Velasco et al., 1987), and tremor in Parkinson's disease (Stefani et al., 2009). Both hemispatial (Watson and Heilman, 1979) and motor neglect (Bogousslavsky et al., 1987) have been described in patients with lesions to the centromedian nucleus.

The mediodorsal thalamic nucleus is relatively larger in humans compared to other species. It receives fibres from the brainstem (superior colliculus, vestibular nuclei), substantia nigra, ventral pallidum, and amygdala, and is reciprocally connected with the orbitofrontal cortex, anterior and dorsolateral prefrontal cortex, ventromedial and anterior cingulate gyri, medial temporal lobe and temporal pole. The mediodorsal thalamic nucleus participates in regulating sleep, drive, motivation, emotion, and is involved in executive functions, working memory, and attention. In leucotomized psychotic patients a retrograde degeneration of the thalamic fibres can be traced back to the mediodorsal thalamic nucleus (Pakkenberg, 1993). A decrease in anxiety, tension, aggression, and obsessive thinking has been described in these patients. The above benefits were often accompanied by reduced psychomotor activity and sometimes marked apathy (Mettler, 1949). A reduced volume and number of neurons in the mediodorsal thalamic nucleus has also been found in non-leucotomized schizophrenic patients (Pakkenberg, 1990). A recent diffusion tensor imaging study suggests that a significant increase in the mean diffusivity of the mediodorsal thalamic nucleus and nucleus accumbens can be measured in schizophrenic patients even before detectable volumetric changes occur. The severity of changes in mean diffusivity has been correlated with the performance in a working memory task (Spoletini et al., 2011). Diffusion tensor imaging tractography can also detect significant white matter changes (i.e. reduced fractional anisotropy and increased mean diffusivity) along thalamo-cortical connections from the mediodorsal thalamic nucleus in chronic schizophrenic patients (Oh et al., 2009) but not in first-episode psychosis (Qiu et al., 2009). Together these studies suggest that in patients with schizophrenia the white matter changes may be secondary to a primary degeneration of the thalamic nuclei. Sporadic and familial fatal insomnias are a group of prion diseases characterized by disrupted sleep, autonomic hyperactivity (e.g. increased blood pressure and heart rate), cognitive deficits, and motor abnormalities such as myoclonus (sudden muscle contraction), ataxia (lack of muscle coordination), and dysphagia (difficulty in swallowing) (Lugaresi et al., 1986; Montagna et al., 2003). The disorder is associated with the accumulation of prion proteins preferentially in the mediodorsal (and anterior) thalamic nuclei (Manetto et al., 1992). The thalamic prionic lesions disconnect subcortical and cortical limbic regions involved in the control of sleep regulation, autonomic functions, and motivation. Such a thalamic disconnection results in a shift to persistent waking behaviour, metabolic hyperactivity (Montagna et al., 2003), and progressive disturbance of memory, planning, attention and vigilance (Galassi et al., 1992).

The ventral thalamic nuclear group is divided into four nuclei; ventral anterior, ventral lateral, ventral posterior, and ventral medial. The ventral anterior (VA) nucleus receives inputs from the pallidum, and substantia nigra, and is reciprocally connected with cingulate and premotor-prefrontal cortices (BA 6, 8, 46). Dystonia, stuttering, and reduced spontaneous speech have been described in association with dysfunction of the ventral anterior nucleus or its connections to frontal areas (Watkins et al., 2008; Bonhila et al., 2009). In subjects with writer's cramp, an idiopathic focal hand dystonia characterized by muscle cramps during writing, deep brain stimulation of the ventral anterior nucleus improves symptoms (Fukaya et al., 2007).

The ventral lateral nucleus consists of anterior (VLa), posterior (VLp), and medial (VLm) divisions. The ventral lateral anterior nucleus receives fibres from the globus pallidus, while the ventral lateral posterior nucleus receives from the contralateral deep cerebellar nuclei. Additional subcortical projections have been reported from the spinothalamic tract and vestibular nuclei. Numerous cortical afferents to both ventral lateral anterior and ventral lateral posterior nuclei originate from precentral motor cortex (BA 4 and posterior 6). The ventral lateral anterior nucleus sends efferents to the supplementary motor area and lateral premotor cortex, while the ventral lateral posterior nucleus projects to the primary motor cortex. Ataxia, mild motor weakness, language and memory difficulties have been described in patients with lesions of the ventral lateral nuclear group. In restless leg syndrome, decreased fractional anisotropy values were observed in the ventral lateral complex and in widespread parietal and frontal white matter areas adjacent to the motor, premotor, and somatosensory cortices (Unrath et al., 2008). Severe drug-resistant tremor in patients with Parkinson's disease, essential tremor, and multiple sclerosis can be alleviated with a neurosurgical procedure consisting of placing a stereotactically targeted lesion in the ventrolateral complex (i.e. thalamotomy) that results in a disconnection between the basal ganglia and the motor and premotor cortices (Hariz, 2003). The ventral posterior nucleus consists of a lateral (VPl), a medial (VPm), and a smaller inferior (VPi) division. The ventral posterior lateral nucleus receives the medial lemniscal and spinothalamic somatosensory (limbs and body) inputs, while the ventral posterior medial nucleus receives trigeminal afferents (from the head). Both nuclei project to the primary somatosensory cortex in the postcentral gyrus. The ventral medial thalamic nucleus (VM) is anterior to the ventral posterior nuclei and is an important relay station of the activating system. It also receives nociceptive and thermoceptive information from the spinothalamic and trigeminal-thalamic tracts and sends projections to the posterior insula. Symptoms associated with pathology of this nucleus include altered (reduced or increased) pain or temperature perception.

The lateral thalamic nuclear group includes the small lateral dorsal (LD) nucleus, the lateral posterior (LP) nucleus, and the large pulvinar. This group receives its main subcortical afferents from the superior colliculus and is reciprocally connected with the parietal, occipital, and temporal lobes. In addition connections to the lateral posterior nucleus and pulvinar originate from motor and premotor areas, including the frontal eye field, and regions of the limbic lobe (i.e. posterior cingulate and parahippocampal cortices). This group of nuclei represents an important relay station for reciprocal cortico-cortical modulation during tasks requiring visual-sensory-motor integration (Sherman and Guillery, 2001). Human lesion studies suggest that the pulvinar, which is the largest of the thalamic nuclei, can be divided into specialized regions. Damage to the medial pulvinar, which carries connections to the amygdala and orbitofrontal cortex, impairs the ability to recognize briefly presented fearful expressions (Ward et al., 2007) and causes delays in processing visual threats (Ward et al., 2005). Similarly selective lesions to the anterior or posterior pulvinar show that spatial and temporal visual attention deficits are dissociable (Arend et al., 2008). Damage to the anterior pulvinar disconnects communications with the posterior parietal and visual cortices, and thereby degrades spatial attention processes, whereas damage to posterior pulvinar disrupts communication with the ventral visual occipitotemporal areas, causing impairment of object recognition and in the top-down control of visual attention towards task-relevant stimuli. A recent volumetric measurement study in subjects with attention deficit and hyperactivity disorder (ADHD) showed that the pulvinar is smaller in these patients compared to controls but relatively enlarged in patients treated with stimulants compared to those untreated (Ivanov et al., 2010).

Posterior and ventral to the thalamic nuclei is the metathalamus, which includes the lateral geniculate nucleus (LGN) and the medial geniculate nucleus (MGN). The lateral geniculate nucleus receives

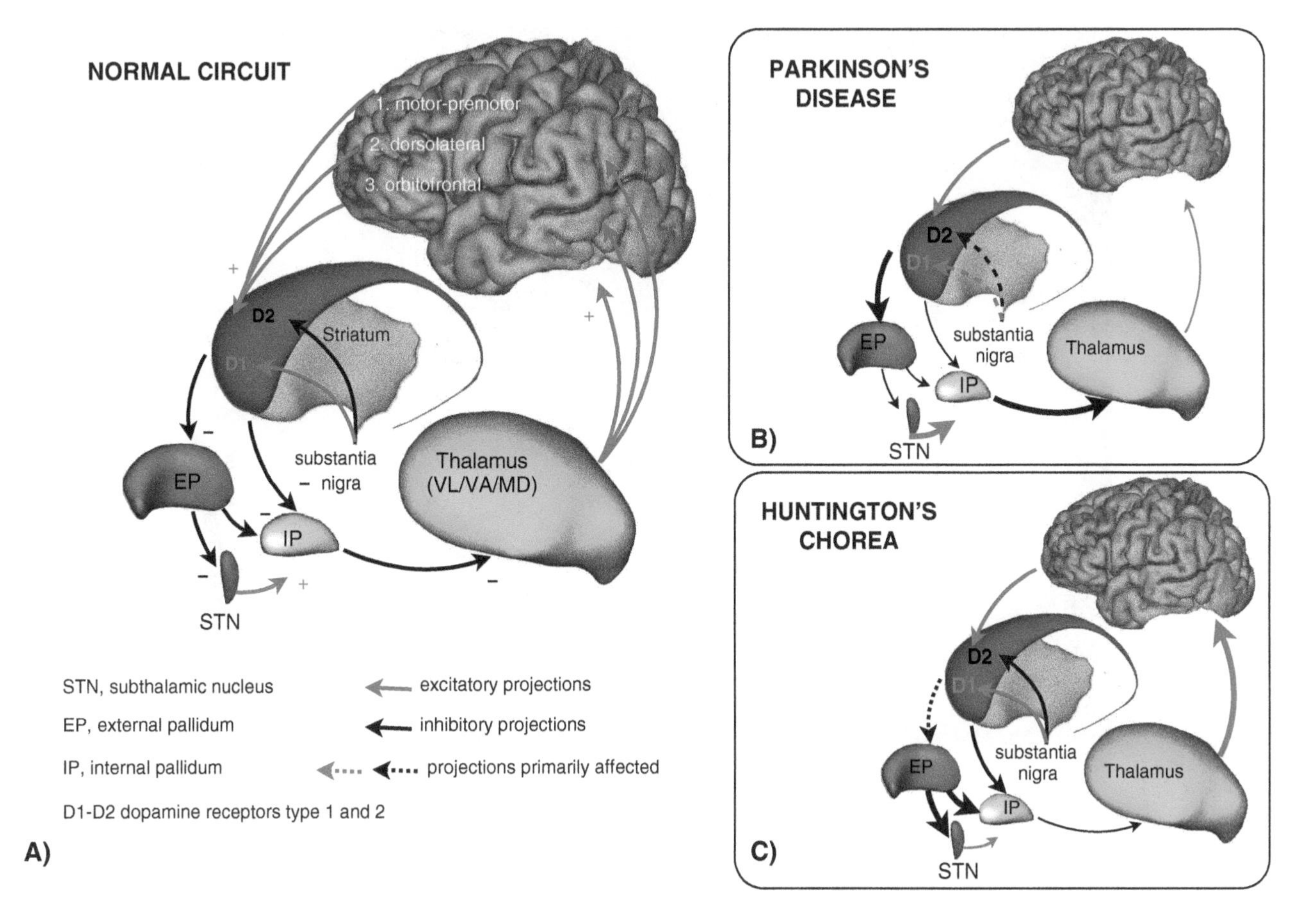

Figure 10.6 A) Diagram of the direct (cortico-striatal-pallido-thalamo-cortical) and indirect (cortico-striatal-pallido-subthalamic-pallido-thalamo-cortical) loops connecting the cerebral cortex to the basal ganglia and thalamus. The activity of the striatal neurons is regulated by dopaminergic fibres from the substantia nigra. The direct and indirect loops are part of three segregated circuits (i.e. 'motor', 'associative', and 'limbic') which originate from distinct cortical regions (i.e. motor-premotor, dorsolateral, and orbitofrontal) and relay through different thalamic nuclei (MD, mediodorsal; VA, ventroanterior; VL, ventrolateral). B, C) Models proposed to explain the neurophysiopathology of Parkinson's disease and Huntington's chorea.

visual inputs mainly from the retina and primary visual cortex (BA 17), to which sends the majority of its efferents through the optic radiations (retrolenticular thalamic peduncle). Minor afferents to the lateral geniculate nucleus come from the superior colliculus. Lesions to the lateral geniculate nucleus or its connections cause hemianopia, the bilateral loss of vision in one hemifield (see also Chapter 8). The medial geniculate nucleus receives auditory fibres form the inferior colliculus and other brainstem nuclei and projects mainly to the primary auditory cortex in the superior temporal lobe (BA 41) and to a minor extent to adjacent temporal areas. The connection from the inferior colliculus to the medial geniculate nucleus is ipsilateral but the inferior colliculus receives auditory inputs is from both cochlear nuclei. Diffusion tensor imaging tractography has been used to automatically segment the medial and lateral geniculate nuclei from surrounding structures based on their distinctive patterns of connectivity to cortical areas (Devlin et al., 2006). Post-mortem dissections of the volumes of the medial geniculate nucleus and auditory radiations show a symmetrical distribution (Rademacher et al., 2002). The redundancy of the central auditory system and its strong bilateral representation above the level of the cochlear nuclei can, in part, explain the observation of gross deficits in hearing only for extensive bilateral lesions. Unilateral lesions that include the auditory system, produce subtle hearing dysfunctions that can only be detected with sophisticated psychoacoustic and electrophysiological testing (Häusler and Levine, 2000).

Cortico-basal ganglia circuit

The basal ganglia represent an important relay station of a group of parallel circuits connecting the cerebral cortex to the thalamus. Anatomically, it is possible to distinguish two main cortico-subcortical loops (Figure 10.6A) (Alexander et al., 1986; Nieuwenhuys et al., 2008). The direct loop includes, in sequence, excitatory corticostriatal, inhibitory striatopallidal (internal pallidum), inhibitory pallidothalamic, and excitatory thalamo-cortical connections. The indirect loop includes, again sequentially, excitatory corticostriatal, inhibitory striatopallidal (external pallidum), inhibitory pallido-subthalamic, excitatory subthalamic-pallidal, inhibitory pallidothalamic, and excitatory thalamocortical connections. The stimulation of the two loops has opposite effects on the cortex: the direct loop increases thalamic activation (positive feedback), while the indirect reduces it (negative feedback). The direct and indirect loops are regulated by dopaminergic fibres from the ventral mesencephalic nuclei (e.g. the substantia nigra, pars compacta). The overall function of these loops is to facilitate initiation and execution of movement (Hauber, 1998), selection

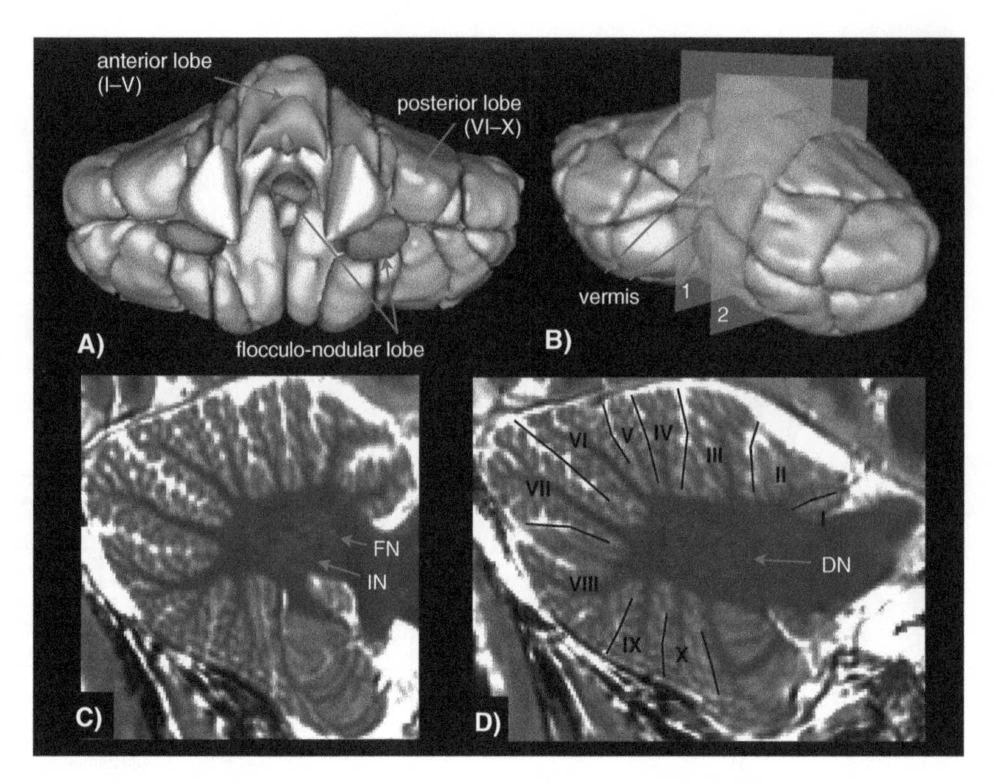

Figure 10.7 A) Anterior and B) posterior view of the human cerebellum, with parcellation according to Dow and Moruzzi (1958) division into an anterior lobe, a posterior lobe, and a flocculo-nodular lobe (Makris et al., 2005). C–D) Right sectional views of parasagittal images of the cerebellar hemispheres; the planes of section pass through the deep cerebellar nuclei (Deoni and Catani, 2007). C) The classical division of the vermis into ten regions (Larsell and Jansen, 1972). DN, dentate nucleus; FN, fastigial nucleus; IN, interposed nucleus.

of purposeful patterns of movement in response to internal and environmental stimuli (Pessiglione et al., 2003), and reward and motivation (Pessiglione et al., 2006).

A further functional subdivision separates the cortico-basal ganglia network into motor, associative, and limbic circuits (Alexander et al., 1986; Lehéricy et al., 2004; Schmahmann and Pandya, 2008).

The motor circuit includes the motor and premotor cortex, the putamen, the external and internal pallidum, the subthalamic nucleus, and the ventrolateral thalamic nuclei. Disorders of the motor circuit manifest with extrapyramidal motor deficits (Bhatia and Marsden, 1994). In Parkinson's disease, for example, the loss of dopaminergic projections from the substantia nigra (pars compacta) results ultimately in a reduced excitatory thalamic output to the cortex (Figure 10.6B). This explains the difficulties in the initiation of movement and the bradykinesia (slowness of movement) typical of parkinsonian patients (Wichmann and DeLong, 1993). In Huntington's chorea the loss of striatal gamma-aminobutyric acid (GABA) neurons that project to the external pallidum results in the increased excitatory thalamic output to the cortex (Figure 10.6C). This explains the abnormal excessive movements observed in choreic patients (DeLong, 1990). Two recent studies suggest that frontostriatal white matter changes can be detected with diffusion tensor imaging not only in symptomatic Huntington's disease patients (Douaud et al., 2009), but also in individuals known to carry the genetic mutation that causes Huntington's disease (Rosas et al., 2006).

The associative circuit is composed of the dorsolateral prefrontal cortex, dorsal caudate nucleus, internal pallidum, and ventral anterior thalamic nuclei. Lesions to the associative circuit impair attention, working memory, strategy formation, and cognitive flexibility (Stuss and Benson, 1984). Lesion studies suggest a functional asymmetry in this circuit: spatial neglect and visuo-spatial disorientation are more frequently described in patients with right caudate lesions, while deficits in verbal encoding, reduced verbal fluency, and aphasia occur more often in left caudate lesions (Caplan et al., 1990; Trépanier et al., 1998; Kumral et al., 1999).

The limbic circuit is composed of the medial and orbitofrontal cortex, ventral striatum (i.e. nucleus accumbens), external and internal pallidum, and mediodorsal thalamic nucleus. The limbic circuit is involved in the neurobiology of neurodevelopmental disorders (e.g. schizophrenia) (Spoletini et al., 2010), addiction (Wise, 1987; Robbins and Everitt, 2002; Kalivas, 2009), and obsessive compulsive disorder (Mataix-Cols et al., 2004; Remijnse et al., 2006). In schizophrenia, for example, functional imaging studies have attributed positive symptoms, such as hallucinations and delusions, to an increased activity within the thalamo-limbic projections. Conversely, behavioural changes that resemble the negative symptoms of schizophrenia, such as disinhibition, irritability, mood lability, and reduced motivation have been reported for acquired lesions of the ventral striatum (Levy and Dubois, 2006; Schoenbaum et al., 2007). In young children with fragile X syndrome, a disorder characterized by reduced or absent expression of the fragile X mental retardation 1

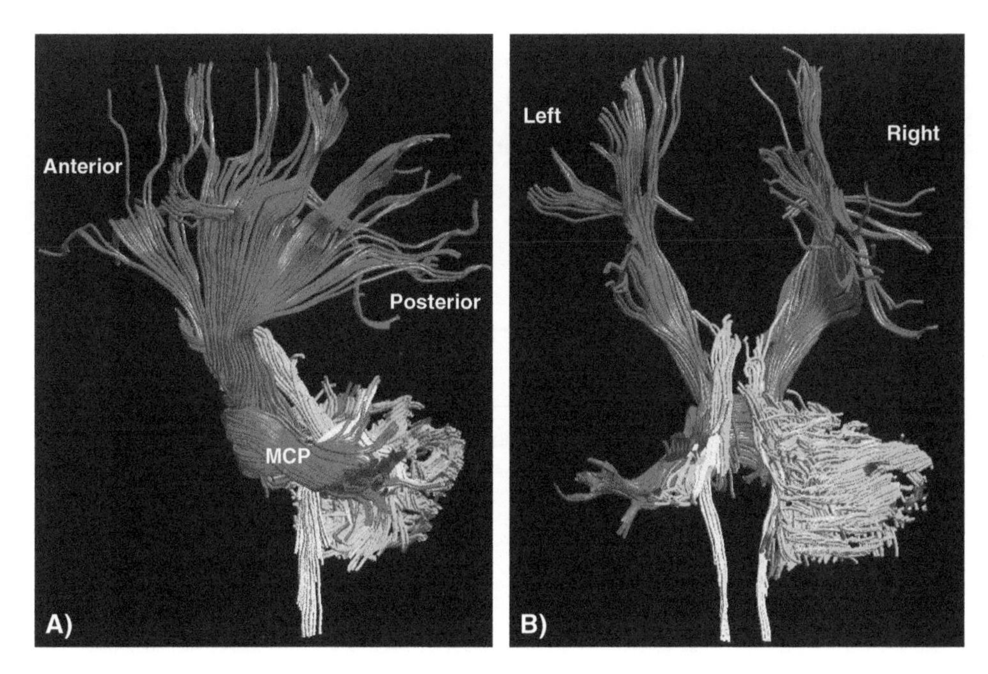

Figure 10.8 A) Left lateral and B) posterior view of the long and short intrinsic cerebellar connections visualized with diffusion tensor tractography. The cortico-ponto-cerebellar pathway (red) carries information from the cerebral cortex to the contralateral cerebellum through the middle cerebellar peduncle (MCP). The inferior cerebellar peduncle (ICP) contains spino-cerebellar fibres (yellow) projecting to the ipsilateral cerebellar hemisphere. The intrinsic cerebellar fibres (cyan) connect the cerebellar cortex to the deep cerebellar nuclei. The cerebellum elaborates information from the cerebral cortex and spinal cord and send feedback projections to the contralateral thalamus and red nucleus through the superior cerebellar peduncle (green) (Catani et al., 2008).

(FMR1) protein, Haas et al. (2009) found an increase in the number of streamlines (a surrogate of tract volume) in the ventral frontostriatal pathway compared to typically developing and developmentally delayed children. The authors concluded that altered FMR1 protein expression can alter white matter anatomy during early brain development. In children and adolescents with Tourette's syndrome, increased diffusivity has been reported in the frontostriatal pathway compared to typically developing subjects (Govindan et al., 2010). These studies suggest that diffusion tensor imaging may be used to detect early white matter abnormalities of the limbic circuit in neurodevelopmental disorders.

Cerebellar pathways

The cerebellum (Latin for 'small brain') occupies the posterior fossa, a space between the posterior aspect of the brainstem and the ventral surface of the occipital lobes. On the external surface it is possible to identify a medial structure, the vermis, and two lateral cerebellar hemispheres. The classical lobar division of the cerebellum distinguishes an anterior lobe, a larger posterior lobe, and a smaller flocculonodular lobe (Figure 10.7) (Dow and Moruzzi, 1958; Makris et al., 2005). A further division of the cerebellum identifies five lobules in the anterior lobe (lobules I–V) and five lobules in the posterior lobe (lobules VI–X). Within the white matter of the cerebellar hemisphere it is possible to identify on each side three pairs of deep cerebellar nuclei located laterally to the midline: the dentate, interposed, and fastigial nuclei. The dentate nucleus is the largest of the cerebellar nuclei and receives connections mainly from the lateral cerebellar hemispheres. The volume of the right dentate nucleus is larger than the left in right-handed subjects, with an opposite asymmetry in left-handed (Deoni and Catani, 2007). The interposed nucleus is composed of two smaller nuclei, the emboliform (anterior interposed) and globose (posterior interposed) nuclei. Both nuclei receive inputs from the medial cerebellar hemispheres. The fastigial is the most medial nucleus and receives projections from the vermis and flocculonodular lobe (Figure 10.7).

The cerebellar cortex and the deep cerebellar nuclei are densely connected to extracerebellar structures through the superior, middle, and inferior cerebellar peduncles (Figure 10.8). The two main cerebellar afferent pathways are the spinocerebellar tract, entering the cerebellum through the inferior and superior cerebellar peduncles, and the pontocerebellar tract (the final limb of the cortico-ponto-cerebellar pathway) entering the cerebellum through the middle cerebellar peduncle. The inferior cerebellar peduncle also carries reciprocal connections with structures of the brainstem, including the inferior olivary nucleus, vestibular nuclei, and reticular nuclei. The superior cerebellar peduncle is the major cerebellar output, originating from the deep cerebellar nuclei and projecting to the contralateral cerebral hemisphere via the thalamus, and to other contralateral brainstem regions including the red nucleus. It also carries some afferent fibres from the brainstem and from the spinal cord (ventral spinocerebellar tract).

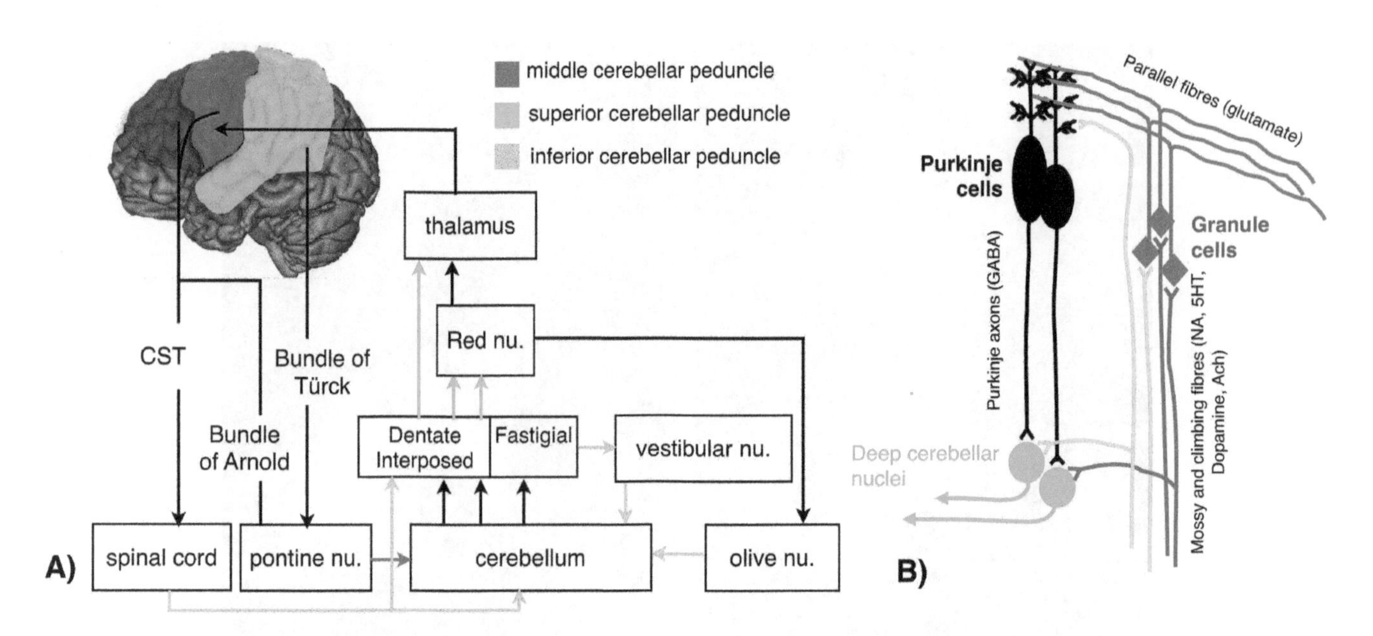

Figure 10.9 A) A simplified diagram of the parallel cerebellar loops. A cerebro-cerebellar parallel neural circuit connects the motor (blue), prefrontal (purple), and parietal-temporal-occipital (cyan) cortex to the cerebellum through the pontine nuclei. The same cerebral areas receive afferents from the deep cerebellar nuclei via the thalamus through the superior cerebellar peduncle (green arrows). Other loops connect the cerebellum to the vestibular nuclei (via the inferior cerebellar peduncle, ICP) and the inferior olivary nucleus (via the interposed, dentate and red nucleus). B) Diagram of the principal intracerebellar neurons and fibres. Note that the colours in both figures correspond to the colours of the tracts in Figure 10.8.

Long-range cerebellar loops and intracerebellar circuits

The long cerebellar tracts form a system organized in parallel but partially overlapping loops (Figure 10.9A). The first of these, the cortico-ponto-cerebellar-thalamo-cortical loop originates from the motor, association, and paralimbic regions of the cerebral cortex (Schmahmann and Pandya 2008). These regions have topographically organized feedforward pontine projections travelling within the bundle of Arnold (fibres from the frontal regions) and the bundle of Türck (fibres from the parieto-temporal-occipital regions) (Brodal, 1978; Wiesendanger et al., 1979; Hartmann-von Monakow et al., 1981; Glickstein et al., 1985; Schmahmann et al., 2004). The pontine nuclei in turn project to distinct cerebellar zones. Those nuclei receiving from the cerebral motor cortex project to the vermis and the medial portion of the cerebellar hemispheres. Those nuclei receiving from the association and paralimbic areas project to the lateral zones of the cerebellar hemispheres. The vermis and the medial cerebellar hemispheres receive inputs also from the vestibular nuclei and the spinocerebellar tracts. Each cortical zone of the cerebellum is connected with a cerebellar output nucleus: the medial cerebellar hemisphere to the interposed nucleus, the lateral cerebellar hemisphere to the dentate nucleus. Each cerebellar nucleus in turn projects back to the contralateral cortical motor and premotor areas through the ventrolateral nucleus of the thalamus (Middleton and Strick, 1997).

The second loop is the cerebellar-rubro-olivo-cerebellar network. The red nucleus is a relatively large nucleus of the midbrain that receives cerebellar projections mainly from the dentate nucleus (and some from the interposed) and gives rise to short rubro-thalamic fibres and long rubro-spinal fibres. Some of the rubro-spinal projections end in the inferior olivary nucleus, which in turn projects back to the cerebellum via the inferior cerebellar peduncle.

A third loop includes feed-forward projections from the vestibular nuclei and reticular nuclei to the vermis and feedback projections from the fastigial nucleus.

The cerebellum has a central role in motor coordination and learning, as suggested by the anatomical features of its long-range connections (Voogd, 2003; Roitman et al., 2005; Ramnani, 2006). Through the spino-cerebellar and vestibulo-cerebellar pathways the cerebellum receives a large amount of information regarding the position of the limbs, body, and head in the space. Information regarding the space of the surrounding environment and the intended motor action reaches the cerebellum through the bundles of Türck and Arnold, respectively. The role of the cerebellum is to compare and integrate the information received and coordinate the correct execution of movements, from the initial preparatory phase to its completion (Dow and Moruzzi, 1958). The initial preparatory stage of a movement involves the anticipatory actions of the proximal axial muscles that control body posture and produce stereotyped movement. The proximal muscles are under control of the red nucleus (rubro-spinal tract) and the olivary complex (olivo-spinal tract), and indirectly the cerebellum through its cerebello-rubro and cerebello-olivo connections. The second stage requires

Table 10.2 Neurological and neuropsychiatric manifestations of cerebellar disorders

Function	Localization	Clinical manifestations
Motor	Anterior lobe and vermis	Gait ataxia, limb ataxia, dysarthric speech, intention tremor, hypotonia
Vestibular	Flocculo-nodular lobe	Vertigo, nausea, nystagmus, gait ataxia
Cognition	Posterior cerebellum	Inattentiveness, distractibility, hyperactivity, perseveration, impulsiveness, disinhibition, difficulty shifting focus of attention, reduced verbal fluency
Thought, personality and social behaviour	Posterior cerebellum	Obsessional thinking, compulsive and ritualistic behaviour, thought disorders and paranoid delusions, oppositional behaviour, disinhibition, passivity, immaturity, childishness, difficulty with social interactions, unawareness of social cues or boundaries, overly gullible and trusting behaviour
Emotion	Flocculo-nodular lobe	Lability, incongruous feelings, pathological laughing and crying, anxiety, agitation, irritability, anger, aggression, flattened affect, emotional blunting, apathy, anhedonia, depression

the facilitation and execution of finely tuned movements of the distal muscles in order to reach a target object or execute an intended action. This requires a synergy between contraction and release of agonistic and antagonistic muscles, and a precise guidance of the limb and head movements according to the dynamic changes of the object and the body position in space. The cerebellum participates in the above functions (but not the initiation of a movement) through its projections to the cortical motor and premotor areas. Finally the cerebellum carries out the adjustments necessary to correct errors occurring during the execution of movements. This process is also fundamental for the automatization of learned movements, which depends on synaptic changes occurring within the intracerebellar circuit.

The principal components of the intracerebellar circuit are the parallel fibres from the granule cells and the axonal projections from the Purkinje neurons (Figure 10.9B). The axons of the granule cells ascend to the most superficial cortical layer (i.e. molecular) of the cerebellum, bifurcate at T-shaped branches into parallel fibres, and project to the Purkinje cells where they exert an excitatory (glutamate) effect. The Purkinje axons have a zonal parallel organization and terminate in the deep cerebellar nuclei where they have an inhibitory effect (GABA). The axonal projections from the deep cerebellar nuclei to the thalamus and red nucleus are excitatory (glutamate). The Purkinje and granule cells, and the neurons of the deep cerebellar nuclei are the first relay station for all afferent cerebellar fibres. The two principal inputs to the cerebellum are: i) mossy fibres from the pontine nuclei, medulla (e.g. vestibular nuclei), and spinal cord; ii) climbing fibres from the contralateral inferior olivary nucleus. Both sets of fibres are excitatory and send collaterals to the deep cerebellar nuclei before projecting to the granule cells and directly to the Purkinje cells. The other cells of the cerebellar cortex (i.e. basket, Golgi, and stellate) connect to the Purkinje and granule cells through relatively short pre- and post-synaptic inhibitory (GABA) projections (Eccles et al., 1967). The ability of the cerebellum to compute a large amount of incoming information and generate a relatively small but timely and topographically precise output signal depends on the summation of the incoming excitatory signal from the extracerebellar regions and the inhibitory effect of the Purkinje cells. The summation occurs on the neuronal membranes of the deep cerebellar nuclei, whose output is only one-fortieth of the overall incoming signal reaching the cerebellum.

Clinical manifestations of cerebellar lesions

The anatomical considerations above allow some conclusions about the correlation between cerebellar syndromes and lesion topography to be drawn (Table 10.2). The most frequent cerebellar deficit is the loss of motor coordination, also known as cerebellar ataxia. Lesions to the vermis present with severe ataxia of axial muscles (i.e. trunk and proximal limbs). The patient is unable to stand or sometimes even sit unaided. If aided he can only stand with his feet widely apart and take small steps with frequent loss of equilibrium and falls (gait ataxia).

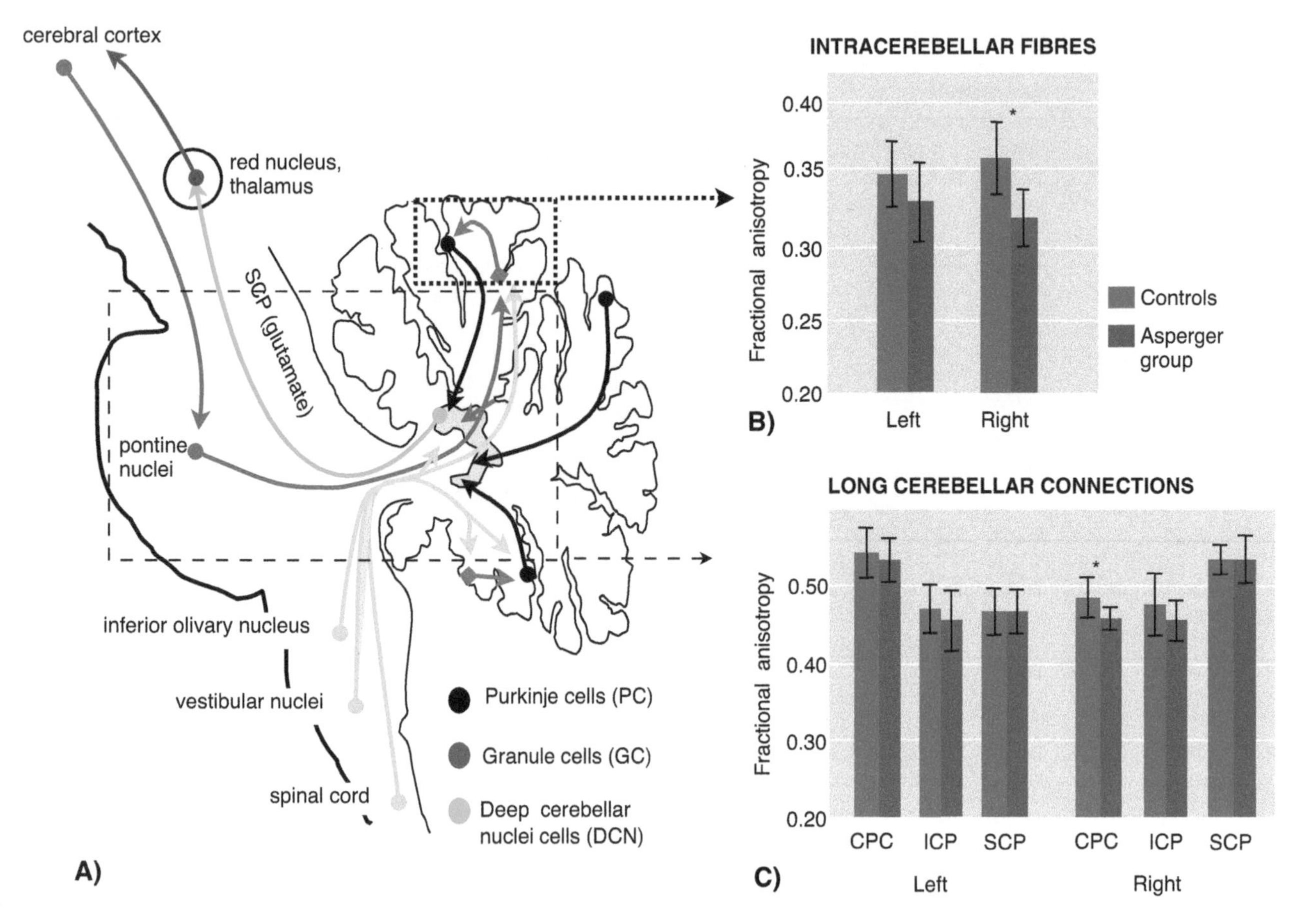

Figure 10.10 A) Diagram of the main long and intrinsic cerebellar connections and B, C) tract-specific measurements of fractional anisotropy in the short intracebellar tracts (upper histogram) and long cerebellar tracts (lower histogram) in subjects with Asperger's syndrome and healthy controls. Compared to controls (red bars) the Asperger's syndrome group (blue bars) shows a statistically significant reduction of fractional anisotropy values in the right short intrinsic cerebellar fibres (*p <0.001). Reduced fractional anisotropy values in the short intrinsic cerebellar fibres are likely to reflect changes in the parallel fibres and Purkinje axons that have been documented in histological studies of autistic brains. Also the Asperger's syndrome group shows reduced fractional anisotropy values in the right superior cerebellar tract (SCP) (*p <0.001) but not in the other tracts, such as the cortico-ponto-cerebellar tract (CPC) and inferior cerebellar tract (ICP) (Catani et al., 2008).

Ataxia of the limbs accompanied by loss of muscle tone (hypotonia) and a tremor that worsens with movement (intention tremor) are often associated with lesions to the anterior cerebellar lobe. If the lesion involves the posterior lobe and flocculus, the patient presents a severe vestibular syndrome characterized by nystagmus, rotation of the head, nausea, and vertigo (Duncan et al., 1975; Lee et al., 2006). Damage to the lateral cerebellar hemispheres manifests with symptoms that have many features of the frontal lobe syndrome, such as attentional, executive, and reasoning dysfunction, personality and behavioural changes, and mood disorder (Schmahmann and Sherman, 1998; Exner et al., 2004). Most of these subjects have mild and transient symptoms. The non-motor cerebellar symptoms are usually described under the umbrella of the cerebellar cognitive affective syndrome (CCAS) (Müller et al.,

1998; Schmahmann and Sherman, 1998), a concept that extends the functions of the cerebellum beyond its classical motor functions (for critiques of the CCAS see Glickstein, 1993). Underlying the CCAS is the concept of the universal cerebellar transform that considers the cerebellum to have an essential role in the automatization and optimization of behavioural responses according to context. The universal cerebellar transform hypothesis suggests that the cerebellum modulates cognition and emotion in the same way that it coordinates movement (Leiner et al., 1991; Schmahmann, 1991; Botez, 1992; Ito, 1993; Middleton and Strick, 1994; Andreasen et al., 1996; Schmahmann and Sherman, 1998). The tractography study by Ramnani et al. (2006) seems to give anatomical support in favour of the cerebellar participation to cognition. Using probabilistic diffusion tensor imaging these authors found in humans a relatively larger contribution of the prefrontal areas to the cortico-ponto-cerebellar tract compared to monkeys. Further evidence in support of the role of the cerebellum in cognition, affect, and behaviour comes from patients with cerebellar lesions. Cognitive and affective deficits are, for example, frequently observed in young patients with the posterior fossa syndrome, a disorder of acute onset following the surgical resection of vermian midline tumours. Children usually develop buccal and lingual apraxia, apathy, and poverty of spontaneous movement that sometimes manifests with severe mutism. Emotional lability is also frequent in these children, with a rapid fluctuation between irritability, agitation, giggling, and easy distractibility (Wisoff and Epstein, 1984; Pollack, 1997; Levisohn et al., 2000; Riva and Giorgi, 2000). Non-motor symptoms have also been described in preterm subjects with altered cerebellar anatomy (Allin et al., 2001), in patients with cerebellar hypoplasia (Limperopoulos et al., 2006), and congenital malformations (Tavano et al., 2007). Similar symptoms have been described in adults with multiple system atrophy (Parvizi et al., 2007), in superficial siderosis (Polidori et al., 2001; van Harskamp et al., 2005), and stroke (Malm et al., 1998; Leggio et al., 2000; Neau et al., 2000; Parvizi et al., 2001). Occasionally patients are significantly impaired in executive functions, working memory, motor shifting, and perseveration. Verbal fluency is impaired to the point of telegraphic speech or mutism but true aphasic disorders are rare (Frank et al., 2010). Anomia, agrammatic speech, and abnormal syntactic structure are also observed, with abnormal prosody (i.e. high pitched, hypophonic whining). Visuospatial impairment in diagram drawing or copying, and in conceptualization of figures have been reported but they are difficult to interpret because of the concomitant presence of attentional deficits and poor coordination. Some patients display simultanagnosia (inability to perceive more than one single object at time).

The behavioural manifestations of cerebellar lesions include personality changes (notable with posterior lobe lesions that involve the vermis and fastigial nucleus), impulsive and disinhibited behaviour (e.g. over-familiarity, flamboyant actions, inappropriate comments) and flattening of affect (Schmahmann and Sherman 1998; Schmahmann et al., 2007; Schmahmann and Pandya, 2008). Regressive, childlike behaviours and obsessive–compulsive traits can also be observed.

Cerebellar pathology and cerebellar neurological signs are very common in psychiatric neurodevelopmental disorders (Haas H. et al., 1996; Katsetos et al., 1997) such as autism (Courchesne, 1997; Harris et al., 1999) and schizophrenia (Andreasen et al., 1996; Schmahmann et al., 2007). Tractography has been recently applied to study cerebellar connections in autism (Catani et al., 2008) and schizophrenia (Magnotta et al., 2008; Kanaan, 2009). Tract-specific measurements of fractional anisotropy showed that compared to controls, subjects with Asperger syndrome (a milder form of autism) have lower fractional anisotropy values in the right short intrinsic cerebellar fibres compared to controls (Figure 10.10). Reduced fractional anisotropy values in the short intrinsic cerebellar fibres are in keeping with post-mortem findings in the disorder (Vargas et al., 2005), and are likely to reflect changes in the parallel fibres and Purkinje axons. In the Asperger's syndrome fractional anisotropy values are also reduced in the right superior cerebellar peduncle compared to controls. These results suggest a specific involvement of intrinsic cerebellar circuitry and main cerebellar outflow within the cerebello-thalamic network (Catani et al., 2008). In schizophrenia a reduction in the fractional anisotropy has also been reported (Magnotta et al., 2008), but this is not specific to a single cerebellar tract and does not correlate with positive or negative symptoms (Kanaan et al., 2009). Others have used tractography to evaluate white matter cerebellar atrophy in phenytoin users with (Lee et al., 2003), and also tract integrity in spinocerebellar degeneration (Taoka et al., 2007) and in congenital malformations (Lee et al., 2005).

Finally cerebellar syndromes can occur with lesions outside the cerebellum, through a mechanism known as 'crossed cerebellar diaschisis' (CCD) (Baron et al., 1980; Pantano et al., 1986). The term diaschisis was introduced by Monakow (1914) at the beginning of the last century to highlight the possible recovery of dysfunctional distant regions connected to the lesioned areas. The concept received empirical support much later, when positron emission tomography studies of patients with cerebral stroke (Baron et al., 1980; Pantano et al., 1986) and tumours (Fulham et al., 1992) showed reduced perfusion of the contralateral cerebellum. In patients treated with intravenous thrombolysis—a treatment that facilitates reperfusion of the ischaemic cerebral areas—the clinical recovery is strongly associated with the normalization of cerebellar perfusion (Sobesky et al., 2005).

References

Aggleton, J.P. (2008). Understanding anterograde amnesia: disconnections and hidden lesions. *Q J Exp Psychol*, **61**(10), 1441–71.

Alexander, G.E., Delong, M.R., and Strick, P.L. (1986). Parallel organization of functionally segregated circuits linking basal ganglia and cortex. *Annu Rev Neurosci*, **9**, 357–81.

Allin, M., Matsumoto, H., Santhouse, A. M., et al. (2001). Cognitive and motor function and the size of the cerebellum in adolescents born very pre-term. *Brain*, **124**(1), 60–6.

Andreasen, N.C., O'Leary, D.S., Cizadlo, T., et al. (1996). Schizophrenia and cognitive dysmetria: a positron-emission tomography study of dysfunctional prefrontal-thalamic-cerebellar circuitry. *Proc Natl Acad Sci USA*, **93**(18), 9985–90.

Arend, I., Rafal, R., and Ward, R. (2008). Spatial and temporal deficits are regionally dissociable in patients with pulvinar lesions. *Brain*, **131**(8), 2140–52.

Baron, J.C., Bousser, M.G., Comar, D., and Castaigne, P. (1980). 'Crossed cerebellar diaschisis' in human suptratentorial brain infarction. *Trans Am Neurol Assess*, **105**, 459–61.

Beevor, C.E. and Horsley, V. (1890). An experimental investigation into the arrangement of the excitable fibres of the internal capsule of the Bonnet monkey (*Macacus sinicus*). *Phil Trans R Soc Lond B*, **181**, 49–88.

Behrens, T.E., Johansen-Berg, H., Woolrich, M.W., et al. (2003). Non-invasive mapping of connections between human thalamus and cortex using diffusion imaging. *Nat Neurosci*, **6**(7), 750–7.

Bhatia, K.P. and Marsden, C.D. (1994). The behavioural and motor consequences of focal lesions of the basal ganglia in man. *Brain*, **117**(4), 859–76.

Bogousslavsky, J., Miklossy, J., Deruaz, J.P., Regli, F., and Assal, G. (1987). Unilateral left paramedian infarction of thalamus and midbrain: a clinico-pathological study. *J Neurol Neurosurg Psychiatry*, **49**(6), 686–94.

Bonilha, L., de Vries, P.M., Hurd, M.W., et al. (2009). Disrupted thalamic prefrontal pathways in patients with idiopathic dystonia. *Parkinsonism Relat Disord*, **15**(1), 64–7.

Botez, M.I. (1992). The neuropsychology of the cerebellum: an emerging concept. *Arch Neurol*, **49**(12), 1229–30.

Brodal, P. (1978). The corticopontine projection in the rhesus monkey. Origin and principles of organization. *Brain*, **101**(2), 251–83.

Caplan, L.R., Schmahmann, J.D., Kase, C.S., et al. (1990). Caudate infarcts. *Arch Neurol*, **47**(2), 133–43.

Catani, M., Jones, D., Daly, E., et al. (2008). Altered cerebellar feedback projections in Asperger syndrome. *NeuroImage*, **41**(4), 1184–91.

Courchesne, E. (1997). Brainstem, cerebellar and limbic neuroanatomical abnormalities in autism. *Curr Opin Neurobiol*, **7**(2), 269–78.

Crosby, E.C., Humphrey, T., and Lauer, E.W. (1962). *Correlative anatomy of the nervous system*. New York: Macmillian Co.

Dejerine, J. (1901). *Anatomie des Centres Nerveux, Vol. 2*. Paris: Rueff et Cie.

DeLong, M.R. (1990). Primate models of movement disorders of basal ganglia origin. *Trends Neurosci*, **13**(7), 281–5.

Deoni, S.C. and Catani, M. (2007). Visualization of the deep cerebellar nuclei using quantitative T1 and rho magnetic resonance imaging at 3 Tesla. *NeuroImage*, **37**(4), 1260–6.

Devlin, J.T., Sillery, E.L., Hall, D.A., et al. (2006). Reliable identification of the auditory thalamus using multi-modal structural analyses. *NeuroImage*, **30**(4), 1112–20.

Douaud, G., Behrens, T.E., Poupon, C., et al. (2009). In vivo evidence for the selective subcortical degeneration in Huntington's disease. *NeuroImage*, **46**(4), 958–66.

Dow, R.S. and Moruzzi, G. (1958). *The Physiology and Pathology of the Cerebellum*. Minneapolis, MN: University of Minnesota Press.

Duncan, G.W., Parker, S.W., and Fisher, C.M. (1975). Acute cerebellar infarction in the PICA territory. *Arch Neurol*, **32**(6), 364–8.

Eccles, J.C., Ito, M., and Szentágothai, J. (1967). *The cerebellum as a neuronal machine*. Berlin: Springer-Verlag.

Eluvathingal, T.J., Hasan, K.M., Kramer, L.A., Fletcher, J.M., and Ewing-Cobbs, L. (2007). Quantitative diffusion tensor tractography of association and projection fibers in normally developing children and adolescents. *Cereb Cortex*, **17**(12), 2760–68.

Exner, C., Weniger, G., and Irle, E. (2004). Cerebellar lesions in the PICA but not SCA territory impair cognition. *Neurology*, **63**(11), 2132–5.

Flechsig, P.E. (1876). *Die Leitungsbahnen im Gehirn und Rückenmark des Menschen auf Grund entwicklungsgeschichtlicher Untersuchungen*. Leipzig: Engelmann.

Fukaya, C., Katayama, Y., Kano, T., et al. (2007). Thalamic deep brain stimulation for writer's cramp. *J Neurosurg*, **107**(5), 977–82.

Fulham, M.J., Brooks, R.A., Hallett M., and Di Chiro, G. (1992). Cerebellar diaschisis revisited: pontine hypometabolism and dentate sparing. *Neurology*, **42**(12), 2267–73.

Frank, B., Maschke, M., Groetschel, H., et al. (2010). Aphasia and neglect are uncommon in cerebellar disease negative findings in a prospective study in acute cerebellar stroke. *Cerebellum*, **9**(4), 556–66.

Gallassi, R., Morreale, A., Montagna, P., Gambetti, P., and Lugaresi, E. (1992). 'Fatal familial insomnia': neuropsychological study of a disease with thalamic degeneration. *Cortex*, **28**(2), 175–87.

Glickstein, M. (1993). Motor skills but not cognitive tasks. *Trends Neurosci*, **16**(11), 450–1; discussion 453–4.

Glickstein, M., May, J.G., and Mercier, B.E. (1985). Corticopontine projection in the macaque: the distribution of labelled cortical cells after large injections of horseradish peroxidase in the pontine nuclei. *J Comp Neurol*, **235**(3), 343–59.

Good, C.D., Johnsrude, I.S., Ashburner, J., et al. (2001). Cerebral asymmetry and the effects of sex and handedness on brain structure: a voxel-based morphometric analysis of 465 normal adult human brains. *NeuroImage*, **14**(3), 685–700.

Govindan, R.M., Makki. M.I., Wilson, B.J., Behen, M.E., and Chugani, H.T. (2010). Abnormal water diffusivity in corticostriatal projections in children with Tourette syndrome. *Hum Brain Mapp*, **31**(11), 1665–74.

Haas, B.W., Barnea-Goraly, N., Lightbody, A.A., et al. (2009). Early white-matter abnormalities of the ventral frontostriatal pathway in fragile X syndrome. *Dev Med Child Neurol*, **51**(8), 593–9.

Haas, R.H., Townsend, J., Courchesne, E., et al. (1996). Neurologic abnormalities in infantile autism. *J Child Neurol*, **11**(2), 84–92.

Hariz, M.I. (2003). From functional neurosurgery to "interventional" neurology: survey of publications on thalamotomy, pallidotomy, and deep brain stimulation for Parkinson's disease from 1966 to 2001. *Mov Disord*, **18**(8), 845–53.

Harris, N.S., Courchesne, E., Townsend, J., Carper, R.A., and Lord, C., (1999). Neuroanatomic contributions to slowed orienting of attention in children with autism. *Brain Res Cogn Brain Res*, **8**(1), 61–71.

Hartman-von Monakow, K., Akert, K., and Künzle H. (1981). Projection of precentral, premotor and prefrontal cortex to the basilar pontine grey and to nucleus reticularis tegmenti pontis in the monkey (Macaca fascicularis). *Schweiz Arch Neurol Neurochir Psychiat*, **129**(2), 189–208.

Hauber, W. (1998). Involvement of basal ganglia transmitter systems in movement initiation. *Prog Neurobiol*, **56**(5), 507–40.

Häusler, R. and Levine, R.A. (2000). Auditory dysfunction in stroke. *Acta Otolaryngol*, **120**(6), 689–703.

Hervé, P.Y., Crivello, F., Perchey, G., Mazoyer, B., and Tzourio-Mazoyer, N. (2006). Handedness and cerebral anatomical asymmetries in young adult males. *NeuroImage*, **29**(4), 1066–79.

Ito, M. (Ed.) (1993). *How Does the Cerebellum Facilitate Thought?* Oxford: Oxford University Press.

Ivanov, I., Bansal, R., and Hao, X. (2010). Morphological abnormalities of the thalamus in youths with attention deficit hyperactivity disorder. *Am J Psychiatry*, **167**(4), 397–408.

Kalivas, P.W. (2009). The glutamate homeostasis hypothesis of addiction. *Nat Rev Neurosci*, **10**(1), 561–72.

Kanaan, R.A., Borgwardt, S., McGuire, P.K., et al. (2009). Microstructural organization of cerebellar tracts in schizophrenia. *Biol Psychiat*, **66**(11), 1067–9.

Katsetos, C.D., Hyde, T.M., and Herman, M.M. (1997). Neuropathology of the cerebellum in schizophrenia-an update: 1996 and future directions. *Biol Psychiat*, **42**(3), 213–24.

Kertesz, A. and Geschwind, N. (1971). Patterns of pyramidal decussation and their relationship to handedness. *Arch Neurol*, **24**(4), 326–32.

Kumral, E., Evyapan, D., and Balkir, K. (1999). Acute caudate vascular lesions. *Stroke*, **30**(1), 100–8.

Larsell, O. and Jansen, J. (1972). *The comparative anatomy and histology of the cerebellum. III. The human cerebellum, cerebellar connections, and cerebellar cortex*. Minneapolis, MN: University of Minnesota Press.

Lee, H., Sohn, S.I., Cho, Y.W., et al. (2006). Cerebellar infarction presenting isolated vertigo: frequency and vascular topographical patterns. *Neurology*, **67**(7), 1178–83.

Lee, S.-K., Mori, S., Kim, D. J., et al. (2003). Diffusion tensor MRI and fiber tractography of cerebellar atrophy in phenytoin users. *Epilepsia*, **44**(12), 1536–40.

Lee, S.-K., Kim, D.I., Kim, J., et al. (2005). Diffusion-tensor MR imaging and fiber tractography: a new method of describing aberrant fiber connections in developmental CNS anomalies. *Radiographics*, **25**(1), 53–65; discussion 66–8.

Leggio, M.G., Silveri, M.C., Petrosini, L., and Molinari, M. (2000). Phonological grouping is specifically affected in cerebellar patients: a verbal fluency study. *J Neurol Neurosurg Psychiat*, **69**, 102–6.

Lehéricy, S., Ducros, M., Van De Moortele, P.-F., et al. (2004). Diffusion tensor fiber tracking shows distinct corticostriatal circuits in humans. *Ann Neurol*, **55**(4), 522–9.

Leiner, H.C., Leiner, A.L., and Dow, R.S. (1991). The human cerebro-cerebellar system: its computing, cognitive, and language skills. *Behav Brain Res*, **44**,113–28.

Levisohn, L., Cronin-Golomb, A., and Schmahmann, J. D. (2000). Neuropsychological consequences of cerebellar tumor resection in children: Cerebellar cognitive affective syndrome in a pediatric population. *Brain*, **123**(5), 1041–50.

Levy, R. and Dubois, B. (2006). Apathy and the functional anatomy of the prefrontal cortex-basal ganglia circuits. *Cereb Cortex*, **16**(7), 916–28.

Limperopoulos, C., Robertson, R. L., Estroff, J.A., et al. (2006). Diagnosis of inferior vermian hypoplasia by fetal magnetic resonance imaging: potential pitfalls and neurodevelopmental outcome. *Am J Obstet Gynecol*, **194**(4), 1070–6.

Lugaresi, E., Medori, R., Montagna, P., et al. (1986). Fatal familial insomnia and dysautonomia with selective degeneration of thalamic nuclei. *N Engl J Med*, **315**(16), 997–1003.

Magnotta, V.A., Adixa, M.L., Caprahanc, A., et al. (2008). Investigating connectivity between the cerebellum and thalamus in schizophrenia using diffusion tensor tractography: a pilot study. *Psychiat Res Neuroimaging*, **163**(3), 193–200.

Makris, N., Schlerf, J.E., Hodge, S.M., et al. (2005). MRI-based surface-assisted parcellation of human cerebellar cortex: an anatomically specified method with estimate of reliability. *NeuroImage*, **25**(4), 1146–60.

Malm, J., Kristensen, B., Karlsson, T., et al. (1998). Cognitive impairment in young adults with infratentorial infarcts. *Neurology*, **51**(2), 433–40.

Manetto, V., Medori, R., Cortelli, P., et al. (1992). Fatal familial insomnia: clinical and pathologic study of five new cases. *Neurology*, **42**(2), 312–9.

Mataix-Cols, D., Wooderson, S., Lawrence, N., et al. (2004). Distinct neural correlates of washing, checking, and hoarding symptom dimensions in obsessive-compulsive disorder. *Arch Gen Psychiat*, **61**(6), 564–76.

Mettler, F.A. (1949). Summary. In Mettler, F.A. (Ed.) *Selective Partial Ablation of the Frontal Cortex*, pp. 477–514. New York: PB Hoeber.

Middleton, F.A. and Strick, P.L. (1994). Anatomical evidence for cerebellar and basal ganglia involvement in higher cognitive function. *Science*, **266**(5184), 458–61.

Middleton, F.A. and Strick, P.L. (1997). Cerebellar output channels. The cerebellum and cognition. *Int Rev Neurobiol*, **41**, 61–82.

Mistichelli, D. (1709). *Trattato dell'Apoplessia*. Roma: A. de Rossi alla Piazza di Ceri.

Monakow, C. von. (1914). *Die Lokalisation im Grosshirn und der Abbau der Funktion durch kortikale Herde*. Wiesbaden: JF Bergmann.

Montagna, P., Gambetti, P., Cortelli, P., and Lugaresi, E. (2003). Familial and sporadic fatal insomnia. *Lancet Neurol*, **2**(3), 167–76.

Morel, A., Magnin, M., and Jenamonod, D. (1997). Multiarchitectonic and stereotactic atlas of the human thalamus. *J Comp Neurol*, **387**(4), 588–630.

Müller, R.-A., Courchesne, E., and Allen, G. (1998). The cerebellum: so much more. *Science*, **282**(5390), 879.

Nathan, P.W., Smith, M.C., and Deacon, P. (1990). The corticospinal tracts in man. Course and location of fibres at different segmental levels. *Brain*, **113**(2), 303–24.

Neau, J.P., Arroyo-Anllo, E., Bonnaud, V., Ingrand, P., and Gil, R. (2000). Neuropsychological disturbances in cerebellar infarcts. *Acta Neurolog Scand*, **102**(6), 363–70.

Newton, J.M., Ward, N.S., Parker G.J., et al. (2006). Non-invasive mapping of corticofugal fibres from multiple motor areas-relevance to stroke recovery. *Brain*, **129**(7), 1844–58.

Nieuwenhuys, R., Voogd, J., and Huijzen, C. (2008). *The human central nervous system*, fourth edition. Berlin: Springer-Verlag.

Oh, J.S., Kubicki, M., and Rosenberger, G. (2009). Thalamo-frontal white matter alterations in chronic schizophrenia: a quantitative diffusion tractography study. *Hum Brain Mapp*, **30**(11), 3812–25.

Pakkenberg, B. (1990). Pronounced reduction of total neuron number in mediodorsal thalamic nucleus and nucleus accumbens in schizophrenics, *Arch Gen Psychiatry*, **47**(11), 1023–28.

Pakkenberg, B. (1993). Leucotomized schizophrenics lose neurons in the mediodorsal thalamic nucleus. *Neuropathol Appl Neurobiol*, **19**(5), 373–80.

Pantano, P., Baron, J.C., Samson, Y., Bousser, M.G., Derouesne, C., and Comar, D. (1986). Crossed cerebellar diaschisis. Further studies. *Brain*, **109**(4), 677–94.

Parvizi, J., Anderson, S.W., Martin, C.O., Damasio, H., and Damasio, A.R. (2001). Pathological laughter and crying: a link to the cerebellum. *Brain*, **124**(9), 1708–19.

Parvizi, J., Joseph, J., Press, D.Z., and Schmahmann, J.D. (2007). Pathological laughter and crying in patients with multiple system atrophy-cerebellar type. *Mov Disord*, **22**(6), 798–803.

Paus, T., Zijdenbos, A., Worsley, K., et al. (1999). Structural maturation of neural pathways in children and adolescents: in vivo study. *Science*, **283**(5409), 1908–11.

Penfield, W. and Boldrey E. (1937). Somatic motor and sensory representation in the cerebral cortex of man as studied by electrical stimulation. *Brain*, **60**(4), 389–440.

Pessiglione, M., Guehl, D., Agid, Y., et al. (2003). Impairment of context-adapted movement selection in a primate model of presymptomatic Parkinson's disease. *Brain*, **126**(6), 1392–408.

Pessiglione, M., Seymour, B., Flandin, G., et al. (2006). Dopamine-dependent prediction errors underpin reward-seeking behaviour in humans. *Nature*, **442**(7106), 1042–5.

Polidori, M.C., Pelliccioli, G., Catani, M., et al. (2001). Superficial siderosis of the central nervous system: a 70-year-old man with ataxia, depression and visual deficits. *Gerontology*, **47**(2), 93–5.

Pollack, I.F. (1997). Posterior fossa syndrome. *Cerebellum Cognit Int Rev Neurobiol*, **41**, 411–32.

Pourfour du Petit, F. (1710). Theory of contralateral innervation Trois lettres d'un médecin des hôpitaux du Roy . . . contient un nouveau système du cerveau, etc. Namur: C.G. Albert.

Qiu, A., Zhong, J., Graham, S., Chia, M.Y., and Sim, K. (2009). Combined analyses of thalamic volume, shape and white matter integrity in first-episode schizophrenia. *NeuroImage*, **47**(4), 1163–71.

Rademacher, J., Bürgel, U., Geyer, S., et al. (2001). Variability and asymmetry in the human precentral motor system. A cytoarchitectonic and myeloarchitectonic brain mapping study. *Brain*, **124**(11), 2232–58.

Rademacher, J., Bürgel, U., and Zilles, K. (2002). Stereotaxic localization, intersubject variability, and interhemispheric differences of the human auditory thalamocortical system. *NeuroImage* **17**(1), 142–60.

Ramnani, N. (2006). The primate cortico-cerebellar system: anatomy and function. *Nat Rev Neurosci*, **7**(7), 511–22.

Ramnani, N., Behrens, T.E., Johansen-Berg, H., et al. (2006). The evolution of prefrontal inputs to the cortico-pontine system: diffusion imaging evidence from Macaque monkeys and humans. *Cereb Cortex*, **16**(6), 811–18.

Remijnse, P.L., Nielen, M.M., van Balkom, A.J., et al. (2006). Reduced orbitofrontal-striatal activity on a reversal learning task in obsessive compulsive disorder. *Arch Gen Psychiat*, **63**(11), 1225–36.

Riva, D. and Giorgi, C. (2000). The cerebellum contributes to higher function during development: evidence from a series of children surgically treated for posterior fossa tumors. *Brain*, **123**(5), 1051–61.

Robbins, T.W. and Everitt, B.J. (2002). Limbic-striatal memory systems and drug addiction. *Neurobiol Learning Memory*, **78**(3), 625–36.

Roitman, A.V., Pasalar, S., Johnson, M.T., and Ebner, T.J. (2005). Position, direction of movement, and speed tuning of cerebellar Purkinje cells during circular manual tracking in monkey. *J Neurosci*, **25**(40), 9244–57.

Rosas, H.D., Tuch, D.S., Hevelone N.D., et al. (2006). Diffusion tensor imaging in presymptomatic and early Huntington's disease: Selective white matter pathology and its relationship to clinical measures. *Mov Disord*, **21**(9), 1317–25.

Schmahmann, J.D. (1991). An emerging concept: The cerebellar contribution to higher function. Arch Neurol, **48**(11), 1178–87.

Schmahmann, J.D., and Pandya, D.N. (2008). Disconnection syndromes of basal ganglia, thalamus, and cerebrocerebellar systems. *Cortex*, **44**(8), 1037–66.

Schmahmann, J.D., and Sherman, J.C. (1998). The cerebellar cognitive affective syndrome. *Brain*, **121**(4), 561–79.

Schmahmann, J.D., Rosene, D.L., and Pandya, D.N. (2004). Motor projections to the basis pontis in rhesus monkey. *J Comp Neurol*, **478**(3), 248–68.

Schmahmann, J.D, Weilburg, J.B., and Sherman, J.C. (2007). The neuropsychiatry of the cerebellum – insights from the clinic. *Cerebellum*, **6**(3), 254–67.

Schoenbaum, G., Gottfried, J., Murray, E.A., and Ramus, S. (2007). Linking affect to action: critical contributions of the obitofrontal cortex. *Ann NY Acad Sci*, **1121**, 6–8.

Sherman, S.M. and Guillery, R.W. (2001). *Exploring the thalamus*. San Diego, CA: Academic Press.

Sobesky, J., Thiel, A., Ghaemi, M., et al. (2005). Crossed cerebellar diaschisis in acute human stroke: a PET study of serial changes and response to supratentorial reperfusion. *J Cereb Blood Flow Metab*, **25**(12), 1685–91.

Spoletini, I., Cherubini, A., Banfi, G., et al. (2011). Hippocampi, thalami, and accumbens microstructural damage in schizophrenia: a volumetry, diffusivity, and neuropsychological study. *Schizophr Bull*, **37**(1), 118–30.

Stefani, A., Peppe, A., Pierantozzi, M., et al. (2009). Multi-target strategy for Parkinsonian patients: the role of deep brain stimulation in the centromedian-parafascicularis complex. *Brain Res Bull*, **78**(2–3), 113–18.

Stuss, D.T. and Benson, D.F. (1984). Neuropsychological studies of the frontal lobes. *Psychol Bull*, **95**(1), 3–28.

Taoka, T., Kin, T., Nakagawa, H., et al. (2007). Diffusivity and diffusion anisotropy of cerebellar peduncles in cases of spinocerebellar degenerative disease. *NeuroImage*, **37**(2), 387–93.

Tavano, A., Grasso, R., Gagliardi, C., et al. (2007). Disorders of cognitive and affective development in cerebellar malformations. *Brain*, **130**(10), 2646–60.

Thiebaut de Schotten, M., ffytche, D., Bizzi, A., et al. (2011). Atlasing location, asymmetry and inter-subject variability of white matter tracts in the human brain with MR diffusion tractography. *Neuroimage*, **54**(1), 49–59.

Trépanier, L.L., Saint-Cyr, J.A., Lozano, A.M., and Lang, A.E. (1998). Neuropsychological consequences of posteroventral pallidotomy for the treatment of Parkinson's disease. *Neurology*, **51**(1), 207–15.

Unrath, A., Müller, H.P., Ludolph, A.C., Riecker, A., and Kassubek, J. (2008). Cerebral white matter alterations in idiopathic restless legs syndrome, as measured by diffusion tensor imaging. *Mov Disord*, **23**(9), 1250–5.

Van der Werf, Y.D., Witter, M.P., and Groenewegen, H.J. (2002). The intralaminar and midline nuclei of the thalamus. Anatomical and functional evidence for participation in processes of arousal and awareness. *Brain Res Rev*, **39**(2–3), 107–40.

van Harskamp, N.J., Rudge, P., and Cipolotti, L. (2005). Cognitive and social impairments in patients with superficial siderosis. *Brain*, **128**(5), 1082–92.

Vargas, D.L., Nascimbene, C., Krishnan, C., Zimmerman, A.W., and Pardo, C.A. (2005). Neuroglial activation and neuroinflammation in the brain of patients with autism. *Ann Neurol*, **57**(1), 67–81.

Velasco, F., Velasco, M., Ogarrio, C., and Fanghanel, G. (1987). Electrical stimulation of the centromedian thalamic nucleus in the treatment of convulsive seizures: a preliminary report. *Epilepsia*, **28**(4), 421–30.

Voogd, J. (2003). The human cerebellum. *J Chem Neuroanat*, **26**(4), 243–52.

Ward, R., Danziger, S., Quirk, R.T., Goodson, L., and Downing, P. (2005). Suppression of involuntary spatial response activation requires selective attention. *Vis Cogn*, **12**(2), 376–94.

Ward, R., Calder, A.J., Parker, M., and Arend, I. (2007). Emotion recognition following human pulvinar damage. *Neuropsychologia*, **45**(8), 1973–8.

Watkins, K.E., Smith, S.M., Davis, S., and Howell, P. (2008). Structural and functional abnormalities of the motor system in developmental stuttering. *Brain*, **131**(1), 50–9.

Watson, R.T. and Heilman, K.M. (1979). Thalamic neglect. *Neurology*, **29**(5), 690–4.

Weigel, R. and Krauss, J.K. (2004). Center median–parafascicular complex and pain control. Reviewer from a neurosurgical perspective, *Stereotact Funct Neurosurg*, **82**(2–3), 115–26.

Wichmann, T. and DeLong, M.R. (1993). Pathophysiology of parkinsonian motor abnormalities. *Adv Neurol*, **60**, 53–61.

Wiesendanger, R., Wiesendanger, M., and Rüegg, D. G. (1979). An anatomical investigation of the corticopontine projection in the primate (*Macaca fascicularis* and *Saimiri sciureus*) – II. The projection from frontal and parietal association areas. *Neuroscience*, **4**(6), 747–65.

Wise, R.A. (1987). The role of reward pathways in the development of drug dependence. *Pharmacol Ther*, **35**(1–2), 227–63.

Wisoff, J.H. and Epstein, F.J. (1984). Pseudobulbar palsy after posterior fossa operation in children. *Neurosurgery*, **15**(5), 707–9.

Xuereb, J.H., Perry, R.H., Candy, J.M., et al. (1991). Nerve cell loss in the thalamus in Alzheimer's disease and Parkinson's disease. *Brain*, **114**(3), 1363–79.

Yakovlev, P.J. and Rakic, P. (1966). Pattern of decussation of bulbar pyramids and distribution of pyramidal tracts on two sides of the spinal cord. *Trans Am Neurol Ass*, **91**, 366–7.

Zarei, M., Patenaude, B., Damoiseaux, J., et al. (2010). Combining shape and connectivity analysis: An MRI study of thalamic degeneration in Alzheimer's disease. *Neuroimage*, **49**(1), 1–8.

Internal capsule (atlas)

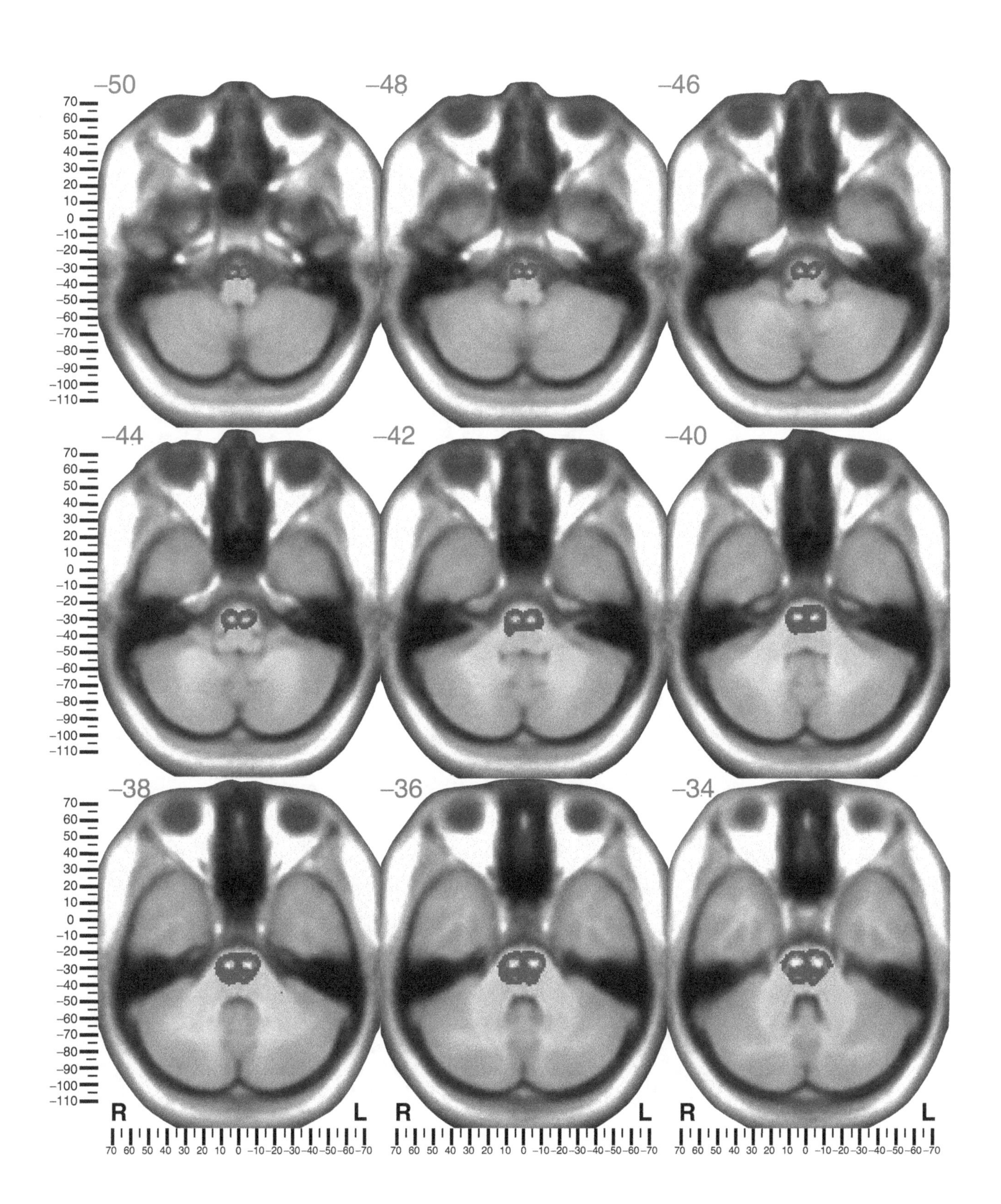

Internal capsule: inferior axial view

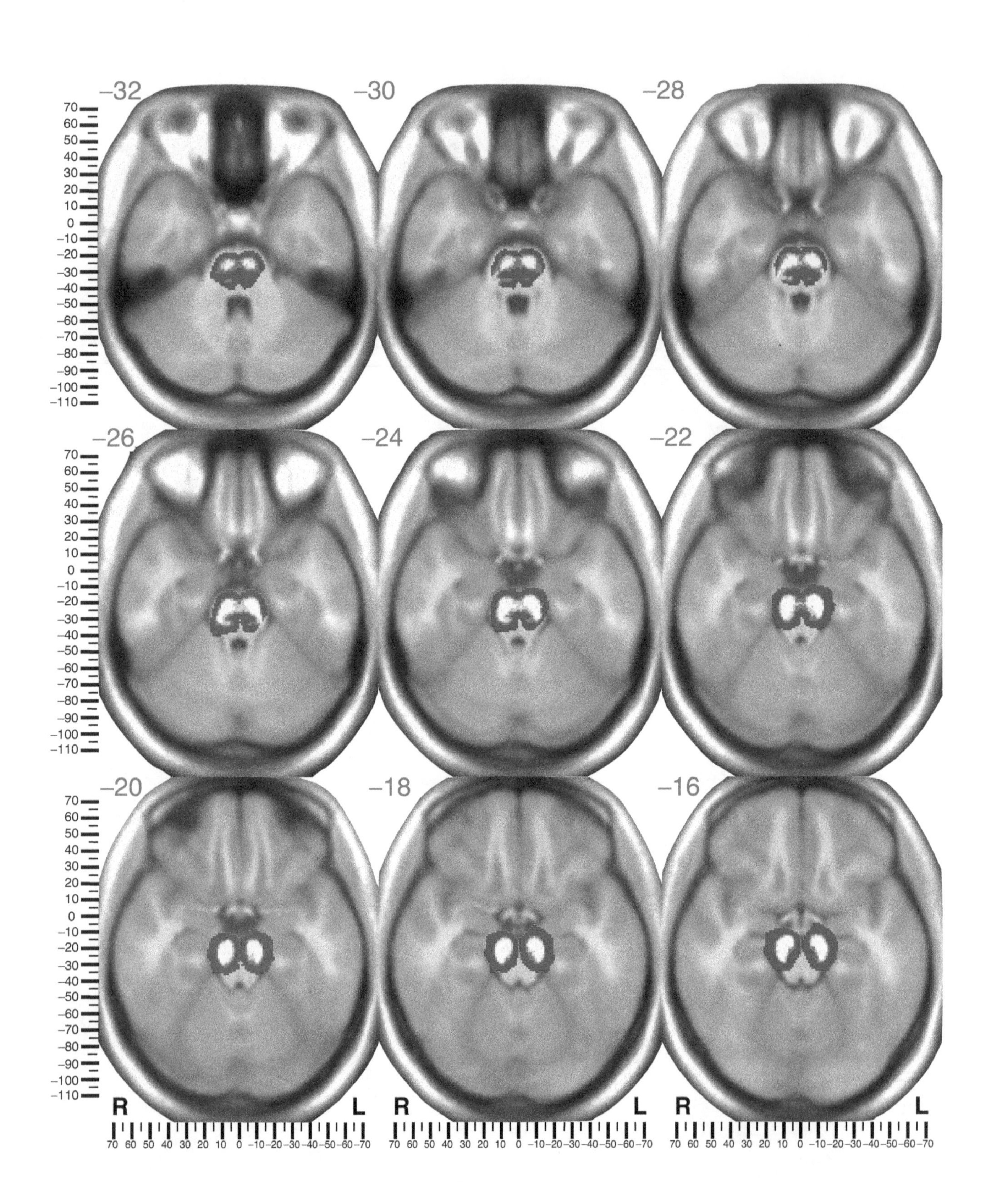

Internal capsule: inferior axial view

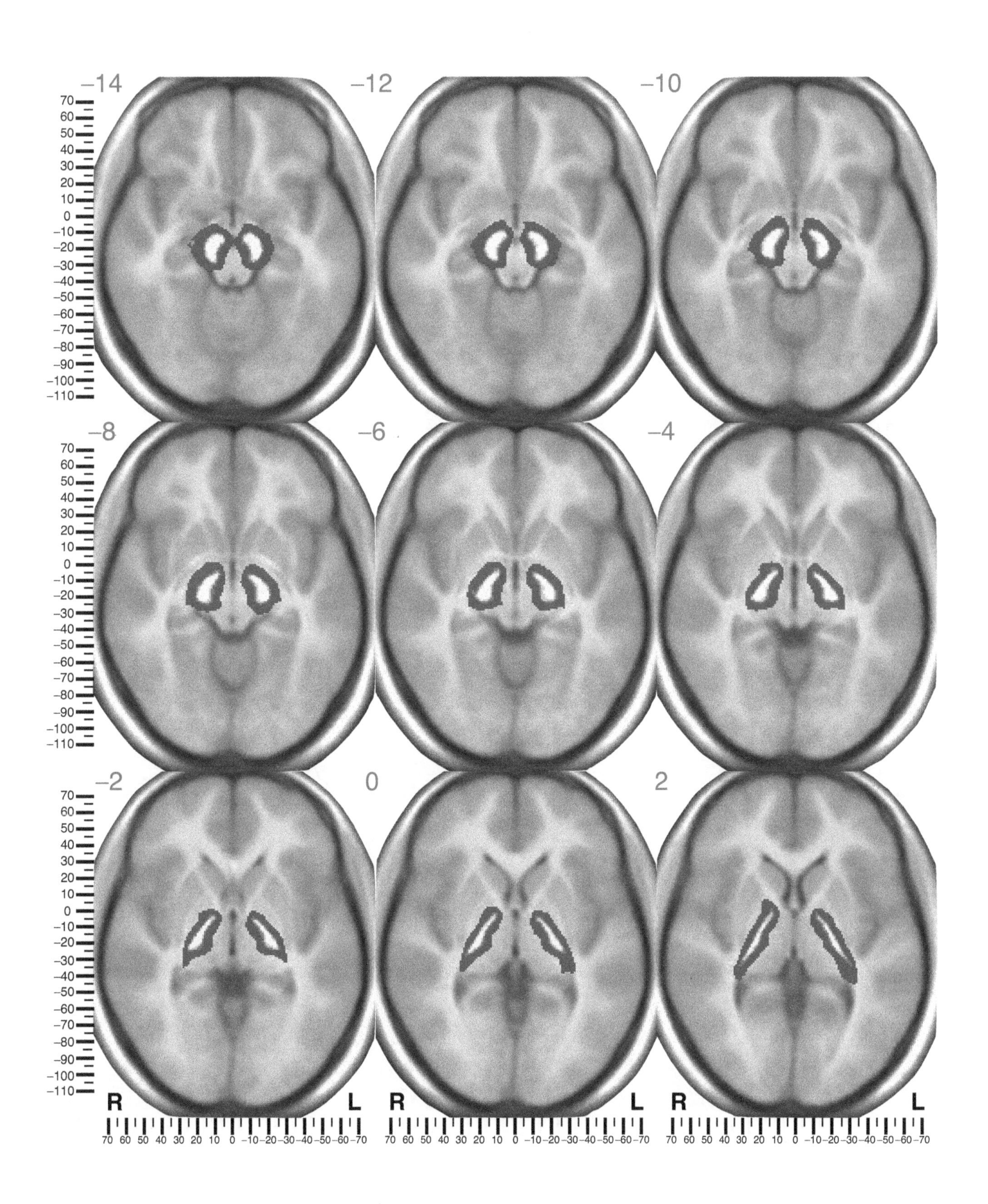

Internal capsule: inferior axial view

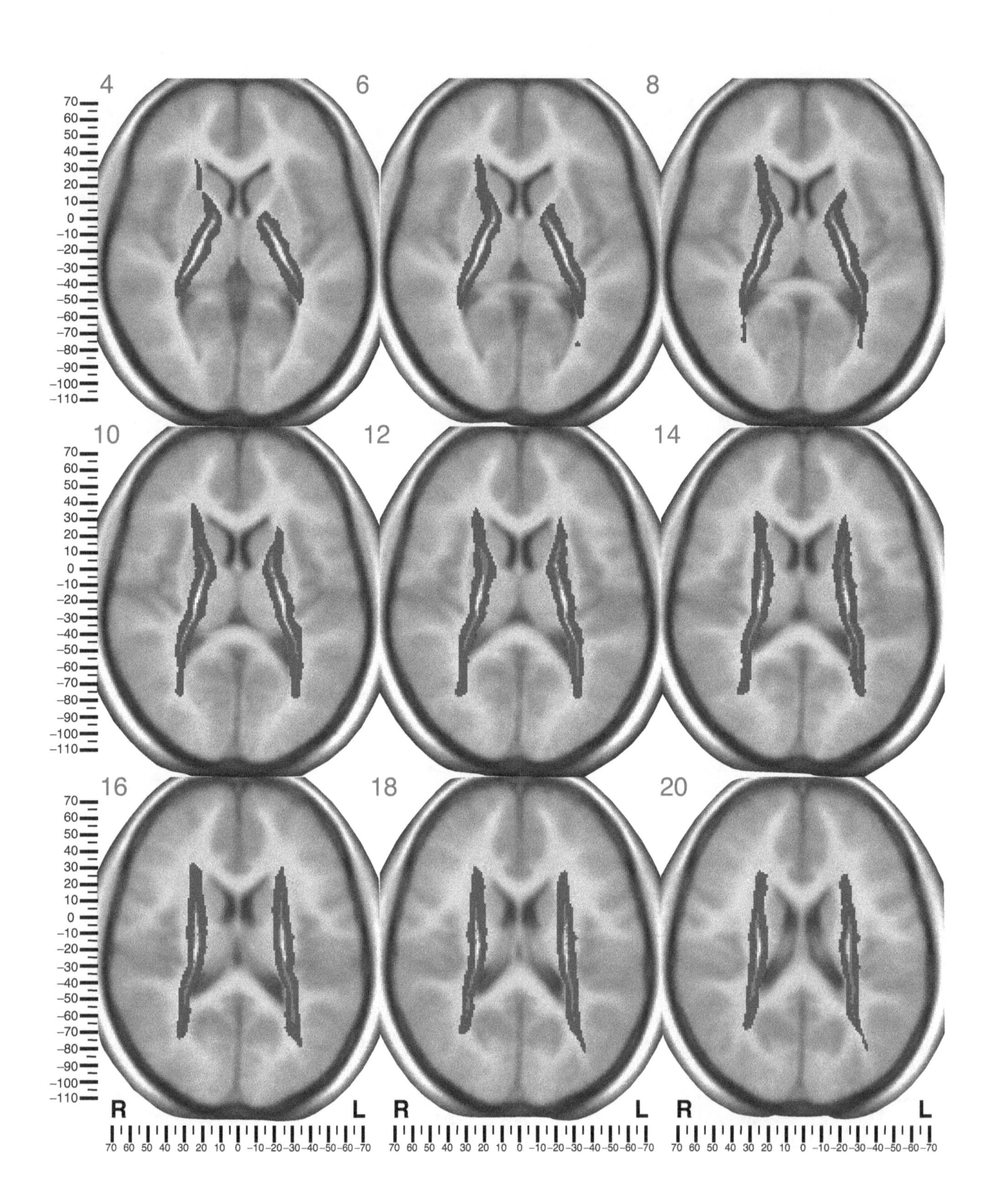

Internal capsule: inferior axial view

p < .05 | > 50% | > 75% | > 90%

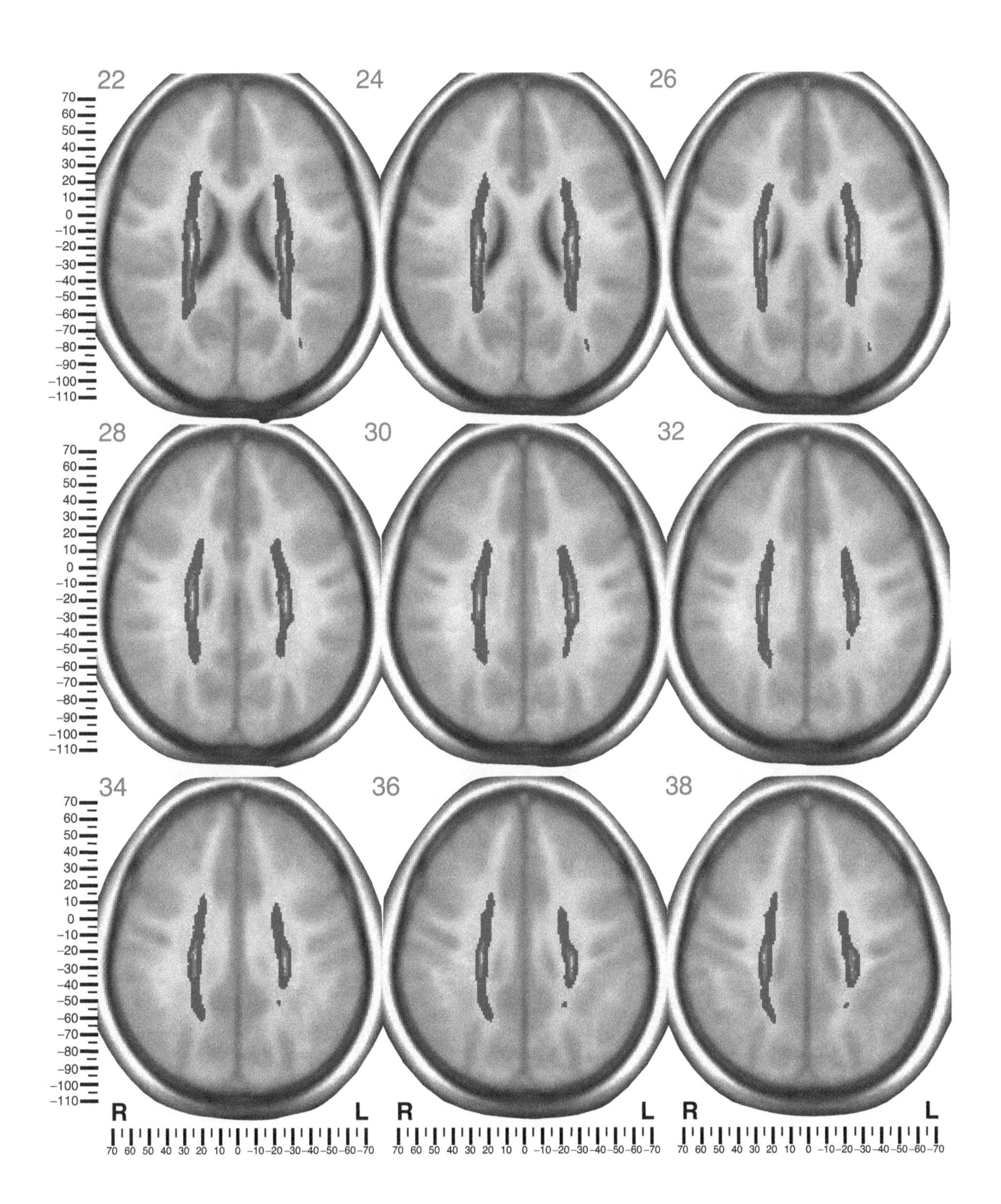

Internal capsule: inferior axial view

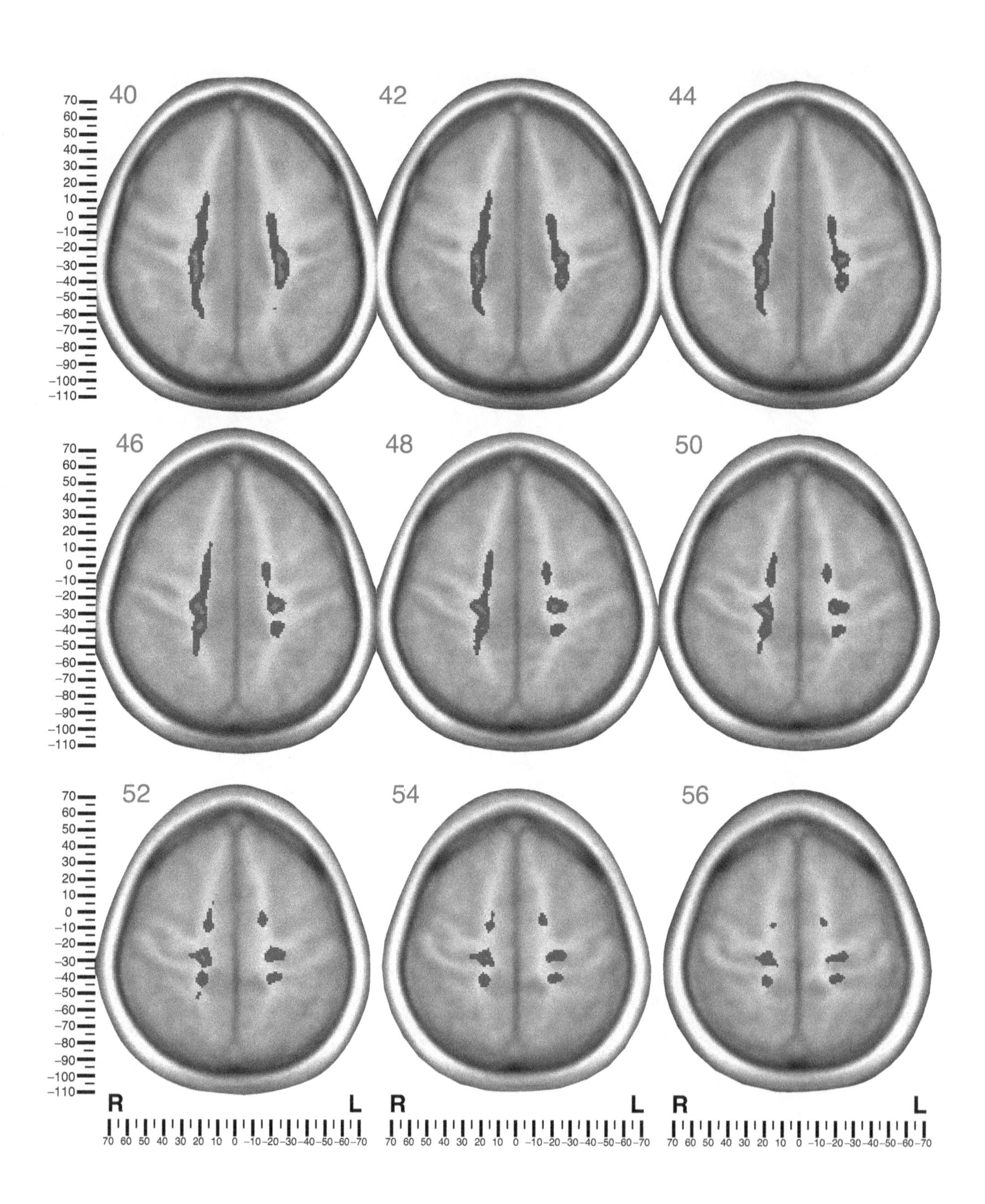

Internal capsule: anterior coronal view

p < .05 | > 50% | > 75% | > 90%

32 30 28 26 24 22 20 18 16 14 12 10

R L

Internal capsule: anterior coronal view

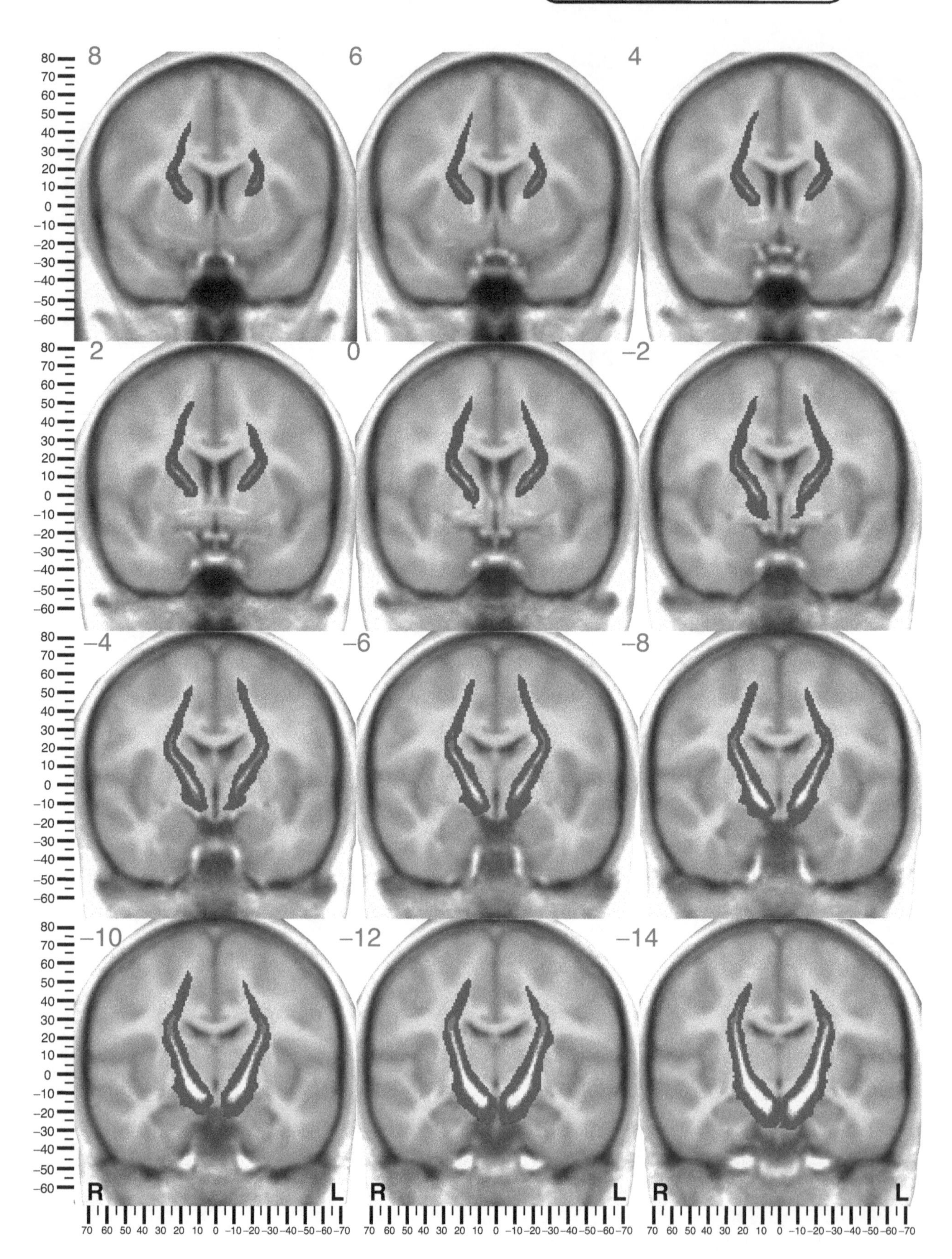

Internal capsule: anterior coronal view

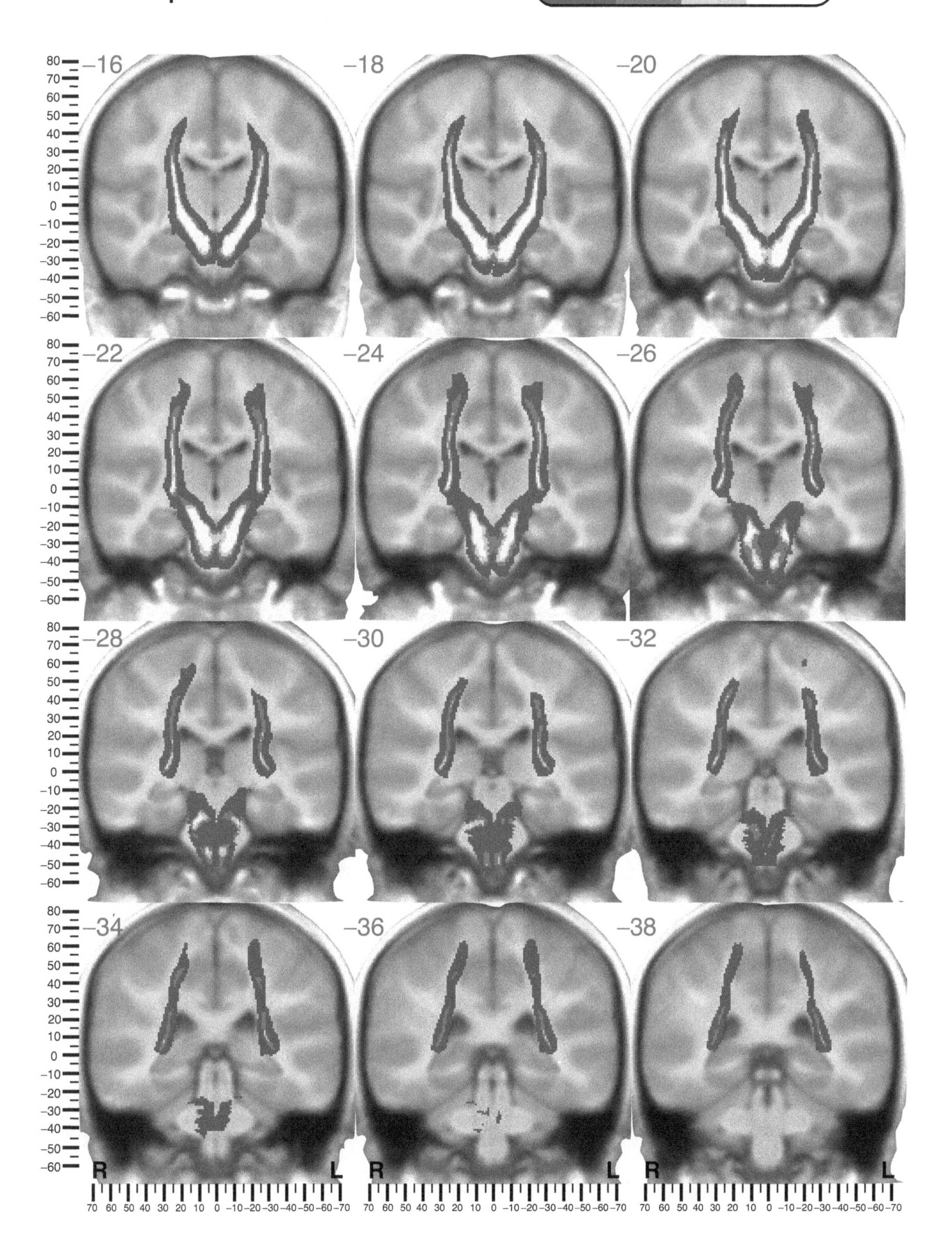

Internal capsule: anterior coronal view

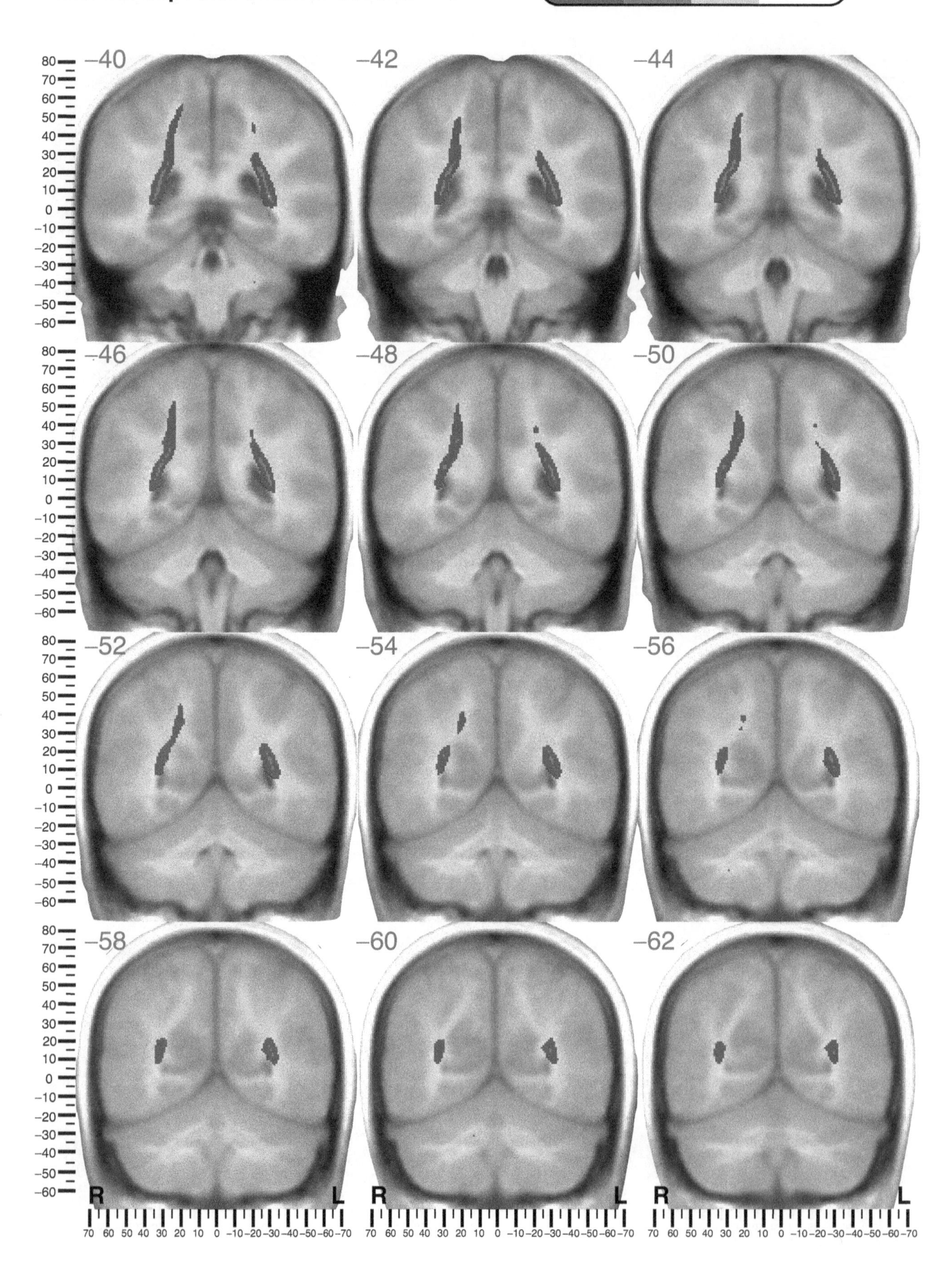

Internal capsule: anterior coronal view

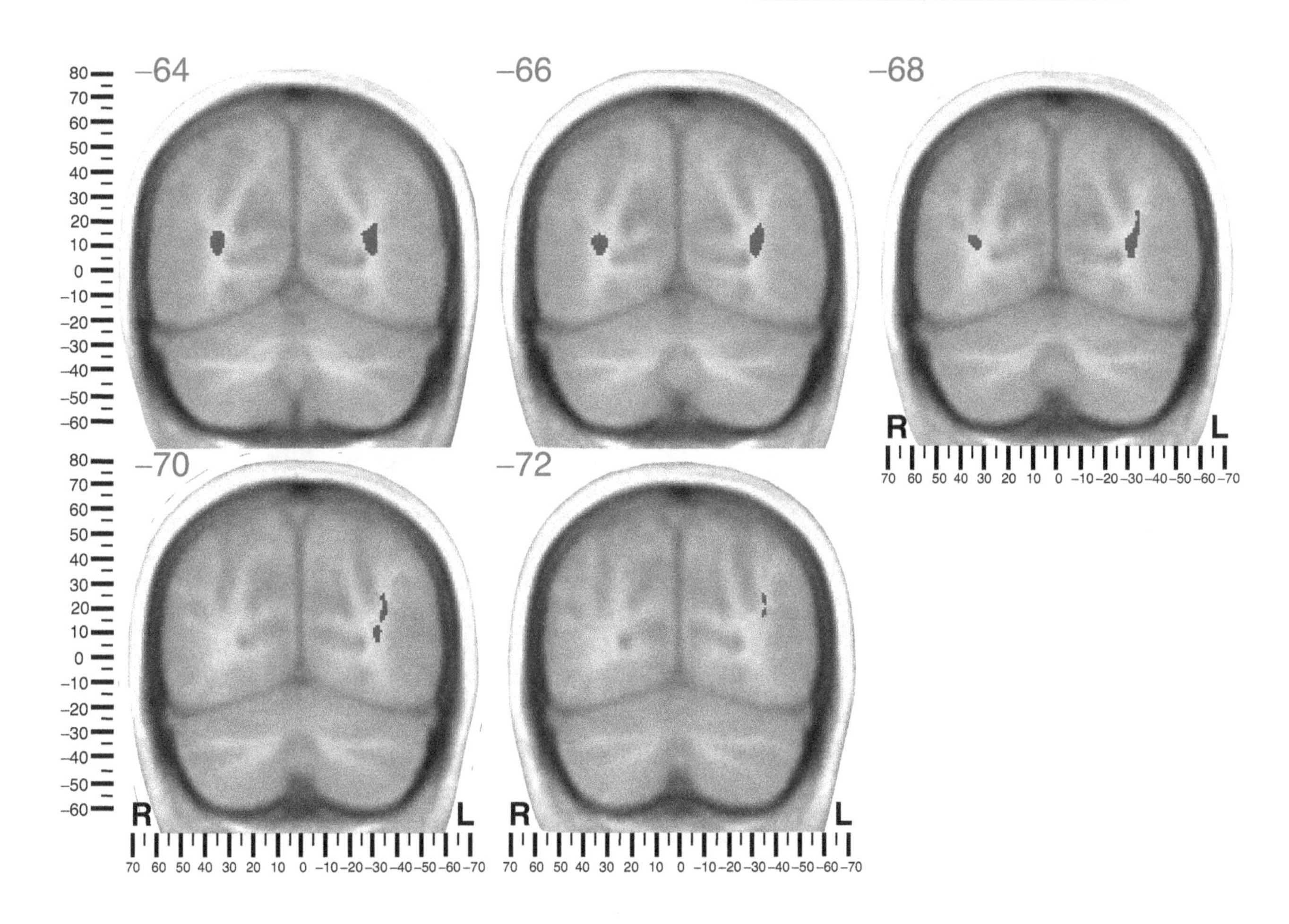

Internal capsule (right): lateral sagittal view

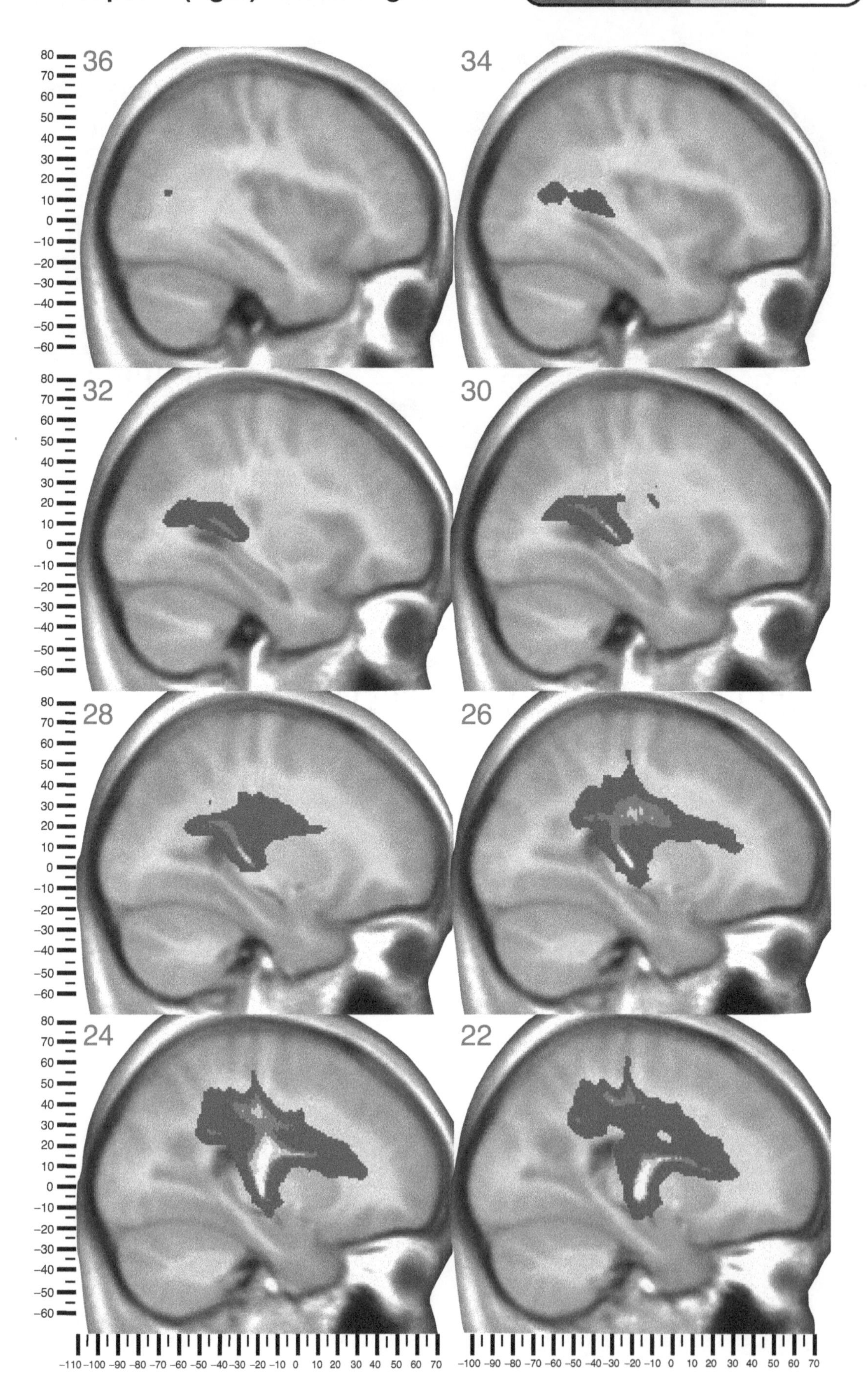

Internal capsule (right): lateral sagittal view

p < .05 | > 50% | > 75% | > 90%

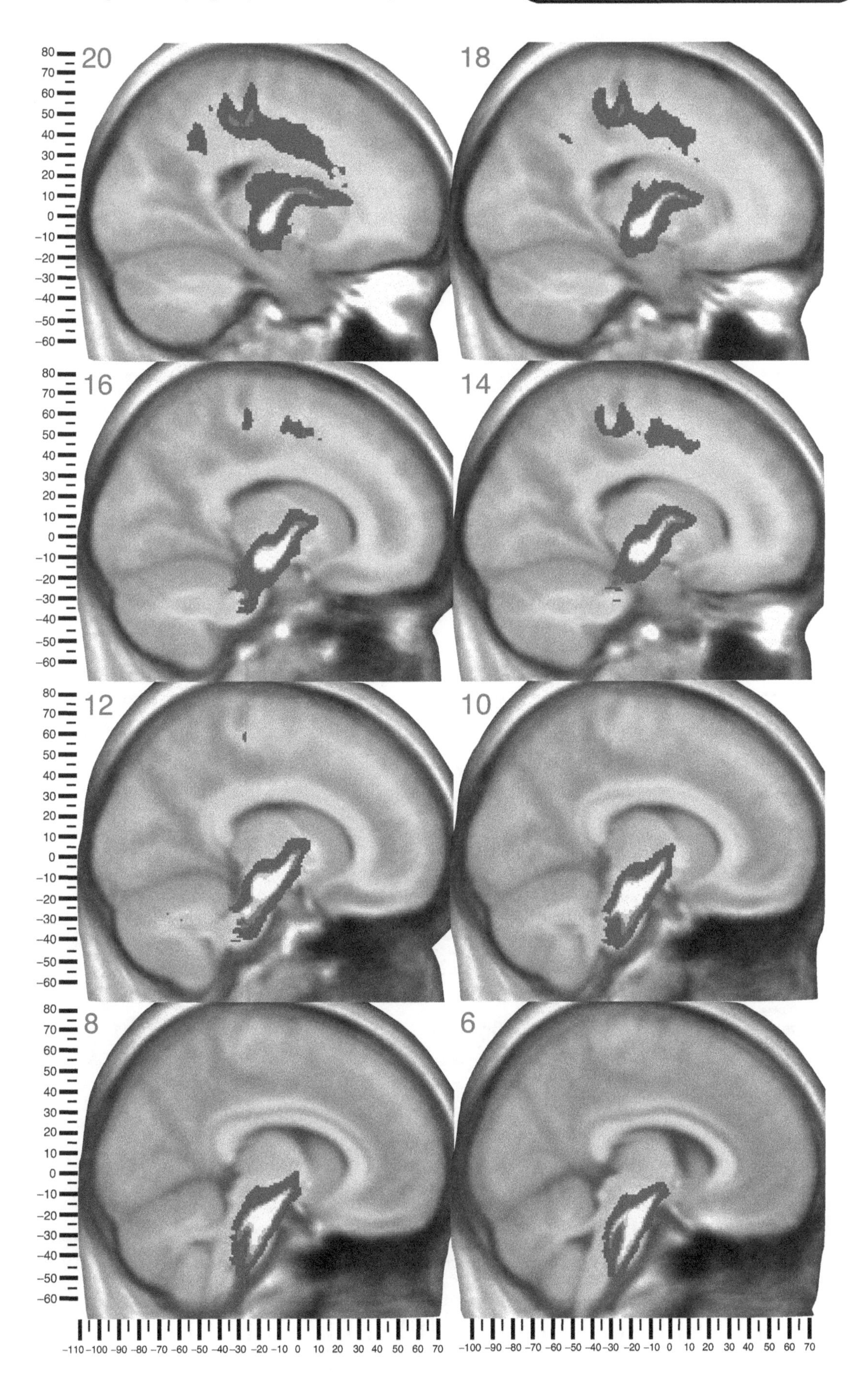

Internal capsule (left): medial sagittal view

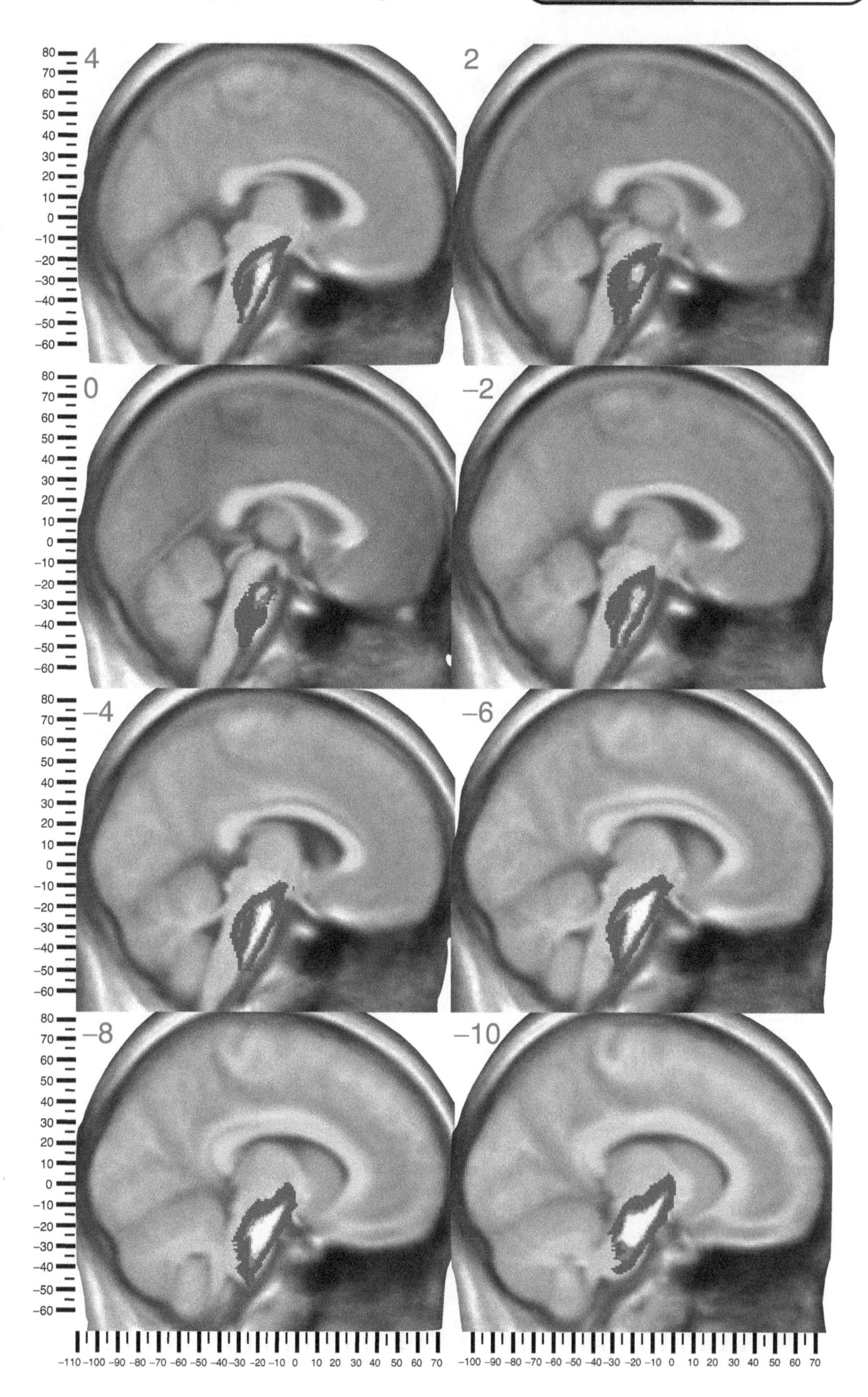

Internal capsule (left): medial sagittal view

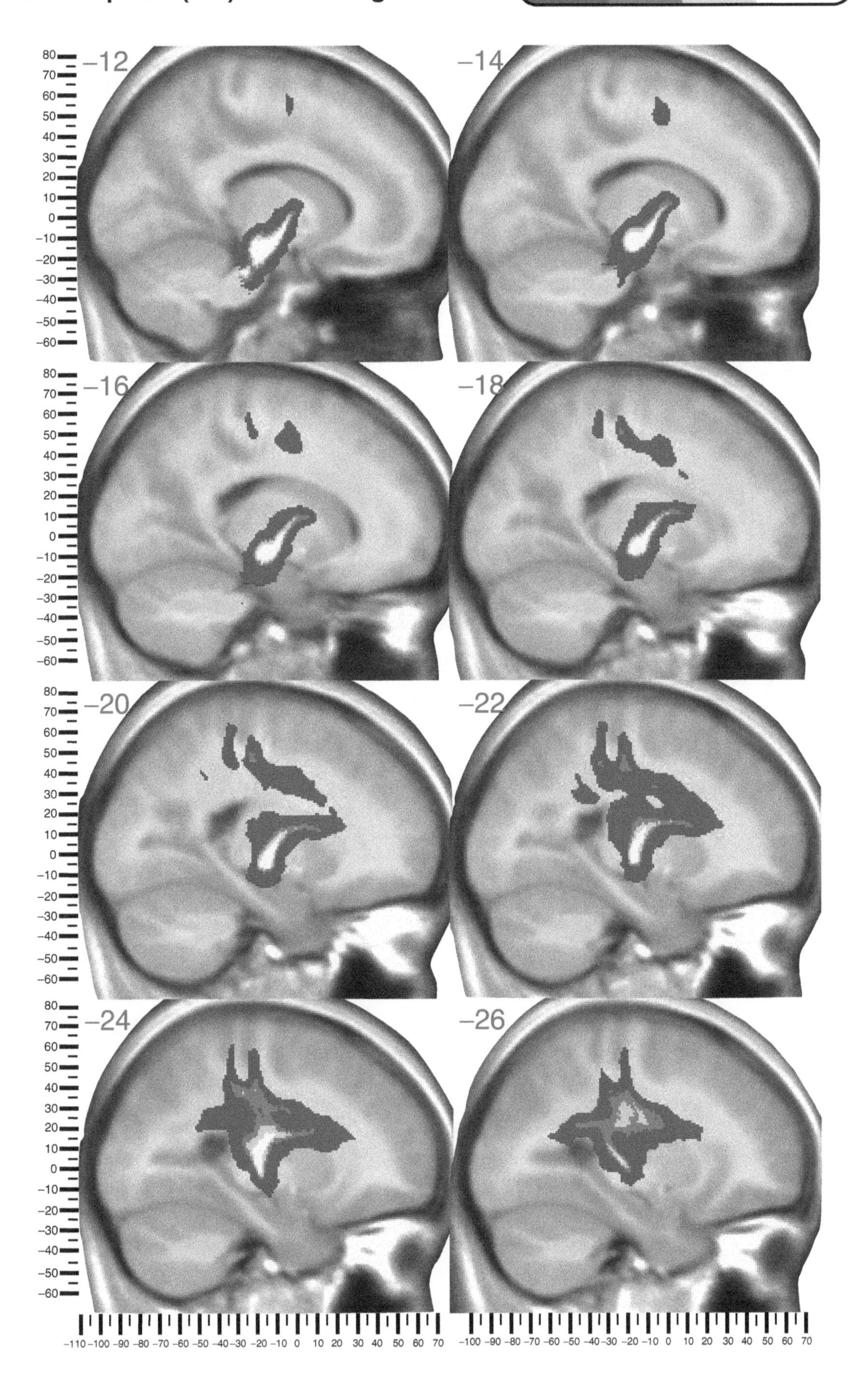

Internal capsule (left): medial sagittal view

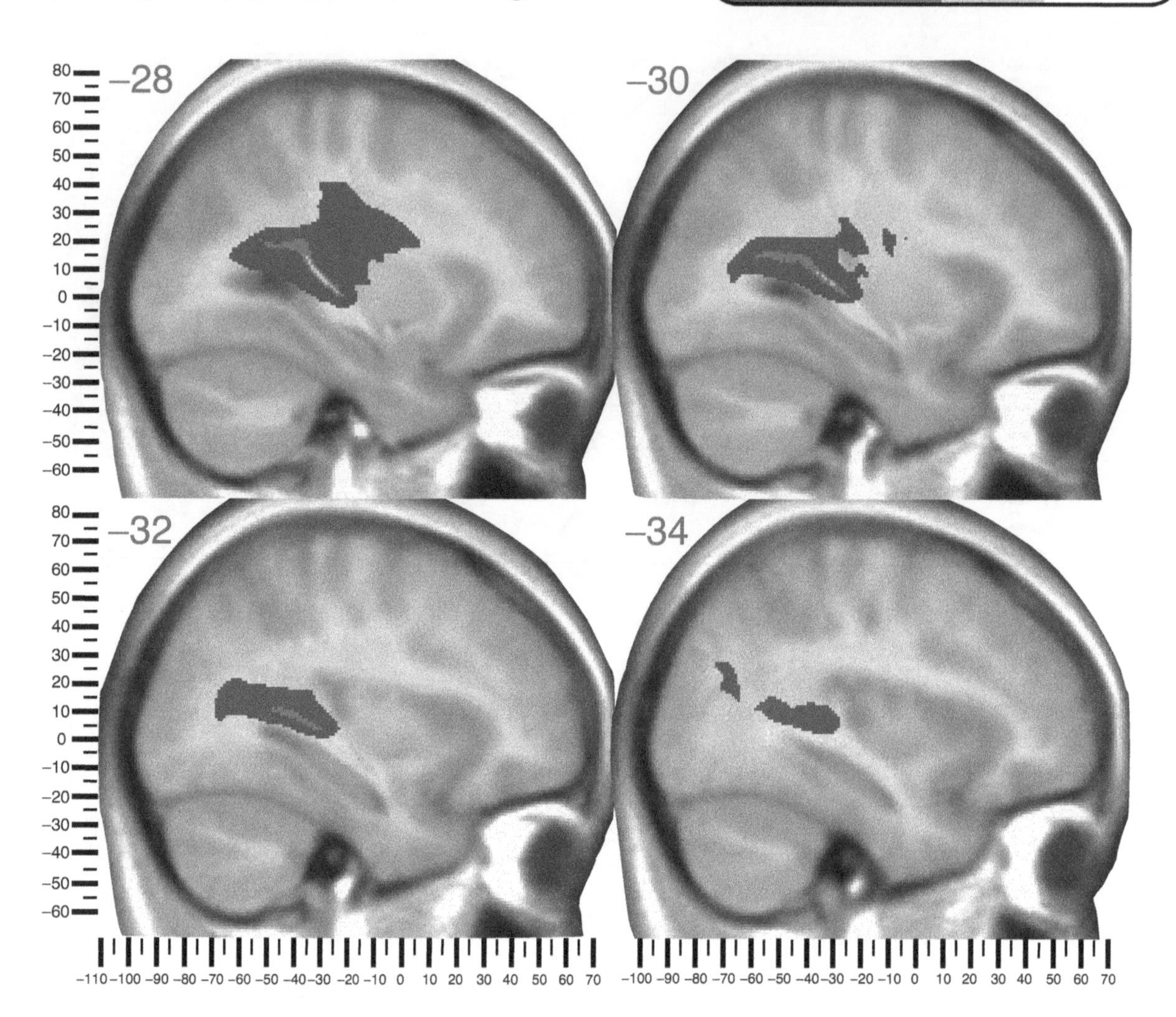

Cortico-ponto-cerebellar tract (atlas)

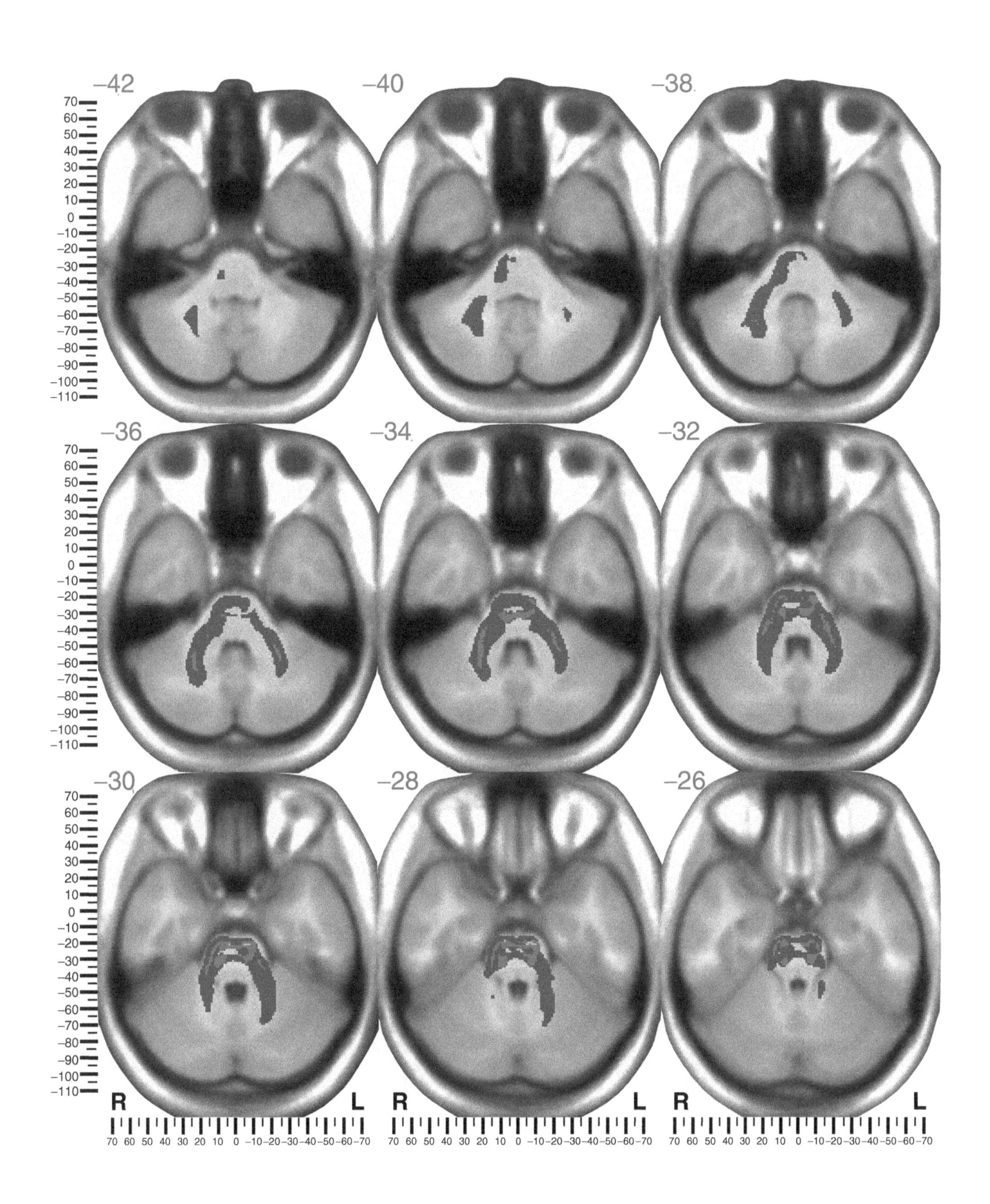

Cortico-ponto-cerebellar: inferior axial view

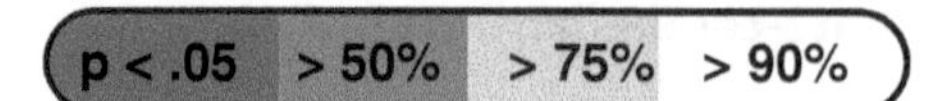

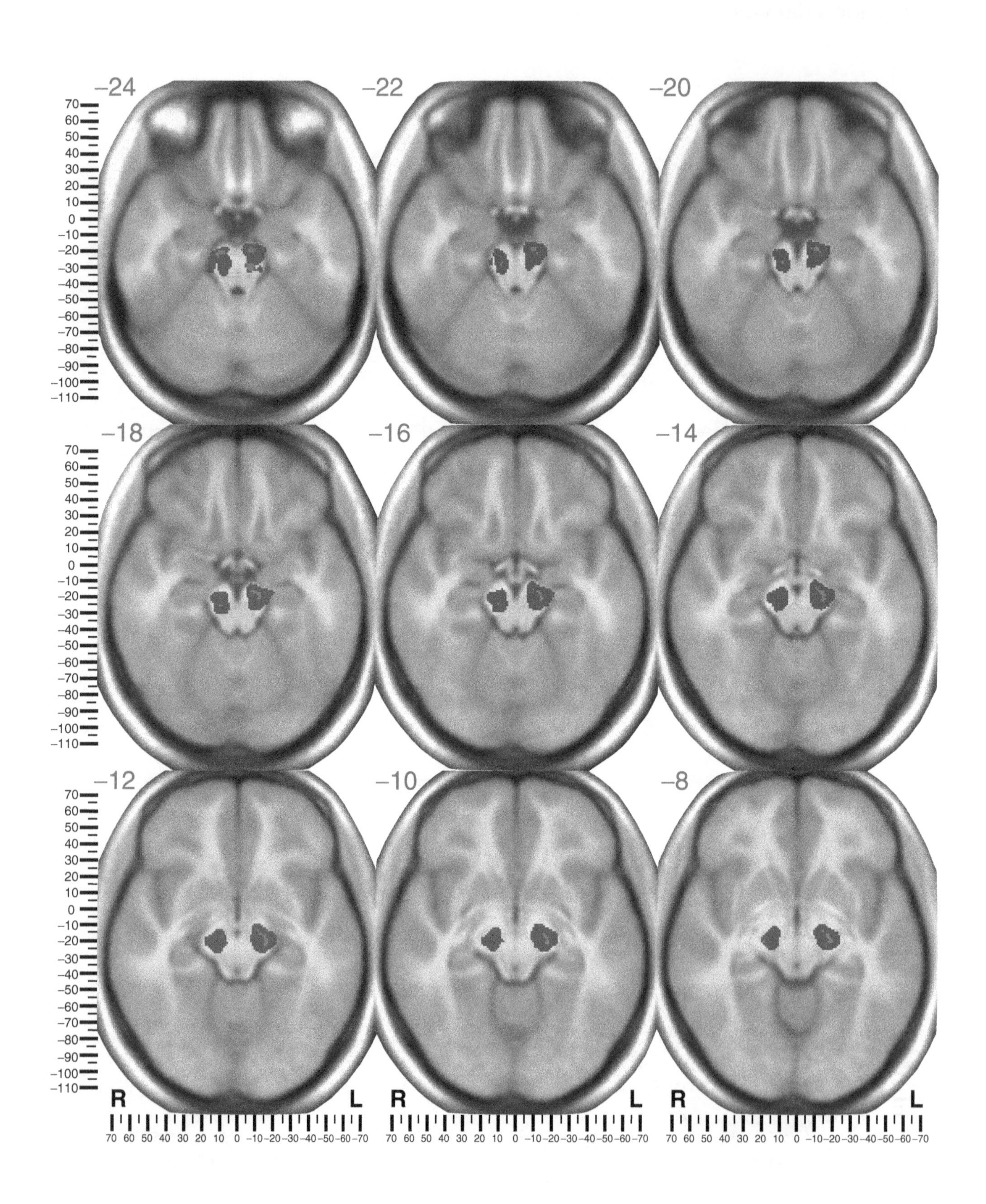

Cortico-ponto-cerebellar: inferior axial view

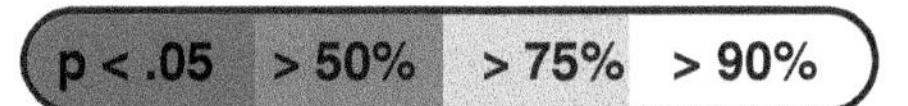

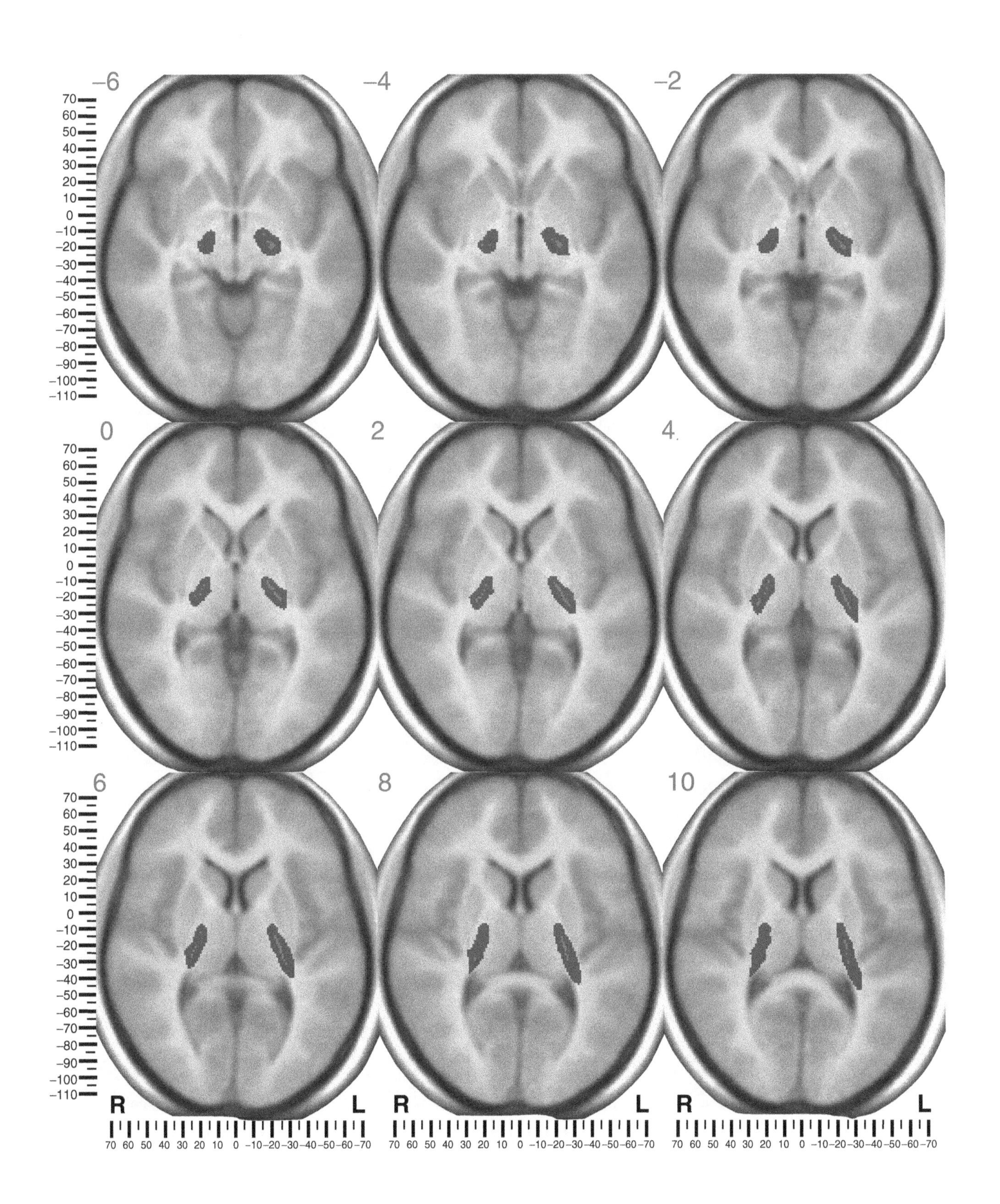

Cortico-ponto-cerebellar: inferior axial view

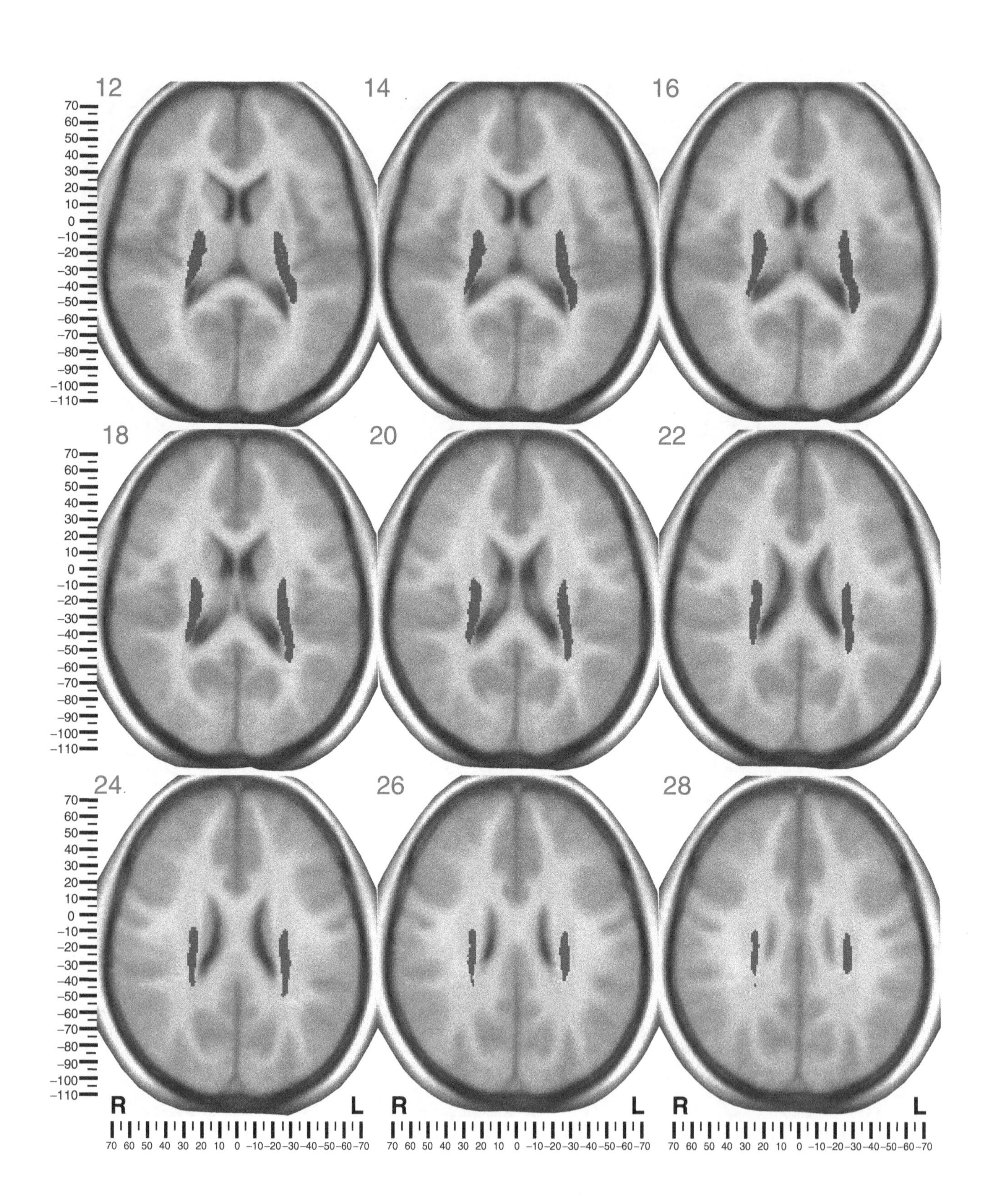

Cortico-ponto-cerebellar: inferior axial view

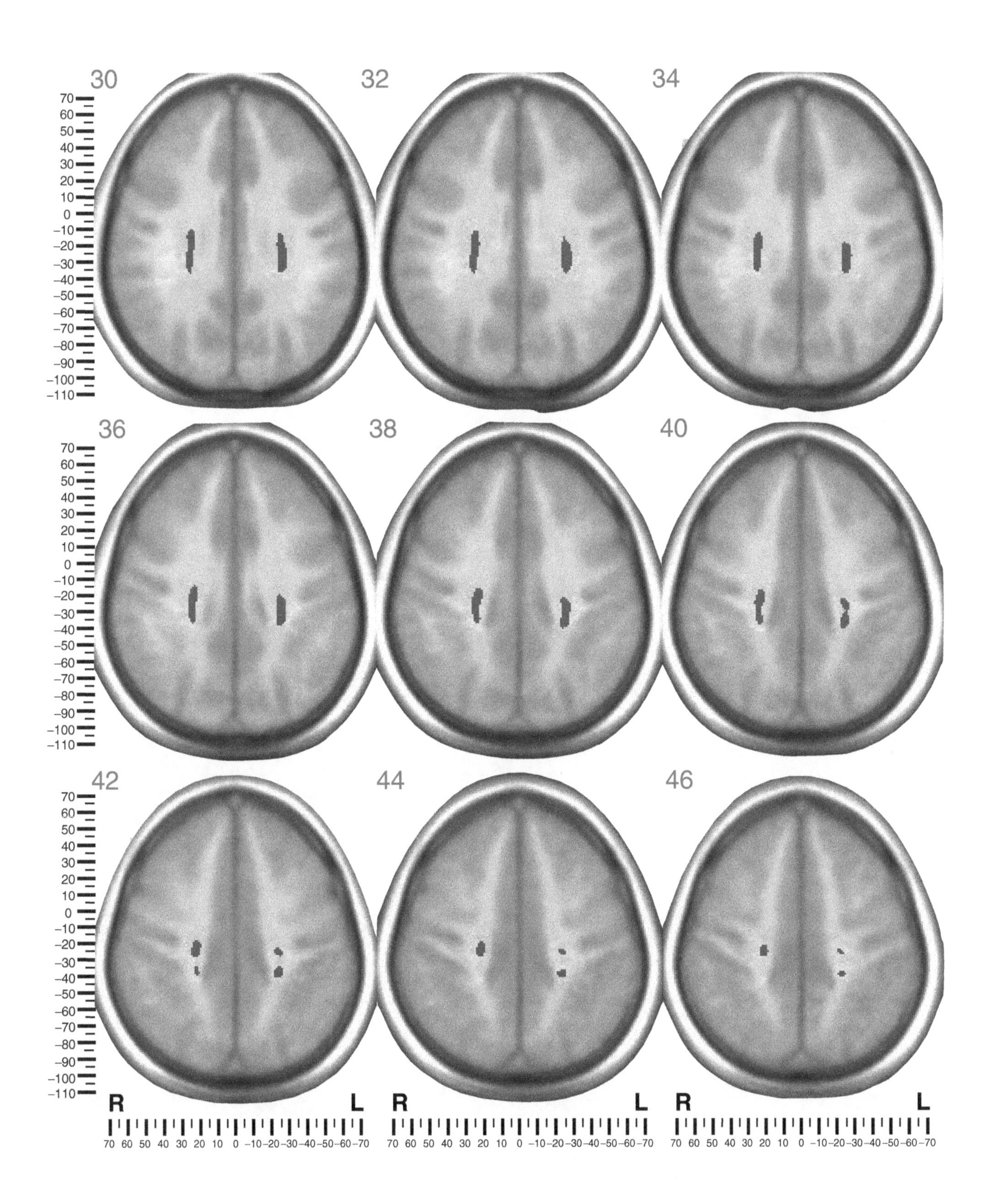

Cortico-ponto-cerebellar: anterior coronal view

p < .05 | > 50% | > 75% | > 90%

−6 −8 −10 −12 −14 −16 −18 −20 −22 −24 −26 −28

80 70 60 50 40 30 20 10 0 −10 −20 −30 −40 −50 −60

R L R L R L

70 60 50 40 30 20 10 0 −10 −20 −30 −40 −50 −60 −70

Cortico-ponto-cerebellar: anterior coronal view

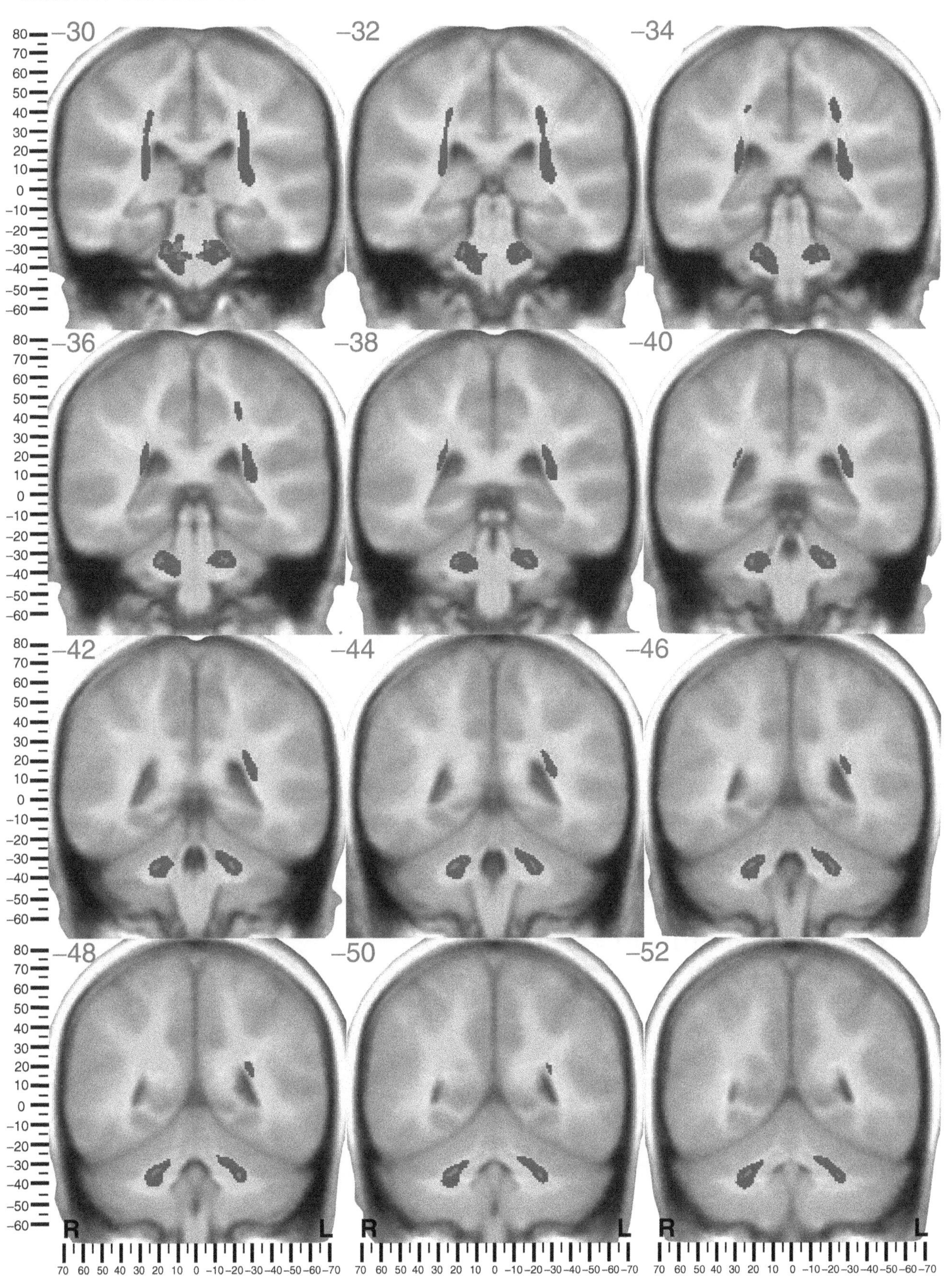

Cortico-ponto-cerebellar: anterior coronal view

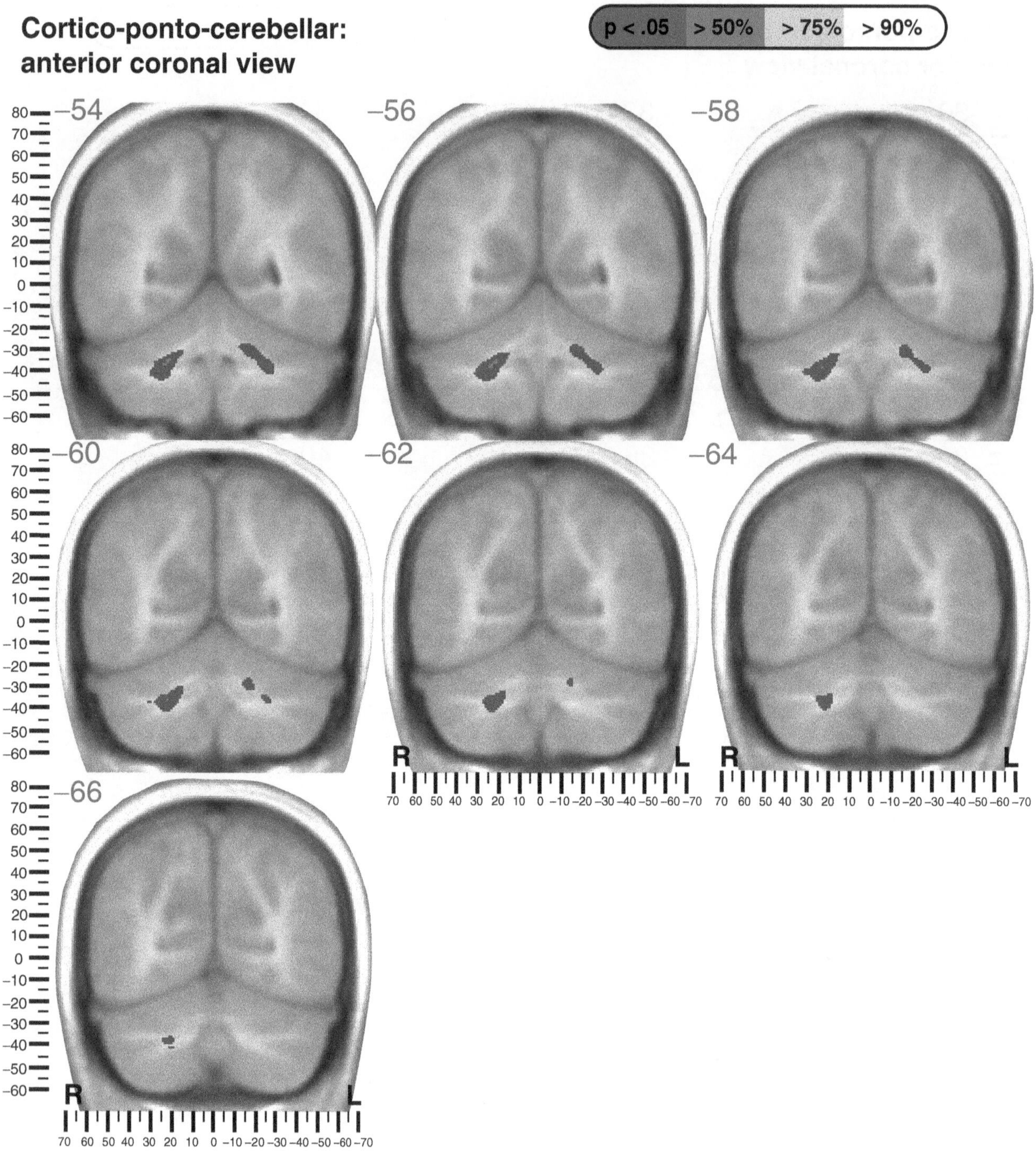

Cortico-ponto-cerebellar: lateral sagittal view

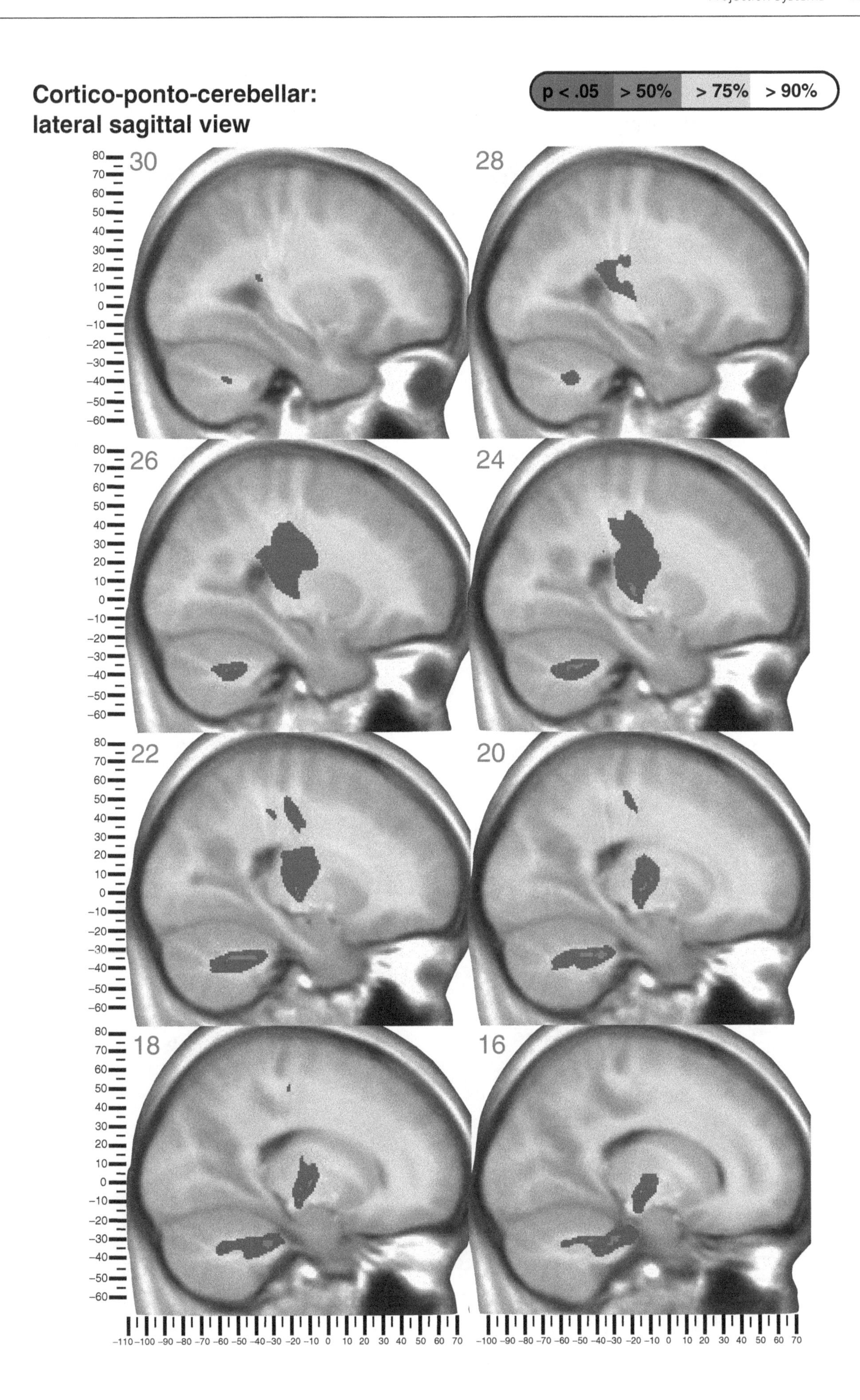

Cortico-ponto-cerebellar: lateral sagittal view

p < .05 | > 50% | > 75% | > 90%

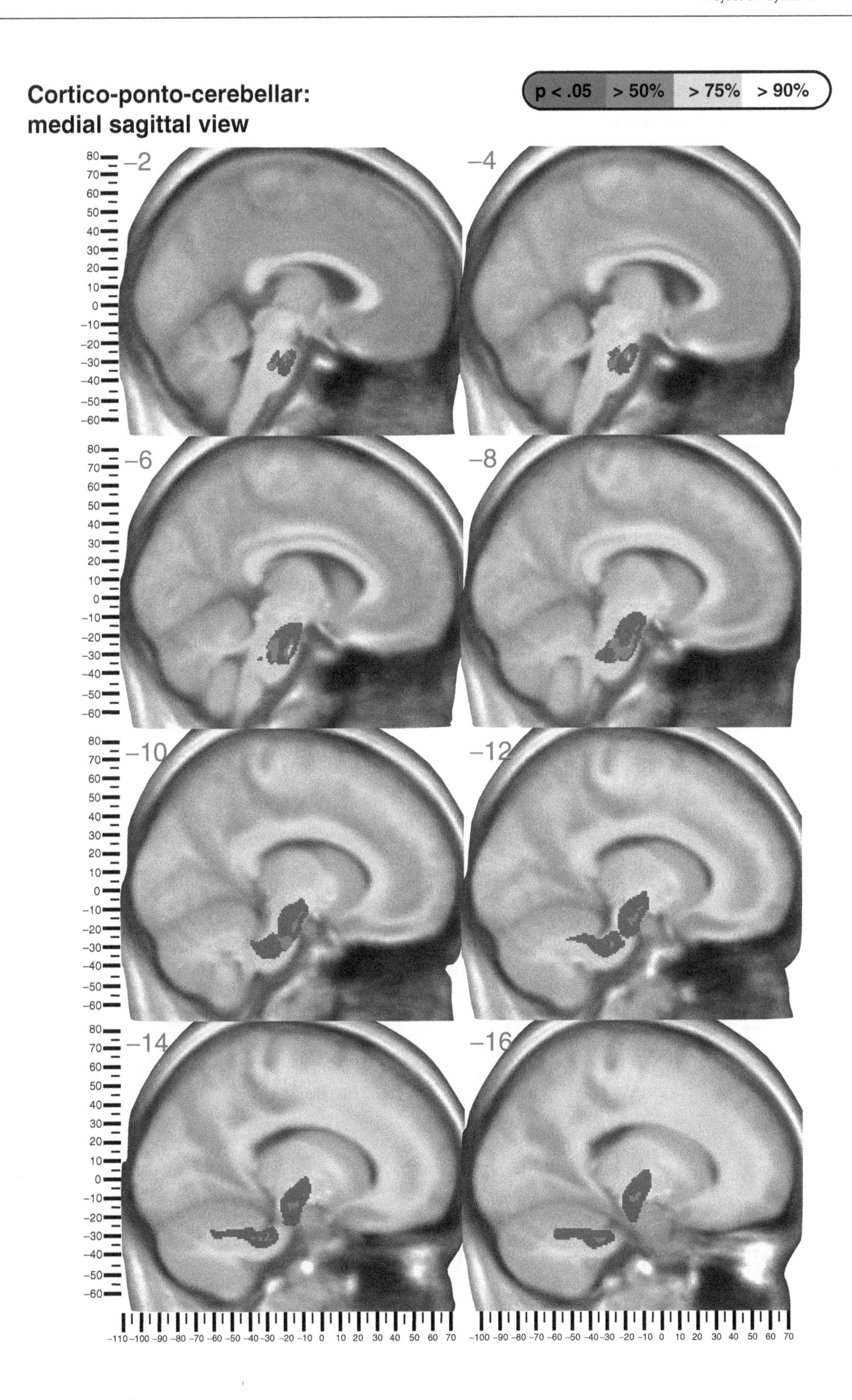
Cortico-ponto-cerebellar:
medial sagittal view
p < .05
> 50%
> 75%
> 90%
−2
−4
−6
−8
−10
−12
−14
−16

Cortico-ponto-cerebellar: medial sagittal view

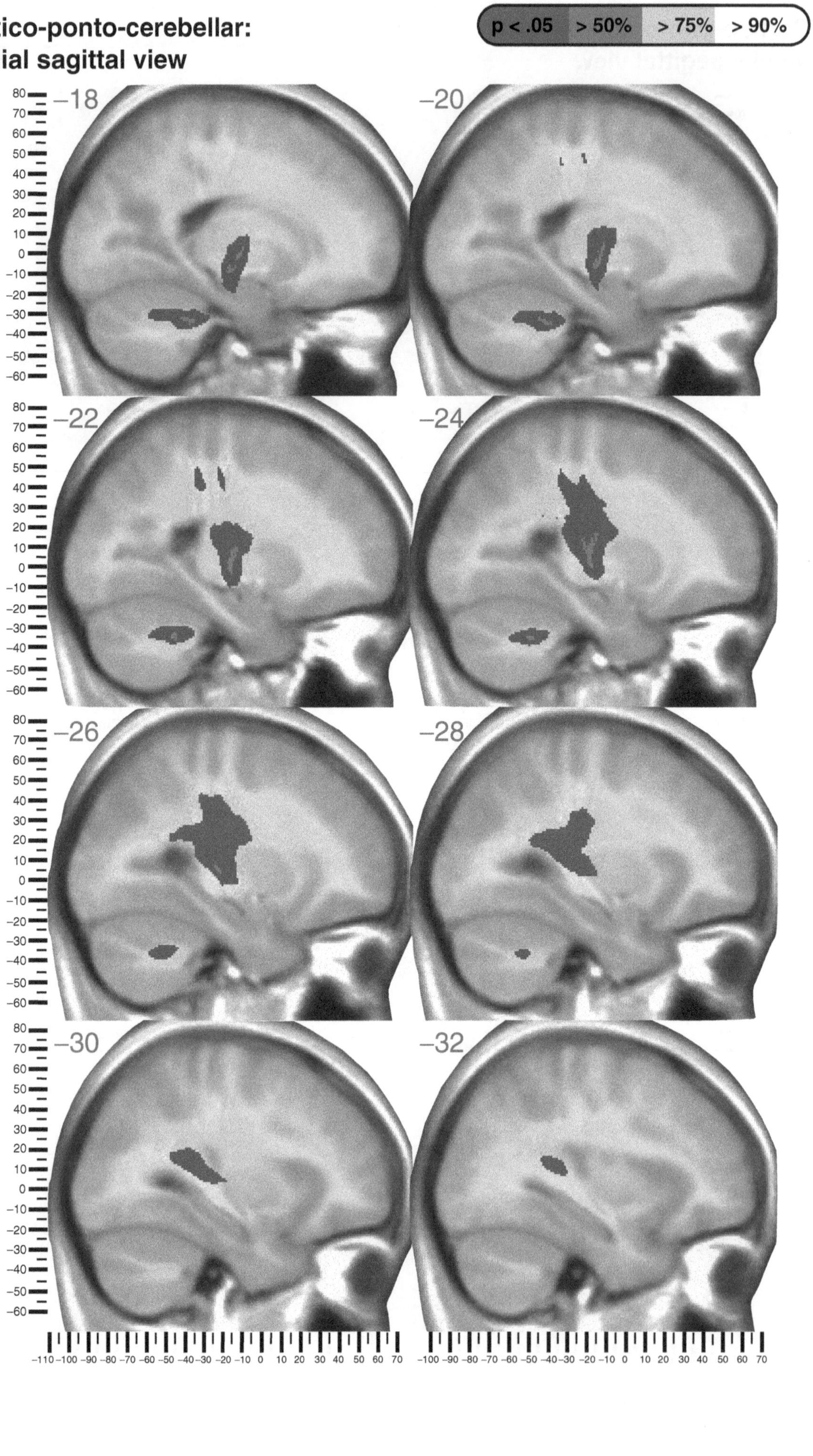

p < .05 | > 50% | > 75% | > 90%

-18 -20 -22 -24 -26 -28 -30 -32

Inferior cerebellar peduncle (atlas)

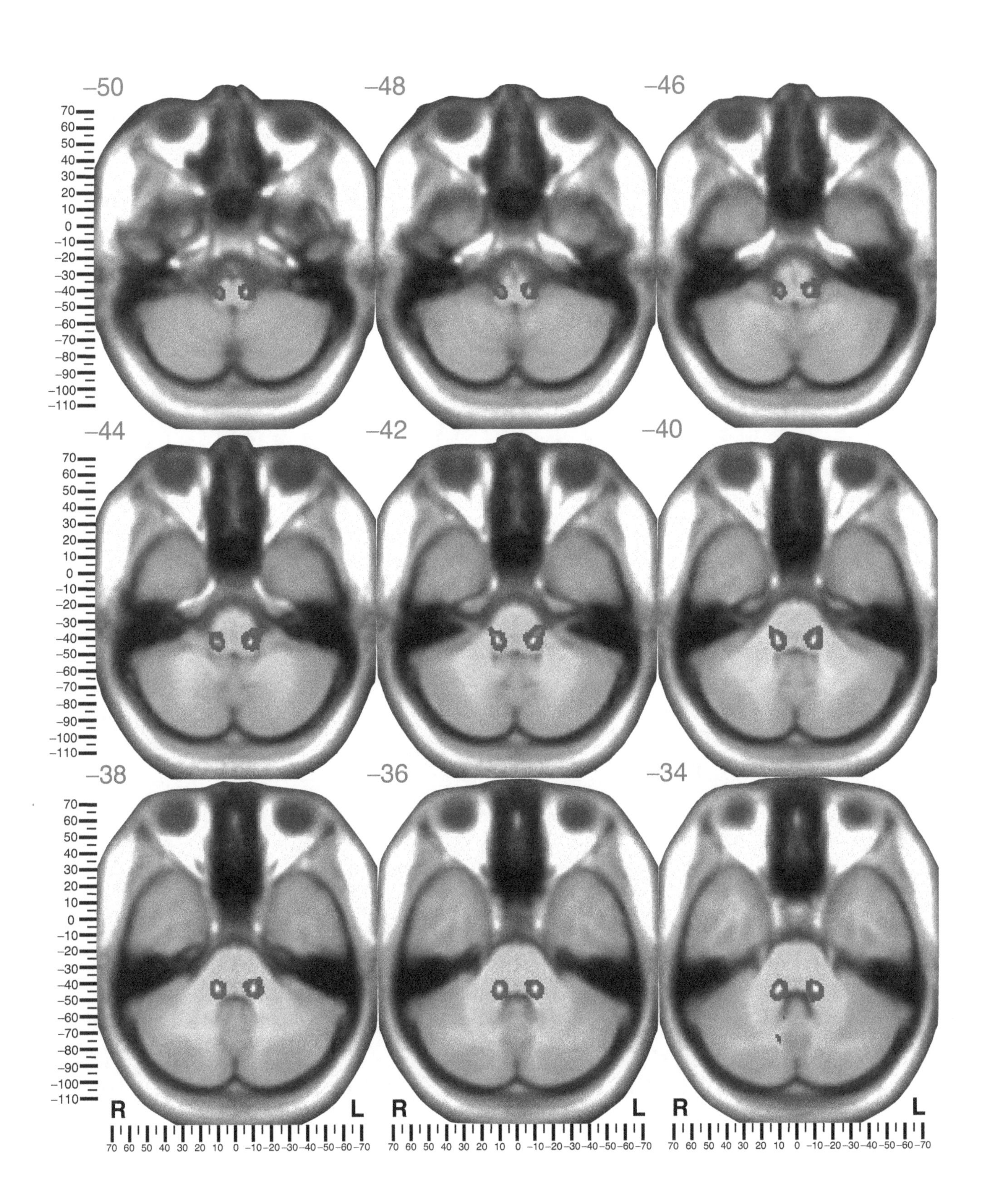

Inferior cerebellar pedunculus: inferior axial view

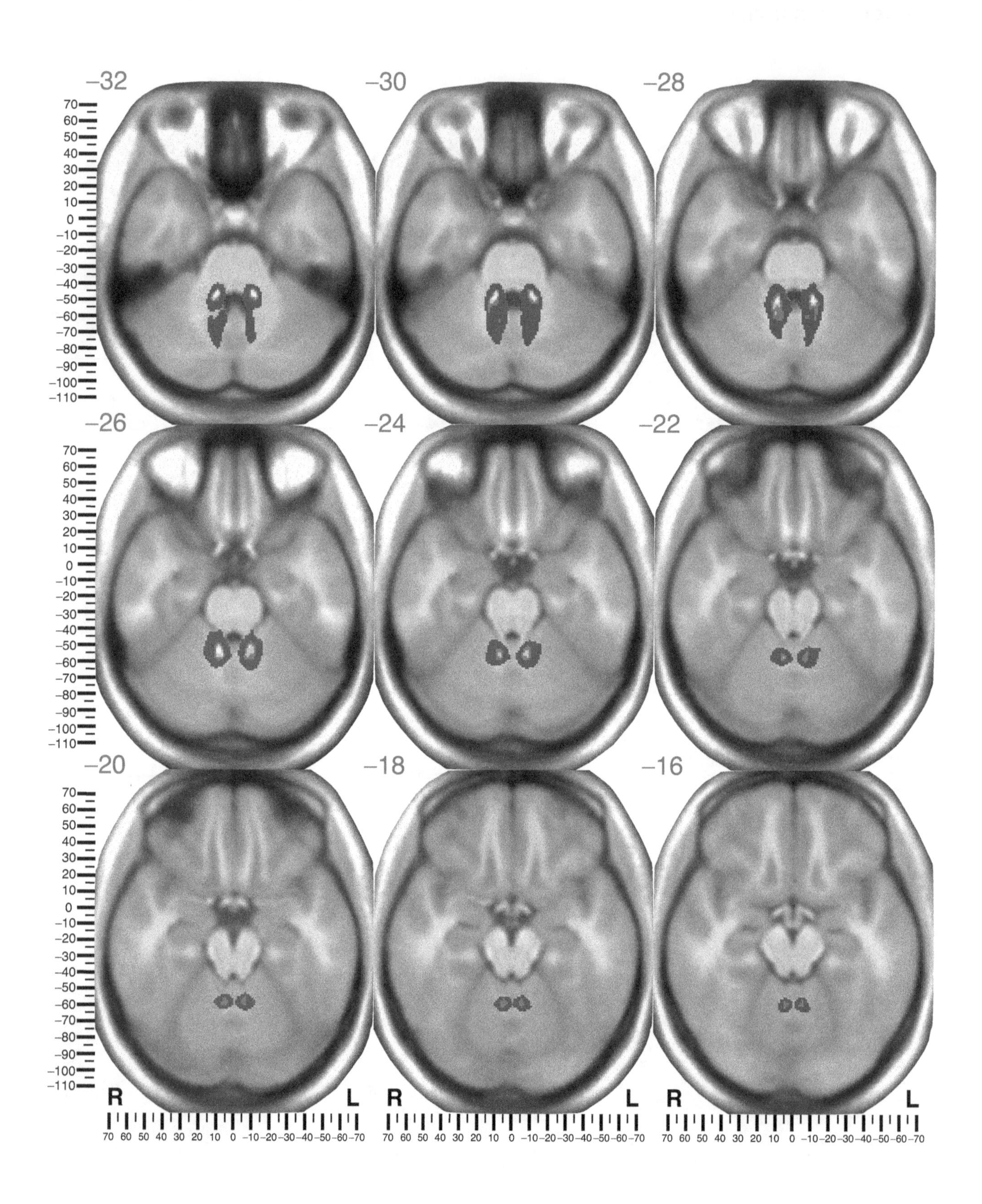

Inferior cerebellar pedunculus: anterior coronal view

p < .05 | > 50% | > 75% | > 90%

−30 −32 −34

−36 −38 −40

−42 −44 −46

−48 −50 −52

R L

70 60 50 40 30 20 10 0 −10 −20 −30 −40 −50 −60 −70

Inferior cerebellar pedunculus: anterior coronal view

p < .05 | > 50% | > 75% | > 90%

Inferior cerebellar pedunculus (right): lateral sagittal view

p < .05 | > 50% | > 75% | > 90%

16 14 12 10 8 6 4 2

Inferior cerebellar pedunculus (left): medial sagittal view

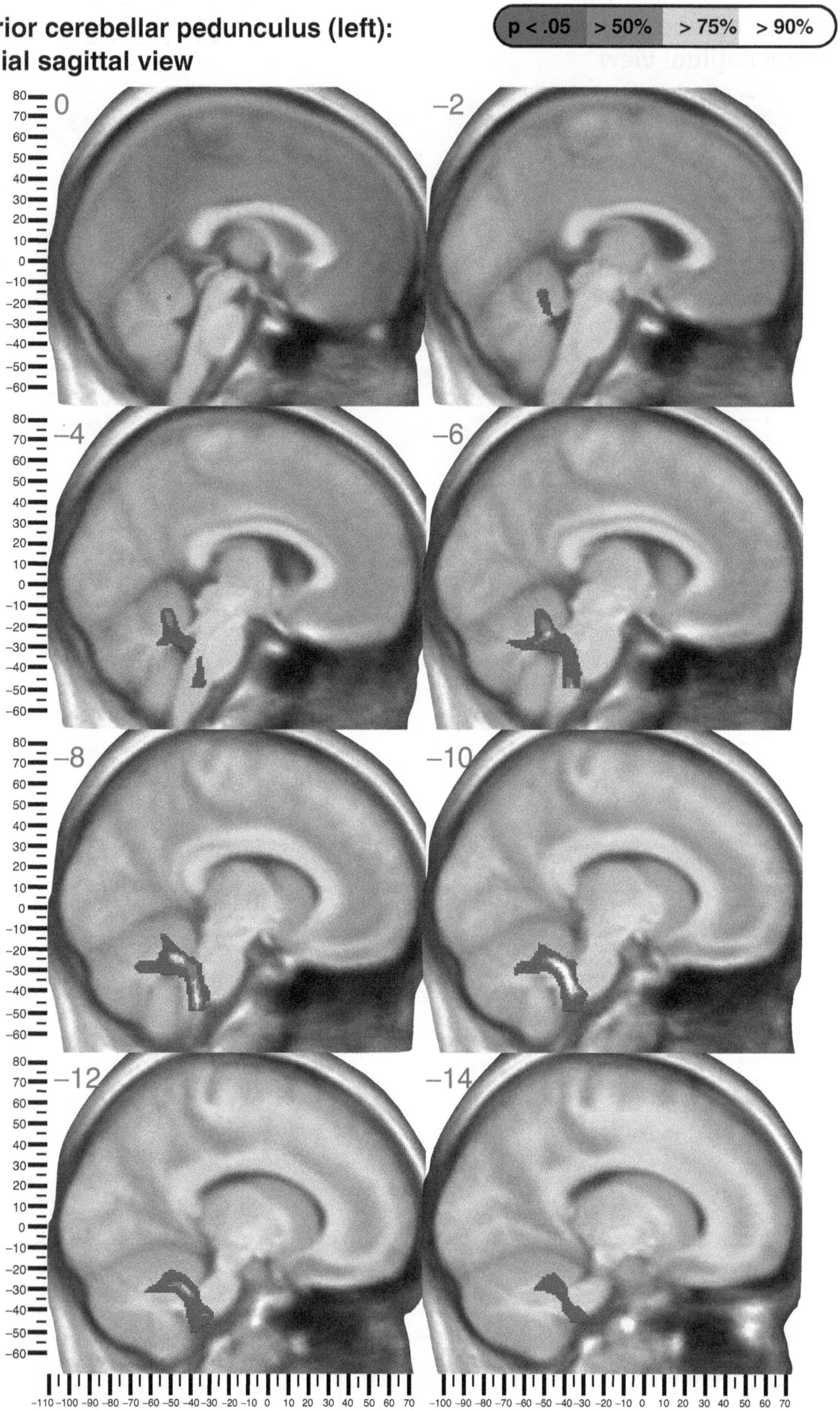

p < .05 | > 50% | > 75% | > 90%

0

−2

−4

−6

−8

−10

−12

−14

Superior cerebellar peduncle (atlas)

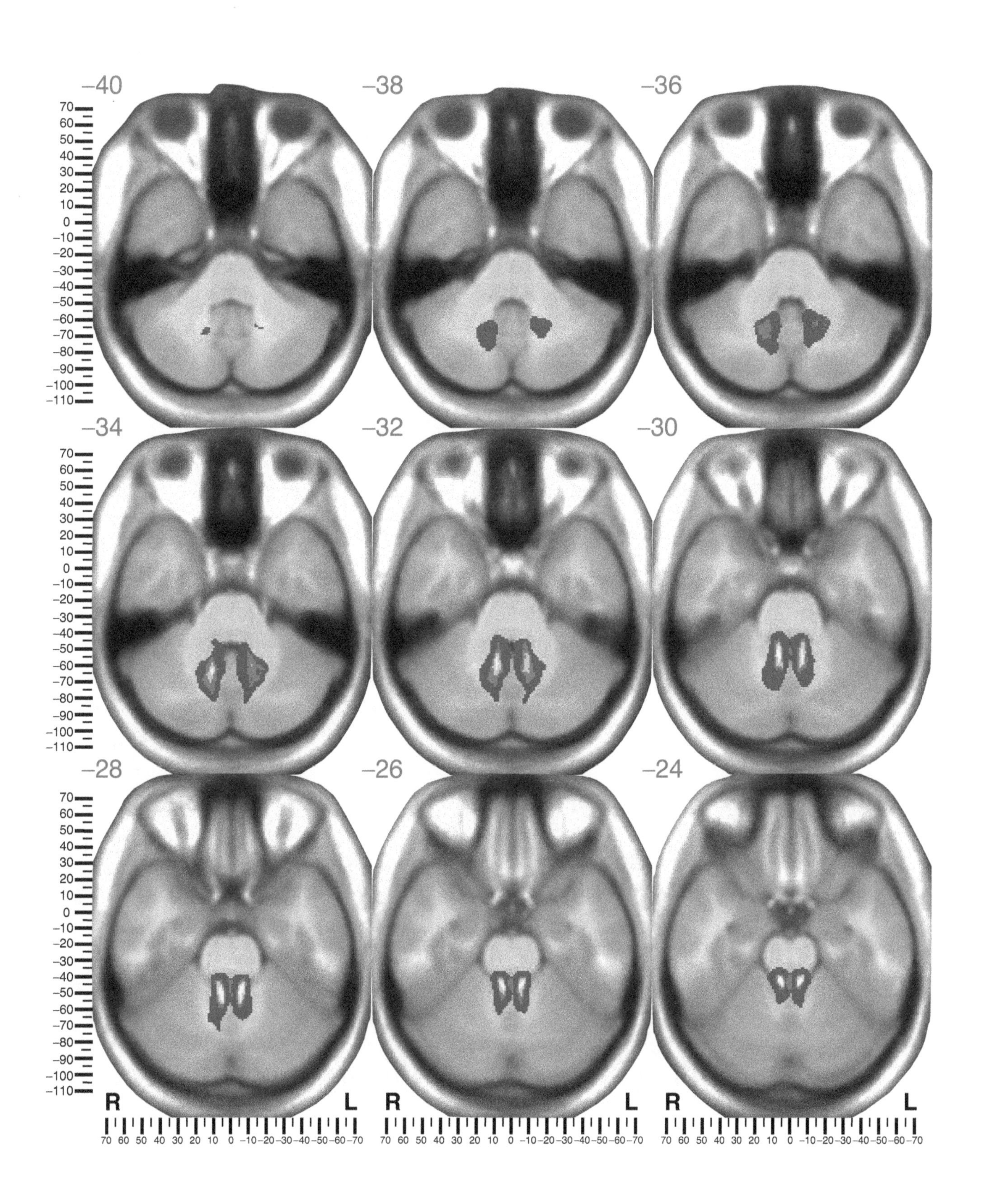

Superior cerebellar pedunculus: inferior axial view

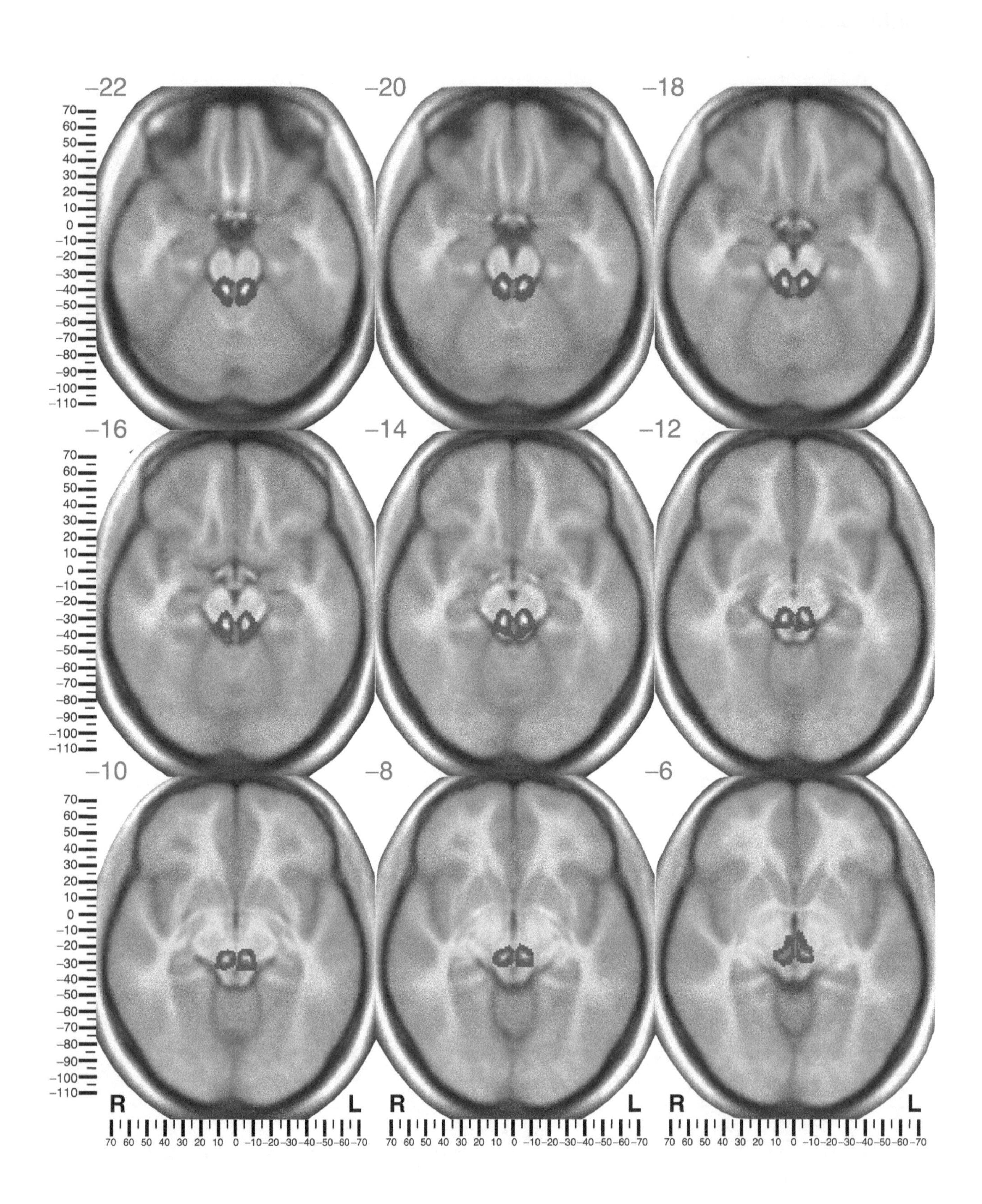

Superior cerebellar pedunculus: anterior coronal view

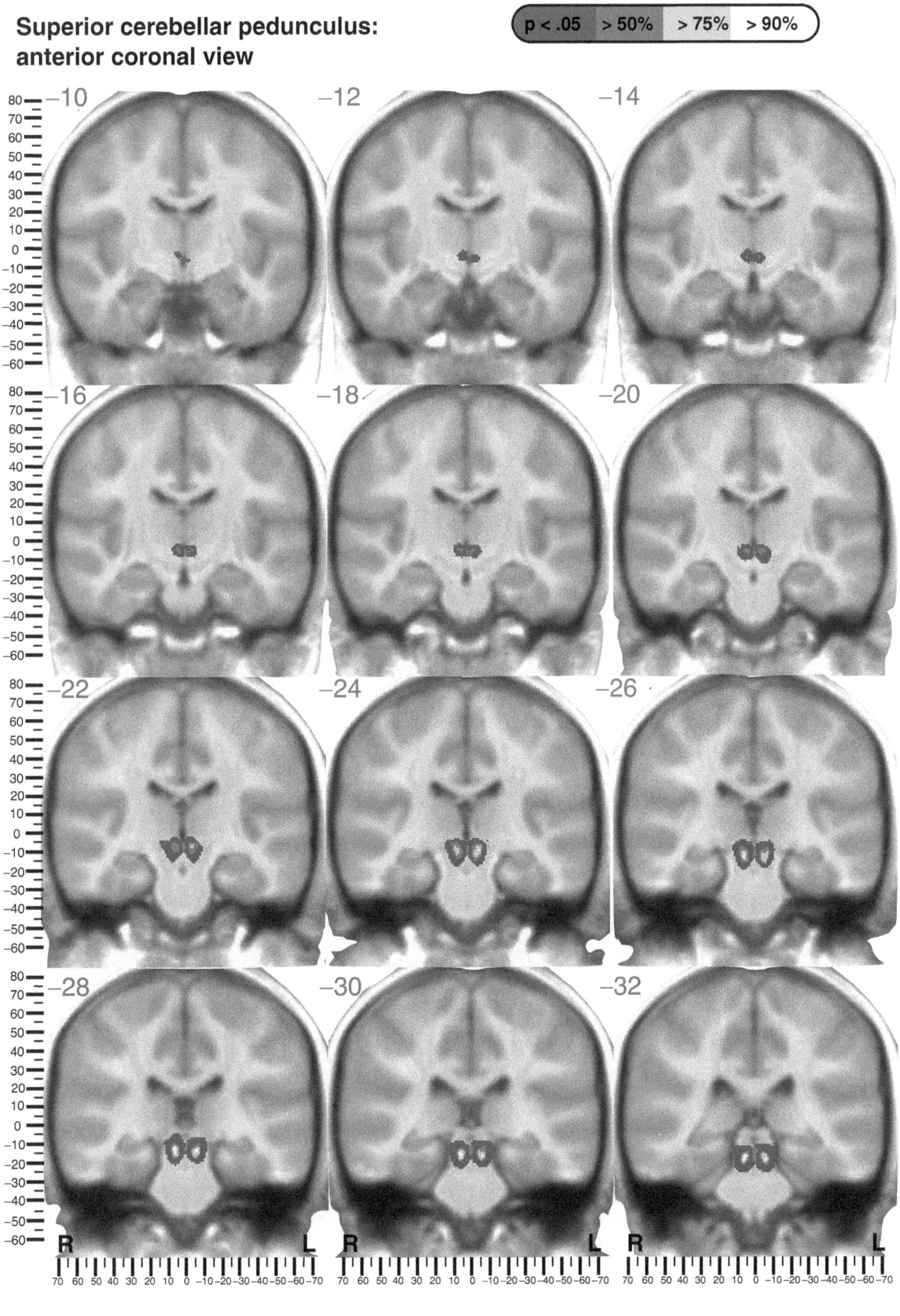

p < .05 | > 50% | > 75% | > 90%

-10 -12 -14 -16 -18 -20 -22 -24 -26 -28 -30 -32

80 70 60 50 40 30 20 10 0 -10 -20 -30 -40 -50 -60

R L

70 60 50 40 30 20 10 0 -10 -20 -30 -40 -50 -60 -70

Superior cerebellar pedunculus: anterior coronal view

p < .05 | > 50% | > 75% | > 90%

Superior cerebellar pedunculus: anterior coronal view

p < .05 | > 50% | > 75% | > 90%

–58 –60 –62 –64 –66 –68 –70 –72

R L

80 70 60 50 40 30 20 10 0 –10 –20 –30 –40 –50 –60

70 60 50 40 30 20 10 0 –10 –20 –30 –40 –50 –60 –70

Superior cerebellar pedunculus (right): lateral sagittal view

p < .05 | > 50% | > 75% | > 90%

20 18 16 14 12 10 8 6

Superior cerebellar pedunculus (right and left): medial sagittal view

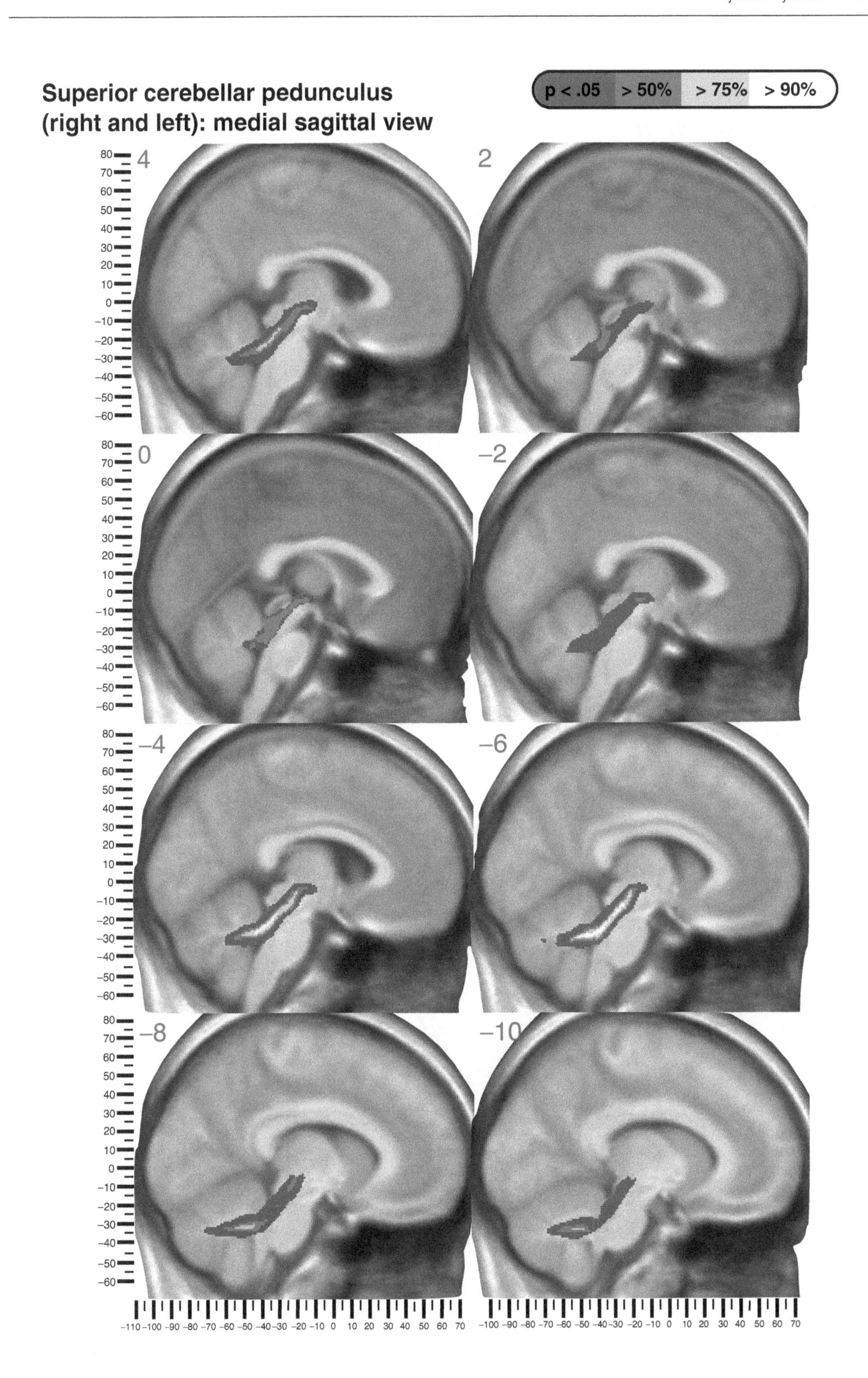

Superior cerebellar pedunculus (left): medial sagittal view

p < .05 | > 50% | > 75% | > 90%

-12

-14

-16

-18

-20

CHAPTER 11

Limbic System

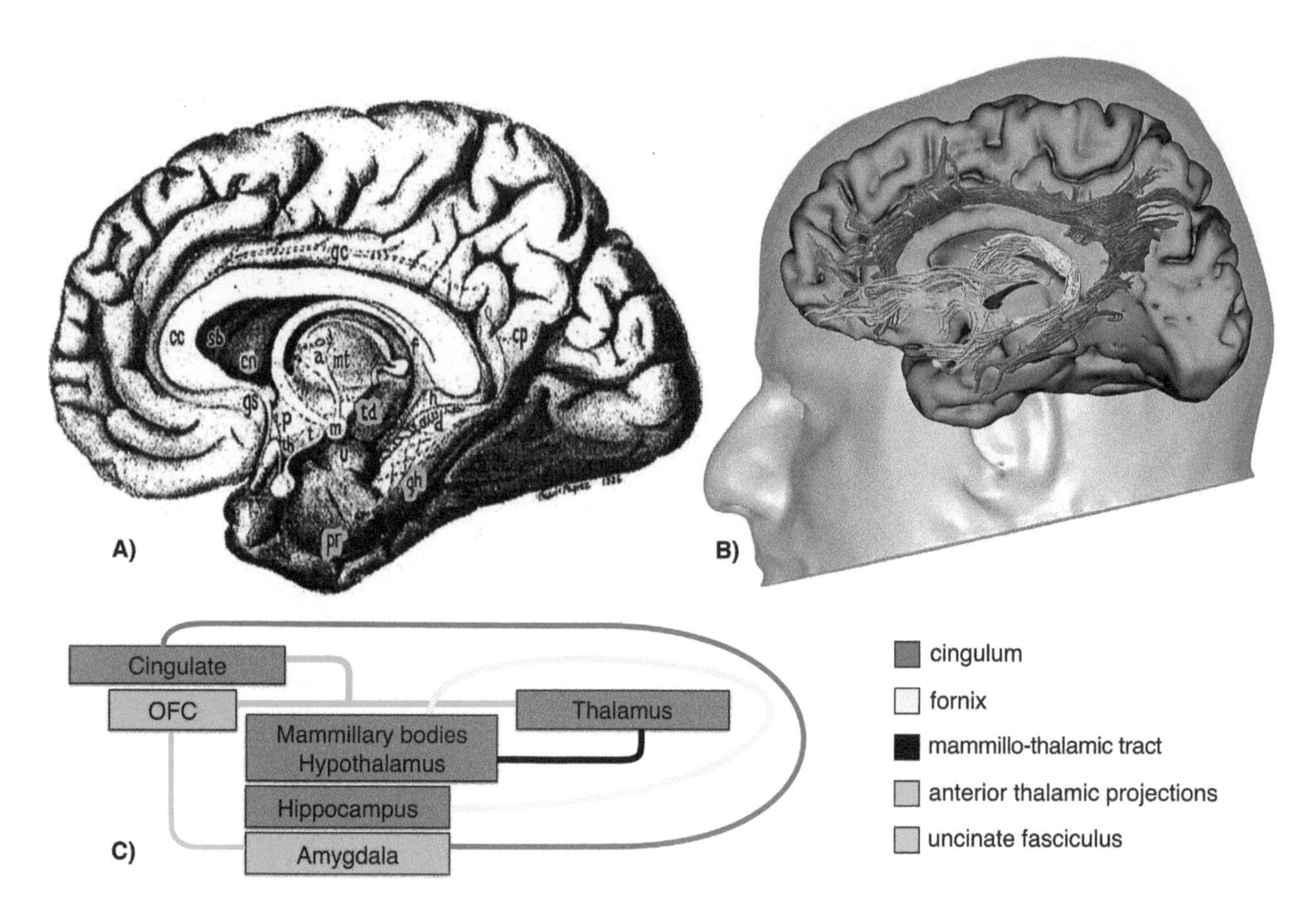

Figure 11.1 Representations of the limbic system, including comparison of (A) the original circuit proposed by Papez in 1937 and (B) current *in vivo* tractography reconstructions. C) A diagram of limbic interconnections based on McLean's 1952 proposal for a unitary model consisting of Papez circuit (grey boxes connected through the fornix, mamillo-thalamic tract, anterior thalamic projections, and cingulum) and Yakovlev's amygdala-orbitofrontal network (orange boxes); OFC, orbitofrontal cortex. The colours in (B) and (C) correspond to the tracts in the legend. a, anterior nucleus; cc, corpus callosum; cn, caudate nucleus; cp, cingulum posterior; d, gyrus dentatus; f, fornix; gc, gyrus cinguli; gh, gyrus hippocampi; gs, gyrus subcallosus; h, hippocampus; m, mammillary body; mt, mammillo-thalamic tract; p, pars optica hipothalami; pr, piriform area; sb, subcallosal bundle; t, tuber cinereum; td, tractus mammillo-tegmentalis; th, tractus hypophyseus; u, uncus.

Introduction

The limbic system is a group of interconnected cortical and subcortical structures dedicated to linking visceral states and emotion to cognition and behaviour (Mesulam, 2000). The use of the term 'limbic' has changed over time. Initially introduced by Thomas Willis (1664) to designate a cortical border encircling the brainstem (*limbus*, Latin 'border') the term has been used in modern neuroscience to indicate a progressively increasing number of regions dedicated to a wide range of functions (Mega et al., 1997; Marshall and Magoun, 1998). Paul Broca (1878) held the view that 'le grand lobe limbique' was mainly an olfactory structure common to all mammalian brains, although he argued that its functions were not limited to olfaction. After Broca's publication the accumulation of experimental evidence from ablation studies in animals broadened the role of the limbic structures to include other aspects of behaviour such as controlling social interactions, regulating predatory behaviour (Brown and Schäfer, 1888), consolidating memories (Bechterew, 1900), and forming emotions (Cannon, 1927). Anatomical and physiological advancements in the field led Christfield Jakob (1906) and James Papez (1937) to formulate the first unified network model for linking action and perception to emotion (Figure 11.1). According to Papez, *'emotion may arise in two ways: as a result of psychic activity and as consequence of hypothalamic activity'*. The psychic activity enters the circuit (later named after Papez) through the hippocampus while visceral and somatic perceptions enter the circuit through the hypothalamus. Thus, according to Papez:

Incitations of cortical origin would pass first to the hippocampal formation and then down by way of the fornix to the mammillary body. From this they would pass upward through the mammillo-thalamic tract, or the fasciculus of Vicq d'Azyr, to the anterior nuclei of the thalamus and thence by the medial thalamocortical radiation [or anterior thalamic projections] to the cortex of the gyrus cinguli [...] The cortex of the cingular gyrus may be looked on as the receptive region for the experiencing of emotion as the result of impulses coming from the hypothalamic region [...] Radiation of the emotive process from the gyrus cinguli to other regions in the cerebral cortex would add emotional coloring to psychic processes occurring elsewhere. (Papez, 1937)

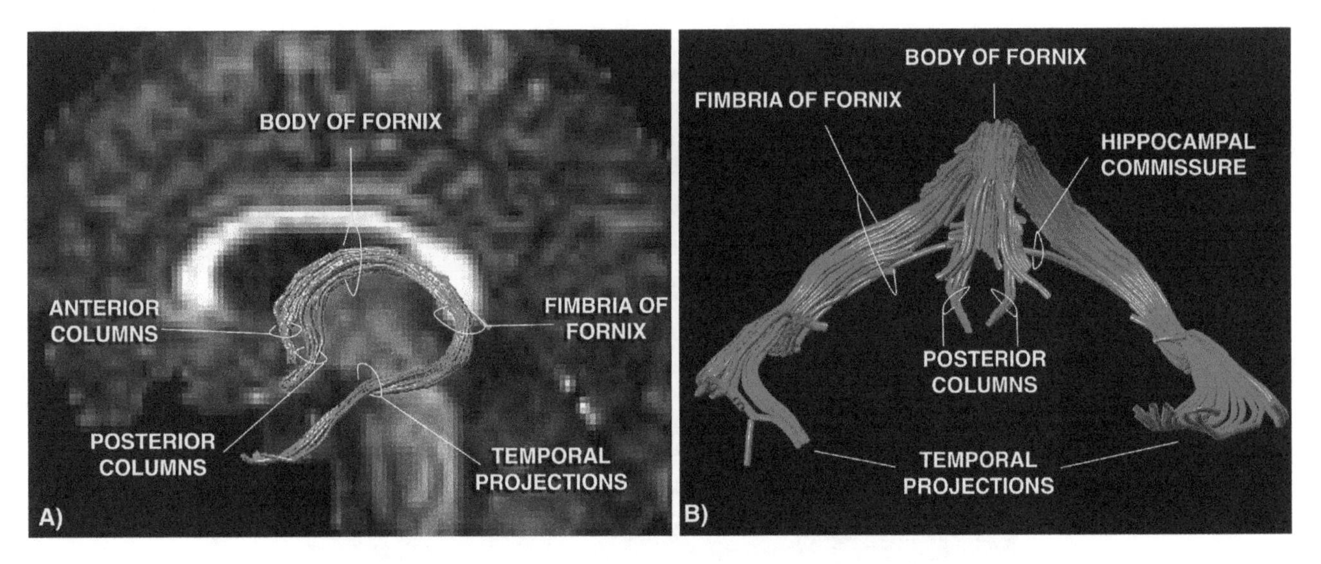

Figure 11.2 A) Lateral and B) anterior views of a tractographic reconstruction of the fornix. Note that the fibres of the hippocampal commissure run together with the temporal fibres of the fornix (Catani et al., 2002).

A decade later, Paul Yakovlev (1948), independently from Papez, proposed that the orbitofrontal cortex, insula, amygdala, and anterior temporal lobe form a network underlying emotion and motivation. In two seminal papers published in 1949 and 1952, Paul D. MacLean crystallized previous works by incorporating both Papez and Yakovlev view into a model of the limbic system that has remained almost unchanged since (MacLean, 1949, 1952). MacLean concluded that the limbic cortex, together with the limbic subcortical structures, is a functionally integrated system interconnected by short- and long-range fibre bundles (Figure 11.1).

Fornix

The fornix is mainly a projection tract connecting the hippocampus with the mammillary body, the anterior thalamic nuclei, and the hypothalamus; it also has a small commissural component known as the hippocampal commissure (Crosby et al., 1962; Aggleton, 2008; Nieuwenhuys et al., 2008). Figure 11.2 shows two views of the fornix reconstructed with *in vivo* diffusion tensor tractography (Catani et al., 2002). Fibres arise from the hippocampus (subiculum and entorhinal cortex) of each side, run through the fimbria, and join beneath the splenium of the corpus callosum to form the body of the fornix. Other fimbrial fibres continue medially, cross the midline, and project to the contralateral hippocampus. Most of the fibres within the body of the fornix run anteriorly beneath the body of the corpus callosum towards the anterior commissure. Above the interventricular foramen, the anterior body of the fornix divides into right and left columns. As each column approaches the anterior commissure it diverges again into two components. One of these, the posterior columns of the fornix, curve ventrally in front of the interventricular foramen of Monroe and posterior to the anterior commissure to enter the mammillary body (post-commissural fornix), adjacent areas of the hypothalamus, and anterior thalamic nucleus. The second component, the anterior columns of the fornix, enter the hypothalamus and project to the septal region and nucleus accumbens (Aggleton, 2008). The fornix also contains some afferent fibres to the hippocampus from septal and hypothalamic nuclei (Niewenhuys et al., 2008).

Mammillo-thalamic tract

The fibres of the mammillo-thalamic tract (bundle of Vicq d'Azyr) originate from the mammillary bodies and after a very short course terminate in the anterior and dorsal nuclei of the thalamus (Figure 11.1). A ventrally directed branch projects from the mammillary bodies to the tegmental nuclei (mammillo-tegmental tract). According to Nauta (1958), the mammillo-tegmental tract together with other fibres of the medial forebrain bundle, forms an important circuit between medial limbic structures of the midbrain and hypothalamus to relate visceral perception to emotion and behaviour.

Anterior thalamic projections

The anterior thalamic nuclei receive projections from the fornix and mammillo-thalamic tract and connect through the anterior thalamic projections to the orbitofrontal and anterior cingulate cortex (Figure 11.1). The anterior thalamic projections run in the anterior limb of the internal capsule (see also Chapter 10).

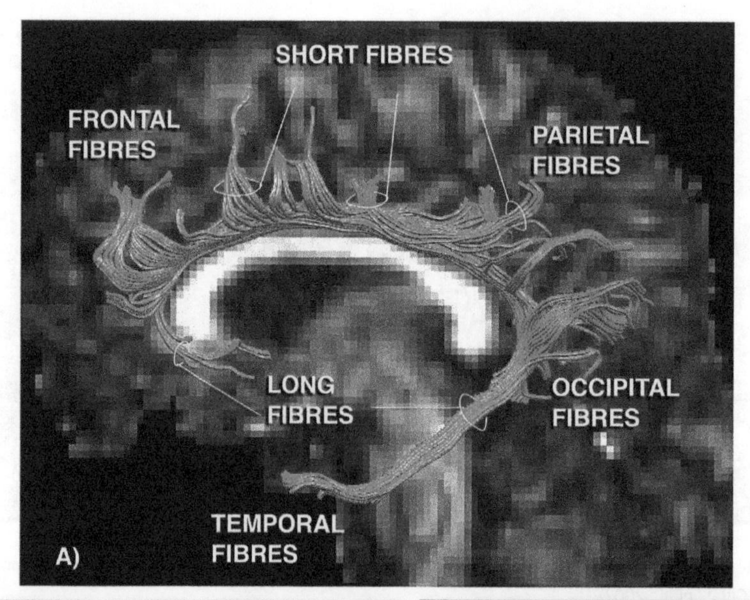

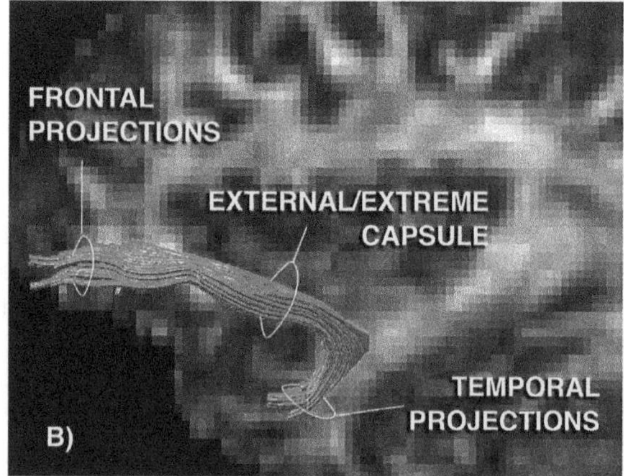

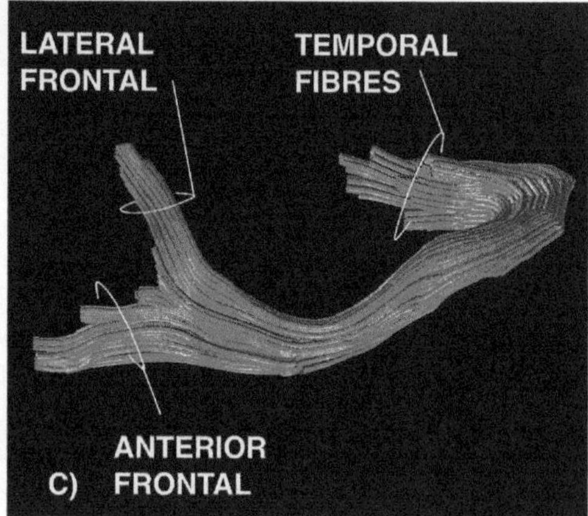

Figure 11.3 Tractographic reconstruction of A) the cingulum in lateral view and the uncinate fasciculus in B) lateral and C) superior view (Catani et al., 2002).

Cingulum

Figure 11.3A shows a lateral view of the trajectory of the cingulum as reconstructed with tractography. The cingulum contains fibres of different lengths, the longest running from the amygdala, uncus, and parahippocampal gyrus to sub-genual areas of the frontal lobe (Crosby et al., 1962; Nieuwenhuys et al., 2008). From the medial temporal lobe, these fibres reach the occipital lobe and arch almost 180 degrees around the splenium to continue anteriorly within the white matter of the cingulate gyrus. The dorsal and anterior fibres of the cingulum follow the shape of the superior aspect of the corpus callosum. After curving around the genu of the corpus callosum, the fibres terminate in the subcallosal gyrus and the paraolfactory area (Crosby et al., 1962). Shorter fibres that join and leave the cingulum along its length, connect adjacent areas of the medial frontal gyrus, paracentral lobule, precuneus, cuneus, cingulate, lingual, and fusiform gyri (Déjérine, 1895; Nieuwenhuys et al., 2008).

The cingulum can be divided into an anterior-dorsal component, which constitutes most of the white matter of the cingulate gyrus, and a posterior-ventral component running within the parahippocampal gyrus, retrosplenial cingulate gyrus, and posterior precuneus. Preliminary data suggest that these subcomponents of the cingulum may have different anatomical features. For example, a higher fractional anisotropy has been found in the left anterior-dorsal segment of the cingulum compared to right, but reduced fractional anisotropy has been reported in the left posterior-ventral component compared to the right (Park et al., 2004; Gong et al., 2005; Wakana et al., 2007). Notwithstanding this, the volume of the cingulum is bilateral and symmetrical in most subjects (Thiebaut de Schotten et al., 2011).

Uncinate fasciculus

The uncinate fasciculus connects the anterior part of the temporal lobe with the orbital and polar frontal cortex. In the tractography reconstruction presented in Figure 11.3B, C, the fibres of the uncinate fasciculus originate from the temporal pole, uncus,

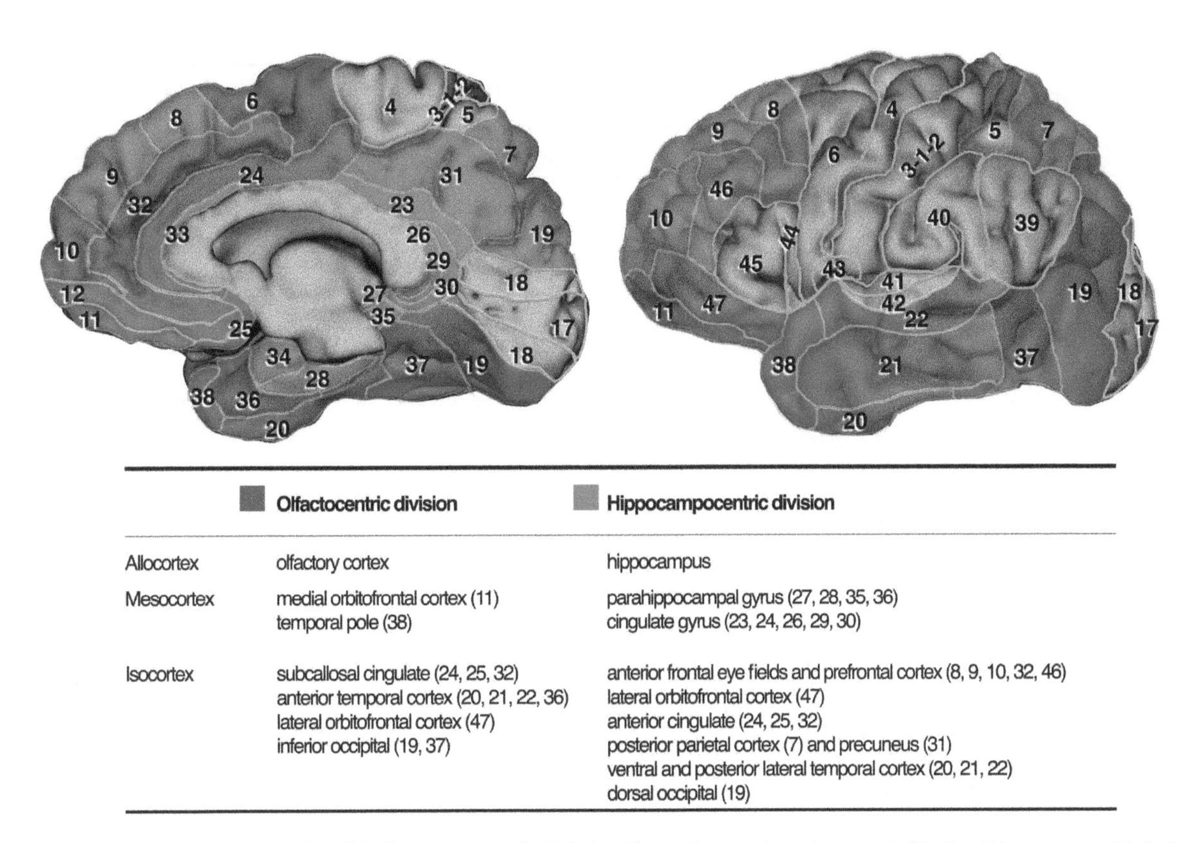

	Olfactocentric division	Hippocampocentric division
Allocortex	olfactory cortex	hippocampus
Mesocortex	medial orbitofrontal cortex (11) temporal pole (38)	parahippocampal gyrus (27, 28, 35, 36) cingulate gyrus (23, 24, 26, 29, 30)
Isocortex	subcallosal cingulate (24, 25, 32) anterior temporal cortex (20, 21, 22, 36) lateral orbitofrontal cortex (47) inferior occipital (19, 37)	anterior frontal eye fields and prefrontal cortex (8, 9, 10, 32, 46) lateral orbitofrontal cortex (47) anterior cingulate (24, 25, 32) posterior parietal cortex (7) and precuneus (31) ventral and posterior lateral temporal cortex (20, 21, 22) dorsal occipital (19)

Figure 11.4 Functional-anatomical separation of the limbic zone, paralimbic belt and isocortical areas into olfactocentric (blue) and hippocampocentric (red) divisions. Some regions (e.g. BA 10, 11, 21, 22, 24, 32, 36, 47) are connected to both divisions.

parahippocampal gyrus, and amygdala, then after a U-turn, enter the floor of the external/extreme capsule. Between the insula and the putamen, the uncinate fasciculus runs inferior to the fronto-occipital fasciculus before entering the orbital region of the frontal lobe. Here, the uncinate splits into a ventro-lateral branch, which terminates in the anterior insula and lateral orbitofrontal cortex, and an antero-medial branch that continues towards the cingulate gyrus and the frontal pole (Déjérine, 1895; Klingler and Gloor, 1960; Crosby et al., 1962). Whether the uncinate fasciculus is a lateralized bundle is still debated. An asymmetry of the volume and density of fibres of this fasciculus has been reported in one post-mortem neurohistological human study in which the uncinate fasciculus was found to be asymmetric in 80% of subjects, containing on average 30% more fibres in the right hemisphere compared to the left (Highley et al., 2002). However, diffusion measurements have also shown higher fractional anisotropy in the left uncinate compared to the right in children and adolescents (Eluvathingal et al., 2007) but not in adults (Thiebaut de Schotten et al., 2011).

Functional anatomy of the limbic system

The limbic system has always been considered as a complex arrangement of transitional structures situated between a visceral 'primitive' subcortical brain and a more evolved cortical one (Yakovlev, 1948; MacLean, 1952).

The subcortical limbic structures include the amygdala, mammillary bodies, hypothalamus, some thalamic nuclei (i.e. anterior, intralaminar, and medial dorsal groups) and the ventral striatum (i.e. nucleus accumbens). The neurons and fibres composing the subcortical limbic structures present a simple arrangement, not dissimilar to other subcortical nuclei of the brainstem regulating basic metabolism, respiration, and circulation.

The cortical components of the limbic system include areas of increasing complexity separated into limbic and paralimbic zones (Mesulam, 2000). At the lower level the corticoid areas of the amygdaloid complex, substantia innominata, together with septal and olfactory nuclei display an anatomical organization that lacks consistent lamination and dendritic orientation. These structures are in part subcortical and in part situated on the ventral and medial surfaces of the cerebral hemispheres. The next level of organization is the allocortex of the olfactory regions and hippocampal complex, where the neurons are well differentiated into layers and their dendrites show an orderly pattern of orientation. The corticoid and allocortical regions are grouped together into the limbic zone of the cerebral cortex as distinct from the paralimbic zone. The latter is mainly composed of 'mesocortex', whose progressive level of structural complexity ranges from a simplified arrangement similar to the allocortex, to the most complex six-layered isocortex.

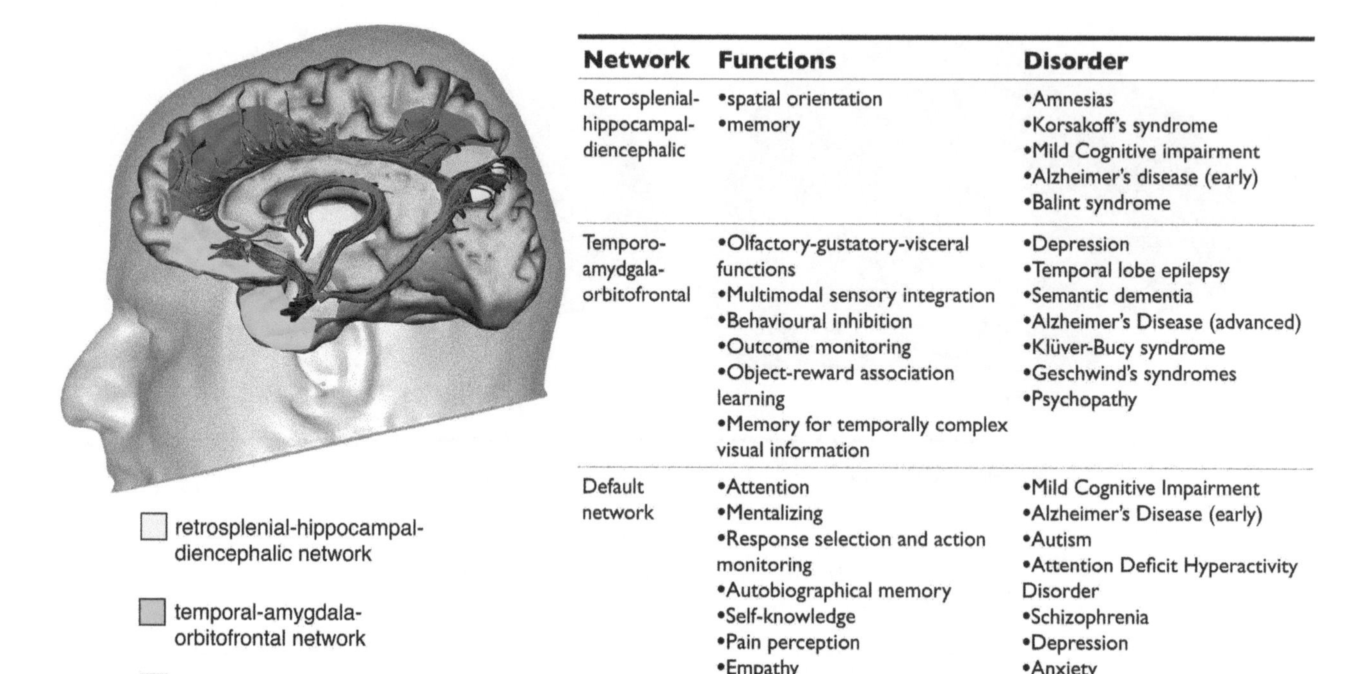

Network	Functions	Disorder
Retrosplenial-hippocampal-diencephalic	•spatial orientation •memory	•Amnesias •Korsakoff's syndrome •Mild Cognitive impairment •Alzheimer's disease (early) •Balint syndrome
Temporo-amydgala-orbitofrontal	•Olfactory-gustatory-visceral functions •Multimodal sensory integration •Behavioural inhibition •Outcome monitoring •Object-reward association learning •Memory for temporally complex visual information	•Depression •Temporal lobe epilepsy •Semantic dementia •Alzheimer's Disease (advanced) •Klüver-Bucy syndrome •Geschwind's syndromes •Psychopathy
Default network	•Attention •Mentalizing •Response selection and action monitoring •Autobiographical memory •Self-knowledge •Pain perception •Empathy •Person perception	•Mild Cognitive Impairment •Alzheimer's Disease (early) •Autism •Attention Deficit Hyperactivity Disorder •Schizophrenia •Depression •Anxiety •Obsessive compulsive disorder

Figure 11.5 The medial limbic structures participating in the activity of three distinct but partially overlapping networks. The main connections of the retrosplenial-hippocampal-diencephalic network are the ventral cingulum, the fornix, and mammillo-thalamic tract; the endstations of this network are indicated in yellow, while its functions and corresponding clinical disorders are listed in the table. The nodes of the temporo-amydgala-orbitofrontal network (indicated in green) are connected by the uncinate fasciculus. The dorsal cingulum is the main connection of the medial default network, whose cortical projections are shown in blue.

The limbic and paralimbic zones can also be divided into olfactocentric and hippocampocentric groups (Figure 11.4) (Mega et al., 1997; Mesulam, 2000). Each division is organized around a central core of allocortex. The olfactocentric division is organized around the primary olfactory piriform cortex and includes the orbitofrontal, insular and temporopolar region. The hippocampocentric division is organized around the hippocampus and includes the parahippocampal and cingulate cortex. Both divisions have reciprocal connections with subcortical limbic structures and surrounding isocortical regions (Figure 11.4). The two divisions overlap in the anterior cingulate cortex.

Functionally the paralimbic areas contribute to the activity of three distinct networks (Figure 11.5).

The network composed of the hippocampal-diencephalic structures (connected through the fornix and mammillo-thalamic tract) and the parahippocampal-retrosplenial circuit (ventral cingulum) is dedicated to memory and spatial orientation (Aggleton, 2008; Vann et al., 2009). Some of the structures composing the network are particularly vulnerable to damage caused by viral infections (e.g. encephalitis) or alcohol (e.g. Korsakoff's syndrome) (Figure 11.5). Imaging studies have documented altered metabolism and reduced functional activation of this network also in age-related neurodegenerative disorders, such as mild cognitive impairment (Nestor et al., 2003; Minoshima et al., 1997) and Alzheimer's disease (Buckner et al., 2005).

The temporo-amygdala-orbitofrontal network (connected through the uncinate fasciculus) is dedicated to the integration of visceral and emotional states, cognition, and behaviour (Mesulam, 2000). In animal studies, disconnection of the uncinate fasciculus causes impairment of object-reward association learning and reduced performances in memory tasks involving temporally complex visual information (Gaffan and Wilson, 2008). Tasks involving the monitoring of outcomes activate the medial orbital cortex (Amodio and Frith, 2006). Damage to this network manifests with cognitive and behavioural symptoms characteristic of temporal lobe epilepsy, mood disorders, traumatic brain injury, and neurodegenerative dementias, including advanced Alzheimer's disease and semantic dementia (Figure 11.5).

The default-mode network consists of a group of medial and lateral regions whose activity decreases during goal-directed tasks (Raichle et al., 2001; Raichle and Snyder, 2007). The anterior cingulate-medial prefrontal cortex and the posterior cingulate-precuneus form the medial portion of the default-mode network and are interconnected through the dorsal cingulum. In functional imaging studies the default-mode network is active during the 'resting state', a condition in which the majority of the subjects engage in an introspective, self-directed stream of thought (i.e. similar to daydreaming). A synchronous deactivation of the default network is observed in the transition between the 'resting state' and the execution of goal directed tasks, irrespective of the nature of the task. The deactivation of the default-mode network has been linked to a number of functions including working memory, focusing attention to sensorially driven activities, understanding other people's intention (mentalizing or theory of mind), prospective thinking (envisioning the future) and memory for personal events (autobiographic memory) (Amodio and Frith, 2006). Altered activation of the default network has been reported in functional imaging studies of patients with neuropsychiatric disorders (Broyd et al., 2009) (Figure 11.5).

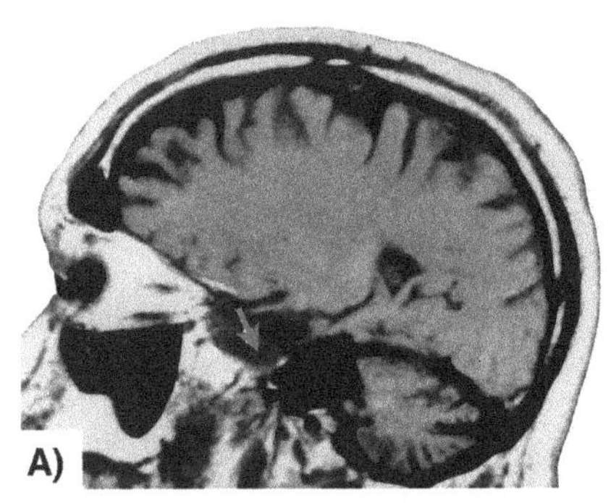

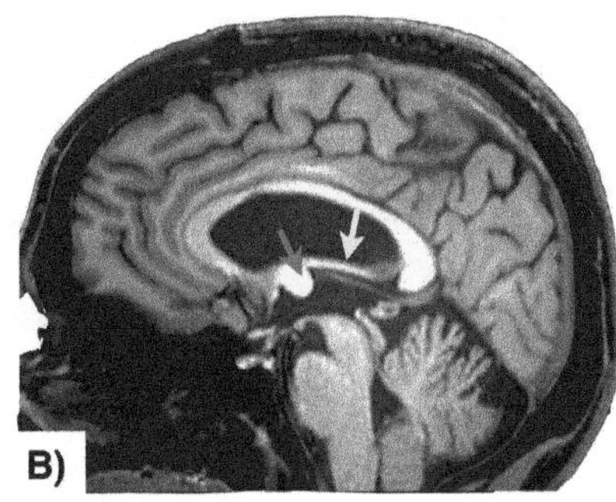

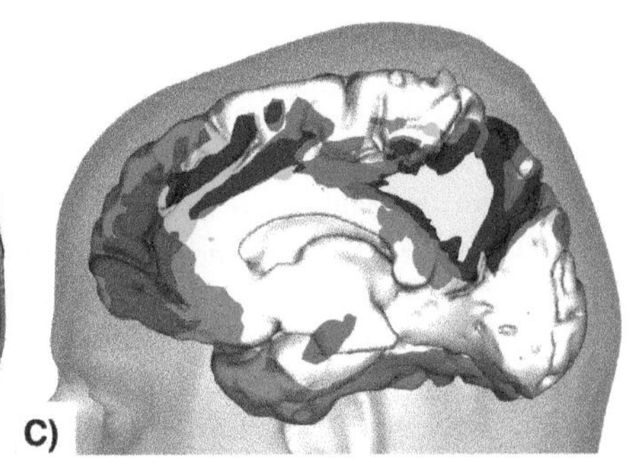

Figure 11.6 Memory deficits associated with lesions to the retrosplenial-hippocampal-diencephalic network. A) Case H.M suffered with severe global amnesia following a bilateral medial temporal lobectomy (red arrow) (Corkin et al., 1997). B) Surgical excision of the colloid cyst of the third ventricle (red arrow) can damage the fibres of the fornix (yellow arrow) and cause global amnesia (Aggleton 2008). C) In patients with Alzheimer's disease the areas with amyloid deposition (revealed by the PIB-compound deposition, green areas), reduced metabolism (blue), and cortical atrophy (purple) overlap with the areas of the default network (red) in the retrosplenial cortex (indicated in yellow) (Buckner et al., 2005).

Limbic syndromes

Amnesias and dementias

Patients with lesions to the hippocampocentric division of the limbic system manifest severe memory deficits (Markowitsch 2000). Some differences in the clinical presentation may be related to the exact location and extension of the damage, and its nature. The common manifestations are those of a global amnesia where the patient is unable to encode, associate, and retrieve new information (anterograde amnesia). In addition there is also some degree of amnesia for events before the brain damage but temporally close to it (retrograde amnesia). The remote memory is well preserved. The patient H.M., who underwent bilateral resection of the medial temporal lobe for pharmacologically intractable epilepsy, is a classical example of a pure global amnesic (Figure 11.6) (Scoville and Milner, 1957). Despite the severity of the short-term memory problems his long-term memory and insight were relatively preserved.

Loss of insight can be associated with confabulation (the spontaneous narrative report of events that never happened) in patients with diencephalic amnesias due to lesions of the mammillary bodies, the thalamic nuclei and their interconnections. Confabulation can be severe in chronic alcoholics with Korsakoff's syndrome, especially if the pathology affects the normal activity of the medial orbitofrontal and anterior cingnulate cortex. In a 32-year-old alcoholic patient that confabulated for 6 weeks, measurement of cerebral perfusion using single photon emission computed tomography, showed hypoperfusion of the anterior and mediodorsal thalamic nuclei, anterior cingulate and orbitofrontal cortex (Benson et al., 1996). A second single photon emission computed tomography repeated after the confabulation stopped showed a 'normalization' of the orbitofrontal and anterior cingulate perfusion. The anterior and mediodorsal thalamic nuclei remained hypoperfused and the patient continued to suffer with profound amnesia (for a review of the anatomy of confabulation see Schneider, 2008).

In patients with vascular thalamic lesions (such as case A.B. reported by Markowitsch et al., 1993) the extension of the damage to the mammillo-thalamic tract is the best predictor of the severity of the memory deficit (Von Cramon et al., 1985).

In patients with colloid cysts of the third ventricle, the surgical removal of the benign tumour can damage the fornix and result in anterograde amnesia, although it is seemingly not as severe as that seen in diencephalic patients (Aggleton, 2008). Another form of hippocampocentric memory dysfunction is associated with lesions to the posterior parahippocampal cortex, retrosplenial cingulate cortex, and posterior precuneus (Valenstein et al., 1987). These patients, in addition to memory deficits, show difficulties in spatial orientation due to the inability to derive directional information from landmark cues in familiar and new environments (Vann et al., 2009). Reduced metabolism of the retrosplenial cortex has also been reported in patients with mild cognitive impairment (Nestor et al., 2003) and early Alzheimer's disease (Minoshima et al., 1997). More recently, a combined cortical morphometry and diffusion imaging study found reduced cortical thickness and white matter abnormalities of these regions (Acosta-Cabronero et al., 2010). Compared to surrounding areas, the parahippocampal, posterior cingulate, and precuneus regions also have a faster rate of atrophy in pre-symptomatic Alzheimer's disease patients (autosomal dominant mutation carriers) (Scahill et al., 2002). Reduced fractional anisotropy has also been found in the cingulum, hippocampus and the posterior corpus callosum of cognitively intact subjects with increased genetic risk of dementia (APOE 4 carriers) (Persson et al., 2006).

Preliminary evidence suggests that diffusion changes in neurodegenerative disorders are likely to reflect severity of underlying white matter pathology. Xie et al. (2005) reported a significant positive correlation between reduced fractional anisotropy values, atrophy of the hippocampus and decline in the mini-mental state examination scores in patients with Alzheimer's disease. In a transgenic mouse model over-expressing beta-amyloid precursor protein, the diffusivity parameters were significantly correlated with the severity of Alzheimer's disease-like pathology in the white matter (Song et al. 2004). In humans, Englund et al. (2004) conducted a parallel post-mortem neuropathological examination and fractional anisotropy quantification of two brains with dementia and reported that the degree of white matter pathology correlated significantly with gradually lower fractional anisotropy values sampled in fifteen regions of interest. Overall, these studies suggest that reduced fractional anisotropy in Alzheimer's disease may reflect white matter axonal degeneration and myelin loss following neuronal degeneration of cortical neurons.

The clinical profile of neurodegenerative disorders varies according to the network affected by the illness. In advanced Alzheimer's disease, for example, the extension of the disease to the olfactory (orbitofrontal-amygdala) division is associated with clinical manifestations such as semantic deficits, language difficulties, personality changes and other behavioural symptoms (e.g. aggression, dishinibition, etc.), which are not present if the pathology is limited to the hippocampocentric division. In the temporal variant of the fronto-temporal dementia and in semantic dementia the white matter abnormalities (Borroni et al., 2007; Agosta et al., 2010) involve the olfactory division first, while damage to the hippocampocentric division occurs later as the disease progresses, involving other cognitive domains such as memory and spatial orientation.

Temporal lobe epilepsy

In patients with unilateral temporal lobe epilepsy the damage to the limbic white matter tracts, such as the fornix (Concha et al., 2005) and the uncinate fasciculus (Diehl et al., 2008), is diffuse and often extends contralaterally from the side of the suspected seizure. In temporal lobe epilepsy patients with mesial hippocampal sclerosis, the decreased fractional anisotropy of the fornix fibres is due to reduced axonal diameter and myelin content (Concha et al., 2010). The diffusion changes in the left uncinate fasciculus correlate with the severity of the deficits in delayed recall (Diehl et al., 2008). Preliminary data also suggest, that in patients with temporal lobe epilepsy undergoing surgery, the pre-operative tractography assessment of the lateralization pattern of the temporal tracts can help to predict naming deficits after the operation (more left lateralized patients showed worse postoperative deficits) (Powell et al., 2008).

In some patients with temporal lobe epilepsy the behavioural symptoms resemble those commonly observed in the Klüver–Bucy syndrome. In the 1930s, Klüver and Bucy conducted a series of experiments in rhesus monkeys that consisted of bilateral surgical removal of the anterior temporal lobe, which include the amygdala and temporal pole (Klüver and Bucy, 1939). After the operation the animals showed a strong tendency to examine objects orally (hyperorality), an irresistible impulse to touch (hypermetamorphosis), loss of normal anger and fear responses, increased sexual activity, and inability to recognize visually presented objects. The first Klüver–Bucy syndrome in humans was described in a patient who received bilateral temporal resection (Terzian and Ore, 1955), but it is usually a condition that clinicians observe in patients with herpes or paraneoplastic encephalitis, tumours, or traumatic brain injury involving the anterior temporal and orbitofrontal cortex (Hayman et al., 1998; Zappala' et al., 2012).

In children with temporal lobe epilepsy, single-photon emission computed tomography reveals hypoperfusion of the basal ganglia and the adjacent frontal and temporal limbic regions. Most of the patients recover after the acute phase, but those with abnormal diffusivity of the temporal and frontal white matter tracts exhibit long-term mental retardation, epilepsy, and persistent oral tendency (Maruyama et al., 2009). Some temporal lobe epilepsy patients present with Geschwind's syndrome, a characteristic change in personality consisting of unusual tendencies to write extensively and in a meticulous manner (hypergraphia), excessive and circumstantial verbal output, deepened cognitive and emotional responses (e.g. excessive moral concerns), viscosity of thought, altered sexuality (usually lack of interest), and hyperreligiosity (Waxman and Geschwind, 1974). The emergence of psychotic symptoms in temporal lobe epilepsy is associated with white matter changes extending to the frontal pathways (Flügel et al., 2006). Behavioural symptoms can respond to surgery. Mitchell et al. (1954) described a case of temporal lobe epilepsy with fetish behaviour. The patient reported highly pleasurable 'thought satisfaction' derived from looking at a safety-pin and sought seclusion in a lavatory to indulge it. Unfortunately the fetish object also triggered severe seizures, which required surgical treatment. Relief not only of the epilepsy but also of the fetishism followed the temporal lobectomy.

Antisocial behaviour

Psychopathic personality disorder (psychopathy) is characterized by features of emotional detachment and antisocial traits (Patrick et al., 1993), and is strongly associated with criminal behaviour and recidivism (Hare et al., 1999). Since the report of the case of Phineas Gage (Figure 11.7) (Harlow, 1848), who displayed 'acquired sociopathy' following frontal lobe injury (Damasio et al., 1994), the orbitofrontal cortex and other regions of the prefrontal cortex have been considered important for personality and social behaviour (Damasio, 2000). For example, the orbitofrontal cortex is crucial to successful reversal learning in which previously rewarded stimuli are associated with punishment. Reversal learning is significantly impaired in adult psychopaths (Budhani et al., 2006) and in young people with psychopathic traits (Budhani, 2005). It has also been reported that violent personality disordered offenders have reduced prefrontal cortex grey matter volume (Raine et al., 2000) and glucose metabolism (Raine et al., 1997), and impaired orbitofrontal cortex activation during aversive conditioning (Veit et al., 2002).

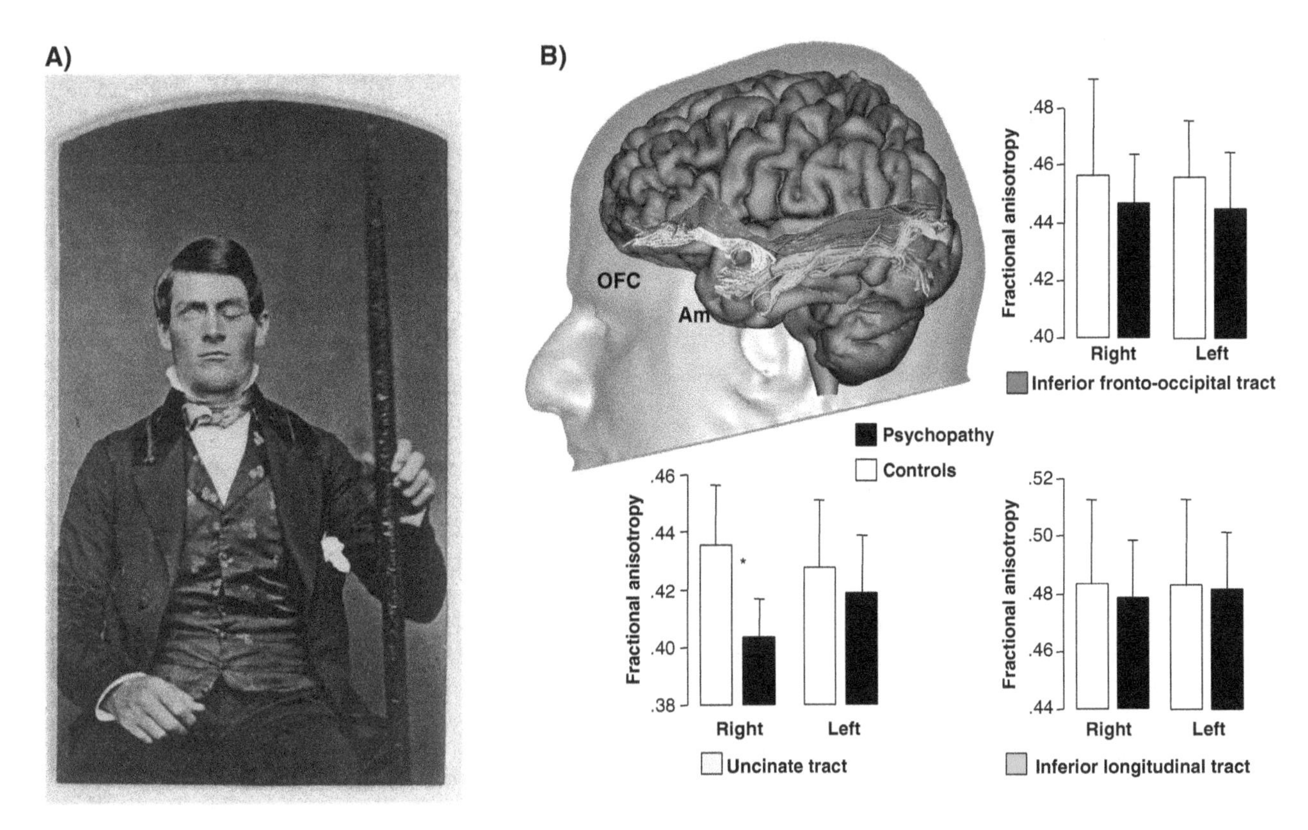

Figure 11.7 A) Phineas Gage photographed with the bar that penetrated his skull through the left orbit and caused frontal damage. B) Tractographic reconstruction of the connections between amygdala, orbitofrontal cortex, and posterior occipital areas. Tract-specific measurements of fractional anisotropy (FA) show that psychopaths have a significantly reduced mean FA in the right uncinate fasciculus (P=0.003) compared to controls. There were no differences in the left uncinate fasciculus (P=0.448) or in the two 'non-limbic' control tracts: the inferior longitudinal fasciculus and inferior fronto-occipital fasiculus (Craig et al., 2009).

In contrast, other researchers have argued that amygdala dysfunction is central to the affective deficits and impairs moral socialization of psychopathy (Blair, 2007a). This latter view is supported by evidence that psychopaths show performance deficits in tasks sensitive to amygdala damage (Levenston et al., 2000; Blair et al., 2001), and have significantly reduced amygdala volume (Tiihonen et al., 2000) and decreased amygdala activation during verbal learning (Kiehl et al., 2001) and facial fear processing (Deeley et al., 2006). Furthermore, stimulation of the amygdala can manifest with irritability, aggression, violent outbursts, and antisocial behaviour.

More recently, the dichotomy between researchers postulating whether orbitofrontal cortex or amygdala dysfunction is central to psychopathy (Abbott, 2001) has narrowed; and it has been suggested instead that the social and emotional deficits of psychopaths may reflect an altered interaction between orbitofrontal cortex and amygdala (van Honk and Schutter, 2006; Blair, 2007b). This view has received support from a recent study that used tractography to measure the volume and integrity of the connections between orbitofrontal cortex and amygdala in psychopaths (Figure 11.7) (Craig et al., 2009). A significantly reduced fractional anisotropy was reported in the uncinate fasciculus of psychopaths compared to healthy subjects with similar age and intelligence. A correlation between measures of antisocial behaviour (as assessed by the Psychopathy Checklist) and anatomical differences in the uncinate fasciculus was also reported. To confirm that these findings were specific to the limbic amygdala-orbitofrontal cortex network, other two 'non-limbic' control tracts connecting the posterior visual areas to either amygdala or orbitofrontal cortex were studied, and no significant between-group differences were found. These results suggest that abnormalities in a specific amygdala-orbitofrontal cortex network underpin the neurobiological basis of psychopathy.

Schizophrenia

The limbic system supports emotion, motivation, and memory functions that are often impaired in schizophrenia. Negative symptoms (e.g. anhedonia) are thought to derive from hypofunctioning of the limbic system, while some positive symptoms (e.g. hallucinations) have been attributed to its hyperactivity. In a recent review of neuroimaging studies in patients with auditory hallucinations, the anterior cingulate cortex and the medial temporal regions are among the limbic structures that were consistently reported to be activated during auditory hallucinations (Allen et al., 2008). Abnormalities of both myelin and oligodendroglial architecture and aberrantly located neurons in myelinated fibre bundles have been found in limbic regions (mainly frontal and anterior temporal) of patients with schizophrenia (Akbarian et al., 1996; Davis et al., 2003). Decrased volume of hippocampus (Bilder et al., 1995) medial temporal lobe, insula, anterior cingulate, and thalamus (Honea et al., 2005) has also been reported.

Voxel-based (Kubicki et al., 2002; Kanaan et al., 2009), region-of-interest, and tractography (Jones et al., 2005; 2006) studies in patients with schizophrenia reported micro-structural changes in the cingulum (Fujiwara et al., 2007), uncinate fasciculus (McIntosh et al., 2008), fornix (Kuroki et al., 2006; Takei et al., 2008), and the anterior thalamic radiations (McIntosh et al., 2008). However, larger studies (Kanaan et al., 2009) and a recent meta-analysis (Ellison-Wright, 2009) suggest that the reported deficits are likely to be part of a wider process rather than be specific to limbic tracts.

Depression and bipolar disorder

Neuroimaging studies show that the subgenual cingulate region (BA 25) is metabolically overactive in depressed patients and its activation reduces with the antidepressant effect of pharmacological treatment (Mayberg et al., 2000), electroconvulsive therapy (Nobler et al., 2001), and transcranial magnetic stimulation of more dorsal frontal regions (Mottaghy et al., 2002). Mayberg et al. (2005) have also shown that chronic direct deep brain stimulation of the white matter fibres adjacent to the subgenual cortex resulted in a significant remission of depression in four of six patients with treatment-resistant depression (Figure 11.8). The antidepressant effect was associated not only with a reduction in the local metabolism of the subgenual region, but also with increased metabolism of dorsal cingulate and other prefrontal areas connected to the subgenual cortex. A recent preliminary tractography study in adolescents with major depressive disorder reported lower fractional anisotropy in the white matter tract connecting the subgenual cingulate region to the amygdala in the right hemisphere (Cullen et al., 2010).

Post-mortem histological studies in patients with bipolar affective disorder have found a reduction in the number and density of glial cells (Ongür et al., 1998; Webster et al., 2005) and decreased neuronal density in the subgenual region of the orbitofrontal cortex (Bouras et al., 2001) and in the dorsolateral prefrontal areas (BA 9) (Rajkowska et al., 2001). Myelin abnormalities in bipolar affective disorder may be related to a decreased expression of genes involved in myelin synthesis and regulation (Aston et al., 2005). These histological findings are supported by neuroimaging studies that reported a general increase in white matter hyperintensities on MRI images (Altshuler et al., 1995) and selective reduced white matter density (Bruno et al., 2004), together with decreased cortical metabolism and volume in the subgenual region (Drevets et al., 1997) of adults with a diagnosis of bipolar affective disorder.

Findings from diffusion imaging studies using region-of-interest or voxel-based approaches have been inconsistent with reports of both decreased (Adler et al., 2004; Versace et al., 2008) and increased (Haznedar, 2005; Versace et al., 2008) fractional anisotropy in bipolar disorder compared to healthy controls. Tractography of the subgenual-amygdala connections showed increased tract volume (Houenou et al., 2007) and reduced fractional anisotropy in the uncinate and anterior thalamic projections (McIntosh et al., 2008).

In a recent single case study, a patient with a history of bipolar disorder and intractable recurrent depression following a right thalamic stroke underwent deep brain stimulation of the subgenual cortex. The patient died 16 months after the implants were positioned without a significant clinical response. A high resolution diffusion imaging dataset acquired post-mortem revealed markedly reduced limbic projections from the thalamus to the subgenual cortex and amygdala in the stroke-affected (right) hemisphere. The authors concluded that reduced limbic connections assessed with diffusion imaging could be a contraindication to deep brain stimulation for depression (McNab et al., 2009).

Obsessive–compulsive disorder

Current anatomical models suggest a network composed of the medial orbitofrontal cortex, the anterior-dorsal cingulate and the striatum underpins obsessive–compulsive disorder. Functional and structural imaging studies support this model (Radua and Mataix-Cols, 2008). A recent study, for example, found increased functional connectivity between the striatum and the orbitofrontal cortex (Harrison et al., 2009). The activity of the regions involved in obsessive–compulsive disorder can be reduced with the surgical severing or the electrophysiological inactivation of their interconnecting fibres. The white matter of the anterior-dorsal cingulate and medial frontal regions are the common targets for the treatment of drug-resistant obsessive–compulsive disorder with deep brain stimulation and psychosurgery (i.e. cingulotomy and capsulotomy) (Figure 11.8). The surgical disconnection causes distant structural changes to the interconnected regions as suggested by a study that showed that individuals undergoing cingulotomy had significant reductions in the volume of the caudate nucleus several months after the operation (Rauch et al., 2000). The anatomical changes in the orbitofrontal cortex are less consistently reported despite clear evidence of its altered activity in obsessive compulsive disorder (Radua and Mataix-Cols, 2008; Harrison et al., 2009). Together, the structural and functional imaging studies suggest that, in obsessive compulsive disorder, the primary anatomical abnormalities occur in the striatum with a distant hodological effect on the paralimbic areas of the anterior cingulate and medial orbitofrontal regions.

Autism spectrum disorder

It has been suggested that some of the social and communication abnormalities typically found in autism spectrum disorder are due to abnormalities in limbic structures (Damasio and Maurer, 1978) and perhaps also in their connectivity (Courchesne and Pierce, 2005; Wickelgren, 2005). Early post-mortem investigations of both

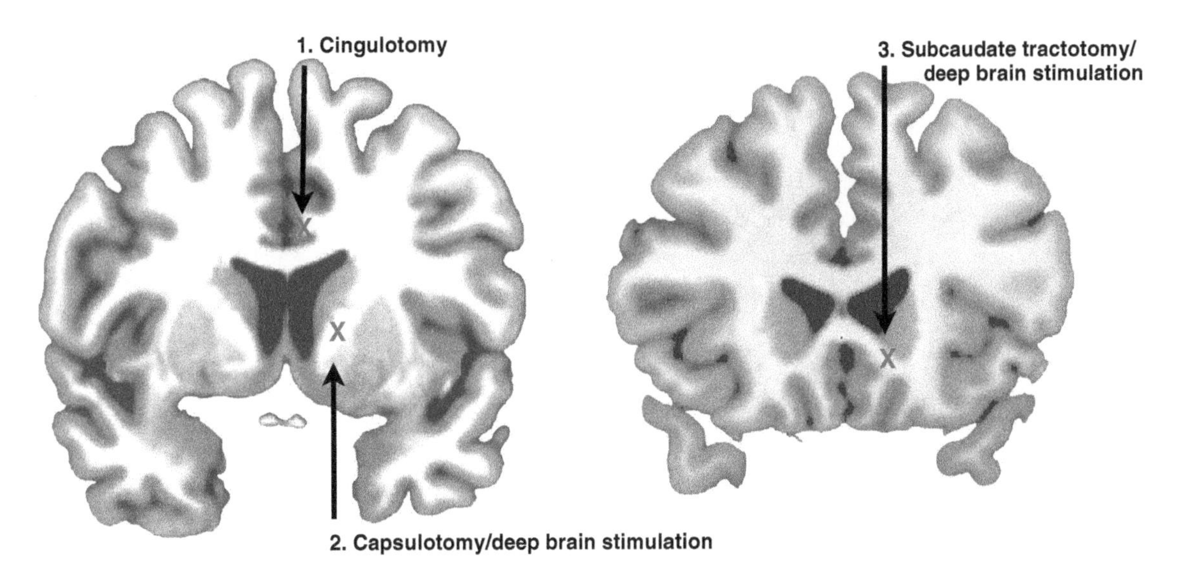

Figure 11.8 Common anatomical targets for the surgical treatment of drug-resistant depression, obsessive–compulsive disorder, and chronic pain. Cingulotomy, capsulotomy, and tractotomy consist of the surgical severing of a fibre bundle, which results in the selective disconnection of two distant regions. An alternative to the surgical disconnection is deep brain stimulation, in which the electrodes are inserted into regions of white matter for the stimulation of selected groups of fibres. For other disorders, like Parkinson disease, the electrodes are implanted in the deep grey matter. Procedures include: 1) Anterior cingulotomy is the bilateral section of the anterior cingulum fibres; this is performed in patients with obsessive–compulsive disorder and chronic pain. 2) The fibres of the anterior internal capsule are the target for capsulotomy or deep brain stimulation; both are used for the treatment of obsessive–compulsive disorder and depression. 3) Subcaudate tractotomy and deep brain stimulation of the frontostriatal fibres are used in patients with depression, and obsessive–compulsive disorder (adapted from Lipsman et al., 2007).

adults and children with autism reported reduced neuronal size and increased cell packing in the hippocampus, amygdala and, to a lesser degree, the enthorinal cortex, mammillary bodies and septal nuclei (Bauman and Kemper, 1985; Raymond et al., 1996; Palmen et al., 2004; Bauman and Kemper, 2005). Recent *in vivo* voxel-based morphometry studies reported significant differences in the anatomy of limbic regions, but with contrasting results with respect to the white matter compartment (Herbert et al., 2003; Barnea-Goraly et al., 2004; Boddaert et al., 2004; Kwon et al., 2004; McAlonan et al., 2005; Salmond et al., 2005; Lee et al., 2007; Lee et al., 2009). For example, several groups reported decreased grey and white matter volumes in the inferior temporal regions and fusiform gyrus in both autism and Asperger's syndrome in young adults (Boddaert et al., 2004; Kwon et al., 2004; McAlonan et al., 2005; Salmond et al., 2005). Herbert et al. (2004) also reported decreased grey matter volume in the same regions in young people with autism but increased white matter volume in regions containing limbic pathways. White matter differences have been reported in a voxel-based diffusion study, which found that children with autism have significant microstructural differences (e.g. reduced fractional anisotropy) in the anterior cingulum and medial temporal lobe (Barnea-Goraly et al., 2004). A recent tractography study showed that compared to healthy controls, adults with Asperger's syndrome had a significantly higher number of streamlines in the cingulum bilaterally and a lower number of streamlines in the right uncinate (Pugliese et al., 2009). Together the post-mortem and *in vivo* studies suggest anatomical changes of the dorsal cingulum and uncinate fasciculus in autism spectrum disorder.

References

Abbott, A. (2001). Into the mind of a killer. *Nature*, **410** (6826), 296–8.

Acosta-Cabronero, J., Williams, G.B., Pengas, G., and Nestor, P.J. (2010). Absolute diffusivities define the landscape of white matter degeneration in Alzheimer's disease. *Brain*, **133**(2), 529–39.

Adler, C.M., Holland, S.K., Schmithorst, V., et al. (2004). Abnormal frontal white matter tracts in bipolar disorder: a diffusion tensor imaging study. *Bipolar Disord*, **6**(3), 197–203.

Aggleton, J.P. (2008). Understanding anterograde amnesia: disconnections and hidden lesions. *Q J Exp Psychol*, **61**(10), 1441–71.

Agosta, F., Henry, R.G., Migliaccio, R., et al. (2010). Language networks in semantic dementia. *Brain*, **133**(1), 286–99.

Akbarian, S., Kim, J.J., Potkin, S.G., et al. (1996). Maldistribution of interstitial neurons in prefrontal white matter of the brains of schizophrenic patients. *Arch Gen Psychiatry*, **53**(5), 425–36.

Allen, P., Larøi, F., McGuire, P.K., and Aleman, A. (2008). The hallucinating brain: A review of structural and functional neuroimaging studies of hallucinations. *Neurosci Biobehav Rev*, **32**(1), 175–91.

Altshuler, L.L., Curran, J.G., Hauser, P., Mintz, J., Denicoff, K., and Post, R. (1995). T2 hyperintensities in bipolar disorder: magnetic resonance imaging comparison and literature meta-analysis. *Am J Psychiatry*, **152**(8), 1139–44.

Amodio, D.M. and Frith, C.D. (2006) Meeting of minds: the medial frontal cortex and social cognition. *Nat Rev Neurosci*, **7**(4), 268–77.

Aston, C., Jiang, L., and Sokolov, B.P. (2005). Transcriptional profiling reveals evidence for signaling and oligodendroglial abnormalities in the temporal cortex from patients with major depressive disorder. *Mol Psychiatry*, **10**(3), 309–22.

Barnea-Goraly, N., Kwon, H., Menon, V., et al. (2004). White matter structure in autism: preliminary evidence from diffusion tensor imaging. *Biol Psychiatry*, **55**(3), 323–6.

Bauman, M. and Kemper, T.L. (1985). Histoanatomic observations of the brain in early infantile autism. *Neurology*, **35**(6), 866–74.

Bauman, M.L. and Kemper, T.L. (2005). Neuroanatomic observations of the brain in autism: a review and future directions. *Int J Dev Neurosci*, **23**(2–3), 183–7.

Bechterew, W. (1900). Demonstration eines gehirns mit Zestörung der vorderen und inneren Theile der Hirnrinde beider Schläfenlappen. *Neurolog Centralbl*, **20**, 990–1.

Benson, D.F., Djenderedjian, A., Miller, B.L., et al. (1996). Neural basis of confabulation. *Neurology*, **46**(5), 1239–43.

Bilder, R.M., Bogerts, B., Ashtari, M., et al. (1995). Anterior hippocampal volume reductions predict frontal lobe dysfunction in first episode schizophrenia. *Schizophr Res*, **17**(1), 47–58.

Blair, R.J. (2007a). The amygdala and ventromedial prefrontal cortex in morality and psychopathy. *Trends Cog Sci*, **11**(9), 387–92.

Blair, R.J. (2007b). Dysfunctions of medial and lateral orbitofrontal cortex in psychopathy. *Ann NY Acad Sci*, **1121**(1), 461–79.

Blair, R.J., Colledge, E., Murray, L., and Mitchell, D.G. (2001). A selective impairment in the processing of sad and fearful expressions in children with psychopathic tendencies. *J Abnorm Child Psychol*, **29**(6), 491–8.

Boddaert, N., Chabane, N., Gervais, H., et al. (2004). Superior temporal sulcus anatomical abnormalities in childhood autism: a voxel-based morphometry MRI study. *NeuroImage*, **23**(1), 364–9.

Borroni, B., Brambati, S.M., Agosti, C., et al. (2007). Evidence of white matter changes on diffusion tensor imaging in frontotemporal dementia. *Arch Neurol*, **64**(2), 246–51.

Bouras, C., Kövari, E., Hof, P.R., Riederer, B.M., and Giannakopoulos, P. (2001). Anterior cingulate cortex pathology in schizophrenia and bipolar disorder. *Acta Neuropathol*, **102**(4), 373–9.

Broca, P. (1878). Anatomie comparée des circonvolutions cérébrales: le grand lobe limbique. *Rev Anthropol*, **1**, 385–498.

Brown, S. and Schäfer, E.A. (1888). An investigation into the functions of the occipital and temporal lobes of the monkey's brain. *Phil Trans Roy Soc Lond B*, **179**, 303–27.

Broyd, S.J., Demanuele, C., Debener, S., Helps, S.K., James, C.J., and Sonuga-Barke, E.J. (2009). Default-mode brain dysfunction in mental disorders: a systematic review. *Neurosci Biobehav Rev*, **33**(3), 279–96.

Bruno, S., Barker, G.J., Cercignani, M., Symms, M., and Ron, M.A. (2004). A study of bipolar disorder using magnetization transfer imaging and voxel-based morphometry. *Brain*, **127**(11), 2433–40.

Buckner, R.L., Snyder, A.Z., Shannon, B.J., et al. (2005). Molecular, structural, and functional characterisation of Alzheimer's disease: evidence for a relationship between default activity, amyloid, and memory. *J Neurosci*, **25**(34), 7709–17.

Budhani, S. and Blair, R.J. (2005). Response reversal and children with psychopathic tendencies: success is a function of salience of contingency change. *J Child Psychol Psychiat Allied Disciplines*, **46**(9), 972–81.

Budhani, S., Richell, R.A., and Blair, R.J. (2006). Impaired reversal but intact acquisition: probabilistic response reversal deficits in adult individuals with psychopathy. *J Abnorm Psychol*, **115**(3), 552–8.

Cannon, W.B. (1927). The James-Lange theory of emotion: a critical examination and an alternate theory. *Am J Psychol*, **39**(1/4), 106–24.

Catani, M., Howard, R.J., Pajevic, S., and Jones, D.K. (2002). Virtual in vivo interactive dissection of white matter fasciculi in the human brain. *NeuroImage*, **17**(1), 77–94.

Concha, L., Beaulieu, C., and Gross, D.W. (2005). Bilateral limbic diffusion abnormalities in unilateral temporal lobe epilepsy. *Ann Neurol*, **57**(2), 188–96.

Concha, L., Livy, D.J., Beaulieu, C., Wheatley, B.M., and Gross, D.W. (2010). In vivo diffusion tensor imaging and histopathology of the fimbria-fornix in temporal lobe epilepsy. *J Neurosci*, **30**(3), 996–1002.

Corkin, S., Amaral D.G., González, R.G., Johnson, K.A., Hyman, B.T. (1997). H.M.'s medial temporal lobe lesion: findings from magnetic resonance imaging. *J Neurosci*, **17**(10), 3964–79.

Courchesne, E. and Pierce, K. (2005). Why the frontal cortex in autism might be talking only to itself: local over-connectivity but long-distance disconnection. *Curr Opin Neurobiol*, **15**(2), 225–30.

Craig, M., Catani, M., Deelay, Q., et al. (2009). Altered connections on the road to psychopathy. *Mol Psychiatry*, **14**(10), 946–53.

Crosby, E.C., Humphery, T., and Lauer, E.W. (1962). *Correlative anatomy of the nervous system*. New York: Oxford University Press.

Cullen, K.R., Klimes-Dougan, B., Muetzel, R., et al. (2010). Altered white matter microstructure in adolescents with major depression: a preliminary study. *J Am Acad Child Adolesc Psychiatry*, **49**(2), 73–83.

Damasio, A.R. (2000). A neural basis for sociopathy. *Arch Gen Psychiatry*, **57**(2), 128–9.

Damasio, A.R. and Maurer, R.G. (1978). A neurological model for childhood autism. *Arch Neurol*, **35**(12), 777–86.

Damasio, H., Grabowski, T., Frank, R., Galaburda, A.M., and Damasio, A.R. (1994). The return of Phineas Gage: clues about the brain from the skull of a famous patient. *Science*, **264**(5162), 1102–5.

Davis, K.L., Stewart, D.G., Friedman, J.I., et al. (2003). White matter changes in schizophrenia: evidence for myelin-related dysfunction. *Arch Gen Psychiatry*, **60**(5), 443–56.

Deeley, Q., Daly, E., Surguladze, S., et al. (2006). Facial emotion processing in criminal psychopathy: Preliminary functional magnetic resonance imaging study. *Br J Psychiatry*, **189**(6), 533–9.

Déjérine, J. (1895). *Anatomie des centres nerveux, Vol 1*. Paris: Ruff et Cie.

Diehl, B., Busch, R.M., Duncan J.S., et al. (2008). Abnormalities in diffusion tensor imaging of the uncinate fasciculus relate to reduced memory in temporal lobe epilepsy. *Epilepsia*, **49**(8), 1409–18.

Drevets, W.C., Price, J.L., Simpson, J.R. Jr, et al. (1997). Subgenual prefrontal cortex abnormalities in mood disorders. *Nature*, **386**(6627), 824–7.

Ellison-Wright, I. and Bullmore, E. (2009). Meta-analysis of diffusion tensor imaging studies in schizophrenia. *Schizoph Res*, **108**(1–3), 3–10.

Englund, E., Sjöbeck, M., Brockstedt, S., Lätt, J., and Larsson, E.M. (2004). Diffusion tensor MRI post-mortem demonstrate pathology. *J Neurol*, **251**(3), 350–2.

Eluvathingal, T.J., Hasan, K.M., Kramer, L.A., Fletcher, J.M., and Ewing-Cobbs, L. (2007). Quantitative diffusion tensor tractography of association and projection fibers in normally developing children and adolescents. *Cereb Cortex*, **17**(12), 2760–8.

Flügel, D., Cercignani, M., Symms, M.R., et al. (2006). Diffusion tensor imaging findings and their correlation with neuropsychological deficits in patients with temporal lobe epilepsy and interictal psychosis. *Epilepsia*, **47**(5), 941–4.

Fujiwara, H., Hirao, K., Namiki, C., et al. (2007). Anterior cingulate pathology and social cognition in schizophrenia: a study of gray matter, white matter and sulcal morphometry. *NeuroImage*, **36**(4), 1236–45.

Gaffan, D. and Wilson, C.R. (2008). Medial temporal and prefrontal function: recent behavioural disconnection studies in the macaque monkey. *Cortex*, **44**(8), 928–35.

Gong, G., Jiang, T., Zhu, C., et al. (2005). Asymmetry analysis of cingulum based on scale-invariant parameterization by diffusion tensor imaging. *Hum Brain Mapp*, **24**(2), 92–8.

Hare, R.D., Cooke, D.J., and Hart, R.D. (1999). Psychopathy and sadistic personality disorder. In Millon, T., Blaney, P.H., and Davies, R.D. (Eds.) *Oxford Textbook of psychopathology*, pp. 555–84. Oxford: Oxford University Press.

Harlow, J.M. (1848). Passage of an iron rod through the head. *Boston Med Surg J*, **39**, 389–93.

Harrison, B.J., Soriano-Mas, C., Pujol, J., et al. (2009). Altered corticostriatal functional connectivity in obsessive-compulsive disorder. *Arch Gen Psychiatry*, **66**(11), 1189–200.

Hayman, L.A., Rexer, J.L., Pavol, M.A., Strite, D., and Meyers, C.A. (1998). Klüver-Bucy syndrome after selective damage of the amygdala and its cortical connections. *J Neuropsychiatry Clin Neurosci*, **10**(3), 354–58.

Haznedar, M.M., Roversi, F., Pallanti, S., et al. (2005). Fronto-thalamo-striatal gray and white matter volumes and anisotropy of their connections in bipolar spectrum illnesses. *Biol Psychiatry*, **57**(7), 733–42.

Herbert, M.R., Ziegler, D.A., Deutsch, C.K., et al. (2003). Dissociations of cerebral cortex, subcortical and cerebral white matter volumes in autistic boys. *Brain*, **126**(5), 1182–92.

Herbert, M.R., Ziegler, D.A., Makris, N., et al. (2004). Localization of white matter volume increase in autism and developmental language disorder. *Ann Neurol*, **55**(4), 530–40.

Highley, J.R., Walker, M.A., Esiri, M.M., Crow, T.J., and Harrison, P.J. (2002). Asymmetry of the uncinate fasciculus: a post-mortem study of normal subjects and patients with schizophrenia. *Cereb Cortex*, **12**(11), 1218–24.

Honea, R., Crow, T.J., Passingham, D., and Mackay, C.E. (2005). Regional deficits in brain volume in schizophrenia: a meta-analysis of voxel-based morphometry studies. *Am J Psychiatry,* **162**(12), 2233–45.

Houenou, J., Wessa, M., Douaud, G., et al. (2007). Increased white matter connectivity in euthymic bipolar patients: diffusion tensor tractography between the subgenual cingulate and the amygdalo-hippocampal complex. *Mol Psychiatry*, **12**(11), 1001–10.

Jakob, C. (1906). Nueva contribución á la fisiopatología de los lóbulos frontales. *La Semana Médica*, **13**(50):1325–9.

Jones, D.K., Catani, M., Pierpaoli, C., et al. (2005). Age effects on diffusion tensor magnetic resonance imaging tractography measures of frontal cortex connections in schizophrenia. Am *J Geriatr Psychiatry,* **13**(12), 1092–9.

Jones, D.K., Catani, M., Pierpaoli, C., et al. (2006). Age effects on diffusion tensor magnetic resonance imaging tractography measures of frontal cortex connections in schizophrenia. *Hum Brain Mapp*, **27**(3), 230–8.

Kanaan, R., Barker, G., Brammer, M., et al. (2009). White matter microstructure in schizophrenia: effects of disorder, duration and medication. *Br J Psychiatry*, **194**(3), 236–42.

Kiehl, K.A., Smith, A.M., Hare, R.D., et al. (2001). Limbic abnormalities in affective processing by criminal psychopaths as revealed by functional magnetic resonance imaging. *Biol Psychiatry*, **50**(9), 677–84.

Klingler, J. and Gloor, P. (1960). The connections of the amygdala and of the anterior temporal cortex in the human brain. *J Comp Neurol*, **115**(3), 333–69.

Klüver, H. and Bucy, P.C. (1939). Preliminary analysis of functions of the temporal lobes in monkeys. *Arch Neurol Psychiatry*, **42**(6), 979–1000.

Kubicki, M., Westin, C-F., Maier, S. E., et al. (2002). Uncinate fasciculus findings in schizophrenia: a magnetic resonance diffusion tensor imaging study. *Am J Psychiatry,* **159**(5), 813–20.

Kuroki, N., Kubicki, M., Nestor, P.G., et al. (2006). Fornix integrity and hippocampal volume in male schizophrenic patients. *Biol Psychiatry*, **60**(1), 22–31.

Kwon, H., Ow, A.W., Pedatella, K.E., Lotspeich, L.J., and Reiss, A.L. (2004). Voxel-based morphometry elucidates structural neuroanatomy of high-functioning autism and Asperger syndrome. *Develop Med Child Neurol*, **46**(11), 760–4.

Lee, J.E., Bigler, E.D., Alexander, A.L., et al. (2007). Diffusion tensor imaging of white matter in the superior temporal gyrus and temporal stem in autism. *Neurosci Lett*, **424**(2), 127–32.

Lee, J.E., Chung, M.K., Lazar, M., et al. (2009). A study of diffusion tensor imaging by tissue-specific, smoothing-compensated voxel-based analysis. *NeuroImage*, **44**(3), 870–83.

Levenston, G.K., Patrick, C.J., Bradley, M.M., and Lang, P.J. (2000). The psychopath as observer: emotion and attention in picture processing. *J Abnorm Psychol*, **109**(3), 373–85.

Lipsman, N., Neimat, J.S., and Lozano, A.M. (2007). Deep brain stimulation for treatment-refractory obsessive-compulsive disorder: the search for a valid target. *Neurosurgery*, **61**(1), 1–11.

MacLean, P.D. (1949). Psychosomatic disease and the 'visceral brain': recent developments bearing on the Papez theory of emotion. *Psychosom Med*, **11**(6), 338–53.

MacLean, P.D. (1952). Some psychiatric implications of physiological studies on frontotemporal portion of limbic system ("visceral brain"). *Electroencephalogr Clin Neurophysiol*, **4**(4), 407–418.

Markowitsch, H.J., von Cramon, D.Y. and Schuri, U. (1993). Mnestic performance profile of a bilateral diencephalic infarct patient with preserved intelligence and severe amnesic disturbances. *J Clin Exp Neuropsychol*, **15**(5), 627–52.

Markowitsch, H.J. (2000). Memory and amnesia. In Mesulam, M. (Ed.) *Principles of Behavioural and Cognitive Neurology*, second edition, pp. 257–93. New York: Oxford University Press.

Marshall, L.H and Magoun, H.W. (1998). *Discoveries in the human brain*. New Jersey: Humana Press.

Maruyama, K., Itomi, S., Sofue, A., and Natsume, J. (2009). Acute encephalitis/encephalopathy associated with transient involuntary movement and limbic dysfunction in the recovery phase. *No To Hattatsu*, **41**(1), 33–6.

Mayberg, H.S., Brannan, S.K., Tekell, J.L., et al. (2000). Regional metabolic effects of fluoxetine in major depression: serial changes and relationship to clinical response. *Biol Psychiatry*, **48**(8), 830–43.

Mayberg, H.S., Lozano, A.M., Voon, V. (2005). Deep brain stimulation for treatment-resistant depression. *Neuron*, **45**(5), 651–60.

McAlonan, G.M., Cheung, V., Cheung, C., et al. (2005). Mapping the brain in autism. A voxel-based MRI study of volumetric differences and intercorrelations in autism. *Brain*, **128**(2), 268–76.

McIntosh, A.M., Muñoz Maniega, S., Lymer, G.K., et al. (2008). White matter tractography in bipolar disorder and schizophrenia. *Biol Psych*, **64**(12), 1088–92.

McNab, J.A., Voets, N.L., Jenkinson, N., et al. (2009). Reduced limbic connections may contraindicate subgenual cingulate deep brain stimulation for intractable depression. *J Neurosurg*, **111**(4), 780–4.

Mega, M.S., Cummings, J.L., Salloway, S., and Maloy, P. (1997). The limbic system: an anatomic, phylogenetic, and clinical perspective. *J Neuropsychiat Clin Neurosci*, **9**(3), 315–30.

Mesulam, M. (2000). Behavioral neuroanatomy: large-scale networks, association cortex, frontal syndromes, the limbic system, and hemispheric specialization. In Mesulam, M. (Ed.) *Principles of Behavioural and Cognitive Neurology*, second edition, pp. 1–119. New York: Oxford University Press.

Minoshima, S., Giordani, B., Berent, S., Frey, K.A., Foster, N.L., and Kuhl, D.E. (1997). Metabolic reduction in the posterior cingulate cortex in very early Alzheimer's disease. *Ann Neurol*, **42**(1), 85–94.

Mitchell, W., Falconer, M.A., and Hill, D. (1954). Epilepsy with fetishism relieved by temporal lobectomy. *Lancet*, **267**(6839), 626–30.

Mottaghy, F.M., Keller, C.E., Gangitano, M., et al. (2002). Correlation of cerebral blood flow and treatment effects of repetitive transcranial magnetic stimulation in depressed patients. *Psychiatry Res*, **115**(1–2), 1–14.

Nauta, W.J. (1958). Hippocampal projections and related neural pathways to the midbrain in the cat. *Brain*, **81**(3), 319–40.

Nestor, P.J., Fryer, T.D., Smielewski P., and Hodges J.R. (2003). Limbic hypometabolism in Alzheimer's disease and mild cognitive impairment. *Ann Neurol*, **54**(3), 343–51.

Nieuwenhuys, R., Vogd, J., and Huijzen, C. (2008). *The human central nervous system*, fourth edition. Berlin: Springer–Verlag.

Nobler, M.S., Oquendo, M.A., Kegeles, L.S., et al. (2001). Decreased regional brain metabolism after ECT. *Am J Psychiatry*, **158**(2), 305–8.

Ongür, D., Drevets, W.C., and Price, J.L. (1998). Glial reduction in the subgenual prefrontal cortex in mood disorders. *Proc Natl Acad Sci USA*, **95**(22), 13290–5.

Palmen, S.J., van Engeland, H., Hof P.R., and Schmitz, C. (2004). Neuropathological findings in autism. *Brain*, **127**(12), 2572–83.

Papez, J.W. (1937). A proposed mechanism of emotion. *Arch Neurol Psychiatry,* **38**(4), 725–43.

Park, H.J., Westin, C.F., Kubicki, M., et al. (2004). White matter hemisphere asymmetries in healthy subjects and in schizophrenia: a diffusion tensor MRI study. *NeuroImage*, **23**(1), 213–23.

Patrick, C.J., Bradley, M.M., and Lang, P.J. (1993). Emotion in the criminal psychopath: Startle reflex modulation. *J Abnorm Psychol*, **102**(1), 82–92.

Persson, J., Lind, J., Larsson, A., et al. (2006). Altered brain white matter integrity in healthy carriers of the APOE epsilon4 allele: a risk for AD? *Neurology*, **66**(7), 1029–33.

Powell, H.W., Parker, G.J., Alexander, D.C., et al. (2008). Imaging language pathways predicts post-operative naming deficits. *J Neurol Neurosurg Psychiatry*, **79**(3), 327–30.

Pugliese, L., Catani, M., Ameis, S., et al. (2009). The anatomy of extended limbic pathways in Asperger syndrome: a preliminary diffusion tensor imaging tractography study. *NeuroImage*, **47**(2), 427–34.

Radua, J. and Mataix-Cols, D. (2009). Voxel-wise meta-analysis of grey matter changes in obsessive–compulsive disorder. *Br J Psychiatry*, **195**(5), 393–402.

Raichle, M.E., MacLeod, A.M., Snyder, A.Z., et al. (2001). A default mode of brain function. *Proc Natl Acad Sci USA*, **98**(2), 676–82.

Raichle, M.E. and Snyder, A.Z. (2007). A default mode of brain function: a brief history. *NeuroImage*, **37**(4), 1083–90.

Raine, A., Buchsbaum, M., and LaCasse, L. (1997). Brain abnormalities in murderers indicated by positron emission tomography. *Biol Psychiatry*, **42**(6), 495–508.

Raine, A., Lencz, T., Bihrle, S., LaCasse, L., and Colletti, P. (2000). Reduced prefrontal gray matter volume and reduced autonomic activity in antisocial personality disorder. *Arch Gen Psychiatry*, **57**(2), 119–27.

Rajkowska, G., Halaris, A., and Selemon, L.D. (2001). Reductions in neuronal and glial density characterize the dorsolateral prefrontal cortex in bipolar disorder. *Biol Psychiatry*, **49**(9),741–52.

Raymond, G.V., Bauman, M.L., and Kemper, T.L. (1996). Hippocampus in autism: a Golgi analysis. *Acta Neuropathol*, **91**(1), 117–9.

Rauch, S.L., Kim, H., Makris, N., et al. (2000). Volume reduction in the caudate nucleus following stereotactic placement of lesions in the anterior cingulate cortex in humans: a morphometric magnetic resonance imaging study. *J Neurosurg*, **93**(6), 1019–25.

Salmond, C.H., Ashburner, J., Connelly, A., et al. (2005). The role of the medial temporal lobe in autistic spectrum disorders. *Eur J Neurosci*, **22**(3), 764–72.

Scahill, R.I., Schott, J.M., Stevens, J.M., Rossor, M.N., and Fox, N.C. (2002). Mapping the evolution of regional atrophy in Alzheimer's disease: unbiased analysis of fluid-registered serial MRI. *Proc Natl Acad Sci USA*, **99**(7), 4703–7.

Schneider, A. (2008). *The confabulating mind: how the brain creates reality*. Oxford: Oxford University Press.

Scoville, W.B. and Milner B. (1957). Loss of recent memory after bilateral hippocampal lesions. *J Neurol Neurosurg Psych*, **20**(1), 11–21.

Song, S.-K., Kim, J.H., Lin, S.-J., Brendza, R.P., and Holtzman, D.M. (2004). Diffusion tensor imaging detects age-dependent white matter changes in a transgenic mouse model with amyloid deposition. *Neurobiol Dis*, **15**(3), 640–7.

Takei, K., Yamasue, H., Abe, O., et al. (2008). Disrupted integrity of the fornix is associated with impaired memory organization in schizophrenia. *Schizoph Res*, **103**(1–3), 52–61.

Terzian, H. and Ore, G.D. (1955). Syndrome of Klüver and Bucy; reproduced in man by bilateral removal of the temporal lobes. *Neurology*, **5**(6), 373–80.

Thiebaut de Schotten, M., ffytche, D., Bizzi, A., et al. (2011). Atlasing location, asymmetry and inter-subject variability of white matter tracts in the human brain with MR diffusion tractography. *NeuroImage*, **54**(1), 49–59.

Tiihonen, J., Hodgins, S., Vaurio, O., et al. (2000). Amygdaloid volume loss in psychopathy. *Soc Neurosci Abst*, **15**, 2017.

Valenstein, E., Bowers, D., Verfaellie, M., et al. (1987). Retrosplenial amnesia. *Brain*, **110**(6), 1631–46.

van Honk, J. and Schutter, D.J.G. (2006). Unmasking feigned sanity: A neurobiological model of emotion processing in primary psychopathy. *Cog Neuropsych*, **11**(3), 285–306.

Vann, S.D., Aggleton, J.P., and Maguire, E.A. (2009). What does the retrosplenial cortex do? *Nat Rev Neurosci*, **10**(11), 792–802.

Versace, A., Almeida, J.R., Hassel, S., et al. (2008). Elevated left and reduced right orbitomedial prefrontal fractional anisotropy in adults with bipolar disorder revealed by tract-based spatial statistics. *Arch Gen Psychiatry*, **65**(9), 1041–52.

Veit, R., Flor, H., Erb, M., Hermann, C., et al. (2002). Brain circuits involved in emotional learning in antisocial behavior and social phobia in humans. *Neurosci Lett*, **328**(3), 233–6.

Von Cramon, D.Y., Hebel, N., and Schuri, U. (1985). A contribution to the anatomical basis of thalamic amnesia. *Brain*, **108**(4), 993–1008.

Wakana, S., Caprihan, A., Panzenboeck, M.M., et al. (2007). Reproducibility of quantitative tractography methods applied to cerebral white matter. *NeuroImage*, **36**(3), 630–44.

Waxman, S.G. and Geschwind, N. (1974). Hypergraphia in temporal lobe epilepsy. *Neurology*, **24**(7), 629–36.

Webster, M.J., O'Grady, J., Kleinman, J.E., and Weickert, C.S. (2005). Glial fibrillary acidic protein mRNA levels in the cingulate cortex of individuals with depression, bipolar disorder and schizophrenia. *Neuroscience*, **133**(2), 453–61.

Wickelgren, I. (2005). Neurology. Autistic brains out of synch? *Science*, **308**(5730), 1856–8.

Willis, T. (1664). *Cerebri anatome*. London: Martyn & Allestry.

Yakovlev, P.I. (1948). Motility, behaviour, and the brain: stereodynamic organization and neural coordinates of behaviour. *J Nerv Ment Dis*, **107**(4), 313–35.

Xie, S., Xiao, J.X., Wang, Y.H., Wu, H.K., Gong, G.L., and Jiang, X.X. (2005). Evaluation of bilateral cingulum with tractography in patients with Alzheimer's disease. *Neuroreport*, **16**(12), 1275–8.

Zappala', G., Thiebaut de Schotten, M., and Eslinger, P.J. (2012). What can we gain with diffusion tensor imaging. *Cortex*, **48**(2), 156–65.

Cingulum (atlas)

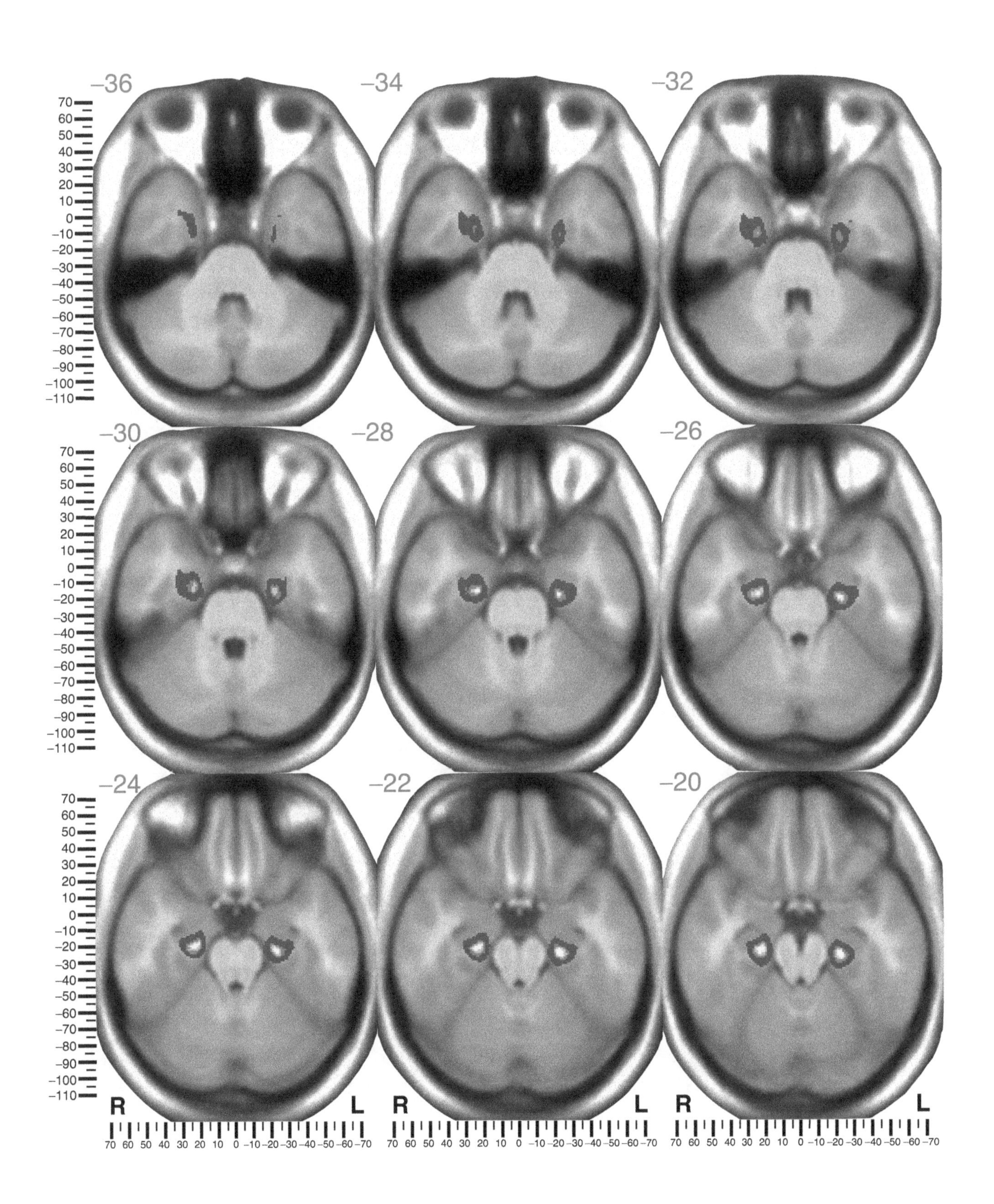

Cingulum: inferior axial view

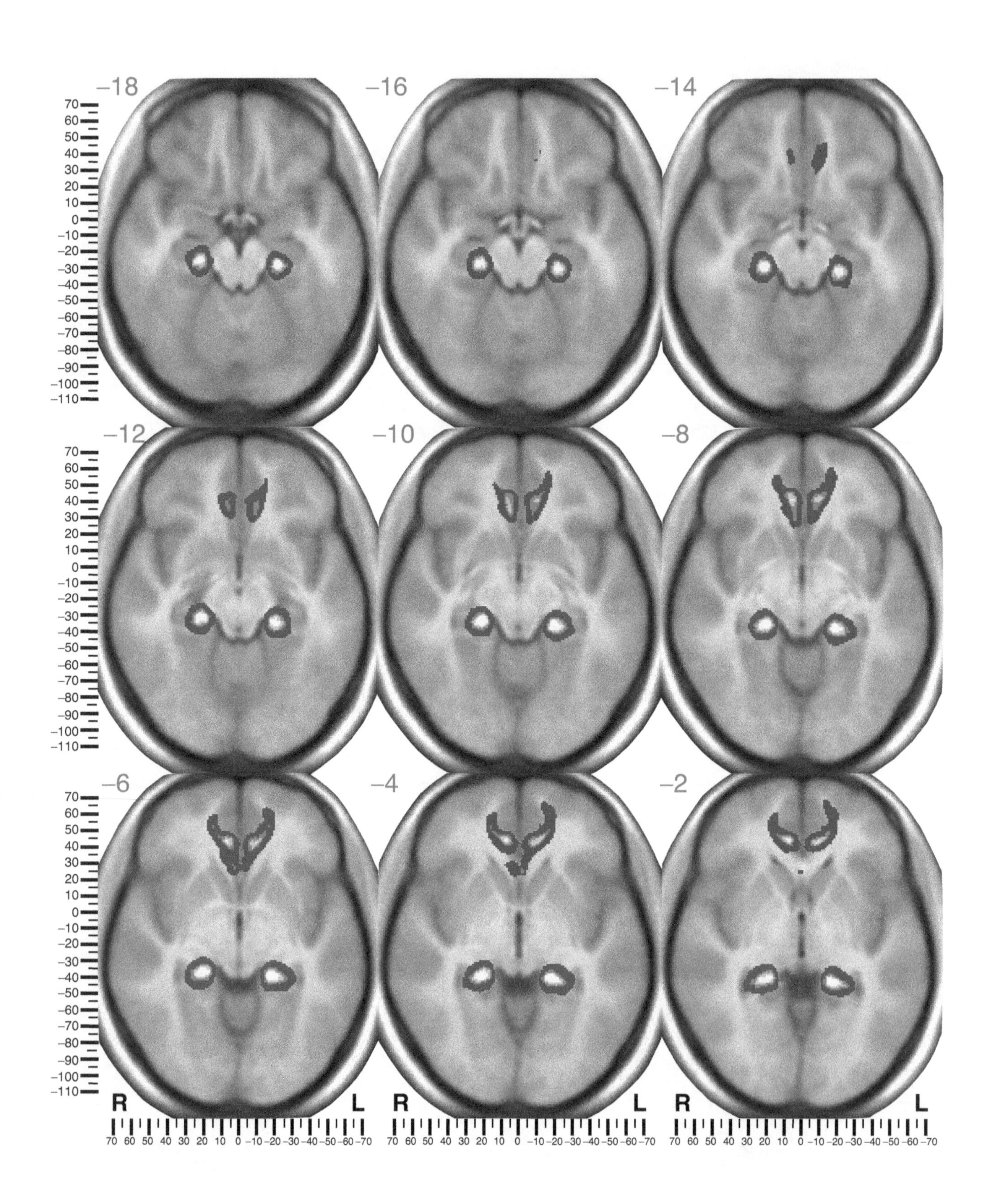

Cingulum: inferior axial view

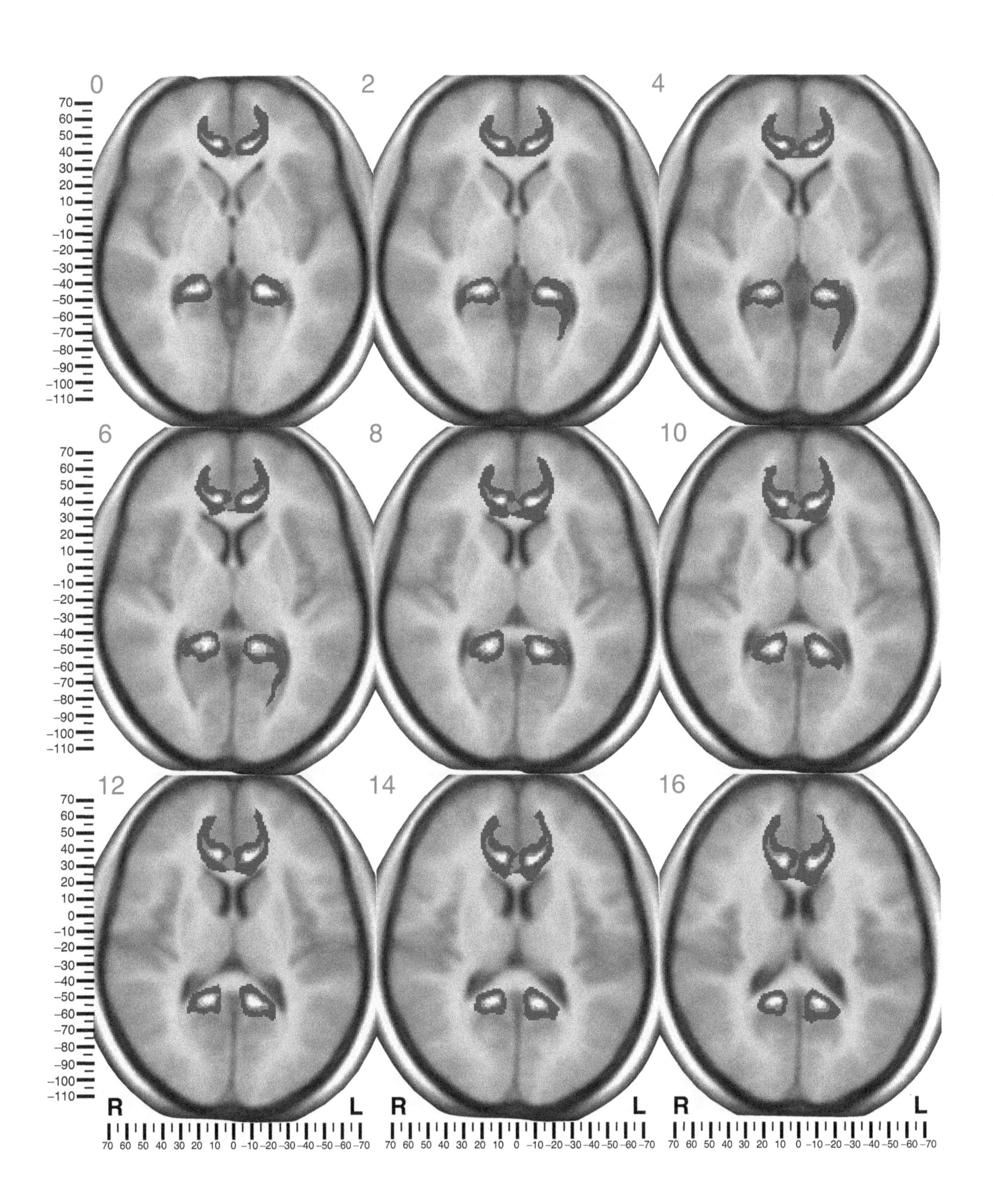

Cingulum: inferior axial view

p < .05 | > 50% | > 75% | > 90%

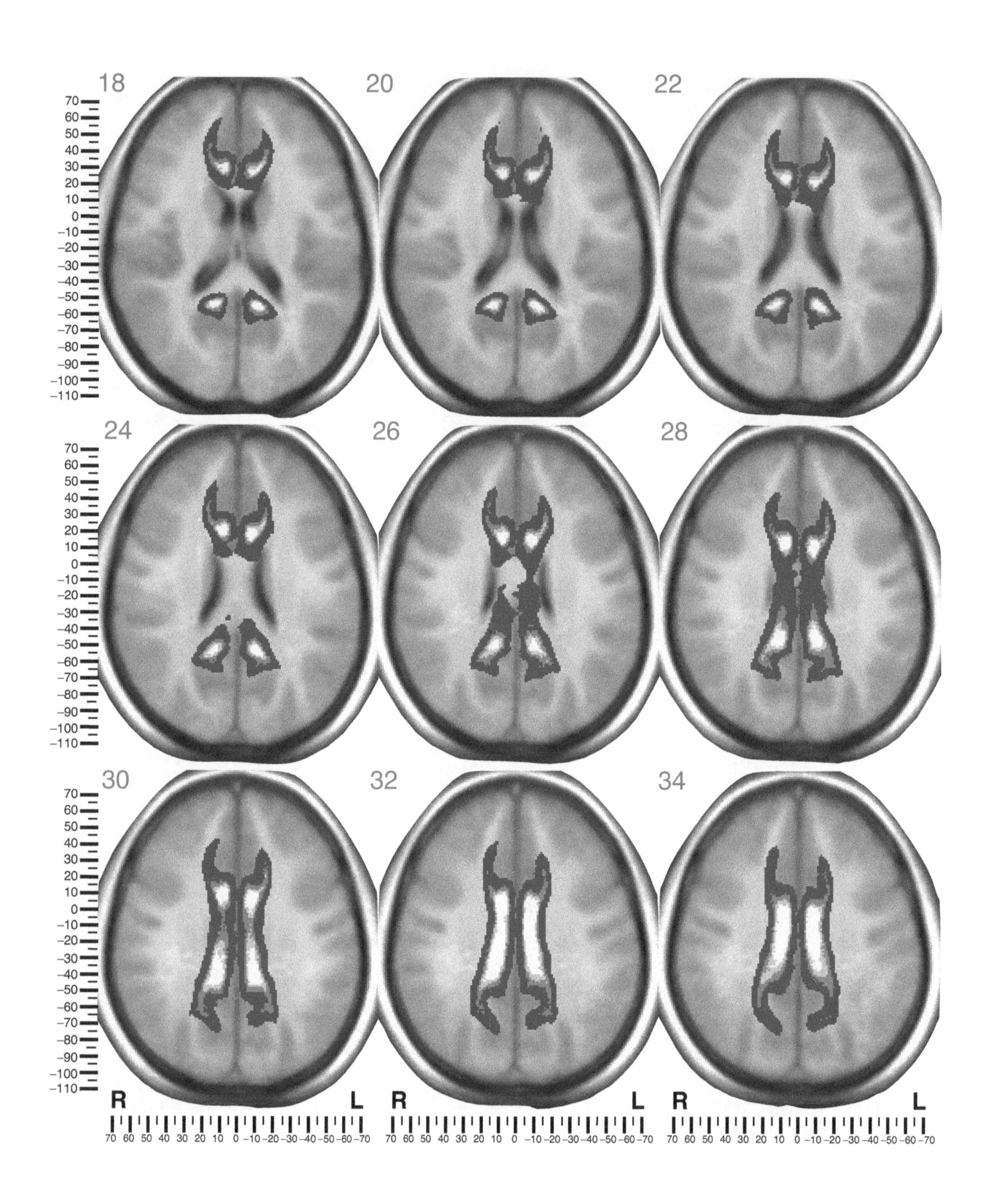

Cingulum: inferior axial view

p < .05 | > 50% | > 75% | > 90%

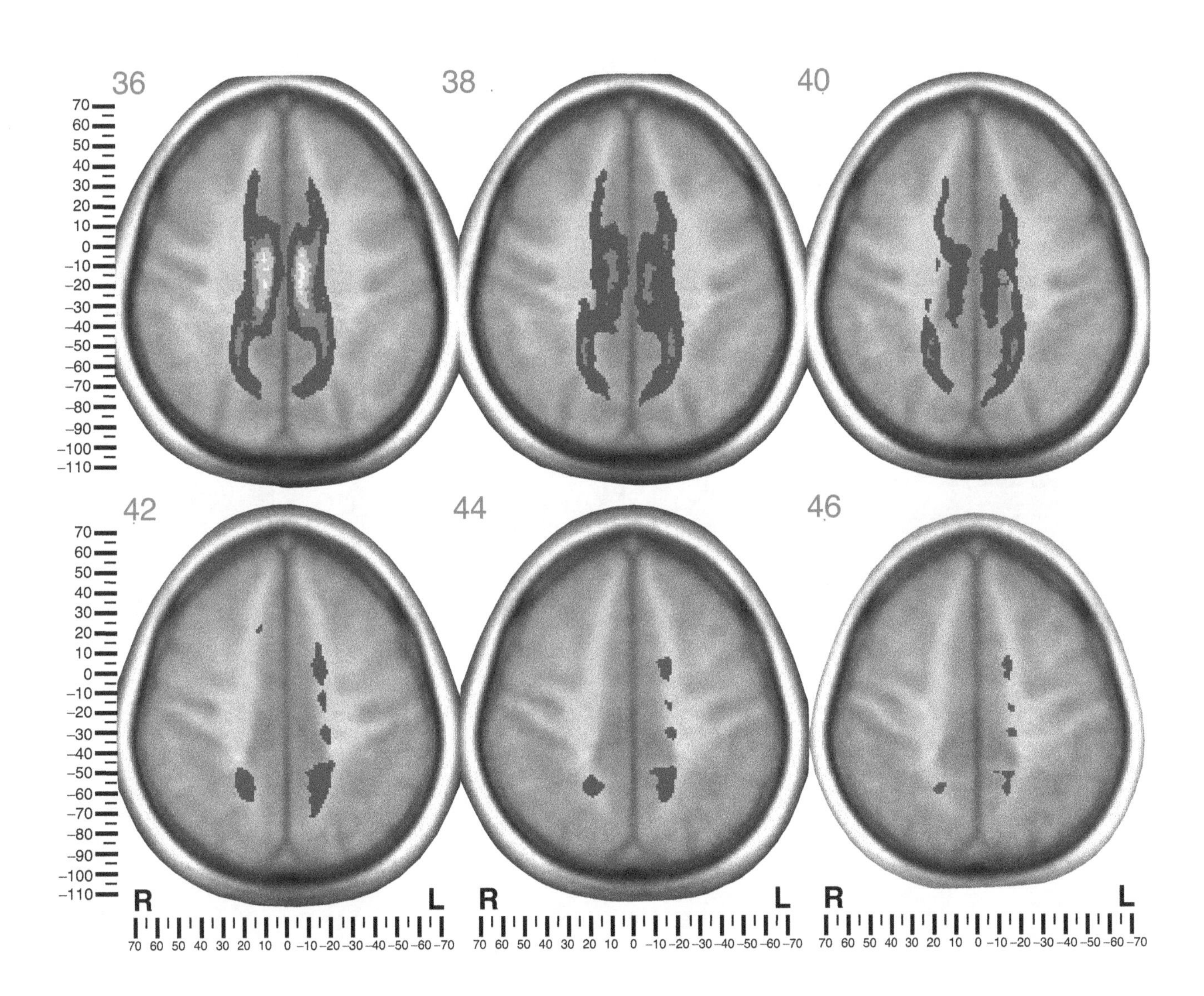

Cingulum: anterior coronal view

p < .05 | > 50% | > 75% | > 90%

80 70 60 50 40 30 20 10 0 −10 −20 −30 −40 −50 −60

46 44 42

80 70 60 50 40 30 20 10 0 −10 −20 −30 −40 −50 −60

40 38 36

80 70 60 50 40 30 20 10 0 −10 −20 −30 −40 −50 −60

34 32 30

80 70 60 50 40 30 20 10 0 −10 −20 −30 −40 −50 −60

28 26 24

R L R L R L

70 60 50 40 30 20 10 0 −10 −20 −30 −40 −50 −60 −70

70 60 50 40 30 20 10 0 −10 −20 −30 −40 −50 −60 −70

70 60 50 40 30 20 10 0 −10 −20 −30 −40 −50 −60 −70

Cingulum: anterior coronal view

p < .05 | > 50% | > 75% | > 90%

22 20 18 16 14 12 10 8 6 4 2 0

80 70 60 50 40 30 20 10 0 −10 −20 −30 −40 −50 −60

R L

70 60 50 40 30 20 10 0 −10 −20 −30 −40 −50 −60 −70

Cingulum: anterior coronal view

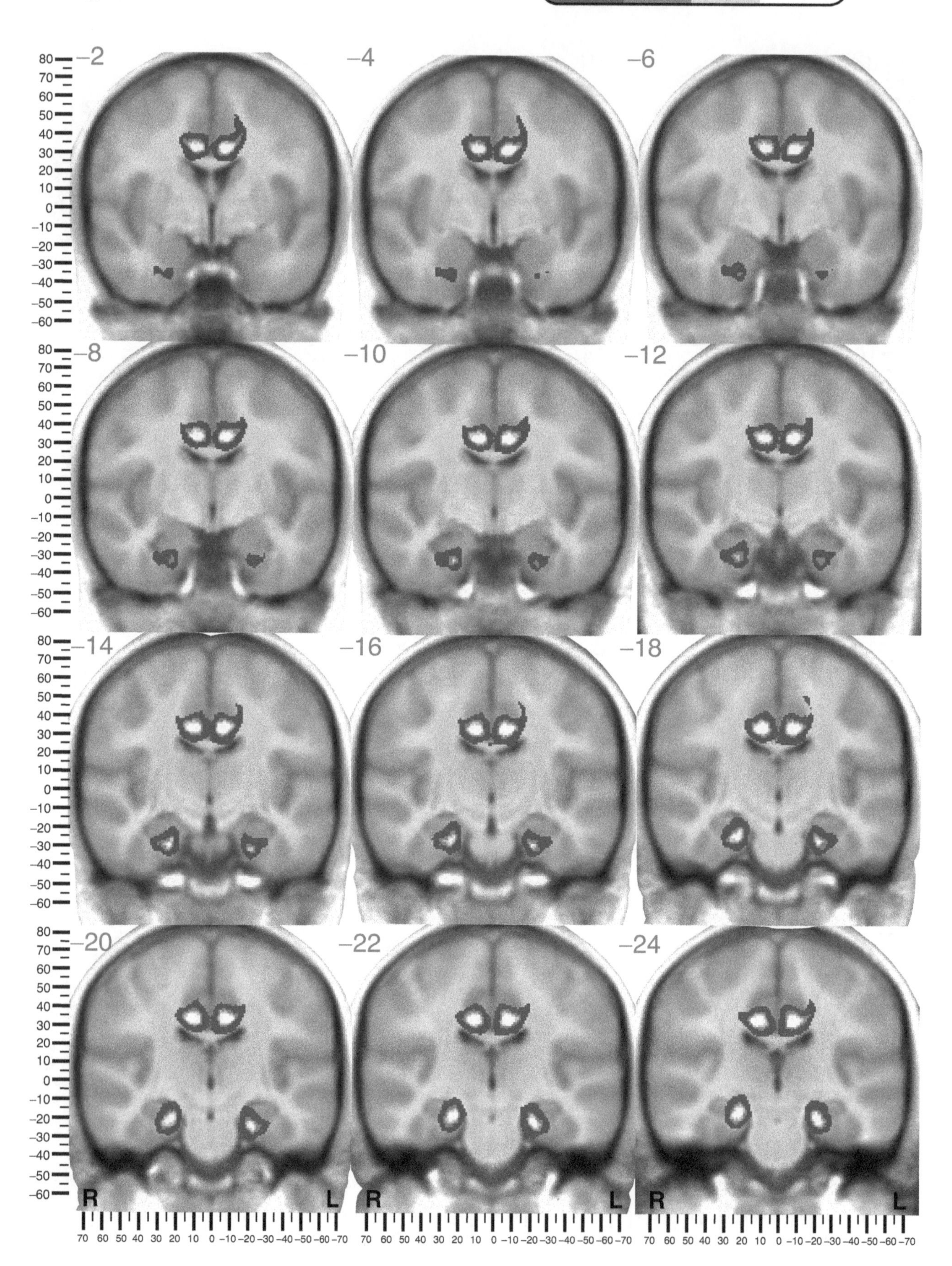

Cingulum: anterior coronal view

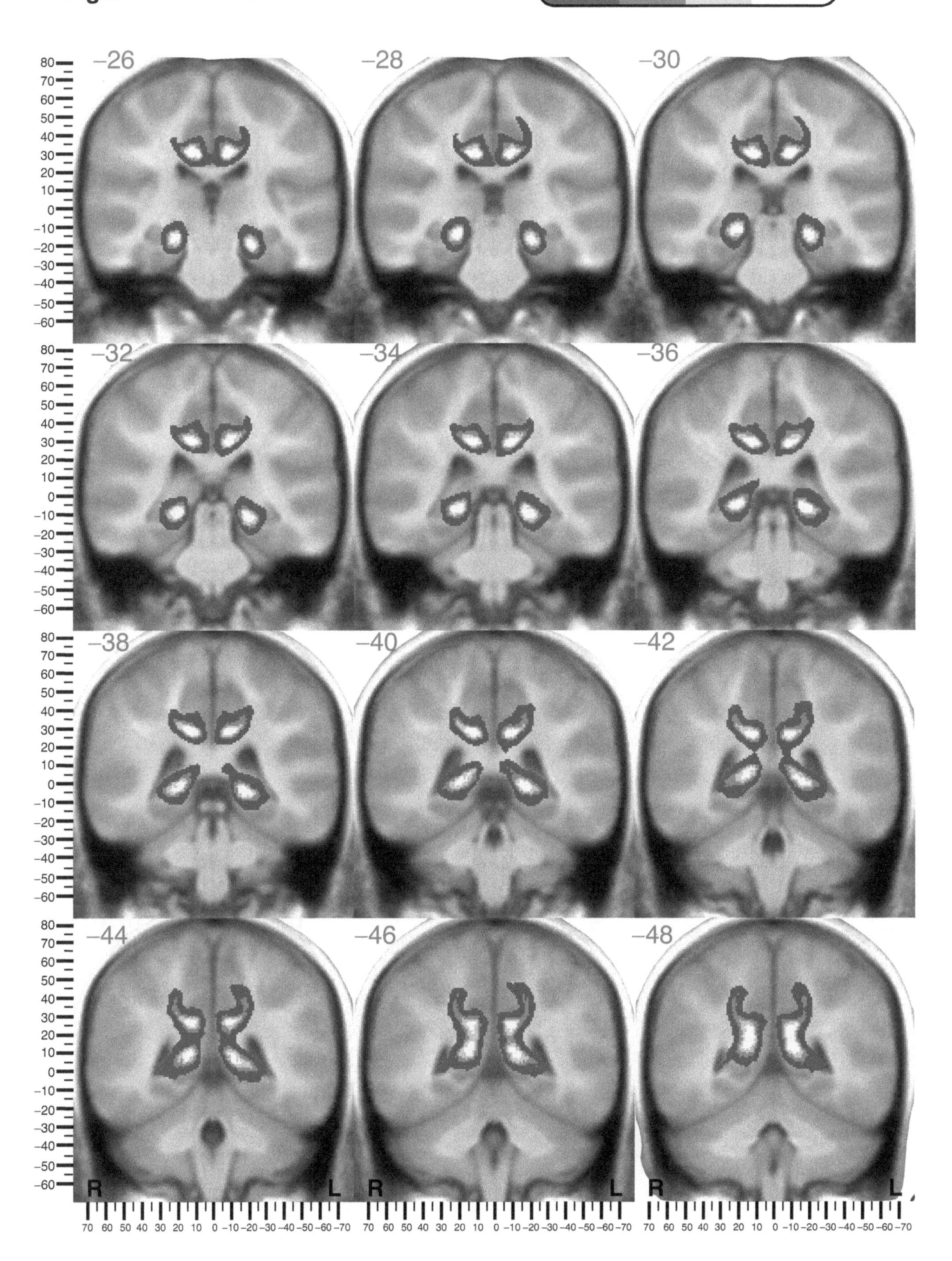

Cingulum: anterior coronal view

p < .05 | > 50% | > 75% | > 90%

−50 −52 −54 −56 −58 −60 −62 −64 −66 −68 −70

R L

Cingulum (right): lateral sagittal view

p < .05 | > 50% | > 75% | > 90%

32 30 28 26 24 22 20 18

80 70 60 50 40 30 20 10 0 −10 −20 −30 −40 −50 −60

−110 −100 −90 −80 −70 −60 −50 −40 −30 −20 −10 0 10 20 30 40 50 60 70

−100 −90 −80 −70 −60 −50 −40 −30 −20 −10 0 10 20 30 40 50 60 70

Cingulum (right): lateral sagittal view

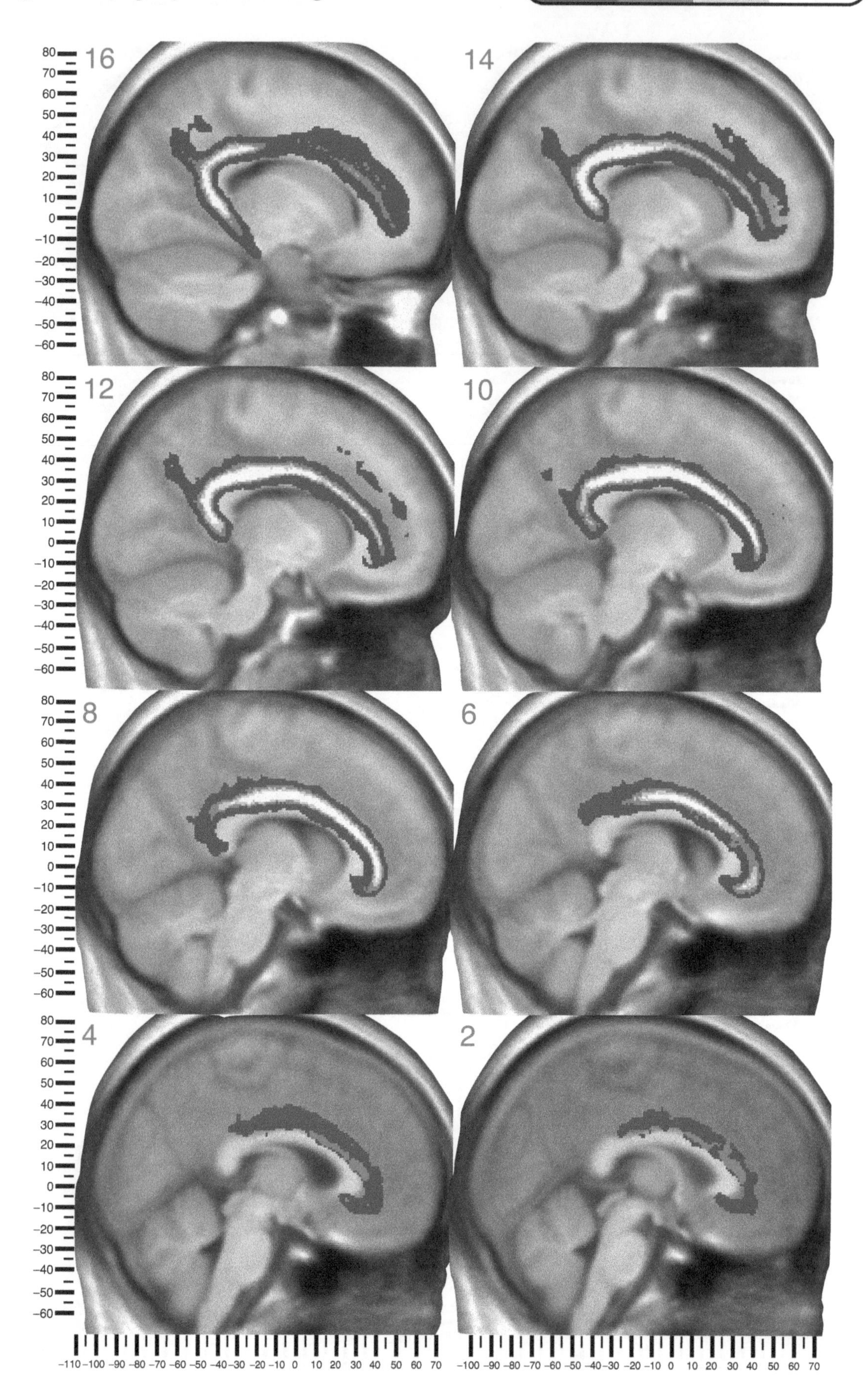

Cingulum (left): medial sagittal view

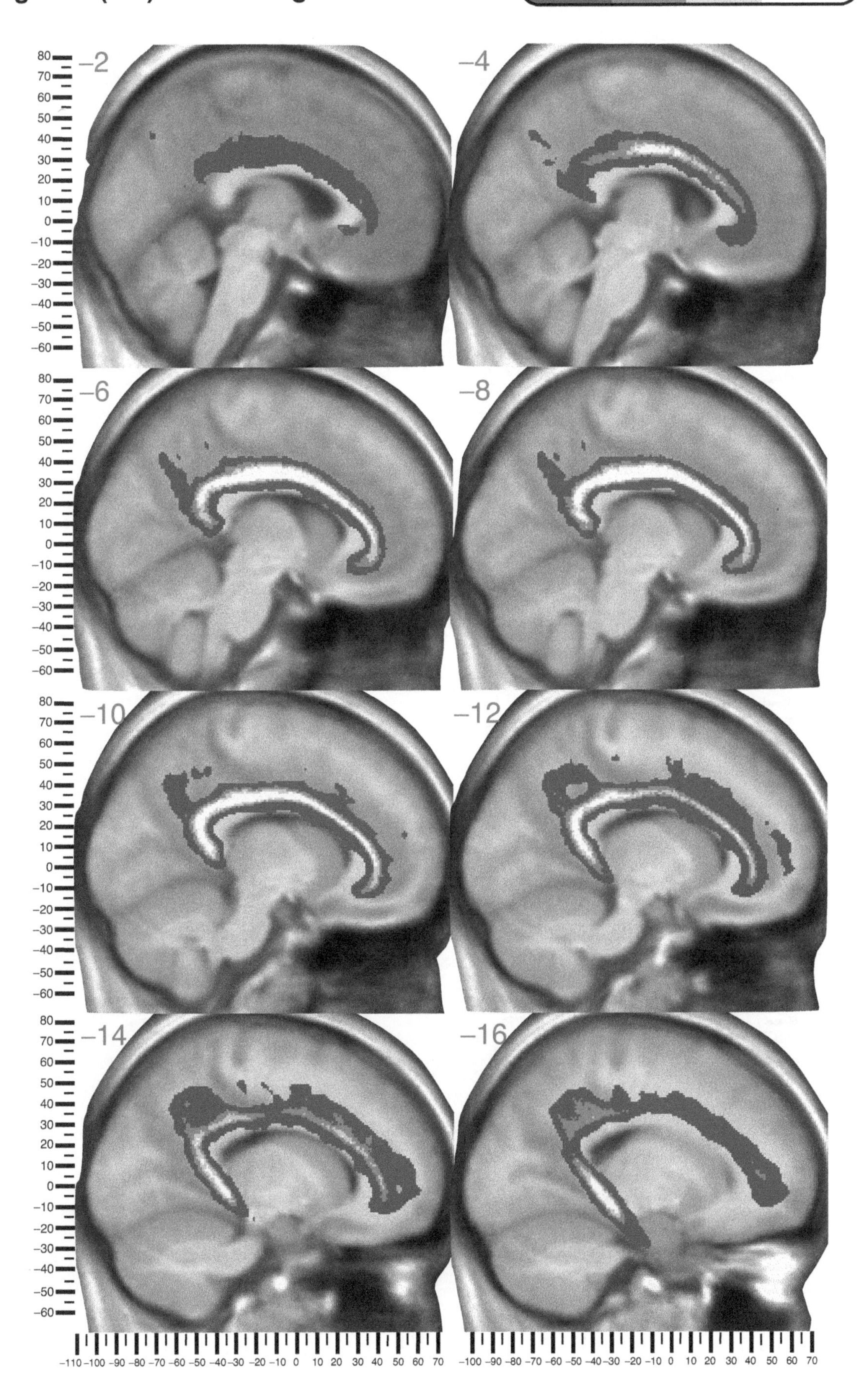

Cingulum (left): medial sagittal view

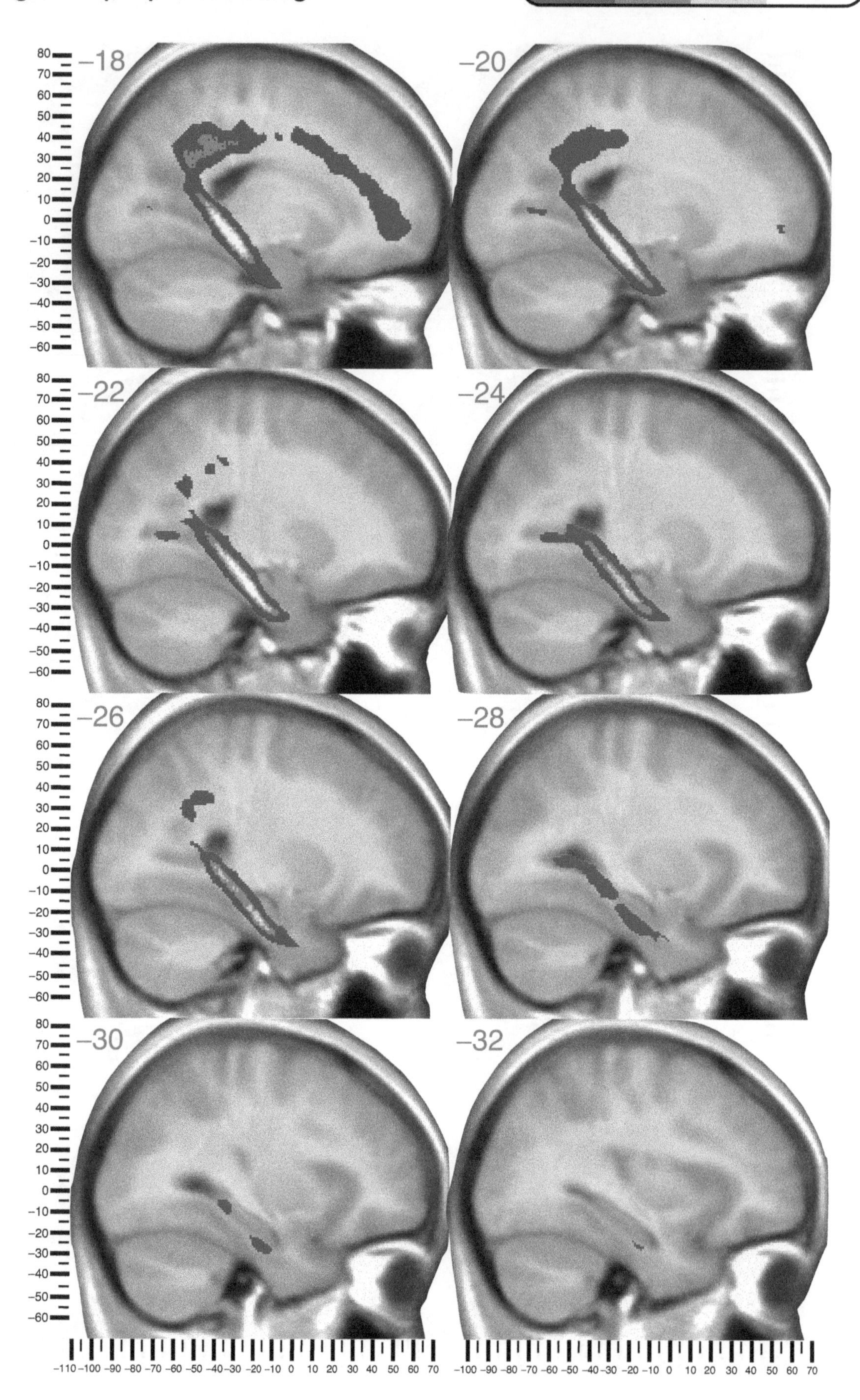

Uncinate fasciculus (atlas)

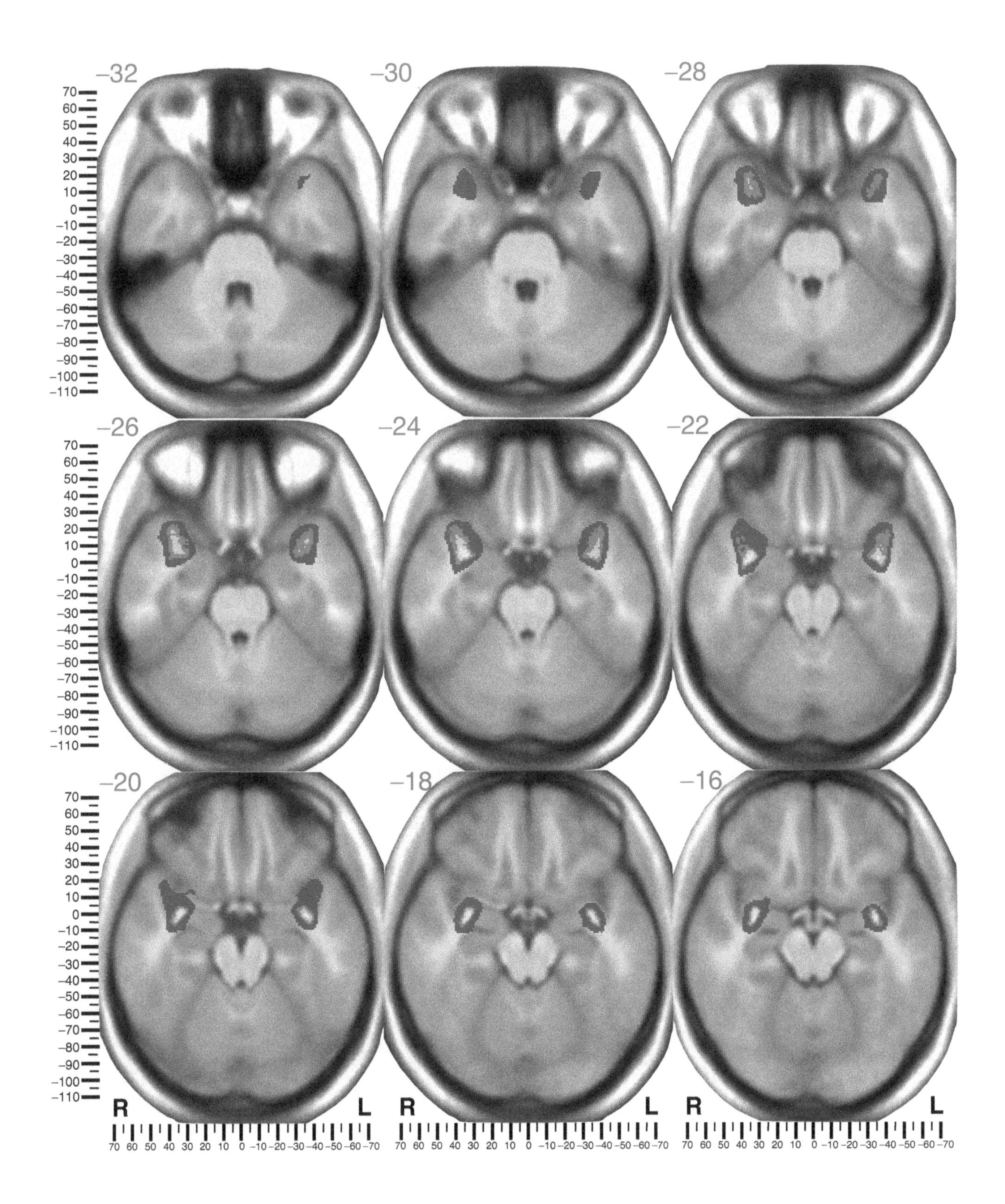

Uncinate: inferior axial view

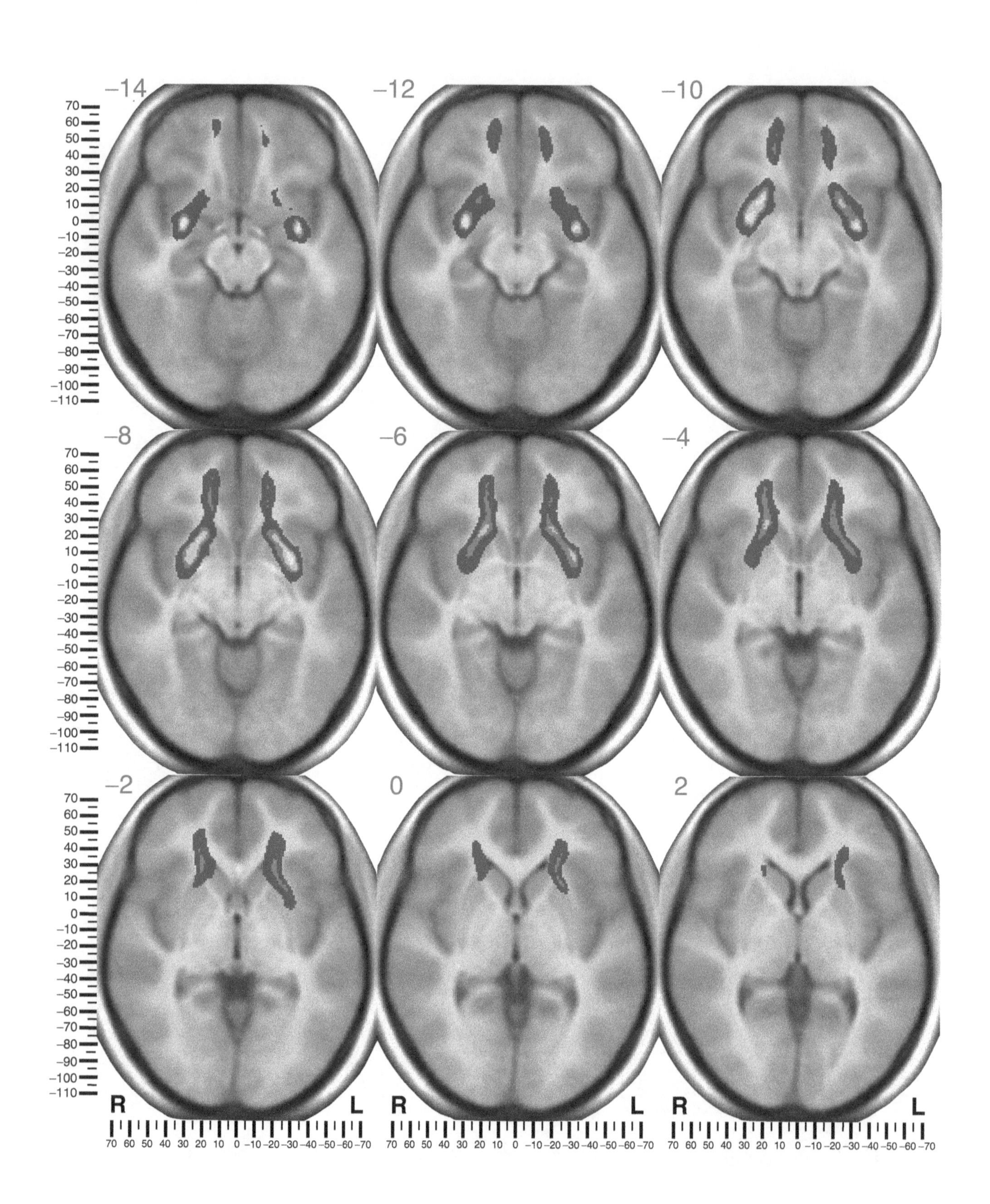

Uncinate: anterior coronal view

p < .05 | > 50% | > 75% | > 90%

80 70 60 50 40 30 20 10 0 –10 –20 –30 –40 –50 –60

54 52 50

80 70 60 50 40 30 20 10 0 –10 –20 –30 –40 –50 –60

48 46 44

80 70 60 50 40 30 20 10 0 –10 –20 –30 –40 –50 –60

42 40 38

80 70 60 50 40 30 20 10 0 –10 –20 –30 –40 –50 –60

36 34 32

R L R L R L

70 60 50 40 30 20 10 0 –10 –20 –30 –40 –50 –60 –70 70 60 50 40 30 20 10 0 –10 –20 –30 –40 –50 –60 –70 70 60 50 40 30 20 10 0 –10 –20 –30 –40 –50 –60 –70

Uncinate: anterior coronal view

Uncinate: anterior coronal view

p < .05 | > 50% | > 75% | > 90%

Uncinate (right): lateral sagittal view

p < .05 | > 50% | > 75% | > 90%

80 70 60 50 40 30 20 10 0 −10 −20 −30 −40 −50 −60

44 42

80 70 60 50 40 30 20 10 0 −10 −20 −30 −40 −50 −60

40 38

80 70 60 50 40 30 20 10 0 −10 −20 −30 −40 −50 −60

36 34

80 70 60 50 40 30 20 10 0 −10 −20 −30 −40 −50 −60

32 30

−110 −100 −90 −80 −70 −60 −50 −40 −30 −20 −10 0 10 20 30 40 50 60 70

−100 −90 −80 −70 −60 −50 −40 −30 −20 −10 0 10 20 30 40 50 60 70

Uncinate (right): lateral sagittal view

p < .05 | > 50% | > 75% | > 90%

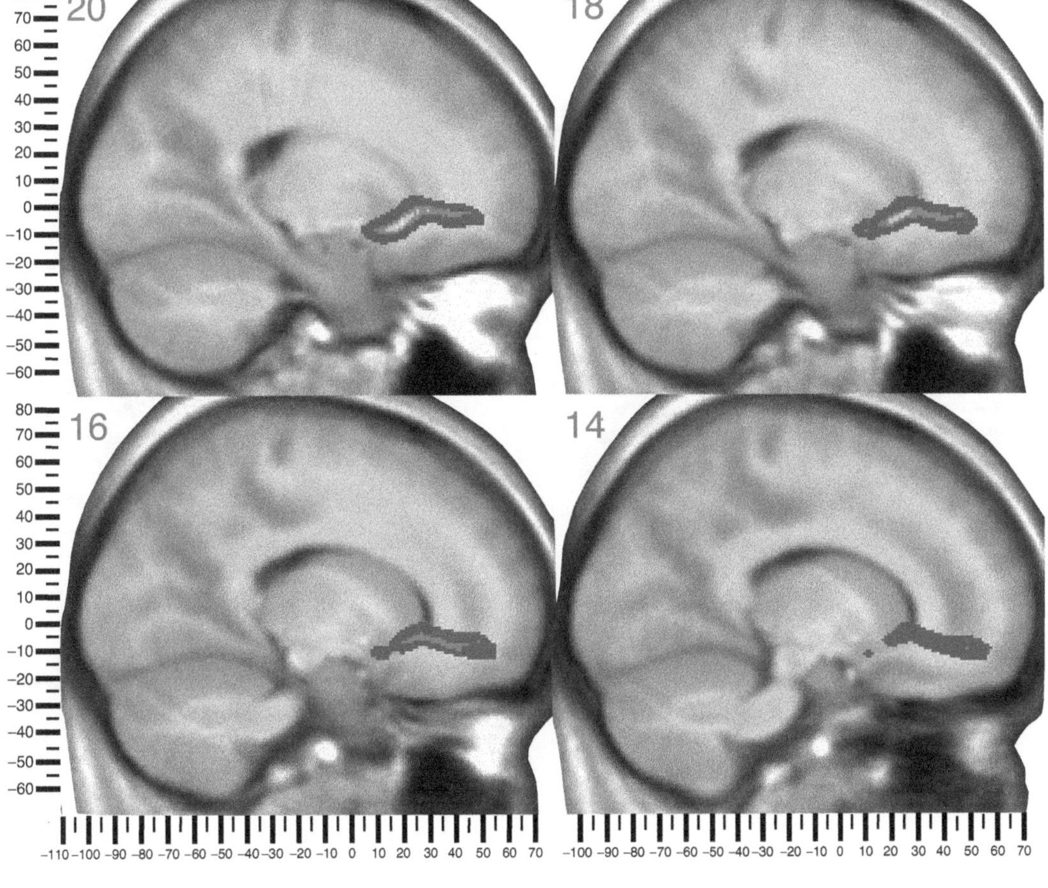

Uncinate (left): medial sagittal view

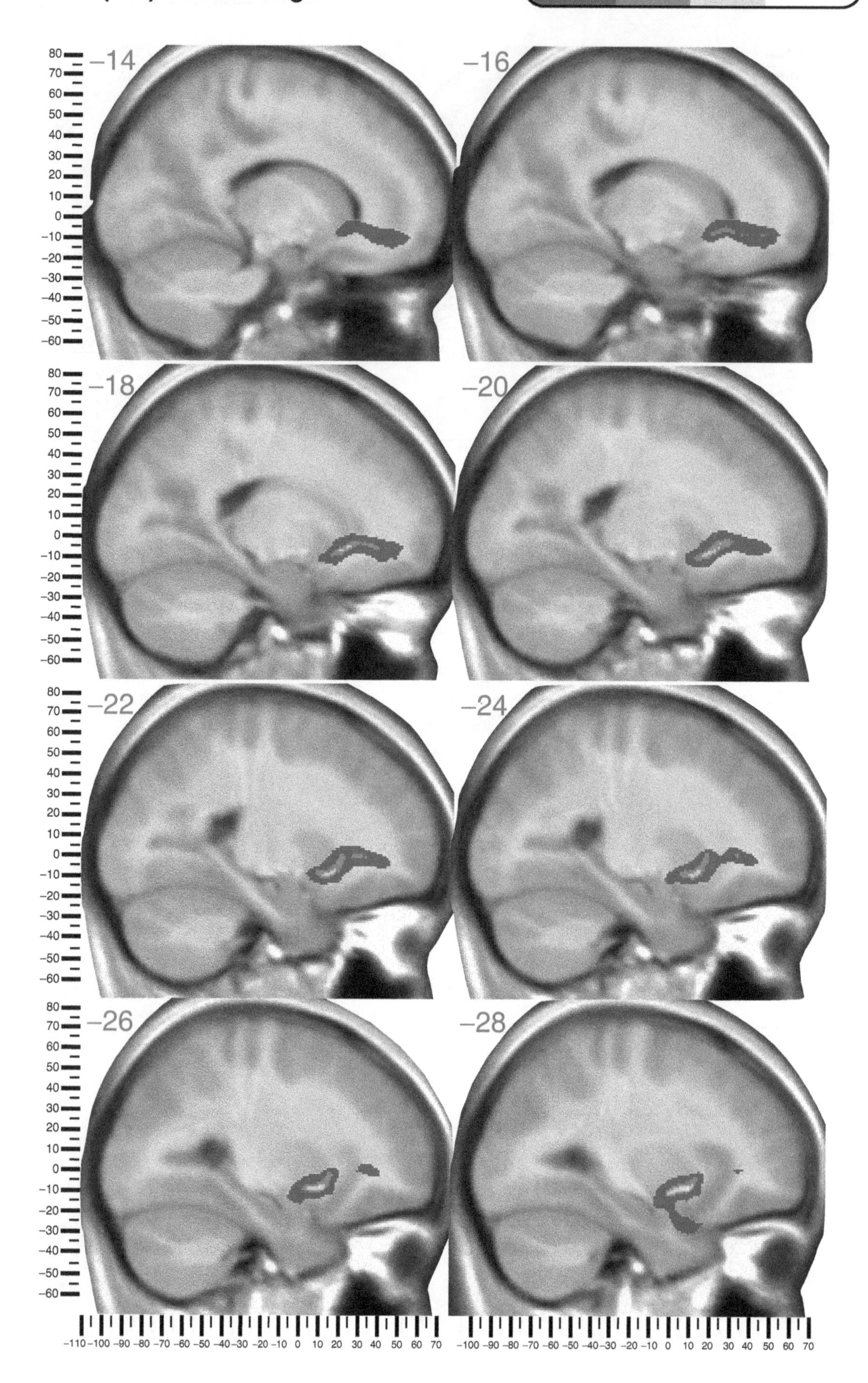

Uncinate (left): medial sagittal view

p < .05 | > 50% | > 75% | > 90%

-30 -32 -34 -36 -38 -40 -42 -44

Fornix (atlas)

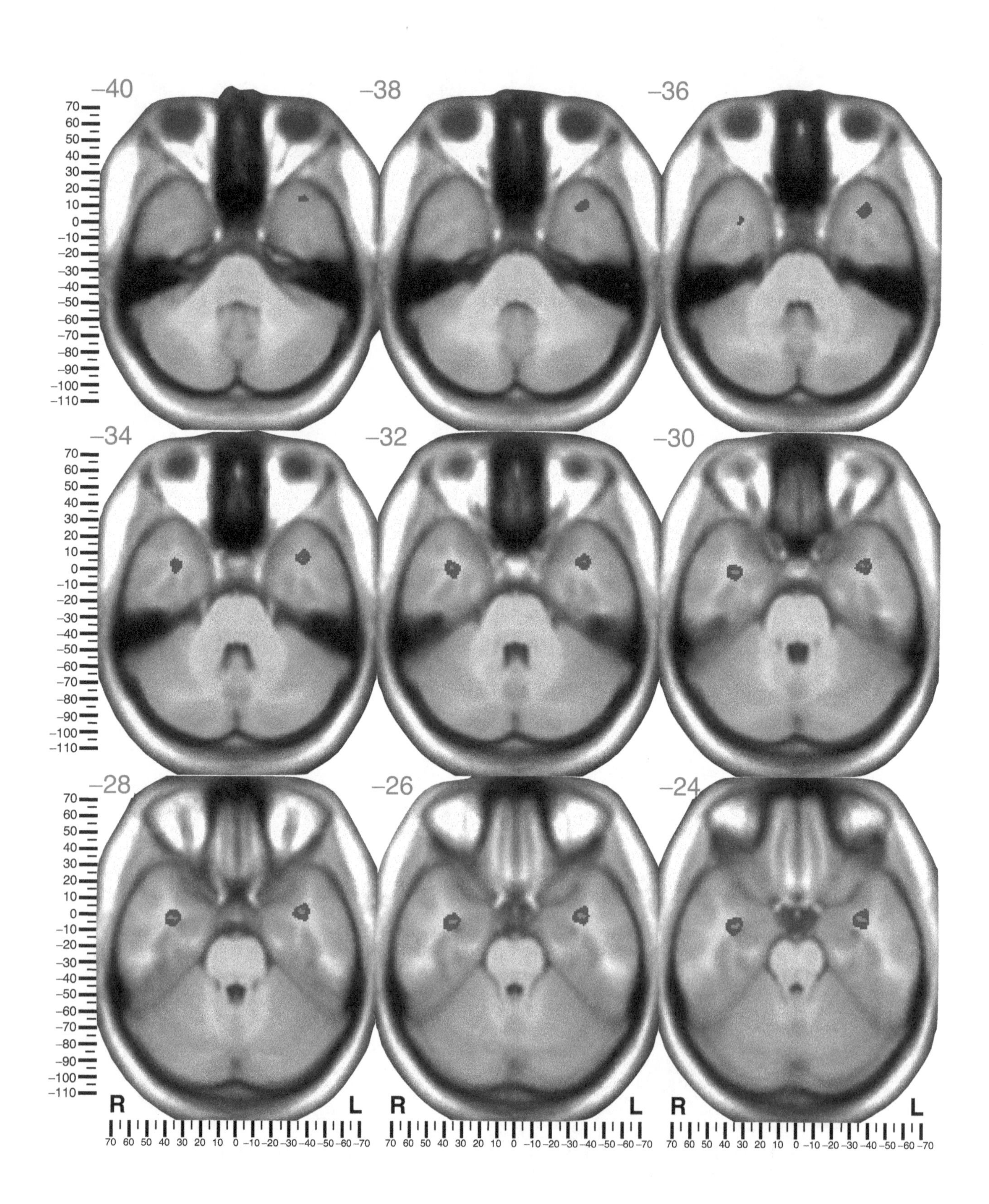

Fornix: inferior axial view

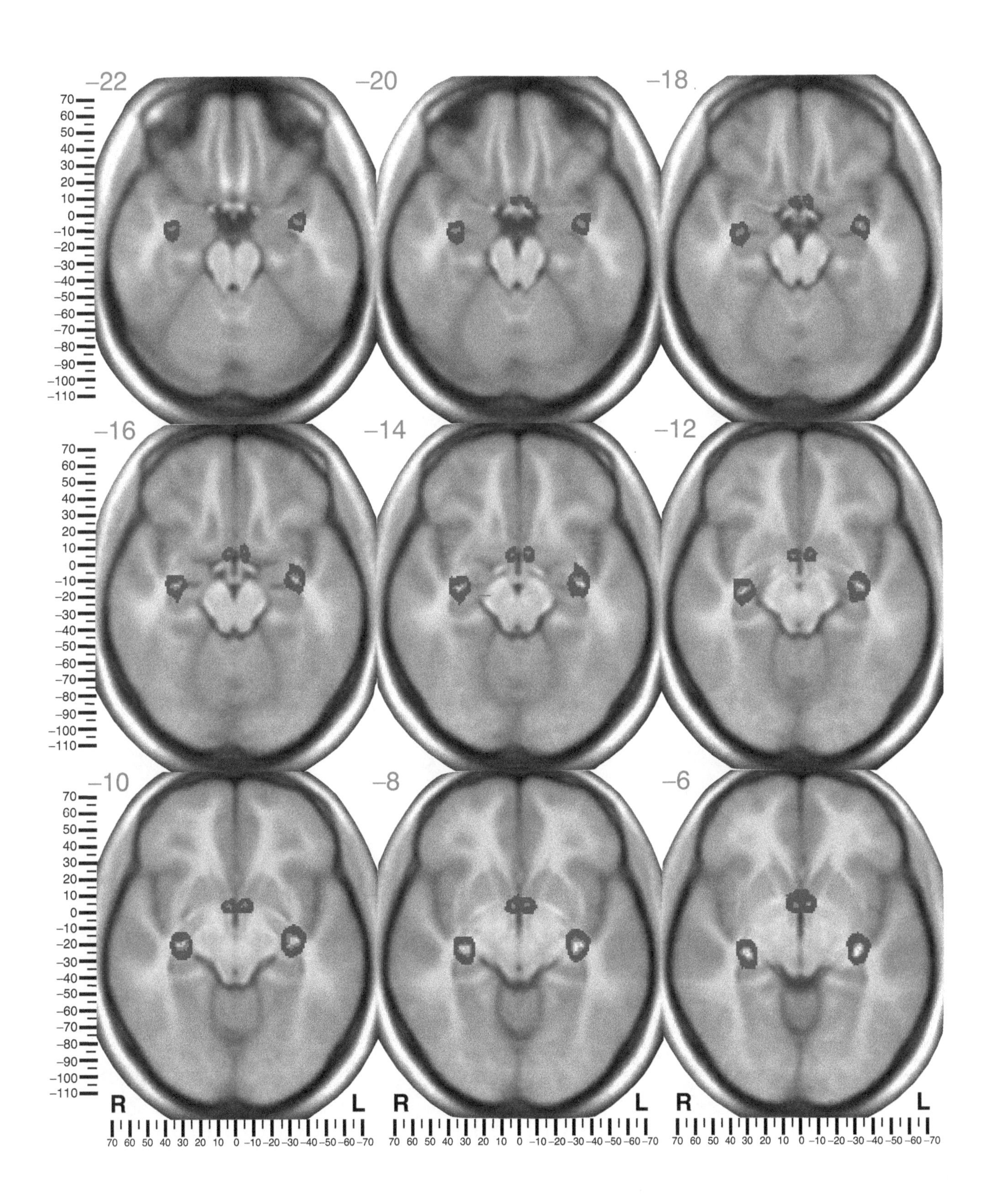

Fornix: inferior axial view

p < .05 | > 50% | > 75% | > 90%

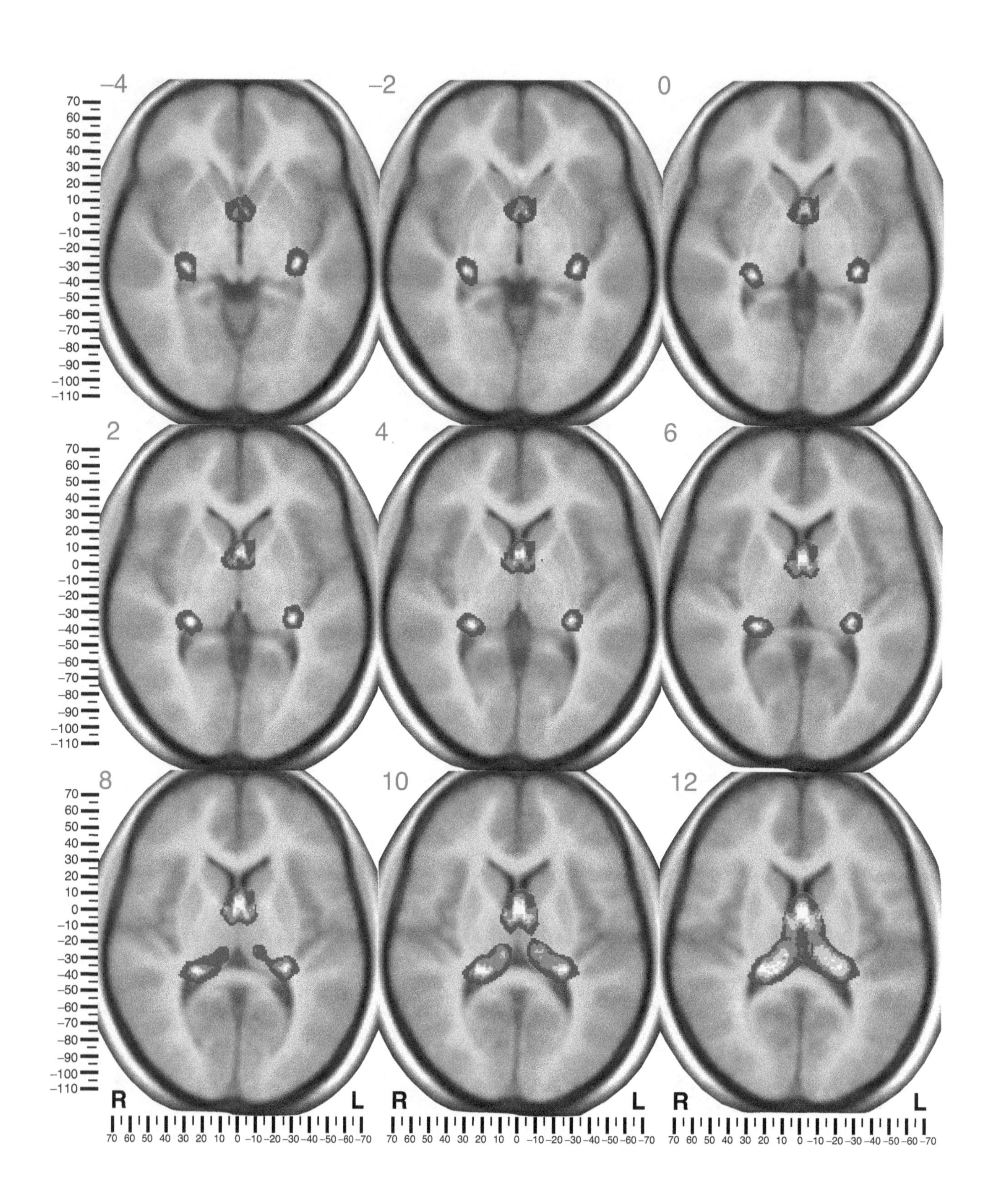

Fornix: inferior axial view

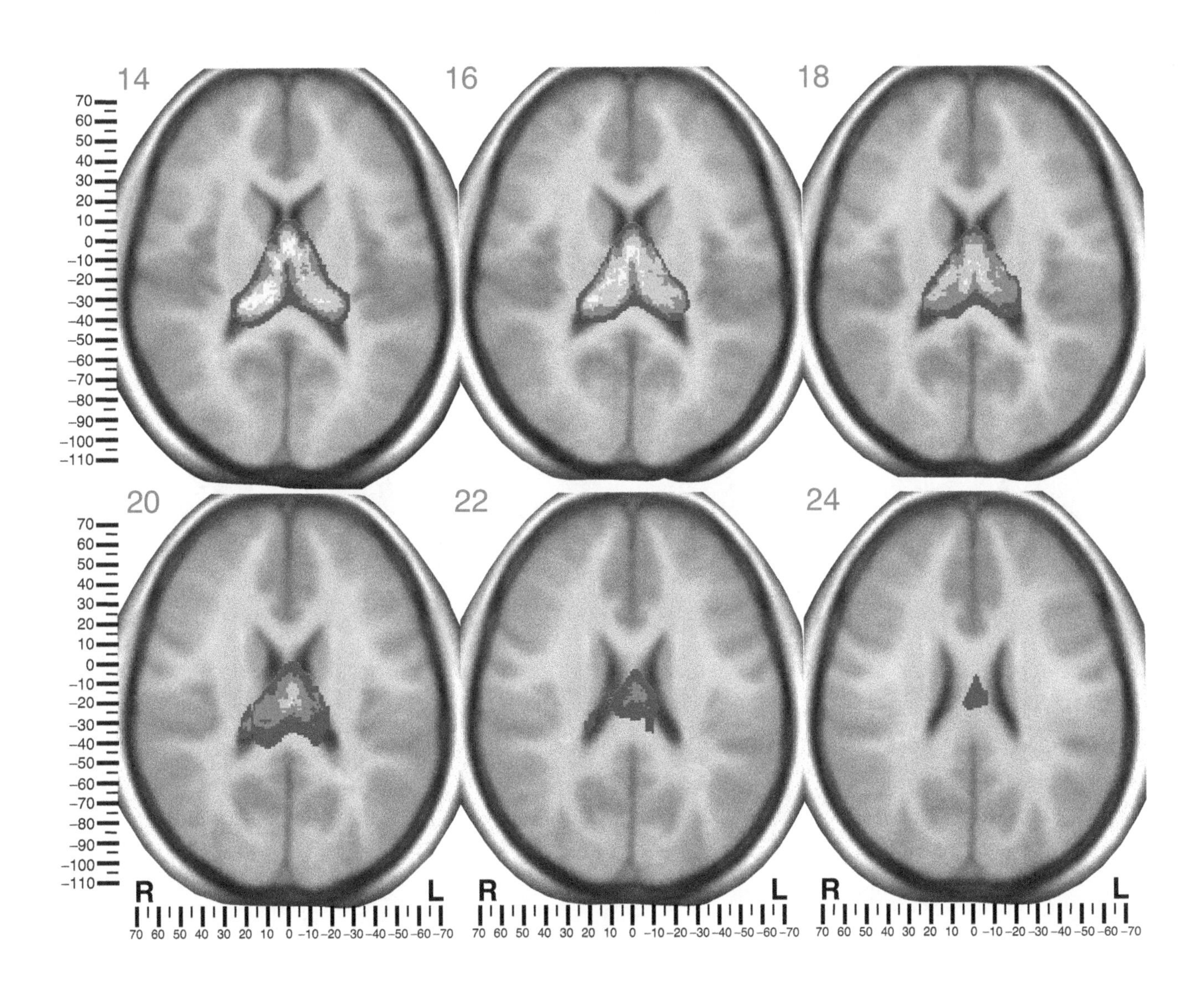

Fornix: anterior coronal view

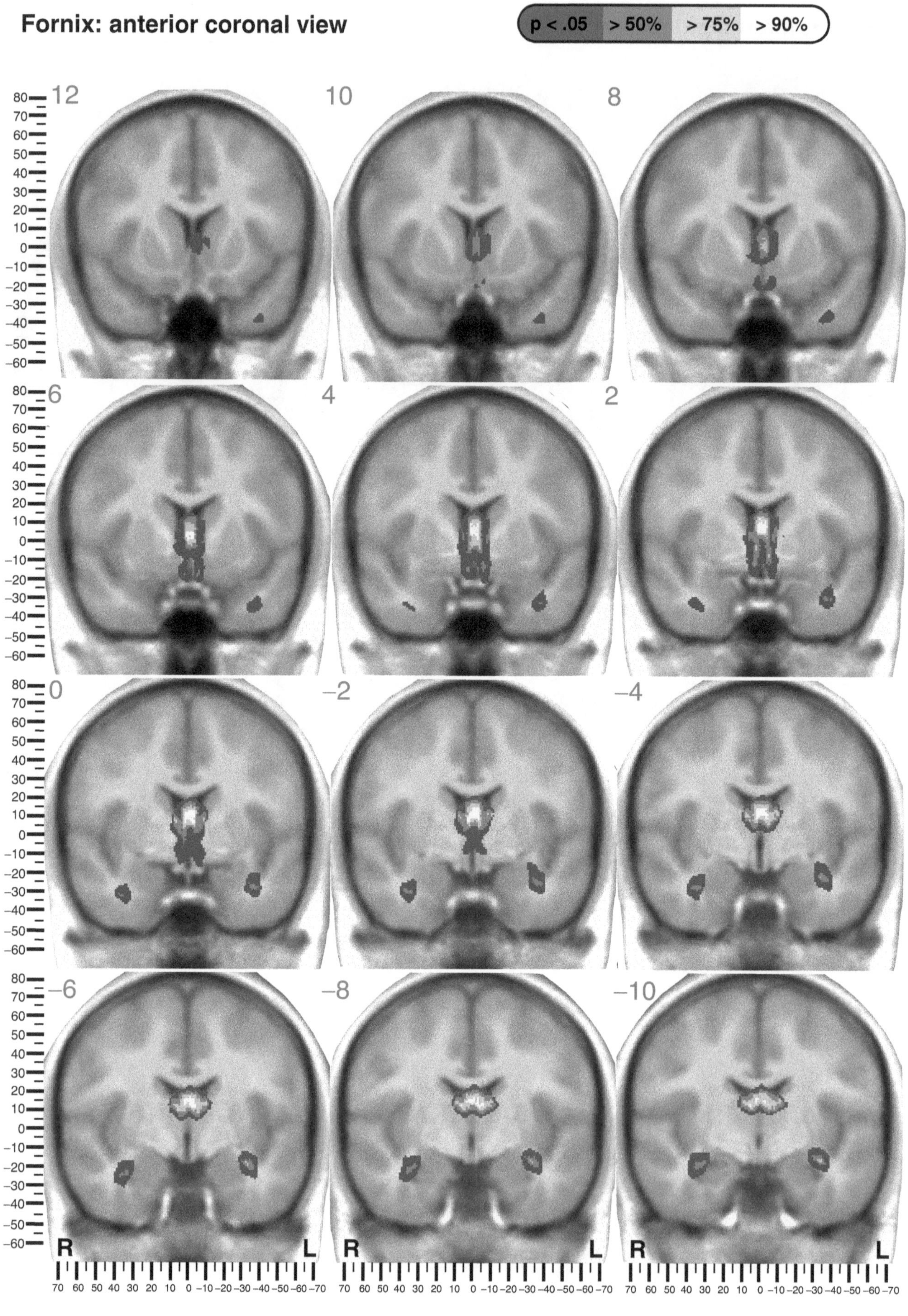

p < .05 | > 50% | > 75% | > 90%

12 10 8

6 4 2

0 −2 −4

−6 −8 −10

R L R L R L

Fornix: anterior coronal view

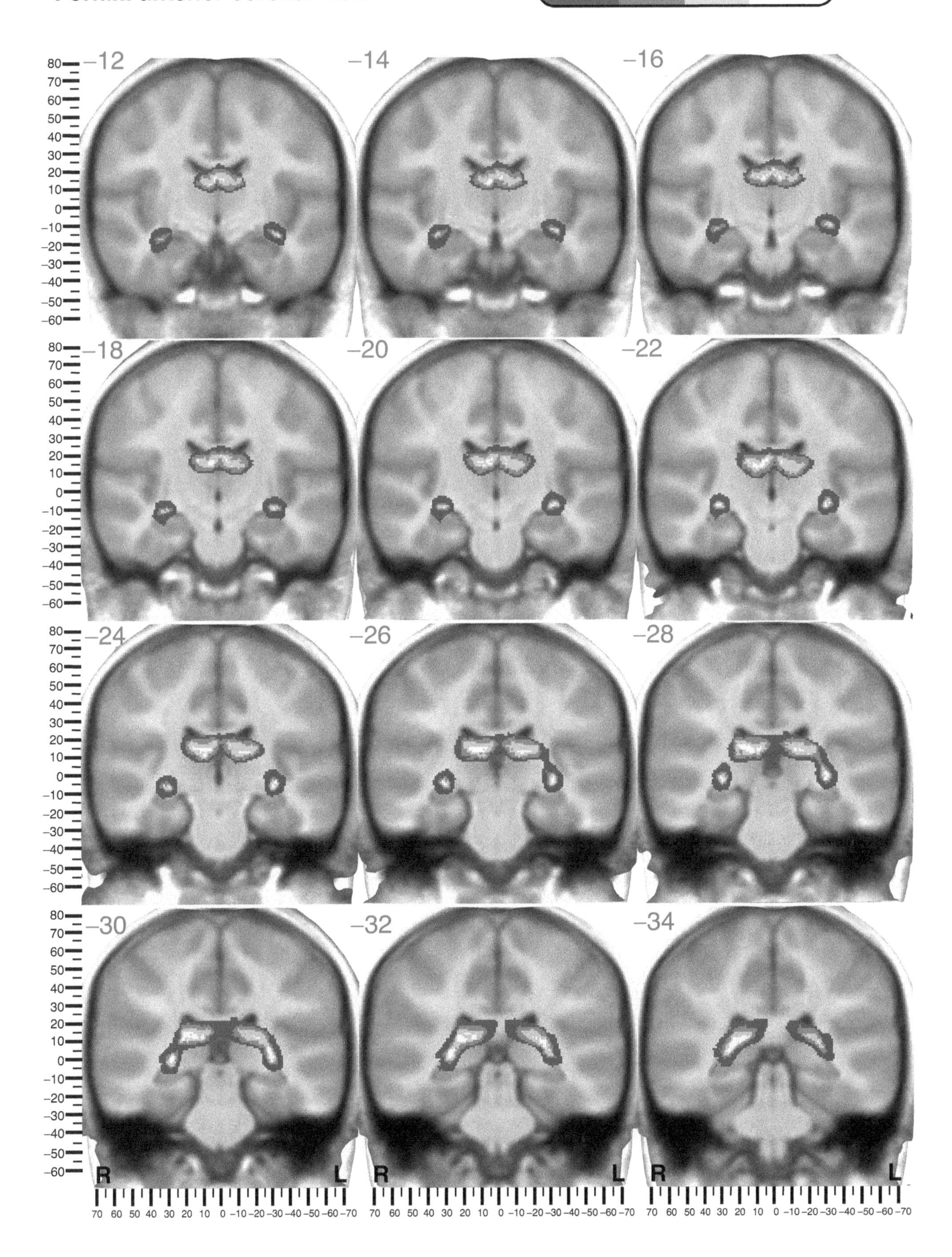

Fornix: anterior coronal view

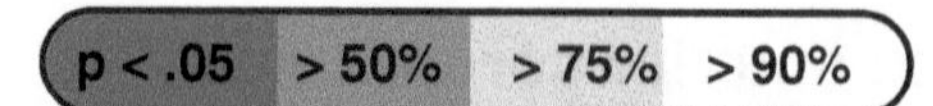

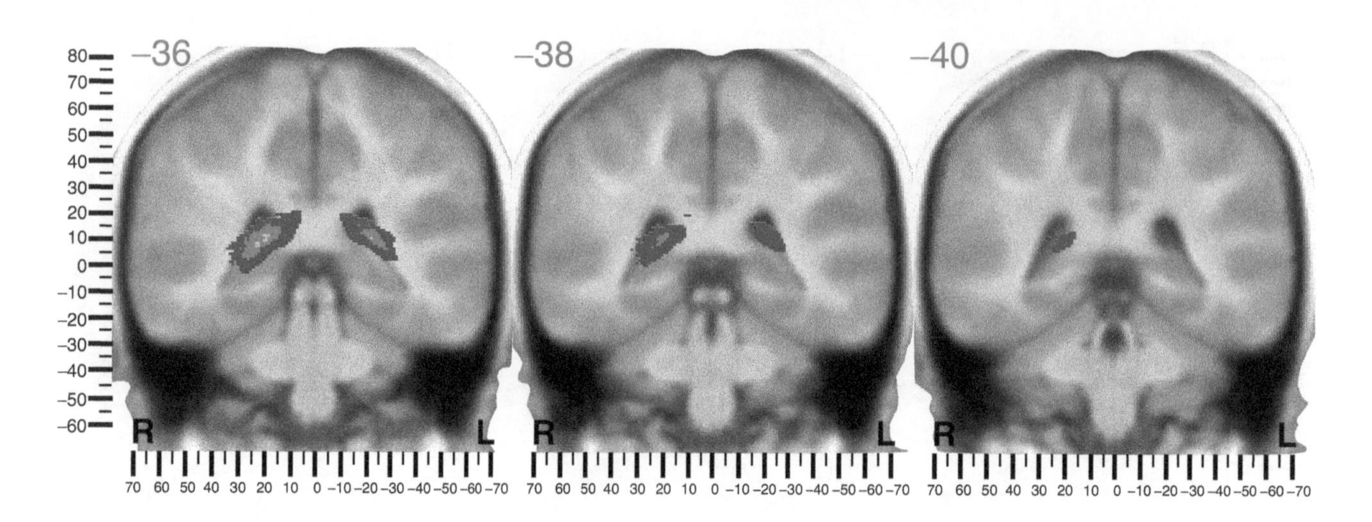

Fornix (right): lateral sagittal view

p < .05 | > 50% | > 75% | > 90%

Fornix (right): lateral sagittal view

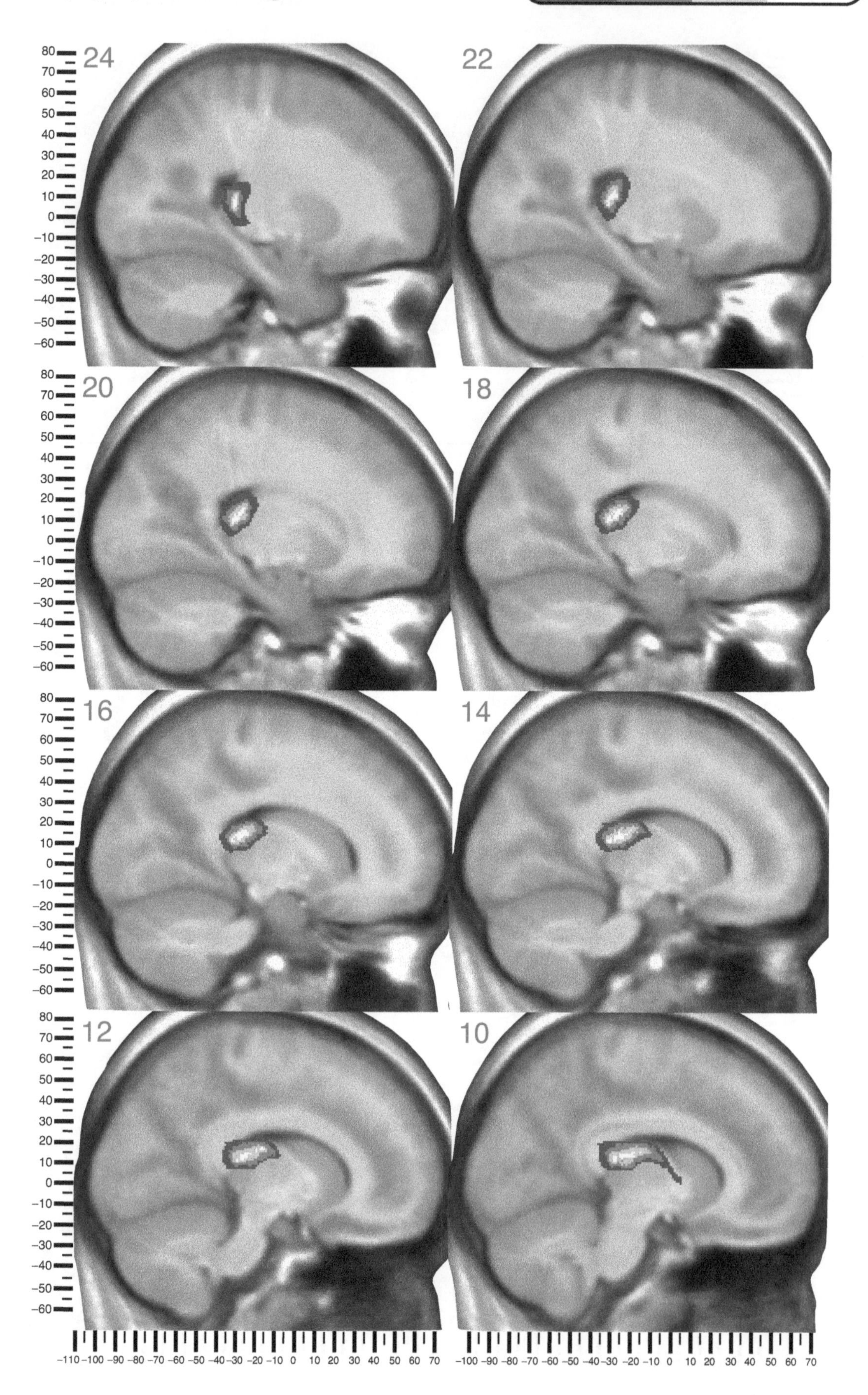

Fornix (right): lateral sagittal view

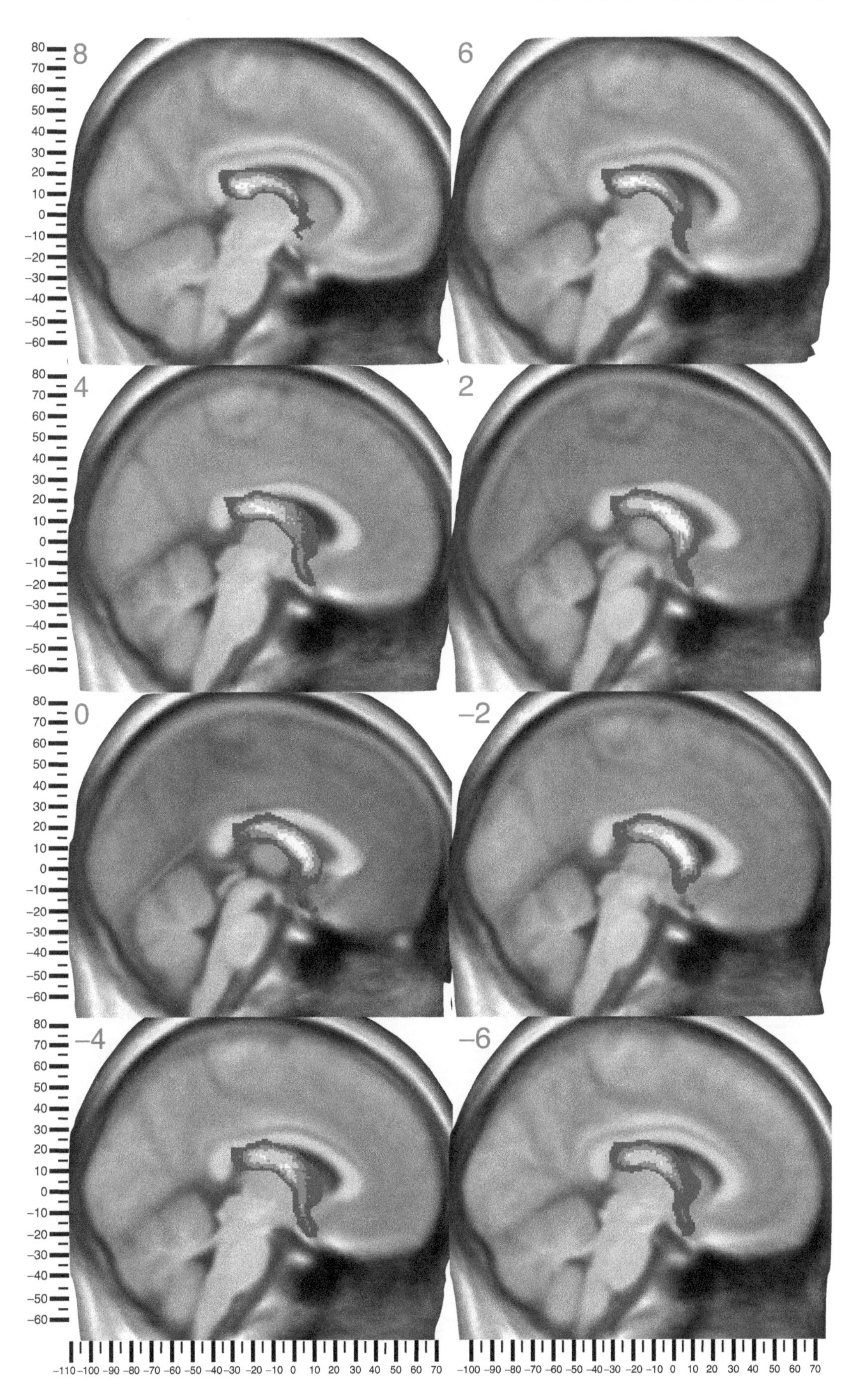

Fornix (left): medial sagittal view

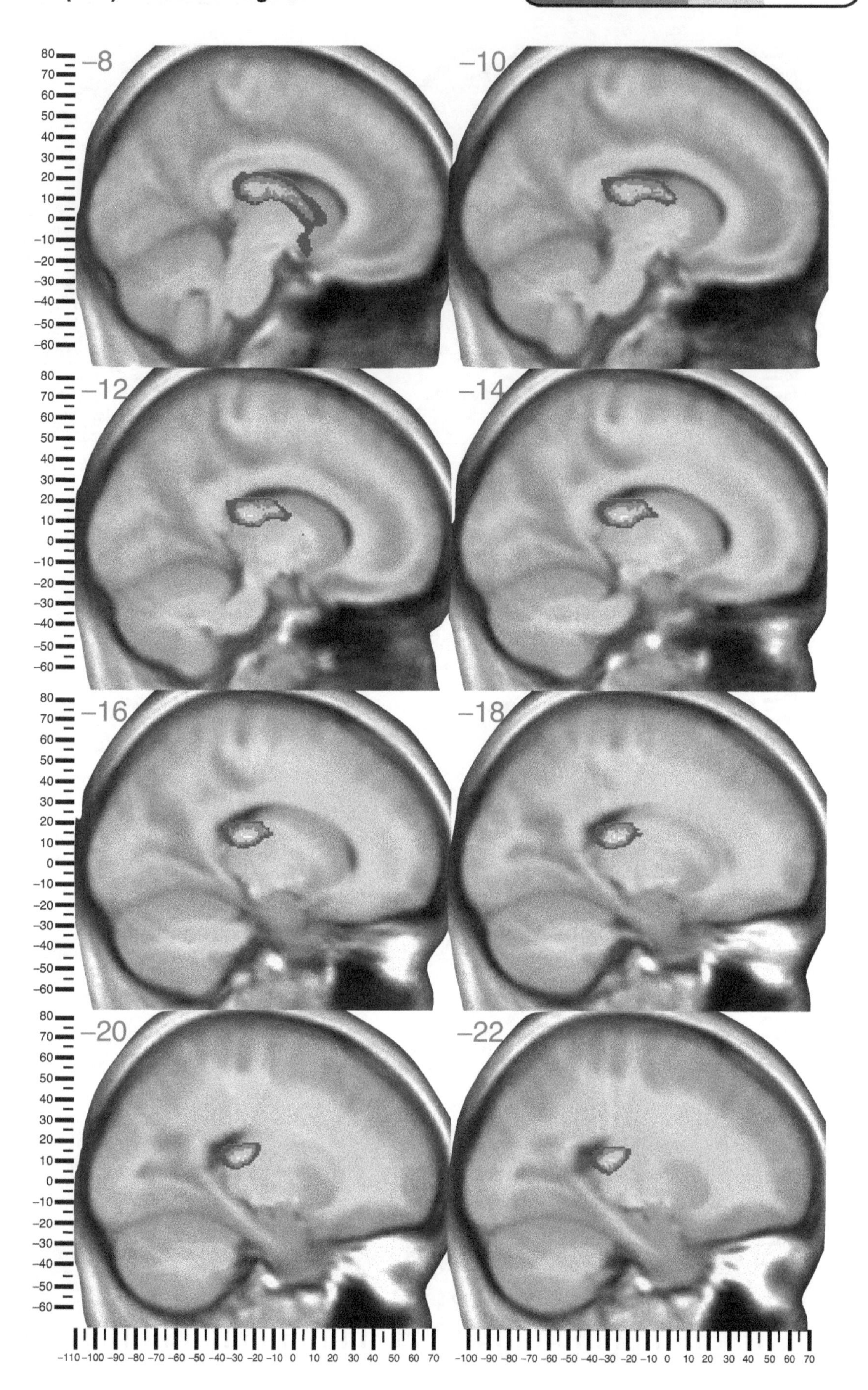

Fornix (left): medial sagittal view

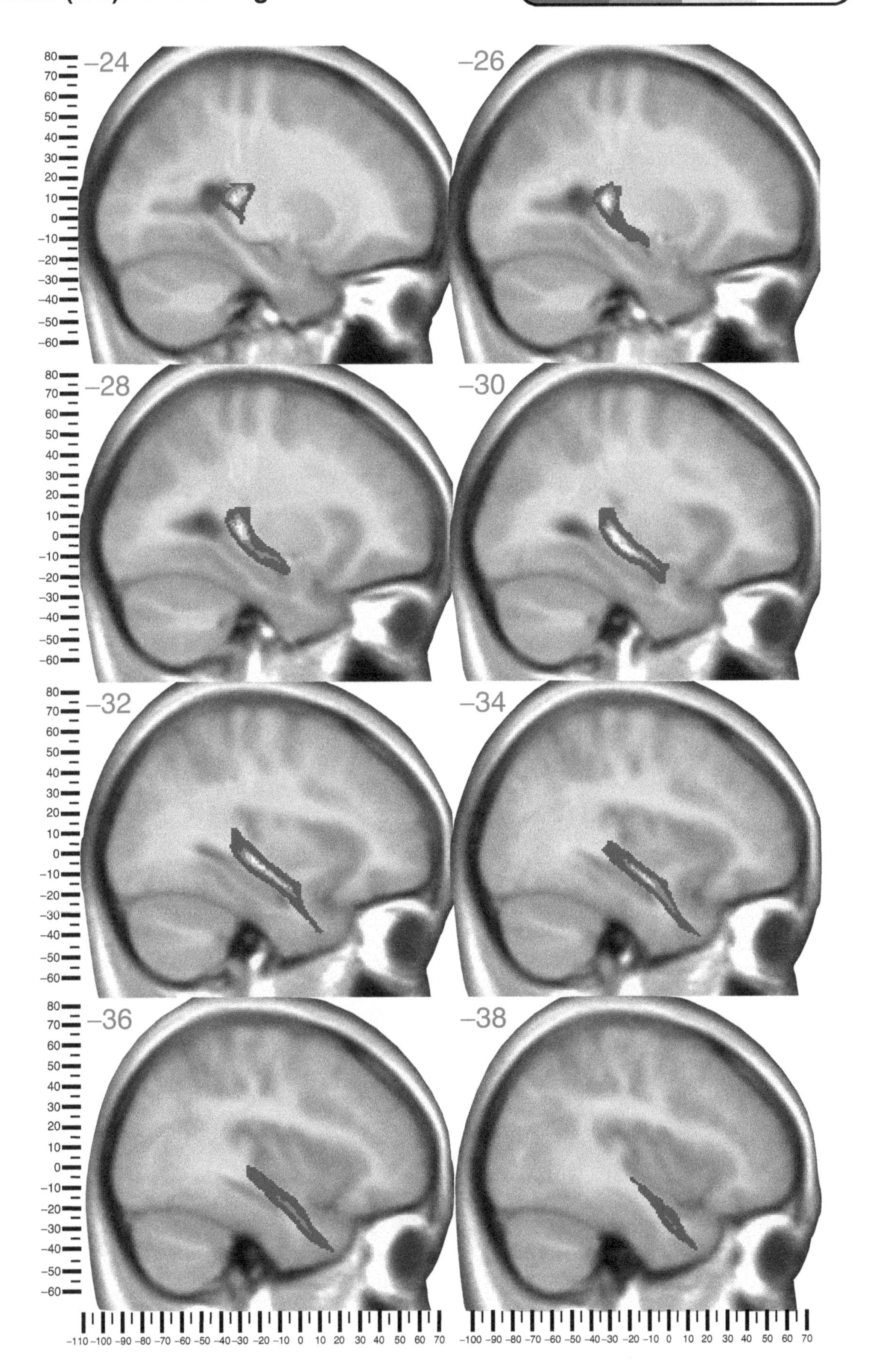

APPENDIX I

Introduction to Diffusion Imaging Tractography

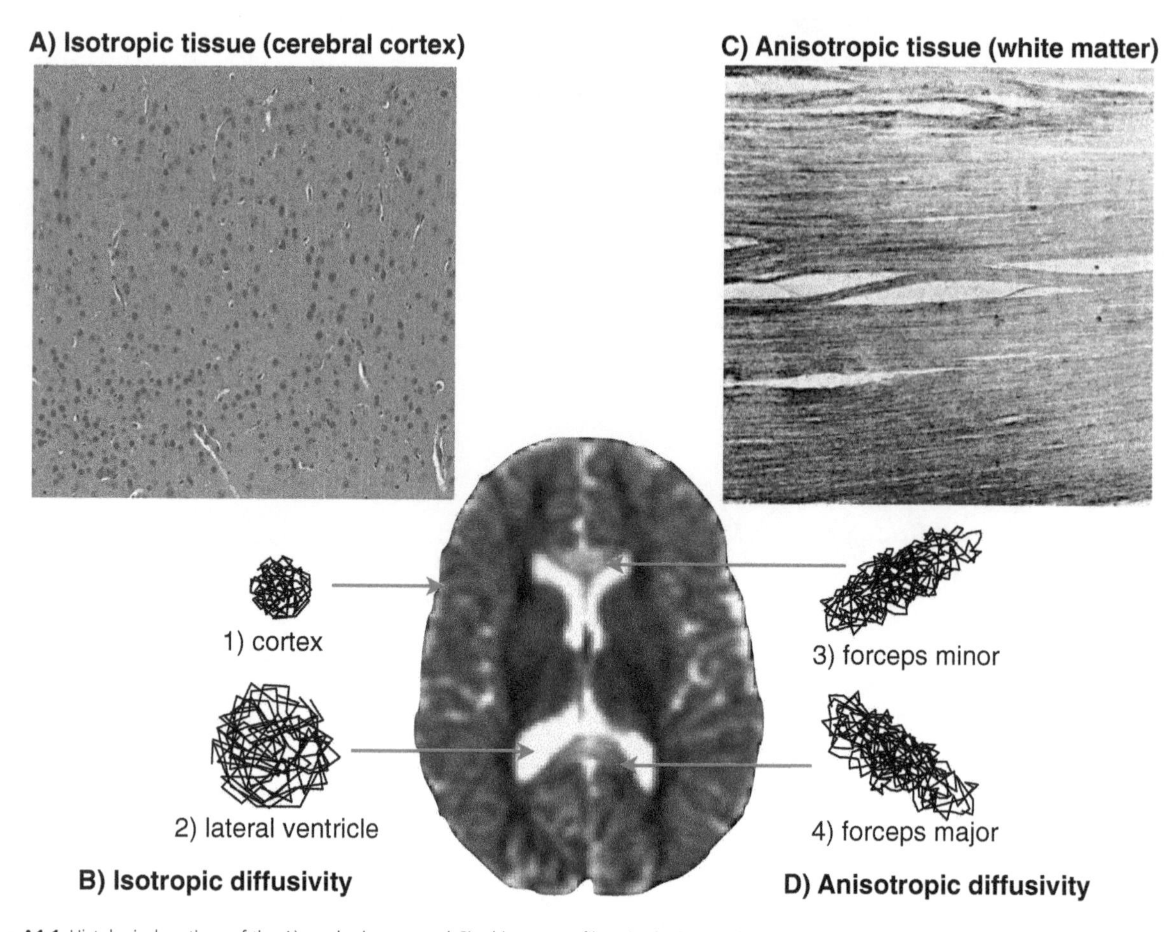

Figure A1.1 Histological sections of the A) cerebral cortex and C) white matter fibres in the human brain. The two tissues differ in their microstructural organization and composition of their biological constituents. In the middle, an axial ADC map of the human brain is shown together with the visualization of the 3D displacement of a water molecule in regions with different degrees of anisotropy and diffusivity: 1) low isotropic diffusivity in the frontal cortex where the overall displacement is equally hindered in all directions; 2) high isotropic diffusivity in the cerebro-spinal fluid of the lateral ventricles; 3) oblique anisotropic diffusivity along the lateral fibres of the genu in the forceps minor; 4) oblique anisotropic diffusivity along the lateral fibres of the splenium in the forceps major. Note that in 1) and 2) the movement of water molecules is random in all directions but in 1) is more restricted than in 2) due to the presence of biological barriers. This is reflected by the difference in grey level between the voxels of the two regions. Also the distinction between grey matter, white matter, and cerebro-spinal fluid is not always possible in this image due to the different orientation of the underlying white matter tissue in some brain regions. For example, the central part of the splenium has bright voxels that are similar to those of the lateral vetnricles containing cerebro-spinal fluid.

The *Atlas of Human Brain Connections* is based on diffusion tensor methods developed in the last 15 years. An overview of these methods is given in this appendix and Appendix II; limitations and future directions are also discussed.

Diffusion tractography is a magnetic resonance imaging (MRI) method for the *in vivo* quantification of certain microstructural characteristics of biological tissues (e.g. axonal composition, degree of myelination, cohesiveness of bundles, etc.) and the virtual reconstruction of white matter trajectories. The diffusion-weighted MRI pulse sequences are sensitive to the displacement of water molecules within biological tissues that follows Einstein's equation, where the mean squared displacement $\langle r^2 \rangle$ is directly proportional to the observation time (t) according to:

$$\langle r^2 \rangle = 6Dt$$

Given a body temperature of 37°C, the diffusion coefficient of free water (D) is about 3×10^{-3} mm^2s^{-1}. Thus, within a voxel containing, for example, cerebrospinal fluid (CSF), where D is very close to the coefficient of free water, the water molecules will displace randomly on an average distance of 20 μm in all directions in about 20 ms. This is true only if the water molecules are allowed to move freely, which is not the case for the water contained in the nervous tissue where the presence of cell membranes, proteins, myelin, intracellular filaments, organelles, and so forth hinders the water displacement. It follows that the displacement and, therefore, the diffusion coefficient in brain tissue are smaller than in free water. This is partly why the term 'apparent diffusion coefficient' (ADC) was coined, to reflect the fact that in the tissue the water diffusion is hindered by several 'biological barriers' (Tanner, 1978; Le Bihan et al., 1986). One of the first applications of diffusion imaging has been in stroke where Moseley et al. (1990) found that the pathological changes that occur in ischaemic tissue affect the diffusivity of water leading to a general reduction of ADC. These changes in diffusivity occur within a few hours from the onset of the stroke, thus offering the possibility of visualizing with diffusion imaging early ischaemic-related modifications that are not visible using other structural MRI sequences.

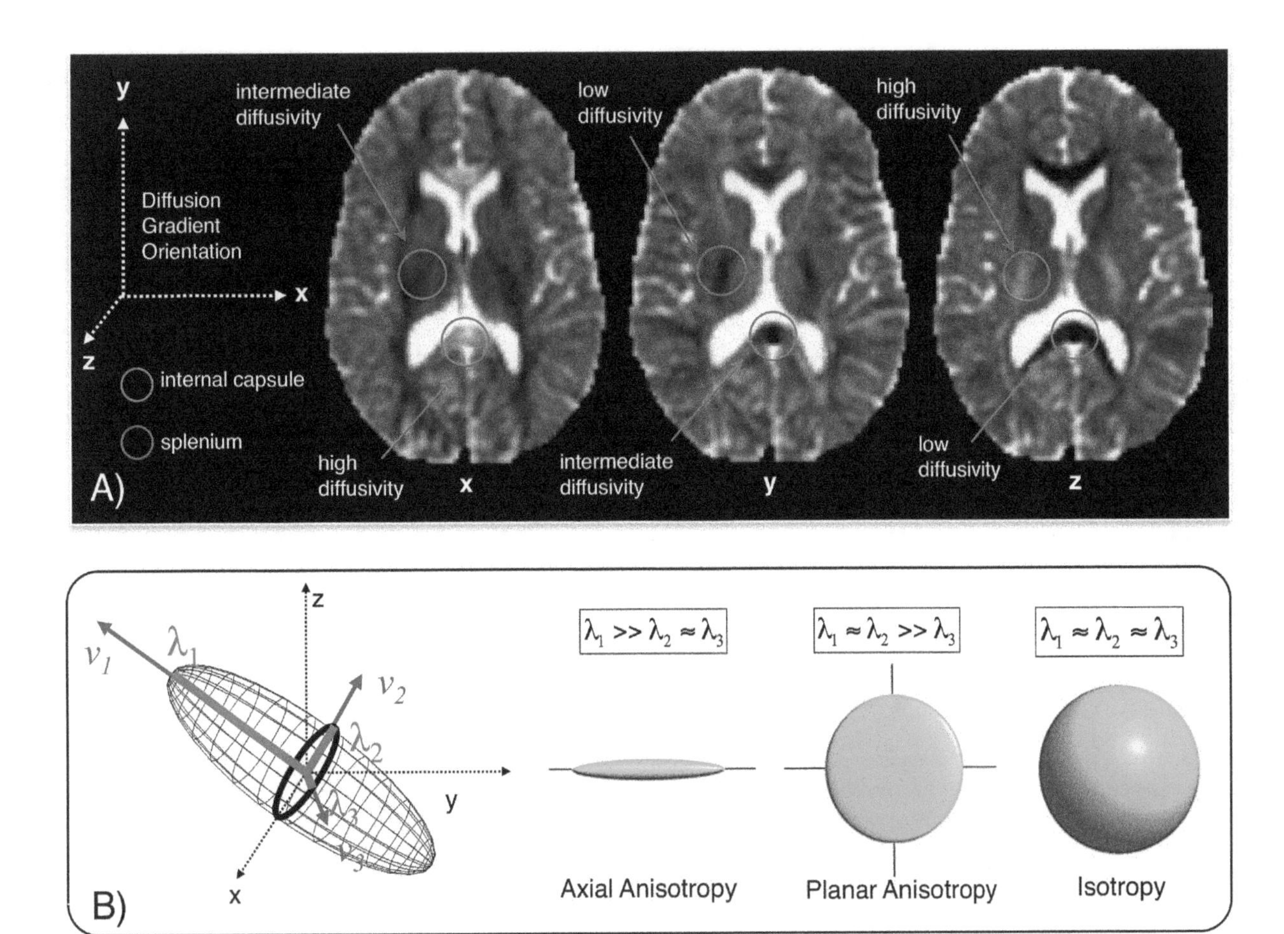

Figure A1.2 A) Monodirectional ADC maps where the signal is sensitized to the displacement of water molecules along the three orthogonal planes (x, latero-lateral direction; y, antero-posterior; z, superior-inferior). The fibres of the internal capsule and splenium have a similar arrangement and composition but different orientation. Hence, the monodirectional ADC values can differ according to the direction of the measured diffusivity. In the first figure (x-direction), the fibres of the internal capsule (blue circle) are perpendicular to the diffusion direction along the x-axis and therefore show a lower diffusivity compared to those of the central portion of the splenium (red circle), which are parallel to the x directions. The diffusivity of the voxels containing fibres of the internal capsule and splenium changes in the other two examples where diffusivity is measured along the y-axis (i.e. anterior-posterior) and z-axis (dorso-ventral) (modified from Jones, 2008). B) Visualization of the diffusion tensor as a diffusion ellipsoid. The size and the shape are defined by the three eigenvalues (λ1, λ2, λ3 in red) while the spatial orientation is described by the three eigenvectors (v1, v2, v3 in blue). In biological tissues the tensor can vary between three possible configurations: i) axial anisotropy (where λ1 is greater than λ2 and λ3) is typical of voxels containing parallel fibres; ii) planar anisotropy (when λ1 is similar to λ2 and greater than λ3) is common in voxels containing, for example, two groups of crossing or diverging fibres; iii) isotropy (when all three eigenvalues are similar) is commonly observed, for example, in the grey matter (Catani and Dell'acqua, 2011).

In the normal isotropic tissue (i.e. its physical properties are identical in all directions), the diffusion of water molecules is reduced equally along all orientations. This is the case, for example, of the grey matter in the subcortical nuclei or the cerebral cortex (Figure A1.1 A, B). Other tissues, like the white matter in the cerebrum and the spinal cord are anisotropic. Here, the fibres are parallelly oriented and the axonal membranes, together with the myelin sheets, represent the greatest biological barrier to the diffusivity of water (Figure A1.1 C, D). Hence, the ADC measured along a direction perpendicular to the fibers is always lower compared to the ADC measured along the direction of the fibres. Diffusion in anisotropic white matter is therefore characterized by having a preferential direction, which varies according to the main orientation of the fibers.

In diffusion MRI the signal is usually sensitized to the displacement of water molecules along a selected direction. The ADC measurement is therefore strictly dependent on the chosen direction, and the ADC values within a voxel of white matter differ according to the selected direction. In other words, the measurement of the ADC inside the white matter is orientationally-variant (Figure A1.2).

The orientationally-variant property of the ADC represents a problem for the exact interpretation of regional changes in the diffusion signal, especially for those lesions localized within specific tracts where the decrease of ADC could be either due to pathological changes or simply be related to a different orientation of the fibres (Moseley and Kucharczyk, 1991; Le Bihan et al., 2001). For this reason the ADC is usually measured along three axes and the mean ADC calculated for clinical purposes. As we will see in the next section, the orientationally-variant characteristic of the diffusion signal can offer additional information about the orientation of the fibres and used to reconstruct virtual trajectories of tracts in the living brain.

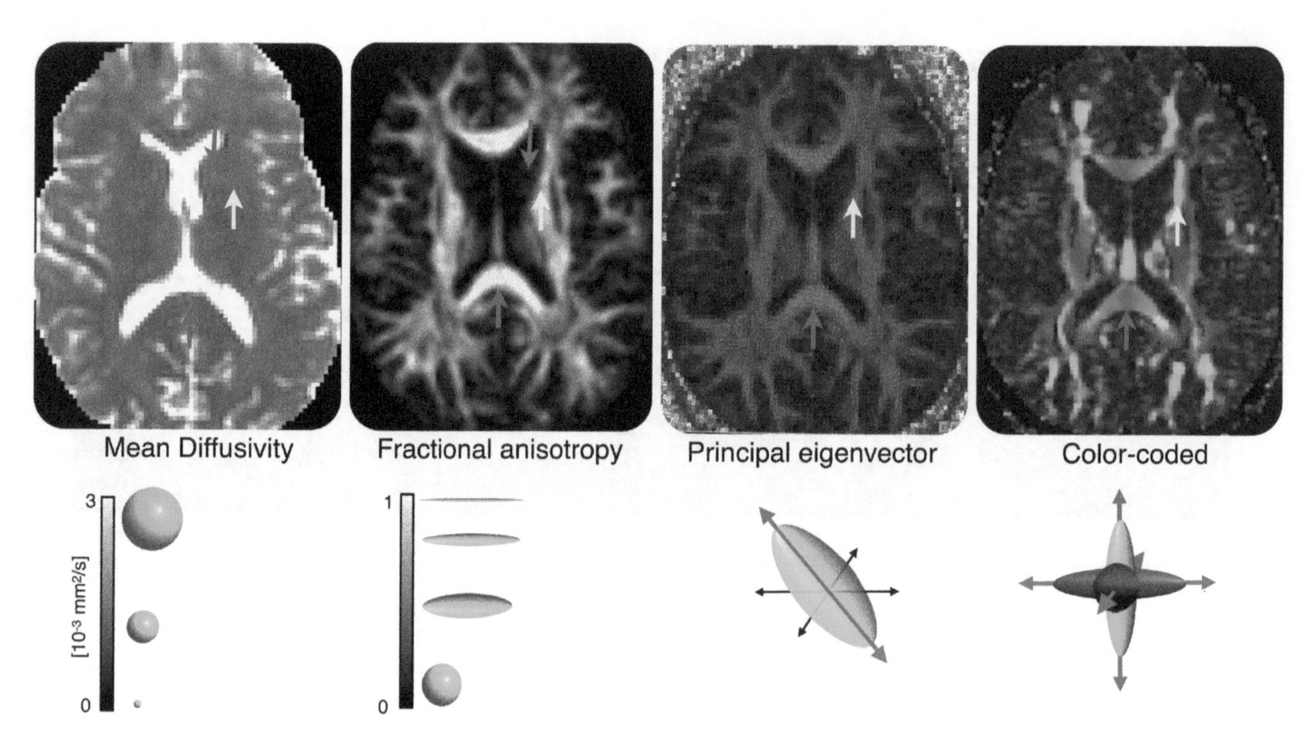

Figure A1.3 The diffusion tensor allows the extraction of quantitative indexes and visualization of maps that provide complementary information about the microstructural properties of the biological tissues and their organization. Thus, for example, two regions such as the anterior limb of the internal capsule (yellow arrow) and the head of the caudate nucleus (red arrow) have similar mean diffusivity (MD) values but different fractional anisotropy (FA) and tensor orientation. This is well described by their respective tensor ellipsoids that have same average 'size' but different shapes (almost spherical for the caudate nucleus) and orientation (the caudate nucleus has no principal orientation being an isotropic tissue). Another example of the complementary information provided by the different diffusion-derived maps are the white matter regions of the anterior limb of the internal capsule and the midpart of the splenium (blue arrow), which have similar MD and FA but different orientation of the direction of maximum diffusivity (visualized either using the principal eigenvector maps or using colour-coded maps).

Diffusion tensor imaging

In 1994, Peter Basser, James Mattiello, and Denis Le Bihan published a seminal paper in diffusion imaging where they showed that if the water diffusion is measured along at least six different directions, it is possible to obtain a mathematical description of the overall displacement of the water molecules, the diffusion tensor (Basser et al., 1994). The tensor provides a synthetic description of the water diffusion in the three-dimensional space and it can be used to extract quantitative indexes that are rotationally invariant (i.e. independent from the orientation of the measurement). The diffusion tensor can also be visualized as a diffusion ellipsoid (Figure A1.2B). The diffusion ellipsoid describes the 3D geometrical profile of the water molecule displacement and it can be calculated from the diffusion coefficient values (eigenvalues) and orientations (eigenvectors) of its three principal axes.

A detailed analysis of the ellipsoids can give precise information about not only the average water molecular displacement (e.g. mean diffusivity, MD) within a voxel but also the degree of tissue anisotropy (e.g. fractional anisotropy, FA) and the main orientation of the underlying white matter fibres (e.g. principal eigenvector or colour-coded maps) (Figure A1.3). These indexes provide complementary information about the microstructural composition and architecture of the brain tissue. Mean diffusivity is a rotational-invariant quantitative index that describes the average mobility of water molecules and is calculated from the three eigenvalues ($\lambda 1$, $\lambda 2$, $\lambda 3$) of the tensor:

$$MD = \frac{Tr(\bar{D})}{3} = \frac{\lambda_1 + \lambda_2 + \lambda_3}{3}$$

Voxels containing grey and white matter tissue show similar MD values (Pierpaoli et al., 1996). In the brain, normal values range from more than 2.0×10^{-3} mm^2/s (in the cerebro-spinal fluid) to 0.6 and 0.9 for grey and white matter, respectively. MD reduces with age within the first years of life and increases in those disorders characterized by demyelination, axonal injury, and oedema.

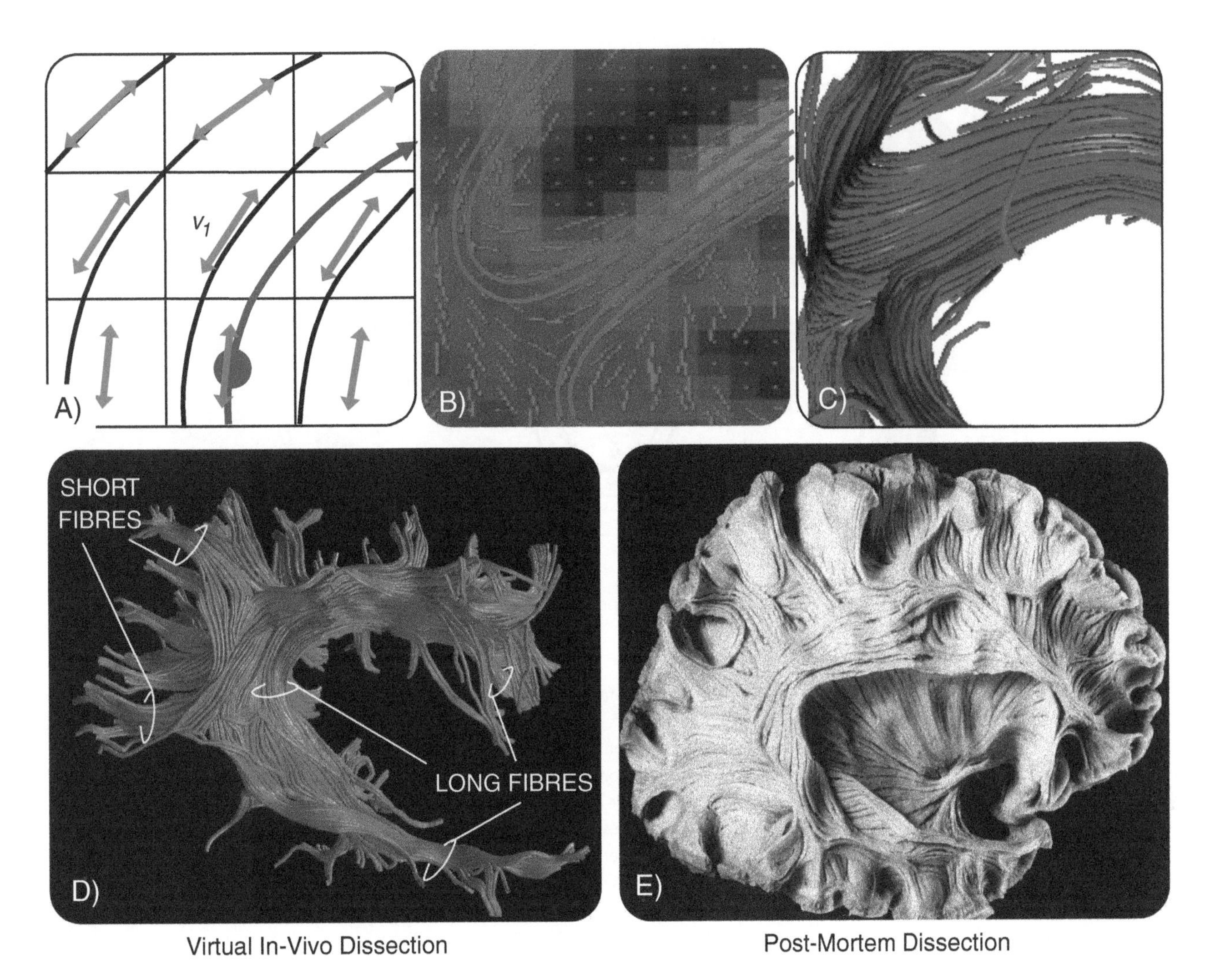

Figure A1.4 Tracking continuous pathways with diffusion tensor imaging. A) Streamline tractography is based on the assumption that in each white matter voxel the principal eigenvector (red arrows) is tangent to the main trajectory of the underlying fibres (black). Starting from a seed voxel (blue circle) the tractography algorithm propagates, voxel by voxel, a streamline (blue) by piecing together neighbouring principal eigenvectors (in this example v1 and so on). B) Axial section of the eigenvector map and streamlines (blue) through the lateral splenium of the corpus callosum. C) Tractography reconstruction of the splenial streamlines visualized as 3D streamtubes. D, E) Comparison between the virtual *in vivo* reconstruction of the arcuate fasciculus (Catani et al., 2002) and the corresponding post-mortem dissection (Gluhbegovic and Williams 1980).

Fractional anisotropy varies from 0 to 1 and represents a quantitative index of the degree of anisotropy of the biological tissue:

$$FA = \frac{\sqrt{3\left((\lambda_1-\bar{\lambda})^2+(\lambda_2-\bar{\lambda})^2+(\lambda_3-\bar{\lambda})^2\right)}}{\sqrt{2\left(\lambda_1^2+\lambda_2^2+\lambda_3^2\right)}}$$

High FA values indicate that one of the three eigenvalues (λ_1) is greater than the other two (λ_2 and λ_3). In the normal brain, FA varies from 0.1 (e.g. in the grey matter) to 0.8 or higher in the white matter. FA gives information about the organization of the tissue (e.g. strongly or weakly anisotropic) and the microarchitecture of the fibres (parallel, crossing, etc.). FA reduces in pathological tissue (e.g. demyelination, oedema, etc.) and is therefore commonly used as an indirect index of microstructural integrity.

Perpendicular and parallel diffusivity are the diffusivity along the principal directions of the diffusion tensor. ADC// = $\lambda 1$ is the longitudinal diffusivity; ADC$\perp$ = ($\lambda 2$+ $\lambda 3$)/2 is the perpendicular diffusivity. The perpendicular diffusivity is generally considered a more sensitive index of axonal or myelin damage, although interpretation of their changes in regions with crossing fibres is not always straightforward (Wheeler-Kingshot and Cercignani, 2009). The principal eigenvector and the colour-coded maps proposed by Pajevic and Pierpaoli (1999) are particularly useful to visualize the principal orientation of the tensor within each voxel. Figure A1.3 shows an example of the colour-coded maps where red is used for latero-lateral orientation, blue for dorso-ventral (and vice versa) orientation, and green for antero-posterior (and vice versa) orientation.

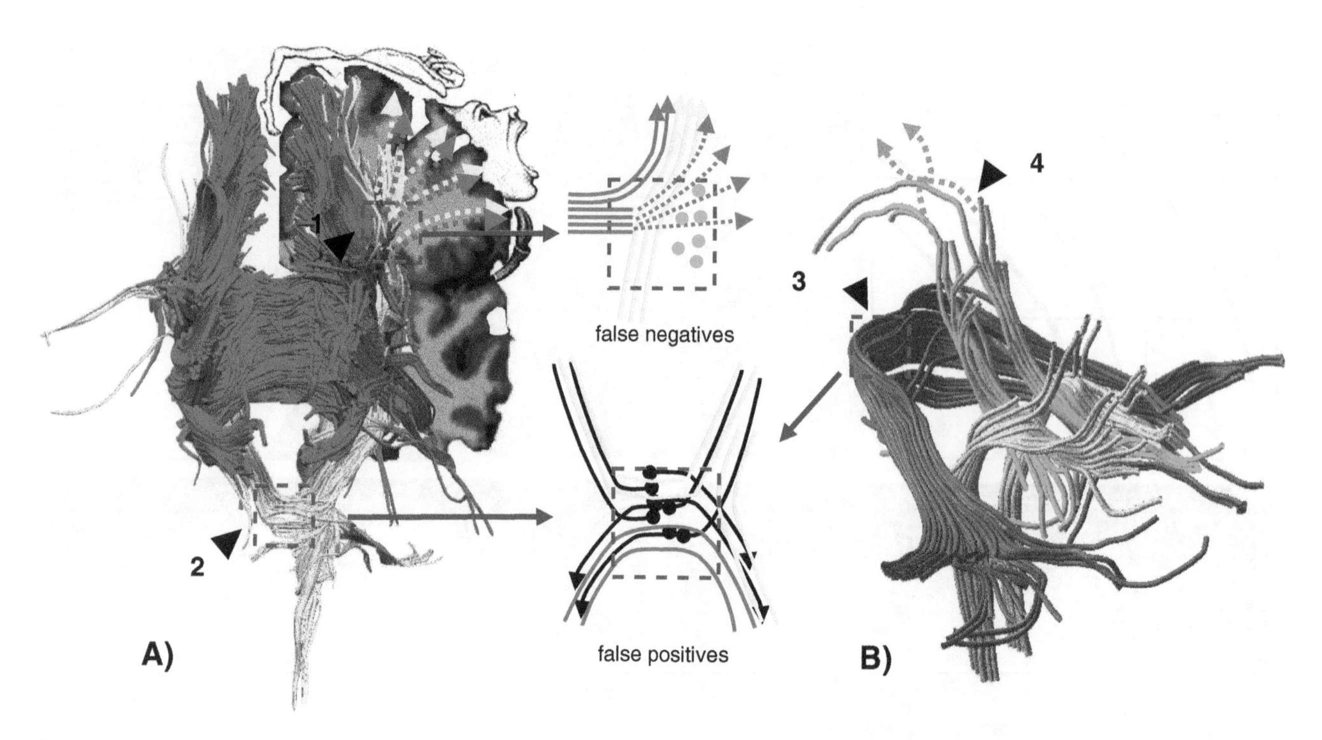

Figure A1.5 Artefactual reconstructions in diffusion tensor tractography generate false positives and false negatives in regions with crossing fibres and partial volumes effects. A) DTI-tractography reconstruction of the corpus callosum (red) and left internal capsule (yellow). (1) Dashed red and yellow lines indicate the expected cortical regions of origin and termination of the internal capsule and callosal tracts, which DTI-tractography is unable to visualize due to the presence of fibre crossing in a region containing the callosal, projection, and association tracts (green dots) (false negatives). (2) Descending streamlines passing through the left internal capsule and ascending into controlateral internal capsule after crossing the pons are artefactual reconstructions (false positive). This false positive is due to the complex anatomy of the pontine region containing crossing fibres together with very small pontine nuclei that reduce the FA levels of the pontine voxels (partial volume effect). B) DTI-tractography representation of the connections of the cerebellum. (3) The red streamlines are artefactual (false positive) reconstructions of a continuous pathway connecting left and right middle cerebellar peduncles through the pons. In this case the tractography algorithm reconstructs continuous streamlines due to the presence of crossing fibres and small pontine nuclei in the pons. (4) The dotted green lines indicate the incomplete reconstruction of the superior cerebellar tracts where streamlines stop before crossing to the controlateral side as expected from known post-mortem neuroanatomy (false negatives) (modified from Catani 2007).

Virtual reconstruction of white matter trajectories

Compared to previously established methods for tracing fibre pathways, such as those used in animal axonal tracing studies, diffusion tensor tractography offers the advantage of being a completely non-invasive technique and therefore its use is not restricted to non-human primates but it can be applied to the living human brain. Furthermore, the data required by the tractography process can be readily obtained on standard clinical MR systems with acquisition times that typically range from 5–20 minutes (depending on data quality required). The main assumption underpinning diffusion tensor tractography is that the diffusion of water molecules inside the brain can be described mathematically by a diffusion tensor whose principal axis aligns with the predominant orientation of the fibres contained within each voxel. Tractography algorithms use these two pieces of information to track the whole white matter pathways by inferring continuity of fibre paths from voxel to voxel. In simple terms, this process is achieved by following the direction of maximum diffusion from a given voxel into a neighbouring voxel (Figure A1.4) (Conturo et al., 1999; Jones et al., 1999; Mori et al., 1999; Basser et al., 2000).

How to piece together discrete estimates of water diffusion within contiguous voxels depends on the algorithm used and the choice of some tracking and stopping parameters. Most of the tractography algorithms adopt angular and/or anisotropy thresholds to avoid unrealistic fibre bending or tracking outside white matter regions (Mori et al., 1999; Basser et al., 2000). The two main applications of tractography are the study of the *in vivo* trajectories of white matter pathways (Catani et al., 2002) and the parcellations of the cortex or subcortical nuclei according to the pattern of streamline connectivity (Behrens et al., 2003). By extracting quantitative diffusion indexes along the dissected tracts, it is also possible to obtain tract-specific measurements indicative of the microstructural organization, composition, and integrity of the tract of interest. The most used indexes are FA and MD. The number of streamlines and number of voxels intersected by the streamlines (i.e. the volume occupied by the streamlines) are commonly used as surrogate measures of tract volume.

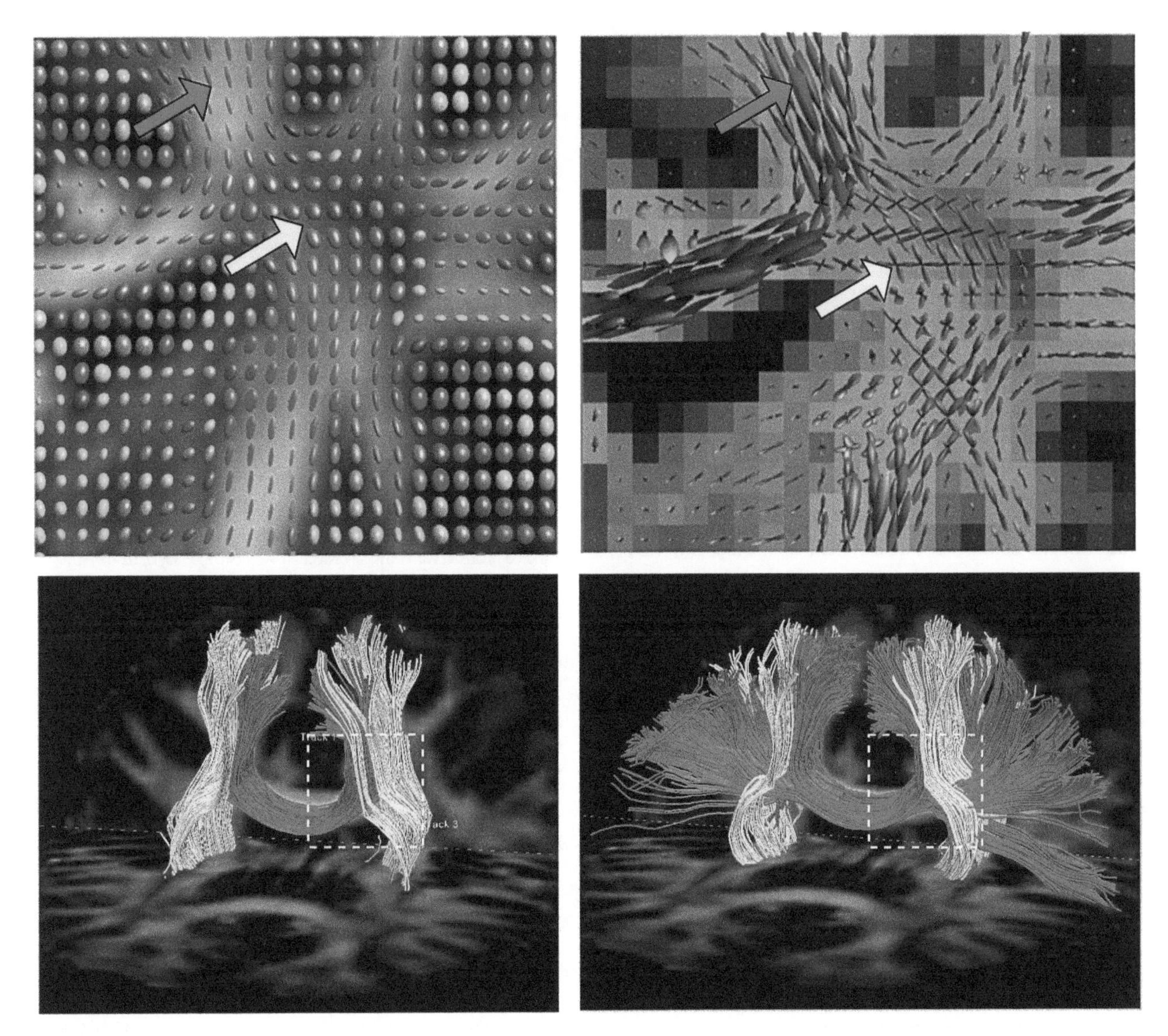

Figure A1.6 Visualization of the white matter organization of the corpus callosum and the corona radiata based on the tensor model (left) and spherical deconvolution (right). In voxels with one fiber population, such as the corpus callosum (red arrow), both models describe orientations that are consistent with the known anatomy. However, in regions with more than one population of crossing fibers (yellow arrow), the tensor model gives an average representation of the water diffusion, whereas the spherical deconvolution model separates different fiber components and describes their individual orientations. The virtual dissections of the corpus callosum based on diffusion tensor tractography reconstruct only the most central part of the corpus callosum (red), while spherical deconvolution tractography shows several streamines of the corpus callosum that cross the streamlines of the corticospinal tract (yellow) and reach the lateral cortex (Catani and Dell'Acqua, 2011).

Limitations

Whilst injected tracers are able to follow the termination of single axons, tractography follows the principal axis of the diffusion tensor, which is obtained by averaging the MRI signal within a voxel. Typically the voxel resolution is too low to identify small fibre bundles. Also the levels of noise in the diffusion data and the intrinsic MR artefacts constitute important factors that affect the precision and accuracy of the diffusion measurements and therefore the tractography reconstruction (Basser et al., 2000; LeBihan et al., 2006). It should be recognized that the results we obtain from tractography are also dependent on a number of factors under the control of the experimenter, such as the angular and anisotropy thresholds and the choice of the tractography algorithm itself. Finally, diffusion tensor tractography assumes that fibers in each voxel are well described by a single orientation estimate, which is a valid assumption for voxels containing only one population of fibres with similar orientation. The majority of white matter voxels, however, contain populations of fibres with multiple orientations; in these regions fibers cross, kiss, merge, or diverge and the tensor model is inadequate to capture this anatomical complexity (Figure A1.5). More recent tractography developments based on HARDI (high angular resolution diffusion imaging) methods (Frank 2001) and appropriate processing techniques are able to resolve in part the fibre crossing problem (Tournier et al., 2004; Tuch, 2004; Wedeen et al., 2005; Alexander, 2005; Behrens et al., 2007).

Preliminary work suggests that it is possible to combine tractography with a spherical deconvolution algorithm to perform dissections of white matter pathways in regions with multiple fiber crossing, like in the corpus callosum (Figure A1.6) and the superior longitudinal fasciculus (Thiebaut de Schotten et al., 2011). One advantage of this approach is that spherical deconvolution datasets can be acquired using clinically feasible protocols (Dell'Acqua et al., 2010).

Compared to classical axonal tracing studies, tractography is unable to differentiate anterograde and retrograde connections, detect the presence of synapses, or determine whether a pathway is functional. All these limitations may lead to tracking pathways that do not exist (false positive) or ineffectively fail to track those that do exist (false negative). A few studies have so far dealt with the issue of validating the tractography results with neuronal tracers (Dauguet et al., 2007; Dyrby et al., 2007) or performing reproducibility analysis on human subjects (Heiervang, et al., 2006; Lawes et al., 2008; Wakana et al., 2007). It is evident from all the considerations above that interpretation of tractography results requires experience and a priori anatomical knowledge.

Finally, in the diseased brain alteration and anatomic distortion due to the presence of pathology, such as brain oedema, bleeding, and compression, generates tissue changes likely to lead to a greater number of artefactual reconstructions (Catani, 2006; Ciccarelli et al., 2008).

References

Alexander, D.C. (2005). Multiple-fiber reconstruction algorithms for diffusion MRI. *Ann NY Acad Sci* **1064**, 113–33.

Basser, P.J., Mattiello, J., and Le Bihan, D. (1994). MR diffusion tensor spectroscopy and imaging. *Biophysical J*, **66**(1), 259–67.

Basser, P.J., Pajevic, S., Pierpaoli C., Duda, J., and Aldroubi, A. (2000). In vivo tractography using DT-MRI data. *Magn Reson Med*, **44**(4), 625–32.

Behrens, T.E.J., Johansen-Berg, H., Woolrichet, M.W., et al. (2003). Non-invasive mapping of connections between human thalamus and cortex using diffusion imaging. *Nat Neurosci*, **6**(7), 750–7.

Behrens, T.E., Johansenn-Berg, H., Jbabdi, S., Rushworth, M.F., and Woolrich, M.W. (2007). Probabilistic diffusion tractography with multiple fibre orientations: What can we gain? *NeuroImage*, **34**(1), 144–55.

Catani, M. (2006). Diffusion tensor magnetic resonance imaging tractography in cognitive disorders. *Curr Opin Neurol*, **19**(6), 599–606.

Catani, M. (2007). From hodology to function. *Brain*, **130**(3), 602–5.

Catani, M., Howard, R.J., Pajevic, S., and Jones, D.K. (2002). Virtual in vivo interactive dissection of white matter fasciculi in the human brain. *NeuroImage*, **17**(1), 77–94.

Catani, M. and Dell'Acqua, F. (2011). DTI tractography and subcortical fasciculi. In Duffau, H. (Ed.) *Brain mapping: from neural basis of cognition to surgical applications*. New York: Springer.

Ciccarelli, O., Catani, M., Johansen-Berg, H., Clark, C., Thompson, A. (2008). Diffusion-based tractography in neurological disorders: concepts, applications, and future developments. *Lancet Neurol* **7**(8), 715–27.

Conturo, T.E., Lori N.F., Cull, T.S., et al. (1999). Tracking neuronal fiber pathways in the living human brain. *Proc Natl Acad Sci USA*, **96**(18), 10422–7.

Dauguet, J., Peled S., Berezovskii V., et al. (2007). Comparison of fiber tracts derived from in-vivo DTI tractography with 3D histological neural tract tracer reconstruction on a macaque brain. *NeuroImage*, **37**(2), 530–8.

Dell'acqua F., Scifo, P., Rizzo, G., et al. (2010). A modified damped Richardson-Lucy algorithm to reduce isotropic background effects in spherical deconvolution. *NeuroImage*, **49**(2), 1446–58.

Dyrby, T.B., Søgaard, L.V., and Parker, G.J. (2007). Validation of in vitro probabilistic tractography. *NeuroImage*, **37**(4), 1267–77.

Frank, L.R. (2001). Anisotropy in high angular resolution diffusion-weighted MRI. *Magn Reson Med* **45**(6), 935–9.

Gluhbegovic, N. and Williams, T.H. (1980). *The human brain*. Hagerstown, MD: Harper and Row.

Heiervang, E., Behrens, T.E., Mackay, C.E., Robson, M.D., and Johansen-Berg, H. (2006). Between session reproducibility and between subject variability of diffusion MR and tractography measures. *NeuroImage*, **33**(3), 867–77.

Jones, D.K., Simmons A., Williams, S.C.R., and Horsfield, M.A. (1999). Non-invasive assessment of axonal fiber connectivity in the human brain via diffusion tensor MRI. *Magn Reson Med*, **42**(1), 37–41.

Lawes, I.N.C., Barrick, T.R., Murugam, V., et al. (2008). Atlas-based segmentation of white matter tracts of the human brain using diffusion tensor tractography and comparison with classical dissection. *NeuroImage*, **39**(1), 62–79.

Le Bihan, D., Breton, E., Lallemand, D., Grenier, P., Cabanis, E., and Laval-Jeantet, M. (1986). MR imaging of intravoxel incoherent motions: application to diffusion and perfusion in neurologic disorders. *Radiology*, **161**(2), 401–7.

Le Bihan, D., Mangin, J.F., Poupon, C., et al. (2001). Diffusion tensor imaging: concepts and applications. *J Magn Reson Imaging*, **13**(4), 534–46.

Le Bihan, D., Poupon, C., Amadon, A., Lethimonnier, F. (2006). Artifacts and pitfalls in diffusion MRI. *J Magn Reson Imaging* **24**(3), 478–88.

Mori, S., Crain, B.J., Chacko, V.P., and Van Zijl, P.C. (1999). Three dimensional tracking of axonal projections in the brain by magnetic resonance imaging. *Ann Neurol*, **45**(2), 265–9.

Moseley, M.E., Cohen, Y., Mintorovitch, J., Chileuitt, L., Shimizu, H. et al. (1990). Early detection of regional cerebral ischemia in cats: comparison of diffusion- and T2-weighted MRI and spectroscopy. *Magn Reson Med*, **14**(2), 330–46.

Moseley, M.E., Kucharczyk, J., Asgari, H.S., and Norman, D. (1991). Anisotropy in diffusion-weighted MRI. *Magn Reson Med*, **19**(2), 321–6.

Pajevic, S., and Pierpaoli, C. (1999). Color schemes to represent the orientation of anisotropic tissues from diffusion tensor data: application to white matter fiber tract mapping in the human brain. *Magn Reson Med*, **42**(3), 526–40. Erratum in: *Magn Reson Med*, **43**(6), 921.

Pierpaoli, C., Jezzard, P., Basser, P.J., et al. (1996). Diffusion tensor MR imaging of the human brain. *Radiology*, **201**(3), 637–48.

Tanner, J.E. (1978). Transient diffusion in a system partitioned by permeable barriers. Application to NMR measurements with a pulsed field gradient. *J Chem Physics*, **69**(4), 1748–54.

Thiebaut de Schotten, M., Dell'Acqua, F., Forkel, S., et al. (2011). A lateralized brain network for visuo-spatial attention. *Nat Neurosci*, **14**(10), 1245–6.

Tournier, J. D., Calamante, F., Gadian, D. G., and Connelly, A. (2004). Direct estimation of the fiber orientation density function from diffusion-weighted MRI data using spherical deconvolution. *NeuroImage*, **23**(3), 1176–1185.

Tuch, D.S. (2004). Q-ball imaging. *Magn Reson Med*, **52**(6), 1358–72.

Wakana, S., Caprihan, A., and Panzenboeck, M. (2007). Reproducibility of quantitative tractography methods applied to cerebral white matter. *NeuroImage*, **36**(3), 630–44.

Wedeen, V.J., Hagmann, P., Tseng, W.Y., Reese, T.G. and Weisskoff, R.M. (2005). Mapping complex tissue architecture with diffusion spectrum magnetic resonance imaging. *Magn Reson Med*, **54**(6), 1377–86.

Wheeler-Kingshott, C.A., and Cercignani, M. (2009). About "axial" and "radial" diffusivities. *Magn Reson Med*, **61**(5), 1255–60.

APPENDIX II

Methods Used to Create the *Atlas of Human Brain Connections**

* This Appendix II is reprinted from *NeuroImage*, **54**(1), Michel Thiebaut de Schotten, Dominic H. ffytche, Alberto Bizzi, Flavio Dell'Acqua, Matthew Allin, Muriel Walshe, Robin Murray, Steven C. Williams, Declan G.M. Murphy, and Marco Catani (2011). Atlasing location, asymmetry and inter-subject variability of white matter tracts in the human brain with MR diffusion tractography, pp. 49–59, Copyright with permission from Elsevier.

Introduction

Until the advent of diffusion tensor imaging (DTI), our knowledge of white matter anatomy was based on a small number of influential 19th- and early 20th-century post-mortem dissection atlases (Burdach, 1819; Déjerine, 1895; Klingler, 1935). In common with their contemporary counterparts (Talairach and Tournoux, 1988), these atlases emphasize the constant or average anatomy of representative subjects at the expense of normal variability between subjects (Amunts and Willmes, 2006). A few post-mortem histological studies addressed the variability of the tracts between the two hemispheres, reporting, for example, asymmetries for the cortico-spinal tract (Flechsig, 1876; Yakovlev and Rakic, 1966; Rademacher et al., 2001), the optic radiations (Bürgel et al., 1999), and the uncinate (Highley et al., 2002). Bürgel et al. (2006) also showed that a significant inter-subject variability exists for each tract within the single hemispheres.

DTI has allowed the study of the *in vivo* anatomy of white matter tracts in the human brain (Catani et al., 2002; Mori et al., 2002, 2005; Wakana et al., 2004; Catani and Thiebaut de Schotten, 2008; Ciccarelli et al., 2008; Lawes et al., 2008) and addressed some of the questions that were difficult to answer with post-mortem dissections such as the location, asymmetry, and inter-subject variability of white matter tracts (Ciccarelli et al., 2003; Barrick et al., 2007; Catani et al., 2007; Wakana et al., 2007; Verhoeven et al., 2010). In recent years several groups have used DTI to produce group atlases of the major white matter tracts (Hua et al., 2008; Lawes et al., 2008; Verhoeven et al., 2010; Wassermann et al., 2010). These atlases contain maps of the major white matter tracts that have a good correspondence with post-mortem blunt dissections and some of them address the inter-subject variability between left and right hemisphere, for example, in relation to gender (Thiebaut de Schotten et al., 2011a).

Our aim was to create a normative atlas of white matter human connections for clinical and research purposes on a large age-matched population of male and female participants. We combined for the first time complementary approaches to visualize group maps based on statistical thresholds and percentage overlap. A comparison between the DTI derived atlas and a previously published post-mortem histological atlas (Bürgel et al., 2006) is also presented and limitations of both approaches discussed.

Subjects

King's College London Review Board approved the study, and informed consent was obtained from all subjects. Forty healthy right-handed volunteers (20 males and 20 females) aged between 18 and 22 years were recruited. To assess the localizing validity of the atlas in patients with brain lesions DTI, was also acquired in a 68-year-old female patient with right hemiplegia due to a glioblastoma multiforme and a 61-year-old male patient with chronic neglect due to right hemisphere stroke.

DTI acquisition and processing

A total of 60 contiguous near-axial slices were acquired on a GE Signa 1.5-T LX MRI system (General Electric, Milwaukee, WI) with 40-mT/m gradients, using an acquisition sequence fully optimized for DTI of white matter, providing isotropic (2.5×2.5×2.5mm) resolution and coverage of the whole head. The acquisition was peripherally-gated to the cardiac cycle with an echo time of 107 ms and an effective repetition time of 15 R–R intervals. At each slice location, seven images were acquired with no diffusion gradient applied, together with 64 diffusion-weighted images in which gradient directions were uniformly distributed in space. The diffusion weighting was equal to a b-value of 1300 s mm^{-2}. Full details of this sequence are given in (Jones et al., 2002). BrainVISA (http://brainvisa.info/) was used to correct for eddy current distortion of raw diffusion-weighted data and calculate the diffusion tensor and the fractional anisotropy (FA) for each voxel. A tractography algorithm based on Euler integration (Jones et al., 2002) (step length 0.5 mm; FA threshold of 0.2, angle threshold of 45°) was used to propagate streamlines from 'seed' regions of interest (ROIs) manually-defined on Anatomist (http://brainvisa.info). To ensure that the operator was blind to hemisphere during the ROIs delineation, and therefore provide protection against subjective bias, half of the datasets were reflected along the midline. Following previously described criteria (Catani et al., 2002, 2003, 2005; Catani and Thiebaut de Schotten, 2008), ROIs were delineated around areas of white matter that represent 'obligatory passages' along the course of each tract (Figure A2.1). This permits to visualize all streamlines of a single tract without constraining its terminal projections, which may vary from subject to subject. A one-ROI approach was used for the anterior commissure, corpus callosum, arcuate, cingulum, internal capsule and fornix. A two-ROI approach was used for the inferior longitudinal fasciculus, inferior fronto-occipital fasciculus, uncinate, cortico-spinal tract, optic radiations, and cerebellar tracts (cortico-ponto-cerebellar, spino-cerebellar, and superior-cerebellar). A two-ROI approach was also used to separate the three segments of the arcuate fasciculus: a direct long segment between Wernicke's and Broca's territories an indirect posterior segment connecting Wernicke's to Geschwind's territories

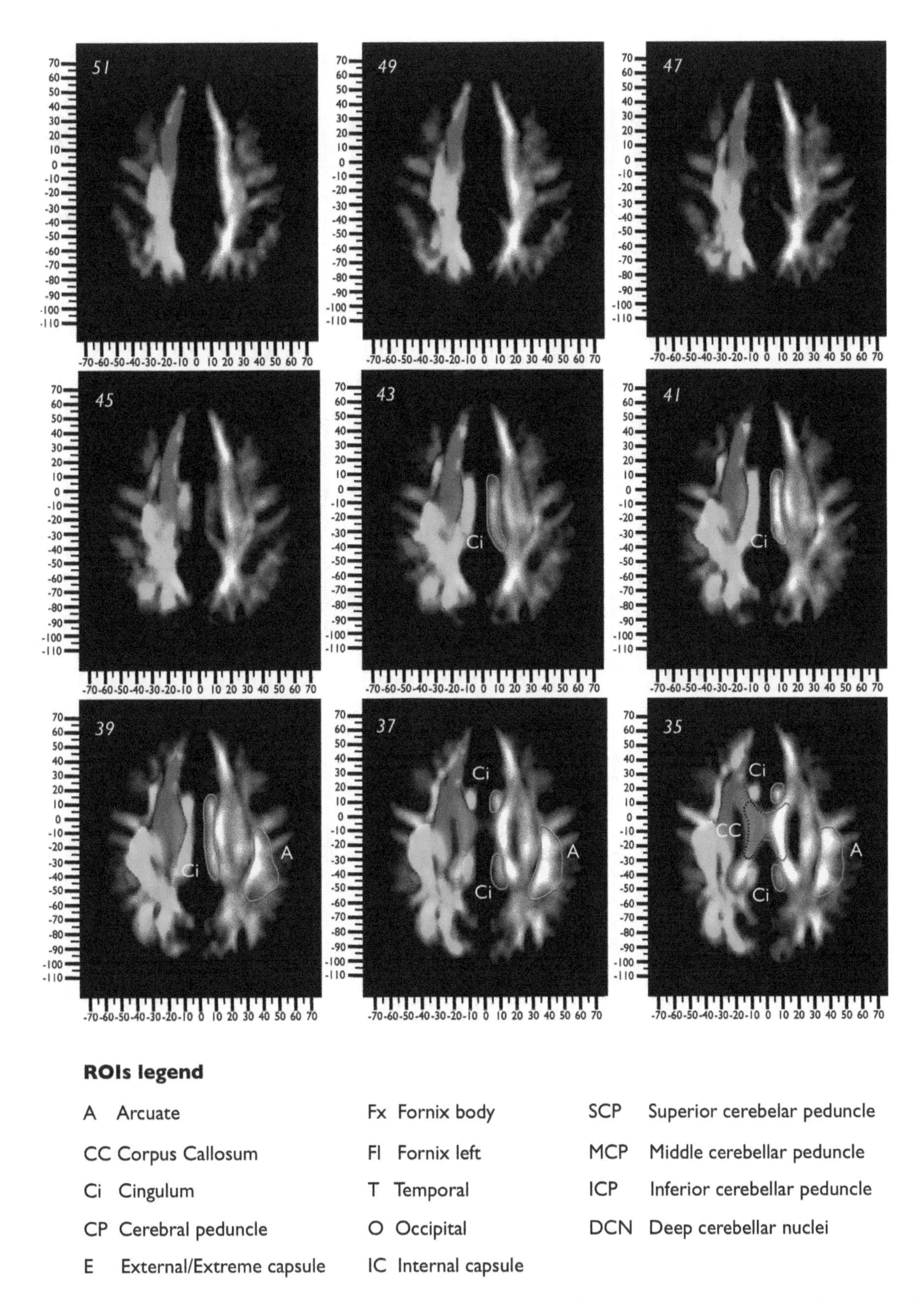

Figure A2.1 Diffusion tensor image template of an average dataset used to delineate the regions of interest. The FA maps on the right provide information about the general anatomy of the major association, commissural and projection white matter tracts. The color maps on the left provide additional information on the local orientation of the tracts, where red color indicates a latero-lateral direction (left to right and right to left), green color an anterior-posterior direction (and vice versa), and blue color a dorsal–ventral direction (and vice versa). Other colors indicate intermediate directions (Catani and Thiebaut de Schotten, 2008).

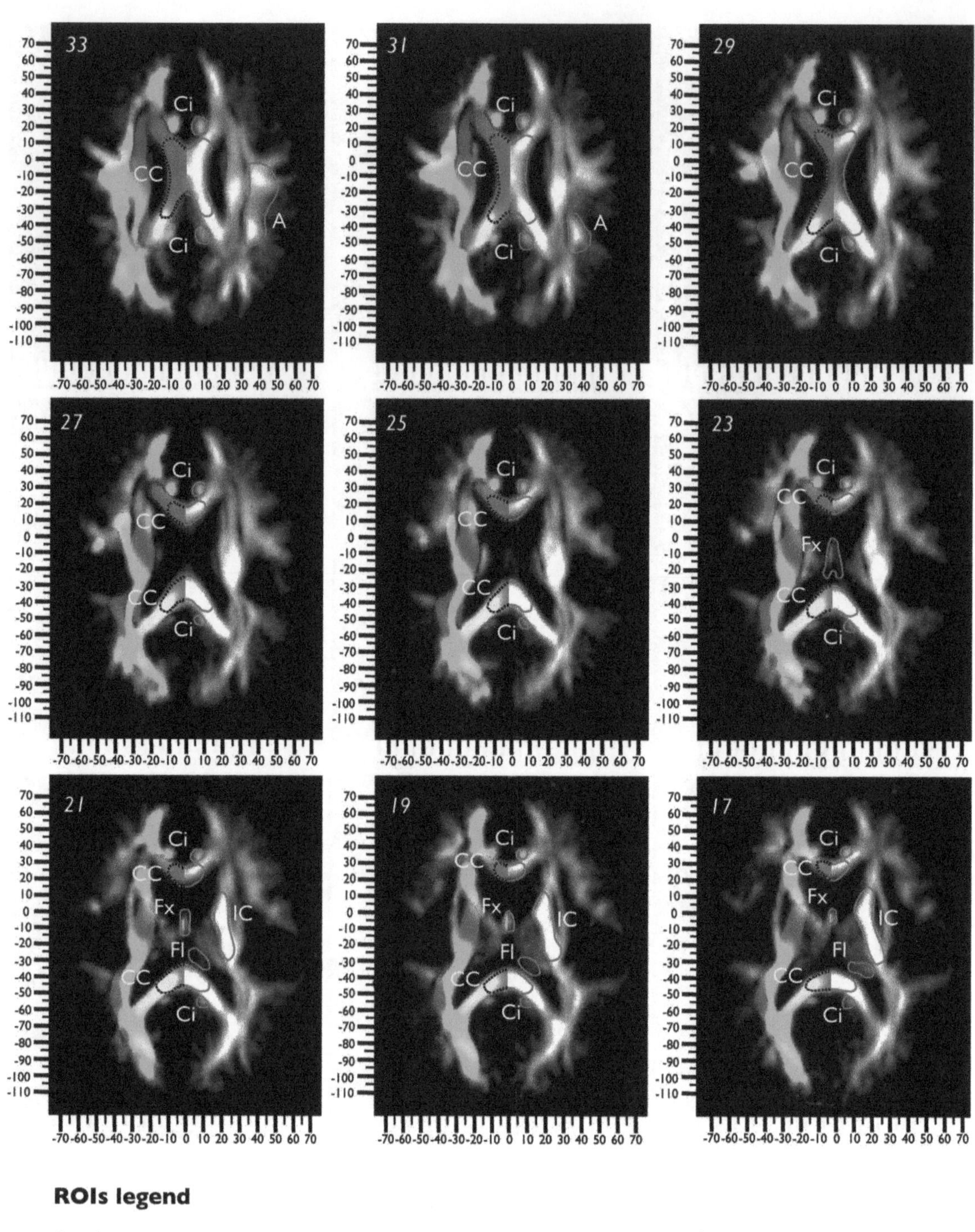

ROIs legend

A Arcuate

CC Corpus Callosum

Ci Cingulum

CP Cerebral peduncle

E External/Extreme capsule

Fx Fornix body

Fl Fornix left

T Temporal

O Occipital

IC Internal capsule

SCP Superior cerebelar peduncle

MCP Middle cerebellar peduncle

ICP Inferior cerebellar peduncle

DCN Deep cerebellar nuclei

Figure A2.1 *Cont'd.*

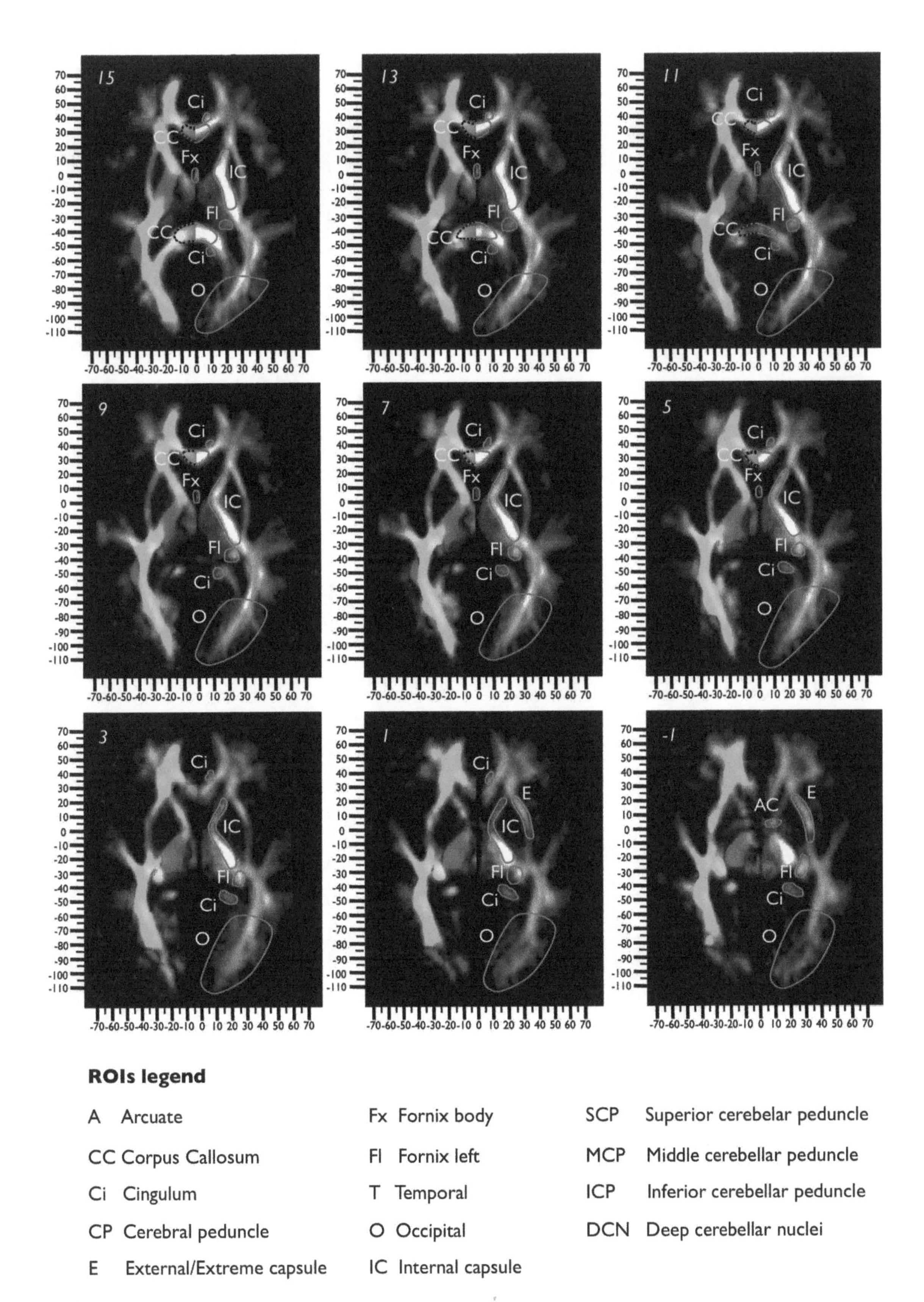

Figure A2.1 *Cont'd.*

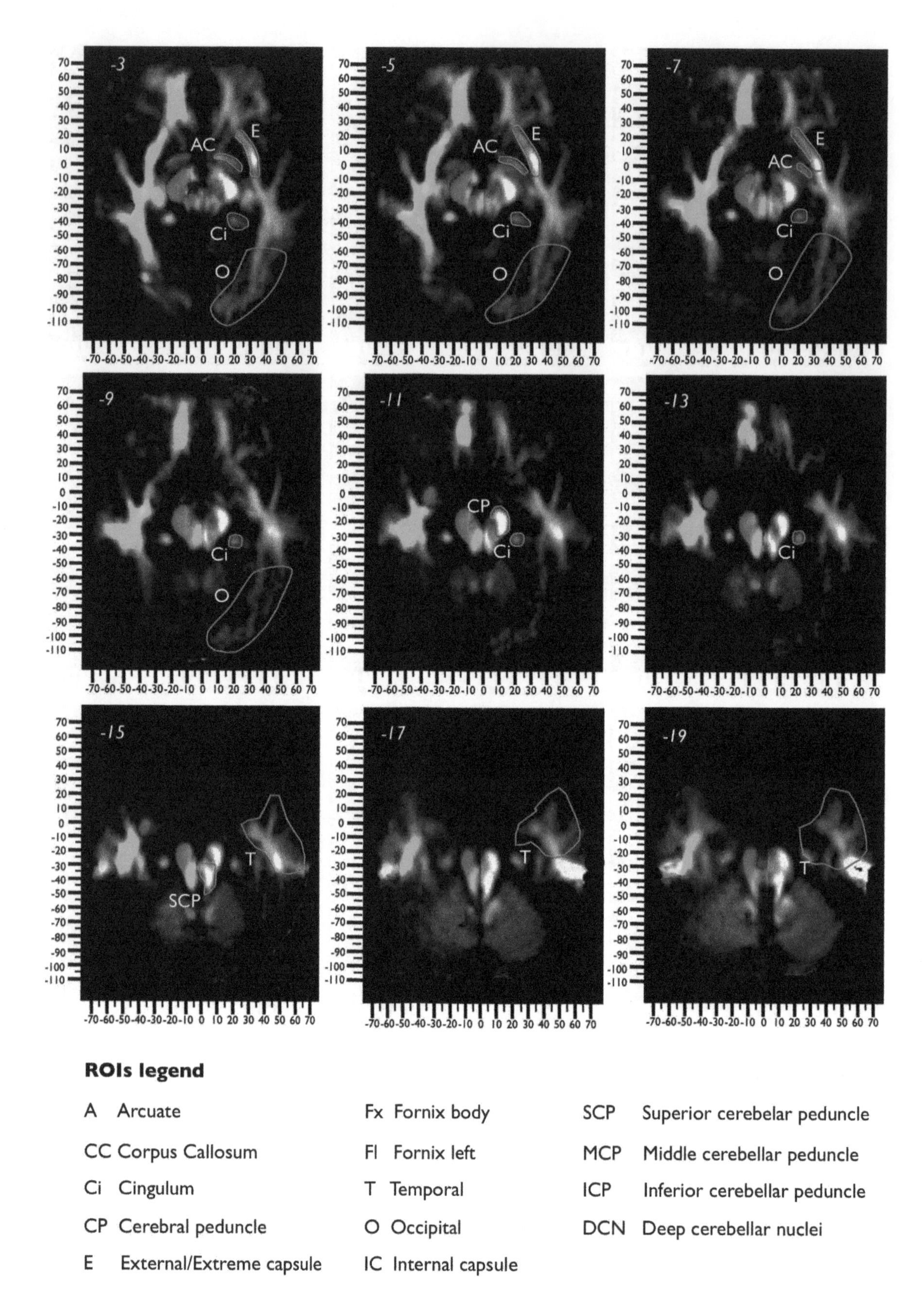

Figure A2.1 *Cont'd.*

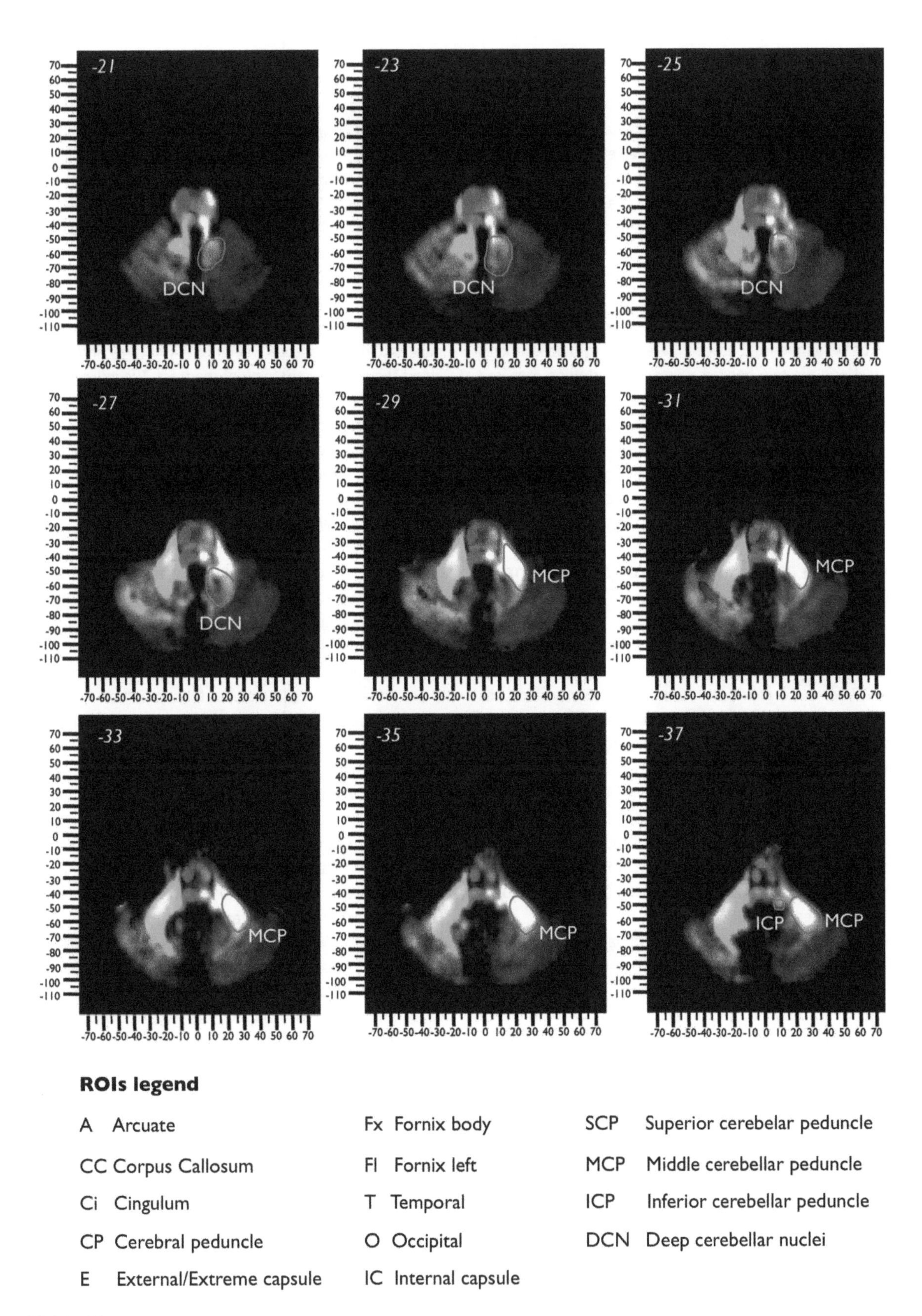

Figure A2.1 *Cont'd.*

(inferior parietal lobule), and an indirect anterior segment connecting Geschwind's to Broca's territories (Catani et al., 2007).

Tract-specific measurements were extracted (i.e. number of streamlines, volume, and fractional anisotropy) and a lateralization indeces calculated (Parker et al., 2005; Powell et al., 2006; Catani et al., 2007; Vernooij et al., 2007; Makris et al., 2008; Matsumoto et al., 2008; Rodrigo et al., 2008; Sivaswamy et al., 2008; Lebel and Beaulieu, 2009) according to the following formula (for example for the number (N.) of streamlines):

(N. streamlines Right – N. streamlines Left)/
(N. streamlines Right + N. streamlines Left)

Analysis of the differences between the left and right hemispheres of each tract and the two genders was performed using repeated measure ANOVA. The lateralization index of each tract was set as the within-subjects factor (14 tracts) and the gender as the between-subjects factor. Statistical significance of the degree of the lateralization was determined using one-sample t test for each tract. Only results that survived Bonferroni correction are presented.

Creation of the atlas based on group effect and variability maps

The B0 images of each subject were normalized to the Montreal Neurological Institute space (MNI, http://www.bic.mni.mcgill.ca/) using the T2 template provided in SPM5 (http://www.fil.ion.ucl.ac.uk/spm/). Then the affine (12 degrees of freedom) and elastic (16 iterations) transformation matrix derived from the normalization of the B0 images was applied to each FA map and an average FA template was created. This average FA template was used to normalize the individual FA maps to generate a normalized FA template. This new FA template was used to normalize again the individual FA maps. This process was repeated five times to generate a sharper final normalized FA template (Ashburner, 2007; Thiebaut de Schotten et al., 2011a).

The binary visitation maps were created for each tract by assigning each voxel a value of 1 or 0 depending on whether the voxel was intersected by the streamlines of the tract (Catani et al., 2007; Lawes et al., 2008; Thiebaut de Schotten et al., 2008). The binary visitation maps of each subject were normalized to the MNI space using the transformation matrix derived from the process of normalization of the FA maps described above. The visitation maps were then analysed following two complementary approaches (Ciccarelli et al., 2003; Thiebaut de Schotten et al., 2011a): i) group effect statistic; ii) percentage overlap. The first analysis aims to create group effect maps following smoothing of the normalized binary maps with a 4-mm (full width at half maximum) isotropic Gaussian kernel. Smoothing has been classically applied to imaging analysis to obtain an approximately Gaussian distribution of the data (Worsley et al., 1992). In our analysis Gaussian distribution of the data was confirmed for three voxels along the inferior fronto-occipital fasciculus using the Shapiro-Wilk test (Shapiro and Wilk, 1965). This allows the use of standard parametric statistics in our dataset to draw statistical inferences (Ashburner and Friston, 2000) about the constant anatomy of the tract. The individual smoothed images were then entered into a design matrix for a one-sample t-test corrected for Family Wise Error (FWE).

The second method creates percentage overlap maps by summing at each point in the MNI space the normalized visitation maps from each subject. In this case the visitation maps are binary and unsmoothed, hence the overlap of the visitation maps varies according to inter-subject variability. This is the method that was also used by Bürgel et al. (2006) to create a histological atlas of white matter tracts. Hence, by following their method we were able to produce percentage maps to compare side-by-side our *in vivo* atlas with their histological atlas (http://www.fz-juelich.de/inb/inb-3//spm_anatomy_toolbox). Only maps with an overlap threshold of more than 50% were used for comparison.

3D rendering of the brains were calculated using the T1 pipeline in BrainVISA. 3D rendering of the maps were obtained using the online command AimsClusterArg in BrainVISA. The overall visualization and screenshots were performed in Anatomist (http://brainvisa.info).

To obtain preliminary data on the localizing validity of the atlas, the T1 or T2 structural images of two patients with brain lesions were normalized in the MNI with SPM5 using affine (12 degrees of freedom) and elastic (16 iterations) transformation and superimposed onto the corresponding slices from the atlas. Then DTI tractography dissections (FA >0.2) of the tract affected by the pathology were performed in the native space of each patient to verify that the atlas localized the tract affected by the lesion correctly.

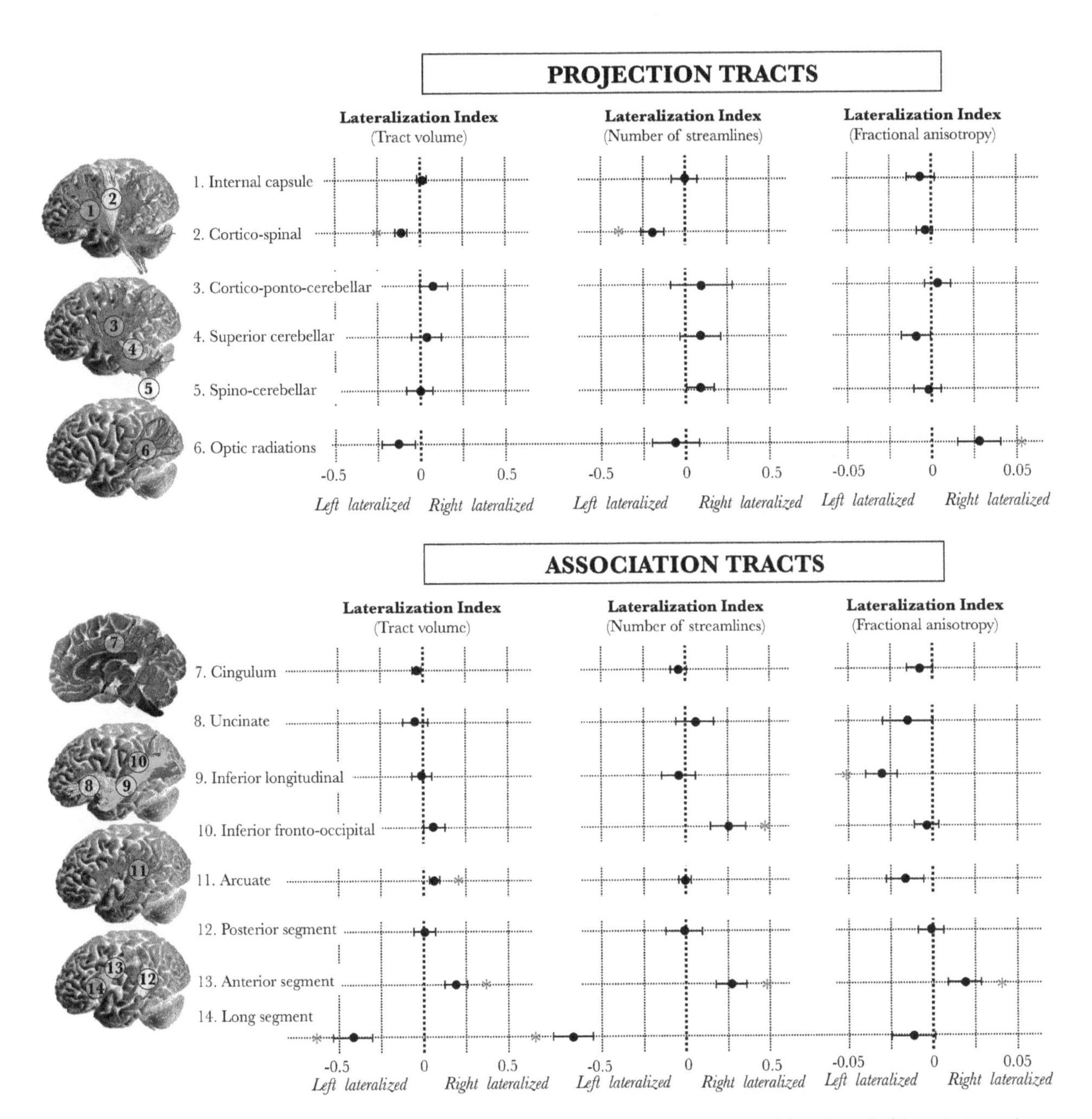

Figure A2.2 Lateralization index of the volume, number of streamlines, and fractional anisotropy (mean ±95% confidence interval) of the projection and association pathways. Red asterisks indicate a statistical significance at p <0.05 after correction for multiple comparisons.

Results

The outcome variables of the tract-specific measurements were used to describe the interhemispheric and gender differences. Both group effect and percentage overlap maps were used to describe the constant and variable features of the dissected tracts and to compare the DTI-derived reconstructions with the post-mortem histology maps.

Hemispheric asymmetries and between gender differences

The lateralization indexes for the volume, number of streamlines, and FA of the dissected tracts are shown in Figure A2.2. A statistically significant leftward asymmetry was found for the volume ($T_{(39)}$ =6 p< 0.001) and number of streamlines ($T_{(39)}$ =5.3 p< 0.001) of the cortico-spinal tract and for the volume ($T_{(39)}$ =7.2 p< 0.001) and number of streamlines ($T_{(39)}$ =10.9 p< 0.001) of the long segment of the arcuate fasciculus. The volume of the optic radiations showed a leftward asymmetry ($T_{(39)}$ =2.7 p <0.01), which did not survive Bonferroni correction.

A statistically significant rightward asymmetry was found for the volume of the arcuate fasciculus ($T_{(39)}$ =3.4 p <0.001) and for the number of streamlines ($T_{(39)}$ =4.4 p <0.001) of the inferior fronto-occipital fasciculus. Of the three segments of the arcuate fasciculus the anterior segment showed a rightward asymmetry in volume ($T_{(39)}$ =5.6 p <0.001) and number of streamlines ($T_{(39)}$ =5.9 p <0.001). Analysis of the FA measurements revealed a leftward asymmetry for the inferior longitudinal fasciculus ($T_{(39)}$ =6.2

COMMISSURAL PATHWAYS

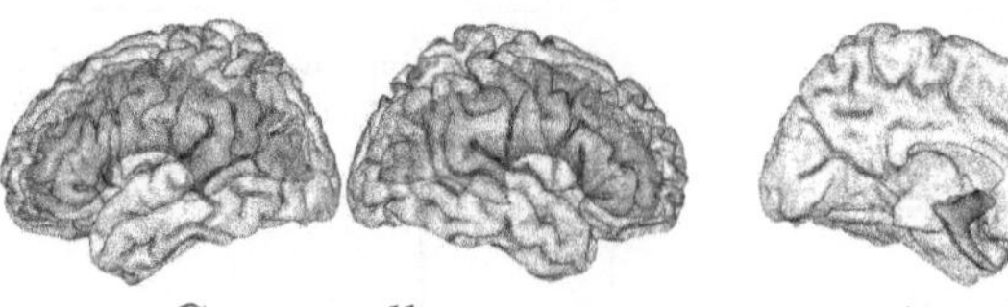

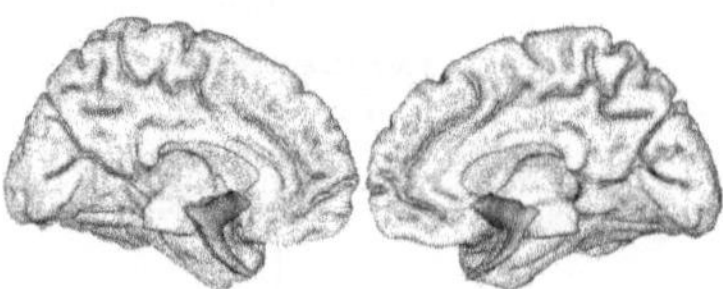

Corpus callosum *Anterior commissure*

ASSOCIATIVE PATHWAYS

MEDIAN NETWORK

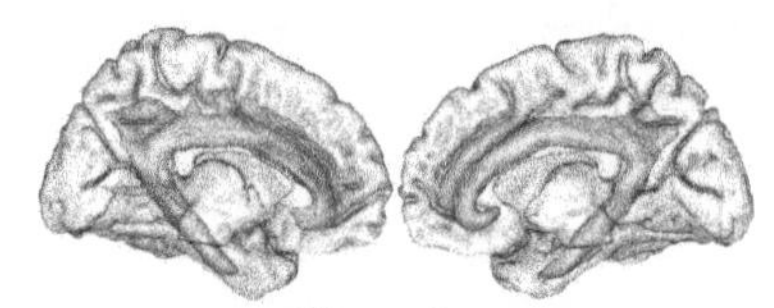

Cingulum

PERISYLVIAN/ARCUATE NETWORK

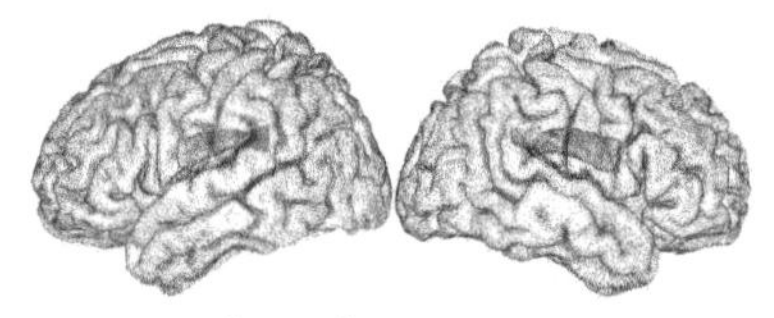

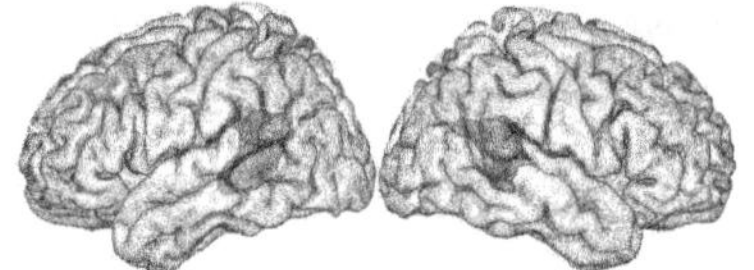

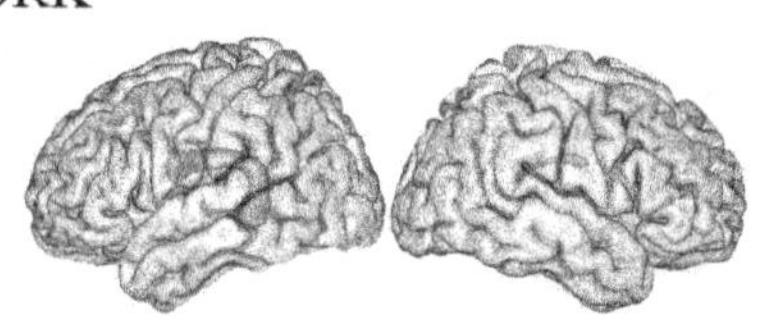

Anterior segment *Posterior segment* *Long segment*

INFERIOR NETWORK

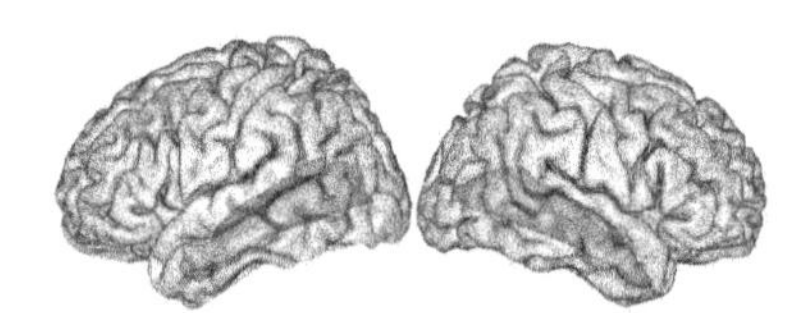

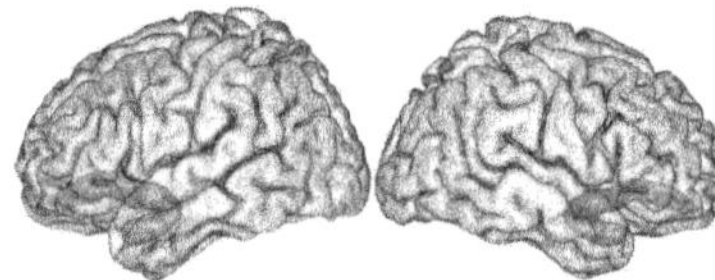

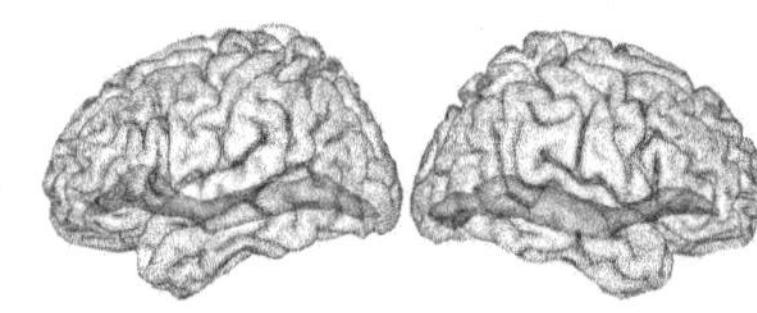

Inferior longitudinal *Uncinate* *Inferior fronto-occipital*

PROJECTION PATHWAYS

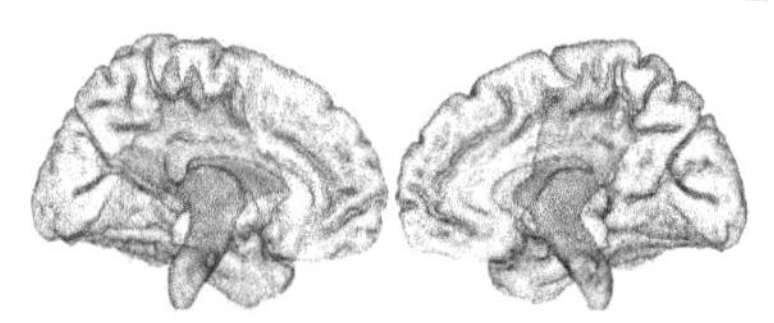

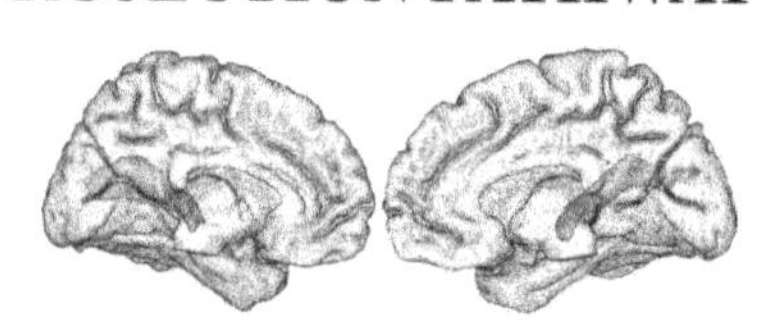

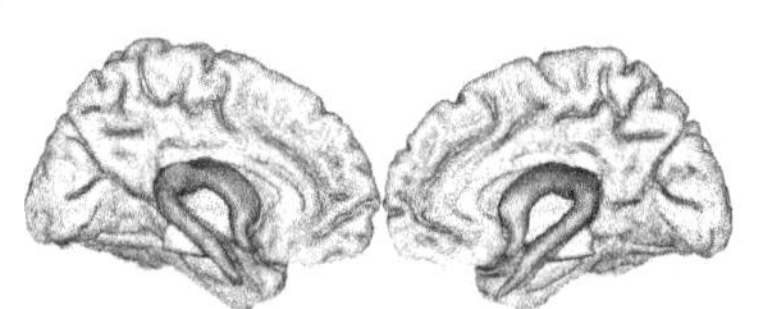

Cortico-spinal *Optic radiations* *Fornix*

CEREBELLAR NETWORK

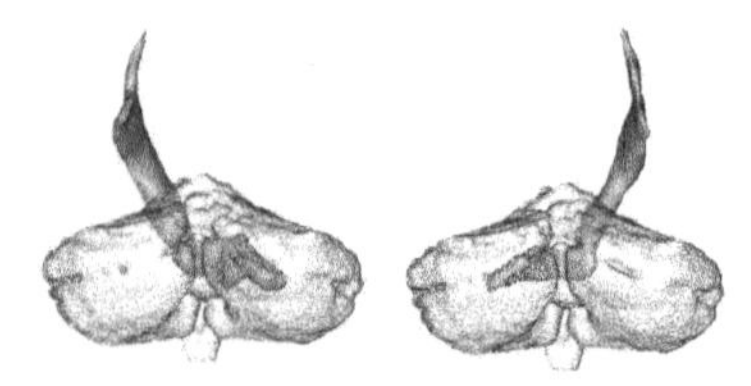

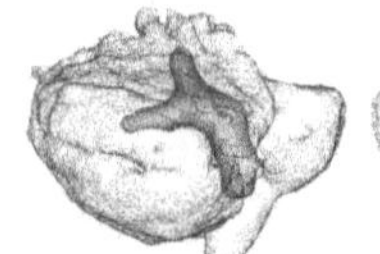

Cortico-ponto-cerebellar *Spino-cerebellar* *Superior cerebellar*

Figure A2.3 3D reconstruction of the major commissural, associative, and projection pathways based on the group effect maps.

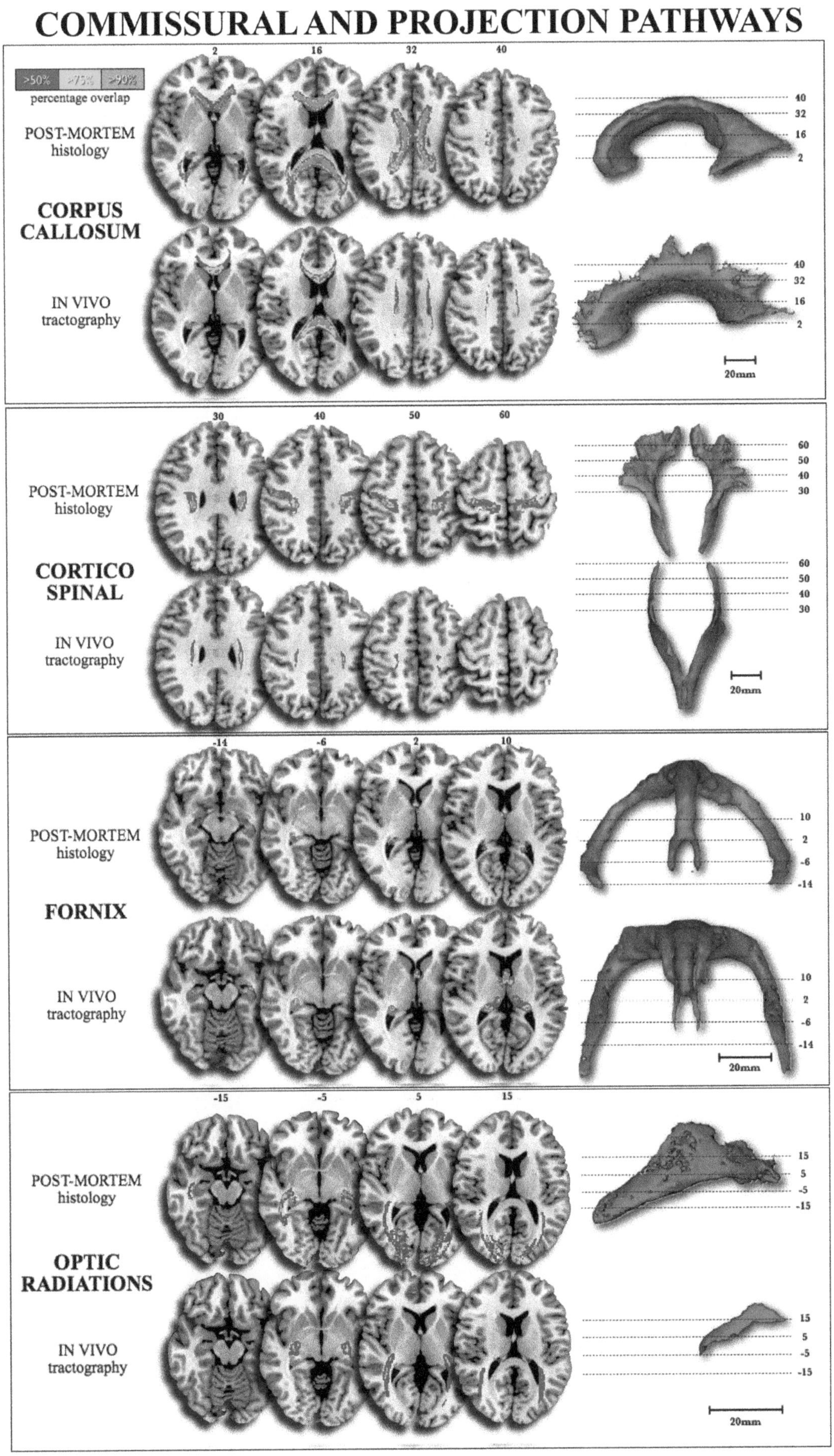

Figure A2.4 Comparison between percentage maps based on post-mortem histology (upper rows) (Bürgel et al., 2006) and DTI tractography (lower rows) of the major commissural and projection tracts (the anterior commissure was not available for the histological maps). Numbers above each slice refer to MNI coordinates. On the right, tridimensional reconstructions of two sets of maps (>50% tract overlap) are shown.

p <0.001). In contrast a rightward lateralization was found for the optic radiations ($T_{(39)}$ =5.1 p <0.001) and the anterior segment of the arcuate fasciculus ($T_{(39)}$ =3.7 p <0.001). The ANOVA analysis showed a significant interaction between the lateralization index of tract volume and the gender ($F_{(13,494)}$ =2.0742 p <0.05), but no interaction for the lateralization index of the number of streamlines and FA. Post hoc analysis revealed a statistically significant gender difference in the lateralization index of the volume of the long segment of the arcuate ($T_{(38)}$ =2.57 p <0.05), with males (lateralization index −0.56±0.36) showing a greater left lateralization compared to females (lateralization index −0.24 ± 0.28). However the difference did not survive Bonferroni correction.

Group effect and overlap maps

The results of the group effect statistic for the commissural, projection, cerebellar, and association tracts are shown in Figure A2.3. All tracts were present in each subject except for the long segment of the arcuate on the right hemisphere. Hence, for the right long segment percentage overlap and group effect maps have been derived only from 16 subjects. The DTI percentage overlap maps of projection, commissural, and associative tracts are shown in Figures A2.4 and A2.5. A concentric distribution of the variability maps was observed for all tracts, with a descending gradient from the central portion (overlap >90%) to the most peripheral zones (overlap <50%).

Finally, composite maps derived from the overlap of the group effect and the percentage overlap maps were created for each dissected tract. The composite maps provide complementary information about the constant anatomy of the tracts and the inter-individual variability as illustrated by the tracts shown in Figure A2.6. The maps of the cingulum, for example, show a symmetrical pattern of both tract volume and percentage overlap maps. For other tracts the lateralization of volume and the degree of overlap differ. This is the case for the inferior fronto-occipital fasciculus where maps show no left–right differences in the tract volume but asymmetrical distribution of the percentage overlap maps. Finally the anterior segment of the arcuate is an example of asymmetry for both volume and percentage maps.

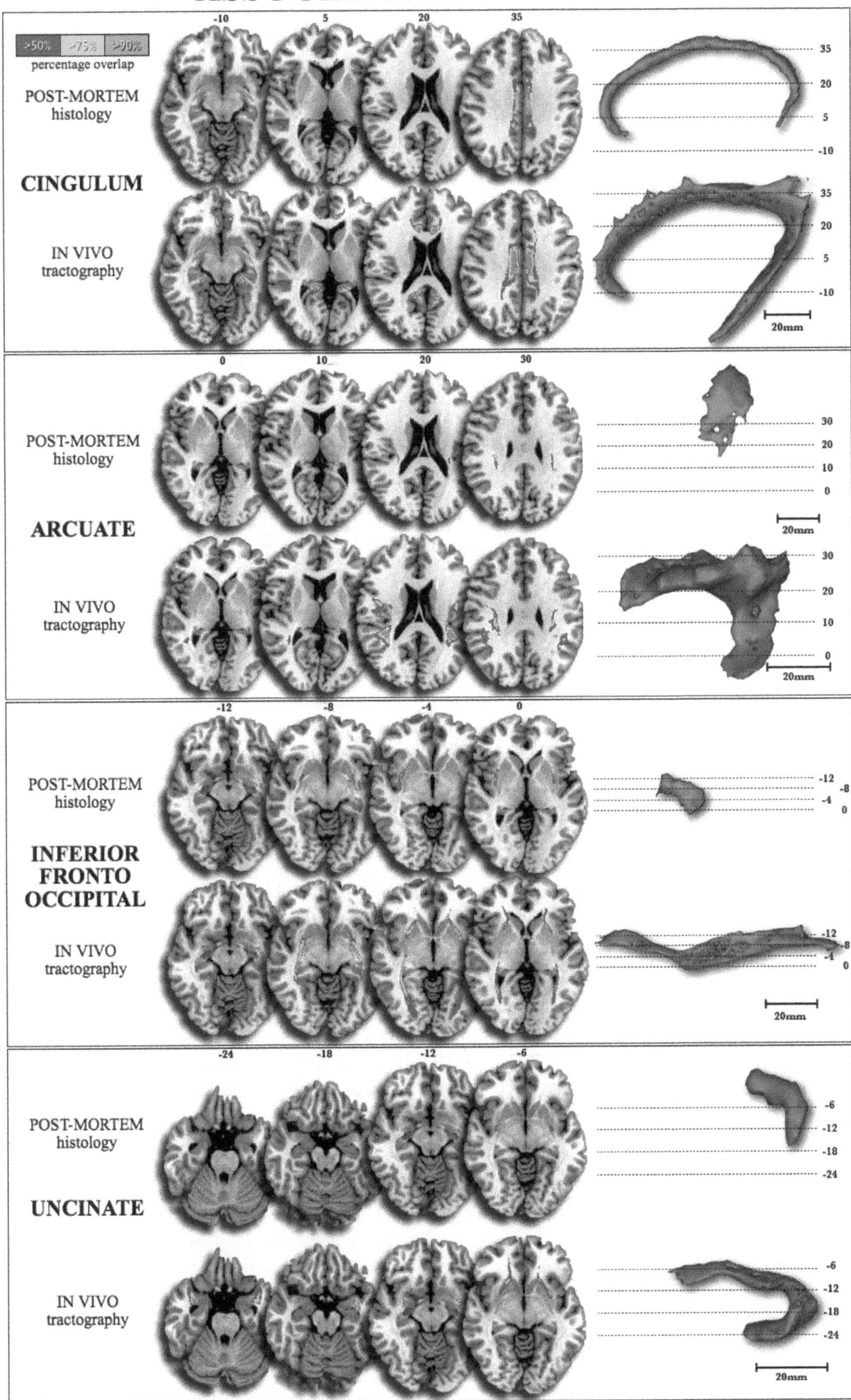

Figure A2.5 Comparison between percentage maps based on post-mortem histology (upper rows) (Bürgel et al., 2006) and DTI tractography (lower rows) of the major association tracts. Numbers above each slice refer to MNI coordinates. On the right, tridimensional reconstructions of two sets of maps (>50% tract overlap) are shown.

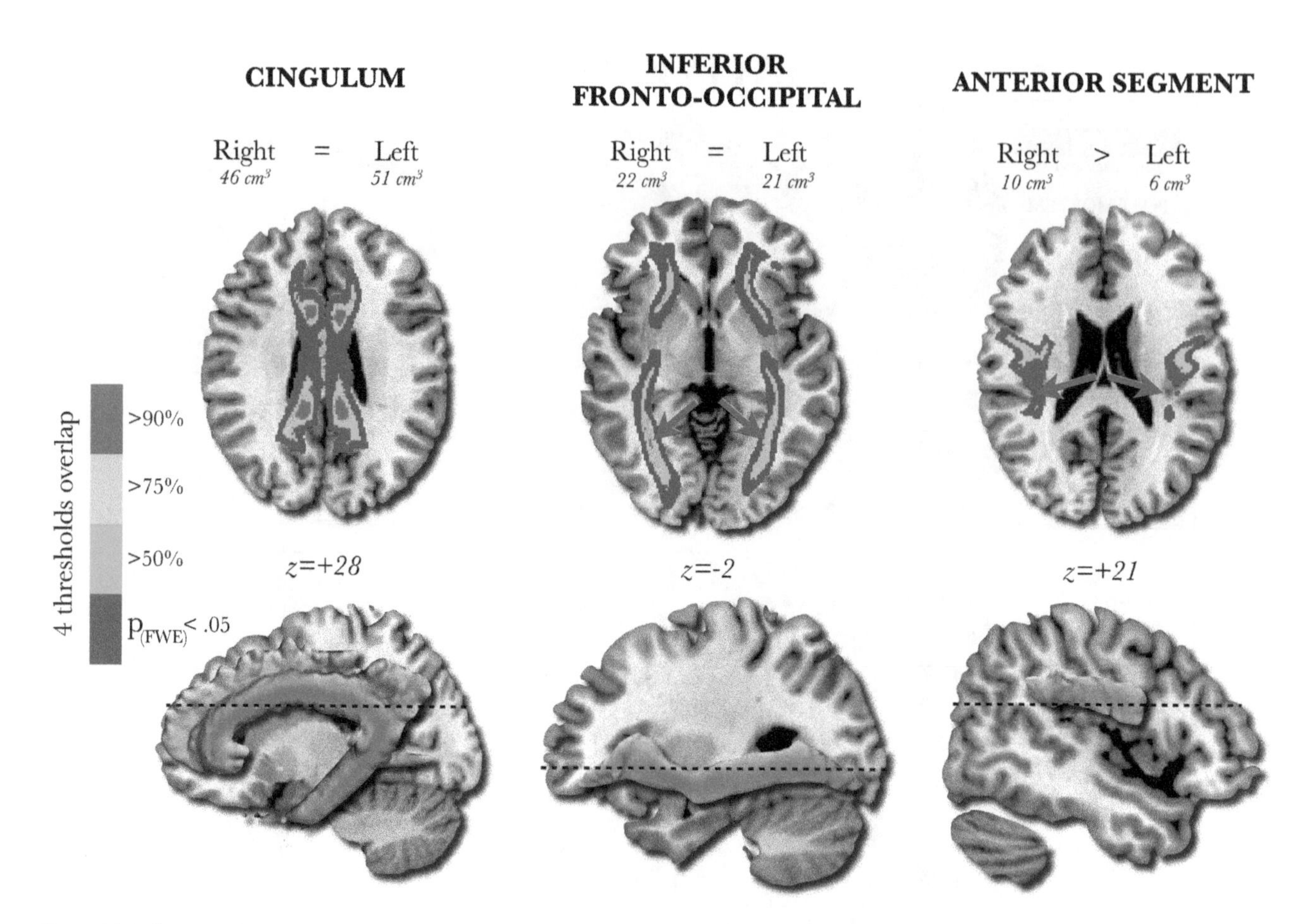

Figure A2.6 The composite maps of the DTI-derived atlas reveal that the left–right differences in volume of the bilateral tracts do not always correlate with the left–right distribution of the percentage overlap maps. The cingulum, for example, shows an almost symmetrical distribution for both volume and percentage overlap maps, while the inferior fronto-occipital fasciculus shows similar left–right volumes but greater overlap in the right side (i.e. lower inter-subject variability) (pink arrows). Finally the anterior segment shows rightward asymmetry for both volume and percentage maps (pink arrows). On the lower panel the dashed lines indicate in the MNI the plane on the z-axis of the axial slice in the row above.

Comparison with histology-based variability maps

The percentage overlap maps derived from DTI-tractography were compared with the histological maps from the atlas of Bürgel et al. (2006). Overall there is a good anatomical correspondence between the DTI-tractography and the histological maps for the central portion of the cortico-spinal tract, corpus callosum, and fornix. Differences between histological and DTI-tractography derived maps were particularly evident for the association tracts and cortical projections of the cortico-spinal tract. In particular the volume of the optic radiations and the cortico-spinal tract was smaller in the DTI maps compared to the histological maps, whereas volume of all association tracts was greater in the DTI maps compared to the histological maps. Differences in tract volume between the two atlases are due to partial reconstructions (false negatives) or overestimation (false positive) of the tracts due to limitations inherent to each technique. Examples of the false negative reconstructions of DTI are the lack of lateral cortico-spinal tract projections to the face/tongue region and the lack of callosal fibres projecting to the ventrolateral regions of the frontal, parietal and occipital lobes and posterior temporal lobe (i.e. tapetum). A false negative result of the histological maps is the lack of terminal projections for all association tracts (Figures A2.4 and A2.5).

Atlas-based tract localization in brains with pathology

Figure A2.7 shows as example, the application of the atlas to two patients presenting with neurological symptoms due to localized brain lesions. The first patient presented with a tumour in the left post-central gyrus and right hemiplegia. The atlas overlapped on the patient's normalized T_2-weighted images indicated that the tumour was infiltrating the posterior portion of the left cortico-spinal tract at the level of the corona radiata. The DTI colour maps and the tractography reconstruction of the projection tracts performed on the patient's own DTI dataset (after placing a ROI in the posterior limb of the internal capsule) confirmed that the tumour involves the streamlines of the cortico-spinal tract at the level of the corona radiata.

The second patient presented with a right middle cerebral artery ischaemic stroke and spatial neglect of the left hemispace. Neglect is a syndrome associated with lesions of the right perisylvian pathways. Overlay of the atlas-derived tract composite maps on the normalized T_1-weighted image suggested involvement of the anterior and long segments of the right arcuate fasciculus. Dissections of the patient's DTI confirmed extension of the lesion to these two tracts.

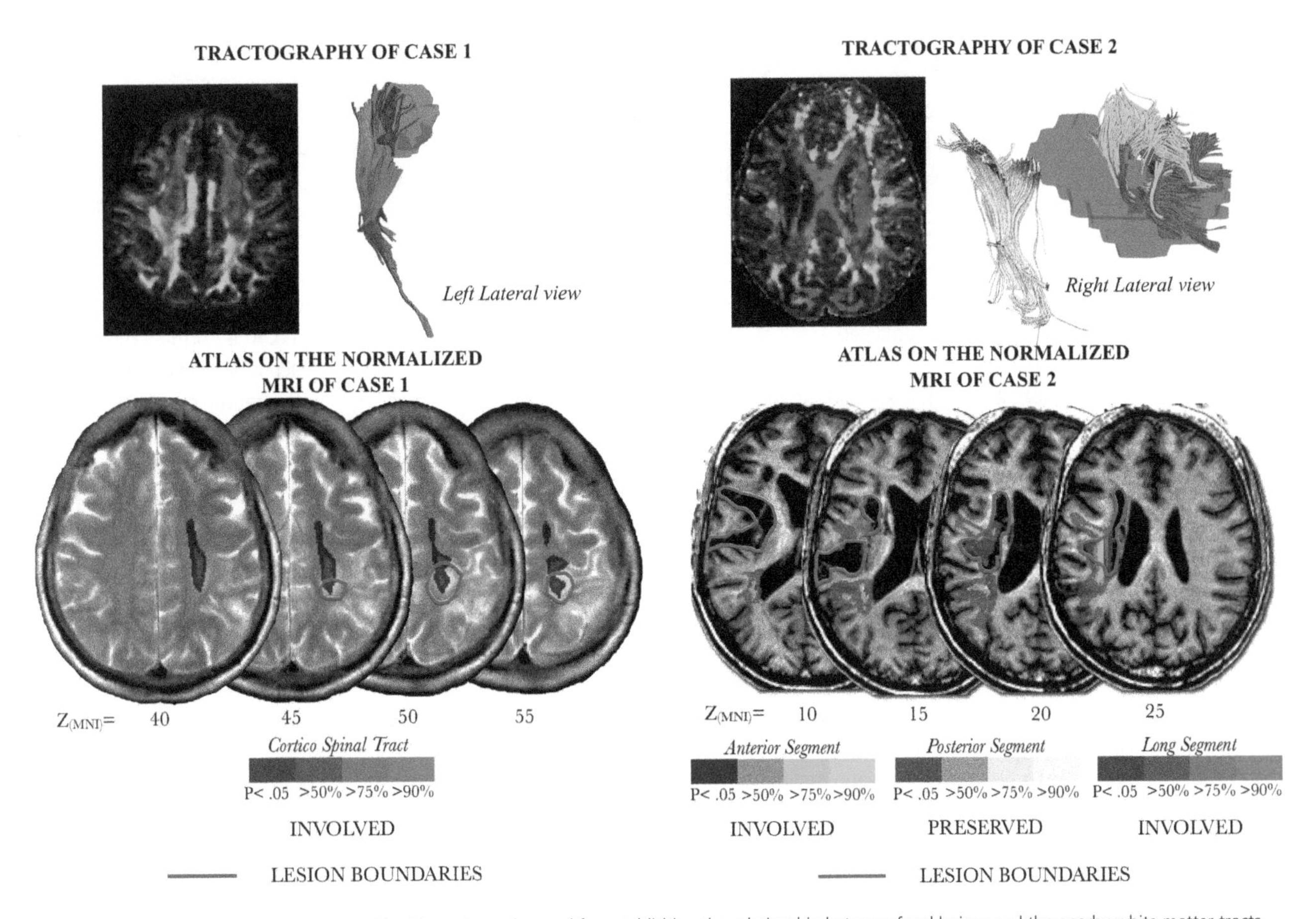

Figure A2.7 The composite maps presented in this study can be used for establishing the relationship between focal lesions and the nearby white matter tracts. In the left upper panel the axial directionality (red, green, and blue, RGB) map of a patient with right lower limb weakness shows an area of reduced FA in the posterior part of the corona radiata. The overlaying of the cortico-spinal tract maps onto the normalized MR images suggests a direct involvement of the posterior portion of the cortico-spinal tract by the tumour. The involvement of the cortico-spinal tract is in this case confirmed by the tractography reconstruction of the cortico-spinal tract that shows an interruption and intersection with the tumour of the posterior cortico-spinal tract streamlines. In the right upper panel the axial directionality (RGB) map of a patient with hemispatial neglect shows an area of reduced FA in the corona radiata (blue) extending into the most lateral voxels (green) containing fibres of the arcuate fasciculus. The exact involvement of the different components of the arcuate fasciculus is difficult to determine on the directionality maps alone. Below, the structural images show a stroke in the hemisphere that according to the atlas involves both anterior and long segments. The tractography analysis confirms involvement of both segments, which are only partially reconstructed in the most anterior regions.

Discussion

We used DTI-tractography to produce a white matter atlas in the MNI space that describes the *in vivo* variability of the major association, commissural, and projection connections. The complete atlas based on the group statistical maps is reproduced in Chapter 6, while the individual percentage maps for each tract are at the end of Chapters 7–11. Three findings emerge from our work (Thiebaut de Schotten et al., 2011). First, tracts like the cortico-spinal tract and the long segment of the arcuate fasciculus are left lateralized, whereas the anterior segment of the arcuate fasciculus and the inferior fronto-occipital fasciculus are right lateralized. Secondly, we confirmed gender differences for the long segment, which is more left lateralized in males as compared to females. Finally, each tract showed an inter-individual variability that could reflect anatomical differences among the healthy population.

The leftward asymmetry found in the cortico-spinal tract and the long segment of the arcuate fasciculus is in agreement with previous findings from post-mortem dissections (Rademacher et al., 2001), *in vivo* DTI-tractography (Nucifora et al., 2005; Hagmann et al., 2006; Eluvathingal et al., 2007; Lebel and Beaulieu, 2009), and voxel-based analysis of T_1-images (Good et al., 2001; Hervé et al., 2006, 2009) (see also Chapter 10). An important question is the extent to which structural differences between the two hemispheres correlate with functional lateralization, and whether the anatomical lateralization of language and sensory-motor pathways reflects differences in language, visuo-spatial processing, and handedness. Preliminary studies combining DTI tractography and fMRI reported no correlation between the lateralization of the arcuate fasciculus volume and the degree of functional lateralization as determined by fMRI during tasks of verbal fluency, verb generation and reading comprehension (Powell et al., 2006). The functional lateralization seems to correlate better with the lateralization of the fractional anisotropy (Powell et al., 2006; Vernooij et al., 2007) and the number of streamlines of the arcuate fasciculus as demonstrated in right-handed healthy individuals (Vernooij et al., 2007) and in patients with temporal lobe epilepsy (Rodrigo et al., 2008).

There are also preliminary findings showing that the extreme left lateralization of the direct long segment (fronto-temporal) is associated with worse performance on a complex verbal memory task that relies on semantic clustering for retrieval (i.e. California Verbal Learning Test, CVLT). The correlation remained significant after splitting the group according to gender, suggesting that the main determinant of CVLT performance is the anatomy (symmetry) of the language pathways, and not the gender. Overall these findings support the notion that lateralization of language to the left hemisphere is an important aspect of human brain organization but paradoxically, a bilateral representation might ultimately be advantageous for certain cognitive functions (see also Chapter 7) (Catani et al., 2007).

Higher FA in the anterior part of the superior cingulum has been previously reported in the left hemisphere using tractography (Gong et al., 2005) and voxel-based statistics (Park et al., 2004). Our analysis of the whole cingulum shows no significant hemispheric asymmetry, suggesting that overall the cingulum has a symmetrical distribution of the FA with only the anterior region showing a leftward asymmetry (see also Chapter 11) (Park et al., 2004; Gong et al., 2005).

In addition to the above findings we report, for the first time, a rightward asymmetry for the volume of the fronto-parietal connections (i.e. anterior segment) of the arcuate fasciculus and the inferior fronto-occipital fasciculus. These tracts convey sensory information from parietal and occipital areas respectively, to a more anterior region in the inferior frontal gyrus. Lesions to the projection areas of the fronto-parietal segment of the arcuate fasciculus and the inferior fronto-occipital fasciculus result in neglect (Doricchi et al., 2008), a syndrome characterized by the inability to process visual sensory information in one hemispace. Neglect occurs mostly with lesions to the right hemisphere and the asymmetry of the fronto-parietal segment of the arcuate fasciculus (Thiebaut de Schotten et al., 2005) and the inferior fronto-occipital fasciculus (Urbanski et al., 2008) may represent the anatomical correlate of right hemisphere dominance for visuo-spatial processing (see also Chapter 7) (Thiebaut de Schotten et al., 2011b).

We found gender differences in the lateralization of the fronto-temporal long segment of the arcuate, with a more left lateralized pattern in males. Previous studies have also reported gender differences in the lateralization of the cortical volume of language regions (Good et al., 2001; Luders et al., 2004), subcortical white matter anatomy (Good et al., 2001; Hagmann et al., 2006), and activation patterns during linguistic tasks (Shaywitz et al., 1995). It has been suggested that gender differences in the lateralization pattern may reflect a different maturational trajectory during development (Paus, 2010; Perrin et al., 2009). Some, for example, documented significant gender differences in the white matter of the left inferior frontal gyrus, a region containing anterior projections of the arcuate fasciculus: boys but not girls showed a linear age-related increase in the white matter volume in this region (Blanton et al., 2004). However, a recent DTI tractography study found no gender differences for the arcuate in the transition from childhood to adolescence (Lebel and Beaulieu, 2009). Hence, the authors suggested that differentiation between the two hemispheres occur early in life and it is maintained during adolescence (see also Chapter 7).

The inter-individual variability we observed for each tract is likely to reflect differences in tract volume, their course, and termination. Other biological factors such as axonal diameter, density, and myelination modulate the diffusion signal and can explain inter-individual variability at a microstructural level (e.g. FA) (see also Appendix 1) (Beaulieu, 2009). However, other methodological explanations should also be considered. The normalization procedure we employed, which deforms each brain to fit with the referential template, may produce misregistration errors (Ashburner, 2007). Even small misregistration errors are likely to generate maps with a higher degree of overlap in the central portion of the tracts, and a lower degree of overlap in the peripheral regions. Misregistration is also frequent in datasets where individual brains show structural differences in relation, for example, to age (Stadlbauer et al., 2008) and handedness (Hagmann et al., 2006). In our sample we recruited only right-handed subjects within a relatively narrow age range to reduce this source of variability. The degree of uncertainty in the estimation of the fibre orientation which is typical of DTI-tractography algorithms may also increase variability across subjects. For example, in regions containing crossing fibres the FA decreases as the confidence intervals in fibre orientation (i.e. 'cone of uncertainty') increases (Jones, 2003). In tractography, accumulated uncertainties in fibre orientation have clear potential for leading to erroneous reconstructions of pathways. For example Jones (2003) demonstrated that the cone of uncertainty is greater in the most terminal regions of the tracts where the reliability of fibre orientation maps decreases. In this study, we set a FA threshold of 0.2 to exclude most of voxels with a high cone of uncertainty and therefore reduce the probability of artefactual reconstructions.

The similarity between our maps and the histology-based atlas (Rademacher et al., 2001; Bürgel et al., 2006) suggests that true anatomical differences contribute to the variability observed in the healthy population. However, compared to the histological maps, our DTI tractography approach was not able to visualize the lateral projection of the cortico-spinal tract and the most anterior part of the optic radiations. This is due to the limitations of current diffusion tensor models in resolving fibre crossing (McNab et al., 2009); in our case, DTI was unable to visualize the fibres of the cortico-spinal tract crossing the arcuate fasciculus. The use of novel algorithms in tractography reconstructions (e.g. spherical deconvolution) may in part overcome these limitations (see also Appendix 1) (Tournier et al., 2004, 2008; Dell'Acqua et al., 2007, 2010). For other tracts, such as the arcuate fasciculus, uncinate, and inferior fronto-occipital fasciculus, DTI tractography was able to show anatomical features that are closer to classical post-mortem blunt dissections than post-mortem histology. This may be due to the difficulty of following anterior to posterior tracts on coronal slices using post-mortem myelin staining methods. Differences in the age range between studies may also explain some of the differences we found between the post-mortem (Bürgel et al., 2006) and the DTI tractography results (Jones et al., 2006; Verhoeven et al., 2010). White matter anatomy changes with age and in the older population white matter changes related to atrophy, reduced myelination, gliosis, and so forth can affect the ability to obtain reliable dissections. The subjects recruited for our study had a relatively young and narrow age range, and therefore the results of our study may not be representative of white matter anatomy across the lifespan. Future atlases will need to take into account the age-related changes in diffusion measurements (Lebel et al., 2008, 2010; Lebel and Beaulieu, 2009; Verhoeven et al., 2010) and how theses changes affect the tracking results.

The atlas could facilitate the identification of the main white matter tracts on T_2-and T_1-weighted MR images in patients with a wide range of pathologies including stroke, multiple sclerosis, acquired and hereditary leucoencephalopathies, neurodegenerative diseases, tumours, and vascular malformations. However, in these patients the overlapping of the atlas maps with the structural data could raise some problems due, for example, to tissue displacement and an altered anatomy secondary to the tumour mass effect (Clark et al., 2003). Hence, the results in these patients should be interpreted with caution. Finally the atlas can also help localize between-group differences derived from voxel based morphometry analysis in psychiatric disorders such as autism (Catani et al., 2008; Pugliese et al., 2009), depression, schizophrenia (Kanaan et al., 2009) and antisocial behaviour (Craig et al., 2009).

In conclusion, a decade ago, radiological information about white matter anatomy could only be inferred from axonal tracing studies in monkeys and a limited number of histological atlases based on human post-mortem dissections. Our results suggest that DTI-tractography maps can complement the information on the constant and variable features of white matter anatomy derived from quantitative histological atlases. Composite maps of interhemispheric asymmetry and interindividual variability have the potential to become a very useful educational and practical tool in establishing the relationship of a lesion with adjacent white matter tracts. This adds a new layer of complexity to the correlation method by linking deficits to inferred dysfunction within distributed networks.

References

Amunts, K. and Willmes, K. (2006). From intersubject variability in clinical syndromes to anatomical variability. *Brain Lang*, **96**(2), 147–50, discussion 157–70.

Ashburner, J. (2007). A fast diffeomorphic image registration algorithm. *NeuroImage*, **38**(1), 95–113.

Ashburner, J. and Friston, K.J., 2000. Voxel-based morphometry–the methods. *NeuroImage*, **11**(6), 805–21.

Barrick, T.R., Lawes, I.N.C., Mackay, C.E., and Clark, C.A. (2007). White matter pathway asymmetry underlies functional lateralization. *Cereb Cortex*, **17** (3), 591–8.

Beaulieu, C. (2009). The biological basis of diffusion anisotropy. In Johansen-Berg, H. and Behrens, T.E. (Eds.) *Diffusion MRI: from quantitative measurement to in vivo neuroanatomy*, pp. 105–26. London: Elsevier.

Blanton, R.E., Levitt, J.G., Peterson, J.R., et al. (2004). Gender differences in the left inferior frontal gyrus in normal children. *NeuroImage*, **22**(2), 626–36.

Burdach, K. (1819). *Vom Bau und Leben des Gehirns und Rückenmarks*. Leipzig: In der Dyk'schen Buchhandlung.

Bürgel, U., Amunts, K., Hoemke, L., et al. (2006). White matter fiber tracts of the human brain: three-dimensional mapping at microscopic resolution, topography and intersubject variability. *NeuroImage*, **29**(4), 1092–105.

Bürgel, U., Schormann, T., Schleicher, A., and Zilles, K. (1999). Mapping of histologically identified long fiber tracts in human cerebral hemispheres to the MRI volume of a reference brain: position and spatial variability of the optic radiation. *NeuroImage*, **10**(5), 489–99.

Catani, M. and Thiebaut de Schotten, M. (2008). A diffusion tensor imaging tractography atlas for virtual in vivo dissections. *Cortex*, **44**(8), 1105–32.

Catani, M., Howard, R.J., Pajevic, S., and Jones, D.K. (2002). Virtual in vivo interactive dissection of white matter fasciculi in the human brain. *NeuroImage*, **17**(1), 77–94.

Catani, M., Jones, D.K., Donato, R., and ffytche, D.H. (2003). Occipito-temporal connections in the human brain. *Brain*, **126**(9), 2093–107.

Catani, M., Jones, D.K., ffytche, D.H. (2005). Perisylvian language networks of the human brain. *Ann Neurol*, **57**(1), 8–16.

Catani, M., Allin, M.P., Husain, M., et al. (2007). Symmetries in human brain language pathways correlate with verbal recall. *Proc Natl Acad Sci USA*, **104**(43), 17163–8.

Catani, M., Jones, D.K., Daly, E., et al. (2008). Altered cerebellar feedback projections in Asperger syndrome. *NeuroImage*, **41**(4), 1184–91.

Ciccarelli, O., Toosy, A.T., Parker, G.J., et al. (2003). Diffusion tractography based group mapping of major white-matter pathways in the human brain. *NeuroImage*, **19**(4), 1545–55.

Ciccarelli, O., Catani, M., Johansen-Berg, H., et al. (2008). Diffusion-based tractography in neurological disorders: concepts, applications, and future developments. *Lancet Neurol*, **7**(8), 715–27.

Clark, C.A., Barrick, T.R., Murphy, M.M., and Bell B.A. (2003). White matter fiber tracking in patients with space-occupying lesions of the brain: a new technique for neurosurgical planning? *NeuroImage*, **20**(3), 1601–8.

Craig, M., Catani, M., Deeley, Q., Latham, R., et al. (2009). Altered connections on the road to psychopathy. *Mol Psychiatry*, **14**(10), 946–53.

Déjerine, J. (1895). *Anatomie des Centres Nerveux*. Paris: Rueff et Cie.

Dell'Acqua, F., Rizzo, G., Scifo, P., et al. (2007). A model-based deconvolution approach to solve fiber crossing in diffusion-weighted MR imaging. *IEEE Trans Biomed Eng*, **54**(3), 462–72.

Dell'Acqua, F., Scifo, P., Rizzo, G., et al. (2010). A modified damped Richardson–Lucy algorithm to reduce isotropic background effects in spherical deconvolution. *NeuroImage*, **49**(2), 1446–58.

Doricchi, F., Thiebaut de Schotten, M., Tomaiuolo, F., and Bartolomeo, P. (2008). White matter (dis)connections and gray matter (dys)functions in visual neglect: gaining insights into the brain networks of spatial awareness. *Cortex*, **44**(8), 983–95.

Eluvathingal, T.J., Hasan, K.M., Kramer, L.A., et al. (2007). Quantitative diffusion tensor tractography of association and projection fibers in normally developing children and adolescents. *Cereb Cortex*, **17**(12), 2760–8.

Flechsig, P.E. (1876). *Die Leitungsbahnen im Gehirn und Rückenmark des Menschen auf Grund entwicklungsgeschichtlicher Untersuchungen*. Leipzig: Engelmann.

Gong, G., Jiang, T., Zhu, C., et al. (2005). Asymmetry analysis of cingulum based on scale-invariant parameterization by diffusion tensor imaging. *Hum Brain Mapp*, **24**(2), 92–8.

Good, C.D., Johnsrude, I.S., Ashburner, J., et al. (2001). Cerebral asymmetry and the effects of sex and handedness on brain structure: a voxel-based morphometric analysis of 465 normal adult human brains. *NeuroImage*, **14**(3), 685–700.

Hagmann, P., Cammoun, L., Martuzzi, R., et al. (2006). Hand preference and sex shape the architecture of language networks. *Hum Brain Mapp*, **27**(10), 828–35.

Hervé, P.Y., Crivello, F., Perchey, G., et al. (2006). Handedness and cerebral anatomical asymmetries in young adult males. *NeuroImage*, **29**(4), 1066–79.

Hervé, P.Y., Leonard, G., Perron, M., et al. (2009). Handedness, motor skills and maturation of the corticospinal tract in the adolescent brain. *Hum Brain Mapp*, **30**(10), 3151–62.

Highley, J.R., Walker, M.A., Esiri, M.M. (2002). Asymmetry of the uncinate fasciculus: a post-mortem study of normal subjects and patients with schizophrenia. *Cereb Cortex*, **12**(11), 1218–24.

Hua, K., Zhang, J., Wakana, S., et al. (2008). Tract probability maps in stereotaxic spaces: analyses of white matter anatomy and tract-specific quantification. *NeuroImage*, **39**(1), 336–47.

Jones, D.K. (2003). Determining and visualizing uncertainty in estimates of fiber orientation from diffusion tensor MRI. *Magn Reson Med*, **49**(1), 7–12.

Jones, D.K., Williams, S.C., Gasston, D., et al. (2002). Isotropic resolution diffusion tensor imaging with whole brain acquisition in a clinically acceptable time. *Hum Brain Mapp*, **15**(4), 216–30.

Jones, D.K., Catani, M., Pierpaoli, C., et al. (2006). Age effects on diffusion tensor magnetic resonance imaging tractography measures of frontal cortex connections in schizophrenia. *Hum Brain Mapp*, **27**(3), 230–8.

Kanaan, R., Borgwardt, S., McGuire, P., et al. (2009). Microstructural organization of cerebellar tracts in schizophrenia. *Biol Psychiatry*, **66**(11), 1067–9.

Klingler, J. (1935). Erleichterung der makroskopischen Präparation des Gehirn durch den Gefrierprozess. *Schweiz Arch Neurol Psychiat*, **36**, 247–56.

Lawes, I.N.C., Barrick, T.R., Murugam, V., et al. (2008). Atlas-based segmentation of white matter tracts of the human brain using diffusion tensor tractography and comparison with classical dissection. *NeuroImage*, **39**(1), 62–79.

Lebel, C., Walker, L., Leemans, A. (2008). Microstructural maturation of the human brain from childhood to adulthood. *NeuroImage*, **40**(3), 1044–55.

Lebel, C. and Beaulieu, C. (2009). Lateralization of the arcuate fasciculus from childhood to adulthood and its relation to cognitive abilities in children. *Hum Brain Mapp*, **30**(11), 3563–73.

Lebel, C., Caverhill-Godkewitsch, S., Beaulieu, C. (2010). Age-related regional variations of the corpus callosum identified by diffusion tensor tractography. *NeuroImage*, **52**(1), 20–31.

Luders, E., Narr, K.L., Thompson, P.M., et al. (2004). Gender differences in cortical complexity. *Nat Neurosci*, **7**(8), 799–800.

Makris, N., Papadimitriou, G.M., Kaiser, J.R., et al. (2008). Delineation of the Middle Longitudinal Fascicle in Humans: A Quantitative, In Vivo, DT-MRI Study. *Cereb Cortex*, **19**(4), 777–785.

Matsumoto, R., Okada, T., Mikuni, N., et al. (2008). Hemispheric asymmetry of the arcuate fasciculus : A preliminary diffusion tensor tractography study in patients with unilateral language dominance defined by Wada test. *J Neurol*, **255**(11), 1703–11.

McNab, J.A., Jbabdi, S., Deoni, S.C., et al. (2009). High resolution diffusion-weighted imaging in fixed human brain using diffusion-weighted steady state free precession. *NeuroImage*, **46**(3), 775–85.

Mori, S., Kaufmann, W., Davatzikos, C., et al. (2002). Imaging cortical association tracts in the human brain using diffusion-tensor-based axonal tracking. *Magn Reson Med*, **47**(2), 215–23.

Mori, S., Crain, B., Wakana, S., et al. (2005). *MRI Atlas of Human White Matter*. Amsterdam: Elsevier.

Nucifora, P.G., Verma, R., Melhem, E.R., Gur, R.E., and Gur, R.C. (2005). Leftward asymmetry in relative fiber density of the arcuate fasciculus. *Neuroreport*, **16**(8), 791–4.

Park, H.J., Westin, C.F., Kubicki, M., et al. (2004). White matter hemisphere asymmetries in healthy subjects and in schizophrenia: a diffusion tensor MRI study. *NeuroImage*, **23**(1), 213–23.

Parker, G.J., Luzzi, S., Alexander, D.C., et al. (2005). Lateralization of ventral and dorsal auditory-language pathways in the human brain. *NeuroImage*, **24**(3), 656–66.

Paus, T. (2010). Growth of white matter in the adolescent brain: myelin or axon? *Brain Cogn*, **72**(1), 26–35.

Perrin, J.S., Leonard, G., Perron, M., et al. (2009). Sex differences in the growth of white matter during adolescence. *NeuroImage*, **45**(4), 1055–66.

Powell, H.W., Parker, G.J., Alexander, D.C., et al. (2006). Hemispheric asymmetries in language-related pathways: a combined functional MRI and tractography study. *NeuroImage*, **32**(1), 388–99.

Pugliese, L., Catani, M., Ameis, S., et al. (2009). The anatomy of extended limbic pathways in Asperger syndrome: a preliminary diffusion tensor imaging tractography study. *NeuroImage*, **47**(2), 427–34.

Rademacher, J., Bürgel, U., Geyer, S., et al. (2001). Variability and asymmetry in the human precentral motor system. A cytoarchitectonic and myeloarchitectonic brain mapping study. *Brain*, **124**(11), 2232–58.

Rodrigo, S., Oppenheim, C., Chassoux, F., et al. (2008). Language lateralization in temporal lobe epilepsy using functional MRI and probabilistic tractography. *Epilepsia*, **49**(8), 1367–76.

Shapiro, S., Wilk, M. (1965). An Analysis of Variance Test for Normality (Complete Samples). *Biometrika*, **52**, 591–611.

Shaywitz, B.A., Shaywitz, S.E., Pugh, K.R., et al. (1995). Sex differences in the functional organization of the brain for language. *Nature*, **373**(6515), 607–9.

Sivaswamy, L., Rajamani, K., Juhasz, C., et al. (2008). The corticospinal tract in Sturge–Weber syndrome: a diffusion tensor tractography study. *Brain Dev*, **30**(7), 447–53.

Stadlbauer, A., Salomonowitz, E., Strunk, G., et al. (2008). Age-related degradation in the central nervous system: assessment with diffusion-tensor imaging and quantitative fiber tracking. *Radiology*, **247**(1), 179–88.

Talairach, J. and Tournoux, P. (1988). *Co-planar Stereotaxic Atlas of the Human Brain: 3Dimensional Proportional System -an Approach to Cerebral Imaging*. New York: Thieme Medical Publishers.

Thiebaut de Schotten, M., Urbanski, M., Duffau, H., et al. (2005). Direct evidence for a parietal-frontal pathway subserving spatial awareness in humans. *Science*, **309**(5744), 2226–8.

Thiebaut de Schotten, M., Kinkingnehun, S., Delmaire, C., et al. (2008). Visualization of disconnection syndromes in humans. *Cortex*, **44**(8), 1097–103.

Thiebaut de Schotten, M., ffytche, D., Bizzi, A., et al. (2011a). Atlasing location, asymmetry and inter-subject variability of white matter tracts in the human brain with MR diffusion tractography. *Neuroimage*, **54**(1), 49–59.

Thiebaut de Schotten, M., Dell'Acqua, F., Forkel, S., et al. (2011b). A lateralized brain network for visuo-spatial attention. *Nat Neurosci*, **14**(10), 1245–46.

Tournier, J.D., Calamante, F., Gadian, D.G., et al. (2004) Direct estimation of the fiber orientation density function from diffusion-weighted MRI data using spherical deconvolution. *NeuroImage*, **23**(2), 1176–85.

Tournier, J.D., Yeh, C.H., Calamante, F., et al. (2008). Resolving crossing fibres using constrained spherical deconvolution: validation using diffusion-weighted imaging phantom data. *NeuroImage*, **42**(5), 617–25.

Urbanski, M., Thiebaut de Schotten, M., Rodrigo, S., et al. (2008). Brain networks of spatial awareness: evidence from diffusion tensor imaging tractography. *J Neurol Neurosurg Psychiat*, **79**(3), 598–601.

Verhoeven, J.S., Sage, C.A., Leemans, A., et al. (2010). Construction of a stereotaxic DTI atlas with full diffusion tensor information for studying white matter maturation from childhood to adolescence using tractography-based segmentations. *Hum Brain Mapp*, **31**(3), 470–86.

Vernooij, M.W., Smits, M., Wielopolski, P.A., et al. (2007). Fiber density asymmetry of the arcuate fasciculus in relation to functional hemispheric language lateralization in both right-and left-handed healthy subjects: a combined fMRI and DTI study. *NeuroImage*, **35**(3), 1064–76.

Wakana, S., Jiang, H., Nagae-Poetscher, L.M., van Zijl, P.C., Mori, S. (2004). Fiber tract-based atlas of human white matter anatomy. *Radiology*, **230**(1), 77–87.

Wakana, S., Caprihan, A., Panzenboeck, M.M., et al. (2007). Reproducibility of quantitative tractography methods applied to cerebral white matter. *NeuroImage*, **36**(3), 630–44.

Wassermann, D.,Bloy, L.,Kanterakis, E., Verma, R., and Deriche, R. (2010). Unsupervised white matter fiber clustering and tract probability map generation: applications of a Gaussian process framework for white matter fibers. *NeuroImage*, **51**(1), 228–41.

Worsley, K.J., Evans, A.C., Marrett, S., et al. (1992). A three-dimensional statistical analysis for CBF activation studies in human brain. *J Cereb Blood Flow Metab*, **12**(6), 900–18.

Yakovlev, P.J. and Rakic, P. (1966). Pattern of decussation of bulbar pyramids and distribution of pyramidal tracts on two sides of the spinal cord. *Trans Am Neurol Assoc*, **91**, 366–7.

Index

abulic syndrome 66
AC *see* anterior commissure
acalculia 68, 250–1
achromatopsia 65, 69, 307
acoustic agnosia 69
ADC (apparent diffusion coefficient) 490, 491, 493
AF *see* arcuate fasciculus
affect *see* emotions
afterimages 309
age-related white matter changes 243–4, 382, 512
aggressive behaviour 48, 446–7
agnosia
  acoustic 69
  prosopagnosia 63, 69, 70, 307
  tactile 350
  visual 69, 307, 309
agraphia 68, 250, 350
akinetopsia 65, 69, 307
Alcmaeon of Croton 12
alcohol abuse 71, 352, 445
alexia 45, 68, 69, 251, 307, 350
  see also dyslexia
alien hand syndrome 350, 352
Alzheimer's disease 71, 250, 251, 385, 444, 445–6
amnesia 70–1, 445
  visual 308
amusia 246, 252
amygdala 10, 22, 23, 24, 304, 443, 444
  lesions 71
anarchic hand syndrome 66, 350
angular gyrus 9, 45, 250
anisotropy 490, 491, 492, 493
anomia 68
  colour 350
  olfactory 71–2, 345
  tactile 350, 352
anoneria 306, 309
anterior commissure (AC)
  atlas 355–64
    with other tracts 80–101, 159–77, 191–231
  description 40, 344–5
  functional anatomy 345
  group effect map 508
anterior thalamic projection(s) 384–5, 441
antisocial behaviour 446–7
Anton's syndrome 69, 306
apathy 66
aphasia 45, 59, 69, 246, 247–9, 350
  see also anomia
apparent diffusion coefficient (ADC) 490, 491, 493
apperceptive agnosia 307
apraxia 45–6, 66, 67, 250–1, 350
aprosodia 69, 252
aqueduct of Sylvius 23
Aranzi, Giulio Cesare 27
archistriatum *see* amygdala; thalamus
arcuate fasciculus (AF)
  anterior segment
    atlas (AF-A only) 278–86
    atlas (with other tracts) 107–17, 124–35, 168–97, 226–28
    diffusion tractography images 66, 67, 241
    group effect map 504
    lateralization 243, 505, 508, 511
    lesions 246, 248, 250
    synonymous with SLF III 255
  atlas 259–95
    with other tracts 96–117, 124–45, 168–97, 225–38
  comparative anatomy 245–6
  description 38–9, 51, 241–2
  diffusion tractography images 51, 240, 241
    compared with post-mortem histology 507
    group effect maps 504, 506
    by lobe 65, 67, 70
    methods used in acquisition 504
  functional anatomy 241–2
  gender differences 243, 511
  historical references 43, 240
  lateralization 242–5, 505, 508, 510
    functional vs. structural 244, 511–2
    right-hemispheric lesions 251–2
  lesions 246–54, 510–1
  long segment 241
    atlas (AF-L only) 269–77
    atlas (with other tracts) 96–117, 124–34, 168–83, 193–6, 227–38
    diffusion tractography images 65, 70, 241
    group effect map 504, 508
    lateralization 242–3, 505, 512
    lesions 250, 252, 257
  posterior segment
    atlas (AF-P only) 287–95
    atlas (with other tracts) 96–117, 126–45, 183, 184–96, 226–38
    diffusion tractography images 67, 70, 241
    group effect map 506
    lateralization 243–4, 505
    lesions 246, 247, 248, 250
Aristotle 2
ascending frontal (precentral) gyrus 8, 66
ascending parietal (postcentral) gyrus 8, 67
Asperger's syndrome 393, 449
associationist models 43–6, 50–1, 57, 58
association pathways 38–9, 43
  see also arcuate fasciculus; cingulum; inferior fronto-occipital fasciculus; inferior longitudinal fasciculus; uncinate fasciculus
associative visual agnosia 307
asymmetries in the brain *see* lateralization
ataxia 67, 69, 350, 391
attention deficit and hyperactivity disorder (ADHD) 352, 386
Auburtin, Simon Alexandre Ernest 16–7
auditory disorders
  cortical deafness 69
  dichotic listening suppression 350
  hallucinations 69, 254, 448
  MGN lesions 386–7
autism spectrum disorder 253, 393, 448–9
autoscopy 70, 310
axial plane 4

Baddeley, Alan David 250
Bailey, Percival Sylvester 33–4
Baillarger, Jules 28
Balint syndrome 68
basal ganglia 22, 23, 24, 387–9, 446
behavioural syndromes
  cerebellar lesions 393
  frontal lobe lesions 48, 67, 445
  psychopathy 446
  temporal lobe lesions 446
Betz, Vladimir Alexewitsch 29
bipolar affective disorder 448
blindness, cerebral 69, 307
Bonin, Gerhardt von 33–4
Bouillaud, Jean-Baptiste 16
brainstem 383, 385, 389
Broca, Paul 16, 17, 18, 65, 440
Broca's aphasia 246, 248
Broca's territory 12, 16, 240–2, 250
Brodmann, Korbinian 31–2
Brodmann's maps 24, 25, 33
bundle of Arnold 390
bundle of Türck 390
Burdach, Karl Friedrich 43, 240

Cajal, Santiago Ramon y 29
calcarine sulcus 9, 10, 302
callosal sulcus 9
callosotomy 49–50, 349, 351
Campbell, Alfred 30, 31
Capgras syndrome (delusion) 69, 309–10
capsulotomy 448
caudate nucleus 22, 23, 24, 388
CC *see* corpus callosum
central (Rolandic) sulcus 8, 11

centromedian (CM) nucleus 384, 385
centrum ovale 42
cephalocentric theory 12
cerebellar cognitive affective syndrome (CCAS) 392–3
cerebellar-rubro-olivocerebellar network 394
cerebellum 389–1
  lesions 391–3
  see also cortico-ponto-cerebellar tract; inferior cerebellar peduncle; superior cerebellar peduncle
cerebral blindness 69, 306–7
cerebral cortex
  historical references 27–34
  isotropy of 491
  localization of function 58–60
  sectional anatomy 24–5
  surface anatomy 8–11
  visual areas 301–3
cerebral peduncles 380, 383
Charcot–Wilbrand syndrome 309
Charles Bonnet syndrome 309
cingulate gyrus 9, 71
cingulate sulcus 9
cingulotomy 448, 449
cingulum
  atlas 453–66
    with other tracts 83–123, 124–45, 148–83, 195–210, 212–27
  description 39, 299, 442
  diffusion tractography images 39, 442
    compared with post-mortem histology 510
    group effect map 508
    by lobe 66, 68, 69, 70, 71
  historical references 43
  lateralization 504, 508, 510
  lesions 448, 449
clinico-anatomical correlation method
  hodotopic framework 62–3
  importance of modern imaging techniques 65–6
  principles 57–62
  in visual disorders 64–6
cognitive syndromes
  arcuate fasciculus lesions 250
  cerebellar lesions 392–3
  cortico-basal ganglia circuit lesions 388
  frontal lobe lesions 67, 388
  limbic system lesions 70, 71, 384–5, 444, 445–6
  occipital lesions 68, 69
  parietal lesions 67, 68
  temporal lesions 69, 70

colour anomia 350
colour vision, achromatopsia 65, 69, 307
commissural pathways 40, 344
  historical references 41, 42–3
  posterior (hippocampal) 344, 441
  see also anterior commissure; corpus callosum
comparative anatomy
  anterior commissure 345
  arcuate fasciculus 245–6
  corpus callosum 345–6, 350
  inferior longitudinal fasciculus (ILF) 304–5
  superior longitudinal fasciculus (SLF) 255
  surface anatomy 12, 15, 245
  visual cortex/pathways 302, 304
conceptual apraxia 250
conduction aphasia 45, 61, 246, 247–9
confabulation 445
connectional neuroanatomy 4, 38–40
  atlas
    axial view 76–123
    coronal view 124–83
    sagittal view 184–238
  historical references 41–52, 59–62, 240, 383
  see also individual tracts
constructional apraxia 67
coronal plane 4
corona radiata 40, 380
  see also internal capsule
corpus callosum (CC)
  atlas 365–78
    with other tracts 94–123, 124–83, 193–229
  callosotomy 49–50, 348, 349–50
  comparative anatomy 345–6, 348
  description 9, 22, 40, 298, 345–6
  diffusion tractography images 40, 51, 344, 346–7, 494, 495
    compared with post-mortem histology 505
    splenium and visual pathways 298
  functional anatomy 349
  historical references 41, 49–50
  lesions 349, 350–2
corpus striatum 22, 23, 24, 388
  historical references 27, 41–2
  see also amygdala
cortico-basal ganglia circuits 387–9
cortico-bulbar tract 380–1, 382–3
cortico-ponto-cerebellar tract (CPC) 389, 390, 498
  atlas 414–24
    with other tracts 80–123, 124–43, 173–83, 196–227
cortico-spinal tract (CST)
  atlas 76–135, 162–83, 195–228
  description 380–2
  group effect map 506
  lateralization 383, 505
  lesions 383, 509, 510
  post-mortem histology vs. *in vivo* tractography 507, 509
CPC *see* cortico-ponto-cerebellar tract
criminal behaviour *see* forensic neuroanatomy
Critchley, Macdonald 57, 59
crossed cerebellar diaschisis (CCD) 393
CST *see* cortico-spinal tract
cuneus 9
Czolgosz, Leon 17

Dandy, Walter 49
da Vinci, Leonardo 2, 25
deafness, cortical 69
decussation of the CST 382
deep brain stimulation 448, 449
deep cerebellar nuclei 389, 390, 391
default-mode network 444
Dejerine, Joseph Jules 16, 17, 45, 52, 240, 381
dementia
  Alzheimer's 71, 250, 251, 385, 444, 445–6
  semantic 70, 446
dentate nucleus 389, 390
depression 49, 448
de-realization 69, 309
descending occipital gyrus 9
diaschisis 48, 60, 62, 393
dichotic listening suppression 350
diencephalon 23–4
diffusion tensor imaging (DTI)
  comparison with histology maps 507, 509, 510
  in diagnosis 383, 511–2
  history of 51–2
  principles of 498–504
  as used to compile this atlas 498–504, 511–3
diffusion tensor value 492
disconnection syndromes 61–2, 307, 350
  historical references 45–6, 50–1
dissection
  history of 2, 26–7, 498
  techniques for fibre dissection 46–7
dorso-lateral surface 8–9
  Vesalius' drawing of 13
dreaming, loss of (anoneria) 306, 309
DTI *see* diffusion tensor imaging
dyscalculia 68, 250–1
dysgenesis of the corpus callosum 349–51
dyslexia 70, 250
  see also alexia
dystonia 386

Economo, Constanin von 33
Egypt, ancient 12
Ehrenberg, Christian Gottfried 29
Einstein, Albert, his brain 17
  diffusion equation 490
emotions
  cerebellar lesions 391, 393
  frontal lobe lesions 48, 66
  limbic lobe lesions 71
  limbic system 440–1, 447
  memory-emotion network 58
  prefrontal cortex 66, 306, 448
  visual hypo-emotionality 65, 309
epilepsy
  split-brain patients 49–50, 345, 349–50
  temporal lobe 70, 446
Erasistratus of Alexandria 12
Essen maps 34
evolution
  of language 245–6
  of specialized areas 348
executive functions 58, 66
Exner's centre 250
Exner, Sigmund 16
extinction 67, 68

face-object identification network 58
faces
  hallucinations 69, 308
  inability to recognize (prosopagnosia) 63, 69, 70, 307
facial weakness 383
false memories 310
fastigial nucleus 390
Fieschi, Giuseppe Marco 15
flashbulb memories 310
Flechsig, Paul Emile 32, 33, 47–8, 298, 382
Flechsig–Meyer's loop 300
Flourens, Marie-Jean Pierre 18
foramen of Luschka 22, 23
foramen of Magendie 22, 23
foramen of Monro 22
forensic neuroanatomy
  antisocial behaviour 447
  historical references 15, 17, 18–19

fornix (Fx)
atlas 476–87
with other tracts 81–113, 124–30, 164–83, 191–231
description 40, 441
diffusion tractography images 40, 51, 71, 441
compared with post-mortem histology 507
group effect map 506
lesions 445, 446
posterior commissure 344, 441
fractional anisotropy 493
fragile X syndrome 388–9
Freeman, Walter 49
Fregoli syndrome 69, 310
frontal eye field 66
frontal lobe 8, 10, 11, 66
Brodmann's map 24, 25
lesions 48, 66, 71, 447, 449
fronto-striatal pathways 301, 389, 449
Fulton, John 48–9
fusiform gyrus 10, 70, 308

Gage, Phineas 446
gait ataxia 391
Galen 13, 26, 41
Gall, Joseph Franz 14–5
gaze abnormalities 67
gender differences 243, 504, 505, 511, 512
Gennari, Francesco 27, 28
Gerstmann's syndrome 68, 250
Geschwind, Norman 19, 50–1, 59, 60, 61, 240
Geschwind's syndrome 446
Geschwind's territory 241–2, 250
global developmental delay 253
globus pallidus 22, 23, 24
Golgi, Camillo 29
granule cells 391
Gratiolet, Louis Pierre 15–6
Greece, ancient 12
group effect maps 504, 505, 506, 508
Gudden, Bernhard Aloys von 47
Guiteau, Charles 17
gyri 8–11, 25–6
see also individual gyri
gyrus rectus 9, 10

habenular commissure 344
hallucinations
auditory 69, 254, 448
visual 65, 71, 307–9, 310
handedness 382, 389
hearing disorders *see* auditory disorders
heautoscopy 70, 310
hemianopsia 65–6, 69, 307
hemiparesis/hemiplegia 66, 383
Henschen, Salomon 301
Herophilus of Alexandria 26
hippocampal-diencephalic-retrosplenial network 71, 444
hippocampal (posterior) commissure 344, 441
hippocampocentric division of the paralimbic zone 24, 443–4, 445
hippocampus 23, 29, 440
history of neuroanatomy 2
connectional anatomy 41–52, 59–62, 240, 381, 440–1, 498
language 12, 16–17, 45, 240, 247
sectional anatomy 25–34
surface anatomy 11–9
vision 301

holistic model of brain function 57, 58
Holmes, Gordon 301
homonymous hemianopia 69
Huntington's chorea 387, 388
Huschke, Emil 16
hyperfunctional syndromes 61, 62, 63, 70, 310
see also hallucinations
hypoemotionality, visual 65, 309
hypothalamus 23, 441

IC *see* internal capsule
ICP *see* inferior cerebellar peduncle
ideomotor apraxia 67, 250
IFOF *see* inferior fronto-occipital fasciculus
ILF *see* inferior longitudinal fasciculus
Imhotep 12
inferior cerebellar peduncle (ICP) 389, 390
atlas 425–30
with other tracts 76–94, 125–44, 203–10, 212–8
inferior colliculus 387
inferior frontal gyrus 8, 66, 241, 512
inferior frontal sulcus 8
inferior fronto-occipital fasciculus (IFOF)
atlas 331–41
with other tracts 92–107, 124–83, 190–205, 217–32
description 39, 299–300
diffusion tractography images 39, 301
compared with post-mortem histology 509
group effect map 506
by lobe 66, 70, 71
functional anatomy 306
lateralization 505, 510, 512
lesions 309
inferior longitudinal fasciculus (ILF)
atlas 320–30
with other tracts 85–111, 124–68, 187–99, 223–38
comparative anatomy 304–5
description 39, 299
diffusion tractography images 39, 29
group effect map 506
by lobe 69, 70, 71
functional anatomy 305
historical references 43
lateralization index 505
lesions 309
inferior occipital gyrus 9, 10
inferior occipital sulcus 9
inferior olivary nucleus 390
inferior temporal gyrus 9, 10
inferior temporal sulcus 9
Inouye, Tatsuji 302
insomnia 386
insula of Reil 9
intelligence 17, 18
inter-individual variability of tracts 34, 508, 511, 512
intermediate frontal sulcus 8
internal capsule (IC)
atlas 331–41
with other tracts 103–17, 124–7, 129–55, 161–77, 193–228
capsulotomy 448
description 40, 380, 381
diffusion tractography images 40, 380, 494
lateralization index 505
internal medullary lamina 384

interposed nucleus 389, 390
inter-subject variability of tracts 34, 498, 504, 508, 512
interventricular foramen (of Monro) 22
intralaminar nuclei 384, 385
intraparietal sulcus 9

Jackson, John Hughlings 17, 57, 58
Jacobsen, Carlyle 48–9

Kleist, Karl 58
Klingler, Josef 46
Klüver–Bucy syndrome 71, 446
Kölliker, Rudolf Albert von 30
Korsakoff's syndrome 71, 445
Koskinas, Georg 33

language
disorders
aphasia 45, 59, 69, 246, 247–9, 350
aprosodia 69, 252
cerebellar lesions 393
mutism 351, 393
evolution 245–6
functional anatomy 58, 240–2
historical references 12, 16–17, 45, 240, 247
lateralization 244, 511–2
Lashley, Karl 58
lateral geniculate nucleus (LGN) 298, 309, 386–7
lateralization 505–8, 510
arcuate fasciculus 242–4, 505, 508, 510
functional vs. structural 244, 511–2
right-hemispheric lesions 251–2
cingulum 498, 508, 510
corpus callosum 346
gender differences 508, 512
IFOF 505, 510, 512
lateralization index 504, 505
uncinate 443, 505
lateral (Sylvian) sulcus 8–9, 11, 14
lateral thalamic nuclei 386, 388
Leborgne (patient of Broca) 17, 64
Leeuwenhoek, Antony van 27
Lelong (patient of Broca), his brain 16
lenticular nucleus 22, 23, 24
Leonardo da Vinci 2, 25
Lettvin, Jerry 60
leucotomy 48–9, 385–6
Leuret, Francois 15–6
LGN (lateral geniculate nucleus) 298, 309, 386–7
Lichtheim, Ludwig 45, 59, 61, 247
Liepmann, Hugo 45–6
Lima, Almeida 48
limbic circuit (cortico-basal ganglia) 387–9
limbic lobe 11, 70, 71
Brodmann's map 24, 25
limbic system
atlas
cingulum 453–66
fornix 476–87
uncinate fasciculus 467–75
description 39, 40, 384, 441–3
functional anatomy 443–4
historical references 47, 440–1
lesions 70, 71–2, 384–5, 444, 445–9
see also cingulum; fornix; uncinate fasciculus
limb-kinetic apraxia 69, 250–1
lingual gyrus 10
lobotomy 48–9

localizationist model of brain function 57, 58
locked-in syndrome 383
Lombroso, Cesare 18–9

MacLean, Paul 441
macropsia 70, 308
magnetic resonance imaging (MRI)
connective anatomy 51–2
sectional anatomy 34
surface anatomy 18, 19
see also diffusion tensor imaging
'main étrangère' 350
Malpighi, Marcello 27, 41
mammillo-tegmental tract 441
mammillo-thalamic tract 440, 441, 445
marginal frontal sulcus 8, 10
marginal sulcus 9
massa intermedia 344, 351
mathematical difficulties (acalculia/dyscalculia) 68, 250
mean diffusivity (MD) 492
medial geniculate nucleus (MGN) 386–7
medial lemniscus 380
medial occipitotemporal (collateral) sulcus 10
medial surface 9–10
mediodorsal thalamic nucleus (MD) 384, 385–6
medulla oblongata 41–2, 382
memory disorders 70, 71, 250, 445–6
memory-emotion network 58
Mesulam, Marsel large-scale networks 58
metamorphopsias 69, 70, 307
metathalamus 385, 386–7
LGN (lateral geniculate nucleus) 298, 309, 386
methods used to compile *Atlas of Human Brain Connections* 497–514
Meyer's loop 298
Meynert, Theodore 30–1, 44
MGN (medial geniculate nucleus) 386–7
Michelangelo Buonarroti 2
micropsia 70, 308
microscopy 27, 28–9, 41, 47–8
middle cerebellar peduncle 389
middle frontal gyrus 8, 241
middle occipital gyrus 9
middle temporal gyrus 9, 69, 241
Mishkin, Mortimer 50
Monakow, Constantin von 47, 48, 59, 61, 240
Mondino de' Luzzi 26–7
Moniz, Antonio Egas 48–9
monkey neuroanatomy *see* comparative anatomy
Montreal Neurological Institute (MNI) co-ordinates 4
motion vision, loss of (akinetopsia) 64, 69, 307
motor aphasia 247, 248
motor disorders
apraxia 45–6, 66, 67, 250–1, 350
ataxia 67, 69, 350, 391
cerebellar lesions 391–3
extrapyramidal defects 388
primary motor cortex lesions 67, 250
pyramidal syndrome 382–3
thalamic nuclei lesions 386
motor neglect 68, 385
motor pathways
cerebellar 389–93
cortico-basal ganglia circuit 387–9
CST (cortico-spinal tract) 380–4
Muratoff bundle 300–1
music, difficulties with (amusia) 246, 252
mutism 351, 393
myelogenetic maps 32, 33, 47–8

neglect 59, 66, 68–9, 251–2, 309, 512
map of patient with 511
neo-associationism 50
neostriatum 22, 23, 24
neuroanatomy 2–4
neurodegenerative disorders *see* dementia; Parkinson's disease
neurodevelopmental disorders 252–4
callosal dysgenesis 349–51
cerebellar lesions 393
cortico-basal ganglia circuit lesions 387–8
limbic system lesions 445
mediodorsal nucleus lesions 385–6
neurosurgery
callosotomy 49–50, 349, 351
capsulotomy 448
cingulotomy 448, 449
deep brain stimulation 448, 449
leucotomy/lobotomy 48–9, 385–6
subcaudate tractotomy 448, 449
Nielsen, J.M. 250
nucleus accumbens 23, 24, 388

object vision (ventral stream) 69, 303, 304
obsessive–compulsive disorder 71, 448
occipital lobe 9, 10, 11, 68
Brodmann's map 24, 25
lesions 69, 306–10
visual pathways 298–306
oculomotor apraxia 67
olfactocentric division of the paralimbic zone 24, 443–4
olfactory disorders 71–2, 345
olfactory frontal sulcus 9, 10
olfactory gyrus 9
optic aphasia 350
optic ataxia 69, 350
optic radiations (OR) 298
atlas 313–19
with other tracts 97–112, 124–57, 193–9, 223–31
diffusion tractography images 298
compared with post-mortem histology 507, 509
group effect map 506
lateralization index 505
orbital gyrus 10
orbital sulcus 10
orbitofrontal cortex 66, 71, 447, 448
orbitofrontal–amygdala network 71, 446
organology 14–5
out-of-body experience 68, 70, 310

paleostriatum (globus pallidus) 22, 23, 24
pallidum 24, 388
Papez, James 47, 440–1
paracentral lobule 9
paracentral sulcus 9
parahippocampal gyrus 9, 10, 304, 442
parahippocampal-retrosplenial network 444, 445
paralimbic area 71, 443–4
parietal lobe 8–9, 11, 66–8, 241
Brodmann's map 24, 25
lesions in 67–8
parieto-occipital sulcus 9, 11
Parkinson's disease 387, 388
pars opercularis 8
pars orbitalis 8
pars triangularis 8
pelopsia 70, 308
perisylvian pathways *see* arcuate fasciculus
personality changes/disorders 66, 71, 393, 446
phosphenes 306, 308
phrenology (organology) 14–5
physiognomy 2
Piccolomini, Arcangelo 26, 27
pineal recess/body 22
polar temporal sulcus 10
pons 27, 380
pontine nuclei 390
postcentral (ascending parietal) gyrus 8, 67
posterior fossa syndrome 393
posterior (hippocampal) commissure 344, 441
posterior intermediate parietal sulcus 9
posterior parietal gyrus 9, 67
post-traumatic stress disorder 69, 310
praxis 45
see also apraxia
precentral (ascending frontal) gyrus 8, 66
precuneus 9, 67, 442
prefrontal cortex 66, 306, 446, 448
preoccipital notch 9, 10, 11
primary progressive aphasia 248, 249
primate neuroanatomy *see* comparative anatomy
projection pathways 38, 40, 380
historical references 47–8, 381
see also individual tracts
proprioception 67
prosencephalon 23–4
prosopagnosia 64, 69, 70, 307
pseudobulbar syndrome 383
pseudothalamic syndrome 68
psychiatric disorders *see* neurodevelopmental disorders
psychopathy 446, 447
pulvinar 298, 386
Purkinje (Purkyně) cells 391
Purkyně (Purkinje), Jan Evangelista 29
putamen 22, 23, 24, 388
pyramidal syndrome 383

quadranopia 69, 306

reading difficulties
alexia 45, 68, 69, 251, 307, 350
dyslexia 70, 250
red nucleus 389, 390
reduplicative phenomena 69, 309–10
Reil, Johann Christian 43, 240
Remak, Robert 29
Renaissance, practise of anatomy 2, 13–14, 25, 26–7, 41–2
restless leg syndrome 386
reticular commissure 344
retinotopy 301–3
retrosplenial cortex 445
Riddoch phenomenon 307
Rolandic (central) sulcus 8, 11
Rolando, Luigi 14
Rosett, J. 46–7
rostral sulcus 9

sagittal plane 4
schizophrenia 252–4, 386, 388, 393, 448
surgery 49, 385–6

SCP *see* superior cerebellar peduncle
sectional neuroanatomy 4, 22–5
historical references 25–34
semantic dementia 70, 446
sensory aphasia 69, 247, 248
sensory pathways 67, 380, 384–7
septum pellucidum 22
SFOF *see* superior fronto-occipital fasciculus
sleep disorders
anoneria 306, 309
insomnia 386
SLF *see* superior longitudinal fasciculus
smell, sense of 71–2, 345
spasticity 383
spatial attention 58, 251–2
see also neglect
spatial vision (dorsal stream) 69, 303, 307
speech *see* language
Sperry, Roger 49
spherical deconvolution (SD) tractography 495
spino-cerebellar tract 389, 505, 506
spino-thalamic tract 380
Spitzka, Edward Anthony 17, 19
Spitzka, Edward Charles 17
splenium of the corpus callosum 40, 298, 345, 346, 349, 350
split-brain patients 49–50, 345, 351–2
Steno, Nicolaus 42
striatum 22, 23, 24, 388, 448
stroke, DTI in 490, 510
subcallosal fasciculus 300
subcallosal gyrus 9
subcaudate tractotomy 449
subparietal sulcus 9
sulci 8–11, 26
see also individual sulci
superior cerebellar peduncle (SCP) 344, 389, 390, 506
atlas 431–8
with other tracts 82–98, 201–21
lateralization index 382, 504
superior frontal gyrus 8, 9, 254
superior frontal sulcus 8
superior fronto-occipital fasciculus (SFOF) 255, 300–1, 351
superior longitudinal fasciculus (SLF) 65, 66, 254–5
branch II atlas 116–20, 136–57, 191–5, 226–8
branch III *see* arcuate fasciculus (AF), anterior segment
superior occipital gyrus 9
superior occipital sulcus 9
superior temporal gyrus 9
superior temporal sulcus 9
supramarginal gyrus 9, 242, 250
supraoptic recess 22
surface neuroanatomy 3–4, 8–11, 19
historical references 11–9
Sylvian (lateral) sulcus 8–9, 11, 14
Sylvius, Franciscus 14
synaesthesia 65, 308, 310
syphilis 17

tactile anomia 350, 352
tactile function 67, 380
Talairach and Tournoux atlas 4, 34
tectal commissure of Forel 344
telencephalon 23
teleopsia 70, 308
temporal lobe 9, 10, 11, 69, 241
Brodmann's map 24, 25
lesions 69–70, 446
temporo-amygdala-orbitofrontal network 71, 444
text hallucinations 308
thalamic (ascending) projections 380, 384–7, 441
thalamus 22, 23, 24, 384
3D rendering procedures 504
touch, sense of 67, 380
Tourette's syndrome 389
tractography *see* diffusion tensor imaging
transverse occipital sulcus 9
tremor 386
Turner, William 16

uncinate fasciculus
atlas 467–75
with other tracts 85–103, 157–83, 189–206, 217–33
description 39, 442–3
diffusion tractography images 39, 442
compared with post-mortem histology 509
group effect map 506
by lobe 66, 70, 71
historical references 43
lateralization 443, 505
lesions 70, 444, 446, 447
uncus 10

Valentin, Gabriel Gustav 29
Variolo, Costanzo 27
ventral surface 10
ventral thalamic nuclei 386
ventricular system 22–3, 26–7
Vesalius, Andreas 13, 26, 27, 41
vestibular syndrome 392
Vicq d'Azyr, Felix 14, 42–3
Vieussens, Raymond 42
visual cortex 301–3
visual disorders 64–6
achromatopsia 65, 69, 307
agnosia 63, 69, 70, 307
akinetopsia 65, 69, 307
cerebral blindness 69, 306–7
corpus callosum lesions 349, 350–2
hallucinations 65, 69, 307–9, 310
hemispatial neglect 68, 251–2, 308, 511
hypoemotionality 65, 309
metamorphopsias 69, 70, 307
occipital lobe lesions 69, 306–10
parietal lobe lesions 67–8
temporal lobe lesions 69–70
thalamic nuclei lesions 386
see also gaze abnormalities
visual pathways
atlas
IFOF 331–41
ILF 320–30
optic radiations 313–9
description 299–301
functional and comparative anatomy 303–6
interhemispheric 348
see also optic radiations
Vogt, Cecile 31, 32
Vogt, Oskart 31, 32
voxel based morphometry (VBM) 34

Wagner, Rudolph 18
Watts, James 49
Wernicke, Carl 16, 17, 44–5, 58, 61, 240
Wernicke's aphasia 69, 246, 248
Wernicke's territory 45, 59, 240–2, 250
Wernicke–Geschwind model 240
white matter
anisotropy of 491
historical references 26, 27, 41–2
Willis, Thomas 13–4, 26, 27, 41–2, 440
working memory 250
Wren, Christopher 13
writer's cramp 386
writing difficulties (agraphia) 68, 250, 350

Yakovlev, Paul Ivan 50, 441